T0281570

PROBLEMS IN SOLID STATE PHYSICS WITH SOLUTIONS

PROBLEMS IN SOLID STATE PHYSICS WITH SOLUTIONS

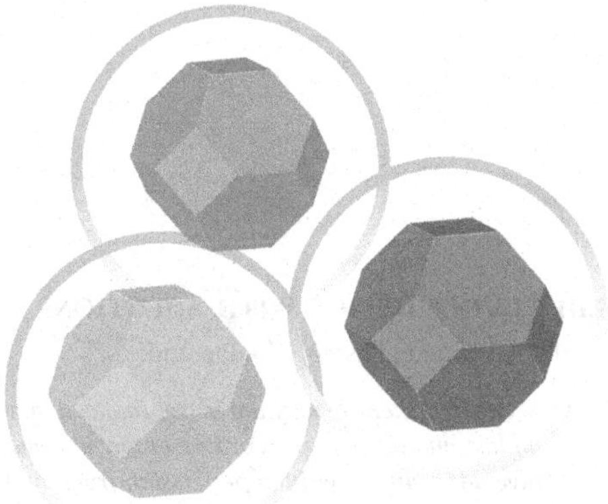

Fuxiang Han
Dalian University of Technology, China

World Scientific

NEW JERSEY • LONDON • SINGAPORE • BEIJING • SHANGHAI • HONG KONG • TAIPEI • CHENNAI

Published by

World Scientific Publishing Co. Pte. Ltd.
5 Toh Tuck Link, Singapore 596224
USA office: 27 Warren Street, Suite 401-402, Hackensack, NJ 07601
UK office: 57 Shelton Street, Covent Garden, London WC2H 9HE

British Library Cataloguing-in-Publication Data
A catalogue record for this book is available from the British Library.

PROBLEMS IN SOLID STATE PHYSICS WITH SOLUTIONS

ISBN-13 978-981-4365-02-4
ISBN-10 981-4365-02-5
ISBN-13 978-981-4366-87-8 (pbk)
ISBN-10 981-4366-87-0 (pbk)

Printed in Singapore by B & Jo Enterprise Pte Ltd

To my wife Pan Yanmei

Without her constant support and encouragement, I could not have finished writing this book.

Preface

This book provides a practical approach to consolidate one's already gained knowledge or to learn new knowledge in solid state physics through solving problems. It contains 300 problems. For the convenience of those using the book, the problems are partitioned into 30 chapters that cover the commonly taught subjects in introductory and advanced solid state physics courses. There are both simple problems that are routine exercises in solid state physics and challenging problems that call for more time and efforts for their solutions. The author has carefully worked out every problem and provided detailed solutions to all 300 problems with special attention paid to the clear presentation of physical ideas and mathematical techniques that are also relevant to research work at the frontiers of condensed matter physics. Many commonly used mathematical methods in condensed matter physics can be used in obtaining solutions to some of the problems.

Although this book *alone* is not suitable for beginners in solid state physics, it can be used in combination with with any book on solid state physics [a list of reference books is given at the end of this book]. The problems presented can be used in homework assignments in an introductory or advanced course on solid state physics for undergraduate or graduate students. Even though being able to solve problems, homework problems in a course in particular, is not the ultimate goal of studying solid state physics (as a matter of fact, it is not the ultimate goal of studying any branch of science), it is exceedingly important for a better understanding of basic concepts and for a quick grasp of methodology in solid state physics. It also helps the reader to get acquainted with the problem–solving paradigm in solid state physics.

This book can also be used as a reference book for typical problems and mathematical techniques in solid state physics. Many problems in this book are not just plain exercises; they are actual targets to which problem-solving ideas and mathematical tools are applied, not just for solving the problems, but also for elucidating the relevant physical ideas and mathematical techniques. In practice, it is perhaps more fascinating and rewarding to learn a new idea or technique through solving a real challenging problem than through reading only. In this aspect, this book is not a plain collection of problems but it presents a large number of problem–solving ideas and procedures, some of which are useful to practitioners in condensed matter physics.

For the convenience of quickly reviewing the relevant materials, a succinct recapitulation of important concepts and formulas is provided at the beginning of each chapter. With the formulas not derived and the notations in them not explained in most cases, the recapitulations are not aimed at complete expositions of relevant subjects. A recapitulation in a chapter can be used only as a place for formulas or as a brief summary of major subjects covered in the chapter. If a desired piece of information cannot be found in the recapitulation, a reference book should be consulted.

To avoid turning pages back and forth, the solution to a problem follows immediately the statement of the problem, with a short rule to indicate the end of the statement of the problem and the beginning of the solution. This way, when the solution to a problem is consulted, the statement of the problem can be conveniently looked up. By just reading the statement of a problem without peeking at the provided solution, the reader can work out a solution to the problem on his/her own.

The writing of this book has been supported by grants for teaching reform from the Teaching Affairs Division and the Graduate School at the Dalian University of Technology.

Dalian, March 2011 *Fuxiang Han*

Contents

Chapter 1

Drude Theory of Metals

(1) *Four assumptions in the Drude theory*

I A given conduction electron in a metal interacts with an ion only when it makes a collision with the ion. Between two successive collisions of the conduction electron, its interaction both with other electrons and with ions is neglected.

II Collisions of conduction electrons with ions in the Drude model are instantaneous events as in kinetic theory. The velocity of the conduction electron that suffers the collision is abruptly changed.

III A conduction electron in a metal makes a collision with an ion with probability $1/\tau$ per unit time.

IV Local thermal equilibrium of conduction electrons with their environment is achieved only through their collisions with ions. Immediately after a collision, the velocity of the conduction electron is not related to its velocity before the collision, with the direction of the velocity random and the magnitude of the velocity determined by the temperature at the collision place.

(2) *Collision probability density*

$$p(t) = \frac{1}{\tau}\mathrm{e}^{-t/\tau}.$$

(3) *Equation of motion*

$$\frac{\mathrm{d}\boldsymbol{p}(t)}{\mathrm{d}t} = -\frac{\boldsymbol{p}(t)}{\tau} + \boldsymbol{f}(t).$$

(4) *Drude DC conductivity*

$$\sigma_0 = \frac{ne^2\tau}{m}.$$

(5) *Drude AC conductivity*

$$\sigma(\omega) = \frac{\sigma_0}{1 - \mathrm{i}\omega\tau}.$$

(6) *Wave equation for* $\boldsymbol{E}$ *in a metal*

$$\nabla^2 \boldsymbol{E} + \frac{\omega^2 \epsilon_r(\omega)}{c^2}\boldsymbol{E} = 0.$$

(7) *Relative dielectric function*

$$\epsilon_r(\omega) = 1 + \frac{\mathrm{i}\sigma}{\epsilon_0 \omega}.$$

(8) *Plasma frequency*

$$\omega_p^2 = \frac{ne^2}{\epsilon_0 m}.$$

(9) *Hall coefficient*

$$R_{\rm H} = -\frac{1}{ne}.$$

(10) *Thermal conductivity*

$$\kappa = \frac{1}{3}\ell v c_v.$$

(11) *Wiedemann-Franz law*

$$\frac{\kappa}{\sigma} = \frac{3}{2}\left(\frac{k_{\rm B}}{e}\right)^2 T.$$

(12) *Thermopower*

$$Q = -\frac{c_v}{3ne} = -\frac{k_{\rm B}}{2e}.$$

1-1 Electron speed distribution. An electron collided with an ion at a place having a temperature T. Find the probability for the electron having emerged with a speed greater than $v_m = \sqrt{2k_{\rm B}T/m}$.

According to *Assumption IV*, the probability for the electron to emerge with a speed falling within the interval between v and $v + {\rm d}v$ is given by

$${\rm d}P = f_{\rm M}(v)dv = 4\pi\left(\frac{m}{2\pi k_{\rm B}T}\right)^{3/2} {\rm e}^{-mv^2/2k_{\rm B}T} v^2 {\rm d}v,$$

where $f_{\rm M}(v)$ is Maxwell's speed distribution function. Thus,

$$\begin{aligned} P_{v>v_m} &= 4\pi\left(\frac{m}{2\pi k_{\rm B}T}\right)^{3/2} \int_{v_m}^{\infty} {\rm e}^{-mv^2/2k_{\rm B}T} v^2 {\rm d}v = \frac{4}{\sqrt{\pi}}\int_1^{\infty} {\rm e}^{-x^2} x^2 {\rm d}x \\ &= \frac{2}{\sqrt{\pi}}\left[-x{\rm e}^{-x^2}\Big|_1^{\infty} + \int_1^{\infty} {\rm e}^{-x^2}{\rm d}x\right] = \frac{2}{\sqrt{\pi}{\rm e}} + \frac{2}{\sqrt{\pi}}\int_1^{\infty} {\rm e}^{-x^2}{\rm d}x \\ &= \frac{2}{\sqrt{\pi}{\rm e}} + {\rm erfc}(1) \approx 0.572, \end{aligned}$$

where ${\rm erfc}(x)$ is the complementary error function.

1-2 Average and standard deviation of the collision time interval. For a randomly picked conduction electron in a metal at time $t = 0$, the probability density function for the conduction electron to have a collision with an ion at time t is given by

$$p(t) = \frac{1}{\tau}\,{\rm e}^{-t/\tau}.$$

(1) Evaluate the average of the time interval for the electron to have its next collision.

(2) Evaluate the average of the square of the time interval for the electron to have its next collision.

(3) Evaluate the standard deviation of the time interval for the electron to have its next collision.

(4) Taking the above standard deviation of the time interval as the time uncertainty in the time-energy uncertainty relation in quantum mechanics, estimate the energy uncertainty in quantum mechanics. Comment on the results of the energy uncertainty given by classical and quantum theories.

---

(1) The average of the time interval for the electron to have its next collision is given by

$$\bar{t} = \int_0^\infty \mathrm{d}t\ tp(t) = \frac{1}{\tau}\int_0^\infty \mathrm{d}t\ te^{-t/\tau} = \tau\int_0^\infty \mathrm{d}x\ xe^{-x} = \tau.$$

(2) The average of the square of the time interval for the electron to have its next collision is given by

$$\overline{t^2} = \int_0^\infty \mathrm{d}t\ t^2 p(t) = \frac{1}{\tau}\int_0^\infty \mathrm{d}t\ t^2 \mathrm{e}^{-t/\tau} = \tau^2\int_0^\infty \mathrm{d}x\ x^2 \mathrm{e}^{-x} = 2\tau^2.$$

(3) The standard deviation of the time interval for the electron to have its next collision is then given by

$$\Delta t = \left[\overline{t^2} - (\bar{t})^2\right]^{1/2} = (2\tau^2 - \tau^2)^{1/2} = \tau.$$

(4) From $\tau\Delta E_q \sim \hbar$, we have $\Delta E_q \sim \hbar/\tau$ in quantum theory. In classical theory, ΔE_c is given by $\Delta E_c = (3/2)^{1/2} k_\mathrm{B} T$ that can be obtained by using Maxwell's distribution. We make the following comments on ΔE_q and ΔE_c.

- ΔE_q in quantum theory is intrinsic in nature and it is the consequence of the wave property of microscopic particles, while ΔE_c in classical theory is statistical in nature.
- While ΔE_c in classical theory depends on temperature, ΔE_q in quantum theory does not.
- For $\tau \sim 10^{-15}\ \mathrm{s}^{-1}$, $\Delta E_q \sim 10^{-12}$ erg. At $T = 273$ K, $\Delta E_c \sim 10^{-14}$ erg. Thus, at not too high temperatures, $\Delta E_c < \Delta E_q$.

1-3 Two successive collisions. Assume that a conduction electron in a metal experiences two successive collisions at times t_1 and t_2, respectively. Let $T = t_2 - t_1$ be the time interval between the two successive collisions.

(1) Find the probability density function for T.

(2) Evaluate the average $\overline{T}$ of the time interval between two successive collisions.

---

(1) The probability density for the occurrence of the first collision is given as usual by $p(t_1) = \tau^{-1}\mathrm{e}^{-t_1/\tau}$. The probability density for the occurrence of the second collision is given by $p(t_2) = \tau^{-1}\mathrm{e}^{-(t_2-t_1)/\tau}$. Let $g(T)$ be the probability density function for T. According to statistics, we have

$$\begin{aligned} g(T) &= \int_0^\infty \mathrm{d}t_1 \int_0^\infty \mathrm{d}t_2\, \delta(t_2 - t_1 - T)p(t_1)p(t_2) \\ &= \frac{1}{\tau^2}\int_0^\infty \mathrm{d}t_1 \int_0^\infty \mathrm{d}t_2\, \delta(t_2 - t_1 - T)\mathrm{e}^{-t_2/\tau} \\ &= \frac{1}{\tau^2}\mathrm{e}^{-T/\tau}\int_0^\infty \mathrm{d}t_1\, \mathrm{e}^{-t_1/\tau} = \frac{1}{\tau}\mathrm{e}^{-T/\tau}. \end{aligned}$$

(2) The average of T is given by

$$\overline{T} = \int_0^\infty \mathrm{d}T\, Tg(T) = \frac{1}{\tau}\int_0^\infty \mathrm{d}T\, T\mathrm{e}^{-T/\tau} = \tau.$$

1-4 Conductivity of a superconductor. Assume that the real part of the conductivity of a superconductor is described by $\mathrm{Re}\,\sigma(\omega) = A\delta(\omega)$ with $\delta(\omega)$ the Dirac δ-function. This expression can be taken as the $\tau \to \infty$ limit of the real part of $\sigma(\omega) = \sigma_0/(1 - \mathrm{i}\omega\tau)$ with $\sigma_0 = ne^2\tau/m$.

(1) Express A in terms of the electron density, mass, and charge.
(2) Evaluate the integral $\int_{-\infty}^{\infty} \mathrm{d}\omega\, \mathrm{Re}\,\sigma(\omega)$.

---

(1) The real part of $\sigma(\omega) = \sigma_0/(1 - \mathrm{i}\omega\tau)$ is given by

$$\mathrm{Re}\,\sigma(\omega) = \frac{\sigma_0}{1 + (\omega\tau)^2} = \frac{ne^2\tau/m}{1 + (\omega\tau)^2}.$$

For the superconductor, we have

$$\begin{aligned} \mathrm{Re}\,\sigma(\omega) = A\delta(\omega) &= \lim_{\tau\to\infty} \frac{ne^2\tau/m}{1 + (\omega\tau)^2} \\ &= \frac{ne^2}{m}\lim_{\tau\to\infty} \frac{1/\tau}{\omega^2 + 1/\tau^2} = \frac{\pi ne^2}{m}\delta(\omega). \end{aligned}$$

Thus,

$$A = \frac{\pi ne^2}{m}.$$

(2) Making use of the property of the δ-function, we have

$$\int_{-\infty}^{\infty} \mathrm{d}\omega\ \mathrm{Re}\,\sigma(\omega) = \frac{\pi n e^2}{m}\int_{-\infty}^{\infty} \mathrm{d}\omega\ \delta(\omega) = \frac{\pi n e^2}{m}.$$

1-5 Relative dielectric function of a metal. In the Drude theory of metals, the complex relative dielectric function of a metal is given by

$$\epsilon_r(\omega) = 1 + \frac{\mathrm{i}\sigma(\omega)}{\epsilon_0\omega},$$

where $\sigma(\omega)$ is the optical conductivity (the AC electrical conductivity)

$$\sigma(\omega) = \frac{\sigma_0}{1 - \mathrm{i}\omega\tau}, \quad \sigma_0 = \frac{ne^2\tau}{m}.$$

(1) Find the real and imaginary parts, $\mathrm{Re}\,\sigma(\omega)$ and $\mathrm{Im}\,\sigma(\omega)$, of $\sigma(\omega)$. Evaluate the integral

$$\frac{1}{\pi}\int_{-\infty}^{\infty} \frac{\mathrm{d}\omega}{\omega}\,\mathrm{Im}\,\sigma(\omega).$$

(2) Find the real and imaginary parts, $\mathrm{Re}\,\epsilon_r(\omega)$ and $\mathrm{Im}\,\epsilon_r(\omega)$, of $\epsilon_r(\omega)$.
(3) For $\sigma_0\tau/\epsilon_0 = 2$, plot $\mathrm{Re}\,\epsilon_r(\omega)$ and $\mathrm{Im}\,\epsilon_r(\omega)$ as functions of $\omega\tau$.
(4) Evaluate the square root of $\epsilon_r(\omega)$, $\sqrt{\epsilon_r(\omega)}$.
(5) The index of refraction $n(\omega)$ and the extinction coefficient $k(\omega)$ are given by the real and imaginary parts of $\sqrt{\epsilon_r(\omega)}$, respectively, $n(\omega) = \mathrm{Re}\,\sqrt{\epsilon_r(\omega)}$ and $k(\omega) = \mathrm{Im}\,\sqrt{\epsilon_r(\omega)}$. Find the explicit expressions for $n(\omega)$ and $k(\omega)$ in terms of the real and imaginary parts of $\epsilon_r(\omega)$. For simplicity in notations, use ϵ_r' for $\mathrm{Re}\,\epsilon_r$ and ϵ_r'' for $\mathrm{Im}\,\epsilon_r$.

---

(1) Separating the real and imaginary parts of $\sigma(\omega)$, we have

$$\sigma(\omega) = \frac{\sigma_0(1 + \mathrm{i}\omega\tau)}{1 + (\omega\tau)^2} = \frac{\sigma_0}{1 + (\omega\tau)^2} + \mathrm{i}\frac{\sigma_0\omega\tau}{1 + (\omega\tau)^2}.$$

Thus,

$$\mathrm{Re}\,\sigma(\omega) = \frac{\sigma_0}{1 + (\omega\tau)^2}, \quad \mathrm{Im}\,\sigma(\omega) = \frac{\sigma_0\omega\tau}{1 + (\omega\tau)^2}.$$

The value of the concerned integral is given by

$$\frac{1}{\pi}\int_{-\infty}^{\infty} \frac{\mathrm{d}\omega}{\omega}\,\mathrm{Im}\,\sigma(\omega) = \frac{\sigma_0\tau}{\pi}\int_{-\infty}^{\infty} \frac{\mathrm{d}\omega}{1 + (\omega\tau)^2} = \sigma_0.$$

(2) Separating the real and imaginary parts of $\epsilon_r(\omega)$, we have

$$\epsilon_r(\omega) = 1 + \frac{\mathrm{i}\sigma(\omega)}{\epsilon_0\omega} = 1 + \frac{\mathrm{i}}{\epsilon_0\omega}\left[\frac{\sigma_0}{1+(\omega\tau)^2} + \mathrm{i}\frac{\sigma_0\omega\tau}{1+(\omega\tau)^2}\right]$$

$$= 1 - \frac{\sigma_0\tau/\epsilon_0}{1+(\omega\tau)^2} + \mathrm{i}\frac{\sigma_0/\epsilon_0}{\omega[\,1+(\omega\tau)^2\,]}.$$

Thus,

$$\mathrm{Re}\,\epsilon_r(\omega) = 1 - \frac{\sigma_0\tau/\epsilon_0}{1+(\omega\tau)^2},\ \ \mathrm{Im}\,\epsilon_r(\omega) = \frac{\sigma_0/\epsilon_0}{\omega[\,1+(\omega\tau)^2\,]}.$$

(3) For $\sigma_0\tau/\epsilon_0 = 2$, we have

$$\mathrm{Re}\,\epsilon_r(\omega) = 1 - \frac{2}{1+(\omega\tau)^2},\ \ \mathrm{Im}\,\epsilon_r(\omega) = \frac{2}{(\omega\tau)[\,1+(\omega\tau)^2\,]}.$$

The plots of $\mathrm{Re}\,\epsilon_r(\omega)$ and $\mathrm{Im}\,\epsilon_r(\omega)$ are given in Fig. 1.1.

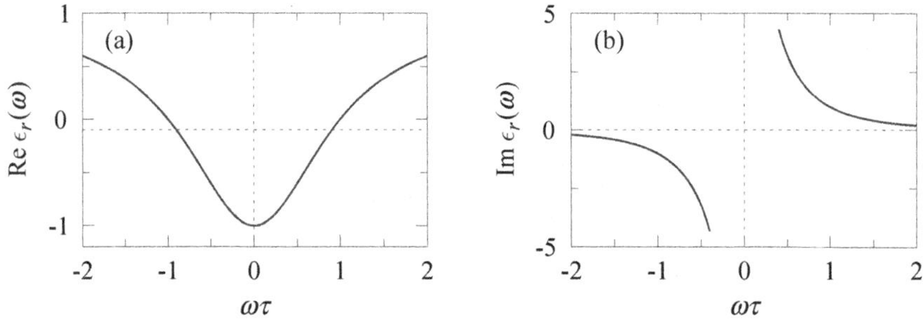

Fig. 1.1 Plots of the real (a) and imaginary (b) parts of the relative dielectric function $\epsilon_r(\omega)$ as functions of $\omega\tau$.

(4) Let $\epsilon_r'(\omega) = \mathrm{Re}\,\epsilon_r(\omega)$ and $\epsilon_r''(\omega) = \mathrm{Im}\,\epsilon_r(\omega)$. We then have

$$\epsilon_r = \epsilon_r' + \mathrm{i}\epsilon_r'' = \left(\epsilon_r'^{\,2} + \epsilon_r''^{2}\right)^{1/2}\mathrm{e}^{\mathrm{i}\tan^{-1}(\epsilon_r''/\epsilon_r')}.$$

Taking the square root of ϵ_r, we have

$$\sqrt{\epsilon_r(\omega)} = \left(\epsilon_r'^{\,2} + \epsilon_r''^{2}\right)^{1/4}\mathrm{e}^{\mathrm{i}\tan^{-1}(\epsilon_r''/\epsilon_r')/2}$$

$$= \left(\epsilon_r'^{\,2} + \epsilon_r''^{2}\right)^{1/4}\cos\left[\tan^{-1}(\epsilon_r''/\epsilon_r')/2\right] + \mathrm{i}\left(\epsilon_r'^{\,2} + \epsilon_r''^{2}\right)^{1/4}$$
$$\times \sin\left[\tan^{-1}(\epsilon_r''/\epsilon_r')/2\right].$$

Doing some algebras, we find that

$$\cos\left[\tan^{-1}(\epsilon_r''/\epsilon_r')/2\right] = \frac{1}{2^{1/2}}\left[\,1 + [1+(\epsilon_r''/\epsilon_r')^2]^{-1/2}\,\right]^{1/2},$$

$$\sin\left[\tan^{-1}(\epsilon_r''/\epsilon_r')/2\right] = \frac{1}{2^{1/2}}\left[\,1 - [1+(\epsilon_r''/\epsilon_r')^2]^{-1/2}\,\right]^{1/2}.$$

We then have

$$\sqrt{\epsilon_r(\omega)} = \frac{1}{2^{1/2}}\left[\left(\epsilon_r'^{\,2}+\epsilon_r''^{2}\right)^{1/2}+\epsilon_r'\right]^{1/2} + \frac{\mathrm{i}}{2^{1/2}}\left[\left(\epsilon_r'^{\,2}+\epsilon_r''^{2}\right)^{1/2}-\epsilon_r'\right]^{1/2}.$$

(5) From the above result for $\sqrt{\epsilon_r(\omega)}$, we obtain the following index of refraction $n(\omega)$ and the extinction coefficient $k(\omega)$

$$n(\omega) = \frac{1}{2^{1/2}}\left[\left(\epsilon_r'^2 + \epsilon_r''^2\right)^{1/2} + \epsilon_r'\right]^{1/2},$$

$$k(\omega) = \frac{1}{2^{1/2}}\left[\left(\epsilon_r'^2 + \epsilon_r''^2\right)^{1/2} - \epsilon_r'\right]^{1/2}.$$

1-6 Propagation of electromagnetic radiation in a metal. A beam of electromagnetic radiation of *high frequency* ω_0 is shed perpendicularly on a flat surface of a metal. It is found experimentally that the radiation can only penetrate a very small distance into the metal, with its strength dropping to 1/e at a small distance λ_0. Using the Drude model, determine

(1) the plasma frequency of the metal;
(2) the electron density of the metal;
(3) the expression of λ at an arbitrary frequency ω within a narrow frequency window below ω_0.

---

(1) The equation for the electric field $\boldsymbol{E}$ inside the metal reads

$$\frac{\mathrm{d}^2\boldsymbol{E}}{\mathrm{d}x^2} + \frac{\omega^2}{c^2}\epsilon_r(\omega)\boldsymbol{E} = 0,$$

where $\epsilon_r(\omega) = 1 - \omega_p^2/\omega^2$ at high frequencies. Since the radiation can only penetrate a very small distance into the metal, $\epsilon_r(\omega)$ must be negative, $\epsilon_r(\omega) < 0$. Then the solution to the above equation decays exponentially as x increases

$$\boldsymbol{E}(x) = \boldsymbol{E}(0)\mathrm{e}^{-[-\epsilon_r(\omega)]^{1/2}\omega x/c}.$$

From the fact that the strength of the radiation of frequency ω_0 drops to 1/e at a small distance λ_0, we have

$$\mathrm{e}^{-[-\epsilon_r(\omega_0)]^{1/2}\omega_0\lambda_0/c} = \frac{1}{\mathrm{e}}.$$

Solving for $\epsilon_r(\omega_0)$, we obtain $\epsilon_r(\omega_0) = -(c/\omega_0\lambda_0)^2$. From $\epsilon_r(\omega_0) = 1 - \omega_p^2/\omega_0^2$, we finally obtain

$$\omega_p = \omega_0\left[1 + \left(\frac{c}{\omega_0\lambda_0}\right)^2\right]^{1/2}.$$

(2) From $\omega_p^2 = ne^2/\epsilon_0 m$, we obtain the electron density of the metal

$$n = \frac{\epsilon_0 m\omega_0^2}{\mathrm{e}^2}\left[1 + \left(\frac{c}{\omega_0\lambda_0}\right)^2\right].$$

(3) To find the expression of λ at an arbitrary frequency ω, we replace λ_0 in ω_p with λ and ω_0 with ω. We then have

$$\omega_p = \omega\left[1 + \left(\frac{c}{\omega\lambda}\right)^2\right]^{1/2}.$$

Since ω_p is an intrinsic property of the metal, it is independent of ω. We thus have $\mathrm{d}\omega_p^2/\mathrm{d}\omega = 0$ from which we obtain

$$0 = \frac{\mathrm{d}\omega_p^2}{\mathrm{d}\omega} = 2\omega\left[1 + \left(\frac{c}{\omega\lambda}\right)^2\right] - \frac{2c^2}{\omega\lambda^3}\left(\lambda + \omega\frac{\mathrm{d}\lambda}{\mathrm{d}\omega}\right) = 2\omega - \frac{2c^2}{\lambda^3}\frac{\mathrm{d}\lambda}{\mathrm{d}\omega}$$

which leads to

$$\mathrm{d}\left(\frac{1}{\lambda^2}\right) = -\frac{1}{c^2}\mathrm{d}\omega^2.$$

Integrating yields

$$\frac{1}{\lambda^2} - \frac{1}{\lambda_0^2} = \frac{1}{c^2}\left(\omega_0^2 - \omega^2\right).$$

Solving for λ, we obtain

$$\lambda = \frac{\lambda_0}{\left[1 + (\omega_0^2 - \omega^2)\lambda_0^2/c^2\right]^{1/2}}.$$

1-7 Thermal conduction of a one-dimensional metal. The constant temperature gradient $\mathrm{d}T/\mathrm{d}x = -\gamma < 0$ exists in a metal bar whose left end is maintained at temperature T_h and is taken to be the origin of the x axis.

(1) Compute the thermal energy difference per electron between $x - v\tau$ and $x + v\tau$.

(2) Evaluate the thermal current density at x.

---

(1) From $\mathrm{d}T/\mathrm{d}x = -\gamma$, we have $\mathrm{d}T = -\gamma\mathrm{d}x$. Integrating yields $T = T_h - \gamma x$. The thermal energy of an electron is given by $\varepsilon(x) = 3k_\mathrm{B}T/2 = 3k_\mathrm{B}(T_h - \gamma x)/2$. The thermal energy difference per electron between $x - v\tau$ and $x + v\tau$ is given by

$$\begin{aligned}\varepsilon(x - v\tau) - \varepsilon(x + v\tau) &= \frac{3}{2}k_\mathrm{B}\left[T_h - \gamma(x - v\tau)\right] \\ &\quad - \frac{3}{2}k_\mathrm{B}\left[T_h - \gamma(x + v\tau)\right] \\ &= 3\gamma v\tau k_\mathrm{B} = 3\gamma\ell k_\mathrm{B}.\end{aligned}$$

(2) From $j_q = (nv/2)\big[\varepsilon(x - v\tau) - \varepsilon(x + v\tau)\big]$, we have

$$j_q = \frac{nv}{2} \cdot 3\gamma\ell k_{\mathrm{B}} = \frac{3}{2} nv\gamma\ell k_{\mathrm{B}}.$$

1-8 Metal in a uniform static electric field. A metal at uniform temperature is placed in a uniform static electric field $\boldsymbol{E}$. A conduction electron in the metal suffers a collision with an ion, and then it suffers a second collision with another ion after a time t.

(1) Find the average energy lost to the ion in the second collision.
(2) Find the average energy lost to the ions per electron per collision.
(3) Suppose that the temperature in the metal is not uniform with a constant temperature gradient given by $\boldsymbol{\nabla}T$. Find the average energy loss to the ions per electron per collision when both the applied static electric field and the constant temperature gradient are taken into account.

---

(1) The speed the electron acquires between two collisions is $v = eEt/m$. Thus, its kinetic energy right before the second collision is

$$E_k = \frac{1}{2}mv^2 = \frac{1}{2m}(eEt)^2.$$

The electron loses its kinetic energy to the ion in its second collision. Therefore, the energy loss is given by

$$E_{\text{loss}} = E_k = \frac{1}{2m}(eEt)^2.$$

(2) To find the average energy lost to the ions per electron per collision, we need to evaluate the average of t^2. Making use of the probability density function for t, $p(t) = \tau^{-1}\mathrm{e}^{-t/\tau}$, we have

$$\overline{t^2} = \int_0^\infty \mathrm{d}t\; t^2 p(t) = \frac{1}{\tau}\int_0^\infty \mathrm{d}t\; t^2 \mathrm{e}^{-t/\tau} = 2\tau^2.$$

Thus, the average energy loss to the ions per electron per collision is given by

$$\overline{E}_{\text{loss}} = \frac{(eE)^2\overline{t^2}}{2m} = \frac{(eE\tau)^2}{m}.$$

(3) Assume that the second collision occurred at $\boldsymbol{r}$. Then, the first collision occurred at $\boldsymbol{r} + e\boldsymbol{E}t^2/2m$, where $-e\boldsymbol{E}t^2/2m$ is the displacement of the electron between two collisions in the uniform electric field $\boldsymbol{E}$. In the presence of the temperature gradient, the energy loss of the electron includes the thermal energy difference between the two collision spots in addition to its kinetic energy. The temperatures at the two collision spots are respectively given by $T(\boldsymbol{r}+e\boldsymbol{E}t^2/2m)$ and $T(\boldsymbol{r})$. The energy loss of the electron is then given by

$$\begin{aligned}
E_{\text{loss}} &= \varepsilon\left[T\left(\boldsymbol{r}+\frac{et^2}{2m}\boldsymbol{E}\right)\right] - \varepsilon[T(\boldsymbol{r})] + \frac{(eEt)^2}{2m} \\
&\approx \varepsilon\left[T(\boldsymbol{r}) + \frac{et^2}{2m}\boldsymbol{\nabla}T\cdot\boldsymbol{E}\right] - \varepsilon[T(\boldsymbol{r})] + \frac{(eEt)^2}{2m} \\
&\approx \varepsilon[T(\boldsymbol{r})] + \frac{et^2}{2m}\frac{\mathrm{d}\varepsilon}{\mathrm{d}T}\boldsymbol{\nabla}T\cdot\boldsymbol{E} - \varepsilon[T(\boldsymbol{r})] + \frac{(eEt)^2}{2m} \\
&= \frac{et^2}{2m}\frac{\mathrm{d}\varepsilon}{\mathrm{d}T}\boldsymbol{\nabla}T\cdot\boldsymbol{E} + \frac{(eEt)^2}{2m}.
\end{aligned}$$

Making use of $\overline{t^2} = 2\tau^2$, we obtain the average energy lost to the ions per electron per collision

$$\overline{E}_{\text{loss}} = \frac{e\tau^2}{m}\frac{\mathrm{d}\varepsilon}{\mathrm{d}T}\boldsymbol{\nabla}T\cdot\boldsymbol{E} + \frac{(eE\tau)^2}{m}.$$

Chapter 2

Sommerfeld Theory of Metals

(1) *Born-von Karman boundary conditions*

$$\psi(x,\, y,\, z+L) = \psi(x,\, y,\, z),$$
$$\psi(x,\, y+L,\, z) = \psi(x,\, y,\, z),$$
$$\psi(x+L,\, y,\, z) = \psi(x,\, y,\, z).$$

(2) *Single-electron levels*

$$\varepsilon_{\boldsymbol{k}} = \frac{\hbar^2 \boldsymbol{k}^2}{2m},\ \psi_{\boldsymbol{k}}(\boldsymbol{r}) = \frac{1}{\sqrt{\mathscr{V}}}\mathrm{e}^{\mathrm{i}\boldsymbol{k}\cdot\boldsymbol{r}},$$
$$\boldsymbol{k} = \left(n_x \boldsymbol{e}_x + n_y \boldsymbol{e}_y + n_z \boldsymbol{e}_z\right)\frac{2\pi}{L},$$
$$n_x, n_y, n_z = 0,\ \pm 1,\ \pm 2,\ \cdots.$$

(3) *k-space density of levels* $\dfrac{\mathscr{V}}{8\pi^3}$.

(4) *Conversion between a* $\boldsymbol{k}$*-summation and a* $\boldsymbol{k}$*-integration*

$$\lim_{\mathscr{V}\to\infty} \frac{1}{\mathscr{V}} \sum_{\boldsymbol{k}} F(\boldsymbol{k}) = \frac{1}{(2\pi)^3}\int \mathrm{d}\boldsymbol{k}\, F(\boldsymbol{k}).$$

(5) *Density of states*

$$g(\varepsilon) = \frac{2}{\mathscr{V}}\sum_{\boldsymbol{k}} \delta(\varepsilon - \varepsilon_{\boldsymbol{k}}) = \frac{1}{4\pi^3}\int_{S(\varepsilon)} \frac{\mathrm{d}S_k}{|\boldsymbol{\nabla}_{\boldsymbol{k}}\varepsilon_{\boldsymbol{k}}|},$$

$$g(\varepsilon) = \frac{1}{2\pi^2}\left(\frac{2m}{\hbar^2}\right)^{3/2} \varepsilon^{1/2}\theta(\varepsilon) = \frac{3n\varepsilon^{1/2}}{2\varepsilon_{\mathrm{F}}^{3/2}}\theta(\varepsilon),$$

$$g'(\varepsilon) = \frac{g(\varepsilon)}{2\varepsilon},\ \varepsilon > 0,$$

$$g(\varepsilon_{\mathrm{F}}) = \frac{3n}{2\varepsilon_{\mathrm{F}}} = \frac{mk_{\mathrm{F}}}{\pi^2\hbar^2},\ g'(\varepsilon_{\mathrm{F}}) = \frac{3n}{4\varepsilon_{\mathrm{F}}^2}.$$

(6) *Fermi wave vector* $k_{\mathrm{F}} = (3\pi^2 n)^{1/3}$.

(7) *Ground-state energy of the electron gas* $\dfrac{E_0}{\mathscr{V}} = \dfrac{3}{5}n\varepsilon_{\mathrm{F}}$.

(8) *Pressure of the electron gas* $P = \dfrac{2E_0}{3\mathscr{V}} = \dfrac{2n\varepsilon_{\mathrm{F}}}{5}$.

(9) *Bulk modulus of the electron gas* $B = \dfrac{2}{3}n\varepsilon_{\mathrm{F}}$.

(10) *Fermi-Dirac distribution*

$$n_{\mathrm{F}}(\varepsilon) = \frac{1}{\mathrm{e}^{\beta(\varepsilon-\mu)}+1}, \quad -\frac{\partial n_{\mathrm{F}}(\varepsilon)}{\partial\varepsilon} \approx \delta(\varepsilon-\mu),\ T \to 0.$$

(11) *Sommerfeld expansion*

$$\int_{-\infty}^{\infty} \mathrm{d}\varepsilon\ \eta(\varepsilon)n_{\mathrm{F}}(\varepsilon) = \int_{-\infty}^{\mu} \mathrm{d}\varepsilon\ \eta(\varepsilon) + \frac{\pi^2}{6}\eta'(\mu)(k_{\mathrm{B}}T)^2 + \frac{7\pi^4}{360}\eta'''(\mu)(k_{\mathrm{B}}T)^4 + O((k_{\mathrm{B}}T)^6).$$

(12) *Internal energy density* $u = \left[\dfrac{3}{5} + \dfrac{\pi^2}{4}\left(\dfrac{k_{\mathrm{B}}T}{\varepsilon_{\mathrm{F}}}\right)^2\right]n\varepsilon_{\mathrm{F}}$.

(13) *Low-temperature electronic specific heat* $c_v = \dfrac{\pi^2}{2}\dfrac{k_{\mathrm{B}}T}{\varepsilon_{\mathrm{F}}}nk_{\mathrm{B}}$.

(14) *Low-temperature specific heat of a metal* $c_v = \gamma T + AT^3$.

2-1 Fermi-Dirac distribution function at low temperatures. The electrons in a metal obey the Fermi-Dirac distribution

$$n_{\mathrm{F}}(\varepsilon) = \frac{1}{\mathrm{e}^{(\varepsilon-\mu)/k_{\mathrm{B}}T}+1}.$$

(1) Find the Fermi-Dirac distribution function $n_{\mathrm{F}}^0(\varepsilon)$ at $T = 0$ by taking explicitly the $T \to 0$ limit of $n_{\mathrm{F}}(\varepsilon)$. Note that $\mu(T=0) \equiv \varepsilon_{\mathrm{F}}$.

(2) Plot $n_{\mathrm{F}}^0(\varepsilon)$ and $n_{\mathrm{F}}(\varepsilon)$ as functions of $\varepsilon/\varepsilon_{\mathrm{F}}$. For $n_{\mathrm{F}}(\varepsilon)$, neglect the temperature dependence of μ and plot it at $T = 0.01\varepsilon_{\mathrm{F}}/k_{\mathrm{B}}$ and $T = 0.1\varepsilon_{\mathrm{F}}/k_{\mathrm{B}}$.

(3) Differentiate $n_{\mathrm{F}}(\varepsilon)$ with respect to ε. Plot $-\partial n_{\mathrm{F}}(\varepsilon)/\partial\varepsilon$ as a function of $\varepsilon/\varepsilon_{\mathrm{F}}$ at $T = 0.01\varepsilon_{\mathrm{F}}/k_{\mathrm{B}}$, neglecting the temperature dependence of μ.

(4) Show that $-\partial n_{\mathrm{F}}(\varepsilon)/\partial\varepsilon \to c\delta(\varepsilon-\varepsilon_{\mathrm{F}})$ in the low-temperature limit. Here c is a constant. Find the value of c.

(5) The approximation $-\partial n_{\mathrm{F}}(\varepsilon)/\partial\varepsilon \to c\delta(\varepsilon-\varepsilon_{\mathrm{F}})$ is good only to the lowest order in T. Making use of this approximation and $u = \int_0^\infty \mathrm{d}\varepsilon\ \varepsilon g(\varepsilon)n_{\mathrm{F}}(\varepsilon)$, compute the ground-state energy per unit volume of the electrons, $u_0 \equiv E_0/V = u|_{T=0}$.

(1) Making use of $\mu(T=0)=\varepsilon_F$, we have

$$\lim_{T\to 0} e^{(\varepsilon-\varepsilon_F)/k_BT} = \begin{cases} 0, & \varepsilon < \varepsilon_F, \\ 1, & \varepsilon = \varepsilon_F, \\ \infty, & \varepsilon > \varepsilon_F \end{cases}$$

$$\Longrightarrow \lim_{T\to 0} \frac{1}{e^{(\varepsilon-\varepsilon_F)/k_BT}+1} = \begin{cases} 1, & \varepsilon < \varepsilon_F, \\ 1/2, & \varepsilon = \varepsilon_F, \\ 0, & \varepsilon > \varepsilon_F. \end{cases}$$

Therefore,

$$n_F^0(\varepsilon) = \lim_{T\to 0} n_F(\varepsilon) = \begin{cases} 1, & \varepsilon < \varepsilon_F, \\ 0, & \varepsilon > \varepsilon_F. \end{cases}$$

In arriving at the above result, we have made use of the fact that the value of a function at a discontinuous point is equal to the average of its limiting values from the left and right of the discontinuous point.

(2) The plots of $n_F^0(\varepsilon)$ and $n_F(\varepsilon)$ are given in Fig. 2.1(a).

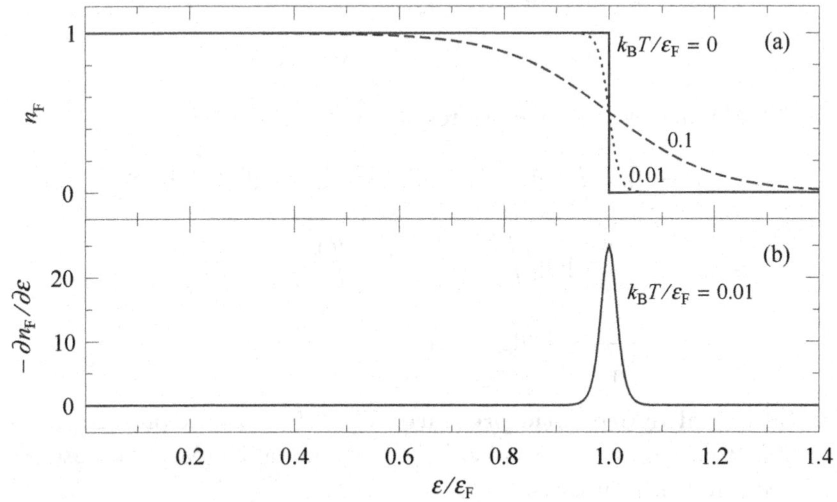

Fig. 2.1 Plots of the Fermi-Dirac distribution function (a) and its derivative (b) as functions of $\varepsilon/\varepsilon_F$. The temperatures are indicated on the curves.

(3) Differentiating $n_F(\varepsilon)$ with respect to ε yields

$$-\frac{\partial n_F}{\partial \varepsilon} = \frac{1}{k_BT}\frac{e^{(\varepsilon-\mu)/k_BT}}{\left[e^{(\varepsilon-\mu)/k_BT}+1\right]^2}.$$

The above result is plotted in Fig. 2.1(b) for $k_BT/\varepsilon_F = 0.01$.

(4) In the $T \to 0$ limit, we have

$$-\frac{\partial n_{\rm F}}{\partial \varepsilon} = \frac{1}{k_{\rm B}T}\frac{{\rm e}^{(\varepsilon-\mu)/k_{\rm B}T}}{\left[{\rm e}^{(\varepsilon-\mu)/k_{\rm B}T}+1\right]^2} \xrightarrow{T\to 0} \frac{1}{k_{\rm B}T}\frac{{\rm e}^{(\varepsilon-\varepsilon_{\rm F})/k_{\rm B}T}}{\left[{\rm e}^{(\varepsilon-\varepsilon_{\rm F})/k_{\rm B}T}+1\right]^2}$$

$$\xrightarrow{T\to 0} \frac{1}{k_{\rm B}T}\begin{cases} 0, & \varepsilon < \varepsilon_{\rm F} \\ 0, & \varepsilon > \varepsilon_{\rm F} \\ 1/4, & \varepsilon = \varepsilon_{\rm F} \end{cases} \xrightarrow{T\to 0} \begin{cases} 0, & \varepsilon \neq \varepsilon_{\rm F} \\ \infty, & \varepsilon = \varepsilon_{\rm F} \end{cases}$$

$$= c\,\delta(\varepsilon - \varepsilon_{\rm F}).$$

The integral of $-\partial n_{\rm F}/\partial\varepsilon$ over ε is given by

$$\int_0^\infty {\rm d}\varepsilon\left(-\frac{\partial n_{\rm F}}{\partial\varepsilon}\right) = \frac{1}{k_{\rm B}T}\int_0^\infty {\rm d}\varepsilon\,\frac{{\rm e}^{(\varepsilon-\mu)/k_{\rm B}T}}{\left[{\rm e}^{(\varepsilon-\mu)/k_{\rm B}T}+1\right]^2}$$

$$= \int_{-\mu/k_{\rm B}T}^\infty {\rm d}x\,\frac{{\rm e}^x}{({\rm e}^x+1)^2} \xrightarrow{T\to 0} \int_{-\infty}^\infty {\rm d}x\,\frac{{\rm e}^x}{({\rm e}^x+1)^2}$$

$$= \int_0^\infty {\rm d}t\,\frac{1}{(t+1)^2} = 1.$$

Thus, $c = 1$ and we have

$$-\frac{\partial n_{\rm F}}{\partial\varepsilon} \xrightarrow{T\to 0} \delta(\varepsilon - \varepsilon_{\rm F}).$$

(5) Making use of the above result, we have

$$u_0 = \lim_{T\to 0}\int_0^\infty {\rm d}\varepsilon\,\varepsilon g(\varepsilon)n_{\rm F}(\varepsilon) = \frac{3n}{2\varepsilon_{\rm F}^{3/2}}\lim_{T\to 0}\int_0^\infty {\rm d}\varepsilon\,\varepsilon^{3/2}n_{\rm F}(\varepsilon)$$

$$= \frac{3n}{5\varepsilon_{\rm F}^{3/2}}\lim_{T\to 0}\int_0^\infty {\rm d}\varepsilon\,\varepsilon^{5/2}\left(-\frac{\partial n_{\rm F}}{\partial\varepsilon}\right)$$

$$= \frac{3n}{5\varepsilon_{\rm F}^{3/2}}\int_0^\infty {\rm d}\varepsilon\,\varepsilon^{5/2}\delta(\varepsilon-\varepsilon_{\rm F}) = \frac{3}{5}n\varepsilon_{\rm F}.$$

2-2 Effects of hydrostatic pressure. When hydrostatic pressure is applied to a metal, the Fermi energy of the metal increases because the electron density increases.

(1) Derive an expression for the change in the Fermi energy within the free electron model.

(2) Compute the pressure which is required to change the Fermi energy by a factor of 1.000 001 for Cu. The electron density in Cu is 8.47×10^{28} m^{-3}.

(3) Evaluate the change in the density of states at the Fermi surface for Cu when its Fermi energy is changed by a factor of 1.000 001.

---

(1) From $\varepsilon_{\rm F} = \hbar^2 k_{\rm F}^2/2m = \hbar^2(3\pi^2 n)^{2/3}/2m$, we have $n = (2m\varepsilon_{\rm F}/\hbar^2)^{3/2}/3\pi^2$. From $P = 2n\varepsilon_{\rm F}/5$, we obtain the following expression for the pressure in terms of $\varepsilon_{\rm F}$

$$P = \frac{2}{15\pi^2}\left(\frac{2m}{\hbar^2}\right)^{3/2}\varepsilon_{\rm F}^{5/2}.$$

Differentiating P with respect to $\varepsilon_{\rm F}$, we obtain

$$\frac{{\rm d}P}{{\rm d}\varepsilon_{\rm F}} = \frac{1}{3\pi^2}\left(\frac{2m\varepsilon_{\rm F}}{\hbar^2}\right)^{3/2} = \frac{1}{3\pi^2}k_{\rm F}^3 = n.$$

We thus have

$$\Delta P \approx n\Delta\varepsilon_{\rm F}.$$

(2) For $n = 8.47 \times 10^{28}\ {\rm m}^{-3}$, the Fermi energy $\varepsilon_{\rm F}$ is given by $\varepsilon_{\rm F} \approx 1.126\,8 \times 10^{-18}$ J. For $\Delta\varepsilon_{\rm F} = 10^{-6}\varepsilon_{\rm F}$, we have

$$\begin{aligned}\Delta P &\approx n\Delta\varepsilon_{\rm F} = 10^{-6}n\varepsilon_{\rm F} = 10^{-6} \times 8.47 \times 10^{28} \times 1.126\,8 \times 10^{-18}\\ &\approx 9.54 \times 10^4\ {\rm N/m^2} \approx 0.94\ {\rm atm}.\end{aligned}$$

(3) From $g(\varepsilon_{\rm F}) = [(2m/\hbar^2)^{3/2}/2\pi^2]\varepsilon_{\rm F}^{1/2}$, we have

$$\frac{{\rm d}g(\varepsilon_{\rm F})}{{\rm d}\varepsilon_{\rm F}} = \frac{1}{4\pi^2\varepsilon_{\rm F}^{1/2}}\left(\frac{2m}{\hbar^2}\right)^{3/2} = \frac{1}{4\pi^2\varepsilon_{\rm F}^2}\left(\frac{2m\varepsilon_{\rm F}}{\hbar^2}\right)^{3/2} = \frac{3n}{4\varepsilon_{\rm F}^2}.$$

Thus,

$$\Delta g(\varepsilon_{\rm F}) \approx \frac{3n}{4\varepsilon_{\rm F}^2}\Delta\varepsilon_{\rm F}.$$

For $\Delta\varepsilon_{\rm F} = 10^{-6}\varepsilon_{\rm F}$, $n = 8.47 \times 10^{28}\ {\rm m}^{-3}$, and $\varepsilon_{\rm F} \approx 1.126\,8 \times 10^{-18}$ J, we have

$$\Delta g(\varepsilon_{\rm F}) \approx \frac{3n}{4\varepsilon_{\rm F}^2} \times 10^{-6}\varepsilon_{\rm F} = \frac{3n}{4\varepsilon_{\rm F}} \times 10^{-6} \approx 5.64 \times 10^{40}\ {\rm J^{-1}\cdot m^{-3}}.$$

2-3 Approximate expression for the Fermi-Dirac distribution function. The internal energy density of the electron gas in a metal at temperature T is given by $u = \int_0^\infty {\rm d}\varepsilon\ \varepsilon g(\varepsilon)n_{\rm F}(\varepsilon)$ with $g(\varepsilon)$ the density of states per unit volume. The Fermi-Dirac distribution function $n_{\rm F}(\varepsilon)$ is here approximated as

$$n_{\rm F}(\varepsilon) = \begin{cases} 1, & \varepsilon < \varepsilon_{\rm F} - 2k_{\rm B}T,\\ 1/2 - (\varepsilon - \varepsilon_{\rm F})/4k_{\rm B}T, & \varepsilon_{\rm F} - 2k_{\rm B}T \le \varepsilon \le \varepsilon_{\rm F} + 2k_{\rm B}T,\\ 0, & \varepsilon > \varepsilon_{\rm F} + 2k_{\rm B}T.\end{cases}$$

Compute the specific heat per unit volume of the electron gas, $c_v = \partial u/\partial T$, to the lowest order in k_BT/ε_F.

Since the approximate expression of $n_F(\varepsilon)$ is continuous in $(0, \infty)$, we can perform the differentiation of u with respect to T before performing the integral in u to save our computational efforts. Doing so, we obtain

$$c_v = \frac{\partial u}{\partial T} = \int_0^\infty \mathrm{d}\varepsilon\ \varepsilon g(\varepsilon)\frac{\partial n_F}{\partial T} = \frac{1}{4k_BT^2}\int_{\varepsilon_F-2k_BT}^{\varepsilon_F+2k_BT} \mathrm{d}\varepsilon\ \varepsilon g(\varepsilon)(\varepsilon-\varepsilon_F)$$
$$= \frac{3n}{8\varepsilon_F^{3/2}k_BT^2}\int_{\varepsilon_F-2k_BT}^{\varepsilon_F+2k_BT} \mathrm{d}\varepsilon\ \varepsilon^{3/2}(\varepsilon-\varepsilon_F).$$

Since the integration interval is very narrow in comparison with ε_F, we see that $|\varepsilon - \varepsilon_F|/\varepsilon_F \ll 1$ in the entire integration interval. Thus, we can approximate the factor $\varepsilon^{3/2}$ in the integrand as follows

$$\varepsilon^{3/2} = \big[\varepsilon_F + (\varepsilon-\varepsilon_F)\big]^{3/2} \approx \varepsilon_F^{3/2}\left[1 + \frac{3}{2\varepsilon_F}(\varepsilon-\varepsilon_F)\right].$$

We then have

$$c_v = \frac{9n}{16\varepsilon_F k_BT^2}\int_{\varepsilon_F-2k_BT}^{\varepsilon_F+2k_BT} \mathrm{d}\varepsilon\ (\varepsilon-\varepsilon_F)^2 = 3\left(\frac{k_BT}{\varepsilon_F}\right)nk_B.$$

Note that the result $c_v = (\pi^2/2)(k_BT/\varepsilon_F)nk_B$ is usually obtained in textbooks without making the approximation here. From the above result, we see that the approximation here yields the correct temperature dependence for c_v although the prefactor (3 here) is quite off the mark ($\pi^2/2 \approx 4.93$ without making the approximation here). The reason for getting the correct temperature dependence for such an approximation is that the most important contribution to the electronic specific heat arises from the electrons in states close to the Fermi surface.

2-4 Uncertainty in the electron kinetic energy. At temperature T, the average kinetic energy of an electron in a metal is given by

$$\langle\varepsilon\rangle \equiv \frac{U}{N} = \frac{1}{n}\int_{-\infty}^{\infty} \mathrm{d}\varepsilon\ \varepsilon g(\varepsilon)n_F(\varepsilon) \approx \frac{3}{5}\varepsilon_F + \frac{\pi^2}{6n}(k_BT)^2 g(\varepsilon_F)$$

up to the second order in T.

(1) Compute the average of the square of the kinetic energy of an electron in the metal

$$\langle\varepsilon^2\rangle = \frac{1}{n}\int_{-\infty}^{\infty} \mathrm{d}\varepsilon\ \varepsilon^2 g(\varepsilon)n_F(\varepsilon)$$

up to the second order in T.

(2) Find the standard deviation of the kinetic energy of an electron.
(3) Making use of the energy-time uncertainty relation, estimate the average time Δt that an electron stays persistently on a single-electron level.
(4) Estimate Δt numerically for Au at zero temperature. The electron density in Au is $5.9 \times 10^{28}\,\mathrm{m}^{-3}$. Compare the estimated Δt with the typical value of the relaxation time τ.
(5) Comment on the estimation of the relaxation time τ through Δt.

---

(1) Making use of the Sommerfeld expansion, we have

$$\begin{aligned}\langle\varepsilon^2\rangle &= \frac{1}{n}\int_{-\infty}^{\infty} \mathrm{d}\varepsilon\ \varepsilon^2 g(\varepsilon) n_\mathrm{F}(\varepsilon) = \frac{3}{2\varepsilon_\mathrm{F}^{3/2}}\int_0^{\infty} \mathrm{d}\varepsilon\ \varepsilon^{5/2} n_\mathrm{F}(\varepsilon)\\ &\approx \frac{3}{2\varepsilon_\mathrm{F}^{3/2}}\left[\int_0^{\mu} \mathrm{d}\varepsilon\ \varepsilon^{5/2} + \frac{5\pi^2}{12}(k_\mathrm{B}T)^2\mu^{3/2}\right]\\ &= \frac{3}{2\varepsilon_\mathrm{F}^{3/2}}\left[\frac{2}{7}\mu^{7/2} + \frac{5\pi^2}{12}(k_\mathrm{B}T)^2\mu^{3/2}\right].\end{aligned}$$

Inserting $\mu = \varepsilon_\mathrm{F}\left[1 - (\pi^2/12)(k_\mathrm{B}T/\varepsilon_\mathrm{F})^2\right]$ into the above equation and keeping only terms up to the second order in T, we have

$$\begin{aligned}\langle\varepsilon^2\rangle &\approx \frac{3}{2\varepsilon_\mathrm{F}^{3/2}}\left\{\frac{2}{7}\varepsilon_\mathrm{F}^{7/2}\left[1 - \frac{7\pi^2}{24}\left(\frac{k_\mathrm{B}T}{\varepsilon_\mathrm{F}}\right)^2\right] + \frac{5\pi^2}{12}(k_\mathrm{B}T)^2\varepsilon_\mathrm{F}^{3/2}\right\}\\ &= \frac{3}{7}\varepsilon_\mathrm{F}^2\left[1 + \frac{7\pi^2}{6}\left(\frac{k_\mathrm{B}T}{\varepsilon_\mathrm{F}}\right)^2\right] = \frac{3}{7}\varepsilon_\mathrm{F}^2 + \frac{\pi^2}{2}(k_\mathrm{B}T)^2.\end{aligned}$$

(2) The variance of the kinetic energy of an electron is given by

$$\begin{aligned}(\Delta\varepsilon)^2 &= \langle\varepsilon^2\rangle - \langle\varepsilon\rangle^2 = \frac{3}{7}\varepsilon_\mathrm{F}^2 + \frac{\pi^2}{2}(k_\mathrm{B}T)^2 - \varepsilon_\mathrm{F}^2\left[\frac{3}{5} + \frac{\pi^2}{4}\left(\frac{k_\mathrm{B}T}{\varepsilon_\mathrm{F}}\right)^2\right]^2\\ &\approx \frac{3}{7}\varepsilon_\mathrm{F}^2 + \frac{\pi^2}{2}(k_\mathrm{B}T)^2 - \frac{9}{25}\varepsilon_\mathrm{F}^2 - \frac{3\pi^2}{10}(k_\mathrm{B}T)^2\\ &= \frac{12}{175}\varepsilon_\mathrm{F}^2 + \frac{\pi^2}{5}(k_\mathrm{B}T)^2.\end{aligned}$$

Then, the standard deviation of the kinetic energy of an electron is given by

$$\Delta\varepsilon = \left[\frac{12}{175}\varepsilon_\mathrm{F}^2 + \frac{\pi^2}{5}(k_\mathrm{B}T)^2\right]^{1/2} \approx \frac{2}{5}\left(\frac{3}{7}\right)^{1/2}\varepsilon_\mathrm{F}\left[1 + \frac{35\pi^2}{24}\left(\frac{k_\mathrm{B}T}{\varepsilon_\mathrm{F}}\right)^2\right].$$

(3) From the energy-time uncertainty relation, the average time Δt that an electron stays persistently on a single-electron level is given by

$$\Delta t \approx \frac{\hbar}{\Delta\varepsilon} \approx \frac{5}{2}\left(\frac{7}{3}\right)^{1/2}\left(\frac{\hbar}{\varepsilon_{\rm F}}\right)\left[1 - \frac{35\pi^2}{24}\left(\frac{k_{\rm B}T}{\varepsilon_{\rm F}}\right)^2\right]$$
$$\approx 3.818\,8\left(\frac{\hbar}{\varepsilon_{\rm F}}\right)\left[1 - 14.393\,2\left(\frac{k_{\rm B}T}{\varepsilon_{\rm F}}\right)^2\right].$$

(4) For Au, the Fermi energy is given by

$$\varepsilon_{\rm F} = \frac{\hbar^2(3\pi^2 n)^{2/3}}{2m} \approx 8.854\,9 \times 10^{-19}\ {\rm J}.$$

Then, the average time Δt that an electron stays persistently on a single-electron level in Au at zero temperature is given by

$$\Delta t \approx 3.818\,8\left(\frac{\hbar}{\varepsilon_{\rm F}}\right) \approx 4.548\,3 \times 10^{-16}\ {\rm s}.$$

In the Drude model, $\tau \sim 10^{-16}$-10^{-13} s. Thus, Δt lies in the lower end of the τ range.

(5) In the free electron model, electrons are treated as identical fermionic particles of spin 1/2, with the direct Coulomb, the exchange, and the correlation interactions neglected. Since electrons obey the Pauli exclusion principle, the repulsive interaction due to the Pauli exclusion principle is taken into account. If τ is estimated through Δt, the effects of all the interactions are neglected except for the effect due to the Pauli exclusion principle.

2-5 Lindhard function. The Lindhard function $L(\boldsymbol{q})$ for a three-dimensional electron gas is defined by

$$L(\boldsymbol{q}) = \frac{1}{g(\varepsilon_{\rm F})}\int \frac{{\rm d}\boldsymbol{k}}{4\pi^3}\,\frac{n_{\rm F}(\varepsilon_{\boldsymbol{k}}) - n_{\rm F}(\varepsilon_{\boldsymbol{k}+\boldsymbol{q}})}{\varepsilon_{\boldsymbol{k}+\boldsymbol{q}} - \varepsilon_{\boldsymbol{k}}}$$

with $\varepsilon_{\boldsymbol{k}} = \hbar^2 k^2/2m$.

(1) Evaluate the integration over $\boldsymbol{k}$ and find an explicit expression for $L(\boldsymbol{q})$ at zero temperature. The Lindhard function at zero temperature is to be denoted by $L_0(\boldsymbol{q})$. **Hints**: (1) First make the change of integration variables $\boldsymbol{k}+\boldsymbol{q} \to \boldsymbol{k}$ in the term involving $n_{\rm F}(\varepsilon_{\boldsymbol{k}+\boldsymbol{q}})$ and then utilize the Fermi-Dirac distribution function at zero temperature. (2) Express the final result in terms of $x = q/2k_{\rm F}$.

(2) Evaluate $\lim_{q\to 0} L_0(\boldsymbol{q})$.
(3) Evaluate $\lim_{q\to 2k_\mathrm{F}} L_0(\boldsymbol{q})$.
(4) Find the temperature dependence of $L(q = 2k_\mathrm{F})$ in the $T \to 0$ limit.

---

(1) Following the hints given, we have

$$\begin{aligned}
L_0(\boldsymbol{q}) &= \frac{1}{g(\varepsilon_\mathrm{F})}\int \frac{\mathrm{d}\boldsymbol{k}}{4\pi^3}\left[\frac{n_\mathrm{F}^0(\varepsilon_{\boldsymbol{k}})}{\varepsilon_{\boldsymbol{k}+\boldsymbol{q}}-\varepsilon_{\boldsymbol{k}}} - \frac{n_\mathrm{F}^0(\varepsilon_{\boldsymbol{k}+\boldsymbol{q}})}{\varepsilon_{\boldsymbol{k}+\boldsymbol{q}}-\varepsilon_{\boldsymbol{k}}}\right]\\
&= \frac{1}{g(\varepsilon_\mathrm{F})}\int \frac{\mathrm{d}\boldsymbol{k}}{4\pi^3}\, n_\mathrm{F}^0(\varepsilon_{\boldsymbol{k}})\left[\frac{1}{\varepsilon_{\boldsymbol{k}+\boldsymbol{q}}-\varepsilon_{\boldsymbol{k}}} - \frac{1}{\varepsilon_{\boldsymbol{k}}-\varepsilon_{\boldsymbol{k}-\boldsymbol{q}}}\right]\\
&= \frac{m\varepsilon_\mathrm{F}}{6\pi^3\hbar^2 nq}\int_0^{k_\mathrm{F}} \mathrm{d}k\, k \int_{-1}^{1}\mathrm{d}\cos\theta \int_0^{2\pi}\mathrm{d}\varphi\\
&\qquad\qquad \times\left[\frac{1}{\cos\theta + q/2k} - \frac{1}{\cos\theta - q/2k}\right].
\end{aligned}$$

Performing the integration over angles, we have

$$L_0(\boldsymbol{q}) = \frac{2m\varepsilon_\mathrm{F}}{3\pi^2\hbar^2 nq}\int_0^{k_\mathrm{F}}\mathrm{d}k\, k\ln\left|\frac{k+q/2}{k-q/2}\right| = \frac{1}{k_\mathrm{F}q}\int_0^{k_\mathrm{F}}\mathrm{d}k\, k\ln\left|\frac{k+q/2}{k-q/2}\right|.$$

Doing once the integration by parts, we can then perform the integration over k

$$\begin{aligned}
L_0(\boldsymbol{q}) &= \frac{1}{2k_\mathrm{F}q}\left[k_\mathrm{F}^2\ln\left|\frac{q+2k_\mathrm{F}}{q-2k_\mathrm{F}}\right| + q\int_0^{k_\mathrm{F}}\mathrm{d}k\,\frac{k^2}{k^2-q^2/4}\right]\\
&= \frac{1}{2k_\mathrm{F}q}\left[k_\mathrm{F}^2\ln\left|\frac{q+2k_\mathrm{F}}{q-2k_\mathrm{F}}\right| + qk_\mathrm{F} + \frac{q^3}{4}\int_0^{k_\mathrm{F}}\frac{\mathrm{d}k}{k^2-q^2/4}\right]\\
&= \frac{1}{2k_\mathrm{F}q}\left[k_\mathrm{F}^2\ln\left|\frac{q+2k_\mathrm{F}}{q-2k_\mathrm{F}}\right| + qk_\mathrm{F} - \frac{q^2}{4}\ln\left|\frac{q+2k_\mathrm{F}}{q-2k_\mathrm{F}}\right|\right]\\
&= \frac{1}{2} + \frac{k_\mathrm{F}^2 - q^2/4}{2k_\mathrm{F}q}\ln\left|\frac{q+2k_\mathrm{F}}{q-2k_\mathrm{F}}\right| = \frac{1}{2} + \frac{1-x^2}{4x}\ln\left|\frac{1+x}{1-x}\right|,
\end{aligned}$$

where $x = q/2k_\mathrm{F}$.

(2) For $q = 0$ $(x = 0)$, making use of L'Hopital's rule we have

$$\begin{aligned}
L_0(q=0) &= \frac{1}{2} + \lim_{x\to 0}\frac{1-x^2}{4x}\ln\left|\frac{1+x}{1-x}\right| = \frac{1}{2} + \lim_{x\to 0}\frac{1}{4x}\ln\left|\frac{1+x}{1-x}\right|\\
&= \frac{1}{2} + \lim_{x\to 0}\frac{1/(x+1) - 1/(x-1)}{4} = \frac{1}{2} + \frac{1}{2} = 1.
\end{aligned}$$

(3) For $q = 2k_\mathrm{F}$ ($x = 1$), making use of L'Hopital's rule we have

$$\begin{aligned}L_0(2k_\mathrm{F}) &= \frac{1}{2} + \lim_{x\to1}\frac{1-x^2}{4x}\ln\left|\frac{1+x}{1-x}\right| \\ &= \frac{1}{2} + \lim_{x\to1}\frac{(1+x)(1-x)}{4x}\ln\left|\frac{1+x}{1-x}\right| \\ &= \frac{1}{2} - \frac{1}{2}\lim_{x\to1}(1-x)\ln|1-x| = \frac{1}{2} - \frac{1}{2}\lim_{x\to1}\frac{\ln|1-x|}{1/(1-x)} \\ &= \frac{1}{2} - \frac{1}{2}\lim_{x\to1}\frac{-1/(1-x)}{1/(1-x)^2} = \frac{1}{2} + \frac{1}{2}\lim_{x\to1}(1-x) = \frac{1}{2}.\end{aligned}$$

(4) For $q = 2k_\mathrm{F}$, we have at finite temperatures

$$L(2k_\mathrm{F}) = \frac{1}{2k_\mathrm{F}^2}\int_0^\infty \mathrm{d}k\; k n_\mathrm{F}(\varepsilon_{\boldsymbol{k}})\ln\left|\frac{k+k_\mathrm{F}}{k-k_\mathrm{F}}\right|.$$

Because of the presence of $\ln|k - k_\mathrm{F}|$ in the integrand, the most important contribution to the integral comes from the region in the vicinity of $|\boldsymbol{k}| = k_\mathrm{F}$. Making a change of integration variables from k to $\varepsilon = \hbar^2k^2/2m$, we have

$$L(2k_\mathrm{F}) = \frac{1}{4\varepsilon_\mathrm{F}}\int_0^\infty \mathrm{d}\varepsilon\; n_\mathrm{F}(\varepsilon)\ln\left|\frac{\sqrt{\varepsilon}+\sqrt{\varepsilon_\mathrm{F}}}{\sqrt{\varepsilon}-\sqrt{\varepsilon_\mathrm{F}}}\right|.$$

Because of the presence of the branch point of the logarithmic function at $\varepsilon = \varepsilon_\mathrm{F}$, the Sommerfeld expansion can not be used. We turn to the fact that $-\partial n_\mathrm{F}(\varepsilon)/\partial\varepsilon$ can be well represented by $\delta(\varepsilon - \mu)$ at low temperatures. To bring $-\partial n_\mathrm{F}(\varepsilon)/\partial\varepsilon$ into the integrand, we perform an integration by parts. For convenience, we introduce

$$G(\varepsilon) = \int_0^\varepsilon \mathrm{d}\varepsilon'\;\ln\left|\frac{\sqrt{\varepsilon'}+\sqrt{\varepsilon_\mathrm{F}}}{\sqrt{\varepsilon'}-\sqrt{\varepsilon_\mathrm{F}}}\right|.$$

We then have

$$\begin{aligned}L(2k_\mathrm{F}) &= \frac{1}{4\varepsilon_\mathrm{F}}\int_{\varepsilon=0}^{\varepsilon=\infty} n_\mathrm{F}(\varepsilon)\,\mathrm{d}G(\varepsilon) = \frac{1}{4\varepsilon_\mathrm{F}}\int_0^\infty \mathrm{d}\varepsilon\; G(\varepsilon)\left[-\frac{\partial n_\mathrm{F}(\varepsilon)}{\partial\varepsilon}\right] \\ &\approx \frac{1}{4\varepsilon_\mathrm{F}}\int_0^\infty \mathrm{d}\varepsilon\; G(\varepsilon)\delta(\varepsilon-\mu) = \frac{1}{4\varepsilon_\mathrm{F}}G(\mu).\end{aligned}$$

Note that the above result is actually the first term in the Sommerfeld expansion. Inserting the expression of $G(\mu)$ into the above equation and evaluating the remaining integral, we obtain

$$\begin{aligned}
L(2k_{\mathrm{F}}) &= \frac{1}{4\varepsilon_{\mathrm{F}}}\int_0^{\mu} \mathrm{d}\varepsilon\, \ln\left|\frac{\sqrt{\varepsilon}+\sqrt{\varepsilon_{\mathrm{F}}}}{\sqrt{\varepsilon}-\sqrt{\varepsilon_{\mathrm{F}}}}\right| \\
&= \frac{1}{4\varepsilon_{\mathrm{F}}}\left[\mu \ln\left|\frac{\sqrt{\mu}+\sqrt{\varepsilon_{\mathrm{F}}}}{\sqrt{\mu}-\sqrt{\varepsilon_{\mathrm{F}}}}\right|\right. \\
&\quad \left. -\frac{1}{2}\int_0^{\mu} \mathrm{d}\varepsilon\, \sqrt{\varepsilon}\left(\frac{1}{\sqrt{\varepsilon}+\sqrt{\varepsilon_{\mathrm{F}}}} - \frac{1}{\sqrt{\varepsilon}-\sqrt{\varepsilon_{\mathrm{F}}}}\right)\right] \\
&= \frac{1}{2}\sqrt{\mu/\varepsilon_{\mathrm{F}}} + \frac{1}{4}(\mu/\varepsilon_{\mathrm{F}}-1)\ln\left|\frac{\sqrt{\mu/\varepsilon_{\mathrm{F}}}+1}{\sqrt{\mu/\varepsilon_{\mathrm{F}}}-1}\right|.
\end{aligned}$$

Making use of

$$\mu \approx \varepsilon_{\mathrm{F}}\left[1 - \frac{\pi^2}{12}\left(\frac{k_{\mathrm{B}}T}{\varepsilon_{\mathrm{F}}}\right)^2\right].$$

we have

$$\begin{aligned}
L(2k_{\mathrm{F}}) &\approx \frac{1}{2}\left[1 - \frac{\pi^2}{12}\left(\frac{k_{\mathrm{B}}T}{\varepsilon_{\mathrm{F}}}\right)^2\right]^{1/2} \\
&\quad - \frac{\pi^2}{48}\left(\frac{k_{\mathrm{B}}T}{\varepsilon_{\mathrm{F}}}\right)^2 \ln\left|\frac{\sqrt{1-(\pi^2/12)(k_{\mathrm{B}}T/\varepsilon_{\mathrm{F}})^2}+1}{\sqrt{1-(\pi^2/12)(k_{\mathrm{B}}T/\varepsilon_{\mathrm{F}})^2}-1}\right| \\
&\approx \frac{1}{2}\left[1 - \frac{\pi^2}{24}\left(\frac{k_{\mathrm{B}}T}{\varepsilon_{\mathrm{F}}}\right)^2\right] \\
&\quad - \frac{\pi^2}{48}\left(\frac{k_{\mathrm{B}}T}{\varepsilon_{\mathrm{F}}}\right)^2 \ln\frac{2-(\pi^2/24)(k_{\mathrm{B}}T/\varepsilon_{\mathrm{F}})^2}{(\pi^2/24)(k_{\mathrm{B}}T/\varepsilon_{\mathrm{F}})^2} \\
&\approx \frac{1}{2} - \frac{\pi^2}{24}\left(\frac{k_{\mathrm{B}}T}{\varepsilon_{\mathrm{F}}}\right)^2 \ln\frac{\varepsilon_{\mathrm{F}}}{k_{\mathrm{B}}T},
\end{aligned}$$

where we have kept only the terms up to the order $T^2 \ln T$.

2-6 Boltzmann equation. When an electric field $\boldsymbol{E}$ is applied to a metal, the Boltzmann equation for the distribution function of electrons takes on the following form

$$\frac{\partial f}{\partial t} + \boldsymbol{v}\cdot\boldsymbol{\nabla}_{\boldsymbol{r}} f - e\boldsymbol{E}\cdot\boldsymbol{\nabla}_{\boldsymbol{p}} f = -\frac{f-f_0}{\tau},$$

where f is the distribution function in the presence of the electric field, f_0 the distribution function in the absence of the electric field, and τ the relaxation time. When the electric field is homogeneous and constant and is directed along the positive x axis, the Boltzmann equation becomes for the steady state

$$eE_x \frac{\partial f}{\partial p_x} = \frac{f - f_0}{\tau}.$$

We shall proceed by making an approximation to the above equation by replacing f on the left hand side with f_0. Then, the DC electrical conductivity of the metal can be computed by making use of $\boldsymbol{j} = -ne\langle \boldsymbol{p} \rangle / m$ and by evaluating the average $\langle \boldsymbol{p} \rangle$ of momentum $\boldsymbol{p}$ over the distribution function f derived from the above equation.

(1) Taking f_0 as the Maxwell distribution

$$f_{\rm M}(\boldsymbol{p})\mathrm{d}\boldsymbol{p} = \frac{1}{(2\pi m k_{\rm B}T)^{3/2}} \mathrm{e}^{-p^2/2mk_{\rm B}T}\mathrm{d}\boldsymbol{p},$$

compute the DC electrical conductivity of the metal.

(2) Taking f_0 as the Fermi-Dirac distribution

$$n_{\rm F}[\varepsilon(\boldsymbol{p})]\mathrm{d}\boldsymbol{p} = \frac{2}{nh^3} \frac{1}{\mathrm{e}^{(\varepsilon(\boldsymbol{p}) - \mu)/k_{\rm B}T} + 1}\mathrm{d}\boldsymbol{p}$$

with $\varepsilon(\boldsymbol{p}) = p^2/2m$, compute the *low-temperature* DC electrical conductivity of the metal.

---

(1) From

$$eE_x \frac{\partial f_0}{\partial p_x} = \frac{f - f_0}{\tau},$$

we obtain

$$f = f_0 + eE_x\tau \frac{\partial f_0}{\partial p_x}.$$

For the Maxwell distribution $f_0 = f_{\rm M}$, we have

$$\frac{\partial f_{\rm M}}{\partial p_x} = -\frac{p_x}{mk_BT} f_{\rm M}.$$

Then,

$$f = \left(1 - \frac{eE_x\tau p_x}{mk_BT}\right) f_{\rm M}.$$

The y and z components of $\langle \boldsymbol{p} \rangle$ are identically zero since the integrand will be an odd function of p_y or p_z

$$\langle p_y \rangle = \langle p_z \rangle = 0.$$

The x component is given by

$$\begin{aligned}
\langle p_x \rangle &= \int \mathrm{d}\boldsymbol{p}\; p_x f(\boldsymbol{p}) = \int \mathrm{d}\boldsymbol{p}\; p_x \left(1 - \frac{eE_x\tau p_x}{mk_BT}\right) f_{\mathrm{M}}(\boldsymbol{p}) \\
&= -\frac{eE_x\tau}{mk_BT} \int \mathrm{d}\boldsymbol{p}\; p_x^2 f_{\mathrm{M}}(\boldsymbol{p}) \\
&= -\frac{eE_x\tau}{mk_BT} \frac{1}{(2\pi mk_BT)^{3/2}} \cdot \frac{1}{2}\pi^{1/2}(2mk_BT)^{3/2} \cdot (2\pi mk_BT) \\
&= -eE_x\tau.
\end{aligned}$$

Using $\boldsymbol{j} = -ne\langle \boldsymbol{p} \rangle / m$, we obtain

$$j_x = \frac{ne^2\tau}{m} E_x, \quad j_y = j_z = 0.$$

From $j_x = \sigma E_x$, we can infer the value of the DC electrical conductivity and find that

$$\sigma = \frac{ne^2\tau}{m}.$$

The above result is the same as that obtained in the Drude theory.

(2) For the Fermi-Dirac distribution $f_0 = n_{\mathrm{F}}$, we have

$$\frac{\partial n_{\mathrm{F}}}{\partial p_x} = \frac{\partial n_{\mathrm{F}}}{\partial \varepsilon}\frac{\partial \varepsilon}{\partial p_x} = \frac{p_x}{m}\frac{\partial n_{\mathrm{F}}}{\partial \varepsilon}.$$

Then,

$$f = n_{\mathrm{F}} + \frac{eE_x\tau p_x}{m}\frac{\partial n_{\mathrm{F}}}{\partial \varepsilon}.$$

As in Part (1), the y and z components of $\langle \boldsymbol{p} \rangle$ are identically zero

$$\langle p_y \rangle = \langle p_z \rangle = 0.$$

The x component of $\langle \boldsymbol{p} \rangle$ can be computed as follows

$$\begin{aligned}
\langle p_x \rangle &= \int \mathrm{d}\boldsymbol{p}\; p_x \left(n_{\mathrm{F}} + \frac{eE_x\tau p_x}{m}\frac{\partial n_{\mathrm{F}}}{\partial \varepsilon}\right) = \frac{eE_x\tau}{m} \int \mathrm{d}\boldsymbol{p}\; p_x^2 \frac{\partial n_{\mathrm{F}}}{\partial \varepsilon} \\
&= \frac{eE_x\tau}{m} \int_0^\infty \mathrm{d}p\; p^4 \frac{\partial n_{\mathrm{F}}}{\partial \varepsilon} \int_{-1}^{1} \mathrm{d}\cos\theta\; \sin^2\theta \int_0^{2\pi} \mathrm{d}\varphi\; \cos^2\varphi \\
&= \frac{4\pi eE_x\tau}{3m} \int_0^\infty \mathrm{d}p\; p^4 \frac{\partial n_{\mathrm{F}}}{\partial \varepsilon} = \frac{8\sqrt{2}\pi m^{3/2} eE_x\tau}{3} \int_0^\infty \mathrm{d}\varepsilon\; \varepsilon^{3/2} \frac{\partial n_{\mathrm{F}}}{\partial \varepsilon} \\
&\approx \frac{8\sqrt{2}\pi m^{3/2} eE_x\tau}{3} \left(-\frac{2}{nh^3}\right) \int_0^\infty \mathrm{d}\varepsilon\; \varepsilon^{3/2} \delta(\varepsilon - \varepsilon_{\mathrm{F}}) \\
&= -\frac{16\sqrt{2}\pi m^{3/2} eE_x\tau}{3nh^3} \varepsilon_{\mathrm{F}}^{3/2} = -\frac{16\sqrt{2}\pi m^{3/2} eE_x\tau}{3nh^3} \frac{3\pi^2 n\hbar^3}{(2m)^{3/2}} \\
&= -eE_x\tau
\end{aligned}$$

which is identical with the result we obtained in Part (1). However, note that the result obtained here is an approximate one, for we used the δ-function approximation to the derivative of $n_{\rm F}(\varepsilon)$ with respect to ε. The exact quantum result shall contain correction terms to the classical result. These correction terms are negligible at low temperatures. Finally, the DC electrical conductivity computed by using the Fermi-Dirac distribution within the above approximation is given by

$$\sigma = \frac{ne^2\tau}{m}.$$

2-7 Two-dimensional electron gas. Consider the free and independent electron gas in two dimensions.

(1) Find the relation between the Fermi wave vector $k_{\rm F}$ and the electron number density n, where n is the number of electrons per unit area.
(2) Compute the free electron density of states $g(\varepsilon)$ in two dimensions.
(3) Show that the chemical potential of the electron gas in two dimensions is given by

$$\mu(T) = k_{\rm B}T \ln\left[e^{\pi n\hbar^2/mk_{\rm B}T} - 1\right].$$

---

(1) Let $\mathscr{A}$ be the area of the two-dimensional electronic gas and let $k_{\rm F}$ be the Fermi wave vector. From the fact that each quantum state occupies a phase space volume h^2 for a two-dimensional electron gas, we have

$$2 \cdot \frac{\mathscr{A} \cdot \pi(\hbar k_{\rm F})^2}{h^2} = N,$$

where the factor of 2 comes from the spin degeneracy. Thus, the Fermi wave vector is given by

$$k_{\rm F} = (2\pi N/\mathscr{A})^{1/2} = (2\pi n)^{1/2},$$

where $n = N/\mathscr{A}$ is the number of electrons per unit area.

(2) [**Method I**] The number of states per unit area in the momentum interval from $\boldsymbol{p}$ to $\boldsymbol{p} + d\boldsymbol{p}$ is given by $2d\boldsymbol{p}/h^2$ with $\boldsymbol{p}$ a two-dimensional vector. The factor of 2 in the above expression comes from the spin degeneracy. Integrating over the angle, we find that the number of states for the magnitude of momentum in the interval from p to $p + \mathrm{d}p$ is given by $4\pi p\mathrm{d}p/h^2$. Let $g(\varepsilon)$ denote the density of states. We must have $g(\varepsilon)\mathrm{d}\varepsilon = 4\pi p\mathrm{d}p/h^2$. Thus,

$$g(\varepsilon) = \frac{4\pi p}{h^2}\frac{1}{\mathrm{d}\varepsilon/\mathrm{d}p} = \frac{4\pi p}{h^2}\frac{m}{p} = \frac{m}{\pi\hbar^2},\ \varepsilon \geqslant 0.$$

[**Method II**] The density of states per unit area can also be computed as follows

$$\begin{aligned}
g(\varepsilon) &= \frac{2}{\mathscr{A}}\sum_{\boldsymbol{k}}\delta(\varepsilon - \hbar^2k^2/2m) = \frac{1}{2\pi^2}\int \mathrm{d}^2k\ \delta(\varepsilon - \hbar^2k^2/2m) \\
&= \frac{1}{2\pi^2}\int_0^\infty \mathrm{d}k\ k\delta(\varepsilon - \hbar^2k^2/2m)\int_0^{2\pi}\mathrm{d}\varphi \\
&= \frac{1}{\pi}\int_0^\infty \mathrm{d}k\ k\delta(\varepsilon - \hbar^2k^2/2m) \\
&= \frac{2m}{\pi\hbar^2}\int_0^\infty \mathrm{d}k\ k\delta(k^2 - 2m\varepsilon/\hbar^2) = \frac{m}{\pi\hbar^2}\int_0^\infty \mathrm{d}k^2\ \delta(k^2 - 2m\varepsilon/\hbar^2) \\
&= \begin{cases} m/\pi\hbar^2, & \varepsilon \geqslant 0 \\ 0, & \varepsilon < 0 \end{cases} = \frac{m}{\pi\hbar^2}\theta(\varepsilon).
\end{aligned}$$

(3) From

$$n = \int_{-\infty}^\infty \mathrm{d}\varepsilon\ g(\varepsilon)n_{\mathrm{F}}(\varepsilon),$$

we have

$$n = \frac{m}{\pi\hbar^2}\int_0^\infty \mathrm{d}\varepsilon\ \frac{1}{\mathrm{e}^{(\varepsilon-\mu)/k_{\mathrm{B}}T} + 1}.$$

Making a change of integration variables from ε to $x = \mathrm{e}^{(\varepsilon-\mu)/k_{\mathrm{B}}T}$, we have $\mathrm{d}\varepsilon = k_{\mathrm{B}}T\mathrm{d}x/x$. The integral is then cast into a more easily manageable one

$$n = \frac{mk_{\mathrm{B}}T}{\pi\hbar^2}\int_{\mathrm{e}^{-\mu/k_{\mathrm{B}}T}}^\infty \frac{\mathrm{d}x}{x(x+1)} = \frac{mk_{\mathrm{B}}T}{\pi\hbar^2}\ln\left(\mathrm{e}^{\mu/k_{\mathrm{B}}T} + 1\right).$$

Solving for μ from the above equation, we obtain

$$\mu = k_{\mathrm{B}}T\ln\left(\mathrm{e}^{\pi n\hbar^2/mk_{\mathrm{B}}T} - 1\right).$$

2-8 Thermodynamics of an electron gas. The Hamiltonian of a free and independent electron gas is given by

$$\hat{H}_0 = \sum_{\boldsymbol{k}\sigma}(\varepsilon_{\boldsymbol{k}} - \mu)\hat{c}^\dagger_{\boldsymbol{k}\sigma}\hat{c}_{\boldsymbol{k}\sigma},$$

where $\varepsilon_{\boldsymbol{k}} = \hbar^2 k^2/2m$ and $\hat{c}^\dagger_{\boldsymbol{k}\sigma}$ and $\hat{c}_{\boldsymbol{k}\sigma}$ are the creation and annihilation operators of electrons on the single-electron state $|\boldsymbol{k}\sigma\rangle$. The eigenvalues of $\hat{H}_0$ are given by

$$E_{\{n_{\boldsymbol{k}\sigma}\}} = \sum_{\boldsymbol{k}\sigma} n_{\boldsymbol{k}\sigma}(\varepsilon_{\boldsymbol{k}} - \mu)$$

with $n_{\boldsymbol{k}\sigma}$ the occupation number on the single-electron state $|\boldsymbol{k}\sigma\rangle$, $n_{\boldsymbol{k}\sigma} = 0, 1$.

(1) Evaluate the grand partition function for the electron gas, $\mathscr{Z} = \mathrm{Tr}\, \mathrm{e}^{-\hat{H}_0/k_{\mathrm{B}}T}$.
(2) Write down the grand potential for the electron gas.
(3) Compute the entropy of the electron gas and express it in terms of the Fermi-Dirac distribution function $n_{\mathrm{F}}(\varepsilon_{\boldsymbol{k}})$.
(4) Write down a general expression for the electronic specific heat and compare the obtained result with that derived from the internal energy.
(5) Neglecting the temperature dependence of the chemical potential and making use of the Sommerfeld expansion, compute the electronic specific heat at low temperatures.

---

(1) The grand partition function for the electron gas is given by

$$\begin{aligned}\mathscr{Z} &= \mathrm{Tr}\, \mathrm{e}^{-\hat{H}_0/k_{\mathrm{B}}T} = \sum_{\{n_{\boldsymbol{k}\sigma}\}} \mathrm{e}^{-E_{\{n_{\boldsymbol{k}\sigma}\}}/k_{\mathrm{B}}T)} \\ &= \sum_{\{n_{\boldsymbol{k}\sigma}\}} \mathrm{e}^{-(k_{\mathrm{B}}T)^{-1}\sum_{\boldsymbol{k}\sigma} n_{\boldsymbol{k}\sigma}(\varepsilon_{\boldsymbol{k}}-\mu)} \\ &= \prod_{\boldsymbol{k}\sigma} \sum_{n_{\boldsymbol{k}\sigma}=0,1} \mathrm{e}^{-n_{\boldsymbol{k}\sigma}(\varepsilon_{\boldsymbol{k}}-\mu)/k_{\mathrm{B}}T} = \prod_{\boldsymbol{k}\sigma}\left[1 + \mathrm{e}^{-(\varepsilon_{\boldsymbol{k}}-\mu)/k_{\mathrm{B}}T}\right].\end{aligned}$$

(2) The grand potential for the electron gas follows from the grand partition function

$$\begin{aligned}\Phi &= -k_{\mathrm{B}}T \ln \mathscr{Z} = -k_{\mathrm{B}}T \sum_{\boldsymbol{k}\sigma} \ln\left[1 + \mathrm{e}^{-(\varepsilon_{\boldsymbol{k}}-\mu)/k_{\mathrm{B}}T}\right] \\ &= -2k_{\mathrm{B}}T \sum_{\boldsymbol{k}} \ln\left[1 + \mathrm{e}^{-(\varepsilon_{\boldsymbol{k}}-\mu)/k_{\mathrm{B}}T}\right].\end{aligned}$$

(3) The entropy of the electron gas is given by

$$\begin{aligned} S &= -\frac{\partial \Phi}{\partial T} \\ &= 2k_{\rm B}\sum_{\boldsymbol{k}} \ln\left[1 + {\rm e}^{-(\varepsilon_{\boldsymbol{k}}-\mu)/k_{\rm B}T}\right] + \frac{2}{T}\sum_{\boldsymbol{k}} \frac{\varepsilon_{\boldsymbol{k}} - \mu}{{\rm e}^{(\varepsilon_{\boldsymbol{k}}-\mu)/k_{\rm B}T} + 1} \\ &= -2k_{\rm B}\sum_{\boldsymbol{k}} \Big\{ n_{\rm F}(\varepsilon_{\boldsymbol{k}}) \ln n_{\rm F}(\varepsilon_{\boldsymbol{k}}) + \left[1 - n_{\rm F}(\varepsilon_{\boldsymbol{k}})\right] \ln\left[1 - n_{\rm F}(\varepsilon_{\boldsymbol{k}})\right] \Big\}. \end{aligned}$$

(4) The electronic specific heat can be obtained from $C_v = T\partial S/\partial T$. We have

$$\begin{aligned} C_v = T\frac{\partial S}{\partial T} &= 2k_{\rm B}T\sum_{\boldsymbol{k}} \ln\frac{1 - n_{\rm F}(\varepsilon_{\boldsymbol{k}})}{n_{\rm F}(\varepsilon_{\boldsymbol{k}})} \frac{\partial n_{\rm F}(\varepsilon_{\boldsymbol{k}})}{\partial T} \\ &= \frac{2}{k_{\rm B}T^2}\sum_{\boldsymbol{k}} \frac{(\varepsilon_{\boldsymbol{k}} - \mu){\rm e}^{(\varepsilon_{\boldsymbol{k}}-\mu)/k_{\rm B}T}}{\left[{\rm e}^{(\varepsilon_{\boldsymbol{k}}-\mu)/k_{\rm B}T} + 1\right]^2} \left[(\varepsilon_{\boldsymbol{k}} - \mu) + T\frac{\partial \mu}{\partial T}\right] \\ &= \frac{2}{T}\sum_{\boldsymbol{k}} \left[-\frac{\partial n_{\rm F}(\varepsilon_{\boldsymbol{k}})}{\partial \varepsilon_{\boldsymbol{k}}}\right] (\varepsilon_{\boldsymbol{k}} - \mu) \left[(\varepsilon_{\boldsymbol{k}} - \mu) + T\frac{\partial \mu}{\partial T}\right]. \end{aligned}$$

The difference between the above result and that derived from the internal energy is in the factor $(\varepsilon_{\boldsymbol{k}} - \mu)$ between the two pairs of square brackets. This factor is simply $\varepsilon_{\boldsymbol{k}}$ in the result derived from the internal energy. This can be reconciled by noting that

$$\sum_{\boldsymbol{k}} \left[-\frac{\partial n_{\rm F}(\varepsilon_{\boldsymbol{k}})}{\partial \varepsilon_{\boldsymbol{k}}}\right] \left[(\varepsilon_{\boldsymbol{k}} - \mu) + T\frac{\partial \mu}{\partial T}\right] = 0$$

which can be obtained by differentiating both sides of $N = 2\sum_{\boldsymbol{k}} n_{\rm F}(\varepsilon_{\boldsymbol{k}})$ with respect to T with N fixed.

(5) Neglecting the temperature dependence of the chemical potential and converting the $\boldsymbol{k}$-summation into an integration over the electron energy ε, we have

$$\begin{aligned} C_v &= \frac{\mathscr{V}}{T}\int_{-\infty}^{\infty} {\rm d}\varepsilon\; g(\varepsilon)(\varepsilon - \mu)^2 \left[-\frac{\partial n_{\rm F}(\varepsilon)}{\partial \varepsilon}\right] \\ &= \frac{3n\mathscr{V}}{2\varepsilon_{\rm F}^{3/2}T}\int_0^{\infty} {\rm d}\varepsilon\; \varepsilon^{1/2}(\varepsilon - \mu)^2 \left[-\frac{\partial n_{\rm F}(\varepsilon)}{\partial \varepsilon}\right]. \end{aligned}$$

Performing an integration by parts yields

$$C_v = \frac{3n\mathscr{V}}{2\varepsilon_{\rm F}^{3/2}T}\int_0^{\infty} {\rm d}\varepsilon\; n_{\rm F}(\varepsilon)\left[\frac{1}{2}\varepsilon^{-1/2}(\varepsilon - \mu)^2 + 2\varepsilon^{1/2}(\varepsilon - \mu)\right].$$

Making use of the Sommerfeld expansion, we have

$$\begin{aligned} C_v &= \frac{3n\mathscr{V}}{2\varepsilon_{\mathrm{F}}^{3/2}T}\left\{\int_0^{\mu} \mathrm{d}\varepsilon \left[\frac{1}{2}\varepsilon^{-1/2}(\varepsilon-\mu)^2 + 2\varepsilon^{1/2}(\varepsilon-\mu)\right]\right. \\ &\qquad \left. + \frac{\pi^2}{3}(k_{\mathrm{B}}T)^2\mu^{1/2}\right\} \\ &= \frac{\pi^2 n\mathscr{V}}{2\varepsilon_{\mathrm{F}}^{3/2}} k_{\mathrm{B}}^2 T\mu^{1/2} \approx \frac{\pi^2}{2}\frac{k_{\mathrm{B}}T}{\varepsilon_{\mathrm{F}}} n\mathscr{V} k_{\mathrm{B}} \end{aligned}$$

in which the value of the integral on the first line is zero. Note that the electronic specific heat per unit volume is given by

$$c_v \equiv \frac{C_v}{\mathscr{V}} = \frac{\pi^2}{2}\frac{k_{\mathrm{B}}T}{\varepsilon_{\mathrm{F}}} nk_{\mathrm{B}}.$$

Chapter 3

Bravais Lattice

(1) *Bravais lattice*

A Bravais lattice is an infinite array of points that are all equivalent in their relation with other points in both arrangement and orientation.

(2) *Primitive unit cell*

A primitive unit cell is the region of the parallelepiped (the parallelogram in two dimensions) formed by a set of primitive vectors of a Bravais lattice.

The primitive unit cell of a Bravais lattice is nonunique.

All the primitive unit cells of a Bravais lattice have the same volume (the same area in two dimensions).

(3) *Wigner-Seitz cell*

The Wigner-Seitz cell is the region in space which is closer to the reference lattice point than to any other lattice points.

The Wigner-Seitz cell of a Bravais lattice possesses all the symmetries of the Bravais lattice.

The Wigner-Seitz cell is a primitive cell.

Construction of the Wigner-Seitz cell:

i. Draw lines to other lattice points from the reference lattice point.

ii. Bisect each line with a plane.

iii. Take the smallest polyhedron that is bounded by the bisecting planes.

(4) *Two-dimensional Bravais lattices*

Oblique, rectangular, centered rectangular, triangular, square.

(5) *Seven crystal systems*

Triclinic,

monoclinic,

orthorhombic,
tetragonal,
trigonal,
hexagonal,
cubic.

(6) *Fourteen Bravais lattices*

Simple triclinic,
simple monoclinic,
centered monoclinic,
simple orthorhombic,
base-centered orthorhombic,
body-centered orthorhombic,
face-centered orthorhombic,
simple tetragonal,
body-centered tetragonal,
simple trigonal,
simple hexagonal,
simple cubic,
body-centered cubic,
face-centered cubic.

(7) *Mathematical description*

The density of lattice points in a Bravais lattice, $\varrho(\boldsymbol{r}) = \sum_{\{n_i\}} \delta\big(\boldsymbol{r} - \sum_i n_i \boldsymbol{a}_i\big)$, is used to describe mathematically the Bravais lattice.

(8) *Lattice planes*

Three noncollinear lattice points of a Bravais lattice determine a lattice plane in the Bravais lattice. Each lattice plane forms a two-dimensional Bravais lattice.

(9) *Miller indices*

Miller indices are used to specify lattice planes and directions of lattice lines.

Miller indices of a lattice plane can be obtained by taking the reciprocals of the intercepts of the lattice plane with axes and removing the common factors.

3-1 Primitive cells of five two-dimensional Bravais lattices. Construct at least three primitive cells for each of the five two-dimensional Bravais lattices.

Primitive cells of the five two-dimensional Bravais lattices are given in Fig. 3.1. For the oblique and rectangular Bravais lattices, three different primitive cells are given for each, while four different primitive cells are given for the centered rectangular, triangular, and square Bravais lattices.

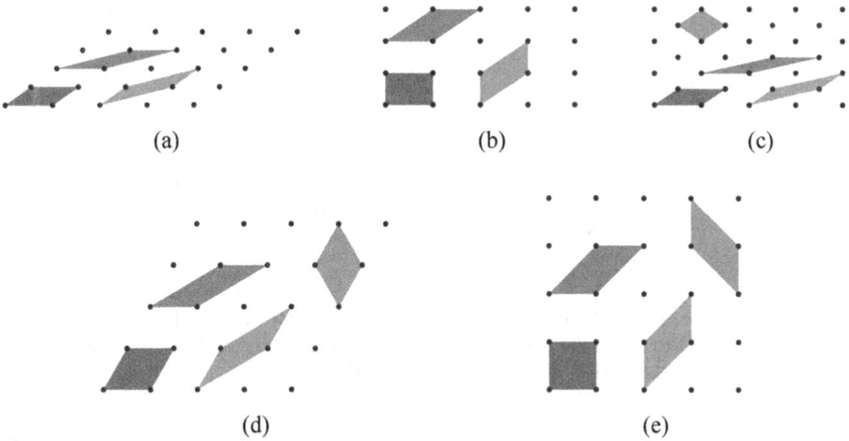

Fig. 3.1 Primitive cells of the five two-dimensional Bravais lattices. (a) Oblique Bravais lattice. (b) Rectangular Bravais lattice. (c) Centered rectangular Bravais lattice. (d) Triangular Bravais lattice. (e) Square Bravais lattice.

3-2 Wigner-Seitz cells of five two-dimensional Bravais lattices. Construct the Wigner-Seitz cells for all the five two-dimensional Bravais lattices.

Wigner-Seitz cells of the five two-dimensional Bravais lattices are shown in Fig. 3.2.

For each Bravais lattice, dotted lines connect the reference lattice point to the nearby lattice points. Solid lines bisect the connecting lines and they also become parts of the boundaries of Wigner-Seitz cells. All the Wigner-Seitz cells are shaded.

3-3 Triangular and centered rectangular Bravais lattices. The triangular Bravais lattice may be viewed as a special case of the centered rectangular Bravais lattice. Find the aspect ratio b/a for which the centered rectangular Bravais lattice becomes the triangular Bravais lattice.

For a centered rectangular Bravais lattice to become a triangular Bravais lattice, the angle α in the centered rectangular Bravais lattice

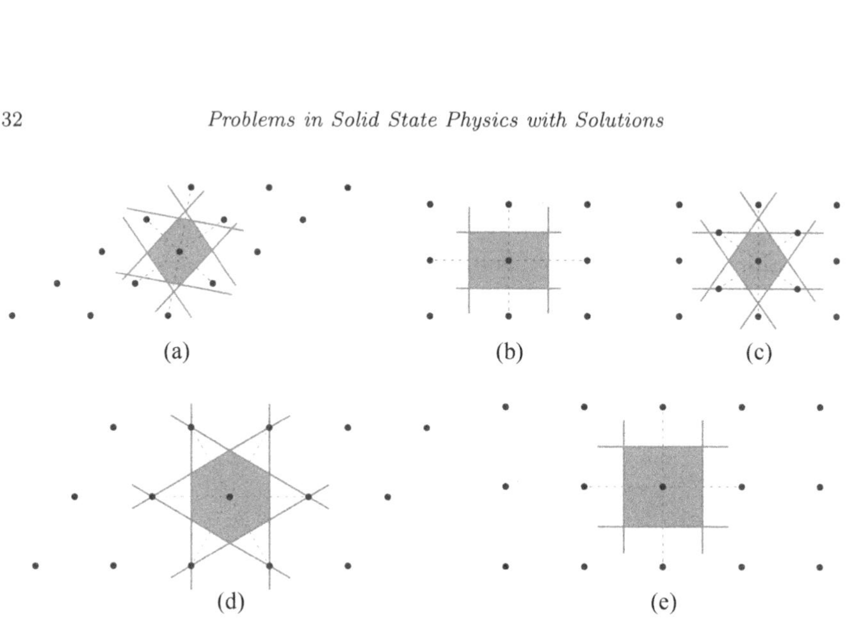

Fig. 3.2 Wigner-Seitz cells of the five two-dimensional Bravais lattices. (a) Oblique Bravais lattice. (b) Rectangular Bravais lattice. (c) Centered rectangular Bravais lattice. (d) Triangular Bravais lattice. (e) Square Bravais lattice.

must be exactly equal to $\pi/3$, which leads to $b = \sqrt{3}\,a$. Therefore, if the aspect ratio b/a is equal to $\sqrt{3}$, the centered rectangular Bravais lattice becomes the triangular Bravais lattice.

3-4 Packing fractions in two dimensions. Imagine that a two-dimensional crystal is constructed by packing circles. The ratio between the area occupied by the circles and the total area is referred to as the packing fraction (or ratio) in two dimensions. Find the packing fractions for the two-dimensional triangular and square Bravais lattices.

The packing is shown for one primitive cell for each of the two lattices in Fig. 3.3. The radius of circles for both the triangular and square Bravais lattices is equal to half of the corresponding lattice constant a. From Fig. 3.3, we see that the total area occupied by circles within a primitive cell is equal to the area of one full circle, $\pi a^2/4$, for both lattices. Making use of the fact that the area of a primitive cell of a triangular lattice is equal to $\sqrt{3}\,a^2/2$, we obtain

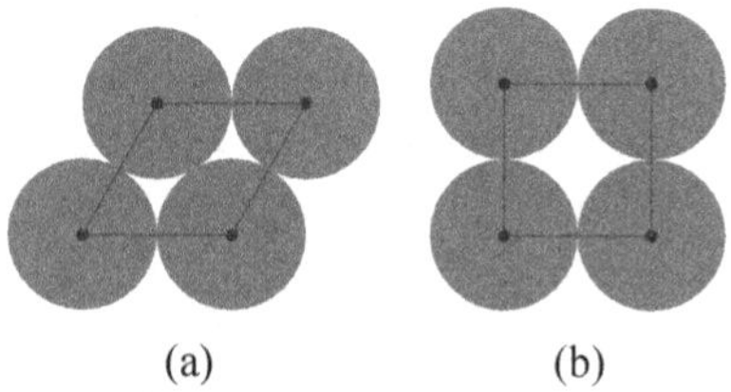

Fig. 3.3 Packing of circles in the two-dimensional triangular (a) and square (b) Bravais lattices.

the following packing factor for the triangular lattice

$$f_{\rm tri} = (\pi a^2/4)/(\sqrt{3}\,a^2/2) = \pi/2\sqrt{3} \approx 0.907.$$

Making use of the fact that the area of a primitive cell of a square lattice is equal to a^2, we obtain the following packing factor for the square lattice

$$f_{\rm sq} = (\pi a^2/4)/a^2 = \pi/4 \approx 0.785.$$

Thus, the triangular lattice is more closely packed than the square lattice.

3-5 Body-centered cubic crystal. A crystal has a basis of one atom per lattice point and a set of primitive vectors (in 10^{-10} m): $\boldsymbol{a} = 3\,\boldsymbol{e}_x$, $\boldsymbol{b} = 3\,\boldsymbol{e}_y$, and $\boldsymbol{c} = 3(\boldsymbol{e}_x + \boldsymbol{e}_y + \boldsymbol{e}_z)/2$, where $\boldsymbol{e}_x$, $\boldsymbol{e}_y$, and $\boldsymbol{e}_z$ are unit vectors in the x, y, and z directions of a Cartesian coordinate system.

(1) What is the Bravais lattice type of this crystal?
(2) Compute the volumes of the primitive and conventional unit cells.

---

(1) The given Bravais lattice is a body-centered cubic Bravais lattice. This can be seen clearly through constructing the symmetrical set of primitive vectors for the body-centered cubic Bravais lattice from the given primitive vectors. Let

$$\boldsymbol{a}_1 = \boldsymbol{c} - \boldsymbol{a} = 3(\boldsymbol{e}_y + \boldsymbol{e}_z - \boldsymbol{e}_x)/2,\ \boldsymbol{a}_2 = \boldsymbol{c}- \\ \boldsymbol{b} = 3(\boldsymbol{e}_z + \boldsymbol{e}_x - \boldsymbol{e}_y)/2, \boldsymbol{a}_3 = \boldsymbol{a} + \boldsymbol{b} - \boldsymbol{c} = 3(\boldsymbol{e}_x + \boldsymbol{e}_y - \boldsymbol{e}_z)/2.$$

If the original primitive vectors can be expressed as linear combinations of the new primitive vectors with integers as coefficients, then the new primitive vectors represent the same Bravais lattice as the original primitive vectors. Expressing the original primitive vectors in terms of the new primitive vectors, we have

$$\boldsymbol{a} = \boldsymbol{a}_2 + \boldsymbol{a}_3,\ \boldsymbol{b} = \boldsymbol{a}_3 + \boldsymbol{a}_1,\ \boldsymbol{c} = \boldsymbol{a}_1 + \boldsymbol{a}_2 + \boldsymbol{a}_3.$$

Thus, the constructed symmetrical set of primitive vectors represents faithfully the original Bravais lattice. Since the symmetrical set of primitive vectors is for a body-centered cubic Bravais lattice, the Bravais lattice type of the crystal is body-centered cubic. The conventional unit cell and two sets of primitive vectors are shown in Fig. 3.4.

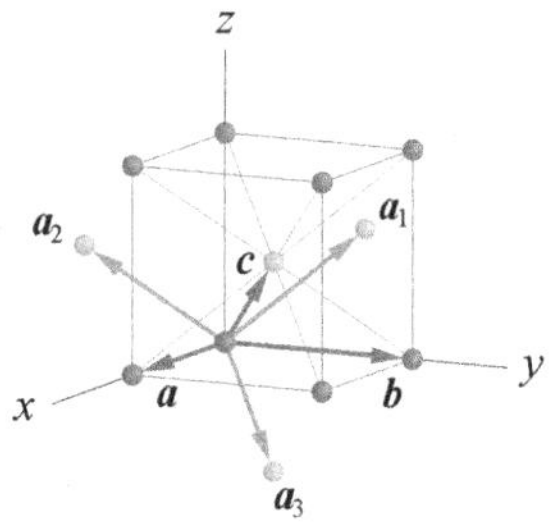

Fig. 3.4 Body-centered cubic Bravais lattice with two sets of primitive vectors.

(2) From the expressions of primitive vectors in the symmetrical set, we see that the lattice constant a is 3×10^{-10} m. Thus, the volume of the primitive unit cell is $a^3/2 = 1.35 \times 10^{-29}$ m^3 and the volume of the conventional unit cell is $a^3 = 2.7 \times 10^{-29}$ m^3.

3-6 Interstices in a face-centered cubic structure. One can place interstitial atoms between the normal crystal atoms in the close packed cubic crystal structure (the face-centered cubic structure). There are two types of interstitial sites in this structure.

(1) State where these two types of interstitial sites are.

(2) Compute the maximum radius that each foreign atom occupying these interstices can have.

---

(1) The FCC structure that is closely-packed with balls is shown in Fig. 3.5. From Fig. 3.5(a), we can clearly see interstices at the middles of edges of conventional cells. When the balls at $(a, a/a, a/2)$ and (a, a, a) are removed as in Fig. 3.5(b), we can clearly see interstices at the centers of conventional

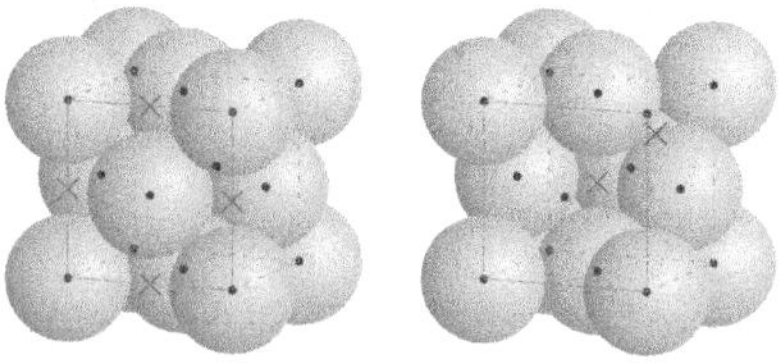

(a) (b)

Fig. 3.5 Face-centered cubic structure. Some of interstices are indicated by crosses (×). (a) All balls in a conventional unit cell. (b) Two balls at $(a, a/a, a/2)$ and (a, a, a) removed.

unit cells and at the centers of the regions each of which is surrounded by three face-center atoms and one corner atom.

If a foreign atom occupies an interstice at the middle of an edge of a conventional unit cell or an interstice at the center of a conventional unit cell as shown in Fig. 3.6(a), it is then surrounded by six host atoms which forman octahedron. Thus, an interstice of this type is referred to as *an octahedral interstice*. Note that the coordination of the interstice at the middle of an edge is identical with that of an interstice at the center of a conventional unit cell (the body center).

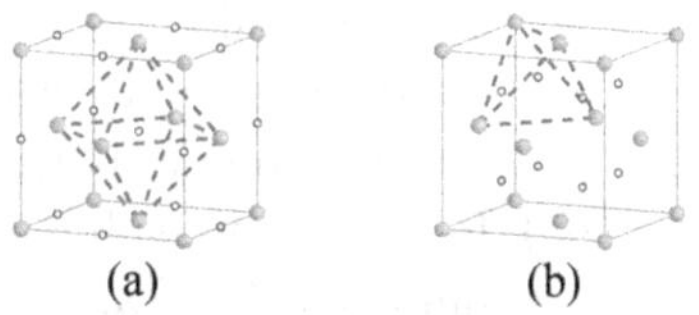

Fig. 3.6 Two types of interstices existing in an FCC structure. (a) Octahedral interstices. (b) Tetrahedral interstices.

If a foreign atom occupies an interstice at the center of the region surrounded by three face-center host atoms and one corner host atom in a conventional unit cell as shown in Fig. 3.6(b), it is then at the center of a tetrahedron formed by these four host atoms. Thus, an interstice of this type is referred to as *a tetrahedral interstice*.

(2) For a foreign atom occupying an octahedral interstice, the maximum radius it can have can be obtained by making use of the fact that the sum of its diameter ($2r$) and that of a host atom ($2R$) is equal to the lattice constant a, $2r + 2R = a$. With $R = a/2\sqrt{2}$ for an FCC structure, we have

$$r = (a - 2R)/2 = (\sqrt{2} - 1)a/2\sqrt{2} \approx 0.146a.$$

For a foreign atom occupying a tetrahedral interstice, the maximum radius it can have can be obtained by making use of the fact that it is located at a distance $\sqrt{3}\,a/4$ from the nearest corner host atom along a body diagonal so that the sum of its radius (r) and that of the host atom (R) is equal to $\sqrt{3}\,a/4$, $r + R = \sqrt{3}\,a/4$. We thus have

$$r = \sqrt{3}\,a/4 - R = \sqrt{3}\,a/4 - a/\sqrt{2} = (\sqrt{3} - \sqrt{2})/4 \approx 0.079a.$$

3-7 Crystal structures and densities.

(1) The crystal structure of iron is BCC at temperatures below 910°C and FCC above. Compute the ratio of the densities of the two structures under the assumption that the atoms can be considered to be close packed spheres the diameters of which are the same in both structures.

(2) In cubic zinc sulphide, each zinc atom is a distance of 0.234 nm from the nearest sulfur atom. Compute the density of zinc sulphide in this phase.

(3) The density of BCC iron is 7,900 kg/m^3. Determine the lattice constant of the cubic unit cell.

---

(1) For a given radius (R) of an iron atom, the lattice constant in the BCC structure is given by $a_{\rm BCC} = 4R/\sqrt{3}$ and the lattice constant in the FCC structure is given by $a_{\rm FCC} = 2\sqrt{2}R$. The densities in the two structures are then given by

$$\rho_{\rm BCC} = 2m_{\rm Fe}/a_{\rm BCC}^3 = 3\sqrt{3}\, m_{\rm Fe}/32R^3,$$
$$\rho_{\rm FCC} = 4m_{\rm Fe}/a_{\rm FCC}^3 = m_{\rm Fe}/4\sqrt{2}\, R^3,$$

where $m_{\rm Fe}$ is the mass of an iron atom. The ratio of the densities of the two structures is thus given by

$$\rho_{\rm FCC}/\rho_{\rm BCC} = 8/3\sqrt{6} \approx 1.089.$$

(2) The distance $d_{\rm ZnS}$ between the nearest neighboring Zn and S atoms in the cubic zinc sulphide is related to the lattice constant a through $d_{\rm ZnS} = \sqrt{3}\, a/4$. From the given value of $d_{\rm ZnS} = 0.234$ nm, we have $a = 4d_{\rm ZnS}/\sqrt{3} \approx 0.540\ 4$ nm. Thus, the density of the cubic zinc sulphide is given by

$$\rho_{\rm ZnS} = 4(m_{\rm Zn} + m_{\rm S})/a^3 \approx 4,100\ {\rm kg/m^3}.$$

(3) From $\rho_{\rm BCC} = 2m_{\rm Fe}/a_{\rm BCC}^3$, we have

$$a_{\rm BCC} = \left(2m_{\rm Fe}/\rho_{\rm Fe}\right)^{1/3} \approx 2.86 \times 10^{-10} = 0.286\ {\rm nm}.$$

3-8 Bond lengths and angles in BCC and FCC structures. The symmetric primitive vectors for the body-centered cubic structure are given by

$(-1/2, 1/2, 1/2)a$, $(1/2, -1/2, 1/2)a$, and $(1/2, 1/2, -1/2)a$;

for the face-centered cubic structure, they are given by

$(0, 1/2, 1/2)a$, $(1/2, 0, 1/2)a$, and $(1/2, 1/2, 0)a$.
Compute the lengths of and angles between the primitive vectors in terms of the lattice constant a for each case.

For the body-centered cubic structure, we set

$$\boldsymbol{a}_1^{\rm BCC} = (-1/2, 1/2, 1/2)a,\ \boldsymbol{a}_2^{\rm BCC} = (1/2, -1/2, 1/2)a,$$
$$\boldsymbol{a}_3^{\rm BCC} = (1/2, 1/2, -1/2)a.$$

We see that the lengths of the three primitive vectors are equal and are given by

$$|\boldsymbol{a}_1^{\rm BCC}| = |\boldsymbol{a}_2^{\rm BCC}| = |\boldsymbol{a}_3^{\rm BCC}| = \sqrt{3}\,a/2.$$

The angles between any two primitive vectors are also equal and are given by

$$\begin{aligned}\angle(\boldsymbol{a}_1^{\rm BCC}, \boldsymbol{a}_2^{\rm BCC}) &= \angle(\boldsymbol{a}_2^{\rm BCC}, \boldsymbol{a}_3^{\rm BCC}) = \angle(\boldsymbol{a}_3^{\rm BCC}, \boldsymbol{a}_1^{\rm BCC}) \\ &= \cos^{-1}\frac{\boldsymbol{a}_1^{\rm BCC}\cdot\boldsymbol{a}_2^{\rm BCC}}{|\boldsymbol{a}_1^{\rm BCC}|^2} \approx 109.47^{\rm o}.\end{aligned}$$

For the face-centered cubic structure, we set

$$\boldsymbol{a}_1^{\rm FCC} = (0, 1/2, 1/2)a,\ \boldsymbol{a}_2^{\rm FCC} = (1/2, 0, 1/2)a,\ \boldsymbol{a}_3^{\rm FCC} = (1/2, 1/2, 0)a.$$

We see that the lengths of the three primitive vectors are equal and are given by

$$|\boldsymbol{a}_1^{\rm FCC}| = |\boldsymbol{a}_2^{\rm FCC}| = |\boldsymbol{a}_3^{\rm FCC}| = a/\sqrt{2}.$$

The angles between any two primitive vectors are also equal and are given by

$$\begin{aligned}\angle(\boldsymbol{a}_1^{\rm FCC}, \boldsymbol{a}_2^{\rm FCC}) &= \angle(\boldsymbol{a}_2^{\rm FCC}, \boldsymbol{a}_3^{\rm FCC}) = \angle(\boldsymbol{a}_3^{\rm FCC}, \boldsymbol{a}_1^{\rm FCC}) \\ &= \cos^{-1}\frac{\boldsymbol{a}_1^{\rm FCC}\cdot\boldsymbol{a}_2^{\rm FCC}}{|\boldsymbol{a}_1^{\rm FCC}|^2} = 60^{\rm o}.\end{aligned}$$

3-9 Neighbors in cubic crystals. Find the number and the relative positions of all the nearest, second-nearest, and third-nearest neighbors of a lattice point in the simple cubic, body-centered cubic, and face-centered cubic Bravais lattices.

---

i. For the simple cubic Bravais lattice, the number of the nearest neighbors is 6. For the lattice point at $(0, 0, 0)$, these nearest neighbors are located at $(\pm 1, 0, 0)a$, $(0, \pm 1, 0)a$, and $(0, 0, \pm 1)a$.

The number of the second nearest neighbors is 12. For the lattice point at $(0,0,0)$, these second nearest neighbors are located at $(0,\pm 1,\pm 1)a$, $(\pm 1,0,\pm 1)a$, and $(\pm 1,\pm 1,0)a$. The number of the third nearest neighbors is 8. For the lattice point at $(0,0,0)$, these third nearest neighbors are located at $(\pm 1,\pm 1,\pm 1)a$.

ii. For the body-centered cubic Bravais lattice, the number of the nearest neighbors is 8. For the lattice point at $(0,0,0)$, these nearest neighbors are located at $(\pm 1/2,\pm 1/2,\pm 1/2)a$. The number of the second nearest neighbors is 6. For the lattice point at $(0,0,0)$, these second nearest neighbors are located at $(\pm 1,0,0)a$, $(0,\pm 1,0)a$, and $(0,0,\pm 1)a$. The number of the third nearest neighbors is 12. For the lattice point at $(0,0,0)$, these third nearest neighbors are located at $(0,\pm 1,\pm 1)a$, $(\pm 1,0,\pm 1)a$, and $(\pm 1,\pm 1,0)a$.

iii. For the face-centered cubic Bravais lattice, the number of the nearest neighbors is 12. For the lattice point at $(0,0,0)$, these nearest neighbors are located at $(0,\pm 1/2,\pm 1/2)a$, $(\pm 1/2,0,\pm 1/2)a$, and $(\pm 1/2,\pm 1/2,0)a$. The number of the second nearest neighbors is 6. For the lattice point at $(0,0,0)$, these second nearest neighbors are located at $(\pm 1,0,0)a$, $(0,\pm 1,0)a$, and $(0,0,\pm 1)a$. The number of the third nearest neighbors is 12. For the lattice point at $(0,0,0)$, these third nearest neighbors are located at $(0,\pm 1,\pm 1)a$, $(\pm 1,0,\pm 1)a$, and $(\pm 1,\pm 1,0)a$.

3-10 Volumes of primitive cells. Compute the volumes of the primitive cells of the simple triclinic and monoclinic Bravais lattices.

A set of primitive vectors for a simple triclinic Bravais lattice are given by

$$\boldsymbol{a}_1 = a\,\boldsymbol{e}_x,\ \boldsymbol{a}_2 = b(\cos\gamma\,\boldsymbol{e}_x + \sin\gamma\,\boldsymbol{e}_y),$$

$$\boldsymbol{a}_3 = c\cos\beta\,\boldsymbol{e}_x + c\big[(\cos\alpha - \cos\beta\cos\gamma)/\sin\gamma\big]\,\boldsymbol{e}_y$$
$$+c\big[\big(1-\cos^2\alpha-\cos^2\beta-\cos^2\gamma+2\cos\alpha\cos\beta\cos\gamma\big)^{1/2}/\sin\gamma\big]\,\boldsymbol{e}_z.$$

The volume of a primitive cell of a simple triclinic Bravais lattice is then given by

$$\begin{aligned} v_c^{\rm tri} &= abc\,\boldsymbol{e}_x\cdot\Big\{\big(1-\cos^2\alpha-\cos^2\beta-\cos^2\gamma+2\cos\alpha\cos\beta\cos\gamma\big)^{1/2}\boldsymbol{e}_x \\ &\quad -\big(1-\cos^2\alpha-\cos^2\beta-\cos^2\gamma+2\cos\alpha\cos\beta\cos\gamma\big)^{1/2}\cot\gamma\,\boldsymbol{e}_y \\ &\quad +\big[(\cos\alpha-\cos\beta\cos\gamma)\cot\gamma-\cos\beta\sin\gamma\big]\boldsymbol{e}_z\Big\} \\ &= \big(1-\cos^2\alpha-\cos^2\beta-\cos^2\gamma+2\cos\alpha\cos\beta\cos\gamma\big)^{1/2}abc. \end{aligned}$$

A set of primitive vectors for a simple monoclinic Bravais lattice are given by

$$\boldsymbol{a}_1 = a\,\boldsymbol{e}_x,\ \ \boldsymbol{a}_2 = b\,\boldsymbol{e}_y,\ \ \boldsymbol{a}_3 = c(\cos\beta\,\boldsymbol{e}_x + \sin\beta\,\boldsymbol{e}_z).$$

The volume of a primitive cell of a simple monoclinic Bravais lattice is then given by

$$v_c^{\text{mon}} = \boldsymbol{a}_1 \cdot \big(\boldsymbol{a}_2 \times \boldsymbol{a}_3\big) = abc\,\boldsymbol{e}_x \cdot \big(\sin\beta\,\boldsymbol{e}_x - \cos\beta\,\boldsymbol{e}_z\big) = abc\sin\beta.$$

3-11 Coulomb interaction energy in an SC structure. We put positive and negative charges, $+q$ and $-q$, alternatively on the sites of a simple cubic Bravais lattice and compute the static Coulomb interaction energy of the charge at the origin with all the other charges.

(1) Derive an analytic expression for the static Coulomb interaction energy. Put the result into the form $-\alpha q^2/4\pi\epsilon_0 a$, where a is the lattice constant and α a numerical constant. Write down an explicit expression for α.

(2) Numerically compute the value of α.

---

(1) Assume that a positive charge, $+q$, is at $(0,0,0)$. Let us consider the Coulomb interaction energy of the charge at $(0,0,0)$ with all the other charges. The charges on the nearest neighbors have the opposite sign, the charges on the second nearest neighbors have the same sign, and *etc.* Thus, the sign of charges changes alternatively among different kinds of neighbors. The position of a general lattice point in a simple cubic Bravais lattice can be expressed as $\boldsymbol{R}_{ijk} = ia\,\boldsymbol{e}_x + ja\,\boldsymbol{e}_y + ka\,\boldsymbol{e}_z$. The charge at $\boldsymbol{R}_{ijk}$ is $(-1)^{i+j+k}q$. Then, the Coulomb interaction energy of the charge at $(0,0,0)$ with all the other charges is given by

$$U_{\text{Coul}} = \frac{q^2}{4\pi\epsilon_0}\sum_{\boldsymbol{R}_{ijk}\neq 0}\frac{(-1)^{i+j+k}}{|\boldsymbol{R}_{ijk}|} = -\frac{\alpha q^2}{4\pi\epsilon_0 a}$$

with

$$\alpha = \sum_{ijk}{}' \frac{(-1)^{i+j+k-1}}{(i^2+j^2+k^2)^{1/2}},$$

where the prime on the summation sign indicates that the term with $i^2+j^2+k^2=0$ is to be excluded in the summation.

(2) Note that the series in the above-derived expression of α converges very slowly. For fast convergence, either the Ewald summation method or the screened Coulomb interaction can be utilized. Here we evaluate the summation directly for several different values of the largest distance to $(0,0,0)$, $\max|\boldsymbol{R}_{ijk}|$ (we will use simply $|\boldsymbol{R}_{ijk}|$ for $\max|\boldsymbol{R}_{ijk}|$ to shorten notations), and then extrapolate the value of the sum to $|\boldsymbol{R}_{ijk}| = \infty$. For convenience, we equivalently extrapolate the value of the sum from finite values of $1/|\boldsymbol{R}_{ijk}|$ to $1/|\boldsymbol{R}_{ijk}| = 0$. The computed values of α at 21 different values of $1/|\boldsymbol{R}_{ijk}|$ corresponding to the maximum value of $|i|$, $|j|$, $|k|$ from 100 to 200 with a spacing of 20 are plotted as a function of $1,000/|\boldsymbol{R}_{ijk}|$ in Fig. 3.7 (solid circles).

From Fig. 3.7, we see that the computed values fall neatly on a straight line. We thus attempt to obtain the value of α for $1/|\boldsymbol{R}_{ijk}| = 0$ by fitting the computed values for it to a straight line and taking the intercept of the fitted straight line with the vertical axis as its final value. For this linear fitting, we write

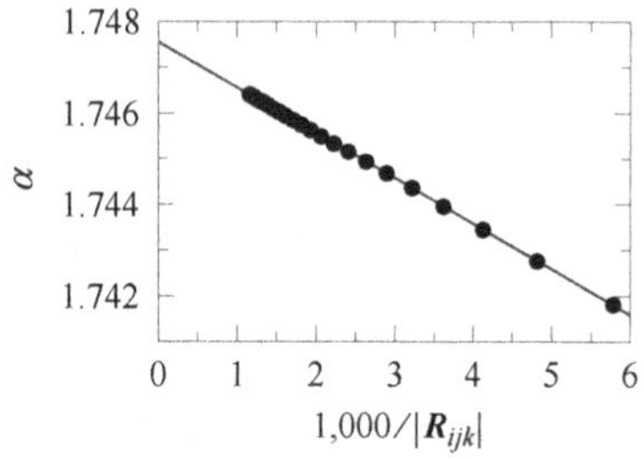

Fig. 3.7 Plot of the parameter α as a function of $1,000/|\boldsymbol{R}_{ijk}|$. The solid line is from the linear least-squares fitting.

$\alpha = ax + b$, where $x = 1,000/|\boldsymbol{R}_{ijk}|$ and a and b are fitting parameters. The quality of the fitting is here gauged by the sum of the squares of the deviations of the fitted values of α from its computed values, denoted by χ^2 and referred to as *the chi-squared function*,

$$\chi^2 = \sum_i (\alpha_i - ax_i - b_i)^2,$$

where α_i and $ax_i + b_i$ are respectively the computed and fitted values of α at x_i. To minimize χ^2 with respect to a and b, we differentiate χ^2 with respect to a and b, respectively, and set the results to zero. We obtain

$$a\sum_i x_i^2 + b\sum_i x_i = \sum_i x_i y_i,$$
$$a\sum_i x_i + nb = \sum_i y_i,$$

where n is the number of computed values of α. Here, $n = 21$. Solving for a and b from the above two equations yields

$$a = \frac{n\sum_i x_i y_i - \sum_i x_i \sum_i y_i}{n\sum_i x_i^2 - (\sum_i x_i)^2},$$
$$b = \frac{\sum_i x_i^2 \sum_i y_i - \sum_i x_i \sum_i x_i y_i}{n\sum_i x_i^2 - (\sum_i x_i)^2}.$$

Such a fitting procedure is known as *the linear least-squares fitting*. Evaluating a and b from the above two expressions using the computed values of α, we obtain

$$a \approx -0.000\,994, \; b \approx 1.747\,558.$$

The fitted straight line $\alpha = ax + b$ is plotted in Fig. 3.7 as a solid line. Extrapolating to $x = 1{,}000/|\boldsymbol{R}_{ijk}| = 0$, we obtain $\alpha = b \approx 1.747\,558$.

Chapter 4

Point Groups

(1) *Point symmetry operation*

A point symmetry operation of a Bravais lattice is a rigid operation that brings the Bravais lattice into itself and leaves at least one lattice point fixed.

(2) *Crystallographic restriction theorem*

The allowed rotation axes of a crystal are limited to 1-, 2-, 3-, 4-, and 6-fold axes.

(3) *Group*

If a set G with a multiplication defined satisfies the following four conditions, then G is a group.

i. [Closure] If a and b are in G, then ab is also in G.

ii. [Associativity] If a, b, and c are in G, then $(ab)c = a(bc)$.

iii. [Identity] There is an element E of G such that, for any element a of G, $aE = Ea = a$.

iv. [Presence of inverses] For any element a of G, there is an element a^{-1} such that $aa^{-1} = a^{-1}a = E$.

(4) *Point group*

A point group is a group whose elements are point symmetry operations that leave at least one lattice point fixed.

(5) *Crystallographic point groups*

C_i, C_1,
C_{2h}, C_2, C_s,
D_{2h}, D_2, C_{2v},
D_{4h}, D_{2d}, D_4, C_{4v}, C_{4h}, S_4, C_4,
D_{3d}, D_3, C_{3v}, S_6, C_3,
D_{6h}, D_{3h}, D_6, C_{6v}, C_{6h}, C_{3h}, C_6,
O_h, T_d, O, T_h, T.

4-1 Identification of groups. Determine which of the following sets are groups. For those sets that are not groups, give at least one axiom (one of the four conditions for a set to be a group) that is violated.

(1) The set of all real numbers under addition.

(2) The set of all real numbers, excluding zero, under division.

(3) The set of all $n \times n$ Hermitian matrices under matrix multiplication.

(4) The set of all $n \times n$ Hermitian matrices under matrix addition.

---

(1) The set is a group.

(2) The set is not a group. The associativity axiom is violated. Note that $(a/b)/c = a/(bc)$ while $a/(b/c) = ac/b$.

(3) The set is not a group. No inverse matrix exists for a singular matrix.

(4) The set is a group.

4-2 Group of Pauli matrices. We now create a group starting from the three Pauli matrices σ_1, σ_2, and σ_3 under the matrix multiplication.

(1) Find other elements in the group using the closure requirement.

(2) Find all the subgroups of the obtained group.

---

(1) Making use of the properties of Pauli matrices, we have

$$\mathrm{i}\sigma_1 = \sigma_2\sigma_3, \quad -\mathrm{i}\sigma_1 = \sigma_3\sigma_2, \quad \mathrm{i}\sigma_2 = \sigma_3\sigma_1, \quad -\mathrm{i}\sigma_2 = \sigma_1\sigma_3,$$
$$\mathrm{i}\sigma_3 = \sigma_1\sigma_2, \quad -\mathrm{i}\sigma_3 = \sigma_2\sigma_1, \quad -\sigma_1 = (\mathrm{i}\sigma_2)\sigma_3, \ -\sigma_2 = (\mathrm{i}\sigma_3)\sigma_1,$$
$$-\sigma_3 = (\mathrm{i}\sigma_1)\sigma_2, \ \mathrm{i} = (\mathrm{i}\sigma_j)\sigma_j = \sigma_j(\mathrm{i}\sigma_j), \ -\mathrm{i} = (-\mathrm{i}\sigma_j)\sigma_j = (\mathrm{i}\sigma_j)(-\sigma_j),$$
$$-1 = (\mathrm{i}\sigma_j)(\mathrm{i}\sigma_j) = (-\mathrm{i}\sigma_j)(-\mathrm{i}\sigma_j), \quad 1 = (\mathrm{i}\sigma_j)(-\mathrm{i}\sigma_j) = (-\mathrm{i}\sigma_j)(\mathrm{i}\sigma_j),$$

where $j = 1$, 2, or 3. Thus, the group is of order 16, with 1 as its identity element. Note that 1 is actually a 2×2 unit matrix.

(2) For the convenience of finding the subgroups, we set up the multiplication table of the group and obtain

	1	-1	i	$-\mathrm{i}$	σ_1	σ_2	σ_3	$-\sigma_1$	$-\sigma_2$	$-\sigma_3$	$\mathrm{i}\sigma_1$	$\mathrm{i}\sigma_2$	$\mathrm{i}\sigma_3$	$-\mathrm{i}\sigma_1$	$-\mathrm{i}\sigma_2$	$-\mathrm{i}\sigma_3$
1	1	-1	i	$-\mathrm{i}$	σ_1	σ_2	σ_3	$-\sigma_1$	$-\sigma_2$	$-\sigma_3$	$\mathrm{i}\sigma_1$	$\mathrm{i}\sigma_2$	$\mathrm{i}\sigma_3$	$-\mathrm{i}\sigma_1$	$-\mathrm{i}\sigma_2$	$-\mathrm{i}\sigma_3$
-1	-1	1	$-\mathrm{i}$	i	$-\sigma_1$	$-\sigma_2$	$-\sigma_3$	σ_1	σ_2	σ_3	$-\mathrm{i}\sigma_1$	$-\mathrm{i}\sigma_2$	$-\mathrm{i}\sigma_3$	$\mathrm{i}\sigma_1$	$\mathrm{i}\sigma_2$	$\mathrm{i}\sigma_3$
i	i	$-\mathrm{i}$	-1	1	$\mathrm{i}\sigma_1$	$\mathrm{i}\sigma_2$	$\mathrm{i}\sigma_3$	$-\mathrm{i}\sigma_1$	$-\mathrm{i}\sigma_2$	$-\mathrm{i}\sigma_3$	$-\sigma_1$	$-\sigma_2$	$-\sigma_3$	σ_1	σ_2	σ_3
$-\mathrm{i}$	$-\mathrm{i}$	i	1	-1	$-\mathrm{i}\sigma_1$	$-\mathrm{i}\sigma_2$	$-\mathrm{i}\sigma_3$	$\mathrm{i}\sigma_1$	$\mathrm{i}\sigma_2$	$\mathrm{i}\sigma_3$	σ_1	σ_2	σ_3	$-\sigma_1$	$-\sigma_2$	$-\sigma_3$
σ_1	σ_1	$-\sigma_1$	$\mathrm{i}\sigma_1$	$-\mathrm{i}\sigma_1$	1	$\mathrm{i}\sigma_3$	$-\mathrm{i}\sigma_2$	-1	$-\mathrm{i}\sigma_3$	$\mathrm{i}\sigma_2$	i	$-\sigma_3$	σ_2	$-\mathrm{i}$	σ_3	$-\sigma_2$
σ_2	σ_2	$-\sigma_2$	$\mathrm{i}\sigma_2$	$-\mathrm{i}\sigma_2$	$-\mathrm{i}\sigma_3$	1	$\mathrm{i}\sigma_1$	$\mathrm{i}\sigma_3$	-1	$-\mathrm{i}\sigma_1$	σ_3	i	$-\sigma_1$	$-\sigma_3$	$-\mathrm{i}$	σ_1
σ_3	σ_3	$-\sigma_3$	$\mathrm{i}\sigma_3$	$-\mathrm{i}\sigma_3$	$\mathrm{i}\sigma_2$	$-\mathrm{i}\sigma_1$	1	$-\mathrm{i}\sigma_2$	$\mathrm{i}\sigma_1$	-1	$-\sigma_2$	σ_1	i	σ_2	$-\sigma_1$	$-\mathrm{i}$
$-\sigma_1$	$-\sigma_1$	σ_1	$-\mathrm{i}\sigma_1$	$\mathrm{i}\sigma_1$	-1	$-\mathrm{i}\sigma_3$	$\mathrm{i}\sigma_2$	1	$\mathrm{i}\sigma_3$	$-\mathrm{i}\sigma_2$	$-\mathrm{i}$	σ_3	$-\sigma_2$	i	$-\sigma_3$	σ_2
$-\sigma_2$	$-\sigma_2$	σ_2	$-\mathrm{i}\sigma_2$	$\mathrm{i}\sigma_2$	$\mathrm{i}\sigma_3$	-1	$-\mathrm{i}\sigma_1$	$-\mathrm{i}\sigma_3$	1	$\mathrm{i}\sigma_1$	$-\sigma_3$	$-\mathrm{i}$	σ_1	σ_3	i	$-\sigma_1$
$-\sigma_3$	$-\sigma_3$	σ_3	$-\mathrm{i}\sigma_3$	$\mathrm{i}\sigma_3$	$-\mathrm{i}\sigma_2$	$\mathrm{i}\sigma_1$	-1	$\mathrm{i}\sigma_2$	$-\mathrm{i}\sigma_1$	1	σ_2	$-\sigma_1$	$-\mathrm{i}$	$-\sigma_2$	σ_1	i
$\mathrm{i}\sigma_1$	$\mathrm{i}\sigma_1$	$-\mathrm{i}\sigma_1$	$-\sigma_1$	σ_1	i	$-\sigma_3$	σ_2	$-\mathrm{i}$	σ_3	$-\sigma_2$	-1	$-\mathrm{i}\sigma_3$	$\mathrm{i}\sigma_2$	1	$\mathrm{i}\sigma_3$	$-\mathrm{i}\sigma_2$
$\mathrm{i}\sigma_2$	$\mathrm{i}\sigma_2$	$-\mathrm{i}\sigma_2$	$-\sigma_2$	σ_2	σ_3	i	$-\sigma_1$	$-\sigma_3$	$-\mathrm{i}$	$-\sigma_1$	$\mathrm{i}\sigma_3$	-1	$-\mathrm{i}\sigma_1$	$-\mathrm{i}\sigma_3$	1	$\mathrm{i}\sigma_1$
$\mathrm{i}\sigma_3$	$\mathrm{i}\sigma_3$	$-\mathrm{i}\sigma_3$	$-\sigma_3$	σ_3	$-\sigma_2$	σ_1	i	σ_2	$-\sigma_1$	$-\mathrm{i}$	$-\mathrm{i}\sigma_2$	$\mathrm{i}\sigma_1$	-1	$\mathrm{i}\sigma_2$	$-\mathrm{i}\sigma_1$	1
$-\mathrm{i}\sigma_1$	$-\mathrm{i}\sigma_1$	$\mathrm{i}\sigma_1$	σ_1	$-\sigma_1$	$-\mathrm{i}$	σ_3	$-\sigma_2$	i	$-\sigma_3$	σ_2	1	$\mathrm{i}\sigma_3$	$-\mathrm{i}\sigma_2$	-1	$-\mathrm{i}\sigma_3$	$\mathrm{i}\sigma_2$
$-\mathrm{i}\sigma_2$	$-\mathrm{i}\sigma_2$	$\mathrm{i}\sigma_2$	σ_2	$-\sigma_2$	$-\sigma_3$	$-\mathrm{i}$	σ_1	σ_3	i	σ_1	$-\mathrm{i}\sigma_3$	1	$\mathrm{i}\sigma_1$	$\mathrm{i}\sigma_3$	-1	$-\mathrm{i}\sigma_1$
$-\mathrm{i}\sigma_3$	$-\mathrm{i}\sigma_3$	$\mathrm{i}\sigma_3$	σ_3	$-\sigma_3$	σ_2	$-\sigma_1$	$-\mathrm{i}$	$-\sigma_2$	σ_1	i	$\mathrm{i}\sigma_2$	$-\mathrm{i}\sigma_1$	1	$-\mathrm{i}\sigma_2$	$\mathrm{i}\sigma_1$	-1

Using the multiplication table, we find the following subgroups of the group.

1. $\{1\}$,
2. $\{1, -1\}$,
3. $\{1, \sigma_1\}$,
4. $\{1, \sigma_2\}$,
5. $\{1, \sigma_3\}$,
6. $\{1, -\sigma_1\}$,
7. $\{1, -\sigma_2\}$,
8. $\{1, -\sigma_3\}$,
9. $\{1, -1, \mathrm{i}, -\mathrm{i}\}$,
10. $\{1, -1, \sigma_1, -\sigma_1\}$,
11. $\{1, -1, \sigma_2, -\sigma_2\}$,
12. $\{1, -1, \sigma_3, -\sigma_3\}$,
13. $\{1, -1, \mathrm{i}\sigma_1, -\mathrm{i}\sigma_1\}$,
14. $\{1, -1, \mathrm{i}\sigma_2, -\mathrm{i}\sigma_2\}$,
15. $\{1, -1, \mathrm{i}\sigma_3, -\mathrm{i}\sigma_3\}$,
16. $\{1, -1, \sigma_1, \sigma_2, -\sigma_1, -\sigma_2, \mathrm{i}\sigma_3, -\mathrm{i}\sigma_3\}$,
17. $\{1, -1, \sigma_2, \sigma_3, -\sigma_2, -\sigma_3, \mathrm{i}\sigma_1, -\mathrm{i}\sigma_1\}$,
18. $\{1, -1, \sigma_3, \sigma_1, -\sigma_3, -\sigma_1, \mathrm{i}\sigma_2, -\mathrm{i}\sigma_2\}$,
19. $\{1, -1, \mathrm{i}, -\mathrm{i}, \sigma_1, -\sigma_1, \mathrm{i}\sigma_1, -\mathrm{i}\sigma_1\}$,
20. $\{1, -1, \mathrm{i}, -\mathrm{i}, \sigma_2, -\sigma_2, \mathrm{i}\sigma_2, -\mathrm{i}\sigma_2\}$,
21. $\{1, -1, \mathrm{i}, -\mathrm{i}, \sigma_3, -\sigma_3, \mathrm{i}\sigma_3, -\mathrm{i}\sigma_3\}$,
22. $\{1, -1, \mathrm{i}, -\mathrm{i}, \sigma_1, -\sigma_1, \mathrm{i}\sigma_1, -\mathrm{i}\sigma_1, \sigma_2, -\sigma_2, \mathrm{i}\sigma_2, -\mathrm{i}\sigma_2, \sigma_3, -\sigma_3, \mathrm{i}\sigma_3, -\mathrm{i}\sigma_3\}$.

4-3 Statements about groups. Let E, a, b, and c be elements of group G with E the identity. Prove the following statements.

(1) If $ab = ac$, then $b = c$; if $ba = ca$, then $b = c$.
(2) $(a^{-1})^{-1} = a$.
(3) If $a^{-1} = b^{-1}$, then $a = b$.
(4) $(ab)^{-1} = b^{-1}a^{-1}$.
(5) G can not have more than one identity.
(6) The inverse of any element of G is unique.
(7) A subgroup H of G contains G's identity.
(8) For a subset H of G to be a subgroup of G, it is necessary and sufficient that H be non-empty and that $ab^{-1} \in H$ for all $a, b \in H$.

---

(1) Multiplying both sides of $ab = ac$ with a^{-1} from the left, we have $a^{-1}ab = a^{-1}ac$, from which it follows that $b = c$ upon making use of $a^{-1}a = E$. Similarly, multiplying both sides of $ba = ca$ with a^{-1} from the right, we have $baa^{-1} = caa^{-1}$, from which it also follows that $b = c$ upon making use of $aa^{-1} = E$.

(2) According to the definition of an inverse, we have $(a^{-1})^{-1}a^{-1} = E$. Multiplying both sides of this equation with a from the right, we have $(a^{-1})^{-1}a^{-1}a = a$. Making use of $a^{-1}a = E$ on the left hand side, we obtain $(a^{-1})^{-1} = a$.

(3) Multiplying both sides of $a^{-1} = b^{-1}$ with a from the left, we have $aa^{-1} = ab^{-1}$. Making use of $aa^{-1} = E$, we have $ab^{-1} = E$. Multiplying both sides of this equation with b from the right and then making use of $b^{-1}b = E$, we obtain $a = b$.

(4) According to the definition of an inverse, we have $(ab)(ab)^{-1} = E$. Multiplying both sides of this equation with a^{-1} from the left and then making use of $a^{-1}a = E$, we have $b(ab)^{-1} = a^{-1}$. Multiplying both sides of this resultant equation with b^{-1} from the left and then making use of $b^{-1}b = E$, we obtain $(ab)^{-1} = b^{-1}a^{-1}$.

(5) If G has another identity E', we then have $EE' = E'$ and $EE' = E$. These two equations imply that $E' = E$. Therefore, G can not have more than one identity.

(6) If the element g of G has two inverses h and h', we then have $h^{-1} = g$ and $h'^{-1} = g$, from which it follows that $h'^{-1} = h^{-1}$. We already proved in the above that, if $a^{-1} = b^{-1}$, then $a = b$. We then have $h' = h$. Therefore, the inverse of any element of G is unique.

(7) Let h be an element of the subgroup H. The inverse h^{-1} of h is in H according to the axiom for the presence of inverses. Invoking the axiom for the closureness, we see that hh^{-1} is in H. However, $hh^{-1} = E$. Thus, the subgroup H of G contains G's identity E.

(8) If H is empty, then it can not be a subgroup. Thus, that H is non-empty is necessary. Let H be a subgroup of G. If $b \in H$, then $b^{-1} \in H$. Let $a \in H$. We then have $ab^{-1} \in H$. Therefore, that H is non-empty and that $ab^{-1} \in H$ are necessary for H to be a subgroup. We now prove that they are also sufficient. For this purpose, we derive the four axioms from them and the fact that $H \subset G$.

(i) Assume that H is non-empty. Let $a \in H$. Setting $b = a$ in $ab^{-1} \in H$, we have $aa^{-1} \in H$. Since $aa^{-1} = E$, we thus have $E \in H$.

(ii) If H has more than one element, then there exits another element b in addition to E. Setting $a = E$ in $ab^{-1} \in H$, we have $b^{-1} \in H$. Thus, for any $b \in H$, we can obtain $b^{-1} \in H$ from $ab^{-1} \in H$.

(iii) We have proved that $b^{-1} \in H$ if $b \in H$. Let $c = b^{-1}$. We then have $ac \in H$ from $ab^{-1} \in H$. Thus, the closureness is guaranteed.

(iv) Let a, b, and c be in H. Since they are elements of G, the multiplication associativity is applicable to them, $(ab)c = a(bc)$. Thus, for a subset H of G to be a subgroup of G, it is necessary and sufficient that H be non-empty and that $ab^{-1} \in H$ for all $a, b \in H$.

4-4 Identification of point groups. Identify the point group that is obtained by combining the following symmetry elements: "a 2-fold rotation axis and an inversion center", "two mirror planes at right angles to each other", and "a 2-fold rotation axis and an intersecting mirror plane".

(i) Noting that $iC_2 = \sigma_h$ with σ_h a mirror plane perpendicular to the 2-fold rotation axis, we see that the point group has four elements E, C_2, i, and σ_h. Thus, the group is C_{2h} $(2/m)$

$$C_{2h} = \{E, C_2, i, \sigma_h\}.$$

(ii) Let σ_v and $\sigma_{v'}$ denote the two mirror planes perpendicular to each other. Noting that $\sigma_v\sigma_{v'}$ gives a result that can be obtained through a pure rotation through π about an axis passing through the intersection of the two mirror planes, we see that the group contains C_2. Thus, the group is C_{2v} $(mm2)$

$$C_{2v} = \{E, C_2, \sigma_v, \sigma_{v'}\} = \{E, C_2, 2\sigma_v\}.$$

(iii) Let σ_h denote the mirror plane intersecting the 2-fold rotation axis. Since $C_2\sigma_h = i$, the group is the same as in (i), *i.e.*, it is C_{2h} $(2/m)$

$$C_{2h} = \{E, C_2, i, \sigma_h\}.$$

4-5 Expressions of rotations and reflections. Let a rotation about an axis passing through the origin and perpendicular to the xOy plane through an angle of θ be represented by the matrix R_θ and a reflection in the line passing through the origin and making an angle of $\theta/2$ with the positive x axis be represented by the matrix S_θ.

(1) Show that R_θ and S_θ can be expressed as

$$R_\theta = \begin{pmatrix} \cos\theta & -\sin\theta \\ \sin\theta & \cos\theta \end{pmatrix}, \quad S_\theta = \begin{pmatrix} \cos\theta & \sin\theta \\ \sin\theta & -\cos\theta \end{pmatrix}.$$

(2) Compute the effect of rotating the vector $2\boldsymbol{e}_x + 3\boldsymbol{e}_y$ counterclockwise about the origin through an angle of $\pi/2$ radians.

(3) Compute the effect of reflecting the vector $\boldsymbol{e}_x + \boldsymbol{e}_y$ through the line $y = 2x$.

(4) Compute the effect of rotating the vector $\boldsymbol{e}_y$ counterclockwise about the origin through an angle of $\pi/3$ radians and then reflecting through the line $y = 2x$.

(5) Show that $S_\theta S_\psi$ is a rotation and find the angle of rotation.

(6) Show that $S_\theta R_\psi S_\theta = R_{-\psi}$.

(7) Let $T_{\boldsymbol{v}}$ be a translation through $\boldsymbol{v}$, $T_{\boldsymbol{v}}\boldsymbol{w} = \boldsymbol{w} + \boldsymbol{v}$. Show that $T_{R_\theta \boldsymbol{v}} R_\theta = R_\theta T_{\boldsymbol{v}}$.

---

(1) We first consider a counterclockwise rotation about the z axis through an angle of θ. Assume that the position vector $\boldsymbol{\rho}$ in the xOy plane makes an angle of φ with the x axis before the rotation. We then have $\boldsymbol{\rho} = x\,\boldsymbol{e}_x + y\,\boldsymbol{e}_y = \rho\cos\varphi\,\boldsymbol{e}_x + \rho\sin\varphi\,\boldsymbol{e}_y$. After the rotation, the position vector makes an angle of $\varphi + \theta$ with the x axis. Thus,

$$\begin{aligned} \boldsymbol{\rho}' &= \rho\cos(\varphi+\theta)\,\boldsymbol{e}_x + \rho\sin(\varphi+\theta)\,\boldsymbol{e}_y \\ &= \big(x\cos\theta - y\sin\theta\big)\boldsymbol{e}_x + \big(x\sin\theta + y\cos\theta\big)\boldsymbol{e}_y. \end{aligned}$$

Comparing the above result with $\boldsymbol{\rho}' = x'\,\boldsymbol{e}_x + y'\,\boldsymbol{e}_y$, we obtain

$$\begin{aligned} x' &= x\cos\theta - y\sin\theta, \\ y' &= x\sin\theta + y\cos\theta. \end{aligned}$$

Writing the above relations in matrix form, we have

$$\begin{pmatrix} x' \\ y' \end{pmatrix} = \begin{pmatrix} \cos\theta & -\sin\theta \\ \sin\theta & \cos\theta \end{pmatrix} \begin{pmatrix} x \\ y \end{pmatrix}$$

Comparing the above equation with $\begin{pmatrix} x' \\ y' \end{pmatrix} = R_\theta \begin{pmatrix} x \\ y \end{pmatrix}$, we obtain

$$R_\theta = \begin{pmatrix} \cos\theta & -\sin\theta \\ \sin\theta & \cos\theta \end{pmatrix}.$$

We now consider the reflection in the line passing through the origin and making an angle of $\theta/2$ with the positive x axis. The direction of the line is given by $\boldsymbol{n} = \cos(\theta/2)\boldsymbol{e}_x + \sin(\theta/2)\boldsymbol{e}_y$. Let us write the position vector as the sum of the component in the direction of $\boldsymbol{n}$, $\boldsymbol{r}_\parallel = (\boldsymbol{r} \cdot \boldsymbol{n})\boldsymbol{n}$, and the component in the direction perpendicular to $\boldsymbol{n}$, $\boldsymbol{r}_\perp = \boldsymbol{r} - \boldsymbol{r}_\parallel = \boldsymbol{r} - (\boldsymbol{r} \cdot \boldsymbol{n})\boldsymbol{n}$,

$$\boldsymbol{r} = \boldsymbol{r}_\parallel + \boldsymbol{r}_\perp.$$

Under the reflection, $\boldsymbol{r}_\perp$ changes sign while $\boldsymbol{r}_\parallel$ remains unchanged. We then have

$$\begin{aligned} \boldsymbol{r}' &= \boldsymbol{r}_\parallel - \boldsymbol{r}_\perp = 2(\boldsymbol{r} \cdot \boldsymbol{n})\boldsymbol{n} - \boldsymbol{r} \\ &= 2\big[\,x\cos(\theta/2) + y\sin(\theta/2)\,\big]\big[\cos(\theta/2)\boldsymbol{e}_x + \sin(\theta/2)\boldsymbol{e}_y\,\big] \\ &\quad - x\,\boldsymbol{e}_x - y\,\boldsymbol{e}_y \\ &= \Big\{x\big[\,2\cos^2(\theta/2) - 1\,\big] + 2y\sin(\theta/2)\cos(\theta/2)\Big\}\,\boldsymbol{e}_x \\ &\quad + \Big\{2x\sin(\theta/2)\cos(\theta/2) - y\big[\,1 - 2\sin^2(\theta/2)\,\big]\Big\}\,\boldsymbol{e}_y \\ &= \big(x\cos\theta + y\sin\theta\big)\boldsymbol{e}_x + \big(x\sin\theta - y\cos\theta\big)\boldsymbol{e}_y. \end{aligned}$$

Writing the above equation in matrix form, we have

$$\begin{pmatrix} x' \\ y' \end{pmatrix} = \begin{pmatrix} \cos\theta & \sin\theta \\ \sin\theta & -\cos\theta \end{pmatrix} \begin{pmatrix} x \\ y \end{pmatrix}$$

Comparing the above equation with $\begin{pmatrix} x' \\ y' \end{pmatrix} = S_\theta \begin{pmatrix} x \\ y \end{pmatrix}$, we obtain

$$S_\theta = \begin{pmatrix} \cos\theta & \sin\theta \\ \sin\theta & -\cos\theta \end{pmatrix}.$$

(2) For $\theta = \pi/2$, we have

$$R_{\pi/2} = \begin{pmatrix} 0 & -1 \\ 1 & 0 \end{pmatrix}.$$

The vector $2\boldsymbol{e}_x + 3\boldsymbol{e}_y$ is expressed in matrix form as

$$2\boldsymbol{e}_x + 3\boldsymbol{e}_y = \begin{pmatrix} 2 \\ 3 \end{pmatrix}.$$

We then have

$$R_{\pi/2}(2\boldsymbol{e}_x + 3\boldsymbol{e}_y) = \begin{pmatrix} 0 & -1 \\ 1 & 0 \end{pmatrix}\begin{pmatrix} 2 \\ 3 \end{pmatrix} = \begin{pmatrix} -3 \\ 2 \end{pmatrix}.$$

Thus, $R_{\pi/2}(2\boldsymbol{e}_x + 3\boldsymbol{e}_y) = -3\boldsymbol{e}_x + 2\boldsymbol{e}_y$.

(3) The slope of the line $y = 2x$ is $\tan^{-1} 2$, that is, $\tan(\theta/2) = 2$ with $\theta/2$ the angle the line makes with the positive x axis. We then have $\sin\theta = 4/5$, $\cos\theta = -3/5$, and

$$S_\theta = \frac{1}{5}\begin{pmatrix} -3 & 4 \\ 4 & 3 \end{pmatrix}.$$

The effect of S_θ on $\boldsymbol{e}_x + \boldsymbol{e}_y$ is then given by

$$S_\theta(\boldsymbol{e}_x + \boldsymbol{e}_y) = \frac{1}{5}\begin{pmatrix} -3 & 4 \\ 4 & 3 \end{pmatrix}\begin{pmatrix} 1 \\ 1 \end{pmatrix} = \frac{1}{5}\begin{pmatrix} 1 \\ 7 \end{pmatrix} = \frac{1}{5}(\boldsymbol{e}_x + 7\boldsymbol{e}_y).$$

(4) The effect of rotating the vector $\boldsymbol{e}_y$ counterclockwise about the origin through an angle of $\pi/3$ radians and then reflecting through the line $y = 2x$ is given by

$$\begin{aligned} S_\theta R_{\pi/3}\boldsymbol{e}_y &= \frac{1}{10}\begin{pmatrix} -3 & 4 \\ 4 & 3 \end{pmatrix}\begin{pmatrix} 1 & -\sqrt{3} \\ \sqrt{3} & 1 \end{pmatrix}\begin{pmatrix} 0 \\ 1 \end{pmatrix} = \frac{1}{10}\begin{pmatrix} -3 & 4 \\ 4 & 3 \end{pmatrix}\begin{pmatrix} -\sqrt{3} \\ 1 \end{pmatrix} \\ &= \frac{1}{10}\begin{pmatrix} 4+3\sqrt{3} \\ 3-4\sqrt{3} \end{pmatrix} = \frac{1}{10}\left[(4+3\sqrt{3})\boldsymbol{e}_x + (3-4\sqrt{3})\boldsymbol{e}_y\right]. \end{aligned}$$

(5) Evaluating the matrix product $S_\theta S_\psi$, we have

$$\begin{aligned} S_\theta S_\psi &= \begin{pmatrix} \cos\theta & \sin\theta \\ \sin\theta & -\cos\theta \end{pmatrix}\begin{pmatrix} \cos\psi & \sin\psi \\ \sin\psi & -\cos\psi \end{pmatrix} \\ &= \begin{pmatrix} \cos\theta\cos\psi + \sin\theta\sin\psi & \cos\theta\sin\psi - \sin\theta\cos\psi \\ \sin\theta\cos\psi - \cos\theta\sin\psi & \sin\theta\sin\psi + \cos\theta\cos\psi \end{pmatrix} \\ &= \begin{pmatrix} \cos(\theta-\psi) & -\sin(\theta-\psi) \\ \sin(\theta-\psi) & \cos(\theta-\psi) \end{pmatrix} = R_{\theta-\psi}. \end{aligned}$$

Thus, $S_\theta S_\psi$ is a rotation through an angle of $\theta - \psi$.

(6) Evaluating the matrix product $S_\theta R_\psi S_\theta$, we have

$$\begin{aligned} S_\theta R_\psi S_\theta &= \begin{pmatrix} \cos\theta & \sin\theta \\ \sin\theta & -\cos\theta \end{pmatrix}\begin{pmatrix} \cos\psi & -\sin\psi \\ \sin\psi & \cos\psi \end{pmatrix}\begin{pmatrix} \cos\theta & \sin\theta \\ \sin\theta & -\cos\theta \end{pmatrix} \\ &= \begin{pmatrix} \cos\theta & \sin\theta \\ \sin\theta & -\cos\theta \end{pmatrix}\begin{pmatrix} \cos(\psi+\theta) & \sin(\psi+\theta) \\ \sin(\psi+\theta) & -\cos(\psi+\theta) \end{pmatrix} \\ &= \begin{pmatrix} \cos\psi & \sin\psi \\ -\sin\psi & \cos\psi \end{pmatrix} = R_{-\psi}. \end{aligned}$$

(7) For $T_{R_\theta \boldsymbol{v}} R_\theta \boldsymbol{w}$ with $\boldsymbol{w}$ an arbitrary vector, we have

$$T_{R_\theta \boldsymbol{v}} R_\theta \boldsymbol{w} = R_\theta \boldsymbol{w} + R_\theta \boldsymbol{v} = R_\theta(\boldsymbol{w} + \boldsymbol{v}) = R_\theta T_{\boldsymbol{v}} \boldsymbol{w}.$$

Thus, $T_{R_\theta \boldsymbol{v}} R_\theta = R_\theta T_{\boldsymbol{v}}$.

4-6 Matrix representation of a point group. The elements of a group of order 6 are represented by the following matrices

$$E = \begin{pmatrix} 1 & 0 \\ 0 & 1 \end{pmatrix}, \qquad A = \begin{pmatrix} -1 & 0 \\ 0 & 1 \end{pmatrix}, \qquad B = \begin{pmatrix} 1/2 & 3/2 \\ 1/2 & -1/2 \end{pmatrix},$$

$$C = \begin{pmatrix} 1/2 & -3/2 \\ -1/2 & -1/2 \end{pmatrix}, D = \begin{pmatrix} -1/2 & -3/2 \\ 1/2 & -1/2 \end{pmatrix}, F = \begin{pmatrix} -1/2 & 3/2 \\ -1/2 & -1/2 \end{pmatrix}.$$

(1) Build the multiplication table for the group.
(2) Find the inverses of all the elements.
(3) Identify the point group(s) whose elements can be represented by the above matrices.

---

(1) Using the matrix multiplication, we can build the following multiplication table for the group

	E	A	B	C	D	F
E	E	A	B	C	D	F
A	A	E	D	F	B	C
B	B	F	E	D	C	A
C	C	D	F	E	A	B
D	D	C	A	B	F	E
F	F	B	C	A	E	D

(2) From the multiplication table, we can easily find the inverse of an element by looking through the row which the element belongs to and identifying the column at which the entry is the identity element E. Then the element which the column corresponds to is the inverse of the concerned element. The inverses we find this way are given by

$$E^{-1} = E,\ A^{-1} = A,\ B^{-1} = B,\ C^{-1} = C,\ D^{-1} = F,\ F^{-1} = D.$$

Thus, the inverses of E, A, B and C are themselves, respectively, and D and F are mutually inverses.

(3) Since there is no matrix that represents the spatial inversion i in the given matrices, the point groups whose elements can be represented by the above matrices must not contain i. Since the matrices do not commute except for E, the point groups that can be represented must be non-Abelian. In addition, the order of point groups must be exactly six. Thus, the crystallographic point groups whose elements can be represented by the above matrices are D_3 and C_{3v}.

4-7 Invariant subgroup. If a subgroup H of group G is such that $g^{-1}hg \in H$ for all $h \in H$ and all $g \in G$, then H is *an invariant subgroup* of G.

(1) Find all invariant subgroups of C_{2v}.
(2) Find all invariant subgroups of D_{2h}.

---

(1) The elements of C_{2v} are E, C_2, σ_v, and σ_d. The multiplication table of C_{2v} is given by

	E	C_2	σ_v	σ_d
E	E	C_2	σ_v	σ_d
C_2	C_2	E	σ_d	σ_v
σ_v	σ_v	σ_d	E	C_2
σ_d	σ_d	σ_v	C_2	E

The multiplication table can be obtained by taking as in Fig. 4.1, for example, the C_2 axis along the z axis, the σ_v mirror plane as the zOx plane, and the σ_d mirror plane as the yOz plane. The effects of the symmetry operations in group C_{2v} on a general point with coordinates (x, y, z) are then given by

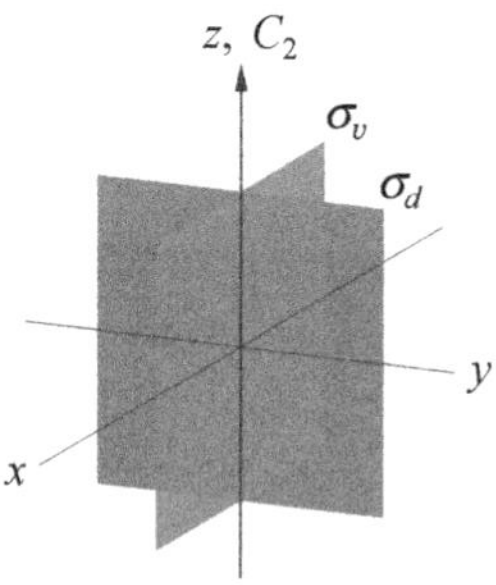

Fig. 4.1 Symmetry operations in point group C_{2v}.

$$E(x, y, z) = (x, y, z),$$
$$C_2(x, y, z) = (-x, -y, z),$$
$$\sigma_v(x, -y, z) = (x, y, z),$$
$$\sigma_d(x, y, z) = (-x, y, z).$$

The result of the product of any two symmetry operations can be obtained by using the above results for individual symmetry operations. For example,

$$C_2\sigma_v(x, y, z) = C_2(x, -y, z) = (-x, y, z) = \sigma_d(x, y, z).$$

Thus, $C_2\sigma_v = \sigma_d$.
From the multiplication table, we see that the inverses of all the elements of C_{2v} are themselves, respectively, since the product of any element with itself is equal to the identity element E. Making use of the multiplication table, we find the following subgroups of C_{2v} with the two trivial subgroups included: C_1, C_s, C_2, C_{2v}, where C_s can represent either $\{E, \sigma_v\}$ or $\{E, \sigma_d\}$ and C_2 represents $\{E, C_2\}$.
Since all the elements of C_{2v} commute with one another, all of its subgroups are invariant subgroups.

(2) Point group D_{2h} is given by $D_{2h} = \{E, C_2, C_2', C_2'', i, \sigma_h, \sigma_v, \sigma_v'\}$ in which σ_h is perpendicular to C_2 (the major rotation axis) and σ_v and σ_v' are parallel to C_2 with σ_v and σ_v' contain respectively C_2' and C_2'' which are perpendicular to each other and are both perpendicular to C_2. The multiplication table of D_{2h} is given by

	E	C_2	C_2'	C_2''	i	σ_h	σ_v	σ_v'
E	E	C_2	C_2'	C_2''	i	σ_h	σ_v	σ_v'
C_2	C_2	E	C_2''	C_2'	σ_h	i	σ_v'	σ_v
C_2'	C_2'	C_2''	E	C_2	σ_v'	σ_v	σ_h	i
C_2''	C_2''	C_2'	C_2	E	σ_v	σ_v'	i	σ_h
i	i	σ_h	σ_v'	σ_v	E	C_2	C_2''	C_2'
σ_h	σ_h	i	σ_v	σ_v'	C_2	E	C_2'	C_2''
σ_v	σ_v	σ_v'	σ_h	i	C_2''	C_2'	E	C_2
σ_v'	σ_v'	σ_v	i	σ_h	C_2'	C_2''	C_2	E

Similarly to what was done regarding point group C_{2v} in the above, the multiplication table for point group D_{2h} can be obtained by taking as in Fig. 4.2, for instance, the C_2 axis along the z axis, the C_2' axis along the x axis, and the C_2'' axis along the y axis and, correspondingly, the σ_h mirror plane as the xOy plane, the σ_v mirror plane as the zOx plane, and the σ_v' mirror plane

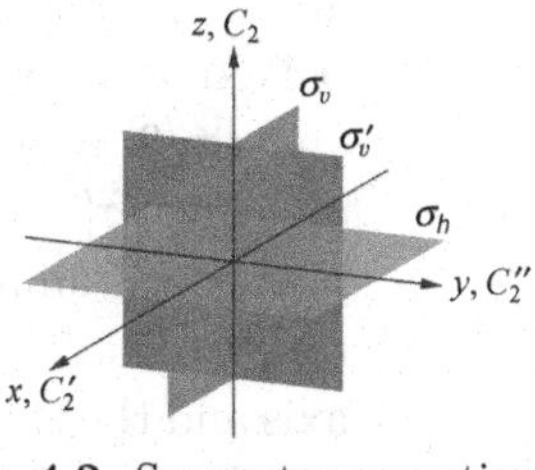

Fig. 4.2 Symmetry operations in point group D_{2h}.

as the yOz plane. The effects of the symmetry operations in point group D_{2h} on a general point with coordinates (x, y, z) are given by

$$\begin{aligned} &E(x,y,z) = (x,y,z), && C_2(x,y,z) = (-x,-y,z),\\ &C_2'(x,y,z) = (x,-y,-z), && C_2''(x,y,z) = (-x,y,-z),\\ &i(x,y,z) = (-x,-y,-z), && \sigma_h(x,y,z) = (x,y,-z),\\ &\sigma_v(x,-y,z) = (x,y,z), && \sigma_v'(x,y,z) = (-x,y,z). \end{aligned}$$

The result of the product of any two symmetry operations can be obtained by using the above results for individual symmetry operations.

Using the multiplication table, we find the following subgroups of D_{2h} with the two trivial subgroups included

$$C_1,\ C_i,\ C_s,\ C_2,\ C_{2h},\ D_2,\ C_{2v},\ D_{2h}.$$

where $C_1 = \{E\}$, $C_i = \{E, i\}$, $C_2 = \{E, C_2\}$, $C_s = \{E, \sigma_h\} = \{E, \sigma_v\} = \{E, \sigma_v'\}$, $D_2 = \{E,\ C_2,\ C_2',\ C_2''\}$, and

$C_{2h} = \{E,\ C_2,\ i,\ \sigma_h\} = \{E,\ C_2',\ i,\ \sigma_v'\} = \{E,\ C_2'',\ i,\ \sigma_v\}$,

$C_{2v} = \{E,\ C_2,\ \sigma_v,\ \sigma_v'\} = \{E,\ C_2',\ \sigma_v,\ \sigma_h\} = \{E,\ C_2'',\ \sigma_v',\ \sigma_h\}$.

Because D_{2h} is an Abelian group and because the inverses of all the elements are themselves, respectively, we have $a^{-1}ba = b$ for any elements a and b of D_{2h}. Therefore, all the subgroups of D_{2h} are its invariant subgroups.

4-8 Subgroups of point group C_{3v}. Consider point group C_{3v}.

(1) Find all the subgroups of this group.

(2) Identify all the invariant subgroups.

---

(1) Point group C_{3v} is given by $C_{3v} = \{E,\ C_3,\ C_3^2,\ \sigma_v,\ \sigma_v',\ \sigma_v''\}$ in which C_3 and C_3^2 are rotations about a 3-fold axis (the principal axis) through angles of $2\pi/3$ and $4\pi/3$, respectively, and σ_v, σ_v', and σ_v'' are three mirror planes containing the 3-fold rotation axis. The 3-fold rotation axis and the three mirror planes are shown in Fig. 4.3.Using the

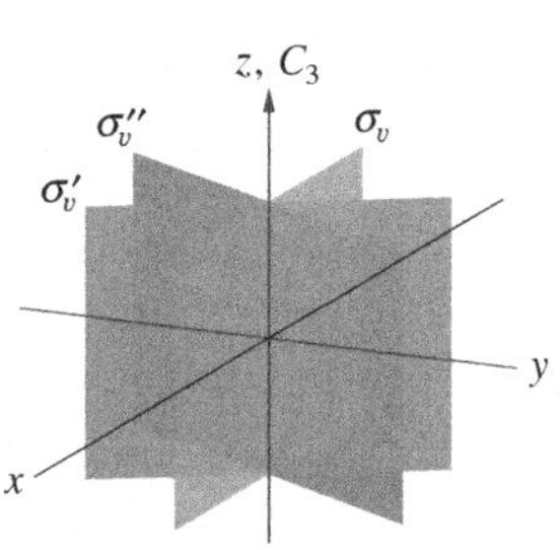

Fig. 4.3 Symmetry operations in point group C_{3v}.

coordinate system given in Fig. 4.3, the elements of C_{3v} are represented by [*cf.* Problem **4-5**]

$$E = \begin{pmatrix} 1 & 0 & 0 \\ 0 & 1 & 0 \\ 0 & 0 & 1 \end{pmatrix}, \quad C_3 = \frac{1}{2}\begin{pmatrix} -1 & -\sqrt{3} & 0 \\ \sqrt{3} & -1 & 0 \\ 0 & 0 & 1 \end{pmatrix}, \quad C_3^2 = \frac{1}{2}\begin{pmatrix} -1 & \sqrt{3} & 0 \\ -\sqrt{3} & -1 & 0 \\ 0 & 0 & 1 \end{pmatrix},$$

$$\sigma_v = \begin{pmatrix} 1 & 0 & 0 \\ 0 & -1 & 0 \\ 0 & 0 & 1 \end{pmatrix}, \quad \sigma_v' = \frac{1}{2}\begin{pmatrix} -1 & -\sqrt{3} & 0 \\ -\sqrt{3} & 1 & 0 \\ 0 & 0 & 1 \end{pmatrix}, \quad \sigma_v'' = \frac{1}{2}\begin{pmatrix} -1 & \sqrt{3} & 0 \\ \sqrt{3} & 1 & 0 \\ 0 & 0 & 1 \end{pmatrix},$$

Since the z coordinate remains unchanged under all the symmetry operations in point group C_{3v}, we can ignore the z coordinate and consider only the two remaining dimensions. With the product of two symmetry operations carried out using the above matrix representations, we can construct the multiplication table for C_{3v} as given below

	E	C_3	C_3^2	σ_v	σ_v'	σ_v''
E	E	C_3	C_3^2	σ_v	σ_v'	σ_v''
C_3	C_3	C_3^2	E	σ_v''	σ_v	σ_v'
C_3^2	C_3^2	E	C_3	σ_v'	σ_v''	σ_v
σ_v	σ_v	σ_v'	σ_v''	E	C_3	C_3^2
σ_v'	σ_v'	σ_v''	σ_v	C_3^2	E	C_3
σ_v''	σ_v''	σ_v	σ_v'	C_3	C_3^2	E

Using the multiplication table, we find the following subgroups of C_{3v} with the two trivial subgroups included C_1, C_s, C_3, C_{3v}, where $C_s = \{E, \sigma_v\} = \{E, \sigma_v'\} = \{E, \sigma_v''\}$ and $C_3 = \{E, C_3, C_3^2\}$.

(2) The two trivial subgroups are invariant subgroups. We now examine the two remaining subgroups. Since

$$C_3^{-1}\{E, \sigma_v\}C_3 = \{E, C_3^2\sigma_v C_3\} = \{E, \sigma_v''\} \neq \{E, \sigma_v\},$$

C_s is not an invariant subgroup. For C_3, we have

$$\sigma_v^{-1}\{E,\ C_3,\ C_3^2\}\sigma_v = \{E,\ \sigma_v C_3 \sigma_v,\ \sigma_v C_3^2 \sigma_v\} = \{E,\ C_3^2,\ C_3\},$$

$$\sigma_v'^{-1}\{E,\ C_3,\ C_3^2\}\sigma_v' = \{E,\ \sigma_v' C_3 \sigma_v',\ \sigma_v' C_3^2 \sigma_v'\} = \{E,\ C_3^2,\ C_3\},$$

$$\sigma_v''^{-1}\{E,\ C_3,\ C_3^2\}\sigma_v'' = \{E,\ \sigma_v'' C_3 \sigma_v'',\ \sigma_v'' C_3^2 \sigma_v''\} = \{E,\ C_3^2,\ C_3\}.$$

Thus, C_3 is an invariant subgroup. In summary, point group C_{3v} has three invariant subgroups, C_1, C_{3v}, and C_3, with the first two being trivial subgroups.

4-9 Abelian groups. *An Abelian group* is the one in which any element commutes with all the other elements.

(1) Show that the cyclic groups are Abelian groups.
(2) Identify all the non-commuting pairs of elements in group T.

---

(1) The elements of a cyclic group of order n can be expressed as C_n^j for $j = 0, 1, 2, \cdots, n-1$ with $C_n^0 \equiv E$ being the identity element. The product of two elements C_n^j and C_n^k can be manipulated as follows

$$C_n^j C_n^k = C_n^{j+k} = C_n^k C_n^j$$

since C_n^j is actually the jth power of C_n. Thus, the cyclic groups are Abelian groups. Note that, if $j + k \geqslant n$, C_n^{j+k} is equivalent to C_n^{j+k-n}.

(2) Denote the elements of group T by

$$E,\ C_3^{(1)},\ C_3^{(1)^2},\ C_3^{(2)},\ C_3^{(2)^2},\ C_3^{(3)},\ C_3^{(3)^2},$$
$$C_3^{(4)},\ C_3^{(4)^2},\ C_2^{(1)},\ C_2^{(2)},\ C_2^{(3)}.$$

The non-commuting pairs of elements in group T are

$$\big(C_3^{(i)}, C_3^{(j)}\big),\ \big(C_3^{(i)}, C_3^{(j)^2}\big),\ \big(C_3^{(i)^2}, C_3^{(j)^2}\big),\ \text{for } i \neq j = 1, 2, 3, 4;$$
$$\big(C_3^{(j)}, C_2^{(\ell)}\big),\ \big(C_3^{(j)^2}, C_2^{(\ell)}\big),\ \text{for } j = 1, 2, 3, 4 \text{ and } \ell = 1, 2, 3.$$

4-10 Equivalence classes. Two elements a and b of group G are said to be *equivalent* if there exists an element $g \in G$ such that $g^{-1}ag = b$. A subset of G consisting of elements of G which are equivalent to one another is referred to as *an equivalence class* of G. Decompose point group C_{3v} into classes.

The multiplication table of point group C_{3v} was obtained in Problem **4-8**. Here we use it to decompose C_{3v} into classes. Let us start with the identity element E. Since E commutes with all the other elements in the group, we always have $g^{-1}Eg = E$ for any $g \in C_{3v}$. Therefore, the identity element E is in an equivalence class by itself. This equivalence class is simply denoted by E, $E = \{E\}$.

We next consider C_3. Since E and C_3^2 commute with C_3, we do not need to use them in finding elements equivalent to C_3. Making use of elements σ_v, σ'_v, and σ''_v, we have with the help of the multiplication table

$$\sigma_v^{-1}C_3\sigma_v = \sigma_v C_3 \sigma_v = \sigma_v\sigma''_v = C_3^2,$$
$$\sigma_v'^{-1}C_3\sigma'_v = \sigma'_v C_3 \sigma'_v = \sigma'_v\sigma_v = C_3^2,$$
$$\sigma_v''^{-1}C_3\sigma_v = \sigma''_v C_3 \sigma''_v = \sigma''_v\sigma'_v = C_3^2.$$

Therefore, C_3^2 is equivalent to C_3 and they two constitute an equivalence class. We denote this equivalence class by $2C_3$, $2C_3 = \{C_3, C_3^2\}$. For the three remaining elements, we start with σ_v. Again, the identity element E does not need to be considered. Making use of elements C_3, C_3^2, σ'_v, and σ''_v, we have with the help of the multiplication table

$$(C_3)^{-1}\sigma_v C_3 = C_3^2\sigma_v C_3 = C_3^2\sigma'_v = \sigma''_v,$$
$$(C_3^2)^{-1}\sigma_v C_3^2 = C_3\sigma_v C_3^2 = C_3\sigma''_v = \sigma'_v,$$
$$(\sigma'_v)^{-1}\sigma_v\sigma'_v = \sigma'_v\sigma_v\sigma'_v = \sigma'_v C_3 = \sigma''_v,$$
$$(\sigma''_v)^{-1}\sigma_v\sigma''_v = \sigma''_v\sigma_v\sigma''_v = \sigma''_v C_3^2 = \sigma'_v.$$

Therefore, σ_v, σ'_v, and σ''_v constitute an equivalence class. We denote this equivalence class by $3\sigma_v$, $3\sigma_v = \{\sigma_v, \sigma'_v, \sigma''_v\}$.
In summary, point group C_{3v} can be decomposed into three equivalence classes, E, $2C_3$, and $3\sigma_v$.

4-11 Point group C_{4v}. Consider the symmetry group of a square (C_{4v}).

(1) Set up the multiplication table of the group.
(2) Decompose the group elements into classes.
(3) Find all subgroups of the group and identify all invariant subgroups.

---

(1) The elements of point group C_{4v} are E, C_4, $C_4^2 = C_2$, C_4^3, σ_v, σ'_v, σ_d, and σ'_d. As shown in Fig. 4.4, we take the 4-fold rotation axis (the principal axis) along the z axis, the zOx plane as the σ_v mirror plane, the yOz plane as the σ'_v mirror plane, the plane $y = x$ as the σ_d mirror plane, and the plane $y = -x$ as the σ'_d mirror plane. Since the z coordinate of a point remains

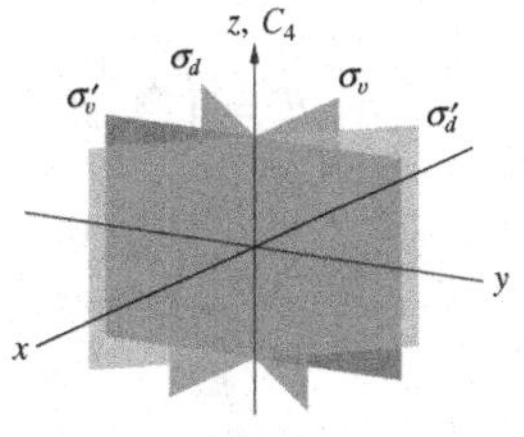

Fig. 4.4 Symmetry operations in point group C_{4v} in three dimensions.

unchanged under the symmetry operations in C_{4v}, we may consider the two-dimensional case. In two dimensions, the mirror planes become mirror lines. The symmetry operations in two dimensions are shown in Fig. 4.5. The z axis is now perpendicular to the paper and points out of the paper. The three rotations C_4, $C_4^2 = C_2$, and C_4^3 are now rotations on the xOy plane through angles $\pi/2$, π, and $3\pi/2$, respectively. The four reflections σ_v, σ_v', σ_d, and σ_d' are now reflections in the x axis, the y axis, the line $y = x$, and the line $y = -x$, respectively. To construct the multiplication table, we can either graphically or algebraically figure out the effects of the symmetry operations and their products. Here we adopt the algebraic approach. For this purpose, we need to represent algebraically the symmetry operations. This can be accomplished by making use of the matrix representations derived in Problem **4-5** for rotations and reflections in two dimensions. The identity element E is simply given by a 2×2 unit matrix. The matrices for the rotations and reflections can be obtained by plugging appropriate angles into R_θ and S_θ, respectively. We have

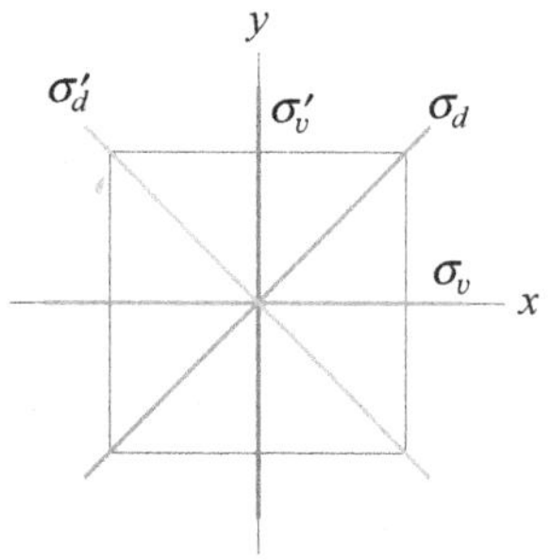

Fig. 4.5 Symmetry operations in point group C_{4v} in two dimensions.

$$E = \begin{pmatrix} 1 & 0 \\ 0 & 1 \end{pmatrix}, \quad C_4 = \begin{pmatrix} 0 & -1 \\ 1 & 0 \end{pmatrix}, C_2 = \begin{pmatrix} -1 & 0 \\ 0 & -1 \end{pmatrix}, C_4^3 = \begin{pmatrix} 0 & 1 \\ -1 & 0 \end{pmatrix},$$

$$\sigma_v = \begin{pmatrix} 1 & 0 \\ 0 & -1 \end{pmatrix}, \sigma_v' = \begin{pmatrix} -1 & 0 \\ 0 & 1 \end{pmatrix}, \sigma_d = \begin{pmatrix} 0 & 1 \\ 1 & 0 \end{pmatrix}, \quad \sigma_d' = \begin{pmatrix} 0 & -1 \\ -1 & 0 \end{pmatrix}.$$

Using the above matrix representations, we obtain the following multiplication table

	E	C_4	C_2	C_4^3	σ_v	σ_v'	σ_d	σ_d'
E	E	C_4	C_2	C_4^3	σ_v	σ_v'	σ_d	σ_d'
C_4	C_4	C_2	C_4^3	E	σ_d	σ_d'	σ_v'	σ_v
C_2	C_2	C_4^3	E	C_4	σ_v'	σ_v	σ_d'	σ_d
C_4^3	C_4^3	E	C_4	C_2	σ_d'	σ_d	σ_v	σ_v'
σ_v	σ_v	σ_d'	σ_v'	σ_d	E	C_2	C_4^3	C_4
σ_v'	σ_v'	σ_d	σ_v	σ_d'	C_2	E	C_4	C_4^3
σ_d	σ_d	σ_v	σ_d'	σ_v'	C_4	C_4^3	E	C_2
σ_d'	σ_d'	σ_v'	σ_d	σ_v	C_4^3	C_4	C_2	E

(2) The identity element E is in its own equivalence class denoted by E, $E = \{E\}$ since it commutes with all other elements. In identifying equivalence classes of other elements, we do not need to use E since $E^{-1}gE = g$ for any $g \in C_{4v}$. We now consider the element C_4 for which C_2 and C_4^3 need not be used. Making use of elements σ_v, σ_v', σ_d, and σ_d', we have with the help of the multiplication table

$$\sigma_v^{-1}C_4\sigma_v = \sigma_v C_4 \sigma_v = \sigma_v\sigma_d = C_4^3,$$
$$\sigma_v'^{-1}C_4\sigma_v' = \sigma_v' C_4 \sigma_v' = \sigma_v'\sigma_d' = C_4^3,$$
$$\sigma_d^{-1}C_4\sigma_d = \sigma_d C_4 \sigma_d = \sigma_d\sigma_v' = C_4^3,$$
$$\sigma_d'^{-1}C_4\sigma_d' = \sigma_d' C_4 \sigma_d' = \sigma_d'\sigma_v = C_4^3.$$

Therefore, C_4 and C_4^3 constitute an equivalence class that is to be denoted by $2C_4$, $2C_4 = \{C_4, C_4^3\}$.
C_2 is in its own equivalence class since it commutes with C_4 and C_4^3 as well as with the reflections. Its own equivalence class is denoted by C_2, $C_2 = \{C_2\}$.
For σ_v, making use of elements C_4, C_4^3, σ_d, and σ_d', we have with the help of the multiplication table

$$(C_4)^{-1}\sigma_v C_4 = C_4^3\sigma_v C_4 = C_4^3\sigma_d' = \sigma_v',$$
$$(C_4^3)^{-1}\sigma_v C_4^3 = C_4\sigma_v C_4^3 = C_4\sigma_d = \sigma_v',$$
$$\sigma_d^{-1}\sigma_v\sigma_d = \sigma_d\sigma_v\sigma_d = \sigma_d C_4^3 = \sigma_v',$$
$$\sigma_d'^{-1}\sigma_v\sigma_d' = \sigma_d'\sigma_v\sigma_d' = \sigma_d' C_4 = \sigma_v'.$$

Therefore, σ_v and σ_v' constitute an equivalence class which is to be denoted by $2\sigma_v, 2\sigma_v = \{\sigma_v, \sigma_v'\}$. In analogy with σ_v and σ_v', we expect that the two remaining elements σ_d and σ_d' are in an equivalence class. It turns out to be true. Making use of elements C_4, C_4^3, σ_v and σ_v', we have with the help of the multiplication table

$$\begin{aligned}
(C_4)^{-1}\sigma_d C_4 &= C_4^3\sigma_d C_4 = C_4^3\sigma_v = \sigma_d', \\
(C_4^3)^{-1}\sigma_d C_4^3 &= C_4\sigma_d C_4^3 = C_4\sigma_v' = \sigma_d', \\
\sigma_v^{-1}\sigma_d\sigma_v &= \sigma_v\sigma_d\sigma_v = \sigma_v C_4 = \sigma_d', \\
\sigma_v'^{-1}\sigma_d\sigma_v' &= \sigma_v'\sigma_d\sigma_v' = \sigma_v' C_4^3 = \sigma_d'.
\end{aligned}$$

Therefore, σ_v and σ_v' are indeed in an equivalence class which is to be denoted by $2\sigma_d$, $2\sigma_d = \{\sigma_d, \sigma_d'\}$. In summary, point group C_{4v} has five equivalence classes: E, $2C_4$, C_2, $2C_v$, $2C_d$.

(3) Using the multiplication table, we see that the subgroups of C_{4v} are

$$C_1, C_s, C_2, C_4, C_{2v}, C_{4v}.$$

Subgroups C_1 and C_{4v} are two trivial subgroups. Subgroup C_s consists of E and one of the four mirror planes. We thus have

$$C_s = \{E, \sigma_v\} = \{E, \sigma_v'\} = \{E, \sigma_d\} = \{E, \sigma_d'\}.$$

Subgroup C_2 consists of elements E and C_2, $C_2 = \{E, C_2\}$. Subgroup C_4 consists of elements E, C_4, C_2, and C_4^3. Subgroup C_{2v} consists of elements E, C_2, σ_v, and σ_v' or E, C_2, σ_d, and σ_d'. We thus have

$$C_{2v} = \{E, C_2, \sigma_v, \sigma_v'\} = \{E, C_2, \sigma_d, \sigma_d'\}.$$

From the definition of an invariant subgroup, $g^{-1}Hg = H$ for $g \in G$, we see that a subgroup can be an invariant subgroup if and only if it contains an equivalence class in its entirety. Since C_s does not contain the equivalence classes $2\sigma_v$ and $2\sigma_d$ in their entireties, it is not an invariant subgroup. Since element C_2 commutes with all other elements of C_{4v}, we conclude that subgroup C_2 is an invariant subgroup. For subgroup C_4, we have

$$\begin{aligned}
\sigma_v^{-1}\{E, C_4, C_2, C_4^3\}\sigma_v &= \{E, C_4^3, C_2, C_4\}, \\
\sigma_v'^{-1}\{E, C_4, C_2, C_4^3\}\sigma_v' &= \{E, C_4^3, C_2, C_4\}, \\
\sigma_d^{-1}\{E, C_4, C_2, C_4^3\}\sigma_d &= \{E, C_4^3, C_2, C_4\}, \\
\sigma_d'^{-1}\{E, C_4, C_2, C_4^3\}\sigma_d' &= \{E, C_4^3, C_2, C_4\}.
\end{aligned}$$

Therefore, C_4 is an invariant subgroup. For subgroup $C_{2v} = \{E, C_2, \sigma_v, \sigma_v'\}$, we need only to consider C_4, C_4^3, σ_d, and σ_d' since the elements of subgroup C_{2v} commute with C_2, σ_v, and σ_v'. With the help of the multiplication table, we have

$$\begin{aligned}
&(C_4)^{-1}\{E, C_2, \sigma_v, \sigma_v'\}C_4 = \{E, C_2, \sigma_v', \sigma_v\}, \\
&(C_4^3)^{-1}\{E, C_2, \sigma_v, \sigma_v'\}C_4^3 = \{E, C_2, \sigma_v', \sigma_v\}, \\
&\sigma_d^{-1}\{E, C_2, \sigma_v, \sigma_v'\}\sigma_d = \{E, C_2, \sigma_v, \sigma_v'\}, \\
&{\sigma_d'}^{-1}\{E, C_2, \sigma_v, \sigma_v'\}\sigma_d' = \{E, C_2, \sigma_v', \sigma_v\}.
\end{aligned}$$

Therefore, subgroup C_{2v} is an invariant subgroup. In summary, the invariant subgroups of C_{4v} are C_1, C_2, C_4, C_{2v}, and C_{4v}. We see that, among all the subgroups of C_{4v}, only subgroup C_s is not an invariant subgroup.

Chapter 5

Classification of Bravais Lattices

(1) *Lattice centerings*
P-, A-, B-, C-, I-, & F-type centerings.

(2) *Triclinic crystal system*
Point group: C_i.
Bravais lattice: Simple.
Crystallographic point groups: C_i, C_1.

(3) *Monoclinic crystal system*
Point group: C_{2h}.
Bravais lattices: Simple, centered.
Crystallographic point groups: C_{2h}, C_2, C_s.

(4) *Orthorhombic crystal system*
Point group: D_{2h}.
Bravais lattices: Simple, base-centered, body-centered, face-centered.
Crystallographic point groups: D_{2h}, D_2, C_{2v}.

(5) *Tetragonal crystal system*
Point group: D_{4h}.
Bravais lattices: Simple, body-centered.
Crystallographic point groups: D_{4h}, D_{2d}, D_4, C_{4v}, C_{4h}, S_4, C_4.

(6) *Trigonal crystal system*
Point group: D_{3d}.
Bravais lattice: Simple.
Crystallographic point groups: D_{3d}, D_3, C_{3v}, S_6, C_3.

(7) *Hexagonal crystal system*
Point group: D_{6h}.
Bravais lattice: Simple.
Crystallographic point groups: D_{6h}, D_{3h}, D_6, C_{6v}, C_{6h}, C_{3h}, C_6.

(8) *Cubic crystal system*
Point group: O_h.
Bravais lattices: Simple, body-centered, face-centered.
Crystallographic point groups: O_h, T_d, O, T_h, T.

5-1 Centerings in the hexagonal crystal system. Show that the centerings of all types in the hexagonal crystal system lead only to the simple hexagonal Bravais lattice.

All the six types of centerings in the hexagonal crystal system are shown in Fig. 5.1 with the crystallographic axes and the lengths of the edges of a primitive cell also given on the primitive Bravais lattice in Fig. 5.1(a).

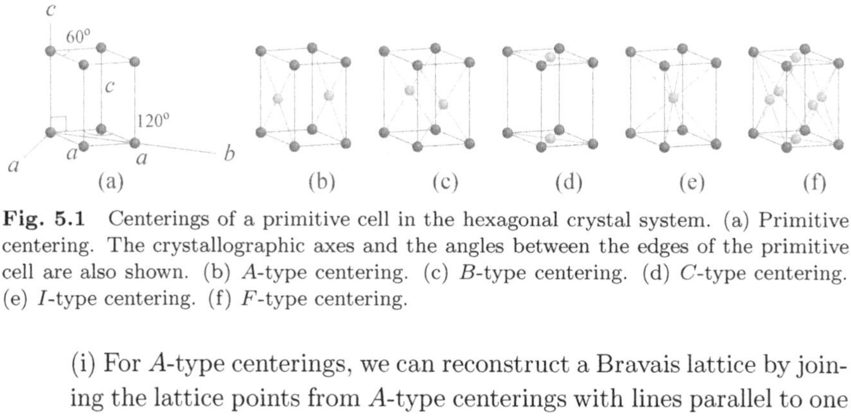

Fig. 5.1 Centerings of a primitive cell in the hexagonal crystal system. (a) Primitive centering. The crystallographic axes and the angles between the edges of the primitive cell are also shown. (b) A-type centering. (c) B-type centering. (d) C-type centering. (e) I-type centering. (f) F-type centering.

(i) For A-type centerings, we can reconstruct a Bravais lattice by joining the lattice points from A-type centerings with lines parallel to one of the two sides of each non-rectangular face of an original primitive cell, by joining them also in the perpendicular directions parallel to the ab plane, and by joining these lattice points with lines parallel to the c axis. The lines joining the lattice points from A-type centerings become edges of the conventional cells of the reconstructed Bravais lattice. The non-joined lattice points from A-type centerings become centers in the faces perpendicular to the c axis and the primitive lattice points become centers in the faces parallel to the c axis. The original and reconstructed Bravais lattices are shown in Figs. 5.2(a) and (b), respectively. We see that the reconstructed Bravais lattice is of lower symmetry than the original Bravais lattice and is a face-centered orthorhombic Bravais lattice with a primitive cell of volume $\sqrt{3}\,a^2c/4$. The origin of the coordinate system for the face-centered orthorhombic Bravais lattice has been shifted through $(\boldsymbol{a}_1 + \boldsymbol{a}_3)/2$.

Therefore, A-type centerings in the hexagonal crystal system do not give rise to any new Bravais lattice.

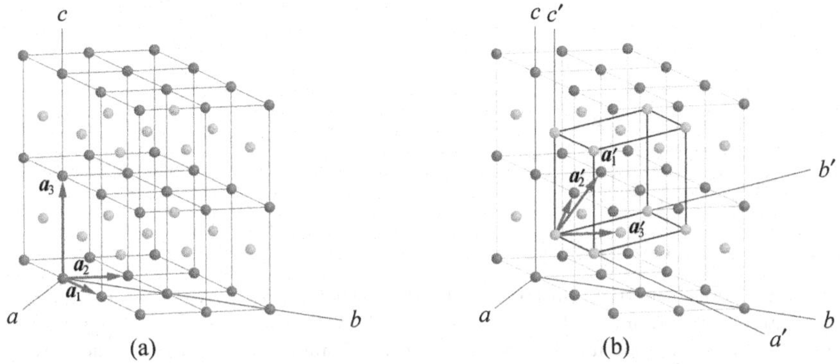

Fig. 5.2 Transformation of a hexagonal Bravais lattice with A-type centerings into a face-centered orthorhombic Bravais lattice. (a) A-type centerings for twelve primitive cells. (b) Transformation into a face-centered orthorhombic Bravais lattice. One conventional cell of the face-centered orthorhombic Bravais lattice is highlighted with its edges drawn in heavy solid lines.

The lattice constants of the face-centered orthorhombic Bravais lattice are $a' = a$, $b' = \sqrt{3}\,a$, and $c' = c$. The expressions for the primitive vectors of the face-centered orthorhombic Bravais lattice shown in Fig. 5.2(b) are

$$\boldsymbol{a}_1' = \frac{1}{2}b'\boldsymbol{e}_y' + \frac{1}{2}c'\boldsymbol{e}_z', \ \boldsymbol{a}_2' = \frac{1}{2}c'\boldsymbol{e}_z' + \frac{1}{2}a'\boldsymbol{e}_x', \ \boldsymbol{a}_3' = \frac{1}{2}a'\boldsymbol{e}_x' + \frac{1}{2}b'\boldsymbol{e}_y',$$

where we have used $\boldsymbol{e}_x'$, $\boldsymbol{e}_y'$, and $\boldsymbol{e}_z'$ as the unit vectors along the three new crystallographic axes a', b', and c', respectively.
(ii) B-type centerings can be treated exactly in the same way as for A-type centerings. A hexagonal Bravais lattice with B-type centerings can be also transformed into a face-centered orthorhombic Bravais lattice as shown in Fig. 5.3. However, the directions of the new crystallographic axes are different from those for A-type centerings. Therefore, B-type centerings in the hexagonal crystal system do not either give rise to any new Bravais lattice.
(iii) For C-type centerings, we can reconstruct a Bravais lattice by joining the lattice points with lines parallel to the two diagonals of each non-rectangular face of an original primitive cell and by joining the lattice points with vertical lines parallel to the c axis. The original and reconstructed Bravais lattices are shown in Figs. 5.4(a) and (b),

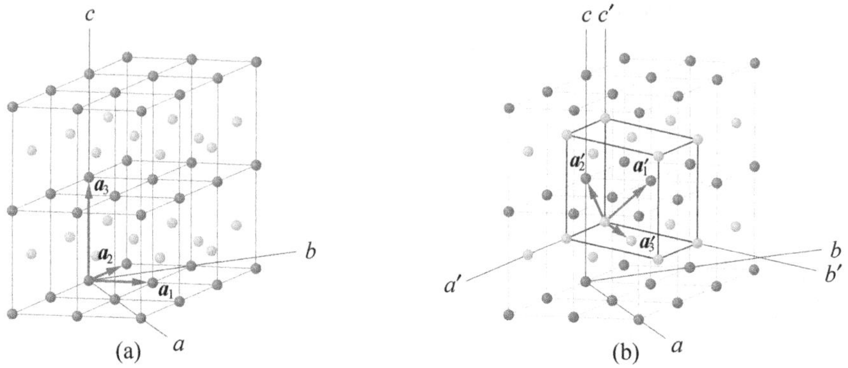

Fig. 5.3 Transformation of a hexagonal Bravais lattice with B-type centerings into a face-centered orthorhombic Bravais lattice. (a) B-type centerings for twelve primitive cells. (b) Transformation into a face-centered orthorhombic Bravais lattice. One conventional cell of the face-centered orthorhombic Bravais lattice is highlighted with its edges drawn in heavy solid lines.

respectively. The reconstructed Bravais lattice is of lower symmetry than the original Bravais lattice and is a simple orthorhombic Bravais lattice with a primitive cell of size $(a/2) \times (\sqrt{3}\,a/2) \times c$. Therefore, C-type centerings in the hexagonal crystal system do not give rise to any new Bravais lattice.

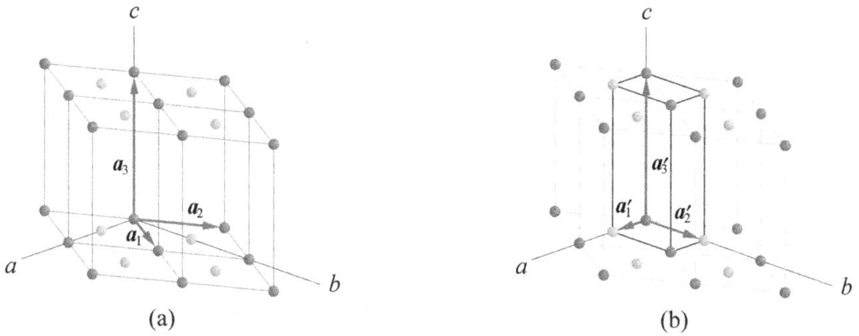

Fig. 5.4 Transformation of a hexagonal Bravais lattice with C-type centerings into a simple orthorhombic Bravais lattice. (a) C-type centerings for four primitive cells. (b) Transformation into a simple orthorhombic Bravais lattice.

(iv) For I-type centerings, we can reconstruct a Bravais lattice by joining the primitive lattice points with lines parallel to the two diagonals of each non-rectangular face of an original primitive cell and by joining these lattice points with vertical lines parallel to the c

axis. The original and reconstructed Bravais lattices are shown in Figs. 5.5(a) and (b), respectively. The reconstructed Bravais lattice is of lower symmetry than the original Bravais lattice and is a face-centered orthorhombic Bravais lattice with a primitive cell of volume $\sqrt{3}\,a^2c/4$. Therefore, I-type centerings in the hexagonal crystal system do not either give rise to any new Bravais lattice.

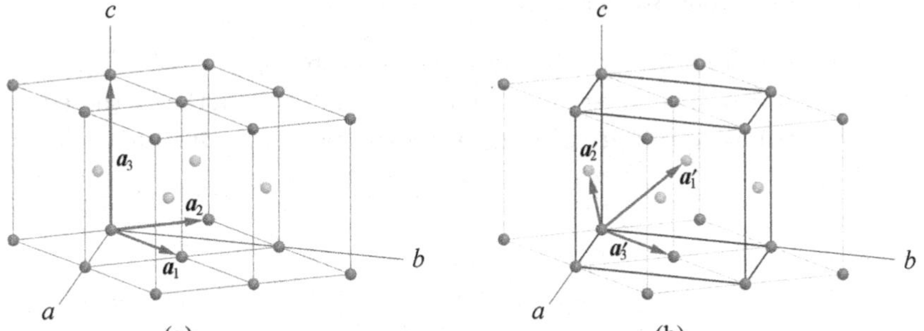

(a) (b)

Fig. 5.5 Transformation of a hexagonal Bravais lattice with I-type centerings into a face-centered orthorhombic Bravais lattice. (a) I-type centerings for four primitive cells. (b) Transformation into a face-centered orthorhombic Bravais lattice.

(v) For F-type centerings, we join lattice points in the same way as for C-type centerings to reconstruct a Bravais lattice. We then obtain a body-centered orthorhombic Bravais lattice as shown in Fig. 5.6(b) with the original Bravais lattice given in Fig. 5.6(a). Therefore, F-type centerings in the hexagonal crystal system do not either give rise to any new Bravais lattice.

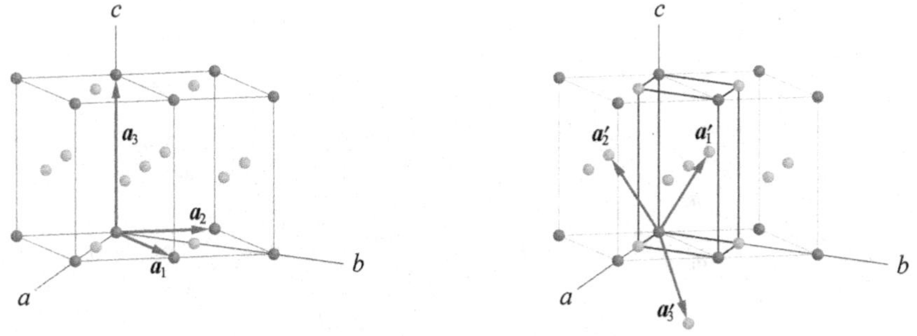

(a) (b)

Fig. 5.6 Transformation of a hexagonal Bravais lattice with F-type centerings into a body-centered orthorhombic Bravais lattice. (a) F-type centerings for two primitive cells. (b) Transformation into a body-centered orthorhombic Bravais lattice.

In summary, all the six types of centerings lead to only one Bravais lattice in the hexagonal crystal system: The simple hexagonal Bravais lattice.

5-2 Centerings in the cubic crystal system. Show that all the six types of centerings in the cubic crystal system lead only to the three different Bravais lattices: The simple, body-centered, and face-centered cubic Bravais lattices.

Shown in Fig. 5.7 are the six types of centerings in the cubic crystal system. The crystallographic axes and the lengths of the edges of a primitive cell are also given on the primitive Bravais lattice in Fig. 5.7(a).

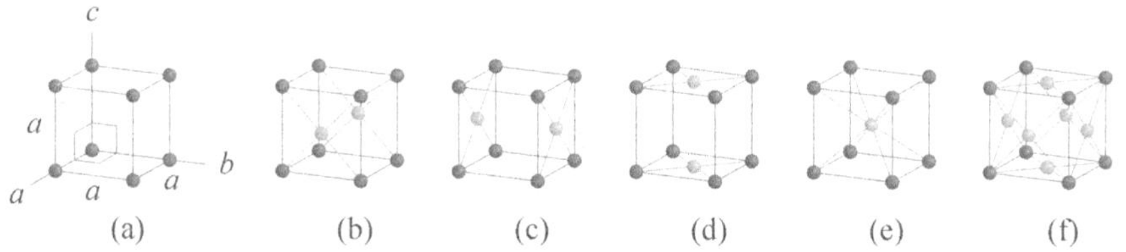

Fig. 5.7 Centerings of a primitive cell in the cubic crystal system. (a) Primitive centering. The crystallographic axes are also shown. (b) A-type centering. (c) B-type centering. (d) C-type centering. (e) I-type centering. (f) F-type centering.

We now examine C-type centerings. Shown in Fig. 5.8 are two primitive cells decorated with C-type centerings. In Fig. 5.8(a), two original cubic primitive cells are given. In Fig. 5.8(b), a primitive cell of size $(a/\sqrt{2}) \times (a/\sqrt{2}) \times a$ for the simple tetragonal Bravais lattice is given.

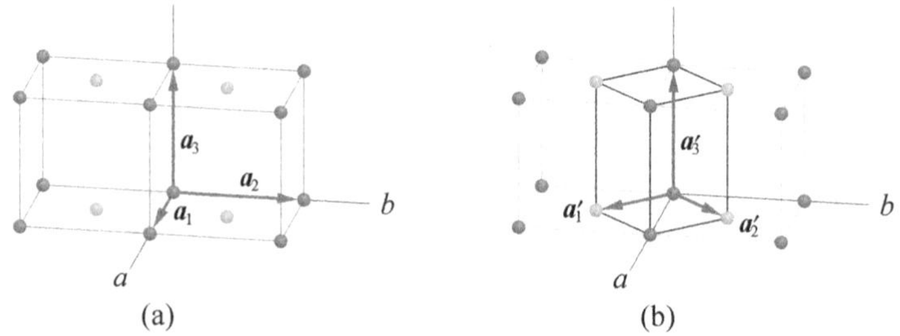

Fig. 5.8 Transformation of a cubic Bravais lattice with C-type centerings into a simple tetragonal Bravais lattice. (a) C-type centerings for two primitive cells. (b) Transformation into a simple tetragonal Bravais lattice.

From Fig. 5.8(b), we see that the cubic Bravais lattice with C-type centerings is actually a simple tetragonal Bravais lattice which is of lower symmetry. Similarly, the cubic Bravais lattices with A- and B-type centerings can both be reduced to simple tetragonal Bravais lattices. Therefore, there are only three different Bravais lattices in the cubic crystal system: The simple, body-centered, and face-centered cubic Bravais lattices.

5-3 Relations between symmetries of crystal systems. We study the relations between the crystal systems.

(1) Starting from the conventional cell of the cubic crystal system (a perfect cube), describe how we can obtain the conventional cells of the tetragonal, orthorhombic, monoclinic, and triclinic crystal systems in turn.

(2) Starting from the conventional cell of the cubic crystal system, describe how we can obtain the conventional cell of the trigonal crystal system.

---

(1) (i) The conventional cell of the tetragonal crystal system can be obtained by reducing the symmetry of the conventional cell of the cubic crystal system through pulling on the two opposite faces. The BCC and FCC Bravais lattices are all reduced to the body-centered tetragonal Bravais lattice.

(ii) The conventional cell of the orthorhombic crystal system can be obtained by reducing the symmetry of the conventional cell of the tetragonal crystal system through deforming the square faces into rectangles.

(iii) The conventional cell of the monoclinic crystal system can be obtained by reducing the symmetry of the conventional cell of the orthorhombic crystal system through distorting the rectangular faces perpendicular to the c-axis into general parallelograms.

(iv) The conventional cell of the triclinic crystal system can be obtained by reducing the symmetry of the conventional cell of the monoclinic crystal system through tilting the c-axis so that it is no longer perpendicular to the other two axes.

(2) The conventional cell of the trigonal crystal system can be obtained by reducing the symmetry of the conventional cell of the cubic crystal system through stretching it along one of its body diagonals.

5-4 Lattice planes and directions in the cubic crystal system. Consider lattice planes and directions in the cubic crystal system.

(1) On a set of cubic conventional unit cells, draw the following directions with two directions on each conventional unit cell

$$[2,-1,2],[4,-1,-2],[1,-1,0],[0,1,-2],[0,0,-2],[1,2,-1].$$

(2) On a set of cubic conventional unit cells, draw the following lattice planes with two lattice planes on each conventional unit cell

$$(1,-1,0),(-1,0,2),(1,1,-1),(2,-1,0),(-2,-1,-2),(2,0,-4).$$

---

(1) The directions are shown in Fig. 5.9.

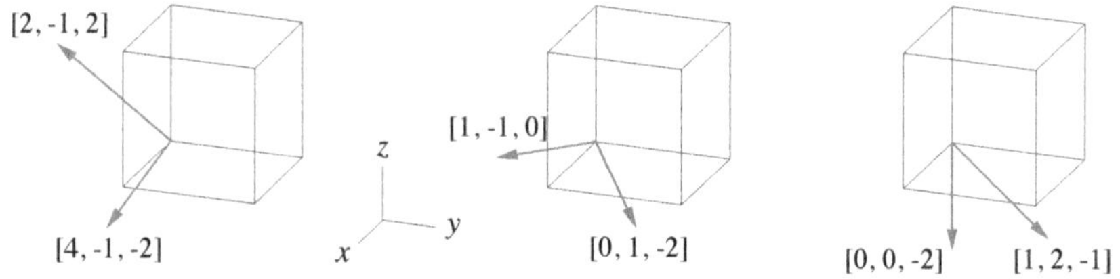

Fig. 5.9 Directions in the cubic crystal system.

(2) The lattice planes are shown in Fig. 5.10.

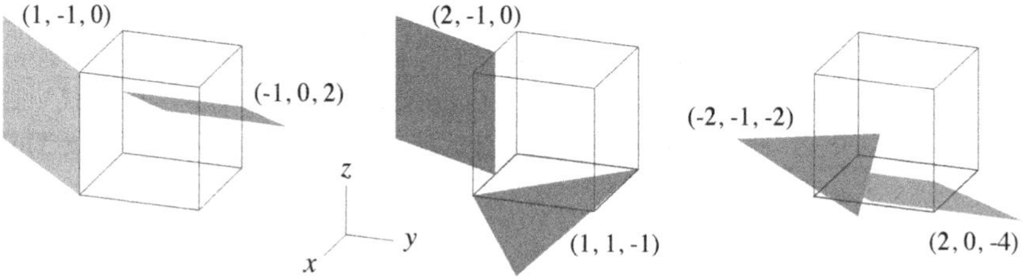

Fig. 5.10 Lattice planes in the cubic crystal system.

Chapter 6

Space Groups of Crystal Structures

(1) *Crystal structure*
A crystal structure = a Bravais lattice + a basis.

(2) *Screw axis*
A screw axis is a symmetry operation that combines a rotation about an axis with a translation parallel to the axis through a distance that is not equal to the length of any lattice vector.
Notation: n_m.
Possible screw axes: 2_1, 3_1, 3_2, 4_1, 4_2, 4_3, 6_1, 6_2, 6_3, 6_4, 6_5.

(3) *Glide plane*
A glide plane is a symmetry operation that combines a translation through a vector not equal to any lattice vector with a reflection in a plane that contains the translation vector.
Types of glide planes: axial, double, diagonal, diamond.
Symbols for glide planes: a, b, c, e, n, d.

(4) *Space group notation $Lijk$.*

(5) *Symmorphic space groups*
The symmetry operations in a symmorphic space group include only lattice translations and point symmetry operations that leave at least one lattice point fixed.
The total number: 73.

(6) *Nonsymmorphic space groups*
The symmorphic space groups become nonsymmorphic space groups when the rotation axes are replaced with the corresponding screw axes and/or the mirror planes are replaced with the corresponding glide planes.
The total number: 157.

(7) *Diamond structure*
It can be described as a face-centered cubic Bravais lattice with a basis consisting of two identical carbon atoms: One at $(0,0,0)a$ and the other at $(1/4,1/4,1/4)a$.
The space group is $Fd\bar{3}m$.

(8) *Sodium chloride structure*
It can be described as a face-centered cubic Bravais lattice with a basis consisting of an Na ion at $(0,0,0)a$ and a Cl ion at $(1/2,0,0)a$.
The space group is $Fm\bar{3}m$.

(9) *Hexagonal close-packed structure*
The stacking sequence is $\cdots ABAB \cdots$.
It can be taken as a simple hexagonal Bravais lattice with a two-atom basis with one atom at $(0,0,0)$ and the other at $(a/2)\boldsymbol{e}_x+(a/2\sqrt{3})\boldsymbol{e}_y+(c/2)\boldsymbol{e}_z$.
The space group is $P6_3/mmc$.
For the ideal HCP structure, $c/a=\sqrt{8/3}\approx 1.633$.

(10) *Cubic close-packed structure*
The stacking sequence is $\cdots ABCABC \cdots$.
The CCP structure is an FCC structure.
The space group is $F\bar{4}3m$.

(11) *Cesium chloride structure*
The CsCl structure can be described as having a simple cubic lattice with a basis consisting of a chlorine ion at $(0,0,0)$ and a cesium ion at $(1/2,1/2,1/2)a$.
The space group is $Pm\bar{3}m$.

(12) *Zincblende structure*
The zincblende structure can be described as having an FCC Bravais lattice with a basis consisting of a sulfur ion at $(0,0,0)$ and a zinc ion at $(1/4,1/4,1/4)a$.
The space group is $F\bar{4}3m$.

(13) *Perovskite structure*
The general chemical formula for the perovskite structure is ABX_3.
The perovskite structure can be described as a simple cubic Bravais lattice with a basis consisting of an A atom at $(0,0,0)$, a B atom at $(1/2,1/2,1/2)a$, and three X atoms at $(1/2,1/2,0)a$, $(1/2,0,1/2)a$, and $(0,1/2,1/2)a$.
The space group is $Pm\bar{3}m$.

6-1 Identification of crystals with symmorphic space groups. For symmorphic space groups $P1$, $P2/m$, $C222$, $Immm$, $P4/mmm$, $I\bar{4}2m$, $P3m1$, $P\bar{6}m2$, $P432$, $I\bar{4}3m$, and $Fm\bar{3}m$, specify their crystal systems, Bravais lattices, and point groups (in both the international and Schönflies symbols).

Space group	Crystal system	Bravais lattice	Point group
$P1$	Triclinic	Simple	$1\ (C_1)$
$P2/m$	Monoclinic	Simple	$2\ (C_2)$
$C222$	Orthorhombic	Centered	$222\ (D_2)$
$Immm$	Orthorhombic	Body-centered	$mmm\ (D_{2h})$
$P4/mmm$	Tetragonal	Simple	$4/mmm\ (D_{4h})$
$I\bar{4}2m$	Tetragonal	Body-centered	$\bar{4}2m\ (D_{2d})$
$P3m1$	Trigonal	Simple	$3m\ (C_{3v})$
$P\bar{6}m2$	Hexagonal	Primitive	$\bar{6}m2\ (D_{3h})$
$P432$	Cubic	Simple	$432\ (O)$
$I\bar{4}3m$	Cubic	Body-centered	$\bar{4}3m\ (T_d)$
$Fm\bar{3}m$	Cubic	Face-centered	$m\bar{3}m\ (O_h)$

6-2 Identification of crystals with nonsymmorphic space groups. For nonsymmorphic space groups $P2_1/m$, $Ccc2$, $Ibca$, $I4_1cd$, $I4_1/acd$, $P3_212$, $R\bar{3}c$, $P6_3/mmc$, $P2_13$, $Fd\bar{3}$, and $Fm\bar{3}c$, specify their crystal systems, Bravais lattices, and point groups (in both the international and Schönflies symbols).

Space group	Crystal system	Bravais lattice	Point group
$P2_1/m$	Monoclinic	Simple	$2/m\ (C_{2h})$
$Ccc2$	Orthorhombic	Centered	$mm2\ (C_{2v})$
$Ibca$	Orthorhombic	Body-centered	$mmm\ (D_{2h})$
$I4_1cd$	Tetragonal	Body-centered	$4mm\ (C_{4v})$
$I4_1/acd$	Tetragonal	Body-centered	$4/mmm\ (D_{4h})$
$P3_212$	Trigonal	Simple	$32\ (D_3)$
$R\bar{3}c$	Trigonal	Simple	$\bar{3}m\ (D_{3d})$
$P6_3/mmc$	Hexagonal	Simple	$6/mmm\ (D_{6h})$
$P2_13$	Cubic	Simple	$23\ (T)$
$Fd\bar{3}$	Cubic	Face-centered	$m\bar{3}\ (T_h)$
$Fm\bar{3}c$	Cubic	Face-centered	$m\bar{3}m\ (O_h)$

6-3 Identification of crystals with symmorphic or nonsymmorphic space groups. For space groups $P23$, $P3_2$, $C2/c$, $B2/m$, $Iba2$, $Fmm2$, $I4cm$, $P6/mmm$, $P6_3mc$, $I222$, and $I2_13$, specify their crystal systems, Bravais lattices, and point groups (in both the international and Schönflies symbols).

Space group	Crystal system	Bravais lattice	Point group
$P23$	Cubic	Simple	23 (T)
$P3_2$	Trigonal	Simple	3 (C_3)
$C2/c$	Monoclinic	Centered	$2/m$ (C_{2h})
$Cmcm$	Orthorhombic	Centered	mmm (D_{2h})
$Iba2$	Orthorhombic	Body-centered	$mm2$ (C_{2v})
$F\overline{4}3c$	Cubic	Face-centered	$\overline{4}3m$ (T_d)
$I4cm$	Tetragonal	Body-centered	$4mm$ (C_{4v})
$P6/mmm$	Hexagonal	Simple	$6/mmm$ (D_{6h})
$P6_3mc$	Hexagonal	Simple	$6mm$ (C_{6v})
$I222$	Orthorhombic	Body-centered	222 (D_2)
$I2_13$	Cubic	Body-centered	23 (T)

6-4 Translation vectors in the diamond structure. The positions of the carbon atoms in diamond take on the form

$$\boldsymbol{r} = (n_1 + x)a\boldsymbol{e}_x + (n_2 + y)a\boldsymbol{e}_y + (n_3 + z)a\boldsymbol{e}_z,$$

where $a = 3.566\ 83 \times 10^{-10}$ m is the lattice constant, n_1, n_2, and n_3 are arbitrary integers, and (x, y, z) takes on one of the following eight values

$$(0,0,0), (1/4,1/4,1/4), (0,1/2,1/2), (1/2,0,1/2),$$
$$(1/2,1/2,0), (1/4,3/4,3/4), (3/4,1/4,3/4), (3/4,3/4,1/4).$$

(1) Consider $\boldsymbol{t} = \xi a\boldsymbol{e}_x + \zeta a\boldsymbol{e}_y + \eta a\boldsymbol{e}_z$, where $0 < \xi,\ \zeta,\ \eta < 1$. Find values of ξ, ζ, and η so that $\boldsymbol{t}$ is a translation vector, that is, if there is a carbon atom at $\boldsymbol{r}$, there will always be another carbon atom at $T_{\boldsymbol{t}}\boldsymbol{r} = \boldsymbol{r} + \boldsymbol{t}$.

(2) Find primitive vectors $\boldsymbol{a}_1$, $\boldsymbol{a}_2$, and $\boldsymbol{a}_3$ such that all translation vectors take on the form $m_1\boldsymbol{a}_1 + m_2\boldsymbol{a}_2 + m_3\boldsymbol{a}_3$ with m_1, m_2, and m_3 integers.

(3) Find the lengths of and the angles between $\boldsymbol{a}_1$, $\boldsymbol{a}_2$, and $\boldsymbol{a}_3$.

---

(1) Writing out $T_{\boldsymbol{t}}\boldsymbol{r} = \boldsymbol{r} + \boldsymbol{t} = \boldsymbol{r}'$ with $\boldsymbol{r}'$ the position of a carbon atom, we have

$$\begin{aligned}(n_1 + x + \xi)a\boldsymbol{e}_x &+ (n_2 + y + \zeta)a\boldsymbol{e}_y + (n_3 + z + \eta)a\boldsymbol{e}_z \\ &= (n_1' + x')a\boldsymbol{e}_x + (n_2' + y')a\boldsymbol{e}_y + (n_3' + z')a\boldsymbol{e}_z\end{aligned}$$

with n_1', n_2', and n_3' integers and (x', y', z') taking on one of the given eight values for (x, y, z). We thus have

$$\xi = n_1' - n_1 + x' - x,$$
$$\zeta = n_2' - n_2 + y' - y,$$
$$\eta = n_3' - n_3 + z' - z.$$

Therefore, the values for ξ can be inferred from those for x by taking the differences of possible pairs with negative values brought into positive ones less than unity through adding suitable positive integers to them. We find that ξ can take on the following four values

$$0,\ 1/4,\ 1/2,\ 3/4.$$

We can similarly find the values for ζ and η. It turns out that ζ and η also take on the above four values. This can be seen from the symmetry in the given eight values for (x, y, z). Thus, including $\boldsymbol{t} = 0$, there are in total 64 translation vectors, $\boldsymbol{t} = \xi a\boldsymbol{e}_x + \zeta a\boldsymbol{e}_y + \eta a\boldsymbol{e}_z$ with $\xi,\ \zeta,\ \eta = 0,\ 1/4,\ 1/2,\ 3/4$.

(2) Since the three nonzero values of ξ, ζ, and η are all multiples of $1/4$, we can simply choose

$$\boldsymbol{a}_1 = (1/4, 0, 0)a,\ \ \boldsymbol{a}_2 = (0, 1/4, 0)a,\ \ \boldsymbol{a}_3 = (0, 0, 1/4)a.$$

Then, all the translation vectors can be expressed in the form $m_1\boldsymbol{a}_1 + m_2\boldsymbol{a}_2 + m_3\boldsymbol{a}_3$ with m_1, m_2, and m_3 integers.

(3) From the above expressions of $\boldsymbol{a}_1$, $\boldsymbol{a}_2$, and $\boldsymbol{a}_3$, we see that their lengths are all equal to $a/4$ and the angles between them are all 90°.

6-5 Conventional and primitive unit cells of a monoclinic lattice. Consider a monoclinic lattice with non-orthogonal angle γ. The conventional basis vectors are given by

$$\boldsymbol{a} = a\boldsymbol{e}_x,\ \ \boldsymbol{b} = b(\cos\gamma\, \boldsymbol{e}_x + \sin\gamma\, \boldsymbol{e}_y),\ \ \boldsymbol{c} = c\boldsymbol{e}_z.$$

Within the conventional cell, the atoms are at $(0, 0, 0)$ and $(\boldsymbol{a} + \boldsymbol{c})/2$.

(1) Show that the same lattice can be generated using the primitive vectors

$$\boldsymbol{a}_1 = (\boldsymbol{a} + \boldsymbol{c})/2,\ \ \boldsymbol{a}_2 = \boldsymbol{b},\ \ \boldsymbol{a}_3 = (\boldsymbol{a} - \boldsymbol{c})/2.$$

(2) Find the volumes of the conventional and primitive unit cells.

---

(1) Expressing $\boldsymbol{a}$, $\boldsymbol{b}$, and $\boldsymbol{c}$ in terms of $\boldsymbol{a}_1$, $\boldsymbol{a}_2$, and $\boldsymbol{a}_3$, we have

$$\boldsymbol{a} = \boldsymbol{a}_1 + \boldsymbol{a}_3,\ \ \boldsymbol{b} = \boldsymbol{a}_2,\ \ \boldsymbol{c} = \boldsymbol{a}_1 - \boldsymbol{a}_3.$$

From the above expressions, we see that $\boldsymbol{a}$, $\boldsymbol{b}$, and $\boldsymbol{c}$ can be expressed in terms of $\boldsymbol{a}_1$, $\boldsymbol{a}_2$, and $\boldsymbol{a}_3$ in the form $m_1\boldsymbol{a}_1 + m_2\boldsymbol{a}_2 + m_3\boldsymbol{a}_3$ with m_1, m_2, and m_3 integers. Therefore, the same lattice can be generated using the primitive vectors $\boldsymbol{a}_1$, $\boldsymbol{a}_2$, and $\boldsymbol{a}_3$.

(2) Assume that γ is an acute angle. The volume of the conventional unit cell is

$$\boldsymbol{a}\cdot(\boldsymbol{b}\times\boldsymbol{c}) = \boldsymbol{e}_x\cdot(-\cos\gamma\boldsymbol{e}_y+\sin\gamma\boldsymbol{e}_x)abc = abc\sin\gamma.$$

The volume of the primitive unit cell is

$$\begin{aligned}\left|\boldsymbol{a}_1\cdot(\boldsymbol{a}_2\times\boldsymbol{a}_3)\right| &= \frac{1}{4}\left|(\boldsymbol{a}+\boldsymbol{c})\cdot\left[\boldsymbol{b}\times(\boldsymbol{a}-\boldsymbol{c})\right]\right| \\ &= \frac{1}{2}\boldsymbol{a}\cdot(\boldsymbol{b}\times\boldsymbol{c}) = \frac{1}{2}abc\sin\gamma.\end{aligned}$$

Note that taking the absolute value of $\boldsymbol{a}_1\cdot(\boldsymbol{a}_2\times\boldsymbol{a}_3)$ is necessary because its value is negative. This is because one of the angles is obtuse.

6-6 BCC and FCC structures of iron. Iron crystallizes in the BCC structure at room temperature with the mass density given by $\rho = 7.86\ \mathrm{g/cm^3}$.

(1) Compute the radius of an iron atom in the BCC structure.

(2) At temperatures above 910°C, iron is in the FCC structure. If we neglect the temperature dependence of the radius of the iron atom, we can then find the density of iron in the FCC structure from that in the BCC structure. Determine whether iron expands or contracts when it undergoes the transformation from the BCC to FCC structure.

---

(1) Take atoms as balls that are packed as closely as allowed by the structure. In the BCC structure, the lattice constant a_{BCC} is related to the radius of an atom through $\sqrt{3}\,a_{\mathrm{BCC}} = 4r$ (along a body diagonal). That is, $r = \sqrt{3}\,a_{\mathrm{BCC}}/4$. In the BCC structure, there are two atoms in a conventional unit cell. Thus, the density is given by

$$\rho_{\mathrm{BCC}} = \frac{2m_{\mathrm{Fe}}}{a^3}.$$

From $\rho_{\mathrm{BCC}} = 7.86\ \mathrm{g/cm^3} = 7.86\times10^3\ \mathrm{kg/m^3}$, we have

$$a_{\mathrm{BCC}} = \left(\frac{2m_{\mathrm{Fe}}}{\rho}\right)^{1/3} \approx 0.286\,8\ \mathrm{nm},$$

and

$$r = \sqrt{3}\,a_{\mathrm{BCC}}/4 \approx 0.124\,2\ \mathrm{nm}.$$

(2) In the FCC structure, a_{FCC} is related to the radius of an atom through $\sqrt{2}\,a_{\text{FCC}} = 4r$ (along a face diagonal). With r unchanged, a_{FCC} is given by

$$a_{\text{FCC}} = 4r/\sqrt{2} \approx 0.351\ 3\ \text{nm}.$$

The density in the FCC structure is then given by

$$\rho_{\text{FCC}} = \frac{4m_{\text{Fe}}}{a_{\text{FCC}}^3} \approx 8,557\ \text{kg/m}^2.$$

Since $\rho_{\text{FCC}} > \rho_{\text{BCC}}$, iron contracts when it undergoes the transformation from the BCC to FCC structure.

6-7 Nearest and second-nearest neighbors in an HCP crystal. Assume that the nearest-neighbor distance in an HCP crystal is $2r$.

(1) Find the numbers of nearest and second-nearest neighbors of an atom in the HCP crystal.

(2) Find the second-nearest-neighbor distance in the HCP crystal.

---

(1) One layer and one atom above the layer in the HCP packing are shown in Fig. 6.1 for the purpose of illustrating the nearest and second-nearest neighbors. From the figure, it is seen that each atom has six nearest neighbors within the layer it belongs to, three in the layer above, and three in the layer below. Therefore, each atom has in total 12 nearest neighbors.

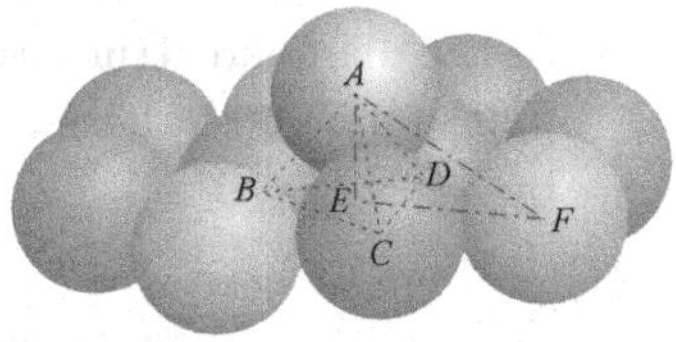

Fig. 6.1 Nearest and second-nearest neighbors in the HCP crystal.

Figure 6.1 shows that the ball located at A has three second nearest neighbors in the layer below. It has three more second nearest neighbors in the layer above. Therefore, each atom has in total 6 second-nearest neighbors.

(2) The triangle in dash-dotted lines is to be used to compute the second-nearest-neighbor distance (given by $\overline{AF}$) in the HCP crystal. The length of the side $\overline{AE}$ is the height of the tetrahedron whose edges are drawn in dotted lines. All the edges of the tetrahedron are of length $2r$. The height of each triangular face of the tetrahedron is $2r\cos 30^\circ = \sqrt{3}\,r$. Thus, the height of the tetrahedron is given by

$$\overline{AE} = \left[\,(2r)^2 - (2\sqrt{3}\,r/3)^2\,\right]^{1/2} = \sqrt{8/3}\,r.$$

The distance between E and F is given by

$$\overline{EF} = \sqrt{3}\,r + \sqrt{3}\,r/3 = 4r/\sqrt{3}.$$

Thus, the second-nearest-neighbor distance is given by

$$\overline{AF} = \left[\,\overline{AE}^2 + \overline{EF}^2\,\right]^{1/2} = 2\sqrt{2}\,r.$$

6-8 Diamond and body-centered tetragonal structures of gray tin. At about 13°C, gray tin (α-Sn) undergoes a phase transition to white tin (β-Sn). α-Sn has a diamond structure with a lattice constant 0.649 nm, while β-Sn has a body-centered tetragonal structure with $a = 0.583$ nm and $c = 0.318$ nm. Evaluate the mass densities in the two phases.

In the diamond structure, there are eight Sn atoms in a conventional unit cell. Thus, the mass density in the diamond structure is given by

$$\rho_{\text{diamond}} = \frac{8m_{\text{Sn}}}{a_{\text{diamond}}^3} \approx 5,768\ \text{kg/m}^3.$$

In the body-centered tetragonal structure, there are two Sn atoms in a conventional unit cell. Thus, the mass density in the body-centered tetragonal structure is given by

$$\rho_{\text{tetra}} = \frac{2m_{\text{Sn}}}{a_{\text{tetra}}^2 c_{\text{tetra}}} \approx 3,647\ \text{kg/m}^3.$$

6-9 Wurtzite and zincblende structures of GaN. GaN can take on both the wurtzite structure that is described with the primitive vectors

$$\boldsymbol{a}_1 = 0.160\,\boldsymbol{e}_x - 0.276\,\boldsymbol{e}_y\,\text{nm},\ \ \boldsymbol{a}_2 = 0.160\,\boldsymbol{e}_x + 0.276\,\boldsymbol{e}_y\,\text{nm},$$
$$\boldsymbol{a}_3 = 0.519\,\boldsymbol{e}_z\,\text{nm}$$

and the zincblende structure of lattice constant $a = 0.450$ nm. In both structures, each Ga atom is coordinated by four N atoms and *vice versa.* But, the next nearest neighbors are arranged differently in the two structures.

(1) Evaluate the volume of a primitive cell in the wurtzite structure.

(2) Compute the number density of Ga-N pairs in the wurtzite structure.

(3) Compute the number density of Ga-N pairs in the zincblende structure.

(4) Can we make GaN transform between the two structures by changing the pressure?

---

(1) The volume of a primitive cell in the wurtzite structure is given by

$$\begin{aligned} v_c^w &= \boldsymbol{a}_1 \cdot (\boldsymbol{a}_2 \times \boldsymbol{a}_3) \\ &= (0.160\,\boldsymbol{e}_x - 0.276\,\boldsymbol{e}_y) \cdot (-0.160\,\boldsymbol{e}_y + 0.276\,\boldsymbol{e}_x) \times 0.519 \\ &\approx 0.045\,84 \text{ nm}^3. \end{aligned}$$

(2) In the wurtzite structure, the number density of Ga-N pairs is given by

$$p_w = 2/v_c^w \approx 4.363 \times 10^{28} \text{ m}^{-3}.$$

(3) In the zincblende structure, the number density of Ga-N pairs is given by

$$p_z = 4/a_z^3 \approx 4.390 \times 10^{28} \text{ m}^{-3}.$$

(4) Since the difference between the number densities of Ga-N pairs in the two structures is very small, we can not make GaN transform between the two structures by changing the pressure. It turns out the two structures both transform into the rocksalt structure under high pressure.

6-10 Crystal structure of diamond. Consider the diamond.

(1) Write down the fractional coordinates of all the atoms within the conventional unit cell of the diamond lattice, *i.e.*, the coordinates of all the atoms within the conventional unit cell in units of the lattice constant a.

(2) Find all the nearest neighbors of an atom and the distance between two nearest-neighboring atoms in units of the lattice constant a.
(3) Find the number of second nearest neighbors of an atom.
(4) Compute the angles between all the pairs of the nearest-neighbor bonds around an atom.

---

(1) The conventional unit cell of the diamond structure is shown in Fig. 6.2. The tetrahedral bonding is also illustrated. The fractional coordinates of the eight atoms are

$$(0,0,0),(1/4,1/4,1/4),$$
$$(0,1/2,1/2),(1/2,0,1/2),$$
$$(1/2,1/2,0),(1/4,3/4,3/4),$$
$$(3/4,1/4,3/4),(3/4,3/4,1/4).$$

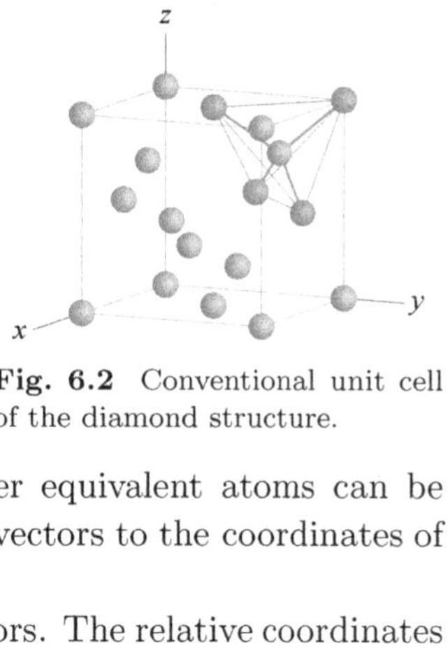

Fig. 6.2 Conventional unit cell of the diamond structure.

Note that the coordinates of other equivalent atoms can be obtained by adding proper lattice vectors to the coordinates of the relevant atoms.

(2) Each atom has four nearest neighbors. The relative coordinates of the nearest neighbors of the atom at $(1/4,3/4,3/4)a$ are

$$(-1/4,-1/4,-1/4)a,\ (-1/4,1/4,1/4)a,$$
$$(1/4,-1/4,1/4)a,\ (1/4,1/4,-1/4)a.$$

In the other set of relative coordinates, all relative coordinates have opposite signs to the above relative coordinates. The nearest-neighbor distance is thus given by $\sqrt{3}\,a/4$.

(3) As can be seen from Fig. 6.2, each atom has twelve second nearest neighbors. These are the nearest neighbors in an FCC structure without a multi-atom basis.

(4) Each atom has four bonds with its nearest neighbors. Let us consider two bonds with atoms at relative coordinates $\boldsymbol{r}_1$ and $\boldsymbol{r}_2$. The angle between these two bonds is given by $\theta = \cos^{-1}(\boldsymbol{r}_1 \cdot \boldsymbol{r}_2/|\boldsymbol{r}_1||\boldsymbol{r}_2|)$.
As can be seen from Fig. 6.2, the angles between different pairs of bonds are all equal. This common angle is found to be 109.47°.

6-11 Crystal structure of CaF_2. Calcium fluoride (CaF_2) has a face-centered cubic lattice with a basis of F^- at $(0,0,0)a$ and $(0,0,1/2)a$ and Ca^{2+} at $(1/4,1/4,1/4)a$ in the conventional unit cell.

(1) Find the coordination numbers of calcium and fluoride ions.

(2) Compute the lattice spacing along the [111] direction between successive planes of calcium and fluoride ions, respectively, in terms of the lattice constant a.

---

(1) The crystal structure of CaF_2 is shown in Fig. 6.3. Emphasized in the figure are the four-coordination of an F ion and the eight coordination of a Ca ion. Thus, the coordination number of an F ion is four while that of a Ca ion is eight.

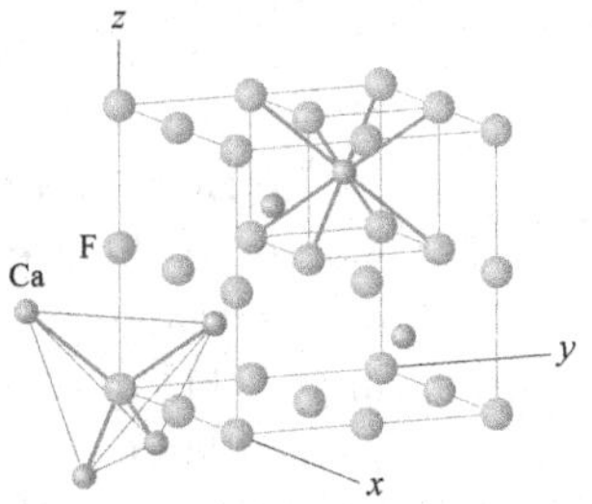

Fig. 6.3 Crystal structure of CaF_2.

(2) Two Ca lattice planes along the [111] direction can be seen in Fig. 6.3, with one containing Ca ions at $(1/4,-1/4,-1/4)a$, $(-1/4,1/4,-1/4)a$, and $(-1/4,-1/4,1/4)a$ and the other containing Ca ions at $(1/4,3/4,3/4)a$, $(3/4,1/4,3/4)a$, and $(3/4,3/4,1/4)a$. However, they are not successive Ca lattice planes because the Ca lattice plane containing the Ca ion at $(1/4,1/4,1/4)a$ is in between. Thus, the lattice spacing along the [111] direction between successive Ca lattice planes is half the sum of the perpendicular distances from the origin to the above-mentioned two Ca lattice planes and we have

$$d_{111}^{\text{Ca}} = \frac{1}{2}\frac{(1,1,1)}{\sqrt{3}}\cdot\big[(3/4,3/4,1/4)a-(1/4,-1/4,-1/4)a\big] = \frac{a}{\sqrt{3}}.$$

On the other hand, all the Ca ions form an FCC Bravais lattice of lattice constant a. The shortest reciprocal lattice vector in the [111] direction in this FCC Bravais lattice is $\boldsymbol{K}_0^{\text{Ca}} = (1,1,1)(2\pi/a)$. Thus, we can also obtain d_{111}^{Ca} in the following way

$$d_{111}^{\text{Ca}} = \frac{2\pi}{|\boldsymbol{K}_0^{\text{Ca}}|} = \frac{a}{\sqrt{3}}.$$

Note that all the F ions form an SC Bravais lattice of lattice constant $a/2$. The shortest reciprocal lattice vector in the [111] direction in this SC Bravais lattice is $\boldsymbol{K}_0^{\mathrm{F}} = (1,1,1)(4\pi/a)$. Thus, the lattice spacing along the [111] direction between successive F lattice planes is given by

$$d_{111}^{\mathrm{F}} = \frac{2\pi}{|\boldsymbol{K}_0^{\mathrm{F}}|} = \frac{a}{2\sqrt{3}}.$$

6-12 Monatomic BCC and FCC crystals. Consider the monatomic BCC and FCC crystals.

(1) Compute the linear packing factor (percentage of a lattice line that is taken up by atoms) for the $\langle 100\rangle$, $\langle 110\rangle$, and $\langle 111\rangle$ directions in each structure. Find the most closely packed directions in the BCC and FCC crystals.

(2) Compute the area packing factor (percentage of the area of a lattice plane that is taken up by atoms) for the $\{100\}$, $\{110\}$, and $\{111\}$ planes in each structure. Find the most closely packed lattice planes in the BCC and FCC crystals.

(3) Compute the volume packing factor (percentage of the volume of the crystal that is taken up by atoms) for the BCC and FCC crystals.

---

(1) (i) BCC crystal.

Shown in Fig. 6.4 is the BCC crystal structure packed most closely with balls. For a monatomic BCC crystal, the radius r of an atom is related to the lattice constant a through $4r = \sqrt{3}\,a$ (along a body diagonal) for most close packing. Thus, $r = \sqrt{3}\,a/4$. Along a $\langle 100\rangle$ direction, the interval occupied by atoms within distance a is $2r = \sqrt{3}\,a/2$. Thus, the linear packing factor along a $\langle 100\rangle$ direction is

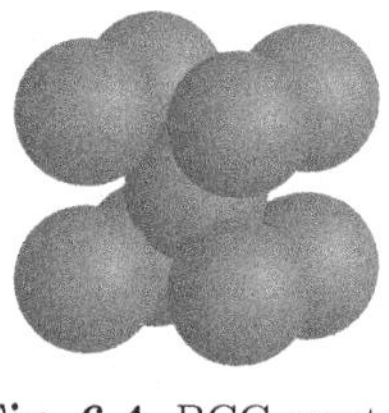

Fig. 6.4 BCC crystal structure packed most closely with balls.

$$f_{\ell,100}^{\mathrm{BCC}} = \frac{2r}{a} = \frac{\sqrt{3}}{2} \approx 0.866.$$

Along a $\langle 110\rangle$ direction, the interval occupied by atoms within distance $\sqrt{2}\,a$ is $2r = \sqrt{3}\,a/2$. Thus, the linear packing factor along a $\langle 110 \rangle$ direction is

$$f_{\ell,110}^{\mathrm{BCC}} = \frac{2r}{\sqrt{2}\,a} = \frac{\sqrt{3}}{2\sqrt{2}} \approx 0.612.$$

Along a $\langle 111 \rangle$ direction, the interval occupied by atoms within distance $\sqrt{3}\,a$ is $4r = \sqrt{3}\,a$. Thus, the linear packing factor along a $\langle 111 \rangle$ direction is

$$f_{\ell,111}^{\mathrm{BCC}} = \frac{4r}{\sqrt{3}\,a} = 1.$$

Thus, the most closely packed directions in a BCC crystal are $\langle 111 \rangle$ directions.

(ii) FCC crystal.
Shown in Fig. 6.5 is the FCC crystal structure packed most closely with balls. For a monatomic FCC crystal, the radius r of an atom is related to the lattice constant a through $4r = \sqrt{2}\,a$ (along a face diagonal) for most close packing. Thus, $r = a/2\sqrt{2}$. Along a $\langle 100 \rangle$ direction, the interval occupied by atoms within distance a is $2r = a/\sqrt{2}$. Thus, the linear packing factor along a $\langle 100 \rangle$ direction is

Fig. 6.5 FCC crystal structure packed most closely with balls.

$$f_{\ell,100}^{\mathrm{FCC}} = \frac{2r}{a} = \frac{1}{\sqrt{2}} \approx 0.707.$$

Along a $\langle 110 \rangle$ direction, the interval occupied by atoms within distance $\sqrt{2}\,a$ is $4r = \sqrt{2}\,a$. Thus, the linear packing factor along a $\langle 110 \rangle$ direction is

$$f_{\ell,110}^{\mathrm{FCC}} = \frac{2r}{a} = 1.$$

Along a $\langle 111 \rangle$ direction, the interval occupied by atoms within distance $\sqrt{3}\,a$ is $2r = a/\sqrt{2}$. Thus, the linear packing factor along a $\langle 110 \rangle$ direction is

$$f_{\ell,111}^{\mathrm{FCC}} = \frac{2r}{\sqrt{3}\,a} = \frac{1}{\sqrt{6}} \approx 0.408.$$

Thus, the most closely packed directions in an FCC crystal are $\langle 110 \rangle$ directions.

(2) (i) BCC crystal.

In a $\{100\}$ plane, the part occupied by atoms within area a^2 is $\pi r^2 = 3\pi a^2/16$. Thus, the area packing factor in a $\{100\}$ plane is

$$f_{a,100}^{\rm BCC} = \frac{\pi r^2}{a^2} = \frac{3\pi}{16} \approx 0.589.$$

In a $\{110\}$ plane, the part occupied by atoms within area $\sqrt{2}\,a^2$ is $2\pi r^2 = 3\pi a^2/8$. Thus, the area packing factor in a $\{110\}$ plane is

$$f_{a,110}^{\rm BCC} = \frac{2\pi r^2}{\sqrt{2}\,a^2} = \frac{3\pi}{8\sqrt{2}} \approx 0.833.$$

A $\{111\}$ plane does not cut a body-center ball through its center. The perpendicular distance of the (111) plane to the center of the ball at $(1/2, 1/2, 1/2)a$ is $\sqrt{3}\,a/2-(1,0,0)a\cdot(1,1,1)/\sqrt{3} = \sqrt{3}\,a/6 = 2r/3$. The area that the ball occupies in the (111) plane is then given by $5\pi r^2/9 = 5\pi a^2/48$. Thus, in a $\{111\}$ plane, the part occupied by atoms within area $\sqrt{3}\,a^2/2$ is $3 \times (\pi r^2/6) + 5\pi r^2/9 = 19\pi r^2/18$. Thus, the area packing factor in a $\{111\}$ plane is

$$f_{a,111}^{\rm BCC} = \frac{19\pi r^2/18}{\sqrt{3}\,a^2/2} = \frac{19\pi}{48\sqrt{3}} \approx 0.718.$$

Thus, the most closely packed lattice planes a BCC crystal are $\{110\}$ planes.

(ii) FCC crystal.

In a $\{100\}$ plane, the part occupied by atoms within area a^2 is $2\pi r^2 = \pi a^2/4$. Thus, the area packing factor in a $\{100\}$ plane is

$$f_{a,100}^{\rm FCC} = \frac{2\pi r^2}{a^2} = \frac{\pi}{4} \approx 0.785.$$

In a $\{110\}$ plane, the part occupied by atoms within area $\sqrt{2}\,a^2$ is $2\pi r^2 = \pi a^2/4$. Thus, the area packing factor in a $\{110\}$ plane is

$$f_{a,110}^{\rm FCC} = \frac{2\pi r^2}{\sqrt{2}\,a^2} = \frac{\pi}{4\sqrt{2}} \approx 0.555.$$

In a $\{111\}$ plane, the part occupied by atoms within area $\sqrt{3}\,a^2/2$ is $\pi r^2/2 + 3\pi r^2/2 = 2\pi r^2 = \pi a^2/4$. Thus, the area packing factor in a $\{111\}$ plane is

$$f_{a,111}^{\rm FCC} = \frac{2\pi r^2}{\sqrt{3}\,a^2/2} = \frac{\pi}{2\sqrt{3}} \approx 0.907.$$

Thus, the most closely packed lattice planes in an FCC crystal are $\{111\}$ planes.

(3) In a BCC crystal, there are two atoms in a conventional cell of volume a^3. With $r = \sqrt{3}\,a/4$, these two atoms occupy a volume of $2 \times 4\pi r^3/3 = \sqrt{3}\,\pi a^3/8$. Thus, the volume packing factor is

$$f_v^{\rm BCC} = \frac{\sqrt{3}\,\pi a^3/8}{a^3} = \frac{\sqrt{3}\,\pi}{8} \approx 0.680.$$

In an FCC crystal, there are four atoms in a conventional cell of volume a^3. With $r = a/2\sqrt{2}$, these four atoms occupy a volume of $4 \times 4\pi r^3/3 = \pi a^3/3\sqrt{2}$. Thus, the volume packing factor is

$$f_v^{\rm FCC} = \frac{\pi a^3/3\sqrt{2}}{a^3} = \frac{\pi}{3\sqrt{2}} \approx 0.740.$$

6-13 Hypothetical ceramic material of AX type. A hypothetical ceramic material of AX type is known to have a mass density of 2.65 g/cm^3 and a unit cell of cubic symmetry with a lattice constant of 0.43 nm. The atomic weights of the A and X elements are 86.6 and 40.3 g/mol, respectively. Determine the possible crystal structure(s) for this material among the rocksalt, cesium chloride, and zincblende structures.

For the CsCl structure, there are one A atom and one X atom in a conventional cell. The density in such a structure is given by

$$\rho_{\rm CsCl} = \frac{m_{\rm A} + m_{\rm X}}{a^3} \approx 2,650 \text{ kg/m}^3 = 2.65 \text{ g/cm}^3.$$

For both the rocksalt and zincblende structures, there are four A atoms and four X atoms in a conventional cell. The density in these structures is given by

$$\rho_{\rm rs,\ zb} = \frac{4(m_{\rm A} + m_{\rm X})}{a^3} \approx 10,601 \text{ kg/m}^3 = 10.601 \text{ g/cm}^3.$$

Thus, the possible crystal structure for this ceramic material is the CsCl structure.

6-14 Crystal structure of CsCl. Consider the CsCl structure.

(1) The ionic radii of Cs and Cl are 0.170 nm and 0.181 nm, respectively. Compute the expected mass density of CsCl in the normal CsCl structure.

(2) Compute the atomic packing factor for CsCl in its normal structure.

(3) Compute the atomic packing factor for a hypothetical form of CsCl in which it adopts the rocksalt (NaCl) structure. Based on this result explain why CsCl adopts its normal structure.

---

(1) For the most closely packed CsCl structure, the lattice constant a is given by

$$a = 2(r_{\rm Cs} + r_{\rm Cl})/\sqrt{3} \approx 0.405 \text{ nm}.$$

Thus, the expected density is

$$\rho = \frac{m_{\rm Cs} + m_{\rm Cl}}{a^3} \approx 4.199 \text{ g/cm}^3.$$

(2) The volume occupied by atoms in a conventional cell is $4\pi(r_{\rm Cs}^3 + r_{\rm Cl}^3)/3$. Thus, the atomic packing factor is

$$f_{\rm CsCl} = \frac{4\pi(r_{\rm Cs}^3 + r_{\rm Cl}^3)}{3a^3} \approx 0.684.$$

(3) The three possible structures for CsCl are given in Fig. 6.6 with Cs and Cl ions are most closely packed.

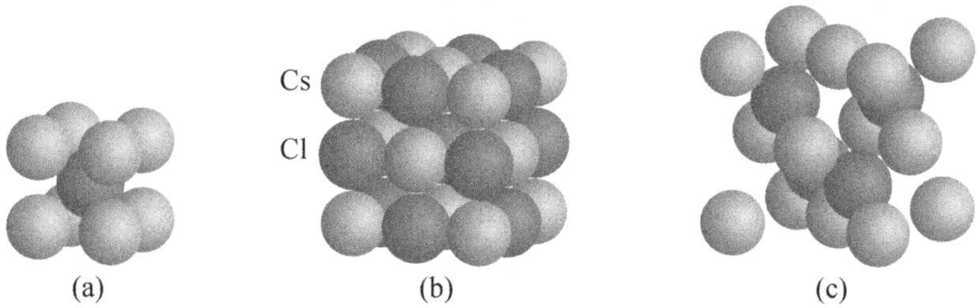

Fig. 6.6 Possible structures of CsCl. (a) Normal CsCl structure. (b) Rocksalt structure. (c) Zincblende structure.

For the rocksalt structure, the lattice constant is

$$a = 2(r_{\rm Cs} + r_{\rm Cl}) = 0.702 \text{ nm}.$$

Since there are four Cs ions and four Cl ions in a conventional cell of the rocksalt structure, the atomic packing factor is given by

$$f_{\rm rs} = \frac{16\pi(r_{\rm Cs}^3 + r_{\rm Cl}^3)}{3a^3} \approx 0.525.$$

For the zincblende structure, the lattice constant is to be determined through $|(1/4, 1/4, 1/4)a| = r_{\rm Cs} + r_{\rm Cl}$ which is obtained from the fact that, if a Cs ion is at $(0, 0, 0)$, then a Cl ion is at $(1/4, 1/4, 1/4)a$. We have

$$a = 4(r_{\rm Cs} + r_{\rm Cl})/\sqrt{3} \approx 0.811 \text{ nm}.$$

Since there are four Cs ions and four Cl ions in a conventional cell of the zincblende structure, the atomic packing factor is given by

$$f_{\rm zb} = \frac{16\pi(r_{\rm Cs}^3 + r_{\rm Cl}^3)}{3a^3} \approx 0.341.$$

A greater atomic packing factor implies that the positive and negative ions can take a greater advantage of the attractive Coulomb interaction between them to lower their energies. From the above results, we see that the atomic packing factor is greater in the normal CsCl structure than those in the rocksalt and zincblende structures. This explains why CsCl adopts its normal structure (the CsCl structure).

Chapter 7

Scattering of X-Rays by a Crystal

(1) *Scattering amplitude* $f(\vartheta, \phi)$.

(2) *Differential scattering cross-section* $\dfrac{\mathrm{d}\sigma}{\mathrm{d}\Omega} = |f(\vartheta, \phi)|^2$.

(3) *Total scattering cross-section* $\sigma = \displaystyle\int \mathrm{d}\Omega\, |f(\vartheta, \phi)|^2$.

(4) *Atomic form factor* $F(\boldsymbol{K}) = \displaystyle\sum_j \alpha_j(\omega)\mathrm{e}^{-\mathrm{i}\boldsymbol{K}\cdot\boldsymbol{r}_j}$.

An approximate expression for $F(\boldsymbol{K})$ in terms of electron charge density $\rho(r)$ is

$$F(\boldsymbol{K}) = -\frac{4\pi}{e}\int_0^\infty \mathrm{d}r\, r^2\rho(r)\frac{\sin(|\boldsymbol{K}|r)}{|\boldsymbol{K}|r}.$$

(5) *Geometric structure factor*

$$S(\boldsymbol{K}) = \sum_\kappa \mathrm{e}^{-\mathrm{i}\boldsymbol{K}\cdot\boldsymbol{d}_\kappa}F_\kappa(\boldsymbol{K}).$$

(6) *Scattering of X-rays by a crystal*

$$\frac{\mathrm{d}\sigma}{\mathrm{d}\Omega} = \frac{1}{2}r_0^2|S(\boldsymbol{K})|^2(1+\cos^2\vartheta)\left|\sum_n \mathrm{e}^{-\mathrm{i}\boldsymbol{K}\cdot\boldsymbol{R}_n}\right|^2,$$

$$\sigma = \frac{8\pi}{3}r_0^2\left|\sum_n \mathrm{e}^{-\mathrm{i}\boldsymbol{K}\cdot\boldsymbol{R}_n}\right|^2|S(\boldsymbol{K})|^2.$$

(7) *Condition for nonzero scattering*

$$\boldsymbol{K}\cdot\boldsymbol{R}_n \equiv (\boldsymbol{k}'-\boldsymbol{k})\cdot\boldsymbol{R}_n = 2m\pi,$$
$$m = 0,\ \pm1,\ \pm2,\ \cdots.$$

7-1 Evaluation of the differential scattering cross-section. A beam of X-rays is incident on a monatomic crystal with 7×10^{11} photons per second hitting perpendicularly a 0.001 m×0.001 m surface of the crystal. The atom number density in the crystal is $5 \times 10^{26}\ \mathrm{m}^{-3}$. A counter, placed at a certain fixed angle θ from the crystal and covering a solid angle of 10^{-5} steradians, registers approximately 140 photons per second. Find the value of the differential scattering cross-section $\mathrm{d}\sigma(\theta)/\mathrm{d}\Omega$.

The number of photons scattered into a unit solid angle per second about the θ direction is given by

$$\mathrm{d}N = \frac{140}{10^{-5}} = 1.4 \times 10^7\ \mathrm{s}^{-1} \cdot \mathrm{rad}^{-1}.$$

The incident photon flux is

$$j_{\mathrm{inc}} = \frac{7 \times 10^{11}}{0.001^2} = 7 \times 10^{17}\ \mathrm{s}^{-1} \cdot \mathrm{m}^{-2}.$$

The differential scattering cross-section is then given by

$$\frac{\mathrm{d}\sigma(\theta)}{\mathrm{d}\Omega} = \frac{\mathrm{d}N}{j_{\mathrm{inc}}} = \frac{1.4 \times 10^7}{7 \times 10^{17}} = 2 \times 10^{-11}\ \mathrm{m}^2/\mathrm{rad} = 2 \times 10^{11}\ \mathrm{Mb/rad}.$$

Here 1 b (barn) $= 10^{-28}\ \mathrm{m}^2$.

7-2 Wave vector transfer in X-ray diffraction. The wave vector transfer in an X-ray diffraction experiment is given by $\boldsymbol{q} = \boldsymbol{k}' - \boldsymbol{k}$. Show that the magnitude of $\boldsymbol{q}$ for elastic scattering is given by $q = 2k\sin\theta$ with $2\theta = \angle(\boldsymbol{k}', \boldsymbol{k})$.

Taking the norm of both sides of $\boldsymbol{q} = \boldsymbol{k}' - \boldsymbol{k}$, we have

$$\begin{aligned} q = |\boldsymbol{k}' - \boldsymbol{k}| &= \left(k^2 + k'^2 - 2\boldsymbol{k}' \cdot \boldsymbol{k}\right)^{1/2} \\ &= \left[\,k^2 + k'^2 - 2kk'\cos(2\theta)\,\right]^{1/2} \\ &= k\left[\,2 - 2\cos(2\theta)\,\right]^{1/2} = 2k\sin\theta, \end{aligned}$$

where we have made use of $k' = k$ for elastic scattering.

7-3 Charged particle in a static magnetic field. A particle of mass m and charge q moves in a plane perpendicular to a uniform static magnetic field B.

(1) Compute the total energy radiated per unit time. Assume that the particle is ultra-relativistic.

(2) If the particle has a total energy $E_0 = \gamma_0 mc^2$ at time $t = 0$, show that its energy is $E = \gamma mc^2 < E_0$ at time t with

$$t \approx \frac{3(4\pi\epsilon_0)m^3c^3}{2q^4B^2}\left(\frac{1}{\gamma} - \frac{1}{\gamma_0}\right).$$

---

(1) Since the particle moves in a plane perpendicular to a uniform static magnetic field, its equation of motion can be written as

$$\frac{\mathrm{d}(m\gamma v)}{\mathrm{d}t} = qBv$$

with m the rest mass of the particle and $\gamma = 1/\sqrt{1 - v^2/c^2}$. Making use of $\mathrm{d}\gamma/\mathrm{d}t = -(2v^2/\gamma^3 c^2)\mathrm{d}v/\mathrm{d}t$, we have

$$\gamma\left(1 - \frac{2v^2}{\gamma^4 c^2}\right)\frac{\mathrm{d}v}{\mathrm{d}t} = \frac{qB}{m}v.$$

The second term in the parentheses on the left hand side is usually much smaller than the first term even in the ultra-relativistic limit unless v is exceedingly close to c. We thus neglect it and write the above equation as

$$a = \frac{\mathrm{d}v}{\mathrm{d}t} = \frac{qB}{\gamma m}v.$$

The energy radiated by the particle per unit time is given by

$$P = \frac{2q^2}{3(4\pi\epsilon_0)c^3}\gamma^4 a^2 = \frac{2q^4 B^2}{3(4\pi\epsilon_0)c^3 m^2}(\gamma v)^2 = \frac{2q^4 B^2}{3(4\pi\epsilon_0)c m^2}(\gamma^2 - 1),$$

where we have made use of $(\gamma v)^2 = c^2(\gamma^2 - 1)$.

(2) On the other hand, the rate of the change of the energy of the particle is given by

$$\frac{\mathrm{d}E}{\mathrm{d}t} = mc^2\frac{\mathrm{d}\gamma}{\mathrm{d}t}.$$

From the fact that the particle looses its energy through electromagnetic radiation, we have

$$\frac{\mathrm{d}E}{\mathrm{d}t} = mc^2\frac{\mathrm{d}\gamma}{\mathrm{d}t} = -P = -\frac{2q^4 B^2}{3(4\pi\epsilon_0)c m^2}(\gamma^2 - 1).$$

We thus have

$$\frac{\mathrm{d}\gamma}{\mathrm{d}t} = -\frac{2q^4 B^2}{3(4\pi\epsilon_0)c^3 m^3}(\gamma^2 - 1).$$

Multiplying both sides with $\mathrm{d}t$ and dividing both sides with $\gamma^2 - 1$, we have

$$\frac{\mathrm{d}\gamma}{\gamma^2 - 1} = -\frac{2q^4 B^2}{3(4\pi\epsilon_0)c^3 m^3}\mathrm{d}t.$$

Integrating both sides from $t = 0$ ($\gamma = \gamma_0$) to t (γ), we obtain

$$\frac{1}{2}\left(\ln\frac{\gamma-1}{\gamma+1} + \ln\frac{\gamma_0+1}{\gamma_0-1}\right) = -\frac{2q^4B^2}{3(4\pi\epsilon_0)c^3m^3}t.$$

For γ, $\gamma_0 \gg 1$, we can make the following approximations

$$\ln\frac{\gamma-1}{\gamma+1} \approx -\frac{2}{\gamma},\ \ln\frac{\gamma_0+1}{\gamma_0-1} \approx \frac{2}{\gamma_0}.$$

We then have

$$t \approx \frac{3(4\pi\epsilon_0)c^3m^3}{2q^4B^2}\left(\frac{1}{\gamma} - \frac{1}{\gamma_0}\right).$$

7-4 Decay of a charged particle through radiation. A particle of mass m and charge q moves in a circle due to the force $\boldsymbol{F} = -(q^2/4\pi\epsilon_0 r^2)\hat{\boldsymbol{r}}$. Assume that the particle moves non-relativistically.

(1) Find the total energy E of the particle as a function of r and evaluate the power P radiated as a function of r.

(2) From $P = -\mathrm{d}E/\mathrm{d}t$, find $\mathrm{d}r/\mathrm{d}t$.

(3) If the particle starts with $r = r_0$ at $t = 0$, compute the time t that the particle takes to reach the $r = 0$ point.

(4) Evaluate t for the particle being an electron. Compare t with the lifetime of a hydrogen atom in the $2p$ state.

---

(1) The potential energy is given by

$$U(r) = -\frac{q^2}{4\pi\epsilon_0 r}.$$

The equation of motion gives us

$$\frac{q^2}{4\pi\epsilon_0 r^2} = m\frac{v^2}{r}$$

from which it follows that

$$\frac{1}{2}mv^2 = \frac{q^2}{2(4\pi\epsilon_0)r},\ a = \frac{v^2}{r} = \frac{q^2}{4\pi\epsilon_0 mr^2}.$$

The total energy of the particle is then given by

$$E = \frac{1}{2}mv^2 + U(r) = -\frac{q^2}{2(4\pi\epsilon_0)r}.$$

The power P radiated by the particle is given by

$$P = \frac{2q^2a^2}{3(4\pi\epsilon_0)c^3} = \frac{2q^6}{3(4\pi\epsilon_0)^3c^3m^2r^4}.$$

(2) The rate of change of the total energy is given by

$$\frac{\mathrm{d}E}{\mathrm{d}t} = \frac{q^2}{2(4\pi\epsilon_0)r^2}\frac{\mathrm{d}r}{\mathrm{d}t}.$$

From $\mathrm{d}E/\mathrm{d}t = -P$, we have

$$r^2\mathrm{d}r = -\frac{4q^4}{3(4\pi\epsilon_0)^2c^3m^2}\mathrm{d}t.$$

(3) Integrating both sides from $t = 0$ $(r = r_0)$ to t $(r = 0)$, we have

$$t = \frac{(4\pi\epsilon_0)^2c^3m^2r_0^3}{4q^4}.$$

(4) For the particle being an electron, we have $q = -e$, $m = m_e$, $r_0 = a_0$. We then have

$$t = \frac{(4\pi\epsilon_0)^2c^3m_e^2a_0^3}{4e^4} = \frac{1}{4}\left(\frac{4\pi\epsilon_0 a_0}{e^2}\right)^2\left(m_ec^2\right)^2\frac{a_0}{c} \approx 1.556\times 10^{-11}\ \mathrm{s}.$$

The lifetime of a hydrogen atom in the $2p$ state is approximately 10^{-8} s. Thus, the electron would fall onto the nucleus much faster than the decay of a hydrogen atom in the $2p$ state.

7-5 Scattering by an atom. A charged particle is scattered elastically on the electric potential $\phi(\boldsymbol{r})$ produced by an atom that has Z electrons and Z protons in the nucleus.

(1) In the Born approximation, derive a general expression for the differential scattering cross-section $\mathrm{d}\sigma/\mathrm{d}\Omega$ in terms of a given charge distribution of electrons $n(\boldsymbol{r})$. The nucleus is treated as a point charge Ze.

(2) Derive a general expression for the differential scattering cross-section for the forward direction $(\theta = 0)$ in terms of $n(\boldsymbol{r})$.

(3) Compute the differential scattering cross-section for the scattering of the particle on a hydrogen atom in the ground state.

(4) Simplify the general expression of $\mathrm{d}\sigma/\mathrm{d}\Omega$ for $|\boldsymbol{q}|a \ll 1$, where $\boldsymbol{q}$ is the wave vector transfer and a the size of the electron charge distribution in the atom.

---

(1) The electric potential due to the electrons and the protons is given by

$$\phi(\boldsymbol{r}) = \frac{Ze}{(4\pi\epsilon_0)r} + \frac{1}{4\pi\epsilon_0}\int \mathrm{d}\boldsymbol{r}'\,\frac{n(\boldsymbol{r}')}{|\boldsymbol{r}-\boldsymbol{r}'|}.$$

The potential energy of the incoming charged particle is given by $Q\phi(\boldsymbol{r})$ with Q its charge

$$V(\boldsymbol{r}) = \frac{ZeQ}{(4\pi\epsilon_0)r} + \frac{Q}{4\pi\epsilon_0}\int \mathrm{d}\boldsymbol{r}'\, \frac{n(\boldsymbol{r}')}{|\boldsymbol{r}-\boldsymbol{r}'|}.$$

In the first-order Born approximation, the scattering amplitude is given by

$$\begin{aligned}
f(\theta) &= -\frac{m}{2\pi\hbar^2}\int \mathrm{d}\boldsymbol{r}'\, e^{\mathrm{i}\boldsymbol{q}\cdot\boldsymbol{r}'}V(\boldsymbol{r}') \\
&= -\frac{m}{2\pi\hbar^2}\frac{Q}{4\pi\epsilon_0}\left[Ze\int \mathrm{d}\boldsymbol{r}'\,\frac{1}{r'}e^{\mathrm{i}\boldsymbol{q}\cdot\boldsymbol{r}'} + \int \mathrm{d}\boldsymbol{r}''\int \mathrm{d}\boldsymbol{r}'\, e^{\mathrm{i}\boldsymbol{q}\cdot\boldsymbol{r}'}\frac{n(\boldsymbol{r}'')}{|\boldsymbol{r}'-\boldsymbol{r}''|}\right] \\
&= -\frac{m}{2\pi\hbar^2}\frac{Q}{4\pi\epsilon_0}\left[Ze\int \mathrm{d}\boldsymbol{r}'\,\frac{1}{r'}e^{\mathrm{i}\boldsymbol{q}\cdot\boldsymbol{r}'}\right. \\
&\qquad\left. + \int \mathrm{d}\boldsymbol{r}''\int \mathrm{d}\boldsymbol{r}'\,\frac{1}{r'}e^{\mathrm{i}\boldsymbol{q}\cdot\boldsymbol{r}'}e^{\mathrm{i}\boldsymbol{q}\cdot\boldsymbol{r}''}n(\boldsymbol{r}'')\right] \\
&= -\frac{m}{2\pi\hbar^2}\frac{Q}{4\pi\epsilon_0}\int \mathrm{d}\boldsymbol{r}'\,\frac{1}{r'}e^{\mathrm{i}\boldsymbol{q}\cdot\boldsymbol{r}'}\left[Ze + \int \mathrm{d}\boldsymbol{r}''\, e^{\mathrm{i}\boldsymbol{q}\cdot\boldsymbol{r}''}n(\boldsymbol{r}'')\right] \\
&= -\frac{2Qm}{(4\pi\epsilon_0)\hbar^2q^2}(Ze + n_{\boldsymbol{q}}),
\end{aligned}$$

where $n_{\boldsymbol{q}}$ is the Fourier components of $n(\boldsymbol{r})$, $n_{\boldsymbol{q}} = \int \mathrm{d}\boldsymbol{r}\, e^{\mathrm{i}\boldsymbol{q}\cdot\boldsymbol{r}}n(\boldsymbol{r})$, and we have made use of the following Fourier transform of $1/r$

$$\begin{aligned}
\int \mathrm{d}\boldsymbol{r}\,\frac{1}{r}\mathrm{e}^{\mathrm{i}\boldsymbol{q}\cdot\boldsymbol{r}} &= \lim_{\delta\to 0^+}\int \mathrm{d}\boldsymbol{r}\,\frac{1}{r}\mathrm{e}^{\mathrm{i}\boldsymbol{q}\cdot\boldsymbol{r}-\delta r} \\
&= 2\pi\lim_{\delta\to 0^+}\int_0^\infty \mathrm{d}r\, r\int_{-1}^{1}\mathrm{d}\cos\theta\, \mathrm{e}^{\mathrm{i}qr\cos\theta-\delta r} \\
&= \frac{2\pi}{\mathrm{i}q}\lim_{\delta\to 0^+}\int_0^\infty \mathrm{d}r\left[e^{(\mathrm{i}q-\delta)r} - e^{-(\mathrm{i}q+\delta)r}\right] \\
&= \frac{2\pi}{\mathrm{i}q}\lim_{\delta\to 0^+}\left(-\frac{1}{\mathrm{i}q-\delta} - \frac{1}{\mathrm{i}q+\delta}\right) = \frac{4\pi}{q^2}.
\end{aligned}$$

In the Born approximation, the differential scattering cross-section is given by

$$\frac{\mathrm{d}\sigma}{\mathrm{d}\Omega} = |f(\theta)|^2 = \frac{4Q^2m^2}{(4\pi\epsilon_0)^2\hbar^4q^4}\left|Ze + n_{\boldsymbol{q}}\right|^2$$

with $q = 2k\sin(\vartheta/2)$.

(2) For forward scattering ($\vartheta = 0$), we have $q = 0$ and $n_{\boldsymbol{q}=0} = -Ze$. Since both the numerator and the denominator of the scattering amplitude tend to zero, we have to resort to L'Hopital's rule to find its limiting value. Assuming a spherically symmetric distribution of electron charges, $n(\boldsymbol{r}) = n(r)$, we have

$$f(0) = -\frac{2Qm}{(4\pi\epsilon_0)\hbar^2} \lim_{q\to 0} \frac{1}{q^2}\left[Ze + \int \mathrm{d}\boldsymbol{r}\; e^{\mathrm{i}qr\cos\theta} n(r) \right]$$
$$= -\frac{2Qm}{(4\pi\epsilon_0)\hbar^2} \lim_{q\to 0} \frac{1}{2q}\left[\mathrm{i}\int \mathrm{d}\boldsymbol{r}\; r\cos\theta e^{\mathrm{i}qr\cos\theta} n(r) \right]$$
$$= \frac{Qm}{(4\pi\epsilon_0)\hbar^2} \int \mathrm{d}\boldsymbol{r}\; r^2 n(r)\cos^2\theta = \frac{4\pi Qm}{3(4\pi\epsilon_0)\hbar^2} \int_0^\infty \mathrm{d}r\; r^4 n(r).$$

The differential scattering cross-section for forward scattering is then given by

$$\left.\frac{\mathrm{d}\sigma}{\mathrm{d}\Omega}\right|_{\theta=0} = |f(0)|^2 = \left(\frac{4\pi Qm}{3(4\pi\epsilon_0)\hbar^2}\right)^2 \left| \int_0^\infty \mathrm{d}r\; r^4 n(r) \right|^2$$

(3) For a hydrogen atom in the ground state, we have $Z = 1$ and $n(\boldsymbol{r}) = n(r) = -e|\psi_{100}|^2 = -(e/\pi a_0^3)\mathrm{e}^{-2r/a_0}$. The Fourier transform of $n(\boldsymbol{r})$ is given by

$$n_{\boldsymbol{q}} = -\frac{e}{\pi a_0^3} \int \mathrm{d}\boldsymbol{r}\; e^{\mathrm{i}\boldsymbol{q}\cdot\boldsymbol{r} - 2r/a_0} = -\frac{e}{(1 + q^2 a_0^2/4)^2}.$$

The differential scattering cross-section for a hydrogen atom in the ground state is then given by

$$\frac{\mathrm{d}\sigma}{\mathrm{d}\Omega} = \frac{e^2 Q^2 m^2 a_0^4}{(4\pi\epsilon_0)^2\hbar^4} \frac{\left(1 + q^2 a_0^2/8\right)^2}{\left(1 + q^2 a_0^2/4\right)^4}.$$

(4) For $|\boldsymbol{q}|a \ll 1$, we can expand the exponential function $\mathrm{e}^{\mathrm{i}\boldsymbol{q}\cdot\boldsymbol{r}}$ in the expression for $n_{\boldsymbol{q}}$. Keeping terms up to the second order, we have for a spherically symmetric distribution of electron charges

$$Ze + n_{\boldsymbol{q}} \approx Ze + \int \mathrm{d}\boldsymbol{r}\; \left[1 + \mathrm{i}\boldsymbol{q}\cdot\boldsymbol{r} - (\boldsymbol{q}\cdot\boldsymbol{r})^2/2 \right] n(r)$$
$$= Ze + \int \mathrm{d}\boldsymbol{r}\; n(r) - \frac{2\pi}{3} q^2 \int_0^\infty \mathrm{d}r\; r^4 n(r)$$
$$= -\frac{2\pi}{3} q^2 \int_0^\infty \mathrm{d}r\; r^4 n(r).$$

The general expression for the differential scattering cross-section then becomes

$$\frac{\mathrm{d}\sigma}{\mathrm{d}\Omega} = \frac{16\pi^2 Q^2 m^2}{9(4\pi\epsilon_0)^2\hbar^4} \left[\int_0^\infty \mathrm{d}r\; r^4 n(r) \right]^2.$$

7-6 Incident and scattered X-ray beams. In an X-ray diffraction experiment, the electric fields in the incident and scattered X-ray beams are given by $\boldsymbol{E}_{\rm inc} = \boldsymbol{\epsilon}_{\boldsymbol{k}} E_0 \mathrm{e}^{\mathrm{i}(\boldsymbol{k}\cdot\boldsymbol{r}-\omega t)}$ and $\boldsymbol{E}_{\rm sc} = \boldsymbol{\epsilon}'_{\boldsymbol{k}'} f(\vartheta,\phi) E_0 \mathrm{e}^{\mathrm{i}(kr-\omega t)}/r$, respectively.

(1) Find the magnetic inductions $\boldsymbol{B}_{\rm inc}$ and $\boldsymbol{B}_{\rm sc}$ in the incident and scattered X-ray beams.

(2) Write down the real parts of $\boldsymbol{E}_{\rm inc}$, $\boldsymbol{B}_{\rm inc}$, $\boldsymbol{E}_{\rm sc}$, and $\boldsymbol{B}_{\rm sc}$, with the polarization vectors $\boldsymbol{\epsilon}_{\boldsymbol{k}}$ and $\boldsymbol{\epsilon}'_{\boldsymbol{k}'}$ taken as complex functions.

(3) Compute the energy densities in the incident and scattered X-ray beams at time t using the real parts of the fields.

(4) Compute the time averages of the above-obtained energy densities.

---

(1) From Maxwell's equations $\boldsymbol{\nabla}\times\boldsymbol{E} = -\partial\boldsymbol{B}/\partial t$, we have

$$\partial\boldsymbol{B}_{\rm inc}/\partial t = -\mathrm{i}(\boldsymbol{k}\times\boldsymbol{\epsilon}_{\boldsymbol{k}})E_0\mathrm{e}^{\mathrm{i}(\boldsymbol{k}\cdot\boldsymbol{r}-\omega t)},$$
$$\partial\boldsymbol{B}_{\rm sc}/\partial t = (\boldsymbol{e}_r\times\boldsymbol{\epsilon}'_{\boldsymbol{k}'})f(\vartheta,\phi)E_0(1-\mathrm{i}kr)\mathrm{e}^{\mathrm{i}(kr-\omega t)}/r^2.$$

The above equations imply that

$$\boldsymbol{B}_{\rm inc} = \omega^{-1}(\boldsymbol{k}\times\boldsymbol{\epsilon}_{\boldsymbol{k}})E_0\mathrm{e}^{\mathrm{i}(\boldsymbol{k}\cdot\boldsymbol{r}-\omega t)},$$
$$\boldsymbol{B}_{\rm sc} = \omega^{-1}(\boldsymbol{e}_r\times\boldsymbol{\epsilon}'_{\boldsymbol{k}'})f(\vartheta,\phi)E_0(\mathrm{i}+kr)\mathrm{e}^{\mathrm{i}(kr-\omega t)}/r^2,$$

where we have set the time-independent parts to zero since no static magnetic fields are present.

(2) The real and imaginary parts of $\boldsymbol{E}_{\rm inc}$ are given by

$$\mathrm{Re}\,\boldsymbol{E}_{\rm inc} = E_0\big[\mathrm{Re}\,\boldsymbol{\epsilon}_{\boldsymbol{k}}\cos(\boldsymbol{k}\cdot\boldsymbol{r}-\omega t) - \mathrm{Im}\,\boldsymbol{\epsilon}_{\boldsymbol{k}}\sin(\boldsymbol{k}\cdot\boldsymbol{r}-\omega t)\big],$$
$$\mathrm{Im}\,\boldsymbol{E}_{\rm inc} = E_0\big[\mathrm{Re}\,\boldsymbol{\epsilon}_{\boldsymbol{k}}\sin(\boldsymbol{k}\cdot\boldsymbol{r}-\omega t) + \mathrm{Im}\,\boldsymbol{\epsilon}_{\boldsymbol{k}}\cos(\boldsymbol{k}\cdot\boldsymbol{r}-\omega t)\big].$$

The real and imaginary parts of $\boldsymbol{B}_{\rm inc}$ are given by

$$\mathrm{Re}\,\boldsymbol{B}_{\rm inc} = (E_0/\omega)\boldsymbol{k}\times\big[\mathrm{Re}\,\boldsymbol{\epsilon}_{\boldsymbol{k}}\cos(\boldsymbol{k}\cdot\boldsymbol{r}-\omega t) - \mathrm{Im}\,\boldsymbol{\epsilon}_{\boldsymbol{k}}\sin(\boldsymbol{k}\cdot\boldsymbol{r}-\omega t)\big],$$
$$\mathrm{Im}\,\boldsymbol{B}_{\rm inc} = (E_0/\omega)\boldsymbol{k}\times\big[\mathrm{Re}\,\boldsymbol{\epsilon}_{\boldsymbol{k}}\sin(\boldsymbol{k}\cdot\boldsymbol{r}-\omega t) + \mathrm{Im}\,\boldsymbol{\epsilon}_{\boldsymbol{k}}\cos(\boldsymbol{k}\cdot\boldsymbol{r}-\omega t)\big].$$

The real and imaginary parts of $\boldsymbol{E}_{\rm sc}$ are given by

$$\begin{aligned}\mathrm{Re}\,\boldsymbol{E}_{\rm sc} = (E_0/r)\big\{&\mathrm{Re}\big[\boldsymbol{\epsilon}'_{\boldsymbol{k}'}f(\vartheta,\phi)\big]\cos(kr-\omega t)\\ &- \mathrm{Im}\big[\boldsymbol{\epsilon}'_{\boldsymbol{k}'}f(\vartheta,\phi)\big]\sin(kr-\omega t)\big\},\\ \mathrm{Im}\,\boldsymbol{E}_{\rm sc} = (E_0/r)\big\{&\mathrm{Re}\big[\boldsymbol{\epsilon}'_{\boldsymbol{k}'}f(\vartheta,\phi)\big]\sin(kr-\omega t)\\ &+ \mathrm{Im}\big[\boldsymbol{\epsilon}'_{\boldsymbol{k}'}f(\vartheta,\phi)\big]\cos(kr-\omega t)\big\}.\end{aligned}$$

The real and imaginary parts of $\boldsymbol{B}_{\rm sc}$ are given by

$$\begin{aligned}
\operatorname{Re}\boldsymbol{B}_{\rm sc} &= (kE_0/\omega r)\boldsymbol{e}_r \times \big\{\operatorname{Re}\big[\,\boldsymbol{\epsilon}'_{\boldsymbol{k}'} f(\vartheta,\phi)\,\big]\cos(kr-\omega t) \\
&\qquad\qquad - \operatorname{Im}\big[\,\boldsymbol{\epsilon}'_{\boldsymbol{k}'} f(\vartheta,\phi)\,\big]\sin(kr-\omega t)\big\} \\
&\quad - (E_0/\omega r^2)\boldsymbol{e}_r \times \big\{\operatorname{Re}\big[\,\boldsymbol{\epsilon}'_{\boldsymbol{k}'} f(\vartheta,\phi)\,\big]\sin(kr-\omega t) \\
&\qquad\qquad + \operatorname{Im}\big[\,\boldsymbol{\epsilon}'_{\boldsymbol{k}'} f(\vartheta,\phi)\,\big]\cos(kr-\omega t)\big\}, \\
\operatorname{Im}\boldsymbol{B}_{\rm sc} &= (kE_0/\omega r)\boldsymbol{e}_r \times \big\{\operatorname{Re}\big[\,\boldsymbol{\epsilon}'_{\boldsymbol{k}'} f(\vartheta,\phi)\,\big]\sin(kr-\omega t) \\
&\qquad\qquad + \operatorname{Im}\big[\,\boldsymbol{\epsilon}'_{\boldsymbol{k}'} f(\vartheta,\phi)\,\big]\cos(kr-\omega t)\big\} \\
&\quad + (E_0/\omega r^2)\boldsymbol{e}_r \times \big\{\operatorname{Re}\big[\,\boldsymbol{\epsilon}'_{\boldsymbol{k}'} f(\vartheta,\phi)\,\big]\cos(kr-\omega t) \\
&\qquad\qquad - \operatorname{Im}\big[\,\boldsymbol{\epsilon}'_{\boldsymbol{k}'} f(\vartheta,\phi)\,\big]\sin(kr-\omega t)\big\}.
\end{aligned}$$

(3) For the incident X-rays, we have

$$\begin{aligned}
u_{\rm inc} &= \epsilon_0\big|\operatorname{Re}\boldsymbol{E}_{\rm inc}\big|^2 \\
&= \epsilon_0 E_0^2\big|\operatorname{Re}\boldsymbol{\epsilon}_{\boldsymbol{k}}\cos(\boldsymbol{k}\cdot\boldsymbol{r}-\omega t) - \operatorname{Im}\boldsymbol{\epsilon}_{\boldsymbol{k}}\sin(\boldsymbol{k}\cdot\boldsymbol{r}-\omega t)\big|^2 \\
&= \epsilon_0 E_0^2\big\{(\operatorname{Re}\boldsymbol{\epsilon}_{\boldsymbol{k}})^2\cos^2(\boldsymbol{k}\cdot\boldsymbol{r}-\omega t) + (\operatorname{Im}\boldsymbol{\epsilon}_{\boldsymbol{k}})^2\sin^2(\boldsymbol{k}\cdot\boldsymbol{r}-\omega t) \\
&\qquad - \operatorname{Re}\boldsymbol{\epsilon}_{\boldsymbol{k}}\cdot\operatorname{Im}\boldsymbol{\epsilon}_{\boldsymbol{k}}\,\sin[2(\boldsymbol{k}\cdot\boldsymbol{r}-\omega t)]\big\}.
\end{aligned}$$

For the scattered X-rays, we have

$$\begin{aligned}
u_{\rm sc} &= \epsilon_0\big|\operatorname{Re}\boldsymbol{E}_{\rm sc}\big|^2 \\
&= \frac{\epsilon_0 E_0^2}{r^2}\big|\operatorname{Re}\big[\,\boldsymbol{\epsilon}'_{\boldsymbol{k}'} f(\vartheta,\phi)\,\big]\cos(kr-\omega t) \\
&\qquad - \operatorname{Im}\big[\,\boldsymbol{\epsilon}'_{\boldsymbol{k}'} f(\vartheta,\phi)\,\big]\sin(kr-\omega t)\big|^2 \\
&= \frac{\epsilon_0 E_0^2}{r^2}\Big\{\big|\operatorname{Re}\big[\,\boldsymbol{\epsilon}'_{\boldsymbol{k}'} f(\vartheta,\phi)\,\big]\big|^2\cos^2(kr-\omega t) \\
&\qquad + \big|\operatorname{Im}\big[\,\boldsymbol{\epsilon}'_{\boldsymbol{k}'} f(\vartheta,\phi)\,\big]\big|^2\sin^2(kr-\omega t) \\
&\qquad - \operatorname{Re}\big[\,\boldsymbol{\epsilon}'_{\boldsymbol{k}'} f(\vartheta,\phi)\,\big]\cdot\operatorname{Im}\big[\,\boldsymbol{\epsilon}'_{\boldsymbol{k}'} f(\vartheta,\phi)\,\big]\sin[2(kr-\omega t)\Big\}.
\end{aligned}$$

(4) Let $T = 2\pi/\omega$ be the period. The time averages of the energy densities are obtained by averaging $u_{\rm inc}$ and $u_{\rm sc}$ in one period by making use of

$$\begin{aligned}
&\frac{1}{T}\int_0^T dt\ \sin^2(\omega t) = \frac{1}{T}\int_0^T dt\ \cos^2(\omega t) = \frac{1}{2}, \\
&\frac{1}{T}\int_0^T dt\ \sin(n\omega t) = \frac{1}{T}\int_0^T dt\ \cos(n\omega t) = 0
\end{aligned}$$

with n an integer. For $u_{\rm inc}$, we have

$$\overline{u}_{\rm inc} = \frac{1}{T}\int_0^T {\rm d}t\ u_{\rm inc} = \frac{1}{2}\epsilon_0 E_0^2\big[\,({\rm Re}\,\boldsymbol{\epsilon_k})^2 + ({\rm Im}\,\boldsymbol{\epsilon_k})^2\,\big]$$
$$= \frac{1}{2}\epsilon_0 E_0^2|\boldsymbol{\epsilon_k}|^2 = \frac{1}{2}\epsilon_0 E_0^2,$$

where we have made use of the fact that the norm of $\boldsymbol{\epsilon_k}$ is unity. The above result is the expected one for the incident X-rays.

For $u_{\rm sc}$, we have

$$\overline{u}_{\rm sc} = \frac{1}{T}\int_0^T {\rm d}t\ u_{\rm sc}$$
$$= \frac{1}{2r^2}\epsilon_0 E_0^2\Big[\,\big|{\rm Re}\big[\,\boldsymbol{\epsilon}'_{\boldsymbol{k}'} f(\vartheta,\phi)\,\big]\big|^2 + \big|{\rm Im}\big[\,\boldsymbol{\epsilon}'_{\boldsymbol{k}'} f(\vartheta,\phi)\,\big]\big|^2\Big]$$
$$= \frac{1}{2r^2}\epsilon_0 E_0^2\Big[\,\big|{\rm Re}\,\boldsymbol{\epsilon}'_{\boldsymbol{k}'}\,{\rm Re}\,f(\vartheta,\phi) - {\rm Im}\,\boldsymbol{\epsilon}'_{\boldsymbol{k}'}\,{\rm Im}\,f(\vartheta,\phi)\big|^2$$
$$+ \big|{\rm Im}\,\boldsymbol{\epsilon}'_{\boldsymbol{k}'}\,{\rm Re}\,f(\vartheta,\phi) + {\rm Re}\,\boldsymbol{\epsilon}'_{\boldsymbol{k}'}\,{\rm Im}\,f(\vartheta,\phi)\big|^2\Big]$$
$$= \frac{1}{2r^2}\epsilon_0 E_0^2\big|f(\vartheta,\phi)\big|^2,$$

where we have made use of the fact that the norm of $\boldsymbol{\epsilon}'_{\boldsymbol{k}'}$ is unity. The above result is the expected one for the scattered X-rays.

7-7 Motion of a bound electron in an atom under the influence of X-rays. The electric field of the incident X-ray beam in an X-ray diffraction experiment is $\boldsymbol{E}_{\rm inc}(\boldsymbol{r},t) = \boldsymbol{\epsilon_k} E_0 {\rm e}^{{\rm i}(\boldsymbol{k}\cdot\boldsymbol{r}-\omega t)}$. The equation of motion of a bound electron in an atom under the influence of $\boldsymbol{E}_{\rm inc}(\boldsymbol{r},t)$ is given by

$$\ddot{\boldsymbol{r}}_j + \gamma\,\dot{\boldsymbol{r}}_j + \omega_j^2\boldsymbol{r}_j = -(e/m)\boldsymbol{\epsilon_k} E_0 {\rm e}^{{\rm i}(\boldsymbol{k}\cdot\boldsymbol{r}_j-\omega t)}.$$

(1) Find the general solution to the equation of motion.
(2) Derive the steady solution to the equation of motion.

---

(1) We first find the general solution to the homogeneous equation, then find a special solution to the inhomogeneous equation, and then obtain the general solution to the inhomogeneous equation by adding up the two previously-derived solutions. For the homogeneous equation

$$\ddot{\boldsymbol{r}}_j + \gamma\,\dot{\boldsymbol{r}}_j + \omega_j^2\boldsymbol{r}_j = 0,$$

the characteristic equation reads

$$\lambda_j^2 + \gamma\lambda_j + \omega_j^2 = 0.$$

The solutions to the above characteristic equation are

$$\lambda_{j,\pm} = -\gamma/2 \pm \mathrm{i}\left(\omega_j^2 - \gamma^2/4\right)^{1/2} = -\gamma/2 \pm \mathrm{i}\omega_{zj},$$

where $\omega_{zj} = \left(\omega_j^2 - \gamma^2/4\right)^{1/2}$. For weak damping, we have $\omega_j > \gamma/2$ for all j's. Thus, the general solution to the homogeneous equation is given by

$$\boldsymbol{r}_j^h = \mathrm{e}^{-\gamma t/2}\left(\boldsymbol{A}_j \mathrm{e}^{\mathrm{i}\omega_{zj}t} + \boldsymbol{B}_j \mathrm{e}^{-\mathrm{i}\omega_{zj}t}\right).$$

For a special solution to the inhomogeneous equation, in consideration of the presence of the exponential function $\mathrm{e}^{\mathrm{i}(\boldsymbol{k}\cdot\boldsymbol{r}_j - \omega t)}$ we set

$$\boldsymbol{r}_j^s = \boldsymbol{C}_j \mathrm{e}^{\mathrm{i}(\boldsymbol{k}\cdot\boldsymbol{r}_j - \omega t)}.$$

Substituting $\boldsymbol{r}_j^s$ into the inhomogeneous equation yields

$$\left(-\omega^2 - \mathrm{i}\gamma\omega + \omega_j^2\right)\boldsymbol{C}_j = -(e/m)\boldsymbol{\epsilon_k} E_0$$

which leads to the following solution to $\boldsymbol{C}_j$

$$\boldsymbol{C}_j = -\frac{(e/m)E_0\boldsymbol{\epsilon_k}}{\omega_j^2 - \omega^2 - \mathrm{i}\gamma\omega}.$$

Thus, the general solution to the equation of motion is given by

$$\begin{aligned}\boldsymbol{r}_j &= \boldsymbol{r}_j^h + \boldsymbol{r}_j^s \\ &= \mathrm{e}^{-\gamma t/2}\left(\boldsymbol{A}_j \mathrm{e}^{\mathrm{i}\omega_{zj}t} + \boldsymbol{B}_j \mathrm{e}^{-\mathrm{i}\omega_{zj}t}\right) - \frac{(e/m)E_0\boldsymbol{\epsilon_k}}{\omega_j^2 - \omega^2 - \mathrm{i}\gamma\omega}\mathrm{e}^{\mathrm{i}(\boldsymbol{k}\cdot\boldsymbol{r}_j - \omega t)}.\end{aligned}$$

(2) The first term in the general solution decays rapidly. After a short period of time, this term becomes undetectable. We then have the following steady-state solution to the equation of motion

$$\boldsymbol{r}_j = -\frac{(e/m)E_0\boldsymbol{\epsilon_k}}{\omega_j^2 - \omega^2 - \mathrm{i}\gamma\omega}\mathrm{e}^{\mathrm{i}(\boldsymbol{k}\cdot\boldsymbol{r}_j - \omega t)}.$$

7-8 Interaction of a bound electron with a plane electromagnetic wave. An electron bound to an atom interacts with the plane electromagnetic wave described by the vector potential $\boldsymbol{A}(\boldsymbol{r}, t) = \boldsymbol{A}_0(\omega)\mathrm{e}^{\mathrm{i}(\boldsymbol{k}\cdot\boldsymbol{r} - \omega t)} + \boldsymbol{A}_0^*(\omega)\mathrm{e}^{-\mathrm{i}(\boldsymbol{k}\cdot\boldsymbol{r} - \omega t)}$. The scalar potential of the electromagnetic wave is equal to zero and the divergence of the vector potential $\boldsymbol{A}(\boldsymbol{r}, t)$ vanishes, $\boldsymbol{\nabla}\cdot\boldsymbol{A}(\boldsymbol{r}, t) = 0$.

(1) The energy flux (the amount of energy crossing a unit area per unit time) of an electromagnetic wave is given by $S(\omega) = \langle|\boldsymbol{E}\times\boldsymbol{B}|\rangle/\mu_0$, where $\langle\cdots\rangle$ denotes the average over time. Compute $S(\omega)$ for the above-given plane electromagnetic wave.

(2) Using Fermi's golden rule, compute the probability per unit time (the transition rate) $R_{i\to j}(\omega)$ for an electron to make a transition from the initial state $|i\rangle$ to the final state $|f\rangle$ induced by the above-given plane electromagnetic wave in terms of the energies and wave functions of $|i\rangle$ and $|f\rangle$.

---

(1) The electric field is given by

$$\boldsymbol{E}(\boldsymbol{r},t) = -\frac{\partial \boldsymbol{A}(\boldsymbol{r},t)}{\partial t} = \mathrm{i}\omega\big[\,\boldsymbol{A}_0(\omega)\mathrm{e}^{\mathrm{i}(\boldsymbol{k}\cdot\boldsymbol{r}-\omega t)} - \boldsymbol{A}_0^*(\omega)\mathrm{e}^{-\mathrm{i}(\boldsymbol{k}\cdot\boldsymbol{r}-\omega t)}\,\big].$$

The magnetic field is given by

$$\boldsymbol{B}(\boldsymbol{r},t) = \boldsymbol{\nabla}\times\boldsymbol{A}(\boldsymbol{r},t) = \mathrm{i}\boldsymbol{k}\times\big[\,\boldsymbol{A}_0(\omega)\mathrm{e}^{\mathrm{i}(\boldsymbol{k}\cdot\boldsymbol{r}-\omega t)} - \boldsymbol{A}_0^*(\omega)\mathrm{e}^{-\mathrm{i}(\boldsymbol{k}\cdot\boldsymbol{r}-\omega t)}\,\big].$$

From the Coulomb gauge condition $\boldsymbol{\nabla}\cdot\boldsymbol{A}(\boldsymbol{r},t) = 0$, we have

$$\boldsymbol{k}\cdot\big[\,\boldsymbol{A}_0(\omega)\mathrm{e}^{\mathrm{i}(\boldsymbol{k}\cdot\boldsymbol{r}-\omega t)} - \boldsymbol{A}_0^*(\omega)\mathrm{e}^{-\mathrm{i}(\boldsymbol{k}\cdot\boldsymbol{r}-\omega t)}\,\big] = 0.$$

The cross product of $\boldsymbol{E}(\boldsymbol{r},t)$ and $\boldsymbol{B}(\boldsymbol{r},t)$ is given by

$$\begin{aligned}\boldsymbol{E}(\boldsymbol{r},t)\times\boldsymbol{B}(\boldsymbol{r},t) &= -\omega\big[\,\boldsymbol{A}_0(\omega)\mathrm{e}^{\mathrm{i}(\boldsymbol{k}\cdot\boldsymbol{r}-\omega t)} - \boldsymbol{A}_0^*(\omega)\mathrm{e}^{-\mathrm{i}(\boldsymbol{k}\cdot\boldsymbol{r}-\omega t)}\,\big]\\
&\quad\times\Big\{\boldsymbol{k}\times\big[\,\boldsymbol{A}_0(\omega)\mathrm{e}^{\mathrm{i}(\boldsymbol{k}\cdot\boldsymbol{r}-\omega t)} - \boldsymbol{A}_0^*(\omega)\mathrm{e}^{-\mathrm{i}(\boldsymbol{k}\cdot\boldsymbol{r}-\omega t)}\,\big]\Big\}\\
&= -\omega\boldsymbol{k}\Big\{\big[\,\boldsymbol{A}_0(\omega)\mathrm{e}^{\mathrm{i}(\boldsymbol{k}\cdot\boldsymbol{r}-\omega t)} - \boldsymbol{A}_0^*(\omega)\mathrm{e}^{-\mathrm{i}(\boldsymbol{k}\cdot\boldsymbol{r}-\omega t)}\,\big]\\
&\quad\cdot\big[\,\boldsymbol{A}_0(\omega)\mathrm{e}^{\mathrm{i}(\boldsymbol{k}\cdot\boldsymbol{r}-\omega t)} - \boldsymbol{A}_0^*(\omega)\mathrm{e}^{-\mathrm{i}(\boldsymbol{k}\cdot\boldsymbol{r}-\omega t)}\,\big]\Big\}\\
&= \omega\boldsymbol{k}\big[\,2\big|\boldsymbol{A}_0(\omega)\big|^2 - \boldsymbol{A}_0(\omega)\cdot\boldsymbol{A}_0(\omega)\mathrm{e}^{2\mathrm{i}(\boldsymbol{k}\cdot\boldsymbol{r}-\omega t)}\\
&\quad - \boldsymbol{A}_0^*(\omega)\cdot\boldsymbol{A}_0^*(\omega)\mathrm{e}^{-2\mathrm{i}(\boldsymbol{k}\cdot\boldsymbol{r}-\omega t)}\,\big].\end{aligned}$$

Taking the time average, we have

$$\langle\boldsymbol{E}(\boldsymbol{r},t)\times\boldsymbol{B}(\boldsymbol{r},t)\rangle = 2\omega\big|\boldsymbol{A}_0(\omega)\big|^2\boldsymbol{k}.$$

We thus have

$$S(\omega) = \frac{1}{\mu_0}\big|\langle\boldsymbol{E}(\boldsymbol{r},t)\times\boldsymbol{B}(\boldsymbol{r},t)\rangle\big| = \frac{2\omega k}{\mu_0}\big|\boldsymbol{A}_0(\omega)\big|^2 = 2c\epsilon_0\omega^2\big|\boldsymbol{A}_0(\omega)\big|^2.$$

(2) In the radiation field, the Hamiltonian of an electron in the atom is given by

$$\hat{H} = \frac{1}{2m}\big(-\mathrm{i}\hbar\boldsymbol{\nabla} + e\boldsymbol{A}\big)^2 = -\frac{\hbar^2}{2m}\nabla^2 - \frac{\mathrm{i}e\hbar}{m}\boldsymbol{A}\cdot\boldsymbol{\nabla} + \frac{e^2}{2m}\boldsymbol{A}^2,$$

where the Coulomb gauge condition $\boldsymbol{\nabla}\cdot\boldsymbol{A} = 0$ has been used. The second term, $-(\mathrm{i}e\hbar/m)\boldsymbol{A}\cdot\boldsymbol{\nabla}$, describes the interaction of an electron with the radiation field. One term in $\boldsymbol{A}$ (with a minus sign in front of ω on the exponential) corresponds to the absorption of a photon and the other term to the emission of a photon. We can consider these two terms separately. Let

$$\hat{H}_{i,-} = -(\mathrm{i}e\hbar/m)\boldsymbol{A}_0(\omega)e^{\mathrm{i}\boldsymbol{k}\cdot\boldsymbol{r}}\cdot\boldsymbol{\nabla}, \quad \hat{H}_{i,+} = -(\mathrm{i}e\hbar/m)\boldsymbol{A}_0^*(\omega)e^{-\mathrm{i}\boldsymbol{k}\cdot\boldsymbol{r}}\cdot\boldsymbol{\nabla}.$$

Note that we have left out the time dependence that is to be integrated to yield the δ-function for the energy conservation. From Fermi's golden rule, we have

$$\begin{aligned} R^{a,e}_{i\to j}(\omega) &= \frac{2\pi}{\hbar}\left|\langle f|\hat{H}_{i,\mp}|i\rangle\right|^2\delta(E_f - E_i \mp \hbar\omega) \\ &= \frac{2\pi e^2\hbar}{m^2}\left|\boldsymbol{A}_0^{(*)}(\omega)\cdot\int \mathrm{d}\boldsymbol{r}\; \mathrm{e}^{\pm\mathrm{i}\boldsymbol{k}\cdot\boldsymbol{r}}\psi_f^*(\boldsymbol{r})\boldsymbol{\nabla}\psi_i(\boldsymbol{r})\right|^2 \\ &\quad\times \delta(E_f - E_i \mp \hbar\omega). \end{aligned}$$

Since $\boldsymbol{k}\cdot\boldsymbol{r}$ is usually very small within the region with the appreciable values of wave functions, we can neglect the exponential function $\mathrm{e}^{\pm\mathrm{i}\boldsymbol{k}\cdot\boldsymbol{r}}$ in the integrand. We then have

$$R^{a,e}_{i\to j}(\omega) = \frac{2\pi e^2\hbar}{m^2}\left|\boldsymbol{A}_0^{(*)}(\omega)\cdot\int \mathrm{d}\boldsymbol{r}\; \psi_f^*(\boldsymbol{r})\boldsymbol{\nabla}\psi_i(\boldsymbol{r})\right|^2\delta(E_f - E_i \mp \hbar\omega).$$

Chapter 8

Reciprocal Lattice

(1) *Primitive vectors of reciprocal lattice*

$$\boldsymbol{b}_1 = 2\pi \frac{\boldsymbol{a}_2 \times \boldsymbol{a}_3}{\boldsymbol{a}_1 \cdot (\boldsymbol{a}_2 \times \boldsymbol{a}_3)},$$

$$\boldsymbol{b}_2 = 2\pi \frac{\boldsymbol{a}_3 \times \boldsymbol{a}_1}{\boldsymbol{a}_1 \cdot (\boldsymbol{a}_2 \times \boldsymbol{a}_3)},$$

$$\boldsymbol{b}_3 = 2\pi \frac{\boldsymbol{a}_1 \times \boldsymbol{a}_2}{\boldsymbol{a}_1 \cdot (\boldsymbol{a}_2 \times \boldsymbol{a}_3)}.$$

(2) *Reciprocal lattice vectors*

$$\boldsymbol{K}_\ell = \sum_j \ell_j \boldsymbol{b}_j = \ell_1 \boldsymbol{b}_1 + \ell_2 \boldsymbol{b}_2 + \ell_3 \boldsymbol{b}_3,$$

$$\ell_j = 0,\ \pm 1,\ \pm 2,\ \cdots .$$

(3) *Formal definition of reciprocal lattice*
The reciprocal lattice of a direct lattice is the set of discrete values of $\boldsymbol{q}$ that satisfy

$$\mathrm{e}^{-\mathrm{i}\boldsymbol{q}\cdot\boldsymbol{R}_n} = 1$$

for all the lattice vectors $\boldsymbol{R}_n$'s of the direct lattice.

(4) *First Brillouin zone*
The Wigner-Seitz cell of the reciprocal lattice of a direct lattice is the first Brillouin zone of the lattice.

(5) *Symbols for high-symmetry points and lines in the interior of the first Brillouin zone*

$$\Gamma,\ \Delta,\ \Sigma,\ \Lambda,\ \cdots .$$

The center of the first Brillouin zone is always denoted by Γ.

(6) *Symbols for high-symmetry points and lines on the surface of the first Brillouin zone*

$$H,\ K,\ L,\ M,\ N,\ P,\ X,\ W,\ \cdots .$$

(7) *Bragg planes*
A Bragg plane is a plane that bisects a lattice vector of the reciprocal lattice. A wave vector on the Bragg plane satisfies

$$\boldsymbol{k} \cdot \hat{\boldsymbol{K}} = |\boldsymbol{K}|/2.$$

(8) *Higher-order Brillouin zones*
The $(n+1)$th Brillouin zone is defined as the region that is not in the $(n-1)$th Brillouin zone and that can be reached from the nth Brillouin zone by crossing only a single Bragg plane.

(9) *Relation between reciprocal lattice vectors and lattice planes*
Corresponding to any reciprocal lattice vector $\boldsymbol{K}$, there exist a family of lattice planes perpendicular to $\boldsymbol{K}$ with the interplanar distance given by $2\pi/|\boldsymbol{K}_0|$, where $\boldsymbol{K}_0$ is the shortest reciprocal lattice vector parallel to $\boldsymbol{K}$. Conversely, for any family of lattice planes, there exist a set of reciprocal lattice vectors perpendicular to the planes and the length of the shortest reciprocal lattice vector is $2\pi/d$ where d is the interplanar distance of the family of lattice planes.

8-1 Reciprocal lattice of the reciprocal lattice. Show that the reciprocal lattice of the reciprocal lattice is the direct lattice.

We now construct the primitive vectors of the reciprocal lattice of the reciprocal lattice using the relations between the primitive vectors of the reciprocal lattice and those of the direct lattice. Let Ω be the volume of a primitive cell of the reciprocal lattice, $\Omega = (2\pi)^3/v_c$. For $\boldsymbol{b}_1 \times \boldsymbol{b}_2$, we have

$$\frac{2\pi}{\Omega}(\boldsymbol{b}_1 \times \boldsymbol{b}_2) = \frac{(2\pi)^3}{\Omega v_c^2}(\boldsymbol{a}_2 \times \boldsymbol{a}_3) \times (\boldsymbol{a}_3 \times \boldsymbol{a}_1).$$

Making use of $\boldsymbol{A} \times (\boldsymbol{B} \times \boldsymbol{C}) = (\boldsymbol{A} \cdot \boldsymbol{C})\boldsymbol{B} - (\boldsymbol{A} \cdot \boldsymbol{B})\boldsymbol{C}$, we have

$$\begin{aligned}\frac{2\pi}{\Omega}(\boldsymbol{b}_1 \times \boldsymbol{b}_2) &= \frac{(2\pi)^3}{\Omega v_c^2}\Big\{\big[(\boldsymbol{a}_2 \times \boldsymbol{a}_3) \cdot \boldsymbol{a}_1\big]\boldsymbol{a}_3 - \big[(\boldsymbol{a}_2 \times \boldsymbol{a}_3) \cdot \boldsymbol{a}_3\big]\boldsymbol{a}_1\Big\} \\ &= \frac{(2\pi)^3}{\Omega v_c}\boldsymbol{a}_3 = \boldsymbol{a}_3,\end{aligned}$$

where we have used $(\boldsymbol{a}_2 \times \boldsymbol{a}_3) \cdot \boldsymbol{a}_1 = v_c$, $(\boldsymbol{a}_2 \times \boldsymbol{a}_3) \cdot \boldsymbol{a}_3 = 0$, and $\Omega = (2\pi)^3/v_c$. We can similarly obtain

$$\frac{2\pi}{\Omega}(\boldsymbol{b}_2 \times \boldsymbol{b}_3) = \boldsymbol{a}_1, \quad \frac{2\pi}{\Omega}(\boldsymbol{b}_3 \times \boldsymbol{b}_1) = \boldsymbol{a}_2.$$

Thus, we have proved that the reciprocal lattice of the reciprocal lattice is the direct lattice.

8-2 Symmetry of the reciprocal lattice. Show that the reciprocal lattice $\{\boldsymbol{K}_\ell\}$ has the same point symmetry as the corresponding direct lattice $\{\boldsymbol{R}_n\}$ for a given crystal.

Let g be an element of the point symmetry group G of the direct lattice $\{\boldsymbol{R}_n\}$. Under the transformation of g or its inverse g^{-1}, any lattice vector $\boldsymbol{R}_n$ either remains unchanged or is transformed into another lattice vector. We have

$$\boldsymbol{R}_{n'} = g^{-1}\boldsymbol{R}_n.$$

By construction, we have

$$\boldsymbol{K}_\ell \cdot \boldsymbol{R}_n = 2\pi m$$

with m an integer. This equality holds for all $\boldsymbol{K}_\ell$'s and $\boldsymbol{R}_n$'s. We also have

$$\boldsymbol{K}_\ell \cdot g^{-1}\boldsymbol{R}_n = 2\pi m'$$

with m' another integer which may or may not be equal to m. Since a scalar product remains unchanged under the action of a point symmetry operation, we have

$$g(\boldsymbol{K}_\ell \cdot g^{-1}\boldsymbol{R}_n) = 2\pi m'.$$

For the left hand side, we have

$$g(\boldsymbol{K}_\ell \cdot g^{-1}\boldsymbol{R}_n) = g\boldsymbol{K}_\ell \cdot (gg^{-1})\boldsymbol{R}_n = g\boldsymbol{K}_\ell \cdot \boldsymbol{R}_n.$$

We thus have

$$g\boldsymbol{K}_\ell \cdot \boldsymbol{R}_n = 2\pi m'$$

which implies that $g\boldsymbol{K}_\ell$ is also a reciprocal lattice vector

$$\boldsymbol{K}'_\ell = g\boldsymbol{K}_\ell.$$

Therefore, the point symmetry group of the direct lattice is also the point symmetry group of the reciprocal lattice.

8-3 Reciprocal lattice of a two-dimensional Bravais lattice. Consider the reciprocal lattice of a two-dimensional Bravais lattice for which a set of primitive vectors are given by $\boldsymbol{a}_1$ and $\boldsymbol{a}_2$.

(1) Show that the primitive vectors of the reciprocal lattice can be expressed as

$$\boldsymbol{b}_1 = \frac{2\pi}{A_c}\big[\,|\boldsymbol{a}_2|^2\boldsymbol{a}_1 - (\boldsymbol{a}_1 \cdot \boldsymbol{a}_2)\boldsymbol{a}_2\,\big],$$

$$\boldsymbol{b}_2 = \frac{2\pi}{A_c}\big[\,|\boldsymbol{a}_1|^2\boldsymbol{a}_2 - (\boldsymbol{a}_2 \cdot \boldsymbol{a}_1)\boldsymbol{a}_1\,\big],$$

where $A_c = |\boldsymbol{a}_1 \times \boldsymbol{a}_2|^2$.

(2) Verify directly that $\boldsymbol{b}_1$ and $\boldsymbol{b}_2$ satisfy the orthogonality relation $\boldsymbol{a}_i \cdot \boldsymbol{b}_j = 2\pi\delta_{ij}$.

---

(1) To use the primitive vector relations between the reciprocal and direct lattices, we set $\boldsymbol{a}_3 = \boldsymbol{e}_z$. For $\boldsymbol{b}_1$, we have

$$\boldsymbol{b}_1 = \frac{2\pi(\boldsymbol{a}_2 \times \boldsymbol{e}_z)}{(\boldsymbol{a}_1 \times \boldsymbol{a}_2) \cdot \boldsymbol{e}_z} = \frac{2\pi(\boldsymbol{a}_2 \times \boldsymbol{e}_z)}{|\boldsymbol{a}_1 \times \boldsymbol{a}_2|},$$

where we have written $(\boldsymbol{a}_1 \times \boldsymbol{a}_2) \cdot \boldsymbol{e}_z$ as $|\boldsymbol{a}_1 \times \boldsymbol{a}_2|$ since $(\boldsymbol{a}_1 \times \boldsymbol{a}_2) \perp \boldsymbol{e}_z$. Note that we have implicitly assumed that $\angle(\boldsymbol{a}_1, \boldsymbol{a}_2)$ is an acute angle. Expressing $\boldsymbol{e}_z$ as $\boldsymbol{e}_z = (\boldsymbol{a}_1 \times \boldsymbol{a}_2)/|\boldsymbol{a}_1 \times \boldsymbol{a}_2|$, we have

$$\begin{aligned}\boldsymbol{b}_1 &= \frac{2\pi(\boldsymbol{a}_2 \times \boldsymbol{e}_z)}{|\boldsymbol{a}_1 \times \boldsymbol{a}_2|} = \frac{2\pi\boldsymbol{a}_2 \times (\boldsymbol{a}_1 \times \boldsymbol{a}_2)}{|\boldsymbol{a}_1 \times \boldsymbol{a}_2|^2} \\ &= \frac{2\pi}{|\boldsymbol{a}_1 \times \boldsymbol{a}_2|^2}\left[\,(\boldsymbol{a}_2 \cdot \boldsymbol{a}_2)\boldsymbol{a}_1 - (\boldsymbol{a}_1 \cdot \boldsymbol{a}_2)\boldsymbol{a}_2\,\right] \\ &= \frac{2\pi}{A_c}\left[\,|\boldsymbol{a}_2|^2\boldsymbol{a}_1 - (\boldsymbol{a}_1 \cdot \boldsymbol{a}_2)\boldsymbol{a}_2\,\right].\end{aligned}$$

We can similarly derive the second expression.

(2) First, we express A_c in another form. Expanding the cross product in A_c, we have

$$\begin{aligned}A_c &= |\boldsymbol{a}_1 \times \boldsymbol{a}_2|^2 = (a_{1x}a_{2y} - a_{1y}a_{2x})^2 \\ &= a_{1x}^2a_{2y}^2 + a_{1y}^2a_{2x}^2 - 2a_{1x}a_{2x}a_{1y}a_{2y} \\ &= |\boldsymbol{a}_1|^2|\boldsymbol{a}_2|^2 - a_{1x}^2a_{2x}^2 - a_{1y}^2a_{2y}^2 - 2a_{1x}a_{2x}a_{1y}a_{2y} \\ &= |\boldsymbol{a}_1|^2|\boldsymbol{a}_2|^2 - (\boldsymbol{a}_1 \cdot \boldsymbol{a}_2)^2.\end{aligned}$$

For $\boldsymbol{a}_1 \cdot \boldsymbol{b}_1$, we have

$$\boldsymbol{a}_1 \cdot \boldsymbol{b}_1 = \frac{2\pi}{A_c}\left[\,|\boldsymbol{a}_1|^2|\boldsymbol{a}_2|^2 - (\boldsymbol{a}_1 \cdot \boldsymbol{a}_2)^2\,\right] = \frac{2\pi}{A_c} \cdot A_c = 2\pi.$$

For $\boldsymbol{a}_1 \cdot \boldsymbol{b}_2$, we have

$$\boldsymbol{a}_1 \cdot \boldsymbol{b}_2 = \frac{2\pi}{A_c}\left[\,|\boldsymbol{a}_1|^2\boldsymbol{a}_1 \cdot \boldsymbol{a}_2 - (\boldsymbol{a}_2 \cdot \boldsymbol{a}_1)|\boldsymbol{a}_1|^2\,\right] = 0.$$

For $\boldsymbol{a}_2 \cdot \boldsymbol{b}_1$, we have

$$\boldsymbol{a}_2 \cdot \boldsymbol{b}_1 = \frac{2\pi}{A_c}\left[\,|\boldsymbol{a}_2|^2\boldsymbol{a}_1 \cdot \boldsymbol{a}_2 - (\boldsymbol{a}_1 \cdot \boldsymbol{a}_2)|\boldsymbol{a}_2|^2\,\right] = 0.$$

For $\boldsymbol{a}_2 \cdot \boldsymbol{b}_2$, we have

$$\boldsymbol{a}_2 \cdot \boldsymbol{b}_2 = \frac{2\pi}{A_c}\left[\,|\boldsymbol{a}_1|^2|\boldsymbol{a}_2|^2 - (\boldsymbol{a}_2 \cdot \boldsymbol{a}_1)^2\,\right] = \frac{2\pi}{A_c} \cdot A_c = 2\pi.$$

We have thus proved that $\boldsymbol{a}_i \cdot \boldsymbol{b}_j = 2\pi\delta_{ij}$.

8-4 Another two-dimensional Bravais lattice. A two-dimensional Bravais lattice has the primitive vectors (in nm)

$$\boldsymbol{a}_1 = 0.4\,\boldsymbol{e}_x, \quad \boldsymbol{a}_2 = 0.1\,\boldsymbol{e}_x + 0.2\,\boldsymbol{e}_y.$$

(1) Compute the primitive vectors of the reciprocal lattice.
(2) Draw the reciprocal lattice and construct the first Brillouin zone.
(3) Draw the planes with the Miller indices (11), (10), and (52).
(4) Compute the distance between the closest (11) planes.

---

(1) We use the matrix formalism here to compute $\boldsymbol{b}_1$ and $\boldsymbol{b}_2$. The matrix forms of $\boldsymbol{a}_1$ and $\boldsymbol{a}_2$ are

$$a_1 = \begin{pmatrix} 0.4 \\ 0 \end{pmatrix}, \quad a_2 = \begin{pmatrix} 0.1 \\ 0.2 \end{pmatrix}.$$

In our computations of $\boldsymbol{b}_1$ and $\boldsymbol{b}_2$, we will need the value of $a_1^t \gamma a_2$. It is evaluated as follows

$$a_1^t \gamma a_2 = (0.4\ 0) \begin{pmatrix} 0 & 1 \\ -1 & 0 \end{pmatrix} \begin{pmatrix} 0.1 \\ 0.2 \end{pmatrix} = 0.08.$$

For the matrix form of $\boldsymbol{b}_1$, we have

$$b_1 = \frac{2\pi \gamma a_2}{a_1^t \gamma a_2} = 25\pi \begin{pmatrix} 0 & 1 \\ -1 & 0 \end{pmatrix} \begin{pmatrix} 0.1 \\ 0.2 \end{pmatrix} = \frac{5\pi}{2} \begin{pmatrix} 2 \\ -1 \end{pmatrix}.$$

For the matrix form of $\boldsymbol{b}_2$, we have

$$b_2 = \frac{2\pi \gamma^t a_1}{a_1^t \gamma a_2} = 25\pi \begin{pmatrix} 0 & -1 \\ 1 & 0 \end{pmatrix} \begin{pmatrix} 0.4 \\ 0 \end{pmatrix} = 10\pi \begin{pmatrix} 0 \\ 1 \end{pmatrix}.$$

In the vector form, we have

$$\boldsymbol{b}_1 = \frac{5\pi}{2}(2\boldsymbol{e}_x - \boldsymbol{e}_y), \quad \boldsymbol{b}_2 = 10\pi\,\boldsymbol{e}_y.$$

(2) The reciprocal lattice and the first Brillouin zone are shown in Fig. 8.1(a). The dotted lines are the lines connecting the concerned point with the lattice points nearby. The solid lines bisect these connecting lines. The smallest region enclosed by the bisecting lines is the first Brillouin zone.

(3) The direct lattice and the (11), (10), and (52) lattice planes are shown in Fig. 8.1(b).

(4) The shortest reciprocal lattice vector perpendicular to the (11) family of lattice planes is $\boldsymbol{K}_0 = \boldsymbol{b}_1 + \boldsymbol{b}_2 = (\boldsymbol{e}_x + 3\boldsymbol{e}_y/2)5\pi$. Thus,

$$d_{11} = \frac{2\pi}{|\boldsymbol{K}_0|} = \frac{4}{5\sqrt{13}} \approx 0.221\,88 \text{ nm}.$$

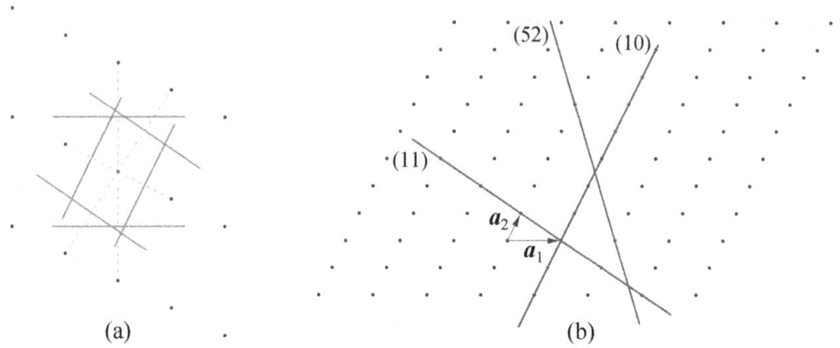

Fig. 8.1 Two-dimensional Bravais lattice and its reciprocal lattice. (a) Reciprocal lattice and first Brillouin zone. (b) Direct lattice and (11), (10), and (52) lattice planes.

8-5 First three Brillouin zones of a two-dimensional triangular lattice. Construct the first three Brillouin zones for a two-dimensional triangular lattice.

We choose the following primitive vectors for the two-dimensional triangular Bravais lattice

$$\boldsymbol{a}_1 = a\,\boldsymbol{e}_x,\ \boldsymbol{a}_2 = -(a/2)\,\boldsymbol{e}_x + (\sqrt{3}\,a/2)\,\boldsymbol{e}_y.$$

The primitive vectors of the reciprocal lattice can be computed straightforwardly using the above primitive vectors of the direct lattice. We find that the primitive vectors of the reciprocal lattice are given by

$$\boldsymbol{b}_1 = (2\pi/a)\,\boldsymbol{e}_x + (2\pi/\sqrt{3}\,a)\,\boldsymbol{e}_y,$$
$$\boldsymbol{b}_2 = (4\pi/\sqrt{3}\,a)\,\boldsymbol{e}_y.$$

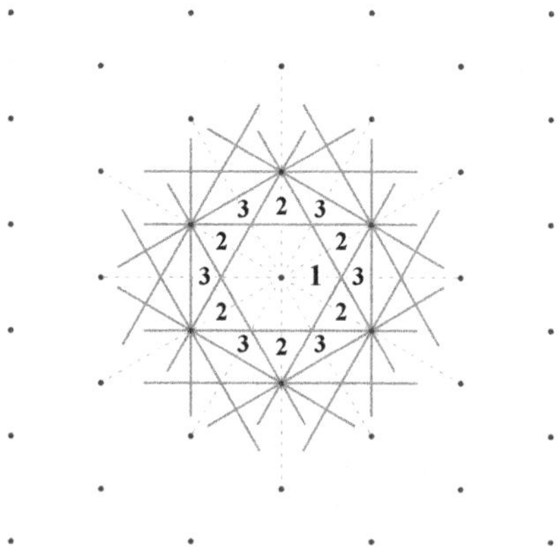

Fig. 8.2 Reciprocal lattice and first three Brillouin zones of the two-dimensional triangular Bravais lattice.

The first three Brillouin zones are constructed in Fig. 8.2. The dotted lines connect the concerned lattice point in the center to its nearby neighbors. The solid lines bisect these connecting lines. The central regular hexagon marked with "1" is the first Brillouin zone. The pieces of the second and third Brillouin zones are marked with "2" and "3", respectively.

8-6 Lengths of first eight reciprocal lattice vectors in SC, BCC, and FCC. Compute the ratios of the lengths of the first eight reciprocal lattice vectors to the length of the shortest reciprocal lattice vector for the SC, BCC, and FCC reciprocal lattices (RLs) and fill the following table.

Ratio	SC RL		BCC RL		FCC RL	
	$hk\ell$	Ratio	$hk\ell$	Ratio	$hk\ell$	Ratio
$\|\boldsymbol{K}_0\|/\|\boldsymbol{K}_0\|$	100	1	111	1	110	1
$\|\boldsymbol{K}_1\|/\|\boldsymbol{K}_0\|$	110	$\sqrt{2}$	200	$2/\sqrt{3}$	200	$\sqrt{2}$
$\|\boldsymbol{K}_2\|/\|\boldsymbol{K}_0\|$	111	$\sqrt{3}$	220	$2\sqrt{2/3}$	211	$\sqrt{3}$
$\|\boldsymbol{K}_3\|/\|\boldsymbol{K}_0\|$	200	2	311	$\sqrt{11/3}$	220	2
$\|\boldsymbol{K}_4\|/\|\boldsymbol{K}_0\|$	210	$\sqrt{5}$	222	2	310	$\sqrt{5}$
$\|\boldsymbol{K}_5\|/\|\boldsymbol{K}_0\|$	211	$\sqrt{6}$	400	$4/\sqrt{3}$	222	$\sqrt{6}$
$\|\boldsymbol{K}_6\|/\|\boldsymbol{K}_0\|$	220	$2\sqrt{2}$	331	$\sqrt{19/3}$	321	$\sqrt{7}$
$\|\boldsymbol{K}_7\|/\|\boldsymbol{K}_0\|$	221	3	420	$2\sqrt{5/3}$	400	$2\sqrt{2}$

8-7 Reciprocal lattice vectors and lattice planes. Suppose that a lattice plane intercepts the $\boldsymbol{a}_1$, $\boldsymbol{a}_2$, and $\boldsymbol{a}_3$ axes at $x_1\boldsymbol{a}_1$, $x_2\boldsymbol{a}_2$, and $x_3\boldsymbol{a}_3$, respectively, and denote $x_1 : x_2 : x_3 = (1/h) : (1/k) : (1/\ell)$.

(1) From the fact that $(-x_1\boldsymbol{a}_1 + x_2\boldsymbol{a}_2) \times (-x_1\boldsymbol{a}_1 + x_3\boldsymbol{a}_3)$ is parallel to the normal of the lattice plane, show that the reciprocal lattice vector $\boldsymbol{K} = h\boldsymbol{b}_1 + k\boldsymbol{b}_2 + \ell\boldsymbol{b}_3$ is perpendicular to this lattice plane.

(2) Assume that $\boldsymbol{K}_0$ is the shortest reciprocal lattice vector in its direction. From the facts that $e^{i\boldsymbol{K}_0 \cdot \boldsymbol{R}} = 1$ and that $\boldsymbol{R} = n_1\boldsymbol{a}_1 + n_2\boldsymbol{a}_2 + n_3\boldsymbol{a}_3 = \boldsymbol{R} - \boldsymbol{0} = \Delta\boldsymbol{R}$, show that the distance between two adjacent parallel lattice planes is given by $d = 2\pi/|\boldsymbol{K}_0|$.

---

(1) To see if $\boldsymbol{K}$ is parallel to $(-x_1\boldsymbol{a}_1 + x_2\boldsymbol{a}_2) \times (-x_1\boldsymbol{a}_1 + x_3\boldsymbol{a}_3)$, we evaluate their cross product. Making use of $\boldsymbol{A} \times (\boldsymbol{B} \times \boldsymbol{C}) =$

$(\boldsymbol{A}\cdot\boldsymbol{C})\boldsymbol{B}-(\boldsymbol{A}\cdot\boldsymbol{B})\boldsymbol{C}$ and $\boldsymbol{b}_i\cdot\boldsymbol{a}_j=2\pi\delta_{ij}$, we have

$$\begin{aligned}\boldsymbol{K}\times\big[(-x_1\boldsymbol{a}_1+x_2\boldsymbol{a}_2)\times(-x_1\boldsymbol{a}_1+x_3\boldsymbol{a}_3)\big]&=\big[\boldsymbol{K}\cdot(-x_1\boldsymbol{a}_1+x_3\boldsymbol{a}_3)\big](-x_1\boldsymbol{a}_1+x_2\boldsymbol{a}_2)\\&\quad-\big[\boldsymbol{K}\cdot(-x_1\boldsymbol{a}_1+x_2\boldsymbol{a}_2)\big](-x_1\boldsymbol{a}_1+x_3\boldsymbol{a}_3)\\&=2\pi(-x_1h+x_3\ell)(-x_1\boldsymbol{a}_1+x_2\boldsymbol{a}_2)\\&\quad-2\pi(-x_1h+x_2k)(-x_1\boldsymbol{a}_1+x_3\boldsymbol{a}_3).\end{aligned}$$

Multiplying out the parentheses yields

$$\begin{aligned}\boldsymbol{K}\times&\big[(-x_1\boldsymbol{a}_1+x_2\boldsymbol{a}_2)\times(-x_1\boldsymbol{a}_1+x_3\boldsymbol{a}_3)\big]\\&=2\pi\big(x_1^2h\boldsymbol{a}_1-x_3x_1\ell\boldsymbol{a}_1-x_1x_2h\boldsymbol{a}_2+x_2x_3\ell\boldsymbol{a}_2\big)\\&\quad-2\pi\big(x_1^2h\boldsymbol{a}_1-x_1x_2k\boldsymbol{a}_1-x_3x_1h\boldsymbol{a}_3+x_2x_3k\boldsymbol{a}_3\big)\\&=2\pi\big[-x_1(x_3\ell-x_2k)\boldsymbol{a}_1-x_2(x_1h-x_3\ell)\boldsymbol{a}_2+x_3(x_1h-x_2k)\boldsymbol{a}_3\big].\end{aligned}$$

Because $x_1:x_2:x_3=(1/h):(1/k):(1/\ell)$, we have

$$x_3\ell-x_2k=0,\ x_1h-x_3\ell=0,\ x_1h-x_2k=0$$

which leads to

$$\boldsymbol{K}\times\big[(-x_1\boldsymbol{a}_1+x_2\boldsymbol{a}_2)\times(-x_1\boldsymbol{a}_1+x_3\boldsymbol{a}_3)\big]=0.$$

We have thus proved that the reciprocal lattice vector $\boldsymbol{K}=h\boldsymbol{b}_1+k\boldsymbol{b}_2+\ell\boldsymbol{b}_3$ is perpendicular to the lattice plane that intercepts the $\boldsymbol{a}_1$, $\boldsymbol{a}_2$, and $\boldsymbol{a}_3$ axes at $x_1\boldsymbol{a}_1$, $x_2\boldsymbol{a}_2$, and $x_3\boldsymbol{a}_3$, respectively.

(2) Let $\boldsymbol{R}$ be a lattice vector on a lattice plane in the family. Since one of the lattice planes in the family passes through the origin, $\boldsymbol{R}$ can be taken as the position difference between two lattice points in two parallel lattice planes in the family. Thus, the absolute value of $\boldsymbol{K}_0\cdot\boldsymbol{R}/|\boldsymbol{K}_0|$ is equal to the distance between these two parallel lattice planes. From $\mathrm{e}^{\mathrm{i}\boldsymbol{K}_0\cdot\boldsymbol{R}}=1$, we have $\boldsymbol{K}_0\cdot\boldsymbol{R}=2\pi m$ with m an integer. Let d be the interplanar distance in the family. Since $|\boldsymbol{K}_0\cdot\boldsymbol{R}|=|\boldsymbol{K}_0|d$ for $m=\pm1$, we have $|\boldsymbol{K}_0|d=2\pi$. Thus,

$$d=\frac{2\pi}{|\boldsymbol{K}_0|}.$$

8-8 Structure factors of BCC and FCC crystals. For the purpose of illustrating the importance of the structure factor in determining the crystal structure, we take a monatomic BCC crystal as an SC crystal with a two-atom basis in which one atom is at $(0,0,0)$ and the other at $(1/2,1/2,1/2)a$ and a monatomic FCC crystal as an SC crystal with a four-atom basis in which the atoms are at $(0,0,0)$, $(1/2,1/2,0)a$, $(1/2,0,1/2)a$, and $(0,1/2,1/2)a$.

(1) Compute the structure factors for the BCC and FCC crystals.

(2) Discuss the consequence of eliminating all the reciprocal lattice vectors at which the structure factor vanishes.

---

(1) The reciprocal lattice vectors for an SC Bravais lattice are of the form $\boldsymbol{K} = (h\boldsymbol{e}_x + k\boldsymbol{e}_y + \ell\boldsymbol{e}_z)(2\pi/a)$. For the BCC crystal, we have

$$S_{\rm BCC}(\boldsymbol{K}) = \sum_{\kappa=1}^{2} {\rm e}^{-{\rm i}\boldsymbol{K}\cdot\boldsymbol{d}_\kappa} = 1 + {\rm e}^{-{\rm i}(h+k+\ell)\pi} = \begin{cases} 2, \ h+k+\ell = 2n, \\ 0, \ h+k+\ell = 2n+1, \end{cases}$$

where n is an integer. For the FCC crystal, we have

$$S_{\rm FCC}(\boldsymbol{K}) = \sum_{\kappa=1}^{2} {\rm e}^{-{\rm i}\boldsymbol{K}\cdot\boldsymbol{d}_\kappa} = 1 + {\rm e}^{-{\rm i}(h+k)\pi} + {\rm e}^{-{\rm i}(k+\ell)\pi} + {\rm e}^{-{\rm i}(\ell+h)\pi} = \begin{cases} 4, \ h,k,\ell \text{ all odd or all even}, \\ 0, \ \text{otherwise}. \end{cases}$$

(2) If all the reciprocal lattice vectors for which the structure factor vanishes are eliminated for the BCC and FCC Bravais lattices, we are then left with the reciprocal lattice vectors only for the BCC and FCC Bravais lattices, respectively.

8-9 First Brillouin zones and interplanar distances.

(1) Derive the relationship between the volume of the primitive cell in the direct lattice and that of the primitive cell in the reciprocal lattice.

(2) Compute the shortest distances from the origin in reciprocal space to the boundary of the first Brillouin zones for BCC and FCC Bravais lattice.

(3) Compute the volumes of the first Brillouin zones for BCC and FCC Bravais lattices.

(4) The Miller indices are generally given for cubic crystals with respect to the basis vectors spanning the conventional cubic unit cell. Thus, the Miller indices of a plane are independent of which type of cubic lattice is being considered. Derive an expression for the distance d_{hkl} between neighboring planes in the family of (hkl) planes.

---

(1) We start with computing the volume Ω of a primitive cell of the reciprocal lattice. Making use of the relations between $\boldsymbol{b}_i$'s and $\boldsymbol{a}_j$'s, we have

$$\begin{aligned}\Omega = \boldsymbol{b}_1 \cdot (\boldsymbol{b}_2 \times \boldsymbol{b}_3) &= (2\pi)^3 \frac{(\boldsymbol{a}_2 \times \boldsymbol{a}_3) \cdot [(\boldsymbol{a}_3 \times \boldsymbol{a}_1) \times (\boldsymbol{a}_1 \times \boldsymbol{a}_2)]}{[\boldsymbol{a}_1 \cdot (\boldsymbol{a}_2 \times \boldsymbol{a}_3)]^3} \\ &= \frac{(2\pi)^3}{v_c^3}(\boldsymbol{a}_2 \times \boldsymbol{a}_3) \cdot \Big\{\big[(\boldsymbol{a}_3 \times \boldsymbol{a}_1) \cdot \boldsymbol{a}_2\big]\boldsymbol{a}_1 - \big[(\boldsymbol{a}_3 \times \boldsymbol{a}_1) \cdot \boldsymbol{a}_1\big]\boldsymbol{a}_2\Big\} \\ &= \frac{(2\pi)^3}{v_c^2}\boldsymbol{a}_1 \cdot (\boldsymbol{a}_2 \times \boldsymbol{a}_3) = \frac{(2\pi)^3}{v_c},\end{aligned}$$

where $v_c = \boldsymbol{a}_1 \cdot (\boldsymbol{a}_2 \times \boldsymbol{a}_3)$ is the volume of a primitive cell of the direct lattice. Thus, the relationship between the volume of a primitive cell in the direct lattice and that of a primitive cell in the reciprocal lattice is given by $\Omega = (2\pi)^3/v_c$.

(2) For a given Bravais lattice, the shortest distance from the origin in reciprocal space to the boundary of the first Brillouin zone is equal to half the length of the shortest nonzero reciprocal lattice vector. With respect to the conventional cell, a reciprocal lattice vector in the cubic crystal system is of the form $\boldsymbol{K} = (h\boldsymbol{e}_x + k\boldsymbol{e}_y + \ell\boldsymbol{e}_z)(2\pi/a)$ with a the lattice constant.

For the BCC Bravais lattice, since $h+k+\ell =$ even, we see that the shortest nonzero reciprocal lattice vector is $\boldsymbol{K} = (\boldsymbol{e}_x + \boldsymbol{e}_y)(2\pi/a)$ or any one of its equivalents. Thus, the shortest distance from the origin in reciprocal space to the boundary of the first Brillouin zone of the BCC Bravais lattice is $\sqrt{2}\pi/a$.

For the FCC Bravais lattice, since h, k, ℓ are all odd or all even, we see that the shortest nonzero reciprocal lattice vector is $\boldsymbol{K}_0 = (\boldsymbol{e}_x + \boldsymbol{e}_y + \boldsymbol{e}_z)(2\pi/a)$ or any one of its equivalents. Thus, the shortest distance from the origin in reciprocal space to the boundary of the first Brillouin zone of the FCC Bravais lattice is $\sqrt{3}\pi/a$.

(3) Since the first Brillouin zone is the Wigner-Seitz cell of the reciprocal lattice, we see that its volume is the volume of a primitive cell of the reciprocal lattice. From the relation $\Omega =$

$(2\pi)^3/v_c$, we can easily find the volume of the first Brillouin zone.

For the BCC Bravais lattice, since $v_c = a^3/2$, we have $\Omega = 16\pi^3/a^3$.

For the FCC Bravais lattice, since $v_c = a^3/4$, we have $\Omega = 32\pi^3/a^3$.

(4) The shortest reciprocal lattice vector parallel to the normal of the family is $\boldsymbol{K}_0 = (h\boldsymbol{e}_x + k\boldsymbol{e}_y + \ell\boldsymbol{e}_z)(2\pi/a)$. Making use of $d = 2\pi/|\boldsymbol{K}_0|$, we have

$$d_{hk\ell} = \frac{2\pi}{|\boldsymbol{K}_0|} = \frac{2}{\sqrt{h^2+k^2+\ell^2}}.$$

8-10 Simple hexagonal lattice and its primitive cell and first Brillouin zone. For a simple hexagonal lattice, we can choose the following primitive vectors

$$\boldsymbol{a}_1 = \frac{\sqrt{3}}{2}a\,\boldsymbol{e}_x + \frac{1}{2}a\,\boldsymbol{e}_y,\quad \boldsymbol{a}_2 = -\frac{\sqrt{3}}{2}a\,\boldsymbol{e}_x + \frac{1}{2}a\,\boldsymbol{e}_y,\quad \boldsymbol{a}_3 = c\,\boldsymbol{e}_z.$$

(1) What volume does the primitive cell have?

(2) Deduce the primitive vectors for the reciprocal lattice. How can the result be described simply?

(3) Sketch the first Brillouin zone.

---

(1) Evaluating the volume of the primitive cell, we have

$$\begin{aligned}
v_c &= \boldsymbol{a}_1\cdot(\boldsymbol{a}_2\times\boldsymbol{a}_3)\\
&= \left(\frac{\sqrt{3}}{2}a\,\boldsymbol{e}_x + \frac{1}{2}a\,\boldsymbol{e}_y\right)\cdot\left[\left(-\frac{\sqrt{3}}{2}a\,\boldsymbol{e}_x + \frac{1}{2}a\,\boldsymbol{e}_y\right)\times(c\,\boldsymbol{e}_z)\right]\\
&= \left(\frac{\sqrt{3}}{2}a\,\boldsymbol{e}_x + \frac{1}{2}a\,\boldsymbol{e}_y\right)\cdot\left(\frac{\sqrt{3}}{2}ac\,\boldsymbol{e}_y + \frac{1}{2}ac\,\boldsymbol{e}_x\right)\\
&= \frac{\sqrt{3}}{2}a\cdot\frac{1}{2}ac + \frac{1}{2}a\cdot\frac{\sqrt{3}}{2}ac = \frac{\sqrt{3}}{2}a^2c.
\end{aligned}$$

(2) For $\boldsymbol{b}_1$, we have

$$\begin{aligned}
\boldsymbol{b}_1 &= 2\pi\frac{\boldsymbol{a}_2\times\boldsymbol{a}_3}{\boldsymbol{a}_1\cdot(\boldsymbol{a}_2\times\boldsymbol{a}_3)}\\
&= \frac{2\pi}{\sqrt{3}a^2c/2}\left(-\frac{\sqrt{3}}{2}a\,\boldsymbol{e}_x + \frac{1}{2}a\,\boldsymbol{e}_y\right)\times c\,\boldsymbol{e}_z\\
&= \frac{4\pi}{\sqrt{3}a}\left(\frac{1}{2}\boldsymbol{e}_x + \frac{\sqrt{3}}{2}\boldsymbol{e}_y\right).
\end{aligned}$$

For $\boldsymbol{b}_2$, we have

$$\boldsymbol{b}_2 = 2\pi\frac{\boldsymbol{a}_3\times\boldsymbol{a}_1}{\boldsymbol{a}_1\cdot(\boldsymbol{a}_2\times\boldsymbol{a}_3)} = \frac{2\pi}{\sqrt{3}a^2c/2}c\,\boldsymbol{e}_z\times\left(\frac{\sqrt{3}}{2}a\,\boldsymbol{e}_x+\frac{1}{2}a\,\boldsymbol{e}_y\right) = \frac{4\pi}{\sqrt{3}a}\left(-\frac{1}{2}\,\boldsymbol{e}_x+\frac{\sqrt{3}}{2}\,\boldsymbol{e}_y\right).$$

For $\boldsymbol{b}_3$, we have

$$\boldsymbol{b}_3 = 2\pi\frac{\boldsymbol{a}_1\times\boldsymbol{a}_2}{\boldsymbol{a}_1\cdot(\boldsymbol{a}_2\times\boldsymbol{a}_3)} = \frac{2\pi}{\sqrt{3}a^2c/2}\left(\frac{\sqrt{3}}{2}a\,\boldsymbol{e}_x+\frac{1}{2}a\,\boldsymbol{e}_y\right)\times\left(-\frac{\sqrt{3}}{2}a\,\boldsymbol{e}_x+\frac{1}{2}a\,\boldsymbol{e}_y\right) = \frac{4\pi}{\sqrt{3}a^2c}\left(\frac{\sqrt{3}}{4}a^2\,\boldsymbol{e}_z+\frac{\sqrt{3}}{4}a^2\,\boldsymbol{e}_z\right) = \frac{2\pi}{c}\,\boldsymbol{e}_z.$$

The first two components (the x and y components) of the above-given primitive vectors of both the reciprocal and direct lattices are shown in Fig. 8.3 on the x-y plane. Solid lines are used for the reciprocal lattice and dashed lines for the direct lattice. Also shown are the Wigner-Seitz cells of the reciprocal and direct lattices.

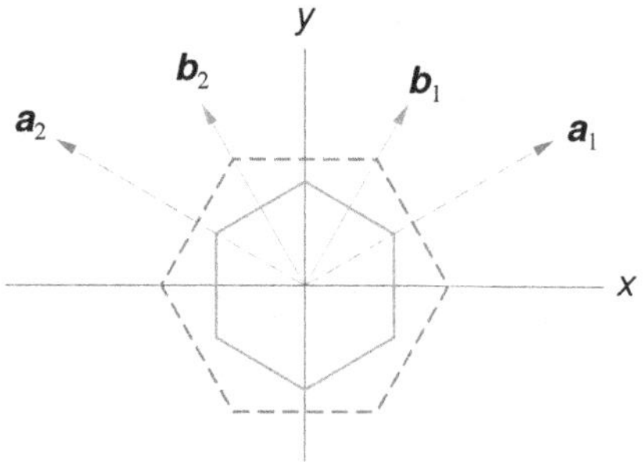

Fig. 8.3 Primitive vectors and *vice versa*. Wigner-Seitz cells of the direct and reciprocal lattices of a two-dimensional triangular Bravais lattice.

From the orientations of the Wigner-Seitz cells, we see that the reciprocal lattice is rotated 30° about the z axis relative to the direct lattice and if $\boldsymbol{b}_1$ ($\boldsymbol{b}_2$) is chosen to be along the positive (negative) x axis, then the primitive vectors of the reciprocal lattice appear to be rotated 30° about the z axis relative to the primitive vectors of the direct lattice.

(3) The first Brillouin zone of the simple hexagonal Bravais lattice is given in Fig. 8.4.

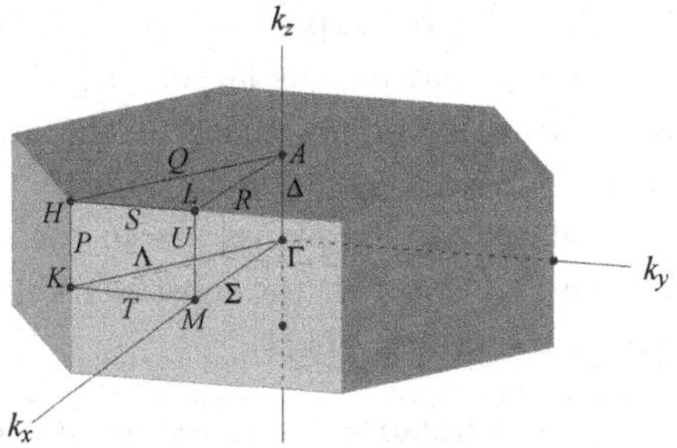

Fig. 8.4 First Brillouin zone of the simple hexagonal Bravais lattice.

8-11 Atom density in a lattice plane. The density of atoms $\sigma_{hk\ell}$ in a lattice plane $(hk\ell)$ of a monatomic crystal is related to the interplanar distance through

$$\sigma_{hkl} = d_{hk\ell}/v_c,$$

where d_{hkl} is the distance between two adjacent lattice planes in the family $(hk\ell)$ and v_c the volume of a primitive cell of the direct lattice.

(1) Prove the above relation.

(2) Find the most densely populated lattice planes in the monatomic BCC and FCC crystals using the above relation.

---

(1) Since each family of lattice planes contains all the lattice points of the Bravais lattice. For a monatomic crystal, the number of atoms is equal to the number of lattice points. Since there is only one lattice point in each primitive cell of volume v_c, the number density of atoms is then given by $n = 1/v_c$. Let $A_{hk\ell}$ be the total area of the lattice planes in the family. Then, the crystal volume can be written as $\mathscr{V} = A_{hk\ell}d_{hk\ell}$. Let N be the total number of atoms in the crystal. Then, the number density of atoms is also given by

$$n = \frac{N}{\mathscr{V}} = \frac{N}{A_{hk\ell}d_{hk\ell}} = \frac{\sigma_{hk\ell}}{d_{hk\ell}},$$

where we have introduced the number of atoms per unit area of the lattice planes, $\sigma_{hk\ell} = N/A_{hk\ell}$. Equating the above expression of n with $n = 1/v_c$ yields

$$\sigma_{hk\ell} = d_{hk\ell}/v_c.$$

(2) From the above-derived expression for the number of atoms per unit area, we see that $\sigma_{hk\ell}$ is larger for a larger value of $d_{hk\ell}$. The family of lattice planes that has the largest $d_{hk\ell}$ has the smallest Miller indices. For a monatomic BCC crystal, the most densely populated lattice planes are in the family $\{110\}$. For a monatomic FCC crystal, the most densely populated lattice planes are in the family $\{111\}$.

8-12 Fourier series of a function with the periodicity of the Bravais lattice. Let the function $f(\boldsymbol{r})$ be a periodic function of a Bravais lattice. Show that the wave vectors ($\boldsymbol{k}$'s) in the Fourier series $f(\boldsymbol{r}) = \sum_{\boldsymbol{k}} F(\boldsymbol{k}) \mathrm{e}^{\mathrm{i}\boldsymbol{k}\cdot\boldsymbol{r}}$ are reciprocal vectors ($\boldsymbol{K}$'s), where $F(\boldsymbol{k})$ is the Fourier transform of $f(\boldsymbol{r})$.

Since $f(\boldsymbol{r})$ is a periodic function of a Bravais lattice, we have for any arbitrary lattice vector $\boldsymbol{R}$ of the direct lattice

$$f(\boldsymbol{r}+\boldsymbol{R}) = f(\boldsymbol{r}).$$

Making use of the Fourier series $f(\boldsymbol{r}) = \sum_{\boldsymbol{k}} F(\boldsymbol{k}) \mathrm{e}^{\mathrm{i}\boldsymbol{k}\cdot\boldsymbol{r}}$, we have

$$f(\boldsymbol{r}+\boldsymbol{R}) = \sum_{\boldsymbol{k}} F(\boldsymbol{k}) \mathrm{e}^{\mathrm{i}\boldsymbol{k}\cdot(\boldsymbol{r}+\boldsymbol{R})} = f(\boldsymbol{r}) = \sum_{\boldsymbol{k}} F(\boldsymbol{k}) \mathrm{e}^{\mathrm{i}\boldsymbol{k}\cdot\boldsymbol{r}},$$

that is,

$$\sum_{\boldsymbol{k}} F(\boldsymbol{k}) \mathrm{e}^{\mathrm{i}\boldsymbol{k}\cdot(\boldsymbol{r}+\boldsymbol{R})} = \sum_{\boldsymbol{k}} F(\boldsymbol{k}) \mathrm{e}^{\mathrm{i}\boldsymbol{k}\cdot\boldsymbol{r}}.$$

To extract $F(\boldsymbol{k})$ from the above equation, we multiply both sides with $\mathrm{e}^{-\mathrm{i}\boldsymbol{k}'\cdot\boldsymbol{r}}$ and then integrate both sides over $\boldsymbol{r}$. Making use of $\mathscr{V}^{-1}\int \mathrm{d}\boldsymbol{r}\ \mathrm{e}^{\mathrm{i}(\boldsymbol{k}-\boldsymbol{k}')\cdot\boldsymbol{r}} = \delta_{\boldsymbol{k}\boldsymbol{k}'}$, we have

$$\sum_{\boldsymbol{k}} F(\boldsymbol{k}) \mathrm{e}^{\mathrm{i}\boldsymbol{k}\cdot\boldsymbol{R}} \int \mathrm{d}\boldsymbol{r}\ \mathrm{e}^{\mathrm{i}(\boldsymbol{k}-\boldsymbol{k}')\cdot\boldsymbol{r}} = \sum_{\boldsymbol{k}} F(\boldsymbol{k}) \int \mathrm{d}\boldsymbol{r}\ \mathrm{e}^{\mathrm{i}(\boldsymbol{k}-\boldsymbol{k}')\cdot\boldsymbol{r}},$$

$$\sum_{\boldsymbol{k}} F(\boldsymbol{k}) \mathrm{e}^{\mathrm{i}\boldsymbol{k}\cdot\boldsymbol{R}} \delta_{\boldsymbol{k}\boldsymbol{k}'} = \sum_{\boldsymbol{k}} F(\boldsymbol{k}) \delta_{\boldsymbol{k}\boldsymbol{k}'},$$

$$F(\boldsymbol{k}) \mathrm{e}^{\mathrm{i}\boldsymbol{k}\cdot\boldsymbol{R}} = F(\boldsymbol{k}),$$

where we have dropped the prime on $\boldsymbol{k}'$ on the last line. To ensure that $f(\boldsymbol{r})$ is a periodic function of the Bravais lattice, the above equality must hold for all lattice vectors, $\boldsymbol{R}$'s. This requires that $\boldsymbol{k}$ satisfy $\mathrm{e}^{\mathrm{i}\boldsymbol{k}\cdot\boldsymbol{R}} = 1$ which is nothing but the equation used in the definition of the reciprocal lattice. Therefore, all $\boldsymbol{k}$'s are reciprocal lattice vectors, $\boldsymbol{K}$'s and the Fourier series can be written as

$$f(\boldsymbol{r}) = \sum_{\boldsymbol{K}} F(\boldsymbol{K}) e^{i\boldsymbol{K}\cdot\boldsymbol{r}}.$$

From the proof given, we see that the reason for the above particular form of the Fourier series is the Bravais lattice periodicity of the function $f(\boldsymbol{r})$.

8-13 Interplanar distances. Find the interplanar distances in the family of lattice planes $(hk\ell)$ in the following Bravais lattices with respect to the given primitive vectors.

(1) The simple triclinic Bravais lattice with

$$\begin{aligned}&\boldsymbol{a}_1 = a\boldsymbol{e}_x,\ \boldsymbol{a}_2 = b(\cos\gamma\boldsymbol{e}_x + \sin\gamma\boldsymbol{e}_y),\\ &\boldsymbol{a}_3 = c\cos\beta\boldsymbol{e}_x + c\big[(\cos\alpha - \cos\beta\cos\gamma)/\sin\gamma\big]\boldsymbol{e}_y\\ &+c\big[\big(1-\cos^2\alpha-\cos^2\beta-\cos^2\gamma+2\cos\alpha\cos\beta\cos\gamma\big)^{1/2}/\sin\gamma\big]\boldsymbol{e}_z.\end{aligned}$$

(2) The centered-monoclinic Bravais lattice with

$$\begin{aligned}&\boldsymbol{a}_1 = (a\,\boldsymbol{e}_x + b\,\boldsymbol{e}_y)/2,\ \boldsymbol{a}_2 = (-a\,\boldsymbol{e}_x + b\,\boldsymbol{e}_y)/2,\\ &\boldsymbol{a}_3 = c(\cos\beta\,\boldsymbol{e}_x + \sin\beta\,\boldsymbol{e}_z).\end{aligned}$$

(3) The face-centered orthorhombic Bravais lattice with

$$\boldsymbol{a}_1 = (b\,\boldsymbol{e}_y + c\,\boldsymbol{e}_z)/2,\ \ \boldsymbol{a}_2 = (c\,\boldsymbol{e}_z + a\,\boldsymbol{e}_x)/2,\ \ \boldsymbol{a}_3 = (a\,\boldsymbol{e}_x + b\,\boldsymbol{e}_y)/2.$$

(4) The body-centered tetragonal Bravais lattice with

$$\begin{aligned}&\boldsymbol{a}_1 = (a\,\boldsymbol{e}_y + c\,\boldsymbol{e}_z - a\,\boldsymbol{e}_x)/2,\ \boldsymbol{a}_2 = (c\,\boldsymbol{e}_z + a\,\boldsymbol{e}_x - a\,\boldsymbol{e}_y)/2,\\ &\boldsymbol{a}_3 = (a\,\boldsymbol{e}_x + a\,\boldsymbol{e}_y - c\,\boldsymbol{e}_z)/2.\end{aligned}$$

The procedure for finding the interplanar distances in families of lattice planes consists of three steps. First, find the expressions for the primitive vectors of the reciprocal lattice $\boldsymbol{b}_1$, $\boldsymbol{b}_2$, and $\boldsymbol{b}_3$ for a given set of primitive vectors of the direct lattice. Then, write down the general expression for the shortest nonzero reciprocal lattice vector perpendicular to the $(hk\ell)$ family of lattice planes, $\boldsymbol{K}_0 = h\boldsymbol{b}_1 + k\boldsymbol{b}_2 + \ell\boldsymbol{b}_3$ with h, k, and ℓ no common factors. And then, obtain the interplanar distance from $d_{hk\ell} = 2\pi/|\boldsymbol{K}_0|$.

(1) For the given set of primitive vectors of the direct lattice, the primitive vectors of the reciprocal lattice are found to be given by

$$\begin{aligned}\boldsymbol{b}_1 &= (2\pi/a)\boldsymbol{e}_x - (2\pi/a)\cot\gamma\,\boldsymbol{e}_y + (2\pi/a\sin\gamma)(\cos\gamma\cos\alpha - \cos\beta)\\ &\qquad\times\left(1-\cos^2\alpha-\cos^2\beta-\cos^2\gamma+2\cos\alpha\cos\beta\cos\gamma\right)^{-1/2}\boldsymbol{e}_z,\\ \boldsymbol{b}_2 &= (2\pi/b\sin\gamma)\boldsymbol{e}_y + (2\pi/b\sin\gamma)(\cos\beta\cos\gamma-\cos\alpha)\\ &\qquad\times\left(1-\cos^2\alpha-\cos^2\beta-\cos^2\gamma+2\cos\alpha\cos\beta\cos\gamma\right)^{-1/2}\boldsymbol{e}_z,\\ \boldsymbol{b}_3 &= (2\pi\sin\gamma/c)\big(1-\cos^2\alpha-\cos^2\beta-\cos^2\gamma\\ &\qquad+2\cos\alpha\cos\beta\cos\gamma\big)^{-1/2}\boldsymbol{e}_z.\end{aligned}$$

From $\boldsymbol{K}_0 = h\boldsymbol{b}_1 + k\boldsymbol{b}_2 + \ell\boldsymbol{b}_3$ and $d_{hk\ell} = 2\pi/|\boldsymbol{K}_0|$, we obtain

$$\begin{aligned}d_{hk\ell} &= \left(1-\cos^2\alpha-\cos^2\beta-\cos^2\gamma+2\cos\alpha\cos\beta\cos\gamma\right)^{1/2}\\ &\quad\times\big[(h^2/a^2)\sin^2\alpha+(k^2/b^2)\sin^2\beta+(\ell^2/c^2)\sin^2\gamma\\ &\qquad+(2hk/ab)(\cos\alpha\cos\beta-\cos\gamma)\\ &\qquad+(2k\ell/bc)(\cos\beta\cos\gamma-\cos\alpha)\\ &\qquad+(2\ell h/ca)(\cos\gamma\cos\alpha-\cos\beta)\big]^{-1/2}.\end{aligned}$$

(2) For the given set of primitive vectors of the direct lattice, the primitive vectors of the reciprocal lattice are found to be given by

$$\begin{aligned}\boldsymbol{b}_1 &= (2\pi/a)\boldsymbol{e}_x + (2\pi/b)\boldsymbol{e}_y - (2\pi/a)\cot\beta\,\boldsymbol{e}_z,\\ \boldsymbol{b}_2 &= -(2\pi/a)\boldsymbol{e}_x + (2\pi/b)\boldsymbol{e}_y + (2\pi/a)\cot\beta\,\boldsymbol{e}_z,\\ \boldsymbol{b}_3 &= (2\pi/c\sin\beta)\,\boldsymbol{e}_z.\end{aligned}$$

From $\boldsymbol{K}_0 = h\boldsymbol{b}_1 + k\boldsymbol{b}_2 + \ell\boldsymbol{b}_3$ and $d_{hk\ell} = 2\pi/|\boldsymbol{K}_0|$, we obtain

$$\begin{aligned}d_{hk\ell} &= 1/\big[(h-k)^2/a^2\sin^2\beta+(h+k)^2/b^2+\ell^2/c^2\sin^2\beta\\ &\qquad-2\ell(h-k)\cos\beta/ca\sin^2\beta\big]^{1/2}.\end{aligned}$$

(3) For the given set of primitive vectors of the direct lattice, the primitive vectors of the reciprocal lattice are found to be given by

$$\begin{aligned}\boldsymbol{b}_1 &= (2\pi/b)\,\boldsymbol{e}_y + (2\pi/c)\,\boldsymbol{e}_z - (2\pi/a)\,\boldsymbol{e}_x,\\ \boldsymbol{b}_2 &= (2\pi/c)\,\boldsymbol{e}_z + (2\pi/a)\,\boldsymbol{e}_x - (2\pi/b)\,\boldsymbol{e}_y,\\ \boldsymbol{b}_3 &= (2\pi/a)\,\boldsymbol{e}_x + (2\pi/b)\,\boldsymbol{e}_y - (2\pi/c)\,\boldsymbol{e}_z.\end{aligned}$$

From $\boldsymbol{K}_0 = h\boldsymbol{b}_1 + k\boldsymbol{b}_2 + \ell\boldsymbol{b}_3$ and $d_{hk\ell} = 2\pi/|\boldsymbol{K}_0|$, we obtain

$$d_{hk\ell} = 1/\big[(k+\ell-h)^2/a^2+(\ell+h-k)^2/b^2+(h+k-\ell)^2/c^2\big]^{1/2}.$$

(4) For the given set of primitive vectors of the direct lattice, the primitive vectors of the reciprocal lattice are found to be given by

$$\begin{aligned} \boldsymbol{b}_1 &= (2\pi/a)\,\boldsymbol{e}_y + (2\pi/c)\,\boldsymbol{e}_z, \\ \boldsymbol{b}_2 &= (2\pi/c)\,\boldsymbol{e}_z + (2\pi/a)\,\boldsymbol{e}_x, \\ \boldsymbol{b}_3 &= (2\pi/a)\,\boldsymbol{e}_x + (2\pi/a)\,\boldsymbol{e}_y. \end{aligned}$$

From $\boldsymbol{K}_0 = h\boldsymbol{b}_1 + k\boldsymbol{b}_2 + \ell\boldsymbol{b}_3$ and $d_{hk\ell} = 2\pi/|\boldsymbol{K}_0|$, we obtain

$$d_{hk\ell} = 1/\left\{[(k+\ell)^2 + (\ell+h)^2]/a^2 + (h+k)^2/c^2\right\}^{1/2}.$$

8-14 Sizes of first Brillouin zones.

(1) Find the ratio of the lengths of the diagonals of each parallelogram face of the first Brillouin zone for the body-centered-cubic Bravais lattice.

(2) Find the ratio of the length of an edge of the polyhedron bounding the first Brillouin zone of the face-centered cubic lattice to the length of the conventional unit cell of the reciprocal lattice.

The first Brillouin zones of the BCC and FCC Bravais lattices are given in Fig. 8.5.

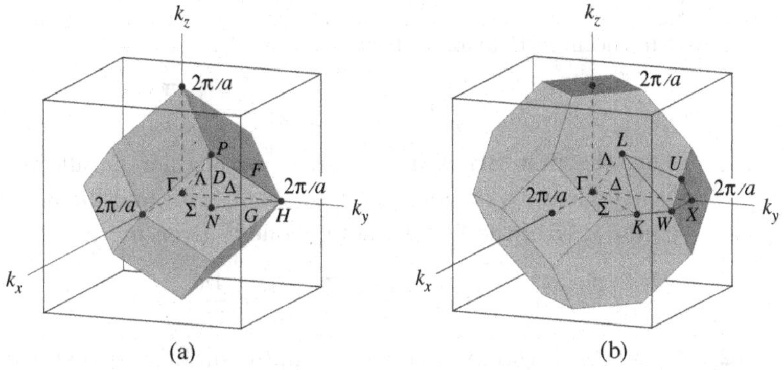

Fig. 8.5 First Brillouin zones of the BCC (a) and FCC (b) Bravais lattices.

(1) From Fig. 8.5(a), we see that the ratio of the lengths of the diagonals of each parallelogram face of the first Brillouin zone for the body-centered-cubic Bravais lattice is given by

$$\frac{\overline{NH}}{\overline{NP}} = \frac{|(0,1,0)(2\pi/a) - (1,1,0)(2\pi/a)|}{|(1,1,1)(2\pi/a) - (1,1,0)(2\pi/a)|} = 1.$$

(2) From Fig. 8.5(b), we see that the length L of an edge of the polyhedron bounding the first Brillouin zone of the face-centered cubic lattice is given by

$$L = 2\overline{KW} = 2|(3/4, 3/4, 0)(2\pi/a) - (1/2, 1, 0)(2\pi/a)| = \sqrt{2}\,\pi/a.$$

Taking into consideration that the length of the conventional unit cell of the reciprocal lattice is $a^* = 4\pi/a$, we have

$$\frac{L}{a^*} = \frac{\sqrt{2}\,\pi/a}{4\pi/a} = \frac{1}{2\sqrt{2}} \approx 0.353\,55.$$

8-15 First Brillouin zone of a simple orthorhombic Bravais lattice. Consider a simple orthorhombic Bravais lattice with lattice constants a, $2a$, and $3a$.

(1) Compute the volume of the first Brillouin zone.

(2) Compute the average number of electrons contributed by each atom when the Fermi surface is an elliptic surface tangent to the boundary of the first Brillouin zone.

---

(1) The primitive vectors of the reciprocal lattice for the simple orthorhombic Bravais lattice are given by

$$\boldsymbol{b}_1 = \frac{2\pi}{a}\boldsymbol{e}_x,\ \boldsymbol{b}_2 = \frac{\pi}{a}\boldsymbol{e}_y,\ \boldsymbol{b}_3 = \frac{2\pi}{3a}\boldsymbol{e}_z.$$

Thus, the first Brillouin zone is a rectangular parallelepiped with the lengths of sides given by $2\pi/a$, π/a, and $2\pi/3a$. The volume $\Omega_{\rm BZ}$ of the first Brillouin zone is given by

$$\Omega_{\rm BZ} = \frac{2\pi}{a}\cdot\frac{\pi}{a}\cdot\frac{2\pi}{3a} = \frac{4\pi^3}{3a^3}.$$

(2) When the Fermi surface is an elliptic surface tangent to the boundary of the first Brillouin zone, the volume of k-space enclosed by the Fermi surface is given by

$$\Omega_{\rm F} = \frac{4\pi}{3}\cdot\frac{\pi}{a}\cdot\frac{\pi}{2a}\cdot\frac{\pi}{3a} = \frac{2\pi^4}{9a^3}.$$

Since the density of states in k-space is $\mathscr{V}/4\pi^3$ (including the spin degeneracy) with $\mathscr{V}$ the volume of the lattice, the number density of electrons is given by

$$n = \frac{\Omega_{\rm F}}{4\pi^3} = \frac{\pi}{18a^3}.$$

The number of electrons contributed by each atom is then given by

$$M_{\rm el} = nv_c = \frac{\pi}{18a^3} \times 6a^3 = \frac{\pi}{3} \approx 1$$

with v_c the volume of a primitive cell of the direct lattice, $v_c = a \cdot (2a) \cdot (3a) = 6a^3$.

8-16 Monatomic monovalent metal with an FCC structure. A monatomic monovalent metal has an FCC structure of lattice constant a. The metal consists of N primitive unit cells (N is very large). Take the conduction electrons in the metal as free independent electrons.

(1) Find the density of levels for electrons.
(2) Find the total number of allowed $\boldsymbol{k}$ values in the first Brillouin zone.
(3) Find the ratio of the Fermi sphere to the volume of the first Brillouin zone.

---

(1) Making use of the fact that a quantum state occupies a volume of h^3 in phase space, we see that the number of states in the phase volume $4\pi\mathscr{V}p^2\mathrm{d}p = 4\pi\hbar^3\mathscr{V}k^2\mathrm{d}k$ is

$$\frac{4\pi\hbar^3\mathscr{V}k^2\mathrm{d}k}{h^3} = \frac{4\pi\mathscr{V}k^2\mathrm{d}k}{(2\pi)^3} = \frac{\mathscr{V}k^2\mathrm{d}k}{2\pi^2}.$$

Taking the spin degeneracy (that contributes a factor of 2) into account, we have the following density of levels for electrons

$$2 \cdot \frac{1}{\mathscr{V}} \cdot \frac{\mathscr{V}k^2dk}{2\pi^2} = \frac{k^2dk}{\pi^2}.$$

(2) Since the total number of allowed $\boldsymbol{k}$ values in the first Brillouin zone is equal to the number of primitive unit cells in the crystal, it is equal to N.

(3) For a monatomic monovalent FCC metal, the electron number density is given by $n = 4/a^3$. The Fermi wave vector is then given by $k_{\rm F} = (3\pi^2 n)^{1/3} = (12\pi^2)^{1/3}/a$. The volume of the Fermi sphere is thus given by

$$\mathscr{V}_{\rm F} = \frac{4\pi}{3}k_{\rm F}^3 = \frac{2(2\pi)^3}{a^3}.$$

The volume of the first Brillouin zone is $\Omega_{\rm BZ} = (2\pi)^3/v_c = 4(2\pi)^3/a^3$. The ratio of $\mathscr{V}_{\rm F}$ to $\Omega_{\rm BZ}$ is then given by

$$\frac{\mathscr{V}_{\rm F}}{\Omega_{\rm BZ}} = \frac{2(2\pi)^3/a^3}{4(2\pi)^3/a^3} = \frac{1}{2}.$$

The shortest distance from the Γ point to the boundary of the first Brillouin zone for an FCC Bravais lattice is the distance between the Γ and N points and is equal to $\sqrt{3}\pi/a$. Since $k_{\rm F}/\overline{\Gamma N} \approx 0.902\,5$, the Fermi sphere can sit quite comfortably within the first Brillouin zone. This provides an explanation for the nearly-spherical Fermi surfaces of monatomic monovalent FCC metals Cu, Ag, and Au.

Chapter 9

Theories and Experiments of X-Ray Diffraction on Crystals

(1) *Commonly used X-ray lines*

K_α lines

K_{α_1}: $L_{\rm III} \to K$.

K_{α_2}: $L_{\rm II} \to K$.

K_β lines

K_{β_1}: $M_{\rm III} \to K$.

K_{β_3}: $M_{\rm II} \to K$.

Quantum numbers of atomic shells.

	K	$L_{\rm II}$	$L_{\rm III}$	$M_{\rm II}$	$M_{\rm III}$
n	1	2	2	3	3
ℓ	0	1	1	1	1
j	1/2	1/2	3/2	1/2	3/2

K_α lines have higher intensity but lower energy than K_β lines.

(2) *Bragg's law*

$$2d_{hk\ell}\sin\theta = n\lambda, \quad n = 1,\, 2,\, 3,\, \cdots,$$

where n is the order of Bragg reflection.

(3) *Selection rules in the cubic crystal system*

Lattice	Selection rules
SC	All h, k, ℓ
BCC	$h + k + \ell = 2m$
FCC	h, k, ℓ: $2m+1$ or $2m$

(4) *Laue condition*

$$\boldsymbol{k} - \boldsymbol{k}' = \boldsymbol{K} \text{ or } \boldsymbol{k}\cdot\hat{\boldsymbol{K}} = |\boldsymbol{K}|/2.$$

(5) *Ewald sphere*
An Ewald sphere is a sphere centered at the tip of the wave vector $\boldsymbol{k}$ of the incident X-rays with its radius equal to the length of $\boldsymbol{k}$.

(6) *Laue method*
A white X-ray source and a stationary single-crystal sample are used.

(7) *Rotating crystal method*
A monochromatic X-ray source is used and a single-crystal sample is kept rotating during the experiment.

(8) *Powder (Debye-Scherrer) method*
A monochromatic X-ray source and a powder sample are used.

(9) *Indexation of diffraction peaks*

$$\theta_1 = \frac{\pi S_1}{2W}, \ \theta_2 = \frac{\pi}{2}\left(1 - \frac{S_2}{W}\right),$$

$$\sin^2\theta = \frac{\lambda^2}{4a^2}(h^2 + k^2 + \ell^2) \ (\text{cubic}),$$

$$h^2 + k^2 + \ell^2 = \frac{4a^2}{\lambda^2}\sin^2\theta = C\sin^2\theta.$$

Procedures:

i. Compute $\sin^2\theta_i$ for all θ_i.
ii. Find C through trial and error.
iii. Round $C\sin^2\theta_i$ to an integer $\mathscr{I}_i$.
iv. Break $\mathscr{I}_i$ into $h^2 + k^2 + \ell^2$.

9-1 Powder diffraction on silicon. A powder diffraction pattern is obtained on silicon using Cu $K\alpha$ radiation ($\lambda = 0.154$ nm). The crystal structure of silicon is cubic with a lattice constant of 0.543 nm. Its unit cell contains 8 silicon atoms located at: $(0, 0, 0)$, $(1/2, 1/2, 0)a$, $(1/2, 0, 1/2)a$, $(0, 1/2, 1/2)a$, $(1/4, 1/4, 1/4)a$, $(3/4, 3/4, 1/4)a$, $(3/4, 1/4, 3/4)a$, and $(1/4, 3/4, 3/4)a$.

(1) Find the Miller indices of the first two lines in the powder diffraction pattern.

(2) Evaluate the ratio of the integrated intensity of the first line to that of the second line.

---

(1) Si has an FCC Bravais lattice. With respect to the conventional unit cell, the reciprocal lattice vectors are of the form

$$\boldsymbol{K} = (h\boldsymbol{e}_x + k\boldsymbol{e}_y + \ell\boldsymbol{e}_z)(2\pi/a),$$

where h, k, and ℓ are integers. The geometric structure factor is given by

$$\begin{aligned}
S(\boldsymbol{K}) &= \sum_{\kappa} \mathrm{e}^{-\mathrm{i}\boldsymbol{K}\cdot\boldsymbol{d}_\kappa} \\
&= 1 + \mathrm{e}^{-\mathrm{i}(h+k)\pi} + \mathrm{e}^{-\mathrm{i}(h+\ell)\pi} + \mathrm{e}^{-\mathrm{i}(k+\ell)\pi} + \mathrm{e}^{-\mathrm{i}(h+k+\ell)\pi/2} \\
&\quad + \mathrm{e}^{-\mathrm{i}(3h+3k+\ell)\pi/2} + \mathrm{e}^{-\mathrm{i}(3h+k+3\ell)\pi/2} + \mathrm{e}^{-\mathrm{i}(h+3k+3\ell)\pi/2} \\
&= \left[1 + (-1)^{h+k} + (-1)^{k+\ell} + (-1)^{\ell+h}\right]\left[1 + (-\mathrm{i})^{h+k+\ell}\right] \\
&= \begin{cases} 4\times 2, & \text{all even or all odd and } h+k+\ell = 4n, \\ 4(1-\mathrm{i}), & \text{all even or all odd and } h+k+\ell = 4n+1, \\ 4(1+\mathrm{i}), & \text{all even or all odd and } h+k+\ell = 4n+3, \\ 0, & \text{otherwise.} \end{cases}
\end{aligned}$$

Note that the first factor on each of the first three lines is for the plain FCC lattice while the second factor is for the two-atom basis. The few sets of Miller indices for reciprocal lattice vectors that give rise to nonzero geometric structure factors are listed below together with the values of the structure factor and the interplanar distance.

$[hk\ell]$	$S(\boldsymbol{K})$	$d_{hk\ell}$ $[a]$	$[hk\ell]$	$S(\boldsymbol{K})$	$d_{hk\ell}$ $[a]$
[111]	$4+4\mathrm{i}$	0.577 350	[422]	8	0.204 124
[220]	8	0.353 553	[333]	$4-4\mathrm{i}$	0.192 450
[311]	$4-4\mathrm{i}$	0.301 511	[440]	8	0.176 777
[400]	8	0.25	[444]	8	0.144 338
[331]	$4+4\mathrm{i}$	0.229 416			

In consideration that, the larger the interplanar distance, the smaller the Bragg angle, we see that the Miller indices correspond to the first two lines are [111] and [220] with the former corresponding to a smaller Bragg angle than the latter.

(2) In consideration that the diffraction intensity is proportional to the absolute value squared of the geometric structure factor, we have the following intensity ratio for the first two lines

$$\frac{I_1}{I_2} = \left|\frac{S([111])}{S[220]}\right|^2 = \left|\frac{4+4\mathrm{i}}{8}\right|^2 = \frac{1}{2}.$$

9-2 Powder diffraction on cubic CaF_2. When a diffraction pattern of a powdered specimen of cubic CaF_2 is taken with X-rays of wavelength 0.154 2 nm, the first four lines occur at the Bragg angles 14.16°, 16.41°, 23.55°, and 27.94°.

(1) Find the Miller indices of the reflections.
(2) Evaluate the lattice constant a.
(3) Show that the result for the lattice constant a is consistent with the measured density of CaF_2, 3,200 kg/m^3.

---

(1) The results of the computations performed in the indexation process are shown in the following table. The predictions for the lattice constant are also given in the last column. The Miller indices of the reflections are given in the second to last column.

θ	$\sin^2\theta$	$C\sin^2\theta$	$h^2+k^2+\ell^2$	hkl	a [nm]
14.16	0.059 844	3.000 00	3	111	0.545 889
16.41	0.079 811	4.000 96	4	200	0.545 824
23.55	0.159 640	8.002 77	8	220	0.545 794
27.94	0.219 536	11.005 40	11	311	0.545 755

(2) We take the average of the predictions for the lattice constant in the last column as an estimate for the lattice constant

$$\overline{a} = \frac{1}{n}\sum_{i=1}^{n} a_i \approx 0.545\ 816 \text{ nm}.$$

The standard error of this estimate is given by

$$\delta a = \left[\frac{1}{n(n-1)}\sum_i (a_i - \overline{a})^2\right]^{1/2} \approx 2.821\ 652 \times 10^{-5} \text{ nm}.$$

Thus, the lattice constant is given by $\overline{a} \pm \delta a \approx 0.545\ 816 \pm 2.821\ 652 \times 10^{-5}$ nm.

(3) Since there are four formula units in a conventional unit cell, the mass density of CaF_2 is given by

$$\rho_m = \frac{1}{a^2}(4A_{\text{Ca}} + 8A_{\text{F}})\text{u} \approx 3,189 \text{ kg/m}^3.$$

Thus, the result for the lattice constant a is consistent with the measured density of CaF_2, 3,200 kg/m^3.

9-3 Debye-Scherrer experiments on two cubic samples. One has made measurements of S_1 from two films, from Debye Scherrer experiments on two different samples. It is known that these two samples are cubic. The wavelength of the X-ray beam used is 0.154 nm and the distance W between the two holes on an unrolled film is 180 mm. The following values of S_1 in mm are obtained for sample 1: 22, 31, 38, 45, 50, 56, 65, 70, 74, 78. For sample 2, the values of S_1 in mm are: 35, 41, 59, 71, 75, 88, 99, 103, 116, 128.

(1) Using the values of S_1 given above and the formulas $\theta = \pi S_1/2W$ and $\sin^2\theta = (\lambda^2/4a^2)(h^2+k^2+\ell^2)$, complete the following table *for each sample*

S_1 (mm)	θ	$\sin^2\theta$	$C\sin^2\theta$	$h^2+k^2+\ell^2$	(hkl)	a
⋮	⋮	⋮	⋮	⋮	⋮	⋮

(2) Determine the cubic structure type of each sample.

(3) Estimate the lattice constant for each sample. Both the mean and the error of the lattice constant must be given.

---

(1) For sample 1, we have ($C = 1/0.036\,408$)

S_1 [mm]	θ	$\sin^2\theta$	$C\sin^2\theta$	$h^2+k^2+\ell^2$	(hkl)	a [nm]
22	0.191 986	0.036 408	1.000 00	1	(100)	0.403 545
31	0.270 526	0.071 416	1.961 55	2	(110)	0.407 481
38	0.331 613	0.105 995	2.911 29	3	(111)	0.409 647
45	0.392 699	0.146 447	4.022 37	4	(200)	0.402 421
50	0.436 332	0.178 606	4.905 68	5	(210)	0.407 406
56	0.488 692	0.220 404	6.053 70	6	(211)	0.401 751
65	0.567 232	0.288 691	7.929 31	8	(220)	0.405 340
70	0.610 865	0.328 990	9.036 18	9	(300)	0.402 736
74	0.645 772	0.362 181	9.947 83	10	(310)	0.404 602
78	0.680 678	0.396 044	10.877 90	11	(311)	0.405 803

For sample 2, we have ($C = 3/0.090424$)

S_1 [mm]	θ	$\sin^2\theta$	$C\sin^2\theta$	$h^2+k^2+\ell^2$	(hkl)	a [nm]
35	0.305 433	0.090 424	3.000 0	3	(111)	0.443 516
41	0.357 792	0.122 645	4.069 0	4	(200)	0.439 739
59	0.514 872	0.242 481	8.044 8	8	(220)	0.442 280
71	0.619 592	0.337 216	11.187 8	11	(311)	0.439 778
75	0.654 498	0.370 590	12.295 1	12	(222)	0.438 162
88	0.767 945	0.482 550	16.009 6	16	(400)	0.443 383
99	0.863 938	0.578 217	19.183 5	19	(331)	0.441 390
103	0.898 845	0.612 476	20.320 1	20	(420)	0.440 009
116	1.012 290	0.719 186	23.860 4	24	(422)	0.444 811
128	1.117 010	0.807 831	26.801 4	27	(511)	0.445 156

(2) Since h, k, and ℓ for sample 1 are neither all odd nor all even and their sum is not even, the underlying lattice of sample 1 is a simple cubic Bravais lattice. Since h, k, and ℓ for sample 2 are either all odd or all even, the underlying lattice of sample 2 is a face-centered cubic Bravais lattice.

(3) Using the values of a in (1) for samples 1 and 2 and the following estimators for the mean and the error of a

$$\overline{a} = \frac{1}{n}\sum_{i=1}^{n} a_i, \quad \delta a = \left[\frac{1}{n(n-1)}\sum_{i=1}^{n}(a_i - \overline{a})^2\right]^{1/2}$$

with $n = 10$, we obtain

$$a \approx 0.4051 \pm 0.0008 \text{ nm for sample 1,}$$

and

$$a \approx 0.4418 \pm 0.0008 \text{ nm for sample 2.}$$

9-4 Crystal structure of $BaTiO_3$. Barium titanate $BaTiO_3$ crystallizes in such a structure that the Ba atoms sit at the corners of a cube with a Ti atom at the cube's center and the O atoms in the centers of its faces.

(1) Describe the structure with a suitable lattice and basis.

(2) Determine the intensity relationship between the first four Bragg reflections with the help of the structure factor. The following form factors apply to the atoms: $f_{\mathrm{Ba}} = 7f_{\mathrm{O}}$ and $f_{\mathrm{Ti}} = 3f_{\mathrm{O}}$.

---

(1) The underlying lattice is a simple cubic Bravais lattice. The basis consists of one Ba atom at $(0, 0, 0)$, one Ti atom at $(1/2, 1/2, 1/2)a$, and three O atoms at $(0, 1/2, 1/2)a$, $(1/2, 0, 1/2)a$, and $(1/2, 1/2, 0)a$.

(2) For a simple cubic Bravais lattice, a reciprocal lattice vector is given by

$$\boldsymbol{K} = \frac{2\pi h}{a}\boldsymbol{e}_x + \frac{2\pi k}{a}\boldsymbol{e}_y + \frac{2\pi \ell}{a}\boldsymbol{e}_z.$$

The structure factor is then given by

$$\begin{aligned} S(\boldsymbol{K}) &= \sum_j f_j(\boldsymbol{K})\mathrm{e}^{\mathrm{i}\boldsymbol{K}\cdot\boldsymbol{d}_j} \\ &= f_{\mathrm{Ba}}(\boldsymbol{K}) + f_{\mathrm{Ti}}(\boldsymbol{K})\mathrm{e}^{\mathrm{i}(h+k+\ell)\pi} + f_{\mathrm{O}}(\boldsymbol{K})\big[\mathrm{e}^{\mathrm{i}(k+\ell)\pi} \\ &\quad + \mathrm{e}^{\mathrm{i}(\ell+h)\pi} + \mathrm{e}^{\mathrm{i}(h+k)\pi}\big] \\ &= \big[7 + 3\mathrm{e}^{\mathrm{i}(h+k+\ell)\pi} + \mathrm{e}^{\mathrm{i}(k+\ell)\pi} + \mathrm{e}^{\mathrm{i}(\ell+h)\pi} + \mathrm{e}^{\mathrm{i}(h+k)\pi}\big] f_{\mathrm{O}}(\boldsymbol{K}). \end{aligned}$$

For the first four reflections, we have

$$S(100) = \left[7 + 3e^{i\pi} + 1 + e^{i\pi} + e^{i\pi}\right] f_O(100) = 3f_O(100)$$
$$S(110) = \left[7 + 3e^{2i\pi} + e^{i\pi} + e^{i\pi} + e^{2i\pi}\right] f_O(110) = 9f_O(110)$$
$$S(111) = \left[7 + 3e^{3i\pi} + e^{2i\pi} + e^{2i\pi} + e^{2i\pi}\right] f_O(111) = 7f_O(111)$$
$$S(200) = \left[7 + 3e^{2i\pi} + 1 + e^{2i\pi} + e^{2i\pi}\right] f_O(200) = 13f_O(200).$$

Making use of the fact that $I(\boldsymbol{K}) \propto |S(\boldsymbol{K})|^2$, we have

$$I(200) : I(111) : I(110) : I(100)$$
$$= 169|f_O(200)|^2 : 49|f_O(111)|^2 : 81|f_O(110)|^2 : 9|f_O(100)|^2$$
$$\approx 18.78|f_O(200)|^2 : 5.44|f_O(111)|^2 : 9|f_O(110)|^2 : |f_O(100)|^2.$$

If the $\boldsymbol{K}$ dependence of $f_O(\boldsymbol{K})$ can be neglected, we then have

$$I(200) : I(111) : I(110) : I(100) = 169 : 49 : 81 : 9$$
$$= 18.78 : 5.44 : 9 : 1.$$

9-5 Temperature dependence of the Bragg angle for Al. At 300 K, Al has an FCC structure with a lattice constant of 0.405 nm. The coefficient of heat expansion for Al is $\alpha = 25 \times 10^{-6}\ \mathrm{K}^{-1}$. Aluminum is studied with Cu $\mathrm{K}_{\alpha 1}$ radiation which has a wavelength of $\lambda = 0.154$ nm. We now investigate the change $\Delta\theta$ in θ due to the change in temperature from 300 to 600 K.

(1) Derive an expression for $\Delta\theta$ in terms of θ.

(2) For the (111) reflection, evaluate $\Delta\theta$.

---

(1) The change in a is given by

$$\Delta a = \alpha \Delta T a = 0.0075a.$$

Differentiating $\sin^2\theta = (h^2 + k^2 + \ell^2)\lambda^2/4a^2$ with respect to a yields

$$2\sin\theta\cos\theta \mathrm{d}\theta = -\frac{\lambda^2}{2a^3}(h^2 + k^2 + \ell^2)\mathrm{d}a,$$
$$2\sin\theta\cos\theta \mathrm{d}\theta = -\frac{2}{a}\sin^2\theta\,\mathrm{d}a,$$
$$\mathrm{d}\theta = -\frac{1}{a}\tan\theta\,\mathrm{d}a,$$
$$\Delta\theta \approx -\frac{1}{a}\tan\theta\,\Delta a,$$
$$\Delta\theta \approx -0.0075\tan\theta.$$

(2) For the (111) reflection, we have

$$\sin^2\theta_{111} = \frac{3\lambda^2}{4a^2} = \frac{3\times 1.54^2}{4\times 4.05^2} \approx 0.11,$$
$$\sin\theta_{111} \approx 0.33,$$
$$\tan\theta_{111} \approx 0.35,$$
$$\Delta\theta_{111} \approx -0.15^{\circ}.$$

9-6 Room-temperature superconductor. Assume that a room-temperature superconductor has been discovered. A theorist predicts that it has a face-centered cubic Bravais lattice with a lattice constant of $a = 1.5$ nm. An experimentalist plans to verify the prediction by performing a powder X-ray diffraction experiment using a Debye-Scherrer camera and X-rays of wavelength 0.15 nm.

(1) Express 2θ of a to-be-observed arc in the diffraction pattern in terms of the Miller indices of the reflecting lattice planes with respect to the cubic conventional cell.

(2) Numerically evaluate the values of 2θ for the first half-dozen arcs in the diffraction pattern.

---

(1) From Bragg's law $2d\sin\theta = n\lambda$ and $d = a/(h^2+k^2+\ell^2)^{1/2}$ in the cubic crystal system, we have

$$2\theta = 2\sin^{-1}\frac{n\lambda}{2d} = 2\sin^{-1}\left[\frac{n\lambda}{2a}(h^2+k^2+\ell^2)^{1/2}\right]$$
$$= 2\sin^{-1}\left[\frac{\lambda}{2a}(\bar{h}^2+\bar{k}^2+\bar{\ell}^2)^{1/2}\right],$$

where $\bar{h} = nh$, $\bar{k} = nk$, and $\bar{\ell} = n\ell$.

(2) The values of $\bar{h}$, $\bar{k}$, and $\bar{\ell}$ for the first half-dozen diffraction peaks for forward reflections are 111, 200, 220, 311, 222, and 400. The corresponding values of 2θ are 9.936°, 11.478°, 16.260°, 19.091°, 19.948°, and 23.074°.

9-7 Powder diffraction on $YBa_2Cu_3O_{6.9}$. It is known that the high-temperature superconductor $YBa_2Cu_3O_{6.9}$ belongs to the orthorhombic crystal system. A powder X-ray diffraction using a diffractometer and the X-rays of wavelength $\lambda = 0.154\,056$ nm has been performed on a powder sample of $YBa_2Cu_3O_{6.9}$ with diffraction peaks observed at the following values of 2θ (in degrees): 15.138, 22.794, 23.184, 27.563, 27.822, 32.533, 32.783, 36.331, 40.341, 46.600, 47.419, 51.463,

52.680, 54.953, 58.189, 58.754, 68.741, 72.945, 73.492, 74.933, 77.569, 78.991, 91.590, 92.892.

(1) Establish selection rules for the orthorhombic crystal system without taking the atom-basis into account.

(2) Index the diffraction peaks observed on $YBa_2Cu_3O_{6.9}$ and identify the underlying Bravais lattice.

(3) Compute the lattice constants a, b, and c for $YBa_2Cu_3O_{6.9}$.

---

(1) As in the simple cubic Bravais lattice, any lattice planes of the simple orthorhombic Bravais lattice can reflect X-rays since the geometric structure factor does not play any role when the atom-basis is not taken into account.

For the base-centered orthorhombic Bravais lattice (the C-type lattice), with respect to the conventional cell, the basis consists of two points: $\boldsymbol{d}_1 = (0, 0, 0)$ and $\boldsymbol{d}_2 = (a/2, b/2, c)$. With the reciprocal lattice vector given by $\boldsymbol{K} = h(2\pi/a)\boldsymbol{e}_x + k(2\pi/b)\boldsymbol{e}_y + \ell(2\pi/c)\boldsymbol{e}_z$, the geometric structure factor with respect to the conventional cell is given by

$$S_{\text{baco}} = \sum_{\kappa=1}^{2} \mathrm{e}^{-\mathrm{i}\boldsymbol{K}\cdot\boldsymbol{d}_\kappa} = 1 + \mathrm{e}^{-\mathrm{i}\pi(h+k)} = \begin{cases} 2, \ h+k = 2n, \\ 0, \ h+k = 2n+1 \end{cases}$$

with n an integer. Thus, the selection rule for the base-centered orthorhombic Bravais lattice (C-type) is that lattice planes with even $h + k$ can reflect X-rays while those with odd $h + k$ can't.

For the body-centered orthorhombic Bravais lattice, with respect to the conventional cell, the basis consists of two points: $\boldsymbol{d}_1 = (0, 0, 0)$ and $\boldsymbol{d}_2 = (a/2, b/2, c/2)$. The geometric structure factor with respect to the conventional cell is given by

$$S_{\text{bco}} = \sum_{\kappa=1}^{2} \mathrm{e}^{-\mathrm{i}\boldsymbol{K}\cdot\boldsymbol{d}_\kappa} = 1 + \mathrm{e}^{-\mathrm{i}\pi(h+k+\ell)} = \begin{cases} 2, \ h+k+\ell = 2n, \\ 0, \ h+k+\ell = 2n+1 \end{cases}$$

with n an integer. Thus, the selection rule for the body-centered orthorhombic Bravais lattice is that lattice planes with even $h + k + \ell$ can reflect X-rays while those with odd $h + k + \ell$ can't. This selection rule is identical in form with that for the body-centered Bravais lattice in the cubic crystal system.

For the face-centered orthorhombic Bravais lattice, with respect to the conventional cell, the basis consists of four points: $\boldsymbol{d}_1 = (0, 0, 0)$, $\boldsymbol{d}_2 = (a/2, b/2, 0)$, $\boldsymbol{d}_3 = (a/2, 0, c/2)$, and

$\boldsymbol{d}_4 = (0, b/2, c/2)$. The geometric structure factor with respect to the conventional cell is given by

$$S_{\text{fco}} = \sum_{\kappa=1}^{2} \mathrm{e}^{-\mathrm{i}\boldsymbol{K}\cdot\boldsymbol{d}_\kappa} = 1 + \mathrm{e}^{-\mathrm{i}\pi(h+k)} + \mathrm{e}^{-\mathrm{i}\pi(h+\ell)} + \mathrm{e}^{-\mathrm{i}\pi(k+\ell)}$$
$$= \begin{cases} 4, & h, k, \ell \text{ all even or all odd,} \\ 0, & \text{otherwise.} \end{cases}$$

Thus, the selection rule for the face-centered orthorhombic Bravais lattice is that lattice planes with h, k, ℓ all even or all odd can reflect X-rays while those with h, k, ℓ neither all even nor all odd can't. This selection rule is identical in form with that for the face-centered Bravais lattice in the cubic crystal system. The above-derived selection rules are summarized in the following table.

Lattice	**Allowed**	**Forbidden**
Simple	All h, k, and ℓ	None
base-centered	$h + k$ even	$h + k$ odd
Body-centered	$h + k + \ell$ even	$h + k + \ell$ odd
Face-centered	h, k, ℓ all even or all odd	h, k, ℓ neither all even nor all odd

(2) Inserting the expression of the interplanar distance $d_{hk\ell} = 1/(h^2/a^2 + k^2/b^2 + \ell^2/c^2)^{1/2}$ into Bragg's law $2d_{hk\ell}\sin\theta = n\lambda$, we obtain

$$\begin{aligned} 4\sin^2\theta &= n^2\lambda^2(h^2/a^2 + k^2/b^2 + \ell^2/c^2) \\ &= (nh)^2/(a/\lambda)^2 + (nk)^2/(b/\lambda)^2 + (n\ell)^2/(c/\lambda)^2 \\ &= \alpha\bar{h}^2 + \beta\bar{k}^2 + \gamma\bar{\ell}^2, \end{aligned}$$

where $\alpha = (\lambda/a)^2$, $\beta = (\lambda/b)^2$, $\gamma = (\lambda/c)^2$, $\bar{h} = nh$, $\bar{k} = nk$, and $\bar{\ell} = n\ell$. Note that the order n of diffraction has been absorbed into the Miller indices and that $\bar{h}$, $\bar{k}$, and $\bar{\ell}$ are no longer restricted to integers without any nontrivial common factors. To index the diffraction peaks, we infer values for α, β, and γ as well as for Miller indices $\bar{h}$, $\bar{k}$, and $\bar{\ell}$ for each diffraction peak from the given diffraction data. The procedure we follow is given as follows.

i. Compute $4\sin^2\theta$ from each given value of 2θ.

ii. Select initial values for lattice constants a, b, and c, and then fit the first 4 or 5 diffraction peaks using the above equation to obtain values for α, β, and γ.

The initial values for the lattice constants can be obtained through trial and error. For the selected initial values for the lattice constants, fit the first 4 or 5 diffraction peaks. If χ^2 of the fitting is large, adjust the initial values for the lattice constants and perform the fitting again. Repeat the process until a satisfactory small value of χ^2 is reached.

iii. Using the obtained values of α, β, and γ and Eq. (9.1), we can find the Miller indices for all the diffraction peaks by picking out, for each diffraction peak, the set of $hk\ell$ that yields the smallest value for $|\alpha\bar{h}^2 + \beta\bar{k}^2 + \gamma\bar{\ell}^2 - 4\sin^2\theta|/4\sin^2\theta$ from all the possible combinations of $hk\ell$.

iv. Using the above-derived Miller indices, refine the values of α, β, and γ by performing a fitting with all the diffraction peaks taken into account.

v. Estimate the values of lattice constants a, b, and c predicted by the diffraction peak at Bragg angle θ_p from

$$a_p = (\lambda/\sqrt{\alpha})f_p,\ b_p = (\lambda/\sqrt{\beta})f_p,\ c_p = (\lambda/\sqrt{\gamma})f_p,$$

where $f_p = 4\sin^2\theta_p/(\alpha\bar{h}_p^2 + \beta\bar{k}_p^2 + \gamma\bar{\ell}_p^2)$.

vi. Finally, evaluate the means and the standard errors of lattice constants from

$$\bar{a} = \frac{1}{n}\sum_{p=1}^{n} a_p,\ \delta a = \left[\frac{1}{n(n-1)}\sum_{p=1}^{n}(a_p - \bar{a})^2\right]^{1/2},$$

and similar expressions for $\bar{b}$ and δb as well as for $\bar{c}$ and δc. The means and standard errors of a lattice constant are often written together in the form $a = \bar{a} \pm \delta a$.

The diffraction peaks can be indexed by following the above-given procedure and the Miller indices of the reflecting lattice planes are given in the following table together with the lattice constants estimated from the diffraction peaks. Since the Miller indices $\bar{h}\bar{k}\bar{\ell}$ in the following table are not subjected to any restrictions put forward by the selection rule for a centered lattice, the underlying Bravais lattice of $YBa_2Cu_3O_{6.9}$ is a simple orthorhombic Bravais lattice.

2θ	$4\sin^2\theta$	$\bar{h}\,\bar{k}\,\bar{\ell}$	a [nm]	b [nm]	c [nm]
15.138	0.069 401	0 0 2	0.382 331	0.388 612	1.166 878
22.794	0.156 193	0 0 3	0.382 381	0.388 663	1.167 031
23.184	0.161 509	1 0 0	0.382 208	0.388 487	1.166 504
27.563	0.226 995	0 1 2	0.383 314	0.389 611	1.169 879
27.822	0.231 196	1 0 2	0.382 482	0.388 765	1.167 340
32.533	0.313 836	0 1 3	0.383 097	0.389 390	1.169 216
32.783	0.318 545	1 1 0	0.382 632	0.388 918	1.167 797
36.331	0.388 784	1 1 2	0.382 991	0.389 283	1.168 893
38.478	0.434 306	0 0 5	0.382 574	0.388 859	1.167 620
40.341	0.475 589	1 1 3	0.382 892	0.389 182	1.168 592
46.600	0.625 825	0 0 6	0.382 704	0.388 991	1.168 016
47.419	0.646 736	2 0 0	0.382 415	0.388 697	1.167 135
51.463	0.753 960	1 1 5	0.382 880	0.389 170	1.168 555
52.680	0.787 468	1 0 6	0.382 634	0.388 920	1.167 804
54.953	0.851 504	1 2 2	0.381 161	0.387 423	1.163 308
58.189	0.945 762	1 1 6	0.382 961	0.389 252	1.168 803
58.754	0.962 573	2 1 3	0.383 036	0.389 328	1.169 029
68.741	1.274 831	2 2 0	0.382 729	0.389 017	1.168 095
72.945	1.413 421	1 2 6	0.382 531	0.388 815	1.167 488
73.492	1.431 702	2 2 3	0.382 782	0.389 070	1.168 255
74.933	1.480 103	0 3 2	0.382 696	0.388 983	1.167 992
77.569	1.569 473	0 3 3	0.382 995	0.389 289	1.168 906
78.991	1.618 074	3 1 0	0.383 176	0.389 470	1.169 456
82.411	1.735 868	1 3 3	0.383 462	0.389 761	1.170 329
91.590	2.055 494	2 3 0	0.3824 39	0.3887 21	1.1672 08
92.892	2.100 907	3 2 1	0.3826 33	0.3889 19	1.1678 00

(3) The values of lattice constants predicted by the diffraction peaks are given in the last three columns in the following table. Evaluating the means and standard errors from these estimates, we obtain $a = 0.382\,7 \pm 0.000\,1$ nm, $b = 0.389\,0 \pm 0.000\,1$ nm, and $c = 1.168\,0 \pm 0.000\,3$ nm. The relative deviations (in percentage) of the estimates of lattice constants from the diffraction peaks are shown in the following table, from which it is seen that the (122) diffraction peak has the largest relative deviation, followed by the (133) and (012) diffraction peaks.

9-8 Three phases of iron. Iron can exist in three different phases that are all cubic, the α phase (below 910°C), the γ phase (between 910°C and 1,403°C), and the δ phase (between 1,403°C and 1,535°C). In powder X-ray diffraction experiments with X-ray beams of wavelength 0.154 183 8 nm, diffraction peaks were observed at $2\theta = 44.67°$, 65.02°, and 82.32° in the α phase at 513 K, at

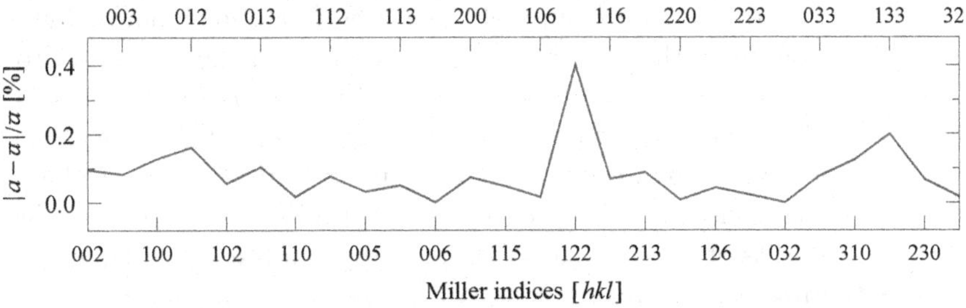

Fig. 9.1 Relative deviations (in percentage) of the estimates of lattice constants from individual diffraction peaks. Note that the relative deviations are the same for all three lattice constants a, b, and c.

$2\theta = 42.66^\circ$, 49.67°, 72.88°, and 88.29° in the γ phase at 1,565 K, and at $2\theta = 43.72^\circ$, 63.55°, and 80.32° in the δ phase at 1,705 K.

(1) Index all the diffraction peaks and determine the structure of each phase.

(2) Evaluate the lattice constants of the three phases at the given temperatures.

---

(1) In the α-phase, because of the number of available diffraction data is small, in order to index the diffraction peaks uniquely, we need to pay attention to the fact that, due to the weak intensity of the diffraction peak at $2\theta = 65.02^\circ$, it is not a first-order diffraction. Taking this fact into account, we know that the Miller indices of lattice planes that give rise to the diffraction peak at $2\theta = 65.02^\circ$ must have a nontrivial common factor. We find that the proper constant C is given by $C = 2/0.144\,42 \approx 13.848\,5$. The results of the indexation are given in the following table.

Phase	2θ	$\sin^2\theta$	$C\sin^2\theta$	$h^2+k^2+\ell^2$	$h\ k\ h$	Lattice	a [nm]	$\bar{a}$ [nm]
	44.67	0.144 42	2.000 00	2	1 1 0		0.286 89	
α	65.02	0.288 85	4.000 23	4	2 0 0	BCC	0.286 88	0.286 90
	82.32	0.433 18	5.999 05	6	2 1 1		0.286 91	
	42.66	0.132 31	3.000 00	3	1 1 1		0.367 10	
γ	49.67	0.176 41	3.999 94	4	2 0 0	FCC	0.36710	0.367 10
	72.88	0.352 81	7.999 93	8	2 2 0		0.367 10	
	88.29	0.485 08	10.999 0	11	3 1 1		0.367 11	
	43.72	0.138 64	2.000 00	2	1 1 0		0.292 81	
δ	63.55	0.277 29	4.000 25	4	2 0 0	BCC	0.292 80	0.292 80
	80.32	0.415 93	6.000 23	6	2 1 1		0.292 80	

In the γ phase, because of the availability of one more diffraction peak, the indexation can be done with the standard procedure. The constant C is given by $C = 3/0.132\ 31 \approx 22.674$. The results of the indexation are also given in the above table. The situation for the δ phase is similar to that for the α phase. The constant C is given by $C = 2/0.138\ 64 \approx 14.425\ 9$. The results of the indexation are also given in the above table.

(2) From the estimates of lattice constants in the second-to-last column in the above table, we can easily obtain the average values of lattice constants, with the results given in the last column in the above table. With more digits retained and the standard errors evaluated, the lattice constants are given by $a = 0.286\ 895\ 1 \pm 0.000\ 000\ 2$ nm in the α phase, $a = 0.367\ 101\ 0 \pm 0.000\ 000\ 1$ nm in the γ phase, and $a = 0.292\ 803\ 8 \pm 0.000\ 000\ 1$ nm in the δ phase. Note that, because of the small number of samples in the population for the lattice constants in each phase, the results for the standard errors may be unreliable.

9-9 Powder diffraction on $La_{1.8}Ba_{0.2}CuO_{4-y}$. In a synchrotron X-ray diffraction study of a powder sample of high temperature superconductor $La_{1.8}Ba_{0.2}CuO_{4-y}$, the diffraction peaks were observed at $2\theta = 34.736°$, $37.442°$, $40.516°$, $45.576°$, $46.493°$, $48.680°$, $54.043°$, $56.369°$, $60.421°$, $60.757°$, $61.598°$, $62.189°$, $63.007°$, and $65.856°$ at 300 K for a tetragonal phase. The wavelength of the used X-ray beam is 0.171 9 nm.

(1) Establish selection rules for the tetragonal crystal system.

(2) Index the diffraction peaks observed and identify the underlying Bravais lattice.

(3) Compute the lattice constants a and c.

---

(1) As in the simple cubic Bravais lattice, any lattice planes of the simple tetragonal Bravais lattice can reflect X-rays since the geometric structure factor does not play any role when the atom-basis is not taken into account.

For the body-centered tetragonal Bravais lattice, with respect to the conventional cell, the basis consists of two points: $\boldsymbol{d}_1 = (0, 0, 0)$ and $\boldsymbol{d}_2 = (a/2, a/2, c)$. With a reciprocal lattice vector given by $\boldsymbol{K} = h(2\pi/a)\boldsymbol{e}_x + k(2\pi/a)\boldsymbol{e}_y + \ell(2\pi/c)\boldsymbol{e}_z$, the geometric structure factor with respect to the conventional cell is given by

$$S_{\rm bco}(\boldsymbol{K}) = \sum_{\kappa=1}^{2} {\rm e}^{-{\rm i}\boldsymbol{K}\cdot\boldsymbol{d}_\kappa} = 1+{\rm e}^{-{\rm i}\pi(h+k+\ell)} = \begin{cases} 2, & h+k+\ell = 2n, \\ 0, & h+k+\ell = 2n+1 \end{cases}$$

with n an integer. Thus, the selection rule for the body-centered tetragonal Bravais lattice is that lattice planes with even $h+k+\ell$ can reflect X-rays while those with odd $h+k+\ell$ can't. This selection rule is identical in form with those for the body-centered Bravais lattices in the cubic and orthorhombic crystal systems.

(2) The indexation procedure is the same as for the orthorhombic crystal system with the results given in the following table together with lattice constants estimated from the diffraction peaks. From the following table, it is seen that the sum of the Miller indices for each diffraction peak is an even number. Therefore, the underlying lattice is a body-centered tetragonal Bravais lattice.

2θ	$4\sin^2\theta$	$\bar{h}\ \bar{k}\ \bar{\ell}$	a [nm]	c [nm]
34.736	0.356 427 6	1 0 3	0.378 404 0	1.331 522 4
37.442	0.412 061 6	1 1 0	0.378 064 6	1.330 328 1
40.516	0.479 550 8	1 1 2	0.378 431 8	1.331 620 3
45.576	0.600 074 9	0 0 6	0.378 393 9	1.331 487 0
46.493	0.623 113 6	1 0 5	0.378 395 4	1.331 492 4
48.680	0.679 472 3	1 1 4	0.378 391 8	1.331 479 6
54.043	0.825 644 1	2 0 0	0.378 413 3	1.331 555 1
56.369	0.892 315 8	2 0 2	0.378 411 1	1.331 547 5
60.421	1.012 754 0	1 1 6	0.378 375 0	1.331 420 6
60.757	1.022 971 0	1 0 7	0.378 359 2	1.331 364 8
61.598	1.048 690 0	2 1 1	0.378 406 9	1.331 532 5
62.189	1.066 887 0	0 0 8	0.378 409 4	1.331 541 5
63.007	1.092 237 0	2 0 4	0.378 389 9	1.331 473 0
65.856	1.181 937 0	2 1 3	0.378 388 9	1.331 469 5

(3) The values of lattice constants predicted by the diffraction peaks are given in the last two columns in the above table. Evaluating the means and standard errors from these estimates, we obtain 0.378 37±0.000 02 nm and $c = 1.331\ 42 \pm 0.000\ 09$ nm. The relative deviations (in ‰) of the estimates from the diffraction peaks are shown in Fig. 9.2, from which it is seen that the (110) diffraction peak has the largest deviation, followed by the (200) and (202) diffraction peaks.

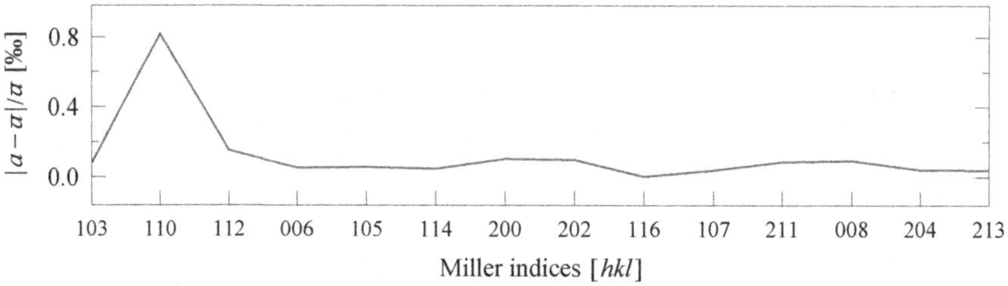

Fig. 9.2 Relative deviations (in part per thousand) of the estimates of lattice constants from the diffraction peaks. Note that the relative deviations are the same for lattice constants a and c.

9-10 X-ray diffraction on Al. An X-ray beam of wavelength 0.158 5 nm is incident parallel to the (010) planes on an aluminum crystal of lattice constant 0.405 nm. When the incident beam makes an angle of 11.31° with the [100] direction, a diffraction beam propagating in the (010) plane is observed. Find the Miller indices for this reflection.

Let φ denote the angle 11.31° that the incident beam makes with the [100] direction. Since the incident and diffracted beams are both in the (010) plane, the normal of the reflecting planes is parallel to the (010) plane and its Miller indices are of the form $[h0\ell]$. The interplanar distance in the family $(h0\ell)$ is $d_{h0\ell} = a/\sqrt{h^2+\ell^2}$. The direction of the incident beam is given to be $\hat{\boldsymbol{n}}_i = (\cos\varphi, 0, \sin\varphi)$. Note that we have chosen a positive z component for $\hat{\boldsymbol{n}}_i$. Let θ be the Bragg angle. We have from Bragg's law and the fact that $\angle(\hat{\boldsymbol{n}}_i, [h0\ell]) = \pi/2 + \theta$

$$\frac{2a}{\sqrt{h^2+\ell^2}}\sin\theta = \lambda, \quad \frac{1}{\sqrt{h^2+\ell^2}}[h0\ell]\cdot\hat{\boldsymbol{n}}_i = -\sin\theta.$$

Eliminating $\sin\theta$ from the above two equations, we obtain

$$h\cos\varphi + \ell\sin\varphi + \frac{\lambda}{2a}(h^2+\ell^2) = 0.$$

We can solve for h and ℓ from the above equation through trial and error. For each chosen pair of two even integral values for h and ℓ, we compute the value of the left hand side of the above equation. The pair that gives the smallest absolute value of the left hand side is the solution. This is most conveniently done using an electronic computer. The solution is given by

$$h = 2, \ \ell = -10$$

which yields a value of $-1.201\,89 \times 10^{-5}$ for the left hand side of the equation for h and ℓ. Therefore, the Miller indices for the reflection are

$$h = 2, \ k = 0, \ \ell = -10.$$

Chapter 10

Determination of Crystal Structure by Neutron Diffraction

(1) *Neutron*

Mass: $1.674\,928 \times 10^{-27}$ kg;
charge: 0;
spin: 1/2;
lifetime: 885 s.

(2) *Fermi pseudopotential*

$$U(\boldsymbol{r}) = \frac{2\pi\hbar^2}{m_\mathrm{n}} b\delta(\boldsymbol{r}).$$

(3) *Geometrical structure factor*

$$F(\boldsymbol{K}) = \sum_\kappa \mathrm{e}^{-W_\kappa}(-b_\kappa)\mathrm{e}^{-\mathrm{i}\boldsymbol{K}\cdot\boldsymbol{d}_\kappa}.$$

(4) *Coherent and incoherent scattering*

$$\frac{\mathrm{d}\sigma}{\mathrm{d}\Omega} = \left.\frac{\mathrm{d}\sigma}{\mathrm{d}\Omega}\right|_{\mathrm{coh}} + \left.\frac{\mathrm{d}\sigma}{\mathrm{d}\Omega}\right|_{\mathrm{inc}},$$

$$\left.\frac{\mathrm{d}\sigma}{\mathrm{d}\Omega}\right|_{\mathrm{coh}} = \sum_{\kappa\lambda} \langle b_\kappa\rangle\langle b_\lambda\rangle\, \mathrm{e}^{-\mathrm{i}\boldsymbol{K}\cdot(\boldsymbol{d}_\kappa - \boldsymbol{d}_\lambda)},$$

$$\left.\frac{\mathrm{d}\sigma}{\mathrm{d}\Omega}\right|_{\mathrm{inc}} = \sum_\kappa \left[\, \langle b_\kappa^2\rangle - \langle b_\kappa\rangle^2 \,\right].$$

(5) *Structure function*

$$S(\boldsymbol{Q}) = \frac{1}{N\langle b\rangle^2} \sum_{j\neq\ell} b_j b_\ell\, \mathrm{e}^{-\mathrm{i}\boldsymbol{Q}\cdot(\boldsymbol{r}_j - \boldsymbol{r}_\ell)} + 1.$$

(6) *Pair distribution function*

$$G(\boldsymbol{r}) = \frac{4\pi r}{N\langle b\rangle^2}\sum_{j\neq\ell} b_j b_\ell\,\delta(\boldsymbol{r}-\boldsymbol{r}_j+\boldsymbol{r}_\ell)$$
$$= 4\pi r\int\frac{\mathrm{d}\boldsymbol{Q}}{(2\pi)^3}\,\mathrm{e}^{\mathrm{i}\boldsymbol{Q}\cdot\boldsymbol{r}}\big[S(\boldsymbol{Q})-1\big],$$
$$G(r) = \frac{2}{\pi}\int_0^\infty \mathrm{d}Q\,Q\big[S(Q)-1\big]\sin(Qr).$$

(7) *Rietveld profile refinement*

The parameters for the unit cell of a crystal are optimized (refined) through fitting the computed diffraction pattern to the experimental data using the least-squares fitting algorithm.

10-1 Probe wavelengths proper for given structures. In order to study structures of size a, it is necessary to have a probe with wavelength λ smaller than or equal to a, $\lambda \leqslant a$.

(1) Find the minimum kinetic energy of a massive particle of mass m that can be used as a probe to resolve such a structure.

(2) For $a = 0.1$ nm, compute the energies of neutrons ($m = 1.675\times 10^{-27}$ kg), electrons ($m = 9.109\times10^{-31}$ kg), and photons required for the structural studies.

(3) Find the wavelength of a neutron with kinetic energy of $k_\mathrm{B}T$ with $T = 300$ K.

(4) A certain crystal has an interplanar spacing of 0.25 nm. Compute the energies that are necessary to observe up to three interference maxima for electrons and neutrons.

---

(1) From the de Bröglie relation $\lambda = h/p$, we have $p = h/\lambda$. The kinetic energy of a particle of mass m is then given by

$$E_\mathrm{kin} = \frac{p^2}{2m} = \frac{h^2}{2m\lambda^2}.$$

(2) For neutrons, we have

$$E_\mathrm{kin}^n = \frac{h^2}{2m_n a^2} = \frac{6.626^2\times10^{-68}}{2\times1.675\times10^{-27}\times0.1^2\times10^{-18}} \approx 81.804\ \mathrm{meV}.$$

For electrons, we have

$$E_\mathrm{kin}^e = \frac{h^2}{2m_e a^2} = \frac{6.626^2\times10^{-68}}{2\times9.109\times10^{-31}\times0.1^2\times10^{-18}} \approx 150.412\ \mathrm{eV}.$$

For photons, we have

$$E^p_{\rm kin} = \frac{hc}{a} = \frac{6.626 \times 10^{-34} \times 2.998 \times 10^8}{0.1 \times 10^{-9}} \approx 12.398 \text{ keV}.$$

(3) For $E^n_{\rm kin} = k_{\rm B}T$ with $T = 300$ K, the wavelength of a neutron is

$$\lambda = \frac{h}{p} = \frac{h}{\sqrt{2m_n k_{\rm B}T}} \approx 0.178 \text{ nm}.$$

(4) From Bragg's law, $2d \sin\theta = n\lambda$, we see that, for $n = 3$, the condition for the appearance of a Bragg peak is $3\lambda \leqslant 2d$. Thus, $\lambda \leqslant 2d/3$ and

$$E_{\rm kin} = \frac{h^2}{2m\lambda^2} \geqslant \frac{9h^2}{8md^2}.$$

For electrons, we have

$$E^e_{\rm kin} \geqslant \frac{9h^2}{8m_e d^2} \approx 54.148 \text{ eV}.$$

For neutrons, we have

$$E^n_{\rm kin} \geqslant \frac{9h^2}{8m_n d^2} \approx 29.450 \text{ meV}.$$

10-2 Reciprocal lattice vectors and families of lattice planes. Bragg's law for scattering of waves by a crystal is given by $n\lambda = 2d \sin\theta$. Any reciprocal lattice vector $\boldsymbol{K}$ is always perpendicular to a family of lattice planes. If the interplanar spacing is given by d, then the magnitude of $\boldsymbol{K}$ is given by $|\boldsymbol{K}| = n(2\pi/d)$ with n a nonnegative integer.

(1) Discuss why n appears in $|\boldsymbol{K}| = n(2\pi/d)$.
(2) Show that Bragg's law can be written as $|\boldsymbol{K}| = 2k \sin\theta$.
(3) Show that Bragg's law can be derived from the crystal momentum conservation law $\hbar\boldsymbol{k} = \hbar\boldsymbol{k}' + \hbar\boldsymbol{K}$.
(4) Determine the scattering angle 2θ for 5 meV neutrons scattering off a pyrolytic graphite monochromator crystal with $|\boldsymbol{K}_{002}| = 18.734 \text{ nm}^{-1}$.

---

(1) For a given family of lattice planes, the first-order diffraction maximum corresponds to the constructive interference between the two successive lattice planes, which corresponds to the shortest reciprocal lattice vector $\boldsymbol{K}_0$ ($|\boldsymbol{K}_0| = 2\pi/d$) perpendicular to the family of lattice planes. The constructive interference between a pair of lattice planes nd apart gives rise to a higher-order diffraction peak with the order given by n, which corresponds to a longer reciprocal lattice vector $\boldsymbol{K} = n\boldsymbol{K}_0$ ($|\boldsymbol{K}| = 2n\pi/d$). In short, the appearance of n in $|\boldsymbol{K}| = n(2\pi/d)$ is reflected in higher-order diffraction.

(2) From $|\boldsymbol{K}| = n(2\pi/d)$, we have $2d = 4n\pi/|\boldsymbol{K}|$. Inserting this relation into Bragg's law, $2d\sin\theta = n\lambda$, we have $(4n\pi/|\boldsymbol{K}|)\sin\theta = n\lambda$. We thus have

$$|\boldsymbol{K}| = 4\pi\sin\theta/\lambda = 2k\sin\theta$$

upon making use of $2\pi/\lambda = k$.

(3) Taking the dot-product of $\boldsymbol{k}$ with both sides of equation $\boldsymbol{k} = \boldsymbol{k}' + \boldsymbol{K}$, we have

$$\begin{aligned} k^2 &= k^2\cos(2\theta) + Kk\sin\theta, \\ 2k\sin^2\theta &= K\sin\theta, \\ 2k\sin\theta &= K, \end{aligned}$$

where we have used $k = k'$ for elastic scattering. We have also made use of the facts that $\angle(\boldsymbol{k}', \boldsymbol{k}) = 2\theta$ and that $\angle(\boldsymbol{k}, \boldsymbol{K}) = \pi/2 - \theta$. Inserting $K = 2n\pi/d$ into the above equation and making use of $k = 2\pi/\lambda$, we immediately obtain Bragg's law, $2d\sin\theta = n\lambda$.

(4) From $2k\sin\theta = K$ and making use of $\hbar k = \sqrt{2m_n E_{\rm kin}}$, we have

$$2\theta = 2\sin^{-1}\frac{K}{2k} = 2\sin^{-1}\frac{\hbar K}{2\sqrt{2m_n E_{\rm kin}}} \approx 1.295 \text{ radians} \approx 74.171^\circ.$$

10-3 Geometric structure factors.

(1) For a simple cubic lattice of lattice constant a, compute the geometric structure factor $F(\boldsymbol{K})$ in units of the nuclear scattering length b with $\boldsymbol{K} = h\boldsymbol{b}_1 + k\boldsymbol{b}_2 + \ell\boldsymbol{b}_3$. Identify the allowed reflections. Do the allowed reflections have different structure factors?

(2) A body-centered cubic (BCC) lattice of lattice constant a can be described as a simple cubic lattice of the same lattice constant with an extra site in the center of the cube. Write down an expression for the nuclear structure factor of a BCC structure using $b = 1$. Identify the allowed reflections. Do the allowed reflections have different structure factors?

(3) A face-centered cubic (FCC) lattice of lattice constant a can be described as a simple cubic lattice of the same lattice constant with extra sites at the centers of all faces of the cube. Write down an expression for the nuclear structure factor of an FCC structure using $b = 1$. Identify the allowed reflections. Do the allowed reflections have different structure factors?

---

(1) In units of the nuclear scattering length b, the geometric structure factor is given by

$$F(\boldsymbol{K}) = -\sum_{\kappa} \mathrm{e}^{-\mathrm{i}\boldsymbol{K}\cdot\boldsymbol{d}_\kappa},$$

where the Debye-Waller factor e^{-W_κ} has been set to unity. For a simple cubic lattice, we have

$$F(\boldsymbol{K}) = -1,$$

since there is only one nucleus in the conventional cell. Therefore, all reflections are allowed and they all have the same structure factor.

(2) For a body-centered cubic (BCC) lattice, we have

$$F(\boldsymbol{K}) = -1 - \mathrm{e}^{-\mathrm{i}(h+k+\ell)\pi} = -2\delta_{h+k+\ell,\mathrm{even}}.$$

Thus, the allowed reflections are those from families of lattice planes with $h+k+\ell$ an even integer. All the allowed reflections have the same structure factor.

(3) For a face-centered cubic (FCC) lattice, we have

$$\begin{aligned} F(\boldsymbol{K}) &= -1 - \mathrm{e}^{-\mathrm{i}(h+k)\pi} - \mathrm{e}^{-\mathrm{i}(k+\ell)\pi} - \mathrm{e}^{-\mathrm{i}(\ell+h)\pi} \\ &= -2\delta_{h,k,\ell,\text{all odd or all even}}. \end{aligned}$$

Thus, the allowed reflections are those from families of lattice planes with h, k, and ℓ all odd or all even. All the allowed reflections have the same structure factor.

10-4 Structure factor of the HCP lattice. Write down an expression for the structure factor of the HCP lattice using $b = 1$. Compute the geometric structure factors for $(hkl) = (100)$, (101), (102), (103), (001), (002), (110), (111), and (112). Identify and explain the regularities in the geometric structure factors.

The HCP structure is a simple hexagonal Bravais lattice with a two-atom basis in which one atom is located at $(0, 0, 0)$ and the other at

$(a/2)\boldsymbol{e}_x+(a/2\sqrt{3})\boldsymbol{e}_y+(c/2)\boldsymbol{e}_z$. Corresponding to the set of primitive vectors of the Bravais lattice,

$$\boldsymbol{a}_1=(a/2)\boldsymbol{e}_x+(\sqrt{3}a/2)\boldsymbol{e}_y,\ \boldsymbol{a}_2=-(a/2)\boldsymbol{e}_x+(\sqrt{3}a/2)\boldsymbol{e}_y,\ \boldsymbol{a}_3=c\boldsymbol{e}_z,$$

the set of primitive vectors of the reciprocal lattice are given by

$$\begin{aligned}\boldsymbol{b}_1&=(2\pi/a)\,\boldsymbol{e}_x+(2\pi/\sqrt{3}a)\,\boldsymbol{e}_y,\\ \boldsymbol{b}_2&=-(2\pi/a)\,\boldsymbol{e}_x+(2\pi/\sqrt{3}a)\,\boldsymbol{e}_y,\\ \boldsymbol{b}_3&=(2\pi/c)\,\boldsymbol{e}_z.\end{aligned}$$

The reciprocal lattice vectors are of the form

$$\boldsymbol{K}=h\boldsymbol{b}_1+k\boldsymbol{b}_2+\ell\boldsymbol{b}_3.$$

The geometric structure factor then has the following form

$$S_{\boldsymbol{K}}=-\sum_{\kappa}\mathrm{e}^{-\mathrm{i}\boldsymbol{K}\cdot\boldsymbol{d}_\kappa}=-1-(-1)^\ell\mathrm{e}^{-\mathrm{i}(4h+2k)\pi/3}.$$

From the above expression of $S_{\boldsymbol{K}}$, we see that, if $h=k$, then the factor containing h and k is equal to unity and the value of $S_{\boldsymbol{K}}$ is real and is determined solely by ℓ. For $h=k$, $S_{\boldsymbol{K}}=0$ if ℓ is odd and $S_{\boldsymbol{K}}=-2$ if ℓ is even. This also occurs for $4h+2k=6n\pi$ with n an integer in addition to the $h=k$ case. These values of h and k include $(h,k)=(0,0)$, $(1,1)$, $(2,2)$, $(3,3)$, $(0,3)$, $(3,0)$, and *etc.*
The values of the structure factor at some reciprocal lattice vectors are given in the following table. The above-summarized regularities in the structure factor can be clearly seen from the table. The vanishing of the structure factor for some reciprocal lattice vectors is because of the destructive interference of scatterings from the two nuclei within the same primitive cell.

$\boldsymbol{K}$	$S_{\boldsymbol{K}}$	$\boldsymbol{K}$	$S_{\boldsymbol{K}}$	$\boldsymbol{K}$	$S_{\boldsymbol{K}}$	$\boldsymbol{K}$	$S_{\boldsymbol{K}}$
(001)	0	(022)	$-\frac{1}{2}-\mathrm{i}\frac{\sqrt{3}}{2}$	(113)	0	(210)	$-\frac{1}{2}-\mathrm{i}\frac{\sqrt{3}}{2}$
(002)	-2	(023)	$-\frac{3}{2}+\mathrm{i}\frac{\sqrt{3}}{2}$	(120)	$-\frac{1}{2}+\mathrm{i}\frac{\sqrt{3}}{2}$	(211)	$-\frac{3}{2}+\mathrm{i}\frac{\sqrt{3}}{2}$
(003)	0	(100)	$-\frac{1}{2}-\mathrm{i}\frac{\sqrt{3}}{2}$	(121)	$-\frac{3}{2}-\mathrm{i}\frac{\sqrt{3}}{2}$	(212)	$-\frac{1}{2}-\mathrm{i}\frac{\sqrt{3}}{2}$
(010)	$-\frac{1}{2}+\mathrm{i}\frac{\sqrt{3}}{2}$	(101)	$-\frac{3}{2}+\mathrm{i}\frac{\sqrt{3}}{2}$	(122)	$-\frac{1}{2}+\mathrm{i}\frac{\sqrt{3}}{2}$	(213)	$-\frac{3}{2}+\mathrm{i}\frac{\sqrt{3}}{2}$
(011)	$-\frac{3}{2}-\mathrm{i}\frac{\sqrt{3}}{2}$	(102)	$-\frac{1}{2}-\mathrm{i}\frac{\sqrt{3}}{2}$	(123)	$-\frac{3}{2}-\mathrm{i}\frac{\sqrt{3}}{2}$	(220)	-2
(012)	$-\frac{1}{2}+\mathrm{i}\frac{\sqrt{3}}{2}$	(103)	$-\frac{3}{2}+\mathrm{i}\frac{\sqrt{3}}{2}$	(200)	$-\frac{1}{2}+\mathrm{i}\frac{\sqrt{3}}{2}$	(221)	0
(013)	$-\frac{3}{2}-\mathrm{i}\frac{\sqrt{3}}{2}$	(110)	-2	(201)	$-\frac{3}{2}-\mathrm{i}\frac{\sqrt{3}}{2}$	(222)	-2
(020)	$-\frac{1}{2}-\mathrm{i}\frac{\sqrt{3}}{2}$	(111)	0	(202)	$-\frac{1}{2}+\mathrm{i}\frac{\sqrt{3}}{2}$	(223)	0
(021)	$-\frac{3}{2}+\mathrm{i}\frac{\sqrt{3}}{2}$	(112)	-2	(203)	$-\frac{3}{2}-\mathrm{i}\frac{\sqrt{3}}{2}$		

10-5 Neutron diffraction on the NaCl structure.

(1) Write down an expression for the structure factor of the NaCl crystal structure. Identify the allowed reflections. Explain why the structure factors of the allowed reflections are different.

(2) CeSb also possesses the NaCl structure. Explain why the structure factors of some reflections nearly vanish.

(3) LiH also possesses the NaCl structure. Compute the structure factors for the three shortest allowed scattering vectors. Find the change in the mean scattering length $\bar{b}$ for H if the naturally occurring H is enriched with 17% D? Examine the influences on the structure factors.

---

(1) The NaCl crystal structure is an FCC crystal structure with a two-ion basis in which Na is located at $(0, 0, 0)$ and Cl at $(a/2, 0, 0)$. The reciprocal lattice vectors are of the form

$$\boldsymbol{K} = (k + \ell - h, \ell + h - k, h + k - \ell)(2\pi/a).$$

Thus, the structure factor is given by

$$S_{\text{NaCl}}(\boldsymbol{K}) = -b_{\text{Na}} - b_{\text{Cl}} e^{-i(k+\ell-h)\pi} = -b_{\text{Na}} - b_{\text{Cl}} e^{-i(k+\ell+h)\pi}$$
$$= \begin{cases} -b_{\text{Na}} + b_{\text{Cl}}, & k + \ell + h \text{ odd}, \\ -b_{\text{Na}} - b_{\text{Cl}}, & k + \ell + h \text{ even}. \end{cases}$$

Note that we have used the primitive cell in the computation. Only reflections from the reciprocal lattice vectors given above are allowed. For these reciprocal lattice vectors, the values of $k + \ell - h$, $\ell + h - k$, and $h + k - \ell$ are all odd or all even, which is commonly stated as the selection rule for FCC structures. The above result implies that the structure factors of allowed reflections are different. This is because the scattering phases from the two ions within the same primitive cell are different.

(2) For CeSb, we have

$$S_{\text{CeSb}}(\boldsymbol{K}) = \begin{cases} -b_{\text{Ce}} + b_{\text{Sb}}, & k + \ell + h \text{ odd}, \\ -b_{\text{Ce}} - b_{\text{Sb}}, & k + \ell + h \text{ even}. \end{cases}$$

For the nuclear scattering lengths, we have $b_{\text{Ce}} = 4.84$ fm and $b_{\text{Sb}} = 5.57$ fm. Since $b_{\text{Ce}} \approx b_{\text{Sb}}$, the structure factors for odd $k + \ell + h$'s nearly vanish. Note that we only consider coherent scattering here.

(3) For LiH, we have

$$S_{\mathrm{LiH}}(\boldsymbol{K}) = \begin{cases} -b_{\mathrm{Li}} + b_{\mathrm{H}},\ k+\ell+h \text{ odd}, \\ -b_{\mathrm{Li}} - b_{\mathrm{H}},\ k+\ell+h \text{ even}. \end{cases}$$

Making use of $b_{\mathrm{Li}} = -1.90$ fm and $b_{\mathrm{H}} = -3.739$ fm, we have for the three shortest allowed scattering vectors

$(hk\ell)$	$S_{\mathrm{LiH}}(\boldsymbol{K})$
(111)	-1.839 fm,
(200)	5.639 fm,
(220)	5.639 fm.

(4) The nuclear scattering length of D is $b_{\mathrm{D}} = 6.671$ fm. If the naturally occurring H is enriched with 17% D, the average value of the nuclear scattering length then becomes

$$\bar{b}_{\mathrm{H}} = 0.83 \times (-3.740\,6) + 0.17 \times 6.671 \approx -1.970\,6 \text{ fm}.$$

Thus, the change in the mean scattering length $\bar{b}$ is

$$\Delta\bar{b}_{\mathrm{H}} = -1.970\,6 + 3.739 = 1.768\,4 \text{ fm}.$$

Making use of the above-derived value for $\bar{b}_{\mathrm{H}}$, we have

$(hk\ell)$	$S_{\mathrm{LiH}}(\boldsymbol{K})$
(111)	-0.070 6 fm,
(200)	3.870 6 fm,
(220)	3.870 6 fm.

We have thus seen that the scattering length has a large effect on the structure factor.

10-6 Be as a neutron attenuator. Beryllium crystallizes in the HCP structure with a lattice constant $a = 0.227$ nm. A thick block of Be powder is often used as a neutron attenuator because it scatters neutrons away from the beam.

(1) Why is the Be powder transparent to long wavelength neutrons?

(2) Compute the critical wavelength for neutron transmission and the corresponding energy.

---

(1) Since the interplanar distances in Be are small due to its small lattice constant. For long wavelength neutrons, $\lambda/2d_{\min}$ is larger than one so that no lattice planes can reflect these neutrons. Therefore, the Be powder is transparent to long wavelength neutrons.

(2) The largest interplanar distance in an ideal HCP structure is in the family (001) with $d_{001} = c = \sqrt{8/3}\,a \approx 0.370\ 7$ nm. From $\lambda_c/2d_{001} = 1$, we have $\lambda_c = 2d_{001} \approx 0.741\ 4$ nm. The corresponding kinetic energy of a neutron is given by

$$E_{\rm kin}^{\rm c} = h^2/2m\lambda_c^2 \approx 1.488 \text{ meV}.$$

10-7 Diffraction on Al of neutrons of energies below 15 meV. A neutron beam contains neutrons with energies below 15 meV. The beam is incident on a single crystal of Al with a lattice constant of 0.405 nm along the [100] direction. Find the direction into which the beam is scattered and the energies that the mono-energetic scattered beams have.

The shortest interplanar distance that can reflect neutrons of energy $E = 15$ meV is given by $\lambda/2 = \sqrt{h^2/8m_nE} \approx 0.116\ 8$ nm. The interplanar distances in the first few families of lattice planes are given below with the Miller indices being with respect to the conventional unit cell.

$\{hk\ell\}$	$\{111\}$	$\{200\}$	$\{220\}$	$\{113\}$	$\{400\}$
$d_{hk\ell}$ [nm]	0.233 8	0.202 5	0.143 2	0.122 1	0.101 25

We see that the (400) family and those with even smaller interplanar distances can not reflect neutrons in the given energy range. Since the neutrons are incident in the [100] direction, the Bragg angle θ for a given reflecting family can be computed from $\sin\theta_{hk\ell} = |h|/\sqrt{h^2+k^2+\ell^2}$. The energy of neutrons in the reflected beam from the $\{hk\ell\}$ family can be obtained from $E_{hk\ell} = h^2/(8m_nd_{hk\ell}^2\sin^2\theta_{hk\ell})$. The direction into which the neutron beam is scattered from a given family can be found as follows. Assume that the direction of the scattered neutron beam from the family $\{hk\ell\}$ is $\hat{\boldsymbol{n}}_s = (x_1, x_2, x_3)$. The scattering direction $\hat{\boldsymbol{n}}_s$ satisfies the following conditions

$$\hat{\boldsymbol{n}}_s \cdot [hk\ell] = -[100]\cdot[hk\ell], \text{ incident angle} = \text{reflection angle},$$
$$\hat{\boldsymbol{n}}_s \cdot ([100]\times[hk\ell]) = 0, \ \hat{\boldsymbol{n}}_s \text{ is in the incident plane},$$
$$|\hat{\boldsymbol{n}}_s|^2 = x_1^2 + x_2^2 + x_3^2 = 1, \ \hat{\boldsymbol{n}}_s \text{ is a unit vector}.$$

From the above equations, we obtain

$$\hat{\boldsymbol{n}}_s = \left(k^2+\ell^2-h^2, -2hk, -2\ell h\right)/\left(k^2+\ell^2+h^2\right),$$

where we have thrown away the other solution with $x_1 = -1$.

With both the energy and interplanar distance limitations taken into consideration, we find that there are only two families of lattice planes that can reflect the given neutrons: The {111} and {200} families. The scattering angles and directions as well as the energies of neutrons in the scattered beams are list below.

Family	**Scattering angle**	**Scattering direction**	**Energy**
{111}	35.26°	$(1,-2,-2)/3$	11.22 meV
{200}	90.00°	$(-1,0,0)$	4.99 meV

10-8 Neutron diffraction peak. The data of a peak collected in neutron diffraction on a powder sample are given in the following table. Fit the peak to a symmetric Gaussian.

2θ	I	2θ	I	2θ	I	2θ	I
123.44	39.634	124.40	649.39	125.57	1868.90	126.84	189.02
123.58	33.537	124.56	865.85	125.85	1621.95	126.99	149.39
123.74	67.073	124.71	1137.20	125.98	1396.34	127.12	73.17
123.85	103.659	124.83	1420.73	126.13	1155.49	127.28	27.44
123.98	164.634	124.98	1652.44	126.27	932.93	127.42	27.44
124.13	289.634	125.28	1917.68	126.44	670.73	127.56	24.39
124.27	408.537	125.42	1954.27	126.57	496.95	127.69	21.34

A symmetric Gaussian shape is given by $y_i = I_{\boldsymbol{K}} \mathrm{e}^{-b_{\boldsymbol{K}}(2\theta_i - 2\theta_{\boldsymbol{K}})^2}$, where $b_{\boldsymbol{K}} = 4\ln 2/H_{\boldsymbol{K}}^2$ with $H_{\boldsymbol{K}}$ the full width at the half maximum. The above expression can be manipulated into a form that is suitable for a linear least-squares fitting. Taking the logarithm of the above expression, we have $\ln y_i = \ln I_{\boldsymbol{K}} - b_{\boldsymbol{K}}(2\theta_i - 2\theta_{\boldsymbol{K}})^2 = \alpha + \beta(2\theta_i) + \gamma(2\theta_i)^2$. where $\alpha = \ln I_{\boldsymbol{K}} - b_{\boldsymbol{K}}(2\theta_{\boldsymbol{K}})^2$, $\beta = 2b_{\boldsymbol{K}}(2\theta_{\boldsymbol{K}})$, $\gamma = -b_{\boldsymbol{K}}$. α, β, and γ are the new fitting parameters. Once they are determined, the original fitting parameters can be inferred from $b_{\boldsymbol{K}} = -\gamma$, $2\theta_{\boldsymbol{K}} = -\beta/2\gamma$, $I_{\boldsymbol{K}} = \mathrm{e}^{\alpha - \beta^2/4\gamma}$. We also set $\eta_i = \ln y_i$ and $x_i = 2\theta_i$. The fitting function is now written as

$$\eta = \alpha + \beta x + \gamma x^2.$$

The sum of the squared deviations of the experimental data from the predicted values of the fitting serves as a measure of the accuracy of the fitting. It is referred to as the chi-squared function and is denoted by χ^2

$$\chi^2 = \sum_i \left(\eta_i - \alpha - \beta x_i - \gamma x_i^2\right)^2.$$

The best values of the fitting parameters are determined by requiring that they minimize χ^2. Differentiating χ^2 with respect to α, β, and γ, respectively, and setting the results to zero, we obtain

$$\alpha n + \beta \sum_i x_i + \gamma \sum_i x_i^2 = \sum_i \eta_i,$$
$$\alpha \sum_i x_i + \beta \sum_i x_i^2 + \gamma \sum_i x_i^3 = \sum_i x_i \eta_i,$$
$$\alpha \sum_i x_i^2 + \beta \sum_i x_i^3 + \gamma \sum_i x_i^4 = \sum_i x_i^2 \eta_i,$$

where n is the number of data points. Solving for α, β, and γ from the above set of linear algebraic equations, we obtain

$$\alpha = \frac{(S_{x^2}S_{x^4} - S_{x^3}^2)S_\eta + (S_{x^2}S_{x^3} - S_x S_{x^4})S_{x\eta} + (S_x S_{x^3} - S_{x^2}^2)S_{x^2\eta}}{n(S_{x^2}S_{x^4} - S_{x^3}^2) + 2S_x S_{x^2} S_{x^3} - (S_{x^2})^3 - S_x^2 S_{x^4}},$$
$$\beta = \frac{(S_{x^2}S_{x^3} - S_x S_{x^4})S_\eta + (nS_{x^4} - S_{x^2}^2)S_{x\eta} + (S_x S_{x^2} - nS_{x^3})S_{x^2\eta}}{n(S_{x^2}S_{x^4} - S_{x^3}^2) + 2S_x S_{x^2} S_{x^3} - (S_{x^2})^3 - S_x^2 S_{x^4}},$$
$$\gamma = \frac{(S_x S_{x^3} - S_{x^2}^2)S_\eta + (S_x S_{x^2} - nS_{x^3})S_{x\eta} + (nS_{x^2} - S_x^2)S_{x^2\eta}}{n(S_{x^2}S_{x^4} - S_{x^3}^2) + 2S_x S_{x^2} S_{x^3} - (S_{x^2})^3 - S_x^2 S_{x^4}},$$

where $S_{x^j} = \sum_i x_i^j$, $S_{x^j\eta} = \sum_i x_i^j \eta_i$. Evaluating the fitting parameters using the above results, we obtain

$$\alpha = -16{,}394.6,\ \beta = 261.481,$$
$$\gamma = -1.042\,13.$$

The values of the original fitting parameters are

$$b_{\boldsymbol{K}} = -1.042\,13,\ 2\theta_{\boldsymbol{K}} = 125.456,$$
$$I_{\boldsymbol{K}} = 1{,}817.72.$$

The experimental data (filled circles) and the fitting result (solid line) are plotted in Fig. 10.1.

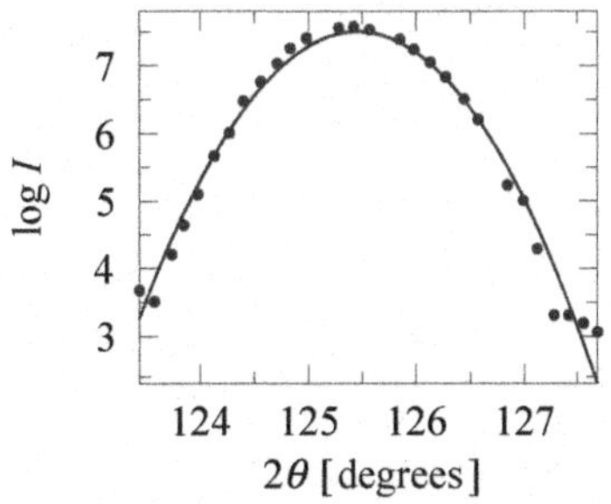

Fig. 10.1 Least-squares fitting for a symmetric Gaussian shape.

Chapter 11

Bonding in Solids

(1) *Ionic bonds*
An ionic bond is formed through the transfer of one or more electrons of one atom to another atom.

(2) *Electronegativity*
The electronegativity is the ability of an atom to attract electrons towards itself.

(3) *Covalent bonds*
A covalent bond is formed through the sharing of a pair of electrons between two atoms.

(4) *Molecular orbitals*
A molecular orbital is a combination of atomic orbitals.
The bonding, antibonding, and nonbonding molecular orbitals are three different kinds of molecular orbitals.

(5) *Metallic bonds*
Metallic bonds are due to the sharing of valence electrons among the positively-charged ions in the entire crystal.

(6) *Van der Waals bonds*
A van der Waals bond is formed through the electron correlation between two atoms or between two molecules.

(7) *Lennard-Jones (12-6) potential*

$$U_{\rm LJ}(R) = 4\epsilon\left[\left(\frac{\sigma}{R}\right)^{12} - \left(\frac{\sigma}{R}\right)^{6}\right].$$

(8) *Van der Waals bonds in molecules*
The bond length is $2^{1/6}\sigma$ and the bond strength is ϵ.

(9) *Hydrogen bonds*
A hydrogen bond is a chemical bond between atomic hydrogen and a strongly electronegative atom.

(10) *Valence electrons*
The valence electron(s) of an atom participate(s) in the bonding of the atom with other atoms.

(11) *Core electrons*
The core electron(s) of an atom do(es) not participate in the bonding of the atom with other atoms in any significant way.

(12) *Ion cores*
The nucleus and the core electron(s) of an atom are collectively referred to as the ion core of the atom.

(13) *Metals*
Metals are crystals with itinerant valence electrons.

(14) *Insulators*
Insulators are crystals with localized valence electrons.
Insulators are classified into
ionic crystals,
covalent crystals,
molecular crystals,
hydrogen-bonded crystals.

Semiconductors are classified as insulators.

11-1 Molecular orbitals for a hydrogen molecule. Verify that the molecular orbitals for a hydrogen molecule

$$\psi_b(\boldsymbol{r}) = c_b\big[\psi_{1s}(\boldsymbol{r}-\boldsymbol{R}_1) + \psi_{1s}(\boldsymbol{r}-\boldsymbol{R}_2)\big],$$
$$\psi_a(\boldsymbol{r}) = c_a\big[\psi_{1s}(\boldsymbol{r}-\boldsymbol{R}_1) - \psi_{1s}(\boldsymbol{r}-\boldsymbol{R}_2)\big],$$

are normalized with the normalization constants given by

$$c_{b,a} = \{2[1 \pm (1 + d/a_0)\mathrm{e}^{-d/a_0}]\}^{-1/2}$$

with $d = |\boldsymbol{R}_2 - \boldsymbol{R}_1|$ the separation between the two nuclei located at $\boldsymbol{R}_1$ and $\boldsymbol{R}_2$, respectively.

From the normalization condition $\int \mathrm{d}\boldsymbol{r}\ |\psi_b(\boldsymbol{r})|^2 = 1$, we have

$$\begin{aligned}\frac{1}{|c_b|^2} &= \frac{1}{\pi a_0^3}\int \mathrm{d}\boldsymbol{r}\ \Big[\mathrm{e}^{-2|\boldsymbol{r}-\boldsymbol{R}_1|/a_0} + \mathrm{e}^{-2|\boldsymbol{r}-\boldsymbol{R}_2|/a_0} \\ &\quad + 2\mathrm{e}^{-|\boldsymbol{r}-\boldsymbol{R}_1|/a_0}\mathrm{e}^{-|\boldsymbol{r}-\boldsymbol{R}_2|/a_0}\Big] \\ &= \frac{2}{\pi a_0^3}\int \mathrm{d}\boldsymbol{r}\ \Big[\mathrm{e}^{-2r/a_0} + \mathrm{e}^{-r/a_0}\mathrm{e}^{-|\boldsymbol{r}-\boldsymbol{d}|/a_0}\Big],\end{aligned}$$

where $\boldsymbol{d} = \boldsymbol{R}_2 - \boldsymbol{R}_1$. The value of the first integral in the square brackets is πa_0^3. The second integral can be evaluated by choosing the polar axis along the direction of $\boldsymbol{d}$. So doing, we have

$$\begin{aligned}\frac{1}{|c_b|^2} &= 2 + \frac{4}{a_0^3}\int_0^\infty \mathrm{d}r\, \mathrm{e}^{-r/a_0}\int_{-1}^{1}\mathrm{d}\cos\theta\, \mathrm{e}^{-(r^2+d^2-2rd\cos\theta)^{1/2}/a_0} \\ &= 2 + \frac{4}{a_0 d}\int_0^\infty \mathrm{d}r\, r\mathrm{e}^{-r/a_0}\int_{|r-d|/a_0}^{(r+d)/a_0}\mathrm{d}t\, \mathrm{e}^{-t} \\ &= 2 + \frac{4}{a_0 d}\int_0^\infty \mathrm{d}r\, r\mathrm{e}^{-r/a_0}\left[\mathrm{e}^{-|r-d|/a_0} - \mathrm{e}^{-(r+d)/a_0}\right] \\ &= 2\left[1 + \left(1 + \frac{d}{a_0}\right)\mathrm{e}^{-d/a_0}\right].\end{aligned}$$

We choose $c_b = \{2[1 + (1 + d/a_0)\mathrm{e}^{-d/a_0}]\}^{-1/2}$. The normalization constant of $\psi_a(\boldsymbol{r})$ can be trivially written down from that for $\psi_b(\boldsymbol{r})$ upon noting that the only difference between them is the sign of the two cross terms in the expansion of the squares of the absolute values of $\psi_a(\boldsymbol{r})$ and $\psi_b(\boldsymbol{r})$ and we have $c_a = \{2[1 - (1 + d/a_0)\mathrm{e}^{-d/a_0}]\}^{-1/2}$.

11-2 Energy of a hydrogen molecule in the bonding molecular orbital. Show that the energy of a hydrogen molecule is given by

$$E_b = \frac{\hbar^2}{ma_0^2}\left\{\frac{1}{\bar{d}} - 2|c_b|^2\left[1 + \frac{2}{\bar{d}} - 2\left(1 + \frac{1}{\bar{d}}\right)\mathrm{e}^{-2\bar{d}} + 3\left(1 + \bar{d} + \frac{\bar{d}^2}{9}\right)\mathrm{e}^{-\bar{d}}\right]\right\}$$

when two electrons are both in the bonding molecular orbital ψ_b with the interaction between them neglected.

Since there exists no spin dependent interaction, we can just concentrate on the spatial part of the wave function. Without consideration of the electron-electron interaction, the energy of the hydrogen atom is given by

$$\begin{aligned}E_b &= \frac{e^2}{d} + \int \mathrm{d}\boldsymbol{r}_1 \int \mathrm{d}\boldsymbol{r}_2\, \Phi^*(\boldsymbol{r}_1, \boldsymbol{r}_2)\left[-\frac{\hbar^2}{2m}\boldsymbol{\nabla}_1^2 - \frac{e^2}{|\boldsymbol{r}_1 - \boldsymbol{R}_1|} - \frac{e^2}{|\boldsymbol{r}_1 - \boldsymbol{R}_2|}\right. \\ &\quad \left. - \frac{\hbar^2}{2m}\boldsymbol{\nabla}_2^2 - \frac{e^2}{|\boldsymbol{r}_2 - \boldsymbol{R}_1|} - \frac{e^2}{|\boldsymbol{r}_2 - \boldsymbol{R}_2|}\right]\Phi(\boldsymbol{r}_1, \boldsymbol{r}_2) \\ &= \frac{e^2}{d} + \int \mathrm{d}\boldsymbol{r}_1 \int \mathrm{d}\boldsymbol{r}_2\, \psi_b^*(\boldsymbol{r}_1)\psi_b^*(\boldsymbol{r}_2)\left[-\frac{\hbar^2}{2m}\boldsymbol{\nabla}_1^2 - \frac{e^2}{|\boldsymbol{r}_1 - \boldsymbol{R}_1|}\right. \\ &\quad \left. - \frac{e^2}{|\boldsymbol{r}_1 - \boldsymbol{R}_2|} - \frac{\hbar^2}{2m}\boldsymbol{\nabla}_2^2 - \frac{e^2}{|\boldsymbol{r}_2 - \boldsymbol{R}_1|} - \frac{e^2}{|\boldsymbol{r}_2 - \boldsymbol{R}_2|}\right]\psi_b(\boldsymbol{r}_1)\psi_b(\boldsymbol{r}_2) \\ &= \frac{e^2}{d} + 2\varepsilon_b,\end{aligned}$$

where

$$\varepsilon_b = \int \mathrm{d}\boldsymbol{r}\ \psi_b^*(\boldsymbol{r})\left[-\frac{\hbar^2}{2m}\boldsymbol{\nabla}^2 - \frac{e^2}{|\boldsymbol{r}-\boldsymbol{R}_1|} - \frac{e^2}{|\boldsymbol{r}-\boldsymbol{R}_2|}\right]\psi_b(\boldsymbol{r})$$

is the energy of a single electron in state ψ_b. The term e^2/d is the Coulomb repulsion of the two nuclei with $d = |\boldsymbol{R}_2 - \boldsymbol{R}_1|$ the separation between them. We now evaluate ε_b. Substituting the expression of ψ_b into ε_b, we obtain

$$\begin{aligned}
\varepsilon_b &= |c_b|^2 \int \mathrm{d}\boldsymbol{r}\ \left[\psi_{100}^*(\boldsymbol{r}-\boldsymbol{R}_1) + \psi_{100}^*(\boldsymbol{r}-\boldsymbol{R}_2)\right]\left[-\frac{\hbar^2}{2m}\boldsymbol{\nabla}^2 - \frac{e^2}{|\boldsymbol{r}-\boldsymbol{R}_1|}\right.\\
&\quad \left. - \frac{e^2}{|\boldsymbol{r}-\boldsymbol{R}_2|}\right]\left[\psi_{100}(\boldsymbol{r}-\boldsymbol{R}_1) + \psi_{100}(\boldsymbol{r}-\boldsymbol{R}_2)\right]\\
&= |c_b|^2 \int \mathrm{d}\boldsymbol{r}\ \left[\psi_{100}^*(\boldsymbol{r}) + \psi_{100}^*(\boldsymbol{r}-\boldsymbol{d})\right]\left[-\frac{\hbar^2}{2m}\boldsymbol{\nabla}^2 - \frac{e^2}{r} - \frac{e^2}{|\boldsymbol{r}-\boldsymbol{d}|}\right]\\
&\quad \times \left[\psi_{100}(\boldsymbol{r}) + \psi_{100}(\boldsymbol{r}-\boldsymbol{d})\right]\\
&= \frac{|c_b|^2}{\pi a_0^3} \int \mathrm{d}\boldsymbol{r}\ \left(\mathrm{e}^{-r/a_0} + \mathrm{e}^{-|\boldsymbol{r}-\boldsymbol{d}|/a_0}\right)\left[-\frac{\hbar^2}{2m}\boldsymbol{\nabla}^2 - \frac{e^2}{r} - \frac{e^2}{|\boldsymbol{r}-\boldsymbol{d}|}\right]\\
&\quad \times \left(\mathrm{e}^{-r/a_0} + \mathrm{e}^{-|\boldsymbol{r}-\boldsymbol{d}|/a_0}\right).
\end{aligned}$$

To proceed further, we need to evaluate $\boldsymbol{\nabla}^2\mathrm{e}^{-r/a_0}$ and $\boldsymbol{\nabla}^2\mathrm{e}^{-|\boldsymbol{r}-\boldsymbol{d}|/a_0}$. The former is easy to evaluate and is given by

$$\begin{aligned}
\boldsymbol{\nabla}^2\mathrm{e}^{-r/a_0} &= \frac{1}{r}\frac{\mathrm{d}^2}{\mathrm{d}r^2}\left(r\mathrm{e}^{-r/a_0}\right) = \frac{1}{r}\frac{\mathrm{d}}{\mathrm{d}r}\left(\mathrm{e}^{-r/a_0} - \frac{r}{a_0}\mathrm{e}^{-r/a_0}\right)\\
&= \frac{1}{r}\left(-\frac{2}{a_0} + \frac{r}{a_0^2}\right)\mathrm{e}^{-r/a_0} = \frac{1}{a_0^2 r}(r - 2a_0)\mathrm{e}^{-r/a_0}.
\end{aligned}$$

Choosing the direction of $\boldsymbol{d}$ as the polar axis, we have $|\boldsymbol{r}-\boldsymbol{d}| = (r^2 + d^2 - 2rd\cos\theta)^{1/2}$. With this in mind, we can write $\boldsymbol{\nabla}^2\mathrm{e}^{-|\boldsymbol{r}-\boldsymbol{d}|/a_0}$ as

$$\begin{aligned}
&\boldsymbol{\nabla}^2\mathrm{e}^{-|\boldsymbol{r}-\boldsymbol{d}|/a_0}\\
&= \frac{1}{a_0^2}\mathrm{e}^{-|\boldsymbol{r}-\boldsymbol{d}|/a_0}\left(\boldsymbol{\nabla}|\boldsymbol{r}-\boldsymbol{d}| \cdot \boldsymbol{\nabla}|\boldsymbol{r}-\boldsymbol{d}| - a_0\boldsymbol{\nabla}^2|\boldsymbol{r}-\boldsymbol{d}|\right)\\
&= \frac{1}{a_0^2}\mathrm{e}^{-|\boldsymbol{r}-\boldsymbol{d}|/a_0}\left[\left(\frac{\partial|\boldsymbol{r}-\boldsymbol{d}|}{\partial r}\right)^2 - \frac{2a_0}{r}\frac{\partial|\boldsymbol{r}-\boldsymbol{d}|}{\partial r} - a_0\frac{\partial^2|\boldsymbol{r}-\boldsymbol{d}|}{\partial r^2}\right.\\
&\quad \left. + \frac{\sin^2\theta}{r^2}\left(\frac{\partial|\boldsymbol{r}-\boldsymbol{d}|}{\partial\cos\theta}\right)^2 + \frac{2a_0\cos\theta}{r^2}\frac{\partial|\boldsymbol{r}-\boldsymbol{d}|}{\partial\cos\theta} - \frac{a_0\sin^2\theta}{r^2}\frac{\partial^2|\boldsymbol{r}-\boldsymbol{d}|}{\partial(\cos\theta)^2}\right].
\end{aligned}$$

We now compute the partial derivatives. For $\partial|\boldsymbol{r}-\boldsymbol{d}|/\partial r$, we have

$$\frac{\partial|\boldsymbol{r}-\boldsymbol{d}|}{\partial r} = \frac{r - d\cos\theta}{(r^2 + d^2 - 2rd\cos\theta)^{1/2}}.$$

For $\partial^2|\boldsymbol{r}-\boldsymbol{d}|/\partial r^2$, we have

$$\frac{\partial^2|\boldsymbol{r}-\boldsymbol{d}|}{\partial r^2}=\frac{1}{(r^2+d^2-2rd\cos\theta)^{1/2}}-\frac{(r-d\cos\theta)^2}{(r^2+d^2-2rd\cos\theta)^{3/2}}.$$

For $\partial|\boldsymbol{r}-\boldsymbol{d}|/\partial\cos\theta$, we have

$$\frac{\partial|\boldsymbol{r}-\boldsymbol{d}|}{\partial\cos\theta}=-\frac{rd}{(r^2+d^2-2rd\cos\theta)^{1/2}}.$$

For $\partial^2|\boldsymbol{r}-\boldsymbol{d}|/\partial(\cos\theta)^2$, we have

$$\frac{\partial^2|\boldsymbol{r}-\boldsymbol{d}|}{\partial(\cos\theta)^2}=-\frac{r^2d^2}{(r^2+d^2-2rd\cos\theta)^{3/2}}.$$

Inserting the above-computed partial derivatives into $\boldsymbol{\nabla}^2\mathrm{e}^{-|\boldsymbol{r}-\boldsymbol{d}|/a_0}$ yields

$$\begin{aligned}
&\boldsymbol{\nabla}^2\mathrm{e}^{-|\boldsymbol{r}-\boldsymbol{d}|/a_0}\\
&=\frac{1}{a_0^2}\mathrm{e}^{-|\boldsymbol{r}-\boldsymbol{d}|/a_0}\left[\frac{(r-d\cos\theta)^2}{r^2+d^2-2rd\cos\theta}-\frac{2a_0}{r}\frac{r-d\cos\theta}{(r^2+d^2-2rd\cos\theta)^{1/2}}\right.\\
&\quad-a_0\left[\frac{1}{(r^2+d^2-2rd\cos\theta)^{1/2}}-\frac{(r-d\cos\theta)^2}{(r^2+d^2-2rd\cos\theta)^{3/2}}\right]\\
&\quad+\frac{\sin^2\theta}{r^2}\frac{r^2d^2}{r^2+d^2-2rd\cos\theta}-\frac{2a_0\cos\theta}{r^2}\frac{rd}{(r^2+d^2-2rd\cos\theta)^{1/2}}\\
&\quad\left.+\frac{a_0\sin^2\theta}{r^2}\frac{r^2d^2}{(r^2+d^2-2rd\cos\theta)^{3/2}}\right]\\
&=\frac{1}{a_0^2}\mathrm{e}^{-|\boldsymbol{r}-\boldsymbol{d}|/a_0}\left(1-\frac{2a_0}{|\boldsymbol{r}-\boldsymbol{d}|}\right).
\end{aligned}$$

Inserting the above results for $\boldsymbol{\nabla}^2\mathrm{e}^{-r/a_0}$ and $\boldsymbol{\nabla}^2\mathrm{e}^{-|\boldsymbol{r}-\boldsymbol{d}|/a_0}$ into ε_b yields

$$\begin{aligned}
\frac{\pi a_0^3}{|c_b|^2}\varepsilon_b=&-\frac{\hbar^2}{ma_0^2}\int\mathrm{d}\boldsymbol{r}\left[\mathrm{e}^{-2r/a_0}+\frac{2a_0}{r}\mathrm{e}^{-2|\boldsymbol{r}-\boldsymbol{d}|/a_0}\right.\\
&\left.+\left(1+\frac{2a_0}{r}\right)\mathrm{e}^{-(r+|\boldsymbol{r}-\boldsymbol{d}|)/a_0}\right].
\end{aligned}$$

Evaluating the integrals in the above equation, we obtain

$$\begin{aligned}
\frac{\pi a_0^3}{|c_b|^2}\varepsilon_b=&-\frac{\hbar^2}{ma_0^2}\left\{\pi a_0^3+\frac{2\pi a_0^3}{d}\left[a_0-(a_0+d)\mathrm{e}^{-2d/a_0}\right]\right.\\
&\left.+\pi a_0^3\left[1+\frac{d}{3a_0^2}(3a_0+d)\right]\mathrm{e}^{-d/a_0}+2\pi a_0^2(a_0+d)\mathrm{e}^{-d/a_0}\right\}\\
=&-\frac{\pi\hbar^2a_0}{m}\left[1+\frac{2}{\bar{d}}-2\left(1+\frac{1}{\bar{d}}\right)\mathrm{e}^{-2\bar{d}}+3\left(1+\bar{d}+\frac{\bar{d}^2}{9}\right)\mathrm{e}^{-\bar{d}}\right],
\end{aligned}$$

where $\bar{d} = d/a_0$ is the separation between the nuclei in units of the Bohr radius a_0. Finally, the energy of the electron in state ψ_b is given by

$$\varepsilon_b = -|c_b|^2 \frac{\hbar^2}{ma_0^2}\left[1+\frac{2}{\bar{d}} - 2\left(1+\frac{1}{\bar{d}}\right)e^{-2\bar{d}} + 3\left(1+\bar{d}+\frac{\bar{d}^2}{9}\right)e^{-\bar{d}}\right].$$

The energy of the hydrogen molecule is then given by

$$E_b = \frac{e^2}{d} + 2\varepsilon_b = \frac{\hbar^2}{ma_0^2}\left\{\frac{1}{\bar{d}} - 2|c_b|^2\left[1+\frac{2}{\bar{d}} - 2\left(1+\frac{1}{\bar{d}}\right)e^{-2\bar{d}}\right.\right.$$
$$\left.\left. +3\left(1+\bar{d}+\frac{\bar{d}^2}{9}\right)e^{-\bar{d}}\right]\right\}.$$

11-3 Energy of a hydrogen molecule in the antibonding molecular orbital. Show that the energy of a hydrogen molecule is given by

$$E_a = \frac{\hbar^2}{ma_0^2}\left\{\frac{1}{\bar{d}} - 2|c_a|^2\left[1+\frac{2}{\bar{d}} - 2\left(1+\frac{1}{\bar{d}}\right)e^{-2\bar{d}} - 3\left(1+\bar{d}+\frac{\bar{d}^2}{9}\right)e^{-\bar{d}}\right]\right\}$$

when two electrons are both in the antibonding molecular orbital ψ_a with the interaction between them neglected.

When the two electrons are both in state ψ_a, the spatial part of the wave function of the system of two electrons is given by

$$\Phi(\boldsymbol{r}_1, \boldsymbol{r}_2) = \psi_a(\boldsymbol{r}_1)\psi_a(\boldsymbol{r}_2).$$

In this case, the energy of the hydrogen molecule is given by

$$E_a = \frac{e^2}{d} + 2\varepsilon_a,$$

where ε_a is the energy of a single electron in state ψ_a and is given by

$$\varepsilon_a = \int d\boldsymbol{r}\ \psi_a^*(\boldsymbol{r})\left[-\frac{\hbar^2}{2m}\boldsymbol{\nabla}^2 - \frac{e^2}{|\boldsymbol{r}-\boldsymbol{R}_1|} - \frac{e^2}{|\boldsymbol{r}-\boldsymbol{R}_2|}\right]\psi_a(\boldsymbol{r})$$
$$= \frac{|c_a|^2}{\pi a_0^3}\int d\boldsymbol{r}\ \left(e^{-r/a_0} - e^{-|\boldsymbol{r}-\boldsymbol{d}|/a_0}\right)\left[-\frac{\hbar^2}{2m}\boldsymbol{\nabla}^2 - \frac{e^2}{r} - \frac{e^2}{|\boldsymbol{r}-\boldsymbol{d}|}\right]$$
$$\times\left(e^{-r/a_0} - e^{-|\boldsymbol{r}-\boldsymbol{d}|/a_0}\right).$$

Changing the signs of the terms containing $e^{-|\boldsymbol{r}-\boldsymbol{d}|/a_0}$ in the result for ψ_b, we obtain

$$\frac{\pi a_0^3}{|c_a|^2}\varepsilon_a = \int d\boldsymbol{r}\ \left(e^{-r/a_0} - e^{-|\boldsymbol{r}-\boldsymbol{d}|/a_0}\right)\left[-\frac{\hbar^2}{2ma_0^2}\frac{1}{r}(r-2a_0)e^{-r/a_0}\right.$$
$$+\frac{\hbar^2}{2ma_0^2}\left(1-\frac{2a_0}{|\boldsymbol{r}-\boldsymbol{d}|}\right)e^{-|\boldsymbol{r}-\boldsymbol{d}|/a_0}$$
$$\left.-\left(\frac{e^2}{r}+\frac{e^2}{|\boldsymbol{r}-\boldsymbol{d}|}\right)\left(e^{-r/a_0} - e^{-|\boldsymbol{r}-\boldsymbol{d}|/a_0}\right)\right]$$
$$= -\frac{\pi\hbar^2 a_0}{m}\left[1+\frac{2}{\bar{d}} - 2\left(1+\frac{1}{\bar{d}}\right)e^{-2\bar{d}} - 3\left(1+\bar{d}+\frac{\bar{d}^2}{9}\right)e^{-\bar{d}}\right].$$

The energy of the electron in state ψ_a is then given by

$$\varepsilon_a = -|c_a|^2 \frac{\hbar^2}{ma_0^2}\left[1 + \frac{2}{\bar{d}} - 2\left(1 + \frac{1}{\bar{d}}\right)e^{-2\bar{d}} - 3\left(1 + \bar{d} + \frac{\bar{d}^2}{9}\right)e^{-\bar{d}}\right].$$

Note that the only difference of this result for ψ_a from that for ψ_b is the opposite sign of the last term in the square brackets except for the normalization constant. The energy of the hydrogen molecule when both electrons are in ψ_a is then given by

$$E_a = \frac{e^2}{d} + 2\varepsilon_a = \frac{\hbar^2}{ma_0^2}\left\{\frac{1}{\bar{d}} - 2|c_a|^2\left[1 + \frac{2}{\bar{d}} - 2\left(1 + \frac{1}{\bar{d}}\right)e^{-2\bar{d}}\right.\right.$$
$$\left.\left. -3\left(1 + \bar{d} + \frac{\bar{d}^2}{9}\right)e^{-\bar{d}}\right]\right\}.$$

11-4 Permanent dipole-permanent dipole interaction. Consider the permanent dipole-permanent dipole interaction.

(1) Show that the thermal average of the permanent dipole-permanent dipole interaction is given by

$$\overline{U}_{pp-pp}(\boldsymbol{r}) = -\frac{\partial}{\partial\beta}\ln\int_{-1}^{1} \mathrm{d}\cos\theta_2\, e^{-\beta U_{pp-pp}(\boldsymbol{r})}$$
$$= \frac{1}{\beta} - \frac{2p_1p_2}{(4\pi\epsilon_0)R^3}\coth\frac{2\beta p_1 p_2}{(4\pi\epsilon_0)R^3}.$$

with $\beta = 1/k_\mathrm{B}T$.

(2) For $2\beta p_1p_2/(4\pi\epsilon_0)R^3 \ll 1$, show that $\overline{U}_{pp-pp}(\boldsymbol{r})$ can be approximated as

$$\overline{U}_{pp-pp}(\boldsymbol{r}) \approx -\frac{4p_1^2p_2^2}{3(4\pi\epsilon_0)^2 k_\mathrm{B}TR^6}.$$

---

(1) Setting both $\boldsymbol{p}_1$ and $\boldsymbol{p}_2$ to be located on the z axis of a Cartesian coordinate system and $\boldsymbol{p}_1$ to be fixed along the z axis, we then have $\boldsymbol{p}_1 \cdot \boldsymbol{p}_2 = p_1 p_2 \cos\theta_2$ and $(\boldsymbol{p}_1 \cdot \hat{\boldsymbol{r}})(\boldsymbol{p}_2 \cdot \hat{\boldsymbol{r}}) = p_1 p_2 \cos\theta_2$ with θ_2 the polar angle of $\boldsymbol{p}_2$. The permanent dipole-permanent dipole interaction is then given by

$$U_{pp-pp}(\boldsymbol{r}) = -\frac{2p_1p_2}{(4\pi\epsilon_0)R^3}\cos\theta_2.$$

Averaging $U_{pp-pp}(\boldsymbol{R})$ over θ_2, we have

$$\begin{aligned}\overline{U}_{pp-pp}(\boldsymbol{r}) &= \int_{-1}^{1} \mathrm{d}\cos\theta_2 U_{pp-pp}(\boldsymbol{R})\mathrm{e}^{-\beta U_{pp-pp}(\boldsymbol{r})} \\ &\quad \times \left[\int_{-1}^{1} \mathrm{d}\cos\theta_2 \mathrm{e}^{-\beta U_{pp-pp}(\boldsymbol{R})}\right]^{-1} \\ &= -\frac{\partial}{\partial\beta}\ln\int_{-1}^{1} \mathrm{d}\cos\theta_2 \mathrm{e}^{-\beta U_{pp-pp}(\boldsymbol{R})} \\ &= -\frac{\partial}{\partial\beta}\ln\int_{-1}^{1} \mathrm{d}\cos\theta_2\, \mathrm{e}^{-2\beta p_1p_2\cos\theta_2/(4\pi\epsilon_0)R^3} \\ &= \frac{1}{\beta} - \frac{2p_1p_2}{(4\pi\epsilon_0)R^3}\frac{\mathrm{e}^{2\beta p_1p_2/(4\pi\epsilon_0)R^3} + \mathrm{e}^{-2\beta p_1p_2/(4\pi\epsilon_0)R^3}}{\mathrm{e}^{2\beta p_1p_2/(4\pi\epsilon_0)R^3} - \mathrm{e}^{-2\beta p_1p_2/(4\pi\epsilon_0)R^3}} \\ &= \frac{1}{\beta} - \frac{2p_1p_2}{(4\pi\epsilon_0)R^3}\coth\frac{2\beta p_1p_2}{(4\pi\epsilon_0)R^3}.\end{aligned}$$

(2) For $2\beta p_1p_2/(4\pi\epsilon_0)R^3 \ll 1$, upon making use of $\coth x \approx 1/x + x/3$ for $x \ll 1$, we have

$$\begin{aligned}\overline{U}_{pp-pp}(\boldsymbol{r}) &\approx \frac{1}{\beta} - \frac{2p_1p_2}{(4\pi\epsilon_0)R^3}\left[\frac{(4\pi\epsilon_0)R^3}{2\beta p_1p_2} + \frac{2\beta p_1p_2}{3(4\pi\epsilon_0)R^3}\right] \\ &= -\frac{4p_1^2p_2^2}{3(4\pi\epsilon_0)^2k_\mathrm{B}TR^6}.\end{aligned}$$

11-5 Van der Waals bond in a diatomic molecule. Consider a diatomic molecule due to the van der Waals bond.

(1) Show that the equilibrium separation is given by $R_0 = 2^{1/6}\sigma$.
(2) Find the strength of the van der Waals bond in the molecule.

---

(1) Differentiating $U_\mathrm{LJ}(R)$ with respect to R, then setting $R = R_0$, and then setting the result to zero, we obtain

$$12\left(\frac{\sigma}{R_0}\right)^{13} - 6\left(\frac{\sigma}{R_0}\right)^{7} = 0$$

which leads to $R_0 = 2^{1/6}\sigma \approx 1.122\,5\sigma$.

(2) At $R = R_0$, we have

$$U_{\rm LJ}(R_0) = 4\epsilon\left(\frac{1}{4} - \frac{1}{2}\right) = -\epsilon.$$

Thus, the bond strength is ϵ.

11-6 Bond in a hypothetical diatomic molecule. The interaction between the two atoms in a hypothetical diatomic molecule is given by

$$U(R) = g\left[\frac{\rho}{R}\,{\rm e}^{\rho/R} - \left(\frac{\rho}{R}\right)^6\right]$$

where g and ρ are positive constants and R is the distance between the two atoms.

(1) Plot $U(R)/g$ against R/ρ for R/ρ in the ranges $[0.076, 0.2]$ and $[0.72, 5]$, respectively.

(2) Find the equation that determines the bond length R_0.

(3) Express the bond strength e_0 in terms of the parameters g, ρ, and R_0. The exponential function should not appear in e_0.

(4) Find R_0 by solving numerically the equation for it. Using the numerical result for R_0, express e_0 in terms of g only.

---

(1) The plots in the two ranges are given in Fig. 11.1.
From the plots, it is seen that the potential has a deep minimum and a local maximum.

(2) Differentiating $U(R)$ with respect to R, then setting $R = R_0$, and then setting the result to zero yields

$$\left(1 + \frac{\rho}{R_0}\right){\rm e}^{\rho/R_0} - 6\left(\frac{\rho}{R_0}\right)^5 = 0.$$

Rearranging, we have

$${\rm e}^{\rho/R_0} = \frac{6}{\rho/R_0 + 1}\left(\frac{\rho}{R_0}\right)^5$$

The above equation is the one that determines the bond length R_0.

(3) At $R = R_0$, the potential energy is given by

$$U(R_0) = g\left[\frac{\rho}{R_0}{\rm e}^{\rho/R_0} - \left(\frac{\rho}{R_0}\right)^6\right] = g\left[\frac{6}{\rho/R_0 + 1}\left(\frac{\rho}{R_0}\right)^6 - \left(\frac{\rho}{R_0}\right)^6\right]$$
$$= -g\left(1 - \frac{6}{\rho/R_0 + 1}\right)\left(\frac{\rho}{R_0}\right)^6 = -g\frac{\rho/R_0 - 5}{\rho/R_0 + 1}\left(\frac{\rho}{R_0}\right)^6.$$

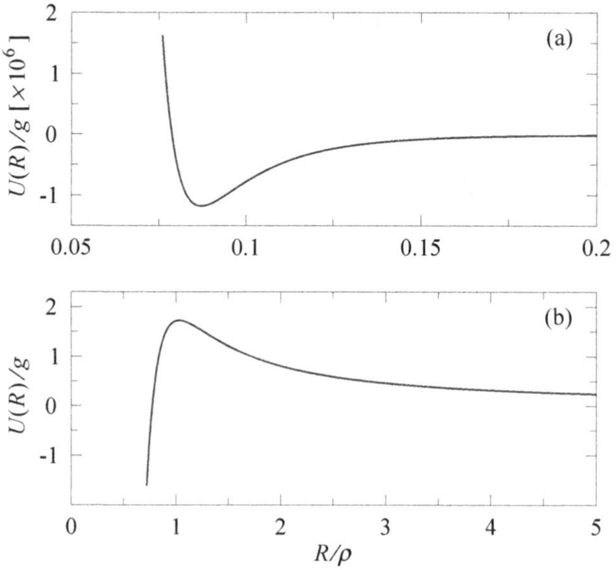

Fig. 11.1 Plots of $U(R)/g$ against R/ρ for R/ρ in the ranges $[0.076, 0.2]$ (a) and $[0.72, 5]$ (b).

Thus, the bond strength is given by

$$e_0 = -U(R_0) = g\frac{\rho/R_0 - 5}{\rho/R_0 + 1}\left(\frac{\rho}{R_0}\right)^6.$$

(4) Numerically solving the equation for R_0, we find that $R_0 \approx 0.087\rho$. The bond strength is then given by $e_0 \approx 1.178 \times 10^6 g$.

Chapter 12

Cohesion of Solids

(1) *Cohesive energy*
The cohesive energy of a crystal is the energy required to break it into free atoms or molecules that are set at rest at infinite separation.

(2) *Lattice energy of an ionic crystal*
The lattice energy of an ionic crystal is the energy required to break it into its free ions that are set at rest at infinite separation.

(3) *Cohesive energy of a molecular crystal*

$$e_0 \equiv -U_{\text{tot}}(R_0)/N \approx 8.61\epsilon.$$

(4) *Equilibrium bulk modulus of a molecular crystal*

$$B_0 \approx 75.19\epsilon/\sigma^3.$$

(5) *Madelung constant*

$$\frac{\alpha}{R} = -\frac{1}{2}\sum_{n\kappa\nu}{}' \frac{(\pm)}{|\boldsymbol{R}_n + \boldsymbol{d}_\nu - \boldsymbol{d}_\kappa|}.$$

Madelung constant for a chain of alternating positive and negative ions is $\alpha = 2\ln 2$.
Madelung constant for NaCl is $\alpha \approx 1.747\,565$.

(6) *Madelung energy*

$$E_{\text{M}} = -\frac{N\alpha q^2}{4\pi\epsilon_0 R_0}.$$

(7) *Lattice energy of an ionic crystal*

$$e_L = -\frac{U_{\text{tot}}(R_0)}{N} = \frac{\alpha q^2}{4\pi\epsilon_0 R_0}\left(1 - \frac{\rho}{R_0}\right).$$

(8) *Ewald summation method*
The idea in the Ewald summation method is to break a slowly-converging series into two. One of the resultant series converges fast in reciprocal space while the other converges fast in real space.

(9) *Ionic radii*
The ionic radius of an element is the effective radius of its ion in a crystal.

(10) *Exchange energy of electrons in an alkali metal*

$$u_{\rm ex} = -\frac{3}{4\pi}\left(\frac{9\pi}{4}\right)^{1/3}\frac{e^2}{4\pi\epsilon_0 r_s} \approx -\frac{12.47}{r_s/a_0}\ {\rm eV/atom}.$$

(11) *Electrostatic energy in an alkali metal*

$$u_{\rm Coul} \approx -\frac{24.38}{r_s/a_0}\ {\rm eV/atom}.$$

(12) *Cohesive energy in an alkali metal*

$$u = u_{\rm kin} + u_{\rm ex} + u_{\rm Coul} \approx \frac{30.07}{(r_s/a_0)^2} - \frac{36.85}{r_s/a_0}\ {\rm eV/atom}.$$

12-1 Morse potential. Consider a fictitious molecular crystal with a simple cubic Bravais lattice. Assume that each atom interacts only with its nearest neighbors through the following Morse potential

$$V(r) = V_0\left[{\rm e}^{-2\kappa(r-\lambda)} - 2{\rm e}^{-\kappa(r-\lambda)}\right],$$

where V_0, κ, and λ are referred to as the Morse parameters.

(1) Find the equilibrium lattice spacing a.

(2) Compute the total binding energy of the crystal. Assume that there are N atoms in the crystal and that the kinetic energy of the atoms is zero.

---

(1) The Morse potential is plotted in Fig. 12.1 for $\kappa a = 1$. Since each atom interacts only with its nearest neighbors, we can determine the equilibrium lattice spacing a using the given $V(r)$. Differentiating $V(r)$ with respect to r and setting the value of the derivative at $r = a$ to zero, we have

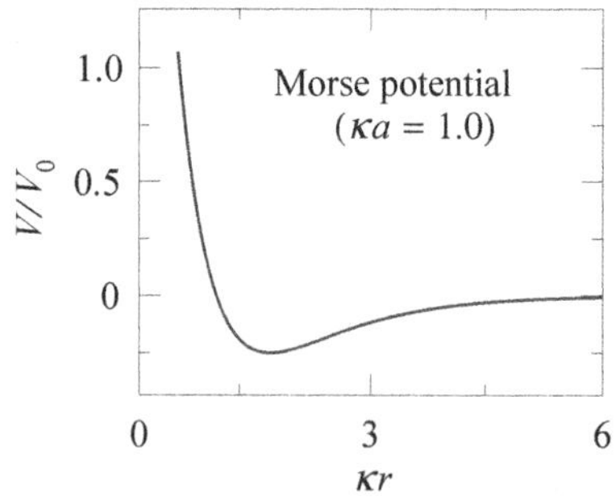

Fig. 12.1 Plot of the Morse potential as a function of κr.

$$0 = \left.\frac{{\rm d}V(r)}{{\rm d}r}\right|_{r=a} = -2\kappa V_0\left[{\rm e}^{-2\kappa(a-\lambda)} - {\rm e}^{-\kappa(a-\lambda)}\right].$$

Thus, $a = \lambda$.

(2) Each atom has six nearest neighbors in a crystal with a simple cubic Bravais lattice. Therefore, the total binding energy of the crystal is given by

$$E_b = -3NV(r=a) = -3NV_0\left[e^{-2\kappa(a-\lambda)} - 2e^{-\kappa(a-\lambda)}\right] = 3NV_0.$$

12-2 One-dimensional crystal of a chain of alternating ions. Consider the cohesion of the familiar one-dimensional crystal of a chain of alternating ions with charges $\pm q$, respectively. Let R be the nearest-neighbor distance and $2N$ the number of ions. In addition to the electrostatic interaction between ions, a repulsive interaction of the form A/R^n also exists between nearest neighbors only.

(1) Derive an expression for the total energy of the crystal.
(2) Find the equilibrium nearest-neighbor distance R_0.
(3) Compute the lattice energy per ion pair.
(4) Compute the work needed to make the crystal shrunken to such an extent that its nearest-neighbor distance is reduced by a fraction of η.

---

(1) The interaction potential between ions i and j is given by

$$U_{ij} = \frac{A}{R^n}\delta_{i=j\pm1} + (-1)^{i-j}\frac{q^2}{4\pi\epsilon_0|i-j|R}.$$

Let N denote the number of ion pairs. Then the total number of ions in the crystal is $2N$. The total energy of the crystal is obtained by summing U_{ij} over i and j with $i \neq j$ and dividing the result by two to remove the double counting. We have

$$\begin{aligned} U_{\text{tot}} &= \frac{1}{2}\sum_{i\neq j} U_{ij} = \frac{1}{2}\sum_{i\neq j}\frac{A}{R^n}\delta_{i=j\pm1} + \frac{1}{2}\sum_{i\neq j}(-1)^{i-j}\frac{q^2}{4\pi\epsilon_0|i-j|R} \\ &= \frac{2NA}{R^n} + \frac{1}{2}\sum_j\sum_{\ell\neq 0}(-1)^\ell\frac{q^2}{4\pi\epsilon_0|\ell|R} = \frac{2NA}{R^n} - \frac{N\alpha q^2}{4\pi\epsilon_0 R}, \end{aligned}$$

where

$$\alpha = \sum_{\ell\neq 0}(-1)^{\ell-1}\frac{1}{|\ell|} = 2\left(1 - \frac{1}{2} + \frac{1}{3} - \frac{1}{4} + \cdots\right) = 2\ln 2$$

is the Madelung constant for this one-dimensional ionic crystal.

(2) At the equilibrium separation, ions do not feel any forces acting on them. Thus, the condition for the determination of the equilibrium nearest-neighbor distance (the equilibrium separation), R_0, is that $\mathrm{d}U_{\mathrm{tot}}/\mathrm{d}R\big|_{R=R_0} = 0$. We then have

$$0 = \left.\frac{\mathrm{d}U_{\mathrm{tot}}}{\mathrm{d}R}\right|_{R=R_0} = -\frac{2nNA}{R_0^{n+1}} + \frac{N\alpha q^2}{4\pi\epsilon_0 R_0^2}.$$

Solving for R_0 from the above equation, we obtain

$$R_0 = \left[\frac{2n(4\pi\epsilon_0)A}{\alpha q^2}\right]^{1/(n-1)}.$$

(3) The total energy at the equilibrium separation is given by

$$U_{\mathrm{tot}}(R_0) = -\frac{(n-1)N\alpha q^2}{n(4\pi\epsilon_0)R_0}.$$

Thus, the total lattice energy is

$$E_L = -U_{\mathrm{tot}}(R_0) = \frac{(n-1)N\alpha q^2}{n(4\pi\epsilon_0)R_0}.$$

The lattice energy per ion pair is given by

$$e_L = \frac{1}{N}E_L = \frac{(n-1)\alpha q^2}{n(4\pi\epsilon_0)R_0}.$$

(4) If the nearest-neighbor distance is reduced, then the total interaction energy will be increased since the crystal has the smallest interaction energy at the equilibrium nearest-neighbor distance R_0. The work W needed to shrink the crystal to the nearest-neighbor distance $(1-\eta)R_0$ is given by $U_{\mathrm{tot}}((1-\eta)R_0) - U_{\mathrm{tot}}(R_0)$. Making use of the general expression for $U_{\mathrm{tot}}(R)$ and the special expression for $U_{\mathrm{tot}}(R_0)$, we have

$$\begin{aligned}
W &= U_{\mathrm{tot}}((1-\eta)R_0) - U_{\mathrm{tot}}(R_0) \\
&= \frac{2NA}{(1-\eta)^n R_0^n} - \frac{N\alpha q^2}{4\pi\epsilon_0(1-\eta)R_0} + \frac{(n-1)N\alpha q^2}{n(4\pi\epsilon_0)R_0} \\
&= \frac{N\alpha q^2}{n(4\pi\epsilon_0)R_0}\left[\frac{1}{(1-\eta)^n} - \frac{n}{1-\eta} + (n-1)\right] \\
&\approx \frac{(n-1)\eta^2 N\alpha q^2}{2(4\pi\epsilon_0)R_0}, \quad \eta \ll 1.
\end{aligned}$$

12-3 Bonding in a three-dimensional ionic crystal. The total energy of a three-dimensional ionic crystal is given by

$$U(R) = \frac{NA}{R^{12}} - \frac{N\alpha Z^2 e^2}{4\pi\epsilon_0 R},$$

where N is the number of ion pairs and R the nearest-neighbor distance.

(1) Derive an expression for the equilibrium value of the nearest-neighbor distance.
(2) Compute the binding energy per ion pair in equilibrium.

---

(1) Differentiating $U(R)$ with respect to R and setting the value of the derivative at $R = R_0$ to zero, we obtain

$$0 = \left.\frac{dU(R)}{dR}\right|_{R=R_0} = -\frac{12NA}{R_0^{13}} + \frac{N\alpha Z^2 e^2}{4\pi\epsilon_0 R_0^2}.$$

Solving for R_0 from the above equation, we have

$$R_0 = \left[\frac{12(4\pi\epsilon_0)A}{\alpha Z^2 e^2}\right]^{1/11}.$$

(2) At the equilibrium separation R_0, the total energy of the crystal is given by

$$U(R_0) = \frac{NA}{R_0}\frac{\alpha Z^2 e^2}{12(4\pi\epsilon_0)A} - \frac{N\alpha Z^2 e^2}{4\pi\epsilon_0 R_0}$$
$$= -\frac{11N\alpha Z^2 e^2}{12(4\pi\epsilon_0)R_0} = -11NA\left[\frac{\alpha Z^2 e^2}{12(4\pi\epsilon_0)A}\right]^{12/11}.$$

Thus, the binding energy per ion pair in equilibrium is given by

$$e_L = -\frac{1}{N}U(R_0) = 11A\left[\frac{\alpha Z^2 e^2}{12(4\pi\epsilon_0)A}\right]^{12/11}.$$

12-4 Born-Meyer theory of bonding in ionic crystals. In the theory of bonding in ionic crystals due to Born and Meyer, the potential energy of an ionic crystal is taken to be

$$U(R) = -\frac{N\alpha e^2}{4\pi\epsilon_0 R} + \frac{NA}{R^n},$$

where N is the number of ion pairs and A and n are constants to be determined from experiment.

(1) Derive an expression for the equilibrium nearest-neighbor distance in terms of A, n, and α.
(2) Compute the lattice energy per ion pair in equilibrium.
(3) Evaluate n and A for NaCl for which the lattice constant is 0.563 nm, the measured lattice energy is 7.95 eV per ion pair, and the Madelung constant α is 1.747 565.

---

(1) Differentiating $U(R)$ with respect to R and setting the value of the derivative at $R = R_0$ to zero, we obtain

$$0 = \left.\frac{\mathrm{d}U(R)}{\mathrm{d}R}\right|_{R=R_0} = \frac{N\alpha e^2}{4\pi\epsilon_0 R_0^2} - \frac{nNA}{R_0^{n+1}}.$$

Solving for R_0 from the above equation, we have

$$R_0 = \left[\frac{n(4\pi\epsilon_0)A}{\alpha e^2}\right]^{1/(n-1)}.$$

(2) The total binding energy in equilibrium is given by

$$U(R_0) = -\frac{N\alpha e^2}{4\pi\epsilon_0 R_0} + \frac{N\alpha e^2}{n(4\pi\epsilon_0)R_0} = -\frac{N(n-1)\alpha e^2}{n(4\pi\epsilon_0)R_0}$$
$$= -N(n-1)\left[\frac{\alpha e^2}{n(4\pi\epsilon_0)A^{1/n}}\right]^{n/(n-1)}.$$

The lattice energy per ion pair in equilibrium is given by

$$e_L = -\frac{1}{N}U(R_0) = (n-1)\left[\frac{\alpha e^2}{n(4\pi\epsilon_0)A^{1/n}}\right]^{n/(n-1)}.$$

(3) Note that e_L can also be written as

$$e_L = \frac{(n-1)\alpha e^2}{n(4\pi\epsilon_0)R_0}$$

from which we can solve for n and obtain

$$n = \frac{1}{1-(4\pi\epsilon_0)R_0 e_L/\alpha e^2} \approx 9.035\,5 \approx 9.$$

From the above-derived expression for R_0, we have

$$A = \frac{\alpha e^2 R_0^{n-1}}{n(4\pi\epsilon_0)} \approx 8.06\times10^{-106}\ \mathrm{J\cdot m}^n \approx 5.03\times10^{-87}\ \mathrm{eV\cdot m}^n.$$

12-5 Bonding in NaCl. We consider the bonding in NaCl again, using its measured bulk modulus B this time. The potential energy per ion pair at distance R is expressed as

$$u(r) = -\frac{\alpha e^2}{4\pi\epsilon_0 R} + \beta\left(\frac{R_0}{R}\right)^n,$$

where R_0 is the nearest-neighbor distance and n and β are constants to be determined below.

(1) Derive a relation between β and n from the equilibrium condition.

(2) Derive an expression for the bulk modulus in equilibrium.

(3) Evaluate n and β using $B = 2.4 \times 10^{10}\ \mathrm{N/m^2}$, $R_0 = 0.282$ nm, and $\alpha = 1.747\,565$.

---

(1) To find the bulk modulus, we first find the pressure

$$P = -\frac{\partial U}{\partial V} = -\frac{\partial u}{\partial v},$$

where v is the volume per ion pair. Noting that there are four ion pairs within a conventional cell and the distance r between two ions in a pair is given by $r = a/2$ with a the lattice constant, we have $v = a^3/4 = 2r^3$. In terms of the derivative with respect to r, we have

$$P = -\frac{1}{6r^2}\frac{\partial u}{\partial r}.$$

The bulk modulus is then given by

$$B = -v\frac{\partial P}{\partial v} = \frac{r}{18}\frac{\partial}{\partial r}\left(\frac{1}{r^2}\frac{\partial u}{\partial r}\right) = \frac{1}{18r}\left(-\frac{2}{r}\frac{\partial u}{\partial r} + \frac{\partial^2 u}{\partial r^2}\right).$$

At the equilibrium distance r_0, we have $\partial u/\partial r\big|_{r=r_0} = 0$. Thus, at the equilibrium distance r_0, the bulk modulus is given by

$$B = \frac{1}{18r_0}\frac{\partial^2 u}{\partial r^2}\bigg|_{r=r_0}.$$

(2) Substituting $u = -\alpha e^2/r + \beta(r_0/r)^n$ into the above equation yields

$$B = \frac{1}{18r_0}\left[-\frac{2\alpha e^2}{r_0^3} + \frac{n(n+1)\beta}{r_0^2}\right].$$

From $\partial u/\partial r\big|_{r=r_0} = 0$, we have

$$\beta = \frac{\alpha e^2}{n r_0}.$$

The expression of the bulk modulus then becomes

$$B = \frac{(n-1)\alpha e^2}{18 r_0^4}.$$

(3) Solving for n from the above expression of B, we obtain

$$n = 1 + \frac{18 r_0^4 B}{\alpha e^2} = 1 + \frac{18 \times 2.82^4 \times 10^{-32} \times 2.4 \times 10^{11}}{1.747565 \times 4.80324^2 \times 10^{-20}} \approx 7.776.$$

The parameter β is given by

$$\beta = \frac{\alpha e^2}{n r_0} = \frac{1.747565 \times 4.80324^2 \times 10^{-20}}{7.776 \times 2.82 \times 10^{-8}} \approx 1.839 \times 10^{-12}\,\text{erg}.$$

12-6 Madelung constant of the CsCl crystal. Consider the Madelung constant of the CsCl crystal.

(1) Derive an expression for the Madelung constant of CsCl using the Ewald summation method.

(2) Numerically evaluate the Madelung constant of CsCl.

---

(1) The general formula for the Madelung constant in the Ewald summation method is given by

$$\frac{\alpha q^2}{R} = -\frac{2\pi}{v_c} \sum_{\boldsymbol{K}}{}' |S_{\boldsymbol{K}}|^2 \frac{\mathrm{e}^{-\boldsymbol{K}^2/4\beta^2}}{|\boldsymbol{K}|^2} + \frac{\beta}{\pi^{1/2}} \sum_{\kappa} q_\kappa^2 - \frac{1}{2} \sum_{n\kappa\nu}{}' q_\kappa q_\nu \frac{\operatorname{erfc}(\beta|\boldsymbol{R}_n + \boldsymbol{d}_\nu - \boldsymbol{d}_\kappa|)}{|\boldsymbol{R}_n + \boldsymbol{d}_\nu - \boldsymbol{d}_\kappa|}$$

The crystal structure of CsCl is an SC Bravais lattice with a two-atom basis. The nearest-neighbor distance R is given by $R = \sqrt{3}\,a/2$ with a the lattice constant. The lattice vectors are of the form

$$\boldsymbol{R}_n = (n_1 \boldsymbol{e}_x + n_2 \boldsymbol{e}_y + n_3 \boldsymbol{e}_z)a$$

with n_1, n_2, and n_3 integers. The volume of a primitive cell of the direct lattice is $v_c = a^3$. The reciprocal lattice vectors are given by

$$\boldsymbol{K} = (h\boldsymbol{e}_x + k\boldsymbol{e}_y + \ell\boldsymbol{e}_z)(2\pi/a)$$

with h, k, and ℓ integers. The structure factor $S_{\boldsymbol{K}}$ for CsCl is given by

$$S_{\boldsymbol{K}} = e\big[1 - \mathrm{e}^{\mathrm{i}(h+k+\ell)\pi}\big] = 2e\delta_{h+k+\ell,\mathrm{odd}},$$

where $q_\kappa = \pm e$ has been used. Inserting the above quantities for CsCl into the general formula for the Madelung constant, we have

$$\alpha_{\mathrm{CsCl}} = \sqrt{3}\left\{-\frac{1}{\pi}\sum_{hk\ell}{}' \frac{\mathrm{e}^{-(h^2+k^2+\ell^2)\pi^2/(\beta a)^2}}{h^2+k^2+\ell^2} - \frac{1}{2}\sum_{ijk}{}' \frac{\mathrm{erfc}\big(\beta a(i^2+j^2+k^2)^{1/2}\big)}{(i^2+j^2+k^2)^{1/2}} + \frac{\beta a}{\pi^{1/2}} + \sum_{ijk} \frac{\mathrm{erfc}\big([(2i+1)^2+(2j+1)^2+(2k+1)^2]^{1/2}\beta a/2\big)}{[(2i+1)^2+(2j+1)^2+(2k+1)^2]^{1/2}}\right\}.$$

(2) Numerically evaluating the above-derived expression for the Madelung constant of CsCl, we find that

$$\alpha_{\mathrm{CsCl}} \approx 1.762\,675.$$

12-7 Alkali metals. A very crude model for alkali metals can be constructed by taking the charge of the valence electron of an atom as uniformly distributed over a sphere of radius r_s about its ion. Assume that the interactions between different atoms are completely screened. Taking into account only the electrostatic interactions of this spherical charge distribution with the ion and with itself, compute the electrostatic energy per atom.

The uniform electron charge density in the sphere of radius r_s is

$$\rho_e = -\frac{3e}{4\pi r_s^3}.$$

The total interaction energy of an atom consists of two parts: The electron-electron interaction energy (U_{ee}) and the electron-ion interaction energy (U_{ei}). We separately evaluate these two parts. The interaction energy of the electron charge distribution with itself is given by

$$\begin{aligned} U_{ee} &= \frac{1}{2}\frac{1}{4\pi\epsilon_0}\rho_e^2 \int \mathrm{d}\boldsymbol{r} \int \mathrm{d}\boldsymbol{r}'\, \frac{1}{|\boldsymbol{r}-\boldsymbol{r}'|} \\ &= \frac{9e^2}{2(4\pi)^2(4\pi\epsilon_0)r_s^6} \int \mathrm{d}\boldsymbol{r} \int \mathrm{d}\boldsymbol{r}'\, \frac{1}{(r^2+r'^2-2rr'\cos\theta)^{1/2}}, \end{aligned}$$

where θ is the angle between $\boldsymbol{r}$ and $\boldsymbol{r}'$. When the integration over $\boldsymbol{r}'$ is performed, the z axis is chosen to point in the direction of $\boldsymbol{r}$. Then θ becomes the polar angle of $\boldsymbol{r}'$. The integration over θ can be easily performed when the integrand is expanded as the following power series with Legendre polynomials as coefficients

$$\frac{1}{(r^2 + r'^2 - 2rr'\cos\theta)^{1/2}} = \sum_{\ell=0}^{\infty} \frac{r_<^\ell}{r_>^{\ell+1}} P_\ell(\cos\theta),$$

where $r_> = \max(r, r')$ and $r_< = \min(r, r')$. The integration over θ can be then performed using the orthogonality relation of Legendre polynomials

$$\int_{-1}^{1} \mathrm{d}x\, P_\ell(x) P_{\ell'}(x) = \frac{2}{2\ell+1}\delta_{\ell\ell'}.$$

We have

$$\begin{aligned}
U_{ee} &= \frac{9e^2}{16\pi(4\pi\epsilon_0)r_s^6} \sum_{\ell=0}^{\infty} \int \mathrm{d}\boldsymbol{r} \int_0^{r_s} \mathrm{d}r'\, r'^2 \frac{r_<^\ell}{r_>^{\ell+1}} \int_{-1}^{1} \mathrm{d}\cos\theta\, P_\ell(\cos\theta) \\
&= \frac{9e^2}{16\pi(4\pi\epsilon_0)r_s^6} \sum_{\ell=0}^{\infty} \int \mathrm{d}\boldsymbol{r} \int_0^{r_s} \mathrm{d}r'\, r'^2 \frac{r_<^\ell}{r_>^{\ell+1}} \cdot 2\delta_{\ell 0} \\
&= \frac{9e^2}{2(4\pi\epsilon_0)r_s^6} \int_0^{r_s} \mathrm{d}r\, r^2 \int_0^{r_s} \mathrm{d}r'\, r'^2 \frac{1}{r_>} \\
&= \frac{9e^2}{2(4\pi\epsilon_0)r_s^6} \int_0^{r_s} \mathrm{d}r\, r^2 \left[\frac{1}{r} \int_0^{r} \mathrm{d}r'\, r'^2 + \int_r^{r_s} \mathrm{d}r'\, r' \right] \\
&= \frac{3e^2}{5(4\pi\epsilon_0)r_s}.
\end{aligned}$$

The electron-ion interaction energy is given by

$$U_{ei} = \frac{e\rho_e}{4\pi\epsilon_0} \int \mathrm{d}\boldsymbol{r}\, \frac{1}{|\boldsymbol{r}|} = -\frac{3e^2}{(4\pi\epsilon_0)r_s^3} \int_0^{r_s} \mathrm{d}r\, r = -\frac{3e^2}{2(4\pi\epsilon_0)r_s}.$$

The interaction energy of the atoms is then given by

$$U = U_{ee} + U_{ei} = \frac{3e^2}{5(4\pi\epsilon_0)r_s} - \frac{3e^2}{2(4\pi\epsilon_0)r_s} = -\frac{9e^2}{10(4\pi\epsilon_0)r_s}.$$

Making use of $e^2/2(4\pi\epsilon_0)a_0 \approx 13.61$ eV, we have

$$U \approx -\frac{24.49}{r_s/a_0}\ \text{eV}.$$

We see that this crude approximation gives a very good result in comparison with the value $U_{\text{Coul}} \approx -24.38/(r_s/a_0)$ eV/atom obtained in a more accurate treatment for the direct Coulomb energy.

12-8 Exchange energy of the electron gas in an alkali metal. Consider the exchange energy of the electron gas in an alkali metal.

(1) Show that

$$\frac{(2\pi)^3}{\mathscr{V}}\sum_{\boldsymbol{k}'}{}' \frac{1}{|\boldsymbol{k}-\boldsymbol{k}'|^2} = 4\pi k_{\mathrm{F}}\eta(k/k_{\mathrm{F}})$$

for the electron gas in the ground state. Here $x = k/k_{\mathrm{F}}$, the prime on the summation sign indicates the restriction that $|\boldsymbol{k}'| \leqslant k_{\mathrm{F}}$, and $\eta(x)$ is given by

$$\eta(x) = \frac{1}{2} + \frac{1-x^2}{4x}\ln\left|\frac{1+x}{1-x}\right|.$$

(2) Show that the exchange energy per atom

$$u_{\mathrm{ex}} = -\frac{e^2}{\epsilon_0 N\mathscr{V}}\sum_{\boldsymbol{k}\boldsymbol{k}'}{}' \frac{1}{|\boldsymbol{k}-\boldsymbol{k}'|^2}$$

is given by

$$u_{\mathrm{ex}} = -\frac{2e^2 k_{\mathrm{F}}}{\pi(4\pi\epsilon_0)N}\sum_{\boldsymbol{k}}{}' \eta(k/k_{\mathrm{F}}) = -\frac{3}{4\pi}\left(\frac{9\pi}{4}\right)^{1/3}\frac{e^2}{4\pi\epsilon_0 r_s}.$$

---

(1) Converting the summation over $\boldsymbol{k}'$ into an integration over $\boldsymbol{k}'$, we have

$$\begin{aligned}
\frac{(2\pi)^3}{\mathscr{V}}\sum_{\boldsymbol{k}'}{}' \frac{1}{|\boldsymbol{k}-\boldsymbol{k}'|^2} &= \int \mathrm{d}\boldsymbol{k}' \, \frac{1}{|\boldsymbol{k}-\boldsymbol{k}'|^2} \\
&= 2\pi\int_0^{k_{\mathrm{F}}} \mathrm{d}k' \, k'^2 \int_{-1}^{1} \frac{\mathrm{d}\cos\theta}{k^2+k'^2-2kk'\cos\theta} \\
&= \frac{2\pi}{k}\int_0^{k_{\mathrm{F}}} \mathrm{d}k' \, k' \ln\left|\frac{k'+k}{k'-k}\right|.
\end{aligned}$$

Performing an integration by parts yields

$$\begin{aligned}
\frac{(2\pi)^3}{\mathscr{V}}\sum_{\boldsymbol{k}'}{}' \frac{1}{|\boldsymbol{k}-\boldsymbol{k}'|^2} &= \frac{\pi}{k}\Bigg[k_{\mathrm{F}}^2 \ln\left|\frac{k+k_{\mathrm{F}}}{k-k_{\mathrm{F}}}\right| \\
&\quad - \int_0^{k_{\mathrm{F}}} \mathrm{d}k' \, k'^2\left(\frac{1}{k'+k} - \frac{1}{k'-k}\right)\Bigg] \\
&= 4\pi k_{\mathrm{F}}\left[\frac{1}{2} + \frac{1-(k/k_{\mathrm{F}})^2}{4(k/k_{\mathrm{F}})}\ln\left|\frac{1+k/k_{\mathrm{F}}}{1-k/k_{\mathrm{F}}}\right|\right] \\
&= 4\pi k_{\mathrm{F}}\eta(k/k_{\mathrm{F}}).
\end{aligned}$$

(2) Making use of the above-derived result, we have

$$u_{\rm ex} = -\frac{2e^2 k_{\rm F}}{\pi(4\pi\epsilon_0)N} {\sum_{\boldsymbol{k}}}' \left[\frac{1}{2} + \frac{k_{\rm F}^2 - k^2}{4kk_{\rm F}} \ln\left|\frac{k+k_{\rm F}}{k-k_{\rm F}}\right|\right].$$

Converting the summation over $\boldsymbol{k}$ into an integration over $\boldsymbol{k}$ yields

$$\begin{aligned} u_{\rm ex} &= -\frac{e^2 k_{\rm F}}{\pi^3(4\pi\epsilon_0)n} \int_0^{k_{\rm F}} {\rm d}k\, k^2 \left[\frac{1}{2} + \frac{k_{\rm F}^2 - k^2}{4kk_{\rm F}} \ln\left|\frac{k+k_{\rm F}}{k-k_{\rm F}}\right|\right] \\ &= -\frac{e^2}{4\pi^3(4\pi\epsilon_0)n} \left[\frac{2}{3}k_{\rm F}^4 + \int_0^{k_{\rm F}} {\rm d}k\, k\big(k_{\rm F}^2 - k^2\big) \ln\left|\frac{k+k_{\rm F}}{k-k_{\rm F}}\right|\right], \end{aligned}$$

where $n = N/\mathscr{V}$ is the electron number density. The integral in the square brackets is an improper integral. To be able to perform an integration by parts, we must ensure that the integrated-out part is finite when it is evaluated at the upper integration limit $k = k_{\rm F}$. For this purpose, we rewrite ${\rm d}k\, k\big(k_{\rm F}^2 - k^2\big)$ as

$${\rm d}k\, k\big(k_{\rm F}^2 - k^2\big) = {\rm d}\big(k_{\rm F}^2 k^2/2 - k^4/4 - kk_{\rm F}^3/4\big) + (k_{\rm F}^3/4){\rm d}k.$$

We then have

$$\begin{aligned} &\int_0^{k_{\rm F}} {\rm d}k\, k\big(k_{\rm F}^2 - k^2\big) \ln\left|\frac{k+k_{\rm F}}{k-k_{\rm F}}\right| \\ &= \frac{1}{4}k_{\rm F}^3 \int_0^{k_{\rm F}} {\rm d}k\, \ln\left|\frac{k+k_{\rm F}}{k-k_{\rm F}}\right| \\ &\quad + \frac{1}{4}\int_0^{k_{\rm F}} {\rm d}k\, \big(k^4 - 2k_{\rm F}^2 k^2 + kk_{\rm F}^3\big)\left(\frac{1}{k+k_{\rm F}} - \frac{1}{k-k_{\rm F}}\right) \\ &= \frac{\ln 2}{2}k_{\rm F}^4 + \frac{1}{4}\int_0^{k_{\rm F}} {\rm d}k\, \big[k^2(k^2 - k_{\rm F}^2) - kk_{\rm F}^2(k - k_{\rm F})\big] \\ &\qquad\qquad \times \left(\frac{1}{k+k_{\rm F}} - \frac{1}{k-k_{\rm F}}\right) \\ &= \frac{\ln 2}{2}k_{\rm F}^4 + \frac{1}{4}\int_0^{k_{\rm F}} {\rm d}k\, k^2(k^2 - k_{\rm F}^2)\left(\frac{1}{k+k_{\rm F}} - \frac{1}{k-k_{\rm F}}\right) \\ &\quad - \frac{1}{4}k_{\rm F}^2 \int_0^{k_{\rm F}} {\rm d}k\, k(k - k_{\rm F})\left(\frac{1}{k+k_{\rm F}} - \frac{1}{k-k_{\rm F}}\right) \\ &= \frac{\ln 2}{2}k_{\rm F}^4 - \frac{1}{2}k_{\rm F}\int_0^{k_{\rm F}} {\rm d}k\, k^2 - \frac{1}{4}k_{\rm F}^2 \int_0^{k_{\rm F}} {\rm d}k\, k\left(\frac{k-k_{\rm F}}{k+k_{\rm F}} - 1\right) \\ &= \frac{\ln 2}{2}k_{\rm F}^4 - \frac{1}{6}k_{\rm F}^4 + \frac{1}{2}k_{\rm F}^3 \int_0^{k_{\rm F}} {\rm d}k\, \frac{k}{k+k_{\rm F}} = \frac{1}{3}k_{\rm F}^4. \end{aligned}$$

Finally,

$$u_{\text{ex}} = -\frac{e^2 k_{\text{F}}^4}{4\pi^3(4\pi\epsilon_0)n}.$$

Making use of $k_{\text{F}} = (3\pi n)^{1/3}$ and $4\pi r_s^3/3 = 1/n$, we can rewrite u_{ex} as

$$u_{\text{ex}} = -\frac{3}{4\pi}\left(\frac{9\pi}{4}\right)^{1/3}\frac{e^2}{(4\pi\epsilon_0)r_s}.$$

Chapter 13

Normal Modes of Lattice Vibrations

(1) *Born-Oppenheimer approximation*
The Born-Oppenheimer approximation can be used to disentangle the motion of electrons from that of nuclei in a solid. In the Born-Oppenheimer approximation, the motion of electrons can be solved while nuclei are taken to be at their equilibrium positions.

(2) *Harmonic approximation*
In the harmonic approximation, the lattice potential energy is Taylor-expanded in the displacements of atoms from their equilibrium positions and only terms up to the second order are kept.

(3) *Properties of the $D(\boldsymbol{R})$ matrix*

i. $D(\boldsymbol{R}_i,\, \boldsymbol{R}_j)$ is a function of $\boldsymbol{R}_i - \boldsymbol{R}_j$.

ii. $D_{\nu\kappa,\,\beta\alpha}(-\boldsymbol{R}) = D_{\kappa\nu,\,\alpha\beta}(\boldsymbol{R})$,

$D_{\nu\kappa,\,\alpha\beta}(-\boldsymbol{R}) = D_{\kappa\nu,\,\alpha\beta}(\boldsymbol{R})$.

iii. $\sum_{j\nu} D_{\kappa\nu,\,\alpha\beta}(\boldsymbol{R}_i - \boldsymbol{R}_j) = 0$.

(4) *Dynamical matrix*

$$\mathscr{D}_{\kappa\nu,\,\alpha\beta}(\boldsymbol{k}) = \frac{1}{\sqrt{m_\kappa m_\nu}} \sum_j \mathrm{e}^{-\mathrm{i}\boldsymbol{k}\cdot(\boldsymbol{R}_i - \boldsymbol{R}_j)} D_{\kappa\nu,\,\alpha\beta}(\boldsymbol{R}_i - \boldsymbol{R}_j).$$

$\mathscr{D}(\boldsymbol{k})$ is a Hermitian matrix.
The eigenvalues of $\mathscr{D}$ yields the dispersions of normal modes, $\omega_s^2 = \lambda_s$.

(5) *Equations for polarization vectors*

$$\sum_{\nu\beta} \big[\mathscr{D}_{\kappa\nu,\,\alpha\beta}(\boldsymbol{k}) - \omega_{\boldsymbol{k}s}^2 \delta_{\kappa\nu}\delta_{\alpha\beta}\big] \epsilon_{\nu\beta}^{(s)}(\boldsymbol{k}) = 0.$$

(6) *Properties of polarization vectors*

$$\boldsymbol{\epsilon}_{\nu}^{(s)*}(\boldsymbol{k}) = \boldsymbol{\epsilon}_{\nu}^{(s)}(-\boldsymbol{k}),$$

$$\sum_{\kappa\alpha} \epsilon_{\kappa\alpha}^{(s)*}(\boldsymbol{k})\epsilon_{\kappa\alpha}^{(s')}(\boldsymbol{k}) = \delta_{ss'},$$

$$\sum_{s} \epsilon_{\kappa\alpha}^{(s)*}(\boldsymbol{k})\epsilon_{\nu\beta}^{(s)}(\boldsymbol{k}) = \delta_{\kappa\nu}\delta_{\alpha\beta}.$$

(7) *Displacements of atoms*

$$u_{j\nu,\,\alpha}(t) = \frac{1}{\sqrt{Nm_\nu}}\sum_{\boldsymbol{k}s} q_{\boldsymbol{k}s}(t)\epsilon_{\nu\alpha}^{(s)}(\boldsymbol{k})\mathrm{e}^{\mathrm{i}\boldsymbol{k}\cdot\boldsymbol{R}_j}.$$

(8) *Hamiltonian of a crystal*

$$H = \frac{1}{2}\sum_{\boldsymbol{k}s} p_{\boldsymbol{k}s}^*(t)p_{\boldsymbol{k}s}(t) + \frac{1}{2}\sum_{\boldsymbol{k}s} \omega_{\boldsymbol{k}s}^2 q_{\boldsymbol{k}s}^*(t)q_{\boldsymbol{k}s}(t).$$

(9) *Acoustical and optical branches*
In a d-dimensional crystal with a p-atom basis, there is (are) d acoustical and $d(p-1)$ optical branch(es).

(10) *Acoustical and optical normal modes*
In a d-dimensional crystal of size of N primitive cells with a p-atom basis, there are dN acoustical and $d(p-1)N$ optical normal modes.

13-1 Normal modes of a linear chain of ions. Consider a linear chain in which alternate ions have masses m_1 and m_2 and only nearest neighbors interact through a spring of force constant K. Find the dispersion relations for the normal modes. Discuss the limiting cases for $m_1 \gg m_2$ and $m_1 = m_2$.

The lattice potential energy is given by

$$\Phi = \frac{1}{2}K\sum_j (u_{j2} - u_{j1})^2 + \frac{1}{2}K\sum_j (u_{j+1,1} - u_{j2})^2,$$

where $u_{j\nu}$ is the displacement of the νth ion ($\nu = 1, 2$) in the jth primitive cell from its equilibrium position. The forces exerting on the two ions in the jth primitive cell can be obtained by differentiating Φ with respect to their displacements. We have

$$\begin{aligned} F_{j1} &= -\frac{\partial\Phi}{\partial u_{j1}} = K(u_{j2} - u_{j1}) - K(u_{j1} - u_{j-1,2}) \\ &= -K(2u_{j1} - u_{j2} - u_{j-1,2}), \end{aligned}$$

and

$$F_{j2} = -\frac{\partial \Phi}{\partial u_{j2}} = -K(u_{j2} - u_{j1}) + K(u_{j+1,1} - u_{j2})$$
$$= -K(2u_{j2} - u_{j1} - u_{j+1,1}).$$

From Newton's second law, we obtain the classical equations of motion for ions

$$m_1 \ddot{u}_{j1} = -K(2u_{j1} - u_{j2} - u_{j-1,2}),$$
$$m_2 \ddot{u}_{j2} = -K(2u_{j2} - u_{j1} - u_{j+1,1}).$$

Let N be the number of primitive cells in the crystal. Then, there are $2N$ equations of motion for $2N$ ions. Let d be the equilibrium distance between two nearest-neighboring ions. The lattice constant is then $a = 2d$ and the lattice vectors can be expressed as $R_j = ja$. To solve the above equations of motion, we make a Fourier transformation to $u_{j\nu}$ with respect to both R_j and time t

$$u_{j\nu} = \sum_{k\omega} Q_\nu(k,\omega) e^{i(kR_j - \omega t)}.$$

The allowed values of k are given by $k_n = 2n\pi/Na$ with $n = 0, \pm 1, \pm 2, \cdots, \pm(N/2-1), N/2$ which follows from the Born-van Karman periodic boundary condition. We now find the allowed values of ω. In the mean time, we also obtain the dispersion relations of the normal modes. Substituting the above Fourier expansion of $u_{j\nu}$ into the equations of motion, we have

$$(2K - m_1\omega^2)Q_1(k,\omega) - K(1 + e^{-ika})Q_2(k,\omega) = 0,$$
$$-K(1 + e^{ika})Q_1(k,\omega) + (2K - m_2\omega^2)Q_2(k,\omega) = 0.$$

The above equations can be regarded as a set of homogeneous linear equations for Fourier coefficients $Q_1(k,\omega)$ and $Q_2(k,\omega)$. According to the theory of linear algebras, the necessary and sufficient condition for the existence of nontrivial solutions is the vanishing of the determinant of coefficients. We then have

$$\det \begin{vmatrix} 2K - m\omega^2 & -K(1 + e^{-ika}) \\ -K(1 + e^{ika}) & 2K - m\omega^2 \end{vmatrix} = 0.$$

Evaluating the determinant, we obtain

$$m_1 m_2 \omega^4 - 2K(m_1 + m_2)\omega^2 + 4K^2 \sin^2(ka/2) = 0.$$

Solving the above equation, we obtain the allowed values of ω

$$\omega = \pm\omega_{k\mathrm{a}}, \pm\omega_{k\mathrm{o}},$$

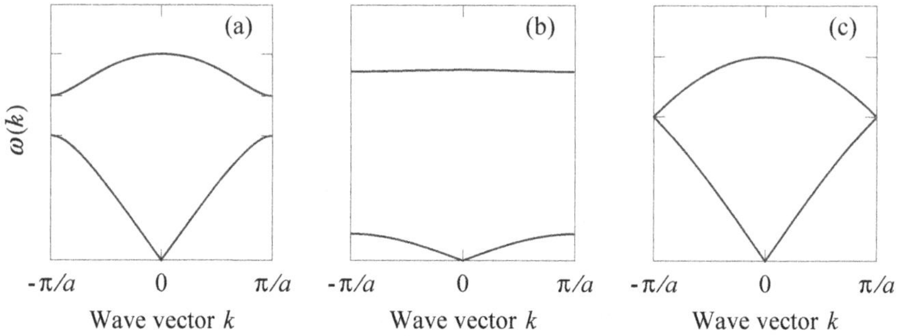

Fig. 13.1 Plots of dispersion relations for a linear chain of alternate positive and negative ions. (a) Comparable but unequal m_1 and m_2. Here $m_1 > m_2$. (b) $m_1 \gg m_2$. (c) $m_1 = m_2 = m$.

where

$$\omega_{k\mathrm{a,o}} = (K/m_1 m_2)^{1/2}\Big\{m_1 + m_2 \mp \big[m_1^2 + m_2^2 + 2m_1 m_2 \cos(ka)\big]^{1/2}\Big\}^{1/2}$$

are the dispersion relations for the acoustical and optical branches, respectively, and they are plotted in Fig. 13.1(a).

For $m_1 \gg m_2$, we have

$$\begin{aligned}\omega_{k\mathrm{a,o}} &\approx (K/m_2)^{1/2}\Big\{1 + m_2/m_1 \mp \big[1 + 2(m_2/m_1)\cos(ka)\big]^{1/2}\Big\}^{1/2} \\ &\approx (K/m_2)^{1/2}\Big\{1 + m_2/m_1 \mp \big[1 + (m_2/m_1)\cos(ka)\big]\Big\}^{1/2} \\ &\approx \begin{cases}(2K/m_1)^{1/2}|\sin(ka/2)|, \\ (2K/m_2)^{1/2}\big[1 + (m_2/2m_1)\cos^2(ka/2)\big] \approx (2K/m_2)^{1/2}.\end{cases}\end{aligned}$$

Thus, in the $m_1 \gg m_2$ limit, the acoustical modes give low-frequency oscillations while the optical modes give high-frequency oscillations with the frequencies nearly constant for all the optical modes. Also note that the acoustical modes are the same as those in a linear chain of identical atoms of mass m_1. $\omega_{k\mathrm{a}}$ and $\omega_{k\mathrm{o}}$ in this limit are plotted in Fig. 13.1(b).

In the $m_1 = m_2$ limit, we have

$$\omega_{k\mathrm{a}} = (4K/m)^{1/2}|\sin(ka/4)|,\ \omega_{k\mathrm{o}} = (4K/m)^{1/2}|\cos(ka/4)|,$$

where $m = m_1 = m_2$. In this limit, the maximum frequency of the acoustical modes is $1/2^{1/2}$ times that of the optical modes. At the boundaries of the first Brillouin zone, the acoustical and optical modes have the same frequency. $\omega_{k\mathrm{a}}$ and $\omega_{k\mathrm{o}}$ in this limit are plotted in Fig. 13.1(c).

13-2 Simple one-dimensional crystal with a two-atom basis. A commonly-seen example of a simple one-dimensional crystal is a line of point masses, each of which has two nearest neighbors: One is distance d away and the other distance $(a - d)$ away $(d \ll a)$ in equilibrium. The two neighboring point masses are connected by springs, with the force constants of the springs between the near-distanced point masses and between the far-distanced point masses given by K and G $(K > G)$, respectively.

(1) Write down the harmonic crystal potential energy in terms of the displacements of point masses from their equilibrium positions.

(2) Set up the classical equations of motion for the point masses.

(3) Solve for the frequencies and polarization vectors of the normal modes of the lattice vibrations from the classical equations of motion.

---

(1) The lattice potential energy is given by

$$\Phi = \frac{1}{2}K\sum_n \big(u_{n1} - u_{n2}\big)^2 + \frac{1}{2}G\sum_n \big(u_{n+1,1} - u_{n2}\big)^2.$$

(2) The forces acting on the two point masses in the nth unit cell are given by

$$F_{n1} = -\frac{\partial \Phi}{\partial u_{n1}} = -K\big(u_{n1} - u_{n2}\big) - G\big(u_{n1} - u_{n-1,2}\big)$$
$$F_{n2} = -\frac{\partial \Phi}{\partial u_{n2}} = -K\big(u_{n2} - u_{n1}\big) - G\big(u_{n2} - u_{n+1,1}\big).$$

From Newton's second law, we obtain the equations of motion for the point masses

$$M\ddot{u}_{n1} = -K\big(u_{n1} - u_{n2}\big) - G\big(u_{n1} - u_{n-1,2}\big),$$
$$M\ddot{u}_{n2} = -K\big(u_{n2} - u_{n1}\big) - G\big(u_{n2} - u_{n+1,1}\big).$$

(3) To solve the above equations of motion, we make a Fourier transformation to $u_{n\alpha}$ with respect to both $R_n = na$ and time t

$$u_{n\alpha} = \sum_{k\omega} Q_\nu(k,\omega)\mathrm{e}^{\mathrm{i}(kR_n - \omega t)}.$$

The allowed values of k are given by $k_j = 2j\pi/Na$ with $j = 0$, ±1, ±2, $\cdots$, $\pm(N/2 - 1)$, $N/2$ which follows from the Born-van Karman periodic boundary condition. We now find the allowed values of ω. In the mean time, we also obtain the dispersion relations of the normal modes. Substituting the above Fourier expansion of $u_{n\alpha}$ into the equations of motion, we have

$$[M\omega^2 - (K+G)]Q_1(k,\omega) + (K + Ge^{-ika})Q_2(k,\omega) = 0,$$
$$(K + Ge^{ika})Q_1(k,\omega) + [M\omega^2 - (K+G)]Q_2(k,\omega) = 0.$$

The above two equations can be taken as a set of homogeneous linear equations for $Q_1(k,\omega)$ and $Q_2(k,\omega)$. The necessary and sufficient condition for the existence of nontrivial solutions is the vanishing determinant of coefficients. We have

$$\begin{vmatrix} M\omega^2 - (K+G) & K + Ge^{-ika} \\ K + Ge^{ika} & M\omega^2 - (K+G) \end{vmatrix} = 0.$$

Evaluating the determinant, we obtain

$$[M\omega^2 - (K+G)]^2 = |K + Ge^{ika}|^2 = K^2 + G^2 + 2KG\cos(ka).$$

Solving for ω from the above equation yields the following allowed values of ω

$$\omega = \pm\omega_{k\mathrm{a}}, \pm\omega_{k\mathrm{o}},$$

where $\omega_{k\mathrm{a}}$ and $\omega_{k\mathrm{o}}$ are the dispersion relations for the acoustical and optical branches and they are given by

$$\omega_{k\mathrm{a,o}} = \left\{ \frac{1}{M}(K+G) \mp \frac{1}{M}\left[K^2 + G^2 + 2KG\cos(ka)\right]^{1/2} \right\}^{1/2}.$$

The equations satisfied by the polarization vectors $\epsilon_1(k)$ and $\epsilon_2(k)$ can be obtained from those for $Q_1(k,\omega)$ and $Q_2(k,\omega)$ by replacing $Q_1(k,\omega)$ and $Q_2(k,\omega)$ with $\epsilon_1(k)$ and $\epsilon_2(k)$, respectively. We have

$$[M\omega^2 - (K+G)]\epsilon_1(k) + (K + Ge^{-ika})\epsilon_2(k) = 0,$$
$$(K + Ge^{ika})\epsilon_1(k) + [M\omega^2 - (K+G)]\epsilon_2(k) = 0.$$

The ratio $\epsilon_2(k)/\epsilon_1(k)$ can be obtained from one of the above equations and we have

$$\frac{\epsilon_2(k)}{\epsilon_1(k)} = -\frac{K + Ge^{ika}}{M\omega^2 - (K+G)}.$$

Making use of the above-derived expressions for $\omega_{ka,o}$, we have

$$\frac{\epsilon_2^{(a,o)}(k)}{\epsilon_1^{(a,o)}(k)} = \pm\frac{K + Ge^{ika}}{|K + Ge^{ika}|} = \pm\frac{K + Ge^{ika}}{[K^2 + G^2 + 2KG\cos(ka)]^{1/2}}.$$

Note that $\left|\epsilon_2^{(a,o)}(k)/\epsilon_1^{(a,o)}(k)\right| = 1$. Making use of the above equation together with the normalization condition $\left|\epsilon_2^{(a,o)}(k)\right|^2 + \left|\epsilon_1^{(a,o)}(k)\right|^2 = 1$, we can easily write down explicit expressions for $\epsilon_1(k)$ and $\epsilon_2(k)$ and have

$$\epsilon_1^{(a,o)}(k) = \frac{1}{2^{1/2}},\ \epsilon_2^{(a,o)}(k) = \pm\frac{K + Ge^{ika}}{2^{1/2}[K^2 + G^2 + 2KG\cos(ka)]^{1/2}},$$

where we have taken the phase of $\epsilon_1^{(a,o)}(k)$ to be zero.

13-3 One-dimensional crystal with next-nearest-neighbor interactions. Consider a one-dimensional crystal of atoms of mass m. Only the interactions up to the next nearest neighbors are taken into account and are modeled by springs with the force constant for the nearest-neighbor interaction given by K and that for the next-nearest-neighbor interaction given by G.

(1) Compute the dispersion relation of the normal modes.

(2) Find the condition on G so that the dispersion curve peaks inside the first Brillouin zone.

(3) Find the expressions for the group and phase velocities and evaluate them at the peak position of the dispersion curve under the condition found in (2).

---

(1) The lattice potential energy is given by

$$\Phi = \frac{1}{2}K\sum_j (u_j - u_{j+1})^2 + \frac{1}{2}G\sum_j (u_j - u_{j+2})^2.$$

The force exerting on the jth atom is given by

$$\begin{aligned} F_j &= -\frac{\partial\Phi}{\partial u_j} \\ &= -2(K+G)u_j + K(u_{j+1} + u_{j-1}) + G(u_{j+2} + u_{j-2}). \end{aligned}$$

The equations of motion for atoms are then given by

$$m\ddot{u}_j = -2(K+G)u_j + K(u_{j+1} + u_{j-1}) + G(u_{j+2} + u_{j-2}).$$

To solve the above equations of motion, we make a Fourier transformation to u_j with respect to both $R_j = ja$ and time t

$$u_j = \sum_{k\omega} Q(k,\omega)e^{i(kR_j - \omega t)}.$$

The allowed values of k are given by $k_n = 2n\pi/Na$ with $n = 0$, ±1, ±2, $\cdots$, $\pm(N/2-1)$, $N/2$ which follows from the Born-van Karman periodic boundary condition. Inserting the above expansion of u_j into the equations of motion, we have

$$-m\omega^2 Q(k,\omega) = \left[-2(K+G)+2K\cos(ka)+2G\cos(2ka)\right]Q(k,\omega).$$

Thus, the allowed values of ω are given by

$$\omega = \pm\omega_a(k),$$

where

$$\begin{aligned}\omega_a(k) &= (2K/m)^{1/2}\Big\{1-\cos(ka)+(G/K)[1-\cos(2ka)\Big\}^{1/2}\\ &= (4K/m)^{1/2}|\sin(ka/2)|\left[1+(4G/K)\cos^2(ka/2)\right]^{1/2}.\end{aligned}$$

It is seen that the presence of the next nearest-neighbor interaction leads to the appearance of the factor

$$\left[1+(4G/K)\cos^2(ka/2)\right]^{1/2}$$

in the dispersion relation which renders the normal-mode frequencies renormalized. The above dispersion is plotted in Fig. 13.2 together with that in the absence of the next nearest-neighbor interaction ($G = 0$).

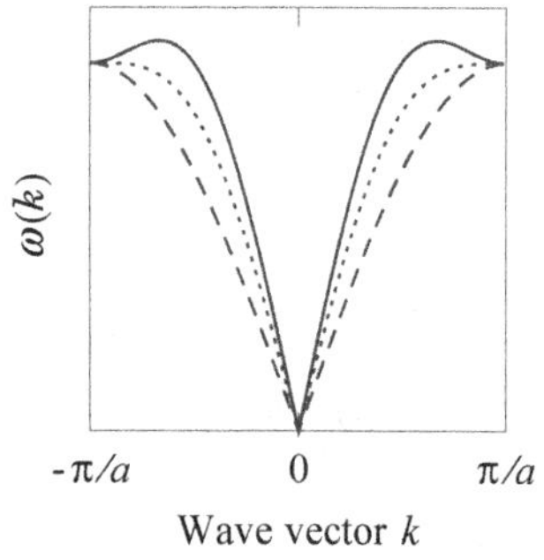

Fig. 13.2 Plots of dispersion relations for a linear chain of alternate positive and negative ions for $G = K/2$. The dotted line is for $G = K/4$ and the dashed line for $G = 0$.

(2) To find the condition on G for the dispersion curve to peak inside the first Brillouin zone, we examine the possible extremal points of $\omega_a^2(k)$. Differentiating $\omega_a^2(k)$ with respect to ka and then setting the result to zero, we have

$$\sin(ka)\left[1+(4G/K)\cos(ka)\right] = 0.$$

The solutions from $\sin(ka) = 0$ can only give the peak in $\omega_a(k)$ at the boundaries of the first Brillouin zone. For the factor $1+(4G/K)\cos(ka)$ to be possibly equal to zero so that $\omega_a(k)$ peaks inside the first Brillouin zone, we must have $4G/K > 1$. This is the condition for $\omega_a(k)$ to peak inside the first Brillouin zone. The peak position is given by $\cos(k_0a) = -K/4G$, where k_0 is the peak position.

(3) The group and phase velocities are given by

$$v_g = \frac{\mathrm{d}\omega_a(k)}{\mathrm{d}k} = \frac{\mathrm{sgn}(k)(Ka^2/m)^{1/2}\cos(ka/2)}{\left[1+(4G/K)\cos^2(ka/2)\right]^{1/2}}\left[1+(4G/K)\cos(ka)\right],$$

$$v_p = \frac{\omega_a(k)}{|k|} = (4K/m)^{1/2}\frac{\sin(ka/2)}{k}\left[1+(4G/K)\cos^2(ka/2)\right]^{1/2}.$$

Since $\mathrm{d}\omega_a(k)/\mathrm{d}k = 0$ at the peak position of $\omega_a(k)$, $v_g(k_0) = 0$. This can be clearly seen from the expression of $\omega_a(k)$ and the condition for $\omega_a(k)$ to peak inside the first Brillouin zone, $1 + (4G/K)\cos(ka) = 0$. The phase velocity at k_0 is given by

$$v_p(k_0) = (Ka^2/m)^{1/2}\frac{1+(K^2+16G^2)/8KG}{\pi - \tan^{-1}[16G^2/K^2-1]^{1/2}}.$$

13-4 Linear chain of atoms with damping. Consider a linear chain of atoms of mass m with the nearest neighboring atoms connected by springs of force constant K. In addition, the motion of each atom is damped, with the damping force $-\Gamma\dot{u}_j$ exerted on the jth atom, where u_j is the displacement of the jth atom from its equilibrium position. Assume that $\Gamma \ll (mK)^{1/2}$.

(1) Write down the equations of motion of atoms with the damping taken into account.

(2) Find the dispersion relation ω_k.

(3) Find the relaxation time of the normal modes.

---

(1) The equations of motion can be obtained from those in the absence of the damping force just by including the term $-\Gamma\dot{u}_j$ for the damping force. We have

$$m\ddot{u}_j = -K(2u_j - u_{j+1} - u_{j-1}) - \Gamma\dot{u}_j.$$

(2) Substituting the Fourier expansion

$$u_j = \sum_{k\omega} Q(k,\omega)\mathrm{e}^{\mathrm{i}(kR_j-\omega t)}$$

with $k_n = 2n\pi/Na$ ($n = 0, \pm1, \pm2, \cdots, \pm(N/2-1), N/2$) into the equations of motion, we obtain

$$-m\omega^2 = -4K\sin^2(ka/2) + \mathrm{i}\Gamma\omega = -\omega_{k0}^2 + \mathrm{i}\Gamma\omega$$

with $\omega_{k0} = (4K/m)^{1/2}|\sin(ka)|$. Solving for ω, we obtain the allowed values of ω

$$\omega = \omega_\pm = -\mathrm{i}\Gamma/2m \pm \left[\omega_{k0}^2 - (\Gamma/2m)^2\right]^{1/2}.$$

The real part of the dispersion relation must be nonnegative. We thus have

$$\omega_k = -\mathrm{i}\Gamma/2m + \left[\omega_{k0}^2 - (\Gamma/2m)^2\right]^{1/2}.$$

Note that ω_k has an imaginary part.

(3) From the time dependence of the atomic displacements due to a normal mode, $\mathrm{e}^{-\mathrm{i}\omega_k t}$, we see that the relaxation time of a normal mode arises from the imaginary part of ω_k. Writing $\omega_k = \omega_k' + \mathrm{i}\omega_k''$ with ω_k' the real part and ω_k'' the imaginary part of ω_k, we have

$$\mathrm{e}^{-\mathrm{i}\omega_k t} = \mathrm{e}^{-(-\omega_k'')t}\mathrm{e}^{-\mathrm{i}\omega_k' t}.$$

The first exponential factor implies the decay of the oscillation amplitude. Note that ω_k'' must be nonpositive to be physically acceptable. Thus, the relaxation time of a normal mode is given by

$$\tau_k = -\frac{1}{\omega_k''}.$$

For $\omega_{k0} > \Gamma/2m$, we have $\omega_k' = \left[\omega_{k0}^2 - (\Gamma/2m)^2\right]^{1/2}$ and $\omega_k'' = -\Gamma/2m$. We see that the frequencies of normal modes are reduced and the normal modes are said to be damped. The relaxation time is given by

$$\tau_k = \frac{2m}{\Gamma}.$$

For $\omega_{k0} < \Gamma/2m$, we have $\omega_k' = 0$ and $\omega_k'' = -\Gamma/2m + \left[(\Gamma/2m)^2 - \omega_{k0}^2\right]^{1/2}$. Since $\omega_k' = 0$, the normal modes are said to be overdamped. The relaxation time in this case is given by

$$\tau_k = \frac{1}{\Gamma/2m - \left[(\Gamma/2m)^2 - \omega_{k0}^2\right]^{1/2}}.$$

13-5 Polarizable molecules with internal degrees of freedom. Consider a linear chain of polarizable molecules with the nearest-neighbor equilibrium distance a. All the molecules are fixed to their positions. However, each molecule has an internal degree of freedom that obeys the equation of motion $\partial^2 p/\partial t^2 = -\omega_0^2 p + E\alpha\omega_0^2$, where p is the electric dipole moment of the molecule (assumed to be parallel to the chain), E the local electric field due to all other molecules, and α the polarizability. Find the dispersion relation $\omega(k)$ for small amplitude polarization waves (optical normal modes). Discuss the dependence of $\omega(0)$ on α.

Since the electric dipole moments on molecules are parallel to the chain, the electric field felt by the ith molecule due to electric dipoles on all other molecules is given by

$$E_i = \frac{1}{4\pi\epsilon_0}\sum_{j(\neq i)}\frac{2p_j}{|R_j - R_i|^3}.$$

The equation of motion for the ith molecule is then given by

$$\frac{\partial^2 p_i}{\partial t^2} = -\omega_0^2 p_i + \frac{2\alpha\omega_0^2}{4\pi\epsilon_0}\sum_{j(\neq i)}\frac{p_j}{|R_j - R_i|^3}.$$

To solve the above equations of motion for all molecules, we make a Fourier transformation to p_i with respect to R_i and t and have

$$p_i = \sum_{k\omega} p(k,\omega)\mathrm{e}^{\mathrm{i}(kR_i - \omega t)}$$

with the allowed values of k given by $k_n = 2\pi n/Na$ ($n = 0, \pm 1, \pm 2, \cdots, \pm(N/2-1), N/2$) as before for this linear chain of N molecules. Inserting the above expansion into the equations of motion yields

$$\sum_{k\omega} p(k,\omega)\left[(\omega^2 - \omega_0^2)\,\mathrm{e}^{\mathrm{i}kR_i} + \frac{2\alpha\omega_0^2}{4\pi\epsilon_0}\sum_{j(\neq i)}\frac{\mathrm{e}^{\mathrm{i}kR_j}}{|R_j - R_i|^3}\right]\mathrm{e}^{-\mathrm{i}\omega t} = 0.$$

Because the basis functions $\mathrm{e}^{-\mathrm{i}\omega t}$'s appear as overall factors and since they are linearly independent, the prefactor of each $\mathrm{e}^{-\mathrm{i}\omega t}$ must vanish for the above equation to hold. We thus have

$$\sum_{k} p(k,\omega)\left[(\omega^2 - \omega_0^2)\,\mathrm{e}^{\mathrm{i}kR_i} + \frac{2\alpha\omega_0^2}{4\pi\epsilon_0}\sum_{j(\neq i)}\frac{\mathrm{e}^{\mathrm{i}kR_j}}{|R_j - R_i|^3}\right] = 0.$$

To remove the summation over k in the above equation, we multiply both sides with $\mathrm{e}^{-\mathrm{i}k'R_i}/N$ and then sum over i. We have

$$\begin{aligned}0 &= \sum_{k} p(k,\omega)\left[(\omega^2 - \omega_0^2)\,\delta_{kk'} + \frac{2\alpha\omega_0^2}{4\pi\epsilon_0}\frac{1}{N}\sum_{i}\sum_{j(\neq i)}\frac{\mathrm{e}^{\mathrm{i}(kR_j - k'R_i)}}{((4\pi\epsilon_0))^3}\right] \\ &= (\omega^2 - \omega_0^2)\,p(k',\omega) + \frac{2\alpha\omega_0^2}{4\pi\epsilon_0}\sum_{k} p(k,\omega)\frac{1}{N}\sum_{i}\sum_{j(\neq i)}\frac{\mathrm{e}^{\mathrm{i}(kR_j - k'R_i)}}{|R_j - R_i|^3}.\end{aligned}$$

Rearranging the term $k'R_i$ on the exponential in the second term into $(k'-k)R_i$ by subtracting and then adding kR_i, we have

$$(\omega^2 - \omega_0^2)\,p(k',\omega) + \frac{2\alpha\omega_0^2}{4\pi\epsilon_0}\sum_{k} p(k,\omega)\frac{1}{N}\sum_{i}\sum_{j(\neq i)}\frac{\mathrm{e}^{\mathrm{i}[k(R_j - R_i) - (k'-k)R_i]}}{|R_j - R_i|^3} = 0.$$

The summation over $j(\neq i)$ can be now rewritten as a summation over $\ell(\neq 0)$ with $R_\ell = R_j - R_i$ because the summand depends on j only through $R_j - R_i$. Then the summation over i and k can be performed

$$0 = (\omega^2 - \omega_0^2)\, p(k', \omega) + \frac{2\alpha\omega_0^2}{4\pi\epsilon_0} \sum_k p(k, \omega) \sum_{\ell(\neq 0)} \frac{e^{ikR_\ell}}{|R_\ell|^3} \left[\frac{1}{N} \sum_i e^{-i(k'-k)R_i} \right]$$

$$= \left[(\omega^2 - \omega_0^2) + \frac{2\alpha\omega_0^2}{4\pi\epsilon_0} \sum_{\ell(\neq 0)} \frac{e^{ik'R_\ell}}{|R_\ell|^3} \right] p(k', \omega).$$

Changing k' to k, we have

$$\omega^2 = \omega_0^2 \left[1 - \frac{2\alpha}{4\pi\epsilon_0} \sum_{\ell(\neq 0)} \frac{e^{ik'R_\ell}}{R_\ell^3} \right] = \omega_0^2 \left[1 - \frac{4\alpha}{4\pi\epsilon_0} \sum_{\ell>0} \frac{\cos(kR_\ell)}{R_\ell^3} \right].$$

Thus, the allowed frequencies are given by

$$\omega = \pm\omega(k),$$

where ω_k is the dispersion relation for polarization waves

$$\omega(k) = \omega_0 \left[1 - \frac{4\alpha}{4\pi\epsilon_0} \sum_{\ell>0} \frac{\cos(kR_\ell)}{R_\ell^3} \right]^{1/2}.$$

For $k = 0$, we have

$$\omega(0) = \omega_0 \left(1 - \frac{4\alpha}{4\pi\epsilon_0} \sum_{\ell>0} \frac{1}{R_\ell^3} \right)^{1/2} \approx \omega_0 \left[1 - \frac{4.808\alpha}{(4\pi\epsilon_0)a^3} \right]^{1/2},$$

where we have made use of $R_\ell = \ell a$ and $\sum_{n=1}^{\infty} n^{-3} = \zeta(3) \approx 1.202$.

From the above expression, we see that $\omega(0)$ depends on α in a square-root fashion. If $\alpha > 0.208(4\pi\epsilon_0)a^3$, $\omega(0)$ then becomes purely imaginary, which implies that the polarization waves of $k = 0$ are over-damped for $\alpha > 0.208(4\pi\epsilon_0)a^3$. The plot of $\omega(0)/\omega_0$ as a function of $\alpha/(4\pi\epsilon_0)a^3$ is given in Fig. 13.3 for $\alpha < 0.208(4\pi\epsilon_0)a^3$.

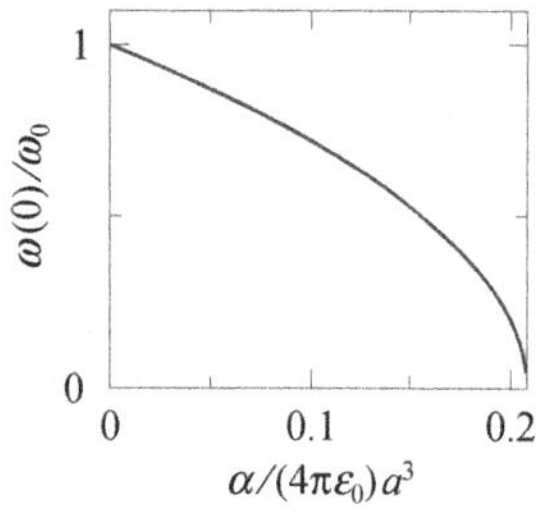

Fig. 13.3 Plot of $\omega(0)/\omega_0$ as a function of $\alpha/(4\pi\epsilon_0)a^3$ for $\alpha < 0.208(4\pi\epsilon_0)a^3$.

13-6 Triatomic linear chain. A triatomic linear chain consists of three different types of atoms of masses m_1, m_2, and m_3, respectively. As usual, it is assumed that only nearest-neighboring atoms interact and the interactions are modeled as mediated through springs of force constants α between atoms of types 2 and 3, β between atoms of types 3 and 1, and γ between atoms of types 1 and 2. Derive an equation that determines the frequencies of normal modes and describe the properties of solutions.

The lattice potential energy is given by

$$\Phi = \frac{1}{2}\gamma\sum_j (u_{j1} - u_{j2})^2 + \frac{1}{2}\alpha\sum_j (u_{j2} - u_3)^2 + \frac{1}{2}\beta\sum_j (u_{j3} - u_{j+1,1})^2.$$

The forces acting on the three atoms in the jth primitive cell are given by

$$\begin{aligned} F_{j1} &= -\frac{\partial\Phi}{\partial u_{j1}} = -(\beta+\gamma)u_{j1} + \gamma u_{j2} + \beta u_{j-1,3}, \\ F_{j2} &= -\frac{\partial\Phi}{\partial u_{j2}} = -(\gamma+\alpha)u_{j2} + \gamma u_{j1} + \alpha u_{j3}, \\ F_{j3} &= -\frac{\partial\Phi}{\partial u_{j3}} = -(\alpha+\beta)u_{j3} + \alpha u_{j2} + \beta u_{j+1,1}. \end{aligned}$$

From Newton's second law, the equations of motions are given by

$$\begin{aligned} m_1\ddot{u}_{j1} &= -(\beta+\gamma)u_{j1} + \gamma u_{j2} + \beta u_{j-1,3}, \\ m_2\ddot{u}_{j2} &= -(\gamma+\alpha)u_{j2} + \gamma u_{j1} + \alpha u_{j3}, \\ m_3\ddot{u}_{j3} &= -(\alpha+\beta)u_{j3} + \alpha u_{j2} + \beta u_{j+1,1}. \end{aligned}$$

To solve the above equations of motion for all atoms, we make a Fourier transformation to $u_{j\nu}$ with respect to R_j and t and have

$$u_{j\nu} = \sum_{k\omega} Q_\nu(k,\omega)\mathrm{e}^{\mathrm{i}(kR_j-\omega t)}$$

with the allowed values of k given by $k_n = 2\pi n/Na$ ($n = 0, \pm 1, \pm 2, \cdots, \pm(N/2-1), N/2$) for a triatomic linear chain of N primitive cells ($3N$ atoms). Inserting the above expansion into the equations of motion yields

$$\begin{aligned} &(\beta+\gamma-m_1\omega^2)Q_1(k,\omega) - \gamma Q_2(k,\omega) - \beta\mathrm{e}^{-\mathrm{i}ka}Q_3(k,\omega) = 0, \\ &-\gamma Q_1(k,\omega) + (\gamma+\alpha-m_2\omega^2)Q_2(k,\omega) - \alpha Q_3(k,\omega) = 0, \\ &-\beta\mathrm{e}^{\mathrm{i}ka}Q_1(k,\omega) - \alpha Q_2(k,\omega) + (\alpha+\beta-m_3\omega^2)Q_3(k,\omega) = 0. \end{aligned}$$

The above equations are homogeneous linear equations for $Q_1(k,\omega)$, $Q_2(k,\omega)$, and $Q_3(k,\omega)$. The necessary and sufficient condition for the existence of nontrivial solutions leads to an equation that determines the allowed values of ω

$$\begin{vmatrix} \beta+\gamma-m_1\omega^2 & -\gamma & -\beta e^{-ika} \\ -\gamma & \gamma+\alpha-m_2\omega^2 & -\alpha \\ -\beta e^{ika} & -\alpha & \alpha+\beta-m_3\omega^2 \end{vmatrix} = 0.$$

Evaluating the determinant on the left hand side, we have

$$m_1 m_2 m_3\,\omega^6 - \left[(\alpha+\beta)m_1m_2 + (\beta+\gamma)m_2m_3 + (\gamma+\alpha)m_3m_1\right]\omega^4 + (\alpha\beta+\beta\gamma+\gamma\alpha)(m_1+m_2+m_3)\omega^2 - 4\alpha\beta\gamma\sin^2(ka/2) = 0.$$

Solving the above equation, we will obtain six allowed values of ω and three dispersion relations for three branches of normal modes if no degeneracy occurs. One branch of normal modes is acoustical and the other two branches are optical.

13-7 Two-dimensional crystal with a square Bravais lattice. Consider a two-dimensional crystal with a square Bravais lattice. With only interactions between the nearest and next nearest neighbors taken into account, the harmonic lattice potential energy of the crystal is given by

$$\Phi = \frac{1}{2a^2}K\sum_{\langle ij\rangle}\left[(\boldsymbol{R}_i-\boldsymbol{R}_j)\cdot(\boldsymbol{u}_i-\boldsymbol{u}_j)\right]^2 + \frac{1}{4a^2}G\sum_{(ij)}\left[(\boldsymbol{R}_i-\boldsymbol{R}_j)\cdot(\boldsymbol{u}_i-\boldsymbol{u}_j)\right]^2,$$

where $\langle ij\rangle$ indicates the summation over the nearest neighbors and (ij) the summation over the next nearest neighbors. Here a is the lattice constant, $\boldsymbol{R}_i$'s are Bravais lattice vectors, and $\boldsymbol{u}_i$'s are deviations of atoms from their equilibrium positions.

(1) Construct the dynamical matrix.

(2) Find the frequencies and polarization vectors of normal modes along the lines Δ, Σ, and Z, respectively.

(3) Plot the dispersion relations along these three high-symmetry lines.

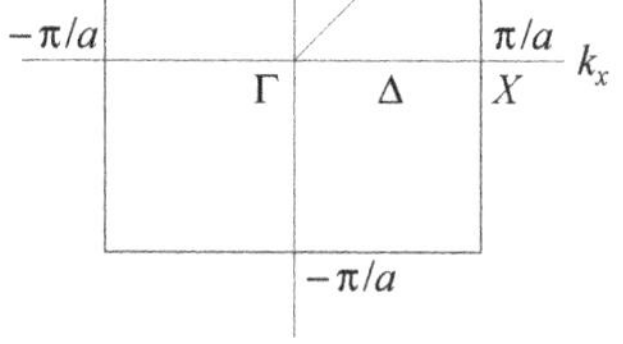

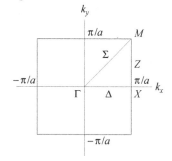

Fig. 13-4 First Brillouin zone of the two-dimensional square lattice.

---

(1) We rewrite the components of the displacement of the jth atom u_{jx} and u_{jy} as x_j and y_j. We also use (m, n) to label a lattice site instead of a single index j. The nearest-neighbor sites of site (m, n) are sites $(m+1, n)$, $(m-1, n)$, $(m, n+1)$, and $(m, n-1)$. The next nearest-neighbor sites of site (m, n) are sites $(m-1, n-1)$, $(m-1, n+1)$, $(m+1, n-1)$, and $(m+1, n+1)$. The harmonic lattice potential energy is then reexpressed as

$$\begin{aligned}\Phi &= \frac{1}{2}K\sum_{mn}\left[(x_{m+1,n} - x_{mn})^2 + (y_{m,n+1} - y_{mn})^2\right] \\ &\quad + \frac{1}{2}G\sum_{mn}\left[(x_{m+1,n-1} - x_{mn} - y_{m+1,n-1} + y_{mn})^2\right. \\ &\qquad\qquad \left. + (x_{m+1,n+1} - x_{mn} + y_{m+1,n+1} - y_{mn})^2\right].\end{aligned}$$

To find the dynamical matrix, we first compute $D_{\alpha\beta}(\boldsymbol{R})$. In our present notations, $\boldsymbol{R}$ is denoted by mn, $\boldsymbol{R} = ma\boldsymbol{e}_x + na\boldsymbol{e}_y$. $D_{\alpha\beta}(\boldsymbol{R} - \boldsymbol{R}')$ is now denoted by $D_{mn\alpha,m'n'\beta}$ which is given by the second-order derivative of Φ

$$D_{mn\alpha,m'n'\beta} = \frac{\partial^2\Phi}{\partial u_{mn,\alpha}\partial u_{m'n',\beta}}.$$

Differentiating Φ with respect to $x_{m'n'}$, we obtain

$$\begin{aligned}\frac{\partial\Phi}{\partial x_{m'n'}} &= K\sum_{mn}(x_{m+1,n} - x_{mn})(\delta_{m+1,m'}\delta_{nn'} - \delta_{m,m'}\delta_{n,n'}) \\ &\quad + G\sum_{mn}\left[(x_{m+1,n-1} - x_{mn} - y_{m+1,n-1} + y_{mn})\right. \\ &\qquad\qquad \times (\delta_{m+1,m'}\delta_{n-1,n'} - \delta_{m,m'}\delta_{n,n'}) \\ &\qquad\qquad + (x_{m+1,n+1} - x_{mn} + y_{m+1,n+1} - y_{mn}) \\ &\qquad\qquad \left.\times (\delta_{m+1,m'}\delta_{n+1,n'} - \delta_{m,m'}\delta_{n,n'})\right] \\ &= K\big(2x_{m',n'} - x_{m'-1,n'} - x_{m'+1,n'}\big) \\ &\quad + G\big(4x_{m'n'} - x_{m'-1,n'-1} - x_{m'-1,n'+1} \\ &\qquad - x_{m'+1,n'-1} - x_{m'+1,n'+1} - y_{m'-1,n'-1} \\ &\qquad + y_{m'-1,n'+1} + y_{m'+1,n'-1} - y_{m'+1,n'+1}\big),\end{aligned}$$

where we have made use of $\partial x_{mn}/\partial x_{m'n'} = \delta_{m,m'}\delta_{n,n'}$. In the second equality, the summation over m and n have been performed by consuming the Kronecker δ-symbols resulting from the derivatives. Differentiating the above result with respect to x_{mn} and y_{mn}, we obtain

$$\begin{aligned}D_{mnx,m'n'x} &= \frac{\partial^2\Phi}{\partial x_{mn}\partial x_{m'n'}}\\ &= K(2\delta_{mm'}\delta_{nn'} - \delta_{m,m'-1}\delta_{nn'} - \delta_{m,m'+1}\delta_{nn'})\\ &+ G(4\delta_{mm'}\delta_{nn'} - \delta_{m,m'-1}\delta_{n,n'-1} - \delta_{m,m'-1}\delta_{n,n'+1}\\ &- \delta_{m,m'+1}\delta_{n,n'-1} - \delta_{m,m'+1}\delta_{n,n'+1}),\end{aligned}$$

and

$$\begin{aligned}D_{mny,m'n'x} &= \frac{\partial^2\Phi}{\partial y_{mn}\partial x_{m'n'}}\\ &= G(-\delta_{m,m'-1}\delta_{n,n'-1} + \delta_{m,m'-1}\delta_{n,n'+1}\\ &\quad + \delta_{m,m'+1}\delta_{n,n'-1} - \delta_{m,m'+1}\delta_{n,n'+1}).\end{aligned}$$

From $D_{mny,m'n'x}$ and the properties of D, we have

$$\begin{aligned}D_{mnx,m'n'y} &= G(-\delta_{m,m'+1}\delta_{n,n'+1} + \delta_{m,m'+1}\delta_{n,n'-1}\\ &\quad + \delta_{m,m'-1}\delta_{n,n'+1} - \delta_{m,m'-1}\delta_{n,n'-1}).\end{aligned}$$

Differentiating Φ with respect to $y_{m'n'}$, we obtain

$$\begin{aligned}\frac{\partial\Phi}{\partial y_{m'n'}} &= K\sum_{mn}(y_{m,n+1} - y_{mn})(\delta_{mm'}\delta_{n+1,n'} - \delta_{mm'}\delta_{nn'})\\ &\quad + G\sum_{mn}\big[(x_{m+1,n-1} - x_{mn} - y_{m+1,n-1} + y_{mn})\\ &\qquad \times(-\delta_{m+1,m'}\delta_{n-1,n'} + \delta_{mm'}\delta_{nn'})\\ &\qquad + (x_{m+1,n+1} - x_{mn} + y_{m+1,n+1} - y_{mn})\\ &\qquad \times(\delta_{m+1,m'}\delta_{n+1,n'} - \delta_{mm'}\delta_{nn'})\big]\\ &= K\big(2y_{m',n'} - y_{m',n'-1} - y_{m',n'+1}\big)\\ &\quad + G\big(4y_{m'n'} - y_{m'-1,n'-1} - y_{m'-1,n'+1}\\ &\qquad - y_{m'+1,n'-1} - y_{m'+1,n'+1} - x_{m'-1,n'-1}\\ &\qquad + x_{m'-1,n'+1} + x_{m'+1,n'-1} - x_{m'+1,n'+1}\big).\end{aligned}$$

Differentiating the above result with respect to x_{mn} and y_{mn}, we obtain

$$\begin{aligned}D_{mnx,m'n'y} &= \frac{\partial^2\Phi}{\partial x_{mn}\partial y_{m'n'}}\\ &= G\big(-\delta_{m,m'-1}\delta_{n,n'-1} + \delta_{m,m'-1}\delta_{n,n'+1}\\ &\quad + \delta_{m,m'+1}\delta_{n,n'-1} - \delta_{m,m'+1}\delta_{n,n'+1}\big),\end{aligned}$$

and

$$\begin{aligned}
D_{mny,m'n'y} = {} & K\big(2\delta_{mm'}\delta_{nn'} - \delta_{mm'}\delta_{n,n'-1} - \delta_{mm'}\delta_{n,n'+1}\big) \\
& + G\big(4\delta_{mm'}\delta_{nn'} - \delta_{m,m'-1}\delta_{n,n'-1} \\
& - \delta_{m,m'-1}\delta_{n,n'+1} - \delta_{m,m'+1}\delta_{n,n'-1} \\
& - \delta_{m,m'+1}\delta_{n,n'+1}\big).
\end{aligned}$$

The result for $D_{mnx,m'n'y}$ is identical with that obtained above for $D_{mny,m'n'x}$. We now compute the elements of the dynamical matrix using its definition. Note that the summation in the definition of the dynamical matrix can be performed by using the Kronecker δ-symbols resulting from the second-order derivatives. For $\mathscr{D}_{11}(\boldsymbol{k})$, we have

$$\begin{aligned}
M\mathscr{D}_{11}(\boldsymbol{k}) &= \sum_{\boldsymbol{R}_{mn}-\boldsymbol{R}_{m'n'}} D_{mnx,m'n'x}\mathrm{e}^{-\mathrm{i}\boldsymbol{k}\cdot(\boldsymbol{R}_{mn}-\boldsymbol{R}_{m'n'})} \\
&= \sum_{\boldsymbol{R}_{mn}-\boldsymbol{R}_{m'n'}} \big[K(2\delta_{mm'}\delta_{nn'} - \delta_{m,m'-1}\delta_{nn'} - \delta_{m,m'+1}\delta_{nn'}) \\
&\quad + G(4\delta_{mm'}\delta_{nn'} - \delta_{m,m'-1}\delta_{n,n'-1} - \delta_{m,m'-1}\delta_{n,n'+1} \\
&\quad - \delta_{m,m'+1}\delta_{n,n'-1} - \delta_{m,m'+1}\delta_{n,n'+1})\big]\mathrm{e}^{-\mathrm{i}\boldsymbol{k}\cdot(\boldsymbol{R}_{mn}-\boldsymbol{R}_{m'n'})} \\
&= K\big(2 - \mathrm{e}^{-\mathrm{i}k_x a} - \mathrm{e}^{\mathrm{i}k_x a}\big) + G\big[4 - \mathrm{e}^{-\mathrm{i}(k_x a+k_y a)} \\
&\quad - \mathrm{e}^{-\mathrm{i}(k_x a-k_y a)} - \mathrm{e}^{\mathrm{i}(k_x a-k_y a)} - \mathrm{e}^{\mathrm{i}(k_x a+k_y a)}\big] \\
&= 2K\big[1 - \cos(k_x a)\big] + 4G\big[1 - \cos(k_x a)\cos(k_y a)\big].
\end{aligned}$$

Note that trigonometric functions appear in an element of the dynamical matrix because only a small number of terms make nonzero contributions to the dynamical matrix due to the short range of the interaction between atoms. For $\mathscr{D}_{12}(\boldsymbol{k})$ and $\mathscr{D}_{21}(\boldsymbol{k})$, we have

$$\begin{aligned}
M\mathscr{D}_{21}(\boldsymbol{k}) = M\mathscr{D}_{12}(\boldsymbol{k}) &= \sum_{\boldsymbol{R}_{mn}-\boldsymbol{R}_{m'n'}} D_{mnx,m'n'y}\mathrm{e}^{-\mathrm{i}\boldsymbol{k}\cdot(\boldsymbol{R}_{mn}-\boldsymbol{R}_{m'n'})} \\
&= G\sum_{\boldsymbol{R}_{mn}-\boldsymbol{R}_{m'n'}} \big(-\delta_{m,m'-1}\delta_{n,n'-1} + \delta_{m,m'-1}\delta_{n,n'+1} \\
&\quad + \delta_{m,m'+1}\delta_{n,n'-1} - \delta_{m,m'+1}\delta_{n,n'+1}\big)\mathrm{e}^{-\mathrm{i}\boldsymbol{k}\cdot(\boldsymbol{R}_{mn}-\boldsymbol{R}_{m'n'})} \\
&= G\big[-\mathrm{e}^{-\mathrm{i}(k_x a+k_y a)} + \mathrm{e}^{-\mathrm{i}(k_x a-k_y a)} - \text{c.c.}\big] \\
&= 4G\sin(k_x a)\sin(k_y a)
\end{aligned}$$

with c.c. standing for the complex conjugate of the expression in front of it. For $D_{22}(\boldsymbol{k})$, we have

$$\begin{aligned}
M\mathscr{D}_{22}(\boldsymbol{k}) &= \sum_{\boldsymbol{R}_{mn}-\boldsymbol{R}_{m'n'}} D_{mny,m'n'y}\mathrm{e}^{-\mathrm{i}\boldsymbol{k}\cdot(\boldsymbol{R}_{mn}-\boldsymbol{R}_{m'n'})} \\
&= \sum_{\boldsymbol{R}_{mn}-\boldsymbol{R}_{m'n'}} \big[K\big(2\delta_{mm'}\delta_{nn'} - \delta_{mm'}\delta_{n,n'-1} - \delta_{mm'}\delta_{n,n'+1}\big) \\
&\quad + G\big(4\delta_{mm'}\delta_{nn'} - \delta_{m,m'-1}\delta_{n,n'-1} - \delta_{m,m'-1}\delta_{n,n'+1} \\
&\quad - \delta_{m,m'+1}\delta_{n,n'-1} - \delta_{m,m'+1}\delta_{n,n'+1}\big)\big]\mathrm{e}^{-\mathrm{i}\boldsymbol{k}\cdot(\boldsymbol{R}_{mn}-\boldsymbol{R}_{m'n'})} \\
&= K\big(2 - \mathrm{e}^{-\mathrm{i}k_y a} - \mathrm{e}^{\mathrm{i}k_y a}\big) + G\big(4 - \mathrm{e}^{-\mathrm{i}(k_x a + k_y a)} \\
&\quad - \mathrm{e}^{\mathrm{i}(k_x a - k_y a)} - \mathrm{e}^{-\mathrm{i}(k_x a - k_y a)} - \mathrm{e}^{\mathrm{i}(k_x a + k_y a)}\big) \\
&= 2K\big[1 - \cos(k_y a)\big] + 4G\big[1 - \cos(k_x a)\cos(k_y a)\big].
\end{aligned}$$

Finally, the dynamical matrix is given by

$$\mathscr{D}(\boldsymbol{k}) = \frac{1}{M}\begin{pmatrix} \begin{matrix} 4G[1-\cos(k_x a)\cos(k_y a)] \\ +2K[1-\cos(k_x a)] \end{matrix} & 4G\sin(k_x a)\sin(k_y a) \\ 4G\sin(k_x a)\sin(k_y a) & \begin{matrix} 4G[1-\cos(k_x a)\cos(k_y a)] \\ +2K[1-\cos(k_y a)] \end{matrix} \end{pmatrix}.$$

Having obtained the dynamical matrix, we now turn to the study of normal modes along the three high-symmetry lines.

(2) *i. Along the* Δ *line*

Along the Δ line, $k_y = 0$. The dynamical matrix reduces to

$$\mathscr{D}^{(\Delta)}(\boldsymbol{k}) = \frac{1}{M}\begin{pmatrix} (2K+4G)[1-\cos(k_x a)] & 0 \\ 0 & 4G[1-\cos(k_x a)] \end{pmatrix}.$$

$\mathscr{D}^{(\Delta)}(\boldsymbol{k})$ is a diagonal matrix whose eigenvalues and corresponding eigenvectors are given by

$$\begin{aligned}
\lambda_1 &= (4K+8G)\sin^2(k_x a/2), & \lambda_2 &= 8G\sin^2(k_x a/2), \\
\epsilon_1 &= \begin{pmatrix} 1 \\ 0 \end{pmatrix}, & \epsilon_2 &= \begin{pmatrix} 0 \\ 1 \end{pmatrix}.
\end{aligned}$$

We thus have two branches of normal modes along the Δ line. The frequencies and polarization vectors of these normal modes are given by

$$\begin{aligned}
\omega_1 &= \big[(4K+8G)/M\big]^{1/2}\big|\sin(k_x a/2)\big|, & \boldsymbol{\epsilon}_1 &= \boldsymbol{e}_x, \\
\omega_2 &= (8G/M)^{1/2}\big|\sin(k_x a/2)\big|, & \boldsymbol{\epsilon}_2 &= \boldsymbol{e}_y.
\end{aligned}$$

ii. Along the Z *line*

Along the Z line, $k_x = \pi/a$. The dynamical matrix reduces to

$$\mathscr{D}^{(\mathrm{Z})}(\boldsymbol{k}) = \frac{1}{M}\begin{pmatrix} 4K + 4G(1+\cos(k_y a)) & 0 \\ 0 & \begin{matrix} 2K(1-\cos(k_y a)) \\ +4G(1+\cos(k_y a)) \end{matrix} \end{pmatrix}.$$

The frequencies and corresponding polarization vectors are given by

$$\omega_1 = \left\{\frac{1}{M}\left[4K + 8G\cos^2\left(\frac{k_y a}{2}\right)\right]\right\}^{1/2}, \qquad \epsilon_1 = \boldsymbol{e}_x,$$

$$\omega_2 = \left\{\frac{1}{M}\left[4K\sin^2\left(\frac{k_y a}{2}\right) + 8G\cos^2\left(\frac{k_y a}{2}\right)\right]\right\}^{1/2}, \ \epsilon_2 = \boldsymbol{e}_y.$$

iii. Along the Σ line

Along the Σ line, $k_x = k_y = k$. $D(\boldsymbol{k})$ reduces to

$$\mathscr{D}^{(\Sigma)}(\boldsymbol{k}) = \frac{1}{M}\begin{pmatrix} \begin{matrix} 2K[1-\cos(ka)] \\ +4G[1-\cos(ka)^2] \end{matrix} & 4G\sin^2(ka) \\ 4G\sin^2(ka) & \begin{matrix} 2K[1-\cos(ka)] \\ +4G[1-\cos^2(ka)] \end{matrix} \end{pmatrix}.$$

The frequencies and corresponding polarization vectors are given by

$$\omega_1 = \left\{\frac{1}{M}\left[4K + 32G\cos^2\left(\frac{ka}{2}\right)\right]\right\}^{1/2}\left|\sin\left(\frac{ka}{2}\right)\right|,$$

$$\epsilon_1 = \frac{1}{2^{1/2}}(\boldsymbol{e}_x + \boldsymbol{e}_y),$$

$$\omega_2 = \left(\frac{4K}{M}\right)^{1/2}\left|\sin\left(\frac{ka}{2}\right)\right|,$$

$$\epsilon_2 = \frac{1}{2^{1/2}}(\boldsymbol{e}_x - \boldsymbol{e}_y).$$

(3) The dispersion curves of normal modes along three high symmetry lines are given in Fig. 13.5.

13-8 Simple cubic crystal. The lattice dynamics of a simple cubic crystal of lattice constant a and atom mass M is studied here with only interactions between nearest-neighboring atoms taken into account. The interactions are modeled as being mediated through springs of force constant γ.

(1) Write down the potential energy of the crystal and construct the dynamical matrix.

(2) Solve for the dispersion relations.

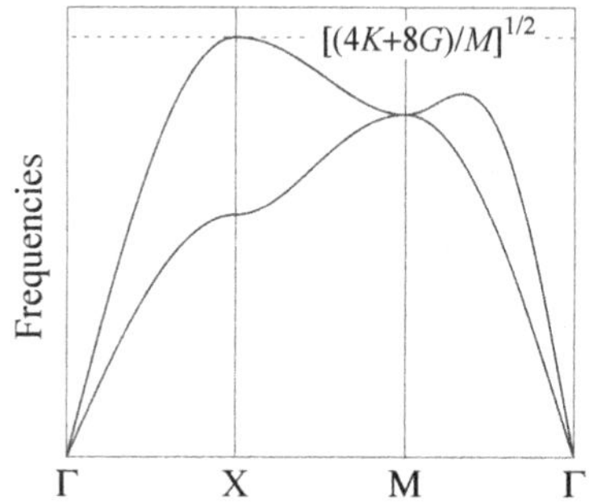

Fig. 13.5 Dispersion curves of normal modes along three high-symmetry lines in a two-dimensional crystal with a square lattice. The value 0.25 for the ratio G/K of the force constants is used here.

(3) Plot the dispersion relations along [100] and label branches properly. Give the physical reason for the zero-frequency normal-mode branches.

(4) Indicate the pattern of displacements of eight atoms in the conventional cell for the mode at $\boldsymbol{k} = (\pi/a, 0, \pi/a)$ with displacements along $\pm\boldsymbol{e}_x$.

---

(1) Let a lattice site be labeled by (ℓmn). The displacement of the atom whose equilibrium position is $\boldsymbol{R}_{\ell mn}$ is denoted by $\boldsymbol{u}_{\ell mn} = x_{\ell mn}\boldsymbol{e}_x + y_{\ell mn}\boldsymbol{e}_y + z_{\ell mn}\boldsymbol{e}_z$. Since the displacements of atoms are small, we take into account only the components of the displacements along the lines connecting nearest-neighboring atoms. This is equivalent to project the displacements along the lines connecting nearest-neighboring atoms, $(\boldsymbol{R}_{\ell mn} - \boldsymbol{R}_{\ell' m' n'})\cdot(\boldsymbol{u}_{\ell mn} - \boldsymbol{u}_{\ell' m' n'})/|\boldsymbol{R}_{\ell mn} - \boldsymbol{R}_{\ell' m' n'}|$. The lattice potential energy is then given by

$$\Phi = \frac{\gamma}{2}\sum_{\ell mn}\Big[\big(x_{\ell mn} - x_{\ell+1,mn}\big)^2 + \big(y_{\ell mn} - y_{\ell,m+1,n}\big)^2 + \big(z_{\ell mn} - z_{\ell m,n+1}\big)^2\Big].$$

Since there are no cross terms between different components of displacements. The $D(\boldsymbol{R})$ matrix and the dynamical matrix are diagonal. We find the $D(\boldsymbol{R})$ matrix by differentiating Φ with respect to displacements. Differentiating Φ with respect to $x_{\ell' m' n'}$, we have

$$\begin{aligned}\frac{\partial \Phi}{\partial x_{\ell' m' n'}} &= \gamma \sum_{\ell mn} (x_{\ell mn} - x_{\ell+1,mn})(\delta_{\ell'\ell}\delta_{m'm}\delta_{n'n} \\ &\qquad -\delta_{\ell',\ell+1}\delta_{m'm}\delta_{n'n}) \\ &= \gamma\big(2x_{\ell' m' n'} - x_{\ell'+1,m'n'} - x_{\ell'-1,m'n'}\big).\end{aligned}$$

Differentiating the above result with respect to $x_{\ell mn}$, $y_{\ell mn}$, and $z_{\ell mn}$, we have

$$\begin{aligned}&\frac{\partial^2 \Phi}{\partial x_{\ell mn}\partial x_{\ell' m' n'}} \\ &\quad = \gamma\big(2\delta_{\ell\ell'}\delta_{mm'}\delta_{nn'} - \delta_{\ell,\ell'+1}\delta_{mm'}\delta_{nn'} - \delta_{\ell,\ell'-1}\delta_{mm'}\delta_{nn'}\big), \\ &\frac{\partial^2 \Phi}{\partial y_{\ell mn}\partial x_{\ell' m' n'}} = \frac{\partial^2 \Phi}{\partial z_{\ell mn}\partial x_{\ell' m' n'}} = 0.\end{aligned}$$

We similarly have

$$\begin{aligned}&\frac{\partial^2 \Phi}{\partial y_{\ell mn}\partial y_{\ell' m' n'}} \\ &\quad = \gamma\big(2\delta_{\ell\ell'}\delta_{mm'}\delta_{nn'} - \delta_{\ell,\ell'}\delta_{m,m'+1}\delta_{nn'} - \delta_{\ell,\ell'}\delta_{m,m'-1}\delta_{nn'}\big), \\ &\frac{\partial^2 \Phi}{\partial x_{\ell mn}\partial y_{\ell' m' n'}} = \frac{\partial^2 \Phi}{\partial z_{\ell mn}\partial y_{\ell' m' n'}} = 0,\end{aligned}$$

and

$$\begin{aligned}&\frac{\partial^2 \Phi}{\partial z_{\ell mn}\partial z_{\ell' m' n'}} \\ &\quad = \gamma\big(2\delta_{\ell\ell'}\delta_{mm'}\delta_{nn'} - \delta_{\ell,\ell'}\delta_{mm'}\delta_{n,n'+1} - \delta_{\ell,\ell'}\delta_{mm'}\delta_{n,n'-1}\big), \\ &\frac{\partial^2 \Phi}{\partial x_{\ell mn}\partial z_{\ell' m' n'}} = \frac{\partial^2 \Phi}{\partial y_{\ell mn}\partial z_{\ell' m' n'}} = 0.\end{aligned}$$

From the definition of the dynamical matrix

$$\mathscr{D}_{\alpha\beta}(\boldsymbol{k}) = \frac{1}{M} \sum_{\boldsymbol{R}_{\ell mn} - \boldsymbol{R}_{\ell' m' n'}} D_{\alpha\beta}(\boldsymbol{R}_{\ell mn} - \boldsymbol{R}_{\ell' m' n'}) \mathrm{e}^{-\mathrm{i}\boldsymbol{k}\cdot(\boldsymbol{R}_{\ell mn} - \boldsymbol{R}_{\ell' m' n'})},$$

we have

$$\begin{aligned}\mathscr{D}_{11}(\boldsymbol{k}) &= \frac{1}{M}\gamma \sum_{\boldsymbol{R}_{\ell mn} - \boldsymbol{R}_{\ell' m' n'}} \big(2\delta_{\ell\ell'}\delta_{mm'}\delta_{nn'} - \delta_{\ell,\ell'+1}\delta_{mm'}\delta_{nn'} \\ &\qquad - \delta_{\ell,\ell'-1}\delta_{mm'}\delta_{nn'}\big)\mathrm{e}^{-\mathrm{i}\boldsymbol{k}\cdot(\boldsymbol{R}_{\ell mn} - \boldsymbol{R}_{\ell' m' n'})} \\ &= \frac{1}{M}\gamma(2 - \mathrm{e}^{\mathrm{i}k_x a} - \mathrm{e}^{-\mathrm{i}k_x a}) = \frac{4\gamma}{M}\sin^2(k_x a).\end{aligned}$$

We similarly have

$$\mathscr{D}_{22}(\boldsymbol{k}) = \frac{4\gamma}{M}\sin^2(k_y a),\ \mathscr{D}_{33}(\boldsymbol{k}) = \frac{4\gamma}{M}\sin^2(k_z a).$$

All other elements of the dynamical matrix are zero. Finally, the dynamical matrix is given by

$$\mathscr{D}(\boldsymbol{k}) = \frac{4\gamma}{M}\begin{pmatrix} \sin^2(k_x a/2) & 0 & 0 \\ 0 & \sin^2(k_y a/2) & 0 \\ 0 & 0 & \sin^2(k_z a/2) \end{pmatrix}.$$

(2) Since the dynamical matrix is diagonal, its eigenvalues and the corresponding eigenfunctions can be trivially written down. We have

$$\omega_1 = \left(\frac{4\gamma}{M}\right)^{1/2} |\sin(k_x a/2)|, \ \boldsymbol{\epsilon}^{(1)} = \begin{pmatrix} 1 \\ 0 \\ 0 \end{pmatrix},$$

$$\omega_2 = \left(\frac{4\gamma}{M}\right)^{1/2} |\sin(k_y a/2)|, \ \boldsymbol{\epsilon}^{(2)} = \begin{pmatrix} 0 \\ 1 \\ 0 \end{pmatrix},$$

$$\omega_3 = \left(\frac{4\gamma}{M}\right)^{1/2} |\sin(k_z a/2)|, \ \boldsymbol{\epsilon}^{(3)} = \begin{pmatrix} 0 \\ 0 \\ 1 \end{pmatrix}.$$

(3) Along [100], $k_x = k$ and $k_y = k_z = 0$. From the polarization vectors in Part (2), for $\boldsymbol{k}$ in the [100] direction we see that ω_1 is the dispersion relation for the longitudinal acoustical branch and that ω_2 and ω_3 are the dispersion relations of two transverse acoustical branches. We thus rename ω_1 as ω_L and ω_2 and ω_3 as ω_{T_1} and ω_{T_2}, respectively. The frequencies of normal modes are then given by

$$\omega_L = (4\gamma/M)^{1/2}|\sin(ka/2)|, \ \omega_{T_1} = \omega_{T_2} = 0.$$

For $\boldsymbol{k}$ in the [100] direction, only the springs parallel to [100] are stretched or compressed so that the displacements of atoms are only along this direction, which leads to the zero frequency for the two transverse branches of acoustical normal modes. The plots of the dispersion relations are given in Fig. 13.6.

(4) The pattern of displacements for $\boldsymbol{k} = (\pi/a, 0, \pi/a)$ can be inferred from the following expression for atomic displacements

$$u_{j\alpha}(t) = \frac{1}{\sqrt{NM}} \sum_{\boldsymbol{k}s} q_{\boldsymbol{k}s}(t)\epsilon_\alpha^{(s)}(\boldsymbol{k})\mathrm{e}^{\mathrm{i}\boldsymbol{k}\cdot\boldsymbol{R}_j}.$$

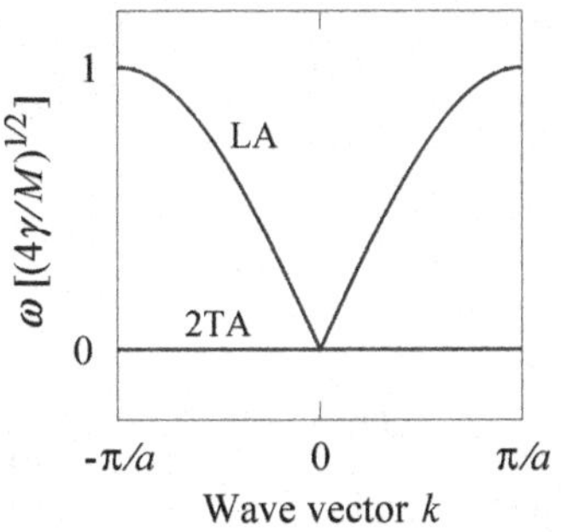

Fig. 13.6 Dispersion relations of normal modes for $\boldsymbol{k}$ in the [100] direction.

The coordinates of the eight atoms at the corners of the conventional cell are

$$(0,0,0)a,\ (1,0,0)a,\ (0,1,0)a,\ (0,0,1)a,$$
$$(1,1,0)a,\ (1,0,1)a,\ (0,1,1)a,\ (1,1,1)a.$$

Note that

$$q_1(t)\epsilon_x^{(1)} + q_2(t)\epsilon_x^{(2)} + q_3(t)\epsilon_x^{(3)} = q_1(t),$$

where we have suppressed the wave vector subscript on q's. We then have the following contributions of the normal mode $\boldsymbol{k} = (\pi/a, 0, \pi/a)$ to the displacements of atoms

$$u_{(0,0,0)x}(t) = \frac{q_1(t)}{\sqrt{NM}},$$
$$u_{(1,0,0)x}(t) = -u_{(0,1,0)x}(t) = u_{(0,0,1)x}(t) = -\frac{q_1(t)}{\sqrt{NM}},$$
$$u_{(1,1,0)x}(t) = -u_{(1,0,1)x}(t) = u_{(0,1,1)x}(t) = -\frac{q_1(t)}{\sqrt{NM}},$$
$$u_{(1,1,1)x}(t) = \frac{q_1(t)}{\sqrt{NM}}.$$

We see that, at any instants, four atoms move in the $+x$ direction and four atoms move in the $-x$ direction. The pattern of displacements is shown in Fig. 13.7 at an instant with $q_1(t) > 0$.

13-9 Three-dimensional monatomic crystal. Consider a three-dimensional monatomic Bravais lattice in which each ion of mass M interacts only with its nearest neighbors with the interaction potential energy given by $\phi(\boldsymbol{r}_i - \boldsymbol{r}_j) = K(|\boldsymbol{r}_i - \boldsymbol{r}_j| - d)^2/2$, where d is the equilibrium spacing between the atoms and K the force constant of the spring connecting the atoms.

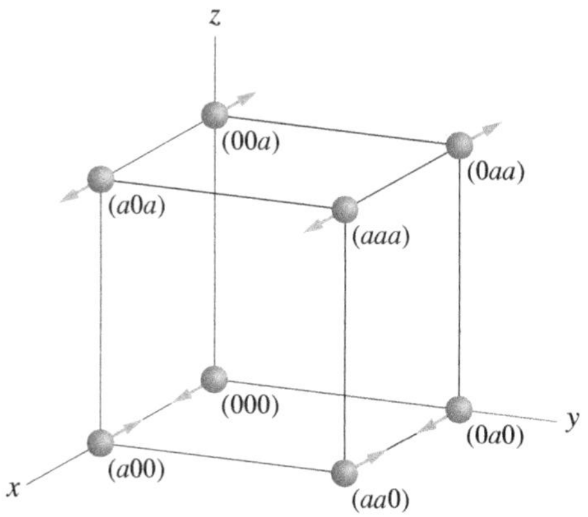

Fig. 13.7 Pattern of displacements of atoms in $\pm e_x$ at the instant with $q_1(t) > 0$ due to the normal mode of wave vector $\boldsymbol{k} = (\pi/a, 0, \pi/a)$.

(1) Show that the frequencies of the three normal modes at each wave vector $\boldsymbol{k}$ are given by $\omega_s(\boldsymbol{k}) = [\lambda_s(\boldsymbol{k})]^{1/2}$, where $\lambda_s(\boldsymbol{k})$'s are the eigenvalues of the 3×3 matrix $\mathscr{D}_{\alpha\beta}(k) = (2K/M)\sum_{\boldsymbol{R}} \hat{R}_\alpha \hat{R}_\beta \sin^2(\boldsymbol{k} \cdot \boldsymbol{R}/2)$ with $\alpha, \beta = x, y, z$. Here the summation over $\boldsymbol{R}$ is only over the nearest neighbors of the atom at the origin.

(2) Now apply the above result to a monatomic FCC crystal. Show that, if $\boldsymbol{k}$ is in the [100] direction, $\boldsymbol{k} = (k, 0, 0)$, then the frequency of the longitudinal acoustical normal mode is given by $\omega_L = 2(2K/M)^{1/2}\sin(ka/4)$ and the frequency of the two degenerate transverse acoustical normal modes is given by $\omega_T = 2(K/M)^{1/2}\sin(ka/4)$. Also consider the cases for $\boldsymbol{k}$ in the [110] and [111] directions

---

(1) The lattice potential energy is given by

$$\Phi = \frac{1}{4}K\sum_{\langle ij\rangle}\left[\,(\hat{\boldsymbol{R}}_i - \hat{\boldsymbol{R}}_j)\cdot(\boldsymbol{u}_i - \boldsymbol{u}_j)\,\right]^2.$$

To derive the dynamical matrix, we first find the elements of the $D(\boldsymbol{R})$ matrix through evaluating the second-order derivatives of the lattice potential energy with respect to atomic displacements. Differentiating Φ with respect to $u_{\ell'\beta}$, we have

$$\begin{aligned}\frac{\partial\Phi}{\partial u_{\ell'\beta}} &= \frac{1}{2}K\sum_{\langle ij\rangle}[(\hat{\boldsymbol{R}}_i-\hat{\boldsymbol{R}}_j)\cdot(\boldsymbol{u}_i-\boldsymbol{u}_j)][(\hat{R}_{i\beta}-\hat{R}_{j\beta})(\delta_{i\ell'}-\delta_{j\ell'})] \\ &= K\sum_{j(\mathrm{nn}\,\ell')}\big[(\hat{\boldsymbol{R}}_{\ell'}-\hat{\boldsymbol{R}}_j)\cdot(\boldsymbol{u}_{\ell'}-\boldsymbol{u}_j)\big](\hat{R}_{\ell'\beta}-\hat{R}_{j\beta}),\end{aligned}$$

where $(\mathrm{nn}\,\ell')$ implies that the summation is limited to the nearest neighbors of ℓ'. Differentiating the above result with respect to $u_{\ell\alpha}$, we have

$$\begin{aligned}\frac{\partial^2\Phi}{\partial u_{\ell\alpha}\partial u_{\ell'\beta}} &= K\sum_{j(\mathrm{nn}\,\ell')}\big[(\hat{R}_{\ell'\alpha}-\hat{R}_{j\alpha})(\delta_{\ell\ell'}-\delta_{\ell j})\big](\hat{R}_{\ell'\beta}-\hat{R}_{j\beta}) \\ &= K\delta_{\ell\ell'}\sum_{j(\mathrm{nn}\,\ell')}(\hat{R}_{\ell'\alpha}-\hat{R}_{j\alpha})(\hat{R}_{\ell'\beta}-\hat{R}_{j\beta}) \\ &\quad - K(\hat{R}_{\ell'\alpha}-\hat{R}_{\ell\alpha})(\hat{R}_{\ell'\beta}-\hat{R}_{\ell\beta})\delta_{\ell\,\mathrm{nn}\,\ell'}.\end{aligned}$$

The elements of the dynamical matrix are then given by

$$\begin{aligned}\mathscr{D}_{\alpha\beta}(\boldsymbol{k}) &= \frac{1}{M}\sum_{\ell}D_{\alpha\beta}(\boldsymbol{R}_\ell-\boldsymbol{R}_{\ell'})\mathrm{e}^{-\mathrm{i}\boldsymbol{k}\cdot(\boldsymbol{R}_\ell-\boldsymbol{R}_{\ell'})} \\ &= \frac{1}{M}\sum_{\ell}\frac{\partial^2\Phi}{\partial u_{\ell\alpha}\partial u_{\ell'\beta}}\mathrm{e}^{-\mathrm{i}\boldsymbol{k}\cdot(\boldsymbol{R}_\ell-\boldsymbol{R}_{\ell'})} \\ &= \frac{K}{M}\sum_{\ell}\bigg[\delta_{\ell\ell'}\sum_{j(\mathrm{nn}\,\ell')}(\hat{R}_{\ell'\alpha}-\hat{R}_{j\alpha})(\hat{R}_{\ell'\beta}-\hat{R}_{j\beta}) \\ &\quad - (\hat{R}_{\ell'\alpha}-\hat{R}_{\ell\alpha})(\hat{R}_{\ell'\beta}-\hat{R}_{\ell\beta})\delta_{\ell\,\mathrm{nn}\,\ell'}\bigg]\mathrm{e}^{-\mathrm{i}\boldsymbol{k}\cdot(\boldsymbol{R}_\ell-\boldsymbol{R}_{\ell'})}.\end{aligned}$$

Simplifying the above expression, we have

$$\begin{aligned}\mathscr{D}_{\alpha\beta}(\boldsymbol{k}) &= \frac{K}{M}\bigg(\sum_{j(\mathrm{nn}\,0)}\hat{R}_{j\alpha}\hat{R}_{j\beta}-\sum_{\ell(\mathrm{nn}\,0)}\hat{R}_{\ell\alpha}\hat{R}_{\ell\beta}\mathrm{e}^{-\mathrm{i}\boldsymbol{k}\cdot\boldsymbol{R}_\ell}\bigg) \\ &= \frac{K}{M}\sum_{j}\hat{R}_{j\alpha}\hat{R}_{j\beta}(1-\mathrm{e}^{-\mathrm{i}\boldsymbol{k}\cdot\boldsymbol{R}_j}) \\ &= \frac{K}{2M}\sum_{j(\mathrm{nn}\,0)}\hat{R}_{j\alpha}\hat{R}_{j\beta}(2-\mathrm{e}^{-\mathrm{i}\boldsymbol{k}\cdot\boldsymbol{R}_j}-\mathrm{e}^{\mathrm{i}\boldsymbol{k}\cdot\boldsymbol{R}_j}) \\ &= \frac{2K}{M}\sum_{j(\mathrm{nn}\,0)}\hat{R}_{j\alpha}\hat{R}_{j\beta}\sin^2(\boldsymbol{k}\cdot\boldsymbol{R}_j/2) \\ &= \frac{2K}{M}\sum_{\boldsymbol{R}(\mathrm{nn}\,0)}\hat{R}_{\alpha}\hat{R}_{\beta}\sin^2(\boldsymbol{k}\cdot\boldsymbol{R}/2).\end{aligned}$$

The secular equation for determining the eigenvalues of matrix $\mathscr{D}(k)$ reads

$$\det\left|\mathscr{D}_{\alpha\beta}(\boldsymbol{k}) - \lambda\delta_{\alpha\beta}\right| = 0.$$

Comparing the above equation with that for the determination of the normal-mode frequencies, we see that λ is identical with ω^2. Therefore, if the eigenvalues of $\mathscr{D}(k)$ are denoted by λ_s's, then the dispersion relations of normal modes are given by $\omega_s = [\lambda_s]^{1/2}$. In the following, the constraint (nn 0) on the summation over $\boldsymbol{R}$ will be suppressed for simplicity in notations.

(2) An atom in a monatomic FCC crystal has 12 nearest neighbors that are located at $(\pm 1/2, \pm 1/2, 0)a$, $(\pm 1/2, 0, \pm 1/2)a$, and $(0, \pm 1/2, \pm 1/2)a$ relative to the atom. For $\boldsymbol{k}$ in the [100] direction, we have

$$\begin{aligned}
\mathscr{D}_{11}(\boldsymbol{k}) &= \frac{2K}{M}\sum_{\boldsymbol{R}} \hat{R}_x^2 \sin^2(\boldsymbol{k}\cdot\boldsymbol{R}/2) \\
&= \frac{2K}{M}\cdot 8\cdot\frac{1}{2}\cdot\sin^2(ka/4) = \frac{8K}{M}\sin^2(ka/4), \\
\mathscr{D}_{22}(\boldsymbol{k}) &= \frac{2K}{M}\sum_{\boldsymbol{R}} \hat{R}_y^2 \sin^2(\boldsymbol{k}\cdot\boldsymbol{R}/2) \\
&= \frac{2K}{M}\cdot 4\cdot\frac{1}{2}\cdot\sin^2(ka/4) = \frac{4K}{M}\sin^2(ka/4), \\
\mathscr{D}_{33}(\boldsymbol{k}) &= \frac{2K}{M}\sum_{\boldsymbol{R}} \hat{R}_z^2 \sin^2(\boldsymbol{k}\cdot\boldsymbol{R}/2) \\
&= \frac{2K}{M}\cdot 4\cdot\frac{1}{2}\cdot\sin^2(ka/4) = \frac{4K}{M}\sin^2(ka/4), \\
\mathscr{D}_{12}(\boldsymbol{k}) &= \frac{2K}{M}\sum_{\boldsymbol{R}} \hat{R}_x\hat{R}_y \sin^2(\boldsymbol{k}\cdot\boldsymbol{R}/2) = 0, \\
\mathscr{D}_{23}(\boldsymbol{k}) &= \frac{2K}{M}\sum_{\boldsymbol{R}} \hat{R}_y\hat{R}_z \sin^2(\boldsymbol{k}\cdot\boldsymbol{R}/2) = 0, \\
\mathscr{D}_{31}(\boldsymbol{k}) &= \frac{2K}{M}\sum_{\boldsymbol{R}} \hat{R}_z\hat{R}_x \sin^2(\boldsymbol{k}\cdot\boldsymbol{R}/2) = 0.
\end{aligned}$$

Thus, the dynamical matrix is given by

$$\mathscr{D}(k) = \frac{4K}{M}\sin^2(ka/4)\begin{pmatrix} 2 & 0 & 0 \\ 0 & 1 & 0 \\ 0 & 0 & 1 \end{pmatrix}.$$

Since $\mathscr{D}(k)$ is diagonal, its eigenvalues are given by its elements on the main diagonal and its eigenvectors (the polarization vectors) are the corresponding basis vectors for the diagonal elements. Thus, the dispersion relations of normal modes are

$$\omega_L = 2\left(\frac{2K}{M}\right)^{1/2} |\sin(ka/4)|, \ \omega_{T_1} = \omega_{T_2} = 2\left(\frac{K}{M}\right)^{1/2} |\sin(ka/4)|,$$

and the corresponding polarization vectors are

$$\boldsymbol{\epsilon}^{(L)} = \boldsymbol{e}_x, \ \boldsymbol{\epsilon}^{(T_1)} = \boldsymbol{e}_y, \ \boldsymbol{\epsilon}^{(T_2)} = \boldsymbol{e}_z.$$

The first dispersion relation is subscripted by "L" to indicate that it is a longitudinal acoustical branch because the direction of the polarization vector is the same as that of $\boldsymbol{k}$. The polarization vectors for the other two branches are perpendicular to $\boldsymbol{k}$. Thus, they are transverse acoustical branches and are denoted by "T_1" and "T_2", respectively.
For $\boldsymbol{k}$ in the [110] direction, we have

$$\begin{aligned}
\mathscr{D}_{11}(\boldsymbol{k}) &= \frac{2K}{M}\sum_{\boldsymbol{R}} \hat{R}_x^2 \sin^2(\boldsymbol{k}\cdot\boldsymbol{R}/2) \\
&= \frac{2K}{M}\left[\sin^2(ka/2) + 2\sin^2(ka/4)\right], \\
\mathscr{D}_{22}(\boldsymbol{k}) &= \frac{2K}{M}\sum_{\boldsymbol{R}} \hat{R}_y^2 \sin^2(\boldsymbol{k}\cdot\boldsymbol{R}/2) \\
&= \frac{2K}{M}\left[\sin^2(ka/2) + 2\sin^2(ka/4)\right], \\
\mathscr{D}_{33}(\boldsymbol{k}) &= \frac{2K}{M}\sum_{\boldsymbol{R}} \hat{R}_z^2 \sin^2(\boldsymbol{k}\cdot\boldsymbol{R}/2) = \frac{8K}{M}\sin^2(ka/4), \\
\mathscr{D}_{12}(\boldsymbol{k}) &= \frac{2K}{M}\sum_{\boldsymbol{R}} \hat{R}_x\hat{R}_y \sin^2(\boldsymbol{k}\cdot\boldsymbol{R}/2) = \frac{2K}{M}\sin^2(ka/2), \\
\mathscr{D}_{23}(\boldsymbol{k}) &= \frac{2K}{M}\sum_{\boldsymbol{R}} \hat{R}_y\hat{R}_z \sin^2(\boldsymbol{k}\cdot\boldsymbol{R}/2) = 0, \\
\mathscr{D}_{31}(\boldsymbol{k}) &= \frac{2K}{M}\sum_{\boldsymbol{R}} \hat{R}_z\hat{R}_x \sin^2(\boldsymbol{k}\cdot\boldsymbol{R}/2) = 0.
\end{aligned}$$

The dynamical matrix is then given by

$$\mathscr{D}(\boldsymbol{k}) = \frac{4K}{M}\sin^2(ka/4)\begin{pmatrix} 1+2\cos^2(ka/4) & 2\cos^2(ka/4) & 0 \\ 2\cos^2(ka/4) & 1+2\cos^2(ka/4) & 0 \\ 0 & 0 & 2 \end{pmatrix}.$$

Since $\mathscr{D}(\boldsymbol{k})$ is a block matrix, one of its eigenvalues is immediately known. It is given by the last diagonal element. Since the corresponding eigenvector (the polarization vector) is perpendicular to $\boldsymbol{k}$, it is a transverse acoustical branch and is referred to as T_2. We have

$$\omega_{T_2} = 2\left(\frac{2K}{M}\right)^{1/2} |\sin(ka/4)|, \ \boldsymbol{\epsilon}^{(T_2)} = \boldsymbol{e}_z = \begin{pmatrix} 0 \\ 0 \\ 1 \end{pmatrix}.$$

To find the dispersion relations and polarization vectors for the other two branches, we now diagonalize the 2×2 submatrix in the upper-left corner. We find that it gives rise to the following dispersion relations and polarization vectors

$$\omega_L = 2\left(\frac{K}{M}\right)^{1/2} \left[\sin^2(ka/4) + \sin^2(ka/2)\right]^{1/2},$$

$$\boldsymbol{\epsilon}^{(L)} = \frac{\boldsymbol{e}_x + \boldsymbol{e}_y}{2^{1/2}} = \frac{1}{2^{1/2}} \begin{pmatrix} 1 \\ 1 \\ 0 \end{pmatrix},$$

$$\omega_{T_1} = 2\left(\frac{K}{M}\right)^{1/2} |\sin(ka/4)|, \ \boldsymbol{\epsilon}^{(T_1)} = \frac{\boldsymbol{e}_x - \boldsymbol{e}_y}{2^{1/2}} = \frac{1}{2^{1/2}} \begin{pmatrix} 1 \\ -1 \\ 0 \end{pmatrix}.$$

As indicated on the frequencies and polarization vectors, we have found the frequencies and polarization vectors for one longitudinal and two transverse acoustical branches for $\boldsymbol{k}$ in the [110] direction.

For $\boldsymbol{k}$ in the [111] direction, we have

$$\mathscr{D}_{11}(\boldsymbol{k}) = \frac{2K}{M} \sum_{\boldsymbol{R}} \hat{R}_x^2 \sin^2(\boldsymbol{k} \cdot \boldsymbol{R}/2) = \frac{4K}{M} \sin^2(ka/2),$$

$$\mathscr{D}_{22}(\boldsymbol{k}) = \frac{2K}{M} \sum_{\boldsymbol{R}} \hat{R}_y^2 \sin^2(\boldsymbol{k} \cdot \boldsymbol{R}/2) = \frac{4K}{M} \sin^2(ka/2),$$

$$\mathscr{D}_{33}(\boldsymbol{k}) = \frac{2K}{M} \sum_{\boldsymbol{R}} \hat{R}_z^2 \sin^2(\boldsymbol{k} \cdot \boldsymbol{R}/2) = \frac{4K}{M} \sin^2(ka/2),$$

$$\mathscr{D}_{12}(\boldsymbol{k}) = \frac{2K}{M} \sum_{\boldsymbol{R}} \hat{R}_x \hat{R}_y \sin^2(\boldsymbol{k} \cdot \boldsymbol{R}/2) = \frac{2K}{M} \sin^2(ka/2),$$

$$\mathscr{D}_{23}(\boldsymbol{k}) = \frac{2K}{M} \sum_{\boldsymbol{R}} \hat{R}_y \hat{R}_z \sin^2(\boldsymbol{k} \cdot \boldsymbol{R}/2) = \frac{2K}{M} \sin^2(ka/2),$$

$$\mathscr{D}_{31}(\boldsymbol{k}) = \frac{2K}{M} \sum_{\boldsymbol{R}} \hat{R}_z \hat{R}_x \sin^2(\boldsymbol{k} \cdot \boldsymbol{R}/2) = \frac{2K}{M} \sin^2(ka/2).$$

The dynamical matrix is then given by

$$\mathscr{D}(\boldsymbol{k}) = \frac{2K}{M} \sin^2(ka/2) \begin{pmatrix} 2 & 1 & 1 \\ 1 & 2 & 1 \\ 1 & 1 & 2 \end{pmatrix}.$$

For convenience of manipulation, we diagonalize the numerical matrix in $\mathscr{D}(\boldsymbol{k})$ instead of $\mathscr{D}(\boldsymbol{k})$ itself. Let η denote an eigenvalue and $(a, b, c)^t$ an eigenvector of this numerical matrix. Its eigenequation reads

$$\begin{pmatrix} 2-\eta & 1 & 1 \\ 1 & 2-\eta & 1 \\ 1 & 1 & 2-\eta \end{pmatrix} \begin{pmatrix} a \\ b \\ c \end{pmatrix} = 0.$$

The secular equation for eigenvalues reads

$$\eta^3 - 6\eta^2 + 9\eta - 4 = 0.$$

The above third-order algebraic equation can be solved through the factorization of the left hand side $(\eta-4)(\eta-1)^2 = 0$. Thus, the eigenvalues are $\eta^{(L)} = 4$ and $\eta^{(T_1)} = \eta^{(T_2)} = 1$. Here we have named the eigenvalue with the value of 4 as if it corresponds to the longitudinal acoustical branch and the two degenerate eigenvalues with the value of 1 as if they correspond to the two transverse acoustical branches. This can be seen to be true from the corresponding eigenvectors that are the polarization vectors of normal modes

$$\boldsymbol{\epsilon}^{(L)} = \frac{1}{3^{1/2}}(\boldsymbol{e}_x + \boldsymbol{e}_y + \boldsymbol{e}_z) = \frac{1}{3^{1/2}} \begin{pmatrix} 1 \\ 1 \\ 1 \end{pmatrix},$$

$$\boldsymbol{\epsilon}^{(T_1)} = \frac{1}{2^{1/2}}(\boldsymbol{e}_x - \boldsymbol{e}_y) = \frac{1}{2^{1/2}} \begin{pmatrix} 1 \\ -1 \\ 0 \end{pmatrix},$$

$$\boldsymbol{\epsilon}^{(T_2)} = \frac{1}{6^{1/2}}(\boldsymbol{e}_x + \boldsymbol{e}_y - 2\boldsymbol{e}_z) = \frac{1}{6^{1/2}} \begin{pmatrix} 1 \\ 1 \\ -2 \end{pmatrix}.$$

We have chosen the two eigenvectors corresponding to the degenerate eigenvalue 1 in accordance with that they are mutually orthogonal and that they are both perpendicular to $\boldsymbol{\epsilon}^{(L)}$. Finally, the frequencies of the three acoustical branches are given by

$$\omega_L = 2\left(\frac{2K}{M}\right)^{1/2} |\sin(ka/2)|,$$

$$\omega_{T_1} = \omega_{T_2} = \left(\frac{2K}{M}\right)^{1/2} |\sin(ka/2)|.$$

13-10 Three-dimensional crystal with a two-atom basis. A three-dimensional crystal has a two-atom basis. The masses of the two atoms in the basis are m_1 and m_2, respectively. Let $\boldsymbol{v}_1$ and $\boldsymbol{v}_2$ be their velocities. Show that, for an optical normal mode at the center of the first Brillouin zone ($\boldsymbol{k} = 0$), $m_1\boldsymbol{v}_1 + m_2\boldsymbol{v}_2 = 0$.

To prove the desired identity, we first prove a property of optical modes. For this purpose, we also need a property of acoustical properties. Let s_a denote an acoustical branch and s_o an optical branch. For an acoustical mode, its frequency and polarization vector satisfy

$$\omega_{s_a}^2(\boldsymbol{k})\epsilon_{\kappa\alpha}^{(s_a)}(\boldsymbol{k}) = \sum_{\nu\beta} \mathscr{D}_{\alpha\beta,\kappa\nu}(\boldsymbol{k})\epsilon_{\nu\beta}^{(s_a)}(\boldsymbol{k}).$$

For $\boldsymbol{k} = 0$, making use of the definition of the dynamical matrix we have

$$\begin{aligned}\omega_{s_a}^2(0)\epsilon_{\kappa\alpha}^{(s_a)}(0) &= \sum_{\nu\beta} \mathscr{D}_{\alpha\beta,\kappa\nu}(0)\epsilon_{\nu\beta}^{(s_a)}(0) \\ &= \sum_{\nu\beta} \frac{1}{\sqrt{m_\kappa m_\nu}} \sum_j D_{\kappa\nu,\alpha\beta}(\boldsymbol{R}_i - \boldsymbol{R}_j)\epsilon_{\nu\beta}^{(s_a)}(0) \\ &= \frac{1}{\sqrt{m_\kappa}} \sum_\beta \sum_{j\nu} D_{\kappa\nu,\alpha\beta}(\boldsymbol{R}_i - \boldsymbol{R}_j)\frac{\epsilon_{\nu\beta}^{(s_a)}(0)}{\sqrt{m_\nu}}\end{aligned}$$

If $\epsilon_{\nu\beta}^{(s_a)}(0)/\sqrt{m_\nu}$ is independent of ν, then

$$\omega_{s_a}^2(0)\epsilon_{\kappa\alpha}^{(s_a)}(0) = \frac{1}{\sqrt{m_\kappa}} \sum_\beta \frac{\epsilon_{\nu\beta}^{(s_a)}(0)}{\sqrt{m_\nu}} \sum_{j\nu} D_{\kappa\nu,\alpha\beta}(\boldsymbol{R}_i - \boldsymbol{R}_j) = 0,$$

where we have made use of the fact that $\sum_{j\nu} D_{\kappa\nu,\alpha\beta}(\boldsymbol{R}_i - \boldsymbol{R}_j) = 0$. Because $\omega_{s_a}^2(0) = 0$ for acoustical modes, we conclude that the ν-independence of $\epsilon_{\nu\beta}^{(s_a)}(0)/\sqrt{m_\nu}$ must be an intrinsic property of acoustical modes. For the two-atom basis in the present problem, we have

$$\frac{\boldsymbol{\epsilon}_1^{(s_a)}(0)}{\sqrt{m_1}} = \frac{\boldsymbol{\epsilon}_2^{(s_a)}(0)}{\sqrt{m_2}}.$$

Making use of the fact that the polarization vector of an optical mode is orthogonal to that of an acoustical mode, we have

$$\boldsymbol{\epsilon}_1^{(s_a)}(0) \cdot \boldsymbol{\epsilon}_1^{(s_o)}(0) + \boldsymbol{\epsilon}_2^{(s_a)}(0) \cdot \boldsymbol{\epsilon}_2^{(s_o)}(0) = 0$$

from which it follows that

$$\sqrt{m_1}\,\boldsymbol{\epsilon}_1^{(s_o)}(0) + \sqrt{m_2}\,\boldsymbol{\epsilon}_2^{(s_o)}(0) = 0.$$

We are now ready to prove the desired identity. From the general expression of atomic displacements

$$u_{j\nu,\,\alpha}(t) = \frac{1}{\sqrt{Nm_\nu}} \sum_{\boldsymbol{k}s} q_{\boldsymbol{k}s}(t)\epsilon^{(s)}_{\nu\alpha}(\boldsymbol{k})\mathrm{e}^{\mathrm{i}\boldsymbol{k}\cdot\boldsymbol{R}_j},$$

we see that the contribution to the atomic displacements from an optical mode at the center of the first Brillouin zone is given by

$$\boldsymbol{u}_{j\nu}(t) = \frac{1}{\sqrt{Nm_\nu}} q_{0s_o}(t)\boldsymbol{\epsilon}^{(s_o)}_\nu(0),$$

where we have written the atomic displacement and polarization vector in vector forms. Since the right hand side is independent of the primitive cell label j, we can drop the subscript j on $\boldsymbol{u}_{j\nu}$. For the combination $m_1\boldsymbol{u}_1 + m_2\boldsymbol{u}_2$, we have

$$m_1\boldsymbol{u}_1 + m_2\boldsymbol{u}_2 = \frac{q_{0s_o}(t)}{\sqrt{N}}\left(\sqrt{m_1}\boldsymbol{\epsilon}^{(s_o)}_1 + \sqrt{m_2}\boldsymbol{\epsilon}^{(s_o)}_2\right) = 0$$

according to the above-proven property of optical modes. Differentiating the above equation with respect to time t, we have

$$m_1\boldsymbol{v}_1 + m_2\boldsymbol{v}_2 = 0,$$

where

$$\boldsymbol{v}_\kappa = \frac{\mathrm{d}\boldsymbol{u}_\kappa}{\mathrm{d}t}.$$

Chapter 14

Quantum Theory of Lattice Vibrations

(1) *Dulong-Petit law* $c_v = 3pnk_{\mathrm{B}}$.

(2) *Phonons*

They are quanta of lattice vibrations. They can be also said to be quanta of the quantum field of atomic displacements. They are bosons of spin zero.

(3) *Annihilation and creation operators of phonons*

$$\hat{a}_{\boldsymbol{k}s} = \left(\frac{\omega_{\boldsymbol{k}s}}{2\hbar}\right)^{1/2}\left(\hat{q}_{\boldsymbol{k}s} + \frac{\mathrm{i}}{\omega_{\boldsymbol{k}s}}\hat{p}_{\boldsymbol{k}s}\right),$$

$$\hat{a}^{\dagger}_{-\boldsymbol{k}s} = \left(\frac{\omega_{\boldsymbol{k}s}}{2\hbar}\right)^{1/2}\left(\hat{q}_{\boldsymbol{k}s} - \frac{\mathrm{i}}{\omega_{\boldsymbol{k}s}}\hat{p}_{\boldsymbol{k}s}\right).$$

(4) *Commutation relations of phonon operators*

$$\left[\hat{a}_{\boldsymbol{k}s},\, \hat{a}^{\dagger}_{\boldsymbol{k}'s'}\right] = \delta_{\boldsymbol{k}\boldsymbol{k}'}\delta_{ss'},$$

$$\left[\hat{a}_{\boldsymbol{k}s},\, \hat{a}_{\boldsymbol{k}'s'}\right] = \left[\hat{a}^{\dagger}_{\boldsymbol{k}s},\, \hat{a}^{\dagger}_{\boldsymbol{k}'s'}\right] = 0.$$

(5) *Quantum field operator of atomic displacements*

$$\hat{u}_{j\nu,\,\alpha} = \sum_{\boldsymbol{k}s}\left(\frac{\hbar}{2Nm_{\nu}\omega_{\boldsymbol{k}s}}\right)^{1/2}\epsilon^{(s)}_{\nu\alpha}(\boldsymbol{k})\big(\hat{a}_{\boldsymbol{k}s} + \hat{a}^{\dagger}_{-\boldsymbol{k}s}\big)\mathrm{e}^{\mathrm{i}\boldsymbol{k}\cdot\boldsymbol{R}_j}.$$

(6) *Hamiltonian of the crystal*

$$\hat{H} = \sum_{\boldsymbol{k}s}\hbar\omega_{\boldsymbol{k}s}\big(\hat{a}^{\dagger}_{\boldsymbol{k}s}\hat{a}_{\boldsymbol{k}s} + 1/2\big).$$

(7) *Eigenvalues and eigenstates of the crystal Hamiltonian*

$$\mathscr{E}_n = \sum_{\boldsymbol{k}s}(n_{\boldsymbol{k}s} + 1/2)\hbar\omega_{\boldsymbol{k}s},\ \ |n\rangle = \prod_{\boldsymbol{k}s}\frac{1}{\sqrt{n_{\boldsymbol{k}s}!}}\big(\hat{a}^{\dagger}_{\boldsymbol{k}s}\big)^{n_{\boldsymbol{k}s}}|0\rangle,$$

$$n = \{n_{\boldsymbol{k}s}\,|\,\forall\,\boldsymbol{k}s\},\ \ n_{\boldsymbol{k}s} = 0,\,1,\,2,\,\cdots.$$

(8) *Time-dependent quantum field operator of atomic displacements*

$$\hat{u}_{j\nu,\alpha}(t) = \sum_{\boldsymbol{k}s}\left(\frac{\hbar}{2Nm_\nu\omega_{\boldsymbol{k}s}}\right)^{1/2}\epsilon^{(s)}_{\nu\alpha}(\boldsymbol{k})\left(\mathrm{e}^{-\mathrm{i}\omega_{\boldsymbol{k}s}t}\hat{a}_{\boldsymbol{k}s} + \mathrm{e}^{\mathrm{i}\omega_{\boldsymbol{k}s}t}\hat{a}^\dagger_{-\boldsymbol{k}s}\right)\mathrm{e}^{\mathrm{i}\boldsymbol{k}\cdot\boldsymbol{R}_j}.$$

(9) *Statistics of phonons*

Phonons obey the Bose-Einstein statistics

$$\langle n_{\boldsymbol{k}s}\rangle = \frac{1}{\mathrm{e}^{\hbar\omega_{\boldsymbol{k}s}/k_\mathrm{B}T} - 1}.$$

(10) *Phonon density of states*

$$g(\omega) = \frac{1}{\mathscr{V}}\sum_{\boldsymbol{k}s}\delta(\omega - \omega_{\boldsymbol{k}s}).$$

$g(\omega)\mathrm{d}\omega$ = the number of phonon states per unit volume in the frequency range from ω to $\omega + \mathrm{d}\omega$.

(11) *Van Hove singularity*

A van Hove singularity implies a peak in the phonon density of states as a function of frequency.

(12) *General expression of the lattice specific heat*

$$c_v = \frac{\partial}{\partial T}\sum_s\int\frac{\mathrm{d}\boldsymbol{k}}{(2\pi)^3}\,\frac{\hbar\omega_{\boldsymbol{k}s}}{\mathrm{e}^{\hbar\omega_{\boldsymbol{k}s}/k_\mathrm{B}T} - 1},$$

$$\lim_{T\to\infty} c_v = 3pnk_\mathrm{B},\quad \lim_{T\to 0} c_v = (2\pi^2/5)\left(k_\mathrm{B}/\hbar c\right)^3 k_\mathrm{B}.$$

(13) *Debye model*

Debye wave vector: $k_\mathrm{D} = \left(6\pi^2 n\right)^{1/3}$.

Debye frequency: $\omega_\mathrm{D} = \left(6\pi^2 n\right)^{1/3}c$.

Debye temperature: $\Theta_\mathrm{D} = \hbar\omega_\mathrm{D}/k_\mathrm{B}$.

Phonon density of states:

$$g_\mathrm{D}(\omega) = (3\omega^2/2\pi^2c^3)\theta(\omega)\,\theta(\omega_\mathrm{D} - \omega).$$

Lattice specific heat:

$$\lim_{T\to\infty} c_v^\mathrm{D} = 3nk_\mathrm{B},$$
$$\lim_{T\to 0} c_v^\mathrm{D} = (12\pi^4/5)\left(T/\Theta_\mathrm{D}\right)^3 nk_\mathrm{B} \approx 234\left(T/\Theta_\mathrm{D}\right)^3 nk_\mathrm{B}.$$

(14) *Einstein model*

The optical normal modes are taken as independent harmonic oscillators with the identical frequency ω_E.

Einstein temperature: $\Theta_\mathrm{E} = \hbar\omega_\mathrm{E}/k_\mathrm{B}$.

Lattice specific heat:

$$c_v^{\mathrm{E}} = \frac{p_{\mathrm{opt}} N k_{\mathrm{B}}}{\mathscr{V}} E(\hbar\omega_{\mathrm{E}}/2k_{\mathrm{B}}T).$$

Einstein function: $E(x) = \dfrac{x^2}{\sinh^2(x)}$.

(15) *Grüneisen parameter*

$$\gamma(T) = \frac{\beta B_T}{c_v},$$

$$\gamma(T) = \frac{\sum_{\boldsymbol{k}s} \gamma_{\boldsymbol{k}s} E(\hbar\omega_{\boldsymbol{k}s}/2k_{\mathrm{B}}T)}{\sum_{\boldsymbol{k}s} E(\hbar\omega_{\boldsymbol{k}s}/2k_{\mathrm{B}}T)},$$

$$\gamma_{ks} = -\frac{\partial \ln \omega_{ks}}{\partial \ln V}.$$

(16) *Specific heat of a metal*

$$c_v = c_v^e + c_v^L = \gamma T + AT^3.$$

14-1 Quantum field operator of atomic momenta. The quantum field operator of atomic displacements for a three-dimensional crystal with a multi-atom basis has been given. We now derive the quantum field operator of atomic momenta. In analogy with the definition of momentum in classical mechanics, let $\hat{P}_{j\nu,\,\alpha}(t) = m_\nu \partial \hat{u}_{j\nu,\,\alpha}(t)/\partial t$.

(1) Write down the explicit expression of $\hat{P}_{j\nu,\,\alpha}(t)$ in terms of operators $\hat{a}_{\boldsymbol{k}s}$ and $\hat{a}_{\boldsymbol{k}s}^\dagger$.

(2) Show that $[\hat{u}_{j\nu,\,\alpha}(t),\, \hat{P}_{\ell\kappa,\,\beta}(t)] = \mathrm{i}\hbar\delta_{j\ell}\delta_{\nu\kappa}\delta_{\alpha\beta}$. Therefore, $\hat{P}_{j\nu,\,\alpha}(t)$ is the momentum field operator conjugate to the displacement field operator $\hat{u}_{j\nu,\,\alpha}(t)$.

(3) Show that $\hat{u}_{j\nu,\,\alpha}(t)$ and $\hat{P}_{j\nu,\,\alpha}(t)$ are Hermitian operators.

---

(1) The time-dependent quantum field operator of atomic displacements is given by

$$\hat{u}_{j\nu,\,\alpha}(t) = \sum_{\boldsymbol{k}s} \left(\frac{\hbar}{2Nm_\nu\omega_{\boldsymbol{k}s}}\right)^{1/2} \epsilon_{\nu\alpha}^{(s)}(\boldsymbol{k}) \left(\mathrm{e}^{-\mathrm{i}\omega_{\boldsymbol{k}s}t}\hat{a}_{\boldsymbol{k}s} + \mathrm{e}^{\mathrm{i}\omega_{\boldsymbol{k}s}t}\hat{a}_{-\boldsymbol{k}s}^\dagger\right)\mathrm{e}^{\mathrm{i}\boldsymbol{k}\cdot\boldsymbol{R}_j}.$$

Differentiating the above equation with respect to t, we have

$$\begin{aligned}\hat{P}_{j\nu,\,\alpha}(t) &= m_\nu \frac{\partial \hat{u}_{j\nu,\,\alpha}(t)}{\partial t} \\ &= -\mathrm{i}\sum_{\boldsymbol{k}s} \left(\frac{m_\nu\hbar\omega_{\boldsymbol{k}s}}{2N}\right)^{1/2} \epsilon_{\nu\alpha}^{(s)}(\boldsymbol{k}) \left(\mathrm{e}^{-\mathrm{i}\omega_{\boldsymbol{k}s}t}\hat{a}_{\boldsymbol{k}s} - \mathrm{e}^{\mathrm{i}\omega_{\boldsymbol{k}s}t}\hat{a}_{-\boldsymbol{k}s}^\dagger\right)\mathrm{e}^{\mathrm{i}\boldsymbol{k}\cdot\boldsymbol{R}_j}.\end{aligned}$$

(2) Making use of the expressions for $\hat{u}_{j\nu,\,\alpha}(t)$ and $\hat{P}_{\ell\kappa,\,\beta}(t)$, we can write the commutator $[\hat{u}_{j\nu,\,\alpha}(t),\, \hat{P}_{\ell\kappa,\,\beta}(t)]$ as

$$\begin{aligned}
&[\hat{u}_{j\nu,\,\alpha}(t),\, \hat{P}_{\ell\kappa,\,\beta}(t)] \\
&\quad = -\mathrm{i}\sum_{\boldsymbol{k}s}\sum_{\boldsymbol{k}'s'}\left(\frac{m_\kappa \hbar^2 \omega_{\boldsymbol{k}'s'}}{4N^2 m_\nu \omega_{\boldsymbol{k}s}}\right)^{1/2} \epsilon_{\nu\alpha}^{(s)}(\boldsymbol{k})\epsilon_{\kappa\beta}^{(s')}(\boldsymbol{k}')\mathrm{e}^{\mathrm{i}(\boldsymbol{k}\cdot\boldsymbol{R}_j+\boldsymbol{k}'\cdot\boldsymbol{R}_\ell)} \\
&\quad\quad \times \left[\mathrm{e}^{-\mathrm{i}\omega_{\boldsymbol{k}s}t}\hat{a}_{\boldsymbol{k}s} + \mathrm{e}^{\mathrm{i}\omega_{\boldsymbol{k}s}t}\hat{a}^\dagger_{-\boldsymbol{k}s},\, \mathrm{e}^{-\mathrm{i}\omega_{\boldsymbol{k}'s'}t}\hat{a}_{\boldsymbol{k}'s'} - \mathrm{e}^{\mathrm{i}\omega_{\boldsymbol{k}'s'}t}\hat{a}^\dagger_{-\boldsymbol{k}'s'}\right].
\end{aligned}$$

The above commutator can be evaluated using $[\hat{a}_{\boldsymbol{k}s}, \hat{a}^\dagger_{\boldsymbol{k}'s'}] = \delta_{\boldsymbol{k}\boldsymbol{k}'}\delta_{ss'}$. We have

$$\begin{aligned}
&[\hat{u}_{j\nu,\,\alpha}(t),\, \hat{P}_{\ell\kappa,\,\beta}(t)] \\
&\quad = \mathrm{i}\sum_{\boldsymbol{k}s}\sum_{\boldsymbol{k}'s'}\left(\frac{m_\kappa \hbar^2 \omega_{\boldsymbol{k}'s'}}{4N^2 m_\nu \omega_{\boldsymbol{k}s}}\right)^{1/2} \epsilon_{\nu\alpha}^{(s)}(\boldsymbol{k})\epsilon_{\kappa\beta}^{(s')}(\boldsymbol{k}')\mathrm{e}^{\mathrm{i}(\boldsymbol{k}\cdot\boldsymbol{R}_j+\boldsymbol{k}'\cdot\boldsymbol{R}_\ell)} \\
&\quad\quad \times \left[\mathrm{e}^{-\mathrm{i}(\omega_{\boldsymbol{k}s}-\omega_{\boldsymbol{k}'s'})t} + \mathrm{e}^{\mathrm{i}(\omega_{\boldsymbol{k}s}-\omega_{\boldsymbol{k}'s'})t}\right]\delta_{\boldsymbol{k},-\boldsymbol{k}'}\delta_{ss'} \\
&\quad = \frac{\mathrm{i}\hbar}{N}\sum_{\boldsymbol{k}s}\left(\frac{m_\kappa}{m_\nu}\right)^{1/2}\epsilon_{\nu\alpha}^{(s)}(\boldsymbol{k})\epsilon_{\kappa\beta}^{(s)*}(\boldsymbol{k})\mathrm{e}^{\mathrm{i}\boldsymbol{k}\cdot(\boldsymbol{R}_j-\boldsymbol{R}_\ell)} \\
&\quad = \mathrm{i}\hbar\delta_{j\ell}\delta_{\nu\kappa}\delta_{\alpha\beta},
\end{aligned}$$

where we have made use of

$$\epsilon_{\kappa\beta}^{(s)}(-\boldsymbol{k}) = \epsilon_{\kappa\beta}^{(s)*}(\boldsymbol{k}),\ \ \sum_s \epsilon_{\nu\alpha}^{(s)}(\boldsymbol{k})\epsilon_{\kappa\beta}^{(s)*}(\boldsymbol{k}) = \delta_{\nu\kappa}\delta_{\alpha\beta},$$

$$\frac{1}{N}\sum_{\boldsymbol{k}}\mathrm{e}^{\mathrm{i}\boldsymbol{k}\cdot(\boldsymbol{R}_j-\boldsymbol{R}_\ell)} = \delta_{j\ell}.$$

(3) Taking the Hermitian conjugation of $\hat{u}_{j\nu,\,\alpha}(t)$, we have

$$\begin{aligned}
&\hat{u}^\dagger_{j\nu,\,\alpha}(t) \\
&\quad = \sum_{\boldsymbol{k}s}\left(\frac{\hbar}{2Nm_\nu\omega_{\boldsymbol{k}s}}\right)^{1/2}\epsilon_{\nu\alpha}^{(s)*}(\boldsymbol{k})\left(\mathrm{e}^{\mathrm{i}\omega_{\boldsymbol{k}s}t}\hat{a}^\dagger_{\boldsymbol{k}s} + \mathrm{e}^{-\mathrm{i}\omega_{\boldsymbol{k}s}t}\hat{a}_{-\boldsymbol{k}s}\right)\mathrm{e}^{-\mathrm{i}\boldsymbol{k}\cdot\boldsymbol{R}_j}.
\end{aligned}$$

Performing a change of dummy summation variables from $\boldsymbol{k}$ to $\boldsymbol{k}' = -\boldsymbol{k}$, then dropping the prime on $\boldsymbol{k}'$, and then making use of $\omega_{-\boldsymbol{k}s} = \omega_{\boldsymbol{k}s}$ and $\epsilon_{\kappa\beta}^{(s)}(-\boldsymbol{k}) = \epsilon_{\kappa\beta}^{(s)*}(\boldsymbol{k})$, we have

$$\begin{aligned}
\hat{u}^\dagger_{j\nu,\,\alpha}(t) &= \sum_{\boldsymbol{k}s}\left(\frac{\hbar}{2Nm_\nu\omega_{\boldsymbol{k}s}}\right)^{1/2}\epsilon_{\nu\alpha}^{(s)}(\boldsymbol{k})\left(\mathrm{e}^{\mathrm{i}\omega_{\boldsymbol{k}s}t}\hat{a}^\dagger_{-\boldsymbol{k}s} + \mathrm{e}^{-\mathrm{i}\omega_{\boldsymbol{k}s}t}\hat{a}_{\boldsymbol{k}s}\right)\mathrm{e}^{\mathrm{i}\boldsymbol{k}\cdot\boldsymbol{R}_j} \\
&= \hat{u}_{j\nu,\,\alpha}(t).
\end{aligned}$$

Thus, $\hat{u}_{j\nu,\,\alpha}(t)$ is Hermitian. We can similarly show that $\hat{P}_{\ell\kappa,\,\beta}(t)$ is also Hermitian as follows

$$\begin{aligned}
&\hat{P}^{\dagger}_{j\nu,\,\alpha}(t)\\
&= \mathrm{i}\sum_{\boldsymbol{k}s}\left(\frac{m_\nu\hbar\omega_{\boldsymbol{k}s}}{2N}\right)^{1/2}\epsilon^{(s)*}_{\nu\alpha}(\boldsymbol{k})\big(\mathrm{e}^{\mathrm{i}\omega_{\boldsymbol{k}s}t}\hat{a}^{\dagger}_{\boldsymbol{k}s}-\mathrm{e}^{-\mathrm{i}\omega_{\boldsymbol{k}s}t}\hat{a}_{-\boldsymbol{k}s}\big)\mathrm{e}^{-\mathrm{i}\boldsymbol{k}\cdot\boldsymbol{R}_j}\\
&= -\mathrm{i}\sum_{\boldsymbol{k}s}\left(\frac{m_\nu\hbar\omega_{\boldsymbol{k}s}}{2N}\right)^{1/2}\epsilon^{(s)}_{\nu\alpha}(\boldsymbol{k})\big(\mathrm{e}^{-\mathrm{i}\omega_{\boldsymbol{k}s}t}\hat{a}_{\boldsymbol{k}s}-\mathrm{e}^{\mathrm{i}\omega_{\boldsymbol{k}s}t}\hat{a}^{\dagger}_{-\boldsymbol{k}s}\big)\mathrm{e}^{\mathrm{i}\boldsymbol{k}\cdot\boldsymbol{R}_j}\\
&= \hat{P}_{j\nu,\,\alpha}(t).
\end{aligned}$$

14-2 Hamiltonian for a 3D crystal with a multi-atom basis. In this problem, the Hamiltonian for a three-dimensional crystal with a multi-atom basis will be derived.

(1) Specializing the above-obtained expression for $\hat{P}_{j\nu,\,\alpha}(t)$ to the time-independent case and making use of the resultant expression, express the kinetic energy of the crystal, $\hat{T}=\sum_{j\nu,\,\alpha}\hat{P}^{\dagger}_{j\nu,\,\alpha}\hat{P}_{j\nu,\,\alpha}/2m_\nu$ in terms of operators $\hat{a}_{\boldsymbol{k}s}$ and $\hat{a}^{\dagger}_{\boldsymbol{k}s}$.

(2) Using the expression of $\hat{u}_{j\nu,\,\alpha}$, express the harmonic lattice potential energy of the crystal, $\hat{\Phi}^{\mathrm{harm}}=(1/2)\sum_{j\ell}\sum_{\nu\kappa,\,\alpha\beta}\hat{u}^{\dagger}_{j\nu,\,\alpha}D_{\nu\kappa,\alpha\beta}(\boldsymbol{R}_j-\boldsymbol{R}_\ell)\hat{u}_{\ell\kappa,\,\beta}$ in terms of operators $\hat{a}_{\boldsymbol{k}s}$ and $\hat{a}^{\dagger}_{\boldsymbol{k}s}$.

(3) Derive the Hamiltonian for a three-dimensional crystal with a multi-atom basis.

---

(1) The time-independent operator $\hat{P}_{j\nu,\,\alpha}$ is given by

$$\hat{P}_{j\nu,\,\alpha}=-\mathrm{i}\sum_{\boldsymbol{k}s}\left(\frac{m_\nu\hbar\omega_{\boldsymbol{k}s}}{2N}\right)^{1/2}\epsilon^{(s)}_{\nu\alpha}(\boldsymbol{k})\big(\hat{a}_{\boldsymbol{k}s}-\hat{a}^{\dagger}_{-\boldsymbol{k}s}\big)\mathrm{e}^{\mathrm{i}\boldsymbol{k}\cdot\boldsymbol{R}_j}.$$

We consider the kinetic and potential energies separately. Inserting the above expression of $\hat{P}_{j\nu,\,\alpha}$ into the kinetic energy operator $\hat{T}=\sum_{j\nu,\,\alpha}\hat{P}^{\dagger}_{j\nu,\,\alpha}\hat{P}_{j\nu,\,\alpha}/2m_\nu$ yields

$$\begin{aligned}
\hat{T}=\frac{1}{4N}\sum_{j\nu,\,\alpha}\sum_{\boldsymbol{k}s}\sum_{\boldsymbol{k}'s'}&\big(\hbar^2\omega_{\boldsymbol{k}s}\omega_{\boldsymbol{k}'s'}\big)^{1/2}\epsilon^{(s)*}_{\nu\alpha}(\boldsymbol{k})\epsilon^{(s')}_{\nu\alpha}(\boldsymbol{k}')\mathrm{e}^{-\mathrm{i}(\boldsymbol{k}-\boldsymbol{k}')\cdot\boldsymbol{R}_j}\\
&\times\big(\hat{a}^{\dagger}_{\boldsymbol{k}s}-\hat{a}_{-\boldsymbol{k}s}\big)\big(\hat{a}_{\boldsymbol{k}'s'}-\hat{a}^{\dagger}_{-\boldsymbol{k}'s'}\big).
\end{aligned}$$

Performing the summation over j, we obtain

$$\hat{T} = \frac{1}{4}\sum_{\nu,\,\alpha}\sum_{\boldsymbol{k}s}\sum_{\boldsymbol{k}'s'}\left(\hbar^2\omega_{\boldsymbol{k}s}\omega_{\boldsymbol{k}'s'}\right)^{1/2}\epsilon_{\nu\alpha}^{(s)*}(\boldsymbol{k})\epsilon_{\nu\alpha}^{(s')}(\boldsymbol{k}')\delta_{\boldsymbol{k}\boldsymbol{k}'}$$
$$\times\left(\hat{a}_{\boldsymbol{k}s}^{\dagger} - \hat{a}_{-\boldsymbol{k}s}\right)\left(\hat{a}_{\boldsymbol{k}'s'} - \hat{a}_{-\boldsymbol{k}'s'}^{\dagger}\right)$$
$$= \frac{1}{4}\sum_{\nu,\,\alpha}\sum_{\boldsymbol{k}ss'}\left(\hbar^2\omega_{\boldsymbol{k}s}\omega_{\boldsymbol{k}s'}\right)^{1/2}\epsilon_{\nu\alpha}^{(s)*}(\boldsymbol{k})\epsilon_{\nu\alpha}^{(s')}(\boldsymbol{k})$$
$$\times\left(\hat{a}_{\boldsymbol{k}s}^{\dagger} - \hat{a}_{-\boldsymbol{k}s}\right)\left(\hat{a}_{\boldsymbol{k}s'} - \hat{a}_{-\boldsymbol{k}s'}^{\dagger}\right).$$

Utilizing the orthonormality relation of the polarization vectors, we have

$$\hat{T} = \frac{1}{4}\sum_{\boldsymbol{k}ss'}\left(\hbar^2\omega_{\boldsymbol{k}s}\omega_{\boldsymbol{k}s'}\right)^{1/2}\delta_{ss'}\left(\hat{a}_{\boldsymbol{k}s}^{\dagger} - \hat{a}_{-\boldsymbol{k}s}\right)\left(\hat{a}_{\boldsymbol{k}s'} - \hat{a}_{-\boldsymbol{k}s'}^{\dagger}\right)$$
$$= \frac{1}{4}\sum_{\boldsymbol{k}s}\hbar\omega_{\boldsymbol{k}s}\left(\hat{a}_{\boldsymbol{k}s}^{\dagger} - \hat{a}_{-\boldsymbol{k}s}\right)\left(\hat{a}_{\boldsymbol{k}s} - \hat{a}_{-\boldsymbol{k}s}^{\dagger}\right).$$

Making use of the commutation relation between $\hat{a}_{\boldsymbol{k}s}$ and $\hat{a}_{\boldsymbol{k}s}^{\dagger}$, we have

$$\hat{T} = \frac{1}{4}\sum_{\boldsymbol{k}s}\hbar\omega_{\boldsymbol{k}s}\left(\hat{a}_{\boldsymbol{k}s}^{\dagger}\hat{a}_{\boldsymbol{k}s} - \hat{a}_{-\boldsymbol{k}s}\hat{a}_{\boldsymbol{k}s} - \hat{a}_{\boldsymbol{k}s}^{\dagger}\hat{a}_{-\boldsymbol{k}s}^{\dagger} + \hat{a}_{-\boldsymbol{k}s}\hat{a}_{-\boldsymbol{k}s}^{\dagger}\right)$$
$$= \frac{1}{4}\sum_{\boldsymbol{k}s}\hbar\omega_{\boldsymbol{k}s}\left(2\hat{a}_{\boldsymbol{k}s}^{\dagger}\hat{a}_{\boldsymbol{k}s} + 1 - \hat{a}_{-\boldsymbol{k}s}\hat{a}_{\boldsymbol{k}s} - \hat{a}_{\boldsymbol{k}s}^{\dagger}\hat{a}_{-\boldsymbol{k}s}^{\dagger}\right).$$

(2) Inserting the expression of $\hat{u}_{j\nu,\,\alpha}$ into the potential energy operator $\hat{\Phi}^{\mathrm{harm}} = (1/2)\sum_{j\ell}\sum_{\nu\kappa,\,\alpha\beta}\hat{u}_{j\nu,\,\alpha}^{\dagger}D_{\nu\kappa,\alpha\beta}(\boldsymbol{R}_j - \boldsymbol{R}_\ell)\hat{u}_{\ell\kappa,\,\beta}$ yields

$$\hat{\Phi}^{\mathrm{harm}} = \frac{\hbar}{4}\sum_{\nu\kappa,\,\alpha\beta}\sum_{\boldsymbol{k}s}\sum_{\boldsymbol{k}'s'}\frac{1}{\sqrt{\omega_{\boldsymbol{k}s}\omega_{\boldsymbol{k}'s'}}}\epsilon_{\nu\alpha}^{(s)*}(\boldsymbol{k})\epsilon_{\kappa\beta}^{(s')}(\boldsymbol{k}')\left(\hat{a}_{\boldsymbol{k}s}^{\dagger} + \hat{a}_{-\boldsymbol{k}s}\right)$$
$$\times\left(\hat{a}_{\boldsymbol{k}'s'} + \hat{a}_{-\boldsymbol{k}'s'}^{\dagger}\right)\frac{1}{N}\sum_{\ell}\mathrm{e}^{-\mathrm{i}(\boldsymbol{k}-\boldsymbol{k}')\cdot\boldsymbol{R}_\ell}$$
$$\times\left[\frac{1}{\sqrt{m_\nu m_\kappa}}\sum_{j}D_{\nu\kappa,\alpha\beta}(\boldsymbol{R}_j - \boldsymbol{R}_\ell)\mathrm{e}^{-\mathrm{i}\boldsymbol{k}\cdot(\boldsymbol{R}_j - \boldsymbol{R}_\ell)}\right]$$
$$= \frac{\hbar}{4}\sum_{\nu\kappa,\,\alpha\beta}\sum_{\boldsymbol{k}s}\sum_{\boldsymbol{k}'s'}\frac{1}{\sqrt{\omega_{\boldsymbol{k}s}\omega_{\boldsymbol{k}'s'}}}\epsilon_{\nu\alpha}^{(s)*}(\boldsymbol{k})\epsilon_{\kappa\beta}^{(s')}(\boldsymbol{k}')\left(\hat{a}_{\boldsymbol{k}s}^{\dagger} + \hat{a}_{-\boldsymbol{k}s}\right)$$
$$\times\left(\hat{a}_{\boldsymbol{k}'s'} + \hat{a}_{-\boldsymbol{k}'s'}^{\dagger}\right)\frac{1}{N}\sum_{\ell}\mathrm{e}^{-\mathrm{i}(\boldsymbol{k}-\boldsymbol{k}')\cdot\boldsymbol{R}_\ell}\mathscr{D}_{\nu\kappa,\alpha\beta}(\boldsymbol{k}),$$

where we have introduced the dynamical matrix $\mathscr{D}$. We next perform the summation over ℓ and obtain a Kronecker δ-symbol $\delta_{\boldsymbol{k}\boldsymbol{k}'}$ that is then used to perform the summation $\boldsymbol{k}'$.

We have

$$\begin{aligned}\hat{\Phi}^{\text{harm}} &= \frac{\hbar}{4}\sum_{\nu\kappa,\,\alpha\beta}\sum_{\boldsymbol{k}ss'}\frac{\epsilon_{\nu\alpha}^{(s)*}(\boldsymbol{k})}{\sqrt{\omega_{\boldsymbol{k}s}\omega_{\boldsymbol{k}s'}}}\big(\hat{a}_{\boldsymbol{k}s}^{\dagger}+\hat{a}_{-\boldsymbol{k}s}\big)\big(\hat{a}_{\boldsymbol{k}s'}+\hat{a}_{-\boldsymbol{k}s'}^{\dagger}\big)\\ &\qquad\times\mathscr{D}_{\nu\kappa,\alpha\beta}(\boldsymbol{k})\epsilon_{\kappa\beta}^{(s')}(\boldsymbol{k})\\ &= \frac{\hbar}{4}\sum_{\nu,\alpha}\sum_{\boldsymbol{k}ss'}\frac{\epsilon_{\nu\alpha}^{(s)*}(\boldsymbol{k})}{\sqrt{\omega_{\boldsymbol{k}s}\omega_{\boldsymbol{k}s'}}}\big(\hat{a}_{\boldsymbol{k}s}^{\dagger}+\hat{a}_{-\boldsymbol{k}s}\big)\big(\hat{a}_{\boldsymbol{k}s'}+\hat{a}_{-\boldsymbol{k}s'}^{\dagger}\big)\omega_{\boldsymbol{k}s'}^{2}\epsilon_{\nu\alpha}^{(s')}(\boldsymbol{k}),\end{aligned}$$

where we have made use of the eigenequation for the dynamical matrix

$$\sum_{\kappa\beta}\mathscr{D}_{\nu\kappa,\alpha\beta}(\boldsymbol{k})\epsilon_{\kappa\beta}^{(s')}(\boldsymbol{k}) = \omega_{\boldsymbol{k}s'}^{2}\epsilon_{\nu\alpha}^{(s')}(\boldsymbol{k}).$$

The summation of $\epsilon_{\nu\alpha}^{(s)*}(\boldsymbol{k})\epsilon_{\nu\alpha}^{(s')}(\boldsymbol{k})$ over ν and α yields $\delta_{ss'}$ which can be utilized to perform the summation over s'. We then have

$$\begin{aligned}\hat{\Phi}^{\text{harm}} &= \frac{1}{4}\sum_{\boldsymbol{k}s}\hbar\omega_{\boldsymbol{k}s}\big(\hat{a}_{\boldsymbol{k}s}^{\dagger}+\hat{a}_{-\boldsymbol{k}s}\big)\big(\hat{a}_{\boldsymbol{k}s}+\hat{a}_{-\boldsymbol{k}s}^{\dagger}\big)\\ &= \frac{1}{4}\sum_{\boldsymbol{k}s}\hbar\omega_{\boldsymbol{k}s}\big(2\hat{a}_{\boldsymbol{k}s}^{\dagger}\hat{a}_{\boldsymbol{k}s}+1+\hat{a}_{-\boldsymbol{k}s}\hat{a}_{\boldsymbol{k}s}+\hat{a}_{\boldsymbol{k}s}^{\dagger}\hat{a}_{-\boldsymbol{k}s}^{\dagger}\big).\end{aligned}$$

(3) Adding up $\hat{T}$ and $\hat{\Phi}^{\text{harm}}$, we obtain the Hamiltonian for the crystal

$$\hat{H} = \hat{T}+\hat{\Phi}^{\text{harm}} = \sum_{\boldsymbol{k}s}\hbar\omega_{\boldsymbol{k}s}\big(\hat{a}_{\boldsymbol{k}s}^{\dagger}\hat{a}_{\boldsymbol{k}s}+1/2\big).$$

14-3 Thermodynamics of a gas of phonons. Take phonons in a crystal as if they move in a box of volume $\mathscr{V}$ and study the thermodynamics of this gas of phonons.

(1) Evaluate the canonical partition function of phonons, $Z = \sum_n \mathrm{e}^{-\mathscr{E}_n/k_{\mathrm{B}}T}$ with $\mathscr{E}_n$ an eigenvalue of the crystal Hamiltonian, and the Helmholtz free energy $F = -k_{\mathrm{B}}T\ln Z$.

(2) From the thermodynamic relation $S = -(\partial F/\partial T)_{\mathscr{V}}$, compute the entropy S.

(3) Express S in terms of the thermal average of the occupation number $n_{\mathrm{B}}(\hbar\omega_{\boldsymbol{k}s}) \equiv \langle n_{\boldsymbol{k}s}\rangle$ of the single-phonon state $|\boldsymbol{k}s\rangle$.

---

(1) The eigenvalues of the crystal Hamiltonian are given by

$$\mathscr{E}_n = \sum_{\boldsymbol{k}s}(n_{\boldsymbol{k}s}+1/2)\hbar\omega_{\boldsymbol{k}s}.$$

Inserting the above expression for $\mathscr{E}_n$ into the definition of the canonical partition function, $Z = \sum_n \mathrm{e}^{-\beta\mathscr{E}_n}$ with $\beta = 1/k_\mathrm{B}T$, we have

$$\begin{aligned} Z &= \sum_n \mathrm{e}^{-\beta\mathscr{E}_n} = \sum_{\{n_{\boldsymbol{k}s}\}} \exp\Big[-\beta\sum_{\boldsymbol{k}s}(n_{\boldsymbol{k}s}+1/2)\hbar\omega_{\boldsymbol{k}s}\Big] \\ &= \sum_{\{n_{\boldsymbol{k}s}\}}\prod_{\boldsymbol{k}s}\exp\big[-\beta(n_{\boldsymbol{k}s}+1/2)\hbar\omega_{\boldsymbol{k}s}\big] \\ &= \prod_{\boldsymbol{k}s}\sum_{n_{\boldsymbol{k}s}=0}^{\infty}\exp\big[-\beta(n_{\boldsymbol{k}s}+1/2)\hbar\omega_{\boldsymbol{k}s}\big] \\ &= \prod_{\boldsymbol{k}s}\frac{\mathrm{e}^{-\beta\hbar\omega_{\boldsymbol{k}s}/2}}{1-\mathrm{e}^{-\beta\hbar\omega_{\boldsymbol{k}s}}} = \prod_{\boldsymbol{k}s}\frac{1}{2\sinh(\beta\hbar\omega_{\boldsymbol{k}s}/2)}. \end{aligned}$$

The Helmholtz free energy then follows from the above expression for the canonical partition function

$$F = -k_\mathrm{B}T\ln Z = k_\mathrm{B}T\sum_{\boldsymbol{k}s}\ln\big[2\sinh(\hbar\omega_{\boldsymbol{k}s}/2k_\mathrm{B}T)\big].$$

(2) Differentiating F with respect to T, we obtain the entropy

$$\begin{aligned} S = -\frac{\partial F}{\partial T} &= -k_\mathrm{B}\sum_{\boldsymbol{k}s}\ln\big[2\sinh(\hbar\omega_{\boldsymbol{k}s}/2k_\mathrm{B}T)\big] \\ &\qquad - k_\mathrm{B}T\sum_{\boldsymbol{k}s}\frac{\cosh(\hbar\omega_{\boldsymbol{k}s}/2k_\mathrm{B}T)}{\sinh(\hbar\omega_{\boldsymbol{k}s}/2k_\mathrm{B}T)}\left(-\frac{\hbar\omega_{\boldsymbol{k}s}}{2k_\mathrm{B}T^2}\right) \\ &= -k_\mathrm{B}\sum_{\boldsymbol{k}s}\ln\big[2\sinh(\hbar\omega_{\boldsymbol{k}s}/2k_\mathrm{B}T)\big] \\ &\quad + \frac{1}{2T}\sum_{\boldsymbol{k}s}\hbar\omega_{\boldsymbol{k}s}\coth(\hbar\omega_{\boldsymbol{k}s}/2k_\mathrm{B}T). \end{aligned}$$

(3) From $n_\mathrm{B}(\hbar\omega_{\boldsymbol{k}s}) = 1/\big(\mathrm{e}^{\beta\hbar\omega_{\boldsymbol{k}s}}-1\big)$, we have $\mathrm{e}^{\beta\hbar\omega_{\boldsymbol{k}s}} = \big[n_\mathrm{B}(\hbar\omega_{\boldsymbol{k}s})+1\big]/n_\mathrm{B}(\hbar\omega_{\boldsymbol{k}s})$ and $\beta\hbar\omega_{\boldsymbol{k}s} = \ln\big\{\big[n_\mathrm{B}(\hbar\omega_{\boldsymbol{k}s})+1\big]/n_\mathrm{B}(\hbar\omega_{\boldsymbol{k}s})\big\}$. Expressing $\sinh(\beta\hbar\omega_{\boldsymbol{k}s}/2)$ and $\coth(\beta\hbar\omega_{\boldsymbol{k}s}/2)$ in terms of $n_\mathrm{B}(\hbar\omega_{\boldsymbol{k}s})$, we have

$$\begin{aligned} \sinh(\beta\hbar\omega_{\boldsymbol{k}s}/2) &= \frac{1}{2n_\mathrm{B}(\hbar\omega_{\boldsymbol{k}s})}\mathrm{e}^{-\beta\hbar\omega_{\boldsymbol{k}s}/2}, \\ \coth(\beta\hbar\omega_{\boldsymbol{k}s}/2) &= 2n_\mathrm{B}(\hbar\omega_{\boldsymbol{k}s})+1. \end{aligned}$$

We can then rewrite S as

$$\begin{aligned}
S &= -k_{\mathrm{B}} \sum_{\boldsymbol{k}s} \ln \frac{\mathrm{e}^{-\beta\hbar\omega_{\boldsymbol{k}s}/2}}{n_{\mathrm{B}}(\hbar\omega_{\boldsymbol{k}s})} + \frac{1}{2T} \sum_{\boldsymbol{k}s} \hbar\omega_{\boldsymbol{k}s} \big[2n_{\mathrm{B}}(\hbar\omega_{\boldsymbol{k}s}) + 1 \big] \\
&= k_{\mathrm{B}} \sum_{\boldsymbol{k}s} \beta\hbar\omega_{\boldsymbol{k}s} \big[n_{\mathrm{B}}(\hbar\omega_{\boldsymbol{k}s}) + 1 \big] + k_{\mathrm{B}} \ln n_{\mathrm{B}}(\hbar\omega_{\boldsymbol{k}s}) \\
&= k_{\mathrm{B}} \sum_{\boldsymbol{k}s} \big[n_{\mathrm{B}}(\hbar\omega_{\boldsymbol{k}s}) + 1 \big] \ln \frac{n_{\mathrm{B}}(\hbar\omega_{\boldsymbol{k}s}) + 1}{n_{\mathrm{B}}(\hbar\omega_{\boldsymbol{k}s})} + k_{\mathrm{B}} \sum_{\boldsymbol{k}s} \ln n_{\mathrm{B}}(\hbar\omega_{\boldsymbol{k}s}) \\
&= k_{\mathrm{B}} \sum_{\boldsymbol{k}s} \Big\{ \big[n_{\mathrm{B}}(\hbar\omega_{\boldsymbol{k}s}) + 1 \big] \ln \big[n_{\mathrm{B}}(\hbar\omega_{\boldsymbol{k}s}) + 1 \big] \\
&\qquad\qquad - n_{\mathrm{B}}(\hbar\omega_{\boldsymbol{k}s}) \ln n_{\mathrm{B}}(\hbar\omega_{\boldsymbol{k}s}) \Big\}.
\end{aligned}$$

14-4 Lattice specific heat of a 1D crystal of inert gas atoms. Consider a one-dimensional crystal of inert gas atoms. Let L be the length of the crystal and N the number of atoms. Let a be the lattice constant. *i.* Evaluate the phonon density of states for this crystal. *ii.* Derive an integral expression for the lattice specific heat of the crystal. *iii.* Evaluate the lattice specific heat in the high- and low-temperature limits.

---

i. The phonon dispersion relation in a one-dimensional crystal of inert gas atoms is given by

$$\omega_k = (4K/m)^{1/2} |\sin(ka/2)| = \omega_m |\sin(ka/2)|$$

with $\omega_m = (4K/m)^{1/2}$. From the definition of the phonon density of states, we have

$$\begin{aligned}
g(\omega) &= \frac{1}{L} \sum_k \delta(\omega - \omega_k) = \frac{1}{2\pi} \int_{-\pi/a}^{\pi/a} \mathrm{d}k\, \delta\big(\omega - \omega_m |\sin(ka/2)|\big) \\
&= \frac{2}{\pi a} \int_0^{\pi/2} \mathrm{d}x\, \delta\big(\omega - \omega_m \sin(x)\big) \\
&= \frac{2}{\pi \omega_m a} \int_0^{\pi/2} \mathrm{d}x\, \frac{\delta(x - \sin^{-1}(\omega/\omega_m))}{|\cos[\sin^{-1}(\omega/\omega_m)]|} \theta(\omega_m - |\omega|) \\
&= \frac{2}{\pi a} \frac{1}{\sqrt{\omega_m^2 - \omega^2}} \theta(\omega)\theta(\omega_m - \omega).
\end{aligned}$$

ii. The internal energy per unit volume is given by

$$\begin{aligned}
u &= u^{\mathrm{eq}} + \int_0^{\infty} \mathrm{d}\omega\, (\hbar\omega) g(\omega) n_{\mathrm{B}}(\hbar\omega) \\
&= u^{\mathrm{eq}} + \frac{2}{\pi a} \int_0^{\omega_m} \mathrm{d}\omega\, \frac{\hbar\omega}{\mathrm{e}^{\hbar\omega/k_{\mathrm{B}}T} - 1} \frac{1}{\sqrt{\omega_m^2 - \omega^2}}.
\end{aligned}$$

Differentiating u with respect to T, we obtain the general expression for the specific heat per unit volume

$$c_v = \frac{2k_B}{\pi a}\int_0^{\omega_m} d\omega\, \frac{1}{\sqrt{\omega_m^2-\omega^2}}\left(\frac{\hbar\omega}{k_B T}\right)^2 \frac{e^{\hbar\omega/k_B T}}{\left(e^{\hbar\omega/k_B T}-1\right)^2}$$

$$= \frac{2k_B t_m}{\pi a}\int_0^{1/t_m} dx\, \frac{1}{\sqrt{1-t_m^2 x^2}}\frac{x^2 e^x}{\left(e^x-1\right)^2},$$

where $t_m = k_B T/\hbar\omega_m = T/\Theta_m$ with $\Theta_m = \hbar\omega_m/k_B$.

iii. In the high-temperature limit, $T \gg \Theta_m$ so that $t_m \gg 1$. Thus, $x \ll 1$ in the entire integration interval. We can then approximate $x^2 e^x/(e^x-1)^2$ as unity and have

$$c_v \approx \frac{2k_B t_m}{\pi a}\int_0^{1/t_m} dx\, \frac{1}{\sqrt{1-t_m^2x^2}} = \frac{2k_B t_m}{\pi a}\cdot\frac{\pi}{2t_m} = \frac{k_B}{a}$$

which is in accordance with the Dulong-Petit law.

In the low-temperature limit, $T \ll \Theta_m$ so that $t_m \ll 1$ and $1/t_m \gg 1$. Because of the presence of $e^x/(e^x-1)^2$, the dominant contribution to the integral comes from the small-x region in which the square root in the denominator, $\sqrt{1-t_m^2x^2}$, can be approximated with unity. We thus have

$$c_v \approx \frac{2k_B t_m}{\pi a}\int_0^{1/t_m} dx\, \frac{x^2 e^x}{\left(e^x-1\right)^2}.$$

Since $1/t_m \gg 1$ and the main contribution is from the small-x region, we can extend the upper integration limit to infinity. We then have

$$c_v \approx \frac{2k_B t_m}{\pi a}\int_0^{\infty} dx\, \frac{x^2 e^x}{\left(e^x-1\right)^2}.$$

The above integral can be evaluated by expanding $e^x/(e^x-1)^2$ as a power series in e^{-x}

$$\frac{e^x}{\left(e^x-1\right)^2} = \frac{e^{-x}}{\left(1-e^{-x}\right)^2} = \sum_{n=1}^{\infty} n\, e^{-nx}.$$

The specific heat is then given by

$$c_v = \frac{2k_B t_m}{\pi a}\sum_{n=1}^{\infty} n\int_0^{\infty} dx\, x^2 e^{-nx} = \frac{2k_B t_m}{\pi a}\sum_{n=1}^{\infty}\frac{2}{n^2}$$

$$= \frac{2k_B t_m}{\pi a}\cdot\frac{\pi^2}{3} = \frac{2\pi}{3a}\frac{T}{\Theta_m}k_B,$$

where we have made use of $\sum_{n=1}^{\infty} n^{-2} = \pi^2/6$. The above result indicates that the lattice specific heat of a one-dimensional crystal at low temperatures is proportional to the temperature.

14-5 Debye model for a 1D crystal of inert gas atoms. Reconsider the above problem within the Debye model. *i.* Find the phonon density of states within the Debye model. *ii.* Determine the Debye frequency. *iii.* Find a general expression for the lattice specific heat. *iv.* Evaluate the lattice specific heat in the high- and low-temperature limits. *v.* Compare the exact and Debye results for the lattice specific heat by plotting them together from zero temperature to the Debye temperature.

---

i. For this one-dimensional crystal, there is only one branch of acoustical phonons. The dispersion relation of the acoustical phonons in the Debye model is given by $\omega = c|k|$. Making use of the dispersion relation $\omega = c|k|$, the phonon density of states in the Debye model is given by

$$g_{\rm D}(\omega) = \int_{-\pi/a}^{\pi/a} \frac{{\rm d}k}{2\pi}\,\delta(\omega - c|k|) = \frac{1}{\pi c}\int_0^{\pi/a} {\rm d}k\;\delta(k - \omega/c)$$
$$= \frac{1}{\pi c}\theta(\omega)\theta(\pi c/a - \omega)$$

which is a constant for ω in $[0, \pi c/a]$.

ii. Let the Debye frequency be denoted by $\omega_{\rm D}$. We now find the value of $\omega_{\rm D}$. For this one-dimensional crystal, there are N acoustical normal modes with N the number of atoms. The length of the crystal is Na. Thus, the number of normal modes (phonon states) per unit length is $N/Na = 1/a$. With $\omega_{\rm D} \leqslant \pi c/a$, we have

$$\frac{1}{a} = \int_0^{\omega_{\rm D}} {\rm d}\omega\; g_{\rm D}(\omega) = \frac{1}{\pi c}\int_0^{\omega_{\rm D}} {\rm d}\omega\;\theta(\pi c/a - \omega) = \frac{1}{\pi c}\omega_{\rm D}.$$

Hence,

$$\omega_{\rm D} = \frac{\pi c}{a}.$$

The above result indicates that $k_{\rm D} = \pi/a$ that is the largest wave vector in the first Brillouin zone. This is because the correct number of normal modes is ensured within the Debye model only if $k_{\rm D} = \pi/a$.

iii. Within the Debye model, the internal energy is given by

$$u = \phi^{\rm eq} + \int_{-\infty}^{\infty} {\rm d}\omega\;\hbar\omega g_D(\omega) n_B(\omega)$$
$$= \phi^{\rm eq} + \frac{1}{\omega_{\rm D} a}\int_0^{\omega_{\rm D}} {\rm d}\omega\;\frac{\hbar\omega}{e^{\hbar\omega/k_{\rm B}T} - 1}$$
$$= \phi^{\rm eq} + \frac{k_{\rm B}T^2}{\Theta_{\rm D} a}\int_0^{\Theta_{\rm D}/T} {\rm d}x\;\frac{x}{e^x - 1},$$

where $\Theta_{\rm D} = \hbar\omega_{\rm D}/k_{\rm B}$ is the Debye temperature.
Differentiating u with respect to T, we obtain the general expression for the lattice specific heat

$$\begin{aligned} c_v &= \frac{\partial u}{\partial T} = \frac{1}{\omega_{\rm D} a k_{\rm B} T^2} \int_0^{\omega_{\rm D}} {\rm d}\omega \, \frac{(\hbar\omega)^2 e^{\hbar\omega/k_{\rm B}T}}{(e^{\hbar\omega/k_{\rm B}T} - 1)^2} \\ &= \frac{k_{\rm B}T}{\Theta_{\rm D} a} \int_0^{\Theta_{\rm D}/T} {\rm d}x \, \frac{x^2 e^x}{(e^x - 1)^2}. \end{aligned}$$

If we perform an integration by parts, the above expression will become the one that can be obtained directly from the expression of u on the second line.

iv. In the high-temperature limit, we have $\Theta_{\rm D}/T \ll 1$ so that $x \ll 1$ in the entire integration interval. We can thus approximate $e^x - 1$ with x and have

$$u = \phi^{\rm eq} + \frac{k_{\rm B}T^2}{\Theta_{\rm D} a} \int_0^{\Theta_{\rm D}/T} {\rm d}x = \phi^{\rm eq} + \frac{1}{a} k_{\rm B} T.$$

The lattice specific heat is then given by

$$c_v = \frac{\partial u}{\partial T} = \frac{k_{\rm B}}{a}$$

which is just the result given by the Dulong-Petit law in one dimension.

In the low-temperature limit, we have $\Theta_{\rm D}/T \gg 1$. Because of the presence of the exponential function in the denominator, the dominant contribution to the integral in u comes from the small-x region. We can thus extend the upper integration limit to infinity and have

$$\begin{aligned} u &= \phi^{\rm eq} + \frac{k_{\rm B}T^2}{\Theta_{\rm D} a} \sum_{n=1}^{\infty} \int_0^{\infty} {\rm d}x \, x e^{-nx} = \phi^{\rm eq} + \frac{k_{\rm B}T^2}{\Theta_{\rm D} a} \sum_{n=1}^{\infty} \frac{1}{n^2} \\ &= \phi^{\rm eq} + \frac{\pi^2 k_{\rm B} T^2}{6\Theta_{\rm D} a}, \end{aligned}$$

where we have made use of $\sum_{n=1}^{\infty} n^{-2} = \pi^2/6$. The lattice specific heat is then given by

$$c_v = \frac{\partial u}{\partial T} = \frac{\pi^2}{3a} \frac{T}{\Theta_{\rm D}} k_{\rm B}.$$

v. The plots of the lattice specific heats computed exactly and by using the Debye model are given in Fig. 14.1 as functions of $T/\Theta_{\rm D}$. We see from the figure that the values of the lattice specific heat from the Debye model are larger than or equal to those of the exact result at all temperatures, which indicates that the Debye model overall overestimates the lattice specific heat for this one-dimensional crystal. At high temperatures, they both tend to the result given by the Dulong-Petit law. At low temperatures, although they both tend to zero as the temperature goes to zero, the exact result goes to zero slower than the result from the Debye model.

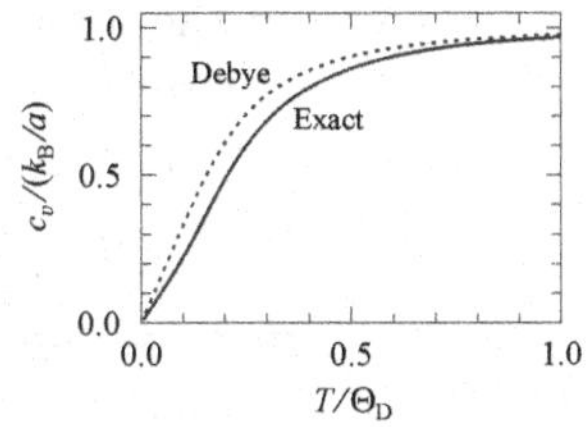

Fig. 14.1: Plots of the lattice specific heats derived exactly and from the Debye model as functions of T/Θ_D.

14-6 Electronic and lattice contributions to the specific heat of a metal. Consider the relative size of the electronic and lattice contributions to the specific heat of a metal using the Sommerfeld theory for the electrons and the Debye model for the phonons. *i.* Find an expression for the ratio c_v^e/c_v^L. *ii.* Determine the temperature T^* at which $c_v^e = c_v^L$. *iii.* Give an estimate on the order of magnitude for T^* in alkali metals.

---

i. From the Sommerfeld theory, we have

$$c_v^e = \frac{\pi^2}{2}\frac{k_B T}{\varepsilon_F} n_e k_B = \frac{\pi^2}{2}\frac{T}{\Theta_F} n_e k_B$$

with n_e the electron number density. From the Debye model, we have

$$c_v^L = \frac{12\pi^4}{5}\left(\frac{T}{\Theta_D}\right)^3 n_a k_B$$

with n_a the number of atoms. We thus have

$$\frac{c_v^e}{c_v^L} = \frac{5 n_e \Theta_D^3}{24\pi^2 \Theta_F n_a T^2}.$$

ii. Setting $c_v^e/c_v^L = 1$ at $T = T^*$, we have

$$T^* = \left(\frac{5 n_e \Theta_D^3}{24\pi^2 \Theta_F n_a}\right)^{1/2}.$$

iii. For an alkali metal, we have $n_e = n_a$. We take $\Theta_D \sim 100$ K and $\Theta_F \sim 10^4$ K. The order of magnitude of T^* is then estimated to be

$$T^* \sim \left(\frac{5 \times 100^3}{24 \times \pi^2 \times 10^4}\right)^{1/2} \sim 1\ \text{K}.$$

For $T \gg 1$ K, the lattice specific heat dominates, while the electronic specific heat dominates for $T \ll 1$ K.

14-7 Specific heat of potassium at low temperatures. A number of values of the specific heat of potassium at low temperatures are given in the following table. *i.* Plot C_v versus T and C_v/T versus T^2. *ii.* Perform a linear least-squares fit of the experimental data for C_v/T to $C_v/T = \gamma + AT^2$ and determine γ and A. *iii.* Estimate the Debye temperature of potassium at low temperatures.

T	C_v	T	C_v	T	C_v	T	C_v
0.2604	0.5852	0.2885	0.6578	0.3644	0.8858	0.4515	1.177
0.2781	0.6306	0.2894	0.6657	0.3734	0.9180	0.4578	1.208
0.2953	0.6786	0.3067	0.7104	0.3935	0.9733	0.4835	1.302
0.2501	0.5592	0.3270	0.7687	0.3994	1.003	0.4969	1.353
0.2650	0.5969	0.3379	0.7962	0.4231	1.021	0.5435	1.551
0.2698	0.6066	0.3478	0.8362	0.4274	1.102	0.5944	1.786

i. The plots of C_v versus T and C_v/T versus T^2 are given in Fig. 14.2.

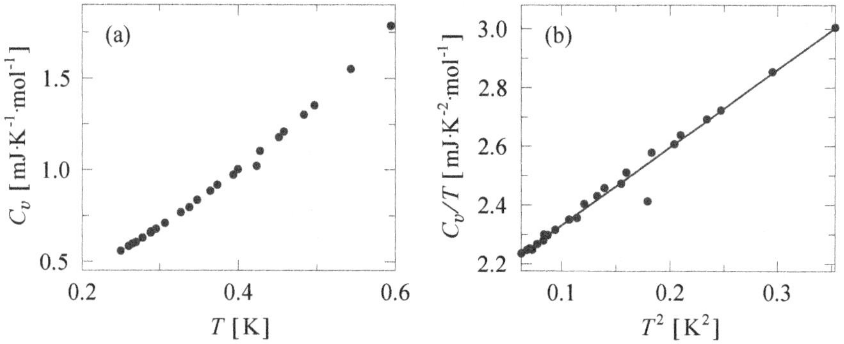

Fig. 14.2 Plots of C_v versus T (a) and C_v/T versus T^2 (b). In (b), the filled circles are experimental data and the solid line is a linear least-squares fit to $C_v/T = \gamma + AT^2$.

ii. Let $c = C_v/T$. At a given temperature T_i, the deviation of the experimentally measured value c_i from the prediction $\gamma + AT_i^2$ by the fitting is $c_i - \gamma - AT_i^2$. There are several different ways to gauge the accuracy of the fitting. Here we use the sum of the squared deviations of data points to gauge the accuracy of the fitting. This sum is referred to as *the chi-squared* function and is denoted by χ^2

$$\chi^2 = \sum_i \left(c_i - \gamma - AT_i^2\right)^2.$$

To get the best fitting, we demand that the values of parameters γ and A minimize χ^2, through which the values of the parameters can be determined. Such a fitting method is known as *the least-squares fitting.* Differentiating χ^2 with respect to γ and A, respectively, and setting the results to zero, we obtain

$$n\gamma + A\sum_i T_i^2 = \sum_i c_i,$$

$$\gamma\sum_i T_i^2 + A\sum_i T_i^4 = \sum_i T_i^2 c_i,$$

where n is the number of data points, $n = 24$. Solving for γ and A from the above two equations, we obtain

$$\gamma = \frac{\sum_i T_i^4 \sum_i c_i - \sum_i T_i^2 \sum_i T_i^2 c_i}{n\sum_i T_i^4 - \left(\sum_i T_i^2\right)^2},$$

$$A = \frac{n\sum_i T_i^2 c_i - \sum_i T_i^2 \sum_i c_i}{n\sum_i T_i^4 - \left(\sum_i T_i^2\right)^2}.$$

Making use of the given experimental data, we can find the values of γ and A using the above two expressions. We find that

$$\gamma \approx 2.066 \text{ mJ} \cdot \text{K}^{-2} \cdot \text{mol}^{-1}, \quad A \approx 2.650 \text{ mJ} \cdot \text{K}^{-4} \cdot \text{mol}^{-1}.$$

The straight line $C_v/T = \gamma + AT^2$ with the above-given values of γ and A is plotted in Fig. 14.2(b).

iii. In the Debye model, A for the specific *per mole* is given by

$$A = \frac{12\pi^4}{5\Theta_\text{D}^3} N_\text{A} k_\text{B},$$

where N_A is the Avogadro constant, $N_\text{A} = 6.022\ 141\ 79 \times 10^{23}$ mol^{-1}. Making use of the above-obtained value for A, we have

$$\frac{12\pi^4}{5\Theta_\text{D}^3} N_\text{A} k_\text{B} = 2.650 \times 10^{-3} \text{ J}.$$

Solving for Θ_D, we obtain

$$\Theta_\text{D} = \left(\frac{12\pi^4 N_\text{A} k_\text{B}}{5 \times 2.650 \times 10^{-3}}\right)^{1/3} \approx 90 \text{ K}.$$

14-8 Phonon density of states for an optical branch. Assume that the dispersion relation of an optical phonon branch in a solid takes on the form $\omega(k) = \omega_0 - Ak^2$ near $\boldsymbol{k} = 0$, where ω_0 and A are positive constants. Find the phonon density of states for $\omega < \omega_0$ and $\omega > \omega_0$, respectively.

According to the definition of the phonon density of states, we have

$$\begin{aligned} g(\omega) &= \frac{1}{\mathscr{V}}\sum_{\boldsymbol{k}}\delta\big(\omega - \omega(\boldsymbol{k})\big) = \frac{1}{2\pi^2}\int_0^\infty dk\, k^2\delta(\omega - \omega_0 + Ak^2) \\ &= \frac{1}{4\pi^2 A}\int_0^\infty dk\, k\big[\,\delta(k - \sqrt{(\omega_0 - \omega)/A}\,) \\ &\quad + \delta(k + \sqrt{(\omega_0 - \omega)/A}\,)\big]\theta(\omega_0 - \omega) \\ &= \frac{(\omega_0 - \omega)^{1/2}}{4\pi^2 A^{3/2}}\theta(\omega_0 - \omega). \end{aligned}$$

Thus, $g(\omega) = 0$ for $\omega > \omega_0$ and

$$g(\omega) = \frac{(\omega_0 - \omega)^{1/2}}{4\pi^2 A^{3/2}} \text{ for } \omega < \omega_0.$$

14-9 Phonon density of states for an acoustical branch. The dispersion relation of the acoustical branch in a one-dimensional crystal with a two-atom basis of identical atoms of mass m is given by

$$\omega_a(k) = \frac{1}{m^{1/2}}\Big\{K + G - \big[K^2 + G^2 + 2KG\cos(ka)\big]^{1/2}\Big\}^{1/2},$$

where K and G are the force constants of the springs connecting the two nearest neighbors of an atom. Evaluate the phonon density of states for this acoustical branch.

According to the definition of the phonon density of states, we have

$$g(\omega) = \frac{1}{\mathscr{L}}\sum_k \delta\big(\omega - \omega_a(k)\big) = \frac{1}{\pi}\int_0^{\pi/a} dk\, \delta\big(\omega - \omega_a(k)\big).$$

Let k_0 denote the root of the equation $\omega - \omega_a(k) = 0$ for k in $[0, \pi/a]$. We have

$$\begin{aligned} \sin(k_0 a) &= (m\omega^2/KG)^{1/2}\big[2(K+G) - m\omega^2\big]^{1/2} \\ &\quad \times \big\{1 - (m\omega^2/4KG)\big[2(K+G) - m\omega^2\big]\big\}^{1/2}. \end{aligned}$$

Note that the above solution exists only if ω satisfies the following two inequalities

$$2(K+G) - m\omega^2 \geqslant 0 \text{ and } 2(K+G) - m\omega^2 \leqslant 4KG/m\omega^2.$$

If any one of the above two inequalities is violated, then the phonon density of states is zero. Since $\min \omega_a(k) = 0$ and $\max \omega_a(k) = \left[2\max(K,G)/m\right]^{1/2}$, the above two inequalities are satisfied for ω taking on any value given by $\omega_a(k)$. Making use of $\delta(\varphi(x)) = \sum_i \delta(x - x_i)/|\varphi'(x_i)|$ with each x_i a first-order zero point of $\varphi(x)$, we have

$$\begin{aligned} g(\omega) &\\ = \frac{1}{\pi|\omega_a'(k_0)|}&\theta(2(K+G) - m\omega^2)\theta(4KG/m\omega^2 - 2(K+G) + m\omega^2) \\ &\times \int_0^{\pi/a} \mathrm{d}k\, \delta(k - k_0) \\ = \frac{1}{\pi|\omega_a'(k_0)|}&\theta(2(K+G) - m\omega^2)\theta(4KG/m\omega^2 - 2(K+G) + m\omega^2). \end{aligned}$$

Evaluating the derivative of $\omega_a(k)$ with respect to k, we have at $k = k_0$

$$\begin{aligned} \omega_a'(k_0) &= \frac{KGa\sin(k_0 a)}{2m\omega\left[K^2 + G^2 + 2KG\cos(k_0 a)\right]^{1/2}} \\ &= \frac{KGa\sin(k_0 a)}{2m\omega(K + G - m\omega^2)} \\ &= \frac{(KG/m)^{1/2} a}{2(K + G - m\omega^2)}\left[2(K+G) - m\omega^2\right]^{1/2} \\ &\quad \times \left\{1 - (m\omega^2/4KG)\left[2(K+G) - m\omega^2\right]\right\}^{1/2}. \end{aligned}$$

Finally, the phonon density of states is given by

$$\begin{aligned} g(\omega) = {}& \frac{2|K + G - m\omega^2|}{\pi a(KG/m)^{1/2}}\left[2(K+G) - m\omega^2\right]^{-1/2} \\ &\times \left\{1 - (m\omega^2/4KG)\left[2(K+G) - m\omega^2\right]\right\}^{-1/2} \\ &\times \theta(2(K+G) - m\omega^2)\theta(4KG/m\omega^2 - 2(K+G) + m\omega^2). \end{aligned}$$

14-10 Number of phonons and its variance. At finite temperatures, phonons in a crystal are constantly created and annihilated so that their number, $n_{\text{ph}} = \sum_{\boldsymbol{k}s} n_{\boldsymbol{k}s}$ with $n_{\boldsymbol{k}s}$ the number of phonons in single-phonon state $|\boldsymbol{k}s\rangle$, fluctuates greatly. *i.* Evaluate the thermal average number of phonons per unit volume $\langle n_{\text{ph}}\rangle = Z^{-1}\sum_n n_{\text{ph}} \mathrm{e}^{-\mathscr{E}_n/k_{\mathrm{B}}T} = Z^{-1}\sum_{\boldsymbol{k}s}\sum_n n_{\boldsymbol{k}s}\mathrm{e}^{-\mathscr{E}_n/k_{\mathrm{B}}T}$ with $Z = \sum_n \mathrm{e}^{-\mathscr{E}_n/k_{\mathrm{B}}T}$ and $\mathscr{E}_n$ the eigenvalues of the harmonic lattice Hamiltonian in a three-dimensional crystal with a multi-atom basis. *ii.* Find the expressions for $\langle n_{\rm ph}\rangle$ in the high- and low-temperature limits. *iii.* Evaluate the variance of the number of phonons, $\mathrm{var}(n_{\rm ph}) = \langle n_{\rm ph}^2\rangle - \langle n_{\rm ph}\rangle^2$.

i. Making up the expression for the partition function on the numerator through multiplying both the numerator and denominator with $\sum_{n_{\boldsymbol{k}s}} \exp\big[-\beta(n_{\boldsymbol{k}s} + 1/2)\hbar\omega_{\boldsymbol{k}s}\big]$, we have

$$\begin{aligned}
\langle n_{\rm ph}\rangle &= \frac{1}{Z}\sum_{\boldsymbol{k}s}\sum_{n} n_{\boldsymbol{k}s} \mathrm{e}^{-\mathscr{E}_n/k_{\rm B}T} \\
&= \frac{1}{Z}\sum_{\boldsymbol{k}s}\sum_{n_{\boldsymbol{k}s}} n_{\boldsymbol{k}s} \exp\big[-\beta(n_{\boldsymbol{k}s} + 1/2)\hbar\omega_{\boldsymbol{k}s}\big] \\
&\quad \frac{\sum_n \mathrm{e}^{-\mathscr{E}_n/k_{\rm B}T}}{\sum_{n_{\boldsymbol{k}s}} \exp\big[-\beta(n_{\boldsymbol{k}s} + 1/2)\hbar\omega_{\boldsymbol{k}s}\big]} \\
&= \sum_{\boldsymbol{k}s} \frac{\sum_{n_{\boldsymbol{k}s}} n_{\boldsymbol{k}s} \exp(-n_{\boldsymbol{k}s}\beta\hbar\omega_{\boldsymbol{k}s})}{\sum_{n_{\boldsymbol{k}s}} \exp(-n_{\boldsymbol{k}s}\beta\hbar\omega_{\boldsymbol{k}s})} \\
&= \sum_{\boldsymbol{k}s} \frac{1}{e^{\beta\hbar\omega_{\boldsymbol{k}s}} - 1} = \sum_{\boldsymbol{k}s} n_{\rm B}(\hbar\omega_{\boldsymbol{k}s}).
\end{aligned}$$

ii. In the high-temperature limit, $\beta\hbar\omega_{\boldsymbol{k}s} \ll 1$. We can then approximate $e^{\beta\hbar\omega_{\boldsymbol{k}s}} - 1$ as $\beta\hbar\omega_{\boldsymbol{k}s} = \hbar\omega_{\boldsymbol{k}s}/k_{\rm B}T$. We thus have

$$\langle n_{\rm ph}\rangle = \sum_{\boldsymbol{k}s} \frac{k_{\rm B}T}{\hbar\omega_{\boldsymbol{k}s}} = AT,$$

with $A = k_{\rm B}\sum_{\boldsymbol{k}s}(\hbar\omega_{\boldsymbol{k}s})^{-1}$. We see that the number of phonons is proportional to the temperature at high temperatures.

In the low-temperature limit, the dominant contribution arises from the normal modes of low frequencies. Since the optical normal modes are of high frequencies, we can neglect them and consider only acoustical normal modes. At low frequencies, the frequencies of acoustical normal modes are linear in wave vectors, $\omega_{\boldsymbol{k}s} \approx c_s(\hat{\boldsymbol{k}})k$. Since the normal modes of high frequencies make negligible contributions, we can also extend to infinity the upper integration limit for the integral over wave vectors. We then have in the high-temperature limit

$$\begin{aligned}\langle n_{\rm ph}\rangle &= \frac{\mathscr{V}}{(2\pi)^3}\int_0^\infty {\rm d}k\, k^2 \int {\rm d}\Omega_{\hat{\boldsymbol{k}}}\, \frac{1}{e^{\beta\hbar c_s(\hat{\boldsymbol{k}})k}-1} \\ &= \frac{\mathscr{V}}{2\pi^2}\left(\frac{k_{\rm B}T}{\hbar}\right)^3 \int \frac{{\rm d}\Omega_{\hat{\boldsymbol{k}}}}{4\pi}\frac{1}{c_s^3(\hat{\boldsymbol{k}})}\int_0^\infty {\rm d}x\, \frac{x^2}{e^x-1} \\ &= \frac{\mathscr{V}}{2\pi^2}\left(\frac{k_{\rm B}T}{\hbar c}\right)^3 \int_0^\infty {\rm d}x\, \frac{x^2}{e^x-1},\end{aligned}$$

where we have introduced the average of $c_s^{-3}(\hat{\boldsymbol{k}})$ over all directions

$$\frac{1}{c^3} = \int \frac{{\rm d}\Omega_{\hat{\boldsymbol{k}}}}{4\pi}\frac{1}{c_s^3(\hat{\boldsymbol{k}})}.$$

Evaluating the integral in $\langle n_{\rm ph}\rangle$, we have

$$\begin{aligned}\langle n_{\rm ph}\rangle &= \frac{\mathscr{V}}{2\pi^2}\left(\frac{k_{\rm B}T}{\hbar c}\right)^3 \sum_{n=1}^\infty \int_0^\infty {\rm d}x\, x^2 {\rm e}^{-nx} \\ &= \frac{\mathscr{V}}{2\pi^2}\left(\frac{k_{\rm B}T}{\hbar c}\right)^3 \sum_{n=1}^\infty \frac{2}{n^3} = \frac{\zeta(3)\mathscr{V}}{\pi^2}\left(\frac{k_{\rm B}T}{\hbar c}\right)^3,\end{aligned}$$

where $\zeta(3) = \sum_{n=1}^\infty n^{-3} \approx 1.202$ is a Riemann zeta-function. The above result indicates that the number of phonons at low temperatures is proportional to the third power of the temperature. It is not surprising at all that the temperature dependence of the number of phonons is the same as that of the lattice specific heat in consideration that the energy absorption and emission by a crystal are through phonons.

iii. To find the variance in the phonon number, we first evaluate $\langle n_{\rm ph}^2\rangle$. From the expression of $n_{\rm ph}$, we have

$$\begin{aligned}\langle n_{\rm ph}^2\rangle &= \frac{1}{Z}\sum_{\boldsymbol{k}s}\sum_{\boldsymbol{k}'s'}\sum_{n} n_{\boldsymbol{k}s}n_{\boldsymbol{k}'s'}{\rm e}^{-\mathscr{E}_n/k_{\rm B}T} \\ &= \frac{1}{Z}\sum_{\boldsymbol{k}s}\sum_{n} n_{\boldsymbol{k}s}^2{\rm e}^{-\mathscr{E}_n/k_{\rm B}T} + \frac{1}{Z}\sum_{\boldsymbol{k}s\neq\boldsymbol{k}'s'}\sum_{n} n_{\boldsymbol{k}s}n_{\boldsymbol{k}'s'}{\rm e}^{-\mathscr{E}_n/k_{\rm B}T}.\end{aligned}$$

The first term can be evaluated as follows

$$
\begin{aligned}
\frac{1}{Z}\sum_{\boldsymbol{k}s}\sum_{n} n_{\boldsymbol{k}s}^2 \mathrm{e}^{-\mathscr{E}_n/k_{\mathrm{B}}T}
&= \frac{1}{Z}\sum_{\boldsymbol{k}s}\sum_{n_{\boldsymbol{k}s}} n_{\boldsymbol{k}s}^2 \exp\big(-n_{\boldsymbol{k}s}\beta\hbar\omega_{\boldsymbol{k}s}\big)\frac{\sum_n \mathrm{e}^{-\mathscr{E}_n/k_{\mathrm{B}}T}}{\sum_{n_{\boldsymbol{k}s}}\exp\big(-n_{\boldsymbol{k}s}\beta\hbar\omega_{\boldsymbol{k}s}\big)}\\
&= \sum_{\boldsymbol{k}s}\frac{\sum_{n_{\boldsymbol{k}s}} n_{\boldsymbol{k}s}^2 \exp\big(-n_{\boldsymbol{k}s}\beta\hbar\omega_{\boldsymbol{k}s}\big)}{\sum_{n_{\boldsymbol{k}s}}\exp\big(-n_{\boldsymbol{k}s}\beta\hbar\omega_{\boldsymbol{k}s}\big)}\\
&= \sum_{\boldsymbol{k}s}\left[\frac{1}{e^{\beta\hbar\omega_{\boldsymbol{k}s}}-1}+\frac{3}{(e^{\beta\hbar\omega_{\boldsymbol{k}s}}-1)^2}+\frac{2}{(e^{\beta\hbar\omega_{\boldsymbol{k}s}}-1)^3}\right]\\
&= \sum_{\boldsymbol{k}s}\big[n_{\mathrm{B}}(\hbar\omega_{\boldsymbol{k}s})+3n_{\mathrm{B}}^2(\hbar\omega_{\boldsymbol{k}s})+2n_{\mathrm{B}}^3(\hbar\omega_{\boldsymbol{k}s})\big].
\end{aligned}
$$

For the second term in $\langle n_{\mathrm{ph}}^2\rangle$, we have

$$
\begin{aligned}
&\frac{1}{Z}\sum_{\boldsymbol{k}s\neq\boldsymbol{k}'s'}\sum_{n} n_{\boldsymbol{k}s}n_{\boldsymbol{k}'s'}\mathrm{e}^{-\mathscr{E}_n/k_{\mathrm{B}}T}\\
&\quad= \sum_{\boldsymbol{k}s\neq\boldsymbol{k}'s'}\frac{\sum_{n_{\boldsymbol{k}s}} n_{\boldsymbol{k}s}\exp\big(-n_{\boldsymbol{k}s}\beta\hbar\omega_{\boldsymbol{k}s}\big)}{\sum_{n_{\boldsymbol{k}s}}\exp\big(-n_{\boldsymbol{k}s}\beta\hbar\omega_{\boldsymbol{k}s}\big)}\\
&\quad\quad \frac{\sum_{n_{\boldsymbol{k}'s'}} n_{\boldsymbol{k}'s'}\exp\big(-n_{\boldsymbol{k}'s'}\beta\hbar\omega_{\boldsymbol{k}'s'}\big)}{\sum_{n_{\boldsymbol{k}'s'}}\exp\big(-n_{\boldsymbol{k}'s'}\beta\hbar\omega_{\boldsymbol{k}'s'}\big)}\\
&\quad= \sum_{\boldsymbol{k}s\neq\boldsymbol{k}'s'}\frac{1}{e^{\beta\hbar\omega_{\boldsymbol{k}s}}-1}\frac{1}{e^{\beta\hbar\omega_{\boldsymbol{k}'s'}}-1} = \sum_{\boldsymbol{k}s\neq\boldsymbol{k}'s'} n_{\mathrm{B}}(\hbar\omega_{\boldsymbol{k}s})n_{\mathrm{B}}(\hbar\omega_{\boldsymbol{k}'s'}).
\end{aligned}
$$

We have thus found the following expression for $\langle n_{\mathrm{ph}}^2\rangle$

$$
\begin{aligned}
\langle n_{\mathrm{ph}}^2\rangle &= \sum_{\boldsymbol{k}s}\big[n_{\mathrm{B}}(\hbar\omega_{\boldsymbol{k}s})+3n_{\mathrm{B}}^2(\hbar\omega_{\boldsymbol{k}s})+2n_{\mathrm{B}}^3(\hbar\omega_{\boldsymbol{k}s})\big]\\
&\quad+\sum_{\boldsymbol{k}s\neq\boldsymbol{k}'s'} n_{\mathrm{B}}(\hbar\omega_{\boldsymbol{k}s})n_{\mathrm{B}}(\hbar\omega_{\boldsymbol{k}'s'}).
\end{aligned}
$$

The variance of the number of phonons is finally given by

$$
\begin{aligned}
\mathrm{var}(n_{\mathrm{ph}}) &= \langle n_{\mathrm{ph}}^2\rangle - \langle n_{\mathrm{ph}}\rangle^2\\
&= \sum_{\boldsymbol{k}s}\big[n_{\mathrm{B}}(\hbar\omega_{\boldsymbol{k}s})+3n_{\mathrm{B}}^2(\hbar\omega_{\boldsymbol{k}s})+2n_{\mathrm{B}}^3(\hbar\omega_{\boldsymbol{k}s})\big]\\
&\quad+\sum_{\boldsymbol{k}s\neq\boldsymbol{k}'s'} n_{\mathrm{B}}(\hbar\omega_{\boldsymbol{k}s})n_{\mathrm{B}}(\hbar\omega_{\boldsymbol{k}'s'}) - \left[\sum_{\boldsymbol{k}s} n_{\mathrm{B}}(\hbar\omega_{\boldsymbol{k}s})\right]^2\\
&= \sum_{\boldsymbol{k}s} n_{\mathrm{B}}(\hbar\omega_{\boldsymbol{k}s})\big[1+2n_{\mathrm{B}}(\hbar\omega_{\boldsymbol{k}s})+2n_{\mathrm{B}}^2(\hbar\omega_{\boldsymbol{k}s})\big].
\end{aligned}
$$

14-11 Grüneisen parameter of a 1D crystal of inert gas atoms. Reexamine the Grüneisen parameter $\gamma(T)$ of a one-dimensional crystal of inert gas atoms with the phonon dispersion relation given by $\omega_k = (4K/m)^{1/2}|\sin(ka/2)|$. *i.* Show algebraically that $\gamma(0) = -1.0$ at $T = 0$. *ii.* Show analytically that $\gamma(T \to \infty) = -\ln 2$ as $T \to \infty$. *iii.* Evaluate numerically $\gamma(T)$ for T/Θ from 0 to 1.5 and plot the results. Here $\Theta = \hbar(4K/m)^{1/2}/k_B$.

---

i. For this one-dimensional crystal, the Grüneisen parameter is given by

$$\gamma(T) = -\frac{\sum_k (ka/2)\cot(ka/2) E(\hbar\omega_k/2k_B T)}{\sum_k E(\hbar\omega_k/2k_B T)},$$

where $E(x)$ is the Einstein function $E(x) = x^2/\sinh^2(x)$. Note that the mode Grüneisen parameter as well as the Grüneisen parameter are all negative. Since $E(x)$ tends to infinity as x goes to infinity, we see that, at $T = 0$, only the acoustical normal mode at $k = 0$ makes a contribution to the Grüneisen parameter. Canceling the Einstein functions from the numerator and denominator and taking the $k \to 0$ limit, we have

$$\begin{aligned}\gamma(0) &= -\lim_{k\to 0}\frac{(ka/2)\cos(ka/2)}{\sin(ka/2)} = -\lim_{k\to 0}\frac{(ka/2)}{\sin(ka/2)} \\ &= -\lim_{k\to 0}\frac{1}{\cos(ka/2)} = -1.\end{aligned}$$

ii. As $T \to \infty$, $\hbar\omega_k/2k_B T \to 0$. Since $E(x) \to 1$ as $x \to 0$, the Grüneisen parameter in the $T \to \infty$ limit is given by

$$\gamma(T \to \infty) = -\frac{a}{2N}\sum_k k\cot(ka/2),$$

where N is the number of atoms in the crystal. The length of the crystal is given by $L = Na$. Converting the summation over k into an integration over k yields

$$\begin{aligned}\gamma(T \to \infty) &= -\frac{a^2}{4\pi}\int_{-\pi/a}^{\pi/a} dk\; k\cot(ka/2) = -\frac{2}{\pi}\int_0^{\pi/2} dx\; x\cot x \\ &= \frac{2}{\pi}\int_0^{\pi/2} dx\; \ln\sin x = -\ln 2.\end{aligned}$$

The above integral is an improper integral. We now give details of its evaluation. Making a change of integration variables from x to $y = \pi/2 - x$, we have

$$I \stackrel{\text{def}}{=} \int_0^{\pi/2} dx\; \ln\sin x = -\int_{\pi/2}^{0} dy\; \ln\cos y = \int_0^{\pi/2} dy\; \ln\cos y.$$

Adding up the two expressions and then dividing the result by two, we have

$$I = -\frac{\pi}{4}\ln 2 + \frac{1}{2}\int_0^{\pi/2} dx\ \ln\sin(2x) = -\frac{\pi}{4}\ln 2 + \frac{1}{4}\int_0^{\pi} dy\ \ln\sin y.$$

Splitting the integration interval into $(0, \pi]$ and $[\pi/2, \pi)$ and then making a change of integration variables from y to $z = y - \pi/2$ in the second integral, we have

$$\begin{aligned} I &= -\frac{\pi}{4}\ln 2 + \frac{1}{4}\int_0^{\pi/2} dy\ \ln\sin y + \frac{1}{4}\int_0^{\pi/2} dz\ \ln\cos z \\ &= -\frac{\pi}{4}\ln 2 + \frac{1}{2}\int_0^{\pi/2} dy\ \ln\sin y. \end{aligned}$$

Thus, the value of the integral is given by $I = -(\pi/2)\ln 2$.

iii. The values of the Grüneisen parameter can be obtained by evaluating numerically the integral in its expression, with the results given in Fig. 14.3 as a plot of the Grüneisen parameter as a function of the temperature. The integration was performed using the adaptive C integration function **int gsl_integration_qags** implemented in the GNU Scientific Library.

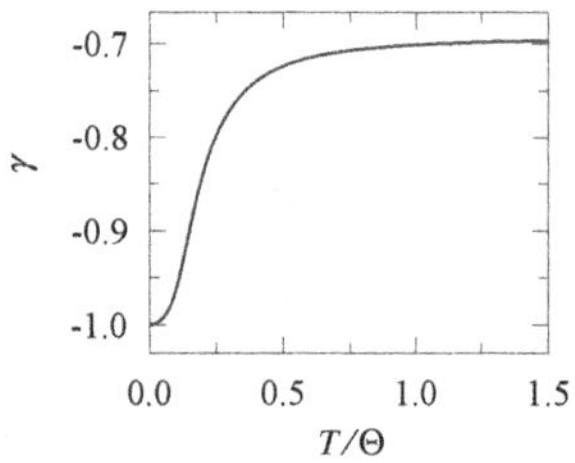

Fig. 14.3 Plot of the Grüneisen parameter as a function of temperature.

Chapter 15

Determination of Phonon Dispersion Relations by Inelastic Neutron Scattering

(1) *Triple-axis spectrometer*
A triple-axis spectrometer allows the measurement of the dynamic structure factor at any point in energy and momentum space physically accessible to the spectrometer.
The three axes are determined by
the monochromator,
the sample,
and the analyzer.
The "constant Q" mode is the one in which the momentum transfer of neutrons is fixed.

(2) *Time-of-flight-spectrometer*
The four major elements in such a spectrometer are *the monochromator* (that selects neutrons), *the chopper* (that controls passing of neutrons), *the monitor* (that keeps time), and *the detectors* (that measure the scattered neutrons).

(3) *Neutron-crystal interaction*

$$\hat{H}_{\text{int}} = \frac{2\pi\hbar^2}{m_n} \sum_{j\nu} b_\nu \delta(\boldsymbol{r} - \boldsymbol{r}_{j\nu}).$$

(4) *Double differential cross-section*

$$\frac{\mathrm{d}^2\sigma}{\mathrm{d}\Omega \mathrm{d}E'} = \frac{q'}{q} \left(\frac{m_n \mathscr{V}}{2\pi\hbar^2} \right)^2 \left| \langle \psi' | \, \hat{H}_{\text{int}} \, | \psi \rangle \right|^2 \delta(E_t' - E_t).$$

(5) *Debye-Waller factor*

$$W_\nu = \sum_{\boldsymbol{k}s} \frac{\hbar}{4N m_\nu \omega_{\boldsymbol{k}s}} \left| \boldsymbol{Q} \cdot \boldsymbol{\epsilon}_\nu^{(s)}(\boldsymbol{k}) \right|^2 \coth\left(\frac{\hbar\omega_{\boldsymbol{k}s}}{2k_\mathrm{B}T} \right).$$

(6) *Geometric structure factor for elastic scattering*

$$S_n(\boldsymbol{K}) = -\sum_\nu b_\nu \mathrm{e}^{-W_\nu - \mathrm{i}\boldsymbol{K}\cdot\boldsymbol{d}_\nu}.$$

(7) *Geometric structure factor for inelastic scattering*

$$\mathscr{S}^{(s)}(\boldsymbol{Q}) = -\sum_\nu \frac{b_\nu}{\sqrt{m_\nu}} \boldsymbol{Q}\cdot\boldsymbol{\epsilon}_\nu^{(s)}(\boldsymbol{Q})\mathrm{e}^{-W_\nu + \mathrm{i}\boldsymbol{Q}\cdot\boldsymbol{d}_\nu}.$$

(8) *Dynamical structure factor*

$$\begin{aligned} S(\boldsymbol{Q},\,\omega) = \frac{(2\pi)^3[n_{\mathrm{B}}(\hbar\omega)+1]}{2pN\langle b^2\rangle v_c\omega}\sum_{\boldsymbol{K}\boldsymbol{k}s}\left|\mathscr{S}^{(s)}(\boldsymbol{Q})\right|^2\big[\delta(\omega-\omega_{\boldsymbol{k}s})\delta(\boldsymbol{Q}-\boldsymbol{k}-\boldsymbol{K}) \\ + \delta(\omega+\omega_{\boldsymbol{k}s})\delta(\boldsymbol{Q}+\boldsymbol{k}-\boldsymbol{K})\big], \end{aligned}$$

$$S(\boldsymbol{Q},\,-\omega) = \mathrm{e}^{-\hbar\omega/k_{\mathrm{B}}T}S(\boldsymbol{Q},\,\omega),$$

$$\frac{\mathrm{d}^2\sigma}{\mathrm{d}\Omega\mathrm{d}E'} = \frac{pNq'}{q}\langle b^2\rangle S(\boldsymbol{Q},\,\omega).$$

15-1 Debye-Waller factors in spaces of different dimensionality. Study the Debye-Waller factor in spaces of different dimensions for thermal neutrons. For phonons, the linear dispersion relation $\omega = ck$ can be used.

(1) Take $\boldsymbol{k}$ as a wave vector in d dimensions and convert the summation over $\boldsymbol{k}$ in the Debye-Waller factor into an integration over $\boldsymbol{k}$.

(2) Show that the Debye-Waller factor W diverges in one and two dimensions.

(3) Evaluate the Debye-Waller factor W in three dimensions.

---

(1) The Debye-Waller factor for the νth atom in a primitive cell is given by

$$W_\nu = \sum_{\boldsymbol{k}s}\frac{\hbar}{4Nm_\nu\omega_{\boldsymbol{k}s}}\left|\boldsymbol{Q}\cdot\boldsymbol{\epsilon}_\nu^{(s)}(\boldsymbol{k})\right|^2\coth\left(\frac{\hbar\omega_{\boldsymbol{k}s}}{2k_{\mathrm{B}}T}\right).$$

Here we consider a monatomic crystal without a multi-atom basis using the Debye model in which only three acoustical phonon branches are included. We then have

$$W = \sum_{\boldsymbol{k}s}\frac{\hbar}{4Nmck}\left|\boldsymbol{Q}\cdot\boldsymbol{\epsilon}^{(s)}(\boldsymbol{k})\right|^2\coth\left(\frac{\hbar ck}{2k_{\mathrm{B}}T}\right).$$

For $\sum_s \left|\boldsymbol{Q}\cdot\boldsymbol{\epsilon}^{(s)}(\boldsymbol{k})\right|^2$, making use of the orthonormality relation of the polarization vectors, we have

$$\sum_s \left|\boldsymbol{Q}\cdot\boldsymbol{\epsilon}^{(s)}(\boldsymbol{k})\right|^2 = \sum_{\alpha\beta} Q_\alpha Q_\beta \sum_s \epsilon_\alpha^{(s)}\epsilon_\beta^{(s)*} = \sum_{\alpha\beta} Q_\alpha Q_\beta \delta_{\alpha\beta} = |\boldsymbol{Q}|^2.$$

We then have

$$W = \frac{\hbar|\boldsymbol{Q}|^2}{4Nmc}\sum_{\boldsymbol{k}}\frac{1}{k}\coth\left(\frac{\hbar ck}{2k_{\mathrm{B}}T}\right),$$

where we have treated the wave vector transfer $\boldsymbol{Q}$ as a constant vector and taken $|\boldsymbol{Q}|^2$ out of the summation over $\boldsymbol{k}$. In d-dimensional k-space, the volume element is given by $\mathrm{d}\boldsymbol{k} = k^{d-1}\mathrm{d}k\mathrm{d}\Omega_d$ with $\mathrm{d}\Omega_d$ given by

$$\mathrm{d}\Omega_d = \mathrm{d}\theta_1 \sin\theta_2 \mathrm{d}\theta_2 \sin^2\theta_3 \mathrm{d}\theta_3 \cdots \sin^{d-2}\theta_{d-1}\mathrm{d}\theta_{d-1}.$$

The surface of a hypersphere of unit radius in d-dimensional space is given by

$$\begin{aligned} S_d = \int \mathrm{d}\Omega_d &= \int_0^{2\pi}\mathrm{d}\theta_1 \int_0^{\pi}\sin\theta_2\mathrm{d}\theta_2 \int_0^{\pi}\sin^2\theta_3\mathrm{d}\theta_3 \cdots \\ &\quad \times \int_0^{\pi}\sin^{d-2}\theta_{d-1}\mathrm{d}\theta_{d-1} \\ &= 2\pi\prod_{j=1}^{d-2}\int_0^{\pi}\sin^j\theta\mathrm{d}\theta = 2^{d-1}\pi\prod_{j=1}^{d-2}\int_0^{\pi/2}\sin^j\theta\mathrm{d}\theta, \end{aligned}$$

where we have taken into consideration that the integral of $\sin^j\theta$ over θ in the interval $[0,\pi/2]$ is equal to that in the interval $[\pi/2,\pi]$. We only consider the case with the dimensionality of the k-space being an integer. We evaluate the integrals over θ separately for even and odd j's. For an even j, we set $j = 2n$ and make a change of integration variables from θ to $x = \tan\theta$. We then have

$$\begin{aligned} \int_0^{\pi/2}\sin^{2n}\theta\,\mathrm{d}\theta &= \int_0^{\infty}\mathrm{d}x\,\frac{x^{2n}}{(x^2+1)^{n+1}} \\ &= \frac{1}{2}\int_{-\infty}^{\infty}\mathrm{d}x\,\frac{x^{2n}}{(x^2+1)^{n+1}} \\ &= \frac{\mathrm{i}\pi}{n!}\lim_{x\to i}\frac{\mathrm{d}^n}{\mathrm{d}x^n}\frac{x^{2n}}{(x+i)^{n+1}} \\ &= \frac{(2n-1)!!}{(2n)!!}\frac{\pi}{2}, \end{aligned}$$

where we have closed the integration contour in the upper-half complex plane and made use of the residue theorem. For an odd j, we set $j = 2n+1$, make a change of integration variables from θ to $x = \sin\theta$, and repeatedly perform integration by parts. We have

$$\begin{aligned}
&\int_0^{\pi/2} \sin^{2n+1}\theta \, \mathrm{d}\theta \\
&= \int_0^1 \mathrm{d}x \, \frac{x^{2n+1}}{\sqrt{1-x^2}} \\
&= -\int_0^1 \mathrm{d}\sqrt{1-x^2}\, x^{2n} \\
&= (2n)\int_0^1 \mathrm{d}x \, x^{2n-1}\sqrt{1-x^2} \\
&= -\frac{1}{3}(2n)\int_0^1 \mathrm{d}[\,(1-x^2)^{3/2}\,]\, x^{2n-2} \\
&= \frac{1}{3}(2n)(2n-2)\int_0^1 \mathrm{d}x \;\; x^{2n-3}(1-x^2)^{3/2} = \cdots \\
&= \frac{1}{3}\frac{1}{5}\cdots\frac{1}{2n-1}(2n)(2n-2)\cdots(2)\int_0^1 \mathrm{d}x \;\; x(1-x^2)^{(2n-1)/2} \\
&= \frac{(2n)!!}{(2n+1)!!}.
\end{aligned}$$

When we insert the above results for integrals into the equation for S_d, we will find that the factorials are all canceled except the last one on the denominator. We then have $S_d = (2\pi)^{d/2}/(d-2)!!$ for an even d and $S_d = (2/\pi)^{1/2}(2\pi)^{d/2}/(d-2)!!$ for an odd d. Making use of $\Gamma(m+1) = m\Gamma(m)$, $\Gamma(m+1/2) = (m-1/2)(m-3/2)\cdots(1/2)\Gamma(1/2)$, and $\Gamma(1/2) = \sqrt{\pi}$, we find that S_d for both even d's and odd d's can be written as $S_d = 2\pi^{d/2}/\Gamma(d/2)$. This result can be also generalized to fractional d's.

Converting the summation over $\boldsymbol{k}$ in W into an integration over $\boldsymbol{k}$ and performing the trivial integration over Ω_d, we have

$$W = \frac{\hbar|\boldsymbol{Q}|^2\mathscr{V}}{2^{d+1}\pi^{d/2}\Gamma(d/2)Nmc}\int_0^{k_\mathrm{D}} \mathrm{d}k\, k^{d-2}\coth\!\left(\frac{\hbar ck}{2k_\mathrm{B}T}\right).$$

(2) Let us examine the behavior of the integrand at the lower limit. For $k \to 0$, $\coth\big(\hbar ck/2k_\mathrm{B}T\big) \to 1/k$. Thus, the integrand tends to k^{d-3} as $k \to 0$, which makes the integral diverge for $d = 1$ and $d = 2$.

(3) For $d = 3$, we have

$$W = \frac{\hbar|\boldsymbol{Q}|^2\mathscr{V}}{8\pi^2 Nmc}\int_0^{k_\mathrm{D}} \mathrm{d}k\, k\coth\left(\frac{\hbar ck}{2k_\mathrm{B}T}\right).$$

For temperatures much higher than the Debye temperature $\Theta_\mathrm{D} = \hbar ck_\mathrm{D}/k_\mathrm{B}$, we have $\hbar ck/2k_\mathrm{B}T \ll 1$ in the entire integration interval [*cf.* Fig. 15.1 for a plot of $\coth x$ as a function of x]. In this case, we can approximate $\coth(\hbar ck/2k_\mathrm{B}T)$ as $1/(\hbar ck/2k_\mathrm{B}T)$. We then have

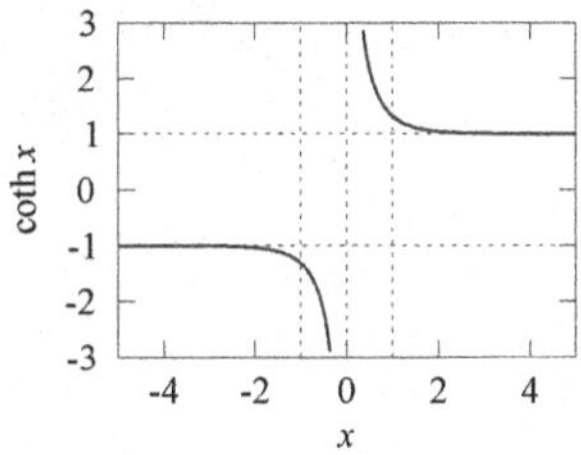

Fig. 15.1 Plot of $\coth x$ as a function of x.

$$W = \frac{|\boldsymbol{Q}|^2\mathscr{V}k_\mathrm{B}T}{4\pi^2 Nmc^2}\int_0^{k_\mathrm{D}}\mathrm{d}k = \frac{|\boldsymbol{Q}|^2\mathscr{V}k_\mathrm{B}Tk_\mathrm{D}}{4\pi^2 Nmc^2}.$$

Making use of $k_\mathrm{D}^3 = 6\pi^3 N/\mathscr{V}$ and taking the value of $\hbar^2|\boldsymbol{Q}|^2/2m$ approximately as the thermal energy of an atom at temperature T, $\hbar^2|\boldsymbol{Q}|^2/2m \approx 3k_\mathrm{B}T/2$, we have

$$W \approx \frac{9}{2}\left(\frac{T}{\Theta_\mathrm{D}}\right)^2, \quad T \gg \Theta_\mathrm{D}.$$

For temperatures much lower than the Debye temperature Θ_D, $T \ll \Theta_\mathrm{D}$, we first cast the integral into a dimensionless form and then make use of $\coth x \approx 1/x$ for $x < 1$ and ≈ 1 for $x > 1$. We have in this case

$$\begin{aligned} W &= \frac{\hbar|\boldsymbol{Q}|^2\mathscr{V}}{8\pi^2 Nmc}\left(\frac{2k_\mathrm{B}T}{\hbar c}\right)^2\int_0^{\Theta_\mathrm{D}/2T}\mathrm{d}x\, x\coth x \\ &\approx 9\left(\frac{T}{\Theta_\mathrm{D}}\right)^3\left(\int_0^1 \mathrm{d}x + \int_1^{\Theta_\mathrm{D}/2T}\mathrm{d}x\, x\right) \\ &\approx \frac{9}{8}\frac{T}{\Theta_\mathrm{D}}, \quad T \ll \Theta_\mathrm{D}. \end{aligned}$$

We now perform the integration over k numerically. For convenience, we introduce the variable $x = \hbar ck/k_\mathrm{B}\Theta_\mathrm{D}$. The expression for W then becomes

$$W = \frac{9}{4}\frac{T}{\Theta_\mathrm{D}}\int_0^1 \mathrm{d}x\, x\coth\left(\frac{\Theta_\mathrm{D}x}{2T}\right).$$

For a given value of T/Θ_D, we numerically evaluate the above integral with the results given in Fig. 15.2 as a plot of W as a function T/Θ_D.

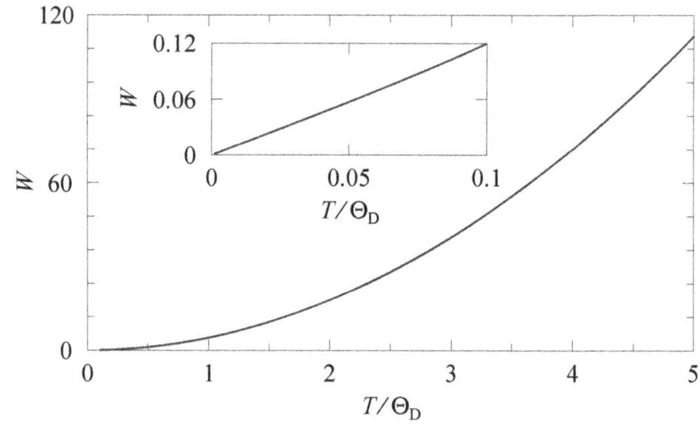

Fig. 15.2 Plot of the Debye-Waller factor W as a function of temperature T. The inset is for low temperatures.

From Fig. 15.2, it is clearly seen that, at high temperatures, W is parabolic in T. The linear temperature dependence of W at low temperatures is obvious from the inset to Fig. 15.2.

15-2 Q-ω region accessible for inelastic neutron scattering. From $\boldsymbol{Q} = \boldsymbol{q} - \boldsymbol{q}'$, we have

$$Q^2 = q^2 + q'^2 - 2qq' \cos\vartheta$$

with ϑ the scattering angle, $\vartheta = \angle(\boldsymbol{q}', \boldsymbol{q})$.

(1) Using the above relation, derive an analytic expression for Q as a function of ω ($\hbar\omega = E - E'$), E, and ϑ.

(2) For $E = 4$ meV, plot ω versus Q for $\vartheta = 0^\circ$ through 180° with a spacing of 20°. Such a plot is the Q-ω region accessible for inelastic neutron scattering at the given incident neutron energy E.

(3) Evaluate $\mathrm{d}\omega/\mathrm{d}Q$ at $\vartheta = 0^\circ$ and $\omega \to 0$ and show that its value is given by $\pm v$ with v the incident neutron velocity.

---

(1) Multiplying both sides of

$$Q^2 = q^2 + q'^2 - 2qq' \cos\vartheta$$

with $\hbar^2/2m$ and then making use of

$$E = \hbar^2 q^2/2m, \ E' = \hbar^2 {q'}^2/2m, \ E' = E - \hbar\omega,$$

we have

$$\frac{\hbar^2 Q^2}{2m} = 2E - \hbar\omega - 2\sqrt{E(E - \hbar\omega)}\,\cos\vartheta.$$

(2) We take Q to be in nm^{-1}. The value of $\hbar^2(1\ \mathrm{nm}^{-1})^2/2m$ is given by $1.964\,897 \times 10^{-5}$ eV. The plot of ω as a function of Q for $\vartheta = 0^\circ$ through 180° with a spacing of 20° is given in Fig. 15.3. The kinetic energy of the incident neutrons is set to be 4 meV.

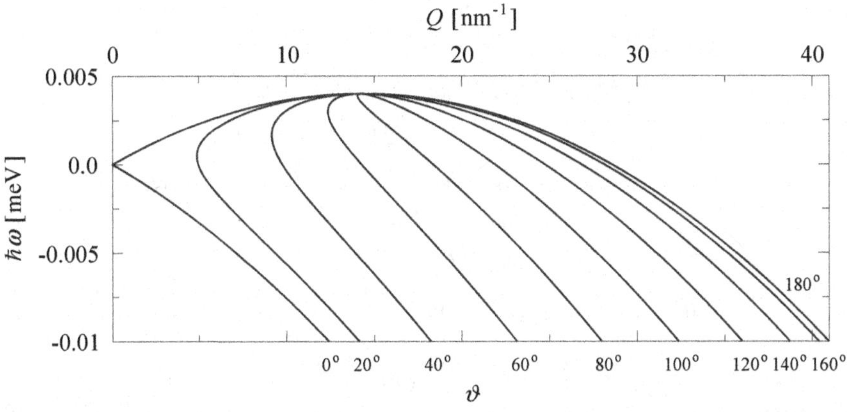

Fig. 15.3 Plot of ω as a function of Q for $\vartheta = 0^\circ$ through 180° with a spacing of 20°. The value of the kinetic energy of the incident neutrons used in this plot is 4 meV.

From Fig. 15.3, we see that the applicable region for measuring phonon dispersion relations is the region with small energy and momentum transfers in the lower-left corner.

(3) Differentiating both sides of the equation for Q^2 with respect to $\hbar\omega$, we obtain

$$\frac{\hbar Q}{m}\frac{\mathrm{d}Q}{\mathrm{d}\omega} = -1 + \left(\frac{E}{E - \hbar\omega}\right)^{1/2} \cos\vartheta.$$

Thus,

$$\frac{\mathrm{d}\omega}{\mathrm{d}Q} = \frac{\hbar Q}{m}\frac{1}{\left[E/(E - \hbar\omega)\right]^{1/2}\cos\vartheta - 1}.$$

At $\vartheta = 0^{\circ}$, we have

$$\frac{\mathrm{d}\omega}{\mathrm{d}Q} = \frac{\hbar Q}{m}\frac{1}{\big[E/(E-\hbar\omega)\big]^{1/2}-1}.$$

For $\omega \to 0$, we can expand the square root and obtain

$$\begin{aligned}\frac{\mathrm{d}\omega}{\mathrm{d}Q} &= \frac{\hbar Q}{m}\frac{1}{\big[1/(1-\hbar\omega/E)\big]^{1/2}-1}\\ &\approx \frac{\hbar Q}{m}\frac{1}{(1+\hbar\omega/2E)-1}\\ &= \frac{2E(\hbar Q)}{m\hbar\omega}\\ &= \frac{\hbar Q}{\hbar\omega}v^2.\end{aligned}$$

For $(\hbar Q)^2/2m$ at $\vartheta = 0^{\circ}$ with $\omega \to 0$, we have

$$\begin{aligned}\hbar^2Q^2/2m &= 2E - \hbar\omega - 2\sqrt{E(E-\hbar\omega)}\\ &= 2E - \hbar\omega - 2E\sqrt{1-\hbar\omega/E}\\ &\approx 2E - \hbar\omega - 2E\big[1-\hbar\omega/2E-(\hbar\omega/E)^2/8\big]\\ &= (\hbar\omega)^2/4E\\ &= (\hbar\omega)^2/2mv^2.\end{aligned}$$

Thus, $\hbar Q/\hbar\omega = \pm v^{-1}$ and

$$\frac{\mathrm{d}\omega}{\mathrm{d}Q} = \pm v.$$

15-3 Cumulant expansion. Prove the following cumulant expansion

$$\langle \mathrm{e}^{\hat{A}}\rangle = \exp\bigg\{\langle\hat{A}\rangle + \frac{1}{2!}\big[\langle\hat{A}^2\rangle - \langle\hat{A}\rangle^2\big] + \frac{1}{3!}\big[\langle\hat{A}^3\rangle - 3\langle A^2\rangle\langle\hat{A}\rangle + 2\langle\hat{A}\rangle^2\big]$$
$$+ \frac{1}{4!}\big[\langle\hat{A}^4\rangle - 4\langle\hat{A}^3\rangle\langle\hat{A}\rangle - 3\langle\hat{A}^2\rangle^2 + 12\langle\hat{A}^2\rangle\langle\hat{A}\rangle^2 - 6\langle\hat{A}\rangle^4\big] + \cdots\bigg\}.$$

For convenience in comparing terms of same orders, we introduce a formal small parameter λ and associate it with $\hat{A}$. We now evaluate $\langle \mathrm{e}^{\lambda\hat{A}}\rangle$ instead of $\langle \mathrm{e}^{\hat{A}}\rangle$. At the end of computations, we set $\lambda = 1$. Expanding $\mathrm{e}^{\lambda\hat{A}}$ as a Taylor series in $\lambda\hat{A}$, we have

$$\langle \mathrm{e}^{\lambda\hat{A}}\rangle = 1 + \lambda\langle\hat{A}\rangle + \frac{1}{2!}\lambda^2\langle\hat{A}^2\rangle + \frac{1}{3!}\lambda^3\langle\hat{A}^3\rangle + \frac{1}{4!}\lambda^4\langle\hat{A}^4\rangle + \cdots.$$

We now write $\langle \mathrm{e}^{\lambda\hat{A}}\rangle$ in the following form

$$\langle \mathrm{e}^{\lambda\hat{A}}\rangle = \exp\bigg[\lambda C_1 + \frac{1}{2!}\lambda^2 C_2 + \frac{1}{3!}\lambda^3 C_3 + \frac{1}{4!}\lambda^4 C_4 + \cdots\bigg].$$

The above expression will be referred to as the second expression of $\langle e^{\lambda\hat{A}}\rangle$. Expanding the above expression as a Taylor series in λ, we have

$$\begin{aligned}\langle e^{\lambda\hat{A}}\rangle &= 1+\lambda C_1+\frac{1}{2!}\lambda^2C_2+\frac{1}{3!}\lambda^3C_3+\frac{1}{4!}\lambda^4C_4\\ &\quad+\frac{1}{2}\lambda^2C_1^2+\frac{1}{8}\lambda^4C_2^2+\frac{1}{2}\lambda^3C_1C_2\\ &\quad+\frac{1}{6}\lambda^4C_1C_3+\frac{1}{3!}\lambda^3C_1^3+\frac{1}{4}\lambda^4C_1^2C_2+\frac{1}{4!}\lambda^4C_1^4+\cdots\\ &= 1+\lambda C_1+\frac{1}{2!}\lambda^2(C_1^2+C_2)\\ &\quad+\frac{1}{3!}\lambda^3(C_3+3C_1C_2+C_1^3)\\ &\quad+\frac{1}{4!}\lambda^4(C_4+3C_2^2+4C_1C_3+6C_1^2C_2+C_1^4).\end{aligned}$$

Equating the coefficients of the terms with the same powers in λ in the two Taylor series of $\langle e^{\lambda\hat{A}}\rangle$ yields

$$\begin{aligned}&C_1=\langle\hat{A}\rangle,\\ &C_1^2+C_2=\langle\hat{A}^2\rangle,\\ &C_3+3C_1C_2+C_1^3=\langle\hat{A}^3\rangle,\\ &C_4+3C_2^2+4C_1C_3+6C_1^2C_2+C_1^4=\langle\hat{A}^4\rangle\,.\end{aligned}$$

The first equation gives us C_1 as $C_1=\langle\hat{A}\rangle$. Inserting this expression for C_1 into the second equation, we immediately obtain the expression for C_2

$$C_2=\langle\hat{A}^2\rangle-\langle\hat{A}\rangle^2\,.$$

Inserting the expressions for C_1 and C_2 into the third equation, we obtain

$$\begin{aligned}C_3 &= \langle\hat{A}^3\rangle-3C_1C_2-C_1^3\\ &= \langle\hat{A}^3\rangle-3\langle\hat{A}\rangle\left[\langle\hat{A}^2\rangle-\langle\hat{A}\rangle^2\right]-\langle\hat{A}\rangle^3\\ &= \langle\hat{A}^3\rangle-3\langle\hat{A}^2\rangle\langle\hat{A}\rangle+2\langle\hat{A}\rangle^3\,.\end{aligned}$$

From the fourth equation, we obtain

$$\begin{aligned}C_4 &= \langle\hat{A}^4\rangle-3C_2^2-4C_1C_3-6C_1^2C_2-C_1^4\\ &= \langle\hat{A}^4\rangle-3(\langle\hat{A}^2\rangle-\langle\hat{A}\rangle^2)^2\\ &\quad-4\langle\hat{A}\rangle(\langle\hat{A}^3\rangle-3\langle\hat{A}^2\rangle\langle\hat{A}\rangle+2\langle\hat{A}\rangle^3)\\ &\quad-6\langle\hat{A}\rangle^2(\langle\hat{A}^2\rangle-\langle\hat{A}\rangle^2)-\langle\hat{A}\rangle^4\\ &= \langle\hat{A}^4\rangle-4\langle\hat{A}^3\rangle\langle\hat{A}\rangle-3\langle\hat{A}^2\rangle^2+12\langle\hat{A}^2\rangle\langle\hat{A}\rangle^2-6\langle\hat{A}\rangle^4\,.\end{aligned}$$

Inserting the above expressions for C_1, C_2, C_3, and C_4 into the second expression of $\langle e^{\lambda\hat{A}}\rangle$, we can obtain the desired cumulant expansion after setting λ to unity. C_1, C_2, C_3, and C_4 are referred to as the first-, second-, third-, and fourth-cumulants.

15-4 Property of the dynamical structure factor. Derive the property of the dynamical structure factor, $S(\boldsymbol{Q},\,-\omega) = e^{-\hbar\omega/k_BT}S(\boldsymbol{Q},\,\omega)$, from its definition.

The dynamical structure factor is defined by

$$S(\boldsymbol{Q},\,\omega) = \frac{(2\pi)^3}{2pN\langle b^2\rangle v_c}\frac{n_B(\hbar\omega)+1}{\omega}\sum_{\boldsymbol{K}\boldsymbol{k}s}\left|\mathscr{S}^{(s)}(\boldsymbol{Q})\right|^2 \times\left[\delta(\omega-\omega_{\boldsymbol{k}s})\delta(\boldsymbol{Q}-\boldsymbol{k}-\boldsymbol{K})+\delta(\omega+\omega_{\boldsymbol{k}s})\delta(\boldsymbol{Q}+\boldsymbol{k}-\boldsymbol{K})\right].$$

Replacing ω in the above expression with $-\omega$, we have

$$S(\boldsymbol{Q},\,-\omega) = -\frac{(2\pi)^3}{2pN\langle b^2\rangle v_c}\frac{n_B(-\hbar\omega)+1}{\omega}\sum_{\boldsymbol{K}\boldsymbol{k}s}\left|\mathscr{S}^{(s)}(\boldsymbol{Q})\right|^2 \times\left[\delta(\omega-\omega_{\boldsymbol{k}s})\delta(\boldsymbol{Q}-\boldsymbol{k}-\boldsymbol{K})+\delta(\omega+\omega_{\boldsymbol{k}s})\delta(\boldsymbol{Q}+\boldsymbol{k}-\boldsymbol{K})\right],$$

where we have made use of the fact that the Dirac δ-function is an even function of its argument. Making use of

$$\begin{aligned} n_B(-\hbar\omega)+1 &= \frac{1}{e^{-\beta\hbar\omega}-1}+1 = \frac{e^{\beta\hbar\omega}}{1-e^{\beta\hbar\omega}}+1 \\ &= -\frac{1}{e^{\beta\hbar\omega}-1} = -e^{-\beta\hbar\omega}\frac{e^{\beta\hbar\omega}}{e^{\beta\hbar\omega}-1} \\ &= -e^{-\beta\hbar\omega}\frac{(e^{\beta\hbar\omega}-1)+1}{e^{\beta\hbar\omega}-1} \\ &= -e^{-\beta\hbar\omega}\left[n_B(\hbar\omega)+1\right], \end{aligned}$$

we have

$$S(\boldsymbol{Q},\,-\omega) = e^{-\beta\hbar\omega}S(\boldsymbol{Q},\,\omega).$$

Chapter 16

Origin of Electronic Energy Bands

(1) *Translation operator*

$$\hat{\mathscr{T}}_{\boldsymbol{R}}\eta(\boldsymbol{r}) = \eta(\boldsymbol{r}+\boldsymbol{R}),$$
$$\hat{\mathscr{T}}_{\boldsymbol{R}}\hat{H}(\boldsymbol{r}) = \hat{H}(\boldsymbol{r}+\boldsymbol{R}).$$

The properties of the translation operator $\hat{\mathscr{T}}_{\boldsymbol{R}}$ are

i. $\hat{\mathscr{T}}_{\boldsymbol{R}}\hat{\mathscr{T}}_{\boldsymbol{R}'} = \hat{\mathscr{T}}_{\boldsymbol{R}+\boldsymbol{R}'}$.

ii. $\hat{\mathscr{T}}^{\dagger}_{\boldsymbol{R}} = \hat{\mathscr{T}}_{-\boldsymbol{R}} = \hat{\mathscr{T}}^{-1}_{\boldsymbol{R}}$.

iii. $[\hat{\mathscr{T}}_{\boldsymbol{R}},\ \hat{H}] = 0$.

(2) *Bloch's theorem*

$$\psi_{n\boldsymbol{k}}(\boldsymbol{r}+\boldsymbol{R}) = \mathrm{e}^{\mathrm{i}\boldsymbol{k}\cdot\boldsymbol{R}}\psi_{n\boldsymbol{k}}(\boldsymbol{r}).$$

The alternative form of Bloch's theorem is given by

$$\psi_{n\boldsymbol{k}}(\boldsymbol{r}) = \mathrm{e}^{\mathrm{i}\boldsymbol{k}\cdot\boldsymbol{r}}u_{n\boldsymbol{k}}(\boldsymbol{r}),$$
$$u_{n\boldsymbol{k}}(\boldsymbol{r}+\boldsymbol{R}) = u_{n\boldsymbol{k}}(\boldsymbol{r}).$$

(3) *Bloch function*

$$\psi_{n\boldsymbol{k}}(\boldsymbol{r}) = \mathrm{e}^{\mathrm{i}\boldsymbol{k}\cdot\boldsymbol{r}}u_{n\boldsymbol{k}}(\boldsymbol{r}).$$

(4) *Allowed Bloch wave vectors*

$$\boldsymbol{k} = \sum_{i=1}^{3}\frac{n_i}{N_i}\boldsymbol{b}_i,\ \ n_i = 0,\ \pm1,\ \pm2,\ \pm(N_i/2-1),\ N_i/2.$$

The number of the allowed Bloch wave vectors in the first Brillouin zone is equal to the number of the primitive cells in the crystal.

(5) $\boldsymbol{k}\cdot\boldsymbol{p}$ *Hamiltonian*

$$\hat{H}_{\boldsymbol{kp}} = \frac{\hbar^2}{2m}\big(-\mathrm{i}\boldsymbol{\nabla}+\boldsymbol{k}\big)^2 + U(\boldsymbol{r}).$$

(6) *Schemes for displaying band structures*

The extended-zone scheme,
the repeated-zone scheme,
the reduced-zone scheme.

(7) *Free-electron band structures*

A free-electron band structure is the zero-periodic-potential limit of the actual band structure with the Bravais lattice kept intact.

A free-electron band structure is also known as an empty-lattice band structure.

(8) *Fermi surface*

The Fermi surface is the manifold, in k-space, of the wave vectors of the occupied single-electron states of the highest energy.

(9) *Density of states in an energy band*

$$g_{n\sigma}(E) = \frac{1}{\mathscr{V}}\sum_{\boldsymbol{k}} \delta(E - E_{n\boldsymbol{k}\sigma}) = \frac{1}{(2\pi)^3}\int \mathrm{d}\boldsymbol{k}\ \delta(E - E_{n\boldsymbol{k}\sigma})$$
$$= \frac{1}{(2\pi)^3}\int_{S_E} \frac{\mathrm{d}S}{|\nabla_{\boldsymbol{k}} E_{n\boldsymbol{k}\sigma}|},$$
$$g(E) = \sum_{n\sigma} g_{n\sigma}(E).$$

(10) *Van Hove singularities*

A van Hove singularity is a point in k-space for which $\nabla_{\boldsymbol{k}} E_{n\boldsymbol{k}\sigma} = 0$.

16-1 Quasi-momentum operator of Bloch electrons. The quantum-mechanical momentum operator $\hat{\boldsymbol{p}} = -\mathrm{i}\hbar\boldsymbol{\nabla}$ does not commute with the effective single-electron Hamiltonian of a solid, $\hat{H} = -\hbar^2\nabla^2/2m + U(\boldsymbol{r})$, where $U(\boldsymbol{r}+\boldsymbol{R}) = U(\boldsymbol{r})$ with $\boldsymbol{R}$ a lattice vector. However, it is possible to construct a quasi-momentum operator $\hat{\boldsymbol{P}}$ that commutes with $\hat{H}$ and whose eigenvalue is the quasi-momentum (or crystal momentum) $\hbar\boldsymbol{k}$

$$[\hat{H}, \hat{\boldsymbol{P}}] = 0, \quad \hat{\boldsymbol{P}}\psi_{n\boldsymbol{k}}(\boldsymbol{r}) = \hbar\boldsymbol{k}\psi_{n\boldsymbol{k}}(\boldsymbol{r}),$$

where $\psi_{n\boldsymbol{k}}(\boldsymbol{r}) = \mathrm{e}^{\mathrm{i}\boldsymbol{k}\cdot\boldsymbol{r}}u_{n\boldsymbol{k}}(\boldsymbol{r})$ is the Bloch function. Since $\hat{\boldsymbol{P}} \to \hat{\boldsymbol{p}}$ for $U(\boldsymbol{r}) = $ const, we can express quasi-momentum operator $\hat{\boldsymbol{P}}$ as $\hat{\boldsymbol{P}} = \hat{\boldsymbol{p}} + \mathrm{i}\hbar\hat{\boldsymbol{F}}$.

(1) Evaluate the commutator $[\hat{H}, \hat{\boldsymbol{p}}]$.

(2) Find the unknown operator $\hat{\boldsymbol{F}}$.

---

(1) Let $\phi(\boldsymbol{r})$ be a well-behaved function of $\boldsymbol{r}$. For the αth component of $\hat{\boldsymbol{p}}$, we have

$$\begin{aligned}[\hat{H},\hat{p}_\alpha]\phi(\boldsymbol{r}) &= [\hat{\boldsymbol{p}}^2/2m+U(\boldsymbol{r}),\hat{p}_\alpha]\phi(\boldsymbol{r}) = [U(\boldsymbol{r}),\hat{p}_\alpha]\phi(\boldsymbol{r})\\ &= U(\boldsymbol{r})\hat{p}_\alpha\phi(\boldsymbol{r}) - \hat{p}_\alpha[U(\boldsymbol{r})\phi(\boldsymbol{r})]\\ &= -\mathrm{i}\hbar U(\boldsymbol{r})\partial\phi(\boldsymbol{r})/\partial x_\alpha + \mathrm{i}\hbar\phi(\boldsymbol{r})\partial U(\boldsymbol{r})/\partial x_\alpha\\ &\quad + \mathrm{i}\hbar U(\boldsymbol{r})\partial\phi(\boldsymbol{r})/\partial x_\alpha\\ &= \mathrm{i}\hbar\phi(\boldsymbol{r})\partial U(\boldsymbol{r})/\partial x_\alpha.\end{aligned}$$

Thus,

$$[\hat{H},\hat{p}_\alpha] = \mathrm{i}\hbar\partial U(\boldsymbol{r})/\partial x_\alpha.$$

For $[\hat{H},\hat{\boldsymbol{p}}]$, we have

$$[\hat{H},\hat{\boldsymbol{p}}] = \mathrm{i}\hbar\boldsymbol{\nabla}U(\boldsymbol{r}).$$

(2) From the above result, $[\hat{H},\hat{\boldsymbol{P}}]=0$, and $\hat{\boldsymbol{P}}=\hat{\boldsymbol{p}}+\mathrm{i}\hbar\hat{\boldsymbol{F}}$, we have

$$[\hat{H},\hat{\boldsymbol{F}}] = -\boldsymbol{\nabla}U(\boldsymbol{r})$$

which must be satisfied by $\hat{\boldsymbol{F}}$ to ensure the commutativity of $\hat{\boldsymbol{P}}$ with $\hat{H}$. From the eigenequation of $\hat{\boldsymbol{P}}$, $\hat{\boldsymbol{P}}\psi_{n\boldsymbol{k}}(\boldsymbol{r}) = \hbar\boldsymbol{k}\psi_{n\boldsymbol{k}}(\boldsymbol{r})$, we have

$$\hat{\boldsymbol{p}}\psi_{n\boldsymbol{k}}(\boldsymbol{r}) + \mathrm{i}\hbar\hat{\boldsymbol{F}}\psi_{n\boldsymbol{k}}(\boldsymbol{r}) = \hbar\boldsymbol{k}\psi_{n\boldsymbol{k}}(\boldsymbol{r}).$$

Making use of

$$\begin{aligned}\hat{\boldsymbol{p}}\psi_{n\boldsymbol{k}}(\boldsymbol{r}) &= \hat{\boldsymbol{p}}[\mathrm{e}^{\mathrm{i}\boldsymbol{k}\cdot\boldsymbol{r}}u_{n\boldsymbol{k}}(\boldsymbol{r})] = \hbar\boldsymbol{k}\psi_{n\boldsymbol{k}}(\boldsymbol{r}) + \mathrm{e}^{\mathrm{i}\boldsymbol{k}\cdot\boldsymbol{r}}\hat{\boldsymbol{p}}u_{n\boldsymbol{k}}(\boldsymbol{r})\\ &= \hbar\boldsymbol{k}\psi_{n\boldsymbol{k}}(\boldsymbol{r}) + \mathrm{e}^{\mathrm{i}\boldsymbol{k}\cdot\boldsymbol{r}}\hat{\boldsymbol{p}}\mathrm{e}^{-\mathrm{i}\boldsymbol{k}\cdot\boldsymbol{r}}\psi_{n\boldsymbol{k}}(\boldsymbol{r}),\end{aligned}$$

we have

$$[\mathrm{i}\hbar\hat{\boldsymbol{F}} + \mathrm{e}^{\mathrm{i}\boldsymbol{k}\cdot\boldsymbol{r}}\hat{\boldsymbol{p}}\mathrm{e}^{-\mathrm{i}\boldsymbol{k}\cdot\boldsymbol{r}}]\psi_{n\boldsymbol{k}}(\boldsymbol{r}) = 0.$$

Hence,

$$\hat{\boldsymbol{F}} = -\frac{1}{\mathrm{i}\hbar}\mathrm{e}^{\mathrm{i}\boldsymbol{k}\cdot\boldsymbol{r}}\hat{\boldsymbol{p}}\mathrm{e}^{-\mathrm{i}\boldsymbol{k}\cdot\boldsymbol{r}}.$$

We can verify directly that the above expression for $\hat{\boldsymbol{F}}$ indeed satisfies $[\hat{H},\hat{\boldsymbol{F}}] = -\boldsymbol{\nabla}U(\boldsymbol{r})$. Making use of the above expression for $\hat{\boldsymbol{F}}$, we have

$$\begin{aligned}[\hat{H},\hat{\boldsymbol{F}}] &= -(\mathrm{i}\hbar)^{-1}[\hat{H},\mathrm{e}^{\mathrm{i}\boldsymbol{k}\cdot\boldsymbol{r}}\hat{\boldsymbol{p}}\mathrm{e}^{-\mathrm{i}\boldsymbol{k}\cdot\boldsymbol{r}}]\\ &= -(\mathrm{i}\hbar)^{-1}\{\mathrm{e}^{\mathrm{i}\boldsymbol{k}\cdot\boldsymbol{r}}[\hat{H},\hat{\boldsymbol{p}}\mathrm{e}^{-\mathrm{i}\boldsymbol{k}\cdot\boldsymbol{r}}] + [\hat{H},\mathrm{e}^{\mathrm{i}\boldsymbol{k}\cdot\boldsymbol{r}}]\hat{\boldsymbol{p}}\mathrm{e}^{-\mathrm{i}\boldsymbol{k}\cdot\boldsymbol{r}}\}\\ &= -(\mathrm{i}\hbar)^{-1}\{\mathrm{e}^{\mathrm{i}\boldsymbol{k}\cdot\boldsymbol{r}}\hat{\boldsymbol{p}}[\hat{H},\mathrm{e}^{-\mathrm{i}\boldsymbol{k}\cdot\boldsymbol{r}}] + \mathrm{e}^{\mathrm{i}\boldsymbol{k}\cdot\boldsymbol{r}}[\hat{H},\hat{\boldsymbol{p}}]\mathrm{e}^{-\mathrm{i}\boldsymbol{k}\cdot\boldsymbol{r}}\\ &\quad + [\hat{H},\mathrm{e}^{\mathrm{i}\boldsymbol{k}\cdot\boldsymbol{r}}]\hat{\boldsymbol{p}}\mathrm{e}^{-\mathrm{i}\boldsymbol{k}\cdot\boldsymbol{r}}\}.\end{aligned}$$

The commutator $\left[\hat{H}, \mathrm{e}^{\pm \mathrm{i}\boldsymbol{k}\cdot\boldsymbol{r}}\right]$ is given by

$$\begin{aligned}
\left[\hat{H}, \mathrm{e}^{\pm \mathrm{i}\boldsymbol{k}\cdot\boldsymbol{r}}\right] &= \frac{1}{2m}\left[\hat{\boldsymbol{p}}^2, \mathrm{e}^{\pm \mathrm{i}\boldsymbol{k}\cdot\boldsymbol{r}}\right] \\
&= \frac{1}{2m}\hat{\boldsymbol{p}}\cdot\left[\hat{\boldsymbol{p}}, \mathrm{e}^{\pm \mathrm{i}\boldsymbol{k}\cdot\boldsymbol{r}}\right] + \frac{1}{2m}\left[\hat{\boldsymbol{p}}, \mathrm{e}^{\pm \mathrm{i}\boldsymbol{k}\cdot\boldsymbol{r}}\right]\cdot\hat{\boldsymbol{p}} \\
&= \frac{1}{2m}\hat{\boldsymbol{p}}\cdot(\pm\hbar\boldsymbol{k})\mathrm{e}^{\pm \mathrm{i}\boldsymbol{k}\cdot\boldsymbol{r}} + \frac{1}{2m}(\pm\hbar\boldsymbol{k})\mathrm{e}^{\pm \mathrm{i}\boldsymbol{k}\cdot\boldsymbol{r}}\cdot\hat{\boldsymbol{p}} \\
&= \frac{\hbar^2\boldsymbol{k}^2}{2m}\mathrm{e}^{\pm \mathrm{i}\boldsymbol{k}\cdot\boldsymbol{r}} \pm \frac{1}{m}\mathrm{e}^{\pm \mathrm{i}\boldsymbol{k}\cdot\boldsymbol{r}}\left(\hbar\boldsymbol{k}\cdot\hat{\boldsymbol{p}}\right).
\end{aligned}$$

Inserting the above result into the right hand side of the equation for $\left[\hat{H}, \hat{\boldsymbol{F}}\right]$ and making use of $\left[\hat{H}, \hat{\boldsymbol{p}}\right] = \mathrm{i}\hbar\boldsymbol{\nabla}U(\boldsymbol{r})$ at the same time, we have

$$\begin{aligned}
\left[\hat{H}, \hat{\boldsymbol{F}}\right] &= -(\mathrm{i}\hbar)^{-1}\left\{\mathrm{e}^{\mathrm{i}\boldsymbol{k}\cdot\boldsymbol{r}}\hat{\boldsymbol{p}}\left[\frac{\hbar^2\boldsymbol{k}^2}{2m}\mathrm{e}^{-\mathrm{i}\boldsymbol{k}\cdot\boldsymbol{r}} - \frac{1}{m}\mathrm{e}^{-\mathrm{i}\boldsymbol{k}\cdot\boldsymbol{r}}\left(\hbar\boldsymbol{k}\cdot\hat{\boldsymbol{p}}\right)\right]\right. \\
&\quad \left. + \mathrm{i}\hbar\boldsymbol{\nabla}U(\boldsymbol{r}) + \left[\frac{\hbar^2\boldsymbol{k}^2}{2m}\mathrm{e}^{\mathrm{i}\boldsymbol{k}\cdot\boldsymbol{r}} + \frac{1}{m}\mathrm{e}^{\mathrm{i}\boldsymbol{k}\cdot\boldsymbol{r}}\left(\hbar\boldsymbol{k}\cdot\hat{\boldsymbol{p}}\right)\right]\hat{\boldsymbol{p}}\mathrm{e}^{-\mathrm{i}\boldsymbol{k}\cdot\boldsymbol{r}}\right\} \\
&= -\boldsymbol{\nabla}U(\boldsymbol{r}) - (\mathrm{i}\hbar)^{-1}\left\{\left[\frac{\hbar^2\boldsymbol{k}^2}{2m} - \frac{1}{m}\left(\hbar\boldsymbol{k}\cdot\hat{\boldsymbol{p}}\right)\right]\left(-\hbar\boldsymbol{k} + \hat{\boldsymbol{p}}\right)\right. \\
&\quad \left. + \left[\frac{\hbar^2\boldsymbol{k}^2}{2m} + \frac{1}{m}\left(\hbar\boldsymbol{k}\right)\cdot\left(-\hbar\boldsymbol{k} + \hat{\boldsymbol{p}}\right)\right]\left(-\hbar\boldsymbol{k} + \hat{\boldsymbol{p}}\right)\right\} \\
&= -\boldsymbol{\nabla}U(\boldsymbol{r}).
\end{aligned}$$

16-2 Free-electron model. Do the following problems using the free electron model.

(1) Show that the Fermi surface lies within the first Brillouin zone for metals with one valence electron that crystallize in the BCC structure.

(2) A monovalent metal crystallizes in a simple cubic lattice. Compute the percentage of bivalent metal atoms that must be added for the Fermi surface to just touch the boundary of the first Brillouin zone.

(3) Compute the needed number of electrons contributed from each atom to ensure that the Fermi sphere is in contact with the boundary of the first Brillouin zone for both FCC and BCC structures.

(4) Consider a two-dimensional crystal with a square lattice and bivalent metal atoms. Construct the Fermi surface in the first and second Brillouin zones. Find the number of valence electrons per atom for which the third Brillouin zone starts to be filled.

---

(1) For a monovalent BCC metal, the electron density is given by $n = 2/a^3$ since there are two electrons in each conventional unit cell. The Fermi wave vector is then given by $k_\mathrm{F} = (3\pi^2 n)^{1/3} = (6\pi^2)^{1/3}/a \approx 3.898/a$. For the first Brillouin zone of a BCC Bravais lattice, see Fig. 16.1(b). The shortest distance from the Γ point to the boundary of the first Brillouin zone of a BCC Bravais lattice with lattice constant a is $\sqrt{2}\pi/a \approx 1.414/a$. Therefore, the Fermi surface lies within the first Brillouin zone.

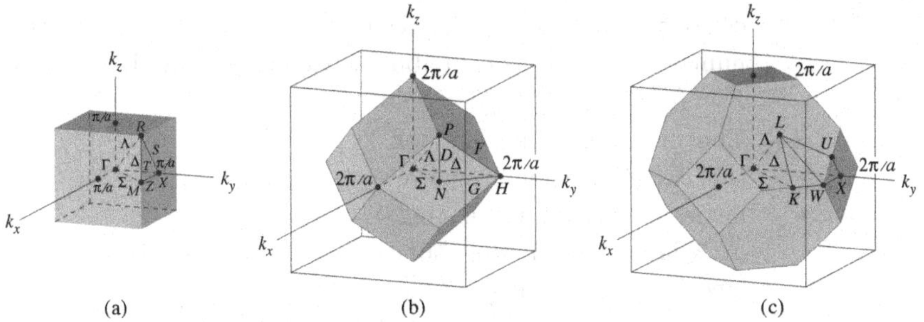

Fig. 16.1 First Brillouin zones for the Bravais lattices in the cubic crystal system. (a) SC Bravais lattice. (b) BCC Bravais lattice. (c) FCC Bravais lattice.

(2) We first compute the electron number density for which the Fermi surface just touches the boundary of the first Brillouin zone. The shortest distance from the Γ point to the boundary of the first Brillouin zone of an SC Bravais lattice with lattice constant a is π/a. Let $k_\mathrm{F} = \pi/a$ [*cf.* Fig. 16.1(a)]. Solving for n from $k_\mathrm{F} = (3\pi^2 n)^{1/3} = \pi/a$, we have $n = \pi/3a^3$. Let f be the percentage of bivalent metal atoms. Then, the number of electrons in a unit cell on average is given by $2x + (1-x) = x+1$. Under this percentage of bivalent atoms, the electron number density is given by $n = (x+1)/a^3$. Equating $n = (x+1)/a^3$ with $n = \pi/3a^3$ yields $(x+1)/a^3 = \pi/3a^3$ from which it follows that $x = \pi/3 - 1 \approx 0.0472 = 4.72\%$.

(3) The shortest distance from the Γ point to the boundary of the first Brillouin zone of an FCC Bravais lattice with lattice constant a is $\sqrt{3}\pi/a$ [*cf.* Fig. 16.1(c)]. Let $k_\mathrm{F} = \sqrt{3}\pi/a$. Solving for n from $k_\mathrm{F} = (3\pi^2 n)^{1/3} = \sqrt{3}\pi/a$, we have $n =$

$\sqrt{3}\pi/a^3$. In consideration that the volume of a primitive cell of the FCC Bravais lattice is $a^3/4$, we see that the needed number of electrons contributed from each atom is given by $Z = (a^3/4)n = \sqrt{3}\pi/4 \approx 1.36$.

The shortest distance from the Γ point to the boundary of the first Brillouin zone of a BCC Bravais lattice with lattice constant a is $\sqrt{2}\pi/a$ [*cf.* Fig. 16.1(b)]. Let $k_F = \sqrt{2}\pi/a$. Solving for n from $k_F = (3\pi^2 n)^{1/3} = \sqrt{2}\pi/a$, we have $n = 2\sqrt{2}\pi/3a^3$. In consideration that the volume of a primitive cell of the BCC Bravais lattice is $a^3/2$, we see that the needed number of electrons contributed from each atom is given by $Z = (a^3/2)n = \sqrt{2}\pi/3 \approx 1.48$.

(4) The first Brillouin zones of the two-dimensional square lattice are shown in Fig. 16.2 in which two Fermi surfaces are also drawn. One Fermi surface lies in the first and second Brillouin zones. The other just touches the third Brillouin zone and has a radius of $k_F = \sqrt{2}\pi/a$. Solving for n from $k_F = (2\pi n)^{1/2} = \sqrt{2}\pi/a$, we have $n = \pi/a^2$. In consideration that the area of a primitive cell of the two-dimensional square Bravais lattice is a^2, we see that the needed num -ber of valence electrons per atom for which the third Brillouin zone starts to be filled is given by $Z = a^2 n = \pi \approx 3.14$.

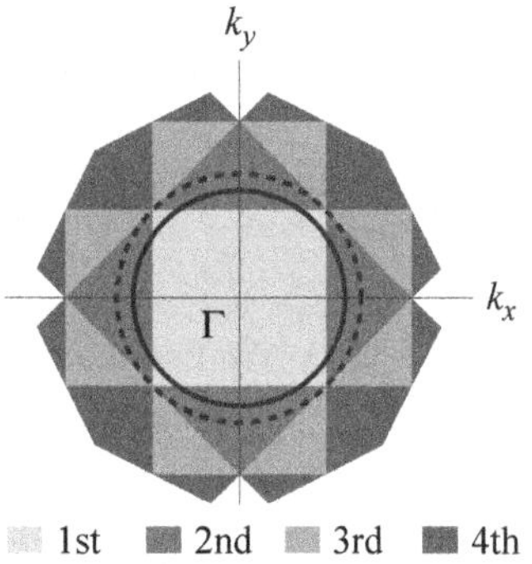

Fig. 16.2 First four Brillouin zones of the two-dimensional square lattice and two Fermi surfaces at different electron densities.

16-3 Infinite chain of atoms with a Peierls distortion. The energy bands for an infinite chain of atoms subject to a periodic distortion (Peierls distortion) are given by

$$E_\pm(k) = \pm\left[(\beta_1 - \beta_2)^2 + 4\beta_1\beta_2\cos^2(ka)\right]^{1/2},$$

where β_1 and β_2 are two different hopping parameters due to the presence of two different atom spacings.

(1) Identify the first Brillouin zone.
(2) Plot the band structure.
(3) Show that the group velocity of an electron at the zone edge is zero.

---

(1) From the periodicity in the given energy bands, we see that the first Brillouin zone is the region in k-space from $-\pi/2a$ to $\pi/2a$.

(2) The band structure is plotted in Fig. 16.3 with the values of the parameters β_1 and β_2 arbitrarily chosen to be 1 eV and 0.8 eV.

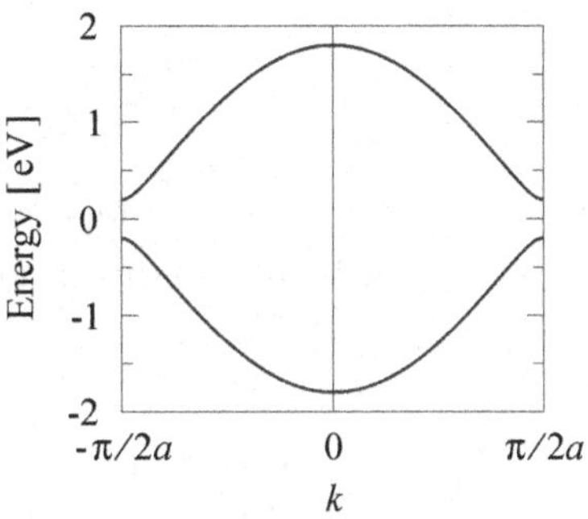

Fig. 16.3 Energy bands for an infinite chain of atoms subject to a Peierls distortion.

(3) Differentiating $E_{\pm}(k)$ with respect to k, we obtain the group velocities for electrons

$$v_{g\pm}(k) = \frac{1}{\hbar}\frac{\partial E_{\pm}(k)}{\partial k} = \frac{\mp(2\beta_1\beta_2 a/\hbar)\sin(2ka)}{\left[(\beta_1 - \beta_2)^2 + 4\beta_1\beta_2\cos^2(ka)\right]^{1/2}}.$$

Since $\sin(2ka) = 0$ at $k = \pm\pi/2a$, the group velocity of an electron at the zone edge is zero.

16-4 Densities of states in 1D and 2D tight-binding energy bands. Compute the densities of states for the following one- and two-dimensional tight-binding energy bands.

(1) A tight-binding energy band $E_k = E_0 - 2t\cos(ka)$ in one dimension.
(2) A tight-binding energy band $E_{\boldsymbol{k}} = E_0 - 2t[\cos(k_x a) + \cos(k_y a)]$ in two dimensions.

---

(1) From the definition of the density of states in one-dimension,

$$g(E) = \frac{2}{\mathscr{L}}\sum_{k\in\mathrm{BZ}} \delta(E - E_k)$$

with $\mathscr{L}$ the length of the one-dimensional crystal, we have

$$g(E) = \frac{2}{\mathscr{L}} \sum_{k \in \mathrm{BZ}} \delta\big(E - E_0 + 2t\cos(ka)\big)$$
$$= \frac{1}{\pi} \int_{-\pi/a}^{\pi/a} \mathrm{d}k\, \delta\big(E - E_0 + 2t\cos(ka)\big)$$
$$= \frac{2}{\pi} \int_{0}^{\pi/a} \mathrm{d}k\, \delta\big(E - E_0 + 2t\cos(ka)\big)$$
$$= \frac{2}{\pi a[\,4t^2 - (E - E_0)^2\,]^{1/2}} \theta(E - E_0 + 2t)\theta(2t - E + E_0).$$

(2) In two dimensions, we have

$$g(E) = \frac{2}{\mathscr{A}} \sum_{\boldsymbol{k} \in \mathrm{BZ}} \delta(E - E_{\boldsymbol{k}})$$

with $\mathscr{A}$ the area of the two-dimensional crystal. Converting the summation over $\boldsymbol{k}$ into an integration over $\boldsymbol{k}$, we have

$$g(E) = \frac{2}{\pi^2} \int_0^{\pi/a} \mathrm{d}k_x \int_0^{\pi/a} \mathrm{d}k_y\, \delta\big(E - E_0 + 2t[\cos(k_x a) + \cos(k_y a)]\big),$$

$$g(e) = \frac{1}{\pi^2 a^2 |t|} \int_0^{\pi} \mathrm{d}x \int_0^{\pi} \mathrm{d}y\, \delta\big(e + \cos x + \cos y\big),$$

where we have made the change of integration variables from k_x and k_y to x and y, $x = k_x a$ and $y = k_y a$, introduced $e = (E - E_0)/2t$, and rewritten $g(E)$ as a function of e, $g(e)$. Through making use of the change of integration variables, $x' = \pi - x$ and $y' = \pi - y$, we can prove that $g(e)$ is an even function of e. Due to this fact, we perform the computation for $e > 0$ in the following. Integrating over y, we have

$$g(e > 0) = \frac{1}{\pi^2 a^2 |t|} \int_0^{\pi} \mathrm{d}x \int_0^{\pi} \mathrm{d}y\, \frac{\delta\big(y - \cos^{-1}(-\cos x - e)\big)}{[1 - (\cos x + e)^2]^{1/2}} \times \theta(1 - \cos x - e)$$
$$= \frac{1}{\pi^2 a^2 |t|} \int_0^{\pi} \mathrm{d}x\, \frac{\theta(1 - \cos x - e)}{[1 - (\cos x + e)^2]^{1/2}}$$
$$= \frac{1}{\pi^2 a^2 t} \int_{-1}^{1} \mathrm{d}u\, \frac{1}{(1 - u^2)^{1/2}} \frac{\theta(1 - u - e)}{[1 - (u + e)^2]^{1/2}}$$
$$= \frac{1}{\pi^2 a^2 |t|} \int_{-1}^{1-e} \mathrm{d}u\, \frac{1}{(1 - u^2)^{1/2}} \frac{1}{[1 - (u + e)^2]^{1/2}}.$$

We now make a change of integration variables so that the lower limit becomes zero while the upper limit becomes unity. Let $v = au + b$. The stated requirements yield

$$0 = -a + b,\ 1 = (1 - e)a + b.$$

Solving for a and b, we have $b = a = 1/(2 - e)$. Thus, $v = (u + 1)/(2 - e)$. The integral now becomes

$$\begin{aligned} g(\varepsilon > 0) &= \frac{1}{\pi^2 a^2 |t|} \int_0^1 \frac{dv}{\{v[2 - (2 - e)v]\}^{1/2}} \frac{1}{\{(1 - v)[e + (2 - e)v]\}^{1/2}} \\ &= \frac{1}{\pi^2 a^2 |t|} \int_0^1 \frac{dv}{[v(1 - v)]^{1/2}} \frac{1}{[2e + (2 - e)^2 v(1 - v)]^{1/2}}. \end{aligned}$$

Making a change of integration variables from v to $\xi = v - 1/2$, we have

$$\begin{aligned} &g(e > 0) \\ &= \frac{4}{\pi^2 a^2 (2 + e)|t|} \int_{-1/2}^{1/2} \frac{d\xi}{(1 - 4\xi^2)^{1/2}} \frac{1}{[1 - 4(2 - e)^2 \xi^2/(2 + e)^2]^{1/2}} \\ &= \frac{8}{\pi^2 a^2 (2 + e)|t|} \int_0^{1/2} \frac{d\xi}{(1 - 4\xi^2)^{1/2}} \frac{1}{[1 - 4(2 - e)^2 \xi^2/(2 + e)^2]^{1/2}} \\ &= \frac{4}{\pi^2 a^2 (2 + e)|t|} \int_0^1 \frac{d\zeta}{(1 - \zeta^2)^{1/2}} \frac{1}{[1 - (2 - e)^2 \zeta^2/(2 + e)^2]^{1/2}}, \end{aligned}$$

where we have made a change of integration variables from ξ to $\zeta = 2\xi$. Comparing the above expression with the complete elliptic integral of the first kind

$$K(k) = \int_0^1 \frac{dx}{[(1 - x^2)(1 - k^2 x^2)]^{1/2}} = \int_0^{\pi/2} \frac{d\varphi}{(1 - k^2 \sin^2 \varphi)^{1/2}},$$

we can write the density of states as

$$\begin{aligned} g(e > 0) &= \frac{4}{\pi^2 a^2 (2 + e)|t|} K((2 - e)/(2 + e))m \\ g(\varepsilon > 0) &= \frac{2}{\pi^2 a^2 (1 + \varepsilon/4t)|t|} K((1 - \varepsilon/4t)/(1 + \varepsilon/4t)), \end{aligned}$$

where $\varepsilon = E - E_0$. Taking into account the fact that $g(\varepsilon)$ is an even function of ε, we have

$$g(\varepsilon) = \frac{2}{\pi^2 a^2 (1 + |\varepsilon/4t|)|t|} K((1 - |\varepsilon/4t|)/(1 + |\varepsilon/4t|)).$$

16-5 Electron energies in a 2D metal with a square lattice. A two-dimensional metal crystallizes in a square lattice. Assume that the free electron model is applicable.

(1) Show that the kinetic energy of an electron at the corner of the first Brillouin zone is twice that of an electron in the middle of the zone boundary.
(2) What is the corresponding factor for a simple cubic lattice in three dimensions?

---

(1) The wave vector of an electron at the corner of the first Brillouin zone of a two-dimensional square lattice is $\boldsymbol{k}_M = (\pm 1, \pm 1)(\pi/a)$. Thus, the kinetic energy of the electron is $E_k(M) = \hbar^2 \boldsymbol{k}_M^2/2m = \pi^2\hbar^2/ma^2$. The wave vector of an electron in the middle of the zone boundary is $\boldsymbol{k}_X = (\pm 1, 0)(\pi/a)$ or $(0, \pm 1)(\pi/a)$. Thus, the kinetic energy of the electron is $E_k(X) = \hbar^2 \boldsymbol{k}_X^2/2m = \pi^2\hbar^2/2ma^2$. We see that $E_k(M) = 2E_k(X)$.
(2) For an SC Bravais lattice in three dimensions, $\boldsymbol{k}_M = (\pm 1, \pm 1, \pm 1)(\pi/a)$ and $E_k(M) = \hbar^2 \boldsymbol{k}_M^2/2m = 3\pi^2\hbar^2/2ma^2$ for an electron at the M point. The wave vector and kinetic energy of an electron at an X point are given by $\boldsymbol{k}_X = (1, 0, 0)(\pi/a)$ and $E_k(X) = \hbar^2 \boldsymbol{k}_X^2/2m = \pi^2\hbar^2/2ma^2$. We thus have $E_k(M) = 3E_k(X)$.

16-6 Allowed wave vectors in a simple cubic crystal. Consider a simple cubic crystal that consists of N^3 primitive unit cells. Show that the number of independent values that the wave vector $\boldsymbol{k}$ can assume within the first Brillouin zone is exactly N^3.

The Born-von-Karman periodic boundary conditions for the Bloch function in a crystal consisting of N^3 primitive unit cells are expressed by

$$\psi_{n\boldsymbol{k}}(\boldsymbol{r} + N\boldsymbol{a}_1) = \psi_{n\boldsymbol{k}}(\boldsymbol{r}),$$
$$\psi_{n\boldsymbol{k}}(\boldsymbol{r} + N\boldsymbol{a}_2) = \psi_{n\boldsymbol{k}}(\boldsymbol{r}),$$
$$\psi_{n\boldsymbol{k}}(\boldsymbol{r} + N\boldsymbol{a}_3) = \psi_{n\boldsymbol{k}}(\boldsymbol{r}),$$

where $\boldsymbol{a}_1$, $\boldsymbol{a}_2$, and $\boldsymbol{a}_3$ are a set of primitive vectors of the direct lattice. By making use of the form of the Bloch function $\psi_{n\boldsymbol{k}}(\boldsymbol{r}) = e^{i\boldsymbol{k}\cdot\boldsymbol{r}} u_{n\boldsymbol{k}}(\boldsymbol{r})$ with $u_{n\boldsymbol{k}}(\boldsymbol{r})$ a periodic function of the direct lattice and expressing $\boldsymbol{k}$ as $\boldsymbol{k} = x_1\boldsymbol{b}_1 + x_2\boldsymbol{b}_2 + x_3\boldsymbol{b}_3$ with $\boldsymbol{b}_1$, $\boldsymbol{b}_2$, and $\boldsymbol{b}_3$ a set of primitive vectors of the reciprocal lattice, we have

$$e^{2\pi i N x_1} = e^{2\pi i N x_2} = e^{2\pi i N x_3} = 1,$$

where we have made use of $\boldsymbol{a}_i \cdot \boldsymbol{b}_j = 2\pi\delta_{ij}$. There exist N different nonequivalent solutions to each of the above three equations. They give rise to N^3 different nonequivalent allowed values to $\boldsymbol{k}$. In the first Brillouin zone, these N^3 different nonequivalent allowed values of $\boldsymbol{k}$ can be written as

$$\boldsymbol{k} = \sum_{i=1}^{3} x_i \boldsymbol{b}_i,\ x_i = \frac{2\pi n_i}{N},$$

$$n_i = 0,\ \pm 1,\ \pm 2,\ \cdots,\ \pm(N/2-1),\ N/2 \text{ for } i = 1,\ 2,\ 3.$$

16-7 Energy bands at the center of the first Brillouin zone for aluminum. Compute the energy interval in eV between the lowest and the next lowest bands at the center of the first Brillouin zone for aluminum with a lattice constant of 0.405 nm if the free electron model is assumed to be valid.

Aluminum has an FCC crystal structure. The lowest energy band at the Γ point (the center of the first Brillouin zone) is from the reciprocal lattice vector $\boldsymbol{K} = 0$. The second lowest energy band is from the reciprocal lattice vectors $\boldsymbol{K} = (\pm 1, \pm 1, \pm 1)(2\pi/a)$ and is thus eight-fold degenerate.
The energy of the lowest energy band at the Γ point is zero, $E_1(\Gamma) = \hbar^2 0^2/2m = 0$. The energy of the second lowest energy band at the Γ point is

$$E_2(\Gamma) = \hbar^2 \left|(\pm 1, \pm 1, \pm 1)(2\pi/a)\right|^2/2m = 6\pi^2\hbar^2/ma^2$$

with a the lattice constant. Thus, the energy interval between the lowest and the next lowest band at the center of the first Brillouin zone is given by

$$\begin{aligned}\Delta E_{12}(\Gamma) &= E_2(\Gamma) - E_1(\Gamma) = \frac{6\pi^2\hbar^2}{ma^2} \\ &= \frac{6 \times \pi^2 \times 1.054\,5716\,28^2 \times 10^{-68}}{9.109\,382\,15 \times 10^{-31} \times 0.405^2 \times 10^{-18}}\ \mathrm{J} \\ &\approx 4.41 \times 10^{-18}\ \mathrm{J} \approx 27.51\ \mathrm{eV}.\end{aligned}$$

16-8 Monovalent metal with an ideal HCP structure. A monovalent metal crystallizes in an ideal HCP structure with $a = 0.32$ nm. The free electron model is assumed to apply.

(1) Find the percentage of the conduction electrons that are in the second band.

(2) Find the largest wavelength a photon can have if it is to be able to excite an electron on the Fermi surface near the Γ point along the $\Gamma \to M$ direction to the next band up.

---

(1) The underlying Bravais lattice of an HCP structure is a simple hexagonal Bravais lattice whose reciprocal lattice is also a simple hexagonal Bravais lattice in reciprocal space. We choose the following primitive vectors for the direct lattice

$$\boldsymbol{a}_1 = (a/2)\,\boldsymbol{e}_x + (\sqrt{3}a/2)\,\boldsymbol{e}_y,\ \boldsymbol{a}_2 = -(a/2)\,\boldsymbol{e}_x + (\sqrt{3}a/2)\,\boldsymbol{e}_y,$$
$$\boldsymbol{a}_3 = c\,\boldsymbol{e}_z.$$

The corresponding primitive vectors of the reciprocal lattice are given by

$$\begin{aligned} \boldsymbol{b}_1 &= (2\pi/a)\,\boldsymbol{e}_x + (2\pi/\sqrt{3}a)\,\boldsymbol{e}_y, \\ \boldsymbol{b}_2 &= -(2\pi/a)\,\boldsymbol{e}_x + (2\pi/\sqrt{3}a)\,\boldsymbol{e}_y, \\ \boldsymbol{b}_3 &= (2\pi/c)\,\boldsymbol{e}_z. \end{aligned}$$

The first Brillouin of the simple hexagonal Bravais lattice is given in Fig. 16.4 with the high-symmetry points and lines labeled with conventionally-used symbols.

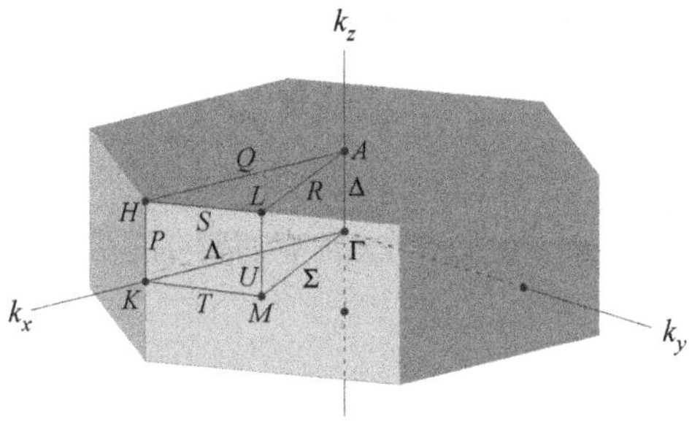

Fig. 16.4 First Brillouin of the simple hexagonal Bravais lattice.

The free-electron (empty-lattice) band structure for the simple hexagonal Bravais lattice is plotted in Fig. 16.5. The HCP structure has a two-atom basis, which implies that there are two electrons in each primitive cell for a monovalent metal. The volume of a primitive cell of the direct lattice is $v_c = \sqrt{3}a^2c/2 = \sqrt{2}a^3$ for an ideal HCP with $c = \sqrt{8/3}\,a$. The electron number density is then given by $n = \sqrt{2}/a^3$.

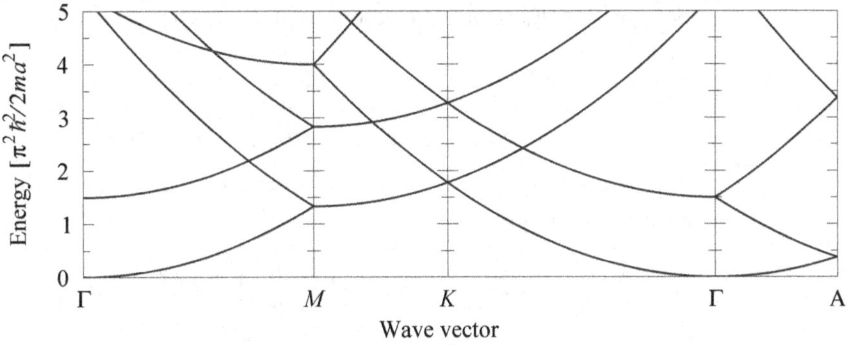

Fig. 16.5 Free-electron band structure of a simple hexagonal Bravais lattice.

If only the free-electron band from $\boldsymbol{K} = 0$ were to be occupied, then the Fermi energy would be given by $(3\pi^2 2^{1/2})^{2/3}\hbar^2/2ma^2 \approx 1.221\ 8\pi^2\hbar^2/2ma^2$. Since this conjectured Fermi energy is higher than the lowest energy of the second band

$$\begin{aligned} E_{2,\min} &= \frac{\hbar^2}{2m}\big[\,\boldsymbol{k}_A - (2\pi/c)\boldsymbol{e}_z\,\big]^2 \\ &= \frac{\hbar^2}{2m}\big[\,(\pi/c)\boldsymbol{e}_z - (2\pi/c)\boldsymbol{e}_z\,\big]^2 = \frac{3}{8}\frac{\pi^2\hbar^2}{2ma^2}, \end{aligned}$$

the second band is also occupied by electrons. Assume that the Fermi energy is $E_{\rm F}$. Making use of the fact that the total number of occupied single-electron states per unit volume with the electron spin taken into consideration is equal to the electron number density, we have

$$\int_0^{E_{\rm F}} {\rm d}E\ g(E) + \int_{E_{2,\min}}^{E_{\rm F}} {\rm d}E\ g(E) = n$$

in which the first term on the left hand side gives the total number of occupied single-electron states in the first band and the second term the total number of occupied single-electron states in the second band. Making use of the explicit expression for the density of states in a free-electron parabolic band

$$g(E) = \frac{1}{2\pi^2}\left(\frac{2m}{\hbar^2}\right)^{3/2} E^{1/2},$$

we have

$$\frac{1}{3\pi^2}\left(\frac{2m}{\hbar^2}\right)^{3/2}\left(2E_{\rm F}^{3/2} - E_{2,\min}^{3/2}\right) = n.$$

Solving for $E_{\rm F}$ from the above equation with $n = \sqrt{2}/a^3$ yields

$$E_{\rm F} = \left(\frac{3}{\sqrt{2}\pi} + \frac{3}{16}\sqrt{\frac{3}{8}}\right)^{2/3}\frac{\pi^2\hbar^2}{2ma^2} \approx 0.85\frac{\pi^2\hbar^2}{2ma^2}.$$

The percentage of the conduction electrons that are in the second band is given by

$$f = \frac{1}{n}\int_{E_{2,\rm min}}^{E_{\rm F}} {\rm d}E\, g(E) = \frac{\pi}{3\sqrt{2}}\left(\frac{2ma^2}{\pi^2\hbar^2}\right)^{3/2}\left(E_{\rm F}^{3/2} - E_{2,\rm min}^{3/2}\right)$$

$$= \frac{\pi}{3\sqrt{2}}\left(\frac{3}{\sqrt{2}\pi} - \frac{3}{16}\sqrt{\frac{3}{8}}\right) = \frac{1}{2} - \frac{\sqrt{3}\pi}{64} \approx 0.415 = 41.5\%.$$

(2) Near the Γ point along the $\Gamma \to M$ direction, the band immediately up is the second occupied band that possesses the following energy dispersion relation

$$E_2(\boldsymbol{k}) = \frac{\hbar^2}{2m}\left(\boldsymbol{k} - \frac{2\pi}{c}\boldsymbol{e}_z\right)^2.$$

Along ΓM, $\boldsymbol{k}$ is of the form $\boldsymbol{k} = \xi(2\pi/a)(\sqrt{3}\,\boldsymbol{e}_x + \boldsymbol{e}_y)/2$ with ξ in $[0, 1/\sqrt{3}]$. The energy dispersion relation in the first band along ΓM is given by

$$E_1(\boldsymbol{k}) = \frac{\hbar^2\boldsymbol{k}^2}{2m}.$$

Taking into account the fact that $\boldsymbol{e}_x$, $\boldsymbol{e}_y$, and $\boldsymbol{e}_z$ are perpendicular to one another, we see that

$$E_2(\boldsymbol{k}) = E_1(\boldsymbol{k}) + \frac{4\pi^2\hbar^2}{2mc^2} = E_1(\boldsymbol{k}) + \frac{3}{2}\frac{\pi^2\hbar^2}{2ma^2}.$$

Thus, the energy difference between the second and first bands is given by

$$\Delta E = E_2(\boldsymbol{k}) - E_1(\boldsymbol{k}) = \frac{3}{2}\frac{\pi^2\hbar^2}{2ma^2}.$$

The wavelength of a photon for the transition from the first to the second band is then given by

$$\lambda = \frac{hc}{\Delta E} = \frac{2}{3}\frac{hc}{\pi^2\hbar^2/2ma^2}$$

$$\approx 2.25 \times 10^{-7}\ {\rm m} = 225\ {\rm nm}.$$

16-9 Energy bands for an FCC lattice in the [111] direction. Compute the lowest lying energy bands in the empty-lattice approximation for an FCC lattice in the [111] direction.

For an FCC lattice of lattice constant a, the reciprocal lattice vectors are of the form

$$\boldsymbol{K} = \left[(k+\ell-h)\boldsymbol{e}_x + (\ell+h-k)\boldsymbol{e}_y + (h+k-\ell)\boldsymbol{e}_z\right]\frac{2\pi}{a}$$

with h, k, and ℓ integers. If we set $\ell_1 = k+\ell-h$, $\ell_2 = \ell+h-k$, and $\ell_3 = h+k-\ell$, we then have

$$\boldsymbol{K} = (\ell_1\boldsymbol{e}_x + \ell_2\boldsymbol{e}_y + \ell_3\boldsymbol{e}_z)\frac{2\pi}{a}$$

in which ℓ_1, and ℓ_2, and ℓ_3 are either all odd or all even. Along the [111] direction, the wave vector $\boldsymbol{k}$ is of the form

$$\boldsymbol{k} = \xi(\boldsymbol{e}_x + \boldsymbol{e}_y + \boldsymbol{e}_z)\frac{2\pi}{a}$$

with ξ in $[0, 1/2]$. The lowest lying bands in the [111] direction in the empty-lattice approximation can be obtained from the first few reciprocal lattice vectors through

$$E^0_{\boldsymbol{k}-\boldsymbol{K}} = \frac{\hbar^2}{2m}(\boldsymbol{k}-\boldsymbol{K})^2.$$

The first few lowest lying empty-lattice bands in the [111] direction are shown in Fig. 16.6 with some of the responsible reciprocal lattice vectors given. If two or more reciprocal lattice vectors yield the same band, then the band is degenerate. In this case, only one reciprocal lattice vector is given. All the reciprocal lattice vectors used to produce the bands in Fig. 16.6 are listed below. The reciprocal lattice vector(s) on each line give(s) rise to a

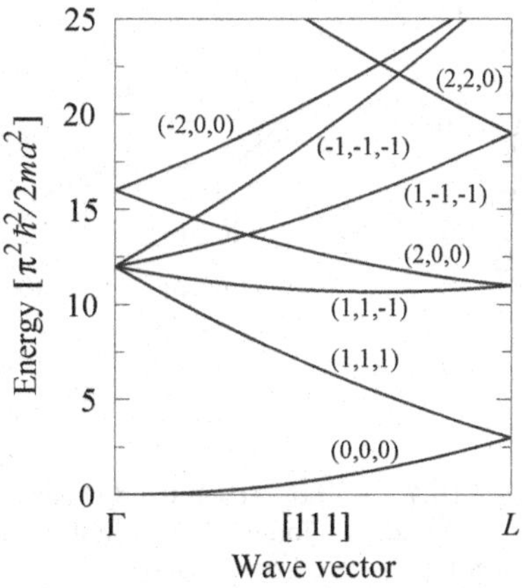

Fig. 16.6 Empty-lattice band structure of an FCC Bravais lattice in the [111] direction.

single band.

$$(0,0,0)(2\pi/a),$$
$$(1,1,1)(2\pi/a),$$
$$(1,1,-1)(2\pi/a),\ (1,-1,1)(2\pi/a),\ (1,-1,1)(2\pi/a),$$
$$(2,0,0)(2\pi/a),\ (0,2,0)(2\pi/a),\ (0,0,2)(2\pi/a),$$
$$(1,-1,-1)(2\pi/a),\ (-1,1,-1)(2\pi/a),\ (-1,-1,1)(2\pi/a),$$
$$(-1,-1,-1)(2\pi/a),$$
$$(-2,0,0)(2\pi/a),\ (0,-2,0)(2\pi/a),\ (0,0,-2)(2\pi/a),$$
$$(2,2,0)(2\pi/a),\ (2,0,2)(2\pi/a),\ (0,2,2)(2\pi/a).$$

16-10 Band structure of a divalent FCC Sr metal. Consider a divalent FCC Sr metal with the lattice constant $a = 0.608$ nm. For this metal, the first Brillouin zone is completely filled. Find the smallest energy gap at the boundary of the first Brillouin zone

---

The empty-lattice band structure of the FCC Bravais lattice is given in Fig. 16.7.

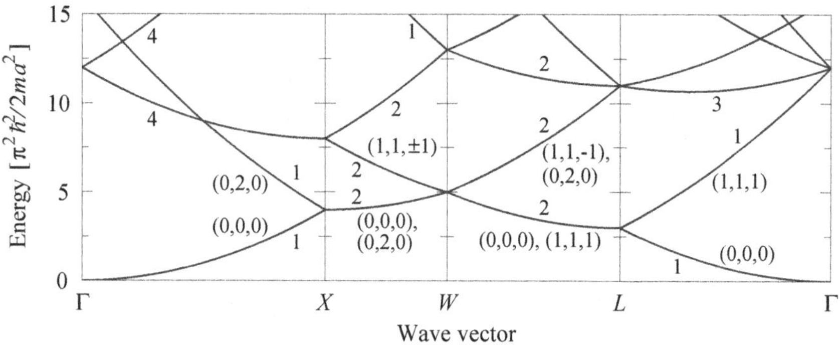

Fig. 16.7 Empty-lattice band structure of the FCC Bravais lattice. The degeneracies of the bands are indicated by numbers and the responsible reciprocal lattice vectors are given for the two lowest bands.

From Fig. 16.7, we can see that the second band has the lowest energy at the L point. For the first Brillouin zone to be completely filled, a band gap must be opened at the L point with its size equal to the difference between the energy of the first band at the W point and the energy of the first band at the L point. Thus, the smallest energy gap at the boundary of the first Brillouin zone is given by

$$\begin{aligned}\Delta E &= E_1(\boldsymbol{k}_W) - E_1(\boldsymbol{k}_L) = \frac{\hbar^2}{2m}(\boldsymbol{k}_W^2 - \boldsymbol{k}_L^2)\\ &= \frac{\hbar^2}{2m}\left[\left|(1/2, 1, 0)\frac{2\pi}{a}\right|^2 - \left|(1/2, 1/2, 1/2)\frac{2\pi}{a}\right|^2\right]\\ &= \frac{\pi^2\hbar^2}{ma^2} \approx 2.03 \text{ eV}.\end{aligned}$$

Chapter 17

Electrons in a Weak Periodic Potential

(1) *Single-electron Bloch functions and energies in an empty-lattice band structure*

$$\psi^0_{\boldsymbol{k}-\boldsymbol{K}}(\boldsymbol{r}) = \frac{1}{\sqrt{\mathscr{V}}}\mathrm{e}^{\mathrm{i}(\boldsymbol{k}-\boldsymbol{K})\cdot\boldsymbol{r}},$$

$$E^0_{\boldsymbol{k}-\boldsymbol{K}} = \frac{\hbar^2(\boldsymbol{k}-\boldsymbol{K})^2}{2m}$$

(2) *Away from Bragg planes*

$$E_{\boldsymbol{k}-\boldsymbol{K}} \approx E^0_{\boldsymbol{k}-\boldsymbol{K}} + {\sum_{\boldsymbol{K}'}}' \frac{|U_{\boldsymbol{K}'-\boldsymbol{K}}|^2}{E^0_{\boldsymbol{k}-\boldsymbol{K}} - E^0_{\boldsymbol{k}-\boldsymbol{K}'}}.$$

(3) *Near Bragg planes*

$$\det\big|(E^0_{\boldsymbol{k}-\boldsymbol{K}_j} - E_{\boldsymbol{k}})\delta_{\boldsymbol{K}_\ell \boldsymbol{K}_j} + U_{\boldsymbol{K}_\ell - \boldsymbol{K}_j}\big| = 0.$$

(4) *Two nearly degenerate single-electron states*

$$E_{\boldsymbol{k}\pm} = \frac{1}{2}\big(E^0_{\boldsymbol{k}-\boldsymbol{K}_2} + E^0_{\boldsymbol{k}-\boldsymbol{K}_1}\big) \pm \left[\frac{1}{4}\big(E^0_{\boldsymbol{k}-\boldsymbol{K}_2} - E^0_{\boldsymbol{k}-\boldsymbol{K}_1}\big)^2 + |U_{\boldsymbol{K}_2-\boldsymbol{K}_1}|^2\right]^{1/2}.$$

On a Bragg plane, we have

$$E_{\boldsymbol{k}\pm} = E^0_{\boldsymbol{K}/2-\boldsymbol{K}_1} \pm |U_{\boldsymbol{K}_2-\boldsymbol{K}_1}|.$$

(5) *Band gap on a Bragg plane*

$$\Delta_{\boldsymbol{K}/2} = 2|U_{\boldsymbol{K}_2-\boldsymbol{K}_1}|.$$

17-1 Electrons in a 2D square lattice. Consider electrons in a two-dimensional square lattice. Atoms are located at $R_{m,n} = m\boldsymbol{a}_1 + n\boldsymbol{a}_2$, where $\boldsymbol{a}_1 = a\boldsymbol{e}_x$ and $\boldsymbol{a}_2 = a\boldsymbol{e}_y$. The electrons are subject to the following weak periodic potential

$$U(\boldsymbol{r}) = V_0 \sum_{j=1,2,3,4} \mathrm{e}^{\mathrm{i}\boldsymbol{K}_j\cdot\boldsymbol{r}} + V_1 \sum_{j=5,6,7,8} \mathrm{e}^{\mathrm{i}\boldsymbol{K}_j\cdot\boldsymbol{r}},$$

where

$$
\begin{aligned}
K_1 &= (2\pi/a)\boldsymbol{e}_x, & K_2 &= -(2\pi/a)\boldsymbol{e}_x, \\
K_3 &= (2\pi/a)\boldsymbol{e}_y, & K_4 &= -(2\pi/a)\boldsymbol{e}_y, \\
K_5 &= (2\pi/a)(\boldsymbol{e}_x + \boldsymbol{e}_y), & K_6 &= (2\pi/a)(-\boldsymbol{e}_x + \boldsymbol{e}_y), \\
K_7 &= (2\pi/a)(\boldsymbol{e}_x - \boldsymbol{e}_y), & K_8 &= -(2\pi/a)(\boldsymbol{e}_x + \boldsymbol{e}_y).
\end{aligned}
$$

(1) Explain why the lowest band at the X point with $\boldsymbol{k}_X = (\pi/a)\boldsymbol{e}_x$ for zero potential is doubly degenerate.

(2) Find the energy splitting of the lowest bands at the X point in the lowest order in the potential if both V_0 and V_1 are small compared to $E_0 = \pi^2\hbar^2/2ma^2$.

(3) Compute the effective mass m_{xx} at the X point for the two lowest bands. m_{xx} is defined by $\hbar^2/m_{xx} = \partial^2 E(\boldsymbol{k})/\partial k_x^2$ with $E(\boldsymbol{k})$ the band energy.

(4) Write down the Hamiltonian matrix for the lowest degenerate space at the M point with $\boldsymbol{k}_M = (\pi/a)(\boldsymbol{e}_x + \boldsymbol{e}_y)$.

---

(1) The first Brillouin zone of the two-dimensional square lattice is shown in Fig. 17.1.

The dispersion relation in the free-electron energy band arising from reciprocal lattice vector $\boldsymbol{K}$ is given by $\hbar^2(\boldsymbol{k} - \boldsymbol{K})^2/2m$. At the X point ($\boldsymbol{k} = (\pi/a)\boldsymbol{e}_x$), the lowest energy is $E_X^0 = \pi^2\hbar^2/2m$. This energy can be given both by the reciprocal vector $\boldsymbol{K} = 0$ and by the reciprocal vector $\boldsymbol{K} = (2\pi/a)\boldsymbol{e}_x$. All other reciprocal lattice vectors lead to energies higher than E_X^0. Thus, the lowest band at the X point for zero potential is doubly degenerate.

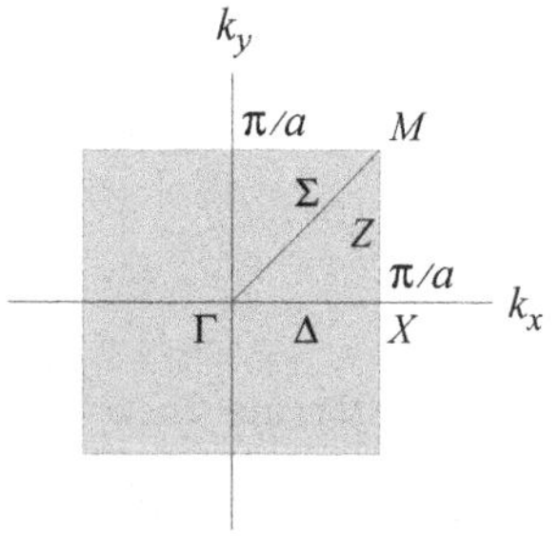

Fig. 17.1 First Brillouin zone of the two-dimensional square lattice.

(2) The Fourier transform of $U(\boldsymbol{r})$ is given by

$$\begin{aligned} U_{\boldsymbol{K}} &= \frac{1}{\mathscr{A}} \int \mathrm{d}\boldsymbol{r}\ \mathrm{e}^{\mathrm{i}\boldsymbol{K}\cdot\boldsymbol{r}} U(\boldsymbol{r}) \\ &= \frac{V_0}{\mathscr{A}} \sum_{j=1,2,3,4} \int \mathrm{d}\boldsymbol{r}\ \mathrm{e}^{\mathrm{i}(\boldsymbol{K}+\boldsymbol{K}_j)\cdot\boldsymbol{r}} + \frac{V_1}{\mathscr{A}} \sum_{j=5,6,7,8} \int \mathrm{d}\boldsymbol{r}\ \mathrm{e}^{\mathrm{i}(\boldsymbol{K}+\boldsymbol{K}_j)\cdot\boldsymbol{r}} \\ &= V_0 \sum_{j=1,2,3,4} \delta_{\boldsymbol{K}\boldsymbol{K}_j} + V_1 \sum_{j=5,6,7,8} \delta_{\boldsymbol{K}\boldsymbol{K}_j}. \end{aligned}$$

In consideration that $\boldsymbol{k}_X = (\pi/a)\boldsymbol{e}_x$, we see that the responsible reciprocal lattice vectors for the double degeneracy of the lowest band at X are $\boldsymbol{K}_1 = 0$ and $\boldsymbol{K}_2 = (2\pi/a)\boldsymbol{e}_x$ with the zero-order energy given by $E_X^0 = \pi^2\hbar^2/2m$. The dispersion relations of the two lowest bands near the X point are given by

$$\begin{aligned} E_{\boldsymbol{k}\pm} &= \big(E^0_{\boldsymbol{k}-\boldsymbol{K}_2} + E^0_{\boldsymbol{k}-\boldsymbol{K}_1}\big)/2 \\ &\quad \pm \Big[\big(E^0_{\boldsymbol{k}-\boldsymbol{K}_2} - E^0_{\boldsymbol{k}-\boldsymbol{K}_1}\big)^2/4 + |U_{\boldsymbol{K}_2-\boldsymbol{K}_1}|^2\Big]^{1/2} \\ &= \hbar^2\boldsymbol{k}^2/2m + \pi\hbar^2(\pi - k_x a)/ma^2 \\ &\quad \pm \big[\pi^2\hbar^4(\pi - k_x a)^2/m^2a^4 + |V_0|^2\big]^{1/2}, \end{aligned}$$

where the above-evaluated Fourier transform of $U(\boldsymbol{r})$ has been used. The energy splitting of the lowest bands at the X point in the lowest order in the potential is given by

$$\begin{aligned} \Delta_X &= E_{\boldsymbol{k}_X+} - E_{\boldsymbol{k}_X-} = 2|U_{\boldsymbol{K}_2-\boldsymbol{K}_1}| \\ &= 2|U_{(2\pi/a)\boldsymbol{e}_x}| = 2|V_0|. \end{aligned}$$

(3) Differentiating $E_{\boldsymbol{k}\pm}$ with respect to k_x, we have

$$\frac{\partial E_{\boldsymbol{k}\pm}}{\partial k_x} = \frac{\hbar^2 k_x}{m} - \frac{\pi\hbar^2}{ma} \mp \frac{\pi^2\hbar^4(\pi - k_x a)}{m^2a^3\big[\pi^2\hbar^4(\pi - k_x a)^2/m^2a^4 + |V_0|^2\big]^{1/2}}.$$

Differentiating the above result with respect to k_x again, we have

$$\begin{aligned} \frac{\partial^2 E_{\boldsymbol{k}\pm}}{\partial k_x^2} &= \frac{\hbar^2}{m} \pm \frac{\pi^2\hbar^4}{m^2a^2\big[\pi^2\hbar^4(\pi - k_x a)^2/m^2a^4 + |V_0|^2\big]^{1/2}} \\ &\quad \mp \frac{\pi^4\hbar^8(\pi - k_x a)^2}{m^4a^6\big[\pi^2\hbar^4(\pi - k_x a)^2/m^2a^4 + |V_0|^2\big]^{3/2}} \\ &= \frac{\hbar^2}{m} \pm \frac{\pi^2\hbar^4|V_0|^2}{m^2a^2\big[\pi^2\hbar^4(\pi - k_x a)^2/m^2a^4 + |V_0|^2\big]^{3/2}}. \end{aligned}$$

At k_X, we have

$$\left.\frac{\partial^2 E_{\boldsymbol{k}\pm}}{\partial k_x^2}\right|_{k_X} = \frac{\hbar^2}{m} \pm \frac{\pi^2\hbar^4}{m^2a^2|V_0|}.$$

Thus, the effective masses at k_X are then given by

$$m_\pm^*(X) = \frac{m}{1 \pm \pi^2\hbar^2/ma^2|V_0|}.$$

Note that the electron effective mass in the lower band at k_X is negative for the value of V_0 smaller than $\pi^2\hbar^2/ma^2$.

(4) At the M point, $\boldsymbol{k}_M = (\pi/a)(\boldsymbol{e}_x + \boldsymbol{e}_y)$, the lowest energy is $E_M^0 = \pi^2\hbar^2/ma^2$. The reciprocal lattice vectors

$$\boldsymbol{K}_1 = 0,\ \boldsymbol{K}_2 = (2\pi/a)\boldsymbol{e}_x,\ \boldsymbol{K}_3 = (2\pi/a)\boldsymbol{e}_y,\ \boldsymbol{K}_4 = (2\pi/a)(\boldsymbol{e}_x+\boldsymbol{e}_y)$$

all give rise to the same energy E_M^0. Thus, the degenerate space at the M point is four-dimensional. In this degenerate space, the matrix representation of the Hamiltonian is given by

$$H = \begin{pmatrix} E^0_{\boldsymbol{k}_M-\boldsymbol{K}_1} & U_{\boldsymbol{K}_1-\boldsymbol{K}_2} & U_{\boldsymbol{K}_1-\boldsymbol{K}_3} & U_{\boldsymbol{K}_1-\boldsymbol{K}_4} \\ U_{\boldsymbol{K}_2-\boldsymbol{K}_1} & E^0_{\boldsymbol{k}_M-\boldsymbol{K}_2} & U_{\boldsymbol{K}_2-\boldsymbol{K}_3} & U_{\boldsymbol{K}_2-\boldsymbol{K}_4} \\ U_{\boldsymbol{K}_3-\boldsymbol{K}_1} & U_{\boldsymbol{K}_3-\boldsymbol{K}_2} & E^0_{\boldsymbol{k}_M-\boldsymbol{K}_3} & U_{\boldsymbol{K}_3-\boldsymbol{K}_4} \\ U_{\boldsymbol{K}_4-\boldsymbol{K}_1} & U_{\boldsymbol{K}_4-\boldsymbol{K}_2} & U_{\boldsymbol{K}_4-\boldsymbol{K}_3} & E^0_{\boldsymbol{k}_M-\boldsymbol{K}_4} \end{pmatrix}.$$

Making use of the explicit expressions of the four reciprocal lattice vectors, we have

$$H = \begin{pmatrix} \pi^2\hbar^2/ma^2 & U_{-(2\pi/a)\boldsymbol{e}_x} & U_{-(2\pi/a)\boldsymbol{e}_y} & U_{-(2\pi/a)(\boldsymbol{e}_x+\boldsymbol{e}_y)} \\ U_{(2\pi/a)\boldsymbol{e}_x} & \pi^2\hbar^2/ma^2 & U_{(2\pi/a)(\boldsymbol{e}_x-\boldsymbol{e}_y)} & U_{-(2\pi/a)\boldsymbol{e}_y} \\ U_{(2\pi/a)\boldsymbol{e}_y} & U_{(2\pi/a)(-\boldsymbol{e}_x+\boldsymbol{e}_y)} & \pi^2\hbar^2/ma^2 & U_{-(2\pi/a)\boldsymbol{e}_x} \\ U_{(2\pi/a)(\boldsymbol{e}_x+\boldsymbol{e}_y)} & U_{(2\pi/a)\boldsymbol{e}_y} & U_{(2\pi/a)\boldsymbol{e}_x} & \pi^2\hbar^2/ma^2 \end{pmatrix}.$$

Making use of the above-evaluated Fourier components of $U(\boldsymbol{r})$, we obtain

$$H = \begin{pmatrix} \pi^2\hbar^2/ma^2 & V_0 & V_0 & V_1 \\ V_0 & \pi^2\hbar^2/ma^2 & V_1 & V_0 \\ V_0 & V_1 & \pi^2\hbar^2/ma^2 & V_0 \\ V_1 & V_0 & V_0 & \pi^2\hbar^2/ma^2 \end{pmatrix}.$$

17-2 Energy bands in a 1D crystal with a two-atom basis. Consider N atoms of type A located at positions $x = R_m = 2mc$ with $m = 1, 2, \cdots, N$, and an equal number of type B atoms located at $x = R_n = (2n+1)c$ with $n = 1, 2, \cdots, N$. Each atom has one electron and one orbital. The orbital for the atom at R_m is denoted by $\psi_m(x - R_m)$

and the orbital for the atom at R_n is denoted by $\phi_n(x-R_n)$. Assume that the orbitals are all orthonormal, that is,

$$(\psi_m, \psi_{m'}) = \delta_{mm'}, \quad (\phi_n, \phi_{n'}) = \delta_{nn'}, \quad (\psi_m, \phi_n) = 0.$$

The single-electron Hamiltonian is given by $\hat{H} = \hat{p}^2/2m + U(x)$ with $U(x)$ a periodic function of the underlying Bravais lattice. The matrix elements of the Hamiltonian are given by

$$(\psi_m, \hat{H}\psi_{m'}) = \varepsilon_{\rm A}\delta_{mm'}, \quad (\phi_n, \hat{H}\phi_{n'}) = \varepsilon_{\rm B}\delta_{nn'},$$

and $(\psi_m, \hat{H}\phi_n) = -t$ if m, n are nearest neighbors, $= 0$ otherwise.

(1) Find the eigenvalues of $\hat{H}$.

(2) Plot the eigenvalues of $\hat{H}$ for $t = 1$ eV, $\varepsilon_{\rm A} - \varepsilon_{\rm B} = t$, and $\varepsilon_{\rm A} + \varepsilon_{\rm B} = 0$.

---

(1) We first construct basis functions satisfying Bloch's theorem from $\psi_m(x)$ and $\phi_n(x)$ by Fourier transforming them with respect to their sites

$$\psi_k(x) = \frac{1}{\sqrt{N}}\sum_m e^{ikR_m}\psi_m(x - R_m),$$

$$\phi_k(x) = \frac{1}{\sqrt{N}}\sum_n e^{ikR_n}\phi_n(x - R_n).$$

The lattice constant is $a = 2c$. From the Born-von Karman periodic boundary conditions $\psi_k(x + Na) = \psi_k(x)$ and $\phi_k(x + Na) = \phi_k(x)$, we find that the allowed values of k are given by $k = 2\pi j/Na$ with $j = 0, \pm 1, \pm 2, \cdots, \pm(N/2-1), N/2$ in the first Brillouin zone. We now compute the matrix elements of $\hat{H}$ between basis functions $\psi_k(x)$ and $\phi_k(x)$ for a given k. For $\langle\psi_k|\hat{H}|\psi_k\rangle$, we have

$$\begin{aligned}\langle\psi_k|\hat{H}|\psi_k\rangle &= \frac{1}{N}\sum_{mm'} e^{ik(R_m - R_{m'})}\langle\psi_{m'}|\hat{H}|\psi_m\rangle \\ &= \frac{1}{N}\varepsilon_{\rm A}\sum_{mm'} e^{ik(R_m - R_{m'})}\delta_{m'm} = \varepsilon_{\rm A}.\end{aligned}$$

For $\langle\phi_k|\hat{H}|\phi_k\rangle$, we have

$$\begin{aligned}\langle\phi_k|\hat{H}|\phi_k\rangle &= \frac{1}{N}\sum_{nn'} e^{ik(R_n - R_{n'})}\langle\phi_{n'}|\hat{H}|\phi_n\rangle \\ &= \frac{1}{N}\varepsilon_{\rm B}\sum_{nn'} e^{ik(R_n - R_{n'})}\delta_{n'n} = \varepsilon_{\rm B}.\end{aligned}$$

For $\langle \psi_k | \hat{H} | \phi_k \rangle$, we have

$$\begin{aligned}\langle \psi_k | \hat{H} | \phi_k \rangle &= \frac{1}{N} \sum_{mn} \mathrm{e}^{\mathrm{i}k(R_m - R_n)} \langle \psi_m | \hat{H} | \phi_n \rangle \\ &= -\frac{1}{N} t \sum_{mn} \mathrm{e}^{\mathrm{i}k(R_m - R_n)} \left(\delta_{R_m, R_n + a/2} + \delta_{R_m, R_n - a/2} \right) \\ &= -\frac{1}{N} t \sum_{n} \left(\mathrm{e}^{\mathrm{i}ka/2} + \mathrm{e}^{-\mathrm{i}ka/2} \right) = -2t \cos(kc).\end{aligned}$$

The matrix representation of $\hat{H}$ for a given k is then given by

$$H = \begin{pmatrix} \varepsilon_{\mathrm{A}} & -2t \cos(ka/2) \\ -2t \cos(ka/2) & \varepsilon_{\mathrm{B}} \end{pmatrix}.$$

The secular equation is

$$\det \begin{vmatrix} \varepsilon_{\mathrm{A}} - E & -2t \cos(ka/2) \\ -2t \cos(ka/2) & \varepsilon_{\mathrm{B}} - E \end{vmatrix} = 0.$$

Solving the above equation for E, we obtain the following eigenvalues for $\hat{H}$

$$E_{\pm} = (\varepsilon_{\mathrm{A}} + \varepsilon_{\mathrm{B}})/2 \pm \left[(\varepsilon_{\mathrm{A}} - \varepsilon_{\mathrm{B}})^2/4 + 4t^2 \cos^2(ka/2) \right]^{1/2}.$$

(2) We now plot the above derived bands. For $\varepsilon_{\mathrm{A}} + \varepsilon_{\mathrm{B}} = 0$ and $\varepsilon_{\mathrm{A}} - \varepsilon_{\mathrm{B}} = t = 1$ eV, the dispersion relations become

$$E_{\pm} = \pm \left[1/4 + 4 \cos^2(ka/2) \right]^{1/2} \text{eV}.$$

The above two energy bands are plotted in Fig. 17.2.

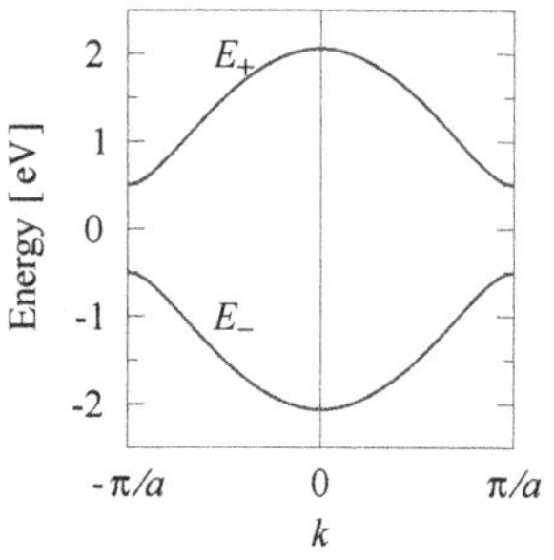

Fig. 17.2 Energy bands of the chain of alternate type-A and type-B atoms.

17-3 Energy gap at the M point in a 2D square lattice. Consider electrons in a square lattice in two dimensions. The periodic potential is given by $U(x, y) = 4U \cos(2\pi x/a) \cos(2\pi y/a)$ with a the lattice constant. Find approximately the energy gap at the corner point M with $\boldsymbol{k}_M = (\pi/a)(\boldsymbol{e}_x + \boldsymbol{e}_y)$ of the first Brillouin zone. It will suffice to solve a 2×2 determinantal equation.

Rewriting $U(x, y)$ in terms of exponential functions, we have

$$\begin{aligned} U(x, y) &= U\left(\mathrm{e}^{\mathrm{i}2\pi x/a} + \mathrm{e}^{-\mathrm{i}2\pi x/a}\right)\left(\mathrm{e}^{\mathrm{i}2\pi y/a} + \mathrm{e}^{-\mathrm{i}2\pi y/a}\right) \\ &= U\left[\mathrm{e}^{\mathrm{i}(2\pi/a)(x+y)} + \mathrm{e}^{-\mathrm{i}(2\pi/a)(x-y)} + \mathrm{e}^{\mathrm{i}(2\pi/a)(x-y)}\right. \\ &\qquad \left. + \mathrm{e}^{-\mathrm{i}(2\pi/a)(x+y)}\right]. \end{aligned}$$

We immediately see that the nonzero Fourier components of $U(x, y)$ are all equal to U, $U_{\boldsymbol{K}} = U$, for $\boldsymbol{K}$ taking on the following values

$$-(2\pi/a)(\boldsymbol{e}_x + \boldsymbol{e}_y),\ (2\pi/a)(\boldsymbol{e}_x - \boldsymbol{e}_y),\ -(2\pi/a)(\boldsymbol{e}_x - \boldsymbol{e}_y),\ (2\pi/a)(\boldsymbol{e}_x + \boldsymbol{e}_y).$$

Since only two reciprocal lattice vectors $\boldsymbol{K}_1 = 0$ and $\boldsymbol{K}_2 = (2\pi/a)(\boldsymbol{e}_x + \boldsymbol{e}_y)$ are connected by $U(x, y)$ among the four degenerate reciprocal lattice vectors 0, $(2\pi/a)\boldsymbol{e}_x$, $(2\pi/a)\boldsymbol{e}_y$, and $(2\pi/a)(\boldsymbol{e}_x + \boldsymbol{e}_y)$ at the M point, we can consider only the subspace spanned by $\boldsymbol{K}_1 = 0$ and $\boldsymbol{K}_2 = (2\pi/a)(\boldsymbol{e}_x + \boldsymbol{e}_y)$. In this subspace, the matrix representation of the Hamiltonian is given by

$$H = \begin{pmatrix} \pi^2\hbar^2/ma^2 & U \\ U & \pi^2\hbar^2/ma^2 \end{pmatrix}.$$

The secular equation reads

$$\det\begin{vmatrix} \pi^2\hbar^2/ma^2 - E & U \\ U & \pi^2\hbar^2/ma^2 - E \end{vmatrix} = 0.$$

Solving E from the above equation, we obtain

$$E_{\pm} = \pi^2\hbar^2/ma^2 \pm U.$$

17-4 Tight-binding band from localized orbitals. Consider the following tight-binding Hamiltonian for a single band of localized orbitals in one dimension

$$\hat{H} = \sum_n \cos(10n\pi/3)\,|n\rangle\langle n| + \frac{1}{2}t\sum_n\left(|n\rangle\langle n+1| + |n\rangle\langle n-1|\right),$$

where n, an integer, labels the atoms along a chain and $|n\rangle$ represents the orbital on the nth atom with $\langle x|n\rangle = \psi(x-n)$. Here the atom separation has been taken as unity. Assume that the orbital on each atom decays fast away from the atom.

(1) Identify the periodicity of the Hamiltonian.

(2) Use Bloch's theorem to reduce the eigenvalue problem of an infinite matrix H obtained by representing the Hamiltonian in the basis of orbitals $|n\rangle$ to that of a small finite matrix of the size of the periodicity of the Hamiltonian.

(3) Compute and plot the energy bands as functions of Bloch wave vector k throughout the first Brillouin zone.

---

(1) From the first term in the Hamiltonian, we see that, each time n changes by three, the value of $\cos(10n\pi/3)$ occurs repeatedly. Thus, the periodicity of the Hamiltonian is three, which implies that the chain has a three-atom basis. The lattice constant is $a = 3$. Let the orbitals on the atoms in a primitive cell be sequentially denoted by $|\xi_m\rangle$, $|\zeta_n\rangle$, and $|\eta_\ell\rangle$. The orbitals along the chain are then given by

$$\cdots,|\xi_{m-1}\rangle,|\zeta_{n-1}\rangle,|\eta_{\ell-1}\rangle,|\xi_m\rangle,|\zeta_n\rangle,|\eta_\ell\rangle,|\xi_{m+1}\rangle,|\zeta_{n+1}\rangle,|\eta_{\ell+1}\rangle,\cdots.$$

(2) In terms of $|\xi_m\rangle$, $|\zeta_n\rangle$, and $|\eta_\ell\rangle$, the Hamiltonian is rewritten as

$$\hat{H} = \sum_m |\xi_m\rangle\langle\xi_m| - \frac{1}{2}\sum_n |\zeta_n\rangle\langle\zeta_n| - \frac{1}{2}\sum_\ell |\eta_\ell\rangle\langle\eta_\ell| + \frac{1}{2}t\sum_n \big(|\xi_n\rangle\langle\zeta_n| + |\zeta_n\rangle\langle\eta_n| + |\eta_n\rangle\langle\xi_{n+1}| + h.c.\big).$$

We now construct basis functions that satisfy Bloch's theorem. For a given wave vector k, we have

$$|\xi_k\rangle = \frac{1}{\sqrt{N}}\sum_m \mathrm{e}^{\mathrm{i}kR_m}\,|\xi_m\rangle,\ |\zeta_k\rangle = \frac{1}{\sqrt{N}}\sum_n \mathrm{e}^{\mathrm{i}kR_n}\,|\zeta_n\rangle,$$
$$|\eta_k\rangle = \frac{1}{\sqrt{N}}\sum_\ell \mathrm{e}^{\mathrm{i}kR_\ell}\,|\eta_\ell\rangle,$$

where N is the number of primitive cells in the chain, $R_m = ma$, $R_n = na + 1$, and $R_\ell = \ell a + 2$. The allowed values of k can be found from the Born-von Karman periodic boundary conditions. We have $k_j = 2\pi j/Na$ with $j = 0, \pm1, \pm2, \cdots, \pm(N/2 - 1), N/2$. We now evaluate matrix elements of the

Hamiltonian between basis functions. For $\langle \xi_k | \hat{H} | \xi_k \rangle$, we have

$$\begin{aligned}\langle \xi_k | \hat{H} | \xi_k \rangle &= \frac{1}{N} \sum_{mm'} \mathrm{e}^{\mathrm{i}k(R_m - R_{m'})} \langle \xi_{m'} | \hat{H} | \xi_m \rangle \\ &= \frac{1}{N} \sum_{mm'm''} \mathrm{e}^{\mathrm{i}k(R_m - R_{m'})} \langle \xi_{m'} | \xi_{m''} \rangle \langle \xi_{m''} | \xi_m \rangle \\ &= \frac{1}{N} \sum_{mm'm''} \mathrm{e}^{\mathrm{i}k(R_m - R_{m'})} \delta_{m'm''} \delta_{m''m} = 1.\end{aligned}$$

Similarly, we have

$$\langle \zeta_k | \hat{H} | \zeta_k \rangle = \langle \eta_k | \hat{H} | \eta_k \rangle = -\frac{1}{2}.$$

For $\langle \xi_k | \hat{H} | \zeta_k \rangle$, we have

$$\langle \xi_k | \hat{H} | \zeta_k \rangle = \frac{1}{N} \sum_{mn} \mathrm{e}^{-\mathrm{i}k(R_m - R_n)} \langle \xi_m | \hat{H} | \zeta_n \rangle = \frac{1}{2} t\mathrm{e}^{\mathrm{i}k}.$$

For $\langle \zeta_k | \hat{H} | \eta_k \rangle$, we have

$$\langle \zeta_k | \hat{H} | \eta_k \rangle = \frac{1}{N} \sum_{n\ell} \mathrm{e}^{-\mathrm{i}k(R_n - R_\ell)} \langle \zeta_n | \hat{H} | \eta_\ell \rangle = \frac{1}{2} t\mathrm{e}^{\mathrm{i}k}.$$

For $\langle \eta_k | \hat{H} | \xi_k \rangle$, we have

$$\langle \eta_k | \hat{H} | \xi_k \rangle = \frac{1}{N} \sum_{\ell m} \mathrm{e}^{-\mathrm{i}k(R_\ell - R_m)} \langle \eta_\ell | \hat{H} | \xi_m \rangle = \frac{1}{2} t\mathrm{e}^{\mathrm{i}k}.$$

Having found all the matrix elements for the Hamiltonian, we can write down its matrix representation and have

$$H = \begin{pmatrix} 1 & t\mathrm{e}^{\mathrm{i}k}/2 & t\mathrm{e}^{-\mathrm{i}k}/2 \\ t\mathrm{e}^{-\mathrm{i}k}/2 & -1/2 & t\mathrm{e}^{\mathrm{i}k}/2 \\ t\mathrm{e}^{\mathrm{i}k}/2 & t\mathrm{e}^{-\mathrm{i}k}/2 & -1/2 \end{pmatrix}.$$

(3) The secular equation is given by

$$\begin{vmatrix} 1 - E & t\mathrm{e}^{\mathrm{i}k}/2 & t\mathrm{e}^{-\mathrm{i}k}/2 \\ t\mathrm{e}^{-\mathrm{i}k}/2 & -1/2 - E & t\mathrm{e}^{\mathrm{i}k}/2 \\ t\mathrm{e}^{\mathrm{i}k}/2 & t\mathrm{e}^{-\mathrm{i}k}/2 & -1/2 - E \end{vmatrix} = 0.$$

Evaluating the determinant, we obtain

$$4E^3 - 3(1 + t^2)E - t^3 \cos(3k) - 1 = 0.$$

The energy bands can be obtained by solving numerically the above equation for various values of k in the first Brillouin zone. Here the Brent method for finding the roots of a nonlinear equation is used. The results are plotted in Fig. 17.3.

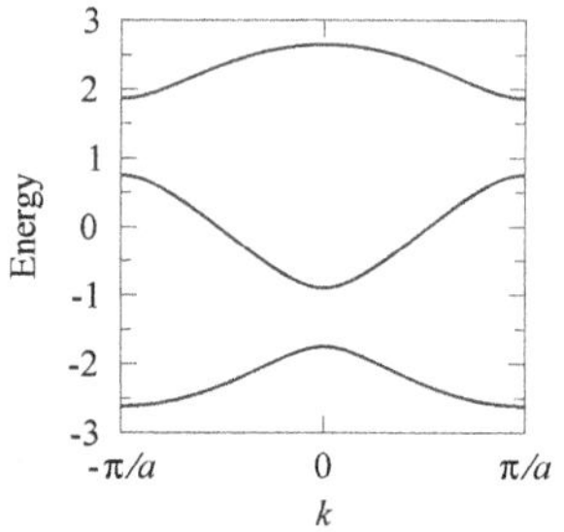

Fig. 17.3 Energy bands of the chain with localized orbitals.

Note that the secular equation is a cubic equation that can be put into the following standard form of Vieta

$$E^3 + pE = q,$$

where $p = -3(1+t^2)/4$ and $q = [t^3\cos(3k)+1]/4$. In terms of $\theta = \cos^{-1}(3q/2\sqrt{-p^3/3})$, the solutions are given by

$$E_1 = 2\sqrt{-p/3}\cos(\theta/3 + 2\pi/3),$$
$$E_2 = 2\sqrt{-p/3}\cos(\theta/3 + 4\pi/3),$$
$$E_3 = 2\sqrt{-p/3}\cos(\theta/3),$$

where the solutions have been put into the ascending order.

17-5 Energy band structure of aluminum. Al is an FCC crystal with the lattice constant $a = 0.405$ nm.

(1) Construct the lowest two bands of the free electron band structure and plot it from Γ to X, from X to W, from W to L, and from L to Γ. Indicate the degeneracies.

(2) Compute the splitting at X and L in eV using the following pseudopotential

$$V(r) = \begin{cases} -Ze^2/4\pi\epsilon_0 r, & r > R_c, \\ 0, & r < R_c, \end{cases}$$

where $Z = 3$ and $R_c = 0.06$ nm.

(3) Write down the Hamiltonian matrix near the W point using the above pseudopotential.

(4) Diagonalize the Hamiltonian matrix near the W point from W to X and from W to L.

---

(1) The reciprocal lattice of the FCC Bravais lattice with a lattice constant a is a BCC Bravais lattice in reciprocal space with a lattice constant $4\pi/a$. The first Brillouin zone of the FCC Bravais lattice is a truncated octahedron shown in Fig. 17.4. In the free-electron (empty-lattice) band structure, the dispersion relation of the band arising from reciprocal lattice vector $\boldsymbol{K}$ is given by

$$E^0_{\boldsymbol{k}} = \frac{\hbar^2}{2m}(\boldsymbol{k} - \boldsymbol{K})^2.$$

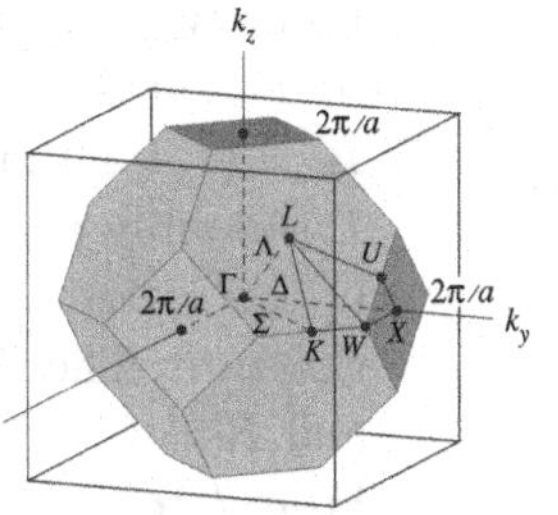

Fig. 17.4 First Brillouin zone of the FCC Bravais lattice.

From Γ to X, $\boldsymbol{k} = k_y \boldsymbol{e}_y$ for k_y from 0 to $2\pi/a$. The lowest energy band is given by $\boldsymbol{K} = 0$. Thus, the first band is nondegenerate. The next energy band is given by $\boldsymbol{K} = (0, 2, 0)(2\pi/a)$. Thus, the second band is also nondegenerate. However, we see from Fig. 17.5 that it intersects the band from $\boldsymbol{K} = (\pm 1, 1, \pm 1)(2\pi/a)$ and the band from $\boldsymbol{K} = (\pm 1, -1, \pm 1)(2\pi/a)$. These two bands are both four-fold degenerate.

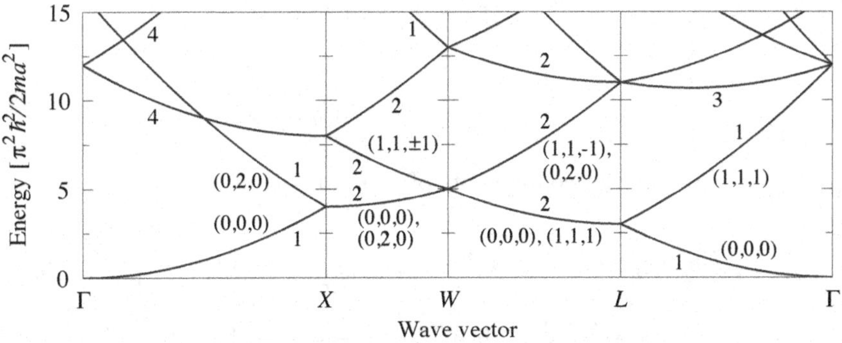

Fig. 17.5 Empty-lattice band structure of the FCC Bravais lattice. The degeneracies of the bands are indicated by numbers and the responsible reciprocal lattice vectors are given for the two lowest bands.

From X to W, $\boldsymbol{k} = (\xi, 1, 0)(2\pi/a)$ with ξ from 0 at X to

1/2 at W. The lowest energy band is from $\boldsymbol{K} = 0$ and $\boldsymbol{K} = (0,2,0)(2\pi/a)$. Thus, the lowest energy band from X to W is doubly degenerate. The second energy band is from $\boldsymbol{K} = (1,1,\pm 1)(2\pi/a)$ and is also doubly degenerate. The third energy band is from $\boldsymbol{K} = (-1,1,\pm 1)(2\pi/a)$ and is also doubly degenerate.

From W to L, $\boldsymbol{k} = (1/2, 1-\zeta, \zeta)(2\pi/a)$ with ζ from 0 at W to 1/2 at L. The lowest energy band is from $\boldsymbol{K} = 0$ and $\boldsymbol{K} = (1,1,1)(2\pi/a)$. Thus, the lowest energy band from W to L is doubly degenerate. The second energy band is from $\boldsymbol{K} = (1,1,-1)(2\pi/a)$ and $\boldsymbol{K} = (0,2,0)(2\pi/a)$ and is also doubly degenerate. The third energy band is from $\boldsymbol{K} = (-1,1,1)(2\pi/a)$ and $\boldsymbol{K} = (2,0,0)(2\pi/a)$ and is also doubly degenerate.

From L to Γ, $\boldsymbol{k} = (\eta,\eta,\eta)(2\pi/a)$ with η from 1/2 at L to 0 at Γ. The lowest energy band is from $\boldsymbol{K} = 0$. Thus, the lowest energy band from L to Γ is nondegenerate. The second energy band is from $\boldsymbol{K} = (1,1,1)(2\pi/a)$ and is also nondegenerate. The third energy band is from $\boldsymbol{K} = (-1,1,1)(2\pi/a)$, $(1,-1,1)(2\pi/a)$, and $(1,1,-1)(2\pi/a)$ and is thus triply degenerate.

(2) The given $V(r)$ is for the potential due to a single atom. In terms of $V(r)$, the periodic potential is given by

$$U(\boldsymbol{r}) = \sum_n V(|\boldsymbol{r} - \boldsymbol{R}_n|) = -\sum_n \frac{Ze^2}{4\pi\epsilon_0|\boldsymbol{r} - \boldsymbol{R}_n|}.$$

The Fourier components of $U(\boldsymbol{r})$ are given by

$$U_{\boldsymbol{K}} = \frac{1}{\mathscr{V}}\int d\boldsymbol{r}\, e^{i\boldsymbol{K}\cdot\boldsymbol{r}} U(\boldsymbol{r}) = -\frac{1}{\mathscr{V}}\sum_n \int_{|\boldsymbol{r}-\boldsymbol{R}_n|>R_0} d\boldsymbol{r}\, e^{i\boldsymbol{K}\cdot\boldsymbol{r}} \frac{Ze^2}{4\pi\epsilon_0|\boldsymbol{r} - \boldsymbol{R}_n|}$$
$$= -\frac{1}{\mathscr{V}}\sum_n \int_{r>R_c} d\boldsymbol{r}\, e^{i\boldsymbol{K}\cdot\boldsymbol{r}} \frac{Ze^2}{4\pi\epsilon_0|\boldsymbol{r}|} = -\frac{1}{v_c}\int_{r>R_c} d\boldsymbol{r}\, e^{i\boldsymbol{K}\cdot\boldsymbol{r}} \frac{Ze^2}{4\pi\epsilon_0|\boldsymbol{r}|},$$

where $v_c = a^3/4$ is the volume of a primitive cell of the direct lattice and $e^{i\boldsymbol{K}\cdot\boldsymbol{R}_n} = 1$ has been used in making the change of integration variables from $\boldsymbol{r}$ to $\boldsymbol{r} - \boldsymbol{R}_n$. To ensure the convergence of the integration, we multiply the integrand with $e^{-\delta r}$ with δ an infinitesimally small positive quantity and to be set to zero in the final result. We then have

$$\begin{aligned}
U_{\boldsymbol{K}} &= -\frac{1}{v_c}\int_{r>R_c} \mathrm{d}\boldsymbol{r}\ \mathrm{e}^{\mathrm{i}\boldsymbol{K}\cdot\boldsymbol{r}-\delta r}\frac{Ze^2}{4\pi\epsilon_0|\boldsymbol{r}|} \\
&= -\frac{2\pi Ze^2}{(4\pi\epsilon_0)v_c}\int_{R_c}^{\infty}\mathrm{d}r\ r\int_{-1}^{1}\mathrm{d}\cos\theta\ \mathrm{e}^{\mathrm{i}Kr\cos\theta-\delta r} \\
&= -\frac{2\pi Ze^2}{(4\pi\epsilon_0)v_c iK}\left[-\frac{\mathrm{e}^{\mathrm{i}(K-\delta)R_c}}{i(K-\delta)}-\frac{\mathrm{e}^{-\mathrm{i}(K+\delta)R_c}}{i(K+\delta)}\right] \\
&\xrightarrow{\delta\to 0^+} -\frac{4\pi Ze^2}{(4\pi\epsilon_0)v_cK^2}\cos(KR_c) = -\frac{16\pi Ze^2}{(4\pi\epsilon_0)K^2a^3}\cos(KR_c).
\end{aligned}$$

Note that $U_{\boldsymbol{K}}$ diverges at $\boldsymbol{K} = 0$. If the Coulomb interaction between electrons is taken into account, the $\boldsymbol{K} = 0$ component can be canceled. We thus ignore this component.

At the X point, the Bragg plane is from $\boldsymbol{K} = (4\pi/a)\boldsymbol{e}_y$. The degeneracy at the X point is due to the reciprocal lattice vectors 0 and $(0,2,0)(2\pi/a)$. Thus, the splitting at X between the two lowest bands is

$$\Delta_X = 2|U_{(0,2,0)(2\pi/a)}| = \frac{Ze^2\cos(4\pi R_c/a)}{\pi(4\pi\epsilon_0)a} \approx 0.974 \text{ eV}.$$

At the L point, the Bragg plane is from $\boldsymbol{K} = (1,1,1)(2\pi/a)$. The degeneracy at the L point is due to the reciprocal lattice vectors 0 and $(1,1,1)(2\pi/a)$. Thus, the splitting at L between the two lowest bands is

$$\Delta_L = 2|U_{(1,1,1)(2\pi/a)}| = \frac{4Ze^2\cos(2\sqrt{3}\pi R_c/a)}{3\pi(4\pi\epsilon_0)a} \approx 0.188 \text{ eV}.$$

(3) Near the W point, the energies are four-fold nearly degenerate with the involved reciprocal lattice vectors being $\boldsymbol{K}_1 = (0,0,0)(2\pi/a)$, $\boldsymbol{K}_2 = (1,1,-1)(2\pi/a)$, $\boldsymbol{K}_3 = (1,1,1)(2\pi/a)$, and $\boldsymbol{K}_4 = (0,2,0)(2\pi/a)$. Thus, the degenerate subspace is four-dimensional. The Hamiltonian matrix in this subspace is given by

$$H = \begin{pmatrix}
\hbar^2(\boldsymbol{k}-\boldsymbol{K}_1)^2/2m & U_{\boldsymbol{K}_1-\boldsymbol{K}_2} & U_{\boldsymbol{K}_1-\boldsymbol{K}_3} & U_{\boldsymbol{K}_1-\boldsymbol{K}_4} \\
U_{\boldsymbol{K}_2-\boldsymbol{K}_1} & \hbar^2(\boldsymbol{k}-\boldsymbol{K}_2)^2/2m & U_{\boldsymbol{K}_2-\boldsymbol{K}_3} & U_{\boldsymbol{K}_2-\boldsymbol{K}_4} \\
U_{\boldsymbol{K}_3-\boldsymbol{K}_1} & U_{\boldsymbol{K}_3-\boldsymbol{K}_2} & \hbar^2(\boldsymbol{k}-\boldsymbol{K}_3)^2/2m & U_{\boldsymbol{K}_3-\boldsymbol{K}_4} \\
U_{\boldsymbol{K}_4-\boldsymbol{K}_1} & U_{\boldsymbol{K}_4-\boldsymbol{K}_2} & U_{\boldsymbol{K}_4-\boldsymbol{K}_3} & \hbar^2(\boldsymbol{k}-\boldsymbol{K}_4)^2/2m
\end{pmatrix}.$$

Evaluating the Fourier components of $U(\boldsymbol{r})$, we have in eV

$$H = \begin{pmatrix}
\hbar^2(\boldsymbol{k}-\boldsymbol{K}_1)^2/2m & -0.0938391 & -0.0938391 & -0.48688 \\
-0.0938391 & \hbar^2(\boldsymbol{k}-\boldsymbol{K}_2)^2/2m & -0.48688 & -0.0938391 \\
-0.0938391 & -0.48688 & \hbar^2(\boldsymbol{k}-\boldsymbol{K}_3)^2/2m & -0.0938391 \\
-0.48688 & -0.0938391 & -0.0938391 & \hbar^2(\boldsymbol{k}-\boldsymbol{K}_4)^2/2m
\end{pmatrix}.$$

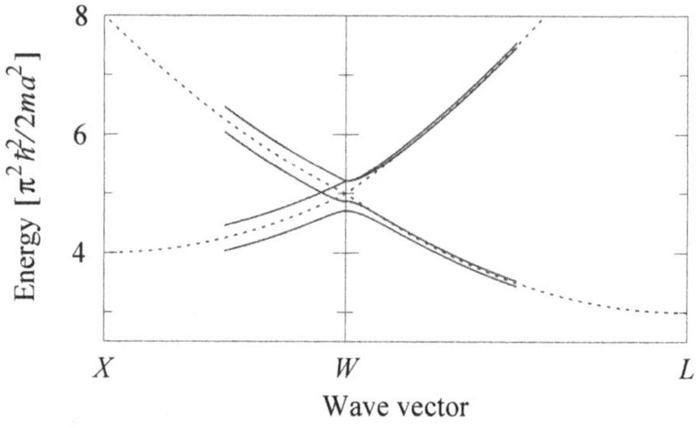

Fig. 17.6 Energy bands of copper near the W point. The solid lines are results from the perturbative computation for the weak potential. The dotted lines are empty-lattice energy bands for the FCC Bravais lattice.

(4) The results obtained through diagonalizing numerically the above matrix near the W point from W to X and from W to L are plotted in Fig. 17.5.

The diagonalization was performed using the linear algebraic functions in the GNU Scientific Library. For this library, see

`http://www.gnu.org/software/gsl/`.

From Fig. 17.5, we see that two bands remain degenerate at the W point and that the splittings along WX are larger than those along WL.

Chapter 18

Methods for Band Structure Computations in Solids

(1) *Single-electron stationary Schrödinger equation*

$$\left[-\frac{\hbar^2}{2m}\nabla^2 + U(\boldsymbol{r})\right]\psi(\boldsymbol{r}) = E\psi(\boldsymbol{r}).$$

(2) *Hartree equations*

$$\left[-\frac{\hbar^2}{2m}\nabla^2 - \frac{1}{4\pi\epsilon_0}\sum_I \frac{Z_I e^2}{|\boldsymbol{r}-\boldsymbol{R}_I|} + \frac{1}{4\pi\epsilon_0}\sum_{i\neq\ell}\int \mathrm{d}\boldsymbol{r}'\, \frac{e^2|\psi_i(\boldsymbol{r}')|^2}{|\boldsymbol{r}-\boldsymbol{r}'|}\right]\psi_\ell(\boldsymbol{r}) = \varepsilon_\ell\psi_\ell(\boldsymbol{r}).$$

(3) *Hartree-Fock equations*

$$\left[-\frac{\hbar^2}{2m}\nabla^2 - \frac{e^2}{4\pi\epsilon_0}\sum_I \frac{Z_I}{|\boldsymbol{r}-\boldsymbol{R}_I|} + \frac{e^2}{4\pi\epsilon_0}\sum_{i\neq\ell}\int \mathrm{d}x'\, \frac{|\psi_{\lambda_i}(x')|^2}{|\boldsymbol{r}-\boldsymbol{r}'|}\right]\psi_{\lambda_\ell}(x) - \frac{e^2}{4\pi\epsilon_0}\sum_{i\neq\ell}\int \mathrm{d}x'\, \frac{\psi^*_{\lambda_i}(x')\psi_{\lambda_\ell}(x')}{|\boldsymbol{r}-\boldsymbol{r}'|}\psi_{\lambda_i}(x) = \varepsilon_{\lambda_\ell}\psi_{\lambda_\ell}(x).$$

(4) *Exchange potential operator*

$$\hat{U}_{\mathrm{ex}}\psi_{\lambda_\ell}(x) = -\frac{e^2}{4\pi\epsilon_0}\sum_{i\neq\ell}\int \mathrm{d}x'\, \frac{\psi^*_{\lambda_i}(x')\psi_{\lambda_\ell}(x')}{|\boldsymbol{r}-\boldsymbol{r}'|}\psi_{\lambda_i}(x).$$

(5) *Exchange energy in a uniform electron gas*

$$E_{\mathrm{ex}}(\boldsymbol{k}) = -\frac{1}{4\pi\epsilon_0}\frac{1}{\mathscr{V}}\sum_{\boldsymbol{k}'\neq\boldsymbol{k}} \frac{4\pi e^2}{|\boldsymbol{k}'-\boldsymbol{k}|^2} = -\frac{e^2}{4\pi\epsilon_0}\frac{2k_{\mathrm{F}}}{\pi}h(k/k_{\mathrm{F}}),$$

$$h(x) = \frac{1}{2} + \frac{1-x^2}{4x}\ln\left|\frac{1+x}{1-x}\right|.$$

(6) *Other computational methods of band structures*

PW	APW	LAPW	LMTO
KKR	OPW	TBA	

(7) *Special points in the first Brillouin zone*
If $\boldsymbol{k}_1$ and $\boldsymbol{k}_2$ are symmetry points, then $\boldsymbol{k}_\ell = \boldsymbol{k}_1 + g_\ell \boldsymbol{k}_2$ for $\ell = 1, 2, \cdots, n_g$ are special points. Weights are given by $w_i = n_i / \sum_i n_i$ with n_i the number of equivalent points of the special point $\boldsymbol{k}_i$.

(8) *Inverse effective mass tensor*

$$\left(\frac{1}{m^*}\right)_{\alpha\beta} = \frac{1}{\hbar^2}\frac{\partial^2 E_{n\boldsymbol{k}}}{\partial k_\alpha \partial k_\beta}.$$

(9) *Effective mass theory*

$$\left[\sum_{\alpha\beta}\left(\frac{\hbar^2}{2m^*}\right)_{\alpha\beta}\left(-\mathrm{i}\frac{\partial}{\partial x_\alpha}\right)\left(-\mathrm{i}\frac{\partial}{\partial x_\beta}\right) + V(\boldsymbol{r})\right]F_n(\boldsymbol{r}) = \left[E - E_n(0)\right]F_n(\boldsymbol{r}).$$

18-1 Plane-wave method for a 1D crystal. For a one-dimensional crystal, we assume that the periodic potential is a repetition of the square potential which is given by

$$U(x) = \begin{cases} V_0, & -a/4 < x < a/4, \\ 0, & -a/2 < x < -a/4 \text{ and } a/4 < x < a/2 \end{cases}$$

in the Wigner-Seitz cell about the lattice point at the origin, where a is the lattice constant and $V_0 > 0$.

(1) Using the plane-wave method, find the two lowest energy eigenvalues at the center and edge of the first Brillouin zone as well as one point halfway between, that is, at $k = 0$, $\pi/2a$, and π/a. As a crude approximation, use only five plane wave basis functions corresponding to reciprocal lattice vectors $K = 0, \pm 2\pi/a, \pm 4\pi/a$.

(2) Find the wave function corresponding to each of these two bands at the three specified k-values.

(3) Plot the electron density as a function of position corresponding to each of these six wave functions.

---

(1) The five basis functions corresponding to $K = 0, \pm 2\pi/a$, and $\pm 4\pi/a$ are

$$\psi_k^0(x) = \frac{1}{L^{1/2}}e^{ikx},\ \psi_{k\mp 2\pi/a}^0(x) = \frac{1}{L^{1/2}}e^{i(k\mp 2\pi/a)x},$$
$$\psi_{k\mp 4\pi/a}^0(x) = \frac{1}{L^{1/2}}e^{i(k\mp 4\pi/a)x}.$$

The Fourier components of the periodic potential are given by

$$U_{K-K'} = \frac{1}{L}\int_{-\infty}^{\infty} dx\, U(x)e^{i(K-K')x} = \frac{1}{a}\int_{-a/2}^{a/2} dx\, U(x)e^{i(K-K')x}$$
$$= \frac{V_0}{a}\int_{-a/4}^{a/4} dx\, e^{i(K-K')x} = \frac{2V_0}{(K-K')a}\sin[(K-K')a/4],$$

where L is the length of the crystal. Within the subspace spanned by the basis functions, the nonzero Fourier components of $U(x)$ are only

$$U_{K-K'=0} = \frac{1}{2}V_0,\ U_{K-K'=\pm 2\pi/a} = \frac{V_0}{\pi},\ U_{K-K'=\pm 6\pi/a} = -\frac{V_0}{3\pi}.$$

Expanding the Bloch function $\psi_k(x)$ in terms of the above basis functions, we have

$$\psi_k(x) = C_k\psi_k^0(x) + C_{k-2\pi/a}\psi_{k-2\pi/a}^0(x) + C_{k+2\pi/a}\psi_{k+2\pi/a}^0(x)$$
$$+ C_{k-4\pi/a}\psi_{k-4\pi/a}^0(x) + C_{k+4\pi/a}\psi_{k+4\pi/a}^0(x).$$

We take the order of the basis functions in the subspace as

$$\psi_{k+4\pi/a}^0(x), \psi_{k+2\pi/a}^0(x), \psi_k^0(x), \psi_{k-2\pi/a}^0(x), \psi_{k-4\pi/a}^0(x).$$

To simplify notations, we introduce

$$\gamma_n(k) = \hbar^2(k - n\pi/a)^2/2m + V_0/2.$$

The equations for the coefficients follow from the stationary Schrödinger equation

$$[\gamma_{-4}(k) - E_k]C_{k+4\pi/a} + (V_0/\pi)C_{k+2\pi/a} - (V_0/3\pi)C_{k-2\pi/a} = 0,$$
$$(V_0/\pi)C_{k+4\pi/a} + [\gamma_{-2}(k) - E_k]C_{k+2\pi/a} + (V_0/\pi)C_k$$
$$- (V_0/3\pi)C_{k-4\pi/a} = 0,$$
$$(V_0/\pi)C_{k+2\pi/a} + [\gamma_0(k) - E_k]C_k + (V_0/\pi)C_{k-2\pi/a} = 0,$$
$$- (V_0/3\pi)C_{k+4\pi/a} + (V_0/\pi)C_k + [\gamma_2(k) - E_k]C_{k-2\pi/a}$$
$$+ (V_0/\pi)C_{k-4\pi/a} = 0,$$
$$- (V_0/3\pi)C_{k+2\pi/a} + (V_0/\pi)C_{k-2\pi/a} + [\gamma_4(k) - E_k]C_{k-4\pi/a} = 0.$$

The secular equation is given by

$$\begin{vmatrix} \gamma_{-4}(k) - E_k & V_0/\pi & 0 & -V_0/3\pi & 0 \\ V_0/\pi & \gamma_{-2}(k) - E_k & V_0/\pi & 0 & -V_0/3\pi \\ 0 & V_0/\pi & \hbar^2 k^2/2m - E_k & V_0/\pi & 0 \\ -V_0/3\pi & 0 & V_0/\pi & \gamma_{-4}(k) - E_k & V_0/\pi \\ 0 & -V_0/3\pi & 0 & V_0/\pi & \gamma_4(k) - E_k \end{vmatrix} = 0.$$

The lowest three bands obtained from solving the above secular equations at various wave vectors in the first Brillouin zone are shown in Fig. 18.1. It is seen clearly from the figure the appearance of the band gaps at the boundaries of the first Brillouin zone. Note that a band gap is also present between the second and third bands at $k = 0$ although the size of this band gap is much smaller than those at $k = \pm\pi/a$. The six points mentioned in the problem are marked with solid circles. The value of V_0 is taken to be $1.5\pi^2\hbar^2/2ma^2$. The C^{++} codes used to obtain the results in Fig. 18.1 were put into three files:

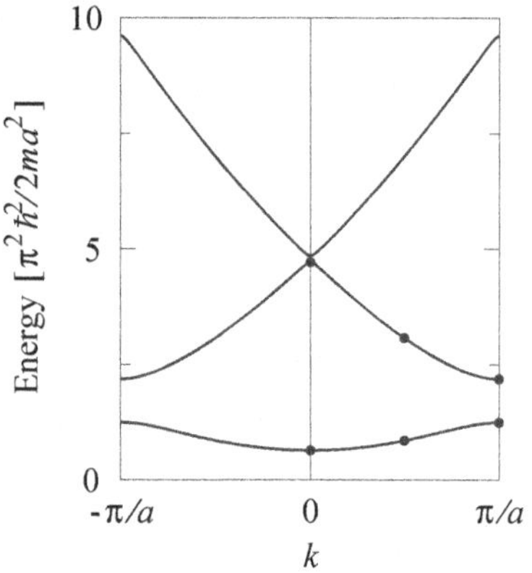

Fig. 18.1 Lowest three bands of the one-dimensional crystal. The six points mentioned in the problem are marked with solid circles.

pwmethod.hpp, **pwmethod.cpp**, and **pw-main.cpp**.
In the first two files, class **PWMethod** is defined and implemented. The third file contains the main program. The file **pwmethod.hpp** reads

```
#ifndef PW_METHOD_HPP
#define PW_METHOD_HPP
#include <gsl/gsl_eigen.h>
#include <gsl/gsl_vector.h>
#include <gsl/gsl_matrix.h>
#include <gsl/gsl_complex.h>
#include <iostream>
class PWMethod {
  public:
    PWMethod(int Kmax, double V0);
    ~PWMethod();
    void initialize(double k)const;
    void diagonalize()const;
    double eigenvalue(int i)const;
    void eigenvector(int i, double* v)const;
    // n: band index;
    // m: number of wave vectors
    //    in $[-\uppi/a, \uppi/a]$;
    // k: wave vectors in $\uppi/a$;
    // e: energies
    //    in $\uppi^{2}\hbar^{2}/2ma^{2}$
    void band(int n, int m, double* k,
              double* e)const;
    gsl_complex planeWave(double k, int K,
                          double x)const;
    double density(int n, double k,
                   double x)const;
    void hamiltonian(std::ostream& os
                     =std::cout)const;
  private:
    int     Kmax_;
    int     dim_;
    double  V0_;
    gsl_matrix* hamil_;
```

```
  gsl_matrix* ident_;
  gsl_vector* eval_;
  gsl_matrix* evec_;
 public:
  static const double PI;
};
#endif
```

To save space, many comments have been stripped off and many white lines have been eliminated. The physical quantities represented by the data members can be easily inferred from their names and the intentions of the member functions are also obvious from their names. Wave vectors (k's) are measured in π/a, reciprocal lattice vectors (K's) in $2\pi/a$, and energies in $\pi^2\hbar^2/2ma^2$. The data member `Kmax_` is the cut-off reciprocal lattice vector used so that all the reciprocal lattice vectors in the basis set are from $-$`Kmax_`$\times 2\pi/a$ through `Kmax_`$\times 2\pi/a$ with a spacing of $2\pi/a$. The class was implemented in such a way that large basis sets can also be used. We performed the diagonalization of the Hamiltonian matrix using the C functions for linear algebras in the GNU Scientific Library (GSL) which can be found at **http://www.gnu.org/software/gsl/**. We also used the matrix, vector, and complex structs in the library.

The file `pwmethod.cpp` reads

```
#include "pwmethod.hpp"
#include <cmath>
#include <gsl/gsl_complex_math.h>
const double PWMethod :: PI = 4.0*atan(1.0);
PWMethod :: PWMethod(int Kmax, double V0)
: Kmax_(Kmax), dim_(Kmax_+Kmax_+1), V0_(V0)
{
  hamil_ = gsl_matrix_alloc(dim_, dim_);
  ident_ = gsl_matrix_alloc(dim_, dim_);
  eval_ = gsl_vector_alloc(dim_);
  evec_ = gsl_matrix_alloc(dim_, dim_);
  gsl_matrix_set_identity(ident_);
}

PWMethod :: ~PWMethod()
{
  gsl_matrix_free(evec_);
  gsl_vector_free(eval_);
  gsl_matrix_free(ident_);
  gsl_matrix_free(hamil_);
}

void PWMethod :: initialize(double k)const
{
  for (int r=-Kmax_; r<=Kmax_; ++r) {
    int Kr = r + r;
    for (int c=-Kmax_; c<=Kmax_; ++c) {
      int Kc = c + c;
      double el = 0.0;
      if (Kr == Kc) {
        el = (k-Kr)*(k-Kr) + 0.5*V0_;
      } else {
        double dK = (Kr-Kc)*PI;
        el = 2.0*V0_*sin(dK*0.25)/dK;
      }
      gsl_matrix_set(hamil_, r+Kmax_,
                     c+Kmax_, el);
    }
  }
}

void PWMethod :: diagonalize()const
{
  gsl_eigen_gensymmv_workspace* w
    = gsl_eigen_gensymmv_alloc(dim_);
  gsl_eigen_gensymmv(hamil_, ident_,
                     eval_, evec_, w);
  gsl_eigen_gensymmv_free(w);

  // sort eigenvalues into the ascending
  // order using the bubble sort
  for (int i=0; i<dim_; ++i) {
    for (int j=i+1; j<dim_; ++j) {
      double ei = gsl_vector_get(eval_, i);
      double ej = gsl_vector_get(eval_, j);
      if (ej < ei) {
        gsl_vector_set(eval_, i, ej);
        gsl_vector_set(eval_, j, ei);
        for (int k=0; k<dim_; ++k) {
          double v_ki
            = gsl_matrix_get(evec_, k, i);
          gsl_matrix_set(evec_, k, i,
            gsl_matrix_get(evec_, k, j));
          gsl_matrix_set(evec_, k, j, v_ki);
        }
      }
    }
  }
}

double PWMethod :: eigenvalue(int i)const
{
  return gsl_vector_get(eval_, i);
}

void PWMethod ::
eigenvector(int i, double* v)const
{
  for (int r = 0; r<dim_; ++r)
    v[r] = gsl_matrix_get(evec_, r, i);
}
```

```
void PWMethod ::
band(int n, int m, double* k, double* e)const
{
  double dwv = 2.0/static_cast<double>(m - 1);
  for (int i=0; i<m; ++i) {
    k[i] = -1.0 + dwv*static_cast<double>(i);
    initialize(k[i]);
    diagonalize();
    e[i] = gsl_vector_get(eval_, n);
  }
}

gsl_complex PWMethod ::
planeWave(double k, int K, double x)const
{
  gsl_complex pw;
  double arg = (k - K - K)*PI*x;
  GSL_SET_COMPLEX(&pw, cos(arg), sin(arg));
  return pw;
}

double PWMethod ::
density(int n, double k, double x)const
{
  gsl_complex amp;
  GSL_SET_COMPLEX(&amp, 0.0, 0.0);
  for (int j=-Kmax_; j<=Kmax_; ++j) {
    amp = gsl_complex_add(amp,
      gsl_complex_mul_real(planeWave(k, j, x),
        gsl_matrix_get(evec_, j+Kmax_, n)));
  }
  return gsl_complex_abs2(amp);
}

void PWMethod ::
hamiltonian(std::ostream& os)const
{
  for (int r=0; r<dim_; ++r) {
    for (int c=0; c<dim_; ++c) {
      os << gsl_matrix_get(hamil_, r, c)
         << (c == dim_-1 ? "\n" : "\t");
    }
  }
}
```

The member function `void initialize(double k)const` constructs the Hamiltonian matrix for the value of the wave vector given by `k`. The Hamiltonian matrix can be output using the member function `void hamiltonian(std::ostream& os=std::cout)const`. The Hamiltonian matrix is used in the member function `void diagonalize()const` to obtain the eigenvalues and eigenfunctions (Bloch functions). Note that the data member `eval_` contains the eigenvalues and that the data member `evec_` contains the values of coefficients C's. The dispersion relation of band `n` can be obtained using the member function `void band(int n, int m, double* k, double* e)const`. The argument `m` is for the number of data points. The wave vectors are contained in `k` and the corresponding energies in `e`.

The file `pw-main.cpp` reads

```
#include "pwmethod.hpp"
#include <iostream>
#include <fstream>
#include <cmath>
#include <cstdlib>
#include <sstream>
#include <string>

using namespace std;

template<typename T>
string toString(T x);

int main()
{
  const int nrlv = 2;
  const int dim = nrlv + nrlv + 1;
  const double V0 = 1.5;

  PWMethod pw(nrlv, V0);

  // compute dispersion relations
  ofstream eos("bands.dat");
  if (!eos) {
    cerr << "Can't open a file to"
         " write energy data!"
      << endl;
    exit(-1);
  }
  const int m = 501;
  double k[m];
  double e[m];
  for (int n=0; n<3; ++n) {
    eos << "Band " << n + 1 << endl;
    pw.band(n, m, k, e);
    for (int i=0; i<m; ++i) {
      eos << k[i] << '\t' << e[i]
          << endl;
    }
    eos << endl;
  }
  eos.close();

  // find Bloch functions and
  // the squares of their amplitudes
  ofstream vos("wavefunctions.dat");
  if (!vos) {
    cerr << "Can't open a file to "
            "write wave function data!"
         << endl;
    exit(-1);
  }
  vos.precision(8);
  for (int k=0; k<=2; ++k) {
    double wv = 0.5*k;
    pw.initialize(wv);
```

```
      pw.diagonalize();
      vos << "At k = " << wv
        << " in $\\uppi/a$." << endl;
      for (int n=0; n<=1; ++n) {
        vos << "In band " << n + 1
          << ":\n  ";
        double wf[dim];
        pw.eigenvector(n, wf);
        for (int j=0; j<dim; ++j)
          vos << wf[j]
            << (j==dim-1 ? '\n' : ' ');
        if ((k == 0 || k == 1 || k == 2)
          && n < 2) {
          string fn = toString(wv);
          fn += "pi_a-band";
          fn += toString(n);
          fn += ".dat";
          ofstream dos(fn.c_str());
          if (!dos) {
            cerr << "Can't open a file to "
                    "write wave function "
                    "data!" << endl;
            exit(-1);
          }
          double dx = 2.0/(m-1);
          for (int j=0; j<m; ++j) {
            double x = -1.0 + dx*j;
            dos << x << '\t'
                << pw.density(n, wv, x)
                << endl;
          }
          dos.close();
        }
      }
      vos << endl;
    }
    vos.close();
    return 0;
  }

  template<typename T>
  string toString(T x)
  {
    stringstream sstrm;
    sstrm << x;
    string str;
    sstrm >> str;
    return str;
  }
```

(2) The coefficients for the wave functions (Bloch functions) at the six k-values are given in the following table. The Bloch function $\psi_{nk}(x)$ can be constructed from the coefficients as follows

$$\psi_{nk}(x) = C_{n,k+4\pi/a}\psi^0_{k+4\pi/a}(x) + C_{n,k+2\pi/a}\psi^0_{k+2\pi/a}(x) + C_{n,0}\psi^0_k(x) + C_{n,k-2\pi/a}\psi^0_{k-2\pi/a}(x) + C_{n,k-4\pi/a}\psi^0_{k-4\pi/a}(x).$$

k	Band (n)	Wave function components $C^0_{n,k+4\pi/a}$	$C^0_{n,k+2\pi/a}$	$C^0_{n,k}$	$C^0_{n,k-2\pi/a}$	$C^0_{n,k-4\pi/a}$
0	1	0.00226765	-0.11477585	0.98673336	-0.11477585	0.00226765
	2	0.03735595	-0.70611935	0.0	0.70611935	-0.03735595
$\pi/2a$	1	6.2448e-05	-0.07541443	0.97292494	-0.21833924	0.00759597
	2	-0.00944213	0.02949086	-0.21679297	-0.97457495	0.04738788
π/a	1	0.00383456	0.03867533	-0.70566115	0.70641045	-0.03893760
	2	0.00569651	-0.04594295	0.70600144	0.70522667	-0.04554237

(3) The plots of $|\psi_{nk}(x)|^2$ for $n = 1$ and 2 and for $k = 0$, $\pi/2a$, and π/a as functions of x are given in Fig. 18.2. The large-amplitude and high-frequency variations as x changes for a given $|\psi_{nk}(x)|^2$ imply that an electron in state $\psi_{nk}(x)$ has a higher kinetic energy.

18-2 Special $\boldsymbol{k}$-points for a simple cubic Bravais lattice. Starting from symmetry points $\boldsymbol{k}_1 = (1/4, 1/4, 1/4)$ and $\boldsymbol{k}_2 = (1/8, 1/8, 1/8)$ in units of $2\pi/a$, derive the four special $\boldsymbol{k}$-points for a simple cubic Bravais lattice.

The point group of the cubic crystal system is O_h ($m\bar{3}m$). All

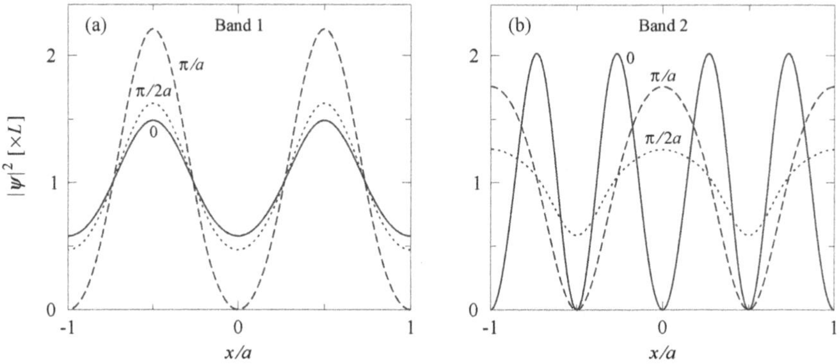

Fig. 18.2 Squares of the amplitudes of Bloch functions of wave vectors $k = 0$, $\pi/2a$, and π/a as functions of coordinate variable x for band 1 (a) and for band 2 (b). The value of the wave vector for each line is given.

the elements of O_h ($m\bar{3}m$) are listed in the following table. Note that the international symbol for an operation is put in a pair of parentheses behind the Schönflies symbol. The symmetry elements are denoted by the directions of rotation axes for rotations and the normals of reflecting planes for reflections.

Description	**Symbol**	**Symmetry elements**
Identity	E (1)	
Three equivalent rotations of angle π	$3[C_2$ (2)]	[100], [010], [001]
Six equivalent rotations of angle $\pi/2$	$6[C_4$ (4)]	[100], [010, [001], $[\bar{1}00]$, $[0\bar{1}0]$, $[00\bar{1}]$
Six equivalent rotations of angle π	$6[C_{2'}$ (2′)]	[110], [101], [011], $[1\bar{1}0]$, $[10\bar{1}]$, $[01\bar{1}]$
Eight equivalent rotations of angle $2\pi/3$	$8[C_3$ (3)]	[111], $[11\bar{1}]$, $[1\bar{1}1]$, $[\bar{1}11]$, $[\bar{1}\bar{1}\bar{1}]$, $[\bar{1}\bar{1}1]$, $[\bar{1}1\bar{1}]$, $[1\bar{1}\bar{1}]$
Inversion	i $(\bar{1})$	
Three equivalent reflections	$3[\sigma_h$ $(m)]$	[100], [010], [001]
Six equivalent rotation-reflections of angle $\pi/2$	$6[S_4$ $(\bar{4})]$	[100], [010, [001], $[\bar{1}00]$, $[0\bar{1}0]$, $[00\bar{1}]$
Six equivalent reflections	$6[\sigma_d$ $(m)]$	[110], [101], [011], $[1\bar{1}0]$, $[10\bar{1}]$, $[01\bar{1}]$
Eight equivalent rotation-reflections of angle $\pi/3$	$8[S_6$ $(\bar{6})]$	[111], $[11\bar{1}]$, $[1\bar{1}1]$, $[\bar{1}11]$, $[\bar{1}\bar{1}\bar{1}]$, $[\bar{1}\bar{1}1]$, $[\bar{1}1\bar{1}]$, $[1\bar{1}\bar{1}]$

We now describe the effects of rotations and reflections. If wave vector $\boldsymbol{k}$ is rotated about the direction $\boldsymbol{n}$ through an angle of θ counterclockwise, then the wave vector becomes

$$\boldsymbol{k}' = \boldsymbol{k}\cos\theta + \boldsymbol{n}(\boldsymbol{n}\cdot\boldsymbol{k})(1-\cos\theta) + (\boldsymbol{n}\times\boldsymbol{k})\sin\theta.$$

If wave vector $\boldsymbol{k}$ is reflected about a plane whose normal is $\boldsymbol{n}$, then the wave vector becomes

$$\boldsymbol{k}' = \boldsymbol{k} - 2\boldsymbol{n}(\boldsymbol{n} \cdot \boldsymbol{k}).$$

The special points are to be found from the two given $\boldsymbol{k}$-points through $\boldsymbol{k}_\ell = \boldsymbol{k}_1 + g_\ell \boldsymbol{k}_2$, where g_ℓ is one of the elements of O_h listed in the above table. The $\boldsymbol{k}$-points obtained for g_ℓ taking in turn all the elements of O_h are listed in the table to be given below. Note that the $\boldsymbol{k}$-points in the table are in $2\pi/a$.

Operation	$\boldsymbol{k}$-point	Operation	$\boldsymbol{k}$-point	Operation	$\boldsymbol{k}$-point
E	(3/8, 3/8, 3/8)	$C_3([111])$	(3/8, 3/8, 3/8)	$S_4([0\bar{1}0])$	(1/8, 1/8, 3/8)
$C_2([100])$	(3/8, 1/8, 1/8)	$C_3([11\bar{1}])$	(3/8, 1/8, 1/8)	$S_4([00\bar{1}])$	(3/8, 1/8, 1/8)
$C_2([010])$	(1/8, 3/8, 1/8)	$C_3([1\bar{1}1])$	(1/8, 1/8, 3/8)	$\sigma_d([110])$	(1/8, 1/8, 3/8)
$C_2([001])$	(1/8, 1/8, 3/8)	$C_3([\bar{1}11])$	(1/8, 3/8, 1/8)	$\sigma_d([101])$	(1/8, 3/8, 1/8)
$C_4([100])$	(3/8, 1/8, 3/8)	$C_3([\bar{1}\bar{1}\bar{1}])$	(3/8, 3/8, 3/8)	$\sigma_d([011])$	(3/8, 1/8, 1/8)
$C_4([010])$	(3/8, 3/8, 1/8)	$C_3([\bar{1}\bar{1}1])$	(1/8, 3/8, 1/8)	$\sigma_d([1\bar{1}0])$	(3/8, 3/8, 3/8)
$C_4([001])$	(1/8, 3/8, 3/8)	$C_3([\bar{1}1\bar{1}])$	(3/8, 1/8, 1/8)	$\sigma_d([10\bar{1}])$	(3/8, 3/8, 3/8)
$C_4([\bar{1}00])$	(3/8, 3/8, 1/8)	$C_3([1\bar{1}\bar{1}])$	(1/8, 1/8, 3/8)	$\sigma_d([01\bar{1}])$	(3/8, 3/8, 3/8)
$C_4([0\bar{1}0])$	(1/8, 3/8, 3/8)	i	(1/8, 1/8, 1/8)	$S_6([111])$	(1/8, 1/8, 1/8)
$C_4([00\bar{1}])$	(3/8, 1/8, 3/8)	$\sigma_h([100])$	(1/8, 3/8, 3/8)	$S_6([11\bar{1}])$	(3/8, 1/8, 3/8)
$C_{2'}([110])$	(3/8, 3/8, 1/8)	$\sigma_h([010])$	(3/8, 1/8, 3/8)	$S_6([1\bar{1}1])$	(1/8, 3/8, 3/8)
$C_{2'}([101])$	(3/8, 1/8, 3/8)	$\sigma_h([001])$	(3/8, 3/8, 1/8)	$S_6([\bar{1}11])$	(3/8, 3/8, 1/8)
$C_{2'}([011])$	(1/8, 3/8, 3/8)	$S_4([100])$	(1/8, 1/8, 3/8)	$S_6([\bar{1}\bar{1}\bar{1}])$	(1/8, 1/8, 1/8)
$C_{2'}([1\bar{1}0])$	(1/8, 1/8, 1/8)	$S_4([010])$	(3/8, 1/8, 1/8)	$S_6([\bar{1}\bar{1}1])$	(1/8, 3/8, 3/8)
$C_{2'}([10\bar{1}])$	(1/8, 1/8, 1/8)	$S_4([001])$	(1/8, 3/8, 1/8)	$S_6([\bar{1}1\bar{1}])$	(3/8, 3/8, 1/8)
$C_{2'}([01\bar{1}])$	(1/8, 1/8, 1/8)	$S_4([\bar{1}00])$	(1/8, 3/8, 1/8)	$S_6([1\bar{1}\bar{1}])$	(3/8, 1/8, 3/8)

The results in the following were obtained using an electronic computer with the effects of symmetry elements handled in the computer codes. We see that repetitions occur. The repetitions are related through the identity operation. There are $\boldsymbol{k}$-points that are related through other point symmetry operations. We find that there are in total four nonequivalent $\boldsymbol{k}$-points: (1/8, 1/8, 1/8), (3/8, 1/8, 1/8), (3/8, 3/8, 1/8), and (3/8, 3/8, 3/8) under the point symmetry. We also find that (1/8, 1/8, 1/8) and (3/8, 3/8, 3/8) have 6 equivalent points each while (3/8, 1/8, 1/8) and (3/8, 3/8, 1/8) have 18 equivalent points each. Thus, their weights are respectively 1/8, 1/8, 3/8, and 3/8. These points are the special $\boldsymbol{k}$-points we are seeking since they are also nonequivalent under the translation symmetry. They are summarized below together with their weights.

$$\boldsymbol{k}_1 = (1/8, 1/8, 1/8),\ w_1 = 1/8; \quad \boldsymbol{k}_2 = (3/8, 1/8, 1/8),\ w_2 = 3/8;$$

$$\boldsymbol{k}_3 = (3/8, 3/8, 1/8),\ w_3 = 3/8; \quad \boldsymbol{k}_4 = (3/8, 3/8, 3/8),\ w_4 = 1/8.$$

18-3 Special $\boldsymbol{k}$-points for a body-centered cubic crystal. The special $\boldsymbol{k}$-points for a body-centered cubic crystal are derived here.

(1) Starting from $\boldsymbol{k}_1 = (1/2, 1/2, 1/2)$ and $\boldsymbol{k}_2 = (1/4, 1/4, 1/4)$ in units of $2\pi/a$, derive the following two special $\boldsymbol{k}$-points for a body-centered cubic Bravais lattice

$$\boldsymbol{k}_1 = (3/4, 1/4, 1/4),\ w_1 = 1/2; \quad \boldsymbol{k}_2 = (1/4, 1/4, 1/4),\ w_2 = 1/2.$$

(2) Using $\boldsymbol{k} = (1/8, 1/8, 1/8)$ and the above-derived two special $\boldsymbol{k}$-points, derive the eight special $\boldsymbol{k}$-points for a body-centered cubic Bravais lattice.

---

(1) The special points are to be found from the two given $\boldsymbol{k}$-points through $\boldsymbol{k}_\ell = \boldsymbol{k}_1 + g_\ell \boldsymbol{k}_2$, where g_ℓ is one of the elements of O_h. The $\boldsymbol{k}$-points obtained for g_ℓ taking in turn all the elements of O_h are listed in the following table. Note that the $\boldsymbol{k}$-points in the table are in $2\pi/a$.

Operation	$\boldsymbol{k}$-point	Operation	$\boldsymbol{k}$-point	Operation	$\boldsymbol{k}$-point
E	(3/4, 3/4, 3/4)	$C_3([111])$	(3/4, 3/4, 3/4)	$S_4([0\bar{1}0])$	(1/4, 1/4, 3/4)
$C_2([100])$	(3/4, 1/4, 1/4)	$C_3([11\bar{1}])$	(3/4, 1/4, 1/4)	$S_4([00\bar{1}])$	(3/4, 1/4, 1/4)
$C_2([010])$	(1/4, 3/4, 1/4)	$C_3([1\bar{1}1])$	(1/4, 1/4, 3/4)	$\sigma_d([110])$	(1/4, 1/4, 3/4)
$C_2([001])$	(1/4, 1/4, 3/4)	$C_3([\bar{1}11])$	(1/4, 3/4, 1/4)	$\sigma_d([101])$	(1/4, 3/4, 1/4)
$C_4([100])$	(3/4, 1/4, 3/4)	$C_3([\bar{1}\bar{1}\bar{1}])$	(3/4, 3/4, 3/4)	$\sigma_d([011])$	(3/4, 1/4, 1/4)
$C_4([010])$	(3/4, 3/4, 1/4)	$C_3([\bar{1}\bar{1}1])$	(1/4, 3/4, 1/4)	$\sigma_d([1\bar{1}0])$	(3/4, 3/4, 3/4)
$C_4([001])$	(1/4, 3/4, 3/4)	$C_3([\bar{1}1\bar{1}])$	(3/4, 1/4, 1/4)	$\sigma_d([10\bar{1}])$	(3/4, 3/4, 3/4)
$C_4([\bar{1}00])$	(3/4, 3/4, 1/4)	$C_3([1\bar{1}\bar{1}])$	(1/4, 1/4, 3/4)	$\sigma_d([01\bar{1}])$	(3/4, 3/4, 3/4)
$C_4([0\bar{1}0])$	(1/4, 3/4, 3/4)	i	(1/4, 1/4, 1/4)	$S_6([111])$	(1/4, 1/4, 1/4)
$C_4([00\bar{1}])$	(3/4, 1/4, 3/4)	$\sigma_h([100])$	(1/4, 3/4, 3/4)	$S_6([11\bar{1}])$	(3/4, 1/4, 3/4)
$C_{2'}([110])$	(3/4, 3/4, 1/4)	$\sigma_h([010])$	(3/4, 1/4, 3/4)	$S_6([1\bar{1}1])$	(1/4, 3/4, 3/4)
$C_{2'}([101])$	(3/4, 1/4, 3/4)	$\sigma_h([001])$	(3/4, 3/4, 1/4)	$S_6([\bar{1}11])$	(3/4, 3/4, 1/4)
$C_{2'}([011])$	(1/4, 3/4, 3/4)	$S_4([100])$	(1/4, 1/4, 3/4)	$S_6([\bar{1}\bar{1}\bar{1}])$	(1/4, 1/4, 1/4)
$C_{2'}([1\bar{1}0])$	(1/4, 1/4, 1/4)	$S_4([010])$	(3/4, 1/4, 1/4)	$S_6([\bar{1}\bar{1}1])$	(1/4, 3/4, 3/4)
$C_{2'}([10\bar{1}])$	(1/4, 1/4, 1/4)	$S_4([001])$	(1/4, 3/4, 1/4)	$S_6([\bar{1}1\bar{1}])$	(3/4, 3/4, 1/4)
$C_{2'}([01\bar{1}])$	(1/4, 1/4, 1/4)	$S_4([\bar{1}00])$	(1/4, 3/4, 1/4)	$S_6([1\bar{1}\bar{1}])$	(3/4, 1/4, 3/4)

With only the point symmetry taken into consideration, we find the following four nonequivalent $\boldsymbol{k}$-points

$$\boldsymbol{k}_1 = (1/4, 1/4, 1/4),\ w_1 = 1/8; \quad \boldsymbol{k}_2 = (3/4, 1/4, 1/4),\ w_2 = 3/8;$$

$$\boldsymbol{k}_3 = (3/4, 3/4, 1/4),\ w_3 = 3/8; \quad \boldsymbol{k}_4 = (3/4, 3/4, 3/4),\ w_4 = 1/8.$$

However, if the translation symmetry is also taken into consideration, not all the four $\boldsymbol{k}$-points are nonequivalent. For the BCC Bravais lattice, the sum of the Miller indices

h, k, and ℓ in the expression of a reciprocal lattice vector, $\boldsymbol{K} = (h\boldsymbol{e}_x + k\boldsymbol{e}_y + \ell\boldsymbol{e}_z)(2\pi/a)$, is an even number, $h + k + \ell =$ even. In consideration of the inversion symmetry, we see that $(1/4, 1/4, 1/4)$ is equivalent to $(-1/4, -1/4, -1/4)$ which is equivalent to $(-1/4, -1/4, 1/4)$, due to $\sigma_h([001])$, which can be transformed into $(3/4, 3/4, 1/4)$ through the translation with $\boldsymbol{K} = (\boldsymbol{e}_x + \boldsymbol{e}_y)(2\pi/a)$. Thus, $\boldsymbol{k}_1$ and $\boldsymbol{k}_3$ are equivalent because of the translation symmetry.
Similarly, $\boldsymbol{k}_2$ and $\boldsymbol{k}_4$ are also equivalent due to $\sigma_h([010])$, $\sigma_h([001])$, and the translation with $\boldsymbol{K} = (\boldsymbol{e}_y + \boldsymbol{e}_z)(2\pi/a)$. Therefore, we have only two nonequivalent special $\boldsymbol{k}$-points

$$\boldsymbol{k}_1 = (3/4, 1/4, 1/4),\ w_1 = 1/2; \quad \boldsymbol{k}_2 = (1/4, 1/4, 1/4),\ w_2 = 1/2.$$

(2) The $\boldsymbol{k}$-points obtained from the above-derived special $\boldsymbol{k}$-point $\boldsymbol{k}_1 = (3/4, 1/4, 1/4)$ and $\boldsymbol{k}_2 = (1/8, 1/8, 1/8)$ are listed in the following table.

Operation	$\boldsymbol{k}$-point	Operation	$\boldsymbol{k}$-point	Operation	$\boldsymbol{k}$-point
E	(7/8, 3/8, 3/8)	$C_3([111])$	(7/8, 3/8, 3/8)	$S_4([0\bar{1}0])$	(5/8, 1/8, 3/8)
$C_2([100])$	(7/8, 1/8, 1/8)	$C_3([11\bar{1}])$	(7/8, 1/8, 1/8)	$S_4([00\bar{1}])$	(7/8, 1/8, 1/8)
$C_2([010])$	(5/8, 3/8, 1/8)	$C_3([1\bar{1}1])$	(5/8, 1/8, 3/8)	$\sigma_d([110])$	(5/8, 1/8, 3/8)
$C_2([001])$	(5/8, 1/8, 3/8)	$C_3([\bar{1}11])$	(5/8, 3/8, 1/8)	$\sigma_d([101])$	(5/8, 3/8, 1/8)
$C_4([100])$	(7/8, 1/8, 3/8)	$C_3([\bar{1}\bar{1}\bar{1}])$	(7/8, 3/8, 3/8)	$\sigma_d([011])$	(7/8, 1/8, 1/8)
$C_4([010])$	(7/8, 3/8, 1/8)	$C_3([\bar{1}\bar{1}1])$	(5/8, 3/8, 1/8)	$\sigma_d([1\bar{1}0])$	(7/8, 3/8, 3/8)
$C_4([001])$	(5/8, 3/8, 3/8)	$C_3([\bar{1}1\bar{1}])$	(7/8, 1/8, 1/8)	$\sigma_d([10\bar{1}])$	(7/8, 3/8, 3/8)
$C_4([\bar{1}00])$	(7/8, 3/8, 1/8)	$C_3([1\bar{1}\bar{1}])$	(5/8, 1/8, 3/8)	$\sigma_d([01\bar{1}])$	(7/8, 3/8, 3/8)
$C_4([0\bar{1}0])$	(5/8, 3/8, 3/8)	i	(5/8, 1/8, 1/8)	$S_6([111])$	(5/8, 1/8, 1/8)
$C_4([00\bar{1}])$	(7/8, 1/8, 3/8)	$\sigma_h([100])$	(5/8, 3/8, 3/8)	$S_6([11\bar{1}])$	(7/8, 1/8, 3/8)
$C_{2'}([110])$	(7/8, 3/8, 1/8)	$\sigma_h([010])$	(7/8, 1/8, 3/8)	$S_6([1\bar{1}1])$	(5/8, 3/8, 3/8)
$C_{2'}([101])$	(7/8, 1/8, 3/8)	$\sigma_h([001])$	(7/8, 3/8, 1/8)	$S_6([\bar{1}11])$	(7/8, 3/8, 1/8)
$C_{2'}([011])$	(5/8, 3/8, 3/8)	$S_4([100])$	(5/8, 1/8, 3/8)	$S_6([\bar{1}\bar{1}\bar{1}])$	(5/8, 1/8, 1/8)
$C_{2'}([1\bar{1}0])$	(5/8, 1/8, 1/8)	$S_4([010])$	(7/8, 1/8, 1/8)	$S_6([\bar{1}\bar{1}1])$	(5/8, 3/8, 3/8)
$C_{2'}([10\bar{1}])$	(5/8, 1/8, 1/8)	$S_4([001])$	(5/8, 3/8, 1/8)	$S_6([\bar{1}1\bar{1}])$	(7/8, 3/8, 1/8)
$C_{2'}([01\bar{1}])$	(5/8, 1/8, 1/8)	$S_4([\bar{1}00])$	(5/8, 3/8, 1/8)	$S_6([1\bar{1}\bar{1}])$	(7/8, 1/8, 3/8)

With only the point symmetry considered, the six nonequivalent special $\boldsymbol{k}$-points are

$$\boldsymbol{k}_1 = (5/8, 1/8, 1/8),\ w_1 = 1/8; \quad \boldsymbol{k}_2 = (5/8, 3/8, 3/8),\ w_2 = 1/8;$$

$$\boldsymbol{k}_3 = (5/8, 3/8, 1/8),\ w_3 = 1/4; \quad \boldsymbol{k}_4 = (7/8, 3/8, 1/8),\ w_4 = 1/4;$$

$$\boldsymbol{k}_5 = (7/8, 1/8, 1/8),\ w_5 = 1/8; \quad \boldsymbol{k}_6 = (7/8, 3/8, 3/8),\ w_6 = 1/8.$$

We see that $\boldsymbol{k}_1 = (5/8, 1/8, 1/8)$ and $\boldsymbol{k}_4 = (7/8, 3/8, 1/8)$ as well as $\boldsymbol{k}_3 = (5/8, 3/8, 1/8)$ and $\boldsymbol{k}_6 = (7/8, 3/8, 3/8)$ are equivalent if the translation symmetry is also taken into consideration. We are now left with only four nonequivalent special $\boldsymbol{k}$-points which are relabeled as

$\boldsymbol{k}_1 = (5/8, 1/8, 1/8)$, $w_1 = 3/8$; $\boldsymbol{k}_2 = (5/8, 3/8, 3/8)$, $w_2 = 1/8$;

$\boldsymbol{k}_3 = (5/8, 3/8, 1/8)$, $w_3 = 3/8$; $\boldsymbol{k}_4 = (7/8, 1/8, 1/8)$, $w_4 = 1/8$.

The $\boldsymbol{k}$-points obtained from the above-derived special $\boldsymbol{k}$-point $(1/4, 1/4, 1/4)$, now referred to as $\boldsymbol{k}_1$, and $\boldsymbol{k}_2 = (1/8, 1/8, 1/8)$ are listed in the following table.

Operation	$\boldsymbol{k}$-point	Operation	$\boldsymbol{k}$-point	Operation	$\boldsymbol{k}$-point
E	(3/8, 3/8, 3/8)	$C_3([111])$	(3/8, 3/8, 3/8)	$S_4([0\bar{1}0])$	(1/8, 1/8, 3/8)
$C_2([100])$	(3/8, 1/8, 1/8)	$C_3([11\bar{1}])$	(3/8, 1/8, 1/8)	$S_4([00\bar{1}])$	(3/8, 1/8, 1/8)
$C_2([010])$	(1/8, 3/8, 1/8)	$C_3([1\bar{1}1])$	(1/8, 1/8, 3/8)	$\sigma_d([110])$	(1/8, 1/8, 3/8)
$C_2([001])$	(1/8, 1/8, 3/8)	$C_3([\bar{1}11])$	(1/8, 3/8, 1/8)	$\sigma_d([101])$	(1/8, 3/8, 1/8)
$C_4([100])$	(3/8, 1/8, 3/8)	$C_3([\bar{1}\bar{1}\bar{1}])$	(3/8, 3/8, 3/8)	$\sigma_d([011])$	(3/8, 1/8, 1/8)
$C_4([010])$	(3/8, 3/8, 1/8)	$C_3([\bar{1}\bar{1}1])$	(1/8, 3/8, 1/8)	$\sigma_d([1\bar{1}0])$	(3/8, 3/8, 3/8)
$C_4([001])$	(1/8, 3/8, 3/8)	$C_3([\bar{1}1\bar{1}])$	(3/8, 1/8, 1/8)	$\sigma_d([10\bar{1}])$	(3/8, 3/8, 3/8)
$C_4([\bar{1}00])$	(3/8, 3/8, 1/8)	$C_3([1\bar{1}\bar{1}])$	(1/8, 1/8, 3/8)	$\sigma_d([01\bar{1}])$	(3/8, 3/8, 3/8)
$C_4([0\bar{1}0])$	(1/8, 3/8, 3/8)	i	(1/8, 1/8, 1/8)	$S_6([111])$	(1/8, 1/8, 1/8)
$C_4([00\bar{1}])$	(3/8, 1/8, 3/8)	$\sigma_h([100])$	(1/8, 3/8, 3/8)	$S_6([11\bar{1}])$	(3/8, 1/8, 3/8)
$C_{2'}([110])$	(3/8, 3/8, 1/8)	$\sigma_h([010])$	(3/8, 1/8, 3/8)	$S_6([1\bar{1}1])$	(1/8, 3/8, 3/8)
$C_{2'}([101])$	(3/8, 1/8, 3/8)	$\sigma_h([001])$	(3/8, 3/8, 1/8)	$S_6([\bar{1}11])$	(3/8, 3/8, 1/8)
$C_{2'}([011])$	(1/8, 3/8, 3/8)	$S_4([100])$	(1/8, 1/8, 3/8)	$S_6([\bar{1}\bar{1}\bar{1}])$	(1/8, 1/8, 1/8)
$C_{2'}([1\bar{1}0])$	(1/8, 1/8, 1/8)	$S_4([010])$	(3/8, 1/8, 1/8)	$S_6([\bar{1}\bar{1}1])$	(1/8, 3/8, 3/8)
$C_{2'}([10\bar{1}])$	(1/8, 1/8, 1/8)	$S_4([001])$	(1/8, 3/8, 1/8)	$S_6([\bar{1}1\bar{1}])$	(3/8, 3/8, 1/8)
$C_{2'}([01\bar{1}])$	(1/8, 1/8, 1/8)	$S_4([\bar{1}00])$	(1/8, 3/8, 1/8)	$S_6([1\bar{1}\bar{1}])$	(3/8, 1/8, 3/8)

Among the $\boldsymbol{k}$-points listed in the following table, only four are nonequivalent. We thus have the following four special $\boldsymbol{k}$-points numbered sequentially after the already-obtained four special $\boldsymbol{k}$-points in the above

$\boldsymbol{k}_6 = (1/8, 1/8, 1/8)$, $w_6 = 1/8$; $\boldsymbol{k}_7 = (3/8, 1/8, 1/8)$, $w_7 = 3/8$;

$\boldsymbol{k}_8 = (3/8, 3/8, 1/8)$, $w_8 = 3/8$; $\boldsymbol{k}_9 = (3/8, 3/8, 3/8)$, $w_9 = 1/8$.

Taking into account the fact that the weights for $(3/4, 1/4, 1/4)$ and $(1/4, 1/4, 1/4)$ are both $1/2$, we have the following eight special $\boldsymbol{k}$-points for the BCC Bravais lattice with their weights summed to unity

$\boldsymbol{k}_1 = (1/8, 1/8, 1/8)$, $w_1 = 1/16$; $\boldsymbol{k}_2 = (3/8, 1/8, 1/8)$, $w_2 = 3/16$;

$\boldsymbol{k}_3 = (3/8, 3/8, 1/8)$, $w_3 = 3/16$; $\boldsymbol{k}_4 = (3/8, 3/8, 3/8)$, $w_4 = 1/16$;

$\boldsymbol{k}_5 = (5/8, 1/8, 1/8)$, $w_5 = 3/16$; $\boldsymbol{k}_6 = (5/8, 3/8, 1/8)$, $w_6 = 3/16$;

$\boldsymbol{k}_7 = (5/8, 3/8, 3/8)$, $w_7 = 1/16$; $\boldsymbol{k}_8 = (7/8, 1/8, 1/8)$, $w_8 = 1/16$.

18-4 Special k-points for a face-centered cubic crystal. The special $\boldsymbol{k}$-points for a face-centered cubic crystal are derived here.

(1) Starting from $\boldsymbol{k}_1 = (1/2, 1/2, 1/2)$ and $\boldsymbol{k}_2 = (1/4, 1/4, 1/4)$ in units of $2\pi/a$, derive the following two special $\boldsymbol{k}$-points for a face-centered cubic Bravais lattice

$$\boldsymbol{k}_1 = (3/4, 1/4, 1/4),\ w_1 = 3/4; \quad \boldsymbol{k}_2 = (1/4, 1/4, 1/4),\ w_2 = 1/4.$$

(2) Using $\boldsymbol{k} = (1/8, 1/8, 1/8)$ and the above-derived two special $\boldsymbol{k}$-points, derive the ten special $\boldsymbol{k}$-points for a face-centered cubic Bravais lattice.

---

(1) In the previous problem, from $\boldsymbol{k}_1 = (1/2, 1/2, 1/2)$ and $\boldsymbol{k}_2 = (1/4, 1/4, 1/4)$ we obtained the following nonequivalent $\boldsymbol{k}$-points under the point symmetry only

$$\boldsymbol{k}_1 = (1/4, 1/4, 1/4),\ w_1 = 1/8; \quad \boldsymbol{k}_2 = (3/4, 1/4, 1/4),\ w_2 = 3/8;$$

$$\boldsymbol{k}_3 = (3/4, 3/4, 1/4),\ w_3 = 3/8; \quad \boldsymbol{k}_4 = (3/4, 3/4, 3/4),\ w_4 = 1/8.$$

For the FCC Bravais lattice, the Miller indices h, k, and ℓ in the expression of a reciprocal lattice vector, $\boldsymbol{K} = (h\boldsymbol{e}_x + k\boldsymbol{e}_y + \ell\boldsymbol{e}_z)(2\pi/a)$, are either all odd or all even. Since $(3/4, 1/4, 1/4)$ is equivalent to $(1/4, 1/4, 3/4)$ which is equivalent to $(-1/4, -1/4, -3/4)$ which can be transformed into $(3/4, 3/4, 1/4)$ through the translation with $\boldsymbol{K} = (\boldsymbol{e}_x + \boldsymbol{e}_y + \boldsymbol{e}_z)(2\pi/a)$, $\boldsymbol{k}_2$ and $\boldsymbol{k}_3$ are equivalent because of the translation symmetry. Similarly, $\boldsymbol{k}_1$ and $\boldsymbol{k}_4$ are also equivalent due to the translation symmetry. Therefore, we have only two nonequivalent special $\boldsymbol{k}$-points

$$\boldsymbol{k}_1 = (3/4, 1/4, 1/4),\ w_1 = 3/4; \quad \boldsymbol{k}_2 = (1/4, 1/4, 1/4),\ w_2 = 1/4.$$

Note that, although the special $\boldsymbol{k}$-points are the same for BCC and FCC, they have different weights.

(2) Using the above-derived special $\boldsymbol{k}$-points and $(1/8, 1/8, 1/8)$, we derived in the previous problem the following ten nonequivalent $\boldsymbol{k}$-points under the point symmetry only

$$\boldsymbol{k}_1 = (5/8, 1/8, 1/8),\ w_1 = 3/32; \quad \boldsymbol{k}_2 = (5/8, 3/8, 3/8),\ w_2 = 3/32;$$

$$\boldsymbol{k}_3 = (5/8, 3/8, 1/8),\ w_3 = 3/16; \quad \boldsymbol{k}_4 = (7/8, 3/8, 1/8),\ w_4 = 3/16;$$

$$\boldsymbol{k}_5 = (7/8, 1/8, 1/8),\ w_5 = 3/32 \quad \boldsymbol{k}_6 = (5/8, 5/8, 1/8),\ w_6 = 3/32;$$

$$\boldsymbol{k}_7 = (1/8, 1/8, 1/8),\ w_7 = 1/32; \quad \boldsymbol{k}_8 = (3/8, 1/8, 1/8),\ w_8 = 3/32;$$

$$\boldsymbol{k}_9 = (3/8, 3/8, 1/8),\ w_9 = 3/32; \quad \boldsymbol{k}_{10} = (3/8, 3/8, 3/8),\ w_{10} = 1/32.$$

Note that the first six $\boldsymbol{k}$-points are from $(3/4, 1/4, 1/4)$ and $(1/8, 1/8, 1/8)$ with a weight of $3/4$ and the last four from $(1/4, 1/4, 1/4)$ and $(1/8, 1/8, 1/8)$ with a weight of $1/4$. These weights have been multiplied to the original weights of the corresponding $\boldsymbol{k}$-points. The $\boldsymbol{k}$-point $(7/8, 3/8, 3/8)$ has been replaced with its equivalent $(5/8, 5/8, 1/8)$. It can be checked that all the above $\boldsymbol{k}$-points are nonequivalent under the translation symmetry for the FCC Bravais lattice. Therefore, they are the special $\boldsymbol{k}$-points for the FCC Bravais lattice.

18-5 Evanescent core potential. Consider the evanescent core potential $w(r) = -\dfrac{z}{R}\left\{\dfrac{1}{x}\left[1 - (1+\beta x)\mathrm{e}^{-\alpha x}\right] - A\mathrm{e}^{-x}\right\}$ with $x = r/R$.

(1) Find the form of $w(r)$ with $\alpha > 1$ in the limit of $r \to \infty$.

(2) Find the form of $w(r)$ in the limit of $r \to 0$ up to the third order in $x = r/R$.

(3) Express A and β in terms of α using the analyticity (cusp-free) conditions that the first- and third-order terms in $\lim_{r\to 0} w(r)$ vanish.

(4) Compute the Fourier transform of $w(r)$, $w(\boldsymbol{q}) = \int \mathrm{d}\boldsymbol{r}\, w(r)\mathrm{e}^{-\mathrm{i}\boldsymbol{q}\cdot\boldsymbol{r}}$.

---

(1) Since the exponential functions $\mathrm{e}^{-\alpha x}$ and e^{-x} decrease very rapidly as x becomes large, we only need to retain the first term in the square brackets. We thus have

$$\lim_{r\to\infty} w(r) = -\frac{z}{Rx} = -\frac{z}{r}.$$

(2) In the $r \to 0$ limit, we expand the exponential functions $\mathrm{e}^{-\alpha x}$ and e^{-x} and keep up to the third-order terms. We have

$$\lim_{r\to 0} w(r) \approx -\frac{z}{R}\Big[\alpha - \beta - A + \frac{1}{2}x(2\alpha\beta - \alpha^2 + 2A) + \frac{1}{6}x^2(\alpha^3 - 3\alpha^2\beta - 3A) + \frac{1}{24}x^3(4\alpha^3\beta - \alpha^4 + 4A)\Big].$$

(3) Setting the coefficients of the first- and third-order terms to zero, we have

$$2\alpha\beta - \alpha^2 + 2A = 0,$$
$$4\alpha^3\beta - \alpha^4 + 4A = 0.$$

Solving for A and β from the above two equations yields

$$\beta = \frac{\alpha(\alpha^2-2)}{4(\alpha^2-1)},\quad A = \frac{\alpha^4}{4(\alpha^2-1)}.$$

(4) The Fourier transform of $w(r)$ is given by

$$\begin{aligned} w_q &= \int \mathrm{d}\boldsymbol{r}\ \mathrm{e}^{-\mathrm{i}\boldsymbol{q}\cdot\boldsymbol{r}} w(r) \\ &= -2\pi z R^2 \int_0^\infty \mathrm{d}x\ x\left[\mathrm{e}^{-\delta x} - (1+\beta x)\mathrm{e}^{-\alpha x} - Ax\mathrm{e}^{-x}\right] \\ &\quad \times \int_{-1}^{1} \mathrm{d}\cos\theta\ \mathrm{e}^{-\mathrm{i}qRx\cos\theta}, \end{aligned}$$

where we have written the first term "1" in the square brackets as $\mathrm{e}^{-\delta x}$ to make its integral converge. In the end, we will set δ to zero. Evaluating the integrals, we obtain

$$\begin{aligned} w_q &= \frac{2\pi \mathrm{i} z R}{q}\int_0^\infty \mathrm{d}x\ \left[\mathrm{e}^{-\delta x} - (1+\beta x)\mathrm{e}^{-\alpha x} - Ax\mathrm{e}^{-x}\right]\left(\mathrm{e}^{\mathrm{i}qRx} - \mathrm{e}^{-\mathrm{i}qRx}\right) \\ &= -4\pi z R^2 \left\{\frac{1}{\delta^2+(qR)^2} - \frac{1}{\alpha^2+(qR)^2} - \frac{2\alpha\beta}{[\alpha^2+(qR)^2]^2} \right. \\ &\qquad \left. - \frac{2A}{[1+(qR)^2]^2}\right\}. \end{aligned}$$

Setting δ to zero, we finally have

$$w_q = -4\pi z R^2 \left\{\frac{1}{(qR)^2} - \frac{1}{\alpha^2+(qR)^2} - \frac{2\alpha\beta}{[\alpha^2+(qR)^2]^2} - \frac{2A}{[1+(qR)^2]^2}\right\}.$$

18-6 Green's function in the KKR method. We derive the Green's function in the original KKR method in this problem.

(1) Prove the identity

$$\sum_{\boldsymbol{R}} \mathrm{e}^{\mathrm{i}\boldsymbol{k}\cdot\boldsymbol{R}}\delta(\boldsymbol{r}-\boldsymbol{r}'-\boldsymbol{R}) = \frac{1}{v_c}\sum_{\boldsymbol{K}} \mathrm{e}^{\mathrm{i}(\boldsymbol{k}-\boldsymbol{K})\cdot(\boldsymbol{r}-\boldsymbol{r}')},$$

where v_c is the volume of a primitive cell, $\boldsymbol{R}$ a lattice vector, and $\boldsymbol{K}$ a reciprocal lattice vector.

(2) Show that the Green's function

$$G(\boldsymbol{k}, E; \boldsymbol{r}, \boldsymbol{r}') = -\frac{1}{\mathscr{V}}\sum_{\boldsymbol{K}} \frac{\mathrm{e}^{\mathrm{i}(\boldsymbol{k}-\boldsymbol{K})\cdot(\boldsymbol{r}-\boldsymbol{r}')}}{(\boldsymbol{k}-\boldsymbol{K})^2 - E}.$$

follows from

$$(\nabla^2 + E)G(\boldsymbol{k}, E; \boldsymbol{r}, \boldsymbol{r}') = \frac{1}{N}\sum_{\boldsymbol{R}} \mathrm{e}^{\mathrm{i}\boldsymbol{k}\cdot\boldsymbol{R}}\delta(\boldsymbol{r}-\boldsymbol{r}'-\boldsymbol{R})$$

with the use of the above identity.

(3) Make a Fourier transformation to the above expression of $G(\boldsymbol{k}, E; \boldsymbol{r}, \boldsymbol{r}')$ in terms of lattice vectors and derive the expression of $G(\boldsymbol{k}, E; \boldsymbol{r}, \boldsymbol{r}')$

$$G(\boldsymbol{k}, E; \boldsymbol{r}, \boldsymbol{r}') = \sum_{\boldsymbol{K}} C_{\boldsymbol{k}-\boldsymbol{K}} \mathrm{e}^{\mathrm{i}(\boldsymbol{k}-\boldsymbol{K})\cdot(\boldsymbol{r}-\boldsymbol{r}')}.$$

---

(1) Integrating both sides of the identity over $\boldsymbol{r}$, we have for the left hand side

$$\int \mathrm{d}\boldsymbol{r} \sum_{\boldsymbol{R}} \mathrm{e}^{\mathrm{i}\boldsymbol{k}\cdot\boldsymbol{R}} \delta(\boldsymbol{r} - \boldsymbol{r}' - \boldsymbol{R}) = \sum_{\boldsymbol{R}} \mathrm{e}^{\mathrm{i}\boldsymbol{k}\cdot\boldsymbol{R}} \int \mathrm{d}\boldsymbol{r}\ \delta(\boldsymbol{r} - \boldsymbol{r}' - \boldsymbol{R})$$
$$= \sum_{\boldsymbol{R}} \mathrm{e}^{\mathrm{i}\boldsymbol{k}\cdot\boldsymbol{R}} = N \sum_{\boldsymbol{K}} \delta_{\boldsymbol{k}\boldsymbol{K}},$$

where we have made use of the fact that $\mathrm{e}^{\mathrm{i}\boldsymbol{k}\cdot\boldsymbol{R}} = 1$ for $\boldsymbol{k}$ equal to any reciprocal lattice vector. If $\boldsymbol{k}$ is indeed equal to one of the reciprocal lattice vectors, we then have $\sum_{\boldsymbol{R}} \mathrm{e}^{\mathrm{i}\boldsymbol{k}\cdot\boldsymbol{R}} = N$. To take all the reciprocal lattice vectors into account, we have made use of $\sum_{\boldsymbol{K}} \delta_{\boldsymbol{k}\boldsymbol{K}}$.

For the right hand side, we have

$$\frac{1}{v_c} \sum_{\boldsymbol{K}} \int \mathrm{d}\boldsymbol{r}\ \mathrm{e}^{\mathrm{i}(\boldsymbol{k}-\boldsymbol{K})\cdot(\boldsymbol{r}-\boldsymbol{r}')} = N \sum_{\boldsymbol{K}} \left[\frac{1}{\mathscr{V}} \int \mathrm{d}\boldsymbol{r}\ \mathrm{e}^{\mathrm{i}(\boldsymbol{k}-\boldsymbol{K})\cdot(\boldsymbol{r}-\boldsymbol{r}')} \right]$$
$$= N \sum_{\boldsymbol{K}} \delta_{\boldsymbol{k}\boldsymbol{K}}.$$

From the above two results, we see that the integrals of the two sides over $\boldsymbol{r}$ are equal. Thus, the identity is valid. This identity can be also proved starting from the use of the integral representation of the Dirac δ-function on the left hand side. We have

$$\sum_{\boldsymbol{R}} \mathrm{e}^{\mathrm{i}\boldsymbol{k}\cdot\boldsymbol{R}} \delta(\boldsymbol{r} - \boldsymbol{r}' - \boldsymbol{R}) = \frac{1}{(2\pi)^3} \sum_{\boldsymbol{R}} \mathrm{e}^{\mathrm{i}\boldsymbol{k}\cdot\boldsymbol{R}} \int \mathrm{d}\boldsymbol{k}'\ \mathrm{e}^{\mathrm{i}\boldsymbol{k}'\cdot(\boldsymbol{r}-\boldsymbol{r}'-\boldsymbol{R})}$$
$$= \frac{1}{\mathscr{V}} \sum_{\boldsymbol{k}'} \mathrm{e}^{\mathrm{i}\boldsymbol{k}'\cdot(\boldsymbol{r}-\boldsymbol{r}')} \sum_{\boldsymbol{R}} \mathrm{e}^{\mathrm{i}(\boldsymbol{k}-\boldsymbol{k}')\cdot\boldsymbol{R}}$$
$$= \frac{N}{\mathscr{V}} \sum_{\boldsymbol{k}'} \mathrm{e}^{\mathrm{i}\boldsymbol{k}'\cdot(\boldsymbol{r}-\boldsymbol{r}')} \sum_{\boldsymbol{K}} \delta_{\boldsymbol{k}-\boldsymbol{k}',\boldsymbol{K}}$$
$$= \frac{1}{v_c} \sum_{\boldsymbol{K}} \mathrm{e}^{\mathrm{i}(\boldsymbol{k}-\boldsymbol{K})\cdot(\boldsymbol{r}-\boldsymbol{r}')},$$

where the above-noted fact that $\mathrm{e}^{\mathrm{i}(\boldsymbol{k}-\boldsymbol{k}')\cdot\boldsymbol{R}} = 1$ for $\boldsymbol{k} - \boldsymbol{k}'$ equal to any reciprocal lattice vector has been utilized.

(2) Making use of the above identity, we have

$$\begin{aligned}
&(\nabla^2 + E)G(\boldsymbol{k}, E; \boldsymbol{r}, \boldsymbol{r}') \\
&\quad = \frac{1}{\mathscr{V}} \sum_{\boldsymbol{K}} \mathrm{e}^{\mathrm{i}(\boldsymbol{k}-\boldsymbol{K})\cdot(\boldsymbol{r}-\boldsymbol{r}')} \\
&\quad = \frac{1}{\mathscr{V}} \sum_{\boldsymbol{K}} \frac{1}{-(\boldsymbol{k}-\boldsymbol{K})^2 + E} (\nabla^2 + E) \mathrm{e}^{\mathrm{i}(\boldsymbol{k}-\boldsymbol{K})\cdot(\boldsymbol{r}-\boldsymbol{r}')} \\
&\quad = (\nabla^2 + E)\bigg[-\frac{1}{\mathscr{V}} \sum_{\boldsymbol{K}} \frac{\mathrm{e}^{\mathrm{i}(\boldsymbol{k}-\boldsymbol{K})\cdot(\boldsymbol{r}-\boldsymbol{r}')}}{(\boldsymbol{k}-\boldsymbol{K})^2 - E} \bigg].
\end{aligned}$$

Comparing the two sides of the above equation, we have

$$G(\boldsymbol{k}, E; \boldsymbol{r}, \boldsymbol{r}') = -\frac{1}{\mathscr{V}} \sum_{\boldsymbol{K}} \frac{\mathrm{e}^{\mathrm{i}(\boldsymbol{k}-\boldsymbol{K})\cdot(\boldsymbol{r}-\boldsymbol{r}')}}{(\boldsymbol{k}-\boldsymbol{K})^2 - E}.$$

(3) Inserting the Fourier transformation of $G(\boldsymbol{k}, E; \boldsymbol{r}, \boldsymbol{r}')$

$$G(\boldsymbol{k}, E; \boldsymbol{r}, \boldsymbol{r}') = \sum_{\boldsymbol{K}} C_{\boldsymbol{k}-\boldsymbol{K}} \mathrm{e}^{\mathrm{i}(\boldsymbol{k}-\boldsymbol{K})\cdot(\boldsymbol{r}-\boldsymbol{r}')}$$

into the equation for it yields

$$\begin{aligned}
&\sum_{\boldsymbol{K}} \big[-(\boldsymbol{k}-\boldsymbol{K})^2 + E \big] C_{\boldsymbol{k}-\boldsymbol{K}} \mathrm{e}^{\mathrm{i}(\boldsymbol{k}-\boldsymbol{K})\cdot(\boldsymbol{r}-\boldsymbol{r}')} \\
&\qquad = \frac{1}{N} \sum_{\boldsymbol{R}} \mathrm{e}^{\mathrm{i}\boldsymbol{k}\cdot\boldsymbol{R}} \delta(\boldsymbol{r}-\boldsymbol{r}'-\boldsymbol{R}) = \frac{1}{\mathscr{V}} \sum_{\boldsymbol{K}} \mathrm{e}^{\mathrm{i}(\boldsymbol{k}-\boldsymbol{K})\cdot(\boldsymbol{r}-\boldsymbol{r}')}.
\end{aligned}$$

To extract $C_{\boldsymbol{k}-\boldsymbol{K}}$ from the above equation, we multiply both sides with $\mathrm{e}^{-\mathrm{i}(\boldsymbol{k}-\boldsymbol{K}')\cdot\boldsymbol{r}}$ and then integrate both sides over $\boldsymbol{r}$. We have

$$\begin{aligned}
&\sum_{\boldsymbol{K}} \big[-(\boldsymbol{k}-\boldsymbol{K})^2 + E \big] C_{\boldsymbol{k}-\boldsymbol{K}} \int \mathrm{d}\boldsymbol{r}\; \mathrm{e}^{-\mathrm{i}(\boldsymbol{K}-\boldsymbol{K}')\cdot\boldsymbol{r}} \\
&\qquad\qquad = \frac{1}{\mathscr{V}} \sum_{\boldsymbol{K}} \int \mathrm{d}\boldsymbol{r}\; \mathrm{e}^{-\mathrm{i}(\boldsymbol{K}-\boldsymbol{K}')\cdot\boldsymbol{r}},
\end{aligned}$$

where we have canceled $\mathrm{e}^{-\mathrm{i}(\boldsymbol{k}-\boldsymbol{K})\cdot\boldsymbol{r}'}$ from the two sides. Making use of the expression for the Kronecker δ-symbol $\delta_{\boldsymbol{K}\boldsymbol{K}'} = \mathscr{V}^{-1} \int \mathrm{d}\boldsymbol{r}\; \mathrm{e}^{-\mathrm{i}(\boldsymbol{K}-\boldsymbol{K}')\cdot\boldsymbol{r}}$, we obtain

$$\mathscr{V} \sum_{\boldsymbol{K}} \big[-(\boldsymbol{k}-\boldsymbol{K})^2 + E \big] C_{\boldsymbol{k}-\boldsymbol{K}} \delta_{\boldsymbol{K}\boldsymbol{K}'} = \sum_{\boldsymbol{K}} \delta_{\boldsymbol{K}\boldsymbol{K}'},$$

$$\mathscr{V} \big[-(\boldsymbol{k}-\boldsymbol{K}')^2 + E \big] C_{\boldsymbol{k}-\boldsymbol{K}'} = 1.$$

Thus,

$$C_{\boldsymbol{k}-\boldsymbol{K}} = -\frac{1}{\mathscr{V}} \frac{1}{(\boldsymbol{k}-\boldsymbol{K})^2 - E},$$

where we have dropped the prime on $\boldsymbol{K}'$. Inserting the above expression of $C_{\boldsymbol{k}-\boldsymbol{K}}$ into the Fourier transformation of $G(\boldsymbol{k}, E; \boldsymbol{r}, \boldsymbol{r}')$, we obtain the following expression for $G(\boldsymbol{k}, E; \boldsymbol{r}, \boldsymbol{r}')$

$$G(\boldsymbol{k}, E; \boldsymbol{r}, \boldsymbol{r}') = -\frac{1}{\mathscr{V}} \sum_{\boldsymbol{K}} \frac{\mathrm{e}^{\mathrm{i}(\boldsymbol{k}-\boldsymbol{K})\cdot(\boldsymbol{r}-\boldsymbol{r}')}}{(\boldsymbol{k}-\boldsymbol{K})^2 - E}.$$

18-7 $\boldsymbol{k} \cdot \boldsymbol{p}$ method for a semiconductor. We now apply the $\boldsymbol{k} \cdot \boldsymbol{p}$ method to a semiconductor. For the semiconductor, we consider its two nondegenerate bands known as the valence (v) and conduction (c) bands. As shown in Fig. 18.3, the energy at the top of the valence band is denoted by $E_v(0)$ and that at the bottom of the conduction band by $E_c(0)$. The band gap is denoted by E_g. For simplicity, we assume that $E_v(0) = 0$ and that $E_c(0) = E_g$. We also assume that the matrix element $\boldsymbol{p}_{cv} = \langle 0_c|\hat{\boldsymbol{p}}|0_v\rangle$ between the valence and conduction bands at $\boldsymbol{k} = 0$ is known.

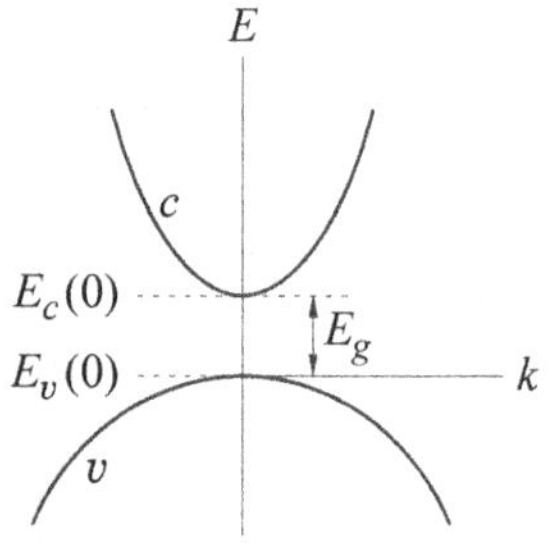

Fig. 18.3 Schematic illustration of the valence band (v), conduction band (c), and band gap (E_g) in a direct-gap semiconductor.

(1) Write down the explicit secular equation for the energy E at $\boldsymbol{k}$.

(2) Solve the secular equation to obtain the energy dispersion relations in the valence and conduction bands for $\boldsymbol{k}$ close to 0.

(3) Derive the approximate expressions for $E_v(\boldsymbol{k})$ and $E_c(\boldsymbol{k})$ for small $|\boldsymbol{k} \cdot \boldsymbol{p}_{cv}|$ and derive the corresponding approximate expressions for the effective masses in the valence and conduction bands.

---

(1) For the present problem, we have only two bands and $\boldsymbol{k}_0 = 0$. The secular equation is given by the determinant of a 2×2

matrix

$$\begin{vmatrix} E_g + \hbar^2\boldsymbol{k}^2/2m - E & (\hbar/m)\boldsymbol{k}\cdot\boldsymbol{p}_{cv} \\ (\hbar/m)\boldsymbol{k}\cdot\boldsymbol{p}_{cv}^* & \hbar^2\boldsymbol{k}^2/2m - E \end{vmatrix} = 0.$$

Evaluating the determinant yields

$$\left(E - \hbar^2\boldsymbol{k}^2/2m\right)^2 - E_g\left(E - \hbar^2\boldsymbol{k}^2/2m\right) - (\hbar/m)^2|\boldsymbol{k}\cdot\boldsymbol{p}_{cv}|^2 = 0.$$

(2) Solving the above secular equation, we obtain

$$E_{c,v}(\boldsymbol{k}) = \frac{\hbar^2\boldsymbol{k}^2}{2m} + \frac{1}{2}E_g\Big\{1 \pm \big[1 + (2\hbar/m)^2|\boldsymbol{k}\cdot\boldsymbol{p}_{cv}|^2\big]^{1/2}\Big\}.$$

(3) For $\boldsymbol{k}$ close to 0, the second term in the square brackets is much smaller than the first term so that we can expand the square root and keep only the first two terms. We then have

$$\begin{aligned} E_{c,v}(\boldsymbol{k}) &\approx \frac{\hbar^2\boldsymbol{k}^2}{2m} + \frac{1}{2}E_g\Big\{1 \pm \big[1 + 2(\hbar/m)^2|\boldsymbol{k}\cdot\boldsymbol{p}_{cv}|^2\big]\Big\} \\ &= \frac{\hbar^2\boldsymbol{k}^2}{2m} + \frac{1}{2}E_g\left\{1 \pm \left[1 + \frac{2\hbar^2}{m^2}\sum_{\alpha\beta} k_\alpha k_\beta p_{cv}^\alpha p_{cv}^{\beta\,*}\right]\right\}. \end{aligned}$$

We thus have

$$E_c(\boldsymbol{k}) = E_g + \frac{\hbar^2\boldsymbol{k}^2}{2m} + \frac{\hbar^2}{m^2E_g}\sum_{\alpha\beta} k_\alpha k_\beta p_{cv}^\alpha p_{cv}^{\beta\,*},$$

$$E_v(\boldsymbol{k}) = \frac{\hbar^2\boldsymbol{k}^2}{2m} - \frac{\hbar^2}{m^2E_g}\sum_{\alpha\beta} k_\alpha k_\beta p_{cv}^\alpha p_{cv}^{\beta\,*}.$$

Combining the two $\boldsymbol{k}$-dependent terms in each expression, we have

$$E_c(\boldsymbol{k}) = E_g + \frac{\hbar^2}{2}\sum_{\alpha\beta}\left(\frac{1}{m}\delta_{\alpha\beta} + \frac{2}{m^2E_g}p_{cv}^\alpha p_{cv}^{\beta\,*}\right)k_\alpha k_\beta,$$

$$E_v(\boldsymbol{k}) = -\frac{\hbar^2}{2}\sum_{\alpha\beta}\left(-\frac{1}{m}\delta_{\alpha\beta} + \frac{2}{m^2E_g}p_{cv}^\alpha p_{cv}^{\beta\,*}\right)k_\alpha k_\beta.$$

From the above expressions, we can infer the inverse effective mass tensors for the conduction and valence bands and have

$$\left(\frac{1}{m_c}\right)_{\alpha\beta} = \frac{1}{m}\delta_{\alpha\beta} + \frac{2}{m^2E_g}p_{cv}^\alpha p_{cv}^{\beta\,*},$$

$$\left(\frac{1}{m_v}\right)_{\alpha\beta} = -\frac{1}{m}\delta_{\alpha\beta} + \frac{2}{m^2E_g}p_{cv}^\alpha p_{cv}^{\beta\,*}.$$

Note that we have written the effective mass in the valence band as that for holes.

18-8 Variational derivation of the tight-binding secular equation. The tight-binding secular equation $\det\big|(E^a_\mu - E)S_{\nu\mu}(\boldsymbol{k}) - \gamma_{\nu\mu}(\boldsymbol{k})\big| = 0$ can be also derived from the variational principle with $\psi_{\boldsymbol{k}}(\boldsymbol{r}) = \sum_\mu C_\mu \phi_{\mu\boldsymbol{k}}(\boldsymbol{r}) = \sum_{\mu j} C_\mu \mathrm{e}^{\mathrm{i}\boldsymbol{k}\cdot\boldsymbol{R}_j}\varphi_\mu(\boldsymbol{r} - \boldsymbol{R}_j)$ taken as the trial wave function.

(1) Evaluate the energy average $E = \langle\psi_{\boldsymbol{k}}|\hat{H}|\psi_{\boldsymbol{k}}\rangle \,/\, \langle\psi_{\boldsymbol{k}}|\psi_{\boldsymbol{k}}\rangle$ with $\hat{H} = -(\hbar^2/2m)\nabla^2 + U(\boldsymbol{r})$.

(2) Differentiate E with respect to C^*_ν and set the result to zero to minimize E. From the resultant equation, infer the secular equation $\det\big|(E^a_\mu - E)S_{\nu\mu}(\boldsymbol{k}) - \gamma_{\nu\mu}(\boldsymbol{k})\big| = 0$.

---

(1) The diagonal matrix element of $\hat{H}$ is given by

$$\begin{aligned}
\langle\psi_{\boldsymbol{k}}|\hat{H}|\psi_{\boldsymbol{k}}\rangle
&= \sum_{\mu\nu j\ell} C^*_\nu C_\mu \int \mathrm{d}\boldsymbol{r}\; \mathrm{e}^{-\mathrm{i}\boldsymbol{k}\cdot\boldsymbol{R}_\ell}\varphi^*_\nu(\boldsymbol{r} - \boldsymbol{R}_\ell) \\
&\qquad \times \big[-(\hbar^2/2m)\nabla^2 + U(\boldsymbol{r})\big]\mathrm{e}^{\mathrm{i}\boldsymbol{k}\cdot\boldsymbol{R}_j}\varphi_\mu(\boldsymbol{r} - \boldsymbol{R}_j) \\
&= \sum_{\mu\nu j\ell} C^*_\nu C_\mu \int \mathrm{d}\boldsymbol{r}\; \mathrm{e}^{\mathrm{i}\boldsymbol{k}\cdot(\boldsymbol{R}_j - \boldsymbol{R}_\ell)}\varphi^*_\nu(\boldsymbol{r} - \boldsymbol{R}_\ell)\big[E^a_\mu + \Delta U(\boldsymbol{r})\big]\varphi_\mu(\boldsymbol{r} - \boldsymbol{R}_j) \\
&= \sum_{\mu\nu j\ell} E^a_\mu C^*_\nu C_\mu \int \mathrm{d}\boldsymbol{r}\; \mathrm{e}^{\mathrm{i}\boldsymbol{k}\cdot(\boldsymbol{R}_j - \boldsymbol{R}_\ell)}\varphi^*_\nu(\boldsymbol{r} - \boldsymbol{R}_\ell)\varphi_\mu(\boldsymbol{r} - \boldsymbol{R}_j) \\
&\quad + \sum_{\mu\nu j\ell} C^*_\nu C_\mu \int \mathrm{d}\boldsymbol{r}\; \mathrm{e}^{\mathrm{i}\boldsymbol{k}\cdot(\boldsymbol{R}_j - \boldsymbol{R}_\ell)}\varphi^*_\nu(\boldsymbol{r} - \boldsymbol{R}_\ell)\Delta U(\boldsymbol{r})\varphi_\mu(\boldsymbol{r} - \boldsymbol{R}_j) \\
&= \sum_{\mu\nu} E^a_\mu C^*_\nu C_\mu S_{\nu\mu}(\boldsymbol{k}) + \sum_{\mu\nu} C^*_\nu C_\mu \gamma_{\nu\mu}(\boldsymbol{k}),
\end{aligned}$$

where we have set $U(\boldsymbol{r}) - V_a(\boldsymbol{r}) = \Delta U(\boldsymbol{r})$ with $V_a(\boldsymbol{r})$ the atomic potential and introduced the overlapping integral $S_{\nu\mu}$ and transfer integral $\gamma_{\nu\mu}$

$$S_{\nu\mu}(\boldsymbol{k}) = \sum_{j\ell} \int \mathrm{d}\boldsymbol{r}\; \mathrm{e}^{\mathrm{i}\boldsymbol{k}\cdot(\boldsymbol{R}_j - \boldsymbol{R}_\ell)}\varphi^*_\nu(\boldsymbol{r} - \boldsymbol{R}_\ell)\varphi_\mu(\boldsymbol{r} - \boldsymbol{R}_j),$$

$$\gamma_{\nu\mu}(\boldsymbol{k}) = \sum_{j\ell} \int \mathrm{d}\boldsymbol{r}\; \mathrm{e}^{\mathrm{i}\boldsymbol{k}\cdot(\boldsymbol{R}_j - \boldsymbol{R}_\ell)}\varphi^*_\nu(\boldsymbol{r} - \boldsymbol{R}_\ell)\Delta U(\boldsymbol{r})\varphi_\mu(\boldsymbol{r} - \boldsymbol{R}_j).$$

The inner product of $\psi_{\boldsymbol{k}}(\boldsymbol{r})$ with itself is given by

$$\begin{aligned}
\langle\psi_{\boldsymbol{k}}|\psi_{\boldsymbol{k}}\rangle &= \sum_{\mu\nu j\ell} C^*_\nu C_\mu \int \mathrm{d}\boldsymbol{r}\; \mathrm{e}^{\mathrm{i}\boldsymbol{k}\cdot(\boldsymbol{R}_j - \boldsymbol{R}_\ell)}\varphi^*_\nu(\boldsymbol{r} - \boldsymbol{R}_\ell)\varphi_\mu(\boldsymbol{r} - \boldsymbol{R}_j) \\
&= \sum_{\mu\nu} C^*_\nu C_\mu S_{\nu\mu}(\boldsymbol{k}).
\end{aligned}$$

The energy average is then given by

$$E = \frac{\langle \psi_{\boldsymbol{k}} | \hat{H} | \psi_{\boldsymbol{k}} \rangle}{\langle \psi_{\boldsymbol{k}} | \psi_{\boldsymbol{k}} \rangle} = \frac{\sum_{\mu\nu} E_\mu^a C_\nu^* C_\mu S_{\nu\mu}(\boldsymbol{k}) + \sum_{\mu\nu} C_\nu^* C_\mu \gamma_{\nu\mu}(\boldsymbol{k})}{\sum_{\mu\nu} C_\nu^* C_\mu S_{\nu\mu}(\boldsymbol{k})}.$$

(2) Differentiating E with respect to C_ν^*, we obtain

$$\begin{aligned} \frac{\partial E}{\partial C_\nu^*} &= \frac{\sum_\mu E_\mu^a C_\mu S_{\nu\mu}(\boldsymbol{k}) + \sum_\mu C_\mu \gamma_{\nu\mu}(\boldsymbol{k})}{\sum_{\mu\nu} C_\nu^* C_\mu S_{\nu\mu}(\boldsymbol{k})} \\ &\quad - \frac{\sum_{\mu\nu} E_\mu^a C_\nu^* C_\mu S_{\nu\mu}(\boldsymbol{k}) + \sum_{\mu\nu} C_\nu^* C_\mu \gamma_{\nu\mu}(\boldsymbol{k})}{\left[\sum_{\mu\nu} C_\nu^* C_\mu S_{\nu\mu}(\boldsymbol{k})\right]^2} \sum_\mu C_\mu S_{\nu\mu}(\boldsymbol{k}). \end{aligned}$$

We can see that the term on the second line can be expressed in terms of E. We have

$$\begin{aligned} \frac{\partial E}{\partial C_\nu^*} &= \frac{\sum_\mu E_\mu^a C_\mu S_{\nu\mu}(\boldsymbol{k}) + \sum_\mu C_\mu \gamma_{\nu\mu}(\boldsymbol{k})}{\sum_{\mu\nu} C_\nu^* C_\mu S_{\nu\mu}(\boldsymbol{k})} \\ &\quad - \frac{E}{\sum_{\mu\nu} C_\nu^* C_\mu S_{\nu\mu}(\boldsymbol{k})} \sum_\mu C_\mu S_{\nu\mu}(\boldsymbol{k}) \\ &= \frac{1}{\sum_{\mu\nu} C_\nu^* C_\mu S_{\nu\mu}(\boldsymbol{k})} \sum_\mu \left[(E_\mu^a - E) S_{\nu\mu}(\boldsymbol{k}) + \gamma_{\nu\mu}(\boldsymbol{k}) \right] C_\mu. \end{aligned}$$

Setting the above result to zero yields

$$\sum_\mu \left[(E_\mu^a - E) S_{\nu\mu}(\boldsymbol{k}) + \gamma_{\nu\mu}(\boldsymbol{k}) \right] C_\mu = 0.$$

The above equations for all ν are a set of homogeneous linear equations for the coefficients C_μ's. The condition for the existence of nontrivial solutions yields the secular equation

$$\det\left| (E_\mu^a - E) S_{\nu\mu}(\boldsymbol{k}) + \gamma_{\nu\mu}(\boldsymbol{k}) \right| = 0.$$

18-9 Tight-binding approximation for a 1D crystal. We now apply the tight-binding approximation to the s band of a one-dimensional crystal of lattice constant a arising from the atomic s orbitals.

(1) Derive the energy dispersion relation $E_s(k)$ for the s band with only the nearest-neighbor overlapping γ taken into account.

(2) Rederive the energy dispersion relation of the s band when the next-nearest-neighbor overlapping γ' and the nearest-neighbor orthogonalization correction α are also taken into consideration.

(3) Plot the above two results in the first Brillouin zone $(-\pi/a, \pi/a)$ for $E_s^a = -6.3$ eV (the energy of the atomic s level), $\beta = 0.7$ eV (the on-site energy), $\gamma = -1.2$ eV, $\gamma' = 0.4$ eV, and $\alpha = 0.15$.

(1) With only the nearest-neighbor overlapping γ taken into account, the energy dispersion relation $E_s(k)$ is given by

$$E_s(k) = E_s^a - \beta - 2\gamma\cos(ka)$$
$$= E_s^0 - 2\gamma\cos(ka).$$

where $E_s^0 = E_s^a - \beta$.

(2) If we also take into consideration the next-nearest-neighbor overlapping γ' and the nearest-neighbor orthogonalization correction α, we then have

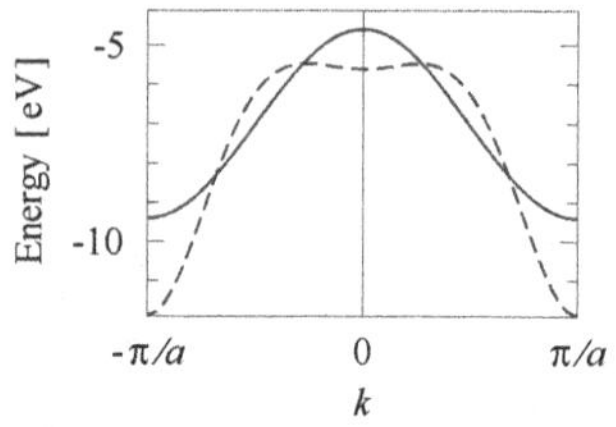

Fig. 18.4 Plots of the one-dimensional tight-binding energy dispersion relations. For the solid line, only the nearest-neighbor overlapping γ is taken into account. For the dashed line, the next-nearest-neighbor overlapping γ' and the nearest-neighbor orthogonalization correction α are also taken into account.

$$E_s(k) = E_s^a - \frac{\beta + 2\gamma\cos(ka) + 2\gamma'\cos(2ka)}{1 + 2\alpha\cos(ka)}.$$

(3) The energy dispersion relations in the above two cases are plotted in Fig. 18.4 in two different line styles.
From Fig. 18.4, we see that the next-nearest-neighbor overlapping γ' and the nearest-neighbor orthogonalization correction α can have a substantial impact on the energy dispersion relation if their values are not too small as given here. They lower the energies at $k = 0$ and $\pm\pi/a$ with the lowering of the energy at $k = \pm\pi/a$ much larger than that at $k = 0$. The lowering of the energy at $k = 0$ breaks the peak there into two.

18-10 Two-dimensional graphite sheet. Consider a two-dimensional graphite sheet. The primitive vectors of the Bravais lattice of the graphite sheet are given by $\boldsymbol{a}_1 = (3a/2)\boldsymbol{e}_x + (\sqrt{3}a/2)\boldsymbol{e}_y$ and $\boldsymbol{a}_2 = (3a/2)\boldsymbol{e}_x - (\sqrt{3}a/2)\boldsymbol{e}_y$, where $a = 0.142$ nm. The energy bands of the graphite sheet are found to be given by

$$E_{c,v}(\boldsymbol{k}) = \pm t\left[1 + 4\cos(\sqrt{3}k_y a/2)\cos(3k_x a/2) + 4\cos^2(\sqrt{3}k_y a/2)\right]^{1/2}$$

with $t = 3$ eV.

(1) Plot $E_{c,v}(\boldsymbol{k})$ along [10], [01], and [11] directions. Indicate the band gap.

(2) Derive a general expression for the effective mass of electrons in the conduction band of the graphite sheet. Find its values at the Γ point and [01] edge of the first Brillouin zone.

---

(1) The primitive vectors of the reciprocal lattice are given by

$$\boldsymbol{b}_1 = \frac{4\pi}{3\sqrt{3}a}\left(\frac{\sqrt{3}}{2}\boldsymbol{e}_x + \frac{3}{2}\boldsymbol{e}_y\right),$$

$$\boldsymbol{b}_2 = \frac{4\pi}{3\sqrt{3}a}\left(\frac{\sqrt{3}}{2}\boldsymbol{e}_x - \frac{3}{2}\boldsymbol{e}_y\right).$$

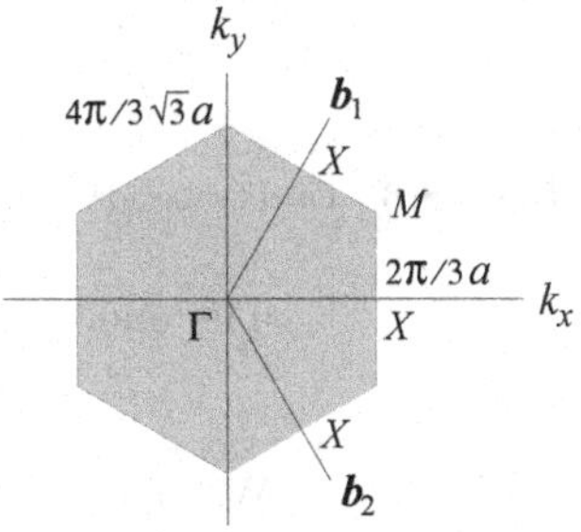

Fig. 18.5 First Brillouin zone for the two-dimensional triangular Bravais lattice.

The lengths of $\boldsymbol{b}_1$ and $\boldsymbol{b}_2$ are both $4\pi/3a$. The first Brillouin zone of the graphite is a regular hexagon of side length $4\pi/3\sqrt{3}a$ as shown in Fig. 18.5. In terms of $\boldsymbol{b}_1$ and $\boldsymbol{b}_2$, a wave vector in the first Brillouin zone is expressed as $\boldsymbol{k} = \xi\boldsymbol{b}_1 + \zeta\boldsymbol{b}_2$. Since, along $\boldsymbol{b}_1$ or $\boldsymbol{b}_2$ the boundary of the first Brillouin zone is at half the length of $\boldsymbol{b}_1$ or $\boldsymbol{b}_2$, we have $\xi = 1/2$, $\zeta = 0$ and $\xi = 0$, $\zeta = 1/2$ at the boundary of the first Brillouin zone along $\boldsymbol{b}_1$ and $\boldsymbol{b}_2$, respectively. Note that [10] ([01]) is the direction of $\boldsymbol{b}_1$ ($\boldsymbol{b}_2$). Along the direction [11], the boundary of the first Brillouin zone is reached at $\xi = \zeta = 1/2$.

From the expression of $\boldsymbol{b}_1$, we see that, along [10], k_x and k_y are given by

$$k_x = \frac{1}{2}\xi\frac{4\pi}{3a},\ k_y = \frac{\sqrt{3}}{2}\xi\frac{4\pi}{3a}.$$

The energies along [10] are given by

$$E_{c,v}(\xi) = \pm t\left[1+4\cos^2(\xi\pi)+4\cos^2(\xi\pi)\right]^{1/2} = \pm t\left[1+8\cos^2(\xi\pi)\right]^{1/2}.$$

The energy bands along the [10] direction are shown in Fig. 18.6.

From the expression of $\boldsymbol{b}_2$, we see that, along [01], k_x and k_y are given by

$$k_x = \frac{1}{2}\zeta\frac{4\pi}{3a},\ k_y = -\frac{\sqrt{3}}{2}\zeta\frac{4\pi}{3a}.$$

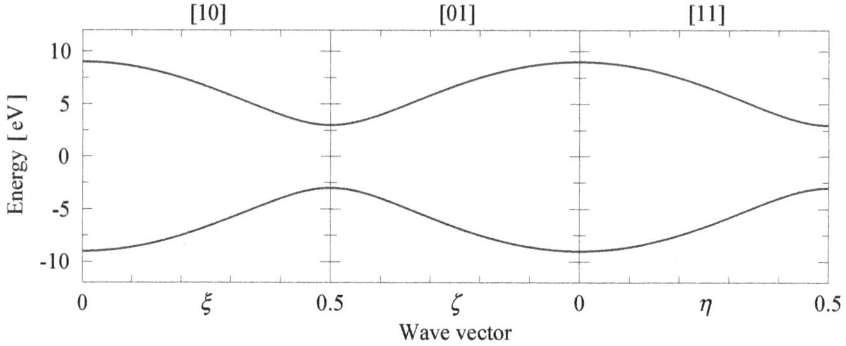

Fig. 18.6 Energy bands of a graphite sheet along the [10], [01], and [11] directions.

The energies along [01] are given by

$$E_{c,v}(\xi) = \pm t\big[\,1 + 8\cos^2(\zeta\pi)\,\big]^{1/2}.$$

We see that the energy dispersion along [01] is identical in form with that along [10]. This is because [01] and [10] are equivalent directions. The energy bands along the [01] direction are also shown in Fig. 18.6.

Along [11], we have $\boldsymbol{k} = \eta(\boldsymbol{b}_1 + \boldsymbol{b}_2)$. We then have

$$k_x = \eta\frac{4\pi}{3a},\ k_y = 0.$$

The energies along [11] are given by

$$E_{c,v}(\xi) = \pm t\big[\,5 + 4\cos(2\eta\pi)\,\big]^{1/2}.$$

The boundary of the first Brillouin zone along $\boldsymbol{e}_x$ is at $\eta = 1/2$. The energy bands along the [11] direction are also shown in Fig. 18.6.

(2) Differentiating $E_c(\boldsymbol{k})$ with respect to k_x, we have

$$\frac{\partial E_c}{\partial k_x} = -\frac{3t^2 a}{E_c}\cos(\sqrt{3}k_y a/2)\sin(3k_x a/2).$$

Differentiating $E_c(\boldsymbol{k})$ with respect to k_y, we have

$$\frac{\partial E_c}{\partial k_y} = -\frac{\sqrt{3}t^2 a}{E_c}\big[\,\sin(\sqrt{3}k_y a/2)\cos(3k_x a/2) + \sin(\sqrt{3}k_y a)\,\big].$$

For $\partial^2 E_c/\partial k_x^2$, we have

$$\frac{\partial E_c^2}{\partial k_x^2} = \frac{3t^2 a}{E_c^2}\frac{\partial E_c}{\partial k_x}\cos(\sqrt{3}k_y a/2)\sin(3k_x a/2) - \frac{9t^2 a^2}{2E_c}\cos(\sqrt{3}k_y a/2)\cos(3k_x a/2)$$

$$= -\frac{9t^4a^2}{E_c^3}\cos^2(\sqrt{3}k_ya/2)\sin^2(3k_xa/2)$$
$$- \frac{9t^2a^2}{2E_c}\cos(\sqrt{3}k_ya/2)\cos(3k_xa/2)$$
$$= -\frac{9t^4a^2}{2E_c^3}\cos(\sqrt{3}k_ya/2)\left[\cos(3k_xa/2)+2\cos(\sqrt{3}k_ya/2)\right]$$
$$\times\left[1+2\cos(3k_xa/2)\cos(\sqrt{3}k_ya/2)\right].$$

For $\partial^2 E_c/\partial k_y \partial k_x$, we have

$$\frac{\partial^2 E_c^2}{\partial k_y \partial k_x} = \frac{3t^2 a}{E_c^2}\frac{\partial E_c}{\partial k_y}\cos(\sqrt{3}k_ya/2)\sin(3k_xa/2)$$
$$+\frac{3\sqrt{3}t^2a^2}{2E_c}\sin(\sqrt{3}k_ya/2)\sin(3k_xa/2)$$
$$= \frac{3\sqrt{3}t^4a^2}{2E_c^3}\sin(\sqrt{3}k_ya/2)\sin(3k_xa/2)$$
$$\times\left[1+2\cos(3k_xa/2)\cos(\sqrt{3}k_ya/2)\right].$$

For $\partial^2 E_c/\partial k_y^2$, we have

$$\frac{\partial^2 E_c}{\partial k_y^2} = \frac{\sqrt{3}t^2 a}{E_c^2}\frac{\partial E_c}{\partial k_y}\left[\sin(\sqrt{3}k_ya/2)\cos(3k_xa/2)+\sin(\sqrt{3}k_ya)\right]$$
$$-\frac{3t^2a^2}{2E_c}\left[\cos(\sqrt{3}k_ya/2)\cos(3k_xa/2)+2\cos(\sqrt{3}k_ya)\right]$$
$$= \frac{3t^4a^2}{2E_c^3}\Big\{2\left[\sin^2(3k_xa/2)-4\cos^2(\sqrt{3}k_ya/2)\right]$$
$$\times\left[1+\cos^2(\sqrt{3}k_ya/2)\right]$$
$$+2\cos^2(\sqrt{3}k_ya/2)-\cos(3k_xa/2)\cos(\sqrt{3}k_ya/2)$$
$$\times\left[1+12\cos^2(\sqrt{3}k_ya/2)\right]\Big\}.$$

The inverse effective mass tensor for electrons in the conduction band is given by

$$\left(\frac{1}{m_e^*}\right)_{\alpha\beta} = \frac{1}{\hbar^2}\begin{pmatrix} \partial^2 E_c/\partial k_x^2 & \partial^2 E_c/\partial k_x \partial k_y \\ \partial^2 E_c/\partial k_y \partial k_x & \partial^2 E_c/\partial k_y^2 \end{pmatrix}_{\alpha\beta}.$$

At the Γ point ($k_x = k_y = 0$), we have

$$\left.\frac{\partial E_c^2}{\partial k_x^2}\right|_\Gamma = \left.\frac{\partial^2 E_c}{\partial k_y^2}\right|_\Gamma = -\frac{3}{2}ta^2,\ \left.\frac{\partial^2 E_c^2}{\partial k_y \partial k_x}\right|_\Gamma = \left.\frac{\partial^2 E_c^2}{\partial k_x \partial k_y}\right|_\Gamma = 0.$$

Thus, the effective mass tensor at the Γ point is a constant times a unit matrix and can be represented by the scalar quantity $m_e^*(\Gamma) = -2\hbar^2/3ta^2$. The negative value of the electron effective mass is because the electron energy dispersion curve near the Γ point is a downward parabola [*cf.* Fig. 18.6]. The [01] edge of the first Brillouin zone is at $X = [\pi/3a, -\pi/\sqrt{3}a]$. At this point, we have

$$\left.\frac{\partial E_c^2}{\partial k_x^2}\right|_X = 0, \quad \left.\frac{\partial^2 E_c}{\partial k_y^2}\right|_X = 3ta^2, \quad \left.\frac{\partial^2 E_c^2}{\partial k_y \partial k_x}\right|_X = \left.\frac{\partial^2 E_c^2}{\partial k_x \partial k_y}\right|_X = -\frac{3\sqrt{3}}{2}ta^2.$$

The inverse effective mass tensor at $X = [\pi/3a, -\pi/\sqrt{3}a]$ is given by

$$\frac{1}{m_e^*(X)} = \begin{pmatrix} 0 & -3\sqrt{3}ta^2/2\hbar^2 \\ -3\sqrt{3}ta^2/2\hbar^2 & 3ta^2/\hbar^2 \end{pmatrix}.$$

Inverting the above matrix, we obtain the effective mass tensor at $X = [\pi/3a, -\pi/\sqrt{3}a]$

$$m_e^*(X) = \begin{pmatrix} -4\hbar^2/9ta^2 & -2\hbar^2/3\sqrt{3}ta^2 \\ -2\hbar^2/3\sqrt{3}ta^2 & 0 \end{pmatrix}.$$

The eigenvalues and eigenvectors of $m_e^*(X)$ are found to be

$$m_1^* = \frac{2\hbar^2}{9ta^2}, \; \phi_1 = \frac{1}{2}\begin{pmatrix} 1 \\ -\sqrt{3} \end{pmatrix}; \quad m_2^* = -\frac{2\hbar^2}{3ta^2}, \; \phi_2 = \frac{1}{2}\begin{pmatrix} \sqrt{3} \\ 1 \end{pmatrix}.$$

The above eigenvalues and eigenvectors indicate that the effective mass is positive in the vicinity of the X point along the direction given by ϕ_1 which is coincident with the direction of $\boldsymbol{b}_2$ ([01]) and that the effective mass is negative in the vicinity of the X point along the direction given by ϕ_2. Note that the value of m_1^* can be directly obtained using the above-given dispersion relation along the [01] direction.

Chapter 19

Dynamics of Bloch Electrons in Electric Fields

(1) *Velocity of an electron in a single-electron state*

$$\boldsymbol{v}_{n\boldsymbol{k}} = \frac{1}{\hbar}\frac{\partial E_{n\boldsymbol{k}}}{\partial \boldsymbol{k}} = \frac{1}{\hbar}\boldsymbol{\nabla}_{\boldsymbol{k}} E_{n\boldsymbol{k}}.$$

(2) *Semi-classical equations of motion*

$$\frac{\mathrm{d}\boldsymbol{r}_{n\boldsymbol{k}}}{\mathrm{d}t} = \boldsymbol{v}_{n\boldsymbol{k}} = \frac{1}{\hbar}\frac{\partial E_{n\boldsymbol{k}}}{\partial \boldsymbol{k}},$$
$$\hbar\frac{\mathrm{d}\boldsymbol{k}}{\mathrm{d}t} = \boldsymbol{F}_{\text{ext}}(\boldsymbol{r}) = -e\boldsymbol{E}(\boldsymbol{r}).$$

(3) *Current density*

$$\boldsymbol{j} = -2\frac{e}{\mathscr{V}}\sum_{\boldsymbol{k},\,\text{occ}} \boldsymbol{v}_{\boldsymbol{k}}.$$

Completely filled energy bands do not contribute to the current.

(4) *Properties of holes*

$$\boldsymbol{k}_h = -\boldsymbol{k}_v,$$
$$E^h_{\boldsymbol{k}_v} = -E^e_{\boldsymbol{k}_v},$$
$$m^*_{h,\alpha\beta}(\boldsymbol{k}_h) = -m^*_{e,\alpha\beta}(\boldsymbol{k}_v),$$
$$\boldsymbol{v}_h = -\boldsymbol{v}_v,$$
$$q_h = q_v = -q_e = e,$$
$$f_h = f_v = 1 - f_e.$$

(5) *Bloch oscillations*

$$x(t) = x_0 + \frac{2\gamma}{eE}\cos(\omega_{\text{BO}}t - k_0 a),$$

$$\omega_{\text{BO}} = \frac{eEa}{\hbar}.$$

19-1 Electrons in an ellipsoidal energy band. Consider an energy band with the following dispersion relation

$$E_{\boldsymbol{k}} = \text{const} + \hbar^2\left(\frac{k_x^2}{2m_x} + \frac{k_y^2}{2m_y} + \frac{k_z^2}{2m_z}\right)$$

which possesses a minimum at $\boldsymbol{k} = 0$. The energy band of this form can occur in an orthorhombic crystal.

(1) Compute the density of states and the electronic specific heat. By comparing the result of the electronic specific heat with that for free electrons, show that the specific heat effective mass is given by $m_s^* = (m_x m_y m_z)^{1/3}$.

(2) Solve the semi-classical equations of motion

$$m_\alpha \frac{\mathrm{d}v_\alpha}{\mathrm{d}t} = -e[E_\alpha + (\boldsymbol{v} \times \boldsymbol{B})_\alpha], \quad \alpha = x,\, y,\, z$$

for $\boldsymbol{E} = 0$ and $\boldsymbol{B} = B\boldsymbol{e}_z$ and show that the cyclotron frequency is given by $\omega_c = eB/m_c^*$, where the cyclotron effective mass is given by $m_c^* = (m_x m_y)^{1/2}$.

---

(1) Let the constant in the dispersion relation be E_0. The density of states is given by

$$\rho(E) = \frac{2}{\mathscr{V}}\sum_{\boldsymbol{k}} \delta(E - E_{\boldsymbol{k}})$$

$$= \frac{1}{4\pi^3}\int \mathrm{d}k_x \mathrm{d}k_y \mathrm{d}k_z\; \delta\left(E - E_0 - \hbar^2\left(\frac{k_x^2}{2m_x} + \frac{k_y^2}{2m_y} + \frac{k_z^2}{2m_z}\right)\right).$$

Note that, since m_x, m_y, and m_z are positive, it is always true that $E_{\boldsymbol{k}} > E_0$. We pull $E - E_0$ out of the δ-function and have

$$g(E) = \frac{\theta(E - E_0)}{4\pi^3(E - E_0)}\iiint_{-\infty}^{\infty} \mathrm{d}k_x \mathrm{d}k_y \mathrm{d}k_z$$

$$\times\, \delta\left(1 - \frac{\hbar^2}{2(E - E_0)}\left(\frac{k_x^2}{m_x} + \frac{k_y^2}{m_y} + \frac{k_z^2}{m_z}\right)\right).$$

We now make a change of integration variables

$$u, v, w = \left(\frac{\hbar^2}{2m_{x,y,z}(E - E_0)}\right)^{1/2} k_{x,y,z}$$

to the above integral and obtain

$$g(E) = \frac{1}{\pi^3\hbar^3}\left(\frac{m_x m_y m_z}{2}\right)^{1/2}(E - E_0)^{1/2}$$

$$\times \iiint_{-\infty}^{\infty} \mathrm{d}u\mathrm{d}v\mathrm{d}w\; \delta(1 - u^2 - v^2 - w^2)$$

$$= C(E - E_0)^{1/2},$$

where

$$C = \frac{1}{\pi^3\hbar^3}\left(\frac{m_x m_y m_z}{2}\right)^{1/2} \iiint_{-\infty}^{\infty} \mathrm{d}u\mathrm{d}v\mathrm{d}w\ \delta(1-u^2-v^2-w^2)$$

$$= \frac{1}{\pi^3\hbar^3}\left(\frac{m_x m_y m_z}{2}\right)^{1/2} \cdot 4\pi \int_0^\infty \mathrm{d}r\ r^2\delta(1-r^2)$$

$$= \frac{1}{\pi^2\hbar^3}(8m_x m_y m_z)^{1/2}\int_0^\infty \mathrm{d}r\ r^2 \cdot \frac{1}{2}[\delta(1+r)+\delta(1-r)]$$

$$= \frac{1}{\pi^2\hbar^3}(2m_x m_y m_z)^{1/2}.$$

Thus, the density of states is given by

$$g(E) = \frac{1}{\pi^2\hbar^3}(2m_x m_y m_z)^{1/2}(E-E_0)^{1/2}\theta(E-E_0)$$

In terms of the Fermi energy E_F, the specific heat is given by

$$c_v = \frac{\pi^2}{3}g(E_\mathrm{F})k_\mathrm{B}^2 T = \frac{1}{3\hbar^3}(2m_x m_y m_z)^{1/2}(E_\mathrm{F}-E_0)^{1/2}k_\mathrm{B}^2 T.$$

Writing the expression of $g(E)$ in the form of the density of states for free electrons, $(1/2\pi^2)(2m_s^*/\hbar^2)^{3/2}(E-E_0)^{1/2}\theta(E-E_0)$ with m_s^* the specific heat (or the DOS) mass, we see that m_s^* is given by

$$m_s^* = (m_x m_y m_z)^{1/3}.$$

(2) The velocity of an electron of wave vector $\boldsymbol{k}$ is

$$\boldsymbol{v} = \frac{1}{\hbar}\boldsymbol{\nabla}_{\boldsymbol{k}}E_{\boldsymbol{k}} = \frac{\hbar k_x}{m_x}\boldsymbol{e}_x + \frac{\hbar k_y}{m_y}\boldsymbol{e}_y + \frac{\hbar k_z}{m_z}\boldsymbol{e}_z$$

We thus have

$$\hbar\frac{\mathrm{d}k_x}{\mathrm{d}t} = -ev_y B = -\frac{eB}{m_y}\hbar k_y,$$

$$\hbar\frac{\mathrm{d}k_y}{\mathrm{d}t} = ev_x B = \frac{eB}{m_x}\hbar k_x,$$

$$\hbar\frac{\mathrm{d}k_z}{\mathrm{d}t} = 0.$$

From the first two equations, we have

$$\frac{\mathrm{d}^2 k_x}{\mathrm{d}t^2} = -\frac{eB}{m_y}\frac{\mathrm{d}k_y}{\mathrm{d}t} = -\frac{(eB)^2}{m_x m_y}k_x.$$

Thus, the cyclotron frequency is $\omega_c = eB/(m_x m_y)^{1/2}$. Comparing this expression with $\omega_c = eB/m_c^*$, we see that the cyclotron mass is given by $m_c^* = /(m_x m_y)^{1/2}$.

19-2 Electrons in the conduction band of a 1D metal. The energy dispersion of the conduction band of a one-dimensional metal is given by

$$E_c(k_x) = E_g + E_1 \sin^2(k_x a/2)$$

with $\hbar k_x$ the crystal momentum, a the lattice constant, and E_g and E_1 constants.

(1) Compute the effective masses of an electron at $k_x = 0$ and π/a.
(2) Derive an expression for the group velocity of an electron using the semi-classical equations of motion.
(3) Plot the group velocity as a function of k_x in the first Brillouin zone.

---

(1) Taking the first- and second-order derivatives of $E_c(k_x)$ with respect to k_x, we have

$$\frac{\partial E_c}{\partial k_x} = \frac{1}{2} E_1 a \sin(k_x a),$$
$$\frac{\partial^2 E_c}{\partial k_x^2} = \frac{1}{2} E_1 a^2 \cos(k_x a).$$

The effective mass is

$$m^* = \frac{\hbar^2}{\partial^2 E_c/\partial k_x^2} = \frac{2\hbar^2}{E_1 a^2 \cos(k_x a)}.$$

Thus, the effective masses of an electron at $k_x = 0$ is $m^*(k_x = 0) = 2\hbar^2/(E_1 a^2)$ and the effective masses of an electron at $k_x = \pi/a$ is $m^*(k_x = \pi/a) = -2\hbar^2/(E_1 a^2)$.

(2) The group velocity of an electron is given by

$$v_x = \frac{1}{\hbar}\frac{\partial E_c}{\partial k_x} = \frac{1}{2\hbar} E_1 a \sin(k_x a).$$

From the above equation, we see that the group velocity vanishes at the center and boundaries of the first Brillouin zone.

(3) The plot of the group velocity as a function of k_x in the first Brillouin zone is given in Fig. 19.1.

19-3 Electrons in a 1D tight-binding conductor. Consider a one-dimensional tight-binding conductor of length L. The dispersion relation of its energy band is given by $E_k = E_0 - (\Delta/2)\cos(ka)$ with a the lattice constant. Assume that the conductor is at zero temperature, that its energy band is partially filled up to the Fermi wave vector k_F with $k_F < \pi/a$, and that there is no scattering of the electrons. The static electric field E is applied to the conductor.

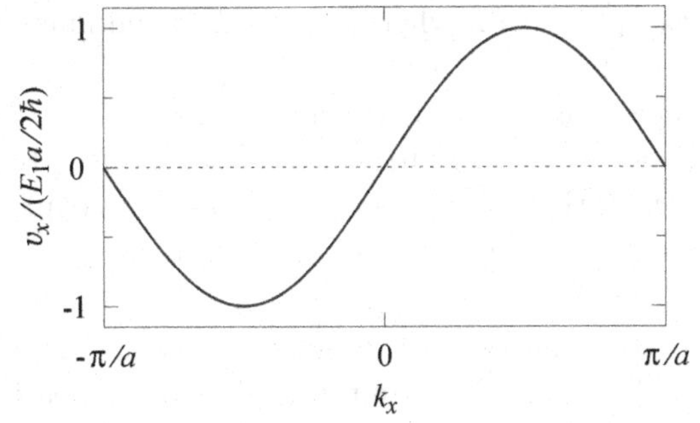

Fig. 19.1 Plot of the group velocity as a function of k_x in the first Brillouin zone.

(1) Show that an electron that has a crystal momentum k_0 at time $t = 0$ has the velocity given by $v(t) = v_{\rm max}\sin(k_0 a - \omega_{\rm BO} t)$ at time t. Give expressions for $\omega_{\rm BO}$ and $v_{\rm max}$.

(2) Assume that, at $t = 0$, all the states below the Fermi energy are occupied by electrons while the states above the Fermi energy are empty. Derive an expression for the current density as a function of time t.

(3) Plot the amplitude of the current density as a function of the electron density from $n = 0$ up to $n = 2/a$.

---

(1) From $\hbar \mathrm{d}k/\mathrm{d}t = -eE$, we have

$$k(t) = k_0 - (eE/\hbar)t.$$

The velocity of an electron of wave vector k at time t is given by

$$v(t) = \frac{1}{\hbar}\frac{\partial E_k}{\partial t} = \frac{1}{2\hbar}a\Delta\sin(ka) = \frac{1}{2\hbar}a\Delta\sin(k_0 a - (eEa/\hbar)t).$$

Reexpressing $v(t)$ in the form

$$v(t) = v_{\rm max}\sin(k_0 a - \omega_{\rm BO} t),$$

we have $v_{\rm max} = a\Delta/2\hbar$ and $\omega_{\rm BO} = eEa/\hbar$.

(2) We first find the Fermi wave vector for a given number density n of electrons. Let N denote the total number of electrons. We then have

$$N = 2\sum_{|k|\leqslant k_{\rm F}} = \frac{L}{\pi}\int_{-k_{\rm F}}^{k_{\rm F}} \mathrm{d}k = \frac{2k_{\rm F}L}{\pi}.$$

Thus, $k_F = \pi n/2$, where $n = N/L$ is the number of electrons per unit length.

We now compute the electric current density. The number of single-electron states from k to $k + \mathrm{d}k$ is given by $2L\mathrm{d}(\hbar k)/h = L\mathrm{d}k/\pi$. If these single-electron states are occupied by electrons, then the contribution to the current density is given by

$$\mathrm{d}j_x = -ev_k \mathrm{d}k/\pi.$$

Note that the current density defined as the charge passing through a point per unit time in one dimension is just the current. We use k to denote the states of electrons at $t = 0$. Then, the velocity of the electron of wave vector k at $t = 0$ is given by the expression in the first part, now written as $v_k(t) = v_{\max} \sin(ka - \omega_{BO}t)$. Making use of this expression for the velocity, we can write $\mathrm{d}j_x$ as

$$\mathrm{d}j_x = -\frac{e}{\pi} v_{\max} \sin(ka - \omega_{BO}t)\mathrm{d}k = -\frac{ea\Delta}{2\pi\hbar} \sin(ka - \omega_{BO}t)\mathrm{d}k.$$

The total current density contributing from all the electrons occupying the states below the Fermi energy can be obtained by integrating the above equation from $k = -k_F$ to $k = k_F$ and we have

$$\begin{aligned} j_x &= -\frac{ea\Delta}{2\pi\hbar} \int_{-k_F}^{k_F} \mathrm{d}k \ \sin(ka - \omega_{BO}t) \\ &= \frac{e\Delta}{2\pi\hbar} \left[\cos(k_F a - \omega_{BO}t) - \cos(k_F a + \omega_{BO}t)\right] \\ &= \frac{e\Delta}{\pi\hbar} \sin(k_F a) \sin(\omega_{BO}t). \end{aligned}$$

Making use of $k_F = \pi n/2$, we have

$$j_x = \frac{e\Delta}{\pi\hbar} \sin(\pi n a/2) \sin(\omega_{BO}t).$$

(3) Let

$$A_j = \frac{e\Delta}{\pi\hbar} \sin(\pi n a/2).$$

The current density is then written as

$$j_x = A_j \sin(\omega_{BO}t).$$

The plot of A_j as a function of the electron density n from $n = 0$ up to $n = 2/a$ is given in Fig. 19-2.

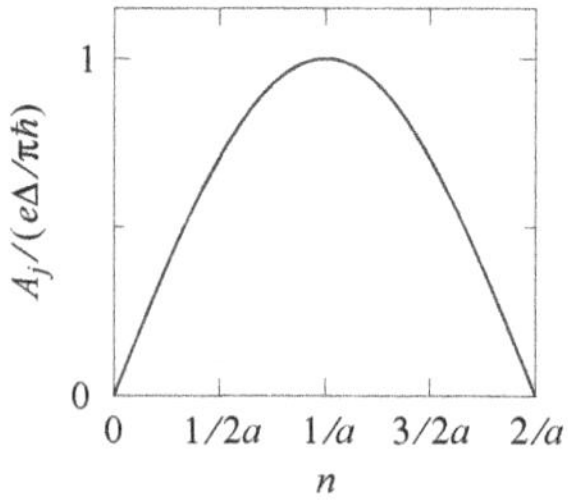

Fig. 19.2 Plot of A_j as a function of n.

19-4 Electrons in a semiconductor superlattice. Consider electrons in a semiconductor superlattice of lattice constant a. The periodic potential in the superlattice is assumed to be weak. The lowest band is filled by electrons of density n. An electric field E is applied in the x-direction. The scattering of electrons is negligible and is hence ignored. Assume that the quasiclassical approximation is valid.

(1) Calculate the fundamental frequency of the oscillation for $a = 5$ nm and $E = 5 \times 10^4$ V/cm. Compare the result to the typical relaxation time of the electrons.

(2) Argue that oscillations appear in the electric polarization. Sketch the electric polarization versus time for various band fillings.

---

(1) From the above problem, the fundamental frequency of the oscillation is given by $\omega_{\rm BO} = eEa/\hbar$. We thus have

$$\begin{aligned}\omega_{\rm BO} &= \frac{1.602\,176\,487 \times 10^{-19} \times 5 \times 10^6 \times 5 \times 10^{-9}}{1.054\,571\,628 \times 10^{-34}} \\ &\approx 3.798 \times 10^{13}\ \text{Hz} = 37.98\ \text{THz}.\end{aligned}$$

The relaxation time is typically $\tau = 10^{-16}$-10^{-14} s. Thus, the Bloch oscillation frequency is near the lower end of relaxation frequencies.

(2) When electrons undergo oscillatory motions, their negative charges deviate from the positive charge background provided by positive ions. This leads to the appearance of the electric polarization. The dipole moment due to the oscillation of an electron of wave vector $\boldsymbol{k}$ is given by $\boldsymbol{p}_{\boldsymbol{k}} = -e\boldsymbol{r}_{\boldsymbol{k}}$. The electric polarization is then given by

$$\boldsymbol{P} = \frac{2}{\mathscr{V}} \sum_{\boldsymbol{k}\,(|\boldsymbol{k}| \leqslant k_{\rm F})} \boldsymbol{p}_{\boldsymbol{k}} = -\frac{2e}{\mathscr{V}} \sum_{\boldsymbol{k}\,(|\boldsymbol{k}| \leqslant k_{\rm F})} \boldsymbol{r}_{\boldsymbol{k}}.$$

Using the results from the above problem, we have

$$P_x = -\frac{2e}{\mathscr{V}} \sum_{\boldsymbol{k}\,(|\boldsymbol{k}|\leqslant k_F)} x_{\boldsymbol{k}} = -\frac{2e}{\mathscr{V}} \sum_{\boldsymbol{k}\,(|\boldsymbol{k}|\leqslant k_F)} \int_0^t \mathrm{d}t'\; v_{\boldsymbol{k},x}(t')$$

$$= -\frac{ea\Delta}{\hbar\mathscr{V}} \sum_{\boldsymbol{k}\,(|\boldsymbol{k}|\leqslant k_F)} \int_0^t \mathrm{d}t'\; \sin\big(k_x a - eEat'/\hbar\big)$$

$$= -\frac{\Delta}{E\mathscr{V}} \sum_{\boldsymbol{k}\,(|\boldsymbol{k}|\leqslant k_F)} \big[\cos\big(k_x a - eEat/\hbar\big) - \cos(k_x a)\big]$$

$$= -\frac{2\Delta}{E\mathscr{V}} \sin(eEat/2\hbar) \sum_{\boldsymbol{k}\,(|\boldsymbol{k}|\leqslant k_F)} \sin\big(k_x a - eEat/2\hbar\big).$$

Note that $P_y = P_z = 0$ since the electric field is in the x-direction. Converting the summation over $\boldsymbol{k}$ into an integration over $\boldsymbol{k}$ yields

$$P_x = -\frac{2\Delta}{(2\pi)^3 E} \sin(eEat/2\hbar) \iiint_{|\boldsymbol{k}|\leqslant k_F} \mathrm{d}k_x \mathrm{d}k_y \mathrm{d}k_z\; \sin\big(k_x a - eEat/2\hbar\big)$$

$$= \frac{2\Delta}{(2\pi)^3 Ea} \sin(eEat/2\hbar) \iint_{k_y^2+k_z^2\leqslant k_F} \mathrm{d}k_y \mathrm{d}k_z$$

$$\times \big[\cos\big((k_F^2 - k_y^2 - k_z^2)^{1/2} a - eEat/2\hbar\big) - \cos(eEat/\hbar)\big]$$

$$= \frac{\Delta}{2\pi^2 Ea} \sin(eEat/2\hbar) \int_0^{k_F} \mathrm{d}k\, k$$

$$\times \big[\cos\big((k_F^2 - k^2)^{1/2} a - eEat/\hbar\big) - \cos(eEat/\hbar)\big].$$

To perform the above integration, we make a change of integration variables from k to $\xi = (k_F^2 - k^2)^{1/2} a$. We have $k\mathrm{d}k = -\xi \mathrm{d}\xi/a^2$ and

$$P_x = \frac{\Delta}{2\pi^2 Ea^3} \sin(eEat/2\hbar) \int_0^{k_F a} \mathrm{d}\xi\, \xi \big[\cos\big(\xi - eEat/\hbar\big) - \cos(eEat/\hbar)\big].$$

Evaluating the integral, we obtain

$$P_x = \frac{\Delta}{2\pi^2 Ea^3} \Big[(k_F a)\sin\big(k_F a - eEat/2\hbar\big) + \cos\big(k_F a - eEat/2\hbar\big)$$

$$- \big((k_F a)^2/2 + 1\big)\cos(eEat/2\hbar)\Big] \sin\big(eEat/2\hbar\big)$$

$$= \frac{\Delta}{4\pi^2 Ea^3} \big[\alpha - (\alpha^2 + \beta^2)^{1/2} \sin\big(eEat/\hbar + \varphi\big)\big],$$

where

$$\alpha = \sin(k_F a) - (k_F a)\cos(k_F a),$$

$$\beta = (k_F a)^2/2 + 1 - (k_F a)\sin(k_F a) - \cos(k_F a),$$

$$\varphi = \tan^{-1}(\alpha/\beta).$$

The electric polarization is plotted against time for $k_{\mathrm{F}}a = 0.2$, 0.3, 0.4, and 0.5 in Fig. 19.3.

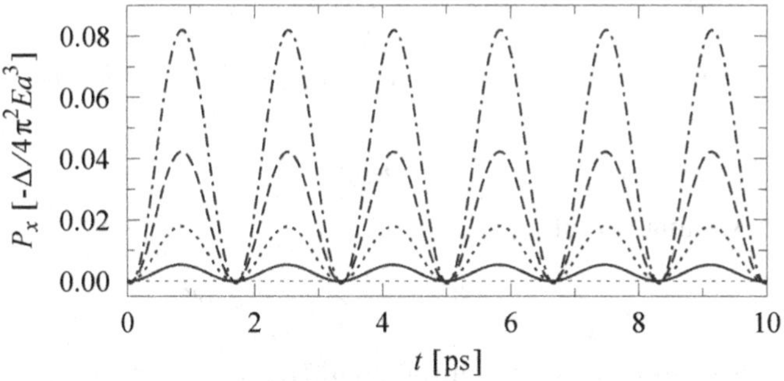

Fig. 19.3 Plot of the electric polarization as a function of time for $k_{\mathrm{F}}a = 0.2$, 0.3, 0.4, and 0.5 from bottom to top.

19-5 Effective mass approximation. Within the effective mass approximation, the Hamiltonian for motion in one dimension is given by $\hat{H} = E(k) - Fx$, where $E(k) = W\cos(ka)$ is the single band energy, F the electric force, and a the lattice constant. The envelope wave function at time $t = 0$ is given by $\psi(x, t = 0) = \delta(x)$.

(1) Express the time-dependent Schrödinger equation and the initial condition in k-space (the momentum representation).

(2) Find the envelope wave function of the form $\psi(k, t) = \mathrm{e}^{\mathrm{i}\varphi(k,t)}$ in k-space at time t.

(3) Express the wave function in real space as the Fourier transform of $\psi(k, t)$. Using the steepest descent method, show the presence of the Bloch oscillations and compare the result with that obtained with the use of the equations of motion.

(4) Find an analytic expression for the Fourier integral. The integral expressions of Bessel functions may be useful. Alternatively, find a numerical solution with a time series of plots.

---

(1) In k-space, the Hamiltonian reads

$$\hat{H} = E(k) - \mathrm{i}F\frac{\partial}{\partial k}.$$

Thus, the time-dependent Schrödinger equation in k-space reads

$$\mathrm{i}\hbar\frac{\partial}{\partial t}\psi(k,t) = \left(E(k) - \mathrm{i}F\frac{\partial}{\partial k}\right)\psi(k,t).$$

In terms of the Fourier component $\psi(k,t)$'s, $\psi(x,t)$ is given by

$$\psi(x,t) = \frac{1}{2\pi}\int_{-\infty}^{\infty} \mathrm{d}k\ \mathrm{e}^{\mathrm{i}kx}\psi(k,t).$$

Making use of

$$\delta(x) = \frac{1}{2\pi}\int_{-\infty}^{\infty} \mathrm{d}k\ \mathrm{e}^{\mathrm{i}kx},$$

we can express the initial condition as

$$\psi(x,t=0) = \frac{1}{2\pi}\int_{-\infty}^{\infty} \mathrm{d}k\ \mathrm{e}^{\mathrm{i}kx}\psi(k,t=0) = \delta(x) = \frac{1}{2\pi}\int_{-\infty}^{\infty} \mathrm{d}k\ \mathrm{e}^{\mathrm{i}kx}.$$

Hence, the initial condition in k-space is

$$\psi(k,t=0) = 1.$$

(2) For $\psi(k,t) = \mathrm{e}^{\mathrm{i}\varphi(k,t)}$, the initial condition $\psi(k,t=0) = 1$ indicates that $\varphi(k,t=0) = 0$. Inserting $\psi(k,t) = \mathrm{e}^{\mathrm{i}\varphi(k,t)}$ into the time-dependent Schrödinger equation in k-space yields

$$-\hbar\frac{\partial}{\partial t}\varphi(k,t) = E(k) + F\frac{\partial}{\partial k}\varphi(k,t).$$

To solve the above equation, we expand $\varphi(k,t)$ as a power series in t. In consideration of the initial condition $\varphi(k,t=0) = 0$, we see that the lowest power in t is one. We thus have

$$\varphi(k,t) = \sum_{\ell=1}^{\infty} a_\ell(k)t^\ell.$$

Inserting the above expansion into the equation for $\varphi(k,t)$ yields

$$-\hbar\sum_{\ell=1}^{\infty} \ell a_\ell(k)t^{\ell-1} = E(k) + F\sum_{\ell=1}^{\infty}\frac{\mathrm{d}a_\ell(k)}{\mathrm{d}k}t^\ell.$$

Rearranging yields

$$\sum_{\ell=0}^{\infty}(\ell+1)a_{\ell+1}(k)t^\ell = -\frac{E(k)}{\hbar} - \frac{F}{\hbar}\sum_{\ell=1}^{\infty}\frac{\mathrm{d}a_\ell(k)}{\mathrm{d}k}t^\ell.$$

Comparing the coefficients of the zeroth-power of t on both sides, we have

$$a_1(k) = -\frac{E(k)}{\hbar}.$$

Comparing the coefficients of the higher powers of t on both sides, we have

$$a_{\ell+1} = -\frac{F}{(\ell+1)\hbar}\frac{\mathrm{d}a_\ell(k)}{\mathrm{d}k}, \ \ell \geqslant 1.$$

Making use of the recurrence relation for $\ell \geqslant 1$, we have

$$\begin{aligned}
a_{\ell+1} &= -\frac{F}{(\ell+1)\hbar}\frac{\mathrm{d}a_\ell(k)}{\mathrm{d}k} = \frac{1}{(\ell+1)\ell}\left(-\frac{F}{\hbar}\right)^2\frac{\mathrm{d}^2 a_{\ell-1}(k)}{\mathrm{d}k^2} = \cdots \\
&= \frac{1}{(\ell+1)\ell\cdots 2}\left(-\frac{F}{\hbar}\right)^\ell \frac{\mathrm{d}^\ell a_1(k)}{\mathrm{d}k^\ell} = \frac{1}{\hbar}\frac{(-1)^{\ell+1}}{(\ell+1)!}\left(\frac{F}{\hbar}\right)^\ell\frac{\mathrm{d}^\ell E(k)}{\mathrm{d}k^\ell} \\
&= \frac{W}{\hbar}\frac{(-1)^{\ell+1}}{(\ell+1)!}\left(\frac{Fa}{\hbar}\right)^\ell\frac{\mathrm{d}^\ell\cos(ka)}{\mathrm{d}(ka)^\ell} \\
&= \begin{cases} \dfrac{W}{\hbar}\dfrac{(-1)^{j+1}}{(2j+1)!}\left(\dfrac{Fa}{\hbar}\right)^{2j}\cos(ka), & \ell = 2j, \ j = 1,2,3,\cdots, \\ \dfrac{W}{\hbar}\dfrac{(-1)^{j+1}}{(2j+2)!}\left(\dfrac{Fa}{\hbar}\right)^{2j+1}\sin(ka), & \ell = 2j+1, \ j = 0,1,2,\cdots. \end{cases}
\end{aligned}$$

Thus,

$$\begin{aligned}
\varphi(k,t) &= \sum_{j=0}^\infty a_{2(j+1)}(k)t^{2(j+1)} + \sum_{j=0}^\infty a_{2j+1}(k)t^{2j+1} \\
&= \frac{W}{Fa}\sin(ka)\sum_{j=0}^\infty\frac{(-1)^{j+1}}{(2j+2)!}\left(\frac{Fa}{\hbar}t\right)^{2j+2} \\
&\quad + \frac{W}{Fa}\cos(ka)\sum_{j=0}^\infty\frac{(-1)^{j+1}}{(2j+1)!}\left(\frac{Fa}{\hbar}t\right)^{2j+1} \\
&= \frac{W}{Fa}\Big\{\sin(ka)\big[\cos(Fat/\hbar)-1\big]-\cos(ka)\sin(Fat/\hbar)\Big\} \\
&= \frac{W}{Fa}\big[\sin(ka - Fat/\hbar) - \sin(ka)\big] \\
&= -\frac{2W}{Fa}\cos(ka - Fat/2\hbar)\sin(Fat/2\hbar).
\end{aligned}$$

(3) In real space, we have

$$\begin{aligned}
\psi(x,t) &= \frac{1}{2\pi}\int_{-\infty}^{\infty}\mathrm{d}k\ \mathrm{e}^{\mathrm{i}kx+\mathrm{i}\varphi(k,t)} \\
&= \frac{1}{2\pi}\int_{-\infty}^{\infty}\mathrm{d}k\ \mathrm{e}^{\mathrm{i}kx-\mathrm{i}(2W/Fa)\cos(ka-Fat/2\hbar)\sin(Fat/2\hbar)}.
\end{aligned}$$

The stationary point k_0 is to be solved from

$$\frac{\partial}{\partial k}\big[kx+\varphi(k,t)\big]\Big|_{k=k_0} = x + \frac{2W}{F}\sin(k_0a - Fat/2\hbar)\sin(Fat/2\hbar) = 0.$$

Thus,

$$\sin(k_0 a - Fat/2\hbar) = -\frac{Fx}{2W\sin(Fat/2\hbar)},$$

$$k_0 a = \frac{Fat}{2\hbar} - \sin^{-1}\frac{Fx}{2W\sin(Fat/2\hbar)}, \quad n = 0 \pm 1, \pm 2, \cdots,$$

$$\cos(k_0 a - Fat/2\hbar) = \frac{\left[4W^2\sin^2(Fat/2\hbar) - (Fx)^2\right]^{1/2}}{2W\sin(Fat/2\hbar)},$$

where we have kept only the solution within the first Brillouin zone and chosen the plus sign in front of the square root in the cosine function. Expanding $kx + \varphi(k,t)$ around the stationary point yields up to the second-order term

$$kx + \varphi(k,t) \approx \kappa(x,t)\frac{x}{a} - \rho(x,t) + \frac{1}{2}|\rho(x,t)|e^{i\alpha(x,t)}\left[ka - \kappa_n(x,t)\right]^2.$$

where

$$\kappa(x,t) = \frac{Fat}{2\hbar} - \sin^{-1}\frac{Fx}{2W\sin(Fat/2\hbar)},$$

$$\rho(x,t) = \left[(2W/Fa)^2\sin^2(Fat/2\hbar) - (x/a)^2\right]^{1/2},$$

$$\alpha(x,t) = \tan^{-1}\frac{\mathrm{Im}\,\rho(x,t)}{\mathrm{Re}\,\rho(x,t)}.$$

Evaluating the integral in the vicinity of the stationary point, we have

$$\psi(x,t) = \frac{1}{\left(-2i\pi|\rho(x,t)|a^2\right)^{1/2}}e^{i\kappa(x,t)x/a - i\rho(x,t)}.$$

Because of the presence of $|\rho(x,t)|$ on the denominator, the value of $\psi(x,t)$ diverges whenever $x = \pm(2W/F)\sin(Fat/2\hbar)$. Thus, as t increases, Bloch oscillations appear in $\psi(x,t)$. From the equations of motion, we can obtain classically the time dependence of the coordinates of individual electrons in given single-electron states. Whereas, we have obtained the time dependence of the envelope wave function in the present method.

(4) To make use of the integral representation of the Bessel function

$$J_\nu(z) = \frac{1}{2\pi}\int_{-\pi+i\infty}^{\pi+i\infty} d\theta\, e^{i\nu\theta - iz\sin\theta},$$

where the path of integration is from $-\pi + i\infty$ to $-\pi \to \pi$ to $\pi + i\infty$, we make the following change of integration variables to the integral for $\psi(x,t)$

$$\pi/2 + k'a = ka - Fat/2\hbar.$$

We then drop the prime on k' and have

$$\psi(x,t) = \frac{1}{2\pi a}\int_{-\infty}^{\infty} \mathrm{d}(ka)\, \mathrm{e}^{\mathrm{i}(ka)(x/a)-\mathrm{i}(2W/Fa)\cos(ka-Fat/2\hbar)\sin(Fat/2\hbar)}$$
$$= \frac{1}{2\pi a}\mathrm{e}^{\mathrm{i}(\pi+Fat/\hbar)(x/2a)}$$
$$\times \int_{-\infty}^{\infty} \mathrm{d}(ka)\, \mathrm{e}^{\mathrm{i}(ka)(x/a)+\mathrm{i}(2W/Fa)\sin(ka)\sin(Fat/2\hbar)}.$$

Since the integrand is analytic in the entire complex plane and its value vanishes on the circle of infinite radius except for two points on the real axis where it oscillates wildly, we can freely change the path of integration with the help of Cauchy's theorem to the one in the integral representation of the Bessel function. We thus have

$$\psi(x,t) = \frac{1}{a}\mathrm{e}^{\mathrm{i}(\pi+Fat/\hbar)(x/2a)} J_{x/a}\big(-(2W/Fa)\sin(Fat/2\hbar)\big).$$

19-6 Semi-classical equations of motion with damping. Taking a damping term into account, we have the following semi-classical equations of motion for a tightly bound electron in one dimension

$$\frac{\mathrm{d}x}{\mathrm{d}t} = \frac{2\gamma a}{\hbar}\sin(ka),\ \ \hbar\frac{\mathrm{d}k}{\mathrm{d}t} = -eE - \frac{m}{\tau}\frac{\mathrm{d}x}{\mathrm{d}t},$$

where the second term on the right hand side of the second equation is the damping term. The values of the parameters in the above equations of motion are assumed to be $\gamma = 1$ eV, $a = 0.2$ nm, $E = 10^6$ V/cm, and $\tau = 10^{-14}$ s.

(1) Rewrite the equations of motion in dimensionless form with distances measured in units of a and time in units of τ.

(2) Integrate the equations of motion and study the effect of the damping on Bloch oscillations.

(3) Analytically describe the final state of the system.

---

(1) We use a bar over a symbol to denote that it is dimensionless. We have

$$\bar{x} = x/a,\ \bar{t} = t/\tau,\ \bar{k} = ka.$$

The dimensionless equations of motion are then given by

$$\frac{\mathrm{d}\bar{x}}{\mathrm{d}\bar{t}} = \alpha\sin(\bar{k}),\ \ \frac{\mathrm{d}\bar{k}}{\mathrm{d}\bar{t}} = -\epsilon - \beta\frac{\mathrm{d}\bar{x}}{\mathrm{d}\bar{t}},$$

where α, ϵ, and β are dimensionless constants and they are given by

$$\alpha = \frac{2\gamma\tau}{\hbar} \approx 30.385\,35,\ \epsilon = \frac{eE\tau a}{\hbar} \approx 0.303\,85,\ \beta = \frac{ma^2}{\hbar\tau} \approx 0.034\,55.$$

(2) Inserting the first equation of motion into the second, we obtain

$$\frac{\mathrm{d}\bar{k}}{\epsilon + \alpha\beta \sin(\bar{k})} = -\mathrm{d}\bar{t}.$$

Let $\bar{k}_0$ be the value of $\bar{k}$ at $\bar{t} = 0$. Integrating both sides of the above equation from $\bar{t} = 0$ to $\bar{t}$ yields

$$\left| \frac{\epsilon \tan(\bar{k}/2) + \alpha\beta - \omega}{\epsilon \tan(\bar{k}/2) + \alpha\beta + \omega} \right| = |C_{\bar{k}_0}| \mathrm{e}^{-\omega \bar{t}},$$

where we have made use of the fact that $\alpha\beta > \epsilon$ and introduced

$$\omega = [\,(\alpha\beta)^2 - \epsilon^2\,]^{1/2},$$

$$C_{\bar{k}_0} = \frac{\epsilon \tan(\bar{k}_0/2) + \alpha\beta - \omega}{\epsilon \tan(\bar{k}_0/2) + \alpha\beta + \omega}.$$

The symbols for the absolute values in the above solution can be eliminated if the signs of $C_{\bar{k}_0}$ and its counterpart at $\bar{t}$ on the left hand side of the solution are known. It turns out that in a certain range of $\bar{k}_0$, they are positive. We now assume that they are positive and write the solution as

$$\frac{\epsilon \tan(\bar{k}/2) + \alpha\beta - \omega}{\epsilon \tan(\bar{k}/2) + \alpha\beta + \omega} = C_{\bar{k}_0} \mathrm{e}^{-\omega \bar{t}}.$$

If $\bar{k}$ does not tend to $\bar{k}_0$ in the $\bar{t} \to 0$ limit, then it is a signal that $C_{\bar{k}_0}$ and its counterpart at $\bar{t}$ are not all positive. Solving for $\bar{k}$ from the above equation, we obtain

$$\bar{k}(\bar{t}) = 2 \tan^{-1}\left[\frac{1}{\epsilon} \frac{(\alpha\beta + \omega) C_{\bar{k}_0} e^{-\omega \bar{t}} - (\alpha\beta - \omega)}{1 - C_{\bar{k}_0} e^{-\omega \bar{t}}} \right].$$

From the first equation of motion, we obtain the time dependence of the dimensionless velocity $\bar{v}$

$$\bar{v}(\bar{t}) = \alpha \sin[\,\bar{k}(\bar{t})\,].$$

$\bar{k}(\bar{t})$ and $\bar{v}(\bar{t})$ are plotted in Fig. 19.4 as functions of $\bar{t}$. From Fig. 19.4(a), it is seen that the wave vector initially decreases at t increases. When the wave vector decreases to zero, it suddenly gains a value equal to a reciprocal lattice vector of the underlying Bravais lattice. However, this sudden gain in the wave vector does not increase the energy of the electron. This is due to the periodicity of the electron band energy in the wave vector. No matter what value the wave vector has at $t = 0$, it tends to the same value (given below) as $t \to \infty$. The jump in the wave vector when its value becomes zero depends on this

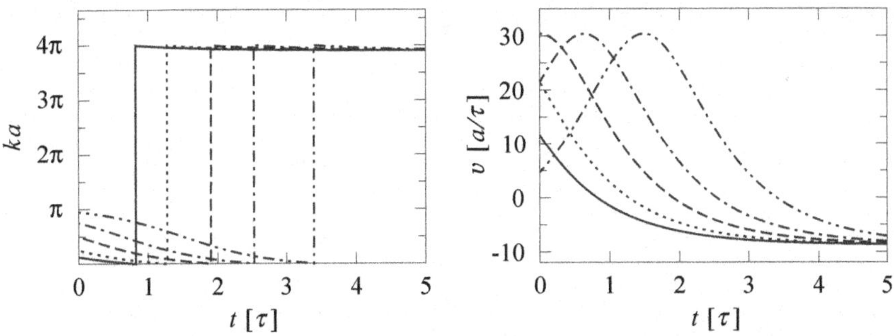

Fig. 19.4 Plots of the wave vector (a) and velocity (b) as functions of time for $k_0a = 0.125\pi$, 0.25π, 0.5π, 0.75π, and 0.95π from left to right in (a) and from bottom to top in (b).

limiting value. This limiting behavior of the wave vector is caused by the damping.

From Fig. 19.4(b), it is seen that the velocities for small initial values of the wave vector start to decrease from $t = 0$ and that the velocities for large initial values of the wave vector first increase, then peaks, and then decrease. They all have the same limiting value. Also note that the values of the velocity are not affected by the jump in the wave vector. Because of the damping, the velocity does not show oscillatory behavior.

(3) From the expression for $\bar{k}(\bar{t})$, we see that, as $\bar{t} \to \infty$, $\bar{k}(\bar{t}) \to 2\tan^{-1}[(\omega - \alpha\beta)/\epsilon]$ no matter what the initial value of the wave vector is. Thus, the final state of the system is the state with the wave vector $2\tan^{-1}[(\omega - \alpha\beta)/\epsilon]$. However, quantum mechanically, no more than two electrons can be in this state because of the restriction due to the Pauli exclusion principle.

Chapter 20

Fundamentals of Semiconductors

(1) *Electronic band structures of semiconductors*
Conduction band, valence band, band gap.

(2) *Parabolic dispersion relation near the conduction band edge for a direct-band-gap semiconductor*

$$\varepsilon_c(\boldsymbol{k}) = \varepsilon_c + \frac{\hbar^2 k^2}{2m_e}.$$

(3) *Parabolic dispersion relation near the conduction band edge for an indirect-band-gap semiconductor*

$$\varepsilon_c(\boldsymbol{k}) = \varepsilon_c + \frac{\hbar^2}{2}\left[\frac{(k_x - k_{x0})^2}{m_e^\ell} + \frac{k_y^2 + k_z^2}{m_e^t}\right].$$

(4) *Parabolic dispersion relation near the valence band edge*

$$\varepsilon_v(\boldsymbol{k}) = \varepsilon_v - \frac{\hbar^2 k^2}{2m_h}.$$

(5) *Temperature dependence of band gaps*

$$E_g(T) = E_g(0) - \frac{\alpha T^2}{T + \beta}.$$

(6) *Doping dependence of band gaps*

$$\Delta E_g(N) = -\frac{3e^2}{16\pi\epsilon}\left(\frac{e^2 N}{4\pi\epsilon k_B T}\right)^{1/2}.$$

(7) *Direct optical absorption*

$$\varepsilon_c(\boldsymbol{k}') = \varepsilon_v(\boldsymbol{k}) + \hbar\omega, \;\; \hbar\boldsymbol{k}' = \hbar\boldsymbol{k} + \hbar\boldsymbol{k}_{\text{photon}} \approx \hbar\boldsymbol{k}.$$

(8) *Indirect optical absorption*

$$\varepsilon_c(\boldsymbol{k}') = \varepsilon_v(\boldsymbol{k}) + \hbar\omega \pm \hbar\Omega_{\boldsymbol{q}}, \;\; \hbar\boldsymbol{k}' = \hbar\boldsymbol{k} \pm \hbar\boldsymbol{q}.$$

(9) *Intrinsic semiconductors*

$$n_i = N_c e^{(\mu-\varepsilon_c)/k_B T},$$
$$p_i = N_v e^{(\varepsilon_v-\mu)/k_B T},$$
$$N_c = 2\left(\frac{m_e k_B T}{2\pi\hbar^2}\right)^{3/2},$$
$$N_v = 2\left(\frac{m_h k_B T}{2\pi\hbar^2}\right)^{3/2},$$
$$\mu = \frac{1}{2}(\varepsilon_v + \varepsilon_c) + \frac{3}{4}k_B T \ln\frac{m_h}{m_e},$$
$$n_i = p_i = 2\left(\frac{k_B T}{2\pi\hbar^2}\right)^{3/2}(m_e m_h)^{3/4} e^{-E_g/2k_B T}.$$

(10) *Donor levels*

$$E_d = -\frac{m_e e^4}{2(4\pi\epsilon)^2\hbar^2},$$
$$r_d = \epsilon_r(m/m_e)a_0.$$

(11) *Numbers of ionized impurities*

$$N_d^+ = \frac{N_d}{g_d e^{(\mu-E_d)/k_B T} + 1},$$
$$N_a^- = \frac{N_a}{g_a e^{(E_a-\mu)/k_B T} + 1}.$$

(12) *Charge neutrality*

$$p + \sum_i Z_i N_i - n = 0.$$

(13) *Law of mass action*

$$np = \frac{1}{2}\left(\frac{k_B T}{\pi\hbar^2}\right)^3 (m_e m_h)^{3/2} e^{-E_g/k_B T}.$$

(14) *n-type semiconductors*

$$n = \frac{1}{2}\left\{(N_d^+ - N_a^-) + \left[(N_d^+ - N_a^-)^2 + 4n_i^2\right]^{1/2}\right\},$$
$$\mu = \varepsilon_c - k_B T \ln(N_c/N_d).$$

(15) *p-type semiconductors*

$$p = \frac{1}{2}\left\{(N_a^- - N_d^+) + \left[(N_a^- - N_d^+)^2 + 4p_i^2\right]^{1/2}\right\},$$
$$\mu = \varepsilon_v + k_B T \ln(N_v/N_a).$$

(16) *Electrical conductivities and mobilities*

$$\sigma_e = \frac{ne^2\tau_e}{m_e}, \ \mu_e = \frac{e\tau_e}{m_e}, \ \sigma_e = ne\mu_e,$$

$$\sigma_h = \frac{pe^2\tau_h}{m_h}, \ \mu_h = \frac{e\tau_h}{m_h}, \ \sigma_h = pe\mu_h.$$

(17) *Matthiessen's rule*

$$1/\mu = 1/\mu_{\rm im} + 1/\mu_{\rm ph}.$$

(18) *Temperature dependence of the electrical conductivity*

$$\sigma = ne(\mu_e + \mu_h) = F(T)\mathrm{e}^{-E_g/2k_BT},$$

$$F(T) = 2e(\mu_e + \mu_h)\left(\frac{k_BT}{2\pi\hbar^2}\right)^{3/2}(m_em_h)^{3/4}\mathrm{e}^{-E_g/2k_BT}.$$

(19) *Energy levels of an exciton*

$$\mathcal{E}_n = E_g - \frac{\mu e^4}{2(4\pi\epsilon)^2\hbar^2 n^2}.$$

(20) *Carrier diffusion in the absence of an applied electric field*

$$n(x) = n_0 + A_e\mathrm{e}^{-x/L_e},$$

$$p(x) = p_0 + A_h\mathrm{e}^{-x/L_h}.$$

(21) *Carrier diffusion in the presence of an applied electric field*

$$n = n_0 + A_e\mathrm{e}^{-\gamma_e x/L_e},$$

$$p = p_0 + A_h\mathrm{e}^{-\gamma_h x/L_h}.$$

20-1 Conduction and valence bands of a hypothetical semiconductor. A hypothetical semiconductor has a conduction band that can be described by $\varepsilon_c(k) = E_1 - E_2\cos(ka)$ and a valence band that can be described by $\varepsilon_v(k) = E_3 - E_4\sin^2(ka)$, where $E_i > 0$ for $i = 1, 2, 3,$ and 4, $E_1 > E_2$, $E_3 < E_1 - E_2$, and $-\pi/a \leqslant k \leqslant \pi/a$.

(1) Find the band widths of the two bands and the band gap between them.

(2) Compute the effective masses of electrons at the conduction band edge and holes at the valence band edge.

---

(1) The band width of the conduction band is given by $W_c = \max[\varepsilon_c(k)] - \min[\varepsilon_c(k)] = (E_1 + E_2) - (E_1 - E_2) = 2E_2.$ The band width of the valence band is given by

$$W_v = \max[\varepsilon_v(k)] - \min[\varepsilon_v(k)] = E_3 - (E_3 - E_4) = E_4.$$

Since the conduction band is above the valence band ($E_3 < E_1 - E_2$), the band gap is given by

$$E_g = \min[\varepsilon_c(k) - \max[\varepsilon_v(k)] = (E_1 - E_2) - E_3 = E_1 - E_2 - E_3.$$

(2) For the conduction band, we have

$$\frac{\partial \varepsilon_c}{\partial k} = E_2 a \sin(ka), \quad \frac{\partial^2 \varepsilon_c}{\partial k^2} = E_2 a^2 \cos(ka).$$

At the conduction band edge, we have

$$\frac{1}{m_e} = \frac{1}{\hbar^2}\left(\frac{\partial^2 \varepsilon_c}{\partial k^2}\right)_{k=0} = \frac{E_2 a^2}{\hbar^2}.$$

Thus, $m_e = \hbar^2 / E_2 a^2$.

For the valence band, we have

$$\frac{\partial \varepsilon_v}{\partial k} = -E_4 a \sin(2ka), \quad \frac{\partial^2 \varepsilon_v}{\partial k^2} = -2E_4 a^2 \cos(2ka).$$

At the valence band edge, we have

$$\frac{1}{m_h} = -\frac{1}{\hbar^2}\left(\frac{\partial^2 \varepsilon_v}{\partial k^2}\right)_{k=0} = \frac{2E_4 a^2}{\hbar^2}.$$

Thus, $m_h = \hbar^2 / 2E_4 a^2$.

20-2 Effective mass, energy, momentum, and velocity of a hole. The electron energy near the top of the valence band in a semiconductor is given by $\varepsilon = -10^{-37}k^2$ J, where $\boldsymbol{k}$ is the wave vector. An electron is removed from the state $\boldsymbol{k} = 10^9 \boldsymbol{e}_x$ m^{-1}. Compute the effective mass, energy, momentum, and velocity of the resulting hole. The sign or direction of each quantity must be given.

The energy of the hole is $\varepsilon_h = -\varepsilon = +10^{-37} \times 10^{18} = +10^{-19}$ J. From $\varepsilon = -10^{-37}k^2$ J, we have

$$\partial^2\varepsilon/\partial k_x^2 = \partial^2\varepsilon/\partial k_y^2 = \partial^2\varepsilon/\partial k_z^2 = -2 \times 10^{-37} \text{ J} \cdot \text{m}^2.$$

Thus, the effective mass of the hole is

$$m_h = -\hbar^2/(\partial^2\varepsilon/\partial k_x^2) = +\hbar^2/(2 \times 10^{-37}) = +2.195 \times 10^{-30} \text{ kg}.$$

The momentum of the hole is

$$\boldsymbol{p}_h = \hbar \boldsymbol{k}_h \approx 6.626 \times 10^{-34} \times (-10^9 \boldsymbol{e}_x) = -6.626 \times 10^{-25} \boldsymbol{e}_x \text{ kg} \cdot \text{m/s}.$$

The velocity of the hole is

$$\boldsymbol{v}_h = \boldsymbol{p}_h / m_h = -6.626 \times 10^{-25} \boldsymbol{e}_x / (2.195 \times 10^{-30}) \approx -3.019 \times 10^5 \boldsymbol{e}_x \text{ m/s}.$$

20-3 Effective mass of electrons at the conduction band edge. The energy of the conduction electrons at the bottom of the conduction band of a semiconductor with $E_g = 0.7$ eV can be approximated by $E = E_g + Ak^2$ where $A = 5 \times 10^{-37}$ J·m^2. Compute the effective mass of electrons at the conduction band edge.

Differentiating E twice with respect to k_x, k_y, and k_z, we have

$$\frac{\partial^2 E}{\partial k_x^2} = \frac{\partial^2 E}{\partial k_y^2} = \frac{\partial^2 E}{\partial k_z^2} = 2A.$$

Thus, the effective mass of electrons at the conduction band edge is given by

$$m_e = \hbar^2/(\partial^2 \varepsilon/\partial k_x^2) = \hbar^2/2A \approx 4.39 \times 10^{-31} \text{ kg}.$$

20-4 Density of states for a single *k*-space ellipsoid in Si. Show that the density of states associated with a single k-space ellipsoid in Si can be written as $g_c(\varepsilon) = (1/2\pi^2)(2m_d/\hbar^2)^{3/2}(\varepsilon - \varepsilon_c)^{1/2}$, where m_d is the density-of-states effective mass given by $m_d = (m_\ell m_t^2)^{1/3}$.

Note that m_ℓ and m_t are respectively the longitudinal and transverse effective masses of electrons. According to the definition of the density of states, we have

$$\begin{aligned} g_c(\varepsilon) &= \frac{2}{\mathscr{V}} \sum_{\boldsymbol{k}} \delta(\varepsilon - \varepsilon_c(\boldsymbol{k})) \\ &= \frac{2}{(2\pi)^3} \int \mathrm{d}k_x \mathrm{d}k_y \mathrm{d}k_z \\ &\quad \times \delta\big(\varepsilon - \varepsilon_c - \hbar^2(k_x - k_{x0})^2/2m_\ell - \hbar^2(k_y^2 + k_z^2)/2m_t\big). \end{aligned}$$

Performing the integration over k_x yields

$$\begin{aligned} g_c(\varepsilon) &= \frac{4m_\ell}{(2\pi)^3\hbar^2} \int \mathrm{d}k_x \mathrm{d}k_y \mathrm{d}k_z \\ &\quad \times \delta\big((k_x - k_{x0})^2 - 2m_\ell(\varepsilon - \varepsilon_c)/\hbar^2 + m_\ell(k_y^2 + k_z^2)/m_t\big) \\ &= \frac{2}{(2\pi)^3}\left(\frac{2m_\ell}{\hbar^2}\right)^{1/2} \int \mathrm{d}k_y \mathrm{d}k_z \, \frac{\theta(\varepsilon - \varepsilon_c - \hbar^2(k_y^2 + k_z^2)/2m_t)}{[\varepsilon - \varepsilon_c - \hbar^2(k_y^2 + k_z^2)/2m_t]^{1/2}}, \end{aligned}$$

where we have used $\delta(x^2 - a^2) = [\delta(x - a) + \delta(x + a)]/2|a|$ in performing the integration over k_x. To perform the integration over k_y and k_z, we use the polar coordinates in the k_y-k_z plane. Since the integrand does not depend on the polar angle, the integration over the polar angle can be trivially performed. We then have

$$g_c(\varepsilon) = \frac{1}{2\pi^2}\left(\frac{2m_\ell}{\hbar^2}\right)^{1/2}\int_0^{[2m_t(\varepsilon-\varepsilon_c)/\hbar^2]^{1/2}} \frac{k\mathrm{d}k\ \theta(\varepsilon-\varepsilon_c)}{(\varepsilon-\varepsilon_c-\hbar^2k^2/2m_t)^{1/2}}$$
$$= \frac{1}{2\pi^2}\left(\frac{2m_\ell}{\hbar^2}\right)^{1/2}\frac{2m_t}{\hbar^2}(\varepsilon-\varepsilon_c)^{1/2}\theta(\varepsilon-\varepsilon_c)$$
$$= \frac{1}{2\pi^2}\left(\frac{2m_d}{\hbar^2}\right)^{3/2}(\varepsilon-\varepsilon_c)^{1/2}\theta(\varepsilon-\varepsilon_c),$$

where $m_d = (m_\ell m_t^2)^{1/3}$.

20-5 Density of states in a nonparabolic conduction band. The band structure of many semiconductors (particularly narrow-band-gap semiconductors) at the conduction band edge is known to be somewhat nonparabolic. A good first approximation to the band structure about $\boldsymbol{k} = 0$ is $\hbar^2\boldsymbol{k}^2/2m_e = \varepsilon(1+\alpha\,\varepsilon)$, where m_e is the effective mass at $\boldsymbol{k} = 0$, ε the energy relative to the conduction band edge, and α the nonparabolicity parameter. Find an expression for the density of states in the conduction band.

The dispersion relation of the conduction band is given by $\varepsilon_c(\boldsymbol{k}) = \varepsilon_c + \varepsilon_{\boldsymbol{k}}$ with $\varepsilon_{\boldsymbol{k}}$ to be solved from $\hbar^2\boldsymbol{k}^2/2m_e = \varepsilon_{\boldsymbol{k}}(1+\alpha\,\varepsilon_{\boldsymbol{k}})$. Note that $\varepsilon_{\boldsymbol{k}}$ is independent of the direction of $\boldsymbol{k}$. According to the definition of the density of states, we have

$$g_c(\varepsilon) = \frac{2}{\mathscr{V}}\sum_{\boldsymbol{k}}\delta\big(\varepsilon-\varepsilon_c(\boldsymbol{k})\big) = \frac{1}{\pi^2}\int \mathrm{d}k\ k^2\delta\big(\varepsilon-\varepsilon_c-\varepsilon_{\boldsymbol{k}}\big).$$

We make a change of integration variables from k to $\varepsilon_{\boldsymbol{k}}$ and have

$$k = \big[(2m_e/\hbar^2)\varepsilon_{\boldsymbol{k}}(1+\alpha\varepsilon_{\boldsymbol{k}})\big]^{1/2},\ k\mathrm{d}k = \frac{m_e}{\hbar^2}\big(1+2\alpha\varepsilon_{\boldsymbol{k}}\big)\mathrm{d}\varepsilon_{\boldsymbol{k}}.$$

The integral can be then easily evaluated

$$g_c(\varepsilon) = \frac{1}{2\pi^2}\left(\frac{2m_e}{\hbar^2}\right)^{3/2}\int \mathrm{d}\varepsilon_{\boldsymbol{k}}\ \big(1+2\alpha\varepsilon_{\boldsymbol{k}}\big)\big[\varepsilon_{\boldsymbol{k}}(1+\alpha\varepsilon_{\boldsymbol{k}})\big]^{1/2} \times \delta\big(\varepsilon-\varepsilon_c-\varepsilon_{\boldsymbol{k}}\big)$$
$$= \frac{1}{2\pi^2}\left(\frac{2m_e}{\hbar^2}\right)^{3/2}\big[1+2\alpha(\varepsilon-\varepsilon_c)\big]\Big\{(\varepsilon-\varepsilon_c)\big[1+\alpha(\varepsilon-\varepsilon_c)\big]\Big\}^{1/2} \times \theta(\varepsilon-\varepsilon_c).$$

20-6 Electron-hole pair excitations. A semiconductor with band gap E_g is exposed to photons of energy $h\nu$. If $h\nu > E_g$, electron-hole pairs are created through direct excitation of electrons from the valence band. Determine the wave vector $\boldsymbol{k}$ and the energy of the electrons and holes. Perform the numerical computation for InSb with $E_g = 0.23$ eV, the effective mass of the electrons $m_e = 0.014m$, that of the holes $m_h = 0.40m$, and $h\nu = 0.50$ eV. Sketch the bands and the transitions.

The energies of the conduction and valence bands are given by

$$\varepsilon_c(\boldsymbol{k}) = \varepsilon_c + \frac{\hbar^2 k^2}{2m_e}, \quad \varepsilon_v(\boldsymbol{k}) = \varepsilon_v - \frac{\hbar^2 k^2}{2m_h}.$$

From the conservation of energy, we have

$$h\nu = \varepsilon_c(\boldsymbol{k}) - \varepsilon_v(\boldsymbol{k}) = \varepsilon_c - \varepsilon_v + \frac{(m_e + m_h)\hbar^2 k^2}{2m_e m_h} = E_g + \frac{(m_e + m_h)\hbar^2 k^2}{2m_e m_h}.$$

Therefore,

$$k = \left[\frac{2(h\nu - E_g)m_e m_h}{\hbar^2(m_e + m_h)}\right]^{1/2}.$$

The energies of the electron and hole are given by

$$\varepsilon_e = \varepsilon_c + \frac{\hbar^2 k^2}{2m_e} = \varepsilon_c + \frac{m_h}{m_e + m_h}(h\nu - E_g),$$

$$\varepsilon_h = -\varepsilon_v + \frac{\hbar^2 k^2}{2m_h} = -\varepsilon_v + \frac{m_e}{m_e + m_h}(h\nu - E_g).$$

Note that $\varepsilon_e + \varepsilon_h = h\nu$. The hole band is the inverted valence band of electrons.

For InSb and $h\nu = 0.50$ eV, we have

$$k = \left[\frac{2 \times (0.5 - 0.23) \times 1.602 \times 10^{-19}}{1.055^2 \times 10^{-68}} \times \frac{0.014 \times 0.4 \times 9.11 \times 10^{-31}}{(0.014 + 0.4)}\right]^{1/2}$$

$$\approx 3.095 \times 10^{10}\ \mathrm{m}^{-1},$$

$$\varepsilon_e = \varepsilon_c + \frac{0.4}{0.014 + 0.4} \times (0.5 - 0.23)$$

$$\approx \varepsilon_c + 0.26\ \mathrm{eV},$$

$$\varepsilon_h = -\varepsilon_v + \frac{0.014}{0.014 + 0.4} \times (0.5 - 0.23)$$

$$\approx -\varepsilon_v + 0.01\ \mathrm{eV}.$$

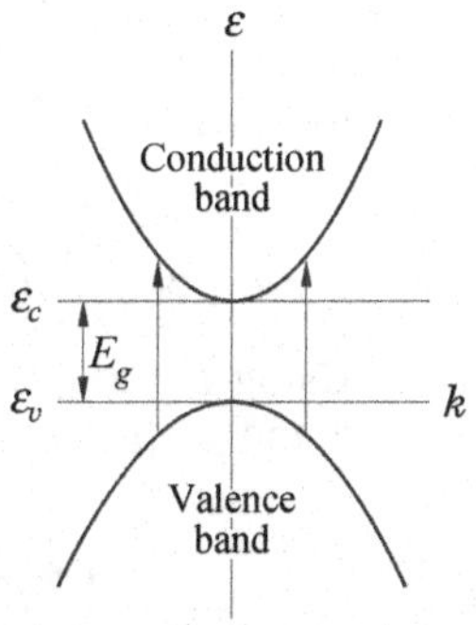

Fig. 20.1 Schematic illustration of bands and direct transitions for a direct-band-gap semiconductor.

For the sketch of the bands and the transitions, see Fig. 20-1.

20-7 Chemical potential of an intrinsic semiconductor. For a particular semiconductor with energy gap $E_g = 0.5$ eV, it is valid to assume that it is an intrinsic semiconductor in the temperature range from 200 to 300 K. How is the chemical potential shifted by this change in temperature? Assume that $m_h = 2m_e$ and the band gap E_g is independent of temperature.

From

$$\mu = \frac{1}{2}(\varepsilon_v + \varepsilon_c) + \frac{3}{4}k_{\rm B}T \ln \frac{m_h}{m_e},$$

we know that when the temperature is raised from T_1 to T_2, the chemical potential μ increases by the mount given by

$$\Delta\mu = \frac{3}{4}k_{\rm B}(T_2 - T_1) \ln \frac{m_h}{m_e}.$$

Making use of the given numerical values and the relation between electron and hole effective masses, we have

$$\Delta\mu = \frac{3}{4} \times 1.381 \times 10^{-23} \times 100 \times \ln 2 \approx 4.48 \times 10^{-3} \text{ eV}.$$

20-8 As-doped silicon crystal. A silicon crystal is doped with 5×10^{20} As atoms/m^3. The donor level lies 0.05 eV from the edge of the conduction band and the band gap is 1.14 eV. Compute the position of the Fermi level at 200 K and the number of unionized arsenic atoms. Assume that $m_e = m_h = m$.

For

$$N_c = 2\left(m_e k_{\rm B}T/2\pi\hbar^2\right)^{3/2},$$

we have

$$N_c = 2\left[9.11\times 10^{-31}\times 1.381\times 10^{-23}\times 200\times (2\times\pi\times 1.055^2\times 10^{-68})^{-1}\right]^{3/2}$$
$$\approx 1.365 \times 10^{25} \text{ m}^{-3}.$$

Using $E_g/2k_{\rm B}T = 1.14 \times 1.602 \times 10^{-19}/(2 \times 1.381 \times 10^{-23} \times 200) \approx 3.306 \times 10^{-3}$ and

$$n_i = 2\left(\frac{k_{\rm B}T}{2\pi\hbar^2}\right)^{3/2}\left(m_e m_h\right)^{3/4} \mathrm{e}^{-E_g/2k_{\rm B}T}$$
$$= 2\left(\frac{mk_{\rm B}T}{2\pi\hbar^2}\right)^{3/2} \mathrm{e}^{-E_g/2k_{\rm B}T} = N_c \mathrm{e}^{-E_g/2k_{\rm B}T},$$

we have $n_i = 1.365 \times 10^{25} \times \mathrm{e}^{-33.06} \approx 5.984 \times 10^{10}$ m^{-3}.

For the concentration of electrons, we have two different expressions

$$n = N_c e^{(\mu-\varepsilon_c)/k_B T},$$

$$n = \frac{1}{2}\left[N_d^+ + \left(N_d^{+2} + 4n_i^2\right)^{1/2}\right]$$

with $N_d^+ = \dfrac{N_d}{g_d e^{(\mu-E_d)/k_B T}+1}$. Equating the two expressions for n, setting $x = e^{(\mu-\varepsilon_c)/k_B T}$, and rearranging, we obtain

$$x^2 - (N_d^+/N_c)x - e^{-E_g/k_B T} = 0.$$

Inserting $N_d^+ = N_d/(\alpha x+1)$ with $\alpha = g_d e^{(\varepsilon_c-E_d)/k_B T}$ into the above equation, we have

$$x^2 - \frac{N_d}{N_c}\frac{x}{\alpha x+1} - e^{-E_g/k_B T} = 0.$$

In consideration that $N_d/N_c \approx 3.663\times 10^{-5}$, $\alpha \approx 36.351$, and $e^{-E_g/k_B T} \approx 1.922\times 10^{-29}$, we see that the solution of x to the above equation is of order 10^{-5}. Therefore, we can neglect $e^{-E_g/k_B T}$ and obtain

$$\alpha x^2 + x - N_d/N_c = 0.$$

The positive solution to the above equation is

$$x_0 = \frac{1}{2\alpha}\left[-1 + (1+4\alpha N_d/N_c)^{1/2}\right] \approx 3.658\times 10^{-5}.$$

The chemical potential is then given by

$$\begin{aligned}\mu &= \varepsilon_c + k_B T \ln x_0 \\ &= \varepsilon_c + \frac{1.381\times 10^{-23}\times 200}{1.602\times 10^{-19}} \times (-5)\times \ln(3.658) \approx \varepsilon_c - 0.112\ \text{eV}.\end{aligned}$$

The percentage of unionized arsenic atoms is given by

$$\begin{aligned}\frac{N_d - N_d^+}{N_d} &= 1 - \frac{1}{\alpha x_0 + 1} = 1 - \frac{1}{366.351\times 3.658\times 10^{-5}+1} \\ &\approx 0.00133 = 0.13\%.\end{aligned}$$

Thus, only a tiny small portion of arsenic atoms are unionized.

20-9 Donors in indium antimonide. Indium antimonide has $E_g = 0.23$ eV, a relative dielectric constant of 18, and an electron effective mass of $m_e = 0.015m$.

(1) Compute the donor ionization energy and the radius of the ground-state orbit.

(2) Find the minimum donor concentration at which the orbits of adjacent impurity atoms overlap appreciably.

---

(1) The donor ionization energy is given by

$$E_d = \frac{m_e e^4}{2(4\pi\epsilon)^2\hbar^2} = \frac{m_e}{m\epsilon_r^2}\frac{\hbar^2}{2ma_0^2} = \frac{0.015}{18^2} \times 13.605$$
$$\approx 6.299 \times 10^{-4}\ \text{eV} = 0.629\,9\ \text{meV}.$$

The radius of the ground-state orbit is

$$r_d = \epsilon_r(m/m_e)a_0 = \frac{18}{0.015} \times 0.529 \times 10^{-10}$$
$$\approx 6.348 \times 10^{-8}\ \text{m} \approx 0.635\ \text{nm}.$$

(2) For the orbits of adjacent impurity atoms to overlap, there must be at least one impurity in the volume $4\pi r_d^3/3$. Thus, the minimum donor concentration must be

$$N_d^{\min} = \frac{1}{4\pi r_d^3/3} = \frac{3}{4\pi \times 6.348^3 \times 10^{-24}} \approx 9.33 \times 10^{20}\ \text{m}^{-3}.$$

20-10 Band gap of InSb. A measurement of the voltage drop U across an InSb sample was made as a function of temperature T with a constant current to determine its band gap, with the measured values of U at several different temperatures given below. During the measurement, the sample was in the temperature range for intrinsic conduction.

T [K]	237.7	245.5	255.5	268.1	275.7	283.7	290.6
U [mV]	6.817	5.862	4.821	3.827	3.358	2.91	2.61

Compute the band gap under the assumption that the variation in the conductivity is partly caused by the increase in the density of charge carriers and partly arises from the mobility μ of the charge carriers being temperature dependent because of the thermal vibrations of atoms. Assume that $\mu \propto T^{-3/2}$ in the relevant temperature range.

For a constant current, we have $U \propto R \propto \rho \propto 1/\sigma$. For $\mu = \mu_e + \mu_h \propto T^{-3/2}$, we have

$$\sigma = 2e(\mu_e + \mu_h)\left(\frac{k_B T}{2\pi\hbar^2}\right)^{3/2}(m_e m_h)^{3/4} e^{-E_g/2k_B T} = C e^{-E_g/2k_B T},$$

where C is a constant independent of T. Hence, $U = U_0 e^{E_g/2k_B T}$ with U_0 independent of T. Taking the logarithm of U, we have

$$\ln U = \ln U_0 + \frac{E_g}{2k_B T}.$$

We see that $\ln U$ is a linear function of the inverse of temperature $(1/T)$, with the slope determined by the band gap. The measured values of $\ln U$ are plotted as a function $1/T$ in Fig. 20-2 together with a linear fitting. From the figure, we see that the experimental data points fall very well on a straight line. To find the band gap E_g, we now fit the data points to a straight line.

Let $\ln U = a + b/T$ with a and b to be determined. To get a best fit, we minimize the sum of the squares of the deviations of the fitted values from the experimental values. For convenience, we set $x = 1/T$ and $y = \ln U$ and use x_i and y_i to denote the corresponding experimental data. The fitting function now reads $y = a + bx$. The sum of the squares of the deviations is given by

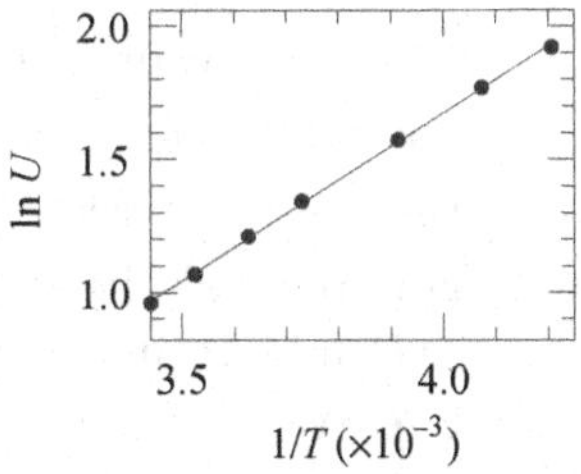

Fig. 20.2 Plot of $\ln U$ as a function of $1/T$. The solid circles represent experimental data and the straight line is a linear fit to the experimental data.

$$\chi^2 = \sum_i (y_i - a - bx_i)^2.$$

To minimize χ^2, we differentiate it with respect to a and b and set the results to zero

$$0 = \frac{1}{2}\frac{\partial \chi^2}{\partial a} = -\sum_i (y_i - a - bx_i) = na + b\sum_i x_i - \sum_i y_i,$$

$$0 = \frac{1}{2}\frac{\partial \chi^2}{\partial b} = -\sum_i x_i(y_i - a - bx_i) = a\sum_i x_i + b\sum_i x_i^2 - \sum_i x_i y_i.$$

Here n is the number of data points. For the present problem, $n = 7$. Solving for a and b from the above two equations, we obtain

$$a = \frac{\sum_i x_i^2 \sum_i y_i - \sum_i x_i \sum_i x_i y_i}{n\sum_i x_i^2 - (\sum_i x_i)^2},$$

$$b = \frac{n\sum_i x_i y_i - \sum_i x_i \sum_i y_i}{n\sum_i x_i^2 - (\sum_i x_i)^2}.$$

We list the values for the sums appearing in the expressions for a and b in the following table.

$\sum_i x_i$	$\sum_i x_i^2$	$\sum_i x_i y_i$	$\sum_i y_i$
0.026 517 3	0.000 100 944	0.037 901	9.841 82

From the above results, we obtain the following values for a and b

$$a = \frac{0.000\,100\,944 \times 9.841\,82 - 0.026\,517\,3 \times 0.037\,901}{7 \times 0.000\,100\,944 - 0.026\,517\,3^2} \approx -3.359\,37,$$
$$b = \frac{7 \times 0.037\,901 - 0.026\,517\,3 \times 9.841\,82}{7 \times 0.000\,100\,944 - 0.026\,517\,3^2} \approx 1,257.95.$$

From the value of b, we obtain the value of E_g

$$E_g = 2k_B b = \frac{2 \times 1.381 \times 10^{-23} \times 1,257.95}{1.602 \times 10^{-19}} \approx 0.217 \text{ eV}.$$

The above value is in consistency with the known values 0.23 eV at 0 K and 0.17 eV at 300 K. The fitting result $\ln U = -3.359\,37 + 1,257.95/T$ is plotted in Fig. 20-2 as the straight line.

20-11 Fermi levels in InP. Compute the position of the intrinsic Fermi level in InP at room temperature at which the band gap is $E_g = 1.26$ eV. Where does the Fermi level lie at $T = 0$ K if E_g is assumed to be temperature independent? The effective masses of electrons and holes are $m_e = 0.073m$ and $m_h = 0.2m$.

The Fermi level at the room temperature (300 K) is given by

$$\begin{aligned}\mu &= \frac{1}{2}(\varepsilon_c + \varepsilon_v) + \frac{3}{4}k_B T \ln \frac{m_h}{m_e} \\ &= \frac{1}{2}(\varepsilon_c + \varepsilon_v) + \frac{3}{4} \times 8.617\,343 \times 10^{-5} \times 300 \times \ln \frac{0.2}{0.073} \\ &\approx \frac{1}{2}(\varepsilon_c + \varepsilon_v) + 0.019\,5 \text{ eV}.\end{aligned}$$

Thus, the Fermi level is 0.019 5 eV above the middle of the band gap. At $T = 0$ K, the Fermi level is at the middle of the band gap and is given by

$$\mu = \varepsilon_F = \frac{1}{2}(\varepsilon_c + \varepsilon_v).$$

20-12 Sb-doped silicon crystal. A silicon crystal is doped with Sb atoms to a concentration of 10^{21} m^{-3}. The impurity level lies 0.04 eV from the nearest band edge. Assume that the band gap at 450 K is $E_g = 1.14$ eV. Compute the conductivity at $T = 450$ K. Approximate the effective masses with the free electron mass and assume that all impurity centers are ionized at room temperature. The mobilities are $\mu_e = 0.13\ \mathrm{m^2/Vs}$ and $\mu_h = 0.05\ \mathrm{m^2/Vs}$.

Sb-doped Si is an n-type semiconductor. For $m_e = m_h = m$, we have

$$\begin{aligned} n_i = p_i &= 2\left(\frac{mk_{\mathrm{B}}T}{2\pi\hbar^2}\right)^{3/2} \mathrm{e}^{-E_g/2k_{\mathrm{B}}T} \\ &= 2 \times \left(\frac{9.109\,382\,15 \times 10^{-31} \times 1.380\,650\,4 \times 10^{-23} \times 450}{2 \times \pi \times 1.054\,571\,628^2 \times 10^{-68}}\right)^{3/2} \\ &\quad \times \mathrm{e}^{-1.14/(2\times 8.617\,343\times 10^{-5}\times 450)} \\ &\approx 1.905\,5 \times 10^{19}\ \mathrm{m^{-3}}. \end{aligned}$$

Since $N_d^+ = N_d$, the concentration of electrons is given by

$$\begin{aligned} n &= \frac{1}{2}\left[N_d + (N_d^2 + 4n_i^2)^{1/2}\right] \\ &= \frac{1}{2}\left[10^{21} + (10^{42} + 4 \times 1.905\,5^2 \times 10^{38})^{1/2}\right] \\ &\approx 1.000\,36 \times 10^{21}\ \mathrm{m^{-3}}. \end{aligned}$$

Note that $n \approx N_d$. The conductivity is given by

$$\begin{aligned} \sigma &= (n\mu_e + p_i\mu_h)e \\ &= \left(1.000\,36 \times 10^{21} \times 0.13 + 1.905\,5 \times 10^{19} \times 0.05\right) \times 1.602\,176 \times 10^{-19} \\ &\approx 20.988\ \mathrm{S/m}, \end{aligned}$$

where S stands for Siemens. Note that $1\ \mathrm{S/m} = 1\ \mathrm{mho/m} = 1\ \Omega^{-1}\cdot\mathrm{m}^{-1}$.

20-13 As-doped germanium. A germanium crystal is doped with 1.0×10^{22} As atoms/m^3, which gives a donor level 13 meV below the conduction band edge. Give the Fermi level and the conductivity at room temperature. How would the Fermi level and the conductivity be influenced (qualitatively) if the crystal were also doped with Al atoms? The Al atoms give rise to an acceptor level 10 meV above the valence band edge. For the crystal, we have $E_g = 0.70$ eV,

$m_e = 0.22m$, $m_h = 0.30m$, $\mu_e = 0.36\ \mathrm{m^2/Vs}$, and $\mu_h = 0.16\ \mathrm{m^2/Vs}$.

N_c is given by

$$\begin{aligned} N_c &= 2\left(\frac{m_e k_B T}{2\pi\hbar^2}\right)^{3/2} \\ &= 2 \times \left(\frac{0.22 \times 9.109\,38 \times 10^{-31} \times 1.380\,65 \times 10^{-23} \times 300}{2 \times \pi \times 1.054\,57^2 \times 10^{-68}}\right)^{3/2} \\ &\approx 2.589 \times 10^{24}\ \mathrm{m^{-3}}. \end{aligned}$$

Assume that $n \approx N_d^+ \approx N_d$. The Fermi level is then given by

$$\begin{aligned} \mu &= \varepsilon_c - k_B T \ln \frac{N_c}{N_d} = \varepsilon_c - k_B T \ln \frac{N_c}{N_d} \\ &= \varepsilon_c - 8.617\,343 \times 10^{-5} \times 300 \times \ln \frac{2.589 \times 10^{24}}{10^{22}} \\ &\approx \varepsilon_c - 0.144\ \mathrm{eV}. \end{aligned}$$

Thus, the Fermi level is 0.144 eV below the conduction band edge. Since acceptors lower the Fermi level while donors raise it, the Fermi level would have been lowered from $\varepsilon_c - 0.144$ eV if the crystal were also doped with Al atoms.

20-14 Concentrations of electrons and holes in a *p*-type semiconductor. Compute the concentrations of electrons and holes in a *p*-type semiconductor if the conductivity is $\sigma = 10\ (\Omega{\cdot}\mathrm{m})^{-1}$, the mobilities are $\mu_e = 0.4\ \mathrm{m^2/Vs}$ and $\mu_h = 0.2\ \mathrm{m^2/Vs}$, and the concentration of intrinsic carriers is $n_i = 2.2 \times 10^{19}\ \mathrm{m^{-3}}$.

Since the semiconductor is a *p*-type semiconductor, the concentration of electrons is given by $n = n_i$. From $\sigma = ne\mu_e + pe\mu_h$ and $n = n_i$, we have

$$\begin{aligned} p &= \frac{\sigma - n_i e \mu_e}{e\mu_h} = \frac{\sigma/e - n_i \mu_e}{\mu_h} \\ &= \frac{10/(1.602\,176\,487 \times 10^{-19}) - 2.2 \times 10^{19} \times 0.4}{0.2} \\ &\approx 2.681 \times 10^{20}\ \mathrm{m^{-3}}. \end{aligned}$$

20-15 Maximum concentration of donor atoms for intrinsic conduction. Determine the maximum concentration (atomic fraction) of donor atoms that can be allowed in the substances in the table below for intrinsic conduction to dominate at room temperature. Is it possible to realize this practically in any of these cases?

Material	Band gap [eV]	Density [$\times 10^3$ kg/m^3]
Diamond	5.33	3.51
Silicon	1.14	2.42
Germanium	0.62	5.35

We first compute the number $\mathscr{N}$ of atoms per unit volume in each material through $\mathscr{N} = \rho_m/A\text{u}$, where A is the atomic weight in atomic mass unit and u is the atomic mass unit. For diamond, we have

$$\mathscr{N}_{\text{diamond}} = \frac{3.51 \times 10^3}{12.011 \times 1.660\,538\,86 \times 10^{-27}} \approx 1.765 \times 10^{29}\ \text{m}^{-3}.$$

For Si, we have

$$\mathscr{N}_{\text{Si}} = \frac{2.42 \times 10^3}{28.085\,5 \times 1.660\,538\,86 \times 10^{-27}} \approx 5.189 \times 10^{28}\ \text{m}^{-3}.$$

For Ge, we have

$$\mathscr{N}_{\text{Ge}} = \frac{5.35 \times 10^3}{72.59 \times 1.660\,538\,86 \times 10^{-27}} \approx 4.438 \times 10^{28}\ \text{m}^{-3}.$$

We now evaluate the value of n_i for each material. For diamond, with the effective mass taken as the free-electron mass, we have

$$\begin{aligned} n_i^{\text{diamond}} &= 2\left(\frac{mk_\text{B}T}{2\pi\hbar^2}\right)^{3/2} \text{e}^{-E_g/2k_\text{B}T} \\ &= 2 \times \left(\frac{9.109\,382 \times 10^{-31} \times 1.380\,650 \times 10^{-23} \times 300}{2 \times \pi \times 1.054\,571^2 \times 10^{-68}}\right)^{3/2} \\ &\quad \times \text{e}^{-5.33/(2\times 8.617\,343\times 10^{-5}\times 300)} \\ &\approx 4.262 \times 10^{-20}\ \text{m}^{-3}. \end{aligned}$$

For Si, we have

$$\begin{aligned} n_i^{\text{Si}} &= 2\left(\frac{m_d k_\text{B}T}{2\pi\hbar^2}\right)^{3/2} \text{e}^{-E_g/2k_\text{B}T} \\ &= 2 \times \left(\frac{1.080 \times 9.109\,382 \times 10^{-31} \times 1.380\,650 \times 10^{-23} \times 300}{2 \times \pi \times 1.054\,571^2 \times 10^{-68}}\right)^{3/2} \\ &\quad \times \text{e}^{-1.14/(2\times 8.617\,343\times 10^{-5}\times 300)} \\ &\approx 7.484 \times 10^{15}\ \text{m}^{-3}. \end{aligned}$$

For Ge, we have

$$n_i^{\rm Ge} = 2\left(\frac{m_d k_{\rm B} T}{2\pi\hbar^2}\right)^{3/2} {\rm e}^{-E_g/2k_{\rm B}T}$$
$$= 2 \times \left(\frac{0.560 \times 9.109\,382 \times 10^{-31} \times 1.380\,650 \times 10^{-23} \times 300}{2 \times \pi \times 1.054\,571^2 \times 10^{-68}}\right)^{3/2}$$
$$\times\, {\rm e}^{-0.62/(2\times 8.617\,343\times 10^{-5}\times 300)}$$
$$\approx 6.518 \times 10^{19}\ {\rm m}^{-3}.$$

We use $n = 2n_i$ to determine the maximum concentration (atomic fraction) of donor atoms that can be allowed in a material for intrinsic conduction to dominate at room temperature. From $n = N_d/2 + (N_d^2/4 + n_i^2)^{1/2}$, we find that, for $n = 2n_i$, N_d is given by $N_d = 3n_i/2$. Let f denote the atomic fraction of donor atoms, $f = N_d/\mathscr{N}$. We have

$$f_{\rm diamond} = \frac{N_d^{\rm diamond}}{\mathscr{N}_{\rm diamond}} = \frac{3n_i^{\rm diamond}}{2\mathscr{N}_{\rm diamond}} = \frac{3 \times 4.262 \times 10^{-20}}{2 \times 1.765 \times 10^{29}}$$
$$\approx 3.622 \times 10^{-49},$$
$$f_{\rm Si} = \frac{N_d^{\rm Si}}{\mathscr{N}_{\rm Si}} = \frac{3n_i^{\rm Si}}{2\mathscr{N}_{\rm Si}} = \frac{3 \times 7.484 \times 10^{15}}{2 \times 5.189 \times 10^{28}} \approx 2.163 \times 10^{-13},$$
$$f_{\rm Ge} = \frac{N_d^{\rm Ge}}{\mathscr{N}_{\rm Ge}} = \frac{3n_i^{\rm Ge}}{2\mathscr{N}_{\rm Ge}} = \frac{3 \times 6.518 \times 10^{19}}{2 \times 4.438 \times 10^{28}} \approx 2.203 \times 10^{-9}.$$

Since the maximum allowed doping levels are all exceedingly low, the materials at these doping levels are essentially pure materials. Thus, this can not be realized practically.

20-16 Conductivity and Fermi level of GaAs. At 300 K, GaAs has a band gap of 1.43 eV, an electron mobility of 1.6 m^2/Vs, a hole mobility of 0.4 m^2/Vs, an electron effective mass of $0.065m$, and a hole effective mass of $0.45m$. All the following questions are for the temperature of 300 K.

(1) What is the intrinsic conductivity? Where is the Fermi level?

(2) In a sample containing only 10^{22} m^{-3} totally ionized donors, where is the Fermi level?

(3) In a sample containing both 10^{22} m^{-3} totally ionized donors and 2×10^{21} m^{-3} totally ionized acceptors, what is the conductivity and where is the Fermi level?

---

(1) The concentrations of electrons and holes for intrinsic conduction are given by

$$n_i = p_i = 2\left(\frac{k_B T}{2\pi\hbar^2}\right)^{3/2} (m_e m_h)^{3/4} e^{-E_g/2k_B T}$$
$$= 2 \times \left(\frac{1.380\,650 \times 10^{-23} \times 300}{2 \times \pi \times 1.054\,571^2 \times 10^{-68}}\right)^{3/2} \times (0.065 \times 0.45)^{3/4}$$
$$\times (9.109\,382\,15 \times 10^{-31})^{3/2} \times e^{-1.43/(2\times 8.617\,343\times 10^{-5}\times 300)}$$
$$\approx 1.728 \times 10^{12}\ \mathrm{m^{-3}}.$$

The intrinsic conductivity is given by

$$\sigma_i = n_i e(\mu_e + \mu_h)$$
$$= 1.728 \times 10^{12} \times 1.602\,176\,487 \times 10^{-19} \times (1.6 + 0.4)$$
$$\approx 5.539 \times 10^{-7}\ \mathrm{S/m}.$$

The chemical potential is given by

$$\mu = \frac{1}{2}(\varepsilon_v + \varepsilon_c) + \frac{3}{4} k_B T \ln \frac{m_h}{m_e}$$
$$= \frac{1}{2}(\varepsilon_v + \varepsilon_c) + \frac{3}{4} \times 8.617\,343 \times 10^{-5} \times 300 \times \ln \frac{0.45}{0.065}$$
$$\approx \frac{1}{2}(\varepsilon_v + \varepsilon_c) + 0.038\ \mathrm{eV}.$$

Thus, the Fermi level is 0.037 5 eV above the middle of the band gap.

(2) N_c is given by

$$N_c = 2\left(\frac{m_e k_B T}{2\pi\hbar^2}\right)^{3/2}$$
$$= 2 \times \left(\frac{0.065 \times 9.109\,38 \times 10^{-31} \times 1.380\,65 \times 10^{-23} \times 300}{2 \times \pi \times 1.054\,57^2 \times 10^{-68}}\right)^{3/2}$$
$$\approx 4.159 \times 10^{23}\ \mathrm{m^{-3}}.$$

The Fermi level is given by

$$\mu = \varepsilon_c - k_B T \ln \frac{N_c}{N_d}$$
$$= \varepsilon_c - 8.617\,343 \times 10^{-5} \times 300 \times \ln \frac{4.159 \times 10^{23}}{10^{22}}$$
$$\approx \varepsilon_c - 0.096\ \mathrm{eV}.$$

Thus, the Fermi level is 0.096 eV below the conduction band edge.

(3) Since $N_d \gg n_i$ and $N_d \gg p_i$, we have $n \approx N_d$ and $p \approx N_a$. Thus, the conductivity is given by

$$\begin{aligned}\sigma &= ne\mu_e + pe\mu_h = (n\mu_e + p\mu_h)e \\ &= (10^{22} \times 1.6 + 2 \times 10^{21} \times 0.4) \times 1.602\,176\,487 \times 10^{-19} \\ &\approx 2,691.656 \text{ S/m}.\end{aligned}$$

The value of N_v is given by

$$\begin{aligned}N_v &= 2\left(\frac{m_h k_B T}{2\pi\hbar^2}\right)^{3/2} \\ &= 2 \times \left(\frac{0.45 \times 9.109\,38 \times 10^{-31} \times 1.380\,65 \times 10^{-23} \times 300}{2 \times \pi \times 1.054\,57^2 \times 10^{-68}}\right)^{3/2} \\ &\approx 7.575 \times 10^{24} \text{ m}^{-3}.\end{aligned}$$

Dividing the equations

$$\begin{aligned}n &= N_c e^{(\mu-\varepsilon_c)/k_B T}, \\ p &= N_v e^{(\varepsilon_v-\mu)/k_B T},\end{aligned}$$

we obtain

$$\frac{n}{p} = \frac{N_c}{N_v} e^{(2\mu-\varepsilon_c-\varepsilon_v)/k_B T}.$$

Solving for μ and making use of $n \approx N_d$ and $p \approx N_a$, we obtain

$$\mu = \frac{1}{2}(\varepsilon_c + \varepsilon_v) + \frac{1}{2}k_B T \ln \frac{N_d N_v}{N_a N_c}.$$

Note that the above result is the average of the following two expressions

$$\begin{aligned}\mu &= \varepsilon_c - k_B T \ln \frac{N_c}{N_d}, \\ \mu &= \varepsilon_v + k_B T \ln \frac{N_v}{N_a}.\end{aligned}$$

Evaluating μ, we have

$$\begin{aligned}\mu &= \frac{1}{2}(\varepsilon_c + \varepsilon_v) \\ &\quad + \frac{1}{2} \times 8.617\,343 \times 10^{-5} \times 300 \times \ln \frac{10^{22} \times 7.575 \times 10^{24}}{2 \times 10^{21} \times 4.159 \times 10^{23}} \\ &\approx \frac{1}{2}(\varepsilon_c + \varepsilon_v) + 0.058.\end{aligned}$$

Thus, the Fermi level is 0.058 eV above the middle of the band gap.

Chapter 21

Density Functional Theory

(1) *Pair density operator*

$$\hat{n}(\boldsymbol{r}, \boldsymbol{r}') = \sum_{i \neq j=1}^{N} \delta(\boldsymbol{r} - \boldsymbol{r}_i)\delta(\boldsymbol{r}' - \boldsymbol{r}_j).$$

(2) *Pair density* $n_{\sigma\sigma'}(\boldsymbol{r}, \boldsymbol{r}')$
The conditional density is denoted by $n_{\sigma'}(\boldsymbol{r}'|\boldsymbol{r}\sigma)$.
In terms of $n_{\sigma'}(\boldsymbol{r}'|\boldsymbol{r}\sigma)$, the pair density is given by

$$n_{\sigma\sigma'}(\boldsymbol{r}, \boldsymbol{r}') = n_{\sigma}(\boldsymbol{r}) n_{\sigma'}(\boldsymbol{r}'|\boldsymbol{r}\sigma).$$

(3) *Pair correlation function* $g_{\sigma\sigma'}(\boldsymbol{r}, \boldsymbol{r}')$

$$n_{\sigma\sigma'}(\boldsymbol{r}, \boldsymbol{r}') = n_{\sigma}(\boldsymbol{r}) g_{\sigma\sigma'}(\boldsymbol{r}, \boldsymbol{r}') n_{\sigma'}(\boldsymbol{r}').$$

(4) *Exchange-correlation hole*

$$n_{\sigma'}(\boldsymbol{r}'|\boldsymbol{r}\sigma) = n_{\sigma'}(\boldsymbol{r}') + n_{\mathrm{xc}}^{\sigma\sigma'}(\boldsymbol{r}, \boldsymbol{r}').,$$

$$n_{\mathrm{xc}}^{\sigma\sigma'}(\boldsymbol{r}, \boldsymbol{r}') = \left[g_{\sigma\sigma'}(\boldsymbol{r}, \boldsymbol{r}') - 1\right] n_{\sigma'}(\boldsymbol{r}').$$

(5) *Hartree energy functional*

$$E_{\mathrm{H}}[n] = \sum_{\sigma\sigma'} \int \mathrm{d}\boldsymbol{r} \mathrm{d}\boldsymbol{r}' \, \frac{e^2 n_{\sigma}(\boldsymbol{r}) n_{\sigma'}(\boldsymbol{r}')}{2(4\pi\epsilon_0)|\boldsymbol{r} - \boldsymbol{r}'|}.$$

(6) *Exchange-correlation energy functional*

$$E_{\mathrm{xc}}[n] = \sum_{\sigma\sigma'} \int \mathrm{d}\boldsymbol{r} \mathrm{d}\boldsymbol{r}' \frac{e^2 n_{\sigma}(\boldsymbol{r}) n_{\mathrm{xc}}^{\sigma\sigma'}(\boldsymbol{r}, \boldsymbol{r}')}{2(4\pi\epsilon_0)|\boldsymbol{r} - \boldsymbol{r}'|}.$$

(7) *Functional derivative*

$$\frac{\delta F[g(x)]}{\delta g(y)} = \lim_{\epsilon \to 0} \frac{1}{\epsilon} \Big\{ F[g(x) + \epsilon\delta(x - y)] - F[g(x)] \Big\}.$$

(8) *Thomas-Fermi kinetic energy functional*

$$T_{\rm TF}[n] = \kappa \int \mathrm{d}\boldsymbol{r}\; n^{5/3}(\boldsymbol{r}),$$

$$\kappa = 3\hbar^2(3\pi^2)^{2/3}/10m.$$

(9) *Thomas-Fermi energy functional*

$$E_{\rm TF}[n] = T_{\rm TF}[n] + \int \mathrm{d}\boldsymbol{r} v_{\rm ext}(\boldsymbol{r})n(\boldsymbol{r}) + \frac{1}{2}\frac{e^2}{4\pi\epsilon_0}\int \mathrm{d}\boldsymbol{r}\mathrm{d}\boldsymbol{r}'\frac{n(\boldsymbol{r})n(\boldsymbol{r}')}{|\boldsymbol{r}-\boldsymbol{r}'|}.$$

(10) *Thomas-Fermi equation*

$$\mu = \frac{5}{3}\kappa n^{2/3}(\boldsymbol{r}) + v_{\rm ext}(\boldsymbol{r}) + \frac{e^2}{4\pi\epsilon_0}\int \mathrm{d}\boldsymbol{r}'\;\frac{n(\boldsymbol{r}')}{|\boldsymbol{r}-\boldsymbol{r}'|}.$$

(11) *Hohenberg-Kohn Theorem I* For any electron system in an external potential $v_{\rm ext}(\boldsymbol{r})$, the potential is uniquely determined, except for an additive constant, by the ground-state electron density $n(\boldsymbol{r})$, and vice versa.

(12) *Hohenberg-Kohn Theorem II* A universal functional of the electron density $n(\boldsymbol{r})$ for the energy functional $E[n]$ can be defined for all electron systems. The exact ground-state energy is the global minimum for a given external potential $v_{\rm ext}(\boldsymbol{r})$, and the electron density $n(\boldsymbol{r})$ that minimizes the energy functional is the exact ground-state electron density.

(13) *Hohenberg-Kohn energy functional*

$$E_{\rm HK}[n] = F_{\rm HK}[n] + \int \mathrm{d}\boldsymbol{r}\; n(\boldsymbol{r})v_{\rm ext}(\boldsymbol{r}).$$

(14) *Universal functional*

$$F_{\rm HK}[n] = \frac{1}{2}\frac{e^2}{4\pi\epsilon_0}\int \mathrm{d}\boldsymbol{r}\mathrm{d}\boldsymbol{r}'\;\frac{n(\boldsymbol{r})n(\boldsymbol{r}')}{|\boldsymbol{r}-\boldsymbol{r}'|} + E_{\rm xc}[n].$$

(15) *Kohn-Sham ansatz*

$$n(\boldsymbol{r}) = \sum_{\sigma} n_\sigma(\boldsymbol{r}) = \sum_{i\sigma}|\psi_{i\sigma}(\boldsymbol{r})|^2.$$

(16) *Kohn-Sham energy functional*

$$E_{\rm KS} = \frac{\hbar^2}{2m}\sum_{i\sigma}\int \mathrm{d}\boldsymbol{r}\,|\boldsymbol{\nabla}\psi_{i\sigma}(\boldsymbol{r})|^2 + \int \mathrm{d}\boldsymbol{r}\, n(\boldsymbol{r})v_{\rm ext}(\boldsymbol{r}) + E_{\rm H}[n] + E_{\rm xc}[n].$$

(17) *Kohn-Sham potential*

$$V_{\rm KS}^{\sigma}(\boldsymbol{r}) = v_{\rm ext}(\boldsymbol{r}) + V_{\rm H}[n] + V_{\rm xc}^{\sigma}[n],$$
$$V_{\rm H}^{\sigma}[n] = \delta E_{\rm H}[n]/\delta n_{\sigma}(\boldsymbol{r}),$$
$$V_{\rm xc}^{\sigma}[n] = \delta E_{\rm xc}[n]/\delta n_{\sigma}(\boldsymbol{r}).$$

(18) *Kohn-Sham equations*

$$\left[-\frac{\hbar^2}{2m}\nabla^2 + V_{\rm KS}^{\sigma}(\boldsymbol{r})\right]\psi_{i\sigma}(\boldsymbol{r}) = \varepsilon_{i\sigma}\psi_{i\sigma}(\boldsymbol{r}).$$

(19) *Local density approximation*

$$\mathscr{E}_{\rm x}^{\rm LDA}(n(\boldsymbol{r})) = -2^{-1/3}\kappa_{\rm x} n^{4/3}(\boldsymbol{r}),$$
$$\kappa_{\rm x} = (3/2)(3/4\pi)^{1/3} \approx 0.9305 \approx 1.$$

(20) *Local spin density approximation*

$$\mathscr{E}_{\rm x}^{\rm LSDA}(n_{\uparrow}(\boldsymbol{r}), n_{\downarrow}(\boldsymbol{r})) = -\kappa_{\rm x}\sum_{\sigma} n_{\sigma}^{4/3}(\boldsymbol{r}).$$

(21) *Generalized gradient approximation*

$$E_{\rm xc}^{\rm GGA}[n_{\uparrow}, n_{\downarrow}, \boldsymbol{\nabla} n_{\uparrow}, \boldsymbol{\nabla} n_{\downarrow}] = \int {\rm d}\boldsymbol{r}\, \mathscr{E}_{\rm xc}^{\rm GGA}(n_{\uparrow}(\boldsymbol{r}), n_{\downarrow}(\boldsymbol{r}), \boldsymbol{\nabla} n_{\uparrow}(\boldsymbol{r}), \boldsymbol{\nabla} n_{\downarrow}(\boldsymbol{r})),$$

Becke88, PW91, mPW, PBE,

Meta-GGA functionals,

Hybrid-GGA functionals,

Hybrid-meta-GGA functionals.

21-1 Electron densities in the Hartree and Fock-Hartree models. The expressions for the electron density of a many-electron system in the Hartree and Hartree-Fock methods will be derived here.

(1) Derive the expression for the electron density of a many-electron system, $n^{\rm H}(\boldsymbol{r}) = \sum_{i=1}^{N} |\psi_i(\boldsymbol{r}_i)|^2$, in the Hartree method.

(2) Derive the expression for the electron density of a many-electron system, $n^{\rm HF}(\boldsymbol{r}) = \sum_{\lambda s_z} \left|\psi_{\lambda}(\boldsymbol{r}, s_z)\right|^2$, in the Hartree-Fock method.

The electron density $n(\boldsymbol{r})$ can be obtained from the expectation value of the electron density operator $\hat{n}(\boldsymbol{r}) = \sum_{i=1}^{N} \delta(\boldsymbol{r} - \boldsymbol{r}_i)$ in the ground-state wave function Ψ

$$n(\boldsymbol{r}) = \sum_{i=1}^{N} \oint \mathrm{d}x_1 \cdots \mathrm{d}x_i \cdots \mathrm{d}x_N \; \Psi^*(x_1, \cdots, x_i, \cdots, x_N)\delta(\boldsymbol{r} - \boldsymbol{r}_i)$$
$$\times \Psi(x_1, \cdots, x_i, \cdots, x_N)$$
$$= \sum_{i=1}^{N} \sum_{s_{zi}} \oint \mathrm{d}x_1 \cdots \mathrm{d}x_{i-1}\mathrm{d}x_{i+1} \cdots \mathrm{d}x_N \; \Psi^*(x_1, \cdots, \boldsymbol{r}s_{zi}, \cdots, x_N)$$
$$\times \Psi(x_1, \cdots, \boldsymbol{r}s_{zi}, \cdots, x_N)$$
$$= N \sum_{s_z} \oint \mathrm{d}x_2 \cdots \mathrm{d}x_N \; \left|\Psi(\boldsymbol{r}s_z, x_2, \cdots, x_N)\right|^2.$$

Note that the expression on the last line is valid only if $|\Psi|^2$ remains unchanged with respect to the exchange of two electrons. Thus, it is not applicable to the Hartree model.

(1) In the Hartree model, the ground-state wave function Ψ is written as a product of single-electron wave functions

$$\Psi(\boldsymbol{r}_1, \boldsymbol{r}_2, \cdots, \boldsymbol{r}_N) = \prod_i \psi_i(\boldsymbol{r}_i),$$

where the electron spins are implicitly taken into account. With the use of the above ground-state wave function, the electron density in the Hartree model is given by

$$n(\boldsymbol{r}) = \sum_{i=1}^{N} \int \mathrm{d}\boldsymbol{r}_1 \cdots \mathrm{d}\boldsymbol{r}_{i-1}\mathrm{d}\boldsymbol{r}_{i+1} \cdots \mathrm{d}\boldsymbol{r}_N$$
$$\times |\Psi(\boldsymbol{r}_1, \cdots, \boldsymbol{r}_{i-1}, \boldsymbol{r}, \boldsymbol{r}_{i+1}, \cdots, \boldsymbol{r}_N)|^2$$
$$= \sum_{i=1}^{N} |\psi_i(\boldsymbol{r})|^2 \int \mathrm{d}\boldsymbol{r}_1 \; |\psi_1(\boldsymbol{r}_1)|^2 \cdots \int \mathrm{d}\boldsymbol{r}_{i-1} \; |\psi_{i-1}(\boldsymbol{r}_{i-1})|^2$$
$$\times \int \mathrm{d}\boldsymbol{r}_{i+1} \; |\psi_{i+1}(\boldsymbol{r}_{i+1})|^2 \cdots \int \mathrm{d}\boldsymbol{r}_N \; |\psi_N(\boldsymbol{r}_N)|^2$$
$$= \sum_{i=1}^{N} \psi_i^*(\boldsymbol{r}_i)\psi_i(\boldsymbol{r}_i) = \sum_{i=1}^{N} \left|\psi_i(\boldsymbol{r}_i)\right|^2,$$

where the normalization of single-electron wave functions has been used.

(2) In the Hartree-Fock model, the ground-state wave function Ψ is written as a Slater determinant

$$\Psi(x_1, x_2, \cdots, x_N) = \frac{1}{\sqrt{N!}} \sum_P (-1)^P \psi_{\lambda_{P_1}}(x_1)\psi_{\lambda_{P_2}}(x_2) \cdots \psi_{\lambda_{P_N}}(x_N).$$

With the use of the above ground-state wave function, the electron density in the Hartree-Fock model is given by

$$n(\boldsymbol{r}) = \frac{1}{N!}\sum_{i=1}^{N}\sum_{s_{zi}}\sum_{PP'}(-1)^{P}(-1)^{P'}\oint dx_1\cdots dx_{i-1}dx_{i+1}\cdots dx_N$$
$$\times\,\psi^*_{\lambda_{P'_1}}(x_1)\cdots\psi^*_{\lambda_{P'_{i-1}}}(x_{i-1})\psi^*_{\lambda_{P'_i}}(\boldsymbol{r}s_{zi})\psi^*_{\lambda_{P'_{i+1}}}(x_{i+1})\cdots\psi^*_{\lambda_{P'_N}}(x_N)$$
$$\times\,\psi_{\lambda_{P_1}}(x_1)\cdots\psi_{\lambda_{P_{i-1}}}(x_{i-1})\psi_{\lambda_{P_i}}(\boldsymbol{r}s_{zi})\psi_{\lambda_{P_{i+1}}}(x_{i+1})\cdots\psi_{\lambda_{P_N}}(x_N),$$

where the permutation in Ψ^* has been denoted by P'. Grouping together $\psi^*_{\lambda_{P'_j}}(x_j)$ and $\psi_{\lambda_{P_j}}(x_j)$ and then making use of the orthonormality relation of single-electron wave functions, we have

$$n(\boldsymbol{r})$$
$$= \frac{1}{N!}\sum_{i=1}^{N}\sum_{s_{zi}}\sum_{PP'}(-1)^{P}(-1)^{P'}\psi^*_{\lambda_{P'_i}}(\boldsymbol{r}s_{zi})\psi_{\lambda_{P_i}}(\boldsymbol{r}s_{zi})$$
$$\times\oint dx_1\,\psi^*_{\lambda_{P'_1}}(x_1)\psi_{\lambda_{P_1}}(x_1)\cdots\oint dx_{i-1}\,\psi^*_{\lambda_{P'_{i-1}}}(x_{i-1})\psi_{\lambda_{P_{i-1}}}(x_{i-1})$$
$$\times\oint dx_{i+1}\,\psi^*_{\lambda_{P'_{i+1}}}(x_{i+1})\psi_{\lambda_{P_{i+1}}}(x_{i+1})\cdots\oint dx_N\,\psi^*_{\lambda_{P'_N}}(x_N)\psi_{\lambda_{P_N}}(x_N)$$
$$= \frac{1}{N!}\sum_{i=1}^{N}\sum_{s_{zi}}\sum_{PP'}(-1)^{P}(-1)^{P'}\psi^*_{\lambda_{P'_i}}(\boldsymbol{r}s_{zi})\psi_{\lambda_{P_i}}(\boldsymbol{r}s_{zi})\prod_{j\,(\neq i)}\delta_{\lambda_{P'_j}\lambda_{P_j}}.$$

The Kronecker δ-symbols in the above expression imply that the two permutations must be exactly identical for nonzero contributions. We thus have

$$n(\boldsymbol{r}) = \frac{1}{N!}\sum_{i=1}^{N}\sum_{s_{zi}}\sum_{P}\psi^*_{\lambda_{P_i}}(\boldsymbol{r}s_{zi})\psi_{\lambda_{P_i}}(\boldsymbol{r}s_{zi}).$$

All the $N!$ permutations yield the same result due to the summation over i. We thus have

$$n(\boldsymbol{r})=\sum_{\lambda s_z}\psi^*_{\lambda}(\boldsymbol{r}s_z)\psi_{\lambda}(\boldsymbol{r}s_z)=\sum_{\lambda s_z}\left|\psi_{\lambda}(\boldsymbol{r}s_z)\right|^2.$$

Inserting $\psi_\lambda(\boldsymbol{r}s_z) = \psi_i(\boldsymbol{r})\chi_\sigma(s_z)$ into the above equation and utilizing the normalization of spin wave functions, we have

$$n(\boldsymbol{r}) = \sum_{i\sigma}\psi^*_i(\boldsymbol{r})\psi_i(\boldsymbol{r})\sum_{s_z}\chi^*_\sigma(s_z)\chi_\sigma(s_z) = \sum_{i\sigma}\psi^*_i(\boldsymbol{r})\psi_i(\boldsymbol{r})$$
$$= 2\sum_{i=1}^{N}\psi^*_i(\boldsymbol{r})\psi_i(\boldsymbol{r}) = 2\sum_{i=1}^{N}\left|\psi_i(\boldsymbol{r})\right|^2.$$

21-2 Pair densities and correlation functions in the Hartree-Fock model. The pair density and correlation function in the Hartree-Fock model have been given in the text.

(1) Evaluate the summations over $\boldsymbol{k}$ and $\boldsymbol{k}'$ in

$$n_{\sigma\sigma}^{\mathrm{HF}}(\boldsymbol{r},\boldsymbol{r}') = \frac{1}{4}n^2 - \frac{1}{\mathscr{V}^2}\sum_{\boldsymbol{k}\boldsymbol{k}'}{}' \mathrm{e}^{-\mathrm{i}(\boldsymbol{k}-\boldsymbol{k}')\cdot(\boldsymbol{r}-\boldsymbol{r}')}$$

to obtain the result for the same-spin pair density.

(2) Compute numerically the same-spin pair density and the pair correlation function as functions of $k_{\mathrm{F}}|\boldsymbol{r}-\boldsymbol{r}'|$ and plot the pair correlation function.

---

(1) Since the exponential function $\mathrm{e}^{-\mathrm{i}(\boldsymbol{k}-\boldsymbol{k}')\cdot(\boldsymbol{r}-\boldsymbol{r}')}$ can be written as a product of two exponential functions, $\mathrm{e}^{-\mathrm{i}(\boldsymbol{k}-\boldsymbol{k}')\cdot(\boldsymbol{r}-\boldsymbol{r}')} = \mathrm{e}^{-\mathrm{i}\boldsymbol{k}\cdot(\boldsymbol{r}-\boldsymbol{r}')}\mathrm{e}^{\mathrm{i}\boldsymbol{k}'\cdot(\boldsymbol{r}-\boldsymbol{r}')}$, only one summation is actually needed to be performed. We have

$$\frac{1}{\mathscr{V}^2}\sum_{\boldsymbol{k}\boldsymbol{k}'}{}' \mathrm{e}^{-\mathrm{i}(\boldsymbol{k}-\boldsymbol{k}')\cdot(\boldsymbol{r}-\boldsymbol{r}')} = \left|\frac{1}{\mathscr{V}}\sum_{\boldsymbol{k}}{}' \mathrm{e}^{-\mathrm{i}\boldsymbol{k}\cdot(\boldsymbol{r}-\boldsymbol{r}')}\right|^2,$$

where we have made use of the fact that $\sum'_{\boldsymbol{k}} \mathrm{e}^{-\mathrm{i}\boldsymbol{k}\cdot(\boldsymbol{r}-\boldsymbol{r}')}$ and $\sum'_{\boldsymbol{k}'} \mathrm{e}^{\mathrm{i}\boldsymbol{k}'\cdot(\boldsymbol{r}-\boldsymbol{r}')}$ are mutually complex conjugates. Converting the summation over $\boldsymbol{k}$ into an integration over $\boldsymbol{k}$, we have

$$\begin{aligned}
&\frac{1}{\mathscr{V}}\sum_{\boldsymbol{k}}{}' \mathrm{e}^{-\mathrm{i}\boldsymbol{k}\cdot(\boldsymbol{r}-\boldsymbol{r}')} \\
&= \frac{1}{(2\pi)^3}\int_{|\boldsymbol{k}|\leqslant k_{\mathrm{F}}} \mathrm{d}\boldsymbol{k}\ \mathrm{e}^{-\mathrm{i}\boldsymbol{k}\cdot(\boldsymbol{r}-\boldsymbol{r}')} \\
&= \frac{1}{(2\pi)^2}\int_0^{k_{\mathrm{F}}} \mathrm{d}k\ k^2 \int_{-1}^{1} \mathrm{d}\cos\theta\ \mathrm{e}^{-\mathrm{i}k|\boldsymbol{r}-\boldsymbol{r}'|\cos\theta} \\
&= \frac{1}{2\pi^2|\boldsymbol{r}-\boldsymbol{r}'|}\int_0^{k_{\mathrm{F}}} \mathrm{d}k\ k\sin(k|\boldsymbol{r}-\boldsymbol{r}'|) \\
&= \frac{1}{2\pi^2|\boldsymbol{r}-\boldsymbol{r}'|^3}\int_0^{k_{\mathrm{F}}|\boldsymbol{r}-\boldsymbol{r}'|} \mathrm{d}x\ x\sin x \\
&= \frac{k_{\mathrm{F}}^3}{2\pi^2}\,\frac{\sin(k_{\mathrm{F}}|\boldsymbol{r}-\boldsymbol{r}'|) - (k_{\mathrm{F}}|\boldsymbol{r}-\boldsymbol{r}'|)\cos(k_{\mathrm{F}}|\boldsymbol{r}-\boldsymbol{r}'|)}{(k_{\mathrm{F}}|\boldsymbol{r}-\boldsymbol{r}'|)^3} \\
&= \frac{3n}{2}\,\frac{j_1(k_{\mathrm{F}}|\boldsymbol{r}-\boldsymbol{r}'|)}{k_{\mathrm{F}}|\boldsymbol{r}-\boldsymbol{r}'|},
\end{aligned}$$

where $j_1(x) = (\sin x - x\cos x)/x^2$ is the first-order spherical Bessel function of the first kind. We finally have

$$n_{\sigma\sigma}^{\mathrm{HF}}(\boldsymbol{r},\boldsymbol{r}') = \frac{1}{4}n^2\left[1 - \frac{9j_1^2(k_{\mathrm{F}}|\boldsymbol{r}-\boldsymbol{r}'|)}{(k_{\mathrm{F}}|\boldsymbol{r}-\boldsymbol{r}'|)^2}\right].$$

(2) From $n_{\sigma\sigma'}(\boldsymbol{r},\boldsymbol{r}') = n_\sigma(\boldsymbol{r})g_{\sigma\sigma'}(\boldsymbol{r},\boldsymbol{r}')n_{\sigma'}(\boldsymbol{r}')$, we obtain the following same-spin pair correlation function for a uniform electron gas

$$g_{\sigma\sigma}^{\mathrm{HF}}(|\boldsymbol{r}-\boldsymbol{r}'|) = 1 - \frac{9j_1^2(k_F|\boldsymbol{r}-\boldsymbol{r}'|)}{(k_F|\boldsymbol{r}-\boldsymbol{r}'|)^2}.$$

The same-spin pair correlation function is plotted in Fig. 21.1. The same-spin pair density has the same distance dependence as the same-spin pair correlation function since they are related through $n_{\sigma\sigma}^{\mathrm{HF}}(\boldsymbol{r},\boldsymbol{r}') = (n^2/4)g_{\sigma\sigma}^{\mathrm{HF}}(|\boldsymbol{r}-\boldsymbol{r}'|)$.

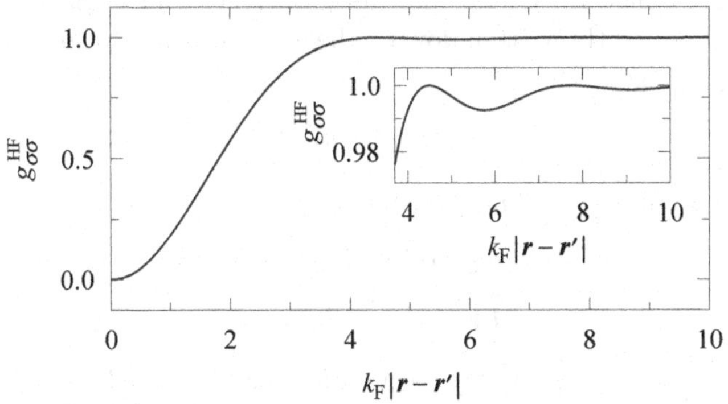

Fig. 21.1 Plot of the pair correlation function $g_{\sigma\sigma}^{\mathrm{HF}}$. Inset: The part of $g_{\sigma\sigma}^{\mathrm{HF}}$ for large values of $k_F|\boldsymbol{r}-\boldsymbol{r}'|$ for displaying the oscillatory behavior of the pair correlation function.

21-3 Jellium model. The Hamiltonian in the jellium model is given by

$$\hat{H} = \hat{H}_0 + \hat{H}_1, \quad \hat{H}_0 = \sum_{\boldsymbol{k}\sigma} \frac{\hbar^2 \boldsymbol{k}^2}{2m} \hat{c}_{\boldsymbol{k}\sigma}^\dagger \hat{c}_{\boldsymbol{k}\sigma},$$

$$\hat{H}_1 = \sum_{\boldsymbol{k}\boldsymbol{k}'\sigma\sigma'} \sum_{\boldsymbol{q}\neq 0} U_{\boldsymbol{q}}\, \hat{c}_{\boldsymbol{k}+\boldsymbol{q},\sigma}^\dagger \hat{c}_{\boldsymbol{k}'-\boldsymbol{q},\sigma'}^\dagger \hat{c}_{\boldsymbol{k}'\sigma'} \hat{c}_{\boldsymbol{k}\sigma}.$$

(1) Write down the eigenvectors of $\hat{H}_0$ for the noninteracting electron gas in terms of creation operators of electrons.
(2) Compute the expectation value of $\hat{H}_1$ in the ground state of the noninteracting electron gas. Write down the total energy per electron.
(3) Show that the pair correlation function in the ground state of the noninteracting electron gas

$$g(\boldsymbol{r}_1,\boldsymbol{r}_2) = \mathcal{V}^2 \sum_{s_{1z}s_{2z}} \not{\int} \mathrm{d}x_3 \cdots \not{\int} \mathrm{d}x_N \left|\Psi_0(x_1,\cdots,x_N)\right|^2$$

can be written as $(1/4)\sum_{\sigma\sigma'} g_{\sigma\sigma'}(\boldsymbol{r}_1,\boldsymbol{r}_2)$ with

$$g_{\sigma\sigma'}(\boldsymbol{r}_1,\boldsymbol{r}_2) = \frac{2\mathcal{V}^2}{N(N-1)}\sum_{\boldsymbol{k}\boldsymbol{k}'}{}' \sum_{s_{1z}s_{2z}} \big|\psi_{\boldsymbol{k}\sigma}(x_1)\psi_{\boldsymbol{k}'\sigma'}(x_2) - \psi_{\boldsymbol{k}\sigma}(x_2)\psi_{\boldsymbol{k}'\sigma'}(x_1)\big|^2.$$

(4) Evaluate $g_{\uparrow\uparrow}(\boldsymbol{r}_1,\boldsymbol{r}_2)$ and $g_{\uparrow\downarrow}(\boldsymbol{r}_1,\boldsymbol{r}_2)$.

(5) Compute the interaction energy per electron through

$$\frac{E_{\rm int}}{N} = \frac{1}{2}\frac{e^2}{4\pi\epsilon_0}\frac{\int \mathrm{d}\boldsymbol{r}\,[g(r)-1]/r}{\int \mathrm{d}\boldsymbol{r}\,[1-g(r)]}.$$

---

(1) Since $\hat{H}_0$ is diagonal, its eigenvectors in Fock space can be directly written down. The eigenvector corresponding to the eigenvalue

$$E_{\{n\}} = \sum_{\boldsymbol{k}\sigma} n_{\boldsymbol{k}\sigma}\frac{\hbar^2\boldsymbol{k}^2}{2m} = \sum_{\boldsymbol{k}\sigma} n_{\boldsymbol{k}\sigma}\varepsilon_{\boldsymbol{k}}$$

with $\varepsilon_{\boldsymbol{k}} = \hbar^2\boldsymbol{k}^2/2m$ is given by

$$\begin{aligned}|\{n_{\boldsymbol{k}\sigma}\}\rangle &= \big(\hat{c}^\dagger_{\boldsymbol{k}_1\uparrow}\big)^{n_{\boldsymbol{k}_1\uparrow}}\big(\hat{c}^\dagger_{\boldsymbol{k}_1\downarrow}\big)^{n_{\boldsymbol{k}_1\downarrow}}\big(\hat{c}^\dagger_{\boldsymbol{k}_2\uparrow}\big)^{n_{\boldsymbol{k}_2\uparrow}}\big(\hat{c}^\dagger_{\boldsymbol{k}_2\downarrow}\big)^{n_{\boldsymbol{k}_2\downarrow}}\cdots|0\rangle \\ &= \prod_{\boldsymbol{k}\sigma}\big(\hat{c}^\dagger_{\boldsymbol{k}\sigma}\big)^{n_{\boldsymbol{k}\sigma}}|0\rangle,\end{aligned}$$

where $n_{\boldsymbol{k}\sigma}$ is the occupation number of the single-electron state $|\boldsymbol{k}\sigma\rangle$ with $\hat{c}^\dagger_{\boldsymbol{k}\sigma}$ the creation operator of the single-electron state and $|0\rangle$ the vacuum state.

(2) Let $k_{\rm F}$ be the Fermi wave vector. The ground state of the non-interacting electron gas is the eigenstate of $\hat{H}_0$ corresponding to the lowest eigenvalue. In the ground state, all the single-electron states within the Fermi sphere are occupied while those outside are empty. The state vector of the ground state is given by

$$|\mathrm{F}\rangle = \prod_{\boldsymbol{k}\sigma\ (|\boldsymbol{k}|\leqslant k_{\rm F})}\hat{c}^\dagger_{\boldsymbol{k}\sigma}|0\rangle,$$

where F represents the filled Fermi sphere. The expectation value of $\hat{H}_1$ in the ground state is given by

$$\begin{aligned}\langle\hat{H}_1\rangle_0 &= \sum_{\boldsymbol{k}\boldsymbol{k}'\sigma\sigma'}\sum_{\boldsymbol{q}\neq 0} U_{\boldsymbol{q}}\,\langle\hat{c}^\dagger_{\boldsymbol{k}+\boldsymbol{q},\sigma}\hat{c}^\dagger_{\boldsymbol{k}'-\boldsymbol{q},\sigma'}\hat{c}_{\boldsymbol{k}'\sigma'}\hat{c}_{\boldsymbol{k}\sigma}\rangle_0 \\ &= \sum_{\boldsymbol{k}\boldsymbol{k}'\sigma\sigma'}\sum_{\boldsymbol{q}\neq 0} U_{\boldsymbol{q}}\,\langle 0|\Bigg(\prod_{\boldsymbol{k}_1\sigma_1\ (|\boldsymbol{k}_1|\leqslant k_{\rm F})}\hat{c}_{\boldsymbol{k}_1\sigma_1}\Bigg)\hat{c}^\dagger_{\boldsymbol{k}+\boldsymbol{q},\sigma}\hat{c}^\dagger_{\boldsymbol{k}'-\boldsymbol{q},\sigma'}\hat{c}_{\boldsymbol{k}'\sigma'}\hat{c}_{\boldsymbol{k}\sigma} \\ &\quad\times\Bigg(\prod_{\boldsymbol{k}_2\sigma_2\ (|\boldsymbol{k}_2|\leqslant k_{\rm F})}\hat{c}^\dagger_{\boldsymbol{k}_2\sigma_2}\Bigg)|0\rangle.\end{aligned}$$

We now move $\hat{c}_{\boldsymbol{k}\sigma}$ across $\prod_{\boldsymbol{k}_2\sigma_2\ (|\boldsymbol{k}_2|\leqslant k_{\rm F})} \hat{c}^\dagger_{\boldsymbol{k}_2\sigma_2}$ to the front of $|0\rangle$ using the anticommutation relation $\{\hat{c}_{\boldsymbol{k}\sigma}, \hat{c}^\dagger_{\boldsymbol{k}'\sigma'}\} = \delta_{\boldsymbol{k}\boldsymbol{k}'}\delta_{\sigma\sigma}$. As soon as $\hat{c}_{\boldsymbol{k}\sigma}$ can act directly on $|0\rangle$, a zero result yields. If $|\boldsymbol{k}| > k_{\rm F}$, we can see immediately that $\langle \hat{H}_1\rangle_0 = 0$ since $\hat{c}_{\boldsymbol{k}\sigma}$ anticommutes with all the operators on its right. If $|\boldsymbol{k}| \leqslant k_{\rm F}$, on its right there exists only a single operator $\hat{c}^\dagger_{\boldsymbol{k}\sigma}$ that does not anticommute with it. Upon making use of the anticommutation relation $\{\hat{c}_{\boldsymbol{k}\sigma}, \hat{c}^\dagger_{\boldsymbol{k}\sigma}\} = 1$, we see that both $\hat{c}_{\boldsymbol{k}\sigma}$ and $\hat{c}^\dagger_{\boldsymbol{k}\sigma}$ are eliminated in the nonzero term. We now consider the overall sign arising from anticommuting $\hat{c}_{\boldsymbol{k}\sigma}$ with $\hat{c}^\dagger_{\boldsymbol{k}_2\sigma_2}$'s to reach the immediate left of $\hat{c}^\dagger_{\boldsymbol{k}\sigma}$. For convenience, we take the order of operators in $\prod_{\boldsymbol{k}_2\sigma_2\ (|\boldsymbol{k}_2|\leqslant k_{\rm F})} \hat{c}^\dagger_{\boldsymbol{k}_2\sigma_2}$ to be

$$\hat{c}^\dagger_{\boldsymbol{k}_1\uparrow}\hat{c}^\dagger_{\boldsymbol{k}_1\downarrow}\hat{c}^\dagger_{\boldsymbol{k}_2\uparrow}\hat{c}^\dagger_{\boldsymbol{k}_2\downarrow}, \cdots \hat{c}^\dagger_{\boldsymbol{k}_{N/2}\uparrow}\hat{c}^\dagger_{\boldsymbol{k}_{N/2}\downarrow},$$

where $|\boldsymbol{k}_{N/2}| = k_{\rm F}$ with N (assumed to be even) the number of electrons. The overall sign arising from anticommuting $\hat{c}_{\boldsymbol{k}\sigma}$ with $\hat{c}^\dagger_{\boldsymbol{k}_2\sigma_2}$'s to reach the immediate left of $\hat{c}^\dagger_{\boldsymbol{k}\sigma}$ is given by

$$(-1)^{\sum_{\boldsymbol{k}_2\sigma_2\ (\boldsymbol{k}_2\to\boldsymbol{k})} n_{\boldsymbol{k}_2\sigma_2} + \delta_{\sigma\downarrow}},$$

where $\boldsymbol{k}_2 \to \boldsymbol{k}$ implies that all the values of $\boldsymbol{k}_2$ up to $\boldsymbol{k}$ with $\boldsymbol{k}$ excluded.

We can make a similar argument to $\hat{c}^\dagger_{\boldsymbol{k}+\boldsymbol{q},\sigma}$ which is to be moved to the left across $\prod_{\boldsymbol{k}_1\sigma_1\ (|\boldsymbol{k}_1|\leqslant k_{\rm F})} \hat{c}_{\boldsymbol{k}_1\sigma_1}$ so that it can act directly on $\langle 0|$. Only if $|\boldsymbol{k}+\boldsymbol{q}| \leqslant k_{\rm F}$, can we obtain a nonzero result. The overall sign arising from anticommuting $\hat{c}^\dagger_{\boldsymbol{k}+\boldsymbol{q},\sigma}$ with $\hat{c}_{\boldsymbol{k}_1\sigma_1}$'s to reach the immediate right of $\hat{c}_{\boldsymbol{k}+\boldsymbol{q},\sigma}$ is given by

$$(-1)^{\sum_{\boldsymbol{k}_1\sigma_1\ (\boldsymbol{k}_1\to\boldsymbol{k}+\boldsymbol{q})} n_{\boldsymbol{k}_1\sigma_1} + \delta_{\sigma\downarrow}}.$$

Taking all the above arguments into consideration, we have

$$\begin{aligned}\langle \hat{H}_1\rangle_0 = &\sum_{\boldsymbol{k}\boldsymbol{k}'\sigma\sigma'}\sum_{\boldsymbol{q}\neq 0} U_{\boldsymbol{q}}\theta(k_{\rm F} - |\boldsymbol{k}+\boldsymbol{q}|)\theta(k_{\rm F} - |\boldsymbol{k}|) \\ &\times (-1)^{\sum_{\boldsymbol{k}_1\sigma_1(\boldsymbol{k}_1\to\boldsymbol{k}+\boldsymbol{q})} n_{\boldsymbol{k}_1\sigma_1} + \sum_{\boldsymbol{k}_2\sigma_2(\boldsymbol{k}_2\to\boldsymbol{k})} n_{\boldsymbol{k}_2\sigma_2}} \\ &\times \langle 0| \Bigg(\prod_{\substack{\boldsymbol{k}_1\sigma_1 \\ (|\boldsymbol{k}_1|\leqslant k_{\rm F}, \\ \boldsymbol{k}_1\sigma_1 \neq \boldsymbol{k}+\boldsymbol{q}\sigma)}} \hat{c}_{\boldsymbol{k}_1\sigma_1}\Bigg) \hat{c}^\dagger_{\boldsymbol{k}'-\boldsymbol{q},\sigma'}\hat{c}_{\boldsymbol{k}'\sigma'} \Bigg(\prod_{\substack{\boldsymbol{k}_2\sigma_2 \\ (|\boldsymbol{k}_2|\leqslant k_{\rm F}, \\ \boldsymbol{k}_2\sigma_2 \neq \boldsymbol{k}\sigma)}} \hat{c}^\dagger_{\boldsymbol{k}_2\sigma_2}\Bigg)|0\rangle.\end{aligned}$$

Making similar arguments to $\hat{c}_{\boldsymbol{k}'\sigma'}$ and $\hat{c}^\dagger_{\boldsymbol{k}'-\boldsymbol{q},\sigma'}$, we obtain

$$\begin{aligned}\langle \hat{H}_1\rangle_0 = &\sum_{\boldsymbol{k}\boldsymbol{k}'\sigma\sigma'}\sum_{\boldsymbol{q}\neq 0} U_{\boldsymbol{q}}\theta(k_{\mathrm{F}}-|\boldsymbol{k}+\boldsymbol{q}|)\theta(k_{\mathrm{F}}-|\boldsymbol{k}|)\theta(k_{\mathrm{F}}-|\boldsymbol{k}'-\boldsymbol{q}|)\\ &\times \theta(k_{\mathrm{F}}-|\boldsymbol{k}'|)(-1)^{\sum_{\boldsymbol{k}_1\sigma_1(\boldsymbol{k}_1\to\boldsymbol{k}+\boldsymbol{q})} n_{\boldsymbol{k}_1\sigma_1}+\sum_{\boldsymbol{k}_2\sigma_2(\boldsymbol{k}_2\to\boldsymbol{k})} n_{\boldsymbol{k}_2\sigma_2}}\\ &\times (-1)^{\sum_{\boldsymbol{k}_1\sigma_1(\boldsymbol{k}_1\to\boldsymbol{k}'-\boldsymbol{q})} n_{\boldsymbol{k}_1\sigma_1}+\sum_{\boldsymbol{k}_2\sigma_2(\boldsymbol{k}_2\to\boldsymbol{k}')} n_{\boldsymbol{k}_2\sigma_2}}\\ &\times (-1)^{\theta(\boldsymbol{k}'\sqsupset\boldsymbol{k})+\theta(\boldsymbol{k}'-\boldsymbol{q}\sqsupset\boldsymbol{k}+\boldsymbol{q})}\\ &\times \langle 0|\Bigg(\prod_{\substack{\boldsymbol{k}_1\sigma_1(|\boldsymbol{k}_1|\leqslant k_{\mathrm{F}},\\ \boldsymbol{k}_1\sigma_1\neq\boldsymbol{k}+\boldsymbol{q}\sigma,\boldsymbol{k}'-\boldsymbol{q}\sigma')}}\hat{c}_{\boldsymbol{k}_1\sigma_1}\Bigg)\Bigg(\prod_{\substack{\boldsymbol{k}_2\sigma_2(|\boldsymbol{k}_2|\leqslant k_{\mathrm{F}},\\ \boldsymbol{k}_2\sigma_2\neq\boldsymbol{k}\sigma,\boldsymbol{k}'\sigma')}}\hat{c}^\dagger_{\boldsymbol{k}_2\sigma_2}\Bigg)|0\rangle,\end{aligned}$$

where $\boldsymbol{k}' \sqsupset \boldsymbol{k}$ implies that $\boldsymbol{k}'$ is beyond $\boldsymbol{k}$ in $\hat{c}^\dagger_{\boldsymbol{k}_1\uparrow}\hat{c}^\dagger_{\boldsymbol{k}_1\downarrow}\hat{c}^\dagger_{\boldsymbol{k}_2\uparrow}\hat{c}^\dagger_{\boldsymbol{k}_2\downarrow},\cdots\hat{c}^\dagger_{\boldsymbol{k}_{N/2}\uparrow}\hat{c}^\dagger_{\boldsymbol{k}_{N/2}\downarrow}$. The average in the above equation is nonzero only for two possible different cases: (1) $\boldsymbol{k}\sigma=\boldsymbol{k}+\boldsymbol{q}\sigma$ and $\boldsymbol{k}'\sigma'=\boldsymbol{k}'-\boldsymbol{q}\sigma'$ or (2) $\boldsymbol{k}\sigma=\boldsymbol{k}'-\boldsymbol{q}\sigma'$ and $\boldsymbol{k}'\sigma'=\boldsymbol{k}+\boldsymbol{q}\sigma$. Since $\boldsymbol{q}\neq 0$, the first case can never be realized. Therefore, the average is nonzero only if $\boldsymbol{k}\sigma=\boldsymbol{k}'-\boldsymbol{q}\sigma'$ and $\boldsymbol{k}'\sigma'=\boldsymbol{k}+\boldsymbol{q}\sigma$. We thus have $\boldsymbol{k}'=\boldsymbol{k}+\boldsymbol{q}$ and $\sigma'=\sigma$ and

$$\begin{aligned}\langle \hat{H}_1\rangle_0 &= -\sum_{\boldsymbol{k}\sigma}\sum_{\boldsymbol{q}\neq 0} U_{\boldsymbol{q}}\theta(k_{\mathrm{F}}-|\boldsymbol{k}|)\theta(k_{\mathrm{F}}-|\boldsymbol{k}+\boldsymbol{q}|)\\ &= -2\sum_{\boldsymbol{k}'\neq\boldsymbol{k}} U_{\boldsymbol{k}'-\boldsymbol{k}}\theta(k_{\mathrm{F}}-|\boldsymbol{k}|)\theta(k_{\mathrm{F}}-|\boldsymbol{k}'|).\end{aligned}$$

Using a prime on the summation sign to denote that $|\boldsymbol{k}|$, $|\boldsymbol{k}'|\leqslant k_{\mathrm{F}}$ and inserting $U_{\boldsymbol{q}}=e^2/2\mathscr{V}\epsilon_0 q^2$ into the above equation, we have

$$\langle \hat{H}_1\rangle_0 = -2{\sum_{\boldsymbol{k}'\neq\boldsymbol{k}}}' U_{\boldsymbol{k}'-\boldsymbol{k}} = -\frac{e^2}{\mathscr{V}\epsilon_0}{\sum_{\boldsymbol{k}'\neq\boldsymbol{k}}}'\frac{1}{|\boldsymbol{k}'-\boldsymbol{k}|^2}.$$

The summations in the above equation can be evaluated by converting them into integrals. We have

$${\sum_{\boldsymbol{k}'\neq\boldsymbol{k}}}'\frac{1}{|\boldsymbol{k}'-\boldsymbol{k}|^2} = \frac{\mathscr{V}k_{\mathrm{F}}}{2\pi^2}{\sum_{\boldsymbol{k}}}'\left[\frac{1}{2}+\frac{k_{\mathrm{F}}^2-k^2}{4kk_{\mathrm{F}}}\ln\left|\frac{k_{\mathrm{F}}+k}{k_{\mathrm{F}}-k}\right|\right] = \frac{\mathscr{V}^2k_{\mathrm{F}}^4}{(2\pi)^4}.$$

We finally have

$$\langle \hat{H}_1\rangle_0 = -\frac{\mathscr{V}e^2k_{\mathrm{F}}^4}{(2\pi)^4\epsilon_0} = -\frac{3Ne^2k_{\mathrm{F}}}{4\pi(4\pi\epsilon_0)}$$

which is just the total exchange energy in the electron gas. Taking into account the kinetic energy per electron, $3E_{\mathrm{F}}/5$, we have the following total energy per electron

$$\frac{E}{N} = \frac{3}{5}E_{\mathrm{F}} - \frac{3e^2k_{\mathrm{F}}}{4\pi(4\pi\epsilon_0)}.$$

(3) Inserting the Slater determinant for the ground-state wave function into $g(\boldsymbol{r}_1, \boldsymbol{r}_2)$, we have

$$
\begin{aligned}
& g(\boldsymbol{r}_1, \boldsymbol{r}_2) \\
& = \frac{\mathcal{V}^2}{N!} \sum_{s_{1z} s_{2z}} \sum_{PP'} (-1)^P (-1)^{P'} \psi^*_{\boldsymbol{k}_{P'_1} \sigma_{P'_1}}(x_1) \psi_{\boldsymbol{k}_{P_1} \sigma_{P_1}}(x_1) \\
& \quad \times \psi^*_{\boldsymbol{k}_{P'_2} \sigma_{P'_2}}(x_2) \psi_{\boldsymbol{k}_{P_2} \sigma_{P_2}}(x_2) \sum\!\!\!\!\!\!\int \mathrm{d}x_3 \psi^*_{\boldsymbol{k}_{P'_3} \sigma_{P'_3}}(x_3) \psi_{\boldsymbol{k}_{P_3} \sigma_{P_3}}(x_3) \cdots \\
& \quad \times \sum\!\!\!\!\!\!\int \mathrm{d}x_N \psi^*_{\boldsymbol{k}_{P'_N} \sigma_{P'_N}}(x_N) \psi_{\boldsymbol{k}_{P_N} \sigma_{P_N}}(x_N) \\
& = \frac{\mathcal{V}^2}{N!} \sum_{s_{1z} s_{2z}} \sum_{PP'} (-1)^P (-1)^{P'} \psi^*_{\boldsymbol{k}_{P'_1} \sigma_{P'_1}}(x_1) \psi_{\boldsymbol{k}_{P_1} \sigma_{P_1}}(x_1) \\
& \quad \times \psi^*_{\boldsymbol{k}_{P'_2} \sigma_{P'_2}}(x_2) \psi_{\boldsymbol{k}_{P_2} \sigma_{P_2}}(x_2) \prod_{i=3}^{N} \delta_{P_i P'_i} \\
& = \frac{(N-2)!\mathcal{V}^2}{N!} \sideset{}{'}\sum_{\boldsymbol{k}\boldsymbol{k}'} \sum_{s_{1z} s_{2z}} \sum_{\sigma\sigma'} \big[\psi^*_{\boldsymbol{k}\sigma}(x_1) \psi_{\boldsymbol{k}\sigma}(x_1) \psi^*_{\boldsymbol{k}'\sigma'}(x_2) \psi_{\boldsymbol{k}'\sigma'}(x_2) \\
& \qquad - \psi^*_{\boldsymbol{k}'\sigma'}(x_1) \psi_{\boldsymbol{k}\sigma}(x_1) \psi^*_{\boldsymbol{k}\sigma}(x_2) \psi_{\boldsymbol{k}'\sigma'}(x_2) \big] \\
& = \frac{\mathcal{V}^2}{N(N-1)} \sideset{}{'}\sum_{\boldsymbol{k}\boldsymbol{k}'} \sum_{s_{1z} s_{2z}} \sum_{\sigma\sigma'} \big[\psi^*_{\boldsymbol{k}\sigma}(x_1) \psi_{\boldsymbol{k}\sigma}(x_1) \psi^*_{\boldsymbol{k}'\sigma'}(x_2) \psi_{\boldsymbol{k}'\sigma'}(x_2) \\
& \qquad - \psi^*_{\boldsymbol{k}'\sigma'}(x_1) \psi_{\boldsymbol{k}\sigma}(x_1) \psi^*_{\boldsymbol{k}\sigma}(x_2) \psi_{\boldsymbol{k}'\sigma'}(x_2) \big],
\end{aligned}
$$

where the prime on the summation sign indicates that $|\boldsymbol{k}|$, $|\boldsymbol{k}'| \leqslant k_{\mathrm{F}}$. Switching the dummy summation variables $\boldsymbol{k}\sigma$ with $\boldsymbol{k}'\sigma'$, adding up the resultant expression and the original one, and dividing the result by two, we obtain

$$
\begin{aligned}
& g(\boldsymbol{r}_1, \boldsymbol{r}_2) \\
& = \frac{\mathcal{V}^2}{2N(N-1)} \sideset{}{'}\sum_{\boldsymbol{k}\boldsymbol{k}'} \sum_{s_{1z} s_{2z}} \sum_{\sigma\sigma'} \big[\psi^*_{\boldsymbol{k}\sigma}(x_1) \psi_{\boldsymbol{k}\sigma}(x_1) \psi^*_{\boldsymbol{k}'\sigma'}(x_2) \psi_{\boldsymbol{k}'\sigma'}(x_2) \\
& \qquad - \psi^*_{\boldsymbol{k}'\sigma'}(x_1) \psi_{\boldsymbol{k}\sigma}(x_1) \psi^*_{\boldsymbol{k}\sigma}(x_2) \psi_{\boldsymbol{k}'\sigma'}(x_2) \\
& \qquad + \psi^*_{\boldsymbol{k}'\sigma'}(x_1) \psi_{\boldsymbol{k}'\sigma'}(x_1) \psi^*_{\boldsymbol{k}\sigma}(x_2) \psi_{\boldsymbol{k}\sigma}(x_2) \\
& \qquad - \psi^*_{\boldsymbol{k}\sigma}(x_1) \psi_{\boldsymbol{k}'\sigma'}(x_1) \psi^*_{\boldsymbol{k}'\sigma'}(x_2) \psi_{\boldsymbol{k}\sigma}(x_2) \big] \\
& = \frac{\mathcal{V}^2}{2N(N-1)} \sideset{}{'}\sum_{\boldsymbol{k}\boldsymbol{k}'} \sum_{s_{1z} s_{2z}} \sum_{\sigma\sigma'} \big| \psi_{\boldsymbol{k}\sigma}(x_1) \psi_{\boldsymbol{k}'\sigma'}(x_2) - \psi_{\boldsymbol{k}\sigma}(x_2) \psi_{\boldsymbol{k}'\sigma'}(x_1) \big|^2 \\
& = \frac{1}{4} \sum_{\sigma\sigma'} g_{\sigma\sigma'}(\boldsymbol{r}_1, \boldsymbol{r}_2),
\end{aligned}
$$

where

$$g_{\sigma\sigma'}(\boldsymbol{r}_1, \boldsymbol{r}_2) = \frac{2\mathscr{V}^2}{N(N-1)} \sum_{\boldsymbol{k}\boldsymbol{k}'}{}' \sum_{s_{1z}s_{2z}} \left|\psi_{\boldsymbol{k}\sigma}(x_1)\psi_{\boldsymbol{k}'\sigma'}(x_2) - \psi_{\boldsymbol{k}\sigma}(x_2)\psi_{\boldsymbol{k}'\sigma'}(x_1)\right|^2.$$

(4) For $g_{\uparrow\uparrow}(\boldsymbol{r}_1, \boldsymbol{r}_2)$, we have

$$g_{\uparrow\uparrow}(\boldsymbol{r}_1, \boldsymbol{r}_2) = \frac{2\mathscr{V}^2}{N(N-1)} \sum_{\boldsymbol{k}\boldsymbol{k}'}{}' \sum_{s_{1z}s_{2z}} \left|\psi_{\boldsymbol{k}\uparrow}(x_1)\psi_{\boldsymbol{k}'\uparrow}(x_2) - \psi_{\boldsymbol{k}\uparrow}(x_2)\psi_{\boldsymbol{k}'\uparrow}(x_1)\right|^2.$$

For noninteracting electrons, we can write $\psi_{\boldsymbol{k}\sigma}(x)$ as

$$\psi_{\boldsymbol{k}}(\boldsymbol{r})\chi_\sigma(s_z) = \mathscr{V}^{-1/2}\mathrm{e}^{\mathrm{i}\boldsymbol{k}\cdot\boldsymbol{r}}\chi_\sigma(s_z).$$

We then have

$$\begin{aligned} g_{\uparrow\uparrow}(\boldsymbol{r}_1, \boldsymbol{r}_2) &= \frac{2}{N(N-1)} \sum_{\boldsymbol{k}\boldsymbol{k}'}{}' \left|\mathrm{e}^{\mathrm{i}(\boldsymbol{k}\cdot\boldsymbol{r}_1 + \boldsymbol{k}'\cdot\boldsymbol{r}_2)} - \mathrm{e}^{\mathrm{i}(\boldsymbol{k}\cdot\boldsymbol{r}_2 + \boldsymbol{k}'\cdot\boldsymbol{r}_1)}\right|^2 \\ &\quad \sum_{s_{1z}s_{2z}} \left|\chi_\uparrow(s_{1z})\chi_\uparrow(s_{2z})\right|^2 \\ &= \frac{2}{N(N-1)} \sum_{\boldsymbol{k}\boldsymbol{k}'}{}' \left|\mathrm{e}^{\mathrm{i}(\boldsymbol{k}\cdot\boldsymbol{r}_1 + \boldsymbol{k}'\cdot\boldsymbol{r}_2)} - \mathrm{e}^{\mathrm{i}(\boldsymbol{k}\cdot\boldsymbol{r}_2 + \boldsymbol{k}'\cdot\boldsymbol{r}_1)}\right|^2 \\ &= \frac{2\mathscr{V}^2}{N(N-1)(2\pi)^6} \int_{|\boldsymbol{k}|\leqslant k_{\mathrm{F}}} \mathrm{d}\boldsymbol{k} \int_{|\boldsymbol{k}'|\leqslant k_{\mathrm{F}}} \mathrm{d}\boldsymbol{k}' \left|\mathrm{e}^{\mathrm{i}(\boldsymbol{k}-\boldsymbol{k}')\cdot(\boldsymbol{r}_1 - \boldsymbol{r}_2)} - 1\right|^2, \end{aligned}$$

where we have converted the summations into integrations. Expanding the absolute value, we have

$$\begin{aligned} g_{\uparrow\uparrow}(\boldsymbol{r}_1, \boldsymbol{r}_2) &= \frac{2\mathscr{V}^2}{N(N-1)(2\pi)^6} \int_{|\boldsymbol{k}|\leqslant k_{\mathrm{F}}} \mathrm{d}\boldsymbol{k} \int_{|\boldsymbol{k}'|\leqslant k_{\mathrm{F}}} \mathrm{d}\boldsymbol{k}' \\ &\quad \times \left[2 - \mathrm{e}^{\mathrm{i}(\boldsymbol{k}-\boldsymbol{k}')\cdot(\boldsymbol{r}_1 - \boldsymbol{r}_2)} - \mathrm{e}^{-\mathrm{i}(\boldsymbol{k}-\boldsymbol{k}')\cdot(\boldsymbol{r}_1 - \boldsymbol{r}_2)}\right] \\ &= \frac{\mathscr{V}^2}{N(N-1)} \left[n^2 - \left|\frac{2}{(2\pi)^3} \int_{|\boldsymbol{k}|\leqslant k_{\mathrm{F}}} \mathrm{d}\boldsymbol{k}\, \mathrm{e}^{\mathrm{i}\boldsymbol{k}\cdot(\boldsymbol{r}_1 - \boldsymbol{r}_2)}\right|^2\right]. \end{aligned}$$

The result for the integral can be expressed in terms of the spherical Bessel function of the first kind, $j_1(x) = (\sin x - x\cos x)/x^2$. We have

$$\begin{aligned} \int_{|\boldsymbol{k}|\leqslant k_{\mathrm{F}}} \mathrm{d}\boldsymbol{k}\, \mathrm{e}^{\mathrm{i}\boldsymbol{k}\cdot(\boldsymbol{r}_1 - \boldsymbol{r}_2)} &= 2\pi \int_0^{k_{\mathrm{F}}} \mathrm{d}k\, k^2 \int_{-1}^{1} \mathrm{d}\cos\theta\, \mathrm{e}^{\mathrm{i}k|\boldsymbol{r}_1 - \boldsymbol{r}_2|\cos\theta} \\ &= \frac{4\pi}{|\boldsymbol{r}_1 - \boldsymbol{r}_2|} \int_0^{k_{\mathrm{F}}} \mathrm{d}k\, k\sin(k|\boldsymbol{r}_1 - \boldsymbol{r}_2|) \\ &= \frac{3}{2}(2\pi)^3 n \frac{j_1(k_{\mathrm{F}}|\boldsymbol{r}_1 - \boldsymbol{r}_2|)}{k_{\mathrm{F}}|\boldsymbol{r}_1 - \boldsymbol{r}_2|}. \end{aligned}$$

Inserting the above result into the expression for $g_{\uparrow\uparrow}(\boldsymbol{r}_1, \boldsymbol{r}_2)$ yields

$$g_{\uparrow\uparrow}(\boldsymbol{r}_1, \boldsymbol{r}_2) = \frac{\mathscr{V}^2 n^2}{N(N-1)}\left[1 - \frac{9j_1^2(k_F|\boldsymbol{r}_1 - \boldsymbol{r}_2|)}{(k_F|\boldsymbol{r}_1 - \boldsymbol{r}_2|)^2}\right].$$

In the $N \to \infty$ limit, we can take $N - 1$ as N so that

$$\lim_{N\to\infty} \frac{\mathscr{V}^2 n^2}{N(N-1)} = \lim_{N\to\infty} \frac{\mathscr{V}^2 n^2}{N^2} = 1.$$

We thus have

$$g_{\uparrow\uparrow}(\boldsymbol{r}_1, \boldsymbol{r}_2) = 1 - \frac{9j_1^2(k_F|\boldsymbol{r}_1 - \boldsymbol{r}_2|)}{(k_F|\boldsymbol{r}_1 - \boldsymbol{r}_2|)^2}.$$

For $g_{\uparrow\downarrow}(\boldsymbol{r}_1, \boldsymbol{r}_2)$, we have

$$g_{\uparrow\downarrow}(\boldsymbol{r}_1, \boldsymbol{r}_2) = \frac{2\mathscr{V}^2}{N(N-1)} {\sum_{\boldsymbol{k}\boldsymbol{k}'}}' \sum_{s_{1z}s_{2z}} \left|\psi_{\boldsymbol{k}\uparrow}(x_1)\psi_{\boldsymbol{k}'\downarrow}(x_2) - \psi_{\boldsymbol{k}\uparrow}(x_2)\psi_{\boldsymbol{k}'\downarrow}(x_1)\right|^2.$$

Inserting $\psi_{\boldsymbol{k}\sigma}(x) = \mathscr{V}^{-1/2} e^{i\boldsymbol{k}\cdot\boldsymbol{r}} \chi_\sigma(s_z)$ into the above expression, we have

$$g_{\uparrow\downarrow}(\boldsymbol{r}_1, \boldsymbol{r}_2) = \frac{2}{N(N-1)} {\sum_{\boldsymbol{k}\boldsymbol{k}'}}' \sum_{s_{1z}s_{2z}} \Big|e^{i(\boldsymbol{k}\cdot\boldsymbol{r}_1 + \boldsymbol{k}'\cdot\boldsymbol{r}_2)}\chi_\uparrow(s_{1z})\chi_\downarrow(s_{2z}) - e^{i(\boldsymbol{k}'\cdot\boldsymbol{r}_1 + \boldsymbol{k}\cdot\boldsymbol{r}_2)}\chi_\uparrow(s_{2z})\chi_\downarrow(s_{1z})\Big|^2.$$

Because of the orthogonality between the spin wave functions of different spin directions, the cross terms from the expansion of the absolute value in the above equation vanish. We then have

$$g_{\uparrow\downarrow}(\boldsymbol{r}_1, \boldsymbol{r}_2) = \frac{2}{N(N-1)} {\sum_{\boldsymbol{k}\boldsymbol{k}'}}' 2 = \frac{N}{(N-1)} \xrightarrow{N\to\infty} 1,$$

where we have made use of the fact that there are $N/2$ allowed values for $\boldsymbol{k}$. From the symmetry consideration, we have

$$g_{\downarrow\downarrow}(\boldsymbol{r}_1, \boldsymbol{r}_2) = g_{\uparrow\uparrow}(\boldsymbol{r}_1, \boldsymbol{r}_2),\ \ g_{\downarrow\uparrow}(\boldsymbol{r}_1, \boldsymbol{r}_2) = g_{\uparrow\downarrow}(\boldsymbol{r}_1, \boldsymbol{r}_2).$$

For $g(\boldsymbol{r}_1, \boldsymbol{r}_2)$, we have

$$g(\boldsymbol{r}_1, \boldsymbol{r}_2) = \frac{1}{4}\sum_{\sigma\sigma'} g_{\sigma\sigma'}(\boldsymbol{r}_1, \boldsymbol{r}_2) = 1 - \frac{9j_1^2(k_F|\boldsymbol{r}_1 - \boldsymbol{r}_2|)}{2(k_F|\boldsymbol{r}_1 - \boldsymbol{r}_2|)^2}.$$

Since $g(\boldsymbol{r}_1, \boldsymbol{r}_2)$ is a function only of $|\boldsymbol{r}_1 - \boldsymbol{r}_2|$, we can write it as

$$g(r) = 1 - \frac{9j_1^2(k_F r)}{2(k_F r)^2}.$$

(5) Making use of the above-obtained expression for $g(r)$, we can evaluate straightforwardly the interaction energy per electron. For the integral $\int \mathrm{d}\boldsymbol{r}\,[\,1-g(r)\,]$, we have

$$\begin{aligned}\int \mathrm{d}\boldsymbol{r}\,[\,1-g(r)\,] &= \frac{9}{2}\int \mathrm{d}\boldsymbol{r}\,\frac{j_1^2(k_\mathrm{F}r)}{(k_\mathrm{F}r)^2}\\ &= \frac{18\pi}{k_\mathrm{F}^3}\int_0^\infty \mathrm{d}x\; j_1^2(x) = \frac{3\pi^2}{k_\mathrm{F}^3} = \frac{1}{n},\end{aligned}$$

where we have made use of

$$\int_0^\infty \mathrm{d}x\; j_\nu^2(x) = \frac{\pi}{2(2\nu+1)},\quad \mathrm{Re}\,\nu > -\frac{1}{2}.$$

For the integral $\int \mathrm{d}\boldsymbol{r}\;(e^2/4\pi\epsilon_0 r)[\,g(r)-1\,]$, we have

$$\frac{e^2}{4\pi\epsilon_0}\int \mathrm{d}\boldsymbol{r}\;\frac{1}{r}[\,g(r)-1\,] = -\frac{9e^2}{2\epsilon_0 k_\mathrm{F}^2}\int_0^\infty \mathrm{d}x\;\frac{j_1^2(x)}{x} = -\frac{9e^2}{8\epsilon_0 k_\mathrm{F}^2},$$

where we have made use of

$$\int_0^\infty \mathrm{d}x\; x^{\alpha-1} j_\nu^2(x) = -\frac{\sqrt{\pi}\,\Gamma(1-\alpha/2)\Gamma(\alpha/2+\nu)}{2(\alpha-1)\Gamma((1-\alpha)/2)\Gamma(-\alpha/2+\nu+2)}$$

for $\mathrm{Re}(\alpha+2\nu) > 0$ and $\mathrm{Re}\,\alpha < 2$. We thus have

$$\frac{E_\mathrm{int}}{N} = -\frac{3e^2 k_\mathrm{F}}{4\pi(4\pi\epsilon_0)}$$

which is the exchange energy per electron. Note that the above result was previously obtained by evaluating the expectation value of the interaction Hamiltonian in the ground state of the noninteracting electron gas.

Note that the above-used integrals of spherical Bessel functions can be deduced from[1]

$$\begin{aligned}&\int_0^\infty \mathrm{d}x\; x^{-\lambda} J_\nu(\alpha x) J_\mu(\alpha x)\\ &\quad= \frac{\alpha^{\lambda-1}\Gamma(\lambda)\Gamma\big(\frac{\nu+\mu-\lambda+1}{2}\big)}{2^\lambda\Gamma\big(\frac{-\nu+\mu+\lambda+1}{2}\big)\Gamma\big(\frac{\nu+\mu+\lambda+1}{2}\big)\Gamma\big(\frac{\nu-\mu+\lambda+1}{2}\big)}\\ &\quad\quad[\,\mathrm{Re}(\nu+\mu+1) > \mathrm{Re}\,\lambda > 0,\ \alpha > 0\,]\end{aligned}$$

using $j_\nu(x) = \sqrt{\pi/2x}\,J_{\nu+1/2}(x)$.

[1]M. Abramowitz and I. A. Stegun, *Handbook of Mathematical Functions* (National Bureau of Standards, Applied Mathematics Series 55, Tenth Printing, 1972).

21-4 Computation of one functional derivative. Evaluate the functional derivative of the functional

$$F[g(x)] = \int_a^b \mathrm{d}x\ \ln\left[1 + \mathrm{e}^{-g(x)}\right].$$

From the definition of the functional derivative, we have

$$\begin{aligned}
\frac{\delta F[g]}{\delta g(y)} &= \lim_{\epsilon\to 0}\frac{1}{\epsilon}\int_a^b \mathrm{d}x\ \left\{\ln\left[1+\mathrm{e}^{-g(x)-\epsilon\delta(x-y)}\right] - \ln\left[1+\mathrm{e}^{-g(x)}\right]\right\} \\
&= \lim_{\epsilon\to 0}\frac{1}{\epsilon}\int_a^b \mathrm{d}x\left\{\ln\left[1+\mathrm{e}^{-g(x)} - \epsilon\mathrm{e}^{-g(x)}\delta(x-y)\right] - \ln\left[1+\mathrm{e}^{-g(x)}\right]\right\} \\
&= \lim_{\epsilon\to 0}\frac{1}{\epsilon}\int_a^b \mathrm{d}x\left\{\ln\left[1+\mathrm{e}^{-g(x)}\right] - \frac{\epsilon\mathrm{e}^{-g(x)}\delta(x-y)}{1+\mathrm{e}^{-g(x)}} - \ln\left[1+\mathrm{e}^{-g(x)}\right]\right\} \\
&= -\int_a^b \mathrm{d}x\ \frac{\mathrm{e}^{-g(x)}\delta(x-y)}{1+\mathrm{e}^{-g(x)}} = -\frac{\mathrm{e}^{-g(y)}}{1+\mathrm{e}^{-g(y)}}\theta(b-y)\theta(y-a) \\
&= -\frac{1}{\mathrm{e}^{g(y)}+1}\theta(b-y)\theta(y-a),
\end{aligned}$$

where we have assumed that $b > a$. From the result in the first equality on the second-to-last line, we see that, if the functional $F[g]$ is expressed in terms of the functional density $\mathscr{F}(g(x))$,

$$F[g] = \int_a^b \mathrm{d}x\ \mathscr{F}(g(x)),$$

then the functional derivative of $F[g]$ can be evaluated as follows

$$\frac{\delta F[g]}{\delta g(y)} = \int_a^b \mathrm{d}x\ \delta(x-y)\frac{\partial \mathscr{F}(g(x))}{\partial g(x)}.$$

This procedure for evaluating functional derivatives can be proved using the definition of the functional derivative from which we have

$$\begin{aligned}
\frac{\delta F[g]}{\delta g(y)} &= \lim_{\epsilon\to 0}\int_a^b \mathrm{d}x\ \frac{\mathscr{F}(g(x)+\epsilon\delta(x-y)) - \mathscr{F}(g(x))}{\epsilon} \\
&= \int_a^b \mathrm{d}x\ \delta(x-y)\lim_{\epsilon\to 0}\frac{\mathscr{F}(g(x)+\epsilon\delta(x-y)) - \mathscr{F}(g(x))}{\epsilon\delta(x-y)} \\
&= \int_a^b \mathrm{d}x\ \delta(x-y)\frac{\partial \mathscr{F}(g(x))}{\partial g(x)},
\end{aligned}$$

where the definition of the derivative of an ordinary function has been used. The above-proved procedure can be used to evaluate functional derivatives whenever appropriate. However, if $\mathscr{F}(g(x))$ contains spatial derivatives of $g(x)$ [*e.g.*, $\mathscr{F}(g(x), \mathrm{d}g(x)/\mathrm{d}x)$], the above procedure is not applicable. In such a case, the definition of the functional derivative must be used. A general rule can be also derived for this case from the definition of the functional derivative [see the next problem]. For the purpose of verifying the above-proved procedure, we apply it to the present problem. We have

$$\begin{aligned}\frac{\delta F[g]}{\delta g(y)} &= \int_a^b \mathrm{d}x\, \delta(x-y) \frac{\partial}{\partial g(x)} \ln\left[1 + \mathrm{e}^{-g(x)}\right] \\ &= \int_a^b \mathrm{d}x\, \delta(x-y) \frac{-\mathrm{e}^{-g(x)}}{1+\mathrm{e}^{-g(x)}} \\ &= -\frac{\mathrm{e}^{-g(y)}}{1+\mathrm{e}^{-g(y)}} \theta(b-y)\theta(y-a) = -\frac{1}{\mathrm{e}^{g(y)}+1}\theta(b-y)\theta(y-a).\end{aligned}$$

We thus obtained the same result as we did starting from the definition of the functional derivative.

21-5 Computation of six functional derivatives. Compute the functional derivatives of the following functionals. It is assumed that all the functions are well-behaved so that all the integrals converge to finite values.

(1) $F[g] = \displaystyle\int_a^b \frac{\mathrm{d}x}{g^2(x)+1}$. **(2)** $F_f[g] = \displaystyle\int_{-\infty}^{\infty} \mathrm{d}x\, \frac{\mathrm{d}g(x)}{\mathrm{d}x} f(x)$.

(3) $F[g] = \displaystyle\int_0^{\infty} \mathrm{d}x\, \mathrm{e}^{-g^2(x)}$. **(4)** $F[g] = \dfrac{1}{2}\displaystyle\iint \mathrm{d}\boldsymbol{r}\mathrm{d}\boldsymbol{r}'\, g(\boldsymbol{r})v(|\boldsymbol{r}-\boldsymbol{r}'|)g(\boldsymbol{r}')$.

(5) $E[n] = c\displaystyle\int \mathrm{d}\boldsymbol{r}\, n^{4/3}(\boldsymbol{r})$. **(6)** $E[n] = \dfrac{\hbar^2}{8m}\displaystyle\int \mathrm{d}\boldsymbol{r}\, \frac{[\boldsymbol{\nabla} n(\boldsymbol{r})]^2}{n(\boldsymbol{r})}$.

---

(1) Making use of the above-proved procedure for evaluating functional derivatives, we have for the first functional

$$\begin{aligned}\frac{\delta F[g]}{\delta g(y)} &= \int_a^b \mathrm{d}x\, \delta(x-y) \frac{\partial}{\partial g(x)} \frac{1}{g^2(x)+1} \\ &= \int_a^b \mathrm{d}x\, \delta(x-y) \frac{-2g(x)}{[\,g^2(x)+1\,]^2} \\ &= -\frac{2g(y)}{[\,g^2(y)+1\,]^2}\theta(b-y)\theta(y-a).\end{aligned}$$

(2) For the second functional, making use of the definition of the functional derivative, we have

$$\begin{aligned}\frac{\delta F_f[g]}{\delta g(y)} &= \lim_{\epsilon\to 0}\frac{1}{\epsilon}\int_{-\infty}^{\infty} \mathrm{d}x \left\{ \frac{\mathrm{d}[\,g(x)+\epsilon\delta(x-y)\,]}{\mathrm{d}x} f(x) - \frac{\mathrm{d}g(x)}{\mathrm{d}x} f(x) \right\} \\ &= \int_{-\infty}^{\infty} \mathrm{d}x\, \frac{\mathrm{d}\delta(x-y)}{\mathrm{d}x} f(x) = -\int_{-\infty}^{\infty} \mathrm{d}x\, \delta(x-y) \frac{\mathrm{d}f(x)}{\mathrm{d}x} \\ &= -\frac{\mathrm{d}f(y)}{\mathrm{d}y}.\end{aligned}$$

Note that it would be easier to take the functional derivative if an integration by parts had been first performed.

(3) For the third functional, we have

$$\frac{\delta F[g]}{\delta g(y)} = \int_0^\infty \mathrm{d}x\ \delta(x-y)\frac{\partial \mathrm{e}^{-g^2(x)}}{\partial g(x)} = -2\int_0^\infty \mathrm{d}x\ \delta(x-y)g(x)\mathrm{e}^{-g^2(x)}$$
$$= -2g(y)\mathrm{e}^{-g^2(y)}\theta(y).$$

(4) For the fourth functional, we have

$$\frac{\delta F[g]}{\delta g(\boldsymbol{r}'')} = \frac{1}{2}\iint \mathrm{d}\boldsymbol{r}\mathrm{d}\boldsymbol{r}'\left[\delta(\boldsymbol{r}-\boldsymbol{r}'')\frac{\partial}{\partial g(\boldsymbol{r})} + \delta(\boldsymbol{r}'-\boldsymbol{r}'')\frac{\partial}{\partial g(\boldsymbol{r}')}\right]$$
$$\times\left[g(\boldsymbol{r})v(|\boldsymbol{r}-\boldsymbol{r}'|)g(\boldsymbol{r}')\right]$$
$$= \frac{1}{2}\iint \mathrm{d}\boldsymbol{r}\mathrm{d}\boldsymbol{r}'\left[\delta(\boldsymbol{r}-\boldsymbol{r}'')v(|\boldsymbol{r}-\boldsymbol{r}'|)g(\boldsymbol{r}')\right.$$
$$\left. + \delta(\boldsymbol{r}'-\boldsymbol{r}'')g(\boldsymbol{r})v(|\boldsymbol{r}-\boldsymbol{r}'|)\right]$$
$$= \frac{1}{2}\left[\int \mathrm{d}\boldsymbol{r}'\ v(|\boldsymbol{r}''-\boldsymbol{r}'|)g(\boldsymbol{r}') + \int \mathrm{d}\boldsymbol{r}\ g(\boldsymbol{r})v(|\boldsymbol{r}-\boldsymbol{r}''|)\right]$$
$$= \int \mathrm{d}\boldsymbol{r}'\ v(|\boldsymbol{r}''-\boldsymbol{r}'|)g(\boldsymbol{r}').$$

Dropping the double prime on the free variable $\boldsymbol{r}''$, we have

$$\frac{\delta F[g]}{\delta g(\boldsymbol{r})} = \int \mathrm{d}\boldsymbol{r}'\ v(|\boldsymbol{r}-\boldsymbol{r}'|)g(\boldsymbol{r}').$$

(5) For the fifth functional, we have

$$\frac{\delta E[n]}{\delta n(\boldsymbol{r}')} = c\int \mathrm{d}\boldsymbol{r}\ \delta(\boldsymbol{r}-\boldsymbol{r}')\frac{\partial}{\partial n(\boldsymbol{r})}n^{4/3}(\boldsymbol{r})$$
$$= \frac{4}{3}c\int \mathrm{d}\boldsymbol{r}\ \delta(\boldsymbol{r}-\boldsymbol{r}')n^{1/3}(\boldsymbol{r}) = \frac{4}{3}cn^{1/3}(\boldsymbol{r}').$$

Dropping the prime on the free variable $\boldsymbol{r}'$, we have

$$\frac{\delta E[n]}{\delta n(\boldsymbol{r})} = \frac{4}{3}cn^{1/3}(\boldsymbol{r}).$$

(6) Since this functional contains $\boldsymbol{\nabla} n(\boldsymbol{r})$, we can not use the procedure proved previously. Instead, we resort to the definition of the functional derivative. We have

$$\frac{\delta E[n]}{\delta n(\boldsymbol{r}')} = \frac{\hbar^2}{8m}\lim_{\epsilon\to 0}\frac{1}{\epsilon}\int \mathrm{d}\boldsymbol{r}\left\{\frac{\left[\boldsymbol{\nabla} n(\boldsymbol{r}) + \epsilon\boldsymbol{\nabla}\delta(\boldsymbol{r}-\boldsymbol{r}')\right]^2}{n(\boldsymbol{r})+\epsilon\delta(\boldsymbol{r}-\boldsymbol{r}')} - \frac{\left[\boldsymbol{\nabla} n(\boldsymbol{r})\right]^2}{n(\boldsymbol{r})}\right\}$$
$$= \frac{\hbar^2}{8m}\int \mathrm{d}\boldsymbol{r}\left\{\frac{2}{n(\boldsymbol{r})}\boldsymbol{\nabla} n(\boldsymbol{r})\cdot\boldsymbol{\nabla}\delta(\boldsymbol{r}-\boldsymbol{r}') - \frac{\left[\boldsymbol{\nabla} n(\boldsymbol{r})\right]^2}{n^2(\boldsymbol{r})}\delta(\boldsymbol{r}-\boldsymbol{r}')\right\}.$$

Performing an integration by parts to the first term in the curly brackets yields

$$\begin{aligned}\frac{\delta E[n]}{\delta n(\boldsymbol{r}')} &= -\frac{\hbar^2}{8m}\int \mathrm{d}\boldsymbol{r}\ \delta(\boldsymbol{r}-\boldsymbol{r}')\left\{2\boldsymbol{\nabla}\cdot\frac{\boldsymbol{\nabla}n(\boldsymbol{r})}{n(\boldsymbol{r})}+\frac{[\boldsymbol{\nabla}n(\boldsymbol{r})]^2}{n^2(\boldsymbol{r})}\right\}\\ &= -\frac{\hbar^2}{8m}\left\{2\boldsymbol{\nabla}'\cdot\frac{\boldsymbol{\nabla}'n(\boldsymbol{r}')}{n(\boldsymbol{r}')}+\frac{[\boldsymbol{\nabla}'n(\boldsymbol{r}')]^2}{n^2(\boldsymbol{r}')}\right\}\\ &= \frac{\hbar^2}{8m}\left\{\frac{[\boldsymbol{\nabla}'n(\boldsymbol{r}')]^2}{n^2(\boldsymbol{r}')}-2\frac{\boldsymbol{\nabla}'^2n(\boldsymbol{r}')}{n(\boldsymbol{r}')}\right\}.\end{aligned}$$

Dropping the prime on the free variable $\boldsymbol{r}'$, we have

$$\frac{\delta E[n]}{\delta n(\boldsymbol{r})} = \frac{\hbar^2}{8m}\left\{\frac{[\boldsymbol{\nabla}n(\boldsymbol{r})]^2}{n^2(\boldsymbol{r})}-2\frac{\boldsymbol{\nabla}^2n(\boldsymbol{r})}{n(\boldsymbol{r})}\right\}.$$

We now find a general procedure for taking the functional derivatives of the functionals of the form

$$F[g(\boldsymbol{r})] = \int \mathrm{d}\boldsymbol{r}\ \mathscr{F}(g(\boldsymbol{r}), \boldsymbol{\nabla}g(\boldsymbol{r})).$$

Making use of the definition of the functional derivative, we have

$$\begin{aligned}\frac{\delta F[g]}{\delta g(\boldsymbol{r})} &= \lim_{\epsilon\to 0}\frac{1}{\epsilon}\int \mathrm{d}\boldsymbol{r}'\ \big[\mathscr{F}\big(g(\boldsymbol{r}')+\epsilon\delta(\boldsymbol{r}'-\boldsymbol{r}), \boldsymbol{\nabla}'g(\boldsymbol{r}')+\epsilon\boldsymbol{\nabla}'\delta(\boldsymbol{r}'-\boldsymbol{r})\big)\\ &\qquad - \mathscr{F}\big(g(\boldsymbol{r}'), \boldsymbol{\nabla}'g(\boldsymbol{r}')\big)\big]\\ &= \lim_{\epsilon\to 0}\frac{1}{\epsilon}\int \mathrm{d}\boldsymbol{r}'\left[\epsilon\frac{\partial\mathscr{F}}{\partial g(\boldsymbol{r}')}\delta(\boldsymbol{r}'-\boldsymbol{r})+\epsilon\frac{\partial\mathscr{F}}{\partial\boldsymbol{\nabla}'g(\boldsymbol{r}')}\boldsymbol{\nabla}'\delta(\boldsymbol{r}'-\boldsymbol{r})\right]\\ &= \int \mathrm{d}\boldsymbol{r}'\left[\frac{\partial\mathscr{F}}{\partial g(\boldsymbol{r}')}\delta(\boldsymbol{r}'-\boldsymbol{r})+\frac{\partial\mathscr{F}}{\partial\boldsymbol{\nabla}'g(\boldsymbol{r}')}\cdot\boldsymbol{\nabla}'\delta(\boldsymbol{r}'-\boldsymbol{r})\right].\end{aligned}$$

Performing an integration by parts to the second term in the square brackets yields

$$\begin{aligned}\frac{\delta F[g]}{\delta g(\boldsymbol{r})} &= \int \mathrm{d}\boldsymbol{r}'\ \delta(\boldsymbol{r}'-\boldsymbol{r})\left[\frac{\partial\mathscr{F}}{\partial g(\boldsymbol{r}')}-\boldsymbol{\nabla}'\cdot\frac{\partial\mathscr{F}}{\partial\boldsymbol{\nabla}'g(\boldsymbol{r}')}\right]\\ &= \frac{\partial\mathscr{F}}{\partial g(\boldsymbol{r})}-\boldsymbol{\nabla}\cdot\frac{\partial\mathscr{F}}{\partial\boldsymbol{\nabla}g(\boldsymbol{r})}.\end{aligned}$$

Thus, for the functionals of the form $F[g(\boldsymbol{r})] = \int \mathrm{d}\boldsymbol{r}\ \mathscr{F}(g(\boldsymbol{r}), \boldsymbol{\nabla}g(\boldsymbol{r}))$, we can make use of the following rule to compute their functional derivatives

$$\frac{\delta F[g]}{\delta g(\boldsymbol{r})} = \frac{\partial\mathscr{F}}{\partial g(\boldsymbol{r})}-\boldsymbol{\nabla}\cdot\frac{\partial\mathscr{F}}{\partial\boldsymbol{\nabla}g(\boldsymbol{r})}.$$

21-6 Action functional and Euler-Lagrange equation. The action functional for a particle in one-dimensional potential $U(x)$ is given by

$$S[x(t), \dot{x}(t)] = \int_{t_a}^{t_b} \mathrm{d}t\; L(x, \dot{x}) = \int_{t_a}^{t_b} \mathrm{d}t \left[\frac{1}{2}m\dot{x}^2 - U(x)\right],$$

where $L(x, \dot{x})$ is the Lagrangian of the particle.

(1) Evaluate the functional derivatives $\delta S/\delta x(t)$ and $\delta S/\delta \dot{x}(t)$.
(2) Find the differential of the action functional, $\delta S[x(t), \dot{x}(t)]$.
(3) Derive the Euler-Lagrange equation by minimizing the action.

---

(1) The functional derivative of $S[x, \dot{x}]$ with respect to $x(t)$ is given by

$$\begin{aligned} \frac{\delta S}{\delta x(t)} &= \frac{\delta}{\delta x(t)} \int_{t_a}^{t_b} \mathrm{d}t'\; L(x(t'), \dot{x}(t')) \\ &= \int_{t_a}^{t_b} \mathrm{d}t'\; \delta(t'-t) \frac{\partial L(x(t'), \dot{x}(t'))}{\partial x(t')} = \frac{\partial L(x, \dot{x})}{\partial x(t)}. \end{aligned}$$

The functional derivative of $S[x, \dot{x}]$ with respect to $\dot{x}(t)$ is given by

$$\begin{aligned} \frac{\delta S}{\delta \dot{x}(t)} &= \frac{\delta}{\delta \dot{x}(t)} \int_{t_a}^{t_b} \mathrm{d}t'\; L(x(t'), \dot{x}(t')) \\ &= \int_{t_a}^{t_b} \mathrm{d}t'\; \delta(t'-t) \frac{\partial L(x(t'), \dot{x}(t'))}{\partial \dot{x}(t')} = \frac{\partial L(x, \dot{x})}{\partial \dot{x}(t)}. \end{aligned}$$

(2) The differential of the action functional is given by

$$\begin{aligned} \delta S &= \int_{t_a}^{t_b} \mathrm{d}t \left[\frac{\delta S}{\delta \dot{x}(t)} \delta \dot{x}(t) + \frac{\delta S}{\delta x(t)} \delta x(t) \right] \\ &= \int_{t_a}^{t_b} \mathrm{d}t \left[\frac{\partial L(x, \dot{x})}{\partial \dot{x}(t)} \delta \dot{x}(t) + \frac{\partial L(x, \dot{x})}{\partial x(t)} \delta x(t) \right]. \end{aligned}$$

Performing an integration by parts to the first term in the square brackets, we have

$$\delta S = -\int_{t_a}^{t_b} \mathrm{d}t \left[\frac{\mathrm{d}}{\mathrm{d}t} \frac{\partial L(x, \dot{x})}{\partial \dot{x}(t)} - \frac{\partial L(x, \dot{x})}{\partial x(t)} \right] \delta x(t),$$

where we have made use of the fact that the integrated part is identically zero.

(3) Minimizing the action S implies that the differential of the action functional, δS, is to be set to zero. From the above-derived expression for δS, we see that, for δS to be zero, the expression in the square brackets in the above equation must be identically zero. We thus have

$$\frac{\mathrm{d}}{\mathrm{d}t}\frac{\partial L(x,\dot{x})}{\partial \dot{x}(t)} - \frac{\partial L(x,\dot{x})}{\partial x(t)} = 0.$$

The above equation is just the Euler-Lagrange equation. If the explicit expression for $L(x,\dot{x})$,

$$L(x,\dot{x}) = \frac{1}{2}m\dot{x}^2 - U(x),$$

had been used in the above algebras, we would have obtained

$$m\ddot{x} = -\frac{\partial U}{\partial x}$$

which is just the statement of Newton's second law. We can also obtain the above result from the Euler-Lagrange equation.

21-7 One-electron system. Consider a one-electron system with the ground-state electron density given by $n(\boldsymbol{r})$. The external potential $v(\boldsymbol{r})$ is uniquely determined within an arbitrary constant.

(1) Show that for a one-electron system the kinetic energy functional $T[n]$ is given by

$$T[n] = \frac{\hbar^2}{8m}\int \mathrm{d}\boldsymbol{r}\, \frac{[\boldsymbol{\nabla} n(\boldsymbol{r})]^2}{n(\boldsymbol{r})}.$$

(2) Apply the variational principle to derive an expression for the potential $v(\boldsymbol{r})$ in terms of the electron density $n(\boldsymbol{r})$.

(3) Use the above-obtained result to find the potentials for a one-dimensional electron density $n(x) = A\mathrm{e}^{-\alpha x^2}$ and for a spherically-symmetric three-dimensional electron density $n(r) = A\mathrm{e}^{-\alpha r}$.

---

(1) In the ground state of a one-electron system described by the wave function $\Psi(\boldsymbol{r})$, the kinetic energy of the electron is given by

$$T = -\frac{\hbar^2}{2m}\int \mathrm{d}\boldsymbol{r}\ \Psi^*(\boldsymbol{r})\nabla^2\Psi(\boldsymbol{r}) = \frac{\hbar^2}{2m}\int \mathrm{d}\boldsymbol{r}\ \boldsymbol{\nabla}\Psi^*(\boldsymbol{r})\cdot\boldsymbol{\nabla}\Psi(\boldsymbol{r})$$

in which the second expression is obtained from the first through an integration by parts. We now consider a one-electron system with a real ground-state wave function, $\Psi^*(\boldsymbol{r}) = \Psi(\boldsymbol{r})$. For such a one-electron system, we can rewrite its kinetic energy in the ground state as

$$T = \frac{\hbar^2}{2m}\int \mathrm{d}\boldsymbol{r}\, [\boldsymbol{\nabla}\Psi(\boldsymbol{r})]^2 = \frac{\hbar^2}{2m}\int \mathrm{d}\boldsymbol{r}\, \frac{[\Psi(\boldsymbol{r})\boldsymbol{\nabla}\Psi(\boldsymbol{r})]^2}{[\Psi(\boldsymbol{r})]^2}$$
$$= \frac{\hbar^2}{8m}\int \mathrm{d}\boldsymbol{r}\, \frac{\{\boldsymbol{\nabla}[\Psi(\boldsymbol{r})]^2\}^2}{[\Psi(\boldsymbol{r})]^2}.$$

For such a one-electron system, the electron density in the ground state is simply given by $n(\boldsymbol{r}) = [\Psi(\boldsymbol{r})]^2$. Inserting $n(\boldsymbol{r}) = [\Psi(\boldsymbol{r})]^2$ into the above equation, we obtain

$$T[n] = \frac{\hbar^2}{8m}\int \mathrm{d}\boldsymbol{r}\, \frac{[\boldsymbol{\nabla}n(\boldsymbol{r})]^2}{n(\boldsymbol{r})}.$$

(2) The exchange-correlation energy is absent in a one-electron system. The energy functional of a one-electron system is then given by

$$E[n] = T[n] + \int \mathrm{d}\boldsymbol{r}\, n(\boldsymbol{r})v(\boldsymbol{r}) = \frac{\hbar^2}{8m}\int \mathrm{d}\boldsymbol{r}\, \frac{[\boldsymbol{\nabla}n(\boldsymbol{r})]^2}{n(\boldsymbol{r})} + \int \mathrm{d}\boldsymbol{r}\, n(\boldsymbol{r})v(\boldsymbol{r}).$$

Taking the functional derivative of the energy functional with respect to $n(\boldsymbol{r})$ and setting the result to zero, we have

$$0 = \frac{\delta E[n]}{\delta n(\boldsymbol{r})} = \frac{\hbar^2}{8m}\left\{\frac{[\boldsymbol{\nabla}n(\boldsymbol{r})]^2}{(n(\boldsymbol{r}))^2} - 2\frac{\nabla^2 n(\boldsymbol{r})}{n(\boldsymbol{r})}\right\} + v(\boldsymbol{r})$$

from which we obtain the expression of $v(\boldsymbol{r})$ for the given electron density $n(\boldsymbol{r})$

$$v(\boldsymbol{r}) = \frac{\hbar^2}{8m}\left\{2\frac{\nabla^2 n(\boldsymbol{r})}{n(\boldsymbol{r})} - \frac{[\boldsymbol{\nabla}n(\boldsymbol{r})]^2}{(n(\boldsymbol{r}))^2}\right\}.$$

(3) For the one-dimensional electron density $n(x) = A\mathrm{e}^{-\alpha x^2}$, we have

$$v(x) = \frac{\hbar^2}{8m}\left\{\frac{2}{n(x)}\frac{\mathrm{d}^2 n(x)}{\mathrm{d}x^2} - \frac{1}{[n(x)]^2}\left[\frac{\mathrm{d}n(x)}{\mathrm{d}x}\right]^2\right\}$$
$$= -\frac{\alpha\hbar^2}{2m} + \frac{\alpha^2\hbar^2}{2m}x^2.$$

We see that the obtained potential is a harmonic potential plus a constant. This is expected since the given electron density is the one that is determined by the wave function of an electron in the ground state of a harmonic potential. Note that $\alpha = m\omega/\hbar$ here instead of the commonly-seen expression $\alpha = \sqrt{m\omega/\hbar}$ in many textbooks on elementary quantum mechanics. Inserting $\alpha = m\omega/\hbar$ into the expression of the potential, we have

$$v(x) = -\frac{1}{2}\hbar\omega + \frac{1}{2}m\omega^2 x^2.$$

Note that the constant term is equal to the negative of the ground-state energy.

For the spherically-symmetric three-dimensional electron density $n(r) = Ae^{-\alpha r}$, we have

$$\begin{aligned} v(r) &= \frac{\hbar^2}{8m}\left\{\frac{2}{rn(r)}\frac{d^2[rn(r)]}{dr^2} - \frac{1}{[n(r)]^2}\left[\frac{dn(r)}{dr}\right]^2\right\} \\ &= \frac{\alpha^2\hbar^2}{8m} - \frac{\alpha\hbar^2}{2mr}. \end{aligned}$$

The potential is actually the Coulomb potential in a hydrogen atom plus a constant. Upon recognizing that $n(r) = Ae^{-\alpha r}$ can be given by the ground-state wave function, $\psi_{100}(r) = (\pi a_0^3)^{-1/2}e^{-r/a_0}$, of the electron in a hydrogen atom, we see that $\alpha = 2/a_0$. Here $a_0 = 4\pi\epsilon_0\hbar^2/me^2$ is the Bohr radius. The potential is then given by

$$v(r) = \frac{me^4}{2(4\pi\epsilon_0)^2\hbar^2} - \frac{e^2}{4\pi\epsilon_0 r}.$$

Note that the constant term, equal to one unit of Rydberg energy, is the negative of the ground-state energy.

21-8 Thomas-Fermi screening. For an electron system in potential $U(\boldsymbol{r})$, the Thomas-Fermi equation is given by

$$\mu = \frac{5}{3}\kappa n^{2/3}(\boldsymbol{r}) + U(\boldsymbol{r}) - \frac{Ze^2}{4\pi\epsilon_0}\int d\boldsymbol{r}'\,\frac{1}{|\boldsymbol{r}-\boldsymbol{r}'|} + \frac{e^2}{4\pi\epsilon_0}\int d\boldsymbol{r}'\,\frac{n(\boldsymbol{r}')}{|\boldsymbol{r}-\boldsymbol{r}'|}.$$

The Thomas-Fermi equation holds for a constant density n_0 if $U(\boldsymbol{r}) = 0$. We now consider an impurity potential $U(r) = -e^2/4\pi\epsilon_0 r$ centered at the origin. Assume that the influence of $U(r)$ on the electron density is small so that the electron density in the presence of $U(r)$ can be written as $n(r) = n_0 + \delta n(r)$ where the change in density, $\delta n(r)$, is small.

(1) Derive a linear equation for $\delta n(\boldsymbol{r})$ by substituting $n(r) = n_0 + \delta n(r)$ into the Thomas-Fermi equation and expanding $n^{2/3}(r)$ in terms of $\delta n(r)$ up to the first order in $\delta n(r)$.

(2) Solve for $\delta n(r)$ through making a Fourier transformation $\delta n(r) = \sum_{\boldsymbol{q}} \delta n_{\boldsymbol{q}} \mathrm{e}^{\mathrm{i}\boldsymbol{q}\cdot\boldsymbol{r}}$ to $\delta n(r)$. Show that $\delta n(r) \sim \mathrm{e}^{-r/\xi}/r$. Find an expression for the "screening length" ξ.

---

(1) Up to the first order in $\delta n(r)$, we have

$$n^{2/3}(r) = \big[n_0 + \delta n(r)\big]^{2/3} \approx n_0^{2/3} + \frac{2}{3n_0^{1/3}}\delta n(r).$$

Inserting the above expansion into the Thomas-Fermi equation, we obtain

$$\begin{aligned}&\frac{10\kappa}{9n_0^{1/3}}\delta n(\boldsymbol{r}) + \frac{e^2}{4\pi\epsilon_0}\int \mathrm{d}\boldsymbol{r}'\, \frac{\delta n(\boldsymbol{r}')}{|\boldsymbol{r}-\boldsymbol{r}'|}\\ &\quad = \mu - \frac{5}{3}\kappa n_0^{2/3} - \frac{n_0 e^2}{4\pi\epsilon_0}\int \frac{\mathrm{d}\boldsymbol{r}'}{|\boldsymbol{r}-\boldsymbol{r}'|} + \frac{Ze^2}{4\pi\epsilon_0}\int \frac{\mathrm{d}\boldsymbol{r}'}{|\boldsymbol{r}-\boldsymbol{r}'|} + \frac{e^2}{4\pi\epsilon_0 r}\\ &\quad = \mu - \frac{5}{3}\kappa n_0^{2/3} + \frac{e^2}{4\pi\epsilon_0 r},\end{aligned}$$

where the third and fourth terms are canceled in the second equality due to the charge neutrality.

(2) Inserting the Fourier transformation

$$\delta n(\boldsymbol{r}) = \sum_{\boldsymbol{q}} \delta n_{\boldsymbol{q}} \mathrm{e}^{\mathrm{i}\boldsymbol{q}\cdot\boldsymbol{r}}$$

into the above equation and making use of

$$\frac{1}{r} = \frac{4\pi}{\mathscr{V}}\sum_{\boldsymbol{q}} \frac{\mathrm{e}^{\mathrm{i}\boldsymbol{q}\cdot\boldsymbol{r}}}{q^2},$$

we have

$$\begin{aligned}&\frac{10\kappa}{9n_0^{1/3}}\sum_{\boldsymbol{q}} \delta n_{\boldsymbol{q}} \mathrm{e}^{\mathrm{i}\boldsymbol{q}\cdot\boldsymbol{r}} + \frac{e^2}{4\pi\epsilon_0}\sum_{\boldsymbol{q}} \delta n_{\boldsymbol{q}} \int \mathrm{d}\boldsymbol{r}'\, \frac{\mathrm{e}^{\mathrm{i}\boldsymbol{q}\cdot\boldsymbol{r}'}}{|\boldsymbol{r}-\boldsymbol{r}'|}\\ &\quad = \mu - \frac{5}{3}\kappa n_0^{2/3} + \frac{e^2}{\epsilon_0 \mathscr{V}}\sum_{\boldsymbol{q}} \frac{\mathrm{e}^{\mathrm{i}\boldsymbol{q}\cdot\boldsymbol{r}}}{q^2}.\end{aligned}$$

Making use of

$$\int \mathrm{d}\boldsymbol{r}'\, \frac{\mathrm{e}^{\mathrm{i}\boldsymbol{q}\cdot\boldsymbol{r}'}}{|\boldsymbol{r}-\boldsymbol{r}'|} = \frac{4\pi}{q^2}\mathrm{e}^{\mathrm{i}\boldsymbol{q}\cdot\boldsymbol{r}},$$

we have

$$\frac{10\kappa}{9n_0^{1/3}}\sum_{\boldsymbol{q}} \delta n_{\boldsymbol{q}} \mathrm{e}^{\mathrm{i}\boldsymbol{q}\cdot\boldsymbol{r}} + \frac{e^2}{\epsilon_0}\sum_{\boldsymbol{q}} \frac{\delta n_{\boldsymbol{q}}}{q^2}\mathrm{e}^{\mathrm{i}\boldsymbol{q}\cdot\boldsymbol{r}} = \mu - \frac{5}{3}\kappa n_0^{2/3} + \frac{e^2}{\epsilon_0 \mathscr{V}}\sum_{\boldsymbol{q}} \frac{\mathrm{e}^{\mathrm{i}\boldsymbol{q}\cdot\boldsymbol{r}}}{q^2}.$$

To extract $\delta n_{\boldsymbol{q}}$ from the above equation, we multiply both sides of the above equation with $\mathscr{V}^{-1}\mathrm{e}^{-\mathrm{i}\boldsymbol{q}'\cdot\boldsymbol{r}}$ and then integrate both sides of the resultant equation over $\boldsymbol{r}$. Making use of

$$\frac{1}{\mathscr{V}}\int \mathrm{d}\boldsymbol{r}\ \mathrm{e}^{\mathrm{i}(\boldsymbol{q}-\boldsymbol{q}')\cdot\boldsymbol{r}} = \delta_{\boldsymbol{q}\boldsymbol{q}'},$$

we have

$$\frac{10\kappa}{9n_0^{1/3}}\delta n_{\boldsymbol{q}} + \frac{e^2\delta n_{\boldsymbol{q}}}{\epsilon_0 q^2} = \left(\mu - \frac{5}{3}\kappa n_0^{2/3}\right)\delta_{\boldsymbol{q}0} + \frac{e^2}{\epsilon_0\mathscr{V}q^2},$$

where we have dropped the prime on $\boldsymbol{q}$. For $\boldsymbol{q} \neq 0$, solving for $n_{\boldsymbol{q}\neq 0}$ from the above equation, we obtain

$$\delta n_{\boldsymbol{q}\neq 0} = \frac{1}{\mathscr{V}}\frac{1}{1+(10\epsilon_0\kappa/9n_0^{1/3}e^2)q^2} = \frac{1}{\mathscr{V}}\frac{1}{1+\xi^2q^2},$$

where

$$\xi = \left(\frac{10\epsilon_0\kappa}{9n_0^{1/3}e^2}\right)^{1/2} = \frac{\hbar k_{\mathrm{F}}}{(3n_0me^2/\epsilon_0)^{1/2}}.$$

For $\boldsymbol{q} = 0$, $\delta n_{\boldsymbol{q}=0}$ must be equal to $1/\mathscr{V}$ so that the two divergent terms on the two sides of the equation for $n_{\boldsymbol{q}}$ cancel. We then obtain the value of the chemical potential

$$\mu = \frac{5}{3}\kappa n_0^{2/3} + \frac{10\kappa}{9n_0^{1/3}\mathscr{V}} = \frac{5}{3}\kappa n_0^{2/3}\left(1+\frac{2}{3n_0\mathscr{V}}\right) = \frac{5}{3}\kappa n_0^{2/3}\left(1+\frac{2}{3N}\right)$$

with $N = n_0\mathscr{V}$ the total number of electrons. Since the value of $\delta n_{\boldsymbol{q}=0}$ can be obtained from the expression of $\delta n_{\boldsymbol{q}\neq 0}$ in the $\boldsymbol{q}\to 0$ limit, the expression for $\delta n_{\boldsymbol{q}}$ valid for all values of $\boldsymbol{q}$ is given by

$$\delta n_{\boldsymbol{q}} = \frac{1}{\mathscr{V}}\frac{1}{1+\xi^2q^2}.$$

$\delta n(\boldsymbol{r})$ is then given by

$$\begin{aligned}\delta n(\boldsymbol{r}) &= \sum_{\boldsymbol{q}}\delta n_{\boldsymbol{q}}\mathrm{e}^{\mathrm{i}\boldsymbol{q}\cdot\boldsymbol{r}} = \frac{\mathscr{V}}{(2\pi)^3}\int \mathrm{d}\boldsymbol{q}\,\delta n_{\boldsymbol{q}}\mathrm{e}^{\mathrm{i}\boldsymbol{q}\cdot\boldsymbol{r}} = \frac{1}{(2\pi)^3}\int \mathrm{d}\boldsymbol{q}\,\frac{\mathrm{e}^{\mathrm{i}\boldsymbol{q}\cdot\boldsymbol{r}}}{1+\xi^2q^2}\\ &= \frac{1}{4\pi^2}\int_0^\infty \mathrm{d}q\,\frac{q^2}{1+\xi^2q^2}\int_{-1}^{1}\mathrm{d}\cos\theta\ \mathrm{e}^{\mathrm{i}qr\cos\theta}\\ &= \frac{1}{4\pi^2\mathrm{i}r}\int_0^\infty \mathrm{d}q\,q\frac{\mathrm{e}^{\mathrm{i}qr}-\mathrm{e}^{-\mathrm{i}qr}}{1+\xi^2q^2} = \frac{1}{4\pi^2\mathrm{i}r}\int_{-\infty}^{\infty}\mathrm{d}q\,\frac{q\mathrm{e}^{\mathrm{i}qr}}{1+\xi^2q^2}.\end{aligned}$$

The integrand of the above integral has two first-order poles at $\pm\mathrm{i}/\xi$. Because of the presence of the exponential factor $\mathrm{e}^{\mathrm{i}qr}$

in the integrand, we can only close the contour in the upper-half complex plane. Making use of the residue theorem, we obtain

$$\delta n(\boldsymbol{r}) = \frac{1}{4\pi\xi^2}\frac{\mathrm{e}^{-r/\xi}}{r}.$$

The length ξ is known as *the Thomas-Fermi screening length*. Note that $\delta n(\boldsymbol{r}) > 0$, which implies that the sign of the screening charge is opposite to that of the charge on the impurity that is positively charged as can be seen from the interaction Hamiltonian $U(r) = -e^2/4\pi\epsilon_0 r$. Note that $\delta n(\boldsymbol{r})$ is the change in the electron density. Integrating $\delta n(\boldsymbol{r})$ over the entire space, we have

$$\int \mathrm{d}\boldsymbol{r}\ \delta n(\boldsymbol{r}) = 1$$

which implies that the total charge in the screening charge cloud is equal to the negative of the charge on the impurity.

21-9 ***N*-representable density.** An N-representable density can be written as the density of an antisymmetric wave function. For a density $n(x)$ in one dimension, we define an auxiliary function $f(x)$ through $\mathrm{d}f(x)/\mathrm{d}x = (2\pi/N)n(x)$ with N the total number of electrons. Define a set of orbitals

$$\psi_k(x) = \sqrt{n(x)/N}\,\mathrm{e}^{\mathrm{i}[kf(x)+\phi(x)]},$$

where k is an integer and $\phi(x)$ a phase factor.

(1) Show that orbitals ψ_k's are orthonormal

$$\int_{-\infty}^{\infty} \mathrm{d}x\ \psi_{k'}^*(x)\psi_k(x) = \delta_{kk'}.$$

(2) Show that orbitals ψ_k's are complete

$$\sum_k \psi_k^*(x)\psi_k(y) = \delta(x-y).$$

(3) Show that a Slater determinant made of N such orbitals yields the density $n(x)$ when the expectation value of the density operator is taken.

---

(1) We first find the size of the space of $f(x)$'s. Integrating both sides of $\mathrm{d}f(x)/\mathrm{d}x = (2\pi/N)n(x)$ over x, we have

$$\int_{f(-\infty)}^{f(\infty)} \mathrm{d}f = \frac{2\pi}{N}\int_{-\infty}^{\infty} \mathrm{d}x\ n(x) = 2\pi.$$

Thus, the size of the space of $f(x)$'s is 2π. If we take $f(-\infty) = 0$, we then have

$$f(x) = \frac{2\pi}{N}\int_{-\infty}^{x} \mathrm{d}x\; n(x),$$

and $f \in [0, 2\pi]$. Making use of the given expression of $\psi_k(x)$, we have

$$\begin{aligned}\int_{-\infty}^{\infty} \mathrm{d}x\; \psi_{k'}^*(x)\psi_k(x) &= \frac{1}{N}\int_{-\infty}^{\infty} \mathrm{d}x\; n(x)\mathrm{e}^{\mathrm{i}(k-k')f(x)} \\ &= \frac{1}{2\pi}\int_0^{2\pi} \mathrm{d}f\; \mathrm{e}^{\mathrm{i}(k-k')f} = \delta_{kk'}.\end{aligned}$$

(2) The completeness relation can be derived as follows

$$\begin{aligned}\sum_k \psi_k^*(x)\psi_k(y) &= \frac{\sqrt{n(x)n(y)}}{N}e^{-\mathrm{i}[\phi(x)-\phi(y)]}\sum_k \mathrm{e}^{-\mathrm{i}k[f(x)-f(y)]} \\ &= \frac{\sqrt{n(x)n(y)}}{N}e^{-\mathrm{i}[\phi(x)-\phi(y)]}\int_{-\infty}^{\infty} \mathrm{d}k\; \mathrm{e}^{-\mathrm{i}k[f(x)-f(y)]} \\ &= \frac{2\pi\sqrt{n(x)n(y)}}{N}e^{-\mathrm{i}[\phi(x)-\phi(y)]}\delta\big(f(x)-f(y)\big) \\ &= \frac{2\pi\sqrt{n(x)n(y)}}{N}e^{-\mathrm{i}[\phi(x)-\phi(y)]}\frac{\delta(x-y)}{|\mathrm{d}f(y)/\mathrm{d}y|} \\ &= \delta(x-y),\end{aligned}$$

where we have made use of $\delta(\varphi(x)) = \sum_i \delta(x-x_i)/|\varphi'(x_i)|$ with a prime on φ denoting its first-order derivative and x_i's its first-order zero points.

(3) The N orbitals are denoted by $\psi_1(x)$, $\psi_2(x)$, $\cdots$, $\psi_N(x)$. The Slater determinant constructed from these N orbitals is given by

$$\Psi(x_1, x_2, \cdots, x_N) = \frac{1}{\sqrt{N!}}\sum_P (-1)^P \psi_{k_{P_1}}(x_1)\psi_{k_{P_2}}(x_2)\cdots\psi_{k_{P_N}}(x_N),$$

where P is a permutation of the set of N integers $\{1, 2, \cdots, N\}$ with $(-1)^P$ its parity. Taking the expectation value of the density operator $\hat{n}(x) = \sum_{i=1}^N \delta(x-x_i)$ in Ψ, we have

$$\begin{aligned}&\int_{-\infty}^{\infty} \mathrm{d}x_1\mathrm{d}x_2\cdots\mathrm{d}x_N\; \Psi^*(x_1, x_2, \cdots, x_N)\hat{n}(x)\Psi(x_1, x_2, \cdots, x_N) \\ &= \frac{1}{N!}\sum_i\sum_{PP'}(-1)^P(-1)^{P'}\int_{-\infty}^{\infty} \mathrm{d}x_1\cdots\mathrm{d}x_i\cdots\mathrm{d}x_N\; \psi_{k_{P_1'}}^*(x_1)\cdots\end{aligned}$$

$$\times\, \psi^*_{k_{P'_2}}(x_i)\cdots\psi^*_{k_{P'_N}}(x_N)\delta(x-x_i)\psi_{k_{P_1}}(x_1)\cdots$$
$$\times\, \psi_{k_{P_2}}(x_i)\cdots\psi_{k_{P_N}}(x_N)$$
$$= \frac{1}{N!}\sum_i\sum_{PP'}(-1)^P(-1)^{P'}\int_{-\infty}^{\infty}\mathrm{d}x_1\,\psi^*_{k_{P'_1}}(x_1)\psi_{k_{P_1}}(x_1)\cdots$$
$$\times\int_{-\infty}^{\infty}\mathrm{d}x_i\,\psi^*_{k_{P'_i}}(x_i)\delta(x-x_i)\psi_{k_{P_i}}(x_i)\cdots$$
$$\times\int_{-\infty}^{\infty}\mathrm{d}x_N\,\psi^*_{k_{P'_N}}(x_N)\psi_{k_{P_N}}(x_N)$$
$$= \frac{1}{N!}\sum_i\sum_{PP'}(-1)^P(-1)^{P'}\psi^*_{k_{P'_i}}(x)\psi_{k_{P_i}}(x)\prod_{j\,(\neq i)}\delta_{k_{P'_j}k_{P_j}}$$
$$= \frac{1}{N!}\sum_i\sum_{PP'}(-1)^P(-1)^{P'}\psi^*_{k_{P'_i}}(x)\psi_{k_{P_i}}(x)\delta_{PP'}$$
$$= \frac{1}{N!}\sum_i\sum_P\frac{n(x)}{N} = n(x).$$

21-10 Electron-ion interaction functional. Show that the electron-ion interaction

$$E_{\text{el-ion}} = \sum_i \sum\!\!\!\!\!\!\int \mathrm{d}x_1\cdots \sum\!\!\!\!\!\!\int \mathrm{d}x_N\;\Psi^*(x_1,\cdots,x_N)v_{\text{ext}}(\boldsymbol{r}_i)\Psi(x_1,\cdots,x_N)$$

with $v_{\text{ext}}(\boldsymbol{r}_i) = -\sum_I Z_I e^2/4\pi\epsilon_0|\boldsymbol{r}_i-\boldsymbol{R}_I|$ in the Hartree-Fock theory can be expressed as a functional of the electron density $n(\boldsymbol{r})$

$$E_{\text{el-ion}}[n] = \int \mathrm{d}\boldsymbol{r}\; n(\boldsymbol{r})v_{\text{ext}}(\boldsymbol{r}).$$

In the Hartree-Fock theory, the trial wave function for the ground state of an electron gas is given by the Slater determinant of single-electron states

$$\Psi(x_1,x_2,\cdots,x_N) = \frac{1}{\sqrt{N!}}\sum_P(-1)^P\psi_{\lambda_{P_1}}(x_1)\psi_{\lambda_{P_2}}(x_2)\cdots\psi_{\lambda_{P_N}}(x_N),$$

where P is a permutation of N integers $\{1,2,\cdots,N\}$ with $(-1)^P$ the parity of the permutation. Making use of the above trial wave function, we have

$$E_{\text{el-ion}} = \frac{1}{N!}\sum_i \sum\!\!\!\!\!\!\int \mathrm{d}x_1\cdots \sum\!\!\!\!\!\!\int \mathrm{d}x_N\;\sum_{PP'}(-1)^P(-1)^{P'}\psi^*_{\lambda_{P'_1}}(x_1)\psi^*_{\lambda_{P'_2}}(x_2)$$
$$\times\cdots\psi^*_{\lambda_{P'_N}}(x_N)v_{\text{ext}}(\boldsymbol{r}_i)\psi_{\lambda_{P_1}}(x_1)\psi_{\lambda_{P_2}}(x_2)\cdots\psi_{\lambda_{P_N}}(x_N),$$

where the permutation in Ψ^* has been written as P'. Grouping together $\psi^*_{\lambda_{P'_j}}(x_j)$ and $\psi_{\lambda_{P_j}}(x_j)$ and then making use of the orthonormality of $\psi_\lambda(x)$'s, we have

$$\begin{aligned}
E_{\text{el-ion}} &= \frac{1}{N!}\sum_i \sum_{PP'}(-1)^P(-1)^{P'} \sumint \mathrm{d}x_1\, \psi^*_{\lambda_{P'_1}}(x_1)\psi_{\lambda_{P_1}}(x_1) \\
&\quad \times \sumint \mathrm{d}x_2\, \psi^*_{\lambda_{P'_2}}(x_2)\psi_{\lambda_{P_2}}(x_2)\cdots \\
&\quad \times \sumint \mathrm{d}x_i\, \psi^*_{\lambda_{P'_i}}(x_i)\psi_{\lambda_{P_i}}(x_i)v_{\text{ext}}(\boldsymbol{r}_i)\cdots \\
&\quad \times \sumint \mathrm{d}x_N\, \psi^*_{\lambda_{P'_N}}(x_N)\psi_{\lambda_{P_N}}(x_N) \\
&= \frac{1}{N!}\sum_i \sum_{PP'}(-1)^P(-1)^{P'}\delta_{P'_1P_1}\delta_{P'_2P_2}\cdots \\
&\quad \times \sumint \mathrm{d}x_i\, \psi^*_{\lambda_{P'_i}}(x_i)\psi_{\lambda_{P_i}}(x_i)v_{\text{ext}}(\boldsymbol{r}_i)\cdots\delta_{P'_NP_N} \\
&= \frac{1}{N!}\sum_i \sum_{PP'}(-1)^P(-1)^{P'}\delta_{P'P} \sumint \mathrm{d}x_i\, \psi^*_{\lambda_{P'_i}}(x_i)\psi_{\lambda_{P_i}}(x_i)v_{\text{ext}}(\boldsymbol{r}_i) \\
&= \frac{1}{N!}\sum_i \sum_{P} \sumint \mathrm{d}x_i\, \psi^*_{\lambda_{P_i}}(x_i)\psi_{\lambda_{P_i}}(x_i)v_{\text{ext}}(\boldsymbol{r}_i).
\end{aligned}$$

Making use of the fact that all the $N!$ permutations yield the same result, we have

$$E_{\text{el-ion}} = \sum_\lambda \sumint \mathrm{d}x\, \psi^*_\lambda(x)\psi_\lambda(x)v_{\text{ext}}(\boldsymbol{r}),$$

where we have used $x = (\boldsymbol{r}, s_z)$ as the integration-summation variable. Inserting $\psi_\lambda(x) = \psi_i(\boldsymbol{r})\chi_\sigma(s_z)$ into the above equation and utilizing the normalization of spin wave functions, we have

$$\begin{aligned}
E_{\text{el-ion}} &= \sum_{i\sigma}\int \mathrm{d}\boldsymbol{r}\, \psi^*_i(\boldsymbol{r})\psi_i(\boldsymbol{r})v_{\text{ext}}(\boldsymbol{r})\sum_{s_z}\chi^*_\sigma(s_z)\chi_\sigma(s_z) \\
&= 2\sum_i \int \mathrm{d}\boldsymbol{r}\, \psi^*_i(\boldsymbol{r})\psi_i(\boldsymbol{r})v_{\text{ext}}(\boldsymbol{r}) = \int \mathrm{d}\boldsymbol{r}\, n(\boldsymbol{r})v_{\text{ext}}(\boldsymbol{r}),
\end{aligned}$$

where

$$n(\boldsymbol{r}) = 2\sum_i \psi^*_i(\boldsymbol{r})\psi_i(\boldsymbol{r}) = 2\sum_i |\psi_i(\boldsymbol{r})|^2.$$

21-11 Janak's Theorem. In this problem, we fill in the steps in Janak's proof of his theorem. The Kohn-Sham equations in atomic units for single-particle energies ε_i's and orbitals $\psi_i(\boldsymbol{r})$'s are

$$[-\nabla^2/2 + V_{\text{H}}(\boldsymbol{r}) + V_{\text{xc}}(\boldsymbol{r})]\psi_i(\boldsymbol{r}) = \varepsilon_i\psi_i(\boldsymbol{r}).$$

Note that the electron-ion interaction is included in $V_{\rm H}(\boldsymbol{r})$ here. To be able to study excited states within the density functional theory, we introduce an occupation number n_i for single-particle orbital $\psi_i(\boldsymbol{r})$ of energy ε_i and define the electron density as $n(\boldsymbol{r}) = \sum_{i=1}^{N} n_i|\psi_i(\boldsymbol{r})|^2$, where N is the number of electrons. With the kinetic energy functional expressed as $T[n] = \sum_i n_i t_i$, where $t_i = \int \mathrm{d}\boldsymbol{r}\ \psi_i^*(-\nabla^2/2)\psi_i = \varepsilon_i - \int \mathrm{d}\boldsymbol{r}\ \psi_i^*(V_{\rm H} + V_{\rm xc})\psi_i$, the energy functional can be written as $E = T[n] + E_{\rm H}[n] + E_{\rm xc}[n]$, where $E_{\rm H}[n]$ includes both the electron-electron and electron-ion interactions. Note that $V_{\rm H}(\boldsymbol{r}) = \delta E_{\rm H}[n]/\delta n(\boldsymbol{r})$ and $V_{\rm xc}(\boldsymbol{r}) = \delta E_{\rm xc}[n]/\delta n(\boldsymbol{r})$. Do not confuse $n(\boldsymbol{r})$ with n_i.

(1) Show that

$$\frac{\partial E}{\partial n_i} = t_i + \sum_j n_j \frac{\partial t_j}{\partial n_i} + \int \mathrm{d}\boldsymbol{r}\ (V_{\rm H} + V_{\rm xc})\left(|\psi_i|^2 + \sum_j n_j \frac{|\psi_j|^2}{\partial n_i}\right).$$

(2) Show that

$$\frac{\partial E}{\partial n_i} = \varepsilon_i.$$

(3) Show that

$$E_{N+1} - E_N = \int_0^1 \mathrm{d}n\ \varepsilon_{\rm lus}(n),$$

where n $(0 \leqslant n \leqslant 1)$ is the number of electrons in the lowest unoccupied state (lus) in the ground state of the N-electron system.

---

(1) Differentiating $E = T[n] + E_{\rm H}[n] + E_{\rm xc}[n]$ with respect to n_i, we have

$$\frac{\partial E}{\partial n_i} = \frac{\partial T[n]}{\partial n_i} + \frac{\partial (E_{\rm H}[n] + E_{\rm xc}[n])}{\partial n_i}.$$

The first term on the right hand side can be straightforwardly evaluated upon making use of $T[n] = \sum_j n_j t_j$, where we have used the dummy summation variable j since the subscript i has been taken by the free variable n_i. We have

$$\begin{aligned}\frac{\partial T[n]}{\partial n_i} &= \frac{\partial}{\partial n_i}\sum_j n_j t_j = \sum_j \left(\frac{\partial n_j}{\partial n_i} t_j + n_j \frac{\partial t_j}{\partial n_i}\right) \\ &= \sum_j \left(\delta_{ij} t_j + n_j \frac{\partial t_j}{\partial n_i}\right) = t_i + \sum_j n_j \frac{\partial t_j}{\partial n_i}.\end{aligned}$$

For the second term in $\partial E/\partial n_i$, we make use of the expression of the differential of a functional. Taking $E_{\rm H}[n]$ as an example, we have

$$\Delta E_{\rm H}[n] = \int d\boldsymbol{r}\, \frac{\delta E_{\rm H}[n]}{\delta n(\boldsymbol{r})} \Delta n(\boldsymbol{r}),$$

where we have used $\Delta E_{\rm H}[n]$ and $\Delta n(\boldsymbol{r})$ to denote infinitesimally small variations of $E_{\rm H}[n]$ and $n(\boldsymbol{r})$, respectively. To find the derivative of $E_{\rm H}[n]$ with respect to n_i, we divide both sides of the above equation with Δn_i and obtain

$$\frac{\Delta E_{\rm H}[n]}{\Delta n_i} = \int d\boldsymbol{r}\, \frac{\delta E_{\rm H}[n]}{\delta n(\boldsymbol{r})} \frac{\Delta n(\boldsymbol{r})}{\Delta n_i}.$$

In the $\Delta n_i \to 0$ limit, we have

$$\begin{aligned}
\frac{\partial E_{\rm H}[n]}{\partial n_i} &= \int d\boldsymbol{r}\, \frac{\delta E_{\rm H}[n]}{\delta n(\boldsymbol{r})} \frac{\partial n(\boldsymbol{r})}{\partial n_i} = \int d\boldsymbol{r}\, V_{\rm H}(\boldsymbol{r}) \frac{\partial n(\boldsymbol{r})}{\partial n_i} \\
&= \int d\boldsymbol{r}\, V_{\rm H}(\boldsymbol{r}) \frac{\partial}{\partial n_i} \sum_j n_j |\psi_j(\boldsymbol{r})|^2 \\
&= \int d\boldsymbol{r}\, V_{\rm H}(\boldsymbol{r}) \sum_j \left(\delta_{ij} |\psi_j(\boldsymbol{r})|^2 + n_j \frac{\partial |\psi_j|^2}{\partial n_i} \right) \\
&= \int d\boldsymbol{r}\, V_{\rm H}(\boldsymbol{r}) \left(|\psi_i(\boldsymbol{r})|^2 + \sum_j n_j \frac{\partial |\psi_j|^2}{\partial n_i} \right).
\end{aligned}$$

We can obtain a similar result for $E_{\rm xc}[n]$. We thus have

$$\frac{\partial E}{\partial n_i} = t_i + \sum_j n_j \frac{\partial t_j}{\partial n_i} + \int d\boldsymbol{r}\, (V_{\rm H} + V_{\rm xc}) \left(|\psi_i|^2 + \sum_j n_j \frac{\partial |\psi_j|^2}{\partial n_i} \right).$$

(2) Inserting the second expression of t_i into the above-derived result, we have

$$\begin{aligned}
\frac{\partial E}{\partial n_i} &= \varepsilon_i - \int d\boldsymbol{r}\, \psi_i^*(V_{\rm H} + V_{\rm xc})\psi_i + \sum_j n_j \frac{\partial \varepsilon_j}{\partial n_i} \\
&\quad - \sum_j n_j \frac{\partial}{\partial n_i} \int d\boldsymbol{r}\, \psi_j^*(V_{\rm H} + V_{\rm xc})\psi_j \\
&\quad + \int d\boldsymbol{r}\, (V_{\rm H} + V_{\rm xc}) \left(|\psi_i|^2 + \sum_j n_j \frac{\partial |\psi_j|^2}{\partial n_i} \right) \\
&= \varepsilon_i + \sum_j n_j \left[\frac{\partial \varepsilon_j}{\partial n_i} - \int d\boldsymbol{r}\, \frac{\partial (V_{\rm H} + V_{\rm xc})}{\partial n_i} |\psi_j|^2 \right].
\end{aligned}$$

Multiplying the Kohn-Sham equation for $\psi_j(\boldsymbol{r})$ from left with $\psi_j^*(\boldsymbol{r})$, then integrating both sides of the resultant equation over $\boldsymbol{r}$, and then differentiating the resultant equation with respect to n_i, we have

$$\int \mathrm{d}\boldsymbol{r}\, \frac{\partial(V_{\mathrm{H}} + V_{\mathrm{xc}})}{\partial n_i} |\psi_j(\boldsymbol{r})|^2 + \int \mathrm{d}\boldsymbol{r}\, \frac{\partial \psi_j^*(\boldsymbol{r})}{\partial n_i} [-\nabla^2/2 + V_{\mathrm{H}}(\boldsymbol{r}) + V_{\mathrm{xc}}(\boldsymbol{r})]\psi_j(\boldsymbol{r}) + \int \mathrm{d}\boldsymbol{r}\, \psi_j^*(\boldsymbol{r})[-\nabla^2/2 + V_{\mathrm{H}}(\boldsymbol{r}) + V_{\mathrm{xc}}(\boldsymbol{r})]\frac{\partial \psi_j(\boldsymbol{r})}{\partial n_i} = \frac{\partial \varepsilon_j}{\partial n_i},$$

where we have made use of the fact that $\psi_j(\boldsymbol{r})$ is normalized to unity. For the second term on the first line, we make use of the Kohn-Sham equation for $\psi_j(\boldsymbol{r})$. For the first term on the second line, we perform integrations by parts twice and then make use of the Kohn-Sham equation for $\psi_j^*(\boldsymbol{r})$. We have

$$\begin{aligned} \frac{\partial \varepsilon_j}{\partial n_i} - \int \mathrm{d}\boldsymbol{r}\, \frac{\partial(V_{\mathrm{H}} + V_{\mathrm{xc}})}{\partial n_i} |\psi_j(\boldsymbol{r})|^2 &= \int \mathrm{d}\boldsymbol{r}\, \frac{\partial \psi_j^*(\boldsymbol{r})}{\partial n_i} \varepsilon_j \psi_j(\boldsymbol{r}) + \int \mathrm{d}\boldsymbol{r}\, \varepsilon_j \psi_j^*(\boldsymbol{r}) \frac{\partial \psi_j(\boldsymbol{r})}{\partial n_i} \\ &= \varepsilon_j \frac{\partial}{\partial n_i} \int \mathrm{d}\boldsymbol{r}\, |\psi_j(\boldsymbol{r})|^2 = 0. \end{aligned}$$

We thus have

$$\frac{\partial E}{\partial n_i} = \varepsilon_i.$$

(3) We consider the case in which the occupation number varies only on the lowest unoccupied state (lus) in the ground state of the N-electron system. Let n denote the occupation number of this state. We have

$$\frac{\partial E}{\partial n} = \varepsilon_{\mathrm{lus}}(n).$$

If $n = 0$, we then have an N-electron system; if $n = 1$, we then have a $(N + 1)$-electron system. Integrating the above equation from $n = 0$ to $n = 1$, we have

$$E_{N+1} - E_N = \int_0^1 \mathrm{d}n\, \varepsilon_{\mathrm{lus}}(n).$$

Chapter 22

Pseudopotentials

(1) *Ashcroft empty-core potential*

$$V_{\rm ps}^{\rm A}(r) = \begin{cases} 0, & r < R_c, \\ -Z/r, & r > R_c. \end{cases}$$

(2) *Abarenkov-Heine potential*

$$V_{\rm ps}^{\rm AH}(r) = \begin{cases} \sum_\ell A_\ell \, |\ell\rangle\langle\ell|, & r < R_c, \\ -Z/r, & r > R_c. \end{cases}$$

(3) *Evanescent core potential*

$$w(r) = -\frac{s}{R}\left\{\frac{1}{x}\left[1-(1+\beta x){\rm e}^{-\alpha x}\right] - A{\rm e}^{-x}\right\}.$$

(4) *Austin pseudopotential*

$$\begin{aligned} \hat{V}_{\rm ps}^{\rm Austin}\,|\phi_n\rangle &= (\hat{U}+\hat{V}_R)\,|\phi_n\rangle \\ &= \hat{U}\,|\phi_n\rangle - \textstyle\sum_c \langle\psi_c|\hat{U}|\phi_n\rangle\,|\psi_c\rangle\,. \end{aligned}$$

(5) *Phillips-Kleinman pseudopotential*

$$\begin{aligned} \hat{V}_{\rm ps}^{\rm PK}\,|\phi_v\rangle &= (\hat{U}+\hat{V}_R)\,|\phi_v\rangle \\ &= \hat{U}\,|\phi_v\rangle + \textstyle\sum_c (E_v - E_c)\,\langle\psi_c|\phi_v\rangle\,|\psi_c\rangle\,. \end{aligned}$$

(6) *Model pseudopotential*

$$(\hat{U}+\hat{V}_R)|\phi_n\rangle = \hat{V}_M\,|\phi_n\rangle + \Big[(\hat{U}-\hat{V}_M)|\phi_n\rangle - \sum_c \langle\psi_c|\hat{U}-\hat{V}_M|\phi_n\rangle|\psi_c\rangle\Big].$$

(7) *Norm conservation*

$$\int_0^{r_\ell} {\rm d}r\, r^2 \big|R_\ell^{\rm PP}(r)\big|^2 = \int_0^{r_\ell} {\rm d}r\, r^2 \big|R_\ell^{\rm AE}(r)\big|^2.$$

(8) *Hamann pseudopotential*

$$\overline{w}_\ell(r) + v_{\rm H}\big([n^{\rm ps}], r\big) + v_{\rm xc}\big([n^{\rm ps}], r\big)$$
$$= v_{\rm xc}^{\rm AE}\big([n^{\rm ps}], r\big)\Big[1 - f\Big(\frac{r}{r_\ell}\Big)\Big] + c_\ell f\Big(\frac{r}{r_\ell}\Big),$$
$$R_\ell^{\rm ps}(r) = \gamma_\ell\big[\overline{R}_\ell(r) + \delta_\ell g_\ell(r)\big],$$
$$\gamma_\ell = R_\ell^{\rm AE}(r)/\overline{R}_\ell(r), \quad r > \tilde{r}_\ell,$$
$$g_\ell(r) = r^{\ell+1} f(r/r_\ell).$$

(9) *Troullier-Martins pseudopotentials*

$$R_\ell^{\rm ps}(r) = \begin{cases} R_{n\ell}^{\rm AE}(r), & r > r_\ell, \\ r^\ell e^{p(r)}, & r < r_\ell, \end{cases}$$
$$p(r) = \sum_{j=0}^{6} c_{2j} r^{2j}.$$

(10) *Ultrasoft pseudopotentials*

$$\Delta Q_{ss'} = \int_0^{R_c} dr\, \Delta Q_{ss'}(r),$$
$$\Delta Q_{ss'}(r) = \phi_s^*(r)\phi_{s'}(r) - \tilde{\phi}_s^*(r)\tilde{\phi}_{s'}(r),$$
$$\chi_s^{\rm ps}(\boldsymbol{r}) = \Big\{E_s - \Big[-\frac{\nabla^2}{2} + V_{\rm L}(r)\Big]\Big\}\psi_s^{\rm ps}(\boldsymbol{r}),$$
$$\delta\hat{V}_{\rm NL}^{\rm US} = \sum_{ss'} D_{ss'}\,|\beta_s\rangle\,\langle\beta_{s'}|,$$
$$D_{ss'} = B_{ss'} + E_{s'}\Delta Q_{ss'},$$
$$B_{ss'} = \langle\psi_s^{\rm ps}|\chi_{s'}^{\rm ps}\rangle,$$
$$|\beta_s\rangle = \sum_{s'} B_{ss'}^{-1}\,|\chi_{s'}^{\rm ps}\rangle,$$
$$\big(\hat{H} - E_s\hat{S}\big)\tilde{\psi}_s = 0,$$
$$\hat{H} = -\frac{1}{2}\nabla^2 + V_{\rm L}(r) + \delta\hat{V}_{\rm NL}^{\rm US},$$
$$\hat{S} = \hat{I} + \sum_{ss'}\Delta Q_{ss'}\,|\beta_s\rangle\langle\beta_{s'}|.$$

22-1 Smooth and oscillatory functions. Consider two functions in one dimension, with one smooth and the other oscillatory,

$$f(x) = C_1 e^{-x^2}, \quad g(x) = C_2 e^{-x^2}\cos(qx),$$

where C_1 and C_2 are normalization constants and q is an arbitrary parameter.

(1) Normalize both functions and then plot them in the same graph to verify that on the same scale one is very smooth while the other is very oscillatory.
(2) Make the Fourier transformation to both functions and derive their Fourier components $F(k)$ and $G(k)$.
(3) Plot $F(k)$ and $G(k)$ as functions of k to verify that, for $q > 1$, the nonzero Fourier components of the oscillatory wave function have a larger domain than the smooth function. Furthermore, show that this domain increases as q increases.

---

(1) We first normalize $g(x)$. The normalization coefficient for $f(x)$ can be obtained from that for $g(x)$ simply by setting $q = 0$. Expressing the cosine function in terms of exponential functions, we have

$$\begin{aligned}
&\int_{-\infty}^{\infty} dx\ |g(x)|^2 \\
&\quad = |C_2|^2 \int_{-\infty}^{\infty} dx\ e^{-2x^2} \cos^2(qx) \\
&\quad = \frac{1}{2}|C_2|^2 \left\{ \sqrt{\frac{\pi}{2}} + \frac{1}{2}\int_{-\infty}^{\infty} dx\ \left[e^{-2(x^2 - iqx)} + e^{-2(x^2 + iqx)} \right] \right\} \\
&\quad = \frac{1}{2}|C_2|^2 \left(\sqrt{\frac{\pi}{2}} + \sqrt{\frac{\pi}{2}}\, e^{-q^2/2} \right) \\
&\quad = \sqrt{\frac{\pi}{8}}\, (1 + e^{-q^2/2})|C_2|^2.
\end{aligned}$$

Thus, $|C_2|^2 = (8/\pi)^{1/2}(1 + e^{-q^2/2})^{-1}$ from the normalization condition for $g(x)$, $\int_{-\infty}^{\infty} dx\ |g(x)|^2 = 1$. We choose $C_2 = (8/\pi)^{1/4}(1 + e^{-q^2/2})^{-1/2}$. Setting $q = 0$, we have $C_1 = (2/\pi)^{1/4}$. The normalized functions are

$$f(x) = \left(\frac{2}{\pi}\right)^{1/4} e^{-x^2},$$

$$g(x) = \left(\frac{8}{\pi}\right)^{1/4} \frac{e^{-x^2}\cos(qx)}{(1 + e^{-q^2/2})^{1/2}}.$$

Functions $f(x)$ and $g(x)$ are plotted in Fig. 22.1 with $q = 15$ for function $g(x)$. From Fig. 22.1, it is clearly seen that function $f(x)$ is a smooth function while function $g(x)$ is an oscillatory function. Function $g(x)$ becomes more wildly oscillatory as q increases.

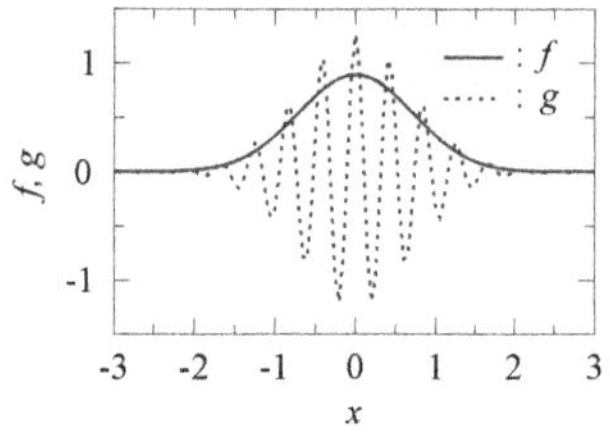

Fig. 22.1 Smooth and oscillatory functions. The value of q is taken to be 15.

(2) The Fourier transform of $g(x)$ is given by

$$\begin{aligned}G(k) &= \frac{1}{(2\pi)^{1/2}}\int_{-\infty}^{\infty} \mathrm{d}x\, g(x)\mathrm{e}^{-\mathrm{i}kx} \\ &= \frac{(2/\pi^3)^{1/4}}{(1+\mathrm{e}^{-q^2/2})^{1/2}}\int_{-\infty}^{\infty} \mathrm{d}x\, \mathrm{e}^{-x^2-\mathrm{i}kx}\cos(qx) \\ &= \frac{1}{(2\pi)^{3/4}(1+\mathrm{e}^{-q^2/2})^{1/2}}\int_{-\infty}^{\infty} \mathrm{d}x\, \mathrm{e}^{-x^2}\left[\mathrm{e}^{-\mathrm{i}(k-q)x}+\mathrm{e}^{-\mathrm{i}(k+q)x}\right] \\ &= \frac{1}{(8\pi)^{1/4}(1+\mathrm{e}^{-q^2/2})^{1/2}}\left[\mathrm{e}^{-(k-q)^2/4}+\mathrm{e}^{-(k+q)^2/4}\right].\end{aligned}$$

The Fourier transform of $f(x)$ can be obtained from that of $g(x)$ by setting $q = 0$. We have

$$F(k) = \frac{1}{(2\pi)^{1/4}}\mathrm{e}^{-k^2/4}.$$

(3) Functions $F(k)$ and $G(k)$ are plotted in Fig. 22.2.

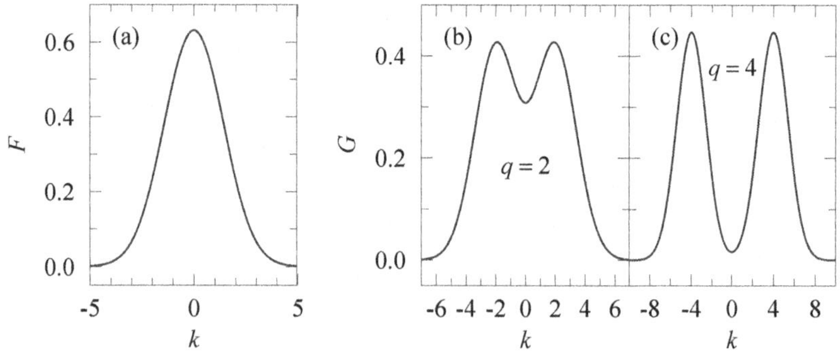

Fig. 22.2 Plots of $F(k)$ and $G(k)$ as functions of k. (a) $F(k)$ versus k. (b) $G(k)$ versus k for $q = 2$. (c) $G(k)$ versus k for $q = 4$.

Comparing Fig. 22.2(a) with Figs. 22.2(b) and (c), we see that the nonzero domain for $G(k)$ is larger than that for $F(k)$ for $q \neq 0$. From Figs. 22.2(b) and (c), we also see that, as q increases, the nonzero domain for $G(k)$ becomes larger. Note that, as q increases from 0, the single peak at $k = 0$ for $q = 0$ splits into two peaks and the separation of the two split-peaks increases as q further increases. Also note that $G(k)$ is even with respect to q as well as with respect to k.

22-2 Atom with a harmonic radial potential. In order to make up an atom for whose electrons the pseudopotentials can be derived analytically, we assume that the electrons in a fictitious atom are subject to a harmonic radial potential $V(r) = m\omega^2 r^2/2$ with the single-electron stationary Schrödinger equation given by

$$\left[-\frac{\hbar^2}{2m}\frac{\mathrm{d}^2}{\mathrm{d}r^2} + \frac{1}{2}m\omega^2 r^2\right]\phi(r) = \varepsilon\phi(r).$$

(1) The wave function of the second excited state is given by $\phi_2(r) = (2\alpha/\sqrt{\pi})^{1/2}(\alpha^2 r^2 - 1/2)\mathrm{e}^{-\alpha^2 r^2/2}$. We now derive a pseudo wave function for this state. Assume that the pseudo wave function takes on the following form

$$\tilde{\phi}_2(r) = \begin{cases} A(\alpha r)^2 \mathrm{e}^{-B(\alpha r)^2}, & r \leqslant r_c, \\ \phi_2(r), & r > r_c. \end{cases}$$

Determine the parameters A and B such that the pseudo wave function and its first-order derivative are continuous at r_c.

(2) Invert the Schrödinger equation for this pseudo wave function and derive the corresponding pseudopotential.

(3) Plot the pseudo wave function and pseudopotential as well as the true wave function and potential with $\alpha r_c = \sqrt{2}$.

---

(1) From the continuity of $\tilde{\phi}_2(r)$ and its first-order derivative at r_c, we have

$$A(\alpha r_c)^2 \mathrm{e}^{-B(\alpha r_c)^2} = (2\alpha/\sqrt{\pi})^{1/2}\left[(\alpha r_c)^2 - 1/2\right]\mathrm{e}^{-(\alpha r_c)^2/2},$$
$$2A\left[1 - B(\alpha r_c)^2\right]\mathrm{e}^{-B(\alpha r_c)^2} = (2\alpha/\sqrt{\pi})^{1/2}\left[5/2 - (\alpha r_c)^2\right]\mathrm{e}^{-(\alpha r_c)^2/2}.$$

Dividing the second equation with the first yields

$$\frac{2}{(\alpha r_c)^2}\left[1 - B(\alpha r_c)^2\right] = \frac{5/2 - (\alpha r_c)^2}{\alpha^2 r_c^2 - 1/2}.$$

Solving for B from the above equation, we obtain

$$B = \frac{2(\alpha r_c)^4 - (\alpha r_c)^2 - 2}{2(\alpha r_c)^2[2(\alpha r_c)^2 - 1]}.$$

From the first equation for A and B, we obtain

$$A = \left(\frac{\alpha}{2\sqrt{\pi}}\right)^{1/2} \frac{2(\alpha r_c)^2 - 1}{(\alpha r_c)^2} \mathrm{e}^{-1/[2(\alpha r_c)^2-1]}.$$

(2) Inverting the Schrödinger equation for the pseudo wave function $\tilde{\phi}_2(r)$ with the pseudopotential $V_{\rm ps}(r)$ as the potential energy

$$\left[-\frac{\hbar^2}{2m}\frac{\mathrm{d}^2}{\mathrm{d}r^2} + V_{\rm ps}(r)\right]\tilde{\phi}_2(r) = \varepsilon_2\tilde{\phi}_2(r)$$

with $\varepsilon_2 = 5\hbar\omega/2$, we obtain the pseudopotential

$$V_{\rm ps}(r) = \varepsilon_2 + \frac{1}{\tilde{\phi}_2(r)}\frac{\hbar^2}{2m}\frac{\mathrm{d}^2\tilde{\phi}_2(r)}{\mathrm{d}r^2}.$$

Inserting the pseudo wave function into the above equation, we obtain the following explicit expression for the pseudopotential

$$V_{\rm ps}(r) = \begin{cases} \dfrac{5}{2}\hbar\omega + \hbar\omega\dfrac{1 - 5B(\alpha r)^2 + 2B^2(\alpha r)^4}{(\alpha r)^2}, & r \leqslant r_c, \\ \dfrac{5}{2}\hbar\omega + \dfrac{1}{2}\hbar\omega\dfrac{5 - 11(\alpha r)^2 + 2(\alpha r)^4}{2(\alpha r)^2 - 1}, & r > r_c. \end{cases}$$

(3) The true and pseudo wave functions as well as the true and pseudo potentials are plotted in Fig. 22.3 against αr.

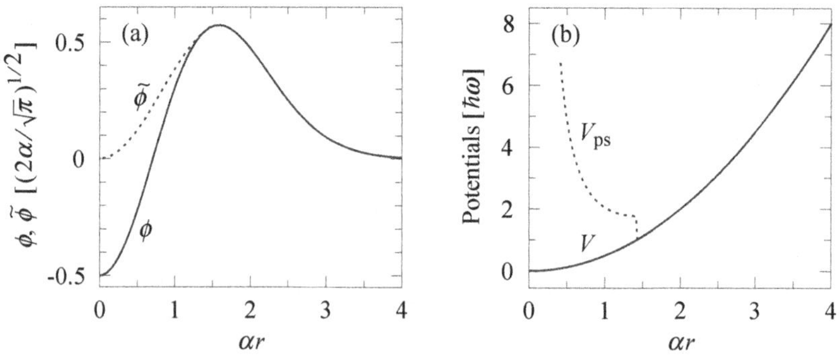

Fig. 22.3 Plots of wave functions (a) and potentials (b) against αr.

From Fig. 22.3, we see that the pseudo wave function is flatter than the true wave function. The pseudopotential is also a bit flatter than the true potential just below $r = r_c$. However, $V_{ps} \to \infty$ as $r \to 0$. This is because the pseudo wave function is zero at $r = 0$. The discontinuity of the pseudopotential at r_c is caused by the discontinuity of the 2nd-order derivative of the pseudo wave function at r_c.

22-3 Numerical solution of the radial Schrödinger equation. When a pseudopotential is generated for a real problem, we need to solve the stationary Schrödinger equation for isolated atoms. The electron density in an atom may be assumed to be spherically symmetric. This implies that the potential experienced by electrons is also spherically symmetric. Strictly speaking, this is valid only in a closed-shell system. In this case, the electron wave function $\psi(\boldsymbol{r})$ can be written as a product of the radial part $R(r)$ and the spherical harmonic function $Y_{\ell m}(\theta, \varphi)$. We express $R(r)$ in terms of a new function $\phi(r)$, $R(r) = \phi(r)/r$. In potential $V(r)$, the radial Schrödinger equation is then transformed into the following equation for $\phi(r)$ in atomic units

$$-\frac{1}{2}\frac{\mathrm{d}^2\phi(r)}{\mathrm{d}r^2} + \left[V(r) + \frac{\ell(\ell+1)}{2r^2}\right]\phi(r) = E\phi(r).$$

(1) For Coulomb potential $V(r) = -1/r$, compute numerically the energy E and wave function $\phi(r)$ of the ground state with $\ell = 0$. For the numerical computation, a uniform or non-uniform grid for r needs to be set up. Take the farthest point $r_{\max}$ on the grid to be at least 10 a.u. from the origin. The correct boundary conditions that $\phi(r) = 0$ at $r = 0$ and $r_{\max}$ must be maintained. Check the numerical results against the exact analytical results for the hydrogen atom.

(2) Replace the Coulomb potential with an Ashcroft-type potential: $V(r) = -1/r$ for $r > R_c$ and 0 for $r \leqslant R_c$. Compute the energies of the ground and first excited states. Take $R_c = 1$ a.u..

(3) We now construct an Abarenkov-Heine-type pseudopotential

$$V_{\mathrm{ps}}(r) = \begin{cases} -1/r, & r > R_c, \\ A, & r \leqslant R_c, \end{cases}$$

where A is a constant to be determined. For $R_c = 1.5$ a.u., find the best value of A to get the correct ground state energy.

(4) With the above-obtained value of A, find the energy and wave function of the first excited state.

---

(1) To solve the stationary Schrödinger equation numerically using the finite difference method, we must first discretize it. For this purpose, we choose a radial point $r_{\max}$ beyond which the wave function $\phi(r)$ is vanishingly small. We thus take $\phi(r) = 0$ for $r \geqslant r_{\max}$. We can then solve the Schrödinger equation only within the interval $[0, r_{\max}]$ under the boundary conditions $\phi(r = 0) = \phi(r = r_{\max}) = 0$. We divide the interval $[0, r_{\max}]$ into $N - 1$ subintervals using the following N points which are also illustrated in Fig. 22.4

$$r_0 = 0, r_1 = h, r_2 = 2h, \cdots, r_j = jh, \cdots, r_{N-1} = r_{\max} = (N-1)h.$$

These N points make up *a uniform grid* with h referred to as *the step size*. A wave function is then approximately specified by giving its value on each of these points. Thus, the wave function $\phi(r)$ is denoted approximately by N real numbers: $\phi_0 = \phi(r_0)$, $\phi_1 = \phi(r_1)$, $\cdots$, $\phi_{N-1} = \phi(r_{N-1})$. We use these N real numbers to form a column matrix that has N rows

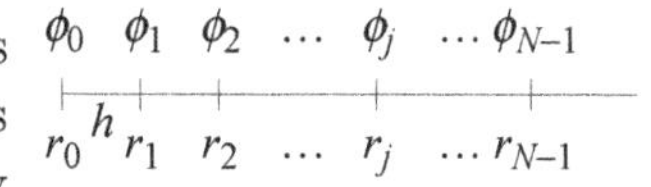

Fig. 22.4 Grid for radial coordinates.

$$\phi(r) = \begin{pmatrix} \phi_0 & \phi_1 & \cdots & \phi_{N-1} \end{pmatrix}^t,$$

where the superscript t on the column matrix represents its transpose. This is actually the radial wave function in the coordinate representation on discretized points. The equation for the wave function $\phi(r)$ on each grid point r_j reads

$$-\frac{1}{2}\frac{\mathrm{d}^2\phi(r)}{\mathrm{d}r^2}\bigg|_{r=r_j} + \left[V(r_j) + \frac{\ell(\ell+1)}{2r_j^2}\right]\phi_j = E\phi_j, \; j = 0, 1, \cdots, N-1.$$

To obtain an expression for the second-order derivative $\mathrm{d}^2\phi(r)/\mathrm{d}r^2\big|_{r=r_j}$ in terms of ϕ_j's, we write down the Taylor expansions of $\phi(r_j + h)$ and $\phi(r_j - h)$ with respect to h that is a small quantity compared to unity in the atomic units

$$\phi(r_j+h) = \phi_j + \left.\frac{\mathrm{d}\phi(r)}{\mathrm{d}r}\right|_{r=r_j} h + \frac{1}{2!}\left.\frac{\mathrm{d}^2\phi(r)}{\mathrm{d}r^2}\right|_{r=r_j} h^2 + \frac{1}{3!}\left.\frac{\mathrm{d}^3\phi(r)}{\mathrm{d}r^3}\right|_{r=r_j} h^3 + O(h^4),$$

$$\phi(r_j-h) = \phi_j - \left.\frac{\mathrm{d}\phi(r)}{\mathrm{d}r}\right|_{r=r_j} h + \frac{1}{2!}\left.\frac{\mathrm{d}^2\phi(r)}{\mathrm{d}r^2}\right|_{r=r_j} h^2 - \frac{1}{3!}\left.\frac{\mathrm{d}^3\phi(r)}{\mathrm{d}r^3}\right|_{r=r_j} h^3 + O(h^4).$$

Adding up the above two equations, we obtain

$$\left.\frac{\mathrm{d}^2\phi(r)}{\mathrm{d}r^2}\right|_{r=r_j} = \frac{1}{h^2}\big[\phi(r_j-h) - 2\phi(r_j) + \phi(r_j+h)\big] + O(h^2)$$
$$= \frac{1}{h^2}\big(\phi_{j-1} - 2\phi_j + \phi_{j+1}\big) + O(h^2).$$

The above expression for $\mathrm{d}^2\phi(r)/\mathrm{d}r^2\big|_{r=r_j}$ is accurate up to the first order in h. Note that we can also write down other higher-order expressions for $\mathrm{d}^2\phi(r)/\mathrm{d}r^2\big|_{r=r_j}$. Inserting the above expression for $\mathrm{d}^2\phi(r)/\mathrm{d}r^2\big|_{r=r_j}$ into the equation for the radial wave function $\phi(r)$ on each grid point r_j , we obtain

$$-\frac{1}{2h^2}\big(\phi_{j-1} - 2\phi_j + \phi_{j+1}\big) + \left[V(r_j) + \frac{\ell(\ell+1)}{2r_j^2}\right]\phi_j = E\phi_j,$$

$j = 0, 1, \cdots, N-1.$

Note that $\phi_{-1} = \phi_N = 0$ since they correspond to points not on the grid. Writing the above equation in the matrix form, we have

$$\begin{pmatrix} 2K+U_0 & -K & 0 & \cdots & 0 \\ -K & 2K+U_1 & -K & \cdots & 0 \\ 0 & -K & 2K+U_2 & \cdots & 0 \\ \vdots & \vdots & \vdots & \ddots & \vdots \\ 0 & 0 & 0 & \cdots & -K \\ 0 & 0 & 0 & \cdots & 2K+U_{N-1} \end{pmatrix} \begin{pmatrix} \phi_0 \\ \phi_1 \\ \phi_2 \\ \vdots \\ \phi_{N-2} \\ \phi_{N-1} \end{pmatrix} = E \begin{pmatrix} \phi_0 \\ \phi_1 \\ \phi_2 \\ \vdots \\ \phi_{N-2} \\ \phi_{N-1} \end{pmatrix},$$

where

$$K = \frac{1}{2h^2}, \; U_j = V(r_j) + \frac{\ell(\ell+1)}{2r_j^2}.$$

We have thus transformed the equation for $\phi(r)$ into a set of linear algebraic equations of the form $A\Phi = E\Phi$, with A the $N \times N$ square matrix on the left hand side and Φ the column matrix that appears on both sides. We expect that E can be obtained through the diagonalization of A. However, to fulfill the boundary conditions $\phi(r = 0) = \phi(r = r_{\max}) = 0$, we can not directly diagonalize this matrix. Instead we diagonalize the matrix obtained after its first and last rows and columns are deleted. We then have a $(N-2) \times (N-2)$ matrix to diagonalize. This also removes the singularity in the potential at $r = 0$. *The smallest eigenvalue is the value for energy* E and *the corresponding eigenvector gives the values of* $\phi(r)$ *on the grid points.* Note that the matrix to be diagonalized is a tridiagonal matrix. Before we turn to the actual computations, we consider the normalization of the wave function. Making use of the extended trapezoidal rule, we have $\int \mathrm{d}r\, \phi^2(r) = h \sum_{j=0}^{N-1} \phi_j^2 = h \sum_{j=1}^{N-2} \phi_j^2$, where we have utilized the boundary conditions $\phi_0 = \phi_{N-1} = 0$. For the ground state with $\ell = 0$ in the Coulomb potential $-1/r$, we use a grid that has a step size of $h = 0.01$ and the point number of $N = 2,000$. The ground-state energy is found to be $E_1 = -0.499\,987$ a.u. which is in excellent agreement with the exact analytical result $E_1 = -0.5$ a.u.. The ground-state radial wave function $R(r)$ obtained numerically is plotted in Fig. 22.5 against r. It has been found to be indistinguishable from the exact radial wave function $R_{10} = 2\mathrm{e}^{-r}$.

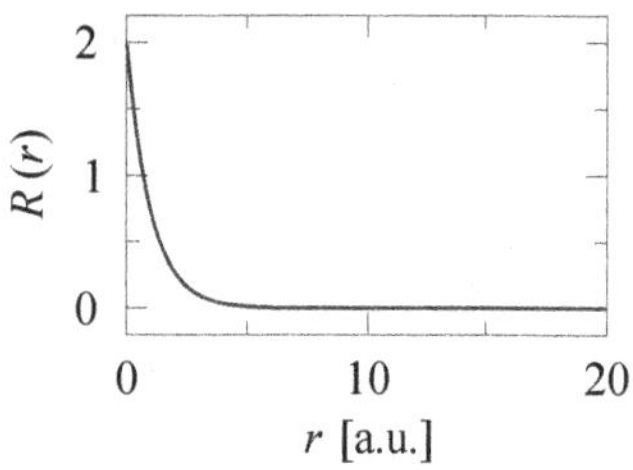

Fig. 22.5 Plot of the ground-state radial wave function $R(r)$ as a function of r for an electron in the Coulomb potential $-1/r$.

(2) For the Ashcroft-type potential with $R_c = 1$ a.u., we have found that the energies of the ground state ($\ell = 0$) and first-excited state ($\ell = 1$) are $-0.273\,151$ a.u. and $-0.121\,843$ a.u., respectively, using a 2,000-point uniform grid for r within the interval $[0, 20]$ a.u..

(3) To find the value of A that gives the exact ground-state energy for the Abarenkov-Heine-type potential, we use the bisection method with respect to the interval $[-2,0]$ a.u. for A. Using a grid of 1,500 points and a value of 15 a.u. for $r_{\max}$, we find that $A \approx -1.252\,118\,11$ gives the ground-state energy $-0.499\,999\,996$ a.u.

(4) Using the above-obtained value of A and the same grid used in obtaining A, we find that the energy of the first excited state with $n = 2$ and $\ell = 0$ is $-0.118\ 044$ a.u. for the Abarenkov-Heine-type potential. The numerically computed radial wave function $\phi(r)/r$ is plotted in Fig. 22.6 as a function of r (the solid line) together with the exact analytical radial wave function $R_{20}(r) = 2^{-3/2}(2-r)\mathrm{e}^{-r/2}$ in a.u. (the dotted line). The numerically obtained radial wave function is the pseudo wave function. From Fig. 22.6, we see that it coincides very well with the exact analytical result for $r > R_c$ and that it is smoother than the exact analytical result for $r < R_c$. Here $r < R_c$ defines the core region. The second-lowest eigenvalue and the corresponding eigenvector of the $(N-2)\times(N-2)$ matrix with $\ell = 0$ defined in the first part of this problem are identified as the energy and radial wave function of the first excited state.

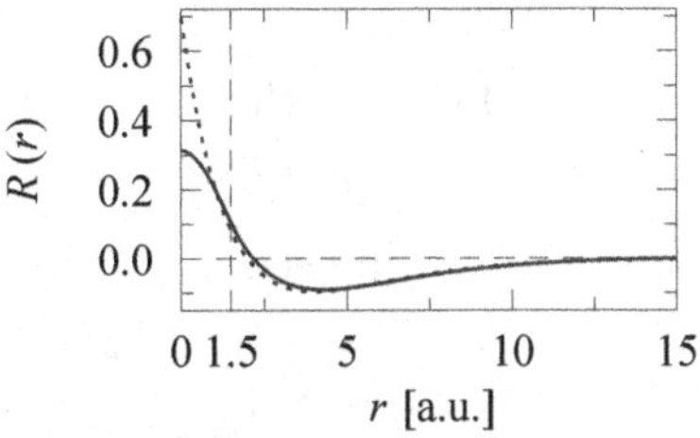

Fig. 22.6 Plot of the radial wave function of the first excited state as a function of r obtained using the Abarenkov-Heine-type potential.

22-4 Kerker scheme for pseudopotentials. In the Kerker scheme for obtaining a pseudopotential, the pseudo radial wave function $R_\ell^{\mathrm{ps}}(r)$ times r is assumed in atomic units (a.u.) to be of the form $F_\ell^{\mathrm{ps}}(r) = rR_\ell^{\mathrm{ps}}(r) = r^{\ell+1}\mathrm{e}^{p(r)}$ for $r \leqslant r_c$, where $p(r) = \alpha\, r^4 + \beta\, r^3 + \gamma\, r^2 + \delta$. The absence of a term linear in r in $p(r)$ is to avoid a singularity in the pseudopotential at $r = 0$. In pseudopotential $V_\ell^{\mathrm{ps}}(r)$, the Schrödinger equation for $F_\ell(r)$ is given in atomic units by

$$\left[-\mathrm{d}^2/\mathrm{d}r^2 + \ell(\ell+1)/r^2 + V_\ell^{\mathrm{ps}}(r) - E\right]F_\ell^{\mathrm{ps}}(r) = 0.$$

The pseudo wave function and pseudopotential are determined from the following four conditions. (1) The real and "pseudo" atom have the same valence eigenvalues for some chosen electronic configuration (usually the atomic ground state). (2) The pseudo wave function

$R_\ell^{\rm ps}(r)$ is nodeless and is identical with the real valence wave function at and beyond a chosen core radius (the cutoff radius) r_c. (3) Both the first and the second derivatives of the wave function $F_\ell^{\rm ps}(r)$ are matched to the real values at r_c. (4) The pseudo-charge contained in the sphere of radius r_c is identical with the real charge in that sphere.

(1) Show that the first three conditions lead to the following three equations

$$\ln[u_\ell(r_c)/r_c^{\ell+1}] = p(r_c),$$
$$r_c u_\ell'(r_c)/u_\ell(r_c) = \ell + 1 + r_c p'(r_c),$$
$$r_c^2 V_\ell^{\rm AE}(r_c) + (\ell+1)^2 - r_c^2\{E + [u_\ell'(r_c)/u_\ell(r_c)]^2\} = r_c^2 p''(r_c),$$

where primes denote derivatives with respect to r, $u_\ell(r)$ is the atomic radial wave function times r, and $V_\ell^{\rm AE}(r)$ is the atomic potential used in the all-electron scheme.

(2) Solve the above three equations for α, β, and γ in terms of δ.

(3) From the fourth condition, derive $2\delta + \ln I - \ln Q = 0$, where Q is the amount of real charge in the core region up to $r = r_c$ and $I = \int_0^{r_c} {\rm d}r\, r^{2(\ell+1)} {\rm e}^{2\alpha r^4 + 2\beta r^3 + 2\gamma r^2}$. This equation can be used to determine δ numerically with α, β, and γ expressed in terms of δ.

(4) Show that the following pseudopotential $V_\ell^{\rm ps}(r)$ follows from the Schrödinger equation given above

$$V_\ell^{\rm ps}(r) = E + \lambda(r)[\,2\ell + 2 + \lambda(r)\, r^2\,] + 12\alpha r^2 + 6\beta r + 2\gamma$$

with $\lambda(r) = 4\alpha r^2 + 3\beta r + 2\gamma$.

---

(1) From $u_\ell(r_c) = F_\ell^{\rm ps}(r_c)$, we have

$$u_\ell(r_c) = r_c^{\ell+1} {\rm e}^{p(r_c)}.$$

Dividing both sides of the above equation with $r_c^{\ell+1}$ and then taking the logarithm of both sides, we have

$$\ln[u_\ell(r_c)/r_c^{\ell+1}] = p(r_c).$$

From $u_\ell'(r_c) = {\rm d}F_\ell^{\rm ps}(r)/{\rm d}r\big|_{r=r_c}$, we have

$$\begin{aligned} u_\ell'(r_c) &= (\ell+1) r_c^\ell {\rm e}^{p(r_c)} + r_c^{\ell+1} p'(r_c) {\rm e}^{p(r_c)} \\ &= \big[\,\ell + 1 + r_c p'(r_c)\,\big] r_c^\ell {\rm e}^{p(r_c)} \\ &= \big[\,\ell + 1 + r_c p'(r_c)\,\big]\big[\,u_\ell(r_c)/r_c\,\big]. \end{aligned}$$

Dividing both sides of the above equation with $u_\ell(r_c)/r_c$ yields

$$r_c u'_\ell(r_c)/u_\ell(r_c) = \ell + 1 + r_c p'(r_c).$$

To make use of the Schrödinger equation for $F_\ell(r)$, we first compute $\mathrm{d}^2 F_\ell^{\mathrm{ps}}(r)/\mathrm{d}r^2$. We have

$$\begin{aligned}
&\frac{\mathrm{d}^2 F_\ell^{\mathrm{ps}}(r)}{\mathrm{d}r^2} \\
&= \frac{\mathrm{d}^2}{\mathrm{d}r^2}\left[r^{\ell+1}\mathrm{e}^{p(r)}\right] = \frac{\mathrm{d}}{\mathrm{d}r}\left[(\ell+1)r^{\ell}\mathrm{e}^{p(r)} + r^{\ell+1}p'(r)\mathrm{e}^{p(r)}\right] \\
&= \left[\ell(\ell+1)r^{-2} + 2(\ell+1)r^{-1}p'(r) + p'^2(r) + p''(r)\right] r^{\ell+1}\mathrm{e}^{p(r)}.
\end{aligned}$$

At $r = r_c$, we have

$$\begin{aligned}
&\left.\frac{\mathrm{d}^2 F_\ell^{\mathrm{ps}}(r)}{\mathrm{d}r^2}\right|_{r=r_c} \\
&= \left[\ell(\ell+1)r_c^{-2} + 2(\ell+1)r_c^{-1}p'(r_c) + p'^2(r_c) + p''(r_c)\right] r_c^{\ell+1}\mathrm{e}^{p(r_c)}.
\end{aligned}$$

Setting $r = r_c$ in the Schrödinger equation for $F_\ell(r)$ and using $V^{\mathrm{ps}}(r_c) = V^{\mathrm{AE}}(r_c)$, we have

$$\begin{aligned}
&-\left[\ell(\ell+1)r_c^{-2} + 2(\ell+1)r_c^{-1}p'(r_c) + p'^2(r_c) + p''(r_c)\right] r_c^{\ell+1}\mathrm{e}^{p(r_c)} \\
&\quad + \ell(\ell+1)r_c^{\ell-1}\mathrm{e}^{p(r_c)} + V^{\mathrm{AE}}(r_c)r_c^{\ell+1}\mathrm{e}^{p(r_c)} - E r_c^{\ell+1}\mathrm{e}^{p(r_c)} = 0.
\end{aligned}$$

Dividing both sides with $r_c^{\ell-1}\mathrm{e}^{p(r_c)}$ yields

$$-\left[2(\ell+1)r_c p'(r_c) + r_c^2 p'^2(r_c) + r_c^2 p''(r_c)\right] + r_c^2\left[V^{\mathrm{AE}}(r_c) - E\right] = 0.$$

Next, we eliminate $p'(r_c)$ from the above equation. From the second equation we obtained in the above, we have

$$r_c p'(r_c) = r_c u'_\ell(r_c)/u_\ell(r_c) - (\ell + 1).$$

We then have

$$\begin{aligned}
r_c^2 p''(r_c) &= r_c^2\left[V_\ell^{\mathrm{AE}}(r_c) - E\right] - 2(\ell+1)r_c p'(r_c) - r_c^2 p'^2(r_c) \\
&= r_c^2\left[V_\ell^{\mathrm{AE}}(r_c) - E\right] - 2(\ell+1)\left[r_c u'_\ell(r_c)/u_\ell(r_c) - (\ell+1)\right] \\
&\quad - \left[r_c u'_\ell(r_c)/u_\ell(r_c) - (\ell+1)\right]^2 \\
&= r_c^2\left[V_\ell^{\mathrm{AE}}(r_c) - E\right] + (\ell+1)^2 - r_c^2\left[u'_\ell(r_c)/u_\ell(r_c)\right]^2 \\
&= r_c^2 V_\ell^{\mathrm{AE}}(r_c) + (\ell+1)^2 - r_c^2\left\{E + \left[u'_\ell(r_c)/u_\ell(r_c)\right]^2\right\}.
\end{aligned}$$

We have thus proved the third equation.

(2) From $p(r) = \alpha r^4 + \beta r^3 + \gamma r^2 + \delta$, we have

$$p'(r_c) = 4\alpha r_c^3 + 3\beta r_c^2 + 2\gamma r_c,$$
$$p''(r_c) = 12\alpha r_c^2 + 6\beta r_c + 2\gamma.$$

From the above-derived equations, we have

$$p(r_c) = \ln[u_\ell(r_c)/r_c^{\ell+1}],$$
$$r_c p'(r_c) = r_c u_\ell'(r_c)/u_\ell(r_c) - (\ell+1),$$
$$r_c^2 p''(r_c) = r_c^2 V_\ell^{\mathrm{AE}}(r_c) + (\ell+1)^2 - r_c^2\{E + [u_\ell'(r_c)/u_\ell(r_c)]^2\}.$$

From now on, $p(r_c)$, $p'(r_c)$, and $p''(r_c)$ are understood to take on the above-given values. Making use of the explicit expressions of $p(r_c)$, $p'(r_c)$, and $p''(r_c)$, we have

$$\alpha r_c^4 + \beta r_c^3 + \gamma r_c^2 + \delta = p(r_c),$$
$$4\alpha r_c^3 + 3\beta r_c^2 + 2\gamma r_c = p'(r_c),$$
$$12\alpha r_c^2 + 6\beta r_c + 2\gamma = p''(r_c).$$

Solving for α, β, and γ from the above three equations yields

$$\alpha = \frac{1}{2r_c^4}\Big\{6\big[p(r_c) - \delta\big] - 4r_c p'(r_c) + r_c^2 p''(r_c)\Big\},$$
$$\beta = \frac{1}{r_c^3}\Big\{-8\big[p(r_c) - \delta\big] + 5r_c p'(r_c) - r_c^2 p''(r_c)\Big\},$$
$$\gamma = \frac{1}{2r_c^3}\Big\{12\big[p(r_c) - \delta\big] - 6r_c p'(r_c) + r_c^2 p''(r_c)\Big\}.$$

With the values of $p(r_c)$, $p'(r_c)$, and $p''(r_c)$ known and independent of α, β, and γ, the above three equations give expressions for α, β, and γ in terms of δ.

(3) Computing the pseudo charge and setting it to be equal to the real charge Q, we have

$$Q = \int_0^{r_c} \mathrm{d}r\, \left|F_\ell^{\mathrm{ps}}(r)\right|^2 = \mathrm{e}^{2\delta} \int_0^{r_c} \mathrm{d}r\, r^{2(\ell+1)} \mathrm{e}^{2\alpha r^4 + 2\beta r^3 + 2\gamma r^2} = \mathrm{e}^{2\delta} I.$$

Taking the logarithm of both sides yields

$$2\delta + \ln I - \ln Q = 0.$$

(4) Making use of $p(r) = \alpha r^4 + \beta r^3 + \gamma r^2 + \delta$ in the above-obtained second-order derivative of $F_\ell^{\rm ps}(r)$, we have

$$\begin{aligned}
&\frac{d^2 F_\ell^{\rm ps}(r)}{dr^2}\\
&= \left[\frac{\ell(\ell+1)}{r^2} + \frac{2(\ell+1)}{r} p'(r) + p'^2(r) + p''(r)\right] F_\ell^{\rm ps}(r)\\
&= \left[\frac{\ell(\ell+1)}{r^2} + 2(\ell+1)\left(4\alpha r^2 + 3\beta r + 2\gamma\right)\right.\\
&\quad \left. + r^2\left(4\alpha r^2 + 3\beta r + 2\gamma\right)^2 + 12\alpha r^2 + 6\beta r + 2\gamma\right] F_\ell^{\rm ps}(r)\\
&= \left[\frac{\ell(\ell+1)}{r^2} + 2(\ell+1)\lambda(r) + r^2\lambda^2(r) + 12\alpha r^2 + 6\beta r + 2\gamma\right] F_\ell^{\rm ps}(r)\\
&= \left\{\frac{\ell(\ell+1)}{r^2} + \lambda(r)\left[2(\ell+1) + r^2\lambda(r)\right] + 12\alpha r^2 + 6\beta r + 2\gamma\right\} F_\ell^{\rm ps}(r).
\end{aligned}$$

From the Schrödinger equation for $F_\ell^{\rm ps}(r)$, we have

$$\begin{aligned}
V_\ell^{\rm ps}(r) &= E + \frac{1}{F_\ell^{\rm ps}(r)} \frac{d^2 F_\ell^{\rm ps}(r)}{dr^2} - \frac{\ell(\ell+1)}{r^2}\\
&= E + \lambda(r)\left[2(\ell+1) + r^2\lambda(r)\right] + 12\alpha r^2 + 6\beta r + 2\gamma.
\end{aligned}$$

22-5 Spherical averages. If the potential $V(\boldsymbol{r})$ is spherically symmetric, the eigenfunction of the single-electron Hamiltonian in an atom can be written as the product of the radial and angular parts, $\psi_{n\ell m}(\boldsymbol{r}) = R_{n\ell}(r) Y_{\ell m}(\theta, \varphi)$. In this case, we may use the spherically averaged electron density. Let $w_{n\ell m}$ be the occupation number of state $\psi_{n\ell m}(\boldsymbol{r})$. The electron density is then given by $n(\boldsymbol{r}) = \sum_{n\ell m} w_{n\ell m} |\psi_{n\ell m}(\boldsymbol{r})|^2$.

(1) The spherical average of $n(\boldsymbol{r})$ can be carried out by assuming that $w_{n\ell m}$ is independent of m, $w_{n\ell m} = w_{n\ell}$. Show that the spherically averaged electron density is given by

$$n(r) = \frac{1}{4\pi} \sum_{n\ell} \omega_{n\ell} R_{n\ell}^2(r).$$

What is the expression of $\omega_{n\ell}$?

(2) The radial electron density $n_s(r)$ is defined by $n_s(r) = 4\pi r^2 n(r)$. Show that

$$\int_0^\infty dr\; n_s(r) = N$$

with N the number of electrons.

(3) The Hartree potential in atomic units is given by $V_{\rm H}(\boldsymbol{r}) = \int {\rm d}\boldsymbol{r}'\ n(\boldsymbol{r}')/|\boldsymbol{r}-\boldsymbol{r}'|$. Show that the spherically averaged Hartree potential is given by

$$V_{\rm H}(r) = \frac{1}{r}\int_0^r {\rm d}r'\ n_s(r') + \int_r^\infty {\rm d}r'\ \frac{n_s(r')}{r'}.$$

---

(1) For $w_{n\ell m} = w_{n\ell}$, making use of $\psi_{n\ell m}(\boldsymbol{r}) = R_{n\ell}(r)Y_{\ell m}(\theta,\varphi)$ we have

$$\begin{aligned} n(r) &= \sum_{n\ell m} w_{n\ell}|\psi_{n\ell m}(\boldsymbol{r})|^2 = \sum_{n\ell} w_{n\ell}R_{n\ell}^2(r)\sum_{m=-\ell}^{\ell}|Y_{\ell m}(\theta,\varphi)|^2 \\ &= \sum_{n\ell} w_{n\ell}R_{n\ell}^2(r)\frac{2\ell+1}{4\pi} = \frac{1}{4\pi}\sum_{n\ell}\omega_{n\ell}R_{n\ell}^2(r), \end{aligned}$$

where we have made use of the addition theorem of spherical harmonic functions, $\sum_{m=-\ell}^{\ell}|Y_{\ell m}(\theta,\varphi)|^2 = (2\ell+1)/4\pi$, and introduced $\omega_{n\ell} = (2\ell+1)w_{n\ell}$.

(2) Making use of the normalization of $R_{n\ell}(r)$, $\int_0^\infty {\rm d}r\ r^2R_{n\ell}^2(r) = 1$, we have

$$\begin{aligned} \int_0^\infty {\rm d}r\ n_s(r) &= 4\pi\int_0^\infty {\rm d}r\ r^2 n(r) = \sum_{n\ell}\omega_{n\ell}\int_0^\infty {\rm d}r\ r^2R_{n\ell}^2(r) \\ &= \sum_{n\ell}\omega_{n\ell} = \sum_{n\ell}(2\ell+1)w_{n\ell} = N \end{aligned}$$

in which $2\ell+1$ is the degeneracy factor for m.

(3) Making use of the spherically averaged electron density in the Hartree potential, we have

$$V_{\rm H}(r) = \int {\rm d}\boldsymbol{r}'\ \frac{n(r')}{|\boldsymbol{r}-\boldsymbol{r}'|}.$$

Inserting the expansion of $1/|\boldsymbol{r}-\boldsymbol{r}'|$ in terms of Legendre polynomials

$$\frac{1}{|\boldsymbol{r}-\boldsymbol{r}'|} = \frac{1}{(r^2+r'^2-2rr'\cos\theta)^{1/2}} = \sum_{\ell=0}^{\infty}\frac{r_<^\ell}{r_>^{\ell+1}}P_\ell(\cos\theta)$$

with $r_< = \min(r,r')$ and $r_> = \max(r,r')$ into the expression for $V_{\rm H}(r)$, we have

$$\begin{aligned} V_{\rm H}(r) &= 2\pi\sum_{\ell=0}^{\infty}\int_0^\infty {\rm d}r'\ r'^2\frac{r_<^\ell}{r_>^{\ell+1}}n(r')\int_{-1}^{1}{\rm d}\cos\theta\ P_\ell(\cos\theta) \\ &= 4\pi\int_0^\infty {\rm d}r'\ \frac{r'^2}{r_>}n(r') \\ &= \frac{1}{r}\int_0^r {\rm d}r'\ n_s(r') + \int_r^\infty {\rm d}r'\ \frac{n_s(r')}{r'}, \end{aligned}$$

where we have made use of $n_s(r) = 4\pi r^2 n(r)$.

22-6 Hedin-Lundqvist interpolation scheme for exchange and correlation. In their numerical computations on a uniform electron gas, Hedin and Lundqvist noticed that the ratio of the derivative of the exchange-correlation potential $V_{xc}(r_s)$ with respect to the electron effective radius r_s to the derivative of the exchange potential $V_x(r_s)$ is almost a linear function of r_s. Let

$$\frac{dV_{xc}(r_s)}{dr_s} = \gamma(r_s)\frac{dV_x(r_s)}{dr_s}.$$

The exchange potential $V_x(r_s)$ is given by $V_x(r_s) = -1/(\pi\alpha r_s)$ a.u. with $\alpha = [4/(9\pi)]^{1/3}$. Hedin and Lundqvist parametrized $\gamma(r_s)$ as

$$\gamma(r_s) = 1 + \frac{Bx}{1+x}, \quad x = \frac{r_s}{A},$$

where A and B are parameters. Through fitting to numerical computations, it was found that $A = 21$ a.u. and $B = 0.773\,4$.

(1) Obtain an expression for $V_{xc}(r_s)$ by integrating the equation that specifies the relation between $dV_{xc}(r_s)/dr_s$ and $dV_x(r_s)/dr_s$.

(2) Integrate again to obtain the exchange-correlation energy per electron $\epsilon_{xc}(r_s) = E_{xc}(r_s)/N$, where N is the number of electrons.

(3) Find the values of the correlation energy per electron $\epsilon_c(r_s)$ in the $r_s \to 0$ and $r_s \to \infty$ limits.

---

(1) Making use of $V_x(r_s) = -1/(\pi\alpha r_s)$ and the expression for $\gamma(r_s)$, we have

$$\frac{dV_{xc}(r_s)}{dr_s} = \frac{1}{\pi\alpha A^2}\left[\frac{1}{x^2} + \frac{B}{x(1+x)}\right].$$

Taking into account the fact that $V_{xc}(r_s) \to 0$ as $r_s \to \infty$, we integrate both sides of the above equation over r_s from r_s to infinity. We have

$$\begin{aligned} V_{xc}(r_s) &= -\frac{1}{\pi\alpha A}\int_x^\infty dx' \left[\frac{1}{x'^2} + \frac{B}{x'(1+x')}\right] \\ &= -\frac{1}{\pi\alpha A}\left[\frac{1}{x} + B\ln\left(1+\frac{1}{x}\right)\right] \\ &= V_x(r_s) - C\ln\left(1+\frac{1}{x}\right), \end{aligned}$$

where $C = B/(\pi\alpha A) \approx 0.022\,5$ a.u.. Note that C is in hartrees here.

(2) In terms of the exchange-correlation energy density $\mathscr{E}_{\rm xc}(\boldsymbol{r})$ and the exchange-correlation energy per electron $\epsilon_{\rm xc}(r_s)$, we can express the exchange-correlation energy $E_{\rm xc}$ as

$$E_{\rm xc}(r_s) = \int \mathrm{d}\boldsymbol{r}\ \mathscr{E}_{\rm xc}(\boldsymbol{r}) = \int \mathrm{d}\boldsymbol{r}\ n(\boldsymbol{r})\epsilon_{\rm xc}(r_s).$$

$V_{\rm xc}(r_s)$ is related to the exchange-correlation energy $E_{\rm xc}(r_s)$ and the exchange-correlation energy density $\mathscr{E}_{\rm xc}$ through

$$V_{\rm xc}(r_s) = \frac{\delta E_{\rm xc}(r_s)}{\delta n(\boldsymbol{r})} = \frac{\partial \mathscr{E}_{\rm xc}(r_s)}{\partial n(\boldsymbol{r})}.$$

For a uniform electron gas, we have $E_{\rm xc}(r_s) = \mathscr{V}\mathscr{E}_{\rm xc} = n\mathscr{V}\epsilon_{\rm xc}(r_s)$. Thus, $\epsilon_{\rm xc}(r_s) = \mathscr{E}_{\rm xc}/n$. From $n = 3/(4\pi r_s^3)$, we have

$$V_{\rm xc}(r_s) = -\frac{4\pi r_s^4}{9}\frac{\partial \mathscr{E}_{\rm xc}(r_s)}{\partial r_s}$$

Integrating yields

$$\begin{aligned}\mathscr{E}_{\rm xc}(r_s) &= \frac{9}{4\pi}\int_{r_s}^{\infty} \mathrm{d}r_s\ \frac{V_{\rm xc}(r_s)}{r_s^4}\\ &= -\frac{9}{4\pi^2\alpha A^4}\int_x^{\infty}\mathrm{d}x'\left[\frac{1}{x'^5}+\frac{B}{x'^4}\ln\left(1+\frac{1}{x'}\right)\right]\\ &= n\epsilon_{\rm x}(r_s) - nC\left[(1+x^3)\ln\left(1+\frac{1}{x}\right)-\frac{1}{3}+\frac{x}{2}-x^2\right],\end{aligned}$$

where $\epsilon_{\rm x}(r_s) = -(3/4)(1/\pi\alpha r_s) = (3/4)V_{\rm x}(r_s)$ is the exchange energy per electron. The exchange-correlation energy per electron $\epsilon_{\rm xc}(r_s)$ is then given by

$$\epsilon_{\rm xc}(r_s) = \epsilon_{\rm x}(r_s) - C\left[(1+x^3)\ln\left(1+\frac{1}{x}\right)-\frac{1}{3}+\frac{x}{2}-x^2\right].$$

(3) The correlation energy per electron $\epsilon_{\rm c}(r_s)$ is given by

$$\epsilon_{\rm c}(r_s) = \epsilon_{\rm xc}(r_s)-\epsilon_{\rm x}(r_s) = -C\left[(1+x^3)\ln\left(1+\frac{1}{x}\right)-\frac{1}{3}+\frac{x}{2}-x^2\right].$$

In the $r_s \to 0$ limit, the most important contribution comes from the term containing $\ln x$. We thus have

$$\epsilon_{\rm c}(r_s \to 0) \to C\ln(r_s/A).$$

In the $r_s \to \infty$ limit, expanding $\ln(1 + 1/x)$ up to the fourth order in $1/x$ and keeping only terms up to the order of $1/x$ for the product $(1 + x^3)\ln(1 + 1/x)$, we have

$$\begin{aligned}\epsilon_c(r_s \to \infty) &\to -C\left[(1+x^3)\left(\frac{1}{x} - \frac{1}{2x^2} + \frac{1}{3x^3}\right) - \frac{1}{3} + \frac{x}{2} - x^2\right] \\ &= -C\left(\frac{1}{x} + x^2 - \frac{x}{2} + \frac{1}{3} - \frac{1}{4x} - \frac{1}{3} + \frac{x}{2} - x^2\right) \\ &= -\frac{3C}{4x} = -\frac{3AC}{4r_s}.\end{aligned}$$

22-7 Simplified OPW pseudopotential. A pseudopotential derived within the orthogonalized plane wave (OPW) formalism is of the form

$$\hat{V}^{\rm ps} = \hat{V} + \sum_\alpha (E_{\boldsymbol{k}} - E_\alpha)|\alpha\rangle\langle\alpha|,$$

where $\hat{V}$ is the nuclear Coulomb potential, $E_{\boldsymbol{k}}$ the band energy to be computed, and $|\alpha\rangle$ an atomic-like core state of energy E_α. We simplify the above expression by retaining only one core state that is taken to be the hydrogen ground state denoted by $|0\rangle$ and demanding that the matrix elements of $\hat{V}^{\rm ps}$ between eigenstates $|\boldsymbol{k}\rangle$'s of the momentum operator be given by

$$\langle\boldsymbol{k}'|\,\hat{V}^{\rm ps}|\boldsymbol{k}\rangle = \langle\boldsymbol{k}'|\,\hat{V}|\boldsymbol{k}\rangle + \beta\,\langle\boldsymbol{k}'|\,E^{(0)}_{\boldsymbol{k}'} - E_0|0\rangle\langle 0|\boldsymbol{k}\rangle,$$

where $E^{(0)}_{\boldsymbol{k}}$ is the free-electron dispersion relation, $E^{(0)}_{\boldsymbol{k}} = \hbar^2\boldsymbol{k}^2/2m$, and β an adjustable parameter of order unity. Here $\hat{V}$ is taken to be the local Coulomb potential in the hydrogen atom, $\langle\boldsymbol{r}'|\,\hat{V}|\boldsymbol{r}\rangle = V(r)\delta(\boldsymbol{r}' - \boldsymbol{r})$ with $V(r) = -e^2/4\pi\epsilon_0 r$.

(1) Show that

$$\langle\boldsymbol{k}'|\,\hat{V}^{\rm ps}|\boldsymbol{k}\rangle = \langle\boldsymbol{k}'|\,\hat{V}|\boldsymbol{k}\rangle - \beta\,\langle\boldsymbol{k}'|\,\hat{V}|0\rangle\langle 0|\boldsymbol{k}\rangle\,.$$

(2) The Fourier transform of $V^{\rm ps}_{\boldsymbol{k}}(\boldsymbol{q}) \stackrel{\rm def}{=} \langle\boldsymbol{k}+\boldsymbol{q}|\,\hat{V}^{\rm ps}\,|\boldsymbol{k}\rangle$ with respect to $\boldsymbol{q}$ is given by $V^{\rm ps}_{\boldsymbol{k}}(\boldsymbol{r}) = \int {\rm d}\boldsymbol{q}\ {\rm e}^{{\rm i}\boldsymbol{q}\cdot\boldsymbol{r}} V^{\rm ps}_{\boldsymbol{k}}(\boldsymbol{q})$. We now consider the average value of $V^{\rm ps}_{\boldsymbol{k}}(\boldsymbol{r})$ on the Fermi surface of the electron system. Set $|\boldsymbol{k}| = k_{\rm F}$ with $k_{\rm F}$ the Fermi wave vector and average $V^{\rm ps}_{\boldsymbol{k}}(\boldsymbol{r})$ over the directions of $\boldsymbol{k}$, with the result denoted by $V^{\rm ps}_{k_{\rm F}}(r)$.

(3) Find the value of the parameter β so that $V^{\rm ps}_{k_{\rm F}}(r) \to$ const for $r \to 0$. Plot $V^{\rm ps}_{k_{\rm F}}(r)$ with β taking on the determined value. Show that $V^{\rm ps}_{k_{\rm F}}(r) \to V(r)$ for $r \to \infty$.

(4) Specify the metal for which this pseudopotential is probably adequate.

---

(1) We first consider the matrix element $\langle \boldsymbol{k}'|\, E_{\boldsymbol{k}'}^{(0)} - E_0|0\rangle$. Making use of the eigenequation for the ground state $|0\rangle$ of the hydrogen atom, $(\hat{\boldsymbol{p}}^2/2m + \hat{V})|0\rangle = E_0|0\rangle$, we have

$$\langle \boldsymbol{k}'|\, E_{\boldsymbol{k}'}^{(0)} - E_0|0\rangle = \langle \boldsymbol{k}'|\, E_{\boldsymbol{k}'}^{(0)} - \hat{\boldsymbol{p}}^2/2m - \hat{V}|0\rangle\,.$$

Since $|\boldsymbol{k}'\rangle$ is an eigenvector of the momentum operator $\hat{\boldsymbol{p}}$, $\hat{\boldsymbol{p}}\,|\boldsymbol{k}'\rangle = \hbar\boldsymbol{k}'\,|\boldsymbol{k}'\rangle$, we have $(\hat{\boldsymbol{p}}^2/2m)|\boldsymbol{k}'\rangle = (\hbar^2\boldsymbol{k}'^2/2m)|\boldsymbol{k}'\rangle$ whose Hermitian conjugate is $\langle \boldsymbol{k}'|(\hat{\boldsymbol{p}}^2/2m) = \langle \boldsymbol{k}'|(\hbar^2\boldsymbol{k}'^2/2m) = \langle \boldsymbol{k}'|\, E_{\boldsymbol{k}'}^{(0)}$. Thus,

$$\langle \boldsymbol{k}'|\, E_{\boldsymbol{k}'}^{(0)} - E_0|0\rangle = \langle \boldsymbol{k}'|\, E_{\boldsymbol{k}'}^{(0)} - E_{\boldsymbol{k}'}^{(0)} - \hat{V}|0\rangle = -\,\langle \boldsymbol{k}'|\,\hat{V}|0\rangle\,.$$

Making use of the above result, we immediately have

$$\langle \boldsymbol{k}'|\,\hat{V}^{\mathrm{ps}}|\boldsymbol{k}\rangle = \langle \boldsymbol{k}'|\,\hat{V}|\boldsymbol{k}\rangle - \beta\,\langle \boldsymbol{k}'|\,\hat{V}|0\rangle\langle 0|\boldsymbol{k}\rangle\,.$$

(2) Setting $\boldsymbol{k}' = \boldsymbol{k} + \boldsymbol{q}$ in $\langle \boldsymbol{k}'|\,\hat{V}^{\mathrm{ps}}|\boldsymbol{k}\rangle$, we have

$$V_{\boldsymbol{k}}^{\mathrm{ps}}(\boldsymbol{q}) \stackrel{\mathrm{def}}{=\!=} \langle \boldsymbol{k}+\boldsymbol{q}|\,\hat{V}^{\mathrm{ps}}\,|\boldsymbol{k}\rangle = \langle \boldsymbol{k}+\boldsymbol{q}|\,\hat{V}|\boldsymbol{k}\rangle - \beta\,\langle \boldsymbol{k}+\boldsymbol{q}|\,\hat{V}|0\rangle\langle 0|\boldsymbol{k}\rangle\,.$$

Since the matrix elements $\langle \boldsymbol{k}+\boldsymbol{q}|\,\hat{V}\,|\boldsymbol{k}\rangle$ and $\langle \boldsymbol{k}+\boldsymbol{q}|\,\hat{V}|0\rangle$ and the inner product $\langle 0|\boldsymbol{k}\rangle$ appear in the above equation, we first evaluate them. For $\langle \boldsymbol{k}+\boldsymbol{q}|\,\hat{V}\,|\boldsymbol{k}\rangle$, we have

$$\langle \boldsymbol{k}+\boldsymbol{q}|\,\hat{V}\,|\boldsymbol{k}\rangle = -\frac{e^2}{(2\pi)^3(4\pi\epsilon_0)}\int \mathrm{d}\boldsymbol{r}\,\frac{\mathrm{e}^{-\mathrm{i}\boldsymbol{q}\cdot\boldsymbol{r}}}{r} = -\frac{e^2}{2\pi^2(4\pi\epsilon_0)q^2}.$$

For $\langle \boldsymbol{k}+\boldsymbol{q}|\,\hat{V}|0\rangle$, we have

$$\begin{aligned}
&\langle \boldsymbol{k}+\boldsymbol{q}|\,\hat{V}|0\rangle \\
&= \int \mathrm{d}\boldsymbol{r}' \int \mathrm{d}\boldsymbol{r}\,\langle \boldsymbol{k}+\boldsymbol{q}|\boldsymbol{r}'\rangle\langle \boldsymbol{r}'|\,\hat{V}|\boldsymbol{r}\rangle\langle \boldsymbol{r}|0\rangle \\
&= -\frac{e^2}{(4\pi\epsilon_0)(2\pi)^{3/2}(\pi a_0^3)^{1/2}}\int \mathrm{d}\boldsymbol{r}' \int \mathrm{d}\boldsymbol{r}\,\frac{\mathrm{e}^{-r/a_0-\mathrm{i}(\boldsymbol{k}+\boldsymbol{q})\cdot\boldsymbol{r}'}}{r}\delta(\boldsymbol{r}'-\boldsymbol{r}) \\
&= -\frac{e^2}{2\pi^2(4\pi\epsilon_0)(2a_0^3)^{1/2}}\int \mathrm{d}\boldsymbol{r}\,\frac{\mathrm{e}^{-r/a_0-\mathrm{i}(\boldsymbol{k}+\boldsymbol{q})\cdot\boldsymbol{r}}}{r} \\
&= -\frac{e^2}{\pi(4\pi\epsilon_0)(2a_0^3)^{1/2}}\int_0^\infty \mathrm{d}r\,r\int_{-1}^1 \mathrm{d}\cos\theta\,\mathrm{e}^{-r/a_0-\mathrm{i}|\boldsymbol{k}+\boldsymbol{q}|r\cos\theta} \\
&= -\frac{e^2}{\mathrm{i}\pi(4\pi\epsilon_0)(2a_0^3)^{1/2}|\boldsymbol{k}+\boldsymbol{q}|}\int_0^\infty \mathrm{d}r\Big[\mathrm{e}^{-(r/a_0)(1-\mathrm{i}|\boldsymbol{k}+\boldsymbol{q}|a_0)} \\
&\qquad\qquad - \mathrm{e}^{-(r/a_0)(1+\mathrm{i}|\boldsymbol{k}+\boldsymbol{q}|a_0)}\Big] \\
&= -\frac{e^2(2a_0)^{1/2}}{\pi(4\pi\epsilon_0)}\frac{1}{1+(|\boldsymbol{k}+\boldsymbol{q}|a_0)^2}.
\end{aligned}$$

For $\langle 0|\boldsymbol{k}\rangle$, we have

$$\begin{aligned}
\langle 0|\boldsymbol{k}\rangle &= \int \mathrm{d}\boldsymbol{r}\ \langle 0|\boldsymbol{r}\rangle\langle \boldsymbol{r}|\boldsymbol{k}\rangle = \frac{1}{(2\pi)^{3/2}(\pi a_0^3)^{1/2}}\int \mathrm{d}\boldsymbol{r}\ \mathrm{e}^{-r/a_0+\mathrm{i}\boldsymbol{k}\cdot\boldsymbol{r}} \\
&= \frac{1}{\pi(2a_0^3)^{1/2}}\int_0^\infty \mathrm{d}r\ r^2\int_{-1}^{1}\mathrm{d}\cos\theta\ \mathrm{e}^{-r/a_0+\mathrm{i}kr\cos\theta} \\
&= \frac{1}{\mathrm{i}\pi(2a_0^3)^{1/2}k}\int_0^\infty \mathrm{d}r\ r\big[\mathrm{e}^{-(r/a_0)(1-\mathrm{i}ka_0)} - \mathrm{e}^{-(r/a_0)(1+\mathrm{i}ka_0)}\big] \\
&= \frac{(2a_0)^{3/2}}{\pi}\frac{1}{[1+(ka_0)^2]^2}.
\end{aligned}$$

Making use of the above results, we have

$$\begin{aligned}
V_{\boldsymbol{k}}^{\mathrm{ps}}(\boldsymbol{q}) &= -\frac{e^2}{2\pi^2(4\pi\epsilon_0)q^2} + \frac{4\beta e^2 a_0^2}{\pi^2(4\pi\epsilon_0)}\frac{1}{1+(|\boldsymbol{k}+\boldsymbol{q}|a_0)^2}\frac{1}{[1+(ka_0)^2]^2} \\
&= -\frac{e^2}{2\pi^2(4\pi\epsilon_0)}\left\{\frac{1}{q^2} - \frac{8\beta a_0^2}{1+(|\boldsymbol{k}+\boldsymbol{q}|a_0)^2}\frac{1}{[1+(ka_0)^2]^2}\right\}.
\end{aligned}$$

The Fourier transform of $V_{\boldsymbol{k}}^{\mathrm{ps}}(\boldsymbol{q}) = \langle \boldsymbol{k}+\boldsymbol{q}|\,\hat{V}^{\mathrm{ps}}\,|\boldsymbol{k}\rangle$ with respect to $\boldsymbol{q}$ is given by

$$\begin{aligned}
V_{\boldsymbol{k}}^{\mathrm{ps}}(\boldsymbol{r}) &= \int \mathrm{d}\boldsymbol{q}\ \mathrm{e}^{\mathrm{i}\boldsymbol{q}\cdot\boldsymbol{r}}V_{\boldsymbol{k}}^{\mathrm{ps}}(\boldsymbol{q}) \\
&= -\frac{e^2}{2\pi^2(4\pi\epsilon_0)}\int \mathrm{d}\boldsymbol{q}\ \mathrm{e}^{\mathrm{i}\boldsymbol{q}\cdot\boldsymbol{r}} \\
&\qquad\times\left\{\frac{1}{q^2} - \frac{8\beta a_0^2}{1+(|\boldsymbol{k}+\boldsymbol{q}|a_0)^2}\frac{1}{[1+(ka_0)^2]^2}\right\} \\
&= V(r)\left\{1 - \frac{8\beta}{[1+(ka_0)^2]^2}\mathrm{e}^{-r/a_0-\mathrm{i}\boldsymbol{k}\cdot\boldsymbol{r}}\right\}.
\end{aligned}$$

Setting $|\boldsymbol{k}| = k_{\mathrm{F}}$ and averaging over the directions of $\boldsymbol{k}$, we have

$$\begin{aligned}
V_{k_{\mathrm{F}}}^{\mathrm{ps}}(r) &= V(r)\bigg\{1 - \frac{8\beta}{[1+(k_{\mathrm{F}}a_0)^2]^2}\mathrm{e}^{-r/a_0} \\
&\qquad\times\frac{1}{4\pi}\int_{-1}^{1}\mathrm{d}\cos\theta\ \mathrm{e}^{-\mathrm{i}k_{\mathrm{F}}r\cos\theta}\int_0^{2\pi}\mathrm{d}\varphi\bigg\} \\
&= V(r)\left\{1 - \frac{8\beta}{[1+(k_{\mathrm{F}}a_0)^2]^2}\frac{\mathrm{e}^{-r/a_0}\sin(k_{\mathrm{F}}r)}{k_{\mathrm{F}}r}\right\}.
\end{aligned}$$

(3) For $V_{k_{\mathrm{F}}}^{\mathrm{ps}}(r) \to \mathrm{const}$ as $r \to 0$, the value inside the curly brackets in the above equation must be of the form $cr+O(r^2)$. Expanding e^{-r/a_0} and $\sin(k_{\mathrm{F}}r)$ with respect to r for $r \to 0$, we have

$$\mathrm{e}^{-r/a_0} \approx 1 - \frac{r}{a_0} + \frac{r^2}{2a_0^2} - \frac{r^3}{6a_0^3},\quad \sin(k_{\mathrm{F}}r) \approx k_{\mathrm{F}}r - \frac{(k_{\mathrm{F}}r)^3}{6}.$$

We then have

$$\frac{e^{-r/a_0}\sin(k_F r)}{k_F r} \approx \left(1 - \frac{r}{a_0} + \frac{r^2}{2a_0^2}\right)\left[1 - \frac{(k_F r)^2}{6}\right] \approx 1 - \frac{r}{a_0}.$$

In the $r \to 0$ limit, $V_{k_F}^{ps}(r)$ becomes

$$V_{k_F}^{ps}(r \to 0) \to V(r)\left\{1 - \frac{8\beta}{[1+(k_F a_0)^2]^2}\left(1 - \frac{r}{a_0}\right)\right\}$$

$$= V(r)\left\{1 - \frac{8\beta}{[1+(k_F a_0)^2]^2} + \frac{8\beta}{[1+(k_F a_0)^2]^2}\frac{r}{a_0}\right\}.$$

For $V_{k_F}^{ps}(r)$ to have a finite value in the $r \to 0$ limit, the following must hold

$$1 - \frac{8\beta}{[1+(k_F a_0)^2]^2} = 0.$$

Thus,

$$\beta = \frac{1}{8}[1+(k_F a_0)^2]^2.$$

At $r = 0$, we have

$$V_{k_F}^{ps}(0) = -\frac{e^2}{4\pi\epsilon_0 a_0}.$$

Since $e^{-r/a_0}/(k_F r) \to 0$ as $r \to \infty$, we see that

$$V_{k_F}^{ps}(r \to \infty) \to V(r) = -\frac{e^2}{4\pi\epsilon_0 r}.$$

For β taking on the above-obtained value, $V_{k_F}^{ps}(r)$ is given by

$$V_{k_F}^{ps}(r) = V(r)\left[1 - \frac{e^{-r/a_0}\sin(k_F r)}{k_F r}\right].$$

The pseudopotential $V_{k_F}^{ps}(r)$ is plotted in Fig. 22.7 (the solid line) together with the nuclear Coulomb potential (the dotted line). The atomic units (a.u.) are used in Fig. 22.7 with the potentials in $e^2/4\pi\epsilon_0 a_0$ and r in a_0. For k_F, the value of $0.59/a_0$ has been used. This is the value for Li. From Fig. 22.7, we see that pseudopotential varies much less than the nuclear Coulomb potential close to the nucleus. The pseudopotential tends to -1 a.u. as $r \to 0$ while the nuclear Coulomb potential tends to $-\infty$ (diverges).

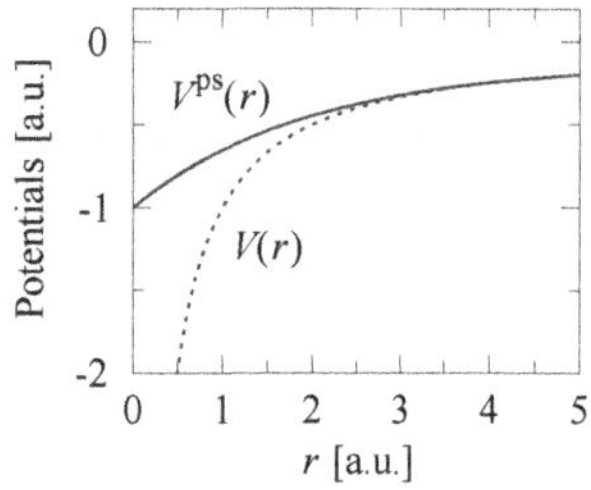

Fig. 22.7 Plot of the pseudopotential $V_{k_F}^{ps}(r)$ as a function of r. The dotted line shows the nuclear Coulomb potential.

(4) Because only one core state, the ground state of the hydrogen atom, is taken into account in constructing the pseudopotential, this pseudopotential is probably adequate for a metal made of two-electron atoms, that is, the lithium metal.

22-8 Equivalence of the norm-conservation condition. To see if the norm-conservation condition gives the correct first-order energy dependence of the phase shift of the scattering from the pseudopotential, we need to check if the energy derivative of the logarithmic derivative of the pseudo wave function is the same as that of the true wave function. Starting from the radial Schrödinger equation in atomic units

$$\left[-\frac{1}{2}\frac{\mathrm{d}^2}{\mathrm{d}r^2}+\frac{\ell(\ell+1)}{2r^2}+V(r)\right]\big[rR_{n\ell}(r)\big]=E_{n\ell}\big[rR_{n\ell}(r)\big],$$

show that

$$-\frac{1}{2}\frac{\partial}{\partial E_{n\ell}}\frac{\partial}{\partial r}\ln R_{n\ell}(r)\bigg|_{r=r_{c\ell}}=\frac{1}{\big[rR_{n\ell}(r)\big]^2}\int_0^{r_{c\ell}}\mathrm{d}r\,\big[rR_{n\ell}(r)\big]^2.$$

In conjunction with the norm-conservation condition, the above equation indicates that the energy derivative of the logarithmic derivative of the pseudo wave function is the same as that of the true wave function so that the first-order energy dependence of the phase shift of the scattering from the pseudopotential is correct.

For convenience, we introduce a short notation for the logarithmic derivative of $rR_{n\ell}(r)$ with respect to r

$$\mathscr{D}(r)=\frac{\partial}{\partial r}\ln\big[rR_{n\ell}(r)\big]=\frac{\big[rR_{n\ell}(r)\big]'}{\big[rR_{n\ell}(r)\big]},$$

where the prime represents the derivative with respect to r. Using $\mathscr{D}(r)$, we can reexpress the radial Schrödinger equation as

$$\mathscr{D}'(r)+\mathscr{D}^2(r)=2\left[\frac{\ell(\ell+1)}{2r^2}+V(r)-E_{n\ell}\right].$$

Differentiating the above equation with respect to $E_{n\ell}$ yields

$$\frac{\partial\mathscr{D}'(r)}{\partial E_{n\ell}}+2\mathscr{D}(r)\frac{\partial\mathscr{D}(r)}{\partial E_{n\ell}}=-2.$$

The two terms on the left hand side of the above equation can be combined by making use of the definition of $\mathscr{D}(r)$. We have

$$\begin{aligned}-2&=\frac{\partial}{\partial r}\frac{\partial\mathscr{D}(r)}{\partial E_{n\ell}}+\frac{1}{\big[rR_{n\ell}(r)\big]^2}\frac{\partial\big[rR_{n\ell}(r)\big]^2}{\partial r}\frac{\partial\mathscr{D}(r)}{\partial E_{n\ell}}\\&=\frac{1}{\big[rR_{n\ell}(r)\big]^2}\frac{\partial}{\partial r}\left\{\big[rR_{n\ell}(r)\big]^2\frac{\partial\mathscr{D}(r)}{\partial E_{n\ell}}\right\}.\end{aligned}$$

Multiplying both sides of the above equation with $\left[rR_{n\ell}(r)\right]^2$ and then integrating both sides from $r = 0$ to $r = r_{c\ell}$, we have

$$\left.\frac{\partial \mathscr{D}(r)}{\partial E_{n\ell}}\right|_{r=r_{c\ell}} = -\frac{2}{\left[r_{c\ell}R_{n\ell}(r_{c\ell})\right]^2}\int_0^{r_{c\ell}} \mathrm{d}r\,\left[rR_{n\ell}(r)\right]^2.$$

Rewriting the derivative of $\mathscr{D}(r)$ on the left hand as that of $R_{n\ell}(r)$, we have

$$\frac{\partial \mathscr{D}(r)}{\partial E_{n\ell}} = \frac{\partial}{\partial E_{n\ell}}\frac{\partial}{\partial r}\ln\left[rR_{n\ell}(r)\right] = \frac{\partial}{\partial E_{n\ell}}\frac{\partial}{\partial r}\ln R_{n\ell}(r).$$

Finally, we have

$$-\frac{1}{2}\frac{\partial}{\partial E_{n\ell}}\frac{\partial}{\partial r}\ln R_{n\ell}(r)\bigg|_{r=r_{c\ell}} = \frac{1}{\left[r_{c\ell}R_{n\ell}(r_{c\ell})\right]^2}\int_0^{r_{c\ell}} \mathrm{d}r\,\left[rR_{n\ell}(r)\right]^2.$$

22-9 Perdew-Zunger parametrization of the correlation energy. Perdew and Zunger parametrized the numerical Monte Carlo results of Ceperley and Alder on a uniform electron gas for the correlation energy and gave the following expression for the correlation energy per electron

$$\epsilon_c = \begin{cases} \dfrac{\gamma}{1+\beta_1\sqrt{r_s}+\beta_2 r_s}, & r_s \geqslant 1, \\ A\ln r_s + B + Cr_s\ln r_s + Dr_s, & r_s < 1, \end{cases}$$

where $\gamma = -0.142\ 3$, $\beta_1 = 1.052\ 9$, $\beta_2 = 0.333\ 4$, $A = 0.031\ 1$, $B = -0.048$, $C = 0.002\ 0$, and $D = -0.011\ 6$ are parameters (all in a.u.). Find the correlation potential V_c.

The total correlation energy $E_c[n]$ is related to the correlation energy per electron $\epsilon_c(n(\boldsymbol{r}))$ through

$$E_c[n] = \int \mathrm{d}\boldsymbol{r}\; n(\boldsymbol{r})\epsilon_c(n(\boldsymbol{r})).$$

Taking the functional derivative of the above expression with respect to $n(\boldsymbol{r})$, we obtain

$$\begin{aligned} V_c(n(\boldsymbol{r})) = \frac{\delta E_c[n]}{\delta n(\boldsymbol{r})} &= \frac{\partial}{\partial n(\boldsymbol{r})}\left[n(\boldsymbol{r})\epsilon_c(n(\boldsymbol{r}))\right] \\ &= \epsilon_c(n(\boldsymbol{r})) + n(\boldsymbol{r})\frac{\partial \epsilon_c(n(\boldsymbol{r}))}{\partial n(\boldsymbol{r})}. \end{aligned}$$

For a uniform gas, $n = 3/(4\pi r_s^3)$. We have

$$\frac{\partial \epsilon_c(n)}{\partial n} = -\frac{4\pi r_s^4}{9}\frac{\partial \epsilon_c(r_s)}{\partial r_s},$$

and

$$V_c(r_s) = \epsilon_c(r_s) - \frac{r_s}{3}\frac{\partial \epsilon_c(r_s)}{\partial r_s}.$$

For $r_s \geqslant 1$, we have

$$V_c(r_s) = \gamma \frac{6 + 7\beta_1\sqrt{r_s} + 8\beta_2 r_s}{(1 + \beta_1\sqrt{r_s} + \beta_2 r_s)^2} = \epsilon_c(r_s)\frac{6 + 7\beta_1\sqrt{r_s} + 8\beta_2 r_s}{1 + \beta_1\sqrt{r_s} + \beta_2 r_s}, \quad r_s \geqslant 1.$$

For $r_s < 1$, we have

$$V_c(r_s) = A \ln r_s + \left(B - \frac{1}{3}A\right) + \frac{2}{3}C r_s \ln r_s + \frac{1}{3}(2D - C)r_s, \quad r_s < 1.$$

Putting together the results for $r_s \geqslant 1$ and $r_s < 1$, we have

$$V_c(r_s) = \begin{cases} \gamma \dfrac{6 + 7\beta_1\sqrt{r_s} + 8\beta_2 r_s}{(1 + \beta_1\sqrt{r_s} + \beta_2 r_s)^2}, & r_s \geqslant 1, \\ A \ln r_s + \left(B - \dfrac{1}{3}A\right) + \dfrac{2}{3}C r_s \ln r_s + \dfrac{1}{3}(2D - C)r_s, & r_s < 1. \end{cases}$$

Chapter 23

Projector-Augmented Plane-Wave Method

(1) *Transformation operator* $\hat{\mathscr{T}}$

$|\psi_\mu\rangle = \hat{\mathscr{T}}\,|\tilde{\psi}_\mu\rangle$,

$|\psi_\mu\rangle$ is the full (true) wave function.

$|\tilde{\psi}_\mu\rangle$ is the auxiliary wave function.

(2) *Determination of* $|\tilde{\psi}_\mu\rangle$

$$\mathscr{T}^\dagger \hat{H} \mathscr{T} |\tilde{\psi}_\mu\rangle = \varepsilon_\mu \mathscr{T}^\dagger \mathscr{T} |\tilde{\psi}_\mu\rangle.$$

(3) *Evaluation of physical quantities*

$$\langle \hat{A} \rangle = \sum_\mu f_\mu \langle \psi_\mu | \hat{A} | \psi_\mu \rangle = \sum_\mu f_\mu \langle \tilde{\psi}_\mu | \mathscr{T}^\dagger A \mathscr{T} | \tilde{\psi}_\mu \rangle.$$

(4) *Projector function*

$$|\tilde{\psi}_\mu\rangle = \sum_i |\tilde{\phi}_{\tau i}\rangle\langle \tilde{p}_{\tau i}|\tilde{\psi}_\mu\rangle,$$

$$\sum_i |\tilde{\phi}_{\tau i}\rangle\langle \tilde{p}_{\tau i}| = \hat{I},$$

$$\langle \tilde{p}_{\tau j}|\tilde{\phi}_{\tau' i}\rangle = \delta_{ij}\delta_{\tau\tau'}.$$

(5) *Expression of* $\hat{\mathscr{T}}$

$$\hat{\mathscr{T}} = \hat{I} + \sum_{\tau i} \big(|\phi_{\tau i}\rangle - |\tilde{\phi}_{\tau i}\rangle\big) \langle \tilde{p}_{\tau i}| .$$

(6) *Decomposition of the true wave function*

$$|\psi_\mu\rangle = |\tilde{\psi}_\mu\rangle + \sum_\tau \big(|\psi^1_{\mu\tau}\rangle - |\tilde{\psi}^1_{\mu\tau}\rangle\big),$$

$$|\psi^1_{\mu\tau}\rangle = \sum_i |\phi_{\tau i}\rangle \langle\tilde{p}_{\tau i}|\tilde{\psi}_\mu\rangle,$$

$$|\tilde{\psi}^1_{\mu\tau}\rangle = \sum_i |\tilde{\phi}_{\tau i}\rangle \langle\tilde{p}_{\tau i}|\tilde{\psi}_\mu\rangle\,.$$

(7) *General formula for local operators*

$$\begin{aligned}\langle\hat{A}\rangle &= \sum_\mu f_\mu\,\langle\tilde{\psi}_\mu|\hat{A}|\tilde{\psi}_\mu\rangle + \sum_{\tau i}\langle\tilde{\varphi}_{\tau i}|\hat{A}|\tilde{\varphi}_{\tau i}\rangle \\ &\quad+\sum_\tau\Big[\sum_{ij}D_{ij}\langle\phi_{\tau j}|\hat{A}|\phi_{\tau i}\rangle+\sum_i\langle\varphi_{\tau i}|\hat{A}|\varphi_{\tau i}\rangle\Big] \\ &\quad-\sum_\tau\Big[\sum_{ij}D_{ij}\langle\tilde{\phi}_{\tau j}|\hat{A}|\tilde{\phi}_{\tau i}\rangle+\sum_i\langle\tilde{\varphi}_{\tau i}|\hat{A}|\tilde{\varphi}_{\tau i}\rangle\Big],\end{aligned}$$

$$D_{ij} = \sum_\mu f_\mu\,\langle\tilde{\psi}_\mu|\tilde{p}_{\tau j}\rangle\,\langle\tilde{p}_{\tau i}|\tilde{\psi}_\mu\rangle\,.$$

(8) *Electron density*

$$n(\boldsymbol{r}) = \tilde{n}(\boldsymbol{r}) + \sum_\tau\left[\,n^1_\tau(\boldsymbol{r}) - \tilde{n}^1_\tau(\boldsymbol{r})\,\right].$$

(9) *Total energy*

$$E([\tilde{\psi}_\mu], R_\tau) = \tilde{E} + \sum_\tau(E^1_\tau - \tilde{E}^1_\tau).$$

23-1 Kohn-Sham equations for auxiliary wave functions. In terms of the auxiliary wave functions, $F([\psi_\mu], \Lambda_{\mu\nu})$ is given by

$$F([\mathscr{T}\tilde{\psi}_\mu], \Lambda_{\mu\nu}) = E[\mathscr{T}\tilde{\psi}_\mu] - \sum_{\mu\nu}[\,\langle\tilde{\psi}_\mu|\mathscr{T}^\dagger\mathscr{T}|\tilde{\psi}_\nu\rangle - \delta_{\mu\nu}\,]\Lambda_{\mu\nu}.$$

Minimize $F([\mathscr{T}\tilde{\psi}_\mu], \Lambda_{\mu\nu})$ with respect to $\tilde{\psi}_\mu(\boldsymbol{r})$ to obtain

$$\hat{\mathscr{T}}^\dagger\hat{H}\hat{\mathscr{T}}|\tilde{\psi}_\mu\rangle = \varepsilon_\mu\hat{\mathscr{T}}^\dagger\hat{\mathscr{T}}|\tilde{\psi}_\mu\rangle.$$

In consideration that $\hat{\mathscr{T}}|\tilde{\psi}_\mu\rangle$ is given by

$$\langle\boldsymbol{r}|\,\hat{\mathscr{T}}|\tilde{\psi}_\mu\rangle = \int d\boldsymbol{r}'\;\mathscr{T}(\boldsymbol{r},\boldsymbol{r}')\tilde{\psi}_\mu(\boldsymbol{r}')$$

in the coordinate representation, the electron number density is expressed as

$$n(\boldsymbol{r}) = \sum_\nu f_\nu\int d\boldsymbol{r}'\int d\boldsymbol{r}''\;\mathscr{T}^*(\boldsymbol{r},\boldsymbol{r}')\tilde{\psi}^*_\nu(\boldsymbol{r}')\mathscr{T}(\boldsymbol{r},\boldsymbol{r}'')\tilde{\psi}_\nu(\boldsymbol{r}'')$$

in terms of $\tilde{\psi}_\mu(\boldsymbol{r})$'s. Taking the functional derivative of $F([\mathscr{T}\tilde{\psi}_\mu], \Lambda_{\mu\nu})$ with respect to $\tilde{\psi}^*_\mu(\boldsymbol{r})$, we have for the first term, $E[\mathscr{T}\tilde{\psi}_\mu]$,

$$
\begin{aligned}
\frac{\delta E[\mathscr{T}\tilde{\psi}_\mu]}{\delta\tilde{\psi}^*_\mu(\boldsymbol{r})} &= \int \mathrm{d}\boldsymbol{r}_1 \, \frac{\delta E[\mathscr{T}\tilde{\psi}_\mu]}{\delta n(\boldsymbol{r}_1)} \frac{\delta n(\boldsymbol{r}_1)}{\delta\tilde{\psi}^*_\mu(\boldsymbol{r})} \\
&= \int \mathrm{d}\boldsymbol{r}_1 \, \frac{\delta E[\mathscr{T}\tilde{\psi}_\mu]}{\delta n(\boldsymbol{r}_1)} \frac{\delta}{\delta\tilde{\psi}^*_\mu(\boldsymbol{r})} \sum_\nu f_\nu \int \mathrm{d}\boldsymbol{r}_2 \int \mathrm{d}\boldsymbol{r}_3 \, \mathscr{T}^*(\boldsymbol{r}_1, \boldsymbol{r}_2)\tilde{\psi}^*_\nu(\boldsymbol{r}_2) \\
&\qquad \times \mathscr{T}(\boldsymbol{r}_1, \boldsymbol{r}_3)\tilde{\psi}_\nu(\boldsymbol{r}_3) \\
&= \int \mathrm{d}\boldsymbol{r}_1 \, \frac{\delta E[\mathscr{T}\tilde{\psi}_\mu]}{\delta n(\boldsymbol{r}_1)} \sum_\nu f_\nu \int \mathrm{d}\boldsymbol{r}_2 \int \mathrm{d}\boldsymbol{r}_3 \, \mathscr{T}^*(\boldsymbol{r}_1, \boldsymbol{r}_2)\delta_{\mu\nu}\delta(\boldsymbol{r} - \boldsymbol{r}_2) \\
&\qquad \times \mathscr{T}(\boldsymbol{r}_1, \boldsymbol{r}_3)\tilde{\psi}_\nu(\boldsymbol{r}_3) \\
&= f_\mu \int \mathrm{d}\boldsymbol{r}_1 \int \mathrm{d}\boldsymbol{r}_3 \, \frac{\delta E[\mathscr{T}\tilde{\psi}_\mu]}{\delta n(\boldsymbol{r}_1)} \mathscr{T}^*(\boldsymbol{r}_1, \boldsymbol{r})\mathscr{T}(\boldsymbol{r}_1, \boldsymbol{r}_3)\tilde{\psi}_\nu(\boldsymbol{r}_3).
\end{aligned}
$$

Making use of $\delta E[\mathscr{T}\tilde{\psi}_\mu]/\delta n(\boldsymbol{r}_1) = \hat{H}(\boldsymbol{r}_1)$ and renaming $\boldsymbol{r}_1$ as $\boldsymbol{r}'$ and $\boldsymbol{r}_3$ as $\boldsymbol{r}''$, we have

$$
\begin{aligned}
\frac{\delta E[\mathscr{T}\tilde{\psi}_\mu]}{\delta\tilde{\psi}^*_\mu(\boldsymbol{r})} &= f_\mu \int \mathrm{d}\boldsymbol{r}' \int \mathrm{d}\boldsymbol{r}'' \, \hat{H}(\boldsymbol{r}')\mathscr{T}^*(\boldsymbol{r}', \boldsymbol{r})\mathscr{T}(\boldsymbol{r}', \boldsymbol{r}'')\tilde{\psi}_\nu(\boldsymbol{r}'') \\
&= f_\mu \int \mathrm{d}\boldsymbol{r}' \int \mathrm{d}\boldsymbol{r}'' \, \mathscr{T}^\dagger(\boldsymbol{r}, \boldsymbol{r}')\hat{H}(\boldsymbol{r}')\mathscr{T}(\boldsymbol{r}', \boldsymbol{r}'')\tilde{\psi}_\nu(\boldsymbol{r}'') \\
&= f_\mu \langle \boldsymbol{r} | \hat{\mathscr{T}}^\dagger \hat{H} \hat{\mathscr{T}} | \tilde{\psi}_\mu \rangle .
\end{aligned}
$$

The functional derivative of the second term in $F([\mathscr{T}\tilde{\psi}_\mu], \Lambda_{\mu\nu})$ with respect to $\tilde{\psi}^*_\mu(\boldsymbol{r})$ is given by

$$
\begin{aligned}
&- \frac{\delta}{\delta\tilde{\psi}^*_\mu(\boldsymbol{r})} \sum_{\mu'\nu} \left[\langle \psi_{\mu'} | \psi_\nu \rangle - \delta_{\mu'\nu} \right] \Lambda_{\mu'\nu} \\
&\quad = - \frac{\delta}{\delta\tilde{\psi}^*_\mu(\boldsymbol{r})} \sum_{\mu'\nu} \Lambda_{\mu'\nu} \int \mathrm{d}\boldsymbol{r}_1 \int \mathrm{d}\boldsymbol{r}_2 \int \mathrm{d}\boldsymbol{r}_3 \, \mathscr{T}^*(\boldsymbol{r}_1, \boldsymbol{r}_2)\tilde{\psi}^*_{\mu'}(\boldsymbol{r}_2) \\
&\qquad \times \mathscr{T}(\boldsymbol{r}_1, \boldsymbol{r}_3)\tilde{\psi}_\nu(\boldsymbol{r}_3) \\
&\quad = - \sum_{\mu'\nu} \Lambda_{\mu'\nu} \int \mathrm{d}\boldsymbol{r}_1 \int \mathrm{d}\boldsymbol{r}_2 \int \mathrm{d}\boldsymbol{r}_3 \, \mathscr{T}^*(\boldsymbol{r}_1, \boldsymbol{r}_2)\delta_{\mu\mu'}\delta(\boldsymbol{r} - \boldsymbol{r}_2) \\
&\qquad \times \mathscr{T}(\boldsymbol{r}_1, \boldsymbol{r}_3)\tilde{\psi}_\nu(\boldsymbol{r}_3).
\end{aligned}
$$

Performing the summation over μ' and integration over $\boldsymbol{r}_2$, we obtain

$$\begin{aligned}
&-\frac{\delta}{\delta\tilde{\psi}_\mu^*(\boldsymbol{r})}\sum_{\mu'\nu}\left[\langle\psi_{\mu'}|\tilde{\psi}_\nu\rangle-\delta_{\mu'\nu}\right]\Lambda_{\mu'\nu}\\
&\quad=-\sum_\nu\Lambda_{\mu\nu}\int d\boldsymbol{r}_1\int d\boldsymbol{r}_3\,\mathscr{T}^*(\boldsymbol{r}_1,\boldsymbol{r})\mathscr{T}(\boldsymbol{r}_1,\boldsymbol{r}_3)\tilde{\psi}_\nu(\boldsymbol{r}_3)\\
&\quad=-\sum_\nu\Lambda_{\mu\nu}\int d\boldsymbol{r}_1\int d\boldsymbol{r}_3\,\mathscr{T}^\dagger(\boldsymbol{r},\boldsymbol{r}_1)\mathscr{T}(\boldsymbol{r}_1,\boldsymbol{r}_3)\tilde{\psi}_\nu(\boldsymbol{r}_3)\\
&\quad=-\sum_\nu\Lambda_{\mu\nu}\langle\boldsymbol{r}|\,\hat{\mathscr{T}}^\dagger\hat{\mathscr{T}}\,|\tilde{\psi}_\nu\rangle\,.
\end{aligned}$$

From $\delta F([\mathscr{T}\tilde{\psi}_\mu],\Lambda_{\mu\nu})/\delta\tilde{\psi}_\mu^*(\boldsymbol{r})=0$, we have

$$\langle\boldsymbol{r}|\,\hat{\mathscr{T}}^\dagger\hat{H}\hat{\mathscr{T}}\,|\tilde{\psi}_\mu\rangle=\sum_\nu(\Lambda_{\mu\nu}/f_\mu)\,\langle\boldsymbol{r}|\,\hat{\mathscr{T}}^\dagger\hat{\mathscr{T}}\,|\tilde{\psi}_\nu\rangle\,.$$

We construct a matrix with $\Lambda_{\mu\nu}/f_\mu$ as its elements and denote this matrix by $\mathscr{L}$. Let S diagonalizes $\mathscr{L}$, $(S^{-1}\mathscr{L}S)_{\mu\nu}=\varepsilon_\mu\delta_{\mu\nu}$. In terms of $\mathscr{L}$, we can rewrite the above equation as

$$\langle\boldsymbol{r}|\,\hat{\mathscr{T}}^\dagger\hat{H}\hat{\mathscr{T}}\,|\tilde{\psi}_\mu\rangle=\langle\boldsymbol{r}|\,\hat{\mathscr{T}}^\dagger\hat{\mathscr{T}}\sum_\nu\mathscr{L}_{\mu\nu}\,|\tilde{\psi}_\nu\rangle\,.$$

We further let $|\tilde{\psi}_\mu\rangle=\sum_\nu S_{\mu\nu}\,|\tilde{\psi}'_\nu\rangle$. We then have

$$\langle\boldsymbol{r}|\,\hat{\mathscr{T}}^\dagger\hat{H}\hat{\mathscr{T}}\sum_\nu S_{\mu\nu}\,|\tilde{\psi}'_\nu\rangle=\langle\boldsymbol{r}|\,\hat{\mathscr{T}}^\dagger\hat{\mathscr{T}}\sum_{\nu\nu'}\mathscr{L}_{\mu\nu}S_{\nu\nu'}\,|\tilde{\psi}'_{\nu'}\rangle\,.$$

Multiplying both side of the above equation with $S^{-1}_{\mu'\mu}$ and then summing both sides over μ, we obtain

$$\langle\boldsymbol{r}|\,\hat{\mathscr{T}}^\dagger\hat{H}\hat{\mathscr{T}}\sum_{\mu\nu}S^{-1}_{\mu'\mu}S_{\mu\nu}\,|\tilde{\psi}'_\nu\rangle=\langle\boldsymbol{r}|\,\hat{\mathscr{T}}^\dagger\hat{\mathscr{T}}\sum_{\mu\nu\nu'}S^{-1}_{\mu'\mu}\mathscr{L}_{\mu\nu}S_{\nu\nu'}\,|\tilde{\psi}'_{\nu'}\rangle,$$

$$\langle\boldsymbol{r}|\,\hat{\mathscr{T}}^\dagger\hat{H}\hat{\mathscr{T}}\sum_\nu\delta_{\mu'\nu}\,|\tilde{\psi}'_\nu\rangle=\langle\boldsymbol{r}|\,\hat{\mathscr{T}}^\dagger\hat{\mathscr{T}}\sum_{\nu'}\varepsilon_{\mu'}\delta_{\mu'\nu'}\,|\tilde{\psi}'_{\nu'}\rangle,$$

$$\langle\boldsymbol{r}|\,\hat{\mathscr{T}}^\dagger\hat{H}\hat{\mathscr{T}}\,|\tilde{\psi}'_{\mu'}\rangle=\varepsilon_{\mu'}\,\langle\boldsymbol{r}|\,\hat{\mathscr{T}}^\dagger\hat{\mathscr{T}}\,|\tilde{\psi}'_{\mu'}\rangle\,.$$

Dropping the primes on both μ' and $\tilde{\psi}'$, we have

$$\langle\boldsymbol{r}|\,\hat{\mathscr{T}}^\dagger\hat{H}\hat{\mathscr{T}}\,|\tilde{\psi}_\mu\rangle=\varepsilon_\mu\,\langle\boldsymbol{r}|\,\hat{\mathscr{T}}^\dagger\hat{\mathscr{T}}\,|\tilde{\psi}_\mu\rangle\,.$$

In the abstract form, we have

$$\hat{\mathscr{T}}^\dagger\hat{H}\hat{\mathscr{T}}\,|\tilde{\psi}_\mu\rangle=\varepsilon_\mu\hat{\mathscr{T}}^\dagger\hat{\mathscr{T}}\,|\tilde{\psi}_\mu\rangle\,.$$

23-2 General formula for local operators. Derive the following general formula for local operators

$$\langle\psi_\mu|\hat{A}|\psi_\mu\rangle = \langle\tilde{\psi}_\mu|\hat{A}|\tilde{\psi}_\mu\rangle + \sum_\tau \left[\langle\psi^1_{\mu\tau}|\hat{A}|\psi^1_{\mu\tau}\rangle - \langle\tilde{\psi}^1_{\mu\tau}|\hat{A}|\tilde{\psi}^1_{\mu\tau}\rangle\right]$$
$$+\sum_\tau \left[\langle\psi^1_{\mu\tau} - \tilde{\psi}^1_{\mu\tau}|\hat{A}|\tilde{\psi}_\mu - \tilde{\psi}^1_{\mu\tau}\rangle + \langle\tilde{\psi}_\mu - \tilde{\psi}^1_{\mu\tau}|\hat{A}|\psi^1_{\mu\tau} - \tilde{\psi}^1_{\mu\tau}\rangle\right]$$
$$+\sum_{\tau\neq\tau'} \langle\psi^1_{\mu\tau} - \tilde{\psi}^1_{\mu\tau}|\hat{A}|\psi^1_{\mu\tau'} - \tilde{\psi}^1_{\mu\tau'}\rangle\,.$$

Inserting

$$|\psi_\mu\rangle = |\tilde{\psi}_\mu\rangle + \sum_\tau (|\psi^1_{\mu\tau}\rangle - |\tilde{\psi}^1_{\mu\tau}\rangle)$$

into $\langle\psi_\mu|\hat{A}|\psi_\mu\rangle$, we have

$$\langle\psi_\mu|\hat{A}|\psi_\mu\rangle$$
$$= [\,\langle\tilde{\psi}_\mu| + \sum_\tau (\langle\psi^1_{\mu\tau}| - \langle\tilde{\psi}^1_{\mu\tau}|)\,]\hat{A}[\,|\tilde{\psi}_\mu\rangle + \sum_{\tau'} (|\psi^1_{\mu\tau'}\rangle - |\tilde{\psi}^1_{\mu\tau'}\rangle)\,]$$
$$= \langle\tilde{\psi}_\mu|\,\hat{A}|\tilde{\psi}_\mu\rangle + \sum_\tau (\langle\psi^1_{\mu\tau}| - \langle\tilde{\psi}^1_{\mu\tau}|)\hat{A}|\tilde{\psi}_\mu\rangle$$
$$+\sum_{\tau'} \langle\tilde{\psi}_\mu|\,\hat{A}(|\psi^1_{\mu\tau'}\rangle - |\tilde{\psi}^1_{\mu\tau'}\rangle)$$
$$+\sum_{\tau\tau'} (\langle\psi^1_{\mu\tau}| - \langle\tilde{\psi}^1_{\mu\tau}|)\hat{A}(|\psi^1_{\mu\tau'}\rangle - |\tilde{\psi}^1_{\mu\tau'}\rangle).$$

Dropping the prime on τ' in the last term on the first line and separating the $\tau = \tau'$ terms from the $\tau \neq \tau'$ ones on the second line, we have

$$\langle\psi_\mu|\hat{A}|\psi_\mu\rangle$$
$$= \langle\tilde{\psi}_\mu|\,\hat{A}|\tilde{\psi}_\mu\rangle + \sum_\tau (\langle\psi^1_{\mu\tau}| - \langle\tilde{\psi}^1_{\mu\tau}|)\hat{A}|\tilde{\psi}_\mu\rangle + \sum_\tau \langle\tilde{\psi}_\mu|\,\hat{A}(|\psi^1_{\mu\tau}\rangle - |\tilde{\psi}^1_{\mu\tau}\rangle)$$
$$+\sum_\tau (\langle\psi^1_{\mu\tau}| - \langle\tilde{\psi}^1_{\mu\tau}|)\hat{A}(|\psi^1_{\mu\tau}\rangle - |\tilde{\psi}^1_{\mu\tau}\rangle)$$
$$+\sum_{\tau\neq\tau'} (\langle\psi^1_{\mu\tau}| - \langle\tilde{\psi}^1_{\mu\tau}|)\hat{A}(|\psi^1_{\mu\tau'}\rangle - |\tilde{\psi}^1_{\mu\tau'}\rangle).$$

Rearranging the second, third, and fourth terms, we obtain

$$\begin{aligned}\langle\psi_\mu|\hat{A}|\psi_\mu\rangle &= \langle\tilde{\psi}_\mu|\,\hat{A}|\tilde{\psi}_\mu\rangle + \sum_\tau\big(\langle\psi^1_{\mu\tau}| - \langle\tilde{\psi}^1_{\mu\tau}|\big)\hat{A}|\tilde{\psi}_\mu - \tilde{\psi}^1_{\mu\tau}\rangle \\
&\quad + \sum_\tau\big(\langle\psi^1_{\mu\tau}| - \langle\tilde{\psi}^1_{\mu\tau}|\big)\hat{A}|\tilde{\psi}^1_{\mu\tau}\rangle \\
&\quad + \sum_\tau\langle\tilde{\psi}_\mu - \tilde{\psi}^1_{\mu\tau}|\,\hat{A}\big(|\psi^1_{\mu\tau}\rangle - |\tilde{\psi}^1_{\mu\tau}\rangle\big) \\
&\quad + \sum_\tau\langle\tilde{\psi}^1_{\mu\tau}|\,\hat{A}\big(|\psi^1_{\mu\tau}\rangle - |\tilde{\psi}^1_{\mu\tau}\rangle\big) \\
&\quad + \sum_\tau\big(\langle\psi^1_{\mu\tau}| - \langle\tilde{\psi}^1_{\mu\tau}|\big)\hat{A}\big(|\psi^1_{\mu\tau}\rangle - |\tilde{\psi}^1_{\mu\tau}\rangle\big) \\
&\quad + \sum_{\tau\neq\tau'}\big(\langle\psi^1_{\mu\tau}| - \langle\tilde{\psi}^1_{\mu\tau}|\big)\hat{A}\big(|\psi^1_{\mu\tau'}\rangle - |\tilde{\psi}^1_{\mu\tau'}\rangle\big) \\
&= \langle\tilde{\psi}_\mu|\,\hat{A}|\tilde{\psi}_\mu\rangle + \sum_\tau\Big[\,\langle\psi^1_{\mu\tau}|\,\hat{A}|\psi^1_{\mu\tau}\rangle - \langle\tilde{\psi}^1_{\mu\tau}|\,\hat{A}|\tilde{\psi}^1_{\mu\tau}\rangle\,\Big] \\
&\quad + \sum_\tau\Big[\,\langle\psi^1_{\mu\tau} - \tilde{\psi}^1_{\mu\tau}|\,\hat{A}|\tilde{\psi}_\mu - \tilde{\psi}^1_{\mu\tau}\rangle \\
&\quad + \sum_\tau\langle\tilde{\psi}_\mu - \tilde{\psi}^1_{\mu\tau}|\,\hat{A}\,|\psi^1_{\mu\tau} - \tilde{\psi}^1_{\mu\tau}\rangle\,\Big] \\
&\quad + \sum_{\tau\neq\tau'}\Big[\,\langle\psi^1_{\mu\tau} - \tilde{\psi}^1_{\mu\tau}|\,\hat{A}\,|\psi^1_{\mu\tau'} - \tilde{\psi}^1_{\mu\tau'}\rangle\,\Big],\end{aligned}$$

where we have made use of the notations $\langle a-b| = \langle a| - \langle b|$ and $|a-b\rangle = |a\rangle - |b\rangle$.

Chapter 24

Experimental Determination of Electronic Band Structures

(1) *Quantized electromagnetic fields*

$$\hat{\boldsymbol{A}}(\boldsymbol{r}) = \sum_{\boldsymbol{k}\lambda}\left(\frac{\hbar}{2\mathscr{V}\epsilon_0\,\omega_{\boldsymbol{k}}}\right)^{1/2}(\hat{a}_{\boldsymbol{k}\lambda} + \hat{a}^\dagger_{-\boldsymbol{k}\lambda})\boldsymbol{\epsilon}^\lambda_{\boldsymbol{k}}\,\mathrm{e}^{\mathrm{i}\boldsymbol{k}\cdot\boldsymbol{r}},$$

$$\hat{H}_{\text{em}} = \sum_{\boldsymbol{k}\lambda}\hbar\omega_{\boldsymbol{k}}\big(\hat{a}^\dagger_{\boldsymbol{k}\lambda}\hat{a}_{\boldsymbol{k}\lambda} + 1/2\big),$$

$$\big[\hat{a}_{\boldsymbol{k}\lambda},\, \hat{a}^\dagger_{\boldsymbol{k}'\lambda'}\big] = \delta_{\boldsymbol{k}\boldsymbol{k}'}\delta_{\lambda\lambda'},$$

$$\big[\hat{a}_{\boldsymbol{k}\lambda},\, \hat{a}_{\boldsymbol{k}'\lambda'}\big] = \big[\hat{a}^\dagger_{\boldsymbol{k}\lambda},\, \hat{a}^\dagger_{\boldsymbol{k}'\lambda'}\big] = 0,$$

$$\hat{a}_{\boldsymbol{k}\lambda}(t) = \mathrm{e}^{-\mathrm{i}\omega_{\boldsymbol{k}}t}\hat{a}_{\boldsymbol{k}\lambda},\quad \hat{a}^\dagger_{\boldsymbol{k}\lambda}(t) = \mathrm{e}^{\mathrm{i}\omega_{\boldsymbol{k}}t}\hat{a}^\dagger_{\boldsymbol{k}\lambda}.$$

(2) *Quantum field operator of electrons*

$$\hat{\Psi}(\boldsymbol{r}, s_z) = \sum_{n\boldsymbol{k}\sigma}\psi_{n\boldsymbol{k}\sigma}(\boldsymbol{r}, s_z)\hat{c}_{n\boldsymbol{k}\sigma},$$

$$\big\{\hat{c}_{n\boldsymbol{k}\sigma},\, \hat{c}^\dagger_{n'\boldsymbol{k}'\sigma'}\big\} = \delta_{nn'}\delta_{\boldsymbol{k}\boldsymbol{k}'}\delta_{\sigma\sigma'},$$

$$\big\{\hat{c}_{n\boldsymbol{k}\sigma},\, \hat{c}_{n'\boldsymbol{k}'\sigma'}\big\} = \big\{\hat{c}^\dagger_{n\boldsymbol{k}\sigma},\, \hat{c}^\dagger_{n'\boldsymbol{k}'\sigma'}\big\} = 0.$$

(3) *Quantum Hamiltonian*

$$\hat{H} = \sum_{n\boldsymbol{k}\sigma}E_{n\boldsymbol{k}\sigma}\hat{c}^\dagger_{n\boldsymbol{k}\sigma}\hat{c}_{n\boldsymbol{k}\sigma} + \sum_{\boldsymbol{q}\lambda}\hbar\omega_{\boldsymbol{q}}\big(\hat{a}^\dagger_{\boldsymbol{q}\lambda}\hat{a}_{\boldsymbol{q}\lambda} + 1/2\big)$$

$$+ \sum_{nn'\boldsymbol{k}\boldsymbol{k}'\sigma\sigma'}\sum_{\boldsymbol{q}\lambda}g_{n'\boldsymbol{k}'\sigma',n\boldsymbol{k}\sigma,\boldsymbol{q}\lambda}\big(\hat{a}_{\boldsymbol{q}\lambda} + \hat{a}^\dagger_{-\boldsymbol{q}\lambda}\big)c^\dagger_{n'\boldsymbol{k}'\sigma'}c_{n\boldsymbol{k}\sigma},$$

$$g_{n'\boldsymbol{k}'\sigma',n\boldsymbol{k}\sigma,\boldsymbol{q}\lambda} = -\frac{\mathrm{i}e\hbar}{m}\left(\frac{\hbar}{2\mathscr{V}\epsilon\,\omega_{\boldsymbol{q}}}\right)^{1/2} \times \sum_{s_z}\int \mathrm{d}\boldsymbol{r}\,\psi^*_{n'\boldsymbol{k}'\sigma'}(\boldsymbol{r}, s_z)\mathrm{e}^{\mathrm{i}\boldsymbol{q}\cdot\boldsymbol{r}}\boldsymbol{\epsilon}^\lambda_{\boldsymbol{q}}\cdot\boldsymbol{\nabla}\psi_{n\boldsymbol{k}\sigma}(\boldsymbol{r}, s_z).$$

(4) *De Haas-van Alphen effect*

The oscillation period is given by

$$P = \Delta\big(1/B\big) = 2\pi e/\hbar(\pi k_{\mathrm{F}}^2).$$

The Lifshits-Kosevich theory gives

$$M_{\mathrm{osc}} = \sum_{k=1}^{\infty} A_k(T,B)\sin(2\pi kF/B + \varphi_k),$$

$$A_k(T,B) = -k_{\mathrm{B}}T\mathscr{V}S_m(\mu)\left(\frac{e}{2\pi^3\hbar B}\right)^{1/2} \times \sum_{k=1}^{\infty}\frac{1}{k^{1/2}}\frac{\cos[(k\hbar^2/2m_0)\partial S_m(\mu)/\partial\mu]}{\left|\partial^2 S_m/\partial k_z^2\right|_0^{1/2}\sinh(\lambda k)},$$

$$F = \frac{\hbar}{2\pi e}S_m(\mu),\quad \varphi_k = \mp\frac{\pi}{4} - 2\pi k\gamma.$$

(5) *Photoemission spectroscopy*

The three-step model consists of
(1) the photo-excitation of the electron,
(2) its propagation towards the surface,
and (3) its escape into the vacuum.

The photocurrent is given by

$$I(E_K,\hbar\omega) \propto w(E_K,\,\hbar\omega)T(E_K)D(E_K).$$

(6) *Response theory of photoemission*

The photocurrent density is given by

$$\mathcal{J}(\boldsymbol{r},s_z,t) \approx \frac{4\sqrt{2}\pi t}{\hbar}\sum_{\nu\kappa j} g_{\nu\kappa j}g_{\kappa\nu j}\boldsymbol{J}_{\nu\nu}n_{\mathrm{F}}(\xi_\kappa)\big[1 - n_{\mathrm{F}}(\xi_\nu)\big] \times n_{\mathrm{B}}(\hbar\omega_j)\delta(E_\nu - E_\kappa - \hbar\omega_j).$$

(7) *Correlated electrons*

$$A_{n\sigma}(\boldsymbol{k},E) = -\frac{1}{\pi}G_{n\sigma}^R(\boldsymbol{k},E) = -\frac{1}{\pi}\frac{\mathrm{Im}\,\Sigma_{n\sigma}(\boldsymbol{k},E)}{(E - E_{n\boldsymbol{k}\sigma} - \Sigma'_{n\sigma})^2 + (\mathrm{Im}\,\Sigma''_{n\sigma})^2},$$

$$\mathcal{J}(\boldsymbol{r},s_z,t) \approx \frac{4\sqrt{2}\pi t}{\hbar}\sum g_{n'\boldsymbol{k}'\sigma',n\boldsymbol{k}\sigma,\boldsymbol{q}\lambda}g_{n\boldsymbol{k}\sigma,n'\boldsymbol{k}'\sigma',\boldsymbol{q}\lambda}\boldsymbol{J}_{n\boldsymbol{k}\sigma,n\boldsymbol{k}\sigma}n_{\mathrm{F}}(\xi_{n\boldsymbol{k}\sigma}) \times \big[1 - n_{\mathrm{F}}(\xi_{n'\boldsymbol{k}'\sigma'})\big]n_{\mathrm{B}}(\hbar\omega_{\boldsymbol{q}\lambda})A_{n\sigma}(\boldsymbol{k},E_{n'\boldsymbol{k}'\sigma'} - \hbar\omega_{\boldsymbol{q}\lambda}).$$

24-1 Quantization of electromagnetic fields. Consider the quantization of electromagnetic fields. The quantized vector potential $\hat{\boldsymbol{A}}$ in vacuum is given by

$$\hat{\boldsymbol{A}}(\boldsymbol{r},t)=\sum_{\boldsymbol{k}\lambda}\left(\frac{\hbar}{2\mathscr{V}\epsilon_0\,\omega_{\boldsymbol{k}}}\right)^{1/2}(\mathrm{e}^{-\mathrm{i}\omega_{\boldsymbol{k}}t}\hat{a}_{\boldsymbol{k}\lambda}+\mathrm{e}^{\mathrm{i}\omega_{\boldsymbol{k}}t}\hat{a}^\dagger_{-\boldsymbol{k}\lambda})\boldsymbol{\epsilon}^\lambda_{\boldsymbol{k}}\mathrm{e}^{\mathrm{i}\boldsymbol{k}\cdot\boldsymbol{r}}$$

in the Coulomb gauge.

(1) Find the quantized electric field $\hat{\boldsymbol{E}}$ and magnetic field $\hat{\boldsymbol{B}}$ from the quantized vector potential $\hat{\boldsymbol{A}}$.

(2) Derive the Hamiltonian of the electromagnetic field from $\hat{H}_{\rm em}=(\epsilon_0/2)\int\mathrm{d}\boldsymbol{r}\;\hat{\boldsymbol{E}}^\dagger\cdot\hat{\boldsymbol{E}}+(1/2\mu_0)\int\mathrm{d}\boldsymbol{r}\;\hat{\boldsymbol{B}}^\dagger\cdot\hat{\boldsymbol{B}}$ using the above derived expressions for $\hat{\boldsymbol{E}}$ and $\hat{\boldsymbol{B}}$.

(3) Evaluate the commutators $[\hat{E}_\alpha(\boldsymbol{r},t),\,\hat{E}^\dagger_\beta(\boldsymbol{r}',t')]$, $[\hat{B}_\alpha(\boldsymbol{r},t),\,\hat{B}^\dagger_\beta(\boldsymbol{r}',t')]$, $[\hat{E}_\alpha(\boldsymbol{r},t),\,\hat{B}_\beta(\boldsymbol{r}',t')]$, where $\alpha,\,\beta=x,\,y,\,z$.

(4) Show that the average of $\hat{\boldsymbol{E}}(\boldsymbol{r},t)$ in the number state $|n_{\boldsymbol{k}\lambda}\rangle$ is identically zero.

(5) Evaluate the average of $\hat{\boldsymbol{E}}(\boldsymbol{r},t)$ in the coherent state $|\gamma_{\boldsymbol{k}\lambda}\rangle=\mathrm{e}^{\gamma_{\boldsymbol{k}\lambda}\hat{a}^\dagger_{\boldsymbol{k}\lambda}-\gamma^*_{\boldsymbol{k}\lambda}\hat{a}_{\boldsymbol{k}\lambda}}|0\rangle$.

---

(1) The electric field is given by

$$\hat{\boldsymbol{E}}(\boldsymbol{r},t)=-\frac{\partial\hat{\boldsymbol{A}}(\boldsymbol{r},t)}{\partial t}$$
$$=\mathrm{i}\sum_{\boldsymbol{k}\lambda}\left(\frac{\hbar\omega_{\boldsymbol{k}}}{2\mathscr{V}\epsilon_0}\right)^{1/2}(\mathrm{e}^{-\mathrm{i}\omega_{\boldsymbol{k}}t}\hat{a}_{\boldsymbol{k}\lambda}-\mathrm{e}^{\mathrm{i}\omega_{\boldsymbol{k}}t}\hat{a}^\dagger_{-\boldsymbol{k}\lambda})\boldsymbol{\epsilon}^\lambda_{\boldsymbol{k}}\mathrm{e}^{\mathrm{i}\boldsymbol{k}\cdot\boldsymbol{r}}.$$

The magnetic field is given by

$$\hat{\boldsymbol{B}}(\boldsymbol{r},t)=\boldsymbol{\nabla}\times\hat{\boldsymbol{A}}(\boldsymbol{r},t)$$
$$=\mathrm{i}\sum_{\boldsymbol{k}\lambda}\left(\frac{\hbar}{2\mathscr{V}\epsilon_0\,\omega_{\boldsymbol{k}}}\right)^{1/2}(\mathrm{e}^{-\mathrm{i}\omega_{\boldsymbol{k}}t}\hat{a}_{\boldsymbol{k}\lambda}+\mathrm{e}^{\mathrm{i}\omega_{\boldsymbol{k}}t}\hat{a}^\dagger_{-\boldsymbol{k}\lambda})(\boldsymbol{k}\times\boldsymbol{\epsilon}^\lambda_{\boldsymbol{k}})\mathrm{e}^{\mathrm{i}\boldsymbol{k}\cdot\boldsymbol{r}}$$

(2) The integral $\int\mathrm{d}\boldsymbol{r}\;\hat{\boldsymbol{E}}^\dagger\cdot\hat{\boldsymbol{E}}$ is given by

$$\int\mathrm{d}\boldsymbol{r}\;\hat{\boldsymbol{E}}^\dagger\cdot\hat{\boldsymbol{E}}=\frac{\hbar}{2\mathscr{V}\epsilon_0}\sum_{\boldsymbol{k}\boldsymbol{k}'\lambda\lambda'}\sqrt{\omega_{\boldsymbol{k}'}\omega_{\boldsymbol{k}}}\,(\mathrm{e}^{\mathrm{i}\omega_{\boldsymbol{k}'}t}\hat{a}^\dagger_{\boldsymbol{k}'\lambda'}-\mathrm{e}^{-\mathrm{i}\omega_{\boldsymbol{k}'}t}\hat{a}_{-\boldsymbol{k}'\lambda'})$$
$$\times\left(\boldsymbol{\epsilon}^{\lambda'*}_{\boldsymbol{k}'}\cdot\boldsymbol{\epsilon}^\lambda_{\boldsymbol{k}}\right)\left(\mathrm{e}^{-\mathrm{i}\omega_{\boldsymbol{k}}t}\hat{a}_{\boldsymbol{k}\lambda}-\mathrm{e}^{\mathrm{i}\omega_{\boldsymbol{k}}t}\hat{a}^\dagger_{-\boldsymbol{k}\lambda}\right)\int\mathrm{d}\boldsymbol{r}\;\mathrm{e}^{-\mathrm{i}(\boldsymbol{k}'-\boldsymbol{k})\cdot\boldsymbol{r}}$$
$$=\frac{1}{2\epsilon_0}\sum_{\boldsymbol{k}\lambda\lambda'}\hbar\omega_{\boldsymbol{k}}\left(\boldsymbol{\epsilon}^{\lambda'*}_{\boldsymbol{k}}\cdot\boldsymbol{\epsilon}^\lambda_{\boldsymbol{k}}\right)\left(\mathrm{e}^{\mathrm{i}\omega_{\boldsymbol{k}}t}\hat{a}^\dagger_{\boldsymbol{k}\lambda'}-\mathrm{e}^{-\mathrm{i}\omega_{\boldsymbol{k}}t}\hat{a}_{-\boldsymbol{k}\lambda'}\right)$$
$$\times\left(\mathrm{e}^{-\mathrm{i}\omega_{\boldsymbol{k}}t}\hat{a}_{\boldsymbol{k}\lambda}-\mathrm{e}^{\mathrm{i}\omega_{\boldsymbol{k}}t}\hat{a}^\dagger_{-\boldsymbol{k}\lambda}\right),$$

where we have made use of $\mathscr{V}^{-1}\int \mathrm{d}\boldsymbol{r}\ \mathrm{e}^{-\mathrm{i}(\boldsymbol{k}'-\boldsymbol{k})\cdot\boldsymbol{r}} = \delta_{\boldsymbol{k}'\boldsymbol{k}}$. Using $\boldsymbol{\epsilon}_{\boldsymbol{k}}^{\lambda'*}\cdot\boldsymbol{\epsilon}_{\boldsymbol{k}}^{\lambda} = \delta_{\lambda'\lambda}$, we have

$$\begin{aligned}\int \mathrm{d}\boldsymbol{r}\ \hat{\boldsymbol{E}}^{\dagger}\cdot\hat{\boldsymbol{E}} &= \frac{1}{2\epsilon_0}\sum_{\boldsymbol{k}\lambda}\hbar\omega_{\boldsymbol{k}}\left(\mathrm{e}^{\mathrm{i}\omega_{\boldsymbol{k}}t}\hat{a}_{\boldsymbol{k}\lambda}^{\dagger} - \mathrm{e}^{-\mathrm{i}\omega_{\boldsymbol{k}}t}\hat{a}_{-\boldsymbol{k}\lambda}\right)\\ &\quad\times\left(\mathrm{e}^{-\mathrm{i}\omega_{\boldsymbol{k}}t}\hat{a}_{\boldsymbol{k}\lambda} - \mathrm{e}^{\mathrm{i}\omega_{\boldsymbol{k}}t}\hat{a}_{-\boldsymbol{k}\lambda}^{\dagger}\right)\\ &= \frac{1}{2\epsilon_0}\sum_{\boldsymbol{k}\lambda}\hbar\omega_{\boldsymbol{k}}\left(\hat{a}_{\boldsymbol{k}\lambda}^{\dagger}\hat{a}_{\boldsymbol{k}\lambda} - \mathrm{e}^{-2\mathrm{i}\omega_{\boldsymbol{k}}t}\hat{a}_{-\boldsymbol{k}\lambda}\hat{a}_{\boldsymbol{k}\lambda}\right.\\ &\quad\left. - \mathrm{e}^{2\mathrm{i}\omega_{\boldsymbol{k}}t}\hat{a}_{\boldsymbol{k}\lambda}^{\dagger}\hat{a}_{-\boldsymbol{k}\lambda}^{\dagger} + \hat{a}_{-\boldsymbol{k}\lambda}\hat{a}_{-\boldsymbol{k}\lambda}^{\dagger}\right).\end{aligned}$$

For the last term in the parentheses, we make a change of dummy summation variables from $\boldsymbol{k}$ to $-\boldsymbol{k}$ with the use of $\omega_{-\boldsymbol{k}} = \omega_{\boldsymbol{k}}$. We then make use of the commutation relation $[\hat{a}_{\boldsymbol{k}\lambda}, \hat{a}_{\boldsymbol{k}\lambda}^{\dagger}] = 1$ to combine it with the first term. We have

$$\begin{aligned}&\int \mathrm{d}\boldsymbol{r}\ \hat{\boldsymbol{E}}^{\dagger}\cdot\hat{\boldsymbol{E}}\\ &= \frac{1}{2\epsilon_0}\sum_{\boldsymbol{k}\lambda}\hbar\omega_{\boldsymbol{k}}\left(2\hat{a}_{\boldsymbol{k}\lambda}^{\dagger}\hat{a}_{\boldsymbol{k}\lambda} + 1 - \mathrm{e}^{-2\mathrm{i}\omega_{\boldsymbol{k}}t}\hat{a}_{-\boldsymbol{k}\lambda}\hat{a}_{\boldsymbol{k}\lambda} - \mathrm{e}^{2\mathrm{i}\omega_{\boldsymbol{k}}t}\hat{a}_{\boldsymbol{k}\lambda}^{\dagger}\hat{a}_{-\boldsymbol{k}\lambda}^{\dagger}\right).\end{aligned}$$

The integral $\int \mathrm{d}\boldsymbol{r}\ \hat{\boldsymbol{B}}^{\dagger}\cdot\hat{\boldsymbol{B}}$ is given by

$$\begin{aligned}\int \mathrm{d}\boldsymbol{r}\ \hat{\boldsymbol{B}}^{\dagger}\cdot\hat{\boldsymbol{B}} &= \frac{\hbar}{2\mathscr{V}\epsilon_0}\sum_{\boldsymbol{k}\boldsymbol{k}'\lambda\lambda'}\frac{1}{\sqrt{\omega_{\boldsymbol{k}'}\omega_{\boldsymbol{k}}}}\left(\mathrm{e}^{\mathrm{i}\omega_{\boldsymbol{k}'}t}\hat{a}_{\boldsymbol{k}'\lambda'}^{\dagger} + \mathrm{e}^{-\mathrm{i}\omega_{\boldsymbol{k}'}t}\hat{a}_{-\boldsymbol{k}'\lambda'}\right)\\ &\quad\times\left(\mathrm{e}^{-\mathrm{i}\omega_{\boldsymbol{k}}t}\hat{a}_{\boldsymbol{k}\lambda} + \mathrm{e}^{\mathrm{i}\omega_{\boldsymbol{k}}t}\hat{a}_{-\boldsymbol{k}\lambda}^{\dagger}\right)\left(\boldsymbol{k}'\times\boldsymbol{\epsilon}_{\boldsymbol{k}'}^{\lambda'*}\right)\cdot\left(\boldsymbol{k}\times\boldsymbol{\epsilon}_{\boldsymbol{k}}^{\lambda}\right)\\ &\quad\times\int \mathrm{d}\boldsymbol{r}\ \mathrm{e}^{-\mathrm{i}(\boldsymbol{k}'-\boldsymbol{k})\cdot\boldsymbol{r}}\\ &= \frac{1}{2\epsilon_0}\sum_{\boldsymbol{k}\lambda\lambda'}\frac{\hbar}{\omega_{\boldsymbol{k}}}\left(\mathrm{e}^{\mathrm{i}\omega_{\boldsymbol{k}}t}\hat{a}_{\boldsymbol{k}\lambda'}^{\dagger} + \mathrm{e}^{-\mathrm{i}\omega_{\boldsymbol{k}}t}\hat{a}_{-\boldsymbol{k}\lambda'}\right)\\ &\quad\times\left(\mathrm{e}^{-\mathrm{i}\omega_{\boldsymbol{k}}t}\hat{a}_{\boldsymbol{k}\lambda} + \mathrm{e}^{\mathrm{i}\omega_{\boldsymbol{k}}t}\hat{a}_{-\boldsymbol{k}\lambda}^{\dagger}\right)\left(\boldsymbol{k}\times\boldsymbol{\epsilon}_{\boldsymbol{k}}^{\lambda'*}\right)\cdot\left(\boldsymbol{k}\times\boldsymbol{\epsilon}_{\boldsymbol{k}}^{\lambda}\right),\end{aligned}$$

where we have made use of $\mathscr{V}^{-1}\int \mathrm{d}\boldsymbol{r}\ \mathrm{e}^{-\mathrm{i}(\boldsymbol{k}'-\boldsymbol{k})\cdot\boldsymbol{r}} = \delta_{\boldsymbol{k}'\boldsymbol{k}}$. The factor $\left(\boldsymbol{k}\times\boldsymbol{\epsilon}_{\boldsymbol{k}}^{\lambda'*}\right)\cdot\left(\boldsymbol{k}\times\boldsymbol{\epsilon}_{\boldsymbol{k}}^{\lambda}\right)$ can be evaluated as follows

$$\begin{aligned}\left(\boldsymbol{k}\times\boldsymbol{\epsilon}_{\boldsymbol{k}}^{\lambda'*}\right)\cdot\left(\boldsymbol{k}\times\boldsymbol{\epsilon}_{\boldsymbol{k}}^{\lambda}\right) &= |\boldsymbol{k}|^2\boldsymbol{\epsilon}_{\boldsymbol{k}}^{\lambda'*}\cdot\boldsymbol{\epsilon}_{\boldsymbol{k}}^{\lambda} - \left(\boldsymbol{k}\cdot\boldsymbol{\epsilon}_{\boldsymbol{k}}^{\lambda'*}\right)\left(\boldsymbol{k}\cdot\boldsymbol{\epsilon}_{\boldsymbol{k}}^{\lambda}\right)\\ &= |\boldsymbol{k}|^2\delta_{\lambda'\lambda} = \frac{1}{c^2}\omega_{\boldsymbol{k}}^2\delta_{\lambda'\lambda},\end{aligned}$$

where we have utilized $\boldsymbol{\epsilon}_{\boldsymbol{k}}^{\lambda'*} \cdot \boldsymbol{\epsilon}_{\boldsymbol{k}}^{\lambda} = \delta_{\lambda'\lambda}$, $\boldsymbol{k} \cdot \boldsymbol{\epsilon}_{\boldsymbol{k}}^{\lambda'*} = \boldsymbol{k} \cdot \boldsymbol{\epsilon}_{\boldsymbol{k}}^{\lambda} = 0$, and $\omega_{\boldsymbol{k}} = c|\boldsymbol{k}|$. We thus have

$$\begin{aligned}
&\int \mathrm{d}\boldsymbol{r}\ \hat{\boldsymbol{B}}^\dagger \cdot \hat{\boldsymbol{B}} \\
&= \frac{1}{2\epsilon_0 c^2} \sum_{\boldsymbol{k}\lambda} \hbar\omega_{\boldsymbol{k}} \big(\mathrm{e}^{\mathrm{i}\omega_{\boldsymbol{k}} t} \hat{a}^\dagger_{\boldsymbol{k}\lambda} + \mathrm{e}^{-\mathrm{i}\omega_{\boldsymbol{k}} t} \hat{a}_{-\boldsymbol{k}\lambda}\big)\big(\mathrm{e}^{-\mathrm{i}\omega_{\boldsymbol{k}} t} \hat{a}_{\boldsymbol{k}\lambda} + \mathrm{e}^{\mathrm{i}\omega_{\boldsymbol{k}} t} \hat{a}^\dagger_{-\boldsymbol{k}\lambda}\big) \\
&= \frac{1}{2\epsilon_0 c^2} \sum_{\boldsymbol{k}\lambda} \hbar\omega_{\boldsymbol{k}} \big(2\hat{a}^\dagger_{\boldsymbol{k}\lambda}\hat{a}_{\boldsymbol{k}\lambda} + 1 + \mathrm{e}^{-2\mathrm{i}\omega_{\boldsymbol{k}} t} \hat{a}_{-\boldsymbol{k}\lambda}\hat{a}_{\boldsymbol{k}\lambda} + \mathrm{e}^{2\mathrm{i}\omega_{\boldsymbol{k}} t} \hat{a}^\dagger_{\boldsymbol{k}\lambda}\hat{a}^\dagger_{-\boldsymbol{k}\lambda}\big).
\end{aligned}$$

The Hamiltonian of the electromagnetic field is then give by

$$\begin{aligned}
\hat{H}_{\mathrm{em}} &= \frac{1}{2}\epsilon_0 \int \mathrm{d}\boldsymbol{r}\ \hat{\boldsymbol{E}}^\dagger \cdot \hat{\boldsymbol{E}} + \frac{1}{2\mu_0} \int \mathrm{d}\boldsymbol{r}\ \hat{\boldsymbol{B}}^\dagger \cdot \hat{\boldsymbol{B}} \\
&= \sum_{\boldsymbol{k}\lambda} \hbar\omega_{\boldsymbol{k}} \big(\hat{a}^\dagger_{\boldsymbol{k}\lambda}\hat{a}_{\boldsymbol{k}\lambda} + 1/2\big).
\end{aligned}$$

(3) Making use of the commutation relations between operators $\hat{a}_{\boldsymbol{k}\lambda}$ and $\hat{a}^\dagger_{\boldsymbol{k}\lambda}$

$$[\hat{a}_{\boldsymbol{k}\lambda}, \hat{a}^\dagger_{\boldsymbol{k}'\lambda'}] = \delta_{\boldsymbol{k}\boldsymbol{k}'}\delta_{\lambda\lambda'},\ [\hat{a}_{\boldsymbol{k}\lambda}, \hat{a}_{\boldsymbol{k}'\lambda'}] = [\hat{a}^\dagger_{\boldsymbol{k}\lambda}, \hat{a}^\dagger_{\boldsymbol{k}'\lambda'}] = 0,$$

we can evaluate the commutator $[\hat{E}_\alpha(\boldsymbol{r},t),\ \hat{E}^\dagger_\beta(\boldsymbol{r}',t')]$ as follows

$$\begin{aligned}
&[\hat{E}_\alpha(\boldsymbol{r},t),\ \hat{E}^\dagger_\beta(\boldsymbol{r}',t')] \\
&\quad = \frac{\hbar}{2\mathscr{V}\epsilon_0} \sum_{\boldsymbol{k}\boldsymbol{k}'\lambda\lambda'} \sqrt{\omega_{\boldsymbol{k}'}\omega_{\boldsymbol{k}}}\, \epsilon^\lambda_{\boldsymbol{k}\alpha}\epsilon^{\lambda'*}_{\boldsymbol{k}'\beta} \mathrm{e}^{-\mathrm{i}(\boldsymbol{k}'\cdot\boldsymbol{r}' - \boldsymbol{k}\cdot\boldsymbol{r})} \\
&\qquad \times \left[\mathrm{e}^{-\mathrm{i}\omega_{\boldsymbol{k}} t}\hat{a}_{\boldsymbol{k}\lambda} - \mathrm{e}^{\mathrm{i}\omega_{\boldsymbol{k}} t}\hat{a}^\dagger_{-\boldsymbol{k}\lambda}, \mathrm{e}^{\mathrm{i}\omega_{\boldsymbol{k}'} t'}\hat{a}^\dagger_{\boldsymbol{k}'\lambda'} - \mathrm{e}^{-\mathrm{i}\omega_{\boldsymbol{k}'} t'}\hat{a}_{-\boldsymbol{k}'\lambda'}\right] \\
&\quad = \frac{\hbar}{2\mathscr{V}\epsilon_0} \sum_{\boldsymbol{k}\boldsymbol{k}'\lambda\lambda'} \sqrt{\omega_{\boldsymbol{k}'}\omega_{\boldsymbol{k}}}\, \epsilon^\lambda_{\boldsymbol{k}\alpha}\epsilon^{\lambda'*}_{\boldsymbol{k}'\beta} \mathrm{e}^{-\mathrm{i}(\boldsymbol{k}'\cdot\boldsymbol{r}' - \boldsymbol{k}\cdot\boldsymbol{r})} \\
&\qquad \times \left[\mathrm{e}^{\mathrm{i}(\omega_{\boldsymbol{k}'} t' - \omega_{\boldsymbol{k}} t)} - \mathrm{e}^{-\mathrm{i}(\omega_{\boldsymbol{k}'} t' - \omega_{\boldsymbol{k}} t)}\right]\delta_{\boldsymbol{k}'\boldsymbol{k}}\delta_{\lambda'\lambda} \\
&\quad = \frac{\mathrm{i}}{\mathscr{V}\epsilon_0} \sum_{\boldsymbol{k}\lambda} \hbar\omega_{\boldsymbol{k}}\, \epsilon^\lambda_{\boldsymbol{k}\alpha}\epsilon^{\lambda*}_{\boldsymbol{k}\beta} \mathrm{e}^{-\mathrm{i}\boldsymbol{k}\cdot(\boldsymbol{r}'-\boldsymbol{r})} \sin[\omega_{\boldsymbol{k}}(t'-t)] \\
&\quad = \delta_{\alpha\beta}\frac{\mathrm{i}}{\mathscr{V}\epsilon_0} \sum_{\boldsymbol{k}} \hbar\omega_{\boldsymbol{k}}\, \mathrm{e}^{-\mathrm{i}\boldsymbol{k}\cdot(\boldsymbol{r}'-\boldsymbol{r})} \sin[\omega_{\boldsymbol{k}}(t'-t)],
\end{aligned}$$

where we have made use of $\sum_\lambda \epsilon^\lambda_{\boldsymbol{k}\alpha}\epsilon^{\lambda*}_{\boldsymbol{k}\beta} = \delta_{\alpha\beta}$. For the commutator $[\hat{B}_\alpha(\boldsymbol{r},t),\ \hat{B}^\dagger_\beta(\boldsymbol{r}',t')]$, we have

$$
\begin{aligned}
&[\hat{B}_\alpha(\boldsymbol{r},t),\ \hat{B}^\dagger_\beta(\boldsymbol{r}',t')] \\
&\quad = \frac{\hbar}{2\mathscr{V}\epsilon_0}\sum_{\boldsymbol{k}\boldsymbol{k}'\lambda\lambda'}\frac{1}{\sqrt{\omega_{\boldsymbol{k}'}\omega_{\boldsymbol{k}}}}\sum_{\mu\nu\gamma\rho}\varepsilon_{\mu\nu\alpha}\varepsilon_{\gamma\rho\beta}k_\mu k'_\gamma \epsilon^{\lambda}_{\boldsymbol{k}\nu}\epsilon^{\lambda'\,*}_{\boldsymbol{k}'\rho}\mathrm{e}^{-\mathrm{i}(\boldsymbol{k}'\cdot\boldsymbol{r}'-\boldsymbol{k}\cdot\boldsymbol{r})} \\
&\qquad \times\left[\mathrm{e}^{-\mathrm{i}\omega_{\boldsymbol{k}}t}\hat{a}_{\boldsymbol{k}\lambda}+\mathrm{e}^{\mathrm{i}\omega_{\boldsymbol{k}}t}\hat{a}^\dagger_{-\boldsymbol{k}\lambda},\ \mathrm{e}^{\mathrm{i}\omega_{\boldsymbol{k}'}t'}\hat{a}^\dagger_{\boldsymbol{k}'\lambda'}+\mathrm{e}^{-\mathrm{i}\omega_{\boldsymbol{k}'}t'}\hat{a}_{-\boldsymbol{k}'\lambda'}\right] \\
&\quad = \frac{\hbar}{2\mathscr{V}\epsilon_0}\sum_{\boldsymbol{k}\boldsymbol{k}'\lambda\lambda'}\frac{1}{\sqrt{\omega_{\boldsymbol{k}'}\omega_{\boldsymbol{k}}}}\sum_{\mu\nu\gamma\rho}\varepsilon_{\mu\nu\alpha}\varepsilon_{\gamma\rho\beta}k_\mu k'_\gamma \epsilon^{\lambda}_{\boldsymbol{k}\nu}\epsilon^{\lambda'\,*}_{\boldsymbol{k}'\rho}\mathrm{e}^{-\mathrm{i}(\boldsymbol{k}'\cdot\boldsymbol{r}'-\boldsymbol{k}\cdot\boldsymbol{r})} \\
&\qquad \times\left[\mathrm{e}^{\mathrm{i}(\omega_{\boldsymbol{k}'}t'-\omega_{\boldsymbol{k}}t)}-\mathrm{e}^{-\mathrm{i}(\omega_{\boldsymbol{k}'}t'-\omega_{\boldsymbol{k}}t)}\right]\delta_{\boldsymbol{k}'\boldsymbol{k}}\delta_{\lambda'\lambda} \\
&\quad = \frac{\mathrm{i}}{\mathscr{V}\epsilon_0}\sum_{\boldsymbol{k}\lambda}\frac{\hbar}{\omega_{\boldsymbol{k}}}\sum_{\mu\nu\gamma\rho}\varepsilon_{\mu\nu\alpha}\varepsilon_{\gamma\rho\beta}k_\mu k_\gamma \epsilon^{\lambda}_{\boldsymbol{k}\nu}\epsilon^{\lambda*}_{\boldsymbol{k}\rho}\mathrm{e}^{-\mathrm{i}\boldsymbol{k}\cdot(\boldsymbol{r}'-\boldsymbol{r})}\sin[\omega_{\boldsymbol{k}}(t'-t)] \\
&\quad = \frac{\mathrm{i}}{\mathscr{V}\epsilon_0}\sum_{\boldsymbol{k}}\frac{\hbar}{\omega_{\boldsymbol{k}}}\sum_{\mu\nu\gamma}\varepsilon_{\mu\nu\alpha}\varepsilon_{\gamma\nu\beta}k_\mu k_\gamma \mathrm{e}^{-\mathrm{i}\boldsymbol{k}\cdot(\boldsymbol{r}'-\boldsymbol{r})}\sin[\omega_{\boldsymbol{k}}(t'-t)] \\
&\quad = \frac{\mathrm{i}}{\mathscr{V}\epsilon_0}\sum_{\boldsymbol{k}}\frac{\hbar}{\omega_{\boldsymbol{k}}}\left(k^2\delta_{\alpha\beta}-k_\alpha k_\beta\right)\mathrm{e}^{-\mathrm{i}\boldsymbol{k}\cdot(\boldsymbol{r}'-\boldsymbol{r})}\sin[\omega_{\boldsymbol{k}}(t'-t)],
\end{aligned}
$$

where we have made use of $\sum_\lambda \epsilon^{\lambda}_{\boldsymbol{k}\nu}\epsilon^{\lambda\,*}_{\boldsymbol{k}\rho} = \delta_{\nu\rho}$ and $\sum_\nu \varepsilon_{\mu\nu\alpha}\varepsilon_{\gamma\nu\beta} = \delta_{\mu\gamma}\delta_{\alpha\beta} - \delta_{\mu\beta}\delta_{\alpha\gamma}$. For the commutator $[\hat{E}_\alpha(\boldsymbol{r},t),\ \hat{B}_\beta(\boldsymbol{r}',t')]$, we have

$$
\begin{aligned}
&[\hat{E}_\alpha(\boldsymbol{r},t),\ \hat{B}_\beta(\boldsymbol{r}',t')] \\
&\quad = -\frac{\hbar}{2\mathscr{V}\epsilon_0}\sum_{\boldsymbol{k}\boldsymbol{k}'\lambda\lambda'}\sqrt{\frac{\omega_{\boldsymbol{k}}}{\omega_{\boldsymbol{k}'}}}\sum_{\mu\nu}\varepsilon_{\mu\nu\beta}k'_\mu \epsilon^{\lambda}_{\boldsymbol{k}\alpha}\epsilon^{\lambda'}_{\boldsymbol{k}'\nu}\mathrm{e}^{\mathrm{i}(\boldsymbol{k}'\cdot\boldsymbol{r}'+\boldsymbol{k}\cdot\boldsymbol{r})} \\
&\qquad \times\left[\mathrm{e}^{-\mathrm{i}\omega_{\boldsymbol{k}}t}\hat{a}_{\boldsymbol{k}\lambda}-\mathrm{e}^{\mathrm{i}\omega_{\boldsymbol{k}}t}\hat{a}^\dagger_{-\boldsymbol{k}\lambda},\ \mathrm{e}^{-\mathrm{i}\omega_{\boldsymbol{k}'}t'}\hat{a}_{\boldsymbol{k}'\lambda'}+\mathrm{e}^{\mathrm{i}\omega_{\boldsymbol{k}'}t'}\hat{a}^\dagger_{-\boldsymbol{k}'\lambda'}\right] \\
&\quad = -\frac{\hbar}{2\mathscr{V}\epsilon_0}\sum_{\boldsymbol{k}\boldsymbol{k}'\lambda\lambda'}\sqrt{\frac{\omega_{\boldsymbol{k}}}{\omega_{\boldsymbol{k}'}}}\sum_{\mu\nu}\varepsilon_{\mu\nu\beta}k'_\mu \epsilon^{\lambda}_{\boldsymbol{k}\alpha}\epsilon^{\lambda'}_{\boldsymbol{k}'\nu}\mathrm{e}^{\mathrm{i}(\boldsymbol{k}'\cdot\boldsymbol{r}'+\boldsymbol{k}\cdot\boldsymbol{r})} \\
&\qquad \times\left[\mathrm{e}^{\mathrm{i}(\omega_{\boldsymbol{k}'}t'-\omega_{\boldsymbol{k}}t)}+\mathrm{e}^{-\mathrm{i}(\omega_{\boldsymbol{k}'}t'-\omega_{\boldsymbol{k}}t)}\right]\delta_{\boldsymbol{k}',-\boldsymbol{k}}\delta_{\lambda'\lambda} \\
&\quad = \frac{\hbar}{\mathscr{V}\epsilon_0}\sum_{\boldsymbol{k}\lambda}\sum_{\mu\nu}\varepsilon_{\mu\nu\beta}k_\mu \epsilon^{\lambda}_{\boldsymbol{k}\alpha}\epsilon^{\lambda\,*}_{\boldsymbol{k}\nu}\mathrm{e}^{-\mathrm{i}\boldsymbol{k}\cdot(\boldsymbol{r}'-\boldsymbol{r})}\cos[\omega_{\boldsymbol{k}}(t'-t)] \\
&\quad = \frac{\hbar}{\mathscr{V}\epsilon_0}\sum_{\boldsymbol{k}\mu}\varepsilon_{\alpha\beta\mu}k_\mu \mathrm{e}^{-\mathrm{i}\boldsymbol{k}\cdot(\boldsymbol{r}'-\boldsymbol{r})}\cos[\omega_{\boldsymbol{k}}(t'-t)],
\end{aligned}
$$

where we have used $\epsilon^{\lambda}_{-\boldsymbol{k},\nu} = \epsilon^{\lambda\,*}_{\boldsymbol{k}\nu}$.

(4) The average of $\hat{\boldsymbol{E}}$ in the number state $|n_{\boldsymbol{k}\lambda}\rangle$ is given by

$$\begin{aligned}
&\langle n_{\boldsymbol{k}\lambda}|\,\hat{\boldsymbol{E}}(\boldsymbol{r},t)|n_{\boldsymbol{k}\lambda}\rangle \\
&= \mathrm{i}\sum_{\boldsymbol{k}'\lambda'}\left(\frac{\hbar\omega_{\boldsymbol{k}'}}{2\mathscr{V}\epsilon_0}\right)^{1/2}\big[\mathrm{e}^{-\mathrm{i}\omega_{\boldsymbol{k}'}t}\langle n_{\boldsymbol{k}\lambda}|\,\hat{a}_{\boldsymbol{k}'\lambda'}|n_{\boldsymbol{k}\lambda}\rangle \\
&\qquad\qquad - \mathrm{e}^{\mathrm{i}\omega_{\boldsymbol{k}'}t}\langle n_{\boldsymbol{k}\lambda}|\,\hat{a}^{\dagger}_{-\boldsymbol{k}'\lambda'}|n_{\boldsymbol{k}\lambda}\rangle\big]\boldsymbol{\epsilon}^{\lambda'}_{\boldsymbol{k}'}\mathrm{e}^{\mathrm{i}\boldsymbol{k}'\cdot\boldsymbol{r}} \\
&= \mathrm{i}\sum_{\boldsymbol{k}'\lambda'}\left(\frac{\hbar\omega_{\boldsymbol{k}'}}{2\mathscr{V}\epsilon_0}\right)^{1/2}\big[\mathrm{e}^{-\mathrm{i}\omega_{\boldsymbol{k}'}t}\cdot 0 - \mathrm{e}^{\mathrm{i}\omega_{\boldsymbol{k}'}t}\cdot 0\big]\boldsymbol{\epsilon}^{\lambda'}_{\boldsymbol{k}'}\mathrm{e}^{\mathrm{i}\boldsymbol{k}'\cdot\boldsymbol{r}} \\
&= 0.
\end{aligned}$$

(5) The average of $\hat{\boldsymbol{E}}(\boldsymbol{r},t)$ in the coherent state $|\gamma_{\boldsymbol{k}\lambda}\rangle$ is given by

$$\begin{aligned}
&\langle \gamma_{\boldsymbol{k}\lambda}|\,\hat{\boldsymbol{E}}(\boldsymbol{r},t)|\gamma_{\boldsymbol{k}\lambda}\rangle \\
&= \mathrm{i}\sum_{\boldsymbol{k}'\lambda'}\left(\frac{\hbar\omega_{\boldsymbol{k}'}}{2\mathscr{V}\epsilon_0}\right)^{1/2}\big[\mathrm{e}^{-\mathrm{i}\omega_{\boldsymbol{k}'}t}\langle \gamma_{\boldsymbol{k}\lambda}|\,\hat{a}_{\boldsymbol{k}'\lambda'}|\gamma_{\boldsymbol{k}\lambda}\rangle \\
&\qquad\qquad - \mathrm{e}^{\mathrm{i}\omega_{\boldsymbol{k}'}t}\langle \gamma_{\boldsymbol{k}\lambda}|\,\hat{a}^{\dagger}_{-\boldsymbol{k}'\lambda'}|\gamma_{\boldsymbol{k}\lambda}\rangle\big]\boldsymbol{\epsilon}^{\lambda'}_{\boldsymbol{k}'}\mathrm{e}^{\mathrm{i}\boldsymbol{k}'\cdot\boldsymbol{r}} \\
&= \mathrm{i}\sum_{\boldsymbol{k}'\lambda'}\left(\frac{\hbar\omega_{\boldsymbol{k}'}}{2\mathscr{V}\epsilon_0}\right)^{1/2}\big[\mathrm{e}^{-\mathrm{i}\omega_{\boldsymbol{k}'}t}\gamma_{\boldsymbol{k}\lambda}\delta_{\boldsymbol{k}'\boldsymbol{k}}\delta_{\lambda'\lambda} \\
&\qquad\qquad - \mathrm{e}^{\mathrm{i}\omega_{\boldsymbol{k}'}t}\gamma^{*}_{\boldsymbol{k}\lambda}\delta_{\boldsymbol{k}',-\boldsymbol{k}}\delta_{\lambda'\lambda}\big]\boldsymbol{\epsilon}^{\lambda'}_{\boldsymbol{k}'}\mathrm{e}^{\mathrm{i}\boldsymbol{k}'\cdot\boldsymbol{r}} \\
&= \mathrm{i}\left(\frac{\hbar\omega_{\boldsymbol{k}}}{2\mathscr{V}\epsilon_0}\right)^{1/2}\big[\mathrm{e}^{\mathrm{i}(\boldsymbol{k}\cdot\boldsymbol{r}-\omega_{\boldsymbol{k}}t)}\gamma_{\boldsymbol{k}\lambda}\boldsymbol{\epsilon}^{\lambda}_{\boldsymbol{k}} - \mathrm{e}^{-\mathrm{i}(\boldsymbol{k}\cdot\boldsymbol{r}-\omega_{\boldsymbol{k}}t)}\gamma^{*}_{\boldsymbol{k}\lambda}\boldsymbol{\epsilon}^{\lambda\,*}_{\boldsymbol{k}}\big].
\end{aligned}$$

For real $\boldsymbol{\epsilon}^{\lambda}_{\boldsymbol{k}}$, we have

$$\begin{aligned}
&\langle \gamma_{\boldsymbol{k}\lambda}|\,\hat{\boldsymbol{E}}(\boldsymbol{r},t)|\gamma_{\boldsymbol{k}\lambda}\rangle \\
&= -\left(\frac{2\hbar\omega_{\boldsymbol{k}}}{\mathscr{V}\epsilon_0}\right)^{1/2}\big[\sin(\boldsymbol{k}\cdot\boldsymbol{r}-\omega_{\boldsymbol{k}}t)\,\mathrm{Re}\,\gamma_{\boldsymbol{k}\lambda} \\
&\qquad\qquad + \cos(\boldsymbol{k}\cdot\boldsymbol{r}-\omega_{\boldsymbol{k}}t)\,\mathrm{Im}\,\gamma_{\boldsymbol{k}\lambda}\big]\boldsymbol{\epsilon}^{\lambda}_{\boldsymbol{k}} \\
&= \left(\frac{2\hbar\omega_{\boldsymbol{k}}}{\mathscr{V}\epsilon_0}\right)^{1/2}\big[(\mathrm{Re}\,\gamma_{\boldsymbol{k}\lambda})^2 \\
&\qquad\qquad + (\mathrm{Im}\,\gamma_{\boldsymbol{k}\lambda})^2\big]^{1/2}\boldsymbol{\epsilon}^{\lambda}_{\boldsymbol{k}}\sin(\boldsymbol{k}\cdot\boldsymbol{r}-\omega_{\boldsymbol{k}}t+\varphi_0) \\
&= \boldsymbol{E}_0\sin(\boldsymbol{k}\cdot\boldsymbol{r}-\omega_{\boldsymbol{k}}t+\varphi_0),
\end{aligned}$$

where

$$\boldsymbol{E}_0 = \left(\frac{2\hbar\omega_{\boldsymbol{k}}}{\mathscr{V}\epsilon_0}\right)^{1/2}\big[(\mathrm{Re}\,\gamma_{\boldsymbol{k}\lambda})^2 + (\mathrm{Im}\,\gamma_{\boldsymbol{k}\lambda})^2\big]^{1/2}\boldsymbol{\epsilon}^{\lambda}_{\boldsymbol{k}},$$

$$\varphi_0 = \tan^{-1}\frac{\mathrm{Im}\,\gamma_{\boldsymbol{k}\lambda}}{\mathrm{Re}\,\gamma_{\boldsymbol{k}\lambda}} + \pi.$$

24-2 Quantum field operator of electrons. In the second quantization formalism and in the independent single-electron approximation, the Hamiltonian $\hat{H}_0$ of the electron subsystem in a solid is given by

$$\hat{H}_0 = \sum_{nk\sigma} E_{nk\sigma} \hat{c}^\dagger_{nk\sigma} \hat{c}_{nk\sigma} = \sum_{nk\sigma} E_{nk\sigma} \hat{\mathcal{N}}_{nk\sigma}.$$

(1) Evaluate the commutators $[\hat{c}_{nk\sigma}, \hat{H}_0]$ and $[\hat{c}^\dagger_{nk\sigma}, \hat{H}_0]$. Find the time dependence of $\hat{c}_{nk\sigma}(t)$ and $\hat{c}^\dagger_{nk\sigma}(t)$ using the results for the commutators and the Heisenberg equation of motion.

(2) Using the above-derived time dependence of $\hat{c}_{nk\sigma}(t)$ and

$$\hat{\Psi}(\boldsymbol{r}, s_z) = \sum_{nk\sigma} \psi_{nk\sigma}(\boldsymbol{r}, s_z) \hat{c}_{nk\sigma},$$

write down an expression for the time-dependent quantum field operator of electrons, $\hat{\Psi}(\boldsymbol{r}, s_z; t)$.

(3) Evaluate the value of the anticommutator

$$\{\hat{\Psi}(\boldsymbol{r}, s_z; t), \hat{\Psi}^\dagger(\boldsymbol{r}', s_z'; t')\}.$$

What is the value of the following equal-time anticommutator?

$$\{\hat{\Psi}(\boldsymbol{r}, s_z; t), \hat{\Psi}^\dagger(\boldsymbol{r}', s_z'; t)\}.$$

---

(1) Evaluating the commutator $[\hat{c}_{nk\sigma}, \hat{H}_0]$, we have

$$\begin{aligned}
&[\hat{c}_{nk\sigma}, \hat{H}_0] \\
&\quad = \sum_{n'k'\sigma'} E_{n'k'\sigma'} [\hat{c}_{nk\sigma}, \hat{c}^\dagger_{n'k'\sigma'} \hat{c}_{n'k'\sigma'}] \\
&\quad = \sum_{n'k'\sigma'} E_{n'k'\sigma'} \big[\{\hat{c}_{nk\sigma}, \hat{c}^\dagger_{n'k'\sigma'}\} \hat{c}_{n'k'\sigma'} - \hat{c}^\dagger_{n'k'\sigma'} \{\hat{c}_{nk\sigma}, \hat{c}_{n'k'\sigma'}\} \big] \\
&\quad = \sum_{n'k'\sigma'} E_{n'k'\sigma'} \hat{c}_{n'k'\sigma'} \delta_{nn'} \delta_{kk'} \delta_{\sigma\sigma'} = E_{nk\sigma} \hat{c}_{nk\sigma}.
\end{aligned}$$

For the commutator $[\hat{c}^\dagger_{nk\sigma}, \hat{H}_0]$, we have

$$\begin{aligned}
&[\hat{c}^\dagger_{nk\sigma}, \hat{H}_0] \\
&\quad = \sum_{n'k'\sigma'} E_{n'k'\sigma'} [\hat{c}^\dagger_{nk\sigma}, \hat{c}^\dagger_{n'k'\sigma'} \hat{c}_{n'k'\sigma'}] \\
&\quad = \sum_{n'k'\sigma'} E_{n'k'\sigma'} \big[\{\hat{c}^\dagger_{nk\sigma}, \hat{c}^\dagger_{n'k'\sigma'}\} \hat{c}_{n'k'\sigma'} - \hat{c}^\dagger_{n'k'\sigma'} \{\hat{c}^\dagger_{nk\sigma}, \hat{c}_{n'k'\sigma'}\} \big] \\
&\quad = - \sum_{n'k'\sigma'} E_{n'k'\sigma'} \hat{c}^\dagger_{n'k'\sigma'} \delta_{nn'} \delta_{kk'} \delta_{\sigma\sigma'} = -E_{nk\sigma} \hat{c}^\dagger_{nk\sigma}.
\end{aligned}$$

Applying the Heisenberg equation of motion to $\hat{c}_{n\boldsymbol{k}\sigma}$ and $\hat{c}^\dagger_{n\boldsymbol{k}\sigma}$, respectively, and making use of the above evaluated commutators, we have

$$\frac{\mathrm{d}\hat{c}_{n\boldsymbol{k}\sigma}}{\mathrm{d}t} = \frac{1}{\mathrm{i}\hbar}[\hat{c}_{n\boldsymbol{k}\sigma}, \hat{H}_0] = -\frac{\mathrm{i}E_{n\boldsymbol{k}\sigma}}{\hbar}\hat{c}_{n\boldsymbol{k}\sigma},$$

$$\frac{\mathrm{d}\hat{c}^\dagger_{n\boldsymbol{k}\sigma}}{\mathrm{d}t} = \frac{1}{\mathrm{i}\hbar}[\hat{c}^\dagger_{n\boldsymbol{k}\sigma}, \hat{H}_0] = \frac{\mathrm{i}E_{n\boldsymbol{k}\sigma}}{\hbar}\hat{c}^\dagger_{n\boldsymbol{k}\sigma}.$$

Integrating the above two equations, respectively, and taking the initial conditions $\hat{c}_{n\boldsymbol{k}\sigma}(t=0) = \hat{c}_{n\boldsymbol{k}\sigma}$ and $\hat{c}^\dagger_{n\boldsymbol{k}\sigma}(t=0) = \hat{c}^\dagger_{n\boldsymbol{k}\sigma}$ into account, we have

$$\hat{c}_{n\boldsymbol{k}\sigma}(t) = \mathrm{e}^{-\mathrm{i}E_{n\boldsymbol{k}\sigma}t/\hbar}\hat{c}_{n\boldsymbol{k}\sigma},$$

$$\hat{c}^\dagger_{n\boldsymbol{k}\sigma}(t) = \mathrm{e}^{\mathrm{i}E_{n\boldsymbol{k}\sigma}t/\hbar}\hat{c}^\dagger_{n\boldsymbol{k}\sigma}.$$

(2) The time-dependent quantum field operator of electrons $\hat{\Psi}(\boldsymbol{r}, s_z; t)$ is given by

$$\hat{\Psi}(\boldsymbol{r}, s_z; t) = \sum_{n\boldsymbol{k}\sigma} \psi_{n\boldsymbol{k}\sigma}(\boldsymbol{r}, s_z)\mathrm{e}^{-\mathrm{i}E_{n\boldsymbol{k}\sigma}t/\hbar}\hat{c}_{n\boldsymbol{k}\sigma}.$$

(3) The Hermitian conjugate of $\hat{\Psi}(\boldsymbol{r}, s_z; t)$ is given by

$$\hat{\Psi}^\dagger(\boldsymbol{r}, s_z; t) = \sum_{n\boldsymbol{k}\sigma} \psi^*_{n\boldsymbol{k}\sigma}(\boldsymbol{r}, s_z)\mathrm{e}^{\mathrm{i}E_{n\boldsymbol{k}\sigma}t/\hbar}\hat{c}^\dagger_{n\boldsymbol{k}\sigma}.$$

Utilizing the anticommutation relations between $\hat{c}_{n\boldsymbol{k}\sigma}$ and $\hat{c}^\dagger_{n\boldsymbol{k}\sigma}$, we can easily evaluate the anticommutator $\{\hat{\Psi}(\boldsymbol{r}, s_z; t), \hat{\Psi}^\dagger(\boldsymbol{r}', s'_z; t')\}$. We have

$$\begin{aligned}
&\{\hat{\Psi}(\boldsymbol{r}, s_z; t), \hat{\Psi}^\dagger(\boldsymbol{r}', s'_z; t')\} \\
&= \sum_{n\boldsymbol{k}\sigma}\sum_{n'\boldsymbol{k}'\sigma'} \psi^*_{n'\boldsymbol{k}'\sigma'}(\boldsymbol{r}', s'_z)\psi_{n\boldsymbol{k}\sigma}(\boldsymbol{r}, s_z) \\
&\qquad \times \mathrm{e}^{\mathrm{i}(E'_{n'\boldsymbol{k}'\sigma}t' - E_{n\boldsymbol{k}\sigma}t)/\hbar}\{\hat{c}_{n\boldsymbol{k}\sigma}, \hat{c}^\dagger_{n'\boldsymbol{k}'\sigma'}\} \\
&= \sum_{n\boldsymbol{k}\sigma}\sum_{n'\boldsymbol{k}'\sigma'} \psi^*_{n'\boldsymbol{k}'\sigma'}(\boldsymbol{r}', s'_z)\psi_{n\boldsymbol{k}\sigma}(\boldsymbol{r}, s_z) \\
&\qquad \times \mathrm{e}^{\mathrm{i}(E'_{n'\boldsymbol{k}'\sigma}t' - E_{n\boldsymbol{k}\sigma}t)/\hbar}\delta_{nn'}\delta_{\boldsymbol{k}\boldsymbol{k}'}\delta_{\sigma\sigma'} \\
&= \sum_{n\boldsymbol{k}\sigma} \psi^*_{n\boldsymbol{k}\sigma}(\boldsymbol{r}', s'_z)\psi_{n\boldsymbol{k}\sigma}(\boldsymbol{r}, s_z)\mathrm{e}^{\mathrm{i}E_{n\boldsymbol{k}\sigma}(t'-t)/\hbar}.
\end{aligned}$$

The value of $\{\hat{\Psi}(\boldsymbol{r}, s_z; t), \hat{\Psi}^\dagger(\boldsymbol{r}', s'_z; t)\}$ can be obtained by setting $t' = t$ in the above equation. Making use of the completeness relation of the single-electron states, we have

$$\begin{aligned}
\{\hat{\Psi}(\boldsymbol{r}, s_z; t), \hat{\Psi}^\dagger(\boldsymbol{r}', s'_z; t)\} &= \sum_{n\boldsymbol{k}\sigma} \psi^*_{n\boldsymbol{k}\sigma}(\boldsymbol{r}', s'_z)\psi_{n\boldsymbol{k}\sigma}(\boldsymbol{r}, s_z) \\
&= \delta(\boldsymbol{r}' - \boldsymbol{r})\delta_{s'_z s_z}.
\end{aligned}$$

24-3 Poisson summation formula. The Poisson summation formula can be expressed in different forms. One of the commonly-seen forms is

$$\sum_{n=-\infty}^{\infty} f(t+nT) = \frac{1}{T}\sum_{k=-\infty}^{\infty} g(k/T)\mathrm{e}^{2\pi\mathrm{i}kt/T},$$

where $g(k/T)$ is the Fourier transform of $f(t)$

$$g(k/T) = \int_{-\infty}^{\infty} \mathrm{d}\tau\ \mathrm{e}^{-2\pi\mathrm{i}k\tau/T} f(\tau).$$

(1) Prove the above form of the Poisson summation formula.
(2) Evaluate $\sum_{n=1}^{\infty}(n^2+1)^{-1}$.
(3) Prove that $\sum_{n=-\infty}^{\infty} \mathrm{e}^{-n^2\pi z} = z^{-1/2}\sum_{k=-\infty}^{\infty} \mathrm{e}^{-k^2\pi/z}$.

---

(1) Let

$$F(t) = \sum_{n=-\infty}^{\infty} f(t+nT).$$

Obviously, $F(t)$ is a periodic function of t with a period of T. We now make a Fourier transformation to $F(t)$ and write

$$F(t) = \frac{1}{T}\sum_{k=-\infty}^{\infty} G(k)\mathrm{e}^{\mathrm{i}2\pi kt/T},$$

where

$$G(k) = \int_{-T/2}^{T/2} \mathrm{d}t\ \mathrm{e}^{-\mathrm{i}2\pi kt/T} F(t).$$

Inserting $F(t) = \sum_{n=-\infty}^{\infty} f(t+nT)$ into the above expression for $G(k)$, we have

$$G(k) = \sum_{n=-\infty}^{\infty}\int_{-T/2}^{T/2} \mathrm{d}t\ \mathrm{e}^{-\mathrm{i}2\pi kt/T} f(t+nT).$$

Making a change of integration variables from t to $\tau = t+nT$ for the integral in the summand yields

$$G(k) = \sum_{n=-\infty}^{\infty}\int_{(2n-1)T/2}^{(2n+1)T/2} \mathrm{d}\tau\ \mathrm{e}^{-\mathrm{i}2\pi k(\tau-nT)/T} f(\tau)$$
$$= \sum_{n=-\infty}^{\infty}\int_{(2n-1)T/2}^{(2n+1)T/2} \mathrm{d}\tau\ \mathrm{e}^{-\mathrm{i}2\pi k\tau/T} f(\tau),$$

where we have made use of $e^{i2\pi kn} = 1$. Directly writing out the summation over n, we see that the summation and integration work together to yield an integral from $\tau = -\infty$ to $\tau = \infty$

$$\begin{aligned} G(k) &= \lim_{n\to\infty}\left[\int_{(-2n-1)T/2}^{(-2n+1)T/2} + \int_{(-2(n-1)-1)T/2}^{(-2(n-1)+1)T/2} + \cdots \right. \\ &\quad \left. + \int_{(2(n-1)-1)T/2}^{(2(n-1)+1)T/2} + \int_{(2n-1)T/2}^{(2n+1)T/2}\right] d\tau\, e^{-i2\pi k\tau/T} f(\tau) \\ &= \lim_{n\to\infty}\left[\int_{(-2n-1)T/2}^{(-2n+1)T/2} + \int_{(-2n+1)T/2}^{(-2n+3)T/2} + \cdots \right. \\ &\quad \left. + \int_{(2n-3)T/2}^{(2n-1)T/2} + \int_{(2n-1)T/2}^{(2n+1)T/2}\right] d\tau\, e^{-i2\pi k\tau/T} f(\tau) \\ &= \lim_{n\to\infty}\int_{(-2n-1)T/2}^{(2n+1)T/2} d\tau\, e^{-i2\pi k\tau/T} f(\tau) \\ &= \int_{-\infty}^{\infty} d\tau\, e^{-i2\pi k\tau/T} f(\tau) = g(k/T), \end{aligned}$$

where

$$g(k/T) = \int_{-\infty}^{\infty} d\tau\, e^{-i2\pi k\tau/T} f(\tau).$$

The form of the variable of $g(k/T)$ is determined by the argument of the exponential function in the integrand. From the form of the above expression, we see that $g(k/T)$ is the Fourier transform of $f(t)$. Substituting $G(k) = g(k/T)$ into the Fourier expansion of $F(t)$ and then plugging $F(t) = \sum_{n=-\infty}^{\infty} f(t+nT)$ into the resultant expression, we obtain

$$\sum_{n=-\infty}^{\infty} f(t+nT) = \frac{1}{T}\sum_{k=-\infty}^{\infty} g(k/T) e^{i2\pi kt/T}.$$

We have thus proved the given form of the Poisson summation formula.

(2) To find the value of the given sum, we take $f(t) = 1/(t^2+1)$ and $T = 1$ in the above Poisson summation formula. For $t = 0$, we have

$$\sum_{n=-\infty}^{\infty} f(n) = \sum_{n=-\infty}^{\infty} \frac{1}{n^2+1} = \sum_{k=-\infty}^{\infty} g(k).$$

Let us now find the Fourier transform of $f(t)$. We have

$$g(k) = \int_{-\infty}^{\infty} dt\, \frac{e^{-i2\pi kt}}{t^2+1}.$$

The integrand has two first-order poles at $t = \pm\mathrm{i}$. For $k \geqslant 0$, we close the integration contour in the lower-half complex plane whereas we close the integration contour in the upper-half complex plane for $k < 0$. We then have

$$\begin{aligned} g(k) &= -2\pi\mathrm{i}\frac{\mathrm{e}^{-2\pi k}}{-2\mathrm{i}}\theta(k) + 2\pi\mathrm{i}\frac{\mathrm{e}^{2\pi k}}{2\mathrm{i}}\theta(-k) \\ &= \pi\mathrm{e}^{-2\pi k}\theta(k) + \pi\mathrm{e}^{2\pi k}\theta(-k) \\ &= \pi\mathrm{e}^{-2\pi|k|}. \end{aligned}$$

We thus have

$$\begin{aligned} \sum_{n=-\infty}^{\infty}\frac{1}{n^2+1} &= \pi\sum_{k=-\infty}^{\infty}\mathrm{e}^{-2\pi|k|} = \pi\left[1 + 2\sum_{k=1}^{\infty}\mathrm{e}^{-2\pi k}\right] \\ &= \pi\left[1 + \frac{2\mathrm{e}^{-2\pi}}{1-\mathrm{e}^{-2\pi}}\right] = \pi\frac{1+\mathrm{e}^{-2\pi}}{1-\mathrm{e}^{-2\pi}} = \pi\frac{\mathrm{e}^{\pi}+\mathrm{e}^{-\pi}}{\mathrm{e}^{\pi}-\mathrm{e}^{-\pi}} \\ &= \pi\coth\pi \approx 3.153\,348, \end{aligned}$$

and

$$\sum_{n=1}^{\infty}\frac{1}{n^2+1} = \frac{\pi\coth\pi - 1}{2} \approx 1.076\,674.$$

(3) To prove the given identity, we take $f(t) = \mathrm{e}^{-t^2}$ and $T = (\pi z)^{1/2}$ in the above Poisson summation formula. For $t = 0$, we have

$$\sum_{n=-\infty}^{\infty} f(n(\pi z)^{1/2}) = \sum_{n=-\infty}^{\infty}\mathrm{e}^{-n^2\pi z} = \frac{1}{(\pi z)^{1/2}}\sum_{k=-\infty}^{\infty} g(k/(\pi z)^{1/2}).$$

We now find the Fourier transform of $f(t)$. We have

$$\begin{aligned} g(k/(\pi z)^{1/2}) &= \int_{-\infty}^{\infty}\mathrm{d}t\,\mathrm{e}^{-2\pi\mathrm{i}kt/(\pi z)^{1/2}}f(t) = \int_{-\infty}^{\infty}\mathrm{d}t\,\mathrm{e}^{-t^2-2\pi^{1/2}\mathrm{i}kt/z^{1/2}} \\ &= \mathrm{e}^{-k^2\pi/z}\int_{-\infty}^{\infty}\mathrm{d}t\;\mathrm{e}^{-(t+\pi^{1/2}\mathrm{i}kt/z^{1/2})^2} = \sqrt{\pi}\,\mathrm{e}^{-k^2\pi/z}. \end{aligned}$$

Thus,

$$\sum_{n=-\infty}^{\infty}\mathrm{e}^{-n^2\pi z} = \frac{1}{z^{1/2}}\sum_{k=-\infty}^{\infty}\mathrm{e}^{-k^2\pi/z}$$

24-4 Application of the Lifshits-Kosevich theory. Consider the Lifshits-Kosevich theory of the de Haas-van Alphen effect for a quadratic dispersion relation $E = \hbar^2k^2/2m = \hbar^2(k_x^2 + k_y^2 + k_z^2)/2m$.

(1) Find the extremal cross-sectional area $S_m(E)$ and its derivatives $\partial S_m/\partial E$ and $\partial^2 S_m/\partial k_z^2$ at energy E.
(2) Evaluate the oscillation frequency F.
(3) Evaluate the oscillation amplitude A_k at zero temperature. Write down an expression for $M_{\rm osc}$ at zero temperature.

---

(1) For free electrons, the dispersion relation is given by $E = \hbar^2 k^2/2m$. At energy E, the extremal cross-section is a circle of radius $k = \sqrt{2mE/\hbar^2}$. Thus, the extremal cross-sectional area $S_m(E)$ is given by

$$S_m(E) = \pi k^2 = \frac{2\pi m}{\hbar^2}E.$$

We then have

$$\frac{\partial S_m(E)}{\partial E} = \frac{2\pi m}{\hbar^2},$$
$$\frac{\partial^2 S_m(E)}{\partial k_z^2} = \pi\frac{\partial^2(k_x^2 + k_y^2 + k_z^2)}{\partial k_z^2} = 2\pi.$$

(2) The oscillation frequency F is given bu

$$F = \frac{\hbar S_m(E)}{2\pi e} = \frac{\hbar \pi k^2}{2\pi e} = \frac{\hbar k^2}{2e}.$$

(3) From the Lifshits-Kosevich theory, the oscillation amplitude A_k is given by

$$A_k(T,B) = k_B T S_m(\mu)\left(\frac{e}{2\pi^3\hbar B}\right)^{1/2}\frac{1}{k^{1/2}}\frac{|\cos[(k\hbar^2/2m)\partial S_m(\mu)/\partial\mu]|}{|\partial^2 S_m/\partial k_z^2|_0^{1/2}|\sinh(\lambda k)|},$$

where $\lambda = (\pi\hbar k_B T/eB)\partial S_m(\mu)/\partial\mu$. Note that $T/\sinh(\lambda k)$ gives a finite result in the $T \to 0$ limit

$$\frac{k_B T}{\sinh(\lambda k)} \xrightarrow{T\to 0} \frac{k_B T}{\lambda k} = \frac{eB}{k\pi\hbar\partial S_m(E_F)/\partial E_F} = \frac{eB\hbar}{2k\pi^2 m}.$$

The value of $\cos[(k\hbar^2/2m)\partial S_m(\mu)/\partial\mu]$ is given by $\cos(k\pi) = (-1)^k$. The value of A_k at zero temperature is then given by

$$A_k(T,B) = \frac{k_F^2}{4\pi^3 m}\left(\frac{\hbar e^3 B}{k^3}\right)^{1/2} = \frac{E_F}{2\pi^3}\left(\frac{e}{k\hbar}\right)^{3/2} B^{1/2}.$$

24-5 Amplitude of dHvA oscillations in a free electron gas. We now derive the temperature dependence of the amplitude of the de Haas-van Alphen oscillations for a free electron gas with a simple spherical Fermi surface.

(1) At a nonzero temperature T, write down an expression for the grand potential as a sum over the Landau levels and an integral over the wave vector component k_z. Assume that the magnetic field $\boldsymbol{B}$ is applied in the z-direction and that the Landau level spacing is much smaller than the Fermi energy E_{F}.

(2) Making use of the Poisson summation formula, perform the summation over the Landau levels. Keep only the oscillatory part of the grand potential. Change the integration variable of the integral arising from the application of the Poisson summation formula to the energy E of an electron. Note that, because the Landau level spacing is much smaller than the Fermi energy E_{F}, the lower limit of the resultant E-integral can be taken to be 0.

(3) Perform the integration over k_z in the oscillatory part of the grand potential.

(4) Integrate E twice by parts and keep only the oscillatory part of the grand potential. Change the integration variable from E to $z = (E - \mu)/k_{\mathrm{B}}T$ with μ the chemical potential. Note that, for $\mu \gg k_{\mathrm{B}}T$, the lower limit of the z-integration can be extended to $-\infty$ with negligible errors. Perform the resultant z-integration by closing the contour in the complex plane of z and making use of the residue theorem.

(5) Obtain the leading oscillatory part of the magnetization from the above-derived oscillatory part of the grand potential.

(6) Express the amplitude of oscillations as a function of temperature. Discuss its behavior at $T = 0$ and $T \gg \hbar\omega_c/k_{\mathrm{B}}$.

---

(1) The single-electron Hamiltonian in the magnetic field $\boldsymbol{B} = B\boldsymbol{e}_z$ reads

$$\hat{H} = \frac{1}{2m}\left(\hat{p}_x^2 + \hat{p}_y^2 + \hat{p}_z^2\right) + \frac{eB}{m}x\hat{p}_y + \frac{e^2B^2}{2m}x^2.$$

The eigenvalues of the above Hamiltonian are given by

$$E_{nk_z} = (n + 1/2)\hbar\omega_c + \hbar^2k_z^2/2m$$

with $n = 0, 1, 2, \cdots$ and $-\infty < k_z < \infty$. Note that each Landau level (each n) is $L^2eB/2\pi\hbar$-fold degenerate. The grand potential is given by

$$\Omega = -\frac{2\mathscr{V}eB}{(2\pi)^2\hbar}k_{\mathrm{B}}T\sum_n\int_{-\infty}^{\infty}\mathrm{d}k_z\ \ln\left[1 + \mathrm{e}^{-\beta(E_{nk_z}-\mu)}\right],$$

where the 2-fold spin degeneracy has been taken into account.

(2) Making use of the Poisson summation formula, we can reexpress the oscillatory part of the grand potential as

$$\Omega_2 = -\frac{4\mathcal{V}eB}{(2\pi)^2\hbar}k_BT\,\mathrm{Re}\sum_{k=1}^{\infty}\int_{-\infty}^{\infty}dk_z\int_{-1/2}^{\infty}dn\,e^{i2\pi kn}\ln\left[1+e^{-\beta(E_{nk_z}-\mu)}\right].$$

Making a change integration variables from n to $E = (n + 1/2)\hbar\omega_c + \hbar^2k_z^2/2m$, we have

$$\Omega_2 = -\frac{4\mathcal{V}mk_BT}{(2\pi)^2\hbar^2}\,\mathrm{Re}\sum_{k=1}^{\infty}(-1)^k\int_{-\infty}^{\infty}dk_z\;e^{-i(\pi k\hbar/m\omega_c)k_z^2}$$
$$\times\int_{\hbar^2k_z^2/2m}^{\infty}dE\;e^{i(2\pi k/\hbar\omega_c)E}\ln\left[1+e^{-\beta(E-\mu)}\right].$$

The most important contribution to the k_z integral comes from the region in which $\hbar^2k_z^2/2m$ is about the size of the Landau level spacing while the most important contribution to the E integral comes from the region close to the chemical potential. Since the Landau level spacing is much smaller than E_F, we can set the lower limit of the E integral to be zero. We thus have

$$\Omega_2 = -\frac{4\mathcal{V}mk_BT}{(2\pi)^2\hbar^2}\,\mathrm{Re}\sum_{k=1}^{\infty}(-1)^k\int_{-\infty}^{\infty}dk_z\;e^{-i(\pi k\hbar/m\omega_c)k_z^2}$$
$$\times\int_{0}^{\infty}dE\;e^{i(2\pi k/\hbar\omega_c)E}\ln\left[1+e^{-\beta(E-\mu)}\right].$$

(3) We can now perform the integration over k_z. Making use of the Gaussian integral formula

$$\int_{-\infty}^{\infty}dz\;e^{-i\alpha z^2} = \sqrt{\frac{\pi}{i\alpha}},$$

we have

$$\Omega_2 = -\frac{4\mathcal{V}mk_BT(eB)^{1/2}}{(2\pi)^2\hbar^{5/2}}\,\mathrm{Re}\,e^{-i\pi/4}\sum_{k=1}^{\infty}\frac{(-1)^k}{k^{1/2}}$$
$$\times\int_{0}^{\infty}dE\;e^{i(2\pi k/\hbar\omega_c)E}\ln\left[1+e^{-\beta(E-\mu)}\right].$$

(4) We now perform integration by parts twice to the E integral to render the integrand into one that is a rapidly decreasing function of E. Since the integrated parts are non-oscillatory, they will be neglected. The oscillatory part of the grand potential is now denoted by $\Omega_{\rm osc}$. We have

$$\Omega_{\rm osc} = \frac{4\mathscr{V}(eB)^{5/2}}{(2\pi)^4\hbar^{1/2}mk_{\rm B}T}\,{\rm Re}\,{\rm e}^{-{\rm i}\pi/4}\sum_{k=1}^{\infty}\frac{(-1)^k}{k^{5/2}} \times \int_0^{\infty}{\rm d}E\,{\rm e}^{{\rm i}(2\pi k/\hbar\omega_c)E}\frac{{\rm e}^{\beta(E-\mu)}}{[{\rm e}^{\beta(E-\mu)}+1]^2}.$$

Since the integrand in the E integral is a rapidly decreasing function E, we can set its lower limit to $-\infty$. With a change of integration variables from E to $z = \beta(E-\mu)$, we have

$$\Omega_{\rm osc} = \frac{4\mathscr{V}(eB)^{5/2}}{(2\pi)^4\hbar^{1/2}m}\,{\rm Re}\,{\rm e}^{-{\rm i}\pi/4}\sum_{k=1}^{\infty}\frac{(-1)^k}{k^{5/2}}{\rm e}^{2\pi{\rm i}k\mu/\hbar\omega_c} \times \int_{-\infty}^{\infty}{\rm d}z\,\frac{{\rm e}^{z+{\rm i}(2\pi kk_{\rm B}T/\hbar\omega_c)z}}{({\rm e}^z+1)^2}.$$

The integrand has second-order poles at $z_n = {\rm i}(2n+1)\pi$ with $n = 0, \pm1, \pm2, \cdots$. In consideration of the sign of the imaginary part of the argument of the exponential in the numerator, we close the integration contour in the upper-half complex plane as shown in Fig. 24.1. Thus, only the poles $z_n = {\rm i}(2n+1)\pi$ for $n = 0, 1, 2, \cdots$ are enclosed by the contour. To obtain the value of the integral using the residue theorem, we need to know the residues at these poles. To simplify notations, we introduce $a = 2\pi kk_{\rm B}T/\hbar\omega_c$. The concerned integral is then written as

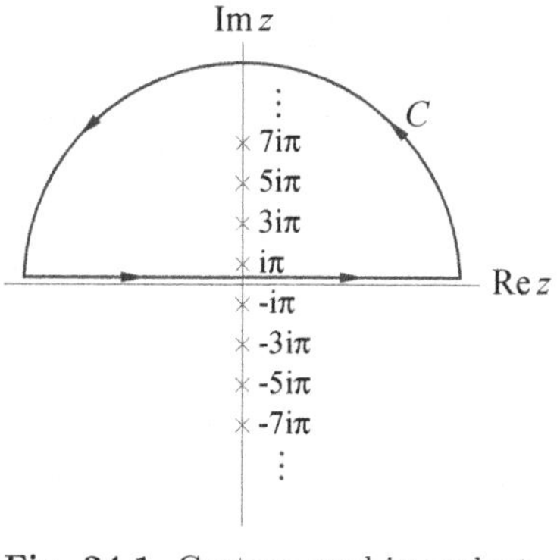

Fig. 24.1 Contour used in evaluating the E integral,

$$I = \int_{-\infty}^{\infty}{\rm d}z\,f(z) = \int_{-\infty}^{\infty}{\rm d}z\,\frac{{\rm e}^{(1+{\rm i}a)z}}{({\rm e}^z+1)^2},$$

where the integrand has been referred to as $f(z)$. For the convenience of finding the residue of the integrand $f(z)$ at the pole

$z_n = \mathrm{i}(2n+1)\pi$, we expand $\mathrm{e}^z + 1$ in the vicinity of the pole as

$$\begin{aligned}\mathrm{e}^z + 1 &= -(z - z_n) - \frac{1}{2!}(z - z_n)^2 - \frac{1}{3!}(z - z_n)^3 - \cdots \\ &= -(z - z_n)\gamma(z),\end{aligned}$$

where $\gamma(z) = 1 + (z - z_n)/2! + (z - z_n)^2/3! + \cdots$. Note that $\gamma(z_n) = 1$ and $\mathrm{d}\gamma/\mathrm{d}z|_{z=z_n} = 1/2$. The residue of $f(z)$ at $z_n = \mathrm{i}(2n+1)\pi$ is given by

$$\begin{aligned}\operatorname*{Res}_{z=z_n} f(z) &= \lim_{z\to z_n} \frac{\mathrm{d}}{\mathrm{d}z}\big[(z - z_n)^2 f(z)\big] \\ &= \lim_{z\to z_n} \frac{\mathrm{d}}{\mathrm{d}z} \frac{(z - z_n)^2 \mathrm{e}^{(1+\mathrm{i}a)z}}{(\mathrm{e}^z + 1)^2} = \lim_{z\to z_n} \frac{\mathrm{d}}{\mathrm{d}z} \frac{\mathrm{e}^{(1+\mathrm{i}a)z}}{\gamma^2(z)} \\ &= \lim_{z\to z_n} \left[-\frac{2\mathrm{e}^{(1+\mathrm{i}a)z}}{\gamma^3(z)} \frac{\mathrm{d}\gamma(z)}{\mathrm{d}z} + \frac{(1 + \mathrm{i}a)\mathrm{e}^{(1+\mathrm{i}a)z}}{\gamma^2(z)} \right] \\ &= -\mathrm{i}a\,\mathrm{e}^{-(2n+1)\pi a}.\end{aligned}$$

The value of integral I is then given by

$$\begin{aligned}I &= 2\pi\mathrm{i} \sum_{n=0}^{\infty} \operatorname*{Res}_{z=z_n} f(z) = 2\pi a \sum_{n=0}^{\infty} \mathrm{e}^{-(2n+1)\pi a} \\ &= 2\pi a\,\mathrm{e}^{-\pi a} \frac{1}{1 - \mathrm{e}^{-2\pi a}} = \frac{\pi a}{\sinh(\pi a)}.\end{aligned}$$

The oscillatory part of the grand potential is then given by

$$\begin{aligned}\Omega_{\mathrm{osc}} &= \frac{\mathscr{V}(eB)^{3/2} k_{\mathrm{B}} T}{2\pi^2 \hbar^{3/2}} \operatorname{Re} \mathrm{e}^{-\mathrm{i}\pi/4} \sum_{k=1}^{\infty} \frac{(-1)^k}{k^{3/2} \sinh(\lambda k)} \mathrm{e}^{2\pi\mathrm{i}k\mu/\hbar\omega_c} \\ &= \frac{\mathscr{V}(eB)^{3/2} k_{\mathrm{B}} T}{2\pi^2 \hbar^{3/2}} \sum_{k=1}^{\infty} \frac{(-1)^k}{k^{3/2} \sinh(\lambda k)} \cos\left(\frac{2\pi k\mu}{\hbar\omega_c} - \frac{\pi}{4}\right),\end{aligned}$$

where we have introduced $\lambda = 2\pi^2 k_{\mathrm{B}} T/\hbar\omega_c = 2\pi^2 m k_{\mathrm{B}} T/\hbar eB$.

(5) Since the cosine factor is the most rapidly varying factor as B changes, we only differentiate this factor with respect to B when Ω_{osc} is differentiated to obtain the oscillatory part of the magnetization M_{osc}. We have

$$\begin{aligned}M_{\mathrm{osc}} &= -\frac{1}{\mathscr{V}} \frac{\partial \Omega_{\mathrm{osc}}}{\partial B} \\ &= \frac{e^{1/2} m k_{\mathrm{B}} T \mu}{\pi \hbar^{5/2} B^{1/2}} \sum_{k=1}^{\infty} \frac{(-1)^{k+1}}{k^{1/2} \sinh(\lambda k)} \sin\left(\frac{2\pi k\mu}{\hbar\omega_c} - \frac{\pi}{4}\right),\end{aligned}$$

where we have made use of $\omega_c = eB/m$ in performing the differentiation.

(6) The amplitude of oscillations for each k is given by

$$A_k(T, B) = \frac{e^{1/2} m k_{\mathrm{B}} T \mu}{\pi \hbar^{5/2} B^{1/2} k^{1/2} \sinh(\lambda k)}.$$

In the $T \to 0$ limit, we have upon making use of $\sinh(\lambda k) \to \lambda k$

$$A_k(T \to 0, B) \to \frac{e^{3/2} B^{1/2} E_{\mathrm{F}}}{2\pi^3 \hbar^{3/2} k^{3/2}} = \frac{(\hbar e^3 B)^{1/2} k_{\mathrm{F}}^2}{4\pi^3 k^{3/2} m} = \frac{E_{\mathrm{F}}}{2\pi^3} \left(\frac{e}{k\hbar}\right)^{3/2} B^{1/2}$$

which is identical with the corresponding result obtained in the previous problem.

In the $T \to \infty$ limit, because $\sinh(\lambda k) \to \infty$, we see that

$$A_k(T \to \infty, B) \to 0.$$

Therefore, the de Haas-van Alphen effect can be observed only at low temperatures.

Chapter 25

Crystal Defects

(1) *Crystal defects*
Point defects (0D defects),
line defects (1D defects),
planar defects (2D defects),
bulk defects (3D defects).

(2) *Point defects*
Vacancies,
self-interstitial atoms,
substitutional impurity atoms,
interstitial impurity atoms.

(3) *Vacancies*
Vacancies are sites which were usually occupied by atoms but which are now unoccupied, also referred to as Schottky defects.

$$x_v = N_v/N \approx e^{s_v/k_B} e^{-E_v/k_B T}.$$

(4) *Schottky defect*
A Schottky defect is a vacancy (or a pair of vacancies in an ionic solid).

$$n^+ = n^- = n e^{-(E^+ + E^-)/2k_B T}.$$

(5) *Frenkel defect*
A Frenkel defect consists of a self-interstitial atom and its vacancy.

$$N_F = \sqrt{NM} e^{-E_F/2k_B T}.$$

(6) *Line defects*
Edge dislocations, screw dislocations, dislocation loops.

(7) *Edge dislocations*
An edge dislocation is caused by the termination of a lattice plane of atoms in the middle of a crystal. It is characterized by the dislocation line, the dislocation plane, and the Burgers vector.

Slip systems in FCC, BCC, and HCP:

Structure	Slip plane	Slip direction
FCC	$\{111\}$	$\langle 1\bar{1}0\rangle$
BCC	$\{110\}$	$\langle \bar{1}11\rangle$
	$\{112\}$	$\langle 11\bar{1}\rangle$
	$\{123\}$	$\langle 111\rangle$
HCP	$\{0001\}$	$\langle 11\bar{2}0\rangle$
	$\{10\bar{1}0\}$	$\langle 11\bar{2}0\rangle$
	$\{10\bar{1}1\}$	$\langle 11\bar{2}0\rangle$

The stress field is given by

$$\sigma_{xx} = -\frac{\mu b}{2\pi(1-\nu)}\frac{y(3x^2+y^2)}{(x^2+y^2)^2},$$

$$\sigma_{yy} = \frac{\mu b}{2\pi(1-\nu)}\frac{y(x^2-y^2)}{(x^2+y^2)^2},$$

$$\sigma_{xy} = \frac{\mu b}{2\pi(1-\nu)}\frac{x(x^2-y^2)}{(x^2+y^2)^2}.$$

The strain energy is given by

$$e_{\text{tot}} = \alpha\mu b^2, \quad \alpha = 0.5\text{–}1.5.$$

(8) *Screw dislocations*
When a shear stress is applied across one end of a crystal, the crystal begins to rip and a screw dislocation forms.
The stress field is given by

$$\sigma_{yz} = \sigma_{zy} = \frac{\mu b}{2\pi}\frac{x}{x^2+y^2},$$

$$\sigma_{zx} = \sigma_{xz} = -\frac{\mu b}{2\pi}\frac{y}{x^2+y^2}.$$

In polar coordinates, the stress field is given by

$$\sigma_{\theta z} = \sigma_{z\theta} = \frac{\mu b}{2\pi r}.$$

The strain energy is given by

$$e_{\text{tot}} = \alpha\mu b^2.$$

(9) *Planar defects*
Stacking faults, grain boundaries, twins.

(10) *Bulk defects*
Voids, precipitates.

25-1 Deduction of the energy of vacancy formation from resistivity data. In the following table, given are the changes in resistivity relative to that measured at 20 K for samples of a metal that have been quenched from various temperatures. By plotting $\ln(\Delta R)$ against $1/T$, find the energy of vacancy formation in this metal. It is known that $\Delta R = CN_v$, where C is a proportional constant and N_v the number of vacancies.

Changes in resistivity relative to that measured at 20 K.

Temperature [K]	800	850	900	950	1,000
Resistivity change [%]	0.46	1.07	2.18	4.14	7.17

The plot of $\ln \Delta R$ as a function of $1,000/T$ is given in Fig. 25.1. Note that ΔR represents the relative change in resistivity. From the figure, we see that $\ln \Delta R$ versus $1,000/T$ is nearly a straight line. The slope of this straight line gives the energy of vacancy formation, E_v. To determine E_v accurately, we now perform a linear least-squares fitting. For simplicity in notations, we set $y = \ln \Delta R$ and $x = 1/T$. We now fit the experimental data to $y = ax + b$. The χ^2 function for this fitting is

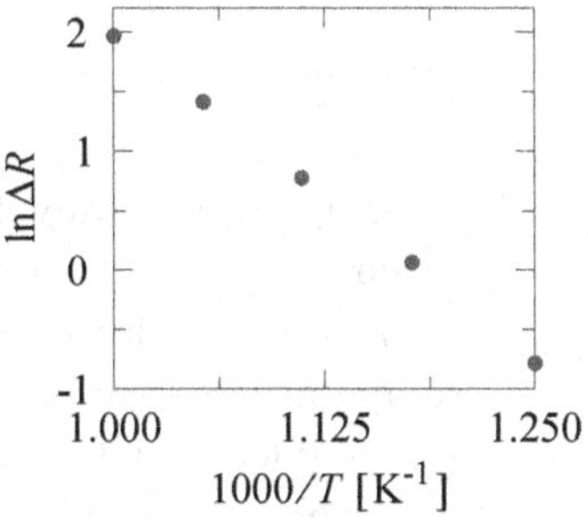

Fig. 25.1 Plot of $\ln \Delta R$ as a function of $1000/T$.

$$\chi^2 = \sum_{i=1}^{n} (y_i - ax_i - b)^2,$$

where n is the number of data points. Here $n = 5$. To minimize χ^2, we differentiate χ^2 with respect to the fitting parameters a and b and set the results to zero. We have

$$a \sum_i x_i^2 + b \sum_i x_i = \sum_i x_i y_i,$$

$$a \sum_i x_i + nb = \sum_i y_i.$$

Solving for a and b from the above two equations, we obtain

$$a = \frac{n\sum_i x_i y_i - \sum_i x_i \sum_i y_i}{n\sum_i x_i^2 - \left(\sum_i x_i\right)^2}, \quad b = \frac{\sum_i x_i^2 \sum_i y_i - \sum_i x_i \sum_i x_i y_i}{n\sum_i x_i^2 - \left(\sum_i x_i\right)^2}.$$

Evaluating a and b using the above expressions and the given data, we obtain

$$a \approx -10,985.1 \text{ K}, \ b \approx 12.974\,1.$$

To infer the energy of vacancy formation, E_v, from the above results, we now write down an analytic expression for ΔR using the established theory for vacancies in crystals. The number of vacant sites is given by $N_v = N e^{s_v/k_B} e^{-E_v/k_B T}$. Making use of $\Delta R = C N_v$, we have

$$\Delta R = A e^{-E_v/k_B T},$$

where $A = C N e^{s_v/k_B}$. Taking the logarithm of the above equation yields

$$\ln \Delta R = \ln A - \frac{E_v}{k_B T}.$$

We thus have

$$E_v = -k_B a = 8.617\,343 \times 10^{-5} \times 10,985.1 \approx 0.95 \text{ eV}.$$

25-2 Energy of vacancy formation in gold. In the following table, given are the changes in resistivity relative to that measured at 78 K for samples of gold that have been quenched from various temperatures. By plotting $\ln(\Delta R)$ against $1/T$, find the energy of vacancy formation in gold. It is known that $\Delta R = C N_v$, where C is a proportional constant and N_v the number of vacancies.

Changes in resistivity relative to that measured at 78 K.

Temperature [K]	920	970	1,020	1,060	1,220
Resistivity change [%]	0.41	0.70	1.40	2.30	9.00

The plot of $\ln \Delta R$ as a function of $1,000/T$ is given in Fig. 25.2. ΔR again represents the relative change in resistivity. From the figure, we see that the data points fall very well on a straight line. As in the previous problem, to determine E_v accurately, we perform a linear least-squares fitting of the experimental data to $y = ax + b$, with $y = \ln \Delta R$ and $x = 1/T$. Using the expressions of a and b ob-

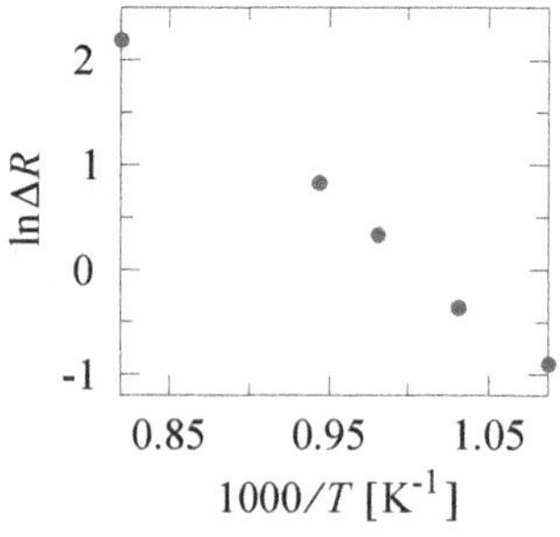

Fig. 25.2 Plot of $\ln \Delta R$ as a function of $1000/T$.

tained in the previous problem, we find that $a \approx -11,765.3$ K and $b \approx 11.862\,7$. From a, we obtain $E_v \approx 1.01$ eV.

25-3 Number of vacant sites. In a given material, it is found that one out of 1,011 atomic sites is vacant at 750 °C and one out of 1,010 sites is vacant at 850 °C. At what temperature will one out of 1,008 atomic sites be vacant?

Let $r(T) = N_v(T)/N$. From the given conditions, we have

$$r(T_1) = \mathrm{e}^{s_v/k_\mathrm{B}}\mathrm{e}^{-E_v/k_\mathrm{B}T_1}, \quad r(T_2) = \mathrm{e}^{s_v/k_\mathrm{B}}\mathrm{e}^{-E_v/k_\mathrm{B}T_2},$$

where

$$T_1 = 750^\circ\mathrm{C} \approx 1,023.15\ \mathrm{K}, \quad r(T_1) = 1/1,011 \approx 9.891\,197 \times 10^{-4},$$
$$T_2 = 850^\circ\mathrm{C} \approx 1,123.15\ \mathrm{K}, \quad r(T_2) = 1/1,010 \approx 9.900\,990 \times 10^{-4}.$$

Solving for E_v and $\mathrm{e}^{s_v/k_\mathrm{B}}$, we obtain

$$E_v = \frac{k_\mathrm{B}T_1T_2}{T_2 - T_1}\ln\frac{r(T_2)}{r(T_1)},$$
$$\mathrm{e}^{s_v/k_\mathrm{B}} = r(T_1)\,\mathrm{e}^{[T_2/(T_2-T_1)]\ln[r(T_2)/r(T_1)]}.$$

At temperature T, we have

$$r_v(T) = r(T_1)\,\mathrm{e}^{[T_2/(T_2-T_1)](1-T_1/T)\ln[r(T_2)/r(T_1)]}.$$

Solving for T from the above equation, we obtain an expression of T for a given value of $r(T)$

$$T = T_1\left\{1 - \frac{T_2 - T_1}{T_2}\frac{\ln[r(T)/r(T_1)]}{\ln[r(T_2)/r(T_1)]}\right\}^{-1}.$$

Plugging the relevant numbers into the above equation, we obtain the temperature at which one out of 1,008 atomic sites is vacant ($r = 1/1,008$)

$$T \approx 1,396.55^\circ\mathrm{C} = 1,669.70\ \mathrm{K}.$$

25-4 Number of occupied lattice sites for every vacancy in Ni. The energy required for vacancy formation in Ni is listed as 1.12 eV. For Ni at 2 °C below its melting point temperature (1,453°C), find the number of occupied lattice sites for every vacancy. The vacancy-formation entropy of Ni is $1.5k_\mathrm{B}$ per vacancy.

For $T = 1,451^\circ\mathrm{C} = 1,724.15$ K, we have

$$N/N_v = \mathrm{e}^{-s_v/k_\mathrm{B}}\mathrm{e}^{E_v/k_\mathrm{B}T} \approx 419/1.$$

Therefore, there are 419 occupied lattice sites for every vacancy.

25-5 Energy of vacancy formation in Al. Given the equilibrium vacancy concentration for Al at 300 °C as 2×10^{17} cm^{-3}, determine the energy of vacancy formation. The vacancy-formation entropy of Al is $2k_B$ per vacancy and the lattice constant of Al is 0.405 nm.

In consideration that Al is a face-centered cubic crystal of lattice constant $a = 0.405$ nm, we see that the number of atoms per unit volume is

$$n = \frac{4}{a^3} \approx 6.021 \times 10^{28}\ \text{m}^{-3}.$$

From $N_v = e^{s_v/k_B} e^{-E_v/k_B T}$, we have

$$E_v = Ts_v + k_B T \ln \frac{N}{N_v} = Ts_v + k_B T \ln \frac{n}{n_v} \approx 0.72\ \text{eV}.$$

25-6 Energy of vacancy formation in Mo. Experimental studies have established that the ratio of the vacancy densities in Mo at 500 °C and 900 °C is 2×10^{-3}. Determine the energy of vacancy formation in Mo.

Using the result

$$E_v = \frac{k_B T_1 T_2}{T_2 - T_1} \ln \frac{r(T_2)}{r(T_1)}$$

derived in **Problem 25-3**, we have

$$E_v = \frac{8.617\,343 \times 10^{-5} \times 773.15 \times 1,173.15}{1,173.15 - 773.15} \ln \frac{1}{2 \times 10^{-3}} \approx 1.21\ \text{eV}.$$

25-7 Schottky defects in copper. The activation energy to form a Schottky defect in copper is 0.89 eV. The vacancy-formation entropy for copper is $1.193k_B$.

(1) Compute the change in density due to Schottky defect formation upon heating copper from 300 K to 1,200 K.

(2) Experimentally the change in density is found to be −6.45%. Attempt to account for any differences between the experimentally determined and theoretically predicted values.

(1) The lattice constant of copper is 0.361 5 nm and the atom number density for a perfect crystal is $n = 8.467\ 1 \times 10^{28}$ m^{-3}. Let $T_1 = 300$ K and $T_2 = 1,200$ K. Let n denote the atom density in a perfect crystal and n_v the density of vacancies. We have

$$n_v(T_1) = n e^{s_v/k_B} e^{-E_v/k_B T_1}, \quad n_v(T_2) = n e^{s_v/k_B} e^{-E_v/k_B T_2}.$$

The change in density from $T = 300$ K to $1,200$ K is given by

$$\begin{aligned}\Delta n &= [n - n(T_2)] - [n - n(T_1)] = n(T_1) - n(T_2) \\ &= n e^{s_v/k_B}\left(e^{-E_v/k_B T_1} - e^{-E_v/k_B T_2}\right).\end{aligned}$$

The relative change is

$$\Delta n/n = e^{s_v/k_B}\left(e^{-E_v/k_B T_1} - e^{-E_v/k_B T_2}\right) \approx -0.06\%.$$

(2) The reason for the large difference between the above-evaluated value and the experimental result is that the experimentally measured density change also includes the lattice expansion associated with increased thermal vibrations of atoms as temperature increases. The lattice expansion results in the increase of the effective atomic radius.

25-8 Schottky defects in an oxide ceramic. Data related to the formation of Schottky defects in some oxide ceramic (having the chemical formula of the form MO with M a metallic element and O the oxygen) are given in the following table.

T [°C]	ρ [g/cm^3]	n_s [m^{-3}]
750	3.50	5.7×10^9
1,000	3.45	?
1,500	3.40	5.8×10^{17}

(1) Find the energy for defect formation (in eV).

(2) Find the equilibrium number of Schottky defects per cubic meter at $1,000$ °C.

(3) Determine the identity of the oxide (*i.e.*, identify the metallic element M).

---

(1) The three temperatures 750 °C, $1,000$ °C, and $1,500$ °C are denoted by T_1, T_2, and T_3, respectively. From $n_s = n\,e^{-E_s/2k_B T}$ where we have used n_s for n_b and E_s for $E^+ + E^- - E_b$, we have at two temperatures T_1 and T_3

$$n_s(T_1) = n\,e^{-E_s/2k_B T_1}, \quad n_s(T_3) = n\,e^{-E_s/2k_B T_3}.$$

Dividing one of the above two equations from the other and solving for E_s from the resultant equation, we obtain

$$E_s = \frac{2k_B T_1 T_3}{T_3 - T_1} \ln \frac{n_s(T_3)}{n_s(T_1)}.$$

Making use of the above-given data at $T_1 = 750°\mathrm{C}$ and $T_3 = 1,500°\mathrm{C}$, we obtain $E_s \approx 7.69$ eV.

(2) From the first of the two equations given in the above, we have

$$n = n_s(T_1)\mathrm{e}^{E_s/2k_\mathrm{B}T_1} \approx 4.87 \times 10^{28}\ \mathrm{m}^{-3}.$$

At $T_2 = 1,000°\mathrm{C} = 1,273.15$ K, we have

$$n_s(T_2) = n\,\mathrm{e}^{-E_s/2k_\mathrm{B}T_2} \approx 2.97 \times 10^{13}\ \mathrm{m}^{-3}.$$

(3) From the number density of MO pairs, n, we can write the density ρ as

$$\rho(T) = \big[\,n - n_s(T)\,\big](m_\mathrm{M} + m_\mathrm{O})\mathrm{u},$$

where m_M and m_O are the atomic weights of M and O, respectively, and u is the atomic mass unit. Solving for m_M from the above equation, we have

$$m_\mathrm{M} = \rho(T)/[\,n - n_s(T)\,]\mathrm{u} - m_\mathrm{O}.$$

Making use of the above results at three temperatures T_1, T_2, and T_3, we obtain the following three values for m_M: 27.30, 26.68, 26.06. The average of the three values for m_M is 26.68. Looking up the atomic weights of the elements in a periodic table of the elements, we find that the element with the atomic weight closest to 26.68 is Al. Therefore, the element represented by M is Al and the oxide is aluminum monoxide (AlO).

25-9 Burgers vectors of dislocations in FCC, BCC, and SC crystals. We study dislocations in FCC, BCC, and SC crystals and identify the Burgers vectors of dislocations in these three kinds of crystals.

(1) Write down the shortest translation vectors in FCC, BCC, and SC structures. Give the expected Burgers vectors of dislocations in these three structures.

(2) Copper has an FCC structure with a lattice constant of $a = 0.361\,48$ nm. Give the Burgers vector $\boldsymbol{b}$ of dislocations and its magnitude $|\boldsymbol{b}|$ in copper.

(3) Consider the BCC phase of iron with a lattice constant of $0.286\,65$ nm. Give the Burgers vector $\boldsymbol{b}$ of dislocations and its magnitude $|\boldsymbol{b}|$ in BCC iron.

(1) The shortest translation vector in a structure is in the direction of the highest linear atom number density. The Burgers vector of dislocations also points in such a direction, with its size equal to the length of the shortest translation vector. In the close packing picture of atoms, atoms touch in the direction of the highest linear atom number density.

In an FCC structure, the [110] direction has the highest linear atom number density which is equal to $1/2R$ with R the atomic radius that is related to the lattice constant a through $R = a/2\sqrt{2}$. The shortest translation vector in an FCC structure is then

$$\boldsymbol{R}_{\min}^{\mathrm{FCC}} = 2R\frac{[110]}{\|[110]\|} = \frac{a}{\sqrt{2}}\frac{[110]}{\sqrt{2}} = \frac{a}{2}[110],$$

where we have used [110] to denote $\boldsymbol{e}_x + \boldsymbol{e}_y$ and $\|[110]\|$ to denote the norm of [110], $\|[110]\| = \sqrt{2}$. Thus, the Burgers vector of dislocations in an FCC structure is

$$\boldsymbol{b}_{\mathrm{FCC}} = \boldsymbol{R}_{\min}^{\mathrm{FCC}} = \frac{a}{2}[110].$$

In a BCC structure, the [111] direction has the highest linear atom number density given by $1/2R$ with $R = \sqrt{3}\,a/4$. The shortest translation vector in a BCC structure is then

$$\boldsymbol{R}_{\min}^{\mathrm{BCC}} = 2R\frac{[111]}{\|[111]\|} = \frac{\sqrt{3}\,a}{2}\frac{[111]}{\sqrt{3}} = \frac{a}{2}[111].$$

The Burgers vector of dislocations in a BCC structure is thus given by

$$\boldsymbol{b}_{\mathrm{BCC}} = \boldsymbol{R}_{\min}^{\mathrm{BCC}} = \frac{a}{2}[111].$$

In an SC structure, the [100] direction has the highest linear atom number density equal to $1/2R$ with $R = a/2$. The shortest translation vector in an SC structure is then

$$\boldsymbol{R}_{\min}^{\mathrm{SC}} = 2R\frac{[100]}{\|[100]\|} = a[100].$$

The Burgers vector of dislocations in an SC structure is thus given by

$$\boldsymbol{b}_{\mathrm{SC}} = \boldsymbol{R}_{\min}^{\mathrm{SC}} = a[100].$$

(2) For Cu, we have

$$\boldsymbol{b}_{\text{Cu}} = \frac{a}{2}[110] = 0.180\,74[110] = 0.180\,74(\boldsymbol{e}_x + \boldsymbol{e}_y)\ \text{nm},$$

$$|\boldsymbol{b}_{\text{Cu}}| = \frac{a}{2}\sqrt{1^2+1^2+0^2} = \frac{a}{\sqrt{2}} \approx 0.255\,60\ \text{nm}.$$

(3) For BCC Fe, we have

$$\boldsymbol{b}_{\text{Fe}} = \frac{a}{2}[111] = 0.143\,325[111] = 0.143\,325(\boldsymbol{e}_x + \boldsymbol{e}_y + \boldsymbol{e}_z)\ \text{nm},$$

$$|\boldsymbol{b}_{\text{Fe}}| = \frac{a}{2}\sqrt{1^2+1^2+1^2} = \frac{\sqrt{3}\,a}{2} \approx 0.248\,25\ \text{nm}.$$

25-10 Two- and three-dimensional defects.

(1) State the orientation relationships between Burgers vectors and dislocation lines for edge and screw dislocations.

(2) An imperfect FCC crystal has the following stacking sequence

$\cdots ABCABCBACBABCABC\cdots.$

Mark with vertical lines the interfaces between the three-dimensional defect in the crystal and the matrix. Identify the three-dimensional defect.

(3) An imperfect FCC crystal has the following stacking sequence

$\cdots ABCABCBCABC\cdots.$

Mark with vertical lines the interfaces between the three-dimensional defect in the crystal and the matrix. Identify the three-dimensional defect.

---

(1) The Burgers vector is perpendicular to the dislocation line for an edge dislocation while it is (anti)parallel to the dislocation line for a screw dislocation.

(2) The interfacial defect that exists for this stacking sequence is a twin boundary whose position is marked on the following sequence

$\cdots ABCAB|CBACB|ABCABC\cdots.$

(3) The interfacial defect that exists within this FCC stacking sequence is a stacking fault whose position is marked on the following sequence

$\cdots ABCAB|CBC|ABC\cdots.$

Within the defect, the stacking sequence is HCP.

25-11 Two perpendicular long edge dislocations. Two infinitely long edge dislocations are placed perpendicularly in an infinite crystal as shown in Fig. 25..3. The crystal is held at a very high temperature.

(1) Derive an expression for the force (per unit length) acting on dislocation 2 due to the stress field of dislocation 1.

(2) Sketch the above-derived force. Assume that the dislocations remain straight.

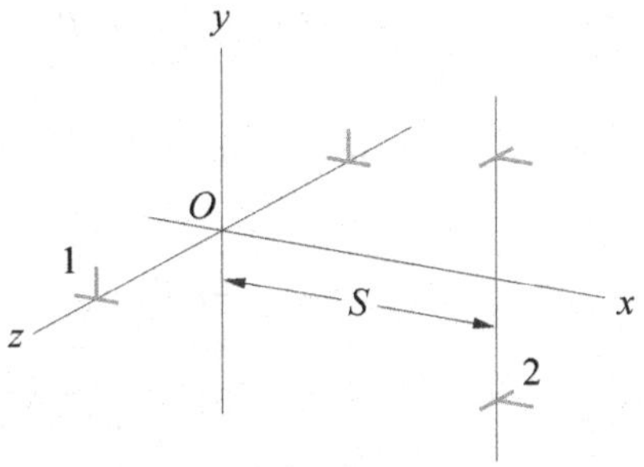

Fig. 25.3 Two perpendicular long edge dislocations.

---

(1) We first outline the derivation of the force (per unit length) acting on dislocation 2 due to the stress field of dislocation 1. Let $\overleftrightarrow{\boldsymbol{\sigma}}_1$ be the shear stress field due to dislocation 1. Let $\boldsymbol{u}$ be the direction of motion and $\boldsymbol{\xi}$ the direction of the dislocation line for dislocation 2. Then, $\boldsymbol{n} = \boldsymbol{\xi} \times \boldsymbol{u}$ is the unit vector of the swept-out area. The shear force acting on dislocation 2 is then given by $\boldsymbol{F} = \overleftrightarrow{\boldsymbol{\sigma}}_1 \cdot \boldsymbol{n}$.

Note that $\boldsymbol{F}$ is the force on a unit area. If dislocation 2 moves, its displacement is given by the Burgers vector $\boldsymbol{b}_2$. The work done by the shear force $\boldsymbol{F}$ is

$$\begin{aligned} W &= \boldsymbol{F} \cdot \boldsymbol{b}_2 = \left(\overleftrightarrow{\boldsymbol{\sigma}}_1 \cdot \boldsymbol{n}\right) \cdot \boldsymbol{b}_2 = \left(\overleftrightarrow{\boldsymbol{\sigma}}_1 \cdot \boldsymbol{b}_2\right) \cdot \boldsymbol{n} \\ &= \left(\overleftrightarrow{\boldsymbol{\sigma}}_1 \cdot \boldsymbol{b}_2\right) \cdot (\boldsymbol{\xi} \times \boldsymbol{u}) = \left[\left(\overleftrightarrow{\boldsymbol{\sigma}}_1 \cdot \boldsymbol{b}_2\right) \times \boldsymbol{\xi}\right] \cdot \boldsymbol{u}, \end{aligned}$$

where we have made use of the fact that $\overleftrightarrow{\boldsymbol{\sigma}}_1$ is a symmetric tensor so that $\left(\overleftrightarrow{\boldsymbol{\sigma}}_1 \cdot \boldsymbol{n}\right) \cdot \boldsymbol{b}_2 = \left(\overleftrightarrow{\boldsymbol{\sigma}}_1 \cdot \boldsymbol{b}_2\right) \cdot \boldsymbol{n}$. Note that the dot product of a tensor $\overleftrightarrow{\boldsymbol{T}}$ and a vector $\boldsymbol{v}$ is given by $\overleftrightarrow{\boldsymbol{T}} \cdot \boldsymbol{v} = \sum_{ij} T_{ij} v_j \boldsymbol{e}_i$. Thus, the force per unit length on dislocation 2 is given by $\boldsymbol{F}_{\mathrm{PK}} = \left(\overleftrightarrow{\boldsymbol{\sigma}}_1 \cdot \boldsymbol{b}_2\right) \times \boldsymbol{\xi}$. The force $\boldsymbol{F}_{\mathrm{PK}}$ is known as *the Peach-Koehler force.*

The nonzero components of $\overleftrightarrow{\boldsymbol{\sigma}}_1$ are given by

$$\begin{aligned} &\left\{\sigma_{1,\,xx}, \sigma_{1,\,yy}, \sigma_{1,\,xy}\right\} \\ &= \frac{\mu b}{2\pi(1-\nu)} \left\{-\frac{y(3x^2+y^2)}{(x^2+y^2)^2}, \frac{y(x^2-y^2)}{(x^2+y^2)^2}, \frac{x(x^2-y^2)}{(x^2+y^2)^2}\right\}. \end{aligned}$$

Note that other nonzero components can be obtained from the symmetry. Specifying to the configuration in the present problem and making use of the above-given $\overleftrightarrow{\sigma}_1$, $\boldsymbol{b}_2 = b\,\boldsymbol{e}_z$, and $\boldsymbol{\xi} = \boldsymbol{e}_y$, we have

$$\begin{aligned}\boldsymbol{F}_{\rm PK} &= b\big(\sigma_{1,zz}\boldsymbol{e}_z\big) \times \boldsymbol{e}_y \\ &= -b\sigma_{1,zz}\boldsymbol{e}_x \\ &= -\nu b\big(\sigma_{1,xx} + \sigma_{1,yy}\big)\boldsymbol{e}_x \\ &= \frac{\mu\nu b^2}{\pi(1-\nu)}\frac{y}{S^2+y^2}\boldsymbol{e}_x.\end{aligned}$$

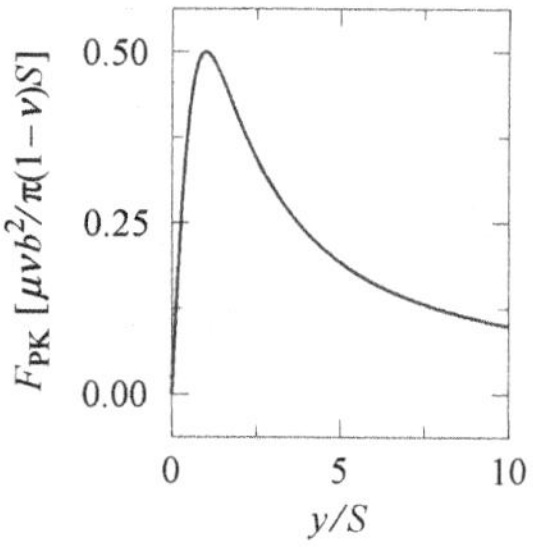

Fig. 25.4 Plot of $F_{\rm PK}$ as a function of y/S.

Thus, the force on dislocation 2 due to the stress field of dislocation 1 is in the x direction and the two dislocations repel each other.

(2) The force $F_{\rm PK}$ is plotted in Fig. 25.4.

Chapter 26

Electron-Phonon Interaction

(1) *Electron-phonon interaction in metals*

$$\hat{H}_{\rm ep} = \sum_{nn'\boldsymbol{k}\boldsymbol{k}'\sigma\sigma'} \sum_{\boldsymbol{q}s} g_{\boldsymbol{q}s}(n'\boldsymbol{k}'\sigma', n\boldsymbol{k}\sigma)\big(\hat{a}_{\boldsymbol{q}s} + \hat{a}^\dagger_{-\boldsymbol{q}s}\big)\hat{c}^\dagger_{n'\boldsymbol{k}'\sigma'}\hat{c}_{n\boldsymbol{k}\sigma},$$

$$g_{\boldsymbol{q}s}(n'\boldsymbol{k}'\sigma', n\boldsymbol{k}\sigma) = \sum_{j\kappa}\left(\frac{\hbar}{2NM_\kappa\omega_{\boldsymbol{q}s}}\right)^{1/2} \mathrm{e}^{\mathrm{i}\boldsymbol{q}\cdot\boldsymbol{R}_j}$$

$$\times \sum_{s_z}\int \mathrm{d}\boldsymbol{r}\; \psi^*_{n'\boldsymbol{k}'\sigma'}(\boldsymbol{r}, s_z)\psi_{n\boldsymbol{k}\sigma}(\boldsymbol{r}, s_z)\boldsymbol{\nabla}U\cdot\boldsymbol{\epsilon}^{(s)}_\kappa(\boldsymbol{q}).$$

(2) *Jellium model*

$$\hat{H}_{\rm ep} = \sum_{\boldsymbol{k}\boldsymbol{q}s\sigma} g_{\boldsymbol{q}s}\big(\hat{a}_{\boldsymbol{q}s} + \hat{a}^\dagger_{-\boldsymbol{q}s}\big)\hat{c}^\dagger_{\boldsymbol{k}+\boldsymbol{q},\sigma}\hat{c}_{\boldsymbol{k}\sigma},$$

$$g_{\boldsymbol{q}s} = \frac{\mathrm{i}}{q^2}\left(\frac{Z^2ne^6\hbar^2}{4M\varepsilon_0^3\mathscr{V}^2}\right)^{1/4}\boldsymbol{q}\cdot\boldsymbol{\varepsilon}^{(s)}(\boldsymbol{q}).$$

(3) *Electron-phonon interaction in ionic crystals*

$$\hat{H}_{\rm ep} = \sum_{\boldsymbol{k}'\boldsymbol{k}\boldsymbol{q}\sigma} g_{\boldsymbol{k}'\boldsymbol{k}\boldsymbol{q}}\big(\hat{a}_{\boldsymbol{q}} + \hat{a}^\dagger_{-\boldsymbol{q}}\big)\hat{c}^\dagger_{\boldsymbol{k}'\sigma}\hat{c}_{\boldsymbol{k}\sigma},$$

$$g_{\boldsymbol{k}'\boldsymbol{k}\boldsymbol{q}} = \frac{\mathrm{i}NZe^2}{\varepsilon_0\mathscr{V}}\left(\frac{\hbar}{2N\omega_{\boldsymbol{q}}}\right)^{1/2}\frac{\boldsymbol{q}}{q^2}\cdot\left[\frac{\boldsymbol{\epsilon}_+(\boldsymbol{q})}{\sqrt{M_+}} - \frac{\boldsymbol{\epsilon}_-(\boldsymbol{q})}{\sqrt{M_-}}\right]\int \mathrm{d}\boldsymbol{r}\psi^*_{\boldsymbol{k}'}(\boldsymbol{r})\psi_{\boldsymbol{k}}(\boldsymbol{r})\mathrm{e}^{\mathrm{i}\boldsymbol{q}\cdot\boldsymbol{r}}.$$

(4) *Deformation potential model*

$$\hat{H}_{\rm ep} = \sum_{\boldsymbol{k}'\boldsymbol{k}\boldsymbol{q}s\sigma} g_{\boldsymbol{k}'\boldsymbol{k}\boldsymbol{q}s}\big(\hat{a}_{\boldsymbol{q}s} + \hat{a}^\dagger_{-\boldsymbol{q}s}\big)\hat{c}^\dagger_{\boldsymbol{k}'\sigma}\hat{c}_{\boldsymbol{k}\sigma},$$

$$g_{\boldsymbol{k}'\boldsymbol{k}\boldsymbol{q}s} = D_c\sum_\kappa\left(\frac{\hbar}{2NM_\kappa\omega_{\boldsymbol{q}s}}\right)^{1/2}\mathrm{i}\boldsymbol{q}\cdot\boldsymbol{\epsilon}_{\kappa s}(\boldsymbol{q})\int \mathrm{d}\boldsymbol{r}\;\psi^*_{\boldsymbol{k}'}(\boldsymbol{r})\psi_{\boldsymbol{k}}(\boldsymbol{r})\mathrm{e}^{\mathrm{i}\boldsymbol{q}\cdot\boldsymbol{r}}.$$

(5) *Electron self-energy in a metal*

$$\Sigma(\boldsymbol{p}, E) = \sum_{\boldsymbol{k}s} |g_{\boldsymbol{k}-\boldsymbol{p},s}|^2 \left[\frac{f_{\boldsymbol{k}} + n_{\boldsymbol{k}-\boldsymbol{p}}}{E - \varepsilon_{\boldsymbol{k}} + \hbar\omega_{\boldsymbol{k}-\boldsymbol{p}}} + \frac{1 - f_{\boldsymbol{k}} + n_{\boldsymbol{k}-\boldsymbol{p}}}{E - \varepsilon_{\boldsymbol{k}} - \hbar\omega_{\boldsymbol{k}-\boldsymbol{p}}} \right].$$

(6) *Electron-phonon coupling function* $\alpha^2 F$

$$\alpha^2 F(\omega) = \frac{\mathscr{V}}{(2\pi)^3 \hbar^2} \left(\int_{S_{\rm F}} \frac{{\rm d}^2 k}{v_{\rm F}} \right)^{-1} \sum_s \int_{S_{\rm F}} \frac{{\rm d}^2 k}{v_{\rm F}} \int_{S_{\rm F}} \frac{{\rm d}^2 k'}{v_{\rm F}} \times |g_{\boldsymbol{k}'-\boldsymbol{k},s}|^2 \delta(\omega - \omega_{\boldsymbol{k}'-\boldsymbol{k},s}).$$

(7) *Effective electron-phonon coupling constant*

$$\lambda = 2 \int_0^\infty {\rm d}\omega \, \frac{\alpha^2 F(\omega)}{\omega}.$$

(8) *Retarded electron self-energy*

$$\Sigma_R(\omega) = \int_{-\infty}^{\infty} {\rm d}\varepsilon \int_0^\infty {\rm d}(\hbar\omega') \, \alpha^2 F(\omega') \times \left[\frac{f(\varepsilon) + n(\hbar\omega')}{\hbar\omega - \varepsilon + \hbar\omega' + {\rm i}\delta} + \frac{1 - f(\varepsilon) + n(\hbar\omega')}{\hbar\omega - \varepsilon - \hbar\omega' + {\rm i}\delta} \right].$$

(9) *Effective electron-electron interaction*

$$\hat{H}_{\rm eff}^{ee} = \frac{1}{2} \sum_{\boldsymbol{k}\sigma\boldsymbol{k}'\sigma'\boldsymbol{q}s} \frac{2|g_{\boldsymbol{q},s}|^2 \hbar\omega_{\boldsymbol{q}s}}{(\varepsilon_{\boldsymbol{k}} - \varepsilon_{\boldsymbol{k}+\boldsymbol{q}})^2 - (\hbar\omega_{\boldsymbol{q}s})^2} \hat{c}^\dagger_{\boldsymbol{k}+\boldsymbol{q},\sigma} \hat{c}^\dagger_{\boldsymbol{k}'-\boldsymbol{q},\sigma'} \hat{c}_{\boldsymbol{k}'\sigma'} \hat{c}_{\boldsymbol{k}\sigma}.$$

(10) *Real-time Green's functions*

$$G_\sigma(\boldsymbol{k}, t - t') = -{\rm i} \, \langle \Omega | \, \hat{T} \big[\hat{\tilde{c}}_{\boldsymbol{k}\sigma}(t) \hat{\tilde{c}}^\dagger_{\boldsymbol{k}\sigma}(t') \big] \, | \Omega \rangle,$$

$$D_s(\boldsymbol{q}, t - t') = -{\rm i} \, \langle \Omega | \, \hat{T} \big[\hat{\tilde{\phi}}_{\boldsymbol{q}s}(t) \hat{\tilde{\phi}}_{-\boldsymbol{q}s}(t') \big] \, | \Omega \rangle \, .$$

(11) *Zeroth-order Green's functions*

$$G_{\sigma 0}(\boldsymbol{k}, \omega) = \frac{1}{\hbar\omega - \varepsilon_{\boldsymbol{k}} + {\rm i}\delta},$$

$$D_{s0}(\boldsymbol{k}, \omega) = \frac{2\hbar\omega_{\boldsymbol{q}s}}{(\hbar\omega)^2 - (\hbar\omega_{\boldsymbol{q}s})^2 + {\rm i}\delta}.$$

(12) *Feynman rules for* $G_{\sigma 0}(\boldsymbol{k}, \omega)$, $D_{s0}(\boldsymbol{q}, \omega)$

$\boldsymbol{k}\sigma, \omega$	$G_{\sigma 0}(\boldsymbol{k}, \omega)$
$\boldsymbol{q}s, \omega$	$D_{s0}(\boldsymbol{q}, \omega)$
$\boldsymbol{q}s$ 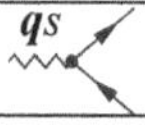	${\rm i}^{1/2} g_{\boldsymbol{q}s}$ or ${\rm i}^{1/2} g^*_{\boldsymbol{q}s}$
	-1
Internal variables	$\sum$ or $\int$

(13) *Dyson equation for electrons*

$$G_\sigma^{-1}(\boldsymbol{k},\omega) = G_{\sigma 0}^{-1}(\boldsymbol{k},\omega) - \Sigma_\sigma(\boldsymbol{k},\omega).$$

(14) *Dyson equation for phonons*

$$D_s^{-1}(\boldsymbol{q},\omega) = D_{s0}^{-1}(\boldsymbol{q},\omega) - \Pi_s(\boldsymbol{q},\omega).$$

(15) *Migdal's theorem*

$$\Gamma = g\left[1 + O\left(\sqrt{m/M}\right)\right].$$

(16) *Retarded and advanced Green's functions*

$$G_{R\sigma}(\boldsymbol{k},t-t') = -\mathrm{i}\,\langle\Omega|\left[\hat{\tilde{c}}_{\boldsymbol{k}\sigma}(t),\hat{\tilde{c}}^\dagger_{\boldsymbol{k}\sigma}(t')\right]_+|\Omega\rangle\,\theta(t-t'),$$

$$G_{A\sigma}(\boldsymbol{k},t-t') = i\,\langle\Omega|\left[\hat{\tilde{c}}_{\boldsymbol{k}\sigma}(t),\hat{\tilde{c}}^\dagger_{\boldsymbol{k}\sigma}(t')\right]_+|\Omega\rangle\,\theta(t'-t),$$

$$D_{Rs}(\boldsymbol{q},t-t') = -\mathrm{i}\,\langle\Omega|\left[\hat{\tilde{\phi}}_{\boldsymbol{q}s}(t),\hat{\tilde{\phi}}_{-\boldsymbol{q}s}(t')\right]|\Omega\rangle\,\theta(t-t'),$$

$$D_{As}(\boldsymbol{q},t-t') = -\mathrm{i}\,\langle\Omega|\left[\hat{\tilde{\phi}}_{-\boldsymbol{q}s}(t'),\hat{\tilde{\phi}}_{\boldsymbol{q}s}(t)\right]|\Omega\rangle\,\theta(t'-t).$$

(17) *Lehmann representations*

$$G_{R\sigma}(\boldsymbol{k},E) = -\frac{1}{\pi}\int_{-\infty}^{\infty}\mathrm{d}E'\,\frac{\mathrm{Im}\,G_{R\sigma}(\boldsymbol{k},E')}{E-E'+\mathrm{i}\delta},$$

$$D_{Rs}(\boldsymbol{q},\omega) = -\frac{1}{\pi}\int_{-\infty}^{\infty}\mathrm{d}\omega'\,\frac{\mathrm{Im}\,D_{Rs}(\boldsymbol{q},\omega')}{\omega-\omega'+\mathrm{i}\delta}.$$

(18) *Kramers-Kronig relations*

$$G'_{R\sigma}(\boldsymbol{k},E) = -\frac{1}{\pi}\,\mathrm{P}\int_{-\infty}^{\infty}\mathrm{d}E'\,\frac{G''_{R\sigma}(\boldsymbol{k},E')}{E-E'},$$

$$G''_{R\sigma}(\boldsymbol{k},E) = \frac{1}{\pi}\,\mathrm{P}\int_{-\infty}^{\infty}\mathrm{d}E'\,\frac{G'_{R\sigma}(\boldsymbol{k},E')}{E-E'},$$

$$D'_{Rs}(\boldsymbol{q},\omega) = -\frac{1}{\pi}\,\mathrm{P}\int_{-\infty}^{\infty}\mathrm{d}\omega'\,\frac{D''_{Rs}(\boldsymbol{q},\omega')}{\omega-\omega'},$$

$$D''_{Rs}(\boldsymbol{q},\omega) = \frac{1}{\pi}\,\mathrm{P}\int_{-\infty}^{\infty}\mathrm{d}\omega'\,\frac{D'_{Rs}(\boldsymbol{q},\omega')}{\omega-\omega'}.$$

(19) *Renormalization constant*

$$Z^{-1} = 1 - \left.\frac{\partial\,\mathrm{Re}\,\Sigma_{R\sigma}(\boldsymbol{k},E)}{\partial E}\right|_{\substack{k\,=\,k_\mathrm{F}\\ E\,=\,E_\mathrm{F}}}.$$

(20) *Effective mass of an electronic excitation*

$$\frac{m}{Zm^*} = 1 + \frac{m}{\hbar^2 k_\mathrm{F}}\left.\frac{\partial\,\mathrm{Re}\,\Sigma_{R\sigma}(\boldsymbol{k},E)}{\partial k}\right|_{\substack{k\,=\,k_\mathrm{F}\\ E\,=\,E_\mathrm{F}}}.$$

(21) *Coherent and incoherent parts*

$$G_{R\sigma}(\boldsymbol{k},E)=\frac{Z}{E-E_{\boldsymbol{k}\sigma}+i\Gamma_{\boldsymbol{k}\sigma}}+(1-Z)G^{\rm inc}_{R\sigma}(\boldsymbol{k},E).$$

(22) *Spectral function*

$$\mathscr{A}_\sigma(\boldsymbol{k},E)=-\frac{1}{\pi}\,{\rm Im}\,G_{R\sigma}(\boldsymbol{k},E)=\frac{1}{\pi}\frac{Z\Gamma_{\boldsymbol{k}\sigma}}{(E-E_{\boldsymbol{k}\sigma})^2+\Gamma^2_{\boldsymbol{k}\sigma}}.$$

(23) *Matsubara Green's functions*

$$\mathscr{G}_\sigma(\boldsymbol{k},\tau-\tau')=-\left\langle\hat{T}_\tau\left[\hat{\tilde{c}}_{\boldsymbol{k}\sigma}(\tau)\hat{\tilde{c}}^\dagger_{\boldsymbol{k}\sigma}(\tau')\right]\right\rangle,$$

$$\mathscr{D}_s(\boldsymbol{q},\tau-\tau')=-\left\langle\hat{T}_\tau\left[\hat{\tilde{\phi}}_{\boldsymbol{q}s}(\tau)\hat{\tilde{\phi}}_{-\boldsymbol{q}s}(\tau')\right]\right\rangle.$$

(24) *Zeroth-order Matsubara Green's functions*

$$\mathscr{G}_{\sigma0}(\boldsymbol{k},{\rm i}\omega_n)=\frac{1}{{\rm i}\hbar\omega_n-\xi_{\boldsymbol{k}\sigma}},$$

$$\mathscr{D}_{s0}(\boldsymbol{q},{\rm i}\omega_m)=\frac{2\hbar\omega_{\boldsymbol{q}s}}{({\rm i}\hbar\omega_m)^2-(\hbar\omega_{\boldsymbol{q}s})^2}.$$

(25) *Feynman rules for* $\mathscr{G}_\sigma(\boldsymbol{k},i\omega_n)$, $\mathscr{D}_s(\boldsymbol{q},i\omega_m)$

$\boldsymbol{k}\sigma, i\omega_n$ 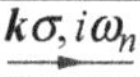	$\mathscr{G}_{\sigma0}(\boldsymbol{k},{\rm i}\omega_n)$
$\boldsymbol{q}s, i\omega_m$ 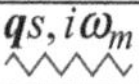	$\mathscr{D}_{s0}(\boldsymbol{q},{\rm i}\omega_m)$
$\boldsymbol{q}s$ 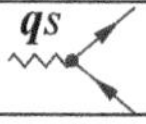	${\rm i}g_{\boldsymbol{q}s}$ or ${\rm i}g^*_{\boldsymbol{q}s}$
	-1
Internal variables	$\sum$

(26) *Dyson equations*

$$\mathscr{G}^{-1}_\sigma(\boldsymbol{k},{\rm i}\omega_n)=\mathscr{G}^{-1}_{\sigma0}(\boldsymbol{k},{\rm i}\omega_n)-\Sigma_\sigma(\boldsymbol{k},{\rm i}\omega_n),$$

$$\mathscr{D}^{-1}_s(\boldsymbol{q},{\rm i}\omega_m)=\mathscr{D}^{-1}_{s0}(\boldsymbol{q},{\rm i}\omega_m)-\Pi_s(\boldsymbol{q},{\rm i}\omega_m).$$

(27) *Spectral representations*

$$\mathscr{G}_\sigma(\boldsymbol{k},{\rm i}\omega_n)=\int_{-\infty}^{\infty}{\rm d}E\,\frac{A_\sigma(\boldsymbol{k},E)}{{\rm i}\hbar\omega_n-E},\quad\mathscr{D}_s(\boldsymbol{q},{\rm i}\omega_m)=\int_{-\infty}^{\infty}{\rm d}(\hbar\omega)\,\frac{B_s(\boldsymbol{q},\omega)}{{\rm i}\hbar\omega_m-\hbar\omega}.$$

26-1 Perturbation computations for the electron-phonon system. Consider an electron-phonon system. The Hamiltonian of the system is given by

$$\hat{H}=\hat{H}_0+\hat{H}_{\rm ep},\ \ \hat{H}_0=\hat{H}_{\rm el}+\hat{H}_{\rm ph},\ \ \hat{H}_{\rm el}=\sum_{\boldsymbol{k}\sigma}\xi_{\boldsymbol{k}}\hat{c}^\dagger_{\boldsymbol{k}\sigma}\hat{c}_{\boldsymbol{k}\sigma},$$

$$\hat{H}_{\rm ph}=\sum_{\boldsymbol{q}s}\hbar\omega_{\boldsymbol{q}s}\hat{a}^\dagger_{\boldsymbol{q}s}\hat{a}_{\boldsymbol{q}s},\ \ \hat{H}_{\rm ep}=\sum_{\boldsymbol{k}\boldsymbol{q}s\sigma}g_{\boldsymbol{q}s}\hat{c}^\dagger_{\boldsymbol{k}+\boldsymbol{q},\sigma}\hat{c}_{\boldsymbol{k}\sigma}(\hat{a}_{\boldsymbol{q}s}+\hat{a}^\dagger_{-\boldsymbol{q}s}),$$

where $\xi_{\boldsymbol{k}} = \varepsilon_{\boldsymbol{k}} - \mu$, with μ the chemical potential, is the electron kinetic energy measured relative to the chemical potential. For brevity, we suppress the electron spin and consider only one longitudinal acoustical branch of phonons. Let $|\Psi\rangle$ be a state of the system characterized by a set of electron occupation numbers $\{n_{\boldsymbol{k}}\}$ and a set of phonon occupation numbers $\{N_{\boldsymbol{q}}\}$,

$$|\Psi\rangle = |n_{\boldsymbol{k}_1}, n_{\boldsymbol{k}_2}, \cdots ; N_{\boldsymbol{q}_1}, N_{\boldsymbol{q}_2}, \cdots\rangle .$$

Up to the second order in the electron-phonon interaction, the total energy of the system in the state $|\Psi\rangle$ is given by

$$E = E_0 + \langle\Psi|\hat{H}_{\rm ep}|\Psi\rangle + \langle\Psi|\hat{H}_{\rm ep}(E_0 - \hat{H}_0)^{-1}\hat{H}_{\rm ep}|\Psi\rangle,\ E_0 = \langle\Psi|\hat{H}_0|\Psi\rangle .$$

where $E_0 = \langle\Psi|\hat{H}_0|\Psi\rangle$.

(1) Show that the energy of the system in the state $|\Psi\rangle$ is given by

$$E = E_0 + \sum_{\boldsymbol{kq}} |g_{\boldsymbol{q}}|^2 n_{\boldsymbol{k}}(1 - n_{\boldsymbol{k}+\boldsymbol{q}}) \times \left(\frac{N_{\boldsymbol{q}}}{\varepsilon_{\boldsymbol{k}} - \varepsilon_{\boldsymbol{k}+\boldsymbol{q}} + \hbar\omega_{\boldsymbol{q}}} + \frac{N_{-\boldsymbol{q}} + 1}{\varepsilon_{\boldsymbol{k}} - \varepsilon_{\boldsymbol{k}+\boldsymbol{q}} - \hbar\omega_{\boldsymbol{q}}}\right).$$

Note that $\omega_{-\boldsymbol{q}} = \omega_{\boldsymbol{q}}$.

(2) Average E in the grand canonical ensemble specified by the Boltzmann distribution $\mathrm{e}^{-\beta\hat{H}_0}$ and show that

$$\langle E\rangle = \langle E_0\rangle + \sum_{\boldsymbol{kq}} |g_{\boldsymbol{q}}|^2 \langle n_{\boldsymbol{k}}\rangle(1 - \langle n_{\boldsymbol{k}+\boldsymbol{q}}\rangle) \times \left(\frac{\langle N_{\boldsymbol{q}}\rangle}{\varepsilon_{\boldsymbol{k}} - \varepsilon_{\boldsymbol{k}+\boldsymbol{q}} + \hbar\omega_{\boldsymbol{q}}} + \frac{\langle N_{\boldsymbol{q}}\rangle + 1}{\varepsilon_{\boldsymbol{k}} - \varepsilon_{\boldsymbol{k}+\boldsymbol{q}} - \hbar\omega_{\boldsymbol{q}}}\right).$$

where $\langle -N_{\boldsymbol{q}}\rangle = \langle N_{\boldsymbol{q}}\rangle$ has been used.

(3) Show that $\langle E\rangle$ can be written as

$$\langle E\rangle = \langle E_0\rangle + \sum_{\boldsymbol{kq}} |g_{\boldsymbol{q}}|^2 \langle n_{\boldsymbol{k}}\rangle \times \left[\frac{2(\varepsilon_{\boldsymbol{k}} - \varepsilon_{\boldsymbol{k}+\boldsymbol{q}})\langle N_{\boldsymbol{q}}\rangle}{(\varepsilon_{\boldsymbol{k}} - \varepsilon_{\boldsymbol{k}+\boldsymbol{q}})^2 - (\hbar\omega_{\boldsymbol{q}})^2} + \frac{1 - \langle n_{\boldsymbol{k}+\boldsymbol{q}}\rangle}{\varepsilon_{\boldsymbol{k}} - \varepsilon_{\boldsymbol{k}+\boldsymbol{q}} - \hbar\omega_{\boldsymbol{q}}}\right].$$

(4) Compute the renormalized phonon energy $\hbar\bar{\omega}_{\boldsymbol{q}}$ from $\hbar\bar{\omega}_{\boldsymbol{q}} = \partial\langle E\rangle/\partial\langle N_{\boldsymbol{q}}\rangle$. Derive an expression for $\hbar\bar{\omega}_{\boldsymbol{q}}$ for $\hbar\omega_{\boldsymbol{q}} \ll |\varepsilon_{\boldsymbol{k}} - \varepsilon_{\boldsymbol{k}+\boldsymbol{q}}|$.

(5) Compute the renormalized electron energy $\bar{\varepsilon}_{\boldsymbol{k}}$ from $\bar{\varepsilon}_{\boldsymbol{k}} = \partial\langle E\rangle/\partial\langle n_{\boldsymbol{k}}\rangle$.

(1) With electron spins suppressed and for only one branch of acoustical phonons, the Hamiltonian of the system is written as

$$\hat{H} = \hat{H}_0 + \hat{H}_{\rm ep}, \quad \hat{H}_0 = \hat{H}_{\rm el} + \hat{H}_{\rm ph},$$
$$\hat{H}_{\rm el} = \sum_{\boldsymbol{k}} \xi_{\boldsymbol{k}} \hat{c}^\dagger_{\boldsymbol{k}} \hat{c}_{\boldsymbol{k}}, \ \hat{H}_{\rm ph} = \sum_{\boldsymbol{q}} \hbar\omega_{\boldsymbol{q}} \hat{a}^\dagger_{\boldsymbol{q}} \hat{a}_{\boldsymbol{q}},$$
$$\hat{H}_{\rm ep} = \sum_{\boldsymbol{kq}} g_{\boldsymbol{q}} \hat{c}^\dagger_{\boldsymbol{k}+\boldsymbol{q}} \hat{c}_{\boldsymbol{k}} (\hat{a}_{\boldsymbol{q}} + \hat{a}^\dagger_{-\boldsymbol{q}}).$$

Since $\hat{H}_{\rm ep}$ contains only one phonon operator in each term, its average in $|\Psi\rangle$ is zero, $\langle\Psi|\hat{H}_{\rm ep}|\Psi\rangle = 0$. To evaluate the second-order term, we first consider $\hat{H}_{\rm ep}|\Psi\rangle$. Making use of $\hat{H}_{\rm ep} = \sum_{\boldsymbol{kq}} g_{\boldsymbol{q}} \hat{c}^\dagger_{\boldsymbol{k}+\boldsymbol{q}} \hat{c}_{\boldsymbol{k}} (\hat{a}_{\boldsymbol{q}} + \hat{a}^\dagger_{-\boldsymbol{q}})$, we have

$$\begin{aligned}\hat{H}_{\rm ep}|\Psi\rangle &= \sum_{\boldsymbol{k}'\boldsymbol{q}'} g_{\boldsymbol{q}'} \hat{c}^\dagger_{\boldsymbol{k}'+\boldsymbol{q}'} \hat{c}_{\boldsymbol{k}'} (\hat{a}_{\boldsymbol{q}'} + \hat{a}^\dagger_{-\boldsymbol{q}'}) |n_{\boldsymbol{k}_1}, n_{\boldsymbol{k}_2}, \cdots; N_{\boldsymbol{q}_1}, N_{\boldsymbol{q}_2}, \cdots\rangle \\ &= \sum_{\ell} \sum_{\boldsymbol{k}'\boldsymbol{q}'} g_{\boldsymbol{q}'} \hat{c}^\dagger_{\boldsymbol{k}'+\boldsymbol{q}'} \hat{c}_{\boldsymbol{k}'} \\ &\quad \times \big[(N_{\boldsymbol{q}_\ell})^{1/2} |n_{\boldsymbol{k}_1}, n_{\boldsymbol{k}_2}, \cdots; \cdots, N_{\boldsymbol{q}_\ell} - 1, \cdots\rangle \, \delta_{\boldsymbol{q}'\boldsymbol{q}_\ell} \\ &\quad + (N_{\boldsymbol{q}_\ell} + 1)^{1/2} |n_{\boldsymbol{k}_1}, n_{\boldsymbol{k}_2}, \cdots; \cdots, N_{\boldsymbol{q}_\ell} + 1, \cdots\rangle \, \delta_{\boldsymbol{q}',-\boldsymbol{q}_\ell} \big],\end{aligned}$$

where we have evaluated the effects of the phonon operators on the state. Evaluating the effects of the electron operators on the state, we have

$$\begin{aligned}\hat{H}_{\rm ep}|\Psi\rangle &= \sum_{i<j} \sum_{\ell} \sum_{\boldsymbol{k}'\boldsymbol{q}'} g_{\boldsymbol{q}'} \big[(N_{\boldsymbol{q}_\ell})^{1/2} (-1)^{\sum_{\alpha=1}^{i-1} n_{\boldsymbol{k}_\alpha}} (-1)^{\sum_{\beta=1}^{j-1} n_{\boldsymbol{k}_\alpha} - 1} \\ &\quad \times n_{\boldsymbol{k}_i} (1 - n_{\boldsymbol{k}_j}) \, | \cdots, n_{\boldsymbol{k}_i} - 1, \cdots, n_{\boldsymbol{k}_j} + 1, \cdots; \cdots, N_{\boldsymbol{q}_\ell} - 1, \cdots\rangle \\ &\quad \times \delta_{\boldsymbol{q}'\boldsymbol{q}_\ell} \delta_{\boldsymbol{k}'\boldsymbol{k}_i} \delta_{\boldsymbol{k}',\boldsymbol{k}_j - \boldsymbol{q}'} \big] \\ &\quad + \big[(N_{\boldsymbol{q}_\ell} + 1)^{1/2} (-1)^{\sum_{\alpha=1}^{j-1} n_{\boldsymbol{k}_\alpha}} (-1)^{\sum_{\beta=1}^{i-1} n_{\boldsymbol{k}_\alpha}} (1 - n_{\boldsymbol{k}_i}) n_{\boldsymbol{k}_j} \\ &\quad \times | \cdots, n_{\boldsymbol{k}_i} + 1, \cdots, n_{\boldsymbol{k}_j} - 1, \cdots; \cdots, N_{\boldsymbol{q}_\ell} + 1, \cdots\rangle \\ &\quad \times \delta_{\boldsymbol{q}',-\boldsymbol{q}_\ell} \delta_{\boldsymbol{k}'\boldsymbol{k}_j} \delta_{\boldsymbol{k}',\boldsymbol{k}_i - \boldsymbol{q}'} \big],\end{aligned}$$

where we have omitted the terms from $\boldsymbol{q}' = 0$ since they are zero due to the fact that $g_{\boldsymbol{q}'=0} = 0$. The phase factors in the above equation can be combined and simplified and the summations over $\boldsymbol{k}'$ and $\boldsymbol{q}'$ can be performed by consuming two δ-symbols. We then have

$$\begin{aligned}\hat{H}_{\rm ep}|\Psi\rangle &= \sum_{i<j} (-1)^{\sum_{\alpha=i}^{j-1} n_{\boldsymbol{k}_\alpha}} \sum_{\ell} \delta_{\boldsymbol{k}_j,\boldsymbol{k}_i+\boldsymbol{q}_\ell} \big[-(N_{\boldsymbol{q}_\ell})^{1/2} n_{\boldsymbol{k}_i} (1 - n_{\boldsymbol{k}_j}) g_{\boldsymbol{q}_\ell} \\ &\qquad \times | \cdots, n_{\boldsymbol{k}_i} - 1, \cdots, n_{\boldsymbol{k}_j} + 1, \cdots; \cdots, N_{\boldsymbol{q}_\ell} - 1, \cdots\rangle\end{aligned}$$

$$+ (N_{\boldsymbol{q}_\ell}+1)^{1/2}(1-n_{\boldsymbol{k}_i})n_{\boldsymbol{k}_j}g_{-\boldsymbol{q}_\ell} \times|\cdots, n_{\boldsymbol{k}_i}+1, \cdots, n_{\boldsymbol{k}_j}-1, \cdots; \cdots, N_{\boldsymbol{q}_\ell}+1, \cdots\rangle\Big].$$

Since the state vectors in the above equation are the eigenvectors of $\hat{H}_0$, the effect of $(E_0 - \hat{H}_0)^{-1}$ on them is to have them multiplied by a factor of the form $1/\Delta E$, where ΔE is the energy difference between $|\Psi\rangle$ and the corresponding state vector in the above equation. For the first state vector, we have $\Delta E_1 = \xi_{\boldsymbol{k}_i} - \xi_{\boldsymbol{k}_j} + \hbar\omega_{\boldsymbol{q}_\ell}$; for the second state vector, we have $\Delta E_2 = -\xi_{\boldsymbol{k}_i} + \xi_{\boldsymbol{k}_j} - \hbar\omega_{\boldsymbol{q}_\ell}$. With the δ-symbol $\delta_{\boldsymbol{k}_j, \boldsymbol{k}_i+\boldsymbol{q}_\ell}$ taken into consideration, we find that $(E_0 - \hat{H}_0)^{-1}\hat{H}_{\rm ep}|\Psi\rangle$ is given by

$$\begin{aligned}
&(E_0 - \hat{H}_0)^{-1}\hat{H}_{\rm ep}|\Psi\rangle \\
&\quad = \sum_{i<j}(-1)^{\sum_{\alpha=i}^{j-1} n_{\boldsymbol{k}_\alpha}} \sum_\ell \delta_{\boldsymbol{k}_j, \boldsymbol{k}_i+\boldsymbol{q}_\ell}\bigg[\frac{(N_{\boldsymbol{q}_\ell})^{1/2} n_{\boldsymbol{k}_i}(1-n_{\boldsymbol{k}_j})g_{\boldsymbol{q}_\ell}}{\xi_{\boldsymbol{k}_i+\boldsymbol{q}_\ell} - \xi_{\boldsymbol{k}_i} - \hbar\omega_{\boldsymbol{q}_\ell}} \\
&\qquad \times|\cdots, n_{\boldsymbol{k}_i}-1, \cdots, n_{\boldsymbol{k}_j}+1, \cdots; \cdots, N_{\boldsymbol{q}_\ell}-1, \cdots\rangle \\
&\qquad + \frac{(N_{\boldsymbol{q}_\ell}+1)^{1/2}(1-n_{\boldsymbol{k}_i})n_{\boldsymbol{k}_j}g_{-\boldsymbol{q}_\ell}}{\xi_{\boldsymbol{k}_i+\boldsymbol{q}_\ell} - \xi_{\boldsymbol{k}_i} - \hbar\omega_{\boldsymbol{q}_\ell}} \\
&\qquad \times|\cdots, n_{\boldsymbol{k}_i}+1, \cdots, n_{\boldsymbol{k}_j}-1, \cdots; \cdots, N_{\boldsymbol{q}_\ell}+1, \cdots\rangle\bigg].
\end{aligned}$$

Taking the inner product of the above equation with $\hat{H}_{\rm ep}|\Psi\rangle$ and making use of the orthonormality relation of the eigenvectors of $\hat{H}_0$, we have

$$\begin{aligned}
&\langle\Psi|\,\hat{H}_{\rm ep}(E_0 - \hat{H}_0)^{-1}\hat{H}_{\rm ep}|\Psi\rangle \\
&\quad = \sum_{i<j}\sum_\ell \delta_{\boldsymbol{k}_j, \boldsymbol{k}_i+\boldsymbol{q}_\ell}\bigg[\frac{N_{\boldsymbol{q}_\ell} n_{\boldsymbol{k}_i}(1-n_{\boldsymbol{k}_j})|g_{\boldsymbol{q}_\ell}|^2}{\xi_{\boldsymbol{k}_i} - \xi_{\boldsymbol{k}_i+\boldsymbol{q}_\ell} + \hbar\omega_{\boldsymbol{q}_\ell}} \\
&\qquad + \frac{(N_{\boldsymbol{q}_\ell}+1)(1-n_{\boldsymbol{k}_i})n_{\boldsymbol{k}_j}|g_{-\boldsymbol{q}_\ell}|^2}{\xi_{\boldsymbol{k}_i+\boldsymbol{q}_\ell} - \xi_{\boldsymbol{k}_i} - \hbar\omega_{\boldsymbol{q}_\ell}}\bigg],
\end{aligned}$$

where we have made use of $n_{\boldsymbol{k}_i}^2 = n_{\boldsymbol{k}_i}$ and $(1 - n_{\boldsymbol{k}_i})^2 = 1 - n_{\boldsymbol{k}_i}$. Performing the summation over j using the remaining δ-symbol and making changes of dummy summation variables, we obtain

$$\begin{aligned}
&\langle\Psi|\,\hat{H}_{\rm ep}(E_0 - \hat{H}_0)^{-1}\hat{H}_{\rm ep}|\Psi\rangle \\
&\quad = \sum_i\sum_\ell \bigg[\frac{N_{\boldsymbol{q}_\ell} n_{\boldsymbol{k}_i}(1-n_{\boldsymbol{k}_i+\boldsymbol{q}_\ell})|g_{\boldsymbol{q}_\ell}|^2}{\xi_{\boldsymbol{k}_i} - \xi_{\boldsymbol{k}_i+\boldsymbol{q}_\ell} + \hbar\omega_{\boldsymbol{q}_\ell}} \\
&\qquad + \frac{(N_{\boldsymbol{q}_\ell}+1)(1-n_{\boldsymbol{k}_i})n_{\boldsymbol{k}_i+\boldsymbol{q}_\ell}|g_{-\boldsymbol{q}_\ell}|^2}{\xi_{\boldsymbol{k}_i+\boldsymbol{q}_\ell} - \xi_{\boldsymbol{k}_i} - \hbar\omega_{\boldsymbol{q}_\ell}}\bigg]
\end{aligned}$$

$$= \sum_{\boldsymbol{kq}} \left[\frac{N_{\boldsymbol{q}} n_{\boldsymbol{k}}(1-n_{\boldsymbol{k}+\boldsymbol{q}})|g_{\boldsymbol{q}}|^2}{\xi_{\boldsymbol{k}} - \xi_{\boldsymbol{k}+\boldsymbol{q}} + \hbar\omega_{\boldsymbol{q}}} + \frac{(N_{\boldsymbol{q}}+1)(1-n_{\boldsymbol{k}})n_{\boldsymbol{k}+\boldsymbol{q}}|g_{-\boldsymbol{q}}|^2}{\xi_{\boldsymbol{k}+\boldsymbol{q}} - \xi_{\boldsymbol{k}} - \hbar\omega_{\boldsymbol{q}}} \right].$$

For the second term in the square brackets, we first let $\boldsymbol{k}' = \boldsymbol{k} + \boldsymbol{q}$, then let $\boldsymbol{q}' = -\boldsymbol{q}$, and then drop the primes on the dummy summation variables. So doing, we can convert the original factor $(1-n_{\boldsymbol{k}})n_{\boldsymbol{k}+\boldsymbol{q}}$ to $(1-n_{\boldsymbol{k}+\boldsymbol{q}})n_{\boldsymbol{k}}$ that also appears in the first term. We then have

$$\langle\Psi|\, \hat{H}_{\rm ep}(E_0 - \hat{H}_0)^{-1}\hat{H}_{\rm ep}|\Psi\rangle$$
$$= \sum_{\boldsymbol{kq}} |g_{\boldsymbol{q}}|^2 n_{\boldsymbol{k}}(1-n_{\boldsymbol{k}+\boldsymbol{q}}) \left(\frac{N_{\boldsymbol{q}}}{\xi_{\boldsymbol{k}} - \xi_{\boldsymbol{k}+\boldsymbol{q}} + \hbar\omega_{\boldsymbol{q}}} + \frac{N_{-\boldsymbol{q}}+1}{\xi_{\boldsymbol{k}} - \xi_{\boldsymbol{k}+\boldsymbol{q}} - \hbar\omega_{\boldsymbol{q}}} \right),$$

where we have made use of $\omega_{-\boldsymbol{q}} = \omega_{\boldsymbol{q}}$. Finally, up to the second order in the electron-phonon interaction, the total energy of the system in the state $|\Psi\rangle$ is given by

$$E = E_0$$
$$+ \sum_{\boldsymbol{kq}} |g_{\boldsymbol{q}}|^2 n_{\boldsymbol{k}}(1-n_{\boldsymbol{k}+\boldsymbol{q}}) \left(\frac{N_{\boldsymbol{q}}}{\xi_{\boldsymbol{k}} - \xi_{\boldsymbol{k}+\boldsymbol{q}} + \hbar\omega_{\boldsymbol{q}}} + \frac{N_{-\boldsymbol{q}}+1}{\xi_{\boldsymbol{k}} - \xi_{\boldsymbol{k}+\boldsymbol{q}} - \hbar\omega_{\boldsymbol{q}}} \right).$$

Since $\xi_{\boldsymbol{k}} - \xi_{\boldsymbol{k}+\boldsymbol{q}} = \varepsilon_{\boldsymbol{k}} - \varepsilon_{\boldsymbol{k}+\boldsymbol{q}}$, we can replace $\xi_{\boldsymbol{k}} - \xi_{\boldsymbol{k}+\boldsymbol{q}}$ with $\varepsilon_{\boldsymbol{k}} - \varepsilon_{\boldsymbol{k}+\boldsymbol{q}}$ in the above equation. We then have

$$E = E_0$$
$$+ \sum_{\boldsymbol{kq}} |g_{\boldsymbol{q}}|^2 n_{\boldsymbol{k}}(1-n_{\boldsymbol{k}+\boldsymbol{q}}) \left(\frac{N_{\boldsymbol{q}}}{\varepsilon_{\boldsymbol{k}} - \varepsilon_{\boldsymbol{k}+\boldsymbol{q}} + \hbar\omega_{\boldsymbol{q}}} + \frac{N_{-\boldsymbol{q}}+1}{\varepsilon_{\boldsymbol{k}} - \varepsilon_{\boldsymbol{k}+\boldsymbol{q}} - \hbar\omega_{\boldsymbol{q}}} \right).$$

(2) The average can be performed separately for the electron and phonon subsystems since there is no coupling between them in $\hat{H}_0$. Note that the contribution from $\boldsymbol{q} = 0$ is zero. For the electron subsystem, with $\boldsymbol{q} \neq 0$ we have

$$\frac{\sum_{\{n_{\boldsymbol{k}}\}} n_{\boldsymbol{k}}(1-n_{\boldsymbol{k}+\boldsymbol{q}}) {\rm e}^{-\beta \sum_{\boldsymbol{k}} n_{\boldsymbol{k}}(\varepsilon_{\boldsymbol{k}}-\mu)}}{\sum_{\{n_{\boldsymbol{k}}\}} {\rm e}^{-\beta \sum_{\boldsymbol{k}} n_{\boldsymbol{k}}(\varepsilon_{\boldsymbol{k}}-\mu)}}$$
$$= \frac{\sum_{n_{\boldsymbol{k}}} n_{\boldsymbol{k}} {\rm e}^{-\beta n_{\boldsymbol{k}}(\varepsilon_{\boldsymbol{k}}-\mu)}}{\sum_{n_{\boldsymbol{k}}} {\rm e}^{-\beta n_{\boldsymbol{k}}(\varepsilon_{\boldsymbol{k}}-\mu)}}$$
$$- \frac{\sum_{n_{\boldsymbol{k}} n_{\boldsymbol{k}+\boldsymbol{q}}} n_{\boldsymbol{k}} n_{\boldsymbol{k}+\boldsymbol{q}} {\rm e}^{-\beta(n_{\boldsymbol{k}}(\varepsilon_{\boldsymbol{k}}-\mu) + n_{\boldsymbol{k}+\boldsymbol{q}}(\varepsilon_{\boldsymbol{k}+\boldsymbol{q}}-\mu))}}{\sum_{n_{\boldsymbol{k}} n_{\boldsymbol{k}+\boldsymbol{q}}} {\rm e}^{-\beta(n_{\boldsymbol{k}}(\varepsilon_{\boldsymbol{k}}-\mu) + n_{\boldsymbol{k}+\boldsymbol{q}}(\varepsilon_{\boldsymbol{k}+\boldsymbol{q}}-\mu))}}$$
$$= \frac{\sum_{n_{\boldsymbol{k}}} n_{\boldsymbol{k}} {\rm e}^{-\beta n_{\boldsymbol{k}}(\varepsilon_{\boldsymbol{k}}-\mu)}}{\sum_{n_{\boldsymbol{k}}} {\rm e}^{-\beta n_{\boldsymbol{k}}(\varepsilon_{\boldsymbol{k}}-\mu)}} \left[1 - \frac{\sum_{n_{\boldsymbol{k}+\boldsymbol{q}}} n_{\boldsymbol{k}+\boldsymbol{q}} {\rm e}^{-\beta n_{\boldsymbol{k}+\boldsymbol{q}}(\varepsilon_{\boldsymbol{k}+\boldsymbol{q}}-\mu)}}{\sum_{n_{\boldsymbol{k}+\boldsymbol{q}}} {\rm e}^{-\beta n_{\boldsymbol{k}+\boldsymbol{q}}(\varepsilon_{\boldsymbol{k}+\boldsymbol{q}}-\mu)}} \right]$$

$$= \frac{e^{-\beta(\varepsilon_{\boldsymbol{k}}-\mu)}}{1+e^{-\beta(\varepsilon_{\boldsymbol{k}}-\mu)}}\left[1-\frac{e^{-\beta(\varepsilon_{\boldsymbol{k}+\boldsymbol{q}}-\mu)}}{1+e^{-\beta(\varepsilon_{\boldsymbol{k}+\boldsymbol{q}}-\mu)}}\right] = \langle n_{\boldsymbol{k}}\rangle(1-\langle n_{\boldsymbol{k}+\boldsymbol{q}}\rangle),$$

where $\langle n_{\boldsymbol{k}}\rangle = 1/[e^{\beta(\varepsilon_{\boldsymbol{k}}-\mu)}+1] = n_{\rm F}(\varepsilon_{\boldsymbol{k}})$ with $n_{\rm F}(\varepsilon_{\boldsymbol{k}})$ the Fermi-Dirac distribution function. For the phonon subsystem, we have

$$\begin{aligned}\frac{\sum_{\{N_{\boldsymbol{q}}\}} N_{\boldsymbol{q}} e^{-\beta\sum_{\boldsymbol{k}} N_{\boldsymbol{q}}\hbar\omega_{\boldsymbol{q}}}}{\sum_{\{N_{\boldsymbol{q}}\}} e^{-\beta\sum_{\boldsymbol{k}} N_{\boldsymbol{q}}\hbar\omega_{\boldsymbol{q}}}} &= \frac{\sum_{N_{\boldsymbol{q}}} N_{\boldsymbol{q}} e^{-\beta N_{\boldsymbol{q}}\hbar\omega_{\boldsymbol{q}}}}{\sum_{N_{\boldsymbol{q}}} e^{-\beta N_{\boldsymbol{q}}\hbar\omega_{\boldsymbol{q}}}} = \frac{e^{-\beta\hbar\omega_{\boldsymbol{q}}}/(1-e^{-\beta\hbar\omega_{\boldsymbol{q}}})^2}{1/(1-e^{-\beta\hbar\omega_{\boldsymbol{q}}})} \\ &= \frac{1}{e^{\beta\hbar\omega_{\boldsymbol{q}}}-1} = \langle N_{\boldsymbol{q}}\rangle,\end{aligned}$$

where $\langle N_{\boldsymbol{q}}\rangle = 1/(e^{\beta\hbar\omega_{\boldsymbol{q}}}-1) = n_{\rm B}(\hbar\omega_{\boldsymbol{q}})$ with $n_{\rm B}(\hbar\omega_{\boldsymbol{q}})$ the Bose-Einstein distribution function. From the expression of $\langle N_{\boldsymbol{q}}\rangle$, we see that $\langle N_{-\boldsymbol{q}}\rangle = \langle N_{\boldsymbol{q}}\rangle$. Thus,

$$E = E_0 + \sum_{\boldsymbol{kq}} |g_{\boldsymbol{q}}|^2 \langle n_{\boldsymbol{k}}\rangle(1-\langle n_{\boldsymbol{k}+\boldsymbol{q}}\rangle)\left(\frac{\langle N_{\boldsymbol{q}}\rangle}{\varepsilon_{\boldsymbol{k}}-\varepsilon_{\boldsymbol{k}+\boldsymbol{q}}+\hbar\omega_{\boldsymbol{q}}} + \frac{\langle N_{\boldsymbol{q}}\rangle+1}{\varepsilon_{\boldsymbol{k}}-\varepsilon_{\boldsymbol{k}+\boldsymbol{q}}-\hbar\omega_{\boldsymbol{q}}}\right).$$

(3) For the convenience of manipulating the above expression, we change the dummy summation variables from $\boldsymbol{q}$ to $\boldsymbol{k}' = \boldsymbol{k}+\boldsymbol{q}$. We then write $\boldsymbol{q}$ as $\boldsymbol{k}'-\boldsymbol{k}$. We have

$$E = E_0 + \sum_{\boldsymbol{kk}'} |g_{\boldsymbol{k}'-\boldsymbol{k}}|^2 \langle n_{\boldsymbol{k}}\rangle(1-\langle n_{\boldsymbol{k}'}\rangle)\left(\frac{\langle N_{\boldsymbol{k}'-\boldsymbol{k}}\rangle}{\varepsilon_{\boldsymbol{k}}-\varepsilon_{\boldsymbol{k}'}+\hbar\omega_{\boldsymbol{k}'-\boldsymbol{k}}} + \frac{\langle N_{\boldsymbol{k}'-\boldsymbol{k}}\rangle+1}{\varepsilon_{\boldsymbol{k}}-\varepsilon_{\boldsymbol{k}'}-\hbar\omega_{\boldsymbol{k}'-\boldsymbol{k}}}\right).$$

We now switch the dummy summation variables $\boldsymbol{k}$ and $\boldsymbol{k}'$. The resultant expression yields the same value as the original one. We add them up and divide the result by two to recover the correct value for E. We have

$$\begin{aligned}E = E_0 + \frac{1}{2}\sum_{\boldsymbol{kk}'}\Bigg\{&|g_{\boldsymbol{k}'-\boldsymbol{k}}|^2 \langle n_{\boldsymbol{k}}\rangle(1-\langle n_{\boldsymbol{k}'}\rangle) \\ &\times\left[\frac{\langle N_{\boldsymbol{k}'-\boldsymbol{k}}\rangle}{\varepsilon_{\boldsymbol{k}}-\varepsilon_{\boldsymbol{k}'}+\hbar\omega_{\boldsymbol{k}'-\boldsymbol{k}}} + \frac{\langle N_{\boldsymbol{k}'-\boldsymbol{k}}\rangle+1}{\varepsilon_{\boldsymbol{k}}-\varepsilon_{\boldsymbol{k}'}-\hbar\omega_{\boldsymbol{k}'-\boldsymbol{k}}}\right] \\ &+ |g_{\boldsymbol{k}-\boldsymbol{k}'}|^2 \langle n_{\boldsymbol{k}'}\rangle(1-\langle n_{\boldsymbol{k}}\rangle) \\ &\times\left[\frac{\langle N_{\boldsymbol{k}-\boldsymbol{k}'}\rangle}{\varepsilon_{\boldsymbol{k}'}-\varepsilon_{\boldsymbol{k}}+\hbar\omega_{\boldsymbol{k}-\boldsymbol{k}'}} + \frac{\langle N_{\boldsymbol{k}-\boldsymbol{k}'}\rangle+1}{\varepsilon_{\boldsymbol{k}'}-\varepsilon_{\boldsymbol{k}}-\hbar\omega_{\boldsymbol{k}-\boldsymbol{k}'}}\right]\Bigg\}.\end{aligned}$$

Making use of $g_{\boldsymbol{k}-\boldsymbol{k}'} = g^*_{\boldsymbol{k}'-\boldsymbol{k}}$ and $\omega_{\boldsymbol{k}-\boldsymbol{k}'} = \omega_{\boldsymbol{k}'-\boldsymbol{k}}$, we have

$$\begin{aligned}
E - E_0 &= \frac{1}{2}\sum_{\boldsymbol{k}\boldsymbol{k}'} |g_{\boldsymbol{k}'-\boldsymbol{k}}|^2 \left\{ \left[\frac{2(\varepsilon_{\boldsymbol{k}} - \varepsilon_{\boldsymbol{k}'})(\langle n_{\boldsymbol{k}}\rangle - \langle n_{\boldsymbol{k}'}\rangle)\langle N_{\boldsymbol{k}'-\boldsymbol{k}}\rangle}{(\varepsilon_{\boldsymbol{k}} - \varepsilon_{\boldsymbol{k}'})^2 - (\hbar\omega_{\boldsymbol{k}'-\boldsymbol{k}})^2} \right] \right. \\
&\qquad \left. + \left[\frac{\langle n_{\boldsymbol{k}}\rangle(1 - \langle n_{\boldsymbol{k}'}\rangle)}{\varepsilon_{\boldsymbol{k}} - \varepsilon_{\boldsymbol{k}'} - \hbar\omega_{\boldsymbol{k}'-\boldsymbol{k}}} + \frac{\langle n_{\boldsymbol{k}'}\rangle(1 - \langle n_{\boldsymbol{k}}\rangle)}{\varepsilon_{\boldsymbol{k}'} - \varepsilon_{\boldsymbol{k}} - \hbar\omega_{\boldsymbol{k}-\boldsymbol{k}'}} \right] \right\} \\
&= \sum_{\boldsymbol{k}\boldsymbol{k}'} |g_{\boldsymbol{k}'-\boldsymbol{k}}|^2 \langle n_{\boldsymbol{k}}\rangle \left[\frac{2(\varepsilon_{\boldsymbol{k}} - \varepsilon_{\boldsymbol{k}'})\langle N_{\boldsymbol{k}'-\boldsymbol{k}}\rangle}{(\varepsilon_{\boldsymbol{k}} - \varepsilon_{\boldsymbol{k}'})^2 - (\hbar\omega_{\boldsymbol{k}'-\boldsymbol{k}})^2} \right. \\
&\qquad \left. + \frac{1 - \langle n_{\boldsymbol{k}'}\rangle}{\varepsilon_{\boldsymbol{k}} - \varepsilon_{\boldsymbol{k}'} - \hbar\omega_{\boldsymbol{k}'-\boldsymbol{k}}} \right] \\
&= \sum_{\boldsymbol{k}\boldsymbol{q}} |g_{\boldsymbol{q}}|^2 \langle n_{\boldsymbol{k}}\rangle \left[\frac{2(\varepsilon_{\boldsymbol{k}} - \varepsilon_{\boldsymbol{k}+\boldsymbol{q}})\langle N_{\boldsymbol{q}}\rangle}{(\varepsilon_{\boldsymbol{k}} - \varepsilon_{\boldsymbol{k}+\boldsymbol{q}})^2 - (\hbar\omega_{\boldsymbol{q}})^2} + \frac{1 - \langle n_{\boldsymbol{k}+\boldsymbol{q}}\rangle}{\varepsilon_{\boldsymbol{k}} - \varepsilon_{\boldsymbol{k}+\boldsymbol{q}} - \hbar\omega_{\boldsymbol{q}}} \right],
\end{aligned}$$

where we have recovered the original dummy summation variable $\boldsymbol{q}$.

(4) The expression for E_0 can be obtained by evaluating $\langle\Psi|\hat{H}_0|\psi\rangle$ and then performing the thermal average. We find that

$$E_0 = \sum_{\boldsymbol{k}} \langle n_{\boldsymbol{k}}\rangle \varepsilon_{\boldsymbol{k}} + \sum_{\boldsymbol{q}} \langle N_{\boldsymbol{q}}\rangle \hbar\omega_{\boldsymbol{q}}.$$

Differentiating E with respect to $\langle N_{\boldsymbol{q}}\rangle$, we have

$$\begin{aligned}
\hbar\bar{\omega}_{\boldsymbol{q}} &= \frac{\partial E}{\partial\langle N_{\boldsymbol{q}}\rangle} \\
&= \hbar\omega_{\boldsymbol{q}} + \frac{\partial}{\partial\langle N_{\boldsymbol{q}}\rangle} \sum_{\boldsymbol{k}\boldsymbol{q}'} |g_{\boldsymbol{q}'}|^2 \langle n_{\boldsymbol{k}}\rangle \frac{2(\varepsilon_{\boldsymbol{k}} - \varepsilon_{\boldsymbol{k}+\boldsymbol{q}'})\langle N_{\boldsymbol{q}'}\rangle}{(\varepsilon_{\boldsymbol{k}} - \varepsilon_{\boldsymbol{k}+\boldsymbol{q}'})^2 - (\hbar\omega_{\boldsymbol{q}'})^2} \\
&= \hbar\omega_{\boldsymbol{q}} + \sum_{\boldsymbol{k}\boldsymbol{q}'} |g_{\boldsymbol{q}'}|^2 \langle n_{\boldsymbol{k}}\rangle \frac{2(\varepsilon_{\boldsymbol{k}} - \varepsilon_{\boldsymbol{k}+\boldsymbol{q}'})}{(\varepsilon_{\boldsymbol{k}} - \varepsilon_{\boldsymbol{k}+\boldsymbol{q}'})^2 - (\hbar\omega_{\boldsymbol{q}'})^2}\delta_{\boldsymbol{q}'\boldsymbol{q}} \\
&= \hbar\omega_{\boldsymbol{q}} + 2|g_{\boldsymbol{q}}|^2 \sum_{\boldsymbol{k}} \frac{(\varepsilon_{\boldsymbol{k}} - \varepsilon_{\boldsymbol{k}+\boldsymbol{q}})\langle n_{\boldsymbol{k}}\rangle}{(\varepsilon_{\boldsymbol{k}} - \varepsilon_{\boldsymbol{k}+\boldsymbol{q}})^2 - (\hbar\omega_{\boldsymbol{q}})^2}.
\end{aligned}$$

For $\hbar\omega_{\boldsymbol{q}} \ll |\varepsilon_{\boldsymbol{k}} - \varepsilon_{\boldsymbol{k}+\boldsymbol{q}}|$, we can neglect the term $(\hbar\omega_{\boldsymbol{q}})^2$ on the denominator and have

$$\hbar\bar{\omega}_{\boldsymbol{q}} \approx \hbar\omega_{\boldsymbol{q}} + 2|g_{\boldsymbol{q}}|^2 \sum_{\boldsymbol{k}} \frac{\langle n_{\boldsymbol{k}}\rangle}{\varepsilon_{\boldsymbol{k}} - \varepsilon_{\boldsymbol{k}+\boldsymbol{q}}}.$$

(5) Differentiating E with respect to $\langle n_{\boldsymbol{k}}\rangle$, we obtain

$$\begin{aligned}\bar{\varepsilon}_{\boldsymbol{k}} &= \frac{\partial E}{\partial\langle n_{\boldsymbol{k}}\rangle} \\ &= \varepsilon_{\boldsymbol{k}} + \sum_{\boldsymbol{q}} |g_{\boldsymbol{q}}|^2 \left[\frac{2(\varepsilon_{\boldsymbol{k}} - \varepsilon_{\boldsymbol{k}+\boldsymbol{q}})\,\langle N_{\boldsymbol{q}}\rangle}{(\varepsilon_{\boldsymbol{k}} - \varepsilon_{\boldsymbol{k}+\boldsymbol{q}})^2 - (\hbar\omega_{\boldsymbol{q}})^2} + \frac{1 - \langle n_{\boldsymbol{k}+\boldsymbol{q}}\rangle}{\varepsilon_{\boldsymbol{k}} - \varepsilon_{\boldsymbol{k}+\boldsymbol{q}} - \hbar\omega_{\boldsymbol{q}}} \right] \\ &\quad - \sum_{\boldsymbol{q}} |g_{\boldsymbol{q}}|^2 \frac{\langle n_{\boldsymbol{k}-\boldsymbol{q}}\rangle}{\varepsilon_{\boldsymbol{k}-\boldsymbol{q}} - \varepsilon_{\boldsymbol{k}} - \hbar\omega_{\boldsymbol{q}}} \\ &= \varepsilon_{\boldsymbol{k}} + \sum_{\boldsymbol{q}} |g_{\boldsymbol{q}}|^2 \left(\frac{\langle N_{\boldsymbol{q}}\rangle + \langle n_{\boldsymbol{k}+\boldsymbol{q}}\rangle}{\varepsilon_{\boldsymbol{k}} - \varepsilon_{\boldsymbol{k}+\boldsymbol{q}} + \hbar\omega_{\boldsymbol{q}}} + \frac{\langle N_{\boldsymbol{q}}\rangle + 1 - \langle n_{\boldsymbol{k}+\boldsymbol{q}}\rangle}{\varepsilon_{\boldsymbol{k}} - \varepsilon_{\boldsymbol{k}+\boldsymbol{q}} - \hbar\omega_{\boldsymbol{q}}} \right).\end{aligned}$$

26-2 Zeroth-order Green's functions from equations of motion. The zeroth-order Green's functions can also be obtained from their equations of motion.

(1) Set up the equation of motion for $\mathscr{G}_{\sigma 0}(\boldsymbol{k}, \tau - \tau')$ defined by

$$\mathscr{G}_{\sigma 0}(\boldsymbol{k}, \tau - \tau') = -\left\langle \hat{T}_\tau \left[\hat{c}_{\boldsymbol{k}\sigma}(\tau) \hat{c}^\dagger_{\boldsymbol{k}\sigma}(\tau') \right] \right\rangle_0$$

by differentiating it with respect to τ and solve it using the Fourier transformation to obtain $\mathscr{G}_{\sigma 0}(\boldsymbol{k}, \mathrm{i}\omega_n) = 1/(\mathrm{i}\hbar\omega_n - \xi_{\boldsymbol{k}\sigma})$.

(2) Set up the equation of motion for $\mathscr{D}_{s0}(\boldsymbol{q}, \tau - \tau')$ defined by

$$\mathscr{D}_{s0}(\boldsymbol{q}, \tau - \tau') = -\left\langle \hat{T}_\tau \left[\hat{\phi}_{\boldsymbol{q}s}(\tau) \hat{\phi}_{-\boldsymbol{q}s}(\tau') \right] \right\rangle_0$$

by differentiating it with respect to τ and solve it using the Fourier transformation to obtain $\mathscr{D}_{s0}(\boldsymbol{q}, \mathrm{i}\omega_m) = 2\hbar\omega_{\boldsymbol{q}s}/[(\mathrm{i}\hbar\omega_m)^2 - (\hbar\omega_{\boldsymbol{q}s})^2]$.

---

(1) Differentiating $\mathscr{G}_{\sigma 0}(\boldsymbol{k}, \tau - \tau')$ with respect to τ, we obtain the equation of motion for $\mathscr{G}_{\sigma 0}(\boldsymbol{k}, \tau - \tau')$

$$\begin{aligned}&\frac{\partial \mathscr{G}_{\sigma 0}(\boldsymbol{k}, \tau - \tau')}{\partial \tau} \\ &= -\frac{\partial}{\partial \tau} \langle \hat{c}_{\boldsymbol{k}\sigma}(\tau)\hat{c}^\dagger_{\boldsymbol{k}\sigma}(\tau')\theta(\tau - \tau') - \hat{c}^\dagger_{\boldsymbol{k}\sigma}(\tau')\hat{c}_{\boldsymbol{k}\sigma}(\tau)\theta(\tau' - \tau)\rangle_0 \\ &= -\langle \partial\hat{c}_{\boldsymbol{k}\sigma}(\tau)/\partial\tau\,\hat{c}^\dagger_{\boldsymbol{k}\sigma}(\tau')\theta(\tau - \tau') - \hat{c}^\dagger_{\boldsymbol{k}\sigma}(\tau')\partial\hat{c}_{\boldsymbol{k}\sigma}(\tau)/\partial\tau\,\theta(\tau' - \tau)\rangle_0 \\ &\quad - \langle \hat{c}_{\boldsymbol{k}\sigma}(\tau)\hat{c}^\dagger_{\boldsymbol{k}\sigma}(\tau') + \hat{c}^\dagger_{\boldsymbol{k}\sigma}(\tau')\hat{c}_{\boldsymbol{k}\sigma}(\tau)\rangle_0\,\delta(\tau - \tau') \\ &= (\xi_{\boldsymbol{k}\sigma}/\hbar)\,\langle \hat{c}_{\boldsymbol{k}\sigma}(\tau)\hat{c}^\dagger_{\boldsymbol{k}\sigma}(\tau')\theta(\tau - \tau') - \hat{c}^\dagger_{\boldsymbol{k}\sigma}(\tau')\hat{c}_{\boldsymbol{k}\sigma}(\tau)\theta(\tau' - \tau)\rangle_0 \\ &\quad - \mathrm{e}^{\hat{H}_0\tau/\hbar}\,\langle \hat{c}_{\boldsymbol{k}\sigma}\hat{c}^\dagger_{\boldsymbol{k}\sigma} + \hat{c}^\dagger_{\boldsymbol{k}\sigma}\hat{c}_{\boldsymbol{k}\sigma}\rangle_0\,\mathrm{e}^{-\hat{H}_0\tau/\hbar}\delta(\tau - \tau') \\ &= -(\xi_{\boldsymbol{k}\sigma}/\hbar)\mathscr{G}_{\sigma 0}(\boldsymbol{k}, \tau - \tau') - \delta(\tau - \tau'),\end{aligned}$$

where we have made use of the imaginary time dependence of $\hat{c}_{\boldsymbol{k}\sigma}(\tau)$, $\hat{c}_{\boldsymbol{k}\sigma}(\tau) = \mathrm{e}^{-\xi_{\boldsymbol{k}\sigma}\tau/\hbar}\hat{c}_{\boldsymbol{k}\sigma}$, and the anticommutation relation $\{\hat{c}_{\boldsymbol{k}\sigma}, \hat{c}^\dagger_{\boldsymbol{k}\sigma}\} = \hat{c}_{\boldsymbol{k}\sigma}\hat{c}^\dagger_{\boldsymbol{k}\sigma} + \hat{c}^\dagger_{\boldsymbol{k}\sigma}\hat{c}_{\boldsymbol{k}\sigma} = 1$. The Fourier transformation of $\mathscr{G}_{\sigma 0}(\boldsymbol{k}, \tau - \tau')$ is given by

$$\mathscr{G}_{\sigma 0}(\boldsymbol{k}, \tau - \tau') = \frac{1}{\beta}\sum_n \mathscr{G}_{\sigma 0}(\boldsymbol{k}, \mathrm{i}\omega_n)\mathrm{e}^{-\mathrm{i}\omega_n(\tau-\tau')},$$

where $\omega_n = (2n+1)\pi/\hbar\beta$ with n an integer. Inserting the above Fourier transformation and the representation of the δ-function, $\delta(\tau-\tau') = (\hbar\beta)^{-1}\sum_n \mathrm{e}^{-\mathrm{i}\omega_n(\tau-\tau')}$, into the equation of motion for $\mathscr{G}_{\sigma 0}(\boldsymbol{k}, \tau - \tau')$, we have

$$-\mathrm{i}\hbar\omega_n\mathscr{G}_{\sigma 0}(\boldsymbol{k}, \mathrm{i}\omega_n) = -\xi_{\boldsymbol{k}}\mathscr{G}_{\sigma 0}(\boldsymbol{k}, \mathrm{i}\omega_n) - 1.$$

Solving for $\mathscr{G}_{\sigma 0}(\boldsymbol{k}, \mathrm{i}\omega_n)$ from the above equation, we obtain

$$\mathscr{G}_{\sigma 0}(\boldsymbol{k}, \mathrm{i}\omega_n) = \frac{1}{\mathrm{i}\hbar\omega_n - \xi_{\boldsymbol{k}\sigma}}.$$

(2) Differentiating $\mathscr{D}_{s0}(\boldsymbol{q}, \tau - \tau')$ with respect to τ, we have

$$\begin{aligned}\frac{\partial\mathscr{D}_{s0}(\boldsymbol{q}, \tau-\tau')}{\partial\tau} &= -\frac{\partial}{\partial\tau}\left\langle\hat{T}_\tau\left[\hat{\phi}_{\boldsymbol{q}s}(\tau)\hat{\phi}_{-\boldsymbol{q}s}(\tau')\right]\right\rangle_0 \\ &= -\left\langle\hat{T}_\tau\left[\partial\hat{\phi}_{\boldsymbol{q}s}(\tau)/\partial\tau\hat{\phi}_{-\boldsymbol{q}s}(\tau')\right]\right\rangle_0 \\ &\quad - \left\langle\left[\hat{\phi}_{\boldsymbol{q}s}(\tau), \hat{\phi}_{-\boldsymbol{q}s}(\tau)\right]\right\rangle_0\delta(\tau-\tau').\end{aligned}$$

Since $[\hat{\phi}_{\boldsymbol{q}s}, \hat{\phi}_{-\boldsymbol{q}s}] = 0$, the second term on the right hand side vanishes. Differentiating again the above equation with respect to τ yields

$$\begin{aligned}\frac{\partial^2\mathscr{D}_{s0}(\boldsymbol{q}, \tau-\tau')}{\partial\tau^2} &= -\left\langle\hat{T}_\tau\left[\partial^2\hat{\phi}_{\boldsymbol{q}s}(\tau)/\partial\tau^2\hat{\phi}_{-\boldsymbol{q}s}(\tau')\right]\right\rangle_0 \\ &\quad - \left\langle\left[\partial\hat{\phi}_{\boldsymbol{q}s}(\tau)/\partial\tau\hat{\phi}_{\boldsymbol{q}s}(\tau), \hat{\phi}_{-\boldsymbol{q}s}(\tau)\right]\right\rangle_0\delta(\tau-\tau').\end{aligned}$$

Making use of $\hat{\phi}_{\boldsymbol{q}s} = \hat{a}_{\boldsymbol{q}s} + \hat{a}^\dagger_{-\boldsymbol{q}s}$, $\hat{a}_{\boldsymbol{q}s}(\tau) = \mathrm{e}^{-\omega_{\boldsymbol{q}s}\tau}\hat{a}_{\boldsymbol{q}s}$, $\hat{a}^\dagger_{\boldsymbol{q}s}(\tau) = \mathrm{e}^{\omega_{\boldsymbol{q}s}\tau}\hat{a}^\dagger_{\boldsymbol{q}s}$, and

$$\begin{aligned}&\left[\partial\hat{\phi}_{\boldsymbol{q}s}(\tau)/\partial\tau\hat{\phi}_{\boldsymbol{q}s}(\tau), \hat{\phi}_{-\boldsymbol{q}s}(\tau)\right] \\ &\quad = \left[-\omega_{\boldsymbol{q}s}\mathrm{e}^{-\omega_{\boldsymbol{q}s}\tau}\hat{a}_{\boldsymbol{q}s} + \omega_{\boldsymbol{q}s}\mathrm{e}^{\omega_{\boldsymbol{q}s}\tau}\hat{a}^\dagger_{-\boldsymbol{q}s}, \mathrm{e}^{-\omega_{\boldsymbol{q}s}\tau}\hat{a}_{-\boldsymbol{q}s} + \mathrm{e}^{\omega_{\boldsymbol{q}s}\tau}\hat{a}^\dagger_{\boldsymbol{q}s}\right] \\ &\quad = -2\omega_{\boldsymbol{q}s},\end{aligned}$$

we obtain the equation of motion for $\mathscr{D}_{s0}(\boldsymbol{q}, \tau - \tau')$

$$\begin{aligned}&\frac{\partial^2\mathscr{D}_{s0}(\boldsymbol{q}, \tau-\tau')}{\partial\tau^2} \\ &\quad = -\left\langle\hat{T}_\tau\left[\partial^2\hat{\phi}_{\boldsymbol{q}s}(\tau)/\partial\tau^2\hat{\phi}_{-\boldsymbol{q}s}(\tau')\right]\right\rangle_0 + 2\omega_{\boldsymbol{q}s}\delta(\tau-\tau') \\ &\quad = \omega^2_{\boldsymbol{q}s}\mathscr{D}_{s0}(\boldsymbol{q}, \tau-\tau') + 2\omega_{\boldsymbol{q}s}\delta(\tau-\tau').\end{aligned}$$

The Fourier transformation of $\mathscr{D}_{s0}(\boldsymbol{q}, \tau-\tau')$ is given by

$$\mathscr{D}_{s0}(\boldsymbol{q}, \tau-\tau') = \frac{1}{\beta}\sum_m \mathscr{D}_{s0}(\boldsymbol{q}, \mathrm{i}\omega_m)\mathrm{e}^{-\mathrm{i}\omega_m(\tau-\tau')},$$

where $\omega_m = 2m\pi/\hbar\beta$ with m an integer. Inserting the above Fourier transformation and the representation of the δ-function, $\delta(\tau-\tau') = (\hbar\beta)^{-1}\sum_m \mathrm{e}^{-\mathrm{i}\omega_m(\tau-\tau')}$, into the equation of motion for $\mathscr{D}_{s0}(\boldsymbol{q}, \tau-\tau')$, we obtain

$$-\hbar\omega_m^2\mathscr{D}_{s0}(\boldsymbol{q}, \mathrm{i}\omega_m) = \hbar\omega_{\boldsymbol{q}s}^2\mathscr{D}_{s0}(\boldsymbol{q}, \mathrm{i}\omega_m) + 2\omega_{\boldsymbol{q}s}.$$

Solving for $\mathscr{D}_s(\boldsymbol{q}, \mathrm{i}\omega_m)$ from the above equation, we obtain

$$\mathscr{D}_{s0}(\boldsymbol{q}, \mathrm{i}\omega_m) = \frac{2\hbar\omega_{\boldsymbol{q}s}}{(\mathrm{i}\hbar\omega_m)^2 - (\hbar\omega_{\boldsymbol{q}s})^2}.$$

26-3 Field operators and single-electron Green's function. In terms of field operators

$$\hat{\psi}(\boldsymbol{r}, t) = \frac{1}{\sqrt{\mathscr{V}}}\sum_{\boldsymbol{k}} \mathrm{e}^{\mathrm{i}\boldsymbol{k}\cdot\boldsymbol{r}}\hat{c}_{\boldsymbol{k}}(t),\ \hat{\psi}^\dagger(\boldsymbol{r}, t) = \frac{1}{\sqrt{\mathscr{V}}}\sum_{\boldsymbol{k}} \mathrm{e}^{-\mathrm{i}\boldsymbol{k}\cdot\boldsymbol{r}}\hat{c}_{\boldsymbol{k}}^\dagger(t),$$

the single-electron Green's function is defined by $G(\boldsymbol{r}, t; \boldsymbol{r}', t') = -\mathrm{i}\,\langle\hat{T}[\hat{\psi}(\boldsymbol{r}, t)\hat{\psi}^\dagger(\boldsymbol{r}', t')]\rangle_0$. We suppress the electron spin here.

(1) Show that $G(\boldsymbol{r}, t; \boldsymbol{r}', t')$ is a function only of the differences $\boldsymbol{r}-\boldsymbol{r}'$ and $t-t'$ for a homogeneous system.

(2) Show that $G_0(\boldsymbol{r}, t; \boldsymbol{r}', t')$ for a noninteracting homogeneous system satisfies the equation $[\mathrm{i}\hbar\mathrm{d}/\mathrm{d}t + (\hbar^2/2m)\mathrm{d}^2/\mathrm{d}x^2]$ $G_0(\boldsymbol{r}, t; \boldsymbol{r}', t') = \delta(\boldsymbol{r}-\boldsymbol{r}')\delta(t-t')$.

(3) Solve for $G_0(\boldsymbol{r}, t; \boldsymbol{r}', t')$ from the above equation for a free electron.

---

(1) Inserting the above-given expansions of field operators into the definition of Green's function $G(\boldsymbol{r}, t; \boldsymbol{r}', t')$ and then making use of $\hat{c}_{\boldsymbol{k}}(t) = \mathrm{e}^{\mathrm{i}\hat{H}t/\hbar}\hat{c}_{\boldsymbol{k}}\mathrm{e}^{-\mathrm{i}\hat{H}t/\hbar}$ and its Hermitian conjugate, we obtain

$$\begin{aligned}
&G(\boldsymbol{r}, t; \boldsymbol{r}', t') \\
&= -\frac{\mathrm{i}}{\mathscr{V}}\sum_{\boldsymbol{k}\boldsymbol{k}'} \mathrm{e}^{\mathrm{i}(\boldsymbol{k}\cdot\boldsymbol{r}-\boldsymbol{k}'\cdot\boldsymbol{r}')}\langle\hat{T}[\hat{c}_{\boldsymbol{k}}(t)\hat{c}_{\boldsymbol{k}'}^\dagger(t')]\rangle_0 \\
&= -\frac{\mathrm{i}}{\mathscr{V}}\sum_{\boldsymbol{k}\boldsymbol{k}'} \mathrm{e}^{\mathrm{i}(\boldsymbol{k}\cdot\boldsymbol{r}-\boldsymbol{k}'\cdot\boldsymbol{r}')}\big[\langle\mathrm{e}^{\mathrm{i}\hat{H}t/\hbar}\hat{c}_{\boldsymbol{k}}\mathrm{e}^{-\mathrm{i}\hat{H}(t-t')/\hbar}\hat{c}_{\boldsymbol{k}'}^\dagger\mathrm{e}^{-\mathrm{i}\hat{H}t'/\hbar}\rangle_0\,\theta(t-t') \\
&\qquad - \langle\mathrm{e}^{\mathrm{i}\hat{H}t'/\hbar}\hat{c}_{\boldsymbol{k}'}^\dagger\mathrm{e}^{\mathrm{i}\hat{H}(t-t')/\hbar}\hat{c}_{\boldsymbol{k}}\mathrm{e}^{-\mathrm{i}\hat{H}t/\hbar}\rangle_0\,\theta(t'-t)\big] \\
&= -\frac{\mathrm{i}}{\mathscr{V}}\sum_{\boldsymbol{k}\boldsymbol{k}'} \mathrm{e}^{\mathrm{i}(\boldsymbol{k}\cdot\boldsymbol{r}-\boldsymbol{k}'\cdot\boldsymbol{r}')}\big[\mathrm{e}^{\mathrm{i}E_0(t-t')}\langle\hat{c}_{\boldsymbol{k}}\mathrm{e}^{-\mathrm{i}\hat{H}(t-t')/\hbar}\hat{c}_{\boldsymbol{k}'}^\dagger\rangle_0\,\theta(t-t') \\
&\qquad - \mathrm{e}^{-\mathrm{i}E_0(t-t')}\langle\hat{c}_{\boldsymbol{k}'}^\dagger\mathrm{e}^{\mathrm{i}\hat{H}(t-t')/\hbar}\hat{c}_{\boldsymbol{k}}\rangle_0\,\theta(t'-t)\big],
\end{aligned}$$

where E_0 is the ground-state energy. The averages in the ground state of a homogeneous system are nonzero only if the creation and annihilation operators of the electron acts on the same plane-wave state, that is, $\boldsymbol{k}' = \boldsymbol{k}$. We then have

$$G(\boldsymbol{r},t;\boldsymbol{r}',t')$$
$$= -\frac{\mathrm{i}}{\mathscr{V}}\sum_{\boldsymbol{k}} \mathrm{e}^{\mathrm{i}\boldsymbol{k}\cdot(\boldsymbol{r}-\boldsymbol{r}')}\big[\mathrm{e}^{\mathrm{i}E_0(t-t')}\langle \hat{c}_{\boldsymbol{k}}\mathrm{e}^{-\mathrm{i}\hat{H}(t-t')/\hbar}\hat{c}^{\dagger}_{\boldsymbol{k}}\rangle_0\,\theta(t-t')$$
$$- \mathrm{e}^{-\mathrm{i}E_0(t-t')}\langle \hat{c}^{\dagger}_{\boldsymbol{k}}\mathrm{e}^{\mathrm{i}\hat{H}(t-t')/\hbar}\hat{c}_{\boldsymbol{k}}\rangle_0\,\theta(t'-t)\big].$$

From the above equation, we see that $G(\boldsymbol{r},t;\boldsymbol{r}',t')$ is a function only of the differences $\boldsymbol{r}-\boldsymbol{r}'$ and $t-t'$ for a homogeneous system.

(2) The Hamiltonian of a free electron is $\hat{H} = \hat{\boldsymbol{p}}^2/2m = -\hbar^2\nabla^2/2m$ and its dispersion relation is given by $\varepsilon_{\boldsymbol{k}} = \hbar^2\boldsymbol{k}^2/2m$. The time dependencies of the annihilation and creation operators are given by $\hat{c}_{\boldsymbol{k}}(t) = \mathrm{e}^{-\mathrm{i}\varepsilon_{\boldsymbol{k}}t/\hbar}\hat{c}_{\boldsymbol{k}}$ and $\hat{c}^{\dagger}_{\boldsymbol{k}}(t) = \mathrm{e}^{\mathrm{i}\varepsilon_{\boldsymbol{k}}t/\hbar}\hat{c}^{\dagger}_{\boldsymbol{k}}$. Differentiating $G_0(\boldsymbol{r},t;\boldsymbol{r}',t')$ with respect to t yields

$$\frac{\partial}{\partial t}G_0(\boldsymbol{r},t;\boldsymbol{r}',t')$$
$$= -\frac{\mathrm{i}}{\mathscr{V}}\sum_{\boldsymbol{k}\boldsymbol{k}'} \mathrm{e}^{\mathrm{i}(\boldsymbol{k}\cdot\boldsymbol{r}-\boldsymbol{k}'\cdot\boldsymbol{r}')}\big[\langle\hat{T}[\partial\hat{c}_{\boldsymbol{k}}(t)/\partial t\hat{c}^{\dagger}_{\boldsymbol{k}'}(t')]\rangle_0$$
$$+ \langle\{\hat{c}_{\boldsymbol{k}}(t),\hat{c}^{\dagger}_{\boldsymbol{k}'}(t)\}\rangle_0\,\delta(t-t')\big]$$
$$= -\frac{\mathrm{i}}{\mathscr{V}}\sum_{\boldsymbol{k}\boldsymbol{k}'} \mathrm{e}^{\mathrm{i}(\boldsymbol{k}\cdot\boldsymbol{r}-\boldsymbol{k}'\cdot\boldsymbol{r}')}\big[-\mathrm{i}(\varepsilon_{\boldsymbol{k}}/\hbar)\langle\hat{T}[\hat{c}_{\boldsymbol{k}}(t)\hat{c}^{\dagger}_{\boldsymbol{k}'}(t')]\rangle_0 + \delta_{\boldsymbol{k}\boldsymbol{k}'}\delta(t-t')\big]$$
$$= \frac{\hbar\boldsymbol{\nabla}^2}{2m\mathscr{V}}\sum_{\boldsymbol{k}\boldsymbol{k}'} \mathrm{e}^{\mathrm{i}(\boldsymbol{k}\cdot\boldsymbol{r}-\boldsymbol{k}'\cdot\boldsymbol{r}')}\langle\hat{T}[\hat{c}_{\boldsymbol{k}}(t)\hat{c}^{\dagger}_{\boldsymbol{k}'}(t')]\rangle_0 - \frac{\mathrm{i}\delta(t-t')}{\mathscr{V}}\sum_{\boldsymbol{k}}\mathrm{e}^{\mathrm{i}\boldsymbol{k}\cdot(\boldsymbol{r}-\boldsymbol{r}')}$$
$$= \mathrm{i}\frac{\hbar}{2m}\boldsymbol{\nabla}^2 G_0(\boldsymbol{r},t;\boldsymbol{r}',t') - \frac{\mathrm{i}\delta(t-t')}{(2\pi)^3}\int \mathrm{d}\boldsymbol{k}\,\mathrm{e}^{\mathrm{i}\boldsymbol{k}\cdot(\boldsymbol{r}-\boldsymbol{r}')}$$
$$= \mathrm{i}\frac{\hbar}{2m}\boldsymbol{\nabla}^2 G_0(\boldsymbol{r},t;\boldsymbol{r}',t') - \mathrm{i}\delta(\boldsymbol{r}-\boldsymbol{r}')\delta(t-t'),$$

that is,

$$\left(\mathrm{i}\hbar\frac{\partial}{\partial t} + \frac{\hbar^2}{2m}\boldsymbol{\nabla}^2\right)G_0(\boldsymbol{r},t;\boldsymbol{r}',t') = \hbar\delta(\boldsymbol{r}-\boldsymbol{r}')\delta(t-t').$$

The above equation is the equation of motion for $G_0(\boldsymbol{r},t;\boldsymbol{r}',t')$. Our next task is to solve for $G_0(\boldsymbol{r},t;\boldsymbol{r}',t')$ from it.

(3) A commonly-used approach for solving the above equation of motion is to utilize the Fourier transformation. Making a

Fourier transformation to $G_0(\boldsymbol{r}, t; \boldsymbol{r}', t')$ with respect to both $\boldsymbol{r} - \boldsymbol{r}'$ and $t - t'$, we have

$$G_0(\boldsymbol{r}, t; \boldsymbol{r}', t') = \frac{1}{(2\pi)^4} \int \mathrm{d}\boldsymbol{k} \int_{-\infty}^{\infty} \mathrm{d}\omega \; \mathrm{e}^{\mathrm{i}[\boldsymbol{k}\cdot(\boldsymbol{r}-\boldsymbol{r}') - \omega(t-t')]} G_0(\boldsymbol{k}, \omega).$$

For the Dirac δ-functions on the right hand side of the equation of motion of $G_0(\boldsymbol{r}, t; \boldsymbol{r}', t')$, we make use of their integral representations in consideration that $G_0(\boldsymbol{r}, t; \boldsymbol{r}', t')$ has been Fourier transformed. The integral representation of the spatial δ-function $\delta(\boldsymbol{r} - \boldsymbol{r}')$ is given by

$$\delta(\boldsymbol{r} - \boldsymbol{r}') = \frac{1}{(2\pi)^3} \int \mathrm{d}\boldsymbol{k} \; \mathrm{e}^{\mathrm{i}\boldsymbol{k}\cdot(\boldsymbol{r}-\boldsymbol{r}')}.$$

The integral representation of the temporal δ-function $\delta(t-t')$ is given by

$$\delta(t - t') = \frac{1}{2\pi} \int_{-\infty}^{\infty} \mathrm{d}\omega \; \mathrm{e}^{-\mathrm{i}\omega(t-t')}$$

Inserting the Fourier transformation of $G_0(\boldsymbol{r}, t; \boldsymbol{r}', t')$ and the integral representations of δ-functions into the equation of motion of $G_0(\boldsymbol{r}, t; \boldsymbol{r}', t')$, we obtain

$$\left(\hbar\omega - \frac{\hbar^2 \boldsymbol{k}^2}{2m}\right) G_0(\boldsymbol{k}, \omega) = \hbar.$$

Thus,

$$G_0(\boldsymbol{k}, \omega) = \frac{\hbar}{\hbar\omega - \hbar^2 \boldsymbol{k}^2/2m}.$$

$G_0(\boldsymbol{r}, t; \boldsymbol{r}', t')$ is then given by

$$G_0(\boldsymbol{r}, t; \boldsymbol{r}', t') = \int \frac{\mathrm{d}\boldsymbol{k}}{(2\pi)^3} \int_{-\infty}^{\infty} \frac{\mathrm{d}\omega}{2\pi} \; \mathrm{e}^{\mathrm{i}[\boldsymbol{k}\cdot(\boldsymbol{r}-\boldsymbol{r}') - \omega(t-t')]} \frac{\hbar}{\hbar\omega - \hbar^2 \boldsymbol{k}^2/2m}.$$

To obtain an explicit expression for $G_0(\boldsymbol{r}, t; \boldsymbol{r}', t')$, we must evaluate the integrations over $\boldsymbol{k}$ and ω in the above equation. Note that the integrand has a first-order pole at $\omega = \hbar \boldsymbol{k}^2/2m$ on the real axis for ω. The integration over ω can be computed using the residue theorem. In consideration of the exponential factor $\mathrm{e}^{-\mathrm{i}\omega(t-t')}$ in the integrand, we close the contour in the lower-half complex plane of ω for $t - t' > 0$ and close the contour in the upper-half complex plane for

$t - t' < 0$. We have

$$\begin{aligned}
&G_0(\boldsymbol{r}, t; \boldsymbol{r}', t') \\
&= -\frac{\mathrm{i}}{2(2\pi)^3} \int \mathrm{d}\boldsymbol{k}\, \mathrm{e}^{\mathrm{i}\boldsymbol{k}\cdot(\boldsymbol{r}-\boldsymbol{r}')} \mathrm{e}^{-\mathrm{i}\hbar k^2 (t-t')/2m} \left[\theta(t-t') - \theta(t'-t)\right] \\
&= -\frac{\theta(t-t') - \theta(t'-t)}{2(2\pi)^2 |\boldsymbol{r}-\boldsymbol{r}'|} \int_{-\infty}^{\infty} \mathrm{d}k\, k\, \mathrm{e}^{\mathrm{i}k|\boldsymbol{r}-\boldsymbol{r}'|} \mathrm{e}^{-\mathrm{i}\hbar k^2 (t-t')/2m} \\
&= -\frac{\mathrm{e}^{\mathrm{i}m|\boldsymbol{r}-\boldsymbol{r}'|^2/2\hbar(t-t')}}{2(2\pi)^2 |\boldsymbol{r}-\boldsymbol{r}'|} \theta(t-t') - \theta(t'-t) \\
&\quad \times \int_{-\infty}^{\infty} \mathrm{d}k\, k\, \mathrm{e}^{-\mathrm{i}[\hbar(t-t')/2m]\,[k - m|\boldsymbol{r}-\boldsymbol{r}'|/\hbar(t-t')]^2} \\
&= -\frac{m\, \mathrm{e}^{\mathrm{i}m|\boldsymbol{r}-\boldsymbol{r}'|^2/2\hbar(t-t')}}{2(2\pi)^2 \hbar(t-t')} \left[\theta(t-t') - \theta(t'-t)\right] \\
&\quad \times \int_{-\infty}^{\infty} \mathrm{d}k'\ \mathrm{e}^{-\mathrm{i}[\hbar(t-t')/2m]k'^2} \\
&= -\frac{\mathrm{i}}{2}\left[\frac{m}{2\pi\mathrm{i}\hbar(t-t')}\right]^{3/2} \mathrm{e}^{\mathrm{i}m|\boldsymbol{r}-\boldsymbol{r}'|^2/2\hbar(t-t')} \left[\theta(t-t') - \theta(t'-t)\right].
\end{aligned}$$

26-4 Feynman rules for the effective electron-electron interaction. Consider the effective electron-electron interaction mediated by phonons in terms of the phonon Green's function and Feynman diagram in Fig. 26.1. The effective electron-electron interaction in Fig. 26.1(b) is obtained by joining the two electron-phonon interaction vertices in Fig. 26.1(a).

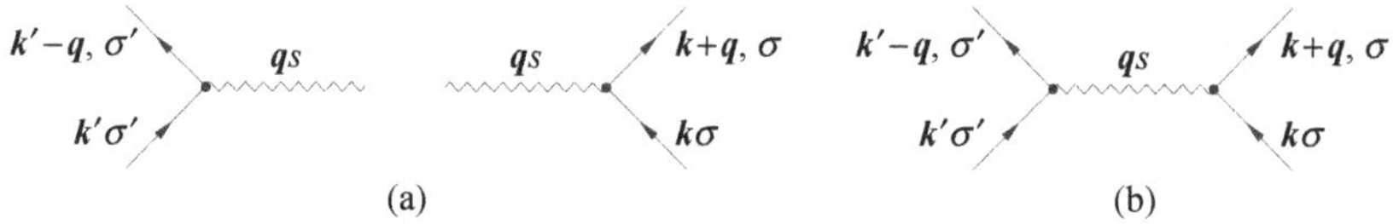

Fig. 26.1 Feynman diagram for the effective electron-electron interaction mediated by phonons.

(1) Write down the analytic expression corresponding to the Feynman diagram in Fig. 26.1(a) for the two vertices multiplied together using the Feynman rules.

(2) Describe how the effective electron-electron interaction

$$\hat{H}^{ee}_{\text{eff}} = \frac{1}{2} \sum_{\boldsymbol{k}\sigma\boldsymbol{k}'\sigma'\boldsymbol{q}s} \frac{2|g_{\boldsymbol{q},s}|^2 \hbar\omega_{\boldsymbol{q}s}}{(\varepsilon_{\boldsymbol{k}} - \varepsilon_{\boldsymbol{k}+\boldsymbol{q}})^2 - (\hbar\omega_{\boldsymbol{q}s})^2} \hat{c}^\dagger_{\boldsymbol{k}+\boldsymbol{q},\sigma} \hat{c}^\dagger_{\boldsymbol{k}'-\boldsymbol{q},\sigma'} \hat{c}_{\boldsymbol{k}'\sigma'} \hat{c}_{\boldsymbol{k}\sigma}$$

can be obtained from the above result. Set up a set of rules that can be used to write down an expression for a higher-order contribution to the effective electron-electron interaction. Provide a method for the verification of the rules.

(3) Write down the corresponding analytic expressions for the effective electron-electron interaction in Figs. 26.2(a) and (b) using the above rules.

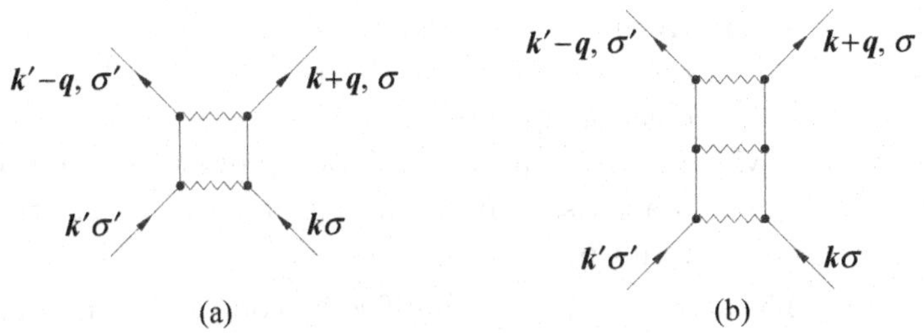

Fig. 26.2 Feynman diagrams for the fourth-order (a) and sixth-order (b) contributions to the effective electron-electron interaction.

---

(1) According to the Feynman rules, we have

$$\mathrm{i}^{1/2}g_{\boldsymbol{q}s}\cdot\mathrm{i}^{1/2}g^*_{\boldsymbol{q}s}D_{s0}(\boldsymbol{q},(\varepsilon_{\boldsymbol{k}}-\varepsilon_{\boldsymbol{k}+\boldsymbol{q}})/\hbar)$$
$$=\frac{2\mathrm{i}|g_{\boldsymbol{q}s}|^2\hbar\omega_{\boldsymbol{q}s}}{(\varepsilon_{\boldsymbol{k}}-\varepsilon_{\boldsymbol{k}+\boldsymbol{q}})^2-(\hbar\omega_{\boldsymbol{q}s})^2+\mathrm{i}\delta}.$$

Since the electron lines are external lines in Fig. 26.1(b), they are not included in the above expression.

(2) Comparing the above expression with the given expression for the effective interaction, we can set up the following rules that can be used to write down an expression for a higher-order contribution to the effective electron-electron interaction.

(i) Assign a factor of $\mathrm{i}^{1/2}|g_{\boldsymbol{q}s}|^2$ to each pair of electron-phonon interaction vertices connected by a phonon line $\boldsymbol{q}s$. Energy and momentum are conserved on each vertex.

(ii) Assign a factor of $D_{s0}(\boldsymbol{q},\omega)$ to each internal phonon line $\boldsymbol{q}s$, with ω to be determined from the electron lines.

(iii) Assign a factor of $G_{\sigma 0}(\boldsymbol{k},\omega)$ to each internal electron line $\boldsymbol{k}\sigma$ with ω being a free variable of the form of the electron energy $\varepsilon_{\boldsymbol{k}'}$ or determined from the conservation of energy.

(iv) Assign an electron creation operator $\hat{c}^\dagger_{\boldsymbol{k}\sigma}$ to each outgoing external electron line $\boldsymbol{k}\sigma$.

(v) Assign an electron annihilation operator $\hat{c}_{\boldsymbol{k}\sigma}$ to each incoming external electron line $\boldsymbol{k}\sigma$.

(vi) Arrange the electron operators counterclockwise as they appear in the diagram, starting from the first creation operator.

(vii) Sum over all wave vectors, spin indices, and phonon branches. Divide the result by two. Multiply the result by $-\mathrm{i}$.

The above rules can be verified by comparing the results derived using these rules with those obtained using the canonical transformation or perturbation method.

(3) According to the above-stated rules, we have for Fig. 26.2(a)

$$\frac{\mathrm{i}}{2}\sum_{\boldsymbol{k}\boldsymbol{k}'\boldsymbol{k}''\sigma\sigma'}\sum_{\boldsymbol{q}\boldsymbol{q}'ss'}|g_{\boldsymbol{q}'s'}|^2|g_{\boldsymbol{q}-\boldsymbol{q}',s'}|^2\frac{2\hbar\omega_{\boldsymbol{q}'s'}}{(\varepsilon_{\boldsymbol{k}+\boldsymbol{q}}-\varepsilon_{\boldsymbol{k}''})^2-(\hbar\omega_{\boldsymbol{q}'s'})^2+\mathrm{i}\delta}$$

$$\times\frac{2\hbar\omega_{\boldsymbol{q}-\boldsymbol{q}',s}}{(\varepsilon_{\boldsymbol{k}''}-\varepsilon_{\boldsymbol{k}})^2-(\hbar\omega_{\boldsymbol{q}-\boldsymbol{q}',s})^2+\mathrm{i}\delta}$$

$$\times\frac{1}{\varepsilon_{\boldsymbol{k}''}-\varepsilon_{\boldsymbol{k}+\boldsymbol{q}-\boldsymbol{q}'}+\mathrm{i}\delta\,\mathrm{sgn}(|\boldsymbol{k}+\boldsymbol{q}-\boldsymbol{q}'|-k_{\mathrm{F}})}$$

$$\times\frac{1}{\varepsilon_{\boldsymbol{k}}+\varepsilon_{\boldsymbol{k}'}-\varepsilon_{\boldsymbol{k}''}-\varepsilon_{\boldsymbol{k}'-\boldsymbol{q}+\boldsymbol{q}'}+\mathrm{i}\delta\,\mathrm{sgn}(|\boldsymbol{k}'-\boldsymbol{q}+\boldsymbol{q}'|-k_{\mathrm{F}})}$$

$$\times\,\hat{c}^\dagger_{\boldsymbol{k}+\boldsymbol{q},\sigma}\hat{c}^\dagger_{\boldsymbol{k}'-\boldsymbol{q},\sigma'}\hat{c}_{\boldsymbol{k}'\sigma'}\hat{c}_{\boldsymbol{k}\sigma}.$$

For Fig. 26.2(b), we have

$$-\frac{1}{2}\sum_{\boldsymbol{k}\boldsymbol{k}'\boldsymbol{k}_1\boldsymbol{k}_2\sigma\sigma'}\sum_{\boldsymbol{q}\boldsymbol{q}'\boldsymbol{q}''ss's''}\frac{2\hbar\omega_{\boldsymbol{q}'s'}|g_{\boldsymbol{q}'s'}|^2|g_{\boldsymbol{q}''s''}|^2|g_{\boldsymbol{q}-\boldsymbol{q}'-\boldsymbol{q}'',s}|^2}{(\varepsilon_{\boldsymbol{k}+\boldsymbol{q}}-\varepsilon_{\boldsymbol{k}_2})^2-(\hbar\omega_{\boldsymbol{q}'s'})^2+\mathrm{i}\delta}$$

$$\times\frac{2\hbar\omega_{\boldsymbol{q}''s''}}{(\varepsilon_{\boldsymbol{k}_2}-\varepsilon_{\boldsymbol{k}_1})^2-(\hbar\omega_{\boldsymbol{q}''s''})^2+\mathrm{i}\delta}\frac{2\hbar\omega_{\boldsymbol{q}-\boldsymbol{q}'-\boldsymbol{q}'',s}}{(\varepsilon_{\boldsymbol{k}_1}-\varepsilon_{\boldsymbol{k}})^2-(\hbar\omega_{\boldsymbol{q}-\boldsymbol{q}'-\boldsymbol{q}'',s})^2+\mathrm{i}\delta}$$

$$\times\frac{1}{\varepsilon_{\boldsymbol{k}_1}-\varepsilon_{\boldsymbol{k}+\boldsymbol{q}-\boldsymbol{q}'-\boldsymbol{q}''}+\mathrm{i}\delta\,\mathrm{sgn}(|\boldsymbol{k}+\boldsymbol{q}-\boldsymbol{q}'-\boldsymbol{q}''|-k_{\mathrm{F}})}$$

$$\times\frac{1}{\varepsilon_{\boldsymbol{k}_2}-\varepsilon_{\boldsymbol{k}'+\boldsymbol{q}-\boldsymbol{q}'}+\mathrm{i}\delta\,\mathrm{sgn}(|\boldsymbol{k}+\boldsymbol{q}-\boldsymbol{q}'|-k_{\mathrm{F}})}$$

$$\times\frac{1}{\varepsilon_{\boldsymbol{k}}+\varepsilon_{\boldsymbol{k}'}-\varepsilon_{\boldsymbol{k}_1}-\varepsilon_{\boldsymbol{k}'-\boldsymbol{q}+\boldsymbol{q}'+\boldsymbol{q}''}+\mathrm{i}\delta\,\mathrm{sgn}(|\boldsymbol{k}'-\boldsymbol{q}+\boldsymbol{q}'+\boldsymbol{q}''|-k_{\mathrm{F}})}$$

$$\times\frac{1}{\varepsilon_{\boldsymbol{k}}+\varepsilon_{\boldsymbol{k}'}-\varepsilon_{\boldsymbol{k}_2}-\varepsilon_{\boldsymbol{k}'-\boldsymbol{q}+\boldsymbol{q}'}+\mathrm{i}\delta\,\mathrm{sgn}(|\boldsymbol{k}'-\boldsymbol{q}+\boldsymbol{q}'|-k_{\mathrm{F}})}$$

$$\times\,\hat{c}^\dagger_{\boldsymbol{k}+\boldsymbol{q},\sigma}\hat{c}^\dagger_{\boldsymbol{k}'-\boldsymbol{q},\sigma'}\hat{c}_{\boldsymbol{k}'\sigma'}\hat{c}_{\boldsymbol{k}\sigma}.$$

26-5 Fourth-order corrections to the phonon Green's function. Consider the fourth-order corrections to the phonon Green's function $D_s(\boldsymbol{q}, \omega)$ at zero temperature.

(1) Write down the contributions corresponding to the Feynman diagrams in Fig. 26.3 using the Feynman rules.

(2) Verify the above results by evaluating explicitly the relevant terms in

$$\begin{aligned} &D_s^{(4)}(\boldsymbol{q}, t-t') \\ &= -\frac{\mathrm{i}}{24\hbar^4}\int_{-\infty}^{\infty} \mathrm{d}t_1 \cdots \mathrm{d}t_4 \,\langle 0|\,\hat{T}\big[\hat{\phi}_{\boldsymbol{q}s}(t)\hat{\phi}_{-\boldsymbol{q}s}(t')\hat{H}_{\mathrm{ep}}(t_1)\cdots\hat{H}_{\mathrm{ep}}(t_4)\big]\,|0\rangle \\ &+ D_s^{(2)}(\boldsymbol{q}, t-t')\frac{1}{2\hbar^2}\int_{-\infty}^{\infty} \mathrm{d}t_1\mathrm{d}t_2 \,\langle 0|\,\hat{T}\big[\hat{H}_{\mathrm{ep}}(t_1)\hat{H}_{\mathrm{ep}}(t_2)\big]\,|0\rangle \\ &- D_{s0}(\boldsymbol{q}, t-t')\frac{1}{24\hbar^2}\int_{-\infty}^{\infty} \mathrm{d}t_1 \cdots \mathrm{d}t_4 \,\langle 0|\,\hat{T}\big[\hat{H}_{\mathrm{ep}}(t_1)\cdots\hat{H}_{\mathrm{ep}}(t_4)\big]\,|0\rangle \,. \end{aligned}$$

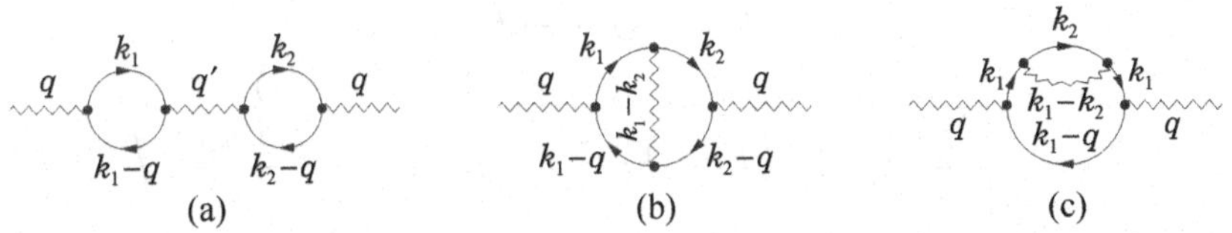

Fig. 26.3 Feynman diagrams for the fourth-order corrections to the phonon Green's function.

---

(1) The electron and phonon lines are labeled in Fig. 26.3. To save space, k has been used to represent collectively the wave vector $\boldsymbol{k}$, the spin index σ, and the frequency ω for electrons. Similarly, q has been used to represent collectively the wave vector $\boldsymbol{q}$, the branch index s, and the frequency ω for phonons. However, $k - q$ implies that only the wave vectors and frequencies are subtracted. Note that, in Fig. 26.3(a), q' differs from q only in the branch indices since the energy and momentum conservation demands that the frequencies and the wave vectors be equal. According to the Feynman rules, we have for Fig. 26.3(a) from left to right

$$
\begin{aligned}
D_s^{(4a)}(\boldsymbol{q},\omega) &= D_{s0}(\boldsymbol{q},\omega)\cdot \mathrm{i}^{1/2} g_{\boldsymbol{q}s} \\
&\quad \cdot(-1)\left[\sum_{\boldsymbol{k}_1\sigma_1}\int \frac{\mathrm{d}(\hbar\omega_1)}{2\pi}\, G_{\sigma_1 0}(\boldsymbol{k}_1,\omega_1)G_{\sigma_1 0}(\boldsymbol{k}_1-\boldsymbol{q},\omega_1-\omega)\right] \\
&\quad \cdot\sum_{s'} \mathrm{i}^{1/2} g_{\boldsymbol{q}s'}\cdot D_{s'0}(\boldsymbol{q},\omega)\cdot \mathrm{i}^{1/2} g^*_{\boldsymbol{q}s'} \\
&\quad \cdot(-1)\left[\sum_{\boldsymbol{k}_2\sigma_2}\int \frac{\mathrm{d}(\hbar\omega_2)}{2\pi}\, G_{\sigma_2 0}(\boldsymbol{k}_2,\omega_2)G_{\sigma_2 0}(\boldsymbol{k}_2-\boldsymbol{q},\omega_2-\omega)\right] \\
&\quad \cdot \mathrm{i}^{1/2} g^*_{\boldsymbol{q}s}\cdot D_{s0}(\boldsymbol{q},\omega) \\
&= -|g_{\boldsymbol{q}s}|^2\left[D_{s0}(\boldsymbol{q},\omega)\right]^2 \sum_{s'} |g_{\boldsymbol{q}s}|^2 D_{s'0}(\boldsymbol{q},\omega) \\
&\quad \times\left[\sum_{\boldsymbol{k}\sigma}\int \frac{\mathrm{d}(\hbar\omega')}{2\pi}\, G_{\sigma 0}(\boldsymbol{k},\omega')G_{\sigma 0}(\boldsymbol{k}-\boldsymbol{q},\omega'-\omega)\right]^2 .
\end{aligned}
$$

For Fig. 26.3(b), we have

$$
\begin{aligned}
&D_s^{(4b)}(\boldsymbol{q},\omega) \\
&\quad = D_{s0}(\boldsymbol{q},\omega)\cdot \mathrm{i}^{1/2} g_{\boldsymbol{q}s}\cdot \sum_{\boldsymbol{k}_1\boldsymbol{k}_2\sigma s'}\int \frac{\mathrm{d}(\hbar\omega_1)}{2\pi}\int \frac{\mathrm{d}(\hbar\omega_2)}{2\pi}\, G_{\sigma 0}(\boldsymbol{k}_1,\omega_1) \\
&\qquad \cdot \mathrm{i}^{1/2} g_{\boldsymbol{k}_1-\boldsymbol{k}_2,s'}\cdot D_{s'0}(\boldsymbol{k}_1-\boldsymbol{k}_2,\omega_1-\omega_2)\cdot G_{\sigma 0}(\boldsymbol{k}_2,\omega_2) \\
&\qquad \cdot \mathrm{i}^{1/2} g^*_{\boldsymbol{q}s}\cdot D_{s0}(\boldsymbol{q},\omega)\cdot G_{\sigma 0}(\boldsymbol{k}_2-\boldsymbol{q},\omega_2-\omega) \\
&\qquad \cdot \mathrm{i}^{1/2} g^*_{\boldsymbol{k}_1-\boldsymbol{k}_2,s'}\cdot G_{\sigma 0}(\boldsymbol{k}_1-\boldsymbol{q},\omega_1-\omega)\cdot(-1) \\
&\quad = |g_{\boldsymbol{q}s}|^2\left[D_{s0}(\boldsymbol{q},\omega)\right]^2 \sum_{\boldsymbol{k}_1\boldsymbol{k}_2\sigma s'} |g_{\boldsymbol{k}_1-\boldsymbol{k}_2,s'}|^2 \int \frac{\mathrm{d}(\hbar\omega_1)}{2\pi}\int \frac{\mathrm{d}(\hbar\omega_2)}{2\pi} \\
&\qquad \times D_{s'0}(\boldsymbol{k}_1-\boldsymbol{k}_2,\omega_1-\omega_2)G_{\sigma 0}(\boldsymbol{k}_1,\omega_1)G_{\sigma 0}(\boldsymbol{k}_1-\boldsymbol{q},\omega_1-\omega) \\
&\qquad \times G_{\sigma 0}(\boldsymbol{k}_2,\omega_2)G_{\sigma 0}(\boldsymbol{k}_2-\boldsymbol{q},\omega_2-\omega).
\end{aligned}
$$

For Fig. 26.3(c), we have

$$
\begin{aligned}
&D_s^{(4c)}(\boldsymbol{q},\omega) \\
&\quad = D_{s0}(\boldsymbol{q},\omega)\cdot \mathrm{i}^{1/2} g_{\boldsymbol{q}s}\cdot \sum_{\boldsymbol{k}_1\boldsymbol{k}_2\sigma s'}\int \frac{\mathrm{d}(\hbar\omega_1)}{2\pi}\int \frac{\mathrm{d}(\hbar\omega_2)}{2\pi}\, G_{\sigma 0}(\boldsymbol{k}_1,\omega_1) \\
&\qquad \cdot \mathrm{i}^{1/2} g_{\boldsymbol{k}_1-\boldsymbol{k}_2,s'}\cdot D_{s'0}(\boldsymbol{k}_1-\boldsymbol{k}_2,\omega_1-\omega_2)\cdot G_{\sigma 0}(\boldsymbol{k}_2,\omega_2) \\
&\qquad \cdot \mathrm{i}^{1/2} g^*_{\boldsymbol{k}_1-\boldsymbol{k}_2,s'}\cdot G_{\sigma 0}(\boldsymbol{k}_1,\omega_1)\cdot \mathrm{i}^{1/2} g^*_{\boldsymbol{q}s}\cdot D_{s0}(\boldsymbol{q},\omega) \\
&\qquad \cdot G_{\sigma 0}(\boldsymbol{k}_1-\boldsymbol{q},\omega_1-\omega)\cdot(-1) \\
&\quad = |g_{\boldsymbol{q}s}|^2\left[D_{s0}(\boldsymbol{q},\omega)\right]^2 \sum_{\boldsymbol{k}_1\boldsymbol{k}_2\sigma s'} |g_{\boldsymbol{k}_1-\boldsymbol{k}_2,s'}|^2 \int \frac{\mathrm{d}(\hbar\omega_1)}{2\pi}\int \frac{\mathrm{d}(\hbar\omega_2)}{2\pi} \\
&\qquad \times D_{s'0}(\boldsymbol{k}_1-\boldsymbol{k}_2,\omega_1-\omega_2)\left[G_{\sigma 0}(\boldsymbol{k}_1,\omega_1)\right]^2 \\
&\qquad \times G_{\sigma 0}(\boldsymbol{k}_1-\boldsymbol{q},\omega_1-\omega)G_{\sigma 0}(\boldsymbol{k}_2,\omega_2).
\end{aligned}
$$

(2) Since the Feynman diagrams in Fig. 26.3 are all connected ones, the terms in $D_s^{(4)}(\boldsymbol{q},\omega)$ can only be obtained from the first term in the given expression for $D_s^{(4)}(\boldsymbol{q},t-t')$. Inserting

$$\hat{H}_{\rm ep} = \sum_{\boldsymbol{k}\boldsymbol{q}s\sigma} g_{\boldsymbol{q}s}\hat{c}^\dagger_{\boldsymbol{k}+\boldsymbol{q},\sigma}\hat{c}_{\boldsymbol{k}\sigma}(\hat{a}_{\boldsymbol{q}s}+\hat{a}^\dagger_{-\boldsymbol{q}s})$$

into the first term in $D_s^{(4)}(\boldsymbol{q},t-t')$ yields

$$-\frac{\mathrm{i}}{24\hbar^4}\int_{-\infty}^{\infty}\mathrm{d}t_1\cdots\mathrm{d}t_4\ \langle 0|\,\hat{T}\big[\hat{\phi}_{\boldsymbol{q}s}(t)\hat{\phi}_{-\boldsymbol{q}s}(t')\hat{H}_{\rm ep}(t_1)\cdots\hat{H}_{\rm ep}(t_4)\big]|0\rangle$$
$$= -\frac{\mathrm{i}}{24\hbar^4}\sum_{\boldsymbol{k}_1\cdots\boldsymbol{k}_4}\sum_{\boldsymbol{q}_1\cdots\boldsymbol{q}_4}\sum_{s_1\cdots s_4}\sum_{\sigma_1\cdots\sigma_4} g_{\boldsymbol{q}_1s_1}g_{\boldsymbol{q}_2s_2}g_{\boldsymbol{q}_3s_3}g_{\boldsymbol{q}_4s_4}\int_{-\infty}^{\infty}\mathrm{d}t_1\cdots\mathrm{d}t_4$$
$$\times\langle 0|\,\hat{T}\big[\hat{\phi}_{\boldsymbol{q}s}(t)\hat{\phi}_{-\boldsymbol{q}s}(t')\hat{\phi}_{\boldsymbol{q}_1s_1}(t_1)\hat{\phi}_{\boldsymbol{q}_2s_2}(t_2)\hat{\phi}_{\boldsymbol{q}_3s_3}(t_3)\hat{\phi}_{\boldsymbol{q}_4s_4}(t_4)\big]|0\rangle$$
$$\times\langle 0|\,\hat{T}\big[\hat{c}^\dagger_{\boldsymbol{k}_1+\boldsymbol{q}_1,\sigma_1}(t_1)\hat{c}_{\boldsymbol{k}_1\sigma_1}(t_1)\hat{c}^\dagger_{\boldsymbol{k}_2+\boldsymbol{q}_2,\sigma_2}(t_2)\hat{c}_{\boldsymbol{k}_2\sigma_2}(t_2)$$
$$\times\,\hat{c}^\dagger_{\boldsymbol{k}_3+\boldsymbol{q}_3,\sigma_3}(t_3)\hat{c}_{\boldsymbol{k}_3\sigma_3}(t_3)\hat{c}^\dagger_{\boldsymbol{k}_4+\boldsymbol{q}_4,\sigma_4}(t_4)\hat{c}_{\boldsymbol{k}_4\sigma_4}(t_4)\big]|0\rangle\,.$$

From the structure of Fig. 26.3(a), we see that it comes from the terms in which the phonon operators are paired up in such a manner: (t,t_1), (t_2,t_3), and (t_4,t') plus those terms obtained through permuting t_1, t_2, t_3, and t_4. In total, there are $4! = 24$ identical terms. The electron operators must be paired up in the manner: (t_1,t_2) and (t_3,t_4) and other permutations in accordance with those for the phonon operators. Note that, because the time sequence is completely fixed through the permutation of time instants for the phonon operators, no additional terms from permutations arise for the electron part. We then have

$$D_s^{(4a)}(\boldsymbol{q},t-t')$$
$$= -\frac{\mathrm{i}^8}{\hbar^4}\sum_{\boldsymbol{k}_1\cdots\boldsymbol{k}_4}\sum_{\boldsymbol{q}_1\cdots\boldsymbol{q}_4}\sum_{s_1\cdots s_4}\sum_{\sigma_1\cdots\sigma_4} g_{\boldsymbol{q}_1s_1}g_{\boldsymbol{q}_2s_2}g_{\boldsymbol{q}_3s_3}g_{\boldsymbol{q}_4s_4}\int_{-\infty}^{\infty}\mathrm{d}t_1\cdots\mathrm{d}t_4$$
$$\times D_{s0}(\boldsymbol{q},t-t_1)\delta_{\boldsymbol{q}_1,-\boldsymbol{q}}\delta_{s_1s}D_{s_20}(\boldsymbol{q}_2,t_2-t_3)\delta_{\boldsymbol{q}_3,-\boldsymbol{q}_2}\delta_{s_3s_2}$$
$$\times D_{s0}(\boldsymbol{q},t_4-t')\delta_{\boldsymbol{q}_4\boldsymbol{q}}\delta_{s_4s}G_{\sigma_10}(\boldsymbol{k}_1,t_1-t_2)\delta_{\boldsymbol{k}_2+\boldsymbol{q}_2,\boldsymbol{k}_1}\delta_{\sigma_2\sigma_1}$$
$$\times(-1)G_{\sigma_10}(\boldsymbol{k}_1+\boldsymbol{q}_1,t_2-t_1)\delta_{\boldsymbol{k}_1+\boldsymbol{q}_1,\boldsymbol{k}_2}\delta_{\sigma_2\sigma_1}G_{\sigma_30}(\boldsymbol{k}_3,t_3-t_4)$$
$$\times\,\delta_{\boldsymbol{k}_4+\boldsymbol{q}_4,\boldsymbol{k}_3}\delta_{\sigma_4\sigma_3}(-1)G_{\sigma_30}(\boldsymbol{k}_3+\boldsymbol{q}_3,t_4-t_3)\delta_{\boldsymbol{k}_3+\boldsymbol{q}_3,\boldsymbol{k}_4}\delta_{\sigma_4\sigma_3}\,,$$

where each Green's function contributes a factor of i and the transpositions of electron creation and annihilation operators lead to the appearance of a factor of -1 for each of two electron Green's functions among the four electron

Green's functions. The presence of δ-symbols in the above equation allows some of the summations over wave vectors, spin indices, and branch indices to be performed. We have

$$D_s^{(4a)}(\boldsymbol{q}, t-t') = -\frac{1}{\hbar^4} \sum_{\boldsymbol{k}_1 \boldsymbol{k}_2 \sigma_1 \sigma_2 s'} |g_{\boldsymbol{q}s}|^2 |g_{\boldsymbol{q}s'}|^2 \int_{-\infty}^{\infty} \mathrm{d}t_1 \cdots \mathrm{d}t_4$$
$$\times D_{s0}(\boldsymbol{q}, t-t_1) D_{s'0}(\boldsymbol{q}, t_2-t_3) D_{s0}(\boldsymbol{q}, t_4-t')$$
$$\times G_{\sigma_1 0}(\boldsymbol{k}_1, t_1-t_2) G_{\sigma_1 0}(\boldsymbol{k}_1-\boldsymbol{q}, t_2-t_1)$$
$$\times G_{\sigma_2 0}(\boldsymbol{k}_2, t_3-t_4) G_{\sigma_2 0}(\boldsymbol{k}_2-\boldsymbol{q}, t_4-t_3),$$

where we have made use of $g_{-\boldsymbol{q}s} = g^*_{\boldsymbol{q}s}$ and renamed some of the dummy summation variables. To obtain $D_s^{(4a)}(\boldsymbol{q}, \omega)$, we insert the Fourier-transformed electron and phonon Green's functions,

$$G_{\sigma 0}(\boldsymbol{k}, t) = \frac{\hbar}{2\pi} \int_{-\infty}^{\infty} \mathrm{d}\omega\, G_{\sigma 0}(\boldsymbol{k}, \omega) \mathrm{e}^{-\mathrm{i}\omega t},$$
$$D_{s0}(\boldsymbol{q}, t) = \frac{\hbar}{2\pi} \int_{-\infty}^{\infty} \mathrm{d}\omega\, D_{s0}(\boldsymbol{q}, \omega) \mathrm{e}^{-\mathrm{i}\omega t},$$

into $D_s^{(4a)}(\boldsymbol{q}, t-t')$ and obtain

$$D_s^{(4a)}(\boldsymbol{q}, t-t')$$
$$= -\frac{\hbar^3}{(2\pi)^7} \sum_{\boldsymbol{k}_1 \boldsymbol{k}_2 \sigma_1 \sigma_2 s'} |g_{\boldsymbol{q}s}|^2 |g_{\boldsymbol{q}s'}|^2 \int_{-\infty}^{\infty} \mathrm{d}\omega_1 \cdots \mathrm{d}\omega_7 \int_{-\infty}^{\infty} \mathrm{d}t_1 \cdots \mathrm{d}t_4$$
$$\times \mathrm{e}^{-\mathrm{i}\omega_1(t-t_1)} \mathrm{e}^{-\mathrm{i}\omega_2(t_2-t_3)} \mathrm{e}^{-\mathrm{i}\omega_3(t_4-t')} \mathrm{e}^{-\mathrm{i}\omega_4(t_1-t_2)} \mathrm{e}^{-\mathrm{i}\omega_5(t_2-t_1)}$$
$$\times \mathrm{e}^{-\mathrm{i}\omega_6(t_3-t_4)} \mathrm{e}^{-\mathrm{i}\omega_7(t_4-t_3)} D_{s0}(\boldsymbol{q}, \omega_1) D_{s'0}(\boldsymbol{q}, \omega_2) D_{s0}(\boldsymbol{q}, \omega_3)$$
$$\times G_{\sigma_1 0}(\boldsymbol{k}_1, \omega_4) G_{\sigma_1 0}(\boldsymbol{k}_1-\boldsymbol{q}, \omega_5) G_{\sigma_2 0}(\boldsymbol{k}_2, \omega_6) G_{\sigma_2 0}(\boldsymbol{k}_2-\boldsymbol{q}, \omega_7).$$

Combining the exponentials with the same time variables for t_1 through t_4 and then performing the integrations over these time variables using $(2\pi)^{-1} \int_{-\infty}^{\infty} \mathrm{d}t\, \mathrm{e}^{\pm \mathrm{i}\Delta\omega t} = \delta(\Delta\omega)$, we obtain

$$D_s^{(4a)}(\boldsymbol{q}, t-t')$$
$$= -\left(\frac{\hbar}{2\pi}\right)^3 \sum_{\boldsymbol{k}_1 \boldsymbol{k}_2 \sigma_1 \sigma_2 s'} |g_{\boldsymbol{q}s}|^2 |g_{\boldsymbol{q}s'}|^2 \int_{-\infty}^{\infty} \mathrm{d}\omega_1 \cdots \mathrm{d}\omega_7\, \mathrm{e}^{-\mathrm{i}(\omega_1 t - \omega_3 t')}$$
$$\times \delta(\omega_1 - \omega_4 + \omega_5) \delta(\omega_2 - \omega_4 + \omega_5) \delta(\omega_2 - \omega_6 + \omega_7)$$
$$\times \delta(\omega_3 - \omega_6 + \omega_7) D_{s0}(\boldsymbol{q}, \omega_1) D_{s'0}(\boldsymbol{q}, \omega_2) D_{s0}(\boldsymbol{q}, \omega_3)$$
$$\times G_{\sigma_1 0}(\boldsymbol{k}_1, \omega_4) G_{\sigma_1 0}(\boldsymbol{k}_1-\boldsymbol{q}, \omega_5) G_{\sigma_2 0}(\boldsymbol{k}_2, \omega_6) G_{\sigma_2 0}(\boldsymbol{k}_2-\boldsymbol{q}, \omega_7)$$

$$
\begin{aligned}
&= -\sum_{\boldsymbol{k}_1\boldsymbol{k}_2\sigma_1\sigma_2 s'} |g_{\boldsymbol{q}s}|^2 |g_{\boldsymbol{q}s'}|^2 \int_{-\infty}^{\infty} \frac{\mathrm{d}(\hbar\omega)}{2\pi}\, \mathrm{e}^{-\mathrm{i}\omega(t-t')} \int_{-\infty}^{\infty} \frac{\mathrm{d}(\hbar\omega_1)}{2\pi} \\
&\quad \times \int_{-\infty}^{\infty} \frac{\mathrm{d}(\hbar\omega_2)}{2\pi} D_{s0}(\boldsymbol{q},\omega) D_{s'0}(\boldsymbol{q},\omega) D_{s0}(\boldsymbol{q},\omega) G_{\sigma_1 0}(\boldsymbol{k}_1,\omega_1) \\
&\quad \times G_{\sigma_1 0}(\boldsymbol{k}_1-\boldsymbol{q},\omega_1-\omega) G_{\sigma_2 0}(\boldsymbol{k}_2,\omega_2) G_{\sigma_2 0}(\boldsymbol{k}_2-\boldsymbol{q},\omega_2-\omega),
\end{aligned}
$$

where we have renamed the frequency variables after the integrations over ω_2, ω_3, ω_5, and ω_7 have been performed by consuming the available δ-functions: $\omega_1 \to \omega$, $\omega_4 \to \omega_1$, $\omega_6 \to \omega_2$. Comparing the integration over ω in the above expression of $D_s^{(4a)}(\boldsymbol{q}, t-t')$ with its Fourier-transformed expression $D_{s0}^{(4a)}(\boldsymbol{q},t) = (\hbar/2\pi)\int_{-\infty}^{\infty} \mathrm{d}\omega\ D_{s0}^{(4a)}(\boldsymbol{q},\omega)\mathrm{e}^{-\mathrm{i}\omega t}$, we see immediately that $D_{s0}^{(4a)}(\boldsymbol{q},\omega)$ is given by

$$
\begin{aligned}
D_s^{(4a)}(\boldsymbol{q},\omega) &= -|g_{\boldsymbol{q}s}|^2 [D_{s0}(\boldsymbol{q},\omega)]^2 \sum_{\boldsymbol{k}_1\boldsymbol{k}_2\sigma_1\sigma_2 s'} |g_{\boldsymbol{q}s'}|^2 D_{s'0}(\boldsymbol{q},\omega) \\
&\quad \times \int_{-\infty}^{\infty} \frac{\mathrm{d}(\hbar\omega_1)}{2\pi} \int_{-\infty}^{\infty} \frac{\mathrm{d}(\hbar\omega_2)}{2\pi} G_{\sigma_1 0}(\boldsymbol{k}_1,\omega_1) G_{\sigma_2 0}(\boldsymbol{k}_2,\omega_2) \\
&\quad \times G_{\sigma_1 0}(\boldsymbol{k}_1-\boldsymbol{q},\omega_1-\omega) G_{\sigma_2 0}(\boldsymbol{k}_2-\boldsymbol{q},\omega_2-\omega) \\
&= -|g_{\boldsymbol{q}s}|^2 [D_{s0}(\boldsymbol{q},\omega)]^2 \sum_{s'} |g_{\boldsymbol{q}s'}|^2 D_{s'0}(\boldsymbol{q},\omega) \\
&\quad \times \left[\sum_{\boldsymbol{k}\sigma} \int_{-\infty}^{\infty} \frac{\mathrm{d}(\hbar\omega')}{2\pi} G_{\sigma 0}(\boldsymbol{k},\omega') G_{\sigma 0}(\boldsymbol{k}-\boldsymbol{q},\omega'-\omega) \right]^2 .
\end{aligned}
$$

We have thus derived the same result as that obtained using the Feynman rules. The verification of $D_s^{(4b)}(\boldsymbol{q},\omega)$ and $D_s^{(4c)}(\boldsymbol{q},\omega)$ will not be given here to save space.

26-6 Spectral function, renormalization constant, and effective mass. The retarded self-energy of quasielectrons in a system is given by $\Sigma_R(\boldsymbol{k}, E) = a(E - E_{\mathrm{F}}) - \mathrm{i}b(E - E_{\mathrm{F}})^2$, where a and b are constants.

(1) Find the spectral function for the quasielectrons.
(2) Evaluate the renormalization constant.
(3) Compute the effective mass.

---

(1) At zero temperature, we have $\mu = E_{\mathrm{F}}$ and $\xi_{\boldsymbol{k}} = \varepsilon_{\boldsymbol{k}} - E_{\mathrm{F}}$. With the electron spin suppressed, the spectral function is

given by

$$\begin{aligned}\mathscr{A}(\boldsymbol{k}, E - E_{\rm F}) &= -\frac{1}{\pi}\operatorname{Im} G_R(\boldsymbol{k}, E) = -\frac{1}{\pi}\operatorname{Im}\frac{1}{E - \varepsilon_{\boldsymbol{k}} - \Sigma_R(\boldsymbol{k}, E)} \\ &= -\frac{1}{\pi}\operatorname{Im}\frac{1}{E - \varepsilon_{\boldsymbol{k}} - a(E - E_{\rm F}) + {\rm i}b(E - E_{\rm F})^2} \\ &= \frac{1}{\pi}\frac{b(E - E_{\rm F})^2}{[(1-a)(E - E_{\rm F}) - \xi_{\boldsymbol{k}}]^2 + b^2(E - E_{\rm F})^4}.\end{aligned}$$

(2) The renormalization constant Z is given by

$$\frac{1}{Z} = 1 - \left.\frac{\partial \operatorname{Re}\Sigma_R(\boldsymbol{k}, E)}{\partial E}\right|_{\substack{k = k_{\rm F} \\ E = E_{\rm F}}} = 1 - a.$$

Thus, $Z = 1/(1-a)$.

(3) The effective mass m^* is given by

$$\frac{1}{m^*} = \frac{Z}{m}\left[1 + \frac{m}{\hbar^2 k_{\rm F}}\left.\frac{\partial \operatorname{Re}\Sigma_{R\sigma}(\boldsymbol{k}, E)}{\partial k}\right|_{\substack{k = k_{\rm F} \\ E = E_{\rm F}}}\right] = \frac{Z}{m}.$$

Thus, $m^* = m/Z = (1-a)m$.

26-7 Real and imaginary parts of the electron retarded self-energy. The imaginary part of the retarded self-energy of an electron in a compound is modeled as $\operatorname{Im}\Sigma_R(\boldsymbol{k}, E) = -\hbar/\tau$ for $|E - E_{\rm F}| > \hbar\omega_c$, $-[(E - E_{\rm F})^2/(\hbar\omega_c)^2](\hbar/\tau)$ for $|E - E_{\rm F}| < \hbar\omega_c$, where ω_c is a characteristic frequency of phonons. The chemical potential for electrons is taken to be equal to the Fermi energy $E_{\rm F}$ at low temperatures of interest here.

(1) Derive the real part of the electron self-energy, $\operatorname{Re}\Sigma_R(\boldsymbol{k}, E)$, for E close to $E_{\rm F}$.

(2) Evaluate the renormalization constant Z.

(3) Estimate the value of Z for $\hbar/\tau \sim \hbar\omega_c$.

---

(1) In consideration that $\Sigma_R(\boldsymbol{k}, E)$ is an analytic function E, we can make use of one of the Kramers-Kronig relations to find the real part of $\Sigma_R(\boldsymbol{k}, E)$ from its imaginary and have

$$\begin{aligned}\operatorname{Re}\Sigma_R(\boldsymbol{k}, E) &= -\frac{1}{\pi}{\rm P}\int_{-\infty}^{\infty} {\rm d}E'\,\frac{\operatorname{Im}\Sigma_R(\boldsymbol{k}, E')}{E - E'} \\ &= \frac{\hbar}{\pi\tau}{\rm P}\int_{-\infty}^{E_{\rm F} - \hbar\omega_c}\frac{{\rm d}E'}{E - E'} + \frac{\hbar}{\pi\tau}{\rm P}\int_{E_{\rm F} + \hbar\omega_c}^{\infty}\frac{{\rm d}E'}{E - E'} \\ &\quad + \frac{\hbar}{\pi\tau(\hbar\omega_c)^2}{\rm P}\int_{E_{\rm F} - \hbar\omega_c}^{E_{\rm F} + \hbar\omega_c}{\rm d}E'\,\frac{(E' - E_{\rm F})^2}{E - E'}.\end{aligned}$$

Since $E \sim E_{\rm F}$, the integrands of the first two integrals have no singularities. Treating the infinities as the $\mathscr{E} \to \infty$ limit, we can perform the first two integrals jointly as follows

$$\int_{-\infty}^{E_{\rm F}-\hbar\omega_c} \frac{{\rm d}E'}{E-E'} + \int_{E_{\rm F}+\hbar\omega_c}^{\infty} \frac{{\rm d}E'}{E-E'}$$

$$= \lim_{\mathscr{E}\to\infty}\left[\int_{-\mathscr{E}}^{E_{\rm F}-\hbar\omega_c} \frac{{\rm d}E'}{E-E'} + \int_{E_{\rm F}+\hbar\omega_c}^{\mathscr{E}} \frac{{\rm d}E'}{E-E'}\right]$$

$$= \lim_{\mathscr{E}\to\infty}\left[\ln\left|\frac{\mathscr{E}+E}{E-E_{\rm F}+\hbar\omega_c}\right| + \ln\left|\frac{E-E_{\rm F}-\hbar\omega_c}{\mathscr{E}-E}\right|\right]$$

$$= \ln\left|\frac{E-E_{\rm F}-\hbar\omega_c}{E-E_{\rm F}+\hbar\omega_c}\right| \approx -\frac{2(E-E_{\rm F})}{\hbar\omega_c},$$

where the fact that $E \sim E_{\rm F}$ has been used. The remaining integral can be evaluated as follows

$${\rm P}\int_{E_{\rm F}-\hbar\omega_c}^{E_{\rm F}+\hbar\omega_c} {\rm d}E' \, \frac{(E'-E_{\rm F})^2}{E-E'}$$

$$= -\int_{E_{\rm F}-\hbar\omega_c}^{E_{\rm F}+\hbar\omega_c} {\rm d}E' \, \big[(E'-E)+2(E-E_{\rm F})\big]$$

$$- (E-E_{\rm F})^2\, {\rm P}\int_{E_{\rm F}-\hbar\omega_c}^{E_{\rm F}+\hbar\omega_c} \frac{{\rm d}E'}{E'-E}$$

$$= -2\hbar\omega_c(E-E_{\rm F})$$

$$- (E-E_{\rm F})^2 \lim_{\delta\to 0^+}\left[\int_{E_{\rm F}-\hbar\omega_c}^{E-\delta} \frac{{\rm d}E'}{E'-E} + \int_{E+\delta}^{E_{\rm F}+\hbar\omega_c} \frac{{\rm d}E'}{E'-E}\right]$$

$$= -2\hbar\omega_c(E-E_{\rm F}) - (E-E_{\rm F})^2 \ln\left|\frac{E-E_{\rm F}-\hbar\omega_c}{E-E_{\rm F}+\hbar\omega_c}\right|$$

$$\approx -2\hbar\omega_c(E-E_{\rm F}) + \frac{2(E-E_{\rm F})^3}{\hbar\omega_c}$$

$$= -\frac{2(E-E_{\rm F})}{\hbar\omega_c}\big[(\hbar\omega_c)^2 - (E-E_{\rm F})^2\big].$$

Thus, ${\rm Re}\,\Sigma_R(\boldsymbol{k}, E)$ for $E \sim E_{\rm F}$ is given by

$${\rm Re}\,\Sigma_R(\boldsymbol{k}, E) \approx -\frac{\hbar}{\pi\tau}\frac{2(E-E_{\rm F})}{\hbar\omega_c}$$

$$-\frac{\hbar}{\pi\tau(\hbar\omega_c)^2}\frac{2(E-E_{\rm F})}{\hbar\omega_c}\big[(\hbar\omega_c)^2 - (E-E_{\rm F})^2\big]$$

$$\approx -\frac{4\hbar}{\pi\tau\hbar\omega_c}(E-E_{\rm F}).$$

(2) The renormalization constant Z is given by

$$\frac{1}{Z} = 1 - \frac{\partial\,{\rm Re}\,\Sigma_R(\boldsymbol{k}, E)}{\partial E}\bigg|_{\substack{k\,=\,k_{\rm F}\\ E\,=\,E_{\rm F}}} = 1 + \frac{4}{\pi\omega_c\tau}.$$

Thus,

$$Z = \frac{1}{1 + 4/\pi\omega_c\tau}.$$

(3) For $\hbar/\tau \sim \hbar\omega_c$, we have

$$Z \sim \frac{1}{1 + 4/\pi} \approx 0.44.$$

26-8 Periodic Anderson model. The Hamiltonian of the periodic Anderson model (PAM) in real space is given by

$$\hat{H} = -t\sum_{\langle ij\rangle\,\sigma}(\hat{c}^\dagger_{i\sigma}\hat{c}_{j\sigma} + \hat{c}^\dagger_{j\sigma}\hat{c}_{i\sigma}) + \varepsilon_d\sum_{j\sigma}\hat{d}^\dagger_{j\sigma}\hat{d}_{j\sigma} + V\sum_{j\sigma}(\hat{c}^\dagger_{j\sigma}\hat{d}_{j\sigma} + \hat{d}^\dagger_{j\sigma}\hat{c}_{j\sigma}),$$

where the subscript $\langle ij\rangle$ on the summation sign implies the summation over nearest neighbors, the $\hat{c}$ operators describe a set of delocalized (conduction) electrons that hop from site to site, the $\hat{d}$ operators describe a set of localized electrons that can not hop from one site to another, and the V term describes the hybridization between the delocalized and localized electrons.

(1) Express the Hamiltonian in k-space (momentum space) and then diagonalize it.

(2) Set up the equation of motion for the Green's function of a conduction electron $G_\uparrow(\boldsymbol{k}, t) = -\mathrm{i}\,\langle\hat{T}[\hat{c}_{\boldsymbol{k}\uparrow}(t)\hat{c}^\dagger_{\boldsymbol{k}\uparrow}(0)]\rangle$ through differentiating $G_\uparrow(\boldsymbol{k}, t)$ with respect to t. Show that the Green's function $F_\uparrow(\boldsymbol{k}, t) = -\mathrm{i}\,\langle\hat{T}[\hat{d}_{\boldsymbol{k}\uparrow}(t)\hat{c}^\dagger_{\boldsymbol{k}\uparrow}(0)]\rangle$ is contained in the equation of motion for $G_\uparrow(\boldsymbol{k}, t)$. Set up the equation of motion for $F_\uparrow(\boldsymbol{k}, t)$.

(3) Solve the coupled equations of $G_\uparrow(\boldsymbol{k}, t)$ and $F_\uparrow(\boldsymbol{k}, t)$ through making Fourier transformations to $G_\uparrow(\boldsymbol{k}, t)$ and $F_\uparrow(\boldsymbol{k}, t)$. Find the self-energy of a conduction electron.

---

(1) The Fourier transformations to $\hat{c}_{j\sigma}$ and $\hat{d}_{j\sigma}$ are given by

$$\hat{c}_{j\sigma} = \frac{1}{\sqrt{N}}\sum_{\boldsymbol{k}}\mathrm{e}^{\mathrm{i}\boldsymbol{k}\cdot\boldsymbol{R}_j}\hat{c}_{\boldsymbol{k}\sigma},\quad \hat{d}_{j\sigma} = \frac{1}{\sqrt{N}}\sum_{\boldsymbol{k}}\mathrm{e}^{\mathrm{i}\boldsymbol{k}\cdot\boldsymbol{R}_j}\hat{d}_{\boldsymbol{k}\sigma},$$

where $\boldsymbol{R}_j$ is the position of the jth lattice site and N the number of lattice sites. For convenience, let $\boldsymbol{\delta}$ denote the relative coordinate of a nearest neighbor of the reference site. We can then write $\sum_{\langle ij\rangle}(\cdots)$ as $(1/2)\sum_{j\boldsymbol{\delta}}(\cdots)$, where the factor $1/2$ is to remove the double counting in the summation since each pair of nearest neighbors is taken into account twice in the summation over j and $\boldsymbol{\delta}$. For the kinetic energy term of delocalized electrons, we have

$$-t\sum_{\langle ij\rangle\,\sigma}\left(\hat{c}^\dagger_{i\sigma}\hat{c}_{j\sigma}+\hat{c}^\dagger_{j\sigma}\hat{c}_{i\sigma}\right)$$

$$=-\frac{t}{2N}\sum_{\boldsymbol{k}\boldsymbol{k}'\sigma}\sum_{\boldsymbol{\delta}}\left(\mathrm{e}^{-\mathrm{i}\boldsymbol{k}'\cdot\boldsymbol{\delta}}+\mathrm{e}^{\mathrm{i}\boldsymbol{k}\cdot\boldsymbol{\delta}}\right)\hat{c}^\dagger_{\boldsymbol{k}'\sigma}\hat{c}_{\boldsymbol{k}\sigma}\sum_j\mathrm{e}^{\mathrm{i}(\boldsymbol{k}-\boldsymbol{k}')\cdot\boldsymbol{R}_j}$$

$$=-\frac{t}{2}\sum_{\boldsymbol{k}\sigma}\sum_{\boldsymbol{\delta}}\left(\mathrm{e}^{-\mathrm{i}\boldsymbol{k}\cdot\boldsymbol{\delta}}+\mathrm{e}^{\mathrm{i}\boldsymbol{k}\cdot\boldsymbol{\delta}}\right)\hat{c}^\dagger_{\boldsymbol{k}\sigma}\hat{c}_{\boldsymbol{k}\sigma}=\sum_{\boldsymbol{k}\sigma}\varepsilon_{\boldsymbol{k}}\hat{c}^\dagger_{\boldsymbol{k}\sigma}\hat{c}_{\boldsymbol{k}\sigma},$$

where we have made use of $N^{-1}\sum_j\mathrm{e}^{\mathrm{i}(\boldsymbol{k}-\boldsymbol{k}')\cdot\boldsymbol{R}_j}=\delta_{\boldsymbol{k}'\boldsymbol{k}}$ and introduced the dispersion relation for delocalized electrons

$$\varepsilon_{\boldsymbol{k}}=-t\sum_{\boldsymbol{\delta}}\cos(\boldsymbol{k}\cdot\boldsymbol{\delta}).$$

For a two-dimensional square lattice, we have $\varepsilon_{\boldsymbol{k}}=-2t[\cos(k_xa)+\cos(k_ya)]$ with a the lattice constant.

For the on-site energy term of localized electrons, we have

$$\varepsilon_d\sum_{j\sigma}\hat{d}^\dagger_{j\sigma}\hat{d}_{j\sigma}=\frac{\varepsilon_d}{N}\sum_{\boldsymbol{k}\boldsymbol{k}'\sigma}\hat{d}^\dagger_{\boldsymbol{k}'\sigma}\hat{d}_{\boldsymbol{k}\sigma}\sum_j\mathrm{e}^{\mathrm{i}(\boldsymbol{k}-\boldsymbol{k}')\cdot\boldsymbol{R}_j}=\varepsilon_d\sum_{\boldsymbol{k}\sigma}\hat{d}^\dagger_{\boldsymbol{k}\sigma}\hat{d}_{\boldsymbol{k}\sigma}.$$

For the hybridization term, we have

$$V\sum_{j\sigma}\left(\hat{c}^\dagger_{j\sigma}\hat{d}_{j\sigma}+\hat{d}^\dagger_{j\sigma}\hat{c}_{j\sigma}\right)=\frac{V}{N}\sum_{\boldsymbol{k}\boldsymbol{k}'\sigma}\left(\hat{c}^\dagger_{\boldsymbol{k}'\sigma}\hat{d}_{\boldsymbol{k}\sigma}+\hat{d}^\dagger_{\boldsymbol{k}'\sigma}\hat{c}_{\boldsymbol{k}\sigma}\right)\sum_j\mathrm{e}^{\mathrm{i}(\boldsymbol{k}-\boldsymbol{k}')\cdot\boldsymbol{R}_j}$$

$$=V\sum_{\boldsymbol{k}\sigma}\left(\hat{c}^\dagger_{\boldsymbol{k}\sigma}\hat{d}_{\boldsymbol{k}\sigma}+\hat{d}^\dagger_{\boldsymbol{k}\sigma}\hat{c}_{\boldsymbol{k}\sigma}\right).$$

Thus, the Hamiltonian in k-space is given by

$$\hat{H}=\sum_{\boldsymbol{k}\sigma}\varepsilon_{\boldsymbol{k}}\hat{c}^\dagger_{\boldsymbol{k}\sigma}\hat{c}_{\boldsymbol{k}\sigma}+\varepsilon_d\sum_{\boldsymbol{k}\sigma}\hat{d}^\dagger_{\boldsymbol{k}\sigma}\hat{d}_{\boldsymbol{k}\sigma}+V\sum_{\boldsymbol{k}\sigma}\left(\hat{c}^\dagger_{\boldsymbol{k}\sigma}\hat{d}_{\boldsymbol{k}\sigma}+\hat{d}^\dagger_{\boldsymbol{k}\sigma}\hat{c}_{\boldsymbol{k}\sigma}\right).$$

Because of the presence of the hybridization term, the Hamiltonian is not in a diagonal form. Since the Hamiltonian is of the quadratic form, it can be diagonalized through a Bogoliubov transformation. To see how the c operators are mixed up with the d operators, we need to turn to Heisenberg's equation of motion. If the commutator $[\hat{c}_{\boldsymbol{k}\sigma},\hat{H}]$ is evaluated for setting up the equation of motion for $\hat{c}_{\boldsymbol{k}\sigma}$, we see that the d operator, $\hat{d}_{\boldsymbol{k}\sigma}$, also appears in the result. Thus, the operators $\hat{c}_{\boldsymbol{k}\sigma}$ and $\hat{d}_{\boldsymbol{k}\sigma}$ as well as $\hat{c}^\dagger_{\boldsymbol{k}\sigma}$ and $\hat{d}^\dagger_{\boldsymbol{k}\sigma}$ are mixed up due to hybridization. In replacement of $\hat{c}_{\boldsymbol{k}\sigma}$ and $\hat{d}_{\boldsymbol{k}\sigma}$, we introduce two new operators $\hat{\alpha}_{\boldsymbol{k}\sigma}$ and $\hat{\beta}_{\boldsymbol{k}\sigma}$ through

$$\hat{\alpha}_{\boldsymbol{k}\sigma}=u_{\boldsymbol{k}}\hat{c}_{\boldsymbol{k}\sigma}+v_{\boldsymbol{k}}\hat{d}_{\boldsymbol{k}\sigma},$$

$$\hat{\beta}_{\boldsymbol{k}\sigma}=p_{\boldsymbol{k}}\hat{c}_{\boldsymbol{k}\sigma}+q_{\boldsymbol{k}}\hat{d}_{\boldsymbol{k}\sigma},$$

where $u_{\boldsymbol{k}}$, $v_{\boldsymbol{k}}$, $p_{\boldsymbol{k}}$, and $q_{\boldsymbol{k}}$ are real coefficients with their values to be determined through requirement that the Hamiltonian is in the diagonal form when expressed in terms of them. To ensure that the physics remains unchanged under the transformation, we require that the new operators satisfy the same anticommutation relations as the original operators. Specifically, the new operators must satisfy the following anticommutation relations

$$\{\hat{\alpha}_{\boldsymbol{k}\sigma}, \hat{\alpha}^\dagger_{\boldsymbol{k}'\sigma'}\} = \delta_{\boldsymbol{k}'\boldsymbol{k}}\delta_{\sigma'\sigma}, \ \{\hat{\alpha}_{\boldsymbol{k}\sigma}, \hat{\alpha}_{\boldsymbol{k}'\sigma'}\} = \{\hat{\alpha}^\dagger_{\boldsymbol{k}\sigma}, \hat{\alpha}^\dagger_{\boldsymbol{k}'\sigma'}\} = 0,$$
$$\{\hat{\beta}_{\boldsymbol{k}\sigma}, \hat{\beta}^\dagger_{\boldsymbol{k}'\sigma'}\} = \delta_{\boldsymbol{k}'\boldsymbol{k}}\delta_{\sigma'\sigma}, \ \{\hat{\beta}_{\boldsymbol{k}\sigma}, \hat{\beta}_{\boldsymbol{k}'\sigma'}\} = \{\hat{\beta}^\dagger_{\boldsymbol{k}\sigma}, \hat{\beta}^\dagger_{\boldsymbol{k}'\sigma'}\} = 0,$$
$$\{\hat{\alpha}_{\boldsymbol{k}\sigma}, \hat{\beta}_{\boldsymbol{k}'\sigma'}\} = \{\hat{\alpha}_{\boldsymbol{k}\sigma}, \hat{\beta}^\dagger_{\boldsymbol{k}'\sigma'}\} = \{\hat{\alpha}^\dagger_{\boldsymbol{k}\sigma}, \hat{\beta}_{\boldsymbol{k}'\sigma'}\} = \{\hat{\alpha}^\dagger_{\boldsymbol{k}\sigma}, \hat{\beta}^\dagger_{\boldsymbol{k}'\sigma'}\} = 0.$$

From the first anticommutation relations on the first and second lines, we have

$$u_{\boldsymbol{k}}^2 + v_{\boldsymbol{k}}^2 = 1, \ p_{\boldsymbol{k}}^2 + q_{\boldsymbol{k}}^2 = 1.$$

Note that the rest anticommutation relations on these lines are satisfied automatically. The first and last anticommutation relations on the third line are also satisfied automatically. The second and third anticommutation relations give the same result since they are Hermitian conjugates to each other. We have

$$u_{\boldsymbol{k}} p_{\boldsymbol{k}} + v_{\boldsymbol{k}} q_{\boldsymbol{k}} = 0.$$

The above equation can be satisfied by choosing $p_{\boldsymbol{k}} = -v_{\boldsymbol{k}}$ and $q_{\boldsymbol{k}} = u_{\boldsymbol{k}}$. Hence, the number of the transformation coefficients has been reduced from four to two and the transformation has been simplified. In terms of these two remaining coefficients, the transformation is written as

$$\hat{\alpha}_{\boldsymbol{k}\sigma} = u_{\boldsymbol{k}}\hat{c}_{\boldsymbol{k}\sigma} + v_{\boldsymbol{k}}\hat{d}_{\boldsymbol{k}\sigma},$$
$$\hat{\beta}_{\boldsymbol{k}\sigma} = -v_{\boldsymbol{k}}\hat{c}_{\boldsymbol{k}\sigma} + u_{\boldsymbol{k}}\hat{d}_{\boldsymbol{k}\sigma}.$$

Note that the remaining two transformation coefficients $u_{\boldsymbol{k}}$ and $v_{\boldsymbol{k}}$ satisfy $u_{\boldsymbol{k}}^2+v_{\boldsymbol{k}}^2 = 1$. We now express the Hamiltonian in terms of the new operators. For this purpose, we first express the original operators in terms of the new ones. Solving for $\hat{c}_{\boldsymbol{k}\sigma}$ and $\hat{d}_{\boldsymbol{k}\sigma}$ from the above transformation with $u_{\boldsymbol{k}}^2 + v_{\boldsymbol{k}}^2 = 1$ taken into account, we obtain

$$\hat{c}_{\boldsymbol{k}\sigma} = u_{\boldsymbol{k}}\hat{\alpha}_{\boldsymbol{k}\sigma} - v_{\boldsymbol{k}}\hat{\beta}_{\boldsymbol{k}\sigma},$$
$$\hat{d}_{\boldsymbol{k}\sigma} = v_{\boldsymbol{k}}\hat{\alpha}_{\boldsymbol{k}\sigma} + u_{\boldsymbol{k}}\hat{\beta}_{\boldsymbol{k}\sigma}.$$

Inserting the above expressions of the original operators into the k-space expression of the Hamiltonian, we can obtain an expression of the Hamiltonian in terms of the new operators, $\hat{\alpha}_{\boldsymbol{k}\sigma}$ and $\hat{\beta}_{\boldsymbol{k}\sigma}$. We find that

$$\begin{aligned}\hat{H} = \sum_{\boldsymbol{k}\sigma}\big[&\big(u_{\boldsymbol{k}}^2\varepsilon_{\boldsymbol{k}} + v_{\boldsymbol{k}}^2\varepsilon_d + 2u_{\boldsymbol{k}}v_{\boldsymbol{k}}V\big)\hat{\alpha}_{\boldsymbol{k}\sigma}^\dagger\hat{\alpha}_{\boldsymbol{k}\sigma} \\ &+ \big(v_{\boldsymbol{k}}^2\varepsilon_{\boldsymbol{k}} + u_{\boldsymbol{k}}^2\varepsilon_d - 2u_{\boldsymbol{k}}v_{\boldsymbol{k}}V\big)\hat{\beta}_{\boldsymbol{k}\sigma}^\dagger\hat{\beta}_{\boldsymbol{k}\sigma}\big] \\ &+ \sum_{\boldsymbol{k}\sigma}\big[u_{\boldsymbol{k}}v_{\boldsymbol{k}}(\varepsilon_d - \varepsilon_{\boldsymbol{k}}) + (u_{\boldsymbol{k}}^2 - v_{\boldsymbol{k}}^2)V\big](\hat{\alpha}_{\boldsymbol{k}\sigma}^\dagger\hat{\beta}_{\boldsymbol{k}\sigma} + \hat{\beta}_{\boldsymbol{k}\sigma}^\dagger\hat{\alpha}_{\boldsymbol{k}\sigma}).\end{aligned}$$

Because $\hat{\alpha}_{\boldsymbol{k}\sigma}^\dagger\hat{\beta}_{\boldsymbol{k}\sigma}$ and $\hat{\beta}_{\boldsymbol{k}\sigma}^\dagger\hat{\alpha}_{\boldsymbol{k}\sigma}$ are mutually Hermitian conjugates, their coefficients are identical as they should be. For $\hat{H}$ to be diagonalized, we demand that the coefficients in front of $\hat{\alpha}_{\boldsymbol{k}\sigma}^\dagger\hat{\beta}_{\boldsymbol{k}\sigma}$ and $\hat{\beta}_{\boldsymbol{k}\sigma}^\dagger\hat{\alpha}_{\boldsymbol{k}\sigma}$ vanish. We then have

$$u_{\boldsymbol{k}}v_{\boldsymbol{k}}(\varepsilon_d - \varepsilon_{\boldsymbol{k}}) + (u_{\boldsymbol{k}}^2 - v_{\boldsymbol{k}}^2)V = 0.$$

Solving for $u_{\boldsymbol{k}}$ and $v_{\boldsymbol{k}}$ from the above equation in conjunction with $u_{\boldsymbol{k}}^2 + v_{\boldsymbol{k}}^2 = 1$, we obtain

$$\begin{aligned}u_{\boldsymbol{k}}^2 &= \frac{1}{2}\left[1 + \frac{\varepsilon_{\boldsymbol{k}} - \varepsilon_d}{\sqrt{(\varepsilon_{\boldsymbol{k}} - \varepsilon_d)^2 + 4V^2}}\right], \\ v_{\boldsymbol{k}}^2 &= \frac{1}{2}\left[1 - \frac{\varepsilon_{\boldsymbol{k}} - \varepsilon_d}{\sqrt{(\varepsilon_{\boldsymbol{k}} - \varepsilon_d)^2 + 4V^2}}\right], \\ u_{\boldsymbol{k}}v_{\boldsymbol{k}} &= \frac{V}{\sqrt{(\varepsilon_{\boldsymbol{k}} - \varepsilon_d)^2 + 4V^2}}.\end{aligned}$$

The diagonalized Hamiltonian is then given by

$$\hat{H} = \sum_{\boldsymbol{k}\sigma}(E_{\boldsymbol{k}}^+\hat{\alpha}_{\boldsymbol{k}\sigma}^\dagger\hat{\alpha}_{\boldsymbol{k}\sigma} + E_{\boldsymbol{k}}^-\hat{\beta}_{\boldsymbol{k}\sigma}^\dagger\hat{\beta}_{\boldsymbol{k}\sigma}),$$

where

$$E_{\boldsymbol{k}}^{\pm} = \frac{1}{2}\big[\varepsilon_{\boldsymbol{k}} + \varepsilon_d \pm \sqrt{(\varepsilon_{\boldsymbol{k}} - \varepsilon_d)^2 + 4V^2}\,\big].$$

(2) Differentiating $G_{\sigma\sigma}(\boldsymbol{k}, t)$ with respect to t yields

$$\begin{aligned}\frac{\partial G_{\sigma\sigma}(\boldsymbol{k},t)}{\partial t} &= -\mathrm{i}\,\langle\hat{T}[\partial\hat{c}_{\boldsymbol{k}\sigma}(t)/\partial t\,\hat{c}_{\boldsymbol{k}\sigma}^\dagger(0)]\rangle - \mathrm{i}\,\langle\{\hat{c}_{\boldsymbol{k}\sigma}(0), \hat{c}_{\boldsymbol{k}\sigma}^\dagger(0)\}\rangle\,\delta(t) \\ &= -\mathrm{i}\,\langle\hat{T}[\partial\hat{c}_{\boldsymbol{k}\sigma}(t)/\partial t\,\hat{c}_{\boldsymbol{k}\sigma}^\dagger(0)]\rangle - \mathrm{i}\delta(t).\end{aligned}$$

The derivative of $\hat{c}_{\boldsymbol{k}\sigma}(t)$ with respect to t can be evaluated by expressing $\hat{c}_{\boldsymbol{k}\sigma}(t)$ in terms of $\hat{\alpha}_{\boldsymbol{k}\sigma}(t)$ and $\hat{\beta}_{\boldsymbol{k}\sigma}(t)$ using the above-derived Bogoliubov transformation. We have

$$\begin{aligned}\frac{\partial\hat{c}_{\boldsymbol{k}\sigma}(t)}{\partial t} &= \frac{\partial}{\partial t}\big[u_{\boldsymbol{k}}\hat{\alpha}_{\boldsymbol{k}\sigma}(t) - v_{\boldsymbol{k}}\hat{\beta}_{\boldsymbol{k}\sigma}(t)\big] \\ &= -\frac{\mathrm{i}}{\hbar}\big[u_{\boldsymbol{k}}E_{\boldsymbol{k}}^+\hat{\alpha}_{\boldsymbol{k}\sigma}(t) - v_{\boldsymbol{k}}E_{\boldsymbol{k}}^-\hat{\beta}_{\boldsymbol{k}\sigma}(t)\big],\end{aligned}$$

where the time dependencies of $\hat{\alpha}_{\boldsymbol{k}\sigma}(t)$ and $\hat{\beta}_{\boldsymbol{k}\sigma}(t)$,

$$\hat{\alpha}_{\boldsymbol{k}\sigma}(t) = \mathrm{e}^{-\mathrm{i}E_{\boldsymbol{k}}^{+}t/\hbar}\hat{\alpha}_{\boldsymbol{k}\sigma}, \quad \hat{\beta}_{\boldsymbol{k}\sigma}(t) = \mathrm{e}^{-\mathrm{i}E_{\boldsymbol{k}}^{-}t/\hbar}\hat{\beta}_{\boldsymbol{k}\sigma}$$

have been used. Reexpressing $\partial\hat{c}_{\boldsymbol{k}\sigma}(t)/\partial t$ in terms of $\hat{c}_{\boldsymbol{k}\sigma}(t)$ and $\hat{d}_{\boldsymbol{k}\sigma}(t)$, we have

$$\begin{aligned}
\mathrm{i}\hbar\frac{\partial\hat{c}_{\boldsymbol{k}\sigma}(t)}{\partial t} &= u_{\boldsymbol{k}}E_{\boldsymbol{k}}^{+}\hat{\alpha}_{\boldsymbol{k}\sigma}(t) - v_{\boldsymbol{k}}E_{\boldsymbol{k}}^{-}\hat{\beta}_{\boldsymbol{k}\sigma}(t) \\
&= u_{\boldsymbol{k}}E_{\boldsymbol{k}}^{+}\left[u_{\boldsymbol{k}}\hat{c}_{\boldsymbol{k}\sigma}(t) + v_{\boldsymbol{k}}\hat{d}_{\boldsymbol{k}\sigma}(t)\right] \\
&\quad - v_{\boldsymbol{k}}E_{\boldsymbol{k}}^{-}\left[-v_{\boldsymbol{k}}\hat{c}_{\boldsymbol{k}\sigma}(t) + u_{\boldsymbol{k}}\hat{d}_{\boldsymbol{k}\sigma}(t)\right] \\
&= \left(u_{\boldsymbol{k}}^{2}E_{\boldsymbol{k}}^{+} + v_{\boldsymbol{k}}^{2}E_{\boldsymbol{k}}^{-}\right)\hat{c}_{\boldsymbol{k}\sigma}(t) + u_{\boldsymbol{k}}v_{\boldsymbol{k}}\left(E_{\boldsymbol{k}}^{+} - E_{\boldsymbol{k}}^{-}\right)\hat{d}_{\boldsymbol{k}\sigma}(t) \\
&= \varepsilon_{\boldsymbol{k}}\hat{c}_{\boldsymbol{k}\sigma}(t) + V\hat{d}_{\boldsymbol{k}\sigma}(t).
\end{aligned}$$

The derivative of $G_{\uparrow}(\boldsymbol{k},t)$ with respect to t is then given by

$$\begin{aligned}
\frac{\partial G_{\sigma\sigma}(\boldsymbol{k},t)}{\partial t} &= -\frac{\varepsilon_{\boldsymbol{k}}}{\hbar}\langle\hat{T}[\hat{c}_{\boldsymbol{k}\sigma}(t)\hat{c}_{\boldsymbol{k}\sigma}^{\dagger}(0)]\rangle - \frac{V}{\hbar}\langle\hat{T}[\hat{d}_{\boldsymbol{k}\sigma}(t)\hat{c}_{\boldsymbol{k}\sigma}^{\dagger}(0)]\rangle - \mathrm{i}\delta(t) \\
&= -\frac{\mathrm{i}\varepsilon_{\boldsymbol{k}}}{\hbar}G_{\sigma\sigma}(\boldsymbol{k},t) - \frac{\mathrm{i}V}{\hbar}F_{\sigma\sigma}(\boldsymbol{k},t) - \mathrm{i}\delta(t).
\end{aligned}$$

Therefore, the Green's function $F_{\sigma\sigma}(\boldsymbol{k},t) = -\mathrm{i}\,\langle\hat{T}[\hat{d}_{\boldsymbol{k}\sigma}(t)\hat{c}_{\boldsymbol{k}\sigma}^{\dagger}(0)]\rangle$ is contained in the equation of motion for $G_{\sigma\sigma}(\boldsymbol{k},t)$. To set up the equation of motion for $F_{\sigma\sigma}(\boldsymbol{k},t)$, we differentiate $F_{\sigma\sigma}(\boldsymbol{k},t)$ with respect to t and obtain

$$\begin{aligned}
\frac{\partial F_{\sigma\sigma}(\boldsymbol{k},t)}{\partial t} &= -\mathrm{i}\,\langle\hat{T}[\partial\hat{d}_{\boldsymbol{k}\sigma}(t)/\partial t\,\hat{c}_{\boldsymbol{k}\sigma}^{\dagger}(0)]\rangle - \mathrm{i}\,\langle\{\hat{d}_{\boldsymbol{k}\sigma}(0),\hat{c}_{\boldsymbol{k}\sigma}^{\dagger}(0)\}\rangle\,\delta(t) \\
&= -\mathrm{i}\,\langle\hat{T}[\partial\hat{d}_{\boldsymbol{k}\sigma}(t)/\partial t\,\hat{c}_{\boldsymbol{k}\sigma}^{\dagger}(0)]\rangle,
\end{aligned}$$

where $\{\hat{d}_{\boldsymbol{k}\sigma}(0),\hat{c}_{\boldsymbol{k}\sigma}^{\dagger}(0)\} = 0$ has been used. We evaluate $\partial\hat{d}_{\boldsymbol{k}\sigma}(t)/\partial t$ in the similar manner as for $\partial\hat{c}_{\boldsymbol{k}\sigma}(t)/\partial t$

$$\begin{aligned}
\mathrm{i}\hbar\frac{\partial\hat{d}_{\boldsymbol{k}\sigma}(t)}{\partial t} &= \mathrm{i}\hbar\frac{\partial}{\partial t}\left[v_{\boldsymbol{k}}\hat{\alpha}_{\boldsymbol{k}\sigma} + u_{\boldsymbol{k}}\hat{\beta}_{\boldsymbol{k}\sigma}\right] = v_{\boldsymbol{k}}E_{\boldsymbol{k}}^{+}\hat{\alpha}_{\boldsymbol{k}\sigma}(t) + u_{\boldsymbol{k}}E_{\boldsymbol{k}}^{-}\hat{\beta}_{\boldsymbol{k}\sigma}(t) \\
&= v_{\boldsymbol{k}}E_{\boldsymbol{k}}^{+}\left[u_{\boldsymbol{k}}\hat{c}_{\boldsymbol{k}\sigma}(t) + v_{\boldsymbol{k}}\hat{d}_{\boldsymbol{k}\sigma}(t)\right] \\
&\quad + u_{\boldsymbol{k}}E_{\boldsymbol{k}}^{-}\left[-v_{\boldsymbol{k}}\hat{c}_{\boldsymbol{k}\sigma}(t) + u_{\boldsymbol{k}}\hat{d}_{\boldsymbol{k}\sigma}(t)\right] \\
&= u_{\boldsymbol{k}}v_{\boldsymbol{k}}\left(E_{\boldsymbol{k}}^{+} - E_{\boldsymbol{k}}^{-}\right)\hat{c}_{\boldsymbol{k}\sigma}(t) + \left(v_{\boldsymbol{k}}^{2}E_{\boldsymbol{k}}^{+} + u_{\boldsymbol{k}}^{2}E_{\boldsymbol{k}}^{-}\right)\hat{d}_{\boldsymbol{k}\sigma}(t) \\
&= V\hat{c}_{\boldsymbol{k}\sigma}(t) + \varepsilon_{d}\hat{d}_{\boldsymbol{k}\sigma}(t).
\end{aligned}$$

Thus,

$$\begin{aligned}
\frac{\partial F_{\sigma\sigma}(\boldsymbol{k},t)}{\partial t} &= -\frac{\varepsilon_{d}}{\hbar}\langle\hat{T}[\hat{d}_{\boldsymbol{k}\sigma}(t)\hat{c}_{\boldsymbol{k}\sigma}^{\dagger}(0)]\rangle - \frac{V}{\hbar}\langle\hat{T}[\hat{c}_{\boldsymbol{k}\sigma}(t)\hat{c}_{\boldsymbol{k}\sigma}^{\dagger}(0)]\rangle \\
&= -\frac{\mathrm{i}\varepsilon_{d}}{\hbar}F_{\sigma\sigma}(\boldsymbol{k},t) - \frac{\mathrm{i}V}{\hbar}G_{\sigma\sigma}(\boldsymbol{k},t).
\end{aligned}$$

Note that we can also evaluate $\partial \hat{c}_{\boldsymbol{k}\sigma}(t)/\partial t$ and $\partial \hat{d}_{\boldsymbol{k}\sigma}(t)/\partial t$ using the Heisenberg equation of motion. Up to now, we have obtained two equations for $G_{\sigma\sigma}(\boldsymbol{k},t)$ and $F_{\sigma\sigma}(\boldsymbol{k},t)$

$$\frac{\partial G_{\sigma\sigma}(\boldsymbol{k},t)}{\partial t} = -\frac{\mathrm{i}\varepsilon_{\boldsymbol{k}}}{\hbar}G_{\sigma\sigma}(\boldsymbol{k},t) - \frac{\mathrm{i}V}{\hbar}F_{\sigma\sigma}(\boldsymbol{k},t) - \mathrm{i}\delta(t),$$

$$\frac{\partial F_{\sigma\sigma}(\boldsymbol{k},t)}{\partial t} = -\frac{\mathrm{i}\varepsilon_d}{\hbar}F_{\sigma\sigma}(\boldsymbol{k},t) - \frac{\mathrm{i}V}{\hbar}G_{\sigma\sigma}(\boldsymbol{k},t).$$

To solve the above set of differential equations, we make the following Fourier transformations to $G_{\sigma\sigma}(\boldsymbol{k},t)$ and $F_{\sigma\sigma}(\boldsymbol{k},t)$ with respect to t

$$G_{\sigma\sigma}(\boldsymbol{k},t) = \int_{-\infty}^{\infty}\frac{\mathrm{d}(\hbar\omega)}{2\pi}\,\mathrm{e}^{-\mathrm{i}\omega t}G_{\sigma\sigma}(\boldsymbol{k},\omega),$$

$$F_{\sigma\sigma}(\boldsymbol{k},t) = \int_{-\infty}^{\infty}\frac{\mathrm{d}(\hbar\omega)}{2\pi}\,\mathrm{e}^{-\mathrm{i}\omega t}F_{\sigma\sigma}(\boldsymbol{k},\omega).$$

Substituting the above Fourier-transformed Green's functions into the equations for $G_{\sigma\sigma}(\boldsymbol{k},t)$ and $F_{\sigma\sigma}(\boldsymbol{k},t)$ and making use of $\delta(t) = (2\pi)^{-1}\int_{-\infty}^{\infty}\mathrm{d}\omega\,\mathrm{e}^{-\mathrm{i}\omega t}$, we have

$$\big(\hbar\omega - \varepsilon_{\boldsymbol{k}}\big)G_{\sigma\sigma}(\boldsymbol{k},\omega) - VF_{\sigma\sigma}(\boldsymbol{k},\omega) = 1,$$

$$-VG_{\sigma\sigma}(\boldsymbol{k},\omega) + \big(\hbar\omega - \varepsilon_d\big)F_{\sigma\sigma}(\boldsymbol{k},\omega) = 0.$$

Solve for $G_{\sigma\sigma}(\boldsymbol{k},\omega)$ and $F_{\sigma\sigma}(\boldsymbol{k},\omega)$ from the above equations, we obtain

$$G_{\sigma\sigma}(\boldsymbol{k},\omega) = \frac{1}{\hbar\omega - \varepsilon_{\boldsymbol{k}} - V^2/(\hbar\omega - \varepsilon_d)},$$

$$F_{\sigma\sigma}(\boldsymbol{k},\omega) = \frac{V}{(\hbar\omega - \varepsilon_{\boldsymbol{k}})(\hbar\omega - \varepsilon_d) - V^2}.$$

(3) From the expression of $G_{\sigma\sigma}(\boldsymbol{k},\omega)$, we see that the self-energy of a conduction electron is

$$\Sigma(\hbar\omega) = \frac{V^2}{\hbar\omega - \varepsilon_d}.$$

26-9 Fourth-order corrections to phonon Green's function at finite temperatures. Consider the fourth-order corrections to the phonon Green's function $\mathscr{D}_s(\boldsymbol{q},\mathrm{i}\omega_m)$ at finite temperatures.

(1) Write down the contributions corresponding to the Feynman diagrams in Fig. 26.3 using the Feynman rules.

(2) Verify the above results by evaluating explicitly the relevant terms in

$$\mathscr{D}_s^{(4)}(\boldsymbol{q},\tau-\tau') = -\int_0^{\hbar\beta}\frac{\mathrm{d}\tau_1}{\hbar}\cdots\int_0^{\hbar\beta}\frac{\mathrm{d}\tau_4}{\hbar}$$

$$\times\big\langle\hat{T}_\tau\big[\hat{\phi}_{\boldsymbol{q}s}(\tau)\hat{\phi}_{-\boldsymbol{q}s}(\tau')\hat{H}_{\mathrm{ep}}(\tau_1)\cdots\hat{H}_{\mathrm{ep}}(\tau_4)\big]\big\rangle_{0,\mathrm{dc}}.$$

(1) According to the Feynman rules, we have for Fig. 26.3(a)

$$\begin{aligned}
&\mathscr{D}_s^{(4a)}(\boldsymbol{q}, \mathrm{i}\omega_m)\\
&= \mathscr{D}_{s0}(\boldsymbol{q}, \mathrm{i}\omega_m) \cdot \mathrm{i}g_{\boldsymbol{q}s} \cdot \frac{1}{\beta} \sum_{\boldsymbol{k}_1 n_1 \sigma_1} \mathscr{G}_{\sigma_1 0}(\boldsymbol{k}_1, \mathrm{i}\omega_{n_1})\\
&\quad \cdot \mathscr{G}_{\sigma_1 0}(\boldsymbol{k}_1 - \boldsymbol{q}, \mathrm{i}\omega_{n_1} - \mathrm{i}\omega_m) \cdot \sum_{s'} \mathrm{i}g_{\boldsymbol{q}s'} \cdot \mathscr{D}_{s'0}(\boldsymbol{q}, \mathrm{i}\omega_m) \cdot \mathrm{i}g^*_{\boldsymbol{q}s'}\\
&\quad \cdot \frac{1}{\beta} \sum_{\boldsymbol{k}_2 n_2 \sigma_2} \mathscr{G}_{\sigma_2 0}(\boldsymbol{k}_2, \mathrm{i}\omega_{n_2}) \cdot \mathscr{G}_{\sigma_2 0}(\boldsymbol{k}_2 - \boldsymbol{q}, \mathrm{i}\omega_{n_2} - \mathrm{i}\omega_m)\\
&\quad \cdot \mathrm{i}g^*_{\boldsymbol{q}s} \cdot \mathscr{D}_{s0}(\boldsymbol{q}, \mathrm{i}\omega_m) \cdot (-1)^2\\
&= |g_{\boldsymbol{q}s}|^2 \left[\mathscr{D}_{s0}(\boldsymbol{q}, \mathrm{i}\omega_m) \right]^2 \sum_{s'} |g_{\boldsymbol{q}s'}|^2 \mathscr{D}_{s'0}(\boldsymbol{q}, \mathrm{i}\omega_m)\\
&\qquad \times \left[\frac{1}{\beta} \sum_{\boldsymbol{k} n \sigma} \mathscr{G}_{\sigma 0}(\boldsymbol{k}, \mathrm{i}\omega_n) \mathscr{G}_{\sigma 0}(\boldsymbol{k} - \boldsymbol{q}, \mathrm{i}\omega_n - \mathrm{i}\omega_m) \right]^2.
\end{aligned}$$

For Fig. 26.3(b), we have

$$\begin{aligned}
\mathscr{D}_s^{(4b)}(\boldsymbol{q}, \mathrm{i}\omega_m) &= \mathscr{D}_{s0}(\boldsymbol{q}, \mathrm{i}\omega_m) \cdot \mathrm{i}g_{\boldsymbol{q}s} \cdot \frac{1}{\beta} \sum_{\boldsymbol{k}_1 n_1 \sigma} \mathscr{G}_{\sigma 0}(\boldsymbol{k}_1, \mathrm{i}\omega_{n_1})\\
&\quad \cdot \mathscr{G}_{\sigma 0}(\boldsymbol{k}_1 - \boldsymbol{q}, \mathrm{i}\omega_{n_1} - \mathrm{i}\omega_m)\\
&\cdot \frac{1}{\beta} \sum_{\boldsymbol{k}_2 n_2} \sum_{s'} \mathrm{i}g_{\boldsymbol{k}_1 - \boldsymbol{k}_2, s'} \cdot \mathscr{D}_{s'0}(\boldsymbol{k}_1 - \boldsymbol{k}_2, \mathrm{i}\omega_m) \cdot \mathrm{i}g^*_{\boldsymbol{k}_1 - \boldsymbol{k}_2, s'}\\
&\cdot \mathscr{G}_{\sigma 0}(\boldsymbol{k}_2, \mathrm{i}\omega_{n_2}) \cdot \mathscr{G}_{\sigma 0}(\boldsymbol{k}_2 - \boldsymbol{q}, \mathrm{i}\omega_{n_2} - \mathrm{i}\omega_m)\\
&\cdot \mathrm{i}g^*_{\boldsymbol{q}s} \cdot \mathscr{D}_{s0}(\boldsymbol{q}, \mathrm{i}\omega_m) \cdot (-1)\\
&= -|g_{\boldsymbol{q}s}|^2 \left[\mathscr{D}_{s0}(\boldsymbol{q}, \mathrm{i}\omega_m) \right]^2 \frac{1}{\beta^2} \sum_{\boldsymbol{k}\boldsymbol{k}' n n' \sigma s'} |g_{\boldsymbol{k}-\boldsymbol{k}', s'}|^2 \mathscr{D}_{s'0}(\boldsymbol{k} - \boldsymbol{k}', \mathrm{i}\omega_m)\\
&\quad \times \mathscr{G}_{\sigma 0}(\boldsymbol{k}, \mathrm{i}\omega_n) \mathscr{G}_{\sigma 0}(\boldsymbol{k} - \boldsymbol{q}, \mathrm{i}\omega_n - \mathrm{i}\omega_m) \mathscr{G}_{\sigma 0}(\boldsymbol{k}', \mathrm{i}\omega_{n'})\\
&\quad \times \mathscr{G}_{\sigma 0}(\boldsymbol{k}' - \boldsymbol{q}, \mathrm{i}\omega_{n'} - \mathrm{i}\omega_m).
\end{aligned}$$

For Fig. 26.3(c), we have

$$\begin{aligned}
&\mathscr{D}_s^{(4c)}(\boldsymbol{q}, \mathrm{i}\omega_m)\\
&= \mathscr{D}_{s0}(\boldsymbol{q}, \mathrm{i}\omega_m) \cdot \mathrm{i}g_{\boldsymbol{q}s} \cdot \frac{1}{\beta} \sum_{\boldsymbol{k}_1 n_1 \sigma} \mathscr{G}_{\sigma 0}(\boldsymbol{k}_1, \mathrm{i}\omega_{n_1})\\
&\quad \cdot \mathscr{G}_{\sigma 0}(\boldsymbol{k}_1 - \boldsymbol{q}, \mathrm{i}\omega_{n_1} - \mathrm{i}\omega_m)\\
&\quad \cdot \frac{1}{\beta} \sum_{\boldsymbol{k}_2 n_2} \sum_{s'} \mathrm{i}g_{\boldsymbol{k}_1 - \boldsymbol{k}_2, s'} \cdot \mathscr{D}_{s'0}(\boldsymbol{k}_1 - \boldsymbol{k}_2, \mathrm{i}\omega_m) \cdot \mathrm{i}g^*_{\boldsymbol{k}_1 - \boldsymbol{k}_2, s'}
\end{aligned}$$

$$\cdot \mathscr{G}_{\sigma 0}(\boldsymbol{k}_2, \mathrm{i}\omega_{n_2}) \cdot \mathscr{G}_{\sigma 0}(\boldsymbol{k}_1, \mathrm{i}\omega_{n_1})$$
$$\cdot \mathrm{i} g^*_{\boldsymbol{q}s} \cdot \mathscr{D}_{s0}(\boldsymbol{q}, \mathrm{i}\omega_m) \cdot (-1)$$
$$= -|g_{\boldsymbol{q}s}|^2 \big[\mathscr{D}_{s0}(\boldsymbol{q}, \mathrm{i}\omega_m)\big]^2 \frac{1}{\beta^2} \sum_{\boldsymbol{k}\boldsymbol{k}'nn'\sigma s'} |g_{\boldsymbol{k}-\boldsymbol{k}',s'}|^2 \mathscr{D}_{s'0}(\boldsymbol{k}-\boldsymbol{k}', \mathrm{i}\omega_m)$$
$$\times \big[\mathscr{G}_{\sigma 0}(\boldsymbol{k}, \mathrm{i}\omega_n)\big]^2 \mathscr{G}_{\sigma 0}(\boldsymbol{k}-\boldsymbol{q}, \mathrm{i}\omega_n - \mathrm{i}\omega_m) \mathscr{G}_{\sigma 0}(\boldsymbol{k}', \mathrm{i}\omega_{n'}).$$

(2) Inserting

$$\hat{H}_{\mathrm{ep}} = \sum_{\boldsymbol{k}\boldsymbol{q}s\sigma} g_{\boldsymbol{q}s} \hat{c}^\dagger_{\boldsymbol{k}+\boldsymbol{q},\sigma} \hat{c}_{\boldsymbol{k}\sigma} (\hat{a}_{\boldsymbol{q}s} + \hat{a}^\dagger_{-\boldsymbol{q}s})$$

into the above-given expression for $\mathscr{D}^{(4)}_s(\boldsymbol{q}, \tau - \tau')$ yields

$$-\frac{1}{\hbar^4} \int_0^{\hbar\beta} \mathrm{d}\tau_1 \cdots \mathrm{d}\tau_4 \, \big\langle \hat{T}_\tau \big[\hat{\phi}_{\boldsymbol{q}s}(\tau) \hat{\phi}_{-\boldsymbol{q}s}(\tau') \hat{H}_{\mathrm{ep}}(\tau_1) \cdots \hat{H}_{\mathrm{ep}}(\tau_4)\big]\big\rangle_{0,\mathrm{dc}}$$
$$= -\frac{1}{\hbar^4} \sum_{\boldsymbol{k}_1 \cdots \boldsymbol{k}_4} \sum_{\boldsymbol{q}_1 \cdots \boldsymbol{q}_4} \sum_{s_1 \cdots s_4} \sum_{\sigma_1 \cdots \sigma_4} g_{\boldsymbol{q}_1 s_1} g_{\boldsymbol{q}_2 s_2} g_{\boldsymbol{q}_3 s_3} g_{\boldsymbol{q}_4 s_4} \int_0^{\hbar\beta} \mathrm{d}\tau_1 \cdots \mathrm{d}\tau_4$$
$$\times \big\langle \hat{T}_\tau \big[\hat{\phi}_{\boldsymbol{q}s}(\tau) \hat{\phi}_{-\boldsymbol{q}s}(\tau') \hat{\phi}_{\boldsymbol{q}_1 s_1}(\tau_1) \hat{\phi}_{\boldsymbol{q}_2 s_2}(\tau_2) \hat{\phi}_{\boldsymbol{q}_3 s_3}(\tau_3) \hat{\phi}_{\boldsymbol{q}_4 s_4}(\tau_4)\big]\big\rangle_{0,\mathrm{dc}}$$
$$\times \big\langle \hat{T}_\tau \big[\hat{c}^\dagger_{\boldsymbol{k}_1+\boldsymbol{q}_1,\sigma_1}(\tau_1) \hat{c}_{\boldsymbol{k}_1\sigma_1}(\tau_1) \hat{c}^\dagger_{\boldsymbol{k}_2+\boldsymbol{q}_2,\sigma_2}(\tau_2) \hat{c}_{\boldsymbol{k}_2\sigma_2}(\tau_2)$$
$$\times \hat{c}^\dagger_{\boldsymbol{k}_3+\boldsymbol{q}_3,\sigma_3}(\tau_3) \hat{c}_{\boldsymbol{k}_3\sigma_3}(\tau_3) \hat{c}^\dagger_{\boldsymbol{k}_4+\boldsymbol{q}_4,\sigma_4}(\tau_4) \hat{c}_{\boldsymbol{k}_4\sigma_4}(\tau_4)\big]\big\rangle_{0,\mathrm{dc}}.$$

From the structure of Fig. 26.3(a), we see that it comes from the term in which the phonon operators are paired up in such a manner: (τ, τ_1), (τ_2, τ_3), and (τ_4, τ'). The electron operators must be paired up in the manner: (τ_1, τ_2) and (τ_3, τ_4). We then have

$$\mathscr{D}^{(4a)}_s(\boldsymbol{q}, \tau - \tau')$$
$$= -\frac{(-1)^7}{\hbar^4} \sum_{\boldsymbol{k}_1 \cdots \boldsymbol{k}_4} \sum_{\boldsymbol{q}_1 \cdots \boldsymbol{q}_4} \sum_{s_1 \cdots s_4} \sum_{\sigma_1 \cdots \sigma_4} g_{\boldsymbol{q}_1 s_1} g_{\boldsymbol{q}_2 s_2} g_{\boldsymbol{q}_3 s_3} g_{\boldsymbol{q}_4 s_4} \int_0^{\hbar\beta} \mathrm{d}\tau_1 \cdots \mathrm{d}\tau_4$$
$$\times \mathscr{D}_{s0}(\boldsymbol{q}, \tau - \tau_1) \delta_{\boldsymbol{q}_1,-\boldsymbol{q}} \delta_{s_1 s} \mathscr{D}_{s_2 0}(\boldsymbol{q}_2, \tau_2 - \tau_3) \delta_{\boldsymbol{q}_3,-\boldsymbol{q}_2} \delta_{s_3 s_2}$$
$$\times \mathscr{D}_{s0}(\boldsymbol{q}, \tau_4 - \tau') \delta_{\boldsymbol{q}_4 \boldsymbol{q}} \delta_{s_4 s} \mathscr{G}_{\sigma_1 0}(\boldsymbol{k}_1, \tau_1 - \tau_2) \delta_{\boldsymbol{k}_2+\boldsymbol{q}_2,\boldsymbol{k}_1} \delta_{\sigma_2\sigma_1}$$
$$\times (-1) \mathscr{G}_{\sigma_1 0}(\boldsymbol{k}_1 + \boldsymbol{q}_1, \tau_2 - \tau_1) \delta_{\boldsymbol{k}_1+\boldsymbol{q}_1,\boldsymbol{k}_2} \delta_{\sigma_2\sigma_1} \mathscr{G}_{\sigma_3 0}(\boldsymbol{k}_3, \tau_3 - \tau_4)$$
$$\times \delta_{\boldsymbol{k}_4+\boldsymbol{q}_4,\boldsymbol{k}_3} \delta_{\sigma_4\sigma_3} (-1) \mathscr{G}_{\sigma_3 0}(\boldsymbol{k}_3 + \boldsymbol{q}_3, \tau_4 - \tau_3) \delta_{\boldsymbol{k}_3+\boldsymbol{q}_3,\boldsymbol{k}_4} \delta_{\sigma_4\sigma_3},$$

where each Green's function contributes a factor of -1 and the transpositions of electron creation and annihilation operators lead to the appearance of a factor of -1 for each of two electron Green's functions among the four electron Green's functions. The presence of δ-symbols in the above equation allows some of the summations over wave vectors, spin indices, and branch indices to be performed. We have

$$\begin{aligned}\mathscr{D}_s^{(4a)}(\boldsymbol{q},\tau-\tau') &= \frac{1}{\hbar^4}\sum_{\boldsymbol{k}_1\boldsymbol{k}_2\sigma_1\sigma_2 s'} |g_{\boldsymbol{q}s}|^2 |g_{\boldsymbol{q}s'}|^2 \int_0^{\hbar\beta} \mathrm{d}\tau_1\cdots\mathrm{d}\tau_4 \\ &\quad\times \mathscr{D}_{s0}(\boldsymbol{q},\tau-\tau_1)\mathscr{D}_{s'0}(\boldsymbol{q},\tau_2-\tau_3)\mathscr{D}_{s0}(\boldsymbol{q},\tau_4-\tau') \\ &\quad\times \mathscr{G}_{\sigma_1 0}(\boldsymbol{k}_1,\tau_1-\tau_2)\mathscr{G}_{\sigma_1 0}(\boldsymbol{k}_1-\boldsymbol{q},\tau_2-\tau_1) \\ &\quad\times \mathscr{G}_{\sigma_2 0}(\boldsymbol{k}_2,\tau_3-\tau_4)\mathscr{G}_{\sigma_2 0}(\boldsymbol{k}_2-\boldsymbol{q},\tau_4-\tau_3),\end{aligned}$$

where we have made use of $g_{-\boldsymbol{q}s} = g^*_{\boldsymbol{q}s}$ and renamed some of the dummy summation variables. To obtain $\mathscr{D}_s^{(4a)}(\boldsymbol{q},\omega)$, we now insert the Fourier-transformed electron and phonon Green's functions,

$$\begin{aligned}\mathscr{G}_{\sigma 0}(\boldsymbol{k},\tau) &= \frac{1}{\beta}\sum_n \mathscr{G}_{\sigma 0}(\boldsymbol{k},\mathrm{i}\omega_n)\mathrm{e}^{-\mathrm{i}\omega_n\tau}, \\ \mathscr{D}_{s0}(\boldsymbol{q},\tau) &= \frac{1}{\beta}\sum_m \mathscr{D}_{s0}(\boldsymbol{q},\mathrm{i}\omega_m)\mathrm{e}^{-\mathrm{i}\omega_m\tau},\end{aligned}$$

into $\mathscr{D}_s^{(4a)}(\boldsymbol{q},\tau-\tau')$ and obtain

$$\begin{aligned}&\mathscr{D}_s^{(4a)}(\boldsymbol{q},\tau-\tau') \\ &= \frac{1}{\beta^7\hbar^4}\sum_{\boldsymbol{k}_1\boldsymbol{k}_2\sigma_1\sigma_2 s'} |g_{\boldsymbol{q}s}|^2 |g_{\boldsymbol{q}s'}|^2 \sum_{m_1 m_2 m_3}\sum_{n_1\cdots n_4}\int_0^{\hbar\beta}\mathrm{d}\tau_1\cdots\mathrm{d}\tau_4 \\ &\times \mathrm{e}^{-\mathrm{i}\omega_{m_1}(t-\tau_1)}\mathrm{e}^{-\mathrm{i}\omega_{m_2}(\tau_2-\tau_3)}\mathrm{e}^{-\mathrm{i}\omega_{m_3}(\tau_4-t')}\mathrm{e}^{-\mathrm{i}\omega_{n_1}(\tau_1-\tau_2)}\mathrm{e}^{-\mathrm{i}\omega_{n_2}(\tau_2-\tau_1)} \\ &\times \mathrm{e}^{-\mathrm{i}\omega_{n_3}(\tau_3-\tau_4)}\mathrm{e}^{-\mathrm{i}\omega_{n_4}(\tau_4-\tau_3)}\mathscr{D}_{s0}(\boldsymbol{q},\mathrm{i}\omega_{m_1})\mathscr{D}_{s'0}(\boldsymbol{q},\mathrm{i}\omega_{m_2})\mathscr{D}_{s0}(\boldsymbol{q},\mathrm{i}\omega_{m_3}) \\ &\times \mathscr{G}_{\sigma_1 0}(\boldsymbol{k}_1,\mathrm{i}\omega_{n_1})\mathscr{G}_{\sigma_1 0}(\boldsymbol{k}_1-\boldsymbol{q},\mathrm{i}\omega_{n_2})\mathscr{G}_{\sigma_2 0}(\boldsymbol{k}_2,\mathrm{i}\omega_{n_3})\mathscr{G}_{\sigma_2 0}(\boldsymbol{k}_2-\boldsymbol{q},\mathrm{i}\omega_{n_4}).\end{aligned}$$

Combining the exponentials with the same time variables for τ_1 through τ_4 and then performing the integrations over these time variables using

$$\begin{aligned}\frac{1}{\hbar\beta}\int_0^{\hbar\beta}\mathrm{d}\tau\,\mathrm{e}^{\pm\mathrm{i}(\omega_m-\omega_{m'})\tau} &= \delta_{\omega_m\omega_{m'}}, \\ \frac{1}{\hbar\beta}\int_0^{\hbar\beta}\mathrm{d}\tau\,\mathrm{e}^{\pm\mathrm{i}(\omega_n-\omega_{n'})\tau} &= \delta_{\omega_n\omega_{n'}},\end{aligned}$$

we obtain

$$\begin{aligned}
&\mathscr{D}_s^{(4a)}(\boldsymbol{q}, \tau - \tau') \\
&= \frac{1}{\beta^3} \sum_{\boldsymbol{k}_1\boldsymbol{k}_2\sigma_1\sigma_2 s'} |g_{\boldsymbol{q}s}|^2 |g_{\boldsymbol{q}s'}|^2 \sum_{m_1 m_2 m_3} \sum_{n_1\cdots n_4} \mathrm{e}^{-\mathrm{i}(\omega_{m_1}\tau - \omega_{m_3}\tau')} \\
&\quad \times \delta_{\omega_{n_2},\omega_{n_2}-\omega_{m_1}} \delta_{\omega_{n_2},\omega_{n_2}-\omega_{m_2}} \delta_{\omega_{n_4},\omega_{n_3}-\omega_{m_2}} \\
&\quad \times \delta_{\omega_{n_4},\omega_{n_3}-\omega_{m_3}} \mathscr{D}_{s0}(\boldsymbol{q}, \mathrm{i}\omega_{m_1}) \mathscr{D}_{s'0}(\boldsymbol{q}, \mathrm{i}\omega_{m_2}) \mathscr{D}_{s0}(\boldsymbol{q}, \mathrm{i}\omega_{m_3}) \\
&\quad \times \mathscr{G}_{\sigma_1 0}(\boldsymbol{k}_1, \mathrm{i}\omega_{n_1}) \mathscr{G}_{\sigma_1 0}(\boldsymbol{k}_1 - \boldsymbol{q}, \mathrm{i}\omega_{n_2})) \\
&\quad \times \mathscr{G}_{\sigma_2 0}(\boldsymbol{k}_2, \mathrm{i}\omega_{n_3}) \mathscr{G}_{\sigma_2 0}(\boldsymbol{k}_2 - \boldsymbol{q}, \mathrm{i}\omega_{n_4}) \\
&= \frac{1}{\beta} \sum_m \mathrm{e}^{-\mathrm{i}\omega_m(\tau-\tau')} \sum_{\boldsymbol{k}_1\boldsymbol{k}_2\sigma_1\sigma_2 s'} |g_{\boldsymbol{q}s}|^2 |g_{\boldsymbol{q}s'}|^2 \\
&\qquad \times \mathscr{D}_{s0}(\boldsymbol{q}, \mathrm{i}\omega_m) \mathscr{D}_{s'0}(\boldsymbol{q}, \mathrm{i}\omega_m) \mathscr{D}_{s0}(\boldsymbol{q}, \mathrm{i}\omega_m) \\
&\qquad \times \frac{1}{\beta^2} \sum_{n_1 n_2} \mathscr{G}_{\sigma_1 0}(\boldsymbol{k}_1, \mathrm{i}\omega_{n_1}) \mathscr{G}_{\sigma_1 0}(\boldsymbol{k}_1 - \boldsymbol{q}, \mathrm{i}\omega_{n_1} - \mathrm{i}\omega_m) \\
&\qquad\qquad \times \mathscr{G}_{\sigma_2 0}(\boldsymbol{k}_2, \mathrm{i}\omega_{n_2}) \mathscr{G}_{\sigma_2 0}(\boldsymbol{k}_2 - \boldsymbol{q}, \mathrm{i}\omega_{n_2} - \mathrm{i}\omega_m),
\end{aligned}$$

where we have renamed the frequency variables after the summations over ω_{m_2}, ω_{m_3}, ω_{n_2}, and ω_{n_4} have been performed by consuming the available δ-symbols: $\omega_{m_1} \to \omega_m$ and $\omega_{n_3} \to \omega_{n_2}$. Comparing the summation over ω_m in the above expression of $\mathscr{D}_s^{(4a)}(\boldsymbol{q}, \tau - \tau')$ with its Fourier-transformed expression $\mathscr{D}_{s0}^{(4a)}(\boldsymbol{q}, \tau) = \beta^{-1} \sum_m \mathscr{D}_{s0}^{(4a)}(\boldsymbol{q}, \mathrm{i}\omega_m) \mathrm{e}^{-\mathrm{i}\omega_m \tau}$, we see immediately that $\mathscr{D}_{s0}^{(4a)}(\boldsymbol{q}, \mathrm{i}\omega_m)$ is given by

$$\begin{aligned}
&\mathscr{D}_s^{(4a)}(\boldsymbol{q}, \mathrm{i}\omega_m) \\
&= \sum_{\boldsymbol{k}_1\boldsymbol{k}_2\sigma_1\sigma_2 s'} |g_{\boldsymbol{q}s}|^2 |g_{\boldsymbol{q}s'}|^2 \mathscr{D}_{s0}(\boldsymbol{q}, \mathrm{i}\omega_m) \mathscr{D}_{s'0}(\boldsymbol{q}, \mathrm{i}\omega_m) \mathscr{D}_{s0}(\boldsymbol{q}, \mathrm{i}\omega_m) \\
&\qquad \times \frac{1}{\beta^2} \sum_{n_1 n_2} \mathscr{G}_{\sigma_1 0}(\boldsymbol{k}_1, \mathrm{i}\omega_{n_1}) \mathscr{G}_{\sigma_1 0}(\boldsymbol{k}_1 - \boldsymbol{q}, \mathrm{i}\omega_{n_1} - \mathrm{i}\omega_m) \\
&\qquad\qquad \times \mathscr{G}_{\sigma_2 0}(\boldsymbol{k}_2, \mathrm{i}\omega_{n_2}) \mathscr{G}_{\sigma_2 0}(\boldsymbol{k}_2 - \boldsymbol{q}, \mathrm{i}\omega_{n_2} - \mathrm{i}\omega_m) \\
&= |g_{\boldsymbol{q}s}|^2 \big[\mathscr{D}_{s0}(\boldsymbol{q}, \mathrm{i}\omega_m) \big]^2 \sum_{s'} |g_{\boldsymbol{q}s'}|^2 \mathscr{D}_{s'0}(\boldsymbol{q}, \mathrm{i}\omega_m) \\
&\qquad \times \left[\frac{1}{\beta} \sum_{\boldsymbol{k}n\sigma} \mathscr{G}_{\sigma 0}(\boldsymbol{k}, \mathrm{i}\omega_n) \mathscr{G}_{\sigma 0}(\boldsymbol{k} - \boldsymbol{q}, \mathrm{i}\omega_n - \mathrm{i}\omega_m) \right]^2 .
\end{aligned}$$

we have thus derived the same result as that obtained using the Feynman rules. The verification of $\mathscr{D}_s^{(4b)}(\boldsymbol{q}, \mathrm{i}\omega_m)$ and $\mathscr{D}_s^{(4c)}(\boldsymbol{q}, \mathrm{i}\omega_m)$ will not be given here to save space.

26-10 Time-ordered product of three operators. Prove explicitly the following identity for the time-ordered product of three operators

$$\int_0^t \mathrm{d}t_1 \int_0^{t_1} \mathrm{d}t_2 \int_0^{t_2} \mathrm{d}t_3\ \hat{V}(t_1)\hat{V}(t_2)\hat{V}(t_3)$$
$$= \frac{1}{3!}\int_0^t \mathrm{d}t_1 \int_0^t \mathrm{d}t_2 \int_0^t \mathrm{d}t_3\ \hat{T}\big[\hat{V}(t_1)\hat{V}(t_2)\hat{V}(t_3)\big],$$

where $\hat{V}$ is the interaction Hamiltonian.

Let us take t_1, t_2, and t_3 as three axes in a three-dimensional space. The integral on the left hand side of the identity is performed within the shaded tetrahedron in Fig. 26.4(a).

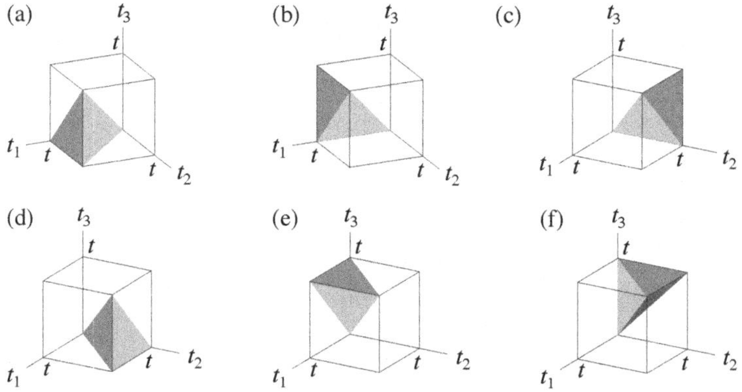

Fig. 26.4 Integration regions for different orders of t_1, t_2, and t_3. (a) $t_1 \geq t_2 \geq t_3$. (b) $t_1 \geq t_3 \geq t_2$. (c) $t_2 \geq t_3 \geq t_1$. (d) $t_2 \geq t_1 \geq t_3$. (e) $t_3 \geq t_1 \geq t_2$. (f) $t_3 \geq t_2 \geq t_1$.

The integration variables in the integral on the left hand side of the identity are in the order $t_1 \geq t_2 \geq t_3$. For the three integration variables, there are in total six different orders. The order of the integration variables in the integral on the left hand side of the identity is just one of them. The other five orders are $t_1 \geq t_3 \geq t_2$, $t_2 \geq t_3 \geq t_1$, $t_2 \geq t_1 \geq t_3$, $t_3 \geq t_1 \geq t_2$, and $t_3 \geq t_2 \geq t_1$ with the corresponding tetrahedron-shaped integration regions shown in Figs. 26.4(b) through (f). These different orders are related through changes of integration variables. However, because the operators $\hat{V}(t_1)$, $\hat{V}(t_2)$, and $\hat{V}(t_3)$ may not commute, their order must be kept unaltered in making changes of integration variables. This can be guaranteed through introducing the time-ordering operator $\hat{T}$ that rearranges the operators so that their correct order with time decreasing from left to right is maintained no matter what changes of integration variables are made. Before we make changes of integration variables, we first write the product of the operators, $\hat{V}(t_1)\hat{V}(t_2)\hat{V}(t_3)$, on the left hand side of the identity as

$$\hat{T}\big[\,\hat{V}(t_1)\hat{V}(t_2)\hat{V}(t_3)\,\big].$$

Since $t_1 \geq t_2 \geq t_3$ for the integral on the left hand side of the identity, the time-ordering operator has no effect at all, that is,

$$\hat{T}\big[\,\hat{V}(t_1)\hat{V}(t_2)\hat{V}(t_3)\,\big] = \hat{V}(t_1)\hat{V}(t_2)\hat{V}(t_3) \text{ for } t_1 \geq t_2 \geq t_3.$$

However, if we use one of the other five orders for t_1, t_2, and t_3 through a change of integration variables, $\hat{T}\big[\,\hat{V}(t_1)\hat{V}(t_2)\hat{V}(t_3)\,\big]$ ensures that the correct order of the operators. Since all the six different orders of the t_1, t_2, and t_3 are here related through changes of integration variables, the values of all the following six integrals corresponding to the six integration regions in Fig. 26.4 are equal

$$\int_0^t \mathrm{d}t_1 \int_0^{t_1} \mathrm{d}t_2 \int_0^{t_2} \mathrm{d}t_3\, \hat{T}\big[\,\hat{V}(t_1)\hat{V}(t_2)\hat{V}(t_3)\,\big],$$
$$\int_0^t \mathrm{d}t_1 \int_0^{t_1} \mathrm{d}t_3 \int_0^{t_3} \mathrm{d}t_2\, \hat{T}\big[\,\hat{V}(t_1)\hat{V}(t_2)\hat{V}(t_3)\,\big],$$
$$\int_0^t \mathrm{d}t_2 \int_0^{t_2} \mathrm{d}t_3 \int_0^{t_3} \mathrm{d}t_1\, \hat{T}\big[\,\hat{V}(t_1)\hat{V}(t_2)\hat{V}(t_3)\,\big],$$
$$\int_0^t \mathrm{d}t_2 \int_0^{t_2} \mathrm{d}t_1 \int_0^{t_1} \mathrm{d}t_3\, \hat{T}\big[\,\hat{V}(t_1)\hat{V}(t_2)\hat{V}(t_3)\,\big],$$
$$\int_0^t \mathrm{d}t_3 \int_0^{t_3} \mathrm{d}t_1 \int_0^{t_1} \mathrm{d}t_2\, \hat{T}\big[\,\hat{V}(t_1)\hat{V}(t_2)\hat{V}(t_3)\,\big],$$
$$\int_0^t \mathrm{d}t_3 \int_0^{t_3} \mathrm{d}t_2 \int_0^{t_2} \mathrm{d}t_1\, \hat{T}\big[\,\hat{V}(t_1)\hat{V}(t_2)\hat{V}(t_3)\,\big].$$

If we add up the above six integrals, we obtain six times of the value of the left hand side of the identity. Thus, the value of the original integral is one sixth of the sum. On the other hand side, the combined integration region is a cube of side t as shown in Fig. 26.4. Hence, the sum of the above integrals can be written as a single integral over the cube and the integral on the left hand side of the identity is equal to one sixth of the integral over the cube

$$\int_0^t \mathrm{d}t_1 \int_0^{t_1} \mathrm{d}t_2 \int_0^{t_2} \mathrm{d}t_3\ \hat{V}(t_1)\hat{V}(t_2)\hat{V}(t_3)$$
$$= \frac{1}{6}\int_0^t \mathrm{d}t_1 \int_0^t \mathrm{d}t_2 \int_0^t \mathrm{d}t_3\ \hat{T}\big[\hat{V}(t_1)\hat{V}(t_2)\hat{V}(t_3)\big]$$
$$= \frac{1}{3!}\int_0^t \mathrm{d}t_1 \int_0^t \mathrm{d}t_2 \int_0^t \mathrm{d}t_3\ \hat{T}\big[\hat{V}(t_1)\hat{V}(t_2)\hat{V}(t_3)\big].$$

26-11 Evaluation of Matsubara sums. Evaluate the following Matsubara sums

(1) $\displaystyle \frac{1}{\beta}\sum_n \mathscr{G}_{\sigma 0}(\boldsymbol{k}, \mathrm{i}\omega_n)\mathscr{G}_{\sigma 0}(\boldsymbol{q}, \mathrm{i}\omega_n + \mathrm{i}\omega_m).$

(2) $\displaystyle \frac{1}{\beta}\sum_n \mathscr{G}_{\sigma 0}(\boldsymbol{k}, \mathrm{i}\omega_n)\mathscr{G}_{\sigma 0}(\boldsymbol{q}, \mathrm{i}\omega_m - \mathrm{i}\omega_n).$

(3) $\displaystyle \frac{1}{\beta}\sum_m \mathscr{D}_{s0}(\boldsymbol{q}, \mathrm{i}\omega_m)\mathscr{G}_{\sigma 0}(\boldsymbol{k}, \mathrm{i}\omega_n + \mathrm{i}\omega_m).$

(4) $\displaystyle \frac{1}{\beta}\sum_m \mathscr{D}_{s0}(\boldsymbol{k}, \mathrm{i}\omega_m)\mathscr{D}_{s0}(\boldsymbol{q}, \mathrm{i}\omega_\ell + \mathrm{i}\omega_m).$

Here $\omega_m = 2m\pi/\hbar\beta$, $\omega_n = (2n+1)\pi/\hbar\beta$, and $\omega_\ell = 2\ell\pi/\hbar\beta$.

---

(1) Making use of the explicit expression of $\mathscr{G}_{\sigma 0}(\boldsymbol{k}, \mathrm{i}\omega_n)$, we have

$$I_1 \equiv \frac{1}{\beta}\sum_n \mathscr{G}_{\sigma 0}(\boldsymbol{k}, \mathrm{i}\omega_n)\mathscr{G}_{\sigma 0}(\boldsymbol{q}, \mathrm{i}\omega_n + \mathrm{i}\omega_m)$$
$$= \frac{1}{\beta}\sum_n \frac{1}{\mathrm{i}\hbar\omega_n - \xi_{\boldsymbol{k}\sigma}}\frac{1}{\mathrm{i}\hbar\omega_n + \mathrm{i}\hbar\omega_m - \xi_{\boldsymbol{q}\sigma}}.$$

Note that $\mathrm{i}\hbar\omega_n$'s are first-order poles of the Fermi-Dirac distribution function $n_{\mathrm{F}}(z) = 1/(\mathrm{e}^{\beta z}+1)$, $z_n = \mathrm{i}\hbar\omega_n$. The residues of $n_{\mathrm{F}}(z)$ at these poles are all equal to $-1/\beta$. We construct a contour, referred to as C, on the complex plane of z to enclose these poles tightly using two parallel lines on the two sides of the imaginary axis and infinitesimally close to it. These two lines are joined at the infinity. The contour encloses all the points $z_n = \mathrm{i}\hbar\omega_n$ in the positive sense. It is shown in Fig. 26.5(a). Note that the contour C encloses *only* the poles of $n_{\mathrm{F}}(z)$.

Writing I_1 as a contour integral along the contour C in Fig. 26.5(a) using the residue theorem, we have

$$I_1 = -\frac{1}{2\pi\mathrm{i}}\oint_C \mathrm{d}z\ n_{\mathrm{F}}(z)\frac{1}{z - \xi_{\boldsymbol{k}\sigma}}\frac{1}{z + \mathrm{i}\hbar\omega_m - \xi_{\boldsymbol{q}\sigma}}.$$

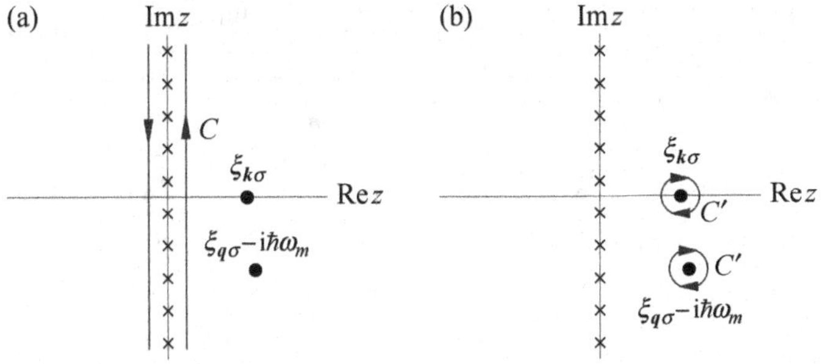

Fig. 26.5 Integration contours for the evaluation of the summation of $\mathscr{G}_{\sigma 0}(\boldsymbol{k}, i\omega_n)\mathscr{G}_{\sigma 0}(\boldsymbol{q}, i\omega_n + i\omega_m)$ over n. (a) Contour C enclosing only the poles of $n_F(z)$. (b) Contour C' enclosing only the poles of $\mathscr{G}_{\sigma 0}(\boldsymbol{k}, z/\hbar)\mathscr{G}_{\sigma 0}(\boldsymbol{q}, z/\hbar + i\omega_m)$.

We now deform the contour through the analytic region of the integrand so that the resultant contour C' encloses only the poles of $\mathscr{G}_{\sigma 0}(\boldsymbol{k}, z/\hbar)\mathscr{G}_{\sigma 0}(\boldsymbol{q}, z/\hbar + i\omega_m)$ clockwise as shown in Fig. 26.5(b). I_1 can be then written as

$$I_1 = -\frac{1}{2\pi i}\oint_{C'} dz\, n_F(z)\frac{1}{z - \xi_{\boldsymbol{k}\sigma}}\,\frac{1}{z + i\hbar\omega_m - \xi_{\boldsymbol{q}\sigma}}.$$

Note that, since the integrand tends to zero faster than $1/|z|$ as $|z| \to \infty$, the integration along any path infinitely far away from the origin is zero. Evaluating the above contour integral using the residue theorem, we have

$$I_1 = \frac{n_F(\xi_{\boldsymbol{k}\sigma})}{i\hbar\omega_m + \xi_{\boldsymbol{k}\sigma} - \xi_{\boldsymbol{q}\sigma}} + \frac{n_F(\xi_{\boldsymbol{q}\sigma} - i\hbar\omega_m)}{\xi_{\boldsymbol{q}\sigma} - i\hbar\omega_m - \xi_{\boldsymbol{k}\sigma}} = \frac{n_F(\xi_{\boldsymbol{k}\sigma}) - n_F(\xi_{\boldsymbol{q}\sigma})}{i\hbar\omega_m + \xi_{\boldsymbol{k}\sigma} - \xi_{\boldsymbol{q}\sigma}},$$

where we have made use of $n_F(\xi_{\boldsymbol{q}\sigma} - i\hbar\omega_m) = n_F(\xi_{\boldsymbol{q}\sigma})$. We thus have

$$\frac{1}{\beta}\sum_n \mathscr{G}_{\sigma 0}(\boldsymbol{k}, i\omega_n)\mathscr{G}_{\sigma 0}(\boldsymbol{q}, i\omega_n + i\omega_m) = \frac{n_F(\xi_{\boldsymbol{k}\sigma}) - n_F(\xi_{\boldsymbol{q}\sigma})}{i\hbar\omega_m + \xi_{\boldsymbol{k}\sigma} - \xi_{\boldsymbol{q}\sigma}}.$$

(2) The sum

$$\begin{aligned} I_2 &\equiv \frac{1}{\beta}\sum_n \mathscr{G}_{\sigma 0}(\boldsymbol{k}, i\omega_n)\mathscr{G}_{\sigma 0}(\boldsymbol{q}, i\omega_m - i\omega_n) \\ &= -\frac{1}{\beta}\sum_n \frac{1}{i\hbar\omega_n - \xi_{\boldsymbol{k}\sigma}}\,\frac{1}{i\hbar\omega_n - i\hbar\omega_m + \xi_{\boldsymbol{q}\sigma}} \end{aligned}$$

can be evaluated similarly to I_1. The difference lies in that the second pole from the Green's functions is now located at $i\hbar\omega_m - \xi_{\boldsymbol{q}\sigma}$. The contour C' now encloses the poles at $\xi_{\boldsymbol{k}\sigma}$ and $i\hbar\omega_m - \xi_{\boldsymbol{q}\sigma}$ as shown in Fig. 26.6(b).

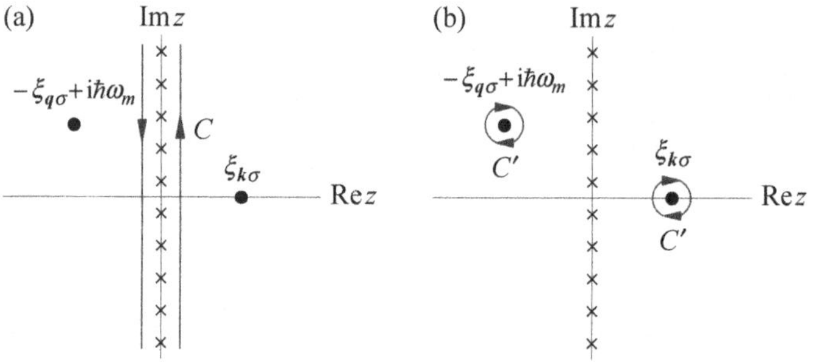

Fig. 26.6 Integration contours for the evaluation of the summation of $\mathscr{G}_{\sigma 0}(\boldsymbol{k}, \mathrm{i}\omega_n)\mathscr{G}_{\sigma 0}(\boldsymbol{q}, \mathrm{i}\omega_m - \mathrm{i}\omega_n)$ over n. (a) Contour C enclosing only the poles of $n_\mathrm{F}(z)$. (b) Contour C' enclosing only the poles of $\mathscr{G}_{\sigma 0}(\boldsymbol{k}, z/\hbar)\mathscr{G}_{\sigma 0}(\boldsymbol{q}, \mathrm{i}\omega_m - z/\hbar)$.

We first write I_2 as a contour integral along C using the residue theorem, then deform C into C', and then evaluate the resultant contour integral using the residue theorem. We have

$$\begin{aligned} I_2 &= \frac{1}{2\pi\mathrm{i}} \oint_C \mathrm{d}z\, n_\mathrm{F}(z) \frac{1}{z - \xi_{\boldsymbol{k}\sigma}} \frac{1}{z - \mathrm{i}\hbar\omega_m + \xi_{\boldsymbol{q}\sigma}} \\ &= \frac{1}{2\pi\mathrm{i}} \oint_{C'} \mathrm{d}z\, n_\mathrm{F}(z) \frac{1}{z - \xi_{\boldsymbol{k}\sigma}} \frac{1}{z - \mathrm{i}\hbar\omega_m + \xi_{\boldsymbol{q}\sigma}} \\ &= -\frac{n_\mathrm{F}(\xi_{\boldsymbol{k}\sigma})}{-\mathrm{i}\hbar\omega_m + \xi_{\boldsymbol{k}\sigma} + \xi_{\boldsymbol{q}\sigma}} - \frac{n_\mathrm{F}(\mathrm{i}\hbar\omega_m - \xi_{\boldsymbol{q}\sigma})}{\mathrm{i}\hbar\omega_m - \xi_{\boldsymbol{k}\sigma} - \xi_{\boldsymbol{q}\sigma}} \\ &= \frac{n_\mathrm{F}(\xi_{\boldsymbol{k}\sigma}) - n_\mathrm{F}(-\xi_{\boldsymbol{q}\sigma})}{\mathrm{i}\hbar\omega_m - \xi_{\boldsymbol{k}\sigma} - \xi_{\boldsymbol{q}\sigma}} = \frac{n_\mathrm{F}(\xi_{\boldsymbol{k}\sigma}) + n_\mathrm{F}(\xi_{\boldsymbol{q}\sigma}) - 1}{\mathrm{i}\hbar\omega_m - \xi_{\boldsymbol{k}\sigma} - \xi_{\boldsymbol{q}\sigma}}, \end{aligned}$$

where we have made use of $n_\mathrm{F}(\mathrm{i}\hbar\omega_m - \xi_{\boldsymbol{q}\sigma}) = n_\mathrm{F}(-\xi_{\boldsymbol{q}\sigma})$ and $n_\mathrm{F}(-\xi_{\boldsymbol{q}\sigma}) = 1 - n_\mathrm{F}(\xi_{\boldsymbol{q}\sigma})$. We thus have

$$\frac{1}{\beta}\sum_n \mathscr{G}_{\sigma 0}(\boldsymbol{k}, \mathrm{i}\omega_n)\mathscr{G}_{\sigma 0}(\boldsymbol{q}, \mathrm{i}\omega_m - \mathrm{i}\omega_n) = \frac{n_\mathrm{F}(\xi_{\boldsymbol{k}\sigma}) + n_\mathrm{F}(\xi_{\boldsymbol{q}\sigma}) - 1}{\mathrm{i}\hbar\omega_m - \xi_{\boldsymbol{k}\sigma} - \xi_{\boldsymbol{q}\sigma}}.$$

(3) Making use of the explicit expressions of $\mathscr{D}_{s0}(\boldsymbol{q}, \mathrm{i}\omega_m)$ and $\mathscr{G}_{\sigma 0}(\boldsymbol{k}, \mathrm{i}\omega_n)$, we can write the third sum as

$$\begin{aligned} I_3 &\equiv \frac{1}{\beta}\sum_m \mathscr{D}_{s0}(\boldsymbol{q}, \mathrm{i}\omega_m)\mathscr{G}_{\sigma 0}(\boldsymbol{k}, \mathrm{i}\omega_n + \mathrm{i}\omega_m) \\ &= \frac{1}{\beta}\sum_m \frac{2\hbar\omega_{\boldsymbol{q}s}}{(\mathrm{i}\hbar\omega_m)^2 - (\hbar\omega_{\boldsymbol{q}s})^2} \frac{1}{\mathrm{i}\omega_n + \mathrm{i}\omega_m - \xi_{\boldsymbol{k}\sigma}}. \end{aligned}$$

The summation over m is a summation over all the first-order poles of the Bose-Einstein distribution function $n_\mathrm{B}(z) = 1/(\mathrm{e}^{\beta z} - 1)$, $z_m = \mathrm{i}\hbar\omega_m$. The residues at these poles are all

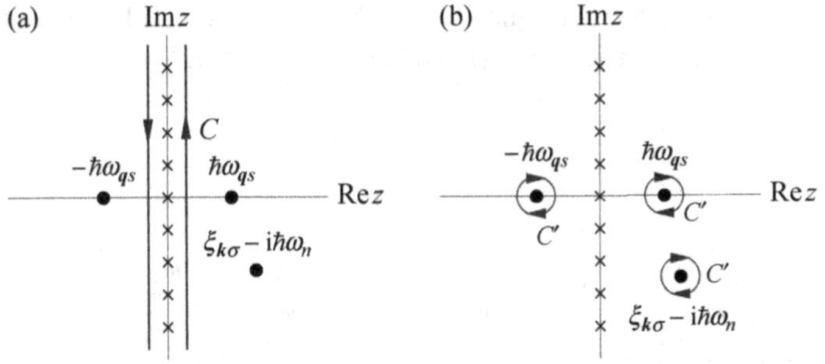

Fig. 26.7 Integration contours for the evaluation of the summation of $\mathscr{D}_{s0}(\boldsymbol{q}, \mathrm{i}\omega_m)\mathscr{G}_{\sigma 0}(\boldsymbol{k}, \mathrm{i}\omega_n + \mathrm{i}\omega_m)$ over m. (a) Contour C enclosing only the poles of $n_{\mathrm{B}}(z)$. (b) Contour C' enclosing only the poles of $\mathscr{D}_{s0}(\boldsymbol{q}, z/\hbar)\mathscr{G}_{\sigma 0}(\boldsymbol{k}, z/\hbar + \mathrm{i}\omega_n)$.

equal to $1/\beta$. These poles and the contour enclosing them are shown in Fig. 26.7(a).

We first write I_3 as a contour integral along the contour C using the residue theorem, then deform the contour C into the contour C', and then evaluate the contour integral along C' using the residue theorem. We have

$$\begin{aligned} I_3 &= \frac{1}{2\pi\mathrm{i}} \oint_C \mathrm{d}z\, n_{\mathrm{B}}(z) \frac{2\hbar\omega_{\boldsymbol{q}s}}{z^2 - (\hbar\omega_{\boldsymbol{q}s})^2} \frac{1}{z + \mathrm{i}\omega_n - \xi_{\boldsymbol{k}\sigma}} \\ &= \frac{1}{2\pi\mathrm{i}} \oint_{C'} \mathrm{d}z\, n_{\mathrm{B}}(z) \frac{2\hbar\omega_{\boldsymbol{q}s}}{z^2 - (\hbar\omega_{\boldsymbol{q}s})^2} \frac{1}{z + \mathrm{i}\omega_n - \xi_{\boldsymbol{k}\sigma}} \\ &= -\frac{n_{\mathrm{B}}(\hbar\omega_{\boldsymbol{q}s}) + n_{\mathrm{F}}(\xi_{\boldsymbol{k}\sigma})}{\mathrm{i}\omega_n - \xi_{\boldsymbol{k}\sigma} + \hbar\omega_{\boldsymbol{q}s}} - \frac{n_{\mathrm{B}}(\hbar\omega_{\boldsymbol{q}s}) - n_{\mathrm{F}}(\xi_{\boldsymbol{k}\sigma}) + 1}{\mathrm{i}\omega_n - \xi_{\boldsymbol{k}\sigma} - \hbar\omega_{\boldsymbol{q}s}}, \end{aligned}$$

where we have made use of $n_{\mathrm{B}}(\xi_{\boldsymbol{k}\sigma} - \mathrm{i}\omega_n) = -n_{\mathrm{F}}(\xi_{\boldsymbol{k}\sigma})$. We thus have

$$\begin{aligned} &\frac{1}{\beta} \sum_m \mathscr{D}_{s0}(\boldsymbol{q}, \mathrm{i}\omega_m)\mathscr{G}_{\sigma 0}(\boldsymbol{k}, \mathrm{i}\omega_n + \mathrm{i}\omega_m) \\ &\qquad = -\frac{n_{\mathrm{B}}(\hbar\omega_{\boldsymbol{q}s}) + n_{\mathrm{F}}(\xi_{\boldsymbol{k}\sigma})}{\mathrm{i}\omega_n - \xi_{\boldsymbol{k}\sigma} + \hbar\omega_{\boldsymbol{q}s}} - \frac{n_{\mathrm{B}}(\hbar\omega_{\boldsymbol{q}s}) - n_{\mathrm{F}}(\xi_{\boldsymbol{k}\sigma}) + 1}{\mathrm{i}\omega_n - \xi_{\boldsymbol{k}\sigma} - \hbar\omega_{\boldsymbol{q}s}}. \end{aligned}$$

(4) Making use of the explicit expression of $\mathscr{D}_{s0}(\boldsymbol{q}, \mathrm{i}\omega_m)$, we can write the fourth sum as

$$\begin{aligned} I_4 &\equiv \frac{1}{\beta} \sum_m \mathscr{D}_{s0}(\boldsymbol{k}, \mathrm{i}\omega_m)\mathscr{D}_{s0}(\boldsymbol{q}, \mathrm{i}\omega_\ell + \mathrm{i}\omega_m) \\ &= \frac{1}{\beta} \sum_m \frac{2\hbar\omega_{\boldsymbol{k}s}}{(\mathrm{i}\hbar\omega_m)^2 - (\hbar\omega_{\boldsymbol{k}s})^2} \frac{2\hbar\omega_{\boldsymbol{q}s}}{(\mathrm{i}\hbar\omega_m + \mathrm{i}\hbar\omega_\ell)^2 - (\hbar\omega_{\boldsymbol{q}s})^2}. \end{aligned}$$

As for I_3, the summation over m in I_4 is a summation over all the first-order poles of the Bose-Einstein distribution function $n_{\mathrm{B}}(z) = 1/(\mathrm{e}^{\beta z} - 1)$, $z_m = \mathrm{i}\hbar\omega_m$. The residues at these poles are all equal to $1/\beta$. These poles and the contour enclosing them are shown in Fig. 26.8(a).

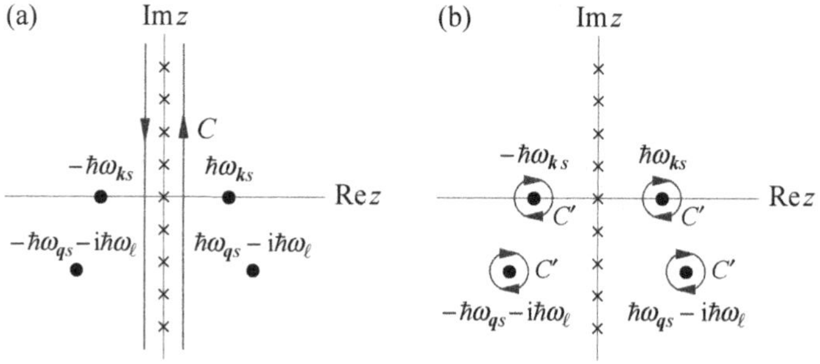

Fig. 26.8 Integration contours for the evaluation of the summation of $\mathscr{D}_{s0}(\boldsymbol{k}, \mathrm{i}\omega_m)\mathscr{D}_{s0}(\boldsymbol{q}, \mathrm{i}\omega_\ell + \mathrm{i}\omega_m)$ over m. (a) Contour C enclosing only the poles of $n_\mathrm{B}(z)$. (b) Contour C' enclosing only the poles of $\mathscr{D}_{s0}(\boldsymbol{k}, z/\hbar)\mathscr{D}_{s0}(\boldsymbol{q}, z/\hbar + \mathrm{i}\omega_\ell)$.

The sum I_4 can be written as a contour integral along C using the residue theorem

$$I_4 = \frac{1}{2\pi \mathrm{i}} \oint_C \mathrm{d}z\, n_\mathrm{B}(z) \frac{2\hbar\omega_{\boldsymbol{k}s}}{z^2 - (\hbar\omega_{\boldsymbol{k}s})^2} \frac{2\hbar\omega_{\boldsymbol{q}s}}{(z + \mathrm{i}\hbar\omega_\ell)^2 - (\hbar\omega_{\boldsymbol{q}s})^2}.$$

In addition to the poles of $n_\mathrm{B}(z)$, the integrand has first-order poles at $\pm\hbar\omega_{\boldsymbol{k}s}$ and $\pm\hbar\omega_{\boldsymbol{q}s} - \mathrm{i}\hbar\omega_\ell$ from the two Green's functions. These poles are enclosed clockwise by the integration contour C' as shown in Fig. 26.8(b). We now deform the integration contour from C to C' and then evaluate the contour integral using the residue theorem. We have

$$\begin{aligned}
I_4 &= \frac{1}{2\pi \mathrm{i}} \oint_{C'} \mathrm{d}z\, n_\mathrm{B}(z) \frac{2\hbar\omega_{\boldsymbol{k}s}}{z^2 - (\hbar\omega_{\boldsymbol{k}s})^2} \frac{2\hbar\omega_{\boldsymbol{q}s}}{(z + \mathrm{i}\hbar\omega_\ell)^2 - (\hbar\omega_{\boldsymbol{q}s})^2} \\
&= -\frac{2\hbar\omega_{\boldsymbol{q}s} n_\mathrm{B}(\hbar\omega_{\boldsymbol{k}s})}{(\mathrm{i}\hbar\omega_\ell + \hbar\omega_{\boldsymbol{k}s})^2 - (\hbar\omega_{\boldsymbol{q}s})^2} + \frac{2\hbar\omega_{\boldsymbol{q}s} n_\mathrm{B}(-\hbar\omega_{\boldsymbol{k}s})}{(\mathrm{i}\hbar\omega_\ell - \hbar\omega_{\boldsymbol{k}s})^2 - (\hbar\omega_{\boldsymbol{q}s})^2} \\
&\quad - \frac{2\hbar\omega_{\boldsymbol{k}s} n_\mathrm{B}(\hbar\omega_{\boldsymbol{q}s})}{(\mathrm{i}\hbar\omega_\ell - \hbar\omega_{\boldsymbol{q}s})^2 - (\hbar\omega_{\boldsymbol{k}s})^2} + \frac{2\hbar\omega_{\boldsymbol{k}s} n_\mathrm{B}(-\hbar\omega_{\boldsymbol{q}s})}{(\mathrm{i}\hbar\omega_\ell + \hbar\omega_{\boldsymbol{q}s})^2 - (\hbar\omega_{\boldsymbol{k}s})^2} \\
&= \left[n_\mathrm{B}(\hbar\omega_{\boldsymbol{q}s}) + n_\mathrm{B}(\hbar\omega_{\boldsymbol{k}s}) + 1\right] \\
&\quad \times \left(\frac{1}{\mathrm{i}\hbar\omega_\ell + \hbar\omega_{\boldsymbol{q}s} + \hbar\omega_{\boldsymbol{k}s}} - \frac{1}{\mathrm{i}\hbar\omega_\ell - \hbar\omega_{\boldsymbol{q}s} - \hbar\omega_{\boldsymbol{k}s}}\right) \\
&\quad + \left[n_\mathrm{B}(\hbar\omega_{\boldsymbol{q}s}) - n_\mathrm{B}(\hbar\omega_{\boldsymbol{k}s})\right] \\
&\quad \times \left(\frac{1}{\mathrm{i}\hbar\omega_\ell - \hbar\omega_{\boldsymbol{q}s} + \hbar\omega_{\boldsymbol{k}s}} - \frac{1}{\mathrm{i}\hbar\omega_\ell + \hbar\omega_{\boldsymbol{q}s} - \hbar\omega_{\boldsymbol{k}s}}\right),
\end{aligned}$$

where we have made use of $n_{\rm B}(-\hbar\omega_{\boldsymbol{q}s}) = -n_{\rm B}(\hbar\omega_{\boldsymbol{q}s}) - 1$. We thus have

$$\begin{aligned}
&\frac{1}{\beta}\sum_m \mathscr{D}_{s0}(\boldsymbol{k}, \mathrm{i}\omega_m)\mathscr{D}_{s0}(\boldsymbol{q}, \mathrm{i}\omega_\ell + \mathrm{i}\omega_m) \\
&\quad = \big[n_{\rm B}(\hbar\omega_{\boldsymbol{q}s}) + n_{\rm B}(\hbar\omega_{\boldsymbol{k}s}) + 1\big] \\
&\qquad \times \left(\frac{1}{\mathrm{i}\hbar\omega_\ell + \hbar\omega_{\boldsymbol{q}s} + \hbar\omega_{\boldsymbol{k}s}} - \frac{1}{\mathrm{i}\hbar\omega_\ell - \hbar\omega_{\boldsymbol{q}s} - \hbar\omega_{\boldsymbol{k}s}}\right) \\
&\qquad + \big[n_{\rm B}(\hbar\omega_{\boldsymbol{q}s}) - n_{\rm B}(\hbar\omega_{\boldsymbol{k}s})\big] \\
&\qquad \times \left(\frac{1}{\mathrm{i}\hbar\omega_\ell - \hbar\omega_{\boldsymbol{q}s} + \hbar\omega_{\boldsymbol{k}s}} - \frac{1}{\mathrm{i}\hbar\omega_\ell + \hbar\omega_{\boldsymbol{q}s} - \hbar\omega_{\boldsymbol{k}s}}\right).
\end{aligned}$$

Another method for evaluating Matsubara sums is to use the integration contour C that is a counterclockwise circle of infinite radius centered at the origin. The summand of the to-be-evaluated Matsubara sum is analytically continued onto the whole complex plane and is then multiplied with the Fermi-Dirac or Bose-Einstein distribution function depending on the frequencies that are to be summed. The resultant function is referred to as $f(z)$. We then consider the contour integral of $f(z)$ along C. Since C encloses all the poles of $f(z)$, the value of the contour integral is equal to $2\pi\mathrm{i}$ times the sum of the residues of $f(z)$ at all its poles. Making use of the fact that the integral of $f(z)$ along C is zero since $f(z)$ tends to zero on C faster than $1/|z|$ as $|z| \to \infty$, we can obtain the value of the sum.

26-12 Generalized spin susceptibility of a free electron gas. In terms of the Matsubara Green's functions for electrons, *the spin susceptibility* for a free electron gas in units of $2\mu_B^2$ is defined by $\chi_s(q) = -\beta^{-1}\sum_k \mathscr{G}_{\sigma 0}(k)\mathscr{G}_{\sigma 0}(k+q)$. The corresponding Feynman diagram is given in Fig. (26.9)(a). Here $k = (\boldsymbol{k}, \mathrm{i}\hbar\omega_n)$ and $q = (\boldsymbol{q}, \mathrm{i}\hbar\omega_m)$ with $\mathrm{i}\hbar\omega_n = \mathrm{i}(2n+1)\pi/\beta$ and $\mathrm{i}\hbar\omega_m = \mathrm{i}2m\pi/\beta$ for integral values of m and n.

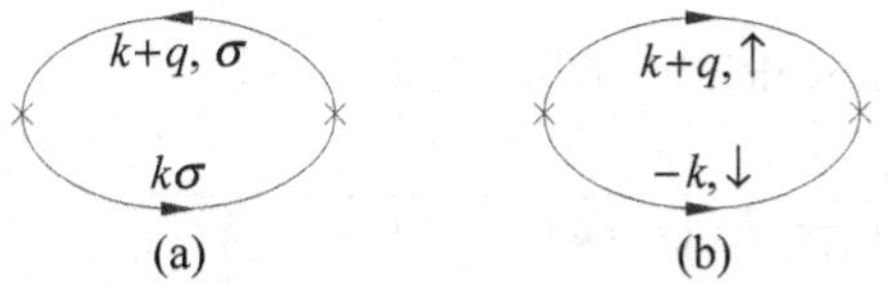

Fig. 26.9 Feynman diagrams for the spin (a) and pairing (b) susceptibilities. Note that a closed electron loop is present in (a) whereas it is absent in (b).

(1) Evaluate the summation over $\mathrm{i}\omega_n$ in $\chi_s(q)$

(2) Derive an expression for the retarded spin susceptibility $\chi_s^R(\boldsymbol{q}, \omega)$ through the analytic continuation $\mathrm{i}\omega_m \to \omega + \mathrm{i}\delta$. $\chi_s^R(\boldsymbol{q}, \omega)$ is also known as *the generalized spin susceptibility*.
(3) Evaluate the imaginary part of $\chi_s^R(\boldsymbol{q}, \omega)$ for $\omega \to 0$ at low temperatures.
(4) Evaluate the real part of the static spin susceptibility $\chi_s^R(\boldsymbol{q}, 0)$ at zero temperature.

---

(1) Writing out explicitly the wave vector and imaginary frequency variables, we have

$$\begin{aligned}\chi_s(\boldsymbol{q}, \mathrm{i}\omega_m) &= -\frac{1}{\beta}\sum_{\boldsymbol{k}n} \mathscr{G}_{\sigma 0}(\boldsymbol{k}, \mathrm{i}\omega_n)\mathscr{G}_{\sigma 0}(\boldsymbol{k}+\boldsymbol{q}, \mathrm{i}\omega_n + \mathrm{i}\omega_m) \\ &= -\frac{1}{\beta}\sum_{\boldsymbol{k}n} \frac{1}{\mathrm{i}\hbar\omega_n - \xi_{\boldsymbol{k}}} \frac{1}{\mathrm{i}\hbar\omega_n + \mathrm{i}\hbar\omega_m - \xi_{\boldsymbol{k}+\boldsymbol{q}}}.\end{aligned}$$

To evaluate the summation over n in the above expression, we consider the contour integral

$$I = \oint_{|z|=R} \mathrm{d}z\, f(z) = \oint_{|z|=R} \mathrm{d}z\, n_{\mathrm{F}}(z) \frac{1}{z - \xi_{\boldsymbol{k}}} \frac{1}{z + \mathrm{i}\hbar\omega_m - \xi_{\boldsymbol{k}+\boldsymbol{q}}}$$

with $R \to \infty$ and $n_{\mathrm{F}}(z) = 1/(\mathrm{e}^{\beta z} + 1)$ the Fermi-Dirac distribution function. Since the integrand goes to zero at least as fast as $1/|z|^2$ as $|z|$ tends to ∞, the value of the contour integral vanishes, $I = 0$. The integrand $f(z)$ has first-order poles at $\mathrm{i}\hbar\omega_n = \mathrm{i}(2n+1)\pi/\beta$, $\xi_{\boldsymbol{k}}$, and $\xi_{\boldsymbol{k}+\boldsymbol{q}} - \mathrm{i}\hbar\omega_m$. Invoking the residue theorem, we have

$$\begin{aligned}0 = &-\frac{1}{\beta}\sum_{n} \frac{1}{\mathrm{i}\hbar\omega_n - \xi_{\boldsymbol{k}}} \frac{1}{\mathrm{i}\hbar\omega_n + \mathrm{i}\hbar\omega_m - \xi_{\boldsymbol{k}+\boldsymbol{q}}} \\ &+ \frac{n_{\mathrm{F}}(\xi_{\boldsymbol{k}})}{\xi_{\boldsymbol{k}} + \mathrm{i}\hbar\omega_m - \xi_{\boldsymbol{k}+\boldsymbol{q}}} + \frac{n_{\mathrm{F}}(\xi_{\boldsymbol{k}+\boldsymbol{q}})}{\xi_{\boldsymbol{k}+\boldsymbol{q}} - \mathrm{i}\hbar\omega_m - \xi_{\boldsymbol{k}}}\end{aligned}$$

from which it follows that

$$-\frac{1}{\beta}\sum_{n} \frac{1}{\mathrm{i}\hbar\omega_n - \xi_{\boldsymbol{k}}} \frac{1}{\mathrm{i}\hbar\omega_n + \mathrm{i}\hbar\omega_m - \xi_{\boldsymbol{k}+\boldsymbol{q}}} = -\frac{n_{\mathrm{F}}(\xi_{\boldsymbol{k}}) - n_{\mathrm{F}}(\xi_{\boldsymbol{k}+\boldsymbol{q}})}{\mathrm{i}\hbar\omega_m + \xi_{\boldsymbol{k}} - \xi_{\boldsymbol{k}+\boldsymbol{q}}}.$$

The spin susceptibility is then given by

$$\chi_s(\boldsymbol{q}, \mathrm{i}\omega_m) = -\sum_{\boldsymbol{k}} \frac{n_{\mathrm{F}}(\xi_{\boldsymbol{k}}) - n_{\mathrm{F}}(\xi_{\boldsymbol{k}+\boldsymbol{q}})}{\mathrm{i}\hbar\omega_m + \xi_{\boldsymbol{k}} - \xi_{\boldsymbol{k}+\boldsymbol{q}}}.$$

(2) Performing the analytic continuation $\mathrm{i}\omega_m \to \omega + \mathrm{i}\delta$, we obtain the retarded spin susceptibility $\chi_s^R(\boldsymbol{q}, \omega)$

$$\chi_s^R(\boldsymbol{q}, \omega) = -\sum_{\boldsymbol{k}} \frac{n_{\mathrm{F}}(\xi_{\boldsymbol{k}}) - n_{\mathrm{F}}(\xi_{\boldsymbol{k}+\boldsymbol{q}})}{\hbar\omega + \xi_{\boldsymbol{k}} - \xi_{\boldsymbol{k}+\boldsymbol{q}} + \mathrm{i}\delta}.$$

The real and imaginary parts of the retarded spin susceptibility are given by

$$\mathrm{Re}\,\chi_s^R(\boldsymbol{q}, \omega) = -\mathrm{P}\sum_{\boldsymbol{k}} \frac{n_{\mathrm{F}}(\xi_{\boldsymbol{k}}) - n_{\mathrm{F}}(\xi_{\boldsymbol{k}+\boldsymbol{q}})}{\hbar\omega + \xi_{\boldsymbol{k}} - \xi_{\boldsymbol{k}+\boldsymbol{q}}},$$

$$\mathrm{Im}\,\chi_s^R(\boldsymbol{q}, \omega) = \pi\sum_{\boldsymbol{k}} \big[n_{\mathrm{F}}(\xi_{\boldsymbol{k}}) - n_{\mathrm{F}}(\xi_{\boldsymbol{k}+\boldsymbol{q}})\big]\delta(\hbar\omega + \xi_{\boldsymbol{k}} - \xi_{\boldsymbol{k}+\boldsymbol{q}}).$$

(3) Converting the summation over $\boldsymbol{k}$ into an integration over $\boldsymbol{k}$, we have for the imaginary part of the retarded spin susceptibility

$$\mathrm{Im}\,\chi_s^R(\boldsymbol{q}, \omega) = \frac{\mathscr{V}}{8\pi^2}\int \mathrm{d}\boldsymbol{k}\,\big[n_{\mathrm{F}}(\xi_{\boldsymbol{k}}) - n_{\mathrm{F}}(\xi_{\boldsymbol{k}+\boldsymbol{q}})\big]\delta(\hbar\omega + \xi_{\boldsymbol{k}} - \xi_{\boldsymbol{k}+\boldsymbol{q}}).$$

Writing $\mathrm{d}\boldsymbol{k}$ as $k^2\mathrm{d}k\,\mathrm{d}\cos\theta\,\mathrm{d}\varphi$, we can trivially perform the integration over φ and obtain the result of 2π since the integrand is independent of φ if the direction of q is chosen as the polar axis. Note that $\xi_{\boldsymbol{k}}$ does not depend on angles, $\xi_{\boldsymbol{k}} = \xi_k$, and that $\xi_{\boldsymbol{k}+\boldsymbol{q}}$ depends only on the polar angle θ. To perform the integrations over k and $\cos\theta$, we make a change of integration variables from k and $\cos\theta$ to $\varepsilon_k = \xi_k + \mu$ and $k' = |\boldsymbol{k} + \boldsymbol{q}|$ through $k^2\mathrm{d}k\,\mathrm{d}\cos\theta = (mk'/\hbar^2 q)\mathrm{d}\varepsilon_k\mathrm{d}k'$. We then have

$$\begin{aligned}
&\mathrm{Im}\,\chi_s^R(\boldsymbol{q}, \omega)\\
&= \frac{m\mathscr{V}}{4\pi\hbar^2 q}\int_0^\infty \mathrm{d}\varepsilon_k \int_{|k-q|}^{k+q} \mathrm{d}k'\,k'\big[n_{\mathrm{F}}(\varepsilon_k) - n_{\mathrm{F}}(\varepsilon_{k'})\big]\delta(\hbar\omega + \varepsilon_k - \varepsilon_{k'})\\
&= \frac{m^2\mathscr{V}}{4\pi\hbar^4 q}\int_0^\infty \mathrm{d}\varepsilon_k \int_{(\varepsilon_k^{1/2}-\varepsilon_q^{1/2})^2}^{(\varepsilon_k^{1/2}+\varepsilon_q^{1/2})^2} \mathrm{d}\varepsilon_{k'}\,[n_{\mathrm{F}}(\varepsilon_k) - n_{\mathrm{F}}(\varepsilon_{k'})]\delta(\hbar\omega + \varepsilon_k - \varepsilon_{k'}),
\end{aligned}$$

where we have written $n_{\mathrm{F}}(\xi_{\boldsymbol{k}})$ as $n_{\mathrm{F}}(\varepsilon_k) = 1/[\mathrm{e}^{\beta(\varepsilon_k - \mu)} + 1]$ and introduced $\varepsilon_q = \hbar^2 q^2/2m$. Performing the integration over $\varepsilon_{k'}$ by consuming the δ-function yields

$$\begin{aligned}
\mathrm{Im}\,\chi_s^R(\boldsymbol{q}, \omega) &= \frac{m^2\mathscr{V}}{4\pi\hbar^4 q}\int_0^\infty \mathrm{d}\varepsilon_k\big[n_{\mathrm{F}}(\varepsilon_k) - n_{\mathrm{F}}(\hbar\omega + \varepsilon_k)\big]\\
&\quad\times \theta\big(\hbar\omega - \varepsilon_q + 2(\varepsilon_k\varepsilon_q)^{1/2}\big)\theta\big(\varepsilon_q + 2(\varepsilon_k\varepsilon_q)^{1/2} - \hbar\omega\big)\\
&= \frac{m^2\mathscr{V}}{4\pi\hbar^4 q}\int_{(\hbar\omega - \varepsilon_q)^2/4\varepsilon_q}^\infty \mathrm{d}\varepsilon_k\big[n_{\mathrm{F}}(\varepsilon_k) - n_{\mathrm{F}}(\hbar\omega + \varepsilon_k)\big].
\end{aligned}$$

For small $\hbar\omega$, we can expand $n_F(\hbar\omega + \varepsilon_k)$ as a Taylor series in $\hbar\omega$. Keeping only the first two terms, we have

$$\begin{aligned}\operatorname{Im}\chi_s^R(\boldsymbol{q},\omega) &= \frac{m^2\mathscr{V}\omega}{4\pi\hbar^3 q}\int_{(\hbar\omega-\varepsilon_q)^2/4\varepsilon_q}^{\infty} \mathrm{d}\varepsilon_k\left[-\frac{\partial n_F(\varepsilon_k)}{\partial\varepsilon_k}\right]\\ &= \frac{m^2\mathscr{V}\omega}{4\pi\hbar^3 q} n_F\big((\hbar\omega-\varepsilon_q)^2/4\varepsilon_q\big).\end{aligned}$$

At low temperatures and for $(\hbar\omega - \varepsilon_q)^2/4\varepsilon_q \ll \varepsilon_F$, the value of the Fermi-Dirac distribution function is essentially unity

$$n_F\big((\hbar\omega-\varepsilon_q)^2/4\varepsilon_q\big) \approx 1 - \mathrm{e}^{(\hbar\omega-\varepsilon_q)^2/4\varepsilon_q k_B T}\mathrm{e}^{-\mu/k_B T}.$$

We then have

$$\operatorname{Im}\chi_s^R(\boldsymbol{q},\omega) \approx \frac{m^2\mathscr{V}\omega}{4\pi\hbar^3 q}\left[1 - \mathrm{e}^{(\hbar\omega-\varepsilon_q)^2/4\varepsilon_q k_B T}\mathrm{e}^{-\mu/k_B T}\right],$$
$$(\hbar\omega-\varepsilon_q)^2/4\varepsilon_q \ll \varepsilon_F.$$

(4) For the real part of the static retarded spin susceptibility, we have

$$\begin{aligned}\operatorname{Re}\chi_s^R(\boldsymbol{q},0) &= -\frac{\mathscr{V}}{(2\pi)^3}\,\mathrm{P}\int \mathrm{d}\boldsymbol{k}\,\frac{n_F(\xi_{\boldsymbol{k}}) - n_F(\xi_{\boldsymbol{k}+\boldsymbol{q}})}{\xi_{\boldsymbol{k}} - \xi_{\boldsymbol{k}+\boldsymbol{q}}}\\ &= -\frac{\mathscr{V}}{(2\pi)^3}\,\mathrm{P}\int \mathrm{d}\boldsymbol{k}\, n_F(\xi_{\boldsymbol{k}})\left[\frac{1}{\xi_{\boldsymbol{k}} - \xi_{\boldsymbol{k}+\boldsymbol{q}}} - \frac{1}{\xi_{\boldsymbol{k}-\boldsymbol{q}} - \xi_{\boldsymbol{k}}}\right].\end{aligned}$$

At zero temperature, we have

$$\begin{aligned}&\operatorname{Re}\chi_s^R(\boldsymbol{q},0;T=0)\\ &\quad= -\frac{\mathscr{V}}{(2\pi)^3}\,\mathrm{P}\int_{|\boldsymbol{k}|\leqslant k_F} \mathrm{d}\boldsymbol{k}\left[\frac{1}{\xi_{\boldsymbol{k}} - \xi_{\boldsymbol{k}+\boldsymbol{q}}} - \frac{1}{\xi_{\boldsymbol{k}-\boldsymbol{q}} - \xi_{\boldsymbol{k}}}\right]\\ &\quad= \frac{m\mathscr{V}}{2\pi^2\hbar^2}\,\mathrm{P}\int_0^{k_F} \mathrm{d}k\, k^2\int_{-1}^{1}\mathrm{d}x\left(\frac{1}{2kqx+q^2} - \frac{1}{2kqx-q^2}\right)\\ &\quad= \frac{m\mathscr{V}}{2\pi^2\hbar^2 q}\,\mathrm{P}\int_0^{k_F} \mathrm{d}k\, k\ln\left|\frac{k+q/2}{k-q/2}\right|.\end{aligned}$$

The integration over k can be completed by first performing an integration by parts

$$\begin{aligned}\operatorname{Re}\chi_s^R(\boldsymbol{q},0;T=0) &= \frac{m\mathscr{V}}{2\pi^2\hbar^2 q}\,\mathrm{P}\int_0^{k_F} \mathrm{d}k\, k\ln\left|\frac{k+q/2}{k-q/2}\right|\\ &= \frac{m\mathscr{V}}{4\pi^2\hbar^2 q}\left[k_F^2\ln\left|\frac{k_F+q/2}{k_F-q/2}\right|\right.\\ &\qquad\left. - \mathrm{P}\int_0^{k_F}\mathrm{d}k\, k^2\left(\frac{1}{k+q/2} - \frac{1}{k-q/2}\right)\right]\\ &= \frac{mk_F\mathscr{V}}{2\pi^2\hbar^2}\left[\frac{1}{2} + \frac{1-x^2}{4x}\ln\left|\frac{x+1}{x-1}\right|\right],\end{aligned}$$

where $x = q/2k_F$.

26-13 Pairing susceptibility. *The pairing susceptibility* of electrons is defined by

$$\chi_p(q) = \frac{1}{\beta}\sum_k \mathscr{G}_{\downarrow 0}(-k)\mathscr{G}_{\uparrow 0}(k+q).$$

The corresponding Feynman diagram is given in Fig. (26.9)(b). The $\boldsymbol{k}$-sum is cut off at $|\xi_{\boldsymbol{k}}| = |\varepsilon_{\boldsymbol{k}} - E_{\rm F}| = \hbar\omega_0$ with $\hbar\omega_0 \ll E_{\rm F}$ because of the nature of the concerned physical processes.

(1) Evaluate the summation over $\mathrm{i}\omega_n$ in $\chi_p(q)$.
(2) Derive an expression for the retarded pairing susceptibility $\chi_p^R(\boldsymbol{q}, \omega)$ through the analytic continuation $\mathrm{i}\hbar\omega_m \to \omega + \mathrm{i}\delta$. Write down an expression for the static pairing susceptibility at zero wave vector transfer $\chi_p^R(0, 0)$.
(3) Evaluate the imaginary part of $\chi_p^R(0, 0)$.
(4) Evaluate approximately the real part of $\chi_p^R(0, 0)$ by approximating $\tanh(x)$ as x for $x < 1$ and 1 for $x > 1$. Find the value of the temperature at which $1 - V_0\,\mathrm{Re}\,\chi_p^R(0, 0) = 0$ with $V_0 \propto 1/\mathscr{V}$ the interaction energy in momentum space with $\mathscr{V}$ the volume of the system. Assume that $\beta\hbar\omega_0 \gg 1$ in the relevant temperature range.

---

(1) Writing out explicitly the wave vector and imaginary frequency variables, we have

$$\begin{aligned}\chi_p(\boldsymbol{q}, \mathrm{i}\omega_m) &= \frac{1}{\beta}\sum_{\boldsymbol{k}n} \mathscr{G}_{\downarrow 0}(-\boldsymbol{k}, -\mathrm{i}\omega_n)\mathscr{G}_{\uparrow 0}(\boldsymbol{k}+\boldsymbol{q}, \mathrm{i}\omega_n + \mathrm{i}\omega_m) \\ &= -\frac{1}{\beta}\sum_{\boldsymbol{k}n} \frac{1}{\mathrm{i}\hbar\omega_n + \xi_{\boldsymbol{k}}}\,\frac{1}{\mathrm{i}\hbar\omega_n + \mathrm{i}\hbar\omega_m - \xi_{\boldsymbol{k}+\boldsymbol{q}}}.\end{aligned}$$

To evaluate the summation over n in the above expression, we consider the contour integral

$$I = \oint_{|z|=R} \mathrm{d}z\, f(z) = \oint_{|z|=R} \mathrm{d}z\, n_{\rm F}(z)\frac{1}{z + \xi_{\boldsymbol{k}}}\,\frac{1}{z + \mathrm{i}\hbar\omega_m - \xi_{\boldsymbol{k}+\boldsymbol{q}}}$$

with $R \to \infty$ and $n_{\rm F}(z) = 1/(\mathrm{e}^{\beta z} + 1)$ the Fermi-Dirac distribution function. Since the integrand goes to zero at least as fast as $1/|z|^2$ as $|z|$ tends to ∞, the value of the contour integral vanishes, $I = 0$. The integrand $f(z)$ has first-order poles at $\mathrm{i}\hbar\omega_n = \mathrm{i}(2n+1)\pi/\beta$, $-\xi_{\boldsymbol{k}}$, and $\xi_{\boldsymbol{k}+\boldsymbol{q}} - \mathrm{i}\hbar\omega_m$. Invoking the residue theorem, we have

$$0 = -\frac{1}{\beta}\sum_n \frac{1}{\mathrm{i}\hbar\omega_n + \xi_{\boldsymbol{k}}}\frac{1}{\mathrm{i}\hbar\omega_n + \mathrm{i}\hbar\omega_m - \xi_{\boldsymbol{k}+\boldsymbol{q}}} + \frac{n_{\mathrm{F}}(-\xi_{\boldsymbol{k}})}{-\xi_{\boldsymbol{k}} + \mathrm{i}\hbar\omega_m - \xi_{\boldsymbol{k}+\boldsymbol{q}}} + \frac{n_{\mathrm{F}}(\xi_{\boldsymbol{k}+\boldsymbol{q}})}{\xi_{\boldsymbol{k}+\boldsymbol{q}} - \mathrm{i}\hbar\omega_m + \xi_{\boldsymbol{k}}}$$

from which it follows that

$$-\frac{1}{\beta}\sum_n \frac{1}{\mathrm{i}\hbar\omega_n + \xi_{\boldsymbol{k}}}\frac{1}{\mathrm{i}\hbar\omega_n + \mathrm{i}\hbar\omega_m - \xi_{\boldsymbol{k}+\boldsymbol{q}}} = -\frac{1 - n_{\mathrm{F}}(\xi_{\boldsymbol{k}}) - n_{\mathrm{F}}(\xi_{\boldsymbol{k}+\boldsymbol{q}})}{\mathrm{i}\hbar\omega_m - \xi_{\boldsymbol{k}} - \xi_{\boldsymbol{k}+\boldsymbol{q}}},$$

where we have made use of $n_{\mathrm{F}}(-\xi_{\boldsymbol{k}}) = 1 - n_{\mathrm{F}}(\xi_{\boldsymbol{k}})$. The pairing susceptibility is then given by

$$\chi_p(\boldsymbol{q}, \mathrm{i}\omega_m) = -\sum_{\boldsymbol{k}} \frac{1 - n_{\mathrm{F}}(\xi_{\boldsymbol{k}}) - n_{\mathrm{F}}(\xi_{\boldsymbol{k}+\boldsymbol{q}})}{\mathrm{i}\hbar\omega_m - \xi_{\boldsymbol{k}} - \xi_{\boldsymbol{k}+\boldsymbol{q}}}.$$

(2) Performing the analytic continuation $\mathrm{i}\omega_m \to \omega + \mathrm{i}\delta$, we obtain the retarded pairing susceptibility $\chi_p^R(\boldsymbol{q}, \omega)$

$$\chi_p^R(\boldsymbol{q}, \omega) = -\sum_{\boldsymbol{k}} \frac{1 - n_{\mathrm{F}}(\xi_{\boldsymbol{k}}) - n_{\mathrm{F}}(\xi_{\boldsymbol{k}+\boldsymbol{q}})}{\hbar\omega - \xi_{\boldsymbol{k}} - \xi_{\boldsymbol{k}+\boldsymbol{q}} + \mathrm{i}\delta}.$$

For $\boldsymbol{q} = 0$ and $\omega = 0$, we have

$$\chi_p^R(0,0) = \sum_{\boldsymbol{k}} \frac{1 - 2n_{\mathrm{F}}(\xi_{\boldsymbol{k}})}{2\xi_{\boldsymbol{k}} - \mathrm{i}\delta} = \frac{\mathscr{V}}{(2\pi)^3}\int \mathrm{d}\boldsymbol{k}\, \frac{1}{2\xi_{\boldsymbol{k}} - \mathrm{i}\delta}\tanh\left(\frac{\xi_{\boldsymbol{k}}}{2k_{\mathrm{B}}T}\right).$$

(3) The imaginary part of the retarded pairing susceptibility $\chi_p^R(0,0)$ is given by

$$\mathrm{Im}\,\chi_p^R(0,0) = \frac{\pi\mathscr{V}}{2(2\pi)^3}\int \mathrm{d}\boldsymbol{k}\,\delta(\xi_{\boldsymbol{k}})\tanh\left(\frac{\xi_{\boldsymbol{k}}}{2k_{\mathrm{B}}T}\right) = 0.$$

(4) The real part of the retarded pairing susceptibility $\chi_p^R(0,0)$ is given by

$$\begin{aligned}
\mathrm{Re}\,\chi_p^R(0,0) &= \frac{\mathscr{V}}{(2\pi)^3}\,\mathrm{P}\!\int \mathrm{d}\boldsymbol{k}\,\frac{1}{2\xi_{\boldsymbol{k}}}\tanh\left(\frac{\xi_{\boldsymbol{k}}}{2k_{\mathrm{B}}T}\right) \\
&= \frac{\mathscr{V}}{8\pi^2}\left(\frac{2m}{\hbar^2}\right)^{3/2}\mathrm{P}\!\int_{-\hbar\omega_0}^{\hbar\omega_0} \mathrm{d}\xi\,(\xi + E_{\mathrm{F}})^{1/2}\frac{1}{\xi}\tanh\left(\frac{\xi}{2k_{\mathrm{B}}T}\right) \\
&\approx \frac{mk_{\mathrm{F}}\mathscr{V}}{4\pi^2\hbar^2}\,\mathrm{P}\!\int_{-\hbar\omega_0}^{\hbar\omega_0} \frac{\mathrm{d}\xi}{\xi}\tanh\left(\frac{\xi}{2k_{\mathrm{B}}T}\right) \\
&= \frac{mk_{\mathrm{F}}\mathscr{V}}{2\pi^2\hbar^2}\int_0^{\hbar\omega_0} \frac{\mathrm{d}\xi}{\xi}\tanh\left(\frac{\xi}{2k_{\mathrm{B}}T}\right),
\end{aligned}$$

where we have approximated $(\xi + E_{\rm F})^{1/2}$ as $E_{\rm F}^{1/2}$ in consideration that $\hbar\omega_0 \ll E_{\rm F}$. Making use of $\tanh(\xi/2k_{\rm B}T) \approx \xi/2k_{\rm B}T$ for $\xi/2k_{\rm B}T < 1$ and 1 for $\xi/2k_{\rm B}T > 1$, we have

$$\begin{aligned}\operatorname{Re}\chi_p^R(0,0) &= \frac{mk_{\rm F}\mathscr{V}}{2\pi^2\hbar^2}\left[\frac{1}{2k_{\rm B}T}\int_0^{2k_{\rm B}T} \mathrm{d}\xi + \int_{2k_{\rm B}T}^{\hbar\omega_0} \frac{\mathrm{d}\xi}{\xi}\right] \\ &= N(0)\left[1 + \ln\left(\frac{\hbar\omega_0}{2k_{\rm B}T}\right)\right],\end{aligned}$$

where we have introduced $N(0) = mk_{\rm F}\mathscr{V}/2\pi^2\hbar^2$ which is the electron density of states for one spin direction at the Fermi surface. From $1 - V_0 \operatorname{Re}\chi_p^R(0,0) = 0$, we have

$$1 - N(0)V_0\left[1 + \ln\left(\frac{\hbar\omega_0}{2k_{\rm B}T_c}\right)\right] = 0,$$

where we have denoted by T_c the temperature at which $1 - V_0 \operatorname{Re}\chi_p^R(0,0) = 0$ holds. Solving for T_c from the above equation, we obtain

$$k_{\rm B}T_c = \frac{\mathrm{e}}{2}\,\hbar\omega_0\,\mathrm{e}^{-1/N(0)V_0} \approx 1.36\,\hbar\omega_0\,\mathrm{e}^{-1/N(0)V_0}.$$

26-14 Localized electrons. The electrons localized on a single site are described by the Hamiltonian

$$\hat{H} = \sum_\sigma \varepsilon_\sigma \hat{c}_\sigma^\dagger \hat{c}_\sigma + U\hat{c}_\uparrow^\dagger \hat{c}_\uparrow \hat{c}_\downarrow^\dagger \hat{c}_\downarrow.$$

(1) Compute the partition function of the system.

(2) Compute the Green's function $G_\sigma(t) = -\mathrm{i}\,\langle \hat{T}\hat{c}_\sigma(t)\hat{c}_\sigma^\dagger(0)\rangle$ and its Fourier transform $G_\sigma(\omega)$, where $\langle\cdots\rangle$ denotes the thermodynamic average, $\langle\cdots\rangle = \operatorname{Tr}\big[\mathrm{e}^{-\beta\hat{H}}(\cdots)\big]$.

(3) Compute $G_\sigma(\omega)$ perturbatively to the second order in U.

---

(1) Since a single site can accommodate at most two electrons, the state space of the system is four dimensional. We use the following four basis vectors

$$|0\rangle,\ |{\uparrow}\rangle,\ |{\downarrow}\rangle,\ |{\uparrow\downarrow}\rangle,$$

where $|0\rangle$ denotes that the site is unoccupied by any electrons, $|{\uparrow}\rangle$ denotes that the site is occupied by only one up-spin electron, $|{\downarrow}\rangle$ denotes that the site is occupied by only one down-spin electron, and $|{\uparrow\downarrow}\rangle$ denotes that the site is doubly occupied by two electrons of opposite spins. These basis vectors are also eigenvectors of the Hamiltonian corresponding respectively to the eigenvalues $E_1 = 0$, $E_2 = \varepsilon_\uparrow$, $E_3 = \varepsilon_\downarrow$, and $E_4 = \varepsilon_\uparrow + \varepsilon_\downarrow + U$. We denote the eigenvectors of the Hamiltonian by $|\phi_1\rangle = |0\rangle$, $|\phi_2\rangle = |\uparrow\rangle$, $|\phi_3\rangle = |\downarrow\rangle$, and $|\phi_4\rangle = |\uparrow\downarrow\rangle$. The partition function is given by

$$Z = \mathrm{Tr}\,\mathrm{e}^{-\beta\hat{H}} = \sum_{n=1}^{4} \mathrm{e}^{-\beta E_n} = 1 + \mathrm{e}^{-\beta\varepsilon_\uparrow} + \mathrm{e}^{-\beta\varepsilon_\downarrow} + \mathrm{e}^{-\beta(\varepsilon_\uparrow+\varepsilon_\downarrow+U)}.$$

(2) The Green's function $G_\sigma(t)$ can be written as

$$G_\sigma(t) = -\mathrm{i}\big[\,\langle \hat{c}_\sigma(t)\hat{c}^\dagger_\sigma(0)\rangle\,\theta(t) - \langle \hat{c}^\dagger_\sigma(0)\hat{c}_\sigma(t)\rangle\,\theta(-t)\,\big].$$

For $\langle \hat{c}_\sigma(t)\hat{c}^\dagger_\sigma(0)\rangle$, we have

$$\begin{aligned} Z\,\langle \hat{c}_\sigma(t)\hat{c}^\dagger_\sigma(0)\rangle &= \mathrm{Tr}\big(\mathrm{e}^{-\beta\hat{H}}\,\hat{c}_\sigma(t)\hat{c}^\dagger_\sigma(0)\big) \\ &= \mathrm{Tr}\big(\mathrm{e}^{-\beta\hat{H}}\mathrm{e}^{\mathrm{i}\hat{H}t/\hbar}\hat{c}_\sigma\mathrm{e}^{-\mathrm{i}\hat{H}t/\hbar}\hat{c}^\dagger_\sigma\big) \\ &= \sum_{n=1}^{4}\langle\phi_n|\,\mathrm{e}^{-\beta\hat{H}}\mathrm{e}^{\mathrm{i}\hat{H}t/\hbar}\hat{c}_\sigma\mathrm{e}^{-\mathrm{i}\hat{H}t/\hbar}\hat{c}^\dagger_\sigma|\phi_n\rangle \\ &= \sum_{n=1}^{4}\mathrm{e}^{(\mathrm{i}t/\hbar-\beta)E_n}\,\langle\phi_n|\,\hat{c}_\sigma\mathrm{e}^{-\mathrm{i}\hat{H}t/\hbar}\hat{c}^\dagger_\sigma|\phi_n\rangle\,. \end{aligned}$$

Writing out the summation over n explicitly, we have

$$\begin{aligned} Z\,\langle \hat{c}_\sigma(t)\hat{c}^\dagger_\sigma(0)\rangle &= \langle 0|\,\hat{c}_\sigma\mathrm{e}^{-\mathrm{i}\hat{H}t/\hbar}\hat{c}^\dagger_\sigma|0\rangle + \mathrm{e}^{(\mathrm{i}t/\hbar-\beta)\varepsilon_\uparrow}\,\langle\uparrow|\,\hat{c}_\sigma\mathrm{e}^{-\mathrm{i}\hat{H}t/\hbar}\hat{c}^\dagger_\sigma|\uparrow\rangle \\ &\quad + \mathrm{e}^{(\mathrm{i}t/\hbar-\beta)\varepsilon_\downarrow}\,\langle\downarrow|\,\hat{c}_\sigma\mathrm{e}^{-\mathrm{i}\hat{H}t/\hbar}\hat{c}^\dagger_\sigma|\downarrow\rangle \\ &= \mathrm{e}^{-\mathrm{i}\varepsilon_\sigma t/\hbar} + \mathrm{e}^{(\mathrm{i}t/\hbar-\beta)\varepsilon_\uparrow}\mathrm{e}^{-\mathrm{i}(\varepsilon_\sigma+\varepsilon_{\bar\sigma}+U)t/\hbar}\delta_{\sigma\downarrow} \\ &\quad + \mathrm{e}^{(\mathrm{i}t/\hbar-\beta)\varepsilon_\downarrow}\mathrm{e}^{-\mathrm{i}(\varepsilon_\sigma+\varepsilon_{\bar\sigma}+U)t/\hbar}\delta_{\sigma\uparrow} \\ &= \mathrm{e}^{-\mathrm{i}\varepsilon_\sigma t/\hbar}\big[1 + \mathrm{e}^{-\mathrm{i}Ut/\hbar-\beta\varepsilon_{\bar\sigma}}(\delta_{\sigma\downarrow}+\delta_{\sigma\uparrow})\big] \\ &= \mathrm{e}^{-\mathrm{i}\varepsilon_\sigma t/\hbar}\big(1 + \mathrm{e}^{-\mathrm{i}Ut/\hbar-\beta\varepsilon_{\bar\sigma}}\big), \end{aligned}$$

where $\bar\sigma = -\sigma$. For $\langle \hat{c}^\dagger_\sigma(0)\hat{c}_\sigma(t)\rangle$, we have

$$\begin{aligned} Z\,\langle \hat{c}^\dagger_\sigma(0)\hat{c}_\sigma(t)\rangle &= \mathrm{e}^{-(\mathrm{i}t/\hbar+\beta)\varepsilon_\uparrow}\,\langle\uparrow|\,\hat{c}^\dagger_\sigma\mathrm{e}^{\mathrm{i}\hat{H}t/\hbar}\hat{c}_\sigma|\uparrow\rangle \\ &\quad + \mathrm{e}^{-(\mathrm{i}t/\hbar+\beta)\varepsilon_\downarrow}\,\langle\downarrow|\,\hat{c}^\dagger_\sigma\mathrm{e}^{\mathrm{i}\hat{H}t/\hbar}\hat{c}_\sigma|\downarrow\rangle \\ &\quad + \mathrm{e}^{-(\mathrm{i}t/\hbar+\beta)(\varepsilon_\uparrow+\varepsilon_\downarrow+U)}\,\langle\uparrow\downarrow|\,\hat{c}^\dagger_\sigma\mathrm{e}^{\mathrm{i}\hat{H}t/\hbar}\hat{c}_\sigma|\uparrow\downarrow\rangle \\ &= \mathrm{e}^{-(\mathrm{i}t/\hbar+\beta)\varepsilon_\uparrow}\delta_{\sigma\uparrow} + \mathrm{e}^{-(\mathrm{i}t/\hbar+\beta)\varepsilon_\downarrow}\delta_{\sigma\downarrow} \\ &\quad + \mathrm{e}^{-(\mathrm{i}t/\hbar+\beta)(\varepsilon_\uparrow+\varepsilon_\downarrow+U)}\mathrm{e}^{\mathrm{i}\varepsilon_{\bar\sigma}t/\hbar} \\ &= \mathrm{e}^{-(\mathrm{i}t/\hbar+\beta)\varepsilon_\sigma} + \mathrm{e}^{-(\mathrm{i}t/\hbar+\beta)(\varepsilon_\sigma+U)-\beta\varepsilon_{\bar\sigma}} \\ &= \mathrm{e}^{-(\mathrm{i}t/\hbar+\beta)\varepsilon_\sigma}\big[1 + \mathrm{e}^{-\mathrm{i}Ut/\hbar-\beta(\varepsilon_{\bar\sigma}+U)}\big]. \end{aligned}$$

The Green's function $G_\sigma(t)$ is then given by

$$G_\sigma(t) = -\mathrm{i}\mathrm{e}^{-\mathrm{i}\varepsilon_\sigma t/\hbar}\{(1+\mathrm{e}^{-\mathrm{i}Ut/\hbar-\beta\varepsilon_{\bar\sigma}})\theta(t) \\ -\mathrm{e}^{-\beta\varepsilon_\sigma}[1+\mathrm{e}^{-\mathrm{i}Ut/\hbar-\beta(\varepsilon_{\bar\sigma}+U)}]\theta(-t)\} \\ \times[1+\mathrm{e}^{-\beta\varepsilon_\uparrow}+\mathrm{e}^{-\beta\varepsilon_\downarrow}+\mathrm{e}^{-\beta(\varepsilon_\uparrow+\varepsilon_\downarrow+U)}]^{-1}.$$

The Fourier transform of $G_\sigma(t)$ is given by

$$G_\sigma(\omega) = \int_{-\infty}^{\infty}\frac{\mathrm{d}t}{\hbar}\,G_\sigma(t)\mathrm{e}^{\mathrm{i}\omega t-\delta|t|} \\ = \frac{-\mathrm{i}}{Z}\left\{\int_0^\infty \frac{\mathrm{d}t}{\hbar}\,[\mathrm{e}^{-\mathrm{i}\varepsilon_\sigma t/\hbar}+\mathrm{e}^{-\beta\varepsilon_{\bar\sigma}}\mathrm{e}^{-\mathrm{i}(\varepsilon_\sigma+U)t/\hbar}]\mathrm{e}^{\mathrm{i}\omega t-\delta t}\right. \\ \left.-\mathrm{e}^{-\beta\varepsilon_\sigma}\int_{-\infty}^0\frac{\mathrm{d}t}{\hbar}[\mathrm{e}^{-\mathrm{i}\varepsilon_\sigma t/\hbar}+\mathrm{e}^{-\beta(\varepsilon_{\bar\sigma}+U)}\mathrm{e}^{-\mathrm{i}(\varepsilon_\sigma+U)t/\hbar}]\mathrm{e}^{\mathrm{i}\omega t+\delta t}\right\} \\ = \frac{1}{Z}\left[\frac{1}{\hbar\omega-\varepsilon_\sigma+\mathrm{i}\delta}+\frac{\mathrm{e}^{-\beta\varepsilon_{\bar\sigma}}}{\hbar\omega-\varepsilon_\sigma-U+\mathrm{i}\delta}\right. \\ \left.+\frac{\mathrm{e}^{-\beta\varepsilon_\sigma}}{\hbar\omega-\varepsilon_\sigma-\mathrm{i}\delta}+\frac{\mathrm{e}^{-\beta(\varepsilon_\sigma+\varepsilon_{\bar\sigma}+U)}}{\hbar\omega-\varepsilon_\sigma-U-\mathrm{i}\delta}\right].$$

(3) To perform perturbation computations, we first find the zeroth-order Green's function which is defined by

$$G_{\sigma 0}(t) = -\mathrm{i}\,\langle\hat T\hat c_\sigma(t)\hat c_\sigma^\dagger(0)\rangle_0,$$

where the subscript "0" on the average sign indicates that the average is with respect to the Hamiltonian

$$\hat H_0 = \sum_\sigma \varepsilon_\sigma\hat c_\sigma^\dagger\hat c_\sigma = \sum_\sigma\varepsilon_\sigma\hat n_\sigma,$$

where $\hat n_\sigma = \hat c_\sigma^\dagger\hat c_\sigma$ is the electron number operator of spin σ. The second part of the total Hamiltonian is now referred to as $\hat H_1$

$$\hat H_1 = U\hat c_\uparrow^\dagger\hat c_\uparrow\hat c_\downarrow^\dagger\hat c_\downarrow$$

which is the interaction Hamiltonian. The total Hamiltonian $\hat H$ is given by

$$\hat H = \hat H_0 + \hat H_1.$$

With respect to $\hat H_0$, the time dependence of $\hat c_\sigma$ and $\hat c_\sigma^\dagger$ is simply given by

$$\hat c_\sigma(t) = \mathrm{e}^{-\mathrm{i}\varepsilon_\sigma t/\hbar}\hat c_\sigma,\ \ \hat c_\sigma^\dagger(t) = \mathrm{e}^{\mathrm{i}\varepsilon_\sigma t/\hbar}\hat c_\sigma^\dagger.$$

$G_{\sigma 0}(t)$ is then given by

$$G_{\sigma 0}(t) = -\mathrm{i}\big[\,\langle \hat{c}_\sigma(t)\hat{c}_\sigma^\dagger(0)\rangle_0\,\theta(t) - \langle \hat{c}_\sigma^\dagger(0)\hat{c}_\sigma(t)\rangle_0\,\theta(-t)\,\big]$$
$$= -\mathrm{i}\,\mathrm{e}^{-\mathrm{i}\varepsilon_\sigma t/\hbar}\big[\,\langle \hat{c}_\sigma\hat{c}_\sigma^\dagger\rangle_0\,\theta(t) - \langle \hat{c}_\sigma^\dagger\hat{c}_\sigma\rangle_0\,\theta(-t)\,\big].$$

We now evaluate the averages in the above expression. For $\langle \hat{c}_\sigma\hat{c}_\sigma^\dagger\rangle_0$, we have

$$Z_0\,\langle \hat{c}_\sigma\hat{c}_\sigma^\dagger\rangle_0 = \mathrm{Tr}\big[\,(1-\hat{n}_\sigma)\mathrm{e}^{-\beta\sum_\sigma \varepsilon_\sigma \hat{n}_\sigma}\,\big] = Z_0 - \mathrm{Tr}\big(\hat{n}_\sigma \mathrm{e}^{-\beta\sum_\sigma \varepsilon_\sigma \hat{n}_\sigma}\big)$$
$$= Z_0 - \sum_{n_\uparrow=0}^{1}\sum_{n_\downarrow=0}^{1} n_\sigma\,\mathrm{e}^{-\beta(\varepsilon_\uparrow n_\uparrow + \varepsilon_\downarrow n_\downarrow)}$$
$$= Z_0 - \mathrm{e}^{-\beta\varepsilon_\sigma}\big(1+\mathrm{e}^{-\beta\varepsilon_{\bar{\sigma}}}\big) = 1 + \mathrm{e}^{-\beta\varepsilon_{\bar{\sigma}}},$$

where

$$Z_0 = \mathrm{Tr}\,\mathrm{e}^{-\beta\hat{H}_0} = \sum_{n_\uparrow=0}^{1}\sum_{n_\downarrow=0}^{1} \mathrm{e}^{-\beta(\varepsilon_\uparrow n_\uparrow + \varepsilon_\downarrow n_\downarrow)}$$
$$= \big(1+\mathrm{e}^{-\beta\varepsilon_\uparrow}\big)\big(1+\mathrm{e}^{-\beta\varepsilon_\downarrow}\big).$$

For $\langle \hat{c}_\sigma^\dagger\hat{c}_\sigma\rangle_0$, we have

$$Z_0\,\langle \hat{c}_\sigma^\dagger\hat{c}_\sigma\rangle_0 = \mathrm{Tr}\big(\hat{n}_\sigma \mathrm{e}^{-\beta\sum_\sigma \varepsilon_\sigma \hat{n}_\sigma}\big) = \mathrm{e}^{-\beta\varepsilon_\sigma}\big(1+\mathrm{e}^{-\beta\varepsilon_{\bar{\sigma}}}\big).$$

We thus have $G_{\sigma 0}(t)$

$$G_{\sigma 0}(t) = -\frac{\mathrm{i}\,\mathrm{e}^{-\mathrm{i}\varepsilon_\sigma t/\hbar}}{1+\mathrm{e}^{-\beta\varepsilon_\sigma}}\big[\,\theta(t) - \mathrm{e}^{-\beta\varepsilon_\sigma}\theta(-t)\,\big].$$

The Fourier transform of $G_{\sigma 0}(t)$ is given by

$$G_{\sigma 0}(\omega) = \int_{-\infty}^{\infty}\frac{\mathrm{d}t}{\hbar}\,G_{\sigma 0}(t)\mathrm{e}^{\mathrm{i}\omega t-\delta|t|}$$
$$= -\frac{\mathrm{i}}{1+\mathrm{e}^{-\beta\varepsilon_\sigma}}\bigg[\int_0^{\infty}\frac{\mathrm{d}t}{\hbar}\,\mathrm{e}^{(\mathrm{i}\omega-\mathrm{i}\varepsilon_\sigma t/\hbar-\delta)t}$$
$$-\,\mathrm{e}^{-\beta\varepsilon_\sigma}\int_{-\infty}^{0}\frac{\mathrm{d}t}{\hbar}\,\mathrm{e}^{(\mathrm{i}\omega-\mathrm{i}\varepsilon_\sigma t/\hbar+\delta)t}\bigg]$$
$$= \frac{1-n_{\mathrm{F}}(\varepsilon_\sigma)}{\hbar\omega-\varepsilon_\sigma+\mathrm{i}\delta} + \frac{n_{\mathrm{F}}(\varepsilon_\sigma)}{\hbar\omega-\varepsilon_\sigma-\mathrm{i}\delta},$$

where $n_{\mathrm{F}}(\varepsilon_\sigma) = 1/(\mathrm{e}^{\beta\varepsilon_\sigma}+1)$. The perturbation series for $G_\sigma(t)$ is

$$G_{\sigma 0}(t) = \sum_{n=0}^{\infty}\frac{(-\mathrm{i})^{n+1}}{\hbar^n}\int_{-\infty}^{\infty}\mathrm{d}t_1\cdots\int_{-\infty}^{\infty}\mathrm{d}t_n$$
$$\times\,\big\langle \hat{T}\hat{c}_\sigma(t)\hat{c}_\sigma^\dagger(0)\hat{H}_1(t_1)\cdots\hat{H}_1(t_n)\big\rangle_{0,\mathrm{dc}},$$

where the subscript "0" indicates that the average is with respect to $\hat{H}_0$ and the subscript "dc" indicates that only those terms corresponding to different connected Feynman diagrams are to be retained. Note that the operators in the above expression are in the interaction picture with the time dependence determined by $\hat{H}_0$. The first-order correction to $G_\sigma(t)$ is given by

$$\begin{aligned}G_\sigma^{(1)}(t) &= -\frac{1}{\hbar}\int_{-\infty}^{\infty} \mathrm{d}t_1 \left\langle \hat{T}\hat{c}_\sigma(t)\hat{c}_\sigma^\dagger(0)\hat{H}_1(t_1)\right\rangle_{0,\mathrm{dc}} \\ &= -\frac{U}{\hbar}\int_{-\infty}^{\infty} \mathrm{d}t_1 \left\langle \hat{T}\hat{c}_\sigma(t)\hat{c}_\sigma^\dagger(0)\hat{c}_\uparrow^\dagger(t_1)\hat{c}_\uparrow(t_1)\hat{c}_\downarrow^\dagger(t_1)\hat{c}_\downarrow(t_1)\right\rangle_{0,\mathrm{dc}}.\end{aligned}$$

Making use of Wick's theorem, we have

$$G_\sigma^{(1)}(t) = \frac{U}{\hbar} n_{\mathrm{F}}(\varepsilon_{\bar{\sigma}}) \int_{-\infty}^{\infty} \mathrm{d}t_1 G_{\sigma 0}(t - t_1) G_{\sigma 0}(t_1).$$

Making use of

$$G_{\sigma 0}(t) = \int_{-\infty}^{\infty} \frac{\mathrm{d}(\hbar\omega)}{2\pi}\, \mathrm{e}^{-\mathrm{i}\omega t} G_{\sigma 0}(\omega),$$

we have

$$\begin{aligned}G_\sigma^{(1)}(t) &= \frac{\hbar U}{(2\pi)^2} n_{\mathrm{F}}(\varepsilon_{\bar{\sigma}}) \int_{-\infty}^{\infty} \mathrm{d}\omega_1 \int_{-\infty}^{\infty} \mathrm{d}\omega_2\, G_{\sigma 0}(\omega_1) G_{\sigma 0}(\omega_2) \\ &\qquad \times \int_{-\infty}^{\infty} \mathrm{d}t_1\, \mathrm{e}^{-\mathrm{i}\omega_1(t - t_1) - \mathrm{i}\omega_2 t_1} \\ &= \frac{\hbar U}{2\pi} n_{\mathrm{F}}(\varepsilon_{\bar{\sigma}}) \int_{-\infty}^{\infty} \mathrm{d}\omega_1 \int_{-\infty}^{\infty} \mathrm{d}\omega_2\, G_{\sigma 0}(\omega_1) G_{\sigma 0}(\omega_2) \mathrm{e}^{-\mathrm{i}\omega_1 t} \delta(\omega_1 - \omega_2) \\ &= U n_{\mathrm{F}}(\varepsilon_{\bar{\sigma}}) \int_{-\infty}^{\infty} \frac{\mathrm{d}(\hbar\omega)}{2\pi} \left[G_{\sigma 0}(\omega)\right]^2 \mathrm{e}^{-\mathrm{i}\omega t}.\end{aligned}$$

Thus,

$$G_\sigma^{(1)}(\omega) = U n_{\mathrm{F}}(\varepsilon_{\bar{\sigma}}) \left[G_{\sigma 0}(\omega)\right]^2.$$

The above first-order correction corresponds to the Feynman diagram in Fig. 26.10(a).

The second-order correction is given by

$$\begin{aligned}G_\sigma^{(2)}(t) &= \frac{\mathrm{i}}{\hbar^2}\int_{-\infty}^{\infty} \mathrm{d}t_1 \int_{-\infty}^{\infty} \mathrm{d}t_2 \left\langle \hat{T}\hat{c}_\sigma(t)\hat{c}_\sigma^\dagger(0)\hat{H}_1(t_1)\hat{H}_1(t_2)\right\rangle_{0,\mathrm{dc}} \\ &= \frac{\mathrm{i}U^2}{\hbar^2}\int_{-\infty}^{\infty} \mathrm{d}t_1 \int_{-\infty}^{\infty} \mathrm{d}t_2 \big\langle \hat{T}\hat{c}_\sigma(t)\hat{c}_\sigma^\dagger(0)\hat{c}_\uparrow^\dagger(t_1)\hat{c}_\uparrow(t_1) \\ &\qquad \times \hat{c}_\downarrow^\dagger(t_1)\hat{c}_\downarrow(t_1)\hat{c}_\uparrow^\dagger(t_2)\hat{c}_\uparrow(t_2)\hat{c}_\downarrow^\dagger(t_2)\hat{c}_\downarrow(t_2)\big\rangle_{0,\mathrm{dc}}.\end{aligned}$$

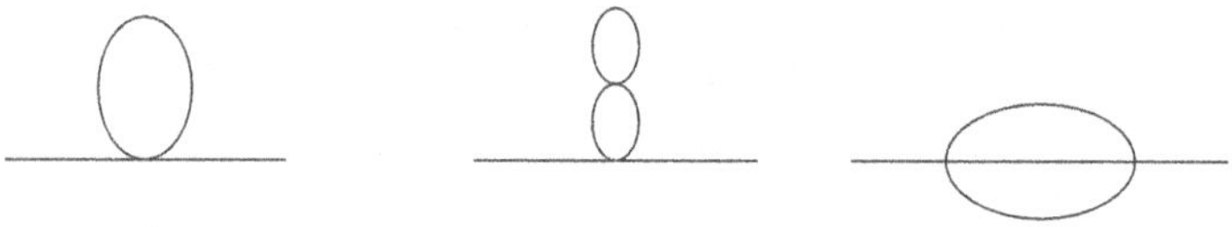

(a) (b)

Fig. 26.10 Feynman diagrams for the Green's function in the single-site Hubbard model. (a) Feynman diagram for the first-order correction. (b) Feynman diagrams for the second-order corrections.

Making use of Wick's theorem, we obtain

$$G_\sigma^{(2)}(t) = \frac{U^2}{\hbar^2}\int_{-\infty}^{\infty} dt_1 \int_{-\infty}^{\infty} dt_2\, G_{\bar\sigma 0}(t_1 - t_2)G_{\bar\sigma 0}(t_2 - t_1) \times \big[-in_F(\varepsilon_\sigma)G_{\sigma 0}(t - t_1)G_{\sigma 0}(t_1) + G_{\sigma 0}(t - t_1)G_{\sigma 0}(t_2)G_{\sigma 0}(t_1 - t_2)\big].$$

Inserting the Fourier transformation of $G_{\sigma 0}(t)$ into the above equation yields

$$\begin{aligned} G_\sigma^{(2)}(t) &= \frac{U^2}{\hbar^2}\int_{-\infty}^{\infty} dt_1 \int_{-\infty}^{\infty} dt_2 \bigg[-in_F(\varepsilon_\sigma)\left(\frac{\hbar}{2\pi}\right)^4 \int_{-\infty}^{\infty} d\omega_1 \cdots \int_{-\infty}^{\infty} d\omega_4\, e^{-i\omega_1 t} \\ &\quad \times e^{i(\omega_1+\omega_4-\omega_2-\omega_3)t_1 + i(\omega_3-\omega_4)t_2} G_{\sigma 0}(\omega_1)G_{\sigma 0}(\omega_2)G_{\bar\sigma 0}(\omega_3)G_{\bar\sigma 0}(\omega_4) \\ &\quad + \left(\frac{\hbar}{2\pi}\right)^5 \int_{-\infty}^{\infty} d\omega_1 \cdots \int_{-\infty}^{\infty} d\omega_5\, e^{-i\omega_1 t} e^{i(\omega_1+\omega_5-\omega_3-\omega_4)t_1} \\ &\quad \times e^{i(\omega_3+\omega_4-\omega_2-\omega_5)t_2} G_{\sigma 0}(\omega_1)G_{\sigma 0}(\omega_2)G_{\sigma 0}(\omega_3)G_{\bar\sigma 0}(\omega_4)G_{\bar\sigma 0}(\omega_5)\bigg] \\ &= U^2 \int_{-\infty}^{\infty} \frac{d(\hbar\omega)}{2\pi}\, e^{-i\omega t}\big[G_{\sigma 0}(\omega)\big]^2 \bigg\{-in_F(\varepsilon_\sigma)\int_{-\infty}^{\infty} \frac{d(\hbar\omega')}{2\pi}\big[G_{\bar\sigma 0}(\omega')\big]^2 \\ &\quad + \int_{-\infty}^{\infty} \frac{d(\hbar\omega')}{2\pi}\int_{-\infty}^{\infty} \frac{d(\hbar\omega'')}{2\pi}\, G_{\sigma 0}(\omega')G_{\bar\sigma 0}(\omega'')G_{\bar\sigma 0}(\omega' + \omega'' - \omega)\bigg\}. \end{aligned}$$

The Fourier transform of $G_\sigma^{(2)}(t)$ is given by

$$\begin{aligned} G_\sigma^{(2)}(\omega) &= U^2\big[G_{\sigma 0}(\omega)\big]^2 \bigg\{-in_F(\varepsilon_\sigma)\int_{-\infty}^{\infty} \frac{d(\hbar\omega')}{2\pi}\big[G_{\bar\sigma 0}(\omega')\big]^2 \\ &\quad + \int_{-\infty}^{\infty} \frac{d(\hbar\omega')}{2\pi}\int_{-\infty}^{\infty} \frac{d(\hbar\omega'')}{2\pi}\, G_{\sigma 0}(\omega')G_{\bar\sigma 0}(\omega'')G_{\bar\sigma 0}(\omega' + \omega'' - \omega)\bigg\}. \end{aligned}$$

The two terms in $G_\sigma^{(2)}(\omega)$ correspond to the Feynman diagrams in Fig. 26.10(b). Making use of the explicit expression of $G_{\sigma 0}(\omega)$, we can perform the integrals in the above equation. We find that

$$\int_{-\infty}^{\infty} \frac{\mathrm{d}(\hbar\omega')}{2\pi}\,\big[G_{\bar\sigma 0}(\omega')\big]^2 = 0,$$

$$\int_{-\infty}^{\infty} \frac{\mathrm{d}(\hbar\omega')}{2\pi}\int_{-\infty}^{\infty} \frac{\mathrm{d}(\hbar\omega'')}{2\pi}\, G_{\sigma 0}(\omega')G_{\bar\sigma 0}(\omega'')G_{\bar\sigma 0}(\omega'+\omega''-\omega)$$
$$= n_{\mathrm{F}}(\varepsilon_{\bar\sigma})\big[1-n_{\mathrm{F}}(\varepsilon_{\bar\sigma})\big]G_{\sigma 0}(\omega).$$

Hence,

$$G_\sigma^{(2)}(\omega) = U^2 n_{\mathrm{F}}(\varepsilon_{\bar\sigma})\big[1-n_{\mathrm{F}}(\varepsilon_{\bar\sigma})\big]\big[G_{\sigma 0}(\omega)\big]^3.$$

26-15 Single impurity. The Hamiltonian describing a single impurity is given by

$$\hat H = \sum_{\boldsymbol k} \varepsilon_{\boldsymbol k}\hat c_{\boldsymbol k}^\dagger \hat c_{\boldsymbol k} + \varepsilon_d \hat d^\dagger \hat d + \sum_{\boldsymbol k}\big(V_{\boldsymbol k}\hat c_{\boldsymbol k}^\dagger \hat d + V_{\boldsymbol k}^* \hat d^\dagger \hat c_{\boldsymbol k}\big).$$

Compute the Green's function

$$G_{\boldsymbol{kk}'}(t) = -\mathrm{i}\theta(t)\langle \hat c_{\boldsymbol k}(t)\hat c_{\boldsymbol k'}^\dagger(0) + \hat c_{\boldsymbol k'}^\dagger(0)\hat c_{\boldsymbol k}(0)\rangle\,.$$

We use the equation-of-motion method to find the Green's function $G_{\boldsymbol{kk}'}(t)$. Differentiating $G_{\boldsymbol{kk}'}(t)$ with respect to t yields

$$\frac{\partial G_{\boldsymbol{kk}'}(t)}{\partial t} = -\mathrm{i}\delta(t)\,\langle\{\hat c_{\boldsymbol k}, \hat c_{\boldsymbol k'}^\dagger\}\rangle - \mathrm{i}\theta(t)\,\langle\{\partial \hat c_{\boldsymbol k}(t)/\partial t, \hat c_{\boldsymbol k'}^\dagger(0)\}\rangle$$
$$= -\mathrm{i}\delta(t)\delta_{\boldsymbol{kk}'} - \mathrm{i}\theta(t)\,\langle\{\partial \hat c_{\boldsymbol k}(t)/\partial t, \hat c_{\boldsymbol k'}^\dagger(0)\}\rangle\,.$$

From the Heisenberg equation of motion, we have

$$\mathrm{i}\hbar\frac{\partial \hat c_{\boldsymbol k}(t)}{\partial t} = \big[\hat c_{\boldsymbol k}(t), \hat H\big] = \varepsilon_{\boldsymbol k}\hat c_{\boldsymbol k}(t) + V_{\boldsymbol k}\hat d(t).$$

We then have

$$\mathrm{i}\hbar\frac{\partial G_{\boldsymbol{kk}'}(t)}{\partial t}$$
$$= \hbar\delta(t)\delta_{\boldsymbol{kk}'} - \mathrm{i}\theta(t)\varepsilon_{\boldsymbol k}\,\langle\{\hat c_{\boldsymbol k}(t), \hat c_{\boldsymbol k'}^\dagger(0)\}\rangle - \mathrm{i}\theta(t)V_{\boldsymbol k}\,\langle\{\hat d(t), \hat c_{\boldsymbol k'}^\dagger(0)\}\rangle$$
$$= \hbar\theta(t)\delta_{\boldsymbol{kk}'} + \varepsilon_{\boldsymbol k}G_{\boldsymbol{kk}'}(t) + V_{\boldsymbol k}F_{\boldsymbol k'}(t),$$

where

$$F_{\boldsymbol k'}(t) = -\mathrm{i}\theta(t)\,\langle\{\hat d(t), \hat c_{\boldsymbol k'}^\dagger(0)\}\rangle\,.$$

Because of the appearance of the new Green's function $F_{\boldsymbol k'}(t)$ in the equation of motion of $G_{\boldsymbol{kk}'}(t)$, we need to set up an additional equation for $F_{\boldsymbol k'}(t)$ to be able to solve for $G_{\boldsymbol{kk}'}(t)$. Differentiating

$F_{\boldsymbol{k}'}(t)$ with respect to t and then making use of the Heisenberg equation of motion, we have

$$\begin{aligned}
\mathrm{i}\hbar\frac{\partial F_{\boldsymbol{k}'}(t)}{\partial t} &= \hbar\delta(t)\,\langle\{\hat{d},\hat{c}^\dagger_{\boldsymbol{k}'}\}\rangle - \mathrm{i}\theta(t)\,\langle\{\mathrm{i}\hbar\partial\hat{d}(t)/\partial t,\hat{c}^\dagger_{\boldsymbol{k}'}(0)\}\rangle \\
&= -\mathrm{i}\theta(t)\,\langle\{\mathrm{i}\hbar\partial\hat{d}(t)/\partial t,\hat{c}^\dagger_{\boldsymbol{k}'}(0)\}\rangle \\
&= -\mathrm{i}\theta(t)\varepsilon_d\,\langle\hat{d}(t),\hat{c}^\dagger_{\boldsymbol{k}'}(0)\}\rangle - \mathrm{i}\theta(t)\sum_{\boldsymbol{k}} V^*_{\boldsymbol{k}}\,\langle\{\hat{c}_{\boldsymbol{k}}(t),\hat{c}^\dagger_{\boldsymbol{k}'}(0)\}\rangle \\
&= \varepsilon_d F_{\boldsymbol{k}'}(t) + \sum_{\boldsymbol{k}} V^*_{\boldsymbol{k}} G_{\boldsymbol{k}\boldsymbol{k}'}(t).
\end{aligned}$$

We have now obtained two linear equations for $G_{\boldsymbol{k}\boldsymbol{k}'}(t)$ and $F_{\boldsymbol{k}'}(t)$. To solve them, we make Fourier transformations to $G_{\boldsymbol{k}\boldsymbol{k}'}(t)$ and $F_{\boldsymbol{k}'}(t)$

$$G_{\boldsymbol{k}\boldsymbol{k}'}(t) = \int_{-\infty}^{\infty}\frac{\mathrm{d}(\hbar\omega)}{2\pi}\,\mathrm{e}^{-\mathrm{i}\omega t}G_{\boldsymbol{k}\boldsymbol{k}'}(\omega),$$
$$F_{\boldsymbol{k}'}(t) = \int_{-\infty}^{\infty}\frac{\mathrm{d}(\hbar\omega)}{2\pi}\,\mathrm{e}^{-\mathrm{i}\omega t}F_{\boldsymbol{k}'}(\omega).$$

Making use of the above Fourier transformations in the equations for $G_{\boldsymbol{k}\boldsymbol{k}'}(t)$ and $F_{\boldsymbol{k}'}(t)$ as well as the following expression for the δ-function

$$\delta(t) = \frac{1}{\hbar}\int_{-\infty}^{\infty}\frac{\mathrm{d}(\hbar\omega)}{2\pi}\,\mathrm{e}^{-\mathrm{i}\omega t},$$

we obtain

$$(\varepsilon_{\boldsymbol{k}} - \hbar\omega)G_{\boldsymbol{k}\boldsymbol{k}'}(\omega) + V_{\boldsymbol{k}}F_{\boldsymbol{k}'}(\omega) = -\delta_{\boldsymbol{k}\boldsymbol{k}'},$$
$$(\varepsilon_d - \hbar\omega)F_{\boldsymbol{k}'}(\omega) + \sum_{\boldsymbol{k}} V^*_{\boldsymbol{k}}G_{\boldsymbol{k}\boldsymbol{k}'}(\omega) = 0.$$

Eliminating $F_{\boldsymbol{k}'}(\omega)$ from the above two equations leads to

$$G_{\boldsymbol{k}\boldsymbol{k}'}(\omega) = \frac{V_{\boldsymbol{k}}}{(\varepsilon_d - \hbar\omega)(\varepsilon_{\boldsymbol{k}} - \hbar\omega)}\sum_{\boldsymbol{k}} V^*_{\boldsymbol{k}}G_{\boldsymbol{k}\boldsymbol{k}'}(\omega) - \frac{\delta_{\boldsymbol{k}\boldsymbol{k}'}}{\varepsilon_{\boldsymbol{k}} - \hbar\omega}.$$

Multiplying both sides of the above equation with $V^*_{\boldsymbol{k}}$ and then summing over $\boldsymbol{k}$, we can solve for $\sum_{\boldsymbol{k}} V^*_{\boldsymbol{k}}G_{\boldsymbol{k}\boldsymbol{k}'}(\omega)$ and obtain

$$\sum_{\boldsymbol{k}} V^*_{\boldsymbol{k}}G_{\boldsymbol{k}\boldsymbol{k}'}(\omega) = -\frac{V^*_{\boldsymbol{k}'}/(\varepsilon_{\boldsymbol{k}'} - \hbar\omega)}{1 - \sum_{\boldsymbol{k}}|V_{\boldsymbol{k}}|^2/[(\varepsilon_d - \hbar\omega)(\varepsilon_{\boldsymbol{k}} - \hbar\omega)]}.$$

The Green's function $G_{\boldsymbol{k}\boldsymbol{k}'}(\omega)$ is finally given by

$$G_{\boldsymbol{k}\boldsymbol{k}'}(\omega) = -\frac{V_{\boldsymbol{k}}V^*_{\boldsymbol{k}'}/[(\varepsilon_d - \hbar\omega)(\varepsilon_{\boldsymbol{k}} - \hbar\omega)(\varepsilon_{\boldsymbol{k}'} - \hbar\omega)]}{1 - \sum_{\boldsymbol{k}}|V_{\boldsymbol{k}}|^2/[(\varepsilon_d - \hbar\omega)(\varepsilon_{\boldsymbol{k}} - \hbar\omega)]} - \frac{\delta_{\boldsymbol{k}\boldsymbol{k}'}}{\varepsilon_{\boldsymbol{k}} - \hbar\omega}.$$

26-16 Discontinuity of the momentum distribution. The momentum distribution of spin-1/2 fermions for a single spin direction is

$$n_{\boldsymbol{k}} = -\mathrm{i}\int_{-\infty}^{\infty} \frac{\mathrm{d}(\hbar\omega)}{2\pi}\, G(\boldsymbol{k},\omega)\mathrm{e}^{\mathrm{i}\eta\omega}$$

with $\eta \to 0^+$. Assume that the Green's function of interacting fermions is given by

$$G(\boldsymbol{k},\omega) = \frac{Z}{\hbar\omega - \varepsilon_{\boldsymbol{k}} + \mathrm{i}\Gamma_{\boldsymbol{k}}\,\mathrm{sgn}(k-k_{\mathrm{F}})} + G_{\mathrm{reg}}(\boldsymbol{k},\omega),$$

where Z is a constant, $\Gamma_{\boldsymbol{k}} > 0$, and $G_{\mathrm{reg}}(\boldsymbol{k},\omega)$ has no poles. Find the contribution of the part of $G(\boldsymbol{k},\omega)$ with poles to the momentum distribution.

Inserting the part of $G(\boldsymbol{k},\omega)$ with poles into the expression for $n_{\boldsymbol{k}}$, we have

$$n_{\boldsymbol{k}} = -\mathrm{i}\int_{-\infty}^{\infty} \frac{\mathrm{d}(\hbar\omega)}{2\pi}\, \frac{Z}{\hbar\omega - \varepsilon_{\boldsymbol{k}} + \mathrm{i}\Gamma_{\boldsymbol{k}}\,\mathrm{sgn}(k-k_{\mathrm{F}})}\mathrm{e}^{\mathrm{i}\eta\omega}.$$

Because of the presence of the factor $\mathrm{e}^{\mathrm{i}\eta\omega}$ with $\eta \to 0^+$, we can only close the contour in the upper-half complex plane. Since the pole is in the lower-half complex plane for $k > k_{\mathrm{F}}$, $n_{\boldsymbol{k}} = 0$ for $k > k_{\mathrm{F}}$. For $k < k_{\mathrm{F}}$, the pole is in the upper-half complex plane and we have

$$n_{\boldsymbol{k}} = -\mathrm{i}\frac{1}{2\pi}\cdot 2\pi\mathrm{i}Z = Z,\ \ k < k_{\mathrm{F}}.$$

Writing the two cases together, we have

$$n_{\boldsymbol{k}} = Z\theta(k_{\mathrm{F}} - k).$$

Thus, Z gives the spectral weight jump at the Fermi surface.

26-17 Multiparticle expectation value. We consider the expectation value of the following product of operators

$$\mathscr{P}_N = \left\langle \prod_{n=1}^{N} \hat{c}_n\hat{c}_n^\dagger \right\rangle_0$$

with $\hat{c}_n$ and $\hat{c}_n^\dagger$ the annihilation and creation operators of a one-dimensional spinless lattice fermion at the lattice site n. Express $\mathscr{P}_N$ in terms of $G_{mn} = \langle \hat{c}_m\hat{c}_n^\dagger\rangle_0$.

The expectation value of the product can be evaluated using Wick's theorem. In applying Wick's theorem, we fix the annihilation operators in the sequence $\hat{c}_1$, $\hat{c}_2$, $\cdots$, $\hat{c}_N$, with only the creation operators permuted to obtain all the possible combinations. In terms of the permutations of the N integers, 1, 2, ..., N, we can write the value of P_N as

$$\mathscr{P}_N = \sum_P (-1)^P G_{1P_1} G_{2P_2} \cdots G_{NP_N},$$

where $(-1)^P$ is the parity of the permutation P.

26-18 Friedel oscillations. The interaction Hamiltonian of a Fermi gas with an impurity located at the origin is

$$\hat{H}_1 = \int \mathrm{d}\boldsymbol{r}\, V(\boldsymbol{r}) \hat{\psi}^\dagger(\boldsymbol{r}) \hat{\psi}(\boldsymbol{r}),$$

where $\hat{\psi}(\boldsymbol{r})$ is the quantum field operator of fermions with the spin variables suppressed and $V(\boldsymbol{r})$ is assumed to be small. Compute the density of fermions as a function of the distance r to the impurity up to the first order in $V(\boldsymbol{r})$. Show that the density of fermions oscillates as a function of r. Find the period of the oscillations.

For this problem, we start from the fermion Green's function in real space. With the fermion spin not suppressed, the fermion Green's function in real space is defined by

$$G(\boldsymbol{r}, t; \boldsymbol{r}', t') = -\mathrm{i} \langle \hat{T} \hat{\psi}(\boldsymbol{r}, t) \hat{\psi}^\dagger(\boldsymbol{r}', t') \rangle .$$

The fermion density at $\boldsymbol{r}$, $n(\boldsymbol{r})$, is given by

$$n(\boldsymbol{r}) = \langle \hat{\psi}^\dagger(\boldsymbol{r}) \hat{\psi}(\boldsymbol{r}) \rangle .$$

From the definition of $G(\boldsymbol{r}, t; \boldsymbol{r}', t')$, we see that, once $G(\boldsymbol{r}, t; \boldsymbol{r}', t')$ is known, we can obtain $n(\boldsymbol{r})$ from

$$n(\boldsymbol{r}) = -\mathrm{i} \lim_{\boldsymbol{r}' \to \boldsymbol{r}} \lim_{t' \to t^+} G(\boldsymbol{r}, t; \boldsymbol{r}', t').$$

For free fermions with the Hamiltonian $\hat{H}_0 = \sum_{\boldsymbol{k}} \xi_{\boldsymbol{k}} \hat{c}_{\boldsymbol{k}}^\dagger \hat{c}_{\boldsymbol{k}}$, making use of

$$\hat{\psi}(\boldsymbol{r}) = \frac{1}{\sqrt{\mathscr{V}}} \sum_{\boldsymbol{k}} \mathrm{e}^{\mathrm{i}\boldsymbol{k}\cdot\boldsymbol{r}} \hat{c}_{\boldsymbol{k}},$$

we have

$$\begin{aligned}
&G_0(\boldsymbol{r}, t; \boldsymbol{r}', t')\\
&\quad = -\frac{\mathrm{i}}{\mathscr{V}}\sum_{\boldsymbol{k}\boldsymbol{k}'} \mathrm{e}^{\mathrm{i}(\boldsymbol{k}\cdot\boldsymbol{r}-\boldsymbol{k}'\cdot\boldsymbol{r}')}\,\langle \hat{T}\hat{c}_{\boldsymbol{k}}(t)\hat{\psi}^\dagger_{\boldsymbol{k}'}(t')\rangle_0\\
&\quad = -\frac{\mathrm{i}}{\mathscr{V}}\sum_{\boldsymbol{k}\boldsymbol{k}'} \mathrm{e}^{\mathrm{i}(\boldsymbol{k}\cdot\boldsymbol{r}-\boldsymbol{k}'\cdot\boldsymbol{r}')}\mathrm{e}^{-\mathrm{i}(\xi_{\boldsymbol{k}}t-\xi_{\boldsymbol{k}'}t')/\hbar}\\
&\qquad\qquad \times\big[\langle \hat{c}_{\boldsymbol{k}}\hat{\psi}^\dagger_{\boldsymbol{k}'}\rangle_0\theta(t-t') - \langle \hat{\psi}^\dagger_{\boldsymbol{k}'}\hat{c}_{\boldsymbol{k}}\rangle_0\theta(t'-t)\big]\\
&\quad = -\frac{\mathrm{i}}{\mathscr{V}}\sum_{\boldsymbol{k}\boldsymbol{k}'} \mathrm{e}^{\mathrm{i}(\boldsymbol{k}\cdot\boldsymbol{r}-\boldsymbol{k}'\cdot\boldsymbol{r}')}\mathrm{e}^{-\mathrm{i}(\xi_{\boldsymbol{k}}t-\xi_{\boldsymbol{k}'}t')/\hbar}\delta_{\boldsymbol{k}\boldsymbol{k}'}\\
&\qquad\qquad \times\big[(1-n_{\boldsymbol{k}})\theta(t-t') - n_{\boldsymbol{k}}\theta(t'-t)\big]\\
&\quad = -\frac{\mathrm{i}}{\mathscr{V}}\sum_{\boldsymbol{k}} \mathrm{e}^{\mathrm{i}\boldsymbol{k}\cdot(\boldsymbol{r}-\boldsymbol{r}')}\mathrm{e}^{-\mathrm{i}\xi_{\boldsymbol{k}}(t-t')/\hbar}\big[(1-n_{\boldsymbol{k}})\theta(t-t') - n_{\boldsymbol{k}}\theta(t'-t)\big].
\end{aligned}$$

The fermion density is given by

$$\begin{aligned}
n(\boldsymbol{r}) &= -\mathrm{i}\lim_{\boldsymbol{r}'\to\boldsymbol{r}}\lim_{t'\to t^+} G(\boldsymbol{r}, t; \boldsymbol{r}', t')\\
&= \frac{1}{\mathscr{V}}\sum_{\boldsymbol{k}} \mathrm{e}^{\mathrm{i}\eta\xi_{\boldsymbol{k}}} n_{\boldsymbol{k}} = \frac{1}{\mathscr{V}}\sum_{\boldsymbol{k}} n_{\boldsymbol{k}} = \frac{N}{\mathscr{V}} = n,
\end{aligned}$$

where $\eta \to 0^+$ and it does not play any significant role in this simple case. We thus see that free fermions have a uniform density given by $N/\mathscr{V}$. We now take the effect of the impurity into account. We first find the first-order correction to the fermion Green's function due to the presence of the impurity. We have

$$\begin{aligned}
&G^{(1)}(\boldsymbol{r}, t; \boldsymbol{r}', t')\\
&= -\int_{-\infty}^{\infty}\frac{\mathrm{d}t_1}{\hbar}\,\langle \hat{T}\hat{\psi}(\boldsymbol{r}, t)\hat{\psi}^\dagger(\boldsymbol{r}', t')\hat{H}_1(t_1)\rangle_{0,\mathrm{dc}}\\
&= -\int_{-\infty}^{\infty}\frac{\mathrm{d}t_1}{\hbar}\int \mathrm{d}\boldsymbol{r}_1\, V(\boldsymbol{r}_1)\,\langle \hat{T}\hat{\psi}(\boldsymbol{r}, t)\hat{\psi}^\dagger(\boldsymbol{r}', t')\hat{\psi}^\dagger(\boldsymbol{r}_1, t_1)\hat{\psi}(\boldsymbol{r}_1, t_1)\rangle_{0;\mathrm{dc}}\\
&= \int_{-\infty}^{\infty}\frac{\mathrm{d}t_1}{\hbar}\int \mathrm{d}\boldsymbol{r}_1\, V(\boldsymbol{r}_1)G_0(\boldsymbol{r}, t; \boldsymbol{r}_1, t_1)G_0(\boldsymbol{r}_1, t_1; \boldsymbol{r}', t')
\end{aligned}$$

$$= -\frac{1}{\mathscr{V}^2}\int_{-\infty}^{\infty}\frac{dt_1}{\hbar}\int d\boldsymbol{r}_1\, V(\boldsymbol{r}_1)$$
$$\times \sum_{\boldsymbol{k}\boldsymbol{k}'} e^{i\boldsymbol{k}\cdot(\boldsymbol{r}-\boldsymbol{r}_1)}e^{-i\xi_{\boldsymbol{k}}(t-t_1)/\hbar}\big[(1-n_{\boldsymbol{k}})\theta(t-t_1)-n_{\boldsymbol{k}}\theta(t_1-t)\big]$$
$$\times e^{i\boldsymbol{k}'\cdot(\boldsymbol{r}_1-\boldsymbol{r}')}e^{-i\xi_{\boldsymbol{k}'}(t_1-t')/\hbar}\big[(1-n_{\boldsymbol{k}'})\theta(t_1-t')-n_{\boldsymbol{k}'}\theta(t'-t_1)\big]$$
$$= -\frac{1}{\mathscr{V}}\sum_{\boldsymbol{k}\boldsymbol{k}'} V(\boldsymbol{k}-\boldsymbol{k}')e^{i(\boldsymbol{k}\cdot\boldsymbol{r}-\boldsymbol{k}'\cdot\boldsymbol{r}')}e^{-i(\xi_{\boldsymbol{k}}t-\xi_{\boldsymbol{k}'}t')/\hbar}\int_{-\infty}^{\infty}\frac{dt_1}{\hbar}\, e^{i(\xi_{\boldsymbol{k}}-\xi_{\boldsymbol{k}'})t_1/\hbar}$$
$$\times\big[(1-n_{\boldsymbol{k}})(1-n_{\boldsymbol{k}'})\theta(t-t_1)\theta(t_1-t')+n_{\boldsymbol{k}}n_{\boldsymbol{k}'}\theta(t_1-t)\theta(t'-t_1)$$
$$-(1-n_{\boldsymbol{k}})n_{\boldsymbol{k}'}\theta(t-t_1)\theta(t'-t_1)-n_{\boldsymbol{k}}(1-n_{\boldsymbol{k}'})\theta(t_1-t)\theta(t_1-t')\big],$$

where we have introduced the Fourier transform of $V(\boldsymbol{r})$, $V(\boldsymbol{k}) = \mathscr{V}^{-1}\int d\boldsymbol{r}\, V(\boldsymbol{r})e^{-i\boldsymbol{k}\cdot\boldsymbol{r}}$. The time step functions in the above equation provide constraints on the t_1-integral. Performing the integration over t_1 with these step functions taken into consideration, we have

$$G^{(1)}(\boldsymbol{r},t;\boldsymbol{r}',t') = \frac{i}{\mathscr{V}}\sum_{\boldsymbol{k}\boldsymbol{k}'} V(\boldsymbol{k}-\boldsymbol{k}')e^{i(\boldsymbol{k}\cdot\boldsymbol{r}-\boldsymbol{k}'\cdot\boldsymbol{r}')}e^{-i(\xi_{\boldsymbol{k}}t-\xi_{\boldsymbol{k}'}t')/\hbar}$$
$$\times\Bigg\{\frac{e^{i(\xi_{\boldsymbol{k}}-\xi_{\boldsymbol{k}'})t}-e^{i(\xi_{\boldsymbol{k}}-\xi_{\boldsymbol{k}'})t'}}{\xi_{\boldsymbol{k}}-\xi_{\boldsymbol{k}'}}\big[(1-n_{\boldsymbol{k}})(1-n_{\boldsymbol{k}'})\theta(t-t')$$
$$-n_{\boldsymbol{k}}n_{\boldsymbol{k}'}\theta(t'-t)\big]-\frac{e^{i(\xi_{\boldsymbol{k}}-\xi_{\boldsymbol{k}'})t_<}}{\xi_{\boldsymbol{k}}-\xi_{\boldsymbol{k}'}-i\delta}(1-n_{\boldsymbol{k}})n_{\boldsymbol{k}'}$$
$$+\frac{e^{i(\xi_{\boldsymbol{k}}-\xi_{\boldsymbol{k}'})t_>}}{\xi_{\boldsymbol{k}}-\xi_{\boldsymbol{k}'}+i\delta}n_{\boldsymbol{k}}(1-n_{\boldsymbol{k}'})\Bigg\},$$

where $t_< = \min(t,t')$ and $t_> = \max(t,t')$. The terms $\pm i\delta$'s on the denominators are introduced to render the integrals to converge at $t_1 = \pm\infty$. Up to the first order in V, the fermion density is given by

$$n(\boldsymbol{r}) = -i\lim_{\boldsymbol{r}'\to\boldsymbol{r}}\lim_{t'\to t^+}\big[G_0(\boldsymbol{r},t;\boldsymbol{r}',t')+G^{(1)}(\boldsymbol{r},t;\boldsymbol{r}',t')\big]$$
$$= n - i\lim_{\boldsymbol{r}'\to\boldsymbol{r}}\lim_{t'\to t^+}G^{(1)}(\boldsymbol{r},t;\boldsymbol{r}',t')$$
$$= n+\frac{1}{\mathscr{V}}\sum_{\boldsymbol{k}\boldsymbol{k}'}V(\boldsymbol{k}-\boldsymbol{k}')e^{i(\boldsymbol{k}-\boldsymbol{k}')\cdot\boldsymbol{r}}\left[\frac{n_{\boldsymbol{k}}(1-n_{\boldsymbol{k}'})}{\xi_{\boldsymbol{k}}-\xi_{\boldsymbol{k}'}+i\delta}-\frac{(1-n_{\boldsymbol{k}})n_{\boldsymbol{k}'}}{\xi_{\boldsymbol{k}}-\xi_{\boldsymbol{k}'}-i\delta}\right].$$

From the denominators of the two fractions in the above equation, we see that fermions on the opposite sides of the Fermi surface with $\xi_{\boldsymbol{k}} \sim \xi_{\boldsymbol{k}'}$ make most important contributions to the density oscillations at low temperatures. Note that the region with $\boldsymbol{k} \sim \boldsymbol{k}'$ does not make a substantial contribution to the density oscillations because $n_{\boldsymbol{k}}(1-n_{\boldsymbol{k}'}) = 0$ for $\boldsymbol{k} = \boldsymbol{k}'$. We can thus make a simplification by setting $V(\boldsymbol{k} - \boldsymbol{k}')$ to $V(2k_{\mathrm{F}})$. We then have

$$n(\boldsymbol{r}) = n + \frac{1}{\mathscr{V}} V(2k_{\mathrm{F}}) \sum_{\boldsymbol{k}\boldsymbol{k}'} \mathrm{e}^{\mathrm{i}(\boldsymbol{k}-\boldsymbol{k}')\cdot\boldsymbol{r}} \left[\frac{n_{\boldsymbol{k}}(1-n_{\boldsymbol{k}'})}{\xi_{\boldsymbol{k}} - \xi_{\boldsymbol{k}'} + \mathrm{i}\delta} - \frac{(1-n_{\boldsymbol{k}})n_{\boldsymbol{k}'}}{\xi_{\boldsymbol{k}} - \xi_{\boldsymbol{k}'} - \mathrm{i}\delta} \right].$$

In the above expression, only the exponential factor $\mathrm{e}^{\mathrm{i}(\boldsymbol{k}-\boldsymbol{k}')\cdot\boldsymbol{r}}$ depends on the directions of $\boldsymbol{k}$ and $\boldsymbol{k}'$. Making use of this fact, we perform the integrations over angles with the summations over $\boldsymbol{k}$ and $\boldsymbol{k}'$ converted into integrations. Converting the integrations over $\boldsymbol{k}$ and $\boldsymbol{k}'$ back into summations after the integrations over angles have been performed, we have

$$n(\boldsymbol{r}) = n + \frac{V(2k_{\mathrm{F}})}{\mathscr{V}} \sum_{\boldsymbol{k}\boldsymbol{k}'} \frac{\sin(kr)\sin(k'r)}{kk'r^2} \left[\frac{n_{\boldsymbol{k}}(1-n_{\boldsymbol{k}'})}{\xi_{\boldsymbol{k}} - \xi_{\boldsymbol{k}'} + \mathrm{i}\delta} - \frac{(1-n_{\boldsymbol{k}})n_{\boldsymbol{k}'}}{\xi_{\boldsymbol{k}} - \xi_{\boldsymbol{k}'} - \mathrm{i}\delta} \right].$$

Because of the presence of the factors $n_{\boldsymbol{k}}(1-n_{\boldsymbol{k}'})$ and $(1-n_{\boldsymbol{k}})n_{\boldsymbol{k}'}$, the major contributions come from the region for both $\xi_{\boldsymbol{k}}$ and $\xi_{\boldsymbol{k}'}$ close to E_{F} with $\boldsymbol{k}$ and $\boldsymbol{k}'$ on the opposite sides of the Fermi surface. Retaining only the real part of the second term and taking the $\xi_{\boldsymbol{k}'} \to \xi_{\boldsymbol{k}}$ limit, we have

$$\begin{aligned} n(\boldsymbol{r}) &\approx n + \frac{1}{\mathscr{V}} V(2k_{\mathrm{F}}) \sum_{\boldsymbol{k}\boldsymbol{k}'} \frac{\sin(kr)\sin(k'r)}{kk'r^2} \delta_{\boldsymbol{k}'\boldsymbol{k}} \lim_{\xi_{\boldsymbol{k}'} \to \xi_{\boldsymbol{k}}} \frac{n_{\boldsymbol{k}} - n_{\boldsymbol{k}'}}{\xi_{\boldsymbol{k}} - \xi_{\boldsymbol{k}'}} \\ &= n + \frac{1}{\mathscr{V}} V(2k_{\mathrm{F}}) \sum_{\boldsymbol{k}} \frac{\sin^2(kr)}{(kr)^2} \frac{\partial n_{\boldsymbol{k}}}{\partial \xi_{\boldsymbol{k}}} \\ &\approx n - \frac{1}{\mathscr{V}} V(2k_{\mathrm{F}}) \sum_{\boldsymbol{k}} \frac{\sin^2(kr)}{(kr)^2} \delta(\xi_{\boldsymbol{k}}) \\ &= n - g(0)V(2k_{\mathrm{F}}) \int_{-E_{\mathrm{F}}}^{\infty} \mathrm{d}\xi\, \frac{\sin^2(kr)}{(kr)^2} \delta(\xi) \\ &= n - g(0)V(2k_{\mathrm{F}}) \frac{\sin^2(k_{\mathrm{F}}r)}{(k_{\mathrm{F}}r)^2} \\ &= n - g(0)V(2k_{\mathrm{F}}) \frac{1-\cos(2k_{\mathrm{F}}r)}{(k_{\mathrm{F}}r)^2}, \end{aligned}$$

where $g(0)$ is the density of states for one spin direction per unit volume at the Fermi energy, $g(0) = mk_{\mathrm{F}}/2\pi^2\hbar^2$. Thus, the fermion density displays oscillations of period π/k_{F}. Such oscillations are known as *Friedel oscillations*.

26-19 Plasma waves. Consider the Feynman diagrams at the finite wave vector $\boldsymbol{k}$ and frequency ω in Fig. 26.11. The sum of these Feynman diagrams describes the effect of dynamic screening of bare interaction $V_{\boldsymbol{k}}$ represented by a wavy line. The dispersion relation of collective excitations, $\omega(\boldsymbol{k})$, is defined by the poles of the screened interaction.

Fig. 26.11 Feynman diagrams for the effect of dynamic screening.

(1) Find the polarization operator $\Pi(\boldsymbol{k},\omega)$ for $k \ll k_{\rm F}$ and $\omega \ll E_{\rm F}/\hbar$.

(2) Sum up the series and find the screened interaction $V_{\rm eff}(\boldsymbol{k},\omega)$ with the bare interaction taken as the pure Coulomb interaction with $V_{\boldsymbol{k}} = e^2/\epsilon_0 k^2$.

(3) Show that, under the above approximation, the dispersion of plasma waves is to be determined from

$$\frac{\omega}{v_{\rm F}k}\ln\frac{\omega+v_{\rm F}k}{\omega-v_{\rm F}k}-\frac{1}{2}=\frac{\epsilon_0 k^2}{e^2 g(0)}.$$

(4) Find the expressions for $\omega(\boldsymbol{k})$ for small and large k.

---

(1) The polarization operator $\Pi(\boldsymbol{k},\mathrm{i}\omega_m)$ per unit volume is given by [*cf.* **Problem 16-12**; however, note that it is per unit volume here.]

$$\begin{aligned}\Pi(\boldsymbol{k},\mathrm{i}\omega_m) &= -\frac{1}{\mathscr{V}\beta}\sum_{\boldsymbol{k}'n}\mathscr{G}_{\sigma 0}(\boldsymbol{k}',\mathrm{i}\omega_n)\mathscr{G}_{\sigma 0}(\boldsymbol{k}'+\boldsymbol{k},\mathrm{i}\omega_n+\mathrm{i}\omega_m)\\ &= -\frac{1}{\mathscr{V}\beta}\sum_{\boldsymbol{k}'n}\frac{1}{\mathrm{i}\hbar\omega_n-\xi_{\boldsymbol{k}'}}\frac{1}{\mathrm{i}\hbar\omega_n+\mathrm{i}\hbar\omega_m-\xi_{\boldsymbol{k}'+\boldsymbol{k}}}\\ &= -\frac{1}{\mathscr{V}}\sum_{\boldsymbol{k}'}\frac{n_{\rm F}(\xi_{\boldsymbol{k}'})-n_{\rm F}(\xi_{\boldsymbol{k}'+\boldsymbol{k}})}{\mathrm{i}\hbar\omega_m+\xi_{\boldsymbol{k}'}-\xi_{\boldsymbol{k}'+\boldsymbol{k}}}.\end{aligned}$$

The retarded polarization operator is

$$\Pi_R(\boldsymbol{k},\omega) = -\frac{1}{\mathscr{V}}\sum_{\boldsymbol{k}'}\frac{n_{\rm F}(\xi_{\boldsymbol{k}'})-n_{\rm F}(\xi_{\boldsymbol{k}'+\boldsymbol{k}})}{\hbar\omega+\xi_{\boldsymbol{k}'}-\xi_{\boldsymbol{k}'+\boldsymbol{k}}+\mathrm{i}\delta}.$$

The real part of $\Pi_R(\boldsymbol{k},\omega)$, $\Pi'_R(\boldsymbol{k},\omega)$, is given by

$$\begin{aligned}
&\Pi'_R(\boldsymbol{k},\omega)\\
&= -\frac{1}{\mathscr{V}}\sum_{\boldsymbol{k}'}\frac{n_{\rm F}(\xi_{\boldsymbol{k}'})-n_{\rm F}(\xi_{\boldsymbol{k}'+\boldsymbol{k}})}{\hbar\omega+\xi_{\boldsymbol{k}'}-\xi_{\boldsymbol{k}'+\boldsymbol{k}}}\\
&= -\frac{1}{\mathscr{V}}\sum_{\boldsymbol{k}'}n_{\rm F}(\xi_{\boldsymbol{k}'})\left[\frac{1}{\hbar\omega+\xi_{\boldsymbol{k}'}-\xi_{\boldsymbol{k}'+\boldsymbol{k}}}-\frac{1}{\hbar\omega+\xi_{\boldsymbol{k}'-\boldsymbol{k}}-\xi_{\boldsymbol{k}'}}\right]\\
&= \frac{mk}{(4\pi\hbar)^2}\int_0^{2k_{\rm F}/k}{\rm d}x\,x\left[\ln\left|\frac{x-\hbar\omega/\varepsilon_k+1}{x+\hbar\omega/\varepsilon_k-1}\right|-\ln\left|\frac{x-\hbar\omega/\varepsilon_k-1}{x+\hbar\omega/\varepsilon_k+1}\right|\right],
\end{aligned}$$

where we have used $2k'/k$ as the integration variable and $\varepsilon_k=\hbar^2k^2/2m$. Performing an integration by parts, we have

$$\begin{aligned}
\Pi'_R(\boldsymbol{k},\omega) &= \frac{mk}{2(4\pi\hbar)^2}\left[\left(\frac{2k_{\rm F}}{k}\right)^2\left(\ln\left|\frac{2k_{\rm F}/k-\hbar\omega/\varepsilon_k+1}{2k_{\rm F}/k+\hbar\omega/\varepsilon_k-1}\right|\right.\right.\\
&\quad\left.-\ln\left|\frac{2k_{\rm F}/k-\hbar\omega/\varepsilon_k-1}{2k_{\rm F}/k+\hbar\omega/\varepsilon_k+1}\right|\right)\\
&\quad-\int_0^{2k_{\rm F}/k}{\rm d}x\,x^2\left(\frac{1}{x-\hbar\omega/\varepsilon_k+1}-\frac{1}{x+\hbar\omega/\varepsilon_k-1}\right.\\
&\quad\left.\left.-\frac{1}{x-\hbar\omega/\varepsilon_k-1}+\frac{1}{x+\hbar\omega/\varepsilon_k+1}\right)\right].
\end{aligned}$$

Performing the remaining integrals, we obtain

$$\begin{aligned}
\Pi'_R(\boldsymbol{k},\omega) &= \frac{mk}{2(4\pi\hbar)^2}\left\{\frac{8k_{\rm F}}{k}\right.\\
&\quad+\left[\left(\frac{2k_{\rm F}}{k}\right)^2-\left(\frac{\hbar\omega}{\varepsilon_k}-1\right)^2\right]\ln\left|\frac{2k_{\rm F}/k-\hbar\omega/\varepsilon_k+1}{2k_{\rm F}/k+\hbar\omega/\varepsilon_k-1}\right|\\
&\quad\left.-\left[\left(\frac{2k_{\rm F}}{k}\right)^2-\left(\frac{\hbar\omega}{\varepsilon_k}+1\right)^2\right]\ln\left|\frac{2k_{\rm F}/k-\hbar\omega/\varepsilon_k-1}{2k_{\rm F}/k+\hbar\omega/\varepsilon_k+1}\right|\right\}.
\end{aligned}$$

For $\hbar\omega\ll E_{\rm F}$ and $k\ll k_{\rm F}$, we can neglect the 1s in the logarithmic functions for which we now have for both of them

$$\begin{aligned}
\ln\left|\frac{2k_{\rm F}/k-\hbar\omega/\varepsilon_k}{2k_{\rm F}/k+\hbar\omega/\varepsilon_k}\right| &= \ln\left|\frac{\hbar k_{\rm F}k/m-\omega}{\hbar k_{\rm F}k/m+\omega}\right|\\
&= \ln\left|\frac{v_{\rm F}k-\omega}{v_{\rm F}k+\omega}\right| = \ln\left|\frac{\omega-v_{\rm F}k}{\omega+v_{\rm F}k}\right|.
\end{aligned}$$

We then have

$$\begin{aligned}
\Pi'_R(\boldsymbol{k},\omega) &\approx \frac{m^2}{(2\pi\hbar)^2\hbar k}\left(v_{\rm F}k+2\omega\ln\left|\frac{\omega-v_{\rm F}k}{\omega+v_{\rm F}k}\right|\right)\\
&= g(0)\left(\frac{1}{2}-\frac{\omega}{v_{\rm F}k}\ln\left|\frac{\omega+v_{\rm F}k}{\omega-v_{\rm F}k}\right|\right),
\end{aligned}$$

where $g(0)$ is the density of states per unit volume for one spin direction at the Fermi energy, $g(0) = mk_{\rm F}/2\pi^2\hbar^2$.

(2) Summing up the series for $V_{\rm eff}(\boldsymbol{k},\omega)$ in Fig. 26.11, we obtain

$$\begin{aligned}
&- V_{\rm eff}(\boldsymbol{k},\omega)\\
&= -V_{\boldsymbol{k}} + V_{\boldsymbol{k}}\Pi'_R(\boldsymbol{k},\omega)V_{\boldsymbol{k}} - V_{\boldsymbol{k}}\Pi'_R(\boldsymbol{k},\omega)V_{\boldsymbol{k}}\Pi'_R(\boldsymbol{k},\omega)V_{\boldsymbol{k}} + \cdots\\
&= -V_{\boldsymbol{k}}\left\{1 - V_{\boldsymbol{k}}\Pi'_R(\boldsymbol{k},\omega) + \left[V_{\boldsymbol{k}}\Pi'_R(\boldsymbol{k},\omega)\right]^2 - \cdots\right\}\\
&= -\frac{V_{\boldsymbol{k}}}{1 + V_{\boldsymbol{k}}\Pi'_R(\boldsymbol{k},\omega)} = -\frac{e^2/\epsilon_0 k^2}{1 + (e^2/\epsilon_0 k^2)\Pi'_R(\boldsymbol{k},\omega)},
\end{aligned}$$

where we have associated with each wavy line the negative of $V_{\boldsymbol{k}}$, $-V_{\boldsymbol{k}}$, according to the Feynman rules for the Coulomb interaction. Thus,

$$V_{\rm eff}(\boldsymbol{k},\omega) = \frac{e^2/\epsilon_0 k^2}{1 + (e^2/\epsilon_0 k^2)\Pi'_R(\boldsymbol{k},\omega)}.$$

(3) The dispersion of plasma waves can be obtained from the equation for determining the poles of $V_{\rm eff}(\boldsymbol{k},\omega)$. Setting the denominator of $V_{\rm eff}(\boldsymbol{k},\omega)$ to zero, we have

$$1 + (e^2/\epsilon_0 k^2)\Pi'_R(\boldsymbol{k},\omega) = 0.$$

Inserting the previously-derived approximate expression for $\Pi'_R(\boldsymbol{k},\omega)$ into the above equation yields

$$\frac{\omega}{v_{\rm F}k}\ln\left|\frac{\omega + v_{\rm F}k}{\omega - v_{\rm F}k}\right| - \frac{1}{2} = \frac{\epsilon_0 k^2}{e^2 g(0)}.$$

(4) For small k with $v_{\rm F}k < \omega$, we can approximate the logarithmic function as follows

$$\ln\left|\frac{\omega + v_{\rm F}k}{\omega - v_{\rm F}k}\right| = \ln\frac{1 + v_{\rm F}k/\omega}{1 - v_{\rm F}k/\omega} \approx \frac{2v_{\rm F}k}{\omega}\left[1 + \frac{1}{3}\left(\frac{v_{\rm F}k}{\omega}\right)^2\right].$$

We have in this case

$$\frac{1}{3}\left(\frac{v_{\rm F}k}{\omega}\right)^2 = \frac{\epsilon_0 k^2}{2e^2 g(0)} - \frac{3}{4}.$$

Thus,

$$\omega = \frac{2v_{\rm F}k/3}{\sqrt{2\epsilon_0 k^2/3e^2 g(0) - 1}}, \quad v_{\rm F}k < \omega.$$

For large k with $v_{\rm F}k > \omega$, we can approximate the logarithmic function as

$$\ln\left|\frac{\omega + v_{\rm F}k}{\omega - v_{\rm F}k}\right| = \ln\frac{1 + \omega/v_{\rm F}k}{1 - \omega/v_{\rm F}k} \approx \frac{2\omega}{v_{\rm F}k}.$$

We then have in this case

$$\left(\frac{\omega}{v_{\mathrm{F}}k}\right)^2 = \frac{\epsilon_0 k^2}{2e^2 g(0)} + \frac{1}{4}.$$

Thus,

$$\omega = \frac{1}{2} v_{\mathrm{F}} k \left[\frac{2\epsilon_0 k^2}{e^2 g(0)} + 1\right]^{1/2}, \quad v_{\mathrm{F}} k > \omega.$$

26-20 Fermi gas in a spin-dependent external field. Consider a non-interacting Fermi gas of spin $s = 1/2$ in the presence of an external potential $V_{\alpha\beta}(\boldsymbol{r})$ which can flip spins. The potential energy of the Fermi gas is given by

$$\hat{H}_1 = \sum_{\alpha\beta} \int \mathrm{d}\boldsymbol{r}\ \hat{\psi}_\alpha^\dagger(\boldsymbol{r}) V_{\alpha\beta}(\boldsymbol{r}) \hat{\psi}_\beta(\boldsymbol{r}),$$

where $\hat{\psi}_\alpha(\boldsymbol{r})$ is the quantum field operator of fermions. In terms of quantum field operators, the Green's function of fermions is defined by

$$G_{\alpha\beta}(\boldsymbol{r}, t; \boldsymbol{r}', t') = -\mathrm{i} \langle \hat{T} \hat{\psi}_\alpha(\boldsymbol{r}, t) \hat{\psi}_\beta^\dagger(\boldsymbol{r}', t') \rangle .$$

(1) Evaluate the derivative of $G_{\alpha\beta}(\boldsymbol{r}, t; \boldsymbol{r}', t')$ with respect to t and establish a partial differential equation for $G_{\alpha\beta}(\boldsymbol{r}, t; \boldsymbol{r}', t')$ by making use of the equation of motion of $\hat{\psi}_\alpha(\boldsymbol{r}, t)$.

(2) Convert the partial differential equation for $G_{\alpha\beta}(\boldsymbol{r}, t; \boldsymbol{r}', t')$ into an integral equation using the Green's function G_0 for the Fermi gas in the absence of an external potential.

(3) Write down Feynman diagrams corresponding to the integral equation.

---

(1) Differentiating $G_{\alpha\beta}(\boldsymbol{r}, t; \boldsymbol{r}', t')$ with respect to t, we have

$$\begin{aligned}
&\frac{\partial}{\partial t} G_{\alpha\beta}(\boldsymbol{r}, t; \boldsymbol{r}', t') \\
&= -\mathrm{i}\delta(t - t') \langle \{\hat{\psi}_\alpha(\boldsymbol{r}, t), \hat{\psi}_\beta^\dagger(\boldsymbol{r}', t)\} \rangle - \mathrm{i} \langle \hat{T} \partial \hat{\psi}_\alpha(\boldsymbol{r}, t)/\partial t\, \hat{\psi}_\beta^\dagger(\boldsymbol{r}', t') \rangle \\
&= -\mathrm{i}\delta_{\alpha\beta}\delta(t - t')\delta(\boldsymbol{r} - \boldsymbol{r}') - \mathrm{i}\langle \hat{T} \partial \hat{\psi}_\alpha(\boldsymbol{r}, t)/\partial t\, \hat{\psi}_\beta^\dagger(\boldsymbol{r}', t') \rangle .
\end{aligned}$$

The derivative of $\hat{\psi}_\alpha(\boldsymbol{r}, t)$ with respect to t can be obtained from the Heisenberg equation of motion. We have

$$\mathrm{i}\hbar \frac{\partial \hat{\psi}_\alpha(\boldsymbol{r}, t)}{\partial t} = \left[\hat{\psi}_\alpha(\boldsymbol{r}, t), \hat{H}\right] = \mathrm{e}^{\mathrm{i}\hat{H}t/\hbar} \left[\hat{\psi}_\alpha(\boldsymbol{r}), \hat{H}\right] \mathrm{e}^{-\mathrm{i}\hat{H}t/\hbar}.$$

The Hamiltonian $\hat{H}$ is given by $\hat{H} = \hat{H}_0 + \hat{H}_1$. We first evaluate $\left[\hat{\psi}_\alpha(\boldsymbol{r}), \hat{H}_0\right]$ and obtain

$$\begin{aligned}
\left[\hat{\psi}_\alpha(\boldsymbol{r}), \hat{H}_0\right] &= -\frac{\hbar^2}{2m}\sum_\beta \int \mathrm{d}\boldsymbol{r}' \left[\hat{\psi}_\alpha(\boldsymbol{r}), \hat{\psi}_\beta^\dagger(\boldsymbol{r}')\nabla'^2\hat{\psi}_\beta(\boldsymbol{r}')\right] \\
&= -\frac{\hbar^2}{2m}\sum_\beta \int \mathrm{d}\boldsymbol{r}' \left\{\hat{\psi}_\alpha(\boldsymbol{r}), \hat{\psi}_\beta^\dagger(\boldsymbol{r}')\right\}\nabla'^2\hat{\psi}_\beta(\boldsymbol{r}') \\
&= -\frac{\hbar^2}{2m}\sum_\beta \int \mathrm{d}\boldsymbol{r}' \, \delta_{\alpha\beta}\delta(\boldsymbol{r}-\boldsymbol{r}')\nabla'^2\hat{\psi}_\beta(\boldsymbol{r}') \\
&= -\frac{\hbar^2}{2m}\nabla^2\hat{\psi}_\alpha(\boldsymbol{r}).
\end{aligned}$$

The commutator $\left[\hat{\psi}_\alpha(\boldsymbol{r}), \hat{H}_1\right]$ is given by

$$\begin{aligned}
\left[\hat{\psi}_\alpha(\boldsymbol{r}), \hat{H}_1\right] &= \sum_{\alpha'\beta} \int \mathrm{d}\boldsymbol{r}' \, V_{\alpha'\beta}(\boldsymbol{r}')\left[\hat{\psi}_\alpha(\boldsymbol{r}), \hat{\psi}_{\alpha'}^\dagger(\boldsymbol{r}')\hat{\psi}_\beta(\boldsymbol{r}')\right] \\
&= \sum_{\alpha'\beta} \int \mathrm{d}\boldsymbol{r}' \, V_{\alpha'\beta}(\boldsymbol{r}')\delta_{\alpha\alpha'}\delta(\boldsymbol{r}-\boldsymbol{r}')\hat{\psi}_\beta(\boldsymbol{r}') \\
&= \sum_\beta V_{\alpha\beta}(\boldsymbol{r})\hat{\psi}_\beta(\boldsymbol{r}).
\end{aligned}$$

We then have

$$\begin{aligned}
&\mathrm{i}\hbar\frac{\partial}{\partial t}G_{\alpha\beta}(\boldsymbol{r}, t; \boldsymbol{r}', t') \\
&\quad = \hbar\delta_{\alpha\beta}\delta(t-t')\delta(\boldsymbol{r}-\boldsymbol{r}') + \mathrm{i}\frac{\hbar^2}{2m}\nabla^2\left\langle \hat{T}\hat{\psi}_\alpha(\boldsymbol{r}, t)\hat{\psi}_\beta^\dagger(\boldsymbol{r}', t')\right\rangle \\
&\qquad - \mathrm{i}\sum_\gamma V_{\alpha\gamma}(\boldsymbol{r})\left\langle \hat{T}\hat{\psi}_\gamma(\boldsymbol{r}, t)\hat{\psi}_\beta^\dagger(\boldsymbol{r}', t')\right\rangle \\
&\quad = \hbar\delta_{\alpha\beta}\delta(t-t')\delta(\boldsymbol{r}-\boldsymbol{r}') - \frac{\hbar^2}{2m}\nabla^2 G_{\alpha\beta}(\boldsymbol{r}, t; \boldsymbol{r}', t') \\
&\qquad + \sum_\gamma V_{\alpha\gamma}(\boldsymbol{r})G_{\gamma\beta}(\boldsymbol{r}, t; \boldsymbol{r}', t').
\end{aligned}$$

We have thus obtained the differential equation for $G_{\alpha\beta}(\boldsymbol{r}, t; \boldsymbol{r}', t')$

$$\begin{aligned}
\left[\mathrm{i}\hbar\frac{\partial}{\partial t} + \frac{\hbar^2}{2m}\nabla^2\right]G_{\alpha\beta}(\boldsymbol{r}, t; \boldsymbol{r}', t') - \sum_\gamma V_{\alpha\gamma}(\boldsymbol{r})G_{\gamma\beta}(\boldsymbol{r}, t; \boldsymbol{r}', t')& \\
= \hbar\delta_{\alpha\beta}\delta(t-t')\delta(\boldsymbol{r}-\boldsymbol{r}')&.
\end{aligned}$$

(2) In the absence of an external potential, G_0 satisfies the following equation

$$\left[\mathrm{i}\hbar\frac{\partial}{\partial t}+\frac{\hbar^2}{2m}\boldsymbol{\nabla}^2\right]G_0(\boldsymbol{r},t;\boldsymbol{r}',t')=\hbar\delta(t-t')\delta(\boldsymbol{r}-\boldsymbol{r}')$$

which can be obtained from the above differential equation for $G_{\alpha\beta}(\boldsymbol{r},t;\boldsymbol{r}',t')$ by removing the term for the external potential and $\delta_{\alpha\beta}$ and replacing $G_{\alpha\beta}(\boldsymbol{r},t;\boldsymbol{r}',t')$ with $G_0(\boldsymbol{r},t;\boldsymbol{r}',t')$. The complex conjugate of $G_0(\boldsymbol{r},t;\boldsymbol{r}',t')$ satisfies the equation

$$\left[-\mathrm{i}\hbar\frac{\partial}{\partial t}+\frac{\hbar^2}{2m}\boldsymbol{\nabla}^2\right]G_0^*(\boldsymbol{r},t;\boldsymbol{r}',t')=\hbar\delta(t-t')\delta(\boldsymbol{r}-\boldsymbol{r}').$$

Changing the variables $\boldsymbol{r}$ and t in the differential equation for $G_{\alpha\beta}(\boldsymbol{r},t;\boldsymbol{r}',t')$ to $\boldsymbol{r}''$ and t'', respectively, then multiplying both sides of the resultant equation with $G_0^*(\boldsymbol{r},t;\boldsymbol{r}'',t'')$ from the left, and then integrating over $\boldsymbol{r}''$ and t'', we obtain

$$\begin{aligned}&\int\mathrm{d}\boldsymbol{r}''\int_{-\infty}^{\infty}\mathrm{d}t''\,G_0^*(\boldsymbol{r},t;\boldsymbol{r}'',t'')\left[\mathrm{i}\hbar\frac{\partial}{\partial t''}+\frac{\hbar^2}{2m}\boldsymbol{\nabla}''^2\right]G_{\alpha\beta}(\boldsymbol{r}'',t'';\boldsymbol{r}',t')\\&\quad-\sum_{\gamma}\int\mathrm{d}\boldsymbol{r}''\int_{-\infty}^{\infty}\mathrm{d}t''\,G_0^*(\boldsymbol{r},t;\boldsymbol{r}'',t'')V_{\alpha\gamma}(\boldsymbol{r}'')G_{\gamma\beta}(\boldsymbol{r}'',t'';\boldsymbol{r}',t')\\&\quad=\hbar\int\mathrm{d}\boldsymbol{r}''\int_{-\infty}^{\infty}\mathrm{d}t''\,G_0^*(\boldsymbol{r},t;\boldsymbol{r}'',t'')\delta(t''-t')\delta(\boldsymbol{r}''-\boldsymbol{r}').\end{aligned}$$

The integrals on the right hand side can be evaluated by making use of the δ-functions. We perform integrations by parts once for t'' and twice for $\boldsymbol{r}''$ to the first term on the left hand side. We have

$$\begin{aligned}&\int\mathrm{d}\boldsymbol{r}''\int_{-\infty}^{\infty}\mathrm{d}t''\,G_{\alpha\beta}(\boldsymbol{r}'',t'';\boldsymbol{r}',t')\left[-\mathrm{i}\hbar\frac{\partial}{\partial t''}+\frac{\hbar^2}{2m}\boldsymbol{\nabla}''^2\right]G_0^*(\boldsymbol{r},t;\boldsymbol{r}'',t'')\\&\quad-\sum_{\gamma}\int\mathrm{d}\boldsymbol{r}''\int_{-\infty}^{\infty}\mathrm{d}t''\,G_0^*(\boldsymbol{r},t;\boldsymbol{r}'',t'')V_{\alpha\gamma}(\boldsymbol{r}'')G_{\gamma\beta}(\boldsymbol{r}'',t'';\boldsymbol{r}',t')\\&\quad=\hbar G_0^*(\boldsymbol{r},t;\boldsymbol{r}',t').\end{aligned}$$

Making use of the equation of $G_0^*(\boldsymbol{r},t;\boldsymbol{r}',t')$, we have

$$\begin{aligned}&\hbar\int\mathrm{d}\boldsymbol{r}''\int_{-\infty}^{\infty}\mathrm{d}t''\,\delta(t-t'')\delta(\boldsymbol{r}-\boldsymbol{r}'')G_{\alpha\beta}(\boldsymbol{r}'',t'';\boldsymbol{r}',t')\\&\quad-\sum_{\gamma}\int\mathrm{d}\boldsymbol{r}''\int_{-\infty}^{\infty}\mathrm{d}t''\,G_0^*(\boldsymbol{r},t;\boldsymbol{r}'',t'')V_{\alpha\gamma}(\boldsymbol{r}'')G_{\gamma\beta}(\boldsymbol{r}'',t'';\boldsymbol{r}',t')\\&\quad=\hbar G_0^*(\boldsymbol{r},t;\boldsymbol{r}',t').\end{aligned}$$

Performing the integrations in the first term on the left hand side by taking advantage of the δ-functions, we obtain the following integral equation for $G_{\alpha\beta}(\boldsymbol{r}, t; \boldsymbol{x}', t')$

$$G_{\alpha\beta}(\boldsymbol{r}, t; \boldsymbol{x}', t') = G_0^*(\boldsymbol{r}, t; \boldsymbol{r}', t') + \frac{1}{\hbar} \sum_{\gamma} \int \mathrm{d}\boldsymbol{r}'' \int_{-\infty}^{\infty} \mathrm{d}t''\, G_0^*(\boldsymbol{r}, t; \boldsymbol{r}'', t'') V_{\alpha\gamma}(\boldsymbol{r}'') G_{\gamma\beta}(\boldsymbol{r}'', t''; \boldsymbol{r}', t').$$

(3) The above differential equation is represented graphically in Fig. 26.12.

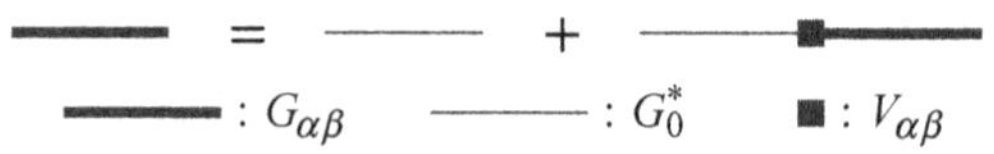

Fig. 26.12 Feynman diagrams corresponding to the integral equation for $G_{\alpha\beta}(\boldsymbol{r}, t; \boldsymbol{r}', t')$. The upper part represents the integral equation. The lower part is the legend for symbols with a thick line for $G_{\alpha\beta}(\boldsymbol{r}, t; \boldsymbol{r}', t')$, a thin line for $G_0^*(\boldsymbol{r}, t; \boldsymbol{r}', t')$, and a solid square for $V_{\alpha\beta}(\boldsymbol{r})$.

Chapter 27

Transport Properties of Solids

(1) *Boltzmann equation*

$$\frac{\partial f(\boldsymbol{r},\boldsymbol{p},t)}{\partial t} + \frac{\boldsymbol{p}}{m}\cdot\boldsymbol{\nabla}_{\boldsymbol{r}} f(\boldsymbol{r},\boldsymbol{p},t) + \boldsymbol{F}\cdot\boldsymbol{\nabla}_{\boldsymbol{p}} f(\boldsymbol{r},\boldsymbol{p},t) = I[f].$$

(2) *Relaxation time approximation*

$$I[f] = -\frac{f(\boldsymbol{r},\boldsymbol{p},t) - f_0(\boldsymbol{r},\boldsymbol{p})}{\tau(\boldsymbol{p})}.$$

(3) *Electrical conductivity tensor of nearly free electrons*

$$\sigma_{\alpha\beta} = 2e^2\!\int \frac{\mathrm{d}\boldsymbol{k}}{(2\pi)^3}\,\tau(\boldsymbol{k})\left[-\frac{\partial n_F(\varepsilon)}{\partial \varepsilon}\right] v_\alpha v_\beta.$$

(4) *Matthiessen's rule*

$$\rho_{\text{tot}} = \rho_{\text{imp}} + \rho_{\text{e-p}} + \rho_{\text{e-e}} + \cdots.$$

(5) *Electron-phonon scattering*

$$\begin{aligned} I[f] = \frac{\pi N Z^2 e^4}{\epsilon_0^2 M \mathscr{V}^2 c\hbar^3}\sum_{\boldsymbol{k}'}(f_{\boldsymbol{k}'} - f_{\boldsymbol{k}})\frac{\left[1-\cos(|\boldsymbol{k}'-\boldsymbol{k}|r_c)\right]^2}{|\boldsymbol{k}'-\boldsymbol{k}|^3} \\ \times\left[n_{\mathrm{B}}(\omega_{\boldsymbol{k}'-\boldsymbol{k}})\delta(\varepsilon_{\boldsymbol{k}'} - \varepsilon_{\boldsymbol{k}} - \hbar\omega_{\boldsymbol{k}'-\boldsymbol{k}}) \right. \\ \left. + \big(n_{\mathrm{B}}(\omega_{\boldsymbol{k}'-\boldsymbol{k}}) + 1\big)\delta(\varepsilon_{\boldsymbol{k}'} - \varepsilon_{\boldsymbol{k}} + \hbar\omega_{\boldsymbol{k}'-\boldsymbol{k}})\right]. \end{aligned}$$

(6) *Electron-impurity scattering*

$$\begin{aligned} I[f] &= -n_i v(\varepsilon)\phi(\varepsilon)\boldsymbol{v}\cdot\boldsymbol{E}\sigma_{\text{tr}}, \\ \sigma_{\text{tr}} &= \int \mathrm{d}\Omega\,(1-\cos\theta)\frac{\mathrm{d}\sigma}{\mathrm{d}\Omega}, \\ \tau_{\text{tr}} &= 1/n_i v(\varepsilon)\sigma_{\text{tr}}. \end{aligned}$$

(7) *Electron-electron scattering*

$$I_1[f] = \frac{e^4(k_BT)^2(\boldsymbol{k}\cdot\boldsymbol{E})\phi(\varepsilon_{\boldsymbol{k}})}{64\pi^2\epsilon_0^2\hbar^2(\varepsilon_{\kappa_s}^3\varepsilon_{\boldsymbol{k}})^{1/2}}[1-n_{\rm F}(\varepsilon_{\boldsymbol{k}})]\left\{1+\frac{\pi^2}{3}n_{\rm F}(\varepsilon_{\boldsymbol{k}}) - \frac{\pi^4}{36}n_{\rm F}^2(\varepsilon_{\boldsymbol{k}})[1-2n_{\rm F}(\varepsilon_{\boldsymbol{k}})]\right\}.$$

(8) *Thermal conductivity tensor of nearly free electrons*

$$\kappa_{\alpha\beta} = -\frac{2}{T}\int\frac{\mathrm{d}\boldsymbol{k}}{(2\pi)^3}v_\alpha v_\beta(\varepsilon-\mu)^2\tau\frac{\partial n_{\rm F}(\varepsilon)}{\partial\varepsilon}.$$

(9) *Wiedemann-Franz law for nearly free electrons*

$$\kappa/\sigma T = (\pi^2/3)(k_{\rm B}/e)^2.$$

(10) *Linear response theory*

$$\hat{H}_{\rm ext}(t) = \frac{1}{\mathscr{V}}\int\mathrm{d}\boldsymbol{r}\ f(\boldsymbol{r},t)\hat{B}(\boldsymbol{r}),$$

$$\langle\hat{A}\rangle(\boldsymbol{q},\omega) = \chi_{AB}(\boldsymbol{q},\omega)f(\boldsymbol{q},\omega),$$

$$\chi_{AB}(\boldsymbol{q},\omega) = \mathrm{i}\int\frac{\mathrm{d}\omega'}{2\pi}\frac{G_{AB}^R(\boldsymbol{q},\omega')}{\omega-\omega'+\mathrm{i}\delta},$$

$$f(\boldsymbol{q},\omega) = \int\frac{\mathrm{d}\boldsymbol{r}\mathrm{d}t}{\mathscr{V}}\,\mathrm{e}^{-\mathrm{i}(\boldsymbol{q}\cdot\boldsymbol{r}-\omega t)}f(\boldsymbol{r},t),$$

$$\langle\hat{A}\rangle(\boldsymbol{q},\omega) = \int\frac{\mathrm{d}\boldsymbol{r}\mathrm{d}t}{\mathscr{V}}\mathrm{e}^{-\mathrm{i}(\boldsymbol{q}\cdot\boldsymbol{r}-\omega t)}\,\langle\hat{A}\rangle(\boldsymbol{r},t),$$

$$G_{AB}^R(\boldsymbol{q},\omega) = \int\frac{\mathrm{d}\boldsymbol{r}\mathrm{d}t}{\mathscr{V}}\mathrm{e}^{-\mathrm{i}(\boldsymbol{q}\cdot\boldsymbol{r}-\omega t)}G_{AB}^R(\boldsymbol{r},t),$$

$$G_{AB}^R(\boldsymbol{r}-\boldsymbol{r}',t-t') = -\mathrm{i}\hbar^{-1}\left\langle[\hat{A}(\boldsymbol{r},t),\hat{B}(\boldsymbol{r}',t')]\right\rangle.$$

(11) *Linear response function*

$$J(\omega) = -2\,\mathrm{Im}\,\chi(\omega),\ \text{spectral function},$$

$$\chi(z) = \int_{-\infty}^{\infty}\frac{\mathrm{d}\omega'}{2\pi}\frac{J(\omega')}{z-\omega'},$$

$$\mathrm{Re}\,\chi(\omega) = -\frac{1}{\pi}\mathrm{P}\int_{-\infty}^{\infty}\mathrm{d}\omega'\,\frac{\mathrm{Im}\,\chi(\omega')}{\omega-\omega'},$$

$$\mathrm{Im}\,\chi(\omega) = \frac{1}{\pi}\mathrm{P}\int_{-\infty}^{\infty}\mathrm{d}\omega'\,\frac{\mathrm{Re}\,\chi(\omega')}{\omega-\omega'}.$$

(12) *Fluctuation-dissipation theorem*

$$\mathrm{Im}\,\chi_{AB}(\boldsymbol{q},\omega) = \frac{1}{2\hbar}\left(\mathrm{e}^{\beta\hbar\omega}-1\right)S_{AB}(\boldsymbol{q},\omega).$$

(13) *Kubo formula for electrical conductivity*

$$\mathrm{Re}\,\sigma_{\alpha\beta} = \frac{\pi\hbar\mathscr{V}}{k_{\mathrm{B}}TZ}\sum_{nm} \mathrm{e}^{-E_n/k_{\mathrm{B}}T}\,\langle n|\hat{J}_\alpha(0)|m\rangle\langle m|\hat{J}^\dagger_\beta(0)|n\rangle\,\delta(E_n - E_m).$$

(14) *Kubo-Greenwood formula*

$$\sigma = \frac{\pi\hbar e^2}{\mathscr{V}}\sum_{\kappa\nu} |\langle\nu|\hat{v}_\alpha|\kappa\rangle|^2\left[-\frac{\partial n_{\mathrm{F}}(\varepsilon_\kappa)}{\partial\varepsilon_\kappa}\right]\delta(\varepsilon_\kappa - \varepsilon_\nu).$$

27-1 Equilibrium distribution function.

(1) Show that the collision integral $I[f]$ vanishes for $f(\boldsymbol{r},\boldsymbol{k},t)$ equal to the equilibrium distribution function, $f(\boldsymbol{r},\boldsymbol{k},t) = h^{-3}n_{\mathrm{F}}(\varepsilon_{\boldsymbol{k}})$.

(2) Show that the electric current density $\boldsymbol{J}$ vanishes for the equilibrium distribution function $h^{-3}n_{\mathrm{F}}(\varepsilon_{\boldsymbol{k}})$.

---

(1) In terms of microscopic transition probabilities and distribution functions, for discretized values of $\boldsymbol{r}$ and $\boldsymbol{k}$ the collision integral can be written as

$$\begin{aligned} I[f] &= \sum_{\boldsymbol{k}'} w_{\boldsymbol{k}'\to\boldsymbol{k}}\left[h^3 f(\boldsymbol{r},\boldsymbol{k}',t) - h^3 f(\boldsymbol{r},\boldsymbol{k},t)\right] \\ &= \sum_{\boldsymbol{k}'} w_{\boldsymbol{k}'\to\boldsymbol{k}} h^3 f(\boldsymbol{r},\boldsymbol{k}',t) - h^3 f(\boldsymbol{r},\boldsymbol{k},t), \end{aligned}$$

where we have made use of $\sum_{\boldsymbol{k}'} w_{\boldsymbol{k}'\to\boldsymbol{k}} = 1$. For $f(\boldsymbol{r},\boldsymbol{k},t) = h^{-3}n_{\mathrm{F}}(\varepsilon_{\boldsymbol{k}})$, we have

$$I[n_{\mathrm{F}}] = \sum_{\boldsymbol{k}'} w_{\boldsymbol{k}'\to\boldsymbol{k}} n_{\mathrm{F}}(\varepsilon_{\boldsymbol{k}'}) - n_{\mathrm{F}}(\varepsilon_{\boldsymbol{k}}).$$

Since $n_{\mathrm{F}}(\varepsilon_{\boldsymbol{k}})$ is an equilibrium distribution function, we must have

$$\sum_{\boldsymbol{k}'} w_{\boldsymbol{k}'\to\boldsymbol{k}} n_{\mathrm{F}}(\varepsilon_{\boldsymbol{k}'}) = n_{\mathrm{F}}(\varepsilon_{\boldsymbol{k}}).$$

Therefore, $I[n_{\mathrm{F}}] = 0$.

(2) The electric current density is given by

$$\boldsymbol{j}(\boldsymbol{r},t) = -\frac{e}{m}\int \mathrm{d}\boldsymbol{p}\,\boldsymbol{p} f(\boldsymbol{r},\boldsymbol{p},t) = -\frac{e\hbar^4}{m}\int \mathrm{d}\boldsymbol{k}\,\boldsymbol{k} f(\boldsymbol{r},\boldsymbol{k},t).$$

For $f(\boldsymbol{r},\boldsymbol{k},t) = h^{-3}n_{\mathrm{F}}(\varepsilon_{\boldsymbol{k}})$, we have

$$\boldsymbol{j} = -\frac{e\hbar}{(2\pi)^3 m}\int \mathrm{d}\boldsymbol{k}\,\boldsymbol{k}\,n_{\mathrm{F}}(\varepsilon_{\boldsymbol{k}}).$$

Since $n_{\rm F}(\varepsilon_{\boldsymbol{k}})$ does not depend on the direction of $\boldsymbol{k}$, we can trivially perform the integrations over angles and obtain a null result

$$\int_{-1}^{1} \mathrm{d}\cos\theta \int_0^{2\pi} \mathrm{d}\varphi\,(\sin\theta\cos\varphi\,\boldsymbol{e}_x + \sin\theta\sin\varphi\,\boldsymbol{e}_y + \cos\theta\,\boldsymbol{e}_z) = 0.$$

Thus,

$$\boldsymbol{j} = -\frac{e\hbar}{(2\pi)^3 m}\int \mathrm{d}\boldsymbol{k}\,\boldsymbol{k}\,n_{\rm F}(\varepsilon_{\boldsymbol{k}}) = 0.$$

27-2 Vector product of mechanical momentum with itself. For an electron in a magnetic field $\boldsymbol{B}$, show that $\hat{\boldsymbol{k}}\times\hat{\boldsymbol{k}} = -\mathrm{i}e\boldsymbol{B}/\hbar$, where $\hat{\boldsymbol{k}}$ is given by $\hat{\boldsymbol{k}} = \hbar^{-1}(\hat{\boldsymbol{p}} + e\boldsymbol{A})$ with $\hat{\boldsymbol{p}}$ the canonical momentum operator of the electron.

We examine the effect of $\hat{\boldsymbol{k}}\times\hat{\boldsymbol{k}}$ on an arbitrary well-behaved function $\psi(\boldsymbol{r})$. Making use of $\boldsymbol{\nabla}\times(\varphi\boldsymbol{f}) = \varphi\boldsymbol{\nabla}\times\boldsymbol{f} + (\boldsymbol{\nabla}\varphi)\times\boldsymbol{f}$, we have

$$\begin{aligned}
\hat{\boldsymbol{k}}\times\hat{\boldsymbol{k}}\,\psi(\boldsymbol{r}) &= \frac{1}{\hbar^2}(\hat{\boldsymbol{p}} + e\boldsymbol{A})\times(\hat{\boldsymbol{p}} + e\boldsymbol{A})\psi(\boldsymbol{r}) \\
&= \frac{e}{\hbar^2}(\hat{\boldsymbol{p}}\times\boldsymbol{A} + e\boldsymbol{A}\times\hat{\boldsymbol{p}})\psi(\boldsymbol{r}) \\
&= -\frac{\mathrm{i}e}{\hbar}\{\boldsymbol{\nabla}\times\big[\boldsymbol{A}\psi(\boldsymbol{r})\big] + \boldsymbol{A}\times\boldsymbol{\nabla}\psi(\boldsymbol{r})\} \\
&= -\frac{\mathrm{i}e}{\hbar}\{\psi(\boldsymbol{r})\boldsymbol{\nabla}\times\boldsymbol{A} + \big[\boldsymbol{\nabla}\psi(\boldsymbol{r})\big]\times\boldsymbol{A} + \boldsymbol{A}\times\boldsymbol{\nabla}\psi(\boldsymbol{r})\} \\
&= -\frac{\mathrm{i}e}{\hbar}\psi(\boldsymbol{r})\boldsymbol{\nabla}\times\boldsymbol{A} = -\frac{\mathrm{i}e}{\hbar}\boldsymbol{B}\psi(\boldsymbol{r}).
\end{aligned}$$

Thus, $\hat{\boldsymbol{k}}\times\hat{\boldsymbol{k}} = -\dfrac{\mathrm{i}e}{\hbar}\boldsymbol{B}$.

27-3 DC conductivity of Bloch electrons in the presence of damping. Consider electrons in the following energy band that is parametrized by anisotropic masses m_x, m_y, and m_z

$$E(\boldsymbol{k}) = \hbar^2k_x^2/2m_x + \hbar^2k_y^2/2m_y + \hbar^2k_z^2/2m_z.$$

Within the relaxation time approximation, the semi-classical equation of motion contains an $\hbar\boldsymbol{k}/\tau$ term

$$\hbar\big(\mathrm{d}\boldsymbol{k}/\mathrm{d}t + \boldsymbol{k}/\tau\big) = -e\big(\boldsymbol{E} + \boldsymbol{v}\times\boldsymbol{B}\big),$$

where the electric and magnetic fields are independent of time. Compute the DC conductivity tensor $\overleftrightarrow{\sigma}$ using $\boldsymbol{J} = -ne\boldsymbol{v}$ and $\boldsymbol{J} = \overleftrightarrow{\sigma}\cdot\boldsymbol{E}$.

Since the equation of motion contains both electron's velocity $\boldsymbol{v}$ and its wave vector $\boldsymbol{k}$, we need a relation between them. This relation can be obtained from the band dispersion relation as follows

$$\boldsymbol{v} = \frac{1}{\hbar}\nabla_{\boldsymbol{k}}E(\boldsymbol{k}) = \frac{\hbar k_x}{m_x}\boldsymbol{e}_x + \frac{\hbar k_y}{m_y}\boldsymbol{e}_y + \frac{\hbar k_z}{m_z}\boldsymbol{e}_z.$$

Inserting the above relation into the equation of motion and setting the time derivative of $\boldsymbol{k}$ to zero for a steady state, we have in the component form

$$\begin{aligned}
\frac{1}{e\tau}k_x + \frac{B_z}{m_y}k_y - \frac{B_y}{m_z}k_z &= -\frac{E_x}{\hbar}, \\
-\frac{B_z}{m_x}k_x + \frac{1}{e\tau}k_y + \frac{B_x}{m_z}k_z &= -\frac{E_y}{\hbar}, \\
\frac{B_y}{m_x}k_x - \frac{B_x}{m_y}k_y + \frac{1}{e\tau}k_z &= -\frac{E_z}{\hbar}.
\end{aligned}$$

We now solve for k_x, k_y, and k_z from the above set of three equations. The coefficient determinant is given by

$$\begin{aligned}
\Delta &= \begin{vmatrix} 1/e\tau & B_z/m_y & -B_y/m_z \\ -B_z/m_x & 1/e\tau & B_x/m_z \\ B_y/m_x & -B_x/m_y & 1/e\tau \end{vmatrix} \\
&= \frac{1}{e^3\tau^3}\left[1 + e^2\tau^2\left(\frac{B_x^2}{m_y m_z} + \frac{B_y^2}{m_z m_x} + \frac{B_z^2}{m_x m_y}\right)\right] = \frac{\delta}{e^3\tau^3},
\end{aligned}$$

where $\delta = 1 + e^2\tau^2(B_x^2/m_y m_z + B_y^2/m_z m_x + B_z^2/m_x m_y)$. The determinant for k_x is

$$\begin{aligned}
\Delta_{k_x} &= \begin{vmatrix} -E_x/\hbar & B_z/m_y & -B_y/m_z \\ -E_y\hbar & e/\tau & B_x/m_z \\ -E_z\hbar & -B_x/m_y & e/\tau \end{vmatrix} \\
&= -\frac{m_x}{e^2\tau^2\hbar}\left[\left(\frac{1}{m_x} + \frac{1}{\mu_{xx}}\right)E_x + \left(\frac{1}{\mu_{xy}} - \frac{1}{\mu_z}\right)E_y \right. \\
&\qquad \left. + \left(\frac{1}{\mu_{zx}} + \frac{1}{\mu_y}\right)E_z\right],
\end{aligned}$$

where

$$\frac{1}{\mu_{\alpha\beta}} = \frac{e^2\tau^2 B_\alpha B_\beta}{m_x m_y m_z}, \quad \alpha, \beta = x, y, z; \quad \frac{1}{\mu_x} = \frac{e\tau B_x}{m_y m_z}, \quad \text{etc.}$$

Note that

$$\frac{1}{\mu_x \mu_y} = \frac{1}{\mu_{xy} m_z}, \quad \text{etc;} \quad \frac{1}{\mu_x^2} = \frac{1}{\mu_{xx}(m_y m_z/m_x)}, \quad \text{etc.}$$

The determinant for k_y is

$$\Delta_{k_y} = \begin{vmatrix} 1/e\tau & -E_x/\hbar & -B_y/m_z \\ -B_z/m_x & -E_y/\hbar & B_x/m_z \\ B_y/m_x & -E_z/\hbar & 1/e\tau \end{vmatrix}$$
$$= -\frac{m_y}{e^2\tau^2\hbar}\left[\left(\frac{1}{\mu_{xy}}+\frac{1}{\mu_z}\right)E_x+\left(\frac{1}{m_y}+\frac{1}{\mu_{yy}}\right)E_y + \left(\frac{1}{\mu_{yz}}-\frac{1}{\mu_x}\right)E_z\right].$$

The determinant for k_z is

$$\Delta_{k_z} = \begin{vmatrix} 1/e\tau & B_z/m_y & -E_x/\hbar \\ -B_z/m_x & 1/e\tau & -E_y/\hbar \\ B_y/m_x & -B_x/m_y & -E_z/\hbar \end{vmatrix}$$
$$= -\frac{m_z}{e^2\tau^2\hbar}\left[\left(\frac{1}{\mu_{zx}}-\frac{1}{\mu_y}\right)E_x+\left(\frac{1}{\mu_{yz}}+\frac{1}{\mu_x}\right)E_y + \left(\frac{1}{m_z}+\frac{1}{\mu_{zz}}\right)E_z\right].$$

The wave vector $\boldsymbol{k}$ is thus given by $\boldsymbol{k} = \frac{\Delta_{k_x}}{\Delta}\boldsymbol{e}_x + \frac{\Delta_{k_y}}{\Delta}\boldsymbol{e}_y + \frac{\Delta_{k_z}}{\Delta}\boldsymbol{e}_z$.

The velocity can be then obtained from $\boldsymbol{k}$ and we have $\boldsymbol{v} = \frac{\hbar\Delta_{k_x}}{m_x\Delta}\boldsymbol{e}_x + \frac{\hbar\Delta_{k_y}}{m_y\Delta}\boldsymbol{e}_y + \frac{\hbar\Delta_{k_z}}{m_z\Delta}\boldsymbol{e}_z$.

The current density follows from $\boldsymbol{J} = -ne\boldsymbol{v}$

$$\boldsymbol{J} = \frac{ne^2\tau}{\delta}\left[\left(\frac{1}{m_x}+\frac{1}{\mu_{xx}}\right)E_x+\left(\frac{1}{\mu_{xy}}-\frac{1}{\mu_z}\right)E_y+\left(\frac{1}{\mu_{zx}}+\frac{1}{\mu_y}\right)E_z\right]\boldsymbol{e}_x$$
$$+\frac{ne^2\tau}{\delta}\left[\left(\frac{1}{\mu_{xy}}+\frac{1}{\mu_z}\right)E_x+\left(\frac{1}{m_y}+\frac{1}{\mu_{yy}}\right)E_y+\left(\frac{1}{\mu_{yz}}-\frac{1}{\mu_x}\right)E_z\right]\boldsymbol{e}_y$$
$$+\frac{ne^2\tau}{\delta}\left[\left(\frac{1}{\mu_{zx}}-\frac{1}{\mu_y}\right)E_x+\left(\frac{1}{\mu_{yz}}+\frac{1}{\mu_x}\right)E_y+\left(\frac{1}{m_z}+\frac{1}{\mu_{zz}}\right)E_z\right]\boldsymbol{e}_z.$$

From $\boldsymbol{J} = \overleftrightarrow{\sigma}\cdot\boldsymbol{E}$, we have

$$\overleftrightarrow{\sigma} = \frac{ne^2\tau}{\delta}\begin{pmatrix} \frac{1}{m_x}+\frac{1}{\mu_{xx}} & \frac{1}{\mu_{xy}}-\frac{1}{\mu_z} & \frac{1}{\mu_{zx}}+\frac{1}{\mu_y} \\ \frac{1}{\mu_{xy}}+\frac{1}{\mu_z} & \frac{1}{m_y}+\frac{1}{\mu_{yy}} & \frac{1}{\mu_{yz}}-\frac{1}{\mu_x} \\ \frac{1}{\mu_{zx}}-\frac{1}{\mu_y} & \frac{1}{\mu_{yz}}+\frac{1}{\mu_x} & \frac{1}{m_z}+\frac{1}{\mu_{zz}} \end{pmatrix}$$

27-4 Matthiessen's rule and its violation. Assume that a metal contains M different kinds of impurities with concentrations n_i and transport times $\tau_i(\varepsilon)$ for $i = 1, 2, \cdots, M$.

(1) Show that the resistivity satisfies Matthiessen's rule at zero temperature, that is, $\rho = \sum_i \rho_i$, where $\rho_i = 1/\sigma_i$ with $\sigma_i = e^2 g(E_{\mathrm{F}}) v_{\mathrm{F}}^2 \tau_i(E_{\mathrm{F}})$.

(2) Show that Matthiessen's rule is not in general valid at finite temperatures unless all τ_i's are independent of energy.

---

(1) In the presence of several different kinds of impurities, the Boltzmann equation can be written as

$$\frac{\mathrm{d}f}{\mathrm{d}t} = \sum_i I_i[f],$$

where $I_i[f]$ is the collision integral for the impurities of the ith kind. In the applied electric field $\boldsymbol{E}$, the distribution function is written as

$$h^3 f = n_{\mathrm{F}}(\varepsilon) + (\boldsymbol{v} \cdot \boldsymbol{E})\phi(\varepsilon).$$

For the impurities of the ith kind, we have

$$h^3 I_i[f] = -n_i v(\varepsilon)\phi(\varepsilon)\boldsymbol{v} \cdot \boldsymbol{E}\, \sigma_{\mathrm{tr}}^i.$$

With $\mathrm{d}f/\mathrm{d}t$ given by $-(e/h^3)(\boldsymbol{v} \cdot \boldsymbol{E})\partial n_{\mathrm{F}}(\varepsilon)/\partial\varepsilon$, we have

$$\begin{aligned} -\frac{e}{h^3}(\boldsymbol{v} \cdot \boldsymbol{E})\frac{\partial n_{\mathrm{F}}(\varepsilon)}{\partial \varepsilon} &= \sum_i I_i[f] = -\frac{1}{h^3}\sum_i n_i v(\varepsilon)\phi(\varepsilon)\boldsymbol{v} \cdot \boldsymbol{E}\, \sigma_{\mathrm{tr}}^i \\ &= -\frac{1}{h^3}\phi(\varepsilon)\boldsymbol{v} \cdot \boldsymbol{E} \sum_i \tau_i^{-1} \\ &= -\frac{1}{h^3}\phi(\varepsilon)\boldsymbol{v} \cdot \boldsymbol{E}\, \bar{\tau}^{-1}(\varepsilon), \end{aligned}$$

where $\tau_i(\varepsilon) = 1/n_i v(\varepsilon)\sigma_{\mathrm{tr}}^i$ and

$$\frac{1}{\bar{\tau}(\varepsilon)} = \sum_i \frac{1}{\tau_i(\varepsilon)}.$$

The conductivity is then given by

$$\sigma = \frac{1}{3}e^2 \int \mathrm{d}\varepsilon \left[-\frac{\partial n_{\mathrm{F}}(\varepsilon)}{\partial \varepsilon}\right] g(\varepsilon) v^2(\varepsilon) \bar{\tau}(E_{\mathrm{F}}).$$

At zero temperature, we have

$$\sigma = \frac{1}{3}e^2 g(E_{\mathrm{F}})\bar{\tau}(E_{\mathrm{F}}) = \frac{ne^2\bar{\tau}(E_{\mathrm{F}})}{m}.$$

For the resistivity, we have

$$\rho = \frac{1}{\sigma} = \frac{m}{ne^2}\frac{1}{\bar{\tau}(E_{\rm F})} = \frac{m}{ne^2}\sum_i \frac{1}{\tau_i(E_{\rm F})} = \sum_i \rho_i.$$

Thus, the resistivity satisfies the Matthiessen's rule at zero temperature

(2) At finite temperatures, in general we can not set $\bar{\tau}(\varepsilon)$ to its value at the Fermi energy $E_{\rm F}$. Instead, we must evaluate σ using

$$\sigma = \frac{1}{3}e^2\int {\rm d}\varepsilon \left[-\frac{\partial n_{\rm F}(\varepsilon)}{\partial \varepsilon}\right] g(\varepsilon)v^2(\varepsilon)\frac{1}{\sum_i [\tau_i(\varepsilon)]^{-1}}.$$

Whereas, for impurities of a single kind, we have

$$\sigma_i = \frac{1}{3}e^2\int {\rm d}\varepsilon \left[-\frac{\partial n_{\rm F}(\varepsilon)}{\partial \varepsilon}\right] g(\varepsilon)v^2(\varepsilon)\tau_i(\varepsilon).$$

From the above two expressions, we see that, in general,

$$\frac{1}{\sigma} \neq \frac{1}{\sigma_1} + \frac{1}{\sigma_2} + \cdots,$$

unless all τ_i's are independent of energy. Therefore, the Matthiessen's rule is not in general valid at finite temperatures unless all τ_is are independent of energy.

27-5 Linear response theory and DC conductivity. The DC conductivity can be derived directly by considering the effect of a uniform static electric field $\boldsymbol{E}$ on the electrons in a metal. The electric potential is given by $\phi(\boldsymbol{r}) = -\boldsymbol{E}\cdot\boldsymbol{r}$.

(1) The energy of electrons in the electric field is given by $e\sum_j \boldsymbol{E}\cdot\boldsymbol{r}_j = -\boldsymbol{E}\cdot\boldsymbol{P}$ with $\boldsymbol{P} = -\sum_j e\boldsymbol{r}_j$. Show that $\boldsymbol{P}$ is related to the current density $\boldsymbol{J} = -\mathscr{V}^{-1}\sum_j e\boldsymbol{v}_j$ through $\boldsymbol{P}(t) = \mathscr{V}\int_{-\infty}^t {\rm d}t'\ \boldsymbol{J}(t')$, where the constant term in $\boldsymbol{P}(t)$ has been neglected.

(2) Quantize $\boldsymbol{P}$ and $\boldsymbol{J}$. Evaluate the commutator $[\hat{P}_\alpha, \hat{J}_\beta]$.

(3) Demonstrate how the average current density $\hat{\boldsymbol{J}}(t)$ is related to the position-dependent current density $\hat{\boldsymbol{J}}(\boldsymbol{r},t)$.

(4) Show that the Hamiltonian of electrons in the electric field can be written as $\hat{H}_{\rm ext}(t) = -\int {\rm d}\boldsymbol{r}\int_{-\infty}^t {\rm d}t'\ \boldsymbol{E}\cdot\hat{\boldsymbol{J}}(\boldsymbol{r},t')$.

(5) Apply the linear response theory to the present problem and derive the DC conductivity.

(1) Differentiating $\boldsymbol{P}(t)$ with respect to t, we have

$$\frac{\mathrm{d}\boldsymbol{P}(t)}{\mathrm{d}t} = -\sum_j e\frac{\mathrm{d}\boldsymbol{r}_j}{\mathrm{d}t} = -\sum_j e\boldsymbol{v}_j = \mathscr{V}\boldsymbol{J}(t).$$

Integrating both sides of the above equation from $t = -\infty$ to t yields

$$\boldsymbol{P}(t) = \boldsymbol{P}(-\infty) + \mathscr{V}\int_{-\infty}^{t} \mathrm{d}t'\ \boldsymbol{J}(t').$$

Dropping the constant term $\boldsymbol{P}(-\infty)$, we have

$$\boldsymbol{P}(t) = \mathscr{V}\int_{-\infty}^{t} \mathrm{d}t'\ \boldsymbol{J}(t').$$

(2) We use the canonical quantization procedure to quantize $\boldsymbol{P}$ and $\boldsymbol{J}$. $\boldsymbol{P}$ can be quantized trivially by taking the electron position $\boldsymbol{r}_j$ in its expression as a quantum operator. As usual, we do not make $\boldsymbol{r}_j$ carry a hat. We have

$$\hat{\boldsymbol{P}} = -\sum_j e\boldsymbol{r}_j.$$

To quantize $\boldsymbol{J}$, we first express it in terms of electron momenta as

$$\boldsymbol{J} = -\frac{1}{\mathscr{V}}\sum_j e\boldsymbol{v}_j = -\frac{e}{m\mathscr{V}}\sum_j \boldsymbol{p}_j.$$

Replacing the electron momentum $\boldsymbol{p}_j$ in the above equation with the electron momentum operator $\hat{\boldsymbol{p}}_j$, we obtain the quantized current density

$$\hat{\boldsymbol{J}} = -\frac{e}{m\mathscr{V}}\sum_j \hat{\boldsymbol{p}}_j.$$

The commutator $[\hat{P}_\alpha, \hat{J}_\beta]$ can be evaluated as follows

$$\begin{aligned}[\hat{P}_\alpha, \hat{J}_\beta] &= \frac{e^2}{m\mathscr{V}}\sum_{jk}[x_j^\alpha, \hat{p}_k^\beta] = \frac{e^2}{m\mathscr{V}}\sum_{jk}\mathrm{i}\hbar\delta_{jk}\delta_{\alpha\beta} \\ &= \frac{\mathrm{i}e^2\hbar}{m\mathscr{V}}\delta_{\alpha\beta}\sum_j 1 = \frac{\mathrm{i}Ne^2\hbar}{m\mathscr{V}}\delta_{\alpha\beta} \\ &= \frac{\mathrm{i}ne^2\hbar}{m}\delta_{\alpha\beta},\end{aligned}$$

where $n = N/\mathscr{V}$ is the electron number density with N the total number of electrons.

(3) The position dependent current density $\hat{\boldsymbol{J}}(\boldsymbol{r},t)$ is given by

$$\hat{\boldsymbol{J}}(\boldsymbol{r},t) = -\sum_j e\boldsymbol{v}_j\,\delta(\boldsymbol{r}-\boldsymbol{r}_j).$$

Averaging $\hat{\boldsymbol{J}}(\boldsymbol{r},t)$ over the entire space, we have

$$\frac{1}{\mathscr{V}}\int \mathrm{d}\boldsymbol{r}\,\hat{\boldsymbol{J}}(\boldsymbol{r},t) = -\frac{1}{\mathscr{V}}\sum_j e\boldsymbol{v}_j\int \mathrm{d}\boldsymbol{r}\,\delta(\boldsymbol{r}-\boldsymbol{r}_j) = -\frac{1}{\mathscr{V}}\sum_j e\boldsymbol{v}_j = \hat{\boldsymbol{J}}(t).$$

Thus,

$$\hat{\boldsymbol{J}}(t) = \frac{1}{\mathscr{V}}\int \mathrm{d}\boldsymbol{r}\,\hat{\boldsymbol{J}}(\boldsymbol{r},t).$$

(4) From the electric potential $\phi(\boldsymbol{r}) = -\boldsymbol{E}\cdot\boldsymbol{r}$, the Hamiltonian of the electron system is given by

$$\begin{aligned} H_{\mathrm{ext}} &= \sum_j (-e)\phi(\boldsymbol{r}_j) = e\sum_j \boldsymbol{E}\cdot\boldsymbol{r}_j(t) \\ &= -\boldsymbol{E}\cdot\boldsymbol{P}(t) = -\mathscr{V}\int_{-\infty}^{t}\mathrm{d}t'\,\boldsymbol{E}\cdot\boldsymbol{J}(t') \end{aligned}$$

Quantizing the above Hamiltonian, we obtain

$$\hat{H}_{\mathrm{ext}} = -\mathscr{V}\int_{-\infty}^{t}\mathrm{d}t'\,\boldsymbol{E}\cdot\hat{\boldsymbol{J}}(t') = -\int \mathrm{d}\boldsymbol{r}\int_{-\infty}^{t}\mathrm{d}t'\,\boldsymbol{E}\cdot\hat{\boldsymbol{J}}(\boldsymbol{r},t').$$

(5) In the linear response theory (lrt), the perturbation due to the external field is usually put into the form

$$\hat{H}_{\mathrm{ext}}^{\mathrm{lrt}}(t) = \frac{1}{\mathscr{V}}\int \mathrm{d}\boldsymbol{r}\,f(\boldsymbol{r},t)\hat{B}(\boldsymbol{r}).$$

Then, the Fourier components of the average of the observable $\hat{A}(t)$ are given by

$$\langle\hat{A}\rangle(\boldsymbol{q},\omega) = \chi_{AB}(\boldsymbol{q},\omega+\mathrm{i}\delta)f(\boldsymbol{q},\omega),$$

where $\chi_{AB}(\boldsymbol{q},\omega+\mathrm{i}\delta)$ is the the linear response function

$$\chi_{AB}(\boldsymbol{q},\omega+\mathrm{i}\delta) = \mathrm{i}\int_{-\infty}^{\infty}\frac{\mathrm{d}\omega'}{2\pi}\,\frac{G_{AB}^{R}(\boldsymbol{q},\omega')}{\omega-\omega'+\mathrm{i}\delta}$$

with

$$\begin{aligned} G_{AB}^{R}(\boldsymbol{q},\omega) &= \int\frac{\mathrm{d}(\boldsymbol{r}-\boldsymbol{r}')}{\mathscr{V}}\int_{-\infty}^{\infty}\mathrm{d}(t-t')\,\mathrm{e}^{-\mathrm{i}[\boldsymbol{q}\cdot(\boldsymbol{r}-\boldsymbol{r}')-\omega(t-t')]} \\ &\quad\times G_{AB}^{R}(\boldsymbol{r}-\boldsymbol{r}',t-t'), \\ G_{AB}^{R}(\boldsymbol{r}-\boldsymbol{r}',t-t') &= -\mathrm{i}\hbar^{-1}\left\langle[\hat{A}(\boldsymbol{r},t),\hat{B}(\boldsymbol{r}',t')]\right\rangle. \end{aligned}$$

If the average of the observable $\hat{A}(t)$ is zero in the absence of disturbance, then $\langle \hat{A} \rangle(\boldsymbol{q}, \omega)$ is simply the response. Otherwise, the difference between the averages of the observable $\hat{A}(t)$ in the presence and absence of disturbance is the response. Here we are interested in the response of the current density to the applied external electric field. Since the current density is zero in the absence of the external electric field, the average of the current density directly gives the response. The response function in this case is the DC conductivity.

In the present problem, we have a scalar product in $\hat{H}_{\text{ext}}$. Writing $\boldsymbol{E} \cdot \hat{\boldsymbol{J}}(\boldsymbol{r}, t')$ as $\sum_\alpha E_\alpha \hat{J}_\alpha(\boldsymbol{r}, t')$, we can treat each term separately. For $E_\alpha \hat{J}_\alpha(\boldsymbol{r}, t')$, we take $f(\boldsymbol{r}, t) = \mathscr{V} E_\alpha$ and $\hat{B}(\boldsymbol{r}) = -\int_{-\infty}^{t} \mathrm{d}t' \; \hat{J}_\alpha(\boldsymbol{r}, t') = -\mathscr{V}^{-1} \hat{P}_\alpha$. Note that $\hat{B}(\boldsymbol{r})$ is actually an operator that does not depend on time explicitly. The time integration in its expression is for obtaining the electron position from its velocity.

We now consider the response of the current density $\hat{J}_\alpha(\boldsymbol{r}, t)$. Since $f(\boldsymbol{r}, t) = \mathscr{V} E_\alpha$ does not depend on $\boldsymbol{r}$ and t, its only nonzero component is that for $\boldsymbol{q} = 0$ and $\omega = 0$

$$f(\boldsymbol{q}, \omega) = \int \frac{\mathrm{d}\boldsymbol{r}}{\mathscr{V}} \int_{-\infty}^{\infty} \mathrm{d}t \; \mathrm{e}^{-\mathrm{i}(\boldsymbol{q}\cdot\boldsymbol{r} - \omega t)} f(\boldsymbol{r}, t) = (2\pi)^4 E_\alpha \delta(\boldsymbol{q}) \delta(\omega).$$

For a uniform current density $\hat{J}_\alpha(\boldsymbol{r}, t) = \hat{J}_\alpha$, the Fourier components of $\hat{A}(\boldsymbol{r}, t) = \hat{J}_\alpha(\boldsymbol{r}, t)$ are given by

$$\begin{aligned} \langle \hat{A} \rangle(\boldsymbol{q}, \omega) &= \int \frac{\mathrm{d}\boldsymbol{r}}{\mathscr{V}} \int_{-\infty}^{\infty} \mathrm{d}t \; \mathrm{e}^{-\mathrm{i}(\boldsymbol{q}\cdot\boldsymbol{r} - \omega t)} \langle \hat{A} \rangle(\boldsymbol{r}, t) \\ &= \frac{(2\pi)^4}{\mathscr{V}} \langle \hat{J}_\alpha \rangle \, \delta(\boldsymbol{q}) \delta(\omega). \end{aligned}$$

From $\langle \hat{A} \rangle(\boldsymbol{q}, \omega) = \chi_{AB}(\boldsymbol{q}, \omega + \mathrm{i}\delta) f(\boldsymbol{q}, \omega)$, we have

$$\langle \hat{J}_\alpha \rangle = \mathscr{V} \sum_\beta \chi_{AB}^{\alpha\beta}(0, 0) E_\beta,$$

where we have taken into account the fact that the disturbance from $\hat{J}_\beta$ may have an effect on a different component of the current density $\hat{J}_\alpha$. Comparing the above equation with $\langle \hat{J}_\alpha \rangle = \sum_\beta \sigma_{\alpha\beta} E_\beta$, we see that the DC conductivity is given by

$$\sigma_{\alpha\beta} = \mathscr{V} \chi_{AB}^{\alpha\beta}(0, 0).$$

We now describe how to evaluate $\chi_{AB}^{\alpha\beta}(0,0)$. Inserting the expressions of $\hat{A}$ and $\hat{B}$ into $G_{AB}^{R}(\boldsymbol{r}-\boldsymbol{r}',t-t') = -\mathrm{i}\hbar^{-1}\langle[\hat{A}(\boldsymbol{r},t),\hat{B}(\boldsymbol{r}',t')]\rangle$, we have

$$\begin{aligned} G_{AB}^{R,\alpha\beta}(\boldsymbol{r}-\boldsymbol{r}',t-t') &= \frac{\mathrm{i}}{\hbar\mathscr{V}}\langle[\hat{J}_\alpha(\boldsymbol{r},t),\hat{P}_\beta(\boldsymbol{r}',t')]\rangle \\ &= \frac{\mathrm{i}}{\hbar\mathscr{V}}\langle[\hat{J}_\alpha(\boldsymbol{r},t),\mathrm{e}^{\mathrm{i}\hat{H}t'/\hbar}\hat{P}_\beta(\boldsymbol{r}')\mathrm{e}^{-\mathrm{i}\hat{H}t'/\hbar}]\rangle, \end{aligned}$$

where $\hat{H}$ is the Hamiltonian of the system in the absence of the applied external electric field. To pursue any further in the valuation of $G_{AB}^{R,\alpha\beta}(\boldsymbol{r}-\boldsymbol{r}',t-t')$, the knowledge of $\hat{H}$ is required. Once $\hat{H}$ is known, we can obtain $G_{AB}^{R,\alpha\beta}(\boldsymbol{r}-\boldsymbol{r}',t-t')$ immediately from the above equation. We can then evaluate the Fourier components of $G_{AB}^{R,\alpha\beta}(\boldsymbol{r}-\boldsymbol{r}')$, $G_{AB}^{R,\alpha\beta}(0,\omega')$'s, from which the response function $\chi_{AB}^{\alpha\beta}(0,0)$ follows. The DC conductivity is then given by $\sigma_{\alpha\beta} = \mathscr{V}\chi_{AB}^{\alpha\beta}(0,0)$.

27-6 Thermopower. The thermopower tensor $S_{\alpha\beta}$ is defined through $E'_\alpha = \sum_\beta S_{\alpha\beta}\partial T/\partial x_\beta$, where E'_α is the αth component of the effective electric field $\boldsymbol{E}'$ and $\partial T/\partial x_\beta$ the βth component of the temperature gradient $\boldsymbol{\nabla}T$. The thermopower $S_{\alpha\beta}$ relates the electric field E_α to the gradient of the temperature under the condition that the net electric current is zero. Derive an expression for $S_{\alpha\beta}$ in terms of

$$R_{\alpha\beta}(\varepsilon) = \int \frac{\mathrm{d}\boldsymbol{k}}{4\pi^3}\,\tau_{\boldsymbol{k}}v_{\boldsymbol{k}\alpha}v_{\boldsymbol{k}\beta}\delta(\varepsilon-\varepsilon_{\boldsymbol{k}}).$$

In the relaxation time approximation, the Boltzmann equation in the steady state reads

$$\frac{\boldsymbol{p}}{m}\cdot\boldsymbol{\nabla}_{\boldsymbol{r}}f^0(\boldsymbol{r},\boldsymbol{p}) + \boldsymbol{F}\cdot\boldsymbol{\nabla}_{\boldsymbol{p}}f^0(\boldsymbol{r},\boldsymbol{p}) = -\frac{f_1}{\tau_{\boldsymbol{p}}}.$$

Substituting $f^0(\boldsymbol{r},\boldsymbol{p}) = h^{-3}n_{\mathrm{F}}(\varepsilon_{\boldsymbol{p}})$ into the left hand side of the above equation, replacing $\boldsymbol{p}$ with $\hbar\boldsymbol{k}$, and using $\boldsymbol{v}_{\boldsymbol{k}} = \hbar\boldsymbol{k}/m$ and $\boldsymbol{F} = -e\boldsymbol{E}$, we have

$$v_{\boldsymbol{k}}\cdot\boldsymbol{\nabla}_{\boldsymbol{r}}n_{\mathrm{F}}(\varepsilon_{\boldsymbol{k}}) - \frac{e}{\hbar}\boldsymbol{E}\cdot\boldsymbol{\nabla}_{\boldsymbol{k}}n_{\mathrm{F}}(\varepsilon_{\boldsymbol{k}}) = -\frac{h^3f_1}{\tau_{\boldsymbol{k}}}.$$

Making use of the explicit expression of $n_{\mathrm{F}}(\varepsilon_{\boldsymbol{k}})$, $n_{\mathrm{F}}(\varepsilon_{\boldsymbol{k}}) = 1/[\mathrm{e}^{(\varepsilon_{\boldsymbol{k}}-\mu)/k_{\mathrm{B}}T}+1]$, we obtain

$$f_1 = -\frac{1}{h^3}\left[-\frac{\partial n_{\mathrm{F}}(\varepsilon_{\boldsymbol{k}})}{\partial\varepsilon_{\boldsymbol{k}}}\right]\boldsymbol{v}_{\boldsymbol{k}}\cdot\left[(e\boldsymbol{E}+\boldsymbol{\nabla}\mu) + \frac{\varepsilon_{\boldsymbol{k}}-\mu}{T}\boldsymbol{\nabla}T\right]\tau_{\boldsymbol{k}}.$$

The current density, with the electron spin degeneracy factor of 2 taken into account, is given by

$$\begin{aligned}
\boldsymbol{j}(\boldsymbol{r}) &= -2e\hbar^3 \int \mathrm{d}\boldsymbol{k}\ \boldsymbol{v}_{\boldsymbol{k}} f(\boldsymbol{r},\boldsymbol{k}) = -2e\hbar^3 \int \mathrm{d}\boldsymbol{k}\ \boldsymbol{v}_{\boldsymbol{k}} f_1(\boldsymbol{r},\boldsymbol{k}) \\
&= \int \frac{\mathrm{d}\boldsymbol{k}}{4\pi^3}\, \boldsymbol{v}_{\boldsymbol{k}} \left[-\frac{\partial n_{\mathrm{F}}(\varepsilon_{\boldsymbol{k}})}{\partial \varepsilon_{\boldsymbol{k}}}\right] \boldsymbol{v}_{\boldsymbol{k}} \cdot \left[(e^2\boldsymbol{E} + e\boldsymbol{\nabla}\mu) + e\frac{\varepsilon_{\boldsymbol{k}}-\mu}{T}\boldsymbol{\nabla}T\right]\tau_{\boldsymbol{k}} \\
&= \int \frac{\mathrm{d}\boldsymbol{k}}{4\pi^3}\, \boldsymbol{v}_{\boldsymbol{k}} \left[-\frac{\partial n_{\mathrm{F}}(\varepsilon_{\boldsymbol{k}})}{\partial \varepsilon_{\boldsymbol{k}}}\right] \boldsymbol{v}_{\boldsymbol{k}} \cdot \left(e^2\boldsymbol{E}' + e\frac{\varepsilon_{\boldsymbol{k}}-\mu}{T}\boldsymbol{\nabla}T\right)\tau_{\boldsymbol{k}},
\end{aligned}$$

where $\boldsymbol{E}' = \boldsymbol{E} + \boldsymbol{\nabla}\mu/e$ is the effective electric field. The components of the current density are given by

$$\begin{aligned}
j_\alpha &= \sum_\beta \int \frac{\mathrm{d}\boldsymbol{k}}{4\pi^3} \left[-\frac{\partial n_{\mathrm{F}}(\varepsilon_{\boldsymbol{k}})}{\partial \varepsilon_{\boldsymbol{k}}}\right] v_{\boldsymbol{k}\alpha} v_{\boldsymbol{k}\beta} \left(e^2 E'_\beta + e\frac{\varepsilon_{\boldsymbol{k}}-\mu}{T}\frac{\partial T}{\partial x_\beta}\right)\tau_{\boldsymbol{k}} \\
&= \sum_\beta \int \mathrm{d}\varepsilon \int \frac{\mathrm{d}\boldsymbol{k}}{4\pi^3}\, \tau_{\boldsymbol{k}} v_{\boldsymbol{k}\alpha} v_{\boldsymbol{k}\beta}\delta(\varepsilon - \varepsilon_{\boldsymbol{k}}) \left[-\frac{\partial n_{\mathrm{F}}(\varepsilon)}{\partial \varepsilon}\right] \\
&\quad \times \left(e^2 E'_\beta + e\frac{\varepsilon-\mu}{T}\frac{\partial T}{\partial x_\beta}\right) \\
&= \sum_\beta \int \mathrm{d}\varepsilon\ R_{\alpha\beta}(\varepsilon) \left[-\frac{\partial n_{\mathrm{F}}(\varepsilon)}{\partial \varepsilon}\right] \left(e^2 E'_\beta + e\frac{\varepsilon-\mu}{T}\frac{\partial T}{\partial x_\beta}\right) \\
&= \sum_\beta \left(e^2 L^{(0)}_{\alpha\beta} E'_\beta + \frac{e}{T} L^{(1)}_{\alpha\beta}\frac{\partial T}{\partial x_\beta}\right),
\end{aligned}$$

where

$$L^{(n)}_{\alpha\beta} = \int \mathrm{d}\varepsilon\ (\varepsilon-\mu)^n R_{\alpha\beta}(\varepsilon)\left[-\frac{\partial n_{\mathrm{F}}(\varepsilon)}{\partial \varepsilon}\right].$$

From $\boldsymbol{j}(\boldsymbol{r}) = 0$, we have

$$\sum_\beta L^{(0)}_{\alpha\beta} E'_\beta = -\frac{1}{eT}\sum_\beta L^{(1)}_{\alpha\beta}\frac{\partial T}{\partial x_\beta}.$$

Taking both the left and right hand sides of the above equation as matrix products and multiplying both sides with the inverse of $L^{(0)}$ from left, we have

$$E'_\alpha = -\frac{1}{eT}\sum_\beta \left(L^{(0)^{-1}} L^{(1)}\right)_{\alpha\beta}\frac{\partial T}{\partial x_\beta}.$$

Thus,

$$S_{\alpha\beta} = -\frac{1}{eT}\left(L^{(0)^{-1}} L^{(1)}\right)_{\alpha\beta}.$$

$L^{(0)}$ and $L^{(1)}$ can be evaluated using the Sommerfeld expansion

$$\int_{-\infty}^{\infty} \mathrm{d}\varepsilon\ \eta(\varepsilon) n_{\mathrm{F}}(\varepsilon) = \int_{-\infty}^{\mu} \mathrm{d}\varepsilon\ \eta(\varepsilon) + \frac{\pi^2}{6}\eta'(\mu)(k_{\mathrm{B}}T)^2 + \frac{7\pi^4}{360}\eta'''(\mu)(k_{\mathrm{B}}T)^4 + O\big((k_{\mathrm{B}}T)^6\big).$$

Up to the second order in $k_{\mathrm{B}}T$, we have

$$L^{(0)} \approx R(\mu) + \frac{\pi^2}{6}(k_{\mathrm{B}}T)^2 R''(\mu), \quad L^{(1)} \approx \frac{\pi^2}{3}(k_{\mathrm{B}}T)^2 R'(\mu).$$

The inverse of $L^{(0)}$ is approximately given by

$$\begin{aligned} {L^{(0)}}^{-1} &\approx \left[R(\mu) + \frac{\pi^2}{6}(k_{\mathrm{B}}T)^2 R''(\mu) \right]^{-1} \\ &= R^{-1}(\mu) \left[1 + \frac{\pi^2}{6}(k_{\mathrm{B}}T)^2 R''(\mu) R^{-1}(\mu) \right]^{-1} \\ &\approx R^{-1}(\mu) \left[1 - \frac{\pi^2}{6}(k_{\mathrm{B}}T)^2 R''(\mu) R^{-1}(\mu) \right]. \end{aligned}$$

Finally, the thermopower is approximately given by

$$\begin{aligned} S &= -\frac{1}{eT}{L^{(0)}}^{-1} L^{(1)} \\ &\approx -\frac{1}{eT} R^{-1}(\mu) \left[1 - \frac{\pi^2}{6}(k_{\mathrm{B}}T)^2 R''(\mu) R^{-1}(\mu) \right] \frac{\pi^2}{3}(k_{\mathrm{B}}T)^2 R'(\mu) \\ &\approx -\frac{\pi^2 k_{\mathrm{B}}^2 T}{3e} R^{-1}(\mu) R'(\mu). \end{aligned}$$

In components, we have

$$S_{\alpha\beta} \approx -\frac{\pi^2 k_{\mathrm{B}}^2 T}{3e} \sum_{\gamma} R_{\alpha\gamma}^{-1}(\mu) R'_{\gamma\beta}(\mu).$$

27-7 Conductivity of a metal. As temperature T tends to zero, the conductivity of a metal saturates at a constant value (*the residual conductivity*) determined by impurities. At $T = 0$, the residual conductivity can be obtained by replacing the derivative of the Fermi function in the general formula for the conductivity

$$\sigma = \frac{1}{3}e^2 \int \mathrm{d}\varepsilon\ g(\varepsilon) \left[-\frac{\partial n_{\mathrm{F}}(\varepsilon)}{\partial \varepsilon} \right] v^2(\varepsilon)\tau_{\mathrm{tr}}(\varepsilon)$$

with the Dirac δ-function, $-\partial n_{\mathrm{F}}(\varepsilon)/\partial\varepsilon = \delta(\varepsilon - E_{\mathrm{F}})$.

(1) Construct an expansion of σ in powers of T for small but finite T with the temperature dependence of the chemical potential μ neglected.

(2) Show that, if $\tau_{\rm tr}(\varepsilon) \propto \varepsilon^\alpha$, then the leading T-dependent correction to the zero-T value of σ increases with T for $\alpha < -3/2$ or $\alpha > -1/2$ and decreases with T for $-3/2 < \alpha < -1/2$.

(3) We now consider the case in which the number density n of carriers is kept constant while the chemical potential is temperature dependent. How would the expansion in the first part change? Derive an expression for the leading term in this case. Show that if $\tau_{\rm tr}$ is independent of ε and the dispersion relation is quadratic, $\varepsilon = \hbar^2 k^2/2m$, then the temperature dependence drops out. Explain this result in terms of the Drude formula $\sigma = ne^2\tau/m$.

(4) Assume that $\tau_{\rm tr}(\varepsilon)$ has a sharp minimum of width much smaller than $k_{\rm B}T$ at energy $\varepsilon_{\rm min} = E_{\rm F} + \varepsilon_0$, $\tau_{\rm tr}(\varepsilon) = \tau_0 - \gamma\delta(\varepsilon - \varepsilon_{\rm min})$. Compute the conductivity in this case.

---

(1) Let $\eta(\varepsilon) = g(\varepsilon)v^2(\varepsilon)\tau_{\rm tr}(\varepsilon)$. We then have

$$\sigma = \frac{1}{3}e^2 \int d\varepsilon\, \eta(\varepsilon)\left[-\frac{\partial n_{\rm F}(\varepsilon)}{\partial \varepsilon}\right].$$

Making use of the Sommerfeld expansion, we have up to the fourth order in $k_{\rm B}T$

$$\begin{aligned}\sigma &\approx \frac{1}{3}e^2\left[\eta(\mu) + \frac{\pi^2}{6}\eta^{(2)}(\mu)(k_{\rm B}T)^2 + \frac{7\pi^4}{360}\eta^{(4)}(\mu)(k_{\rm B}T)^4\right] \\ &= \sigma_0\left[1 + \frac{\pi^2\eta^{(2)}(\mu)}{6\eta(\mu)}(k_{\rm B}T)^2 + \frac{7\pi^4\eta^{(4)}(\mu)}{360\eta(\mu)}(k_{\rm B}T)^4\right],\end{aligned}$$

where $\sigma_0 = e^2\eta(\mu)/3$.

(2) For $\tau_{\rm tr}(\varepsilon) \propto \varepsilon^\alpha$, we gave $\eta(\varepsilon) = c\varepsilon^{3/2+\alpha}$ with $c > 0$. We then have

$$\eta^{(2)}(\mu) = c(3/2 + \alpha)(1/2 + \alpha)\mu^{\alpha - 1/2}.$$

For $\alpha < -3/2$ or $\alpha > -1/2$, $\eta^{(2)}(\mu) > 0$. Thus, the leading T-dependent correction to the zero-T value of σ increases with T in this case. For $-3/2 < \alpha < -1/2$, $\eta^{(2)}(\mu) < 0$. Thus, the leading T-dependent correction to the zero-T value of σ decreases with T in this case.

(3) Up to the second order in $k_{\rm B}T/E_{\rm F}$, we have

$$\mu = E_{\rm F}\left[1 - \frac{\pi^2}{12}\left(\frac{k_{\rm B}T}{E_{\rm F}}\right)^2\right].$$

Since the temperature dependence of μ is now taken into account, the first term in the expansion of σ also depends on the temperature through $\eta(\mu)$. To obtain the T-dependent correction in the second order of k_BT/E_F, we expand $\eta(\mu)$ around E_F and obtain

$$\eta(\mu) \approx \eta(E_F) + \eta'(E_F)(\mu - E_F) \approx \eta(E_F) - \frac{\pi^2\eta'(E_F)}{12E_F}(k_BT)^2.$$

We now have up to the second order in k_BT

$$\sigma = \sigma_0\left\{1 + \frac{\pi^2}{6}\left[\frac{\eta''(E_F)}{\eta(E_F)} - \frac{\eta'(E_F)}{2E_F\eta(E_F)}\right](k_BT)^2\right\},$$

where σ_0 is now given by $\sigma_0 = e^2\eta(E_F)/3$. For τ_{tr} independent of ε and $\varepsilon = \hbar^2k^2/2m$, we have $\eta(\varepsilon) = c\varepsilon^{3/2}$. We then have

$$\eta'(E_F) = \frac{3}{2}cE_F^{1/2},\ \ \eta''(E_F) = \frac{3}{4}cE_F^{-1/2},$$

and

$$\frac{\eta''(E_F)}{\eta(E_F)} - \frac{\eta'(E_F)}{2E_F\eta(E_F)} = \frac{3}{4E_F^2} - \frac{3}{4E_F^2} = 0.$$

Therefore, the leading correction term vanishes and the temperature dependence drops out up to the second order in k_BT. This is in consistency with the Drude conductivity formula $\sigma = ne^2\tau/m$. The present result shows that the Drude conductivity formula is applicable to nearly free electrons.

(4) For $\tau_{tr}(\varepsilon) = \tau_0 - \gamma\delta(\varepsilon - \varepsilon_{min})$, we have

$$\begin{aligned}
\sigma &= \frac{e^2\tau_0}{3}\int d\varepsilon\ g(\varepsilon)v^2(\varepsilon)\left[-\frac{\partial n_F(\varepsilon)}{\partial\varepsilon}\right] \\
&\quad - \frac{e^2\gamma}{3}\int d\varepsilon\ g(\varepsilon)v^2(\varepsilon)\left[-\frac{\partial n_F(\varepsilon)}{\partial\varepsilon}\right]\delta(\varepsilon - \varepsilon_{min}) \\
&= \sigma_0 - \frac{e^2\gamma}{3k_BT}\int d\varepsilon\ g(\varepsilon)v^2(\varepsilon)\frac{e^{(\varepsilon-\mu)/k_BT}}{[e^{(\varepsilon-\mu)/k_BT}+1]^2}\delta(\varepsilon - \varepsilon_{min}) \\
&= \sigma_0 - \frac{e^2\gamma g(\varepsilon_{min})v^2(\varepsilon_{min})e^{(\varepsilon_{min}-\mu)/k_BT}}{3k_BT[e^{(\varepsilon_{min}-\mu)/k_BT}+1]^2} \\
&= \sigma_0 - \frac{e^2\gamma g(\varepsilon_{min})v^2(\varepsilon_{min})}{12k_BT\cosh^2[(\varepsilon_{min}-\mu)/2k_BT]},
\end{aligned}$$

where we have denoted the term containing τ_0 by σ_0 which has a much weaker temperature dependence than the second term.

27-8 Conductivity in terms of the effective mass. Show that the first expression below can be written as the second expression

$$\sigma_{\alpha\beta} = \frac{e^2\tau}{4\pi^3}\int_{\mathrm{BZ}} \mathrm{d}\boldsymbol{k} \left[-\frac{\partial n_{\mathrm{F}}(\varepsilon)}{\partial\varepsilon}\right] v_\alpha(\varepsilon(k))v_\beta(\varepsilon(k)),$$

$$\sigma_{\alpha\beta} = \frac{e^2\tau}{4\pi^3}\int_{\mathrm{BZ}} \mathrm{d}\boldsymbol{k}\, n_{\mathrm{F}}(\varepsilon)\left(\frac{1}{m^*}\right)_{\alpha\beta}.$$

Making use of

$$v_\alpha = \frac{1}{\hbar}\frac{\partial\varepsilon}{\partial k_\alpha}, \quad v_\beta = \frac{1}{\hbar}\frac{\partial\varepsilon}{\partial k_\beta},$$

we have

$$\begin{aligned}\sigma_{\alpha\beta} &= -\frac{e^2\tau}{4\pi^3\hbar^2}\int_{\mathrm{BZ}} \mathrm{d}\boldsymbol{k}\, \frac{\partial n_{\mathrm{F}}(\varepsilon)}{\partial\varepsilon}\frac{\partial\varepsilon}{\partial k_\alpha}\frac{\partial\varepsilon}{\partial k_\beta} \\ &= -\frac{e^2\tau}{4\pi^3\hbar^2}\int_{\mathrm{BZ}} \mathrm{d}\boldsymbol{k}\, \frac{\partial n_{\mathrm{F}}(\varepsilon)}{\partial k_\alpha}\frac{\partial\varepsilon}{\partial k_\beta}.\end{aligned}$$

Performing an integration by parts and making use of the fact that the group velocity of an electron vanishes at the boundaries of the Brillouin zones, we have

$$\sigma_{\alpha\beta} = \frac{e^2\tau}{4\pi^3\hbar^2}\int_{\mathrm{BZ}} \mathrm{d}\boldsymbol{k}\, n_{\mathrm{F}}(\varepsilon)\frac{\partial^2\varepsilon}{\partial k_\alpha\partial k_\beta} = \frac{e^2\tau}{4\pi^3}\int_{\mathrm{BZ}} \mathrm{d}\boldsymbol{k}\, n_{\mathrm{F}}(\varepsilon)\left(\frac{1}{m^*}\right)_{\alpha\beta},$$

where we have made use of the definition of the inverse effective mass tensor

$$\left(\frac{1}{m^*}\right)_{\alpha\beta} = \frac{1}{\hbar^2}\frac{\partial^2\varepsilon}{\partial k_\alpha\partial k_\beta}.$$

Chapter 28

Magnetic Properties of Solids

(1) *Classification of magnetic solids*

diamagnetic, paramagnetic, ferromagnetic, ferrimagnetic, antiferromagnetic.

(2) *Magnetization*

$$\boldsymbol{M} = \lim_{\Delta\mathscr{V}\to 0} \frac{1}{\Delta\mathscr{V}} \int_{\Delta\mathscr{V}} \mathrm{d}\boldsymbol{r}\, \boldsymbol{\mu},$$

$$\boldsymbol{M} = -\boldsymbol{\nabla}_{\boldsymbol{B}} U, \text{ or } M_\alpha = -\partial U/\partial B_\alpha.$$

(3) *Magnetic susceptibility*

$$\chi_{\alpha\beta} = \frac{\partial M_\alpha}{\partial H_\beta}.$$

(4) *Langevin diamagnetism*

$$\chi = -\frac{nZe^2\hbar^2\mu_0}{6m}\langle r^2\rangle .$$

(5) *Paramagnetism of insulators*

$$\chi = \frac{n\mu_0(g\mu_B J)^2}{k_{\mathrm{B}}T} B_J'(g\mu_{\mathrm{B}}BJ/k_{\mathrm{B}}T),$$

$B_J(x)$ is the Brillouin function,

$$B_J(x) = \frac{2J+1}{2J}\coth\left(\frac{2J+1}{2J}x\right) - \frac{1}{2J}\coth\frac{x}{2J}.$$

(6) *Pauli paramagnetism*

Pauli paramagnetic susceptibility reads

$$\chi_{\text{Pauli}} = 2\mu_0\mu_B^2 g(E_F).$$

For free electrons, we have

$$\chi_{\text{Pauli}} = 3n\mu_0\mu_{\mathrm{B}}^2/2k_{\mathrm{B}}T_F.$$

(7) *Landau diamagnetism*

$$\chi_{\text{Landau}} \approx -n\mu_0\mu_{\text{B}}^2/2k_{\text{B}}T_F.$$

(8) *Heisenberg model*

$$\hat{H}_{\text{Heis}} = -J\hbar^{-2}\sum_{\langle j\ell\rangle}\hat{\boldsymbol{S}}_j\cdot\hat{\boldsymbol{S}}_\ell.$$

(9) *Curie-Weiss law*

$$\chi = C/(T - T_c).$$

(10) *Spontaneous magnetization*

$$m = B_J(m/t).$$

(11) *Holstein-Primakoff transformation*

$$\begin{aligned}\hat{S}^+ &= \hbar(2S - \hat{a}^\dagger\hat{a})^{1/2}\hat{a},\\ \hat{S}^- &= \hbar\hat{a}^\dagger(2S - \hat{a}^\dagger\hat{a})^{1/2},\\ \hat{S}^z &= \hbar(S - \hat{a}^\dagger\hat{a}).\end{aligned}$$

(12) *Ferromagnetic magnons*

$$\hbar\omega_{\boldsymbol{k}} = zJS\big(1 - \gamma_{\boldsymbol{k}}\big)\big[1 - \eta(T)/S\big].$$

(13) *Magnetic susceptibility of an antiferromagnet in the paramagnetic phase*

$$\chi = 2C/(T + T_N).$$

(14) *Antiferromagnetic spin waves (antiferromagnons)*

$$\hbar\omega_{\boldsymbol{k}} = zJS\big(1 - \gamma_{\boldsymbol{k}}^2\big)^{1/2}.$$

(15) *Bogoliubov transformation for bosons*

$$\begin{aligned}\hat{\alpha}_{\boldsymbol{k}} &= u_{\boldsymbol{k}}\hat{a}_{\boldsymbol{k}} + v_{\boldsymbol{k}}\hat{b}_{\boldsymbol{k}}^\dagger,\\ \hat{\beta}_{\boldsymbol{k}}^\dagger &= v_{\boldsymbol{k}}\hat{a}_{\boldsymbol{k}} + u_{\boldsymbol{k}}\hat{b}_{\boldsymbol{k}}^\dagger.\end{aligned}$$

(16) *Stoner model*

$$\begin{aligned}\hat{H}^{\text{HF}} &= \sum_{\boldsymbol{k}\sigma}\varepsilon_{\boldsymbol{k}\sigma}\hat{c}_{\boldsymbol{k}\sigma}^\dagger\hat{c}_{\boldsymbol{k}\sigma} + \frac{\Delta^2}{4U} - \frac{U}{4}N(N-2),\\ (NU/E_{F0})\zeta &= (1+\zeta)^{2/3} - (1-\zeta)^{2/3}.\end{aligned}$$

(17) *Transverse dynamic spin susceptibility*

$$\chi^{+-}(\boldsymbol{q},\omega) = \mathrm{i}\frac{\mu_0 g_s^2\mu_B^2}{\hbar^3\mathscr{V}}\int_{-\infty}^{t}\mathrm{d}t'\mathrm{e}^{\mathrm{i}(\omega+i\delta)(t-t')}\left\langle\big[\hat{S}_{\boldsymbol{q}}^+(t), \hat{S}_{-\boldsymbol{q}}^-(t')\big]\right\rangle.$$

(18) *Longitudinal dynamic spin susceptibility*

$$\chi^{zz}(\boldsymbol{q},\omega) = \mathrm{i}\frac{\mu_0 g_s^2 \mu_B^2}{\hbar^3 \mathscr{V}} \int_{-\infty}^{t} \mathrm{d}t' \mathrm{e}^{\mathrm{i}(\omega+i\delta)(t-t')} \left\langle \left[\hat{S}^z_{\boldsymbol{q}}(t), \hat{S}^z_{-\boldsymbol{q}}(t')\right]\right\rangle .$$

(19) *Random phase approximation (RPA)*

$$\chi^{+-}_{\mathrm{RPA}}(\boldsymbol{q},\omega) = \frac{\chi_0^{+-}(\boldsymbol{q},\omega)}{1 - U\Gamma^{+-}(\boldsymbol{q},\omega)},$$

$$\Gamma^{+-}(\boldsymbol{q},\omega) = \sum_{\boldsymbol{k}} \frac{n_{\mathrm{F}}(\varepsilon_{\boldsymbol{k}+\boldsymbol{q},\downarrow}) - n_{\mathrm{F}}(\varepsilon_{\boldsymbol{k}\uparrow})}{\hbar\omega + \varepsilon_{\boldsymbol{k}\uparrow} - \varepsilon_{\boldsymbol{k}+\boldsymbol{q},\downarrow} + \mathrm{i}\delta}.$$

(20) *RKKY interaction*

$$\hat{H}_{\mathrm{eff}} = \frac{1}{\hbar^2 \mathscr{V}^2} \int \mathrm{d}\boldsymbol{r}\, \mathrm{d}\boldsymbol{r}'\, J_{\mathrm{eff}}(\boldsymbol{r} - \boldsymbol{r}') \hat{\boldsymbol{S}}(\boldsymbol{r}) \cdot \hat{\boldsymbol{S}}(\boldsymbol{r}'),$$

$$J_{\mathrm{eff}}(\boldsymbol{r}) = \frac{9\pi (NJ)^2}{4E_F} \frac{1}{(2k_F r)^4} \left[(2k_F r)\cos(2k_F r) - \sin(2k_F r) \right].$$

(21) *Superexchange interaction*

$$\hat{H}_{\mathrm{eff}} = \hat{P}\left(J\hbar^{-2} \sum_{\langle \ell\ell' \rangle} \hat{\boldsymbol{S}}_\ell \cdot \hat{\boldsymbol{S}}_{\ell'} \right) \hat{P},$$

$$J = \frac{8t^4}{(U-\varepsilon)^2} \left(\frac{1}{U} + \frac{1}{U-\varepsilon} \right).$$

(22) *Giant magnetoresistance*
The giant magnetoresistance is a quantum mechanical magnetoresistance effect observed in thin film structures composed of alternating ferromagnetic and nonmagnetic layers.

(23) *Colossal magnetoresistance*
In certain manganese perovskites, the resistance change in an applied magnetic field could be several orders of magnitude higher than that for the giant magnetoresistance effect.

28-1 Localized magnetic moments in a weak magnetic field. Consider a paramagnetic solid that consists of noninteracting magnetic ions with magnetic moment $\boldsymbol{\mu}$. A weak magnetic field $\boldsymbol{B} = B\boldsymbol{e}_z$ has been applied to the solid.

(1) Compute the average of the z-component of the magnetic moments using classical statistical mechanics. Derive the Curie law, $M = C/T$. Provide an expression for the Curie constant C.

(2) Compute again the average of the z-component of the magnetic moments using quantum statistical mechanics. Derive the Curie law, $M = C/T$. Provide an expression for the Curie constant C.

---

(1) In classical mechanics, the energy of a magnetic ion is given by

$$E = -\mu B \cos\theta$$

with $\theta = \angle(\boldsymbol{\mu}, \boldsymbol{B})$. The partition function is given by

$$\begin{aligned} Z &= \int_{-1}^{1} \mathrm{d}\cos\theta\, \mathrm{e}^{\beta\mu B\cos\theta} \int_0^{2\pi} \mathrm{d}\varphi = \frac{2\pi}{\beta\mu B}\left(\mathrm{e}^{\beta\mu B} - \mathrm{e}^{-\beta\mu B}\right) \\ &= \frac{4\pi k_{\mathrm{B}}T}{\mu B}\sinh\left(\frac{\mu B}{k_{\mathrm{B}}T}\right). \end{aligned}$$

The average of the z-component of the magnetic moment is given by

$$\begin{aligned} \langle \mu_z \rangle &= \frac{1}{Z}\int_{-1}^{1} \mathrm{d}\cos\theta \int_0^{2\pi} \mathrm{d}\varphi\ \mu_z \mathrm{e}^{\beta\mu B\cos\theta} \\ &= \frac{2\pi}{Z}\int_{-1}^{1} \mathrm{d}\cos\theta\ \mu\cos\theta\, \mathrm{e}^{\beta\mu B\cos\theta} \\ &= \frac{2\pi}{ZB}\frac{\partial}{\partial\beta}\int_{-1}^{1} \mathrm{d}\cos\theta\ \mathrm{e}^{\beta\mu B\cos\theta} \\ &= \frac{1}{B}\frac{\partial \ln Z}{\partial\beta} = -\frac{k_{\mathrm{B}}T^2}{B}\frac{\partial \ln Z}{\partial T} = \mu\left[\coth\left(\frac{\mu B}{k_{\mathrm{B}}T}\right) - \frac{k_{\mathrm{B}}T}{\mu B}\right] \\ &= \mu L(\mu B/k_{\mathrm{B}}T), \end{aligned}$$

where $L(x)$ is the Langevin function

$$L(x) = \coth x - \frac{1}{x}.$$

For $\mu B/k_{\mathrm{B}}T \ll 1$, making use of $\coth(x) \approx 1/x + x/3$ for $|x| \ll 1$, we have

$$\langle \mu_z \rangle \approx \frac{\mu^2 B}{3k_{\mathrm{B}}T} = \frac{C}{T},$$

where $C = \mu^2 B/3k_{\mathrm{B}}$.

(2) The quantum Hamiltonian is given by

$$\hat{H} = -\hat{\boldsymbol{\mu}} \cdot \boldsymbol{B} = -\hat{\mu}_z B = -g\mu_{\rm B}\hat{J}_z B,$$

where $\hat{J}_z$ is the z-component spin operator of an ion, g the Landé factor, and $\mu_{\rm B}$ the Bohr magneton. Let J be the spin quantum number of an ion. The eigenvalues of $\hat{H}$ are

$$E_M = -g\mu_{\rm B}BM, \ \ M = -J, -J+1, \cdots, J.$$

The partition function is given by

$$\begin{aligned} Z &= \sum_{M=-J}^{J} \mathrm{e}^{-\beta E_M} = \sum_{M=-J}^{J} \mathrm{e}^{\beta g\mu_{\rm B}BM} \\ &= \mathrm{e}^{-\beta g\mu_{\rm B}BJ} \sum_{n=0}^{2J} \mathrm{e}^{\beta g\mu_{\rm B}Bn} \\ &= \mathrm{e}^{-\beta g\mu_{\rm B}BJ} \frac{1 - \mathrm{e}^{\beta g\mu_{\rm B}B(2J+1)}}{1 - \mathrm{e}^{\beta g\mu_{\rm B}B}} \\ &= \frac{\mathrm{e}^{\beta g\mu_{\rm B}B(J+1/2)} - \mathrm{e}^{-\beta g\mu_{\rm B}B(J+1/2)}}{\mathrm{e}^{\beta g\mu_{\rm B}B/2} - \mathrm{e}^{-\beta g\mu_{\rm B}B/2}} \\ &= \frac{\sinh\left[\beta g\mu_{\rm B}B(J+1/2)\right]}{\sinh(\beta g\mu_{\rm B}B/2)}. \end{aligned}$$

The average of the z-component of the magnetic moment is given by

$$\begin{aligned} \langle\mu_z\rangle &= \mathrm{Tr}\left(\hat{\mu}_z \mathrm{e}^{-\beta\hat{H}}\right) = \frac{1}{Z}\sum_{M=-J}^{J} g\mu_{\rm B}M\mathrm{e}^{\beta g\mu_{\rm B}BM} = \frac{1}{B}\frac{\partial \ln Z}{\partial\beta} \\ &= g\mu_{\rm B}\left[\frac{2J+1}{2}\coth\left(\frac{g\mu_{\rm B}B(2J+1)}{2k_{\rm B}T}\right) - \frac{1}{2}\coth\left(\frac{g\mu_{\rm B}B}{2k_{\rm B}T}\right)\right] \\ &= g\mu_{\rm B}JB_J\left(\mu_{\rm B}BJ/k_{\rm B}T\right), \end{aligned}$$

where $B_J(x)$ is the Brillouin function

$$B_J(x) = \frac{2J+1}{2J}\coth\left(\frac{2J+1}{2J}x\right) - \frac{1}{2J}\coth\left(\frac{1}{2J}x\right).$$

For $\mu_{\rm B}BJ/k_{\rm B}T \ll 1$, making use of $\coth x \approx 1/x + x/3$, we have

$$B_J(x) \approx \frac{J+1}{3J}x.$$

We thus have

$$\langle\mu_z\rangle \approx \frac{J(J+1)(g\mu_{\rm B})^2}{3k_{\rm B}T} = \frac{C}{T},$$

where $C = J(J+1)(g\mu_{\rm B})^2/3k_{\rm B}$.

28-2 Holstein-Primakoff transformation. Prove that the spin operator as expressed in the Holstein-Primakoff transformation

$$\hat{S}^+ = \hbar(2S - \hat{a}^\dagger\hat{a})^{1/2}\hat{a}, \quad \hat{S}^- = \hbar\hat{a}^\dagger(2S - \hat{a}^\dagger\hat{a})^{1/2}, \quad \hat{S}^z = \hbar(S - \hat{a}^\dagger\hat{a})$$

satisfies the usual commutation relations

$$[\hat{S}^+, \hat{S}^-] = 2\hbar\hat{S}^z, \quad [\hat{S}^z, \hat{S}^\pm] = \pm\hbar\hat{S}^\pm.$$

For $[\hat{S}^+, \hat{S}^-]$, we have

$$\begin{aligned}
\hbar^{-2}[\hat{S}^+, \hat{S}^-] &= [(2S - \hat{a}^\dagger\hat{a})^{1/2}\hat{a}, \hat{a}^\dagger(2S - \hat{a}^\dagger\hat{a})^{1/2}] \\
&= (2S - \hat{a}^\dagger\hat{a})^{1/2}\hat{a}\hat{a}^\dagger(2S - \hat{a}^\dagger\hat{a})^{1/2} \\
&\quad - \hat{a}^\dagger(2S - \hat{a}^\dagger\hat{a})^{1/2}(2S - \hat{a}^\dagger\hat{a})^{1/2}\hat{a} \\
&= \hat{a}\hat{a}^\dagger(2S - \hat{a}^\dagger\hat{a}) - \hat{a}^\dagger(2S - \hat{a}^\dagger\hat{a})\hat{a} \\
&= 2S(\hat{a}\hat{a}^\dagger - \hat{a}^\dagger\hat{a}) - \hat{a}\hat{a}^\dagger\hat{a}^\dagger\hat{a} + \hat{a}^\dagger\hat{a}^\dagger\hat{a}\hat{a} \\
&= 2S - (1 + \hat{a}^\dagger a)\hat{a}^\dagger\hat{a} + \hat{a}^\dagger(\hat{a}\hat{a}^\dagger - 1)\hat{a} \\
&= 2(S - \hat{a}^\dagger a) = (2/\hbar)\hat{S}_z.
\end{aligned}$$

Thus, $[\hat{S}^+, \hat{S}^-] = 2\hbar\hat{S}_z$. For $[\hat{S}^z, \hat{S}^+]$, we have

$$\begin{aligned}
\hbar^{-2}[\hat{S}^z, \hat{S}^+] &= -[\hat{a}^\dagger\hat{a}, (2S - \hat{a}^\dagger\hat{a})^{1/2}\hat{a}] = -(2S - \hat{a}^\dagger\hat{a})^{1/2}[\hat{a}^\dagger\hat{a}, \hat{a}] \\
&= (2S - \hat{a}^\dagger\hat{a})^{1/2}\hat{a} = \hbar^{-1}\hat{S}^+.
\end{aligned}$$

Thus, $[\hat{S}^z, \hat{S}^+] = \hbar\hat{S}^+$. For $[\hat{S}^z, \hat{S}^-]$, we have

$$\begin{aligned}
\hbar^{-2}[\hat{S}^z, \hat{S}^-] &= -[\hat{a}^\dagger\hat{a}, \hat{a}^\dagger(2S - \hat{a}^\dagger\hat{a})^{1/2}] = -[\hat{a}^\dagger\hat{a}, \hat{a}^\dagger](2S - \hat{a}^\dagger\hat{a})^{1/2} \\
&= -\hat{a}^\dagger(2S - \hat{a}^\dagger\hat{a})^{1/2} = -\hbar^{-1}\hat{S}^-.
\end{aligned}$$

Thus, $[\hat{S}^z, \hat{S}^-] = -\hbar\hat{S}^-$.

28-3 Heisenberg-Weiss model for a ferromagnet. Consider a ferromagnet made of N spin-1/2 particles with magnetic moments $\boldsymbol{m} = m_0\hat{\boldsymbol{S}}/\hbar$. The ferromagnet can have a nonzero magnetization $M(T, V, N)$ even when no external magnetic field is present, provided that T is less than the Curie temperature T_c. In the Heisenberg-Weiss model, $M(T, V, N)$ is implicitly given by

$$M = (Nm_0/2V)\tanh(\lambda\mu_0 m_0 M/2k_B T),$$

where λ is a positive dimensionless parameter. The Curie temperature T_c is given by $T_c = \lambda N\mu_0 m_0^2/4Vk_B$.

(1) Find an expression for $(\partial M/\partial T)_{V,N}$ in terms of T, V, N, and M.

(2) The single-particle partition function for this ferromagnet is given by

$$Z_1(T, V, N) = \cosh(\lambda\mu_0 m_0 M/2k_B T).$$

Compute the internal energy of the ferromagnet.

(3) Evaluate the specific heat C_v of the ferromagnet and give the value of C_v in the $T \to 0$ limit.

---

(1) Differentiating both sides of the above-given expression of M with respect to T, we obtain

$$\frac{\partial M}{\partial T} = \frac{\lambda N\mu_0 m_0^2}{4Vk_B T\cosh^2(\lambda\mu_0 m_0 M/2k_B T)}\left(\frac{\partial M}{\partial T} - \frac{M}{T}\right).$$

Solving for $\partial M/\partial T$ from the above equation yields

$$\frac{\partial M}{\partial T} = \frac{M/T}{1 - 4Vk_B T\cosh^2(\lambda\mu_0 m_0 M/2k_B T)/(\lambda N\mu_0 m_0^2)}.$$

(2) To find the internal energy, we first compute the entropy from the Helmholtz free energy F which is given by

$$F = -Nk_B T\ln Z_1 = -Nk_B T\ln\cosh\left(\frac{\lambda\mu_0 m_0 M}{2k_B T}\right).$$

Differentiating F with respect to T, we have

$$\begin{aligned}S &= -\frac{\partial F}{\partial T}\\ &= Nk_B\ln\cosh\left(\frac{\lambda\mu_0 m_0 M}{2k_B T}\right)\\ &\quad + \frac{N\lambda\mu_0 m_0}{2}\tanh\left(\frac{\lambda\mu_0 m_0 M}{2k_B T}\right)\left(\frac{\partial M}{\partial T} - \frac{M}{T}\right).\end{aligned}$$

Thus,

$$\begin{aligned}U = F + TS &= \frac{N\lambda\mu_0 m_0 T}{2}\tanh\left(\frac{\lambda\mu_0 m_0 M}{2k_B T}\right)\left(\frac{\partial M}{\partial T} - \frac{M}{T}\right)\\ &= \frac{\lambda\mu_0 M^2 V}{(\lambda N\mu_0 m_0^2/4Vk_B T)\cosh^{-2}(\lambda\mu_0 m_0 M/2k_B T) - 1}\\ &= \frac{\lambda\mu_0 M^2 V}{(\lambda N\mu_0 m_0^2/4Vk_B T)\left[1 - \tanh^2(\lambda\mu_0 m_0 M/2k_B T)\right] - 1}\\ &= \frac{\lambda\mu_0 M^2 V}{(T_c/T)(1 - 4M^2V^2/N^2 m_0^2) - 1}.\end{aligned}$$

(3) Differentiating F with respect to T, we obtain the specific heat

$$
\begin{aligned}
C_v &= \frac{\partial U}{\partial T} \\
&= \frac{\lambda \mu_0 M^2 V T_c (1 - 4M^2V^2/N^2m_0^2)}{[T_c(1 - 4M^2V^2/N^2m_0^2) - T]^2} \\
&\quad \times \left[3 + \frac{8T_c M^2 V^2/N^2 m_0^2}{T_c(1 - 4M^2V^2/N^2m_0^2) - T} \right].
\end{aligned}
$$

From the expression of M and that of $\partial M/\partial T$, we can see that, as T tends to zero, M tends to $Nm_0/2V$ exponentially. Therefore, $1 - 4M^2V^2/N^2m_0^2$ goes to zero faster than $[T_c(1 - 4M^2V^2/N^2m_0^2) - T]^3$ that is the term going to zero fastest on the denominator, which implies than $C_v \to 0$ as $T \to 0$.

28-4 Two-site Hubbard model. Consider the Hubbard Hamiltonian for two lattice sites

$$
\hat{H} = -t \sum_{\sigma} \left(\hat{c}_{1\sigma}^{\dagger} \hat{c}_{2\sigma} + \hat{c}_{2\sigma}^{\dagger} \hat{c}_{1\sigma} \right) + U \left(\hat{n}_{1\uparrow} \hat{n}_{1\downarrow} + \hat{n}_{2\uparrow} \hat{n}_{2\downarrow} \right),
$$

where $\hat{n}_{i\sigma} = \hat{c}_{i\sigma}^{\dagger} \hat{c}_{i\sigma}$ is the electron number operator of spin σ on site i, t the hopping matrix element, and $U > 0$ the on-site Coulomb interaction energy. Assume that there are two electrons in the system. For this problem, we use the basis set: $|\uparrow\uparrow\rangle$, $|\downarrow\downarrow\rangle$, $|\uparrow\downarrow\rangle$, $|\downarrow\uparrow\rangle$, $|0 \updownarrow\rangle$, and $|\updownarrow 0\rangle$, where the symbols in the first and second positions in a ket denote the occupations of the first and second sites, respectively, with 0 for not being occupied and $\uparrow$, $\downarrow$, and $\updownarrow$ for being occupied by one up-spin electron, one down-spin electron, and two electrons of opposite spins.

(1) Find the eigenvalues and eigenvectors of the Hamiltonian. Plot the eigenvalues.

(2) Guess the band structure if the two sites are replaced with an infinite lattice with one electron per site for $U/t \gg 1$.

---

(1) In the basis set $\{|\uparrow\uparrow\rangle, |\downarrow\downarrow\rangle, |\uparrow\downarrow\rangle, |\downarrow\uparrow\rangle, |0 \updownarrow\rangle, |\updownarrow 0\rangle\}$, the matrix representation of the Hamiltonian is given by

$$H = \begin{pmatrix} 0 & 0 & 0 & 0 & 0 & 0 \\ 0 & 0 & 0 & 0 & 0 & 0 \\ 0 & 0 & 0 & 0 & -t & -t \\ 0 & 0 & 0 & 0 & -t & -t \\ 0 & 0 & -t & -t & U & 0 \\ 0 & 0 & -t & -t & 0 & U \end{pmatrix}$$

From the above matrix representation, we see that the energy eigenvalue 0 is doubly degenerate with the degenerate space spanned by $|\uparrow\uparrow\rangle$ and $|\downarrow\downarrow\rangle$. The corresponding eigenvectors can be taken as $|\uparrow\uparrow\rangle$ and $|\downarrow\downarrow\rangle$. To obtain the rest energy eigenvalues, we need to diagonalize the following 4×4 matrix

$$\begin{pmatrix} 0 & 0 & -t & -t \\ 0 & 0 & -t & -t \\ -t & -t & U & 0 \\ -t & -t & 0 & U \end{pmatrix}$$

in the subspace spanned by $|\uparrow\downarrow\rangle$, $|\downarrow\uparrow\rangle$, $|0 \updownarrow\rangle$, $|\updownarrow 0\rangle$. Performing the diagonalization, we find the following energy eigenvalues and corresponding eigenvectors

Eigenvalue	**Eigenvector**
0	$\frac{1}{\sqrt{2}}\big[\,|\uparrow\downarrow\rangle - |\downarrow\uparrow\rangle\,\big]$
U	$\frac{1}{\sqrt{2}}\big[\,|0 \updownarrow\rangle - |\updownarrow 0\rangle\,\big]$
$\frac{1}{2}\big[\,U - (U^2+16t^2)^{1/2}\,\big]$	$a_-\big[\,|\uparrow\downarrow\rangle + |\downarrow\uparrow\rangle\,\big] + b_-\big[\,|0 \updownarrow\rangle + |\updownarrow 0\rangle\,\big]$
$\frac{1}{2}\big[\,U + (U^2+16t^2)^{1/2}\,\big]$	$a_+\big[\,|\uparrow\downarrow\rangle + |\downarrow\uparrow\rangle\,\big] - b_+\big[\,|0 \updownarrow\rangle + |\updownarrow 0\rangle\,\big]$

The coefficients $a_\pm$ and $b_\pm$ are given by

$$a_\pm = \frac{2t}{(U^2+16t^2)^{1/4}\big[\,\big(U^2+16t^2\big)^{1/2} \pm U\,\big]^{1/2}},$$

$$b_\pm = \frac{\big[\,\big(U^2+16t^2\big)^{1/2} \pm U\,\big]^{1/2}}{2\big(U^2+16t^2\big)^{1/4}}.$$

The energy eigenvalues are plotted in Fig. 28.1 against U/t. Note that the energy eigenvalue with a value of 0 (the horizontal line) is threefold degenerate. We can see from the figure that, as U/t increases, the two upper lines as well as the two lower lines get closer and closer, respectively. But the separation between the upper and lower lines becomes larger as U/t increases, with the separation proportional to the on-site Coulomb energy U. This is a characteristic feature of the Hubbard model.

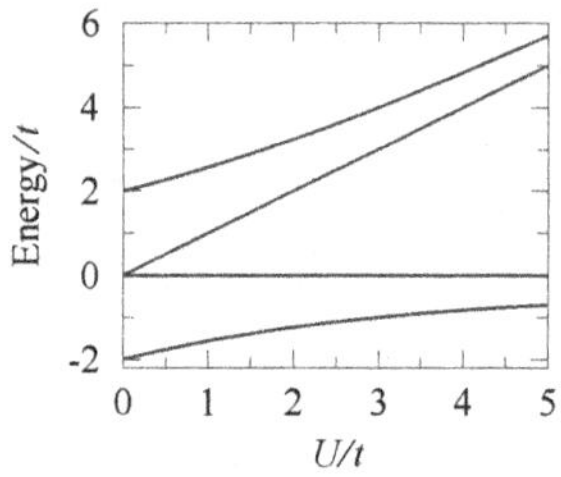

Fig. 28.1 Plots of energy eigenvalues against U/t.

(2) From Fig. 28.1, we can see that two energy bands will form for an infinite lattice. The band gap between these two energy bands is of the order of U. These bands are referred to as *the lower* and *upper Hubbard bands*.

28-5 Spin waves in a two-dimensional triangular ferromagnet. A two-dimensional ferromagnet with a triangular lattice in a magnetic field is described by the Hamiltonian

$$\hat{H} = -\frac{1}{2}J\hbar^{-2}\sum_{j\delta}\hat{\boldsymbol{S}}_j \cdot \hat{\boldsymbol{S}}_{j+\delta} - (g\mu_{\rm B}B/\hbar)\sum_j \hat{S}_j^z,$$

where j labels the lattice sites of the triangular lattice and δ denotes the nearest neighbors of site j. Assume that $S \gg 1$.

(1) Find the spectrum of the spin waves at zero temperature using the Holstein-Primakoff transformation.

(2) Compute the spin-wave contributions to the specific heat per lattice site in the two limits: $k_{\rm B}T \ll g\mu_{\rm B}B \ll J$ and $g\mu_{\rm B}B \ll k_{\rm B}T \ll J$.

---

(1) Introducing a Holstein-Primakoff transformation to the spin operator on each lattice site and approximating $(2S - \hat{a}_j^\dagger\hat{a}_j)^{1/2}$ with $(2S)^{1/2}$ and ignoring the quartic term in magnon operators in $\hat{S}_j^z \hat{S}_{j+\delta}^z$, we have

$$\hat{H} = -\frac{J\hbar^2}{4}\sum_{j\delta}(\hat{S}_j^+ \hat{S}_{j+\delta}^- + \hat{S}_j^- \hat{S}_{j+\delta}^+ + 2\hat{S}_j^z \hat{S}_{j+\delta}^z) - \frac{g\mu_B B}{\hbar}\sum_j \hat{S}_j^z$$
$$\approx -\frac{1}{2}JS\sum_{j\delta}[(\hat{a}_j\hat{a}_{j+\delta}^\dagger + \hat{a}_j^\dagger \hat{a}_{j+\delta} + 2(S - \hat{a}_j^\dagger \hat{a}_j)(S - \hat{a}_{j+\delta}^\dagger \hat{a}_{j+\delta})]$$
$$- g\mu_B B\sum_j (S - \hat{a}_j^\dagger \hat{a}_j)$$
$$\approx E_0 - \frac{1}{2}JS\sum_{j\delta}(\hat{a}_j^\dagger \hat{a}_{j+\delta} + \hat{a}_{j+\delta}^\dagger \hat{a}_j) + (6JS + g\mu_B B)\sum_j \hat{a}_j^\dagger \hat{a}_j,$$

where $E_0 = -3NJS^2 - NSg\mu_B B$. Making a Fourier transformation to $\hat{a}_j$

$$\hat{a}_j = \frac{1}{\sqrt{N}}\sum_{\boldsymbol{q}} \mathrm{e}^{\mathrm{i}\boldsymbol{q}\cdot\boldsymbol{R}_j}\hat{a}_{\boldsymbol{q}},$$

we have

$$\sum_j \hat{a}_j^\dagger \hat{a}_j = \sum_{\boldsymbol{q}\boldsymbol{q}'} \hat{a}_{\boldsymbol{q}'}^\dagger \hat{a}_{\boldsymbol{q}} \frac{1}{N}\sum_j \mathrm{e}^{-\mathrm{i}(\boldsymbol{q}'-\boldsymbol{q})\cdot\boldsymbol{R}_j} = \sum_{\boldsymbol{q}\boldsymbol{q}'} \hat{a}_{\boldsymbol{q}'}^\dagger \hat{a}_{\boldsymbol{q}}\delta_{\boldsymbol{q}'\boldsymbol{q}} = \sum_{\boldsymbol{q}} \hat{a}_{\boldsymbol{q}}^\dagger \hat{a}_{\boldsymbol{q}},$$
$$\sum_{j\delta} \hat{a}_j^\dagger \hat{a}_{j+\delta} = \sum_{\boldsymbol{q}\boldsymbol{q}'\delta} \mathrm{e}^{\mathrm{i}\boldsymbol{q}\cdot\boldsymbol{\delta}}\hat{a}_{\boldsymbol{q}'}^\dagger \hat{a}_{\boldsymbol{q}} \frac{1}{N}\sum_j \mathrm{e}^{-\mathrm{i}(\boldsymbol{q}'-\boldsymbol{q})\cdot\boldsymbol{R}_j} = 6\sum_{\boldsymbol{q}} \gamma_{\boldsymbol{q}}\, \hat{a}_{\boldsymbol{q}}^\dagger \hat{a}_{\boldsymbol{q}},$$
$$\sum_{j\delta} \hat{a}_{j+\delta}^\dagger \hat{a}_j = \sum_{\boldsymbol{q}\boldsymbol{q}'\delta} \mathrm{e}^{-\mathrm{i}\boldsymbol{q}'\cdot\boldsymbol{\delta}}\hat{a}_{\boldsymbol{q}'}^\dagger \hat{a}_{\boldsymbol{q}} \frac{1}{N}\sum_j \mathrm{e}^{-\mathrm{i}(\boldsymbol{q}'-\boldsymbol{q})\cdot\boldsymbol{R}_j} = 6\sum_{\boldsymbol{q}} \gamma_{\boldsymbol{q}}\, \hat{a}_{\boldsymbol{q}}^\dagger \hat{a}_{\boldsymbol{q}},$$

where

$$\gamma_{\boldsymbol{q}} = \frac{1}{6}\sum_{\boldsymbol{\delta}} \mathrm{e}^{\mathrm{i}\boldsymbol{q}\cdot\boldsymbol{\delta}} = \frac{1}{6}\sum_{\boldsymbol{\delta}} \mathrm{e}^{-\mathrm{i}\boldsymbol{q}\cdot\boldsymbol{\delta}}.$$

The Hamiltonian is then given by

$$\hat{H} = \sum_{\boldsymbol{q}}\left[6JS(1-\gamma_{\boldsymbol{q}}) + g\mu_B B)\right]\hat{a}_{\boldsymbol{q}}^\dagger \hat{a}_{\boldsymbol{q}}$$

Thus, the spectrum of spin waves at zero temperature is given by

$$\hbar\omega_{\boldsymbol{q}} = 6JS(1-\gamma_{\boldsymbol{q}}) + g\mu_B B.$$

(2) Since magnons are bosonic particles, they obey the Bose-Einstein distribution. The internal energy of the system per lattice site, u, is then given by

$$u = e_0 + \frac{1}{N}\sum_{\boldsymbol{q}} \frac{\hbar\omega_{\boldsymbol{q}}}{\mathrm{e}^{\beta\hbar\omega_{\boldsymbol{q}}} - 1},$$

where $e_0 = E_0/N$. Differentiating the internal energy U with respect to T, we obtain the contribution of magnons to the specific heat per lattice site

$$c_v = \frac{\mathrm{d}U}{\mathrm{d}T} = \frac{1}{N}\frac{\mathrm{d}}{\mathrm{d}T}\sum_{\boldsymbol{q}} \frac{\hbar\omega_{\boldsymbol{q}}}{\mathrm{e}^{\beta\hbar\omega_{\boldsymbol{q}}} - 1}.$$

The limits $k_{\mathrm{B}}T \ll g\mu_{\mathrm{B}}B \ll J$ and $g\mu_{\mathrm{B}}B \ll k_{\mathrm{B}}T \ll J$ are both low-temperature limits because $k_{\mathrm{B}}T \ll J$ holds in both of them. In these low-temperature limits, most of the excited magnons are of low energy. We can thus use the dispersion relation in the $q \to 0$ limit. To obtain the dispersion relation in the $q \to 0$ limit, we first write down an explicit expression for $\hbar\omega_{\boldsymbol{q}}$ and then take the limit. The coordinates of the six nearest neighbors of the site at the origin are given in Fig. 28.2. The primitive vectors $\boldsymbol{a}_1$ and $\boldsymbol{a}_2$ are given by

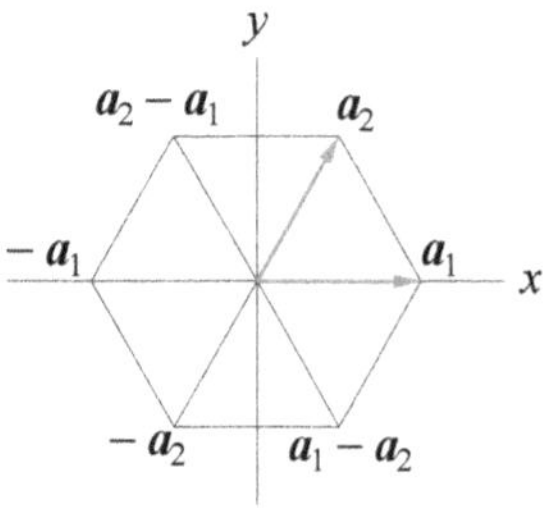

Fig. 28.2 Coordinates of the six nearest neighbors of the lattice site at the origin for a two-dimensional triangular Bravais lattice.

$$\boldsymbol{a}_1 = a\boldsymbol{e}_x, \;\; \boldsymbol{a}_2 = (a/2)\boldsymbol{e}_x + (\sqrt{3}\,a/2)\boldsymbol{e}_y.$$

$\gamma_{\boldsymbol{q}}$ is then given by

$$\begin{aligned}\gamma_{\boldsymbol{q}} &= \frac{1}{6}\big[\mathrm{e}^{\mathrm{i}\boldsymbol{q}\cdot\boldsymbol{a}_1} + \mathrm{e}^{-\mathrm{i}\boldsymbol{q}\cdot\boldsymbol{a}_1} + \mathrm{e}^{\mathrm{i}\boldsymbol{q}\cdot\boldsymbol{a}_2} + \mathrm{e}^{-\mathrm{i}\boldsymbol{q}\cdot\boldsymbol{a}_2} + \mathrm{e}^{\mathrm{i}\boldsymbol{q}\cdot(\boldsymbol{a}_1-\boldsymbol{a}_2)} \\ &\qquad + \mathrm{e}^{-\mathrm{i}\boldsymbol{q}\cdot(\boldsymbol{a}_1-\boldsymbol{a}_2)}\big] \\ &= \frac{1}{3}\big[\cos(q_x a) + \cos(q_x a/2 + \sqrt{3}\,q_y a/2) \\ &\qquad + \cos(q_x a/2 - \sqrt{3}\,q_y a/2)\big] \\ &= \frac{1}{3}\big[\cos(q_x a) + 2\cos(q_x a/2)\cos(\sqrt{3}\,q_y a/2)\big].\end{aligned}$$

Expanding the cosine functions, we obtain up to the second order in qa

$$\begin{aligned}\gamma_{\boldsymbol{q}} &\approx \frac{1}{3}\big\{1 - (q_x a)^2/2 + 2\big[1 - (q_x a)^2/8\big]\big[1 - 3(q_y a)^2/8\big]\big\} \\ &\approx 1 - \frac{1}{4}q^2 a^2.\end{aligned}$$

The dispersion relation then becomes

$$\hbar\omega_{\boldsymbol{q}} \approx 3JSq^2a^2/2 + g\mu_{\mathrm{B}}B.$$

The specific heat c_v is then given by

$$c_v \approx \frac{1}{N}\frac{\mathrm{d}}{\mathrm{d}T}\sum_{\boldsymbol{q}}\frac{3JSq^2a^2/2 + g\mu_{\mathrm{B}}B}{\mathrm{e}^{3\beta JSq^2a^2/2+\beta g\mu_{\mathrm{B}}B} - 1}.$$

For $k_{\mathrm{B}}T \ll g\mu_{\mathrm{B}}B \ll J$, we can ignore the -1 term on the denominator of the integrand in consideration that $\mathrm{e}^{g\mu_{\mathrm{B}}B/k_{\mathrm{B}}T} \gg 1$ and $\mathrm{e}^{3\beta JSq^2a^2/2} \geqslant 1$. We then have

$$\begin{aligned} c_v &\approx \frac{1}{N}\frac{\mathrm{d}}{\mathrm{d}T}\sum_{\boldsymbol{q}}(3JSq^2a^2/2 + g\mu_{\mathrm{B}}B)\mathrm{e}^{-3\beta JSq^2a^2/2-\beta g\mu_{\mathrm{B}}B} \\ &= \frac{\sqrt{3}\,a^2}{4\pi}\frac{\mathrm{d}}{\mathrm{d}T}\left[\mathrm{e}^{-\beta g\mu_{\mathrm{B}}B}\int_0^\infty \mathrm{d}q\, q(3JSq^2a^2/2+g\mu_{\mathrm{B}}B) \right. \\ &\qquad\qquad \left. \times\, \mathrm{e}^{-3\beta JSq^2a^2/2}\right] \\ &= \frac{1}{4\sqrt{3}\,\pi JS}\frac{\mathrm{d}}{\mathrm{d}T}\left\{\left[(k_{\mathrm{B}}T)^2 + g\mu_B Bk_{\mathrm{B}}T\right]\mathrm{e}^{-g\mu_{\mathrm{B}}B/k_{\mathrm{B}}T}\right\} \\ &= \frac{1}{4\sqrt{3}\,\pi JS}\left[2k_{\mathrm{B}}T + 2g\mu_B B + \frac{(g\mu_B B)^2}{k_{\mathrm{B}}T}\right]\mathrm{e}^{-g\mu_{\mathrm{B}}B/k_{\mathrm{B}}T}k_{\mathrm{B}} \\ &\approx \frac{1}{4\sqrt{3}\,\pi}\frac{(g\mu_B B)^2}{JS(k_{\mathrm{B}}T)}\mathrm{e}^{-g\mu_{\mathrm{B}}B/k_{\mathrm{B}}T}k_{\mathrm{B}}. \end{aligned}$$

For $g\mu_{\mathrm{B}}B \ll k_{\mathrm{B}}T \ll J$, we can ignore the term $g\mu_{\mathrm{B}}B$ on the numerator of the integrand and that in the argument of the exponential function on the denominator. We have in this case

$$\begin{aligned} c_v &\approx \frac{\sqrt{3}\,a^2}{4\pi}\frac{\mathrm{d}}{\mathrm{d}T}\int_0^\infty \mathrm{d}q\, q\frac{3JSq^2a^2/2}{\mathrm{e}^{3\beta JSq^2a^2/2} - 1} = \frac{k_{\mathrm{B}}^2T}{2\sqrt{3}\,\pi JS}\int_0^\infty \mathrm{d}x\,\frac{x}{\mathrm{e}^x - 1} \\ &= \frac{k_{\mathrm{B}}^2T}{2\sqrt{3}\,\pi JS}\sum_{n=1}^\infty\int_0^\infty \mathrm{d}x\, x\,\mathrm{e}^{-nx} = \frac{k_{\mathrm{B}}^2T}{2\sqrt{3}\,\pi JS}\sum_{n=1}^\infty\frac{1}{n^2} = \frac{\pi k_{\mathrm{B}}T}{12\sqrt{3}\,JS}k_{\mathrm{B}}. \end{aligned}$$

28-6 Double exchange model on two lattice sites. Consider the double exchange model on two lattice sites ($k = 1, 2$)

$$\hat{H} = t\sum_\alpha(\hat{c}_{1\alpha}^\dagger\hat{c}_{2\alpha} + \hat{c}_{2\alpha}^\dagger\hat{c}_{1\alpha}) - \frac{1}{2}J\sum_{k\alpha\beta}\boldsymbol{S}_k\cdot\hat{c}_{k\alpha}^\dagger\boldsymbol{\sigma}_{\alpha\beta}\hat{c}_{k\beta},$$

where $\boldsymbol{S}_{1,2}$ are classical ($S \gg 1$) core spins on lattice sites 1 and 2, respectively. A conduction electron can hop from one lattice site to another (the first term in the Hamiltonian) and interacts with the core spins through the ferromagnetic exchange interaction (the second term). Assume that there is just one electron on the two lattice sites, $\sum_{k\alpha} \hat{c}^\dagger_{k\alpha}\hat{c}_{k\alpha} = 1$.

(1) Find the lowest eigenvalue $E_0(S_1, S_2)$ of the electron residing on the two lattice sites. Assume that the core spins do not change in time.

(2) Write down an effective Hamiltonian H_{eff} for just the core spins using $E_0(S_1, S_2)$ and the adiabatic approximation which is justified because of the smallness of the parameter $1/S$. This interaction is said to be induced (or mediated) by the electron.

(3) Is H_{eff} ferromagnetic or antiferromagnetic? When can H_{eff} be replaced by the Heisenberg interaction?

---

(1) Making use of the explicit expressions of the Pauli matrices, we can write the Hamiltonian as

$$\hat{H} = t\sum_\alpha (\hat{c}^\dagger_{1\alpha}\hat{c}_{2\alpha} + \hat{c}^\dagger_{2\alpha}\hat{c}_{1\alpha}) - \frac{J}{2}\sum_k \left[S_k^+ \hat{c}^\dagger_{k\downarrow}\hat{c}_{k\uparrow} + S_k^- \hat{c}^\dagger_{k\uparrow}\hat{c}_{k\downarrow} + S_k^z(\hat{c}^\dagger_{k\uparrow}\hat{c}_{k\uparrow} - \hat{c}^\dagger_{k\downarrow}\hat{c}_{k\downarrow})\right],$$

where $S_k^\pm = S_k^x \pm \mathrm{i}S_k^y$. To diagonalize the Hamiltonian, we first rewrite it in the matrix form and have

$$\hat{H} = \begin{pmatrix}\hat{c}^\dagger_{1\uparrow} & \hat{c}^\dagger_{1\downarrow} & \hat{c}^\dagger_{2\uparrow} & \hat{c}^\dagger_{2\downarrow}\end{pmatrix}\begin{pmatrix} -\frac{JS_1^z}{2} & -\frac{JS_1^-}{2} & t & 0 \\ -\frac{JS_1^+}{2} & \frac{JS_1^z}{2} & 0 & t \\ t & 0 & -\frac{JS_2^z}{2} & -\frac{JS_2^-}{2} \\ 0 & t & -\frac{JS_2^+}{2} & \frac{JS_2^z}{2}\end{pmatrix}\begin{pmatrix}\hat{c}_{1\uparrow} \\ \hat{c}_{1\downarrow} \\ \hat{c}_{2\uparrow} \\ \hat{c}_{2\downarrow}\end{pmatrix}.$$

For brevity in notations, we denote the column matrix of annihilation operators by $\hat{\gamma}$ with $\hat{\gamma}_1, \hat{\gamma}_2, \hat{\gamma}_3, \hat{\gamma}_4 = \hat{c}_{1\uparrow}, \hat{c}_{1\downarrow}, \hat{c}_{2\uparrow}, \hat{c}_{2\downarrow}$, respectively. We denote the square matrix by M. We then have $\hat{H} = \hat{\gamma}^\dagger M\hat{\gamma}$. Let $\hat{\varphi} = U^{-1}\hat{\gamma}$ denote new operators in terms of which the Hamiltonian is of the diagonal form. We have

$$\hat{H} = \hat{\varphi}^\dagger U^\dagger M U\hat{\varphi}.$$

We thus see that, if $U^\dagger M U$ is a diagonal matrix, then $\hat{H}$ is diagonalized. To ensure that $U^\dagger M U$ is a diagonal matrix, we can construct U from the eigenvectors of M. Since M is a Hermitian matrix, the matrix constructed this way is a unitary matrix, $U^\dagger = U^{-1}$. The eigenvalues of $\hat{H}$ are then given by the eigenvalues of M. The diagonalization of $\hat{H}$ then reduces to the diagonalization of M. However, for this to work, the algebraic properties of the original operators must be maintained, that is, $\hat{\varphi}$ must satisfy the following anticommutation relations

$$[\hat{\varphi}_m, \hat{\varphi}_n^\dagger] = \delta_{mn},\ \ [\hat{\varphi}_m, \hat{\varphi}_n] = [\hat{\varphi}_m^\dagger, \hat{\varphi}_n^\dagger] = 0,\ \ m, n = 1, 2, 3, 4.$$

We now check to see if the above anticommutation relations are satisfied for the matrix U in $\hat{\varphi} = U^{-1}\hat{\gamma}$ being a unitary matrix. Evaluating $[\hat{\varphi}_m, \hat{\varphi}_n^\dagger]$ using the anticommutation relations between the original operators,

$$\{\hat{\gamma}_\ell, \hat{\gamma}_{\ell'}^\dagger\} = \delta_{\ell\ell'},\ \ \{\hat{\gamma}_\ell, \hat{\gamma}_{\ell'}\} = \{\hat{\gamma}_\ell^\dagger, \hat{\gamma}_{\ell'}^\dagger\} = 0,$$

we have

$$\begin{aligned}\{\hat{\varphi}_m, \hat{\varphi}_n^\dagger\} &= \sum_{\ell\ell'} U_{\ell m}^* U_{\ell' n}\{\hat{\gamma}_\ell, \hat{\gamma}_{\ell'}^\dagger\} = \sum_{\ell\ell'} U_{\ell m}^* U_{\ell' n}\delta_{\ell\ell'} = \sum_{\ell} U_{\ell m}^* U_{\ell n} \\ &= \sum_\ell \left(U^\dagger\right)_{m\ell} U_{\ell n} = \left(U^\dagger U\right)_{mn} = \delta_{mn},\end{aligned}$$

$$\{\hat{\varphi}_m, \hat{\varphi}_n\} = \sum_{\ell\ell'} U_{\ell m}^* U_{\ell' n}^* \{\hat{\gamma}_\ell, \hat{\gamma}_{\ell'}\} = 0,$$

$$\{\hat{\varphi}_m^\dagger, \hat{\varphi}_n^\dagger\} = \sum_{\ell\ell'} U_{\ell m} U_{\ell' n} \{\hat{\gamma}_\ell^\dagger, \hat{\gamma}_{\ell'}^\dagger\} = 0.$$

Therefore, we can identify the eigenvalues of the matrix M as those of the Hamiltonian $\hat{H}$. The secular equation for M reads

$$\begin{vmatrix} -JS_1^z/2 - \lambda & -JS_1^-/2 & t & 0 \\ -JS_1^+/2 & JS_1^z/2 - \lambda & 0 & t \\ t & 0 & -JS_2^z/2 - \lambda & -JS_2^-/2 \\ 0 & t & -JS_2^+/2 & JS_2^z/2 - \lambda \end{vmatrix} = 0$$

with λ an eigenvalue of M. Evaluating the determinant, we have

$$\lambda^4 - \frac{1}{2}\left(J^2S^2 + 4t^2\right)\lambda^2 + \frac{1}{16}J^4S^4 - \frac{1}{2}t^2J^2\boldsymbol{S}_1 \cdot \boldsymbol{S}_2 + t^4 = 0.$$

Solving for λ from the above equation, we obtain the following four energy eigenvalues

$$\pm\frac{JS}{2}\left[1+\frac{4t^2}{J^2S^2}\pm\frac{2^{3/2}t}{JS}\left(1+\frac{\boldsymbol{S}_1\cdot\boldsymbol{S}_2}{S^2}\right)^{1/2}\right]^{1/2}.$$

Denoting the lowest energy eigenvalue by $E_0(S_1, S_2)$, we have

$$E_0(S_1,S_2)=-\frac{JS}{2}\left[1+\frac{4t^2}{J^2S^2}+\frac{2^{3/2}t}{JS}\left(1+\frac{\boldsymbol{S}_1\cdot\boldsymbol{S}_2}{S^2}\right)^{1/2}\right]^{1/2}.$$

(2) From $E_0(S_1, S_2)$, we see that the effective Hamiltonian for the core spins is

$$H_{\text{eff}}=-\frac{JS}{2}\left[1+\frac{4t^2}{J^2S^2}+\frac{2^{3/2}t}{JS}\left(1+\frac{\boldsymbol{S}_1\cdot\boldsymbol{S}_2}{S^2}\right)^{1/2}\right]^{1/2}.$$

(3) Because the value of H_{eff} for $\boldsymbol{S}_1\cdot\boldsymbol{S}_2>0$ is smaller than that for $\boldsymbol{S}_1\cdot\boldsymbol{S}_2<0$, H_{eff} is ferromagnetic. For $|\boldsymbol{S}_1\cdot\boldsymbol{S}_2|\ll S^2$ and $|t|/JS\ll 1$, we can expand the square root functions and reduce H_{eff} to the following form

$$\begin{aligned}H_{\text{eff}}&\approx-\frac{JS}{2}\left[1+\frac{4t^2}{J^2S^2}+\frac{2^{3/2}t}{JS}\left(1+\frac{\boldsymbol{S}_1\cdot\boldsymbol{S}_2}{2S^2}\right)\right]^{1/2}\\&\approx-\frac{JS}{2}\left[1+\frac{2^{1/2}t}{JS}\left(1+\frac{\boldsymbol{S}_1\cdot\boldsymbol{S}_2}{2S^2}\right)\right]\\&=-J_{\text{eff}}\,\boldsymbol{S}_1\cdot\boldsymbol{S}_2+c,\end{aligned}$$

where $J_{\text{eff}}=t/(2^{3/2}S^2)$ and $c=-JS/2-t/2^{1/2}$. The term of the order of $(t/JS)^2$ was neglected in arriving at the approximate expression on the second line. Besides the constant term, the above expression is just the Hamiltonian in the Heisenberg model. Thus, H_{eff} can be replaced by the Heisenberg interaction for $|\boldsymbol{S}_1\cdot\boldsymbol{S}_2|\ll S^2$ and $|t|/JS\ll 1$.

28-7 Specific heat of N noninteracting 1/2-spins in a magnetic field. N noninteracting 1/2-spins are placed in a magnetic field of strength B. Compute the heat capacity of the system $C_m = (\partial U/\partial T)_{\mathcal{V}}$ with U the internal energy of the system.

Assume that $\boldsymbol{B}=B\boldsymbol{e}_z$. The Hamiltonian of a single spin in the magnetic field is given by

$$\hat{h}=2\mu_{\text{B}}\hat{s}_zB/\hbar,$$

where g_s has been taken approximately to be -2. The two eigenvalues of $\hat{h}$ are $e_{m_s} = 2m_s\mu_B B$ for $m_s = -1/2$ and $1/2$. The partition function of a single spin is given by

$$z = \sum_{m_s} e^{-\beta e_{m_s}} = e^{\beta\mu_B B} + e^{-\beta\mu_B B} = 2\cosh(\beta\mu_B B).$$

The free energy of the system is then given by

$$F = -Nk_B T \ln z = -Nk_B T \ln\big[\, 2\cosh(\beta\mu_B B)\, \big].$$

The entropy of the system is

$$S = -\frac{\partial F}{\partial T} = Nk_B \ln\big[\, 2\cosh(\beta\mu_B B)\, \big] - \frac{N\mu_B B}{T}\tanh(\beta\mu_B B).$$

We can obtain the internal energy from F and S as follows

$$U = F + TS = -N\mu_B B \tanh(\beta\mu_B B).$$

The heat capacity is then given by

$$C_m = \frac{\partial U}{\partial T} = -N\mu_B B\frac{\partial \tanh(\beta\mu_B B)}{\partial T} = \frac{N(\mu_B B/k_B T)^2}{\cosh^2(\mu_B B/k_B T)}k_B.$$

28-8 Linear chain of three 1/2-spins. Consider a linear chain of three 1/2-spins with the periodic boundary conditions. With both the spin-spin interaction and a Zeeman term with a magnetic field in the z direction, the Hamiltonian of the system is given by

$$\hat{H} = J\hbar^{-2}\sum_{\langle j\ell\rangle} \hat{\boldsymbol{S}}_j \cdot \hat{\boldsymbol{S}}_\ell - (g_s\mu_B B/\hbar)\sum_j \hat{S}_j^z,$$

where $\langle j\ell\rangle$ indicates the summation over nearest neighbors only.

(1) Show that $\hat{\boldsymbol{S}}_j \cdot \hat{\boldsymbol{S}}_\ell = (\hat{S}_j^+\hat{S}_\ell^- + \hat{S}_j^-\hat{S}_\ell^+)/2 + \hat{S}_j^z\hat{S}_\ell^z$, where $\hat{S}_j^\pm = \hat{S}_j^x \pm i\hat{S}_j^y$.

(2) Write down the matrix representation of the Hamiltonian in the basis consisting of the following vectors

$$|1\rangle = |\uparrow\uparrow\uparrow\rangle,\ \ |2\rangle = |\uparrow\uparrow\downarrow\rangle,\ \ |3\rangle = |\uparrow\downarrow\uparrow\rangle,\ \ |4\rangle = |\downarrow\uparrow\uparrow\rangle,$$
$$|5\rangle = |\uparrow\downarrow\downarrow\rangle,\ \ |6\rangle = |\downarrow\uparrow\downarrow\rangle,\ \ |7\rangle = |\downarrow\downarrow\uparrow\rangle,\ \ |8\rangle = |\downarrow\downarrow\downarrow\rangle.$$

(3) Diagonalize the Hamiltonian.

(4) The system is prepared in state $|5\rangle$ at time $t = 0$. Find the probabilities of the system to be in states $|5\rangle$ and $|6\rangle$ as functions of t.

---

(1) From $\hat{S}_j^{\pm} = \hat{S}_j^x \pm \mathrm{i}\hat{S}_j^y$, we have

$$\hat{S}_j^x = (\hat{S}_j^+ + \hat{S}_j^-)/2, \;\; \hat{S}_j^y = -\mathrm{i}(\hat{S}_j^+ - \hat{S}_j^-)/2.$$

Inserting the above expressions into $\hat{\boldsymbol{S}}_j \cdot \hat{\boldsymbol{S}}_\ell = \hat{S}_j^x \hat{S}_\ell^x + \hat{S}_j^y \hat{S}_\ell^y + \hat{S}_j^z \hat{S}_\ell^z$, we obtain

$$\begin{aligned}\hat{\boldsymbol{S}}_j \cdot \hat{\boldsymbol{S}}_\ell &= \frac{1}{4}[(\hat{S}_j^+ + \hat{S}_j^-)(\hat{S}_\ell^+ + \hat{S}_\ell^-) - (\hat{S}_j^+ - \hat{S}_j^-)(\hat{S}_\ell^+ - \hat{S}_\ell^-)] + \hat{S}_j^z \hat{S}_\ell^z \\ &= \frac{1}{4}[\,(\hat{S}_j^+ \hat{S}_\ell^+ + \hat{S}_j^+ \hat{S}_\ell^- + \hat{S}_j^- \hat{S}_\ell^+ + \hat{S}_j^- \hat{S}_\ell^-) \\ &\quad - (\hat{S}_j^+ \hat{S}_\ell^+ - \hat{S}_j^+ \hat{S}_\ell^- - \hat{S}_j^- \hat{S}_\ell^+ + \hat{S}_j^- \hat{S}_\ell^-)\,] + \hat{S}_j^z \hat{S}_\ell^z \\ &= \frac{1}{2}(\hat{S}_j^+ \hat{S}_\ell^- + \hat{S}_j^- \hat{S}_\ell^+) + \hat{S}_j^z \hat{S}_\ell^z.\end{aligned}$$

(2) Writing out the Hamiltonian, we have

$$\begin{aligned}\hat{H} = J\hbar^{-2}\big[\,&(\hat{S}_1^+ \hat{S}_2^- + \hat{S}_1^- \hat{S}_2^+)/2 + \hat{S}_1^z \hat{S}_2^z + (\hat{S}_2^+ \hat{S}_3^- + \hat{S}_2^- \hat{S}_3^+)/2 \\ &+ \hat{S}_2^z \hat{S}_3^z + (\hat{S}_3^+ \hat{S}_1^- + \hat{S}_3^- \hat{S}_1^+)/2 + \hat{S}_3^z \hat{S}_1^z\,\big] \\ &- (g_s \mu_{\mathrm{B}} B/\hbar)(\hat{S}_1^z + \hat{S}_2^z + \hat{S}_3^z).\end{aligned}$$

Let $h = g_s \mu_{\mathrm{B}} B/J$. We can now write down the matrix representation of the Hamiltonian

$$H = J\begin{pmatrix} \frac{3}{4} - \frac{3h}{2} & 0 & 0 & 0 & 0 & 0 & 0 & 0 \\ 0 & -\frac{1}{4} - \frac{h}{2} & \frac{1}{2} & \frac{1}{2} & 0 & 0 & 0 & 0 \\ 0 & \frac{1}{2} & -\frac{1}{4} - \frac{h}{2} & \frac{1}{2} & 0 & 0 & 0 & 0 \\ 0 & \frac{1}{2} & \frac{1}{2} & -\frac{1}{4} - \frac{h}{2} & 0 & 0 & 0 & 0 \\ 0 & 0 & 0 & 0 & -\frac{1}{4} + \frac{h}{2} & \frac{1}{2} & \frac{1}{2} & 0 \\ 0 & 0 & 0 & 0 & \frac{1}{2} & -\frac{1}{4} + \frac{h}{2} & \frac{1}{2} & 0 \\ 0 & 0 & 0 & 0 & \frac{1}{2} & \frac{1}{2} & -\frac{1}{4} + \frac{h}{2} & 0 \\ 0 & 0 & 0 & 0 & 0 & 0 & 0 & \frac{3}{4} + \frac{3h}{2} \end{pmatrix}.$$

We see that H is a block matrix which consists of four blocks: 1×1, 3×3, 3×3, and 1×1 matrices along the main diagonal. The second and third blocks have similar structures and they differ only in their diagonal elements. From the first block, we can directly write down the first eigenvalue and the corresponding eigenvector

$$E_1 = \Big(\frac{3}{4} - \frac{3h}{2}\Big)J, \;\; |\varphi_1\rangle = |\uparrow\uparrow\uparrow\rangle\,.$$

The second block reads

$$J\begin{pmatrix} -\frac{1}{4}-\frac{h}{2} & \frac{1}{2} & \frac{1}{2} \\ \frac{1}{2} & -\frac{1}{4}-\frac{h}{2} & \frac{1}{2} \\ \frac{1}{2} & \frac{1}{2} & -\frac{1}{4}-\frac{h}{2} \end{pmatrix}$$

whose eigenvalues and corresponding eigenvectors are given by

$$E_2=\left(\frac{3}{4}-\frac{h}{2}\right)J, \qquad |\varphi_2\rangle=\frac{1}{\sqrt{3}}\big[\,|\uparrow\uparrow\downarrow\rangle+|\uparrow\downarrow\uparrow\rangle+|\downarrow\uparrow\uparrow\rangle\,\big],$$

$$E_3=E_4=-\left(\frac{3}{4}+\frac{h}{2}\right)J, \ |\varphi_3\rangle=\frac{1}{\sqrt{6}}\big[\,2\,|\uparrow\uparrow\downarrow\rangle-|\uparrow\downarrow\uparrow\rangle-|\downarrow\uparrow\uparrow\rangle\,\big],$$

$$|\varphi_4\rangle=\frac{1}{\sqrt{2}}\big[\,|\uparrow\downarrow\uparrow\rangle-|\downarrow\uparrow\uparrow\rangle\,\big].$$

Note that the eigenvalue $-(3/4+h/2)J$ is doubly degenerate. The corresponding eigenvectors have been chosen so that they are mutually orthogonal and they are both orthogonal to $|\varphi_2\rangle$. The third block reads

$$J\begin{pmatrix} -\frac{1}{4}+\frac{h}{2} & \frac{1}{2} & \frac{1}{2} \\ \frac{1}{2} & -\frac{1}{4}+\frac{h}{2} & \frac{1}{2} \\ \frac{1}{2} & \frac{1}{2} & -\frac{1}{4}+\frac{h}{2} \end{pmatrix}$$

whose eigenvalues and corresponding eigenvectors are given by

$$E_5=\left(\frac{3}{4}+\frac{h}{2}\right)J, \qquad |\varphi_5\rangle=\frac{1}{\sqrt{3}}\big[\,|\uparrow\downarrow\downarrow\rangle+|\downarrow\uparrow\downarrow\rangle+|\downarrow\downarrow\uparrow\rangle\,\big],$$

$$E_6=E_7=-\left(\frac{3}{4}-\frac{h}{2}\right)J, \ |\varphi_6\rangle=\frac{1}{\sqrt{6}}\big[\,2\,|\uparrow\downarrow\downarrow\rangle-|\downarrow\uparrow\downarrow\rangle-|\downarrow\downarrow\uparrow\rangle\,\big],$$

$$|\varphi_7\rangle=\frac{1}{\sqrt{2}}\big[\,|\downarrow\uparrow\downarrow\rangle-|\downarrow\downarrow\uparrow\rangle\,\big].$$

Note that the eigenvalue $-(3/4-h/2)J$ is doubly degenerate. Since the fourth block is a 1×1 matrix, we can directly write down the last eigenvalue and the corresponding eigenvector

$$E_8=\left(\frac{3}{4}+\frac{3h}{2}\right)J, \ |\varphi_8\rangle=|\downarrow\downarrow\downarrow\rangle\,.$$

(3) Expanding $|\psi(t=0)\rangle = |5\rangle = |\uparrow\downarrow\downarrow\rangle$ in terms of the eigenvectors of the Hamiltonian, we have

$$|\psi(t=0)\rangle = |5\rangle = |\uparrow\downarrow\downarrow\rangle = \frac{1}{\sqrt{3}}\,|\varphi_5\rangle + \sqrt{\frac{2}{3}}\,|\varphi_6\rangle\,.$$

At time t, we have

$$\begin{aligned}|\psi(t)\rangle &= \frac{1}{\sqrt{3}}\,\mathrm{e}^{-\mathrm{i}E_5 t/\hbar}\,|\varphi_5\rangle + \sqrt{\frac{2}{3}}\mathrm{e}^{-\mathrm{i}E_6 t/\hbar}\,|\varphi_6\rangle \\ &= \mathrm{e}^{-\mathrm{i}hJt/2\hbar}\left(\frac{1}{\sqrt{3}}\,\mathrm{e}^{-3\mathrm{i}Jt/4\hbar}\,|\varphi_5\rangle + \sqrt{\frac{2}{3}}\,\mathrm{e}^{3\mathrm{i}Jt/\hbar}\,|\varphi_6\rangle\right).\end{aligned}$$

The probability for the system to be in state $|5\rangle$ at time t is given by

$$\begin{aligned}P_{|5\rangle}(t) &= \left|\langle 5|\psi(t)\rangle\right|^2 = \left|\frac{1}{3}\,\mathrm{e}^{-3\mathrm{i}Jt/4\hbar} + \frac{2}{3}\,\mathrm{e}^{3\mathrm{i}Jt/\hbar}\right|^2 \\ &= \left|\cos(3Jt/4\hbar) + \frac{1}{3}\mathrm{i}\sin(3Jt/4\hbar)\right|^2 = 1 - \frac{8}{9}\sin^2(3Jt/4\hbar).\end{aligned}$$

The probability for the system to be in state $|6\rangle = |\downarrow\uparrow\downarrow\rangle$ at time t is given by

$$P_{|6\rangle}(t) = \left|\langle 6|\psi(t)\rangle\right|^2 = \left|\frac{1}{3}\mathrm{e}^{-3\mathrm{i}Jt/4\hbar} - \frac{1}{3}\,\mathrm{e}^{3\mathrm{i}Jt/\hbar}\right|^2 = \frac{4}{9}\sin^2(3Jt/4\hbar).$$

28-9 Landau's theory for the ferromagnetic phase transition. Consider Landau's theory for the ferromagnetic phase transition. We know that the magnetization M is small near the Curie temperature T_c. Therefore, we can express the free energy $F(M)$ as a power series in M

$$F(M) = a_1 M + a_2 M^2 + a_3 M^3 + a_4 M^4.$$

(1) Explain why we can keep only the terms that contain even powers of M. With only the even-power terms retained, the free energy is given by

$$F(M) = \frac{1}{2}a(T)M^2 + \frac{1}{4}b(T)M^4.$$

(2) Minimize $F(M)$ to obtain the equilibrium values for M.

(3) Assume that $b(T) = b$ is a constant. Draw $F(M)$ schematically for $a > 0$ and $a < 0$. Describe the fundamental difference between the two cases. Explain why the ferromagnetic phase transition occurs.

(4) Expand $a(T)$ to the first order in T about T_c. Find the equilibrium values for M. Express M as $M \propto (T_c - T)^\beta$. Find the value of β.
(5) Find the equilibrium value of $F(M)$ and compute the heat capacity C_v using $C_v = -T\partial^2 F/\partial T^2$.
(6) Add an external magnetic field H to the free energy. Compute the magnetic susceptibility χ and put it into the form

$$\chi \propto \begin{cases} (T-T_c)^\gamma, & T > T_c, \\ (T_c - T)^{\gamma'}, & T < T_c. \end{cases}$$

Find γ and γ'.

---

(1) The spontaneous magnetization $\boldsymbol{M}$ in the absence of a magnetic field can point in any direction. Thus, $F(M)$ and $F(-M)$ must both take the same minimum value in equilibrium, which implies that the coefficients in the terms with odd powers of M are identically zero. Hence, we can keep only the terms that contain even powers of M.
(2) Differentiating $F(M)$ with respect to M and setting the result to zero, we have

$$a(T)M + b(T)M^3 = 0$$

from which we obtain $M = \pm[-a(T)/b(T)]^{1/2}$. We have thrown away the $M = 0$ solution.
(3) The plots of $F(M)$ as functions of M are given in Fig. 28.3 for both $a > 0$ and $a < 0$. The fundamental difference between the $a > 0$ and $a < 0$ cases is that a minimum of $F(M)$ appears at $M \neq 0$ for $a < 0$ while the minimum appears only at $M = 0$ for $a > 0$. Note that the minimum of $F(M)$ at $M = 0$ indicates that no spontaneous magnetization can be present. Therefore, for $F(M)$ to be capable of describing a ferro-

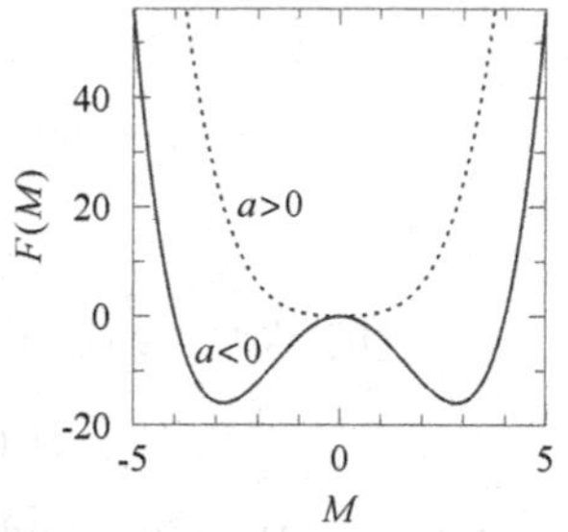

Fig. 28.3 Plots of $F(M)$ as functions of M for $a = 1$ (the dotted line) and $a = -8$ (the solid line) with $b = 1$.

magnetic phase transition, a must be negative. Since $M \neq 0$ in thermodynamic equilibrium for $a < 0$, a new phase appears whenever $M \neq 0$ and the ferromagnetic phase transition has occurred.

(4) We now expand $a(T)$ in T about T_c. Up to the first order in T, we have $a(T) = a(T_c) + a'(T_c)(T - T_c)$. The first order term is larger than zero for $T > T_c$ and is smaller than zero for $T < T_c$. For T_c to be the temperature below which a spontaneous magnetization appears, we set $a(T_c) = 0$. We then have

$$a(T) = a'(T_c)(T - T_c)$$

with $a'(T_c) > 0$. Inserting the above expression into the equilibrium value of M, $M = \pm[-a(T)/b(T)]^{1/2}$ derived previously, we obtain

$$M = \pm\big[\,a'(T_c)/b\,\big]^{1/2}(T_c - T)^{1/2}.$$

Thus,

$$M \propto (T_c - T)^{\beta}$$

with $\beta = 1/2$.

(5) At $M = \pm\big[\,a'(T_c)/b\,\big]^{1/2}(T_c - T)^{1/2}$, the value of $F(M)$, now written as $F(T)$, is given by

$$\begin{aligned} F(T) &= \frac{a'(T_c)}{2}(T - T_c)\left[\frac{a'(T_c)}{b}(T_c - T)\right] \\ &\quad + \frac{1}{4}b\left[\,\frac{a'(T_c)}{b}(T_c - T)\,\right]^2 \\ &= -\frac{[a'(T_c)T_c]^2}{4b}\left(1 - \frac{T}{T_c}\right)^2, \quad T < T_c. \end{aligned}$$

The specific heat for $T < T_c$ is given by

$$C_v = -T\frac{\partial^2 F(T)}{\partial T^2} = \frac{[a'(T_c)]^2}{2b} \quad T < T_c.$$

(6) Adding the term $-MB$ to $F(M)$, we have

$$F(T, B) = \frac{1}{2}a(T)M^2 + \frac{1}{4}bM^4 - MB.$$

We must find the equilibrium value of M again. Differentiating $F(M, B)$ with respect to M for a fixed value of B and setting the result to zero, we have

$$a(T)M + bM^3 - B = 0.$$

From the above equation, we can obtain M as a function of B, $M = M(B)$. The derivative of M with respect to B can be obtained from the above equation. We have

$$\frac{\partial M}{\partial B} = \frac{1}{a(T) + 3bM^2}.$$

For $B = 0$, we have $a(T)M + bM^3 = 0$ from which we obtain the previous result $M = [-a(T)/b]^{1/2}$ for positive M. The magnetic susceptibility is then given by

$$\chi = \left.\frac{\partial M}{\partial H}\right|_{H=0} = \mu_0 \left.\frac{\partial M}{\partial B}\right|_{B=0} = \left.\frac{\mu_0}{a(T) + 3bM^2}\right|_{B=0}$$
$$= \begin{cases} \dfrac{\mu_0}{a'(T_c)(T - T_c)}, & T > T_c \\ \dfrac{\mu_0}{2a'(T_c)(T_c - T)}, & T < T_c \end{cases} \propto \begin{cases} (T - T_c)^{\gamma}, & T > T_c, \\ (T_c - T)^{\gamma'}, & T < T_c \end{cases}$$

with $\gamma = \gamma' = -1$.

28-10 Spin operator in terms of electron annihilation and creation operators. In terms of electron annihilation and creation operators $\hat{c}_\sigma$ and $\hat{c}_\sigma^\dagger$, the components of the spin operator are given by

$$\hat{S}^+ = \hbar \hat{c}_\uparrow^\dagger \hat{c}_\downarrow, \quad \hat{S}^- = \hbar \hat{c}_\downarrow^\dagger \hat{c}_\uparrow, \quad \hat{S}^z = (\hbar/2)(\hat{c}_\uparrow^\dagger \hat{c}_\uparrow - \hat{c}_\downarrow^\dagger \hat{c}_\downarrow).$$

Show directly using the anti-communication relations between electron annihilation and creation operators that the the commutation relations between the components of the spin operator, $[\hat{S}^+, \hat{S}^-] = 2\hbar \hat{S}_z$ and $[\hat{S}^z, \hat{S}^\pm] = \pm\hbar \hat{S}_\pm$, are satisfied.

---

(1) For $[\hat{S}^+, \hat{S}^-]$, we have

$$\hbar^{-2}[\hat{S}^+, \hat{S}^-] = [\hat{c}_\uparrow^\dagger \hat{c}_\downarrow, \hat{c}_\downarrow^\dagger \hat{c}_\uparrow] = -\hat{c}_\downarrow^\dagger \{[\hat{c}_\uparrow^\dagger, \hat{c}_\uparrow\}\hat{c}_\downarrow + \hat{c}_\uparrow^\dagger \{\hat{c}_\downarrow, \hat{c}_\downarrow^\dagger\}\hat{c}_\uparrow$$
$$= -\hat{c}_\downarrow^\dagger \hat{c}_\downarrow + \hat{c}_\uparrow^\dagger \hat{c}_\uparrow = (2/\hbar)\hat{S}^z.$$

Thus, $[\hat{S}^+, \hat{S}^-] = 2\hbar \hat{S}_z$. For $[\hat{S}^z, \hat{S}^+]$, we have

$$2\hbar^{-2}[\hat{S}^z, \hat{S}^+] = [\hat{c}_\uparrow^\dagger \hat{c}_\uparrow - \hat{c}_\downarrow^\dagger \hat{c}_\downarrow, \hat{c}_\uparrow^\dagger \hat{c}_\downarrow] = \hat{c}_\uparrow^\dagger \{\hat{c}_\uparrow, \hat{c}_\uparrow^\dagger\}\hat{c}_\downarrow + \hat{c}_\uparrow^\dagger \{\hat{c}_\downarrow^\dagger, \hat{c}_\downarrow\}\hat{c}_\downarrow$$
$$= \hat{c}_\uparrow^\dagger \hat{c}_\downarrow + \hat{c}_\uparrow^\dagger \hat{c}_\downarrow = (2/\hbar)\hat{S}^+.$$

Thus, $[\hat{S}^z, \hat{S}^+] = \hbar \hat{S}^+$.

(2) For $[\hat{S}^z, \hat{S}^-]$, we have

$$2\hbar^{-2}[\hat{S}^z, \hat{S}^-] = [\hat{c}_\uparrow^\dagger \hat{c}_\uparrow - \hat{c}_\downarrow^\dagger \hat{c}_\downarrow, \hat{c}_\downarrow^\dagger \hat{c}_\uparrow] = -\hat{c}_\downarrow^\dagger \{\hat{c}_\uparrow^\dagger, \hat{c}_\uparrow\}\hat{c}_\uparrow - \hat{c}_\downarrow^\dagger \{\hat{c}_\downarrow, \hat{c}_\downarrow^\dagger\}\hat{c}_\uparrow$$
$$= -\hat{c}_\downarrow^\dagger \hat{c}_\uparrow - \hat{c}_\downarrow^\dagger \hat{c}_\uparrow = -(2/\hbar)\hat{S}^-.$$

Thus, $[\hat{S}^z, \hat{S}^+] = -\hbar \hat{S}^-$.

28-11 Dot product of spin operators in terms of electron operators. The spin operator $\hat{\boldsymbol{S}}_i$ on site i due to electrons is expressed as $\hat{\boldsymbol{S}}_i = (\hbar/2)\sum_{\alpha\beta}\hat{c}^\dagger_{i\alpha}\boldsymbol{\sigma}_{\alpha\beta}\hat{c}_{i\beta}$, where $\boldsymbol{\sigma} = \sigma_x\boldsymbol{e}_x + \sigma_y\boldsymbol{e}_y + \sigma_z\boldsymbol{e}_z$ with σ_x, σ_y, and σ_z the Pauli matrices. Show that, for $i \neq j$,

$$\hat{\boldsymbol{S}}_i \cdot \hat{\boldsymbol{S}}_j = -\frac{\hbar^2}{2}\sum_{\alpha\beta}\hat{c}^\dagger_{i\alpha}\hat{c}^\dagger_{j\beta}\hat{c}_{i\beta}\hat{c}_{j\alpha} - \frac{\hbar^2}{4}\hat{n}_i\hat{n}_j.$$

From $\hat{\boldsymbol{S}}_i = (\hbar/2)\sum_{\alpha\beta}\hat{c}^\dagger_{i\alpha}\boldsymbol{\sigma}_{\alpha\beta}\hat{c}_{i\beta}$, we have

$$\hat{\boldsymbol{S}}_i \cdot \hat{\boldsymbol{S}}_j = \frac{\hbar^2}{4}\sum_{\gamma}\sum_{\alpha\beta\alpha'\beta'}\hat{c}^\dagger_{i\alpha}\hat{c}_{i\beta}\hat{c}^\dagger_{j\alpha'}\hat{c}_{j\beta'}\sigma^\gamma_{\alpha\beta}\sigma^\gamma_{\alpha'\beta'}.$$

Making use of the following properties of the Pauli matrices

$$\begin{aligned}
\sigma^x_{\alpha\beta}\sigma^x_{\alpha'\beta'} &= \big(\delta_{\alpha\beta'}\delta_{\beta\alpha'} + \delta_{\alpha\alpha'}\delta_{\beta\beta'}\big)\big(1-\delta_{\alpha\beta}\big),\\
\sigma^y_{\alpha\beta}\sigma^y_{\alpha'\beta'} &= \big(\delta_{\alpha\beta'}\delta_{\beta\alpha'} - \delta_{\alpha\alpha'}\delta_{\beta\beta'}\big)\big(1-\delta_{\alpha\beta}\big),\\
\sigma^z_{\alpha\beta}\sigma^z_{\alpha'\beta'} &= \delta_{\alpha\beta}\delta_{\alpha'\beta'}\big(2\delta_{\alpha\alpha'} - 1\big),
\end{aligned}$$

we have

$$\begin{aligned}
\hat{\boldsymbol{S}}_i \cdot \hat{\boldsymbol{S}}_j &= \frac{\hbar^2}{4}\sum_{\alpha\beta\alpha'\beta'}\hat{c}^\dagger_{i\alpha}\hat{c}_{i\beta}\hat{c}^\dagger_{j\alpha'}\hat{c}_{j\beta'}\big[2\delta_{\alpha\beta'}\delta_{\beta\alpha'}\big(1-\delta_{\alpha\beta}\big)\\
&\qquad\qquad + \delta_{\alpha\beta}\delta_{\alpha'\beta'}\big(2\delta_{\alpha\alpha'} - 1\big)\big]\\
&= \frac{\hbar^2}{4}\sum_{\alpha\beta\alpha'\beta'}\hat{c}^\dagger_{i\alpha}\hat{c}_{i\beta}\hat{c}^\dagger_{j\alpha'}\hat{c}_{j\beta'}\big(2\delta_{\alpha\beta'}\delta_{\beta\alpha'} - \delta_{\alpha\beta}\delta_{\alpha'\beta'}\big).
\end{aligned}$$

Performing two of the four summations over spin indices using the δ-symbols, we have

$$\begin{aligned}
\hat{\boldsymbol{S}}_i \cdot \hat{\boldsymbol{S}}_j &= \frac{\hbar^2}{2}\sum_{\alpha\beta}\hat{c}^\dagger_{i\alpha}\hat{c}_{i\beta}\hat{c}^\dagger_{j\beta}\hat{c}_{j\alpha} - \frac{\hbar^2}{4}\sum_{\alpha\alpha'}\hat{c}^\dagger_{i\alpha}\hat{c}_{i\alpha}\hat{c}^\dagger_{j\alpha'}\hat{c}_{j\alpha'}\\
&= -\frac{\hbar^2}{2}\sum_{\alpha\beta}\hat{c}^\dagger_{i\alpha}\hat{c}^\dagger_{j\beta}\hat{c}_{i\beta}\hat{c}_{j\alpha} - \frac{\hbar^2}{4}\hat{n}_i\hat{n}_j,
\end{aligned}$$

where we have made use of $\hat{c}_{i\beta}\hat{c}^\dagger_{j\beta} = -\hat{c}^\dagger_{j\beta}\hat{c}_{i\beta}$ for $i \neq j$ and $\hat{n}_i = \sum_\alpha \hat{c}^\dagger_{i\alpha}\hat{c}_{i\alpha}$.

28-12 One-dimensional spin-S Heisenberg quantum ferromagnet. The Hamiltonian of a one-dimensional spin-S Heisenberg quantum ferromagnet is given by

$$\hat{H} = -J\hbar^{-2}\sum_{n=1}^{N}\hat{\boldsymbol{S}}_n \cdot \hat{\boldsymbol{S}}_{n+1}$$

with $J > 0$. The periodic boundary conditions, $\hat{\boldsymbol{S}}_{N+1} = \hat{\boldsymbol{S}}_1$ and $\hat{\boldsymbol{S}}_0 = \hat{\boldsymbol{S}}_N$, are imposed on spins.

(1) Apply the Holstein-Primakoff transformation to the spin operators and expand the Hamiltonian to the quadratic order in boson operators for $S \gg 1$.

(2) Diagonalize the Hamiltonian and show that the spin-wave spectrum is given by

$$\hbar\omega_k = 2JS\big[1 - \cos(ka)\big],$$

where $k = 2\pi n/Na$ with a the lattice constant and $n = 0$, $\pm 1, \cdots, \pm(N-1)/2, N/2$.

---

(1) Making use of the Holstein-Primakoff transformation, we have to the quadratic order in boson operators

$$\begin{aligned}
\hat{H} &= -\frac{J}{2\hbar^2}\sum_{n=1}^{N}\big[\big(\hat{S}_n^+\hat{S}_{n+1}^- + \hat{S}_n^-\hat{S}_{n+1}^+\big) + 2\hat{S}_n^z\hat{S}_{n+1}^z\big] \\
&\approx -\frac{J}{2}\sum_{n=1}^{N}\big[2S\big(\hat{a}_n\hat{a}_{n+1}^\dagger + \hat{a}_n^\dagger\hat{a}_{n+1}\big) \\
&\qquad + 2\big(S - \hat{a}_n^\dagger\hat{a}_n\big)\big(S - \hat{a}_{n+1}^\dagger\hat{a}_{n+1}\big)\big] \\
&\approx -NJS^2 + JS\sum_{n=1}^{N}\big(\hat{a}_{n+1}^\dagger\hat{a}_{n+1} + \hat{a}_n^\dagger\hat{a}_n - \hat{a}_n\hat{a}_{n+1}^\dagger - \hat{a}_n^\dagger\hat{a}_{n+1}\big) \\
&= -NJS^2 + JS\sum_{n=1}^{N}\big(2\hat{a}_n^\dagger\hat{a}_n - \hat{a}_{n+1}^\dagger\hat{a}_n - \hat{a}_n^\dagger\hat{a}_{n+1}\big).
\end{aligned}$$

Making a Fourier transformation to $\hat{a}_n$, we have

$$\hat{a}_n = \frac{1}{\sqrt{N}}\sum_k \mathrm{e}^{\mathrm{i}kR_n}\hat{a}_k,$$

where $R_n = na$ is the position of the nth site. From the periodic boundary condition $\hat{\boldsymbol{S}}_{N+1} = \hat{\boldsymbol{S}}_1$, we have

$$\mathrm{e}^{\mathrm{i}kNa} = 1.$$

Therefore, there are only N allowed values for k. We choose them to be $k = 0, \pm 2\pi/Na, \cdots, \pm(N-1)\pi/Na, \pi/a$.

(2) In terms of $\hat{a}_k$'s, the Hamiltonian is given by

$$\hat{H} \approx -NJS^2 + \frac{JS}{N}\sum_{kk'}\sum_{n=1}^{N}(2 - \mathrm{e}^{-ika} - \mathrm{e}^{ik'a})\mathrm{e}^{\mathrm{i}(k'-k)R_n}\hat{a}_k^\dagger \hat{a}_{k'}$$

$$= -NJS^2 + JS\sum_{kk'}(2 - \mathrm{e}^{-ika} - \mathrm{e}^{ik'a})\delta_{k'k}\hat{a}_k^\dagger \hat{a}_{k'}$$

$$= -NJS^2 + JS\sum_{k}(2 - \mathrm{e}^{-ika} - \mathrm{e}^{ika})\hat{a}_k^\dagger \hat{a}_k$$

$$= -NJS^2 + 2JS\sum_{k}[(1 - \cos(ka)]\hat{a}_k^\dagger \hat{a}_k.$$

Thus, the spin-wave spectrum is given by

$$\hbar\omega_k = 2JS[(1 - \cos(ka)].$$

28-13 Pauli susceptibility of a non-interacting Fermi gas. Consider a non-interacting Fermi gas in a uniform external magnetic field B at $T = 0$. Keeping only the Zeeman part of the interaction with the magnetic field and neglecting orbital effects of the magnetic field, we write the interaction part of the Hamiltonian as

$$\hat{H}_B = \hbar\omega_B \sum_{\boldsymbol{k}}(\hat{c}_{\boldsymbol{k}\uparrow}^\dagger \hat{c}_{\boldsymbol{k}\uparrow} - \hat{c}_{\boldsymbol{k}\downarrow}^\dagger \hat{c}_{\boldsymbol{k}\downarrow}),$$

where $\hbar\omega_B = \mu_B B$.

(1) Describe the exact ground state of non-interacting fermions of spin 1/2 in a uniform magnetic field.

(2) Derive the exact Green's function $G_{\alpha\beta}(\boldsymbol{k}, \omega)$ of such a system given that the chemical potential is μ.

(3) Find the magnetization M in the ground state using the derived Green's function $G_{\alpha\beta}(\boldsymbol{k}, \omega)$, with the result expressed as a function of the density of the Fermi gas.

(4) Find the Pauli susceptibility χ_{Pauli} of the non-interacting Fermi gas.

---

(1) In the presence of the magnetic field, the total Hamiltonian is given by

$$\hat{H} = \hat{H}_0 + \hat{H}_B = \sum_{\boldsymbol{k}\sigma}\varepsilon_{\boldsymbol{k}}\hat{c}_{\boldsymbol{k}\sigma}^\dagger \hat{c}_{\boldsymbol{k}\sigma} + \hbar\omega_B\sum_{\boldsymbol{k}}(\hat{c}_{\boldsymbol{k}\uparrow}^\dagger \hat{c}_{\boldsymbol{k}\uparrow} - \hat{c}_{\boldsymbol{k}\downarrow}^\dagger \hat{c}_{\boldsymbol{k}\downarrow})$$

$$= \sum_{\boldsymbol{k}\sigma}\xi_{\boldsymbol{k}\sigma}\hat{c}_{\boldsymbol{k}\sigma}^\dagger \hat{c}_{\boldsymbol{k}\sigma},$$

where $\xi_{\boldsymbol{k}\sigma} = \varepsilon_{\boldsymbol{k}} + \sigma\hbar\omega_B$ with $\sigma = +1$ corresponding to $\uparrow$ and $\sigma = -1$ to $\downarrow$. In the presence of the magnetic field, let the Fermi wave vector for up-spin fermions be $k_{\mathrm{F}\uparrow}$ and that for down-spin fermions be $k_{\mathrm{F}\downarrow}$. From the fact that the highest energies for up- and down-spin fermions are equal, we have $\hbar^2 k_{\mathrm{F}\uparrow}^2/2m + \hbar\omega_B = \hbar^2 k_{\mathrm{F}\downarrow}^2/2m - \hbar\omega_B$, which relates $k_{\mathrm{F}\uparrow}$ and $k_{\mathrm{F}\downarrow}$. The ground state of the non-interacting Fermi gas in the presence of the magnetic field can be then described as follows: In momentum space, up-spin fermions occupy all the up-spin single-fermion states in the Fermi sphere of radius $k_{\mathrm{F}\uparrow}$ and down-spin fermions occupy all the down-spin single-fermion states in the Fermi sphere of radius $k_{\mathrm{F}\downarrow}$. In terms of the number densities of up- and down-spin fermions, $n_\uparrow$ and $n_\downarrow$, we have $k_{\mathrm{F}\uparrow} = (6\pi^2 n_\uparrow)^{1/3}$ and $k_{\mathrm{F}\downarrow} = (6\pi^2 n_\downarrow)^{1/3}$.

(2) Since the total Hamiltonian is free-particle like, we can write down the exact Green's function $G_{\alpha\beta}(\boldsymbol{k}, \omega)$ immediately as

$$G_{\alpha\beta}(\boldsymbol{k}, \omega) = \frac{\delta_{\alpha\beta}}{\hbar\omega - \xi_{\boldsymbol{k}\alpha} + \mu + \mathrm{i}\delta\,\mathrm{sgn}(\xi_{\boldsymbol{k}\alpha} - \mu)}.$$

(3) The magnetization M is given by

$$M = -\frac{\mu_\mathrm{B}}{\mathscr{V}} \sum_{|\boldsymbol{k}|<k_{\mathrm{F}\uparrow}} \langle \hat{c}_{\boldsymbol{k}\uparrow}^\dagger \hat{c}_{\boldsymbol{k}\uparrow} \rangle_0 + \frac{\mu_\mathrm{B}}{\mathscr{V}} \sum_{|\boldsymbol{k}|<k_{\mathrm{F}\downarrow}} \langle \hat{c}_{\boldsymbol{k}\downarrow}^\dagger \hat{c}_{\boldsymbol{k}\downarrow} \rangle_0 .$$

In terms of $G_{\alpha\beta}(\boldsymbol{k}, t)$, we have

$$M = \frac{\mathrm{i}\mu_\mathrm{B}}{\mathscr{V}} \sum_{|\boldsymbol{k}|<k_{\mathrm{F}\uparrow}} \lim_{t\to 0^-} G_{\uparrow\uparrow}(\boldsymbol{k}, t) - \frac{\mathrm{i}\mu_\mathrm{B}}{\mathscr{V}} \sum_{|\boldsymbol{k}|<k_{\mathrm{F}\downarrow}} \lim_{t\to 0^-} G_{\downarrow\downarrow}(\boldsymbol{k}, t).$$

From $G_{\alpha\beta}(\boldsymbol{k}, \omega)$, we have

$$\begin{aligned} G_{\alpha\beta}(\boldsymbol{k}, t) &= \int_{-\infty}^{\infty} \frac{\mathrm{d}(\hbar\omega)}{2\pi}\, \mathrm{e}^{-\mathrm{i}\omega t} G_{\alpha\beta}(\boldsymbol{k}, \omega) \\ &= \delta_{\alpha\beta} \int_{-\infty}^{\infty} \frac{\mathrm{d}(\hbar\omega)}{2\pi}\, \frac{\mathrm{e}^{-\mathrm{i}\omega t}}{\hbar\omega - \xi_{\boldsymbol{k}\alpha} + \mu + \mathrm{i}\delta\,\mathrm{sgn}(\xi_{\boldsymbol{k}\alpha} - \mu)}. \end{aligned}$$

For $t < 0$, we must close the contour in the upper-half complex plane and obtain upon invoking the residue theorem

$$G_{\alpha\beta}(\boldsymbol{k}, t) = \mathrm{i}\delta_{\alpha\beta}\mathrm{e}^{-\mathrm{i}(\xi_{\boldsymbol{k}\alpha} - \mu)t}\theta(\xi_{\boldsymbol{k}\alpha} - \mu).$$

Note that $\mu = E_\mathrm{F}$ at $T = 0$. The magnetization is then given by

$$M = -\frac{\mu_\mathrm{B}}{\mathscr{V}} \sum_{|\boldsymbol{k}|<k_{\mathrm{F}\uparrow}} 1 + \frac{\mu_\mathrm{B}}{\mathscr{V}} \sum_{|\boldsymbol{k}|<k_{\mathrm{F}\downarrow}} 1 = -\mu_\mathrm{B}(n_\uparrow - n_\downarrow),$$

as expected. We now relate $n_\uparrow$ and $n_\downarrow$ to the electron density n. We can obtain two equations for $n_\uparrow$ and $n_\downarrow$ from the facts that the highest energies for up- and down-spin fermions are equal and that the sum of $n_\uparrow$ and $n_\downarrow$ is equal to n

$$\frac{\hbar^2(6\pi^2 n_\uparrow)^{2/3}}{2m} + \mu_B B = \frac{\hbar^2(6\pi^2 n_\downarrow)^{2/3}}{2m} - \mu_B B,$$
$$n_\uparrow + n_\downarrow = n.$$

We can obtain $n_\uparrow$ and $n_\downarrow$ by solving the above equations. However, for very small B ($\mu_B B/E_F \ll 1$), the values of the two sides of the first equation are very close to $E_F = \hbar^2(3\pi^2 n)^{2/3}/2m$, the Fermi energy in the absence of the external magnetic field. We thus have approximately

$$\frac{\hbar^2(6\pi^2 n_\uparrow)^{2/3}}{2m} + \mu_B B = \frac{\hbar^2(6\pi^2 n_\downarrow)^{2/3}}{2m} - \mu_B B \approx \frac{\hbar^2(3\pi^2 n)^{2/3}}{2m}.$$

Hence,

$$n_\uparrow \approx \frac{n}{2}\left[1 - \frac{3m\mu_B B}{(3\pi^2 n)^{2/3}\hbar^2}\right] = \frac{n}{2}\left(1 - \frac{3\mu_B B}{2E_F}\right),$$
$$n_\downarrow \approx \frac{n}{2}\left[1 + \frac{3m\mu_B B}{(3\pi^2 n)^{2/3}\hbar^2}\right] = \frac{n}{2}\left(1 + \frac{3\mu_B B}{2E_F}\right).$$

The magnetization is then given by

$$M \approx \frac{3\mu_B^2 B n}{2E_F}.$$

(4) The Pauli susceptibility follows from $\partial M/\partial H\big|_{H=0}$

$$\chi_{\text{Pauli}} = \left.\frac{\partial M}{\partial H}\right|_{H=0} = \frac{3\mu_0\mu_B^2 n}{2E_F} = \mu_0\mu_B^2\, g(E_F).$$

Note that $g(E_F)$ in the above expression is for both spin directions.

28-14 Instability of electron gas to ferromagnetism. Consider an electron gas with $n_\uparrow \neq n_\downarrow$. In this case, the contributions to the kinetic energy per electron are different for two spin directions and they are $3\hbar^2 k_{F\uparrow}^2/10m$ and $3\hbar^2 k_{F\downarrow}^2/10m$, respectively, for up- and down-spin electrons. The exchange energies, $-3e^2 k_{F\uparrow}/4\pi(4\pi\epsilon_0)$ and $-3e^2 k_{F\downarrow}/4\pi(4\pi\epsilon_0)$, for two spin directions are also taken into account.

(1) Find the overall energy E per electron in terms of $n_\uparrow$ and $n_\downarrow$.

(2) To study the stability of the unpolarized electron gas, we introduce an order parameter s in terms of which $n_\uparrow$ and $n_\downarrow$ are given by $n_\uparrow = (n+s)/2$ and $n_\downarrow = (n-s)/2$. Evaluate $\mathrm{d}^2E/\mathrm{d}s^2\big|_{s=0}$. Find out whether there exists a critical density n_c above which the unpolarized electron gas becomes unstable to the formation of a spin-polarized (*i.e.*, ferromagnetic) state.

(3) Find the critical value of the dimensionless parameter $r_s = (3/4\pi n)^{1/3}/a_0$ at which the instability occurs.

---

(1) Summing up contributions to the kinetic energy from both spin directions and taking into account the exchange energy, we have for the energy per electron

$$\begin{aligned} E &= \frac{3\hbar^2 k_{\mathrm{F}\uparrow}^2}{10m} + \frac{3\hbar^2 k_{\mathrm{F}\downarrow}^2}{10m} - \frac{3e^2 k_{\mathrm{F}\uparrow}}{4\pi(4\pi\epsilon_0)} - \frac{3e^2 k_{\mathrm{F}\downarrow}}{4\pi(4\pi\epsilon_0)} \\ &= \frac{3\hbar^2(6\pi^2 n_\uparrow)^{2/3}}{10m} + \frac{3\hbar^2(6\pi^2 n_\downarrow)^{2/3}}{10m} - \frac{3e^2(6\pi^2 n_\uparrow)^{1/3}}{4\pi(4\pi\epsilon_0)} \\ &\quad - \frac{3e^2(6\pi^2 n_\downarrow)^{1/3}}{4\pi(4\pi\epsilon_0)} \\ &= \frac{3\hbar^2(6\pi^2)^{2/3}}{10m}\left(n_\uparrow^{2/3} + n_\downarrow^{2/3}\right) - \frac{3e^2(6\pi^2)^{1/3}}{4\pi(4\pi\epsilon_0)}\left(n_\downarrow^{1/3} + n_\downarrow^{1/3}\right). \end{aligned}$$

(2) Inserting $n_\uparrow = (n+s)/2$ and $n_\downarrow = (n-s)/2$ into the expression of E yields

$$\begin{aligned} E &= \frac{3\hbar^2(3\pi^2)^{2/3}}{10m}\left[(n+s)^{2/3} + (n-s)^{2/3}\right] \\ &\quad - \frac{3e^2(3\pi^2)^{1/3}}{4\pi(4\pi\epsilon_0)}\left[(n+s)^{1/3} + (n-s)^{1/3}\right]. \end{aligned}$$

Differentiating E twice with respect to s, we obtain

$$\begin{aligned} \left.\frac{\mathrm{d}^2E}{\mathrm{d}s^2}\right|_{s=0} &= \frac{3\hbar^2(3\pi^2)^{2/3}}{10m}\left(-\frac{4}{9n^{4/3}}\right) - \frac{3e^2(3\pi^2)^{1/3}}{4\pi(4\pi\epsilon_0)}\left(-\frac{4}{9n^{5/3}}\right) \\ &= -\frac{2\hbar^2(3\pi^2)^{2/3}}{15mn^{4/3}} + \frac{e^2(3\pi^2)^{1/3}}{3\pi(4\pi\epsilon_0)n^{5/3}} \\ &= -\frac{2\hbar^2(3\pi^2)^{2/3}}{15mn^{4/3}}\left[1 - \left(\frac{n_c}{n}\right)^{1/3}\right], \end{aligned}$$

where n_c is given by $n_c = 125/24\pi^5 a_0^3$. At $n = n_c$, we have $\mathrm{d}^2E/\mathrm{d}s^2\big|_{s=0} = 0$. Since $\mathrm{d}^2E/\mathrm{d}s^2\big|_{s=0} < 0$ for $n > n_c$, $s = 0$ is a local maximum point of $E(s)$ as can be seen from Fig. 28.4(a). For $n < n_c$, $s = 0$ is the minimum point of $E(s)$ as can be seen from Fig. 28.4(b). Therefore, the ground-state energy is lower for $s \neq 0$ than for $s = 0$ if $n > n_c$. Hence, n_c is the critical density above which the unpolarized electron gas becomes unstable to the formation of a spin-polarized state.

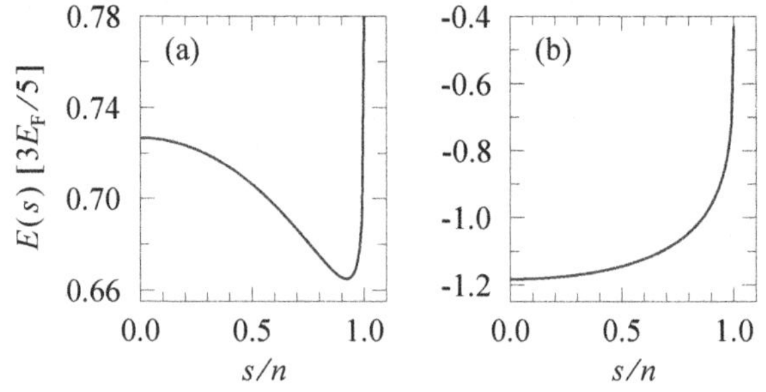

Fig. 28.4 Plots of $E(s)$ against s/n. (a) $k_\mathrm{F}a_0 = 1.25$ ($n \approx 1.162n_c$). (b) $k_\mathrm{F}a_0 = 0.5$ ($n \approx 0.856n_c$). Note that $(n_c/n)^{1/3} = 5/2\pi k_\mathrm{F}a_0$.

(3) For r_s, we have at $n = n_c$

$$r_s^c = \left(\frac{3}{4\pi n_c a_0^3}\right)^{1/3} = \frac{1}{5}(18\pi^4)^{1/3} \approx 2.412.$$

Note that, because the direct Coulomb and correlation interactions are not taken into account, the above estimation of n_c is not accurate.

28-15 Schwinger boson representation. In the Schwinger boson representation, the spin operator is expressed in terms of two bosonic operators $\hat{a}$ and $\hat{b}$

$$\hat{S}^+ = \hbar\hat{a}^\dagger\hat{b},\ \ \hat{S}^- = \hbar\hat{b}^\dagger\hat{a},\ \ \hat{S}^z = \frac{1}{2}\hbar(\hat{a}^\dagger\hat{a} - \hat{b}^\dagger\hat{b}).$$

(1) Show that the above expression of the spin operator satisfies the commutation relations $[\hat{S}^+, \hat{S}^-] = 2\hbar\hat{S}_z$ and $[\hat{S}^z, \hat{S}^\pm] = \pm\hbar\hat{S}_\pm$.

(2) Show that

$$|Sm\rangle = \frac{1}{[(S+m)!(S-m)!]^{1/2}}\left(\hat{a}^\dagger\right)^{S+m}\left(\hat{b}^\dagger\right)^{S-m}|0\rangle$$

is a common eigenstate of $\hat{\boldsymbol{S}}^2$ and $\hat{S}^z$. Note that the physical state space is given by $\{|n_a, n_b\rangle\}$ with $n_a + n_b = 2S$.

---

(1) For $[\hat{S}^+, \hat{S}^-]$, we have

$$\begin{aligned}[\hat{S}^+, \hat{S}^-] &= \hbar^2[\hat{a}^\dagger\hat{b}, \hat{b}^\dagger\hat{a}] = \hbar^2\left(\hat{a}^\dagger[\hat{b}, \hat{b}^\dagger]\hat{a} + \hat{b}^\dagger[\hat{a}^\dagger, \hat{a}]\hat{b}\right) \\ &= \hbar^2\left(\hat{a}^\dagger\hat{a} - \hat{b}^\dagger\hat{b}\right) = 2\hbar\hat{S}^z.\end{aligned}$$

For $[\hat{S}^z, \hat{S}^+]$, we have

$$\begin{aligned}[\hat{S}^z, \hat{S}^+] &= \frac{1}{2}\hbar^2[\hat{a}^\dagger\hat{a} - \hat{b}^\dagger\hat{b}, \hat{a}^\dagger\hat{b}] = \frac{1}{2}\hbar^2\left(\hat{a}^\dagger[\hat{a}, \hat{a}^\dagger]\hat{b} - \hat{a}^\dagger[\hat{b}^\dagger, \hat{b}]\hat{b}\right) \\ &= \hbar^2\hat{a}^\dagger\hat{b} = \hbar\hat{S}^+.\end{aligned}$$

For $[\hat{S}^z, \hat{S}^-]$, we have

$$\begin{aligned}[\hat{S}^z, \hat{S}^-] &= \frac{1}{2}\hbar^2[\hat{a}^\dagger\hat{a} - \hat{b}^\dagger\hat{b}, \hat{b}^\dagger\hat{a}] = \frac{1}{2}\hbar^2\left(\hat{b}^\dagger[\hat{a}^\dagger, \hat{a}]\hat{a} - \hat{b}^\dagger[\hat{b}, \hat{b}^\dagger]\hat{a}\right) \\ &= -\hbar^2\hat{b}^\dagger\hat{a} = -\hbar\hat{S}^-.\end{aligned}$$

(2) We rewrite $\hat{\boldsymbol{S}}^2$ as

$$\hat{\boldsymbol{S}}^2 = \frac{1}{2}\left(\hat{S}^+\hat{S}^- + \hat{S}^-\hat{S}^+\right) + \hat{S}^z\hat{S}^z.$$

Inserting the Schwinger boson representation into the above equation yields

$$\begin{aligned}\hat{\boldsymbol{S}}^2 &= \frac{1}{4}\hbar^2\left(2\hat{a}^\dagger\hat{b}\hat{b}^\dagger\hat{a} + 2\hat{b}^\dagger\hat{a}\hat{a}^\dagger\hat{b} + \hat{a}^\dagger\hat{a}\hat{a}^\dagger\hat{a} - 2\hat{a}^\dagger\hat{a}\hat{b}^\dagger\hat{b} + \hat{b}^\dagger\hat{b}\hat{b}^\dagger\hat{b}\right) \\ &= \frac{1}{4}\hbar^2\left(\hat{a}^\dagger\hat{a} + \hat{b}^\dagger\hat{b}\right)\left(\hat{a}^\dagger\hat{a} + \hat{b}^\dagger\hat{b} + 2\right).\end{aligned}$$

For $\hat{\boldsymbol{S}}^2|Sm\rangle$, we have

$$\hat{\boldsymbol{S}}^2|Sm\rangle = \frac{\hbar^2\left(\hat{a}^\dagger\hat{a} + \hat{b}^\dagger\hat{b}\right)\left(\hat{a}^\dagger\hat{a} + \hat{b}^\dagger\hat{b} + 2\right)}{4[(S+m)!(S-m)!]^{1/2}}\left(\hat{a}^\dagger\right)^{S+m}\left(\hat{b}^\dagger\right)^{S-m}|0\rangle.$$

Making use of $[\hat{a}, (\hat{a}^\dagger)^n] = n(\hat{a}^\dagger)^{n-1}$ and $[\hat{b}, (\hat{b}^\dagger)^n] = n(\hat{b}^\dagger)^{n-1}$, we have

$$\begin{aligned}\hat{\boldsymbol{S}}^2|Sm\rangle &= \frac{\hbar^2}{4[(S+m)!(S-m)!]^{1/2}}(S+m+S-m) \\ &\quad\times(S+m+S-m+2)\left(\hat{a}^\dagger\right)^{S+m}\left(\hat{b}^\dagger\right)^{S-m}|0\rangle \\ &= S(S+1)\hbar^2|Sm\rangle.\end{aligned}$$

Note that, if we make use of the constraint $n_a + n_b = 2S$ for physical states, we have

$$\hat{\boldsymbol{S}}^2 = \frac{1}{4}\hbar^2\big(\hat{a}^\dagger\hat{a} + \hat{b}^\dagger\hat{b}\big)\big(\hat{a}^\dagger\hat{a} + \hat{b}^\dagger\hat{b} + 2\big)$$
$$= \frac{1}{4}\hbar^2(2S)(2S+2) = S(S+1)\hbar^2$$

which implies that all the physical states are eigenstates of $\hat{\boldsymbol{S}}^2$ corresponding to the eigenvalue $S(S+1)\hbar^2$. For $\hat{S}_z\,|Sm\rangle$, we have

$$\hat{S}_z\,|Sm\rangle = \frac{\hbar}{2[\,(S+m)!(S-m)!\,]^{1/2}}\big(\hat{a}^\dagger\hat{a} - \hat{b}^\dagger\hat{b}\big)\big(\hat{a}^\dagger\big)^{S+m}\big(\hat{b}^\dagger\big)^{S-m}|0\rangle$$
$$= \frac{\hbar}{2[\,(S+m)!(S-m)!\,]^{1/2}}(S+m-S+m)$$
$$\times\big(\hat{a}^\dagger\big)^{S+m}\big(\hat{b}^\dagger\big)^{S-m}|0\rangle$$
$$= m\hbar\,|Sm\rangle\,.$$

28-16 Jordan-Wigner transformation. In the Jordan-Wigner transformation, the spin operator in one-dimensional real space is expressed as

$$\hat{S}_\ell^+ = \hat{f}_\ell^\dagger e^{i\pi\sum_{j<\ell}\hat{n}_j},\ \ \hat{S}_\ell^- = e^{-i\pi\sum_{j<\ell}\hat{n}_j}\hat{f}_\ell,\ \ \hat{S}_\ell^z = \hat{f}_\ell^\dagger\hat{f}_\ell - \frac{1}{2},$$

where $\hat{f}_\ell$ and $\hat{f}_\ell^\dagger$ are the annihilation and creation operators of spinless fermions.

(1) Show that

$$\hat{S}_\ell^+\hat{S}_{\ell+1}^- = \hat{f}_\ell^\dagger\hat{f}_{\ell+1}.$$

(2) The Hamiltonian of an anisotropic quantum Heisenberg spin-1/2 chain is given by

$$\hat{H} = -\sum_\ell\Big[\,\frac{1}{2}J_\perp\Big(\hat{S}_\ell^+\hat{S}_{\ell+1}^- + \hat{S}_\ell^-\hat{S}_{\ell+1}^+\Big) + J_z\hat{S}_\ell^z\hat{S}_{\ell+1}^z\,\Big].$$

Show that the Hamiltonian can be written as

$$\hat{H} = -\sum_\ell\left[\frac{1}{2}J_\perp(\hat{f}_\ell^\dagger\hat{f}_{\ell+1} + \hat{f}_{\ell+1}^\dagger\hat{f}_\ell) + J_z\left(\frac{1}{4} - \hat{f}_\ell^\dagger\hat{f}_\ell + \hat{f}_\ell^\dagger\hat{f}_\ell\hat{f}_{\ell+1}^\dagger\hat{f}_{\ell+1}\right)\right].$$

(3) Show that, for one-dimensional spin-1/2 XY-model, that is, $J_z = 0$, the eigenvalues of the Hamiltonian $\hat{H}$ are given by

$$\hbar\omega_k = -J_\perp\cos(ka).$$

---

(1) From the Jordan-Wigner transformation, we have

$$\begin{aligned}\hat{S}^+_\ell \hat{S}^-_{\ell+1} &= \hat{f}^\dagger_\ell \mathrm{e}^{\mathrm{i}\pi\sum_{j<\ell}\hat{n}_j}\mathrm{e}^{-\mathrm{i}\pi\sum_{j<\ell+1}\hat{n}_j}\hat{f}_{\ell+1}\\ &= \hat{f}^\dagger_\ell \mathrm{e}^{-\mathrm{i}\pi\hat{n}_\ell}\hat{f}_{\ell+1},\end{aligned}$$

where we have made use of $[\hat{n}_j, \hat{n}_{j'}] = 0$. Expanding the exponential $\mathrm{e}^{-\mathrm{i}\pi\hat{n}_\ell}$, we have

$$\begin{aligned}\hat{S}^+_\ell \hat{S}^-_{\ell+1} &= \hat{f}^\dagger_\ell\left[1 + \sum_{j=1}^{\infty}\frac{(-\mathrm{i}\pi)^j}{j!}(\hat{n}_\ell)^j\right]\hat{f}_{\ell+1}\\ &= \left[\hat{f}^\dagger_\ell + (\hat{f}^\dagger_\ell)^2\sum_{j=1}^{\infty}\frac{(-\mathrm{i}\pi)^j}{j!}(\hat{f}^\dagger_\ell)^{j-1}(\hat{f}_\ell)^j\right]\hat{f}_{\ell+1}.\end{aligned}$$

Making use of $(\hat{f}^\dagger_\ell)^2 = 0$, we have

$$\hat{S}^+_\ell \hat{S}^-_{\ell+1} = \hat{f}^\dagger_\ell \hat{f}_{\ell+1}.$$

(2) To recast the Hamiltonian, we first express $\hat{S}^+_\ell \hat{S}^-_{\ell+1} + \hat{S}^-_\ell \hat{S}^+_{\ell+1}$ and $\hat{S}^z_\ell \hat{S}^z_{\ell+1}$ in terms of operators $\hat{f}^\dagger_\ell$ and $\hat{f}_\ell$. For $\hat{S}^+_\ell \hat{S}^-_{\ell+1} + \hat{S}^-_\ell \hat{S}^+_{\ell+1}$, we have

$$\begin{aligned}\hat{S}^+_\ell \hat{S}^-_{\ell+1} + \hat{S}^-_\ell \hat{S}^+_{\ell+1} &= \hat{S}^+_\ell \hat{S}^-_{\ell+1} + (\hat{S}^+_\ell \hat{S}^-_{\ell+1})^\dagger = \hat{f}^\dagger_\ell \hat{f}_{\ell+1} + (\hat{f}^\dagger_\ell \hat{f}_{\ell+1})^\dagger\\ &= \hat{f}^\dagger_\ell \hat{f}_{\ell+1} + \hat{f}^\dagger_{\ell+1}\hat{f}_\ell.\end{aligned}$$

For $\hat{S}^z_\ell \hat{S}^z_{\ell+1}$, we have

$$\begin{aligned}\hat{S}^z_\ell \hat{S}^z_{\ell+1} &= \left(\hat{f}^\dagger_\ell \hat{f}_\ell - \frac{1}{2}\right)\left(\hat{f}^\dagger_{\ell+1}\hat{f}_{\ell+1} - \frac{1}{2}\right)\\ &= \frac{1}{4} - \frac{1}{2}\hat{f}^\dagger_\ell \hat{f}_\ell - \frac{1}{2}\hat{f}^\dagger_{\ell+1}\hat{f}_{\ell+1} + \hat{f}^\dagger_\ell \hat{f}_\ell \hat{f}^\dagger_{\ell+1}\hat{f}_{\ell+1}.\end{aligned}$$

The summation of $\hat{S}^z_\ell \hat{S}^z_{\ell+1}$ over ℓ is given by

$$\begin{aligned}\sum_\ell \hat{S}^z_\ell \hat{S}^z_{\ell+1} &= \sum_\ell\left(\frac{1}{4} - \frac{1}{2}\hat{f}^\dagger_\ell \hat{f}_\ell - \frac{1}{2}\hat{f}^\dagger_{\ell+1}\hat{f}_{\ell+1} + \hat{f}^\dagger_\ell \hat{f}_\ell \hat{f}^\dagger_{\ell+1}\hat{f}_{\ell+1}\right)\\ &= \sum_\ell\left(\frac{1}{4} - \hat{f}^\dagger_\ell \hat{f}_\ell + \hat{f}^\dagger_\ell \hat{f}_\ell \hat{f}^\dagger_{\ell+1}\hat{f}_{\ell+1}\right).\end{aligned}$$

Thus, the Hamiltonian can be written as

$$\hat{H} = -\sum_\ell\left[\frac{1}{2}J_\perp(\hat{f}^\dagger_\ell \hat{f}_{\ell+1} + \hat{f}^\dagger_{\ell+1}\hat{f}_\ell) + J_z\left(\frac{1}{4} - \hat{f}^\dagger_\ell \hat{f}_\ell + \hat{f}^\dagger_\ell \hat{f}_\ell \hat{f}^\dagger_{\ell+1}\hat{f}_{\ell+1}\right)\right].$$

(3) In the case that $J_z = 0$, the Hamiltonian becomes

$$\hat{H} = -\frac{1}{2}J_\perp \sum_\ell (\hat{f}_\ell^\dagger \hat{f}_{\ell+1} + \hat{f}_{\ell+1}^\dagger \hat{f}_\ell).$$

To obtain the eigenvalues of $\hat{H}$, we make a Fourier transformation to $\hat{f}_\ell$

$$\hat{f}_\ell = \frac{1}{\sqrt{N}} \sum_k \mathrm{e}^{\mathrm{i}kR_\ell} \hat{f}_k.$$

The Hermitian conjugate of $\hat{f}_\ell$ is given by

$$\hat{f}_\ell^\dagger = \frac{1}{\sqrt{N}} \sum_k \mathrm{e}^{-\mathrm{i}kR_\ell} \hat{f}_k^\dagger.$$

In terms of $\hat{f}_k$ and $\hat{f}_k^\dagger$, we have

$$\begin{aligned}
\hat{H} &= -\frac{1}{2}J_\perp \sum_{kk'} \left(\mathrm{e}^{\mathrm{i}k'a} + \mathrm{e}^{-\mathrm{i}ka}\right) \hat{f}_k^\dagger \hat{f}_{k'} \left[\frac{1}{N}\sum_\ell \mathrm{e}^{-\mathrm{i}(k-k')R_\ell}\right] \\
&= -\frac{1}{2}J_\perp \sum_{kk'} \left(\mathrm{e}^{\mathrm{i}k'a} + \mathrm{e}^{-\mathrm{i}ka}\right) \hat{f}_k^\dagger \hat{f}_{k'} \delta_{kk'} \\
&= -\frac{1}{2}J_\perp \sum_k \left(\mathrm{e}^{\mathrm{i}ka} + \mathrm{e}^{-\mathrm{i}ka}\right) \hat{f}_k^\dagger \hat{f}_k = -J_\perp \sum_k \cos(ka) \hat{f}_k^\dagger \hat{f}_k.
\end{aligned}$$

Therefore, the eigenvalues of $\hat{H}$ are given by $\hbar\omega_k = -J_\perp \cos(ka)$.

28-17 Matsubara susceptibility and Curie's law. Consider a single free spin $S = 1/2$ at temperature T. The Hamiltonian of such a system is zero. The transverse Matsubara susceptibility with respect to a weak external magnetic field is defined as

$$\chi_\mathrm{M}(\mathrm{i}\omega_m) = \left(\frac{\mu}{\hbar}\right)^2 \int_0^{\hbar\beta} \frac{\mathrm{d}\tau}{\hbar}\, \mathrm{e}^{\mathrm{i}\omega_m\tau} \langle \hat{T}_\tau [\, \hat{S}^+(\tau)\hat{S}^-(0)\,] \rangle$$

with μ the magnetic moment of the spin.

(1) Compute $\chi_\mathrm{M}(\mathrm{i}\omega_m)$ for the single free spin at temperature T in a weak external magnetic field along the z direction.

(2) Continue $\chi_\mathrm{M}(\mathrm{i}\omega_m)$ analytically to the real axis of ω and derive the Kubo susceptibility $\chi_\mathrm{K}(\omega)$ so that $\chi_\mathrm{K}(\omega) = \chi_\mathrm{M}(\mathrm{i}\omega_m)\big|_{\omega=\mathrm{i}\omega_m+\mathrm{i}\delta}$.

(3) Derive Curie's law in the static limit.

(1) In a weak external magnetic field along the z direction, the Hamiltonian of the spin reads

$$\hat{H} = -(2\mu B/\hbar)\hat{S}^z.$$

In the common representation of $\hat{\boldsymbol{S}}^2$ and $\hat{S}_z$, the eigenvector of $\hat{H}$ corresponding to the eigenvalue $E_m = -2m\mu B$ with $m = \pm 1/2$ is denoted by $|Sm\rangle$. Here $S = 1/2$. The imaginary time dependence of $\hat{S}^+(\tau)$ can be found as follows

$$\begin{aligned}\hat{S}^+(\tau) &= \mathrm{e}^{\tau\hat{H}/\hbar}\hat{S}^+\mathrm{e}^{-\tau\hat{H}/\hbar}\\ &= \hat{S}^+ + \frac{\tau}{\hbar}[\hat{H},\hat{S}^+] + \frac{1}{2!}\left(\frac{\tau}{\hbar}\right)^2[\hat{H},[\hat{H},\hat{S}^+]] + \cdots\\ &= \hat{S}^+ - \frac{2\mu B\tau}{\hbar^2}[\hat{S}^z,\hat{S}^+] + \frac{1}{2!}\left(-\frac{2\mu B\tau}{\hbar^2}\right)^2[\hat{S}^z,[\hat{S}^z,\hat{S}^+]] + \cdots\\ &= \hat{S}^+\left[1 + \left(-\frac{2\mu B\tau}{\hbar}\right) + \frac{1}{2!}\left(-\frac{2\mu B\tau}{\hbar}\right)^2 + \cdots\right]\\ &= \mathrm{e}^{-2\mu B\tau/\hbar}\hat{S}^+.\end{aligned}$$

The average $\langle\hat{T}_\tau[\,\hat{S}_z(\tau)\hat{S}_z(0)\,]\rangle$ can be evaluated as follows

$$\begin{aligned}&\langle\hat{T}_\tau[\,\hat{S}^+(\tau)\hat{S}^-(0)\,]\rangle\\ &= \frac{1}{Z}\sum_m \mathrm{e}^{-\beta E_m}\langle Sm|\,\hat{T}_\tau[\,\hat{S}^+(\tau)\hat{S}^-(0)\,]|Sm\rangle\\ &= \frac{1}{Z}\mathrm{e}^{-2\mu B\tau/\hbar}\sum_m \mathrm{e}^{-\beta E_m}\langle Sm|\,\hat{S}^+\hat{S}^-\theta(\tau) + \hat{S}^-\hat{S}^+\theta(-\tau)|Sm\rangle\\ &= \frac{\hbar^2}{Z}\mathrm{e}^{-2\mu B\tau/\hbar}\sum_m \mathrm{e}^{-\beta E_m}\{[\,S(S+1) - m(m-1)\,]\theta(\tau)\\ &\quad + [\,S(S+1) - m(m+1)\,]\theta(-\tau)\},\end{aligned}$$

where $Z = \sum_m \mathrm{e}^{-\beta E_m} = \mathrm{e}^{\beta\mu B} + \mathrm{e}^{-\beta\mu B} = 2\cosh(\beta\mu B)$. Evaluating the summation over m in the above equation, we obtain

$$\langle\hat{T}_\tau[\,\hat{S}^+(\tau)\hat{S}^-(0)\,]\rangle = \frac{\hbar^2}{2}\mathrm{e}^{-2\mu B\tau/\hbar}\{1 + \tanh(\beta\mu B)[\,\theta(\tau) - \theta(-\tau)\,]\}.$$

$\chi_{\mathrm{M}}(\mathrm{i}\omega_m)$ is then given by

$$\begin{aligned}\chi_{\mathrm{M}}(\mathrm{i}\omega_m) &= \frac{1}{2}\mu^2[\,1 + \tanh(\beta\mu B)\,]\int_0^{\hbar\beta}\frac{\mathrm{d}\tau}{\hbar}\,\mathrm{e}^{(\mathrm{i}\omega_m - 2\mu B/\hbar)\tau}\\ &= \frac{1}{2}\mu^2[\,1 + \tanh(\beta\mu B)\,]\frac{\mathrm{e}^{-2\beta\mu B} - 1}{\mathrm{i}\hbar\omega_m - 2\mu B}.\end{aligned}$$

(2) Analytically continue $\chi_{\rm M}(\mathrm{i}\omega_m)$ to the real frequency axis yields

$$\chi_{\rm K}(\omega) = \frac{1}{2}\mu^2\big[\,1+\tanh(\beta\mu B)\,\big]\frac{\mathrm{e}^{-2\beta\mu B}-1}{\hbar\omega - 2\mu B + \mathrm{i}\delta}.$$

(3) In the static limit, we have

$$\chi_{\rm K}(0) = \frac{1}{2}\mu^2\big[\,1+\tanh(\beta\mu B)\,\big]\frac{1-\mathrm{e}^{-2\beta\mu B}}{2\mu B - \mathrm{i}\delta}.$$

In the $B \to 0$ limit, the imaginary part vanishes and $\chi_{\rm K}(0)$ is given by

$$\chi_{\rm K}(0) = \frac{1}{2}\mu^2 \lim_{B\to 0}\frac{1-\mathrm{e}^{-2\beta\mu B}}{2\mu B} = \frac{\mu^2}{2k_{\rm B}T}$$

which is Curie's law.

28-18 One-dimensional anisotropic *XY*-model. The Hamiltonian in the one-dimensional anisotropic XY-model is given by

$$\hat{H} = \sum_n \big(J_1\hat{S}_n^x\hat{S}_{n+1}^x + J_2\hat{S}_n^y\hat{S}_{n+1}^y\big),$$

where J_1 and J_2 are coupling constants. Making the Jordan-Wigner transformation to the spin operator, show that the Hamiltonian can be written as

$$\hat{H} = \sum_n \big[\,t(\hat{f}_n^\dagger\hat{f}_{n+1} + \hat{f}_{n+1}^\dagger\hat{f}_n) + \Delta(\hat{f}_n\hat{f}_{n+1} + \hat{f}_{n+1}^\dagger\hat{f}_n^\dagger)\,\big],$$

where $\hat{f}_n$ and $\hat{f}_n^\dagger$ are respectively the annihilation and creation operators of spinless fermions on site n. Provide expressions for t and Δ.

Expressing $\hat{S}_n^x$ and $\hat{S}_n^y$ in terms of $\hat{S}_n^+$ and $\hat{S}_n^-$, we have

$$\hat{S}_n^x\hat{S}_{n+1}^x = \frac{1}{4}\big(\hat{S}_n^+\hat{S}_{n+1}^- + \hat{S}_n^-\hat{S}_{n+1}^+ + \hat{S}_n^+\hat{S}_{n+1}^+ + \hat{S}_n^-\hat{S}_{n+1}^-\big),$$

$$\hat{S}_n^y\hat{S}_{n+1}^y = \frac{1}{4}\big(\hat{S}_n^+\hat{S}_{n+1}^- + \hat{S}_n^-\hat{S}_{n+1}^+ - \hat{S}_n^+\hat{S}_{n+1}^+ - \hat{S}_n^-\hat{S}_{n+1}^-\big).$$

We have shown in **Problem 28-16** that $\hat{S}_n^+\hat{S}_{n+1}^- = \hat{f}_n^\dagger\hat{f}_{n+1}$ from which we can obtain by taking the Hermitian conjugation that $\hat{S}_n^-\hat{S}_{n+1}^+ = \hat{f}_{n+1}^\dagger\hat{f}_n$, where we have made use of $[\hat{S}_{n+1}^+, \hat{S}_n^-] = 0$. For $\hat{S}_n^+\hat{S}_{n+1}^+$, we have

$$\begin{aligned}\hat{S}_n^+\hat{S}_{n+1}^+ &= \hat{f}_n^\dagger\mathrm{e}^{i\pi\sum_{j<n}\hat{n}_j}\hat{f}_{n+1}^\dagger\mathrm{e}^{i\pi\sum_{j<n+1}\hat{n}_j}\\ &= \hat{f}_n^\dagger\mathrm{e}^{i\pi\sum_{j<n}\hat{n}_j}\mathrm{e}^{i\pi\sum_{j<n+1}\hat{n}_j}\hat{f}_{n+1}^\dagger = \hat{f}_n^\dagger\mathrm{e}^{i\pi\hat{n}_n}\hat{f}_{n+1}^\dagger\\ &= \bigg[\,\hat{f}_n^\dagger + (\hat{f}_n^\dagger)^2\sum_{j=1}^{\infty}\frac{(\mathrm{i}\pi)^j}{j!}(\hat{f}_n^\dagger)^{j-1}\hat{f}_n^j\,\bigg]\hat{f}_{n+1}^\dagger = \hat{f}_n^\dagger\hat{f}_{n+1}^\dagger\end{aligned}$$

from which we also have

$$\hat{S}_n^- \hat{S}_{n+1}^- = \hat{f}_{n+1} \hat{f}_n.$$

In the above manipulations, we have made use of $[\hat{n}_j, \hat{n}_\ell] = 0$ and $e^{2i\pi\hat{n}_j} = 1$. The second identity holds because the eigenvalues of $\hat{n}_j$ are 0 and 1. In terms of the spinless fermion operators, the Hamiltonian becomes

$$\begin{aligned}
\hat{H} &= \frac{1}{4}\sum_n \big[(J_1 + J_2)(\hat{S}_n^+ \hat{S}_{n+1}^- + \hat{S}_n^- \hat{S}_{n+1}^+) \\
&\quad + (J_1 - J_2)(\hat{S}_n^+ \hat{S}_{n+1}^+ + \hat{S}_n^- \hat{S}_{n+1}^-) \big] \\
&= \frac{1}{4}\sum_n \big[(J_1 + J_2)(\hat{f}_n^\dagger \hat{f}_{n+1} + \hat{f}_{n+1}^\dagger \hat{f}_n) \\
&\quad + (J_1 - J_2)(\hat{f}_n^\dagger \hat{f}_{n+1}^\dagger + \hat{f}_{n+1} \hat{f}_n) \big] \\
&= \sum_n \big[t(\hat{f}_n^\dagger \hat{f}_{n+1} + \hat{f}_{n+1}^\dagger \hat{f}_n) + \Delta(\hat{f}_n^\dagger \hat{f}_{n+1}^\dagger + \hat{f}_{n+1} \hat{f}_n) \big],
\end{aligned}$$

where

$$t = \frac{1}{4}(J_1 + J_2), \ \ \Delta = \frac{1}{4}(J_1 - J_2).$$

28-19 Phonon-ferromagnon interaction. The interaction between phonons and ferromagnons in a ferromagnet is described by the Hamiltonian

$$\hat{H} = \sum_{\boldsymbol{k}} \big[\hbar\omega_{\boldsymbol{k}}^p \hat{a}_{\boldsymbol{k}}^\dagger \hat{a}_{\boldsymbol{k}} + \hbar\omega_{\boldsymbol{k}}^m \hat{b}_{\boldsymbol{k}}^\dagger \hat{b}_{\boldsymbol{k}} + g_{\boldsymbol{k}}(\hat{a}_{\boldsymbol{k}}^\dagger \hat{b}_{\boldsymbol{k}} + \hat{b}_{\boldsymbol{k}}^\dagger \hat{a}_{\boldsymbol{k}})\big],$$

where $\hat{a}_{\boldsymbol{k}}^\dagger$ and $\hat{a}_{\boldsymbol{k}}$ are phonon operators with the dispersion relation given by $\hbar\omega_{\boldsymbol{k}}^p$, $\hat{b}_{\boldsymbol{k}}^\dagger$ and $\hat{b}_{\boldsymbol{k}}$ are ferromagnon operators with the dispersion relation given by $\hbar\omega_{\boldsymbol{k}}^m$, and $g_{\boldsymbol{k}}$ is the phonon-ferromagnon coupling constant.

(1) Diagonalize the Hamiltonian using the Bogoliubov transformation.

(2) Give the dispersion relations of the new excitations.

(3) Describe how the new excitations arise from the original phonons and ferromagnons.

---

(1) In making a Bogoliubov transformation, we would like to make the linear combinations of the original operators to have as few operators as possible. The minimum number of operators as well as the particular operators participating in a certain linear combination can be determined from Heisenberg's equation of motion. Since the time-dependence of an operator is given by the commutator of the operator with the Hamiltonian according to Heisenberg's equation of motion, we can identify the operators that participate in linear combinations using this commutator. Let us first consider operator $\hat{a}_{\boldsymbol{k}}$. From the expression of the Hamiltonian, we see that the value of $[\hat{a}_{\boldsymbol{k}}, \hat{H}]$ contains $\hat{a}_{\boldsymbol{k}}$ and $\hat{b}_{\boldsymbol{k}}$

$$\begin{aligned}[\hat{a}_{\boldsymbol{k}}, \hat{H}] &= \sum_{\boldsymbol{k}'}[\hat{a}_{\boldsymbol{k}}, \hbar\omega^p_{\boldsymbol{k}'}\hat{a}^\dagger_{\boldsymbol{k}'}\hat{a}_{\boldsymbol{k}'} + \hbar\omega^m_{\boldsymbol{k}'}\hat{b}^\dagger_{\boldsymbol{k}'}\hat{b}_{\boldsymbol{k}'} + g_{\boldsymbol{k}'}(\hat{a}^\dagger_{\boldsymbol{k}'}\hat{b}_{\boldsymbol{k}'} + \hat{b}^\dagger_{\boldsymbol{k}'}\hat{a}_{\boldsymbol{k}'})] \\ &= \sum_{\boldsymbol{k}'}\hbar\omega^p_{\boldsymbol{k}'}[\hat{a}_{\boldsymbol{k}}, \hat{a}^\dagger_{\boldsymbol{k}'}\hat{a}_{\boldsymbol{k}'}] + \sum_{\boldsymbol{k}'} g_{\boldsymbol{k}'}[\hat{a}_{\boldsymbol{k}}, \hat{a}^\dagger_{\boldsymbol{k}'}\hat{b}_{\boldsymbol{k}'}] \\ &= \hbar\omega^p_{\boldsymbol{k}}\hat{a}_{\boldsymbol{k}} + g_{\boldsymbol{k}}\hat{b}_{\boldsymbol{k}}.\end{aligned}$$

We thus make the following Bogoliubov transformation

$$\begin{aligned}\hat{\alpha}_{\boldsymbol{k}} &= u_{\boldsymbol{k}}\hat{a}_{\boldsymbol{k}} + v_{\boldsymbol{k}}\hat{b}_{\boldsymbol{k}}, \\ \hat{\beta}_{\boldsymbol{k}} &= p_{\boldsymbol{k}}\hat{a}_{\boldsymbol{k}} + q_{\boldsymbol{k}}\hat{b}_{\boldsymbol{k}},\end{aligned}$$

where $p_{\boldsymbol{k}}$, $q_{\boldsymbol{k}}$, $u_{\boldsymbol{k}}$, and $v_{\boldsymbol{k}}$ are real coefficients to be determined. To retain all the algebraic properties of the original operators, we demand that the new operators satisfy the following commutation relations

$$\begin{aligned}&[\hat{\alpha}_{\boldsymbol{k}}, \hat{\alpha}^\dagger_{\boldsymbol{k}'}] = \delta_{\boldsymbol{k}\boldsymbol{k}'}, \ \ [\hat{\alpha}_{\boldsymbol{k}}, \hat{\alpha}_{\boldsymbol{k}'}] = [\hat{\alpha}^\dagger_{\boldsymbol{k}}, \hat{\alpha}^\dagger_{\boldsymbol{k}'}] = 0, \\ &[\hat{\beta}_{\boldsymbol{k}}, \hat{\beta}^\dagger_{\boldsymbol{k}'}] = \delta_{\boldsymbol{k}\boldsymbol{k}'}, \ \ [\hat{\beta}_{\boldsymbol{k}}, \hat{\beta}_{\boldsymbol{k}'}] = [\hat{\beta}^\dagger_{\boldsymbol{k}}, \hat{\beta}^\dagger_{\boldsymbol{k}'}] = 0, \\ &[\hat{\alpha}_{\boldsymbol{k}}, \hat{\beta}_{\boldsymbol{k}'}] = [\hat{\alpha}_{\boldsymbol{k}}, \hat{\beta}^\dagger_{\boldsymbol{k}'}] = [\hat{\alpha}^\dagger_{\boldsymbol{k}}, \hat{\beta}_{\boldsymbol{k}'}] = [\hat{\alpha}^\dagger_{\boldsymbol{k}}, \hat{\beta}^\dagger_{\boldsymbol{k}'}] = 0.\end{aligned}$$

Note that the second and third commutation relations on the first two lines are satisfied automatically and so are the first and last ones on the third line. From the first commutation relations on the first and second lines, we have

$$u^2_{\boldsymbol{k}} + v^2_{\boldsymbol{k}} = 1, \ \ p^2_{\boldsymbol{k}} + q^2_{\boldsymbol{k}} = 1.$$

The second and third commutation relations on the third line both give the following result because they are mutually Hermitian conjugates

$$u_{\boldsymbol{k}}p_{\boldsymbol{k}} + v_{\boldsymbol{k}}q_{\boldsymbol{k}} = 0.$$

The above relation can be satisfied by choosing $p_{\boldsymbol{k}} = -v_{\boldsymbol{k}}$ and $q_{\boldsymbol{k}} = u_{\boldsymbol{k}}$. We are then left with only two coefficients $u_{\boldsymbol{k}}$ and $v_{\boldsymbol{k}}$ that satisfy

$$u_{\boldsymbol{k}}^2 + v_{\boldsymbol{k}}^2 = 1.$$

The transformation is now written as

$$\hat{\alpha}_{\boldsymbol{k}} = u_{\boldsymbol{k}}\hat{a}_{\boldsymbol{k}} + v_{\boldsymbol{k}}\hat{b}_{\boldsymbol{k}},$$

$$\hat{\beta}_{\boldsymbol{k}} = -v_{\boldsymbol{k}}\hat{a}_{\boldsymbol{k}} + u_{\boldsymbol{k}}\hat{b}_{\boldsymbol{k}},$$

For the purpose of diagonalizing the Hamiltonian, we now express the Hamiltonian in terms of the new operators. First, expressing the original operators in terms of the new operators, we have

$$\hat{a}_{\boldsymbol{k}} = u_{\boldsymbol{k}}\hat{\alpha}_{\boldsymbol{k}} - v_{\boldsymbol{k}}\hat{\beta}_{\boldsymbol{k}},$$

$$\hat{b}_{\boldsymbol{k}} = v_{\boldsymbol{k}}\hat{\alpha}_{\boldsymbol{k}} + u_{\boldsymbol{k}}\hat{\beta}_{\boldsymbol{k}},$$

Inserting the above expressions of $\hat{a}_{\boldsymbol{k}}$ and $\hat{b}_{\boldsymbol{k}}$ into the Hamiltonian, we obtain

$$\begin{aligned}
\hat{H} = \sum_{\boldsymbol{k}} \{ &\hbar\omega_{\boldsymbol{k}}^p (u_{\boldsymbol{k}}\hat{\alpha}_{\boldsymbol{k}}^\dagger - v_{\boldsymbol{k}}\hat{\beta}_{\boldsymbol{k}}^\dagger)(u_{\boldsymbol{k}}\hat{\alpha}_{\boldsymbol{k}} - v_{\boldsymbol{k}}\hat{\beta}_{\boldsymbol{k}}) \\
&+ \hbar\omega_{\boldsymbol{k}}^m (v_{\boldsymbol{k}}\hat{\alpha}_{\boldsymbol{k}}^\dagger + u_{\boldsymbol{k}}\hat{\beta}_{\boldsymbol{k}}^\dagger)(v_{\boldsymbol{k}}\hat{\alpha}_{\boldsymbol{k}} + u_{\boldsymbol{k}}\hat{\beta}_{\boldsymbol{k}}) \\
&\quad + g_{\boldsymbol{k}} \big[(u_{\boldsymbol{k}}\hat{\alpha}_{\boldsymbol{k}}^\dagger - v_{\boldsymbol{k}}\hat{\beta}_{\boldsymbol{k}}^\dagger)(v_{\boldsymbol{k}}\hat{\alpha}_{\boldsymbol{k}} + u_{\boldsymbol{k}}\hat{\beta}_{\boldsymbol{k}}) \\
&+ (v_{\boldsymbol{k}}\hat{\alpha}_{\boldsymbol{k}}^\dagger + u_{\boldsymbol{k}}\hat{\beta}_{\boldsymbol{k}}^\dagger)(u_{\boldsymbol{k}}\hat{\alpha}_{\boldsymbol{k}} - v_{\boldsymbol{k}}\hat{\beta}_{\boldsymbol{k}}) \big] \} \\
= \sum_{\boldsymbol{k}} \big[&(u_{\boldsymbol{k}}^2 \hbar\omega_{\boldsymbol{k}}^p + v_{\boldsymbol{k}}^2 \hbar\omega_{\boldsymbol{k}}^m + 2u_{\boldsymbol{k}}v_{\boldsymbol{k}}g_{\boldsymbol{k}})\hat{\alpha}_{\boldsymbol{k}}^\dagger \hat{\alpha}_{\boldsymbol{k}} \\
&\quad + (v_{\boldsymbol{k}}^2 \hbar\omega_{\boldsymbol{k}}^p + u_{\boldsymbol{k}}^2 \hbar\omega_{\boldsymbol{k}}^m - 2u_{\boldsymbol{k}}v_{\boldsymbol{k}}g_{\boldsymbol{k}})\hat{\beta}_{\boldsymbol{k}}^\dagger \hat{\beta}_{\boldsymbol{k}} \big] \\
&+ \sum_{\boldsymbol{k}} \big[-u_{\boldsymbol{k}}v_{\boldsymbol{k}} (\hbar\omega_{\boldsymbol{k}}^p - \hbar\omega_{\boldsymbol{k}}^m) + (u_{\boldsymbol{k}}^2 - v_{\boldsymbol{k}}^2) g_{\boldsymbol{k}} \big] (\hat{\alpha}_{\boldsymbol{k}}^\dagger \hat{\beta}_{\boldsymbol{k}} + \hat{\beta}_{\boldsymbol{k}}^\dagger \hat{\alpha}_{\boldsymbol{k}}).
\end{aligned}$$

The Hamiltonian is diagonalized if the coefficient of $\hat{\alpha}_{\boldsymbol{k}}^\dagger \hat{\beta}_{\boldsymbol{k}} + \hat{\beta}_{\boldsymbol{k}}^\dagger \hat{\alpha}_{\boldsymbol{k}}$ on the last line vanishes

$$-u_{\boldsymbol{k}}v_{\boldsymbol{k}} (\hbar\omega_{\boldsymbol{k}}^p - \hbar\omega_{\boldsymbol{k}}^m) + (u_{\boldsymbol{k}}^2 - v_{\boldsymbol{k}}^2) g_{\boldsymbol{k}} = 0.$$

Solving for $u_{\boldsymbol{k}}$ and $v_{\boldsymbol{k}}$ from the above equation together with the condition $u_{\boldsymbol{k}}^2 + v_{\boldsymbol{k}}^2 = 1$, we obtain

$$u_{\boldsymbol{k}}^2 = \frac{1}{2}\left[1 + \frac{\hbar\omega_{\boldsymbol{k}}^p - \hbar\omega_{\boldsymbol{k}}^m}{\sqrt{(\hbar\omega_{\boldsymbol{k}}^p - \hbar\omega_{\boldsymbol{k}}^m)^2 + 4g_{\boldsymbol{k}}^2}} \right],$$

$$v_{\boldsymbol{k}}^2 = \frac{1}{2}\left[1 - \frac{\hbar\omega_{\boldsymbol{k}}^p - \hbar\omega_{\boldsymbol{k}}^m}{\sqrt{(\hbar\omega_{\boldsymbol{k}}^p - \hbar\omega_{\boldsymbol{k}}^m)^2 + 4g_{\boldsymbol{k}}^2}} \right],$$

$$u_{\boldsymbol{k}}v_{\boldsymbol{k}} = \frac{g_{\boldsymbol{k}}}{\sqrt{(\hbar\omega_{\boldsymbol{k}}^p - \hbar\omega_{\boldsymbol{k}}^m)^2 + 4g_{\boldsymbol{k}}^2}}.$$

(2) Evaluating the coefficients of $\hat{\alpha}_{\boldsymbol{k}}^\dagger \hat{\alpha}_{\boldsymbol{k}}$ and $\hat{\beta}_{\boldsymbol{k}}^\dagger \hat{\beta}_{\boldsymbol{k}}$ in the above Hamiltonian expressed in terms of the new operators, we have

$$u_{\boldsymbol{k}}^2 \hbar\omega_{\boldsymbol{k}}^p + v_{\boldsymbol{k}}^2 \hbar\omega_{\boldsymbol{k}}^m + 2u_{\boldsymbol{k}} v_{\boldsymbol{k}} g_{\boldsymbol{k}} = \frac{1}{2}\left[\hbar\omega_{\boldsymbol{k}}^p + \hbar\omega_{\boldsymbol{k}}^m + \sqrt{(\hbar\omega_{\boldsymbol{k}}^p - \hbar\omega_{\boldsymbol{k}}^m)^2 + 4g_{\boldsymbol{k}}^2}\,\right],$$

$$v_{\boldsymbol{k}}^2 \hbar\omega_{\boldsymbol{k}}^p + u_{\boldsymbol{k}}^2 \hbar\omega_{\boldsymbol{k}}^m - 2u_{\boldsymbol{k}} v_{\boldsymbol{k}} g_{\boldsymbol{k}} = \frac{1}{2}\left[\hbar\omega_{\boldsymbol{k}}^p + \hbar\omega_{\boldsymbol{k}}^m - \sqrt{(\hbar\omega_{\boldsymbol{k}}^p - \hbar\omega_{\boldsymbol{k}}^m)^2 + 4g_{\boldsymbol{k}}^2}\,\right].$$

Thus, the diagonalized Hamiltonian is given by

$$\hat{H} = \sum_{\boldsymbol{k}} \left(E_{\boldsymbol{k}}^+ \hat{\alpha}_{\boldsymbol{k}}^\dagger \hat{\alpha}_{\boldsymbol{k}} + E_{\boldsymbol{k}}^- \hat{\beta}_{\boldsymbol{k}}^\dagger \hat{\beta}_{\boldsymbol{k}}\right),$$

where

$$E_{\boldsymbol{k}}^\pm = \frac{1}{2}\left[\hbar\omega_{\boldsymbol{k}}^p + \hbar\omega_{\boldsymbol{k}}^m \pm \sqrt{(\hbar\omega_{\boldsymbol{k}}^p - \hbar\omega_{\boldsymbol{k}}^m)^2 + 4g_{\boldsymbol{k}}^2}\,\right].$$

(3) From the expressions of the new operators in terms of the original operators, $\hat{\alpha}_{\boldsymbol{k}} = u_{\boldsymbol{k}}\hat{a}_{\boldsymbol{k}} + v_{\boldsymbol{k}}\hat{b}_{\boldsymbol{k}}$ and $\hat{\beta}_{\boldsymbol{k}} = -v_{\boldsymbol{k}}\hat{a}_{\boldsymbol{k}} + u_{\boldsymbol{k}}\hat{b}_{\boldsymbol{k}}$, we see that each excitation described by $\hat{\alpha}_{\boldsymbol{k}}$ or $\hat{\beta}_{\boldsymbol{k}}$ is a linear superposition of the original excitations. This is because the interaction between phonons and ferromagnons makes them mix up. We can also say that phonons and ferromagnons are hybridized because of the interaction. Thus, an interaction in a system can bring about new excitations and hence change the properties of the system.

28-20 Weakly interacting dilute Bose gas. The approximate Bogoliubov Hamiltonian for a weakly interacting dilute Bose gas is given by

$$\hat{H} = E_0 + \frac{1}{2}\sum_{\boldsymbol{k}}{}' (\varepsilon_{\boldsymbol{k}} + \Delta)\left(\hat{b}_{\boldsymbol{k}}^\dagger \hat{b}_{\boldsymbol{k}} + \hat{b}_{-\boldsymbol{k}}^\dagger \hat{b}_{-\boldsymbol{k}}\right) + \frac{1}{2}\Delta \sum_{\boldsymbol{k}}{}' \left(\hat{b}_{\boldsymbol{k}}^\dagger \hat{b}_{-\boldsymbol{k}}^\dagger + \hat{b}_{-\boldsymbol{k}}\hat{b}_{\boldsymbol{k}}\right),$$

where the prime on a summation sign indicates that $\boldsymbol{k} \neq 0$, $\hat{b}_{\boldsymbol{k}}$ and $\hat{b}_{\boldsymbol{k}}^\dagger$ are annihilation and creation operators of bosons, and Δ is a small coupling constant.

(1) Diagonalize the Hamiltonian using the Bogoliubov transformation and find the spectrum of collective modes.

(2) Find out whether the spectrum of excitations contains sound modes.

(1) To determine the forms of the linear combinations in the Bogoliubov transformation, we utilize Heisenberg's equation of motion. For the present purpose, the evaluation of relevant commutators suffices. We first consider $\hat{b}_{\boldsymbol{k}}$. The commutator $[\hat{b}_{\boldsymbol{k}}, \hat{H}]$ is given by

$$\begin{aligned}[\hat{b}_{\boldsymbol{k}}, \hat{H}] &= \sum_{\boldsymbol{k}'}{}' [\hat{b}_{\boldsymbol{k}}, (\varepsilon_{\boldsymbol{k}'} + \Delta)\hat{b}^\dagger_{\boldsymbol{k}'}\hat{b}_{\boldsymbol{k}'} + (\Delta/2)(\hat{b}^\dagger_{\boldsymbol{k}'}\hat{b}^\dagger_{-\boldsymbol{k}'} + \hat{b}_{-\boldsymbol{k}'}\hat{b}_{\boldsymbol{k}'})] \\ &= (\varepsilon_{\boldsymbol{k}} + \Delta)\hat{b}_{\boldsymbol{k}} + \Delta\hat{b}^\dagger_{-\boldsymbol{k}}.\end{aligned}$$

Thus, it is mandatory for us to make linear combinations of $\hat{b}_{\boldsymbol{k}}$ and $\hat{b}^\dagger_{-\boldsymbol{k}}$. This can be further confirmed by evaluating the commutation relation between $\hat{b}^\dagger_{-\boldsymbol{k}}$ and $\hat{H}$

$$\begin{aligned}[\hat{b}^\dagger_{-\boldsymbol{k}}, \hat{H}] &= \sum_{\boldsymbol{k}'}{}' [\hat{b}^\dagger_{-\boldsymbol{k}}, (\varepsilon_{\boldsymbol{k}'} + \Delta)\hat{b}^\dagger_{\boldsymbol{k}'}\hat{b}_{\boldsymbol{k}'} + (\Delta/2)(\hat{b}^\dagger_{\boldsymbol{k}'}\hat{b}^\dagger_{-\boldsymbol{k}'} + \hat{b}_{-\boldsymbol{k}'}\hat{b}_{\boldsymbol{k}'})] \\ &= -(\varepsilon_{\boldsymbol{k}} + \Delta)\hat{b}^\dagger_{-\boldsymbol{k}} - \Delta\hat{b}_{\boldsymbol{k}}.\end{aligned}$$

We thus write down the following Bogoliubov transformation

$$\begin{aligned}\hat{\alpha}_{\boldsymbol{k}} &= u_{\boldsymbol{k}}\hat{b}_{\boldsymbol{k}} + v_{\boldsymbol{k}}\hat{b}^\dagger_{-\boldsymbol{k}}, \\ \hat{\beta}^\dagger_{\boldsymbol{k}} &= p_{\boldsymbol{k}}\hat{b}_{\boldsymbol{k}} + q_{\boldsymbol{k}}\hat{b}^\dagger_{-\boldsymbol{k}}.\end{aligned}$$

The new operators must satisfy the following commutation relations to preserve the algebraic properties of the original operators

$$\begin{aligned}&[\hat{\alpha}_{\boldsymbol{k}}, \hat{\alpha}^\dagger_{\boldsymbol{k}'}] = \delta_{\boldsymbol{k}\boldsymbol{k}'}, \ [\hat{\alpha}_{\boldsymbol{k}}, \hat{\alpha}_{\boldsymbol{k}'}] = [\hat{\alpha}^\dagger_{\boldsymbol{k}}, \hat{\alpha}^\dagger_{\boldsymbol{k}'}] = 0, \\ &[\hat{\beta}_{\boldsymbol{k}}, \hat{\beta}^\dagger_{\boldsymbol{k}'}] = \delta_{\boldsymbol{k}\boldsymbol{k}'}, \ [\hat{\beta}_{\boldsymbol{k}}, \hat{\beta}_{\boldsymbol{k}'}] = [\hat{\beta}^\dagger_{\boldsymbol{k}}, \hat{\beta}^\dagger_{\boldsymbol{k}'}] = 0, \\ &[\hat{\alpha}_{\boldsymbol{k}}, \hat{\beta}_{\boldsymbol{k}'}] = [\hat{\alpha}_{\boldsymbol{k}}, \hat{\beta}_{\boldsymbol{k}'}] = [\hat{\alpha}^\dagger_{\boldsymbol{k}}, \hat{\beta}_{\boldsymbol{k}'}] = [\hat{\alpha}^\dagger_{\boldsymbol{k}}, \hat{\beta}^\dagger_{\boldsymbol{k}'}] = 0.\end{aligned}$$

Some of the above commutation relations are satisfied automatically, others yield constraints on the transformation coefficients. We have

$$u_{\boldsymbol{k}}^2 - v_{\boldsymbol{k}}^2 = 1, \ q_{\boldsymbol{k}}^2 - p_{\boldsymbol{k}}^2 = 1, \ u_{\boldsymbol{k}}p_{\boldsymbol{k}} - v_{\boldsymbol{k}}q_{\boldsymbol{k}} = 0.$$

To satisfy the third constraint, we choose $p_{\boldsymbol{k}} = v_{\boldsymbol{k}}$ and $q_{\boldsymbol{k}} = u_{\boldsymbol{k}}$. We are then left with a single constraint

$$u_{\boldsymbol{k}}^2 - v_{\boldsymbol{k}}^2 = 1,$$

and the Bogoliubov transformation becomes

$$\hat{\alpha}_{\boldsymbol{k}} = u_{\boldsymbol{k}}\hat{b}_{\boldsymbol{k}} + v_{\boldsymbol{k}}\hat{b}^\dagger_{-\boldsymbol{k}},$$

$$\hat{\beta}^\dagger_{\boldsymbol{k}} = v_{\boldsymbol{k}}\hat{b}_{\boldsymbol{k}} + u_{\boldsymbol{k}}\hat{b}^\dagger_{-\boldsymbol{k}}.$$

Inverting the above Bogoliubov transformation, we have

$$\hat{b}_{\boldsymbol{k}} = u_{\boldsymbol{k}}\hat{\alpha}_{\boldsymbol{k}} - v_{\boldsymbol{k}}\hat{\beta}^\dagger_{\boldsymbol{k}},$$

$$\hat{b}^\dagger_{-\boldsymbol{k}} = -v_{\boldsymbol{k}}\hat{\alpha}_{\boldsymbol{k}} + u_{\boldsymbol{k}}\hat{\beta}^\dagger_{\boldsymbol{k}}.$$

Expressing the Hamiltonian in terms of the new operators $\hat{\alpha}_{\boldsymbol{k}}$ and $\hat{\beta}^\dagger_{\boldsymbol{k}}$ using the above inverse transformation, we have

$$\begin{aligned}\hat{H} = \sum_{\boldsymbol{k}}{}' \Big\{ &\big[(\varepsilon_{\boldsymbol{k}}+\Delta)\big(u_{\boldsymbol{k}}^2 + v_{\boldsymbol{k}}^2\big)/2 - \Delta u_{\boldsymbol{k}} v_{\boldsymbol{k}}\big]\big(\hat{\alpha}^\dagger_{\boldsymbol{k}}\hat{\alpha}_{\boldsymbol{k}} + \hat{\beta}^\dagger_{\boldsymbol{k}}\hat{\beta}_{\boldsymbol{k}}\big) \\ &+ \big[-(\varepsilon_{\boldsymbol{k}} + \Delta)u_{\boldsymbol{k}}v_{\boldsymbol{k}} + \Delta\big(u_{\boldsymbol{k}}^2 + v_{\boldsymbol{k}}^2\big)/2\big]\big(\hat{\alpha}_{\boldsymbol{k}}\hat{\beta}_{\boldsymbol{k}} + \hat{\beta}^\dagger_{\boldsymbol{k}}\hat{\alpha}^\dagger_{\boldsymbol{k}}\big) \\ &+ (\varepsilon_{\boldsymbol{k}} + \Delta)v_{\boldsymbol{k}}^2 - \Delta u_{\boldsymbol{k}}v_{\boldsymbol{k}}\Big\} + E_0.\end{aligned}$$

From the above expression of $\hat{H}$, we see that $\hat{H}$ is diagonalized if the term on the second line can be eliminated. This can be achieved by setting the coefficient of the term to zero using the freedom that the transformation coefficients are yet to be completely fixed. We thus have

$$-(\varepsilon_{\boldsymbol{k}} + \Delta)u_{\boldsymbol{k}}v_{\boldsymbol{k}} + \Delta\big(u_{\boldsymbol{k}}^2 + v_{\boldsymbol{k}}^2\big)/2 = 0.$$

Solving the above equation in conjunction with the constraint $u_{\boldsymbol{k}}^2 - v_{\boldsymbol{k}}^2 = 1$, we obtain

$$u_{\boldsymbol{k}}^2 = \frac{1}{2}\bigg\{1 + \frac{\varepsilon_{\boldsymbol{k}} + \Delta}{[\varepsilon_{\boldsymbol{k}}(\varepsilon_{\boldsymbol{k}} + 2\Delta)]^{1/2}}\bigg\},$$

$$v_{\boldsymbol{k}}^2 = \frac{1}{2}\bigg\{-1 + \frac{\varepsilon_{\boldsymbol{k}} + \Delta}{[\varepsilon_{\boldsymbol{k}}(\varepsilon_{\boldsymbol{k}} + 2\Delta)]^{1/2}}\bigg\},$$

$$u_{\boldsymbol{k}}v_{\boldsymbol{k}} = \frac{\Delta}{2[\varepsilon_{\boldsymbol{k}}(\varepsilon_{\boldsymbol{k}} + 2\Delta)]^{1/2}}.$$

We see that the two branches of collective excitations are degenerate. The spectrum is given by

$$\begin{aligned}\hbar\omega_{\boldsymbol{k}} &= \frac{1}{2}(\varepsilon_{\boldsymbol{k}} + \Delta)\big(u_{\boldsymbol{k}}^2 + v_{\boldsymbol{k}}^2\big) - \Delta u_{\boldsymbol{k}}v_{\boldsymbol{k}} \\ &= \frac{1}{2}\big[\varepsilon_{\boldsymbol{k}}(\varepsilon_{\boldsymbol{k}} + 2\Delta)\big]^{1/2}.\end{aligned}$$

The diagonalized Hamiltonian reads

$$\hat{H} = \sum_{\boldsymbol{k}}{}' \hbar\omega_{\boldsymbol{k}}\big(\hat{\alpha}^\dagger_{\boldsymbol{k}}\hat{\alpha}_{\boldsymbol{k}} + \hat{\beta}^\dagger_{\boldsymbol{k}}\hat{\beta}_{\boldsymbol{k}}\big) + \frac{1}{2}\sum_{\boldsymbol{k}}{}' \big(2\hbar\omega_{\boldsymbol{k}} - \varepsilon_{\boldsymbol{k}} - \Delta\big) + E_0.$$

(2) For $\varepsilon_{\boldsymbol{k}} \ll \Delta$ the dispersion relation becomes

$$\hbar\omega_{\boldsymbol{k}} \approx (\Delta/2)^{1/2}\varepsilon_{\boldsymbol{k}}^{1/2} \propto k,$$

where we have made use of $\varepsilon_{\boldsymbol{k}} = \hbar^2 k^2/2m$. Therefore, the spectrum of excitations contains sound modes.

28-21 ***XY* model for a one-dimensional quantum magnet.** The Hamiltonian of a one-dimensional quantum magnet in the XY model is given by

$$\hat{H} = \sum_{n=-\infty}^{\infty} \left(J_1 \hat{c}_n^\dagger \hat{c}_{n+1} + J_2 \hat{c}_n \hat{c}_{n+1} - B\hat{c}_n^\dagger \hat{c}_n\right) + \text{h.c.},$$

where J_1, J_2, and B are constants and $\hat{c}_n$ and $\hat{c}_n^\dagger$ are annihilation and creation operators of spin-zero fermions.

(1) Diagonalize the Hamiltonian using the Bogoliubov transformation and find the spectrum of quasiparticles $\hbar\omega_{\boldsymbol{k}}$.
(2) Show that the dispersion disappears if $J_1 = J_2$ and $B = 0$.

---

(1) We first transform the Hamiltonian into momentum space through

$$\hat{c}_n = \frac{1}{N^{1/2}} \sum_k e^{ikR_n} c_k, \quad \hat{c}_n^\dagger = \frac{1}{N^{1/2}} \sum_k e^{-ikR_n} c_k^\dagger,$$

where N is the number of lattice sites and R_n the position of the nth lattice site, $R_n = na$ with a the lattice constant. For $\sum_n \hat{c}_n^\dagger \hat{c}_{n+1} + \text{h.c.}$, we have

$$\begin{aligned}
\sum_n \hat{c}_n^\dagger \hat{c}_{n+1} + \text{h.c.} &= \sum_{kk'} e^{ika} \hat{c}_{k'}^\dagger \hat{c}_k \frac{1}{N} \sum_n e^{-i(k'-k)R_n} + \text{h.c.} \\
&= \sum_k e^{ika} \hat{c}_k^\dagger \hat{c}_k + \text{h.c.} = \sum_k \left(e^{ika} + e^{-ika}\right) \hat{c}_k^\dagger \hat{c}_k \\
&= 2\sum_k \cos(ka) \hat{c}_k^\dagger \hat{c}_k.
\end{aligned}$$

For $\sum_n \hat{c}_n\hat{c}_{n+1} + \text{h.c.}$, we have

$$\begin{aligned}
&\sum_n \hat{c}_n\hat{c}_{n+1} + \text{h.c.} \\
&\quad = \sum_{kk'} \mathrm{e}^{\mathrm{i}ka}\hat{c}_{k'}\hat{c}_k \frac{1}{N}\sum_n \mathrm{e}^{\mathrm{i}(k'+k)R_n} + \text{h.c.} \\
&\quad = \sum_k \mathrm{e}^{\mathrm{i}ka}\hat{c}_{-k}\hat{c}_k + \text{h.c.} = \sum_k \big(\mathrm{e}^{\mathrm{i}ka}\hat{c}_{-k}\hat{c}_k + \mathrm{e}^{-\mathrm{i}ka}\hat{c}_k^\dagger\hat{c}_{-k}^\dagger\big) \\
&\quad = \frac{1}{2}\sum_k \big(\mathrm{e}^{\mathrm{i}ka}\hat{c}_{-k}\hat{c}_k + \mathrm{e}^{-\mathrm{i}ka}\hat{c}_k\hat{c}_{-k} + \mathrm{e}^{-\mathrm{i}ka}\hat{c}_k^\dagger\hat{c}_{-k}^\dagger + \mathrm{e}^{\mathrm{i}ka}\hat{c}_{-k}^\dagger\hat{c}_k^\dagger\big) \\
&\quad = \frac{1}{2}\sum_k \big[\big(\mathrm{e}^{\mathrm{i}ka} - \mathrm{e}^{-\mathrm{i}ka}\big)\hat{c}_{-k}\hat{c}_k + \big(\mathrm{e}^{-\mathrm{i}ka} - \mathrm{e}^{\mathrm{i}ka}\big)\hat{c}_k^\dagger\hat{c}_{-k}^\dagger\big] \\
&\quad = \mathrm{i}\sum_k \sin(ka)\big(\hat{c}_{-k}\hat{c}_k - \hat{c}_k^\dagger\hat{c}_{-k}^\dagger\big).
\end{aligned}$$

For $\sum_n \hat{c}_n^\dagger\hat{c}_n + \text{h.c.}$, we have

$$\begin{aligned}
\sum_n \hat{c}_n^\dagger\hat{c}_n + \text{h.c.} &= \sum_{kk'} \hat{c}_{k'}^\dagger\hat{c}_k \frac{1}{N}\sum_n \mathrm{e}^{-\mathrm{i}(k'-k)R_n} + \text{h.c.} \\
&= \sum_k \hat{c}_k^\dagger\hat{c}_k + \text{h.c.} = 2\sum_k \hat{c}_k^\dagger\hat{c}_k.
\end{aligned}$$

Thus, the Hamiltonian in momentum space is given by

$$\hat{H} = \sum_k \varepsilon_k\big(\hat{c}_k^\dagger\hat{c}_k + \hat{c}_{-k}^\dagger\hat{c}_{-k}\big) + \mathrm{i}J_2\sum_k \sin(ka)\big(\hat{c}_{-k}\hat{c}_k - \hat{c}_k^\dagger\hat{c}_{-k}^\dagger\big),$$

where $\varepsilon_k = J_1\cos(ka) - B$. To find out how to make linear combinations of operators, we evaluate the commutator $[\,\hat{c}_k, \hat{H}\,]$ which appears in Heisenberg's equation of motion for $\hat{c}_k$. We have

$$\begin{aligned}
[\,\hat{c}_k, \hat{H}\,] &= \varepsilon_k\hat{c}_k + \mathrm{i}J_2\sum_{k'} \sin(k'a)[\,\hat{c}_{k'}, \hat{c}_{-k'}\hat{c}_{k'} - \hat{c}_{k'}^\dagger\hat{c}_{-k'}^\dagger\,] \\
&= \varepsilon_k\hat{c}_k + \mathrm{i}J_2\sum_{k'} \sin(k'a)\big(-\hat{c}_{-k'}^\dagger\delta_{k'k} + \hat{c}_{k'}^\dagger\delta_{k',-k}\big) \\
&= \varepsilon_k\hat{c}_k - 2\mathrm{i}J_2\sin(ka)\,\hat{c}_{-k}^\dagger.
\end{aligned}$$

Therefore, we make the following Bogoliubov transformation

$$\begin{aligned}
\hat{\alpha}_k &= u_k\hat{c}_k + \mathrm{i}v_k\hat{c}_{-k}^\dagger, \\
\hat{\beta}_k^\dagger &= p_k\hat{c}_k + \mathrm{i}q_k\hat{c}_{-k}^\dagger
\end{aligned}$$

in which all the coefficients, u_k, v_k, p_k, and q_k, are real. The new operators must satisfy the following anticommutation relations to preserve the algebraic properties of the original operators

$$\begin{aligned}
&\{\hat{\alpha}_k, \hat{\alpha}^\dagger_{k'}\} = \delta_{kk'}, \ \ \{\hat{\alpha}_k, \hat{\alpha}_{k'}\} = \{\hat{\alpha}^\dagger_k, \hat{\alpha}^\dagger_{k'}\} = 0,\\
&\{\hat{\beta}_k, \hat{\beta}^\dagger_{k'}\} = \delta_{kk'}, \ \ \{\hat{\beta}_k, \hat{\beta}_{k'}\} = \{\hat{\beta}^\dagger_k, \hat{\beta}^\dagger_{k'}\} = 0,\\
&\{\hat{\alpha}_k, \hat{\beta}_{k'}\} = \{\hat{\alpha}_k, \hat{\beta}^\dagger_{k'}\} = \{\hat{\alpha}^\dagger_k, \hat{\beta}_{k'}\} = \{\hat{\alpha}^\dagger_k, \hat{\beta}^\dagger_{k'}\} = 0.
\end{aligned}$$

As usual, some of the above anticommutation relations are satisfied automatically and others yield constraints on the transformation coefficients. We obtain the following three constraints from the above anticommutation relations

$$u_k^2 + v_k^2 = 1, \ \ p_k^2 + q_k^2 = 1, \ \ u_k p_k + v_k q_k = 0.$$

From the third constraint, we choose $p_k = -v_k$ and $q_k = u_k$. The Bogoliubov transformation then becomes

$$\begin{aligned}
\hat{\alpha}_k &= u_k \hat{c}_k + \mathrm{i} v_k \hat{c}^\dagger_{-k},\\
\hat{\beta}^\dagger_k &= -v_k \hat{c}_k + \mathrm{i} u_k \hat{c}^\dagger_{-k}.
\end{aligned}$$

The inverse transformation reads

$$\begin{aligned}
\hat{c}_k &= u_k \hat{\alpha}_k - v_k \hat{\beta}^\dagger_k,\\
\hat{c}^\dagger_{-k} &= -\mathrm{i} v_k \hat{\alpha}_k - \mathrm{i} u_k \hat{\beta}^\dagger_k.
\end{aligned}$$

In terms of the new operators $\hat{\alpha}_k$ and $\hat{\beta}^\dagger_k$, the Hamiltonian is given by

$$\begin{aligned}
\hat{H} = \sum_k \Big\{ &\big[\varepsilon_k (u_k^2 - v_k^2) - 2J_2 \sin(ka) u_k v_k\big] (\hat{\alpha}^\dagger_k \hat{\alpha}_k + \hat{\beta}^\dagger_k \hat{\beta}_k)\\
&+ \big[2\varepsilon_k u_k v_k + J_2 \sin(ka)(u_k^2 - v_k^2)\big](\hat{\alpha}_k \hat{\beta}_k + \hat{\beta}^\dagger_k \hat{\alpha}^\dagger_k)\\
&+ 2\varepsilon_k v_k^2 + 2J_2 \sin(ka) u_k v_k \Big\}.
\end{aligned}$$

The Hamiltonian is diagonalized if the term on the second line can be eliminated. This can be accomplished by setting the coefficient of this term to zero. We have

$$2\varepsilon_k u_k v_k + J_2 \sin(ka)(u_k^2 - v_k^2) = 0.$$

Solving for u_k and v_k in conjunction with the constraint $u_k^2 + v_k^2 = 1$, we obtain

$$u_k^2 = \frac{1}{2}\left[1 + \frac{\varepsilon_k}{[\,\varepsilon_k^2 + J_2^2 \sin^2(ka)\,]^{1/2}}\right],$$

$$v_k^2 = \frac{1}{2}\left[1 - \frac{\varepsilon_k}{[\,\varepsilon_k^2 + J_2^2 \sin^2(ka)\,]^{1/2}}\right],$$

$$u_k v_k = -\frac{J_2 \sin(ka)}{2[\,\varepsilon_k^2 + J_2^2 \sin^2(ka)\,]^{1/2}}.$$

The diagonalized Hamiltonian is then given by

$$\hat{H} = \sum_k \hbar\omega_k \big(\hat{\alpha}_k^\dagger \hat{\alpha}_k + \hat{\beta}_k^\dagger \hat{\beta}_k\big) + \sum_k \big(\varepsilon_k - \hbar\omega_k\big),$$

where $\hbar\omega_k = [\,\varepsilon_k^2 + J_2^2 \sin^2(ka)\,]^{1/2}$ is the spectrum of quasiparticles.

(2) If $J_1 = J_2 = J$ and $B = 0$, we have

$$\hbar\omega_k = [\,J_1^2 \cos^2(ka) + J_2^2 \sin^2(ka)\,]^{1/2} = J.$$

Therefore, the dispersion disappears.

28-22 Schwinger-boson mean-field theory for a one-dimensional ferrimagnet. In the Schwinger-boson mean-field theory, the Hamiltonian of a one-dimensional ferrimagnet is given by

$$\hat{H} = E_0 + 2J \sum_k \sum_{m=\pm 1} \big(\lambda \hat{a}_{km}^\dagger \hat{a}_{km} + \mu \hat{b}_{km}^\dagger \hat{b}_{km}\big)$$
$$- 2\zeta J \sum_k \cos(ka)\big(\hat{a}_{k1}\hat{b}_{k,-1} - \hat{a}_{k,-1}\hat{b}_{k1} + \text{h.c.}\big),$$

where J, λ, μ, and ζ are constants and $\hat{a}_{km}$, $\hat{a}_{km}^\dagger$ and $\hat{b}_{km}$, $\hat{b}_{km}^\dagger$ are respectively annihilation and creation operators of two different kinds of spin-one bosons with $-\pi/a < k \leqslant \pi/a$. Diagonalize the Hamiltonian using the Bogoliubov transformation.

We first find out how to make linear combinations. The commutation relation between $\hat{b}_{k1}$ and $\hat{H}$ is

$$[\,\hat{b}_{k1}, \hat{H}\,] = 2J\mu \hat{b}_{k1} + 2\zeta J \cos(ka) \hat{a}_{k,-1}^\dagger.$$

Thus, $\hat{b}_{k1}$ and $\hat{a}_{k,-1}^\dagger$ are to be mixed up. For $\hat{b}_{k,-1}$, we have

$$[\,\hat{b}_{k,-1}, \hat{H}\,] = 2J\mu \hat{b}_{k1} + 2\zeta J \cos(ka) \hat{a}_{k1}^\dagger.$$

Thus, $\hat{b}_{k,-1}$ and $\hat{a}^\dagger_{k1}$ are to be mixed up. To solve the problem, we must make two sets of Bogoliubov transformations, which indicates that the Hamiltonian can be broken into two parts. We rewrite the Hamiltonian as

$$\hat{H} = E_0 + \hat{H}_1 + \hat{H}_2,$$

where

$$\begin{aligned}\hat{H}_1 &= 2J\sum_k \left(\lambda \hat{a}^\dagger_{k,-1}\hat{a}_{k,-1} + \mu \hat{b}^\dagger_{k1}\hat{b}_{k1}\right)\\ &\quad + 2\zeta J\sum_k \cos(ka)\left(\hat{a}_{k,-1}\hat{b}_{k1} + \hat{b}^\dagger_{k1}\hat{a}^\dagger_{k,-1}\right),\\ \hat{H}_2 &= 2J\sum_k \left(\lambda \hat{a}^\dagger_{k1}\hat{a}_{k1} + \mu \hat{b}^\dagger_{k,-1}\hat{b}_{k,-1}\right)\\ &\quad - 2\zeta J\sum_k \cos(ka)\left(\hat{a}_{k1}\hat{b}_{k,-1} + \hat{b}^\dagger_{k,-1}\hat{a}^\dagger_{k1}\right).\end{aligned}$$

We can diagonalize $\hat{H}_1$ and $\hat{H}_2$ separately.

(1) Diagonalization of $\hat{H}_1$

To diagonalize $\hat{H}_1$, we make the following Bogoliubov transformation

$$\begin{aligned}\hat{\alpha}_k &= u_k\hat{b}_{k1} + v_k\hat{a}^\dagger_{k,-1},\\ \hat{\beta}^\dagger_k &= p_k\hat{b}_{k1} + q_k\hat{a}^\dagger_{k,-1}.\end{aligned}$$

The following are the commutation relations that $\hat{\alpha}_k$ and $\hat{\beta}_k$ must satisfy

$$\begin{aligned}&[\hat{\alpha}_{\boldsymbol{k}}, \hat{\alpha}^\dagger_{\boldsymbol{k}'}] = \delta_{\boldsymbol{kk}'},\ \ [\hat{\alpha}_{\boldsymbol{k}}, \hat{\alpha}_{\boldsymbol{k}'}] = [\hat{\alpha}^\dagger_{\boldsymbol{k}}, \hat{\alpha}^\dagger_{\boldsymbol{k}'}] = 0,\\ &[\hat{\beta}_{\boldsymbol{k}}, \hat{\beta}^\dagger_{\boldsymbol{k}'}] = \delta_{\boldsymbol{kk}'},\ \ [\hat{\beta}_{\boldsymbol{k}}, \hat{\beta}_{\boldsymbol{k}'}] = [\hat{\beta}^\dagger_{\boldsymbol{k}}, \hat{\beta}^\dagger_{\boldsymbol{k}'}] = 0,\\ &[\hat{\alpha}_{\boldsymbol{k}}, \hat{\beta}_{\boldsymbol{k}'}] = [\hat{\alpha}_{\boldsymbol{k}}, \hat{\beta}_{\boldsymbol{k}'}] = [\hat{\alpha}^\dagger_{\boldsymbol{k}}, \hat{\beta}_{\boldsymbol{k}'}] = [\hat{\alpha}^\dagger_{\boldsymbol{k}}, \hat{\beta}^\dagger_{\boldsymbol{k}'}] = 0.\end{aligned}$$

From the nontrivial commutation relations, we obtain the following three constraints on the transformation coefficients

$$u_k^2 - v_k^2 = 1,\ \ -p_k^2 + q_k^2 = 1,\ \ u_kp_k - v_kq_k = 0.$$

From the third constraint, we choose $p_k = v_k$ and $q_k = u_k$. We are then left with only a single constraint

$$u_k^2 - v_k^2 = 1.$$

and the Bogoliubov transformation becomes

$$\hat{\alpha}_k = u_k \hat{b}_{k1} + v_k \hat{a}^\dagger_{k,-1},$$
$$\hat{\beta}^\dagger_k = v_k \hat{b}_{k1} + u_k \hat{a}^\dagger_{k,-1}.$$

The inverse transformation is given by

$$\hat{b}_{k1} = u_k \hat{\alpha}_k - v_k \hat{\beta}^\dagger_k,$$
$$\hat{a}^\dagger_{k,-1} = -v_k \hat{\alpha}_k + u_k \hat{\beta}^\dagger_k.$$

In terms of the new operators, $\hat{H}_1$ becomes

$$\begin{aligned}\hat{H}_1 = 2J\sum_k \{ & \left[\lambda v_k^2 + \mu u_k^2 - 2\zeta\cos(ka) u_k v_k\right] \hat{\alpha}^\dagger_k \hat{\alpha}_k \\ & + \left[\lambda u_k^2 + \mu v_k^2 - 2\zeta \cos(ka) u_k v_k\right] \hat{\beta}^\dagger_k \hat{\beta}_k \\ & + \left[-(\lambda+\mu) u_k v_k + \zeta\cos(ka)\left(u_k^2 + v_k^2\right)\right]\left(\hat{\alpha}_k \hat{\beta}_k + \hat{\beta}^\dagger_k \hat{\alpha}^\dagger_k\right) \\ & \quad + (\lambda+\mu) v_k^2 - 2\zeta\cos(ka) u_k v_k \}.\end{aligned}$$

Setting the coefficient of the term on the third line to zero diagonalizes $\hat{H}_1$

$$-(\lambda+\mu) u_k v_k + \zeta\cos(ka)\left(u_k^2 + v_k^2\right)\,\big]\left(\hat{\alpha}_k \hat{\beta}_k + \hat{\beta}_k = 0.\right.$$

Solving for u_k and v_k from the above equation together with $u_k^2 - v_k^2 = 1$, we obtain

$$u_k^2 = \frac{1}{2}\left[1 + \frac{\lambda+\mu}{[(\lambda+\mu)^2 - 4\zeta^2\cos^2(ka)]^{1/2}}\right],$$
$$v_k^2 = \frac{1}{2}\left[-1 + \frac{\lambda+\mu}{[(\lambda+\mu)^2 - 4\zeta^2\cos^2(ka)]^{1/2}}\right],$$
$$u_k v_k = \frac{\zeta\cos(ka)}{[(\lambda+\mu)^2 - 4\zeta^2\cos^2(ka)]^{1/2}}.$$

The diagonalized $\hat{H}_1$ is given by

$$\hat{H}_1 = \sum_k \left(\hbar\omega_k^+ \hat{\alpha}^\dagger_k \hat{\alpha}_k + \hbar\omega_k^- \hat{\beta}^\dagger_k \hat{\beta}_k\right) + \sum_k \left(\hbar\omega_k^+ - 2\mu\right),$$

where the dispersion relations $\hbar\omega_k^\pm$ are given by

$$\hbar\omega_k^\pm = J\left\{[(\lambda+\mu)^2 - 4\zeta^2\cos^2(ka)]^{1/2} \pm (\mu - \lambda)\right\}.$$

(2) Diagonalization of $\hat{H}_2$

Note that, except for the spins of the annihilation and creation operators, $\hat{H}_2$ differs from $\hat{H}_1$ only in the sign of the term containing ζ. In consideration that all the factors and constant terms in the diagonalized $\hat{H}_1$ do not depend on this sign, we will obtain the same values for the corresponding factors and constant terms when $\hat{H}_2$ is diagonalized through the following Bogoliubov transformation

$$\hat{\alpha}'_k = \mu_k \hat{b}_{k,-1} + \nu_k \hat{a}^\dagger_{k1},$$
$$\hat{\beta}'^\dagger_k = \nu_k \hat{b}_{k,-1} + \mu_k \hat{a}^\dagger_{k1}.$$

The diagonalized $\hat{H}_2$ is given by

$$\hat{H}_2 = \sum_k \left(\hbar\omega_k^+ \hat{\alpha}'^\dagger_k \hat{\alpha}'_k + \hbar\omega_k^- \hat{\beta}'^\dagger_k \hat{\beta}'_k\right) + \sum_k \left(\hbar\omega_k^+ - 2\mu\right),$$

To simplify notations, we rename $\hat{\alpha}_k$ as $\hat{\alpha}_{k1}$, $\hat{\beta}_k$ as $\hat{\beta}_{k1}$, $\hat{\alpha}'_k$ as $\hat{\alpha}_{k,-1}$, and $\hat{\beta}'_k$ as $\hat{\beta}_{k,-1}$. Then, the diagonalized total Hamiltonian $\hat{H}$ is given by

$$\hat{H} = \sum_k \sum_{m=\pm 1} \left(\hbar\omega_k^+ \hat{\alpha}^\dagger_{km} \hat{\alpha}_{km} + \hbar\omega_k^- \hat{\beta}^\dagger_{km} \hat{\beta}_{km}\right) + 2\sum_k \left(\hbar\omega_k^+ - 2\mu\right) + E_0.$$

Chapter 29

Optical Properties of Solids

(1) *Complex index of refraction*

$$N = \sqrt{\epsilon_r} = n + \mathrm{i}k.$$

n is the refractive index.

k is the extinction coefficient.

(2) *Absorption coefficient*

$$\alpha = \omega k/c.$$

(3) *Reflectivity at normal incidence*

$$R = \left|\frac{1-N}{1+N}\right|^2 = \frac{(1-n)^2+k^2}{(1+n)^2+k^2}.$$

Reflectivity is also called *reflectance.*

(4) *Kramers-Kronig relations*

$$\epsilon_r'(\omega) = \epsilon_r'(\infty) + \frac{2}{\pi}\mathrm{P}\int_0^\infty \mathrm{d}\omega'\,\frac{\omega'\epsilon_r''(\omega')}{\omega'^2-\omega^2},$$

$$\epsilon_r''(\omega) = -\frac{2\omega}{\pi}\mathrm{P}\int_0^\infty \mathrm{d}\omega'\,\frac{\epsilon_r'(\omega)-\epsilon_r'(\infty)}{\omega'^2-\omega^2},$$

where $\epsilon_r' = \mathrm{Re}\,\epsilon_r$ and $\epsilon_r'' = \mathrm{Im}\,\epsilon_r$.

(5) *Sum rules for the dielectric function*

$$\frac{2}{\pi}\int_0^\infty \mathrm{d}\omega\,\frac{1}{\omega}\epsilon_r''(\omega) = \epsilon_r(0) - 1,$$

$$\int_0^\infty \mathrm{d}\omega\,\omega\epsilon_r''(\omega) = \frac{1}{2}\pi\omega_p^2,$$

$$\int_0^\infty \mathrm{d}\omega\,\omega\frac{1}{\epsilon_r''(\omega)} = -\frac{1}{2}\pi\omega_p^2.$$

(6) *Determination of* n, k, *and* ϵ_r

$$r = (n + \mathrm{i}k - 1)/(n + \mathrm{i}k + 1) = |r|\mathrm{e}^{\mathrm{i}\varphi} \text{ with } |r| = \sqrt{R},$$

$$\varphi(\omega) = -\frac{2\omega}{\pi}\,\mathrm{P}\int_0^\infty \mathrm{d}\omega' \,\frac{\ln(|r(\omega')|/|r(\omega)|)}{\omega'^2 - \omega^2},$$

$$n = \frac{1-R}{1 - 2\sqrt{R}\cos\varphi + R}, \quad k = \frac{2\sqrt{R}\sin\varphi}{1 - 2\sqrt{R}\cos\varphi + R},$$

$$\epsilon_r' = n^2 - k^2, \ \epsilon_r'' = 2nk.$$

(7) *Optical properties of free electrons*

The Drude optical conductivity is

$$\sigma(\omega) = \frac{\sigma_0}{1 - \mathrm{i}\omega\tau}$$

with $\sigma_0 = ne^2\tau/m$.

The dielectric function is

$$\epsilon = \left[\epsilon_L - \frac{\sigma_0\tau}{1+\omega^2\tau^2}\right] + \frac{\mathrm{i}\sigma_0}{\omega(1+\omega^2\tau^2)}.$$

(8) *Plasma frequency*

$$\omega_p^2 = ne^2/\epsilon_0 m.$$

(9) *Lorentz oscillator*

$$\epsilon_r = 1 + \frac{ne^2/\epsilon_0 m}{\omega_0^2 - \omega^2 - \mathrm{i}\gamma\omega}.$$

(10) *Oscillator strength*

$$\epsilon_r = 1 + \frac{ne^2}{\epsilon_0 m}\sum_j \frac{f_j}{\omega_{0j}^2 - \omega^2 - \mathrm{i}\gamma_j\omega}$$

with f_j's the oscillator strengths.

(11) *Absorption edge in a semiconductor*

$$\alpha(\omega) \propto \sqrt{\hbar\omega - E_g}/\hbar\omega.$$

(12) *Critical points*

Defined through $\nabla_{\boldsymbol{k}}(\varepsilon_{\boldsymbol{k}}^c - \varepsilon_{\boldsymbol{k}}^v) = 0$.

Classified according to the number of negative coefficients b_i's in

$$\Delta\varepsilon_{\boldsymbol{k}} \equiv \varepsilon_{\boldsymbol{k}}^c - \varepsilon_{\boldsymbol{k}}^v \approx \Delta\varepsilon_{\boldsymbol{k}_0} + \textstyle\sum_i b_i\kappa_i^2.$$

(13) *Energy of an exciton*

$$\mathcal{E}_{n\boldsymbol{K}}^{\mathrm{ex}} = E_g + \frac{\hbar^2\boldsymbol{K}^2}{2(m_e^* + m_h^*)} - \frac{R_{\mathrm{ex}}}{n^2}.$$

(14) *Absorption of light through indirect interband transitions*
Conservation of energy:

$$\hbar\omega = E_g - \hbar\omega_{\boldsymbol{q}} + (\varepsilon_n - \varepsilon_c) + \varepsilon_p.$$

Conservation of crystal momentum:

$$\boldsymbol{q} = \boldsymbol{k}_n - \boldsymbol{k}_c.$$

Absorption of light with the absorption of a phonon:

$$\alpha_a \propto \frac{(\hbar\omega + \hbar\omega_{\boldsymbol{q}} - E_g)^2}{\mathrm{e}^{\hbar\omega_{\boldsymbol{q}}/k_\mathrm{B}T} - 1}.$$

Absorption of light with the emission of a phonon:

$$\alpha_e \propto \frac{(\hbar\omega - \hbar\omega_{\boldsymbol{q}} - E_g)^2}{1 - \mathrm{e}^{-\hbar\omega_{\boldsymbol{q}}/k_\mathrm{B}T}}.$$

(15) *Raman scattering*
Conservation of momentum and energy:

$$\hbar\boldsymbol{q}_2 = \hbar\boldsymbol{q}_1 \pm \hbar\boldsymbol{Q},$$
$$\hbar\omega_{\boldsymbol{q}_2} = \hbar\omega_{\boldsymbol{q}_1} \pm \hbar\Omega_{\boldsymbol{Q}s}.$$

Stokes and anti-Stokes scattering:

$$I_{\mathrm{anti-Stokes}}/I_{\mathrm{Stokes}} = \mathrm{e}^{-\hbar\Omega_{\boldsymbol{Q}s}/k_\mathrm{B}T}.$$

(16) *Polaritons*
In the long-wavelength limit, the frequencies of polaritons are given by

$$\omega_+^2 \approx (\epsilon_r(0)/\epsilon_r(\infty))\omega_\mathrm{T}^2 = \omega_\mathrm{L}^2,$$
$$\omega_-^2 \approx c^2k^2/\epsilon_r(0).$$

In the short-wavelength limit, the frequencies of polaritons are given by

$$\omega_+^2 \approx c^2k^2/\epsilon_r(\infty),\ \omega_-^2 \approx \omega_\mathrm{T}^2.$$

The dielectric function is given by

$$\epsilon(\omega) = \epsilon(\infty) - \frac{\epsilon(0) - \epsilon(\infty)}{\omega^2/\omega_\mathrm{T}^2 - 1}.$$

29-1 Properties of the electric susceptibility tensor. The time-domain electric field $\boldsymbol{E}(\boldsymbol{r}, t)$ and the polarization $\boldsymbol{P}(\boldsymbol{r}, t)$ are real quantities. Their relation is $P_{\alpha\beta}(\boldsymbol{r}, t) = \epsilon_0 \sum_\beta \chi_{\alpha\beta}(\boldsymbol{r}, t)E_\beta(\boldsymbol{r}, t)$, where $\chi_{\alpha\beta}(\boldsymbol{r}, t)$ is a component of the electric susceptibility tensor.

(1) Show that the complex conjugate of the wave vector and frequency-domain electric susceptibility is given by

$\chi^*_{\alpha\beta}(\boldsymbol{q}, \omega) = \chi_{\alpha\beta}(-\boldsymbol{q}, -\omega)$.

(2) Determine the symmetry of the real and imaginary parts of $\chi^*_{\alpha\beta}(\boldsymbol{q}, \omega)$.

---

(1) We write the Fourier expansions of $\boldsymbol{E}(\boldsymbol{r}, t)$, $\boldsymbol{P}(\boldsymbol{r}, t)$, and $\chi_{\alpha\beta}(\boldsymbol{r}, t)$ as

$$\boldsymbol{E}(\boldsymbol{r}, t) = \int \mathrm{d}\boldsymbol{q} \int_{-\infty}^{\infty} \mathrm{d}\omega \; \mathrm{e}^{\mathrm{i}(\boldsymbol{q}\cdot\boldsymbol{r}-\omega t)} \boldsymbol{E}(\boldsymbol{q}, \omega),$$

$$\boldsymbol{P}(\boldsymbol{r}, t) = \int \mathrm{d}\boldsymbol{q} \int_{-\infty}^{\infty} \mathrm{d}\omega \; \mathrm{e}^{\mathrm{i}(\boldsymbol{q}\cdot\boldsymbol{r}-\omega t)} \boldsymbol{P}(\boldsymbol{q}, \omega),$$

$$\chi_{\alpha\beta}(\boldsymbol{r}, t) = \int \mathrm{d}\boldsymbol{q} \int_{-\infty}^{\infty} \mathrm{d}\omega \; \mathrm{e}^{\mathrm{i}(\boldsymbol{q}\cdot\boldsymbol{r}-\omega t)} \chi_{\alpha\beta}(\boldsymbol{q}, \omega).$$

We then have

$$P_\alpha(\boldsymbol{q}, \omega) = \epsilon_0 \sum_\beta \int \mathrm{d}\boldsymbol{q}' \int_{-\infty}^{\infty} \mathrm{d}\omega' \; \chi_{\alpha\beta}(\boldsymbol{q}-\boldsymbol{q}', \omega-\omega') E_\beta(\boldsymbol{q}', \omega').$$

Taking the complex conjugation of $\boldsymbol{E}(\boldsymbol{r}, t)$ and then making a change of integration variables from $\boldsymbol{q}$ to $-\boldsymbol{q}$ and from ω to $-\omega$, we have

$$\begin{aligned}
\boldsymbol{E}^*(\boldsymbol{r}, t) &= \int \mathrm{d}\boldsymbol{q} \int_{-\infty}^{\infty} \mathrm{d}\omega \; \mathrm{e}^{-\mathrm{i}(\boldsymbol{q}\cdot\boldsymbol{r}-\omega t)} \boldsymbol{E}^*(\boldsymbol{q}, \omega) \\
&= \int \mathrm{d}\boldsymbol{q} \int_{-\infty}^{\infty} \mathrm{d}\omega \; \mathrm{e}^{\mathrm{i}(\boldsymbol{q}\cdot\boldsymbol{r}-\omega t)} \boldsymbol{E}^*(-\boldsymbol{q}, -\omega) \\
&= \boldsymbol{E}(\boldsymbol{r}, t) = \int \mathrm{d}\boldsymbol{q} \int_{-\infty}^{\infty} \mathrm{d}\omega \; \mathrm{e}^{\mathrm{i}(\boldsymbol{q}\cdot\boldsymbol{r}-\omega t)} \boldsymbol{E}(\boldsymbol{q}, \omega).
\end{aligned}$$

We thus have $\boldsymbol{E}^*(-\boldsymbol{q}, -\omega) = \boldsymbol{E}(\boldsymbol{q}, \omega)$. That is,

$$\boldsymbol{E}^*(\boldsymbol{q}, \omega) = \boldsymbol{E}(-\boldsymbol{q}, -\omega).$$

We similarly have

$$\boldsymbol{P}^*(\boldsymbol{q}, \omega) = \boldsymbol{P}(-\boldsymbol{q}, -\omega).$$

Taking the complex conjugation of the expression of $\boldsymbol{P}(\boldsymbol{q}, \omega)$ in terms of $\boldsymbol{E}(\boldsymbol{q}, \omega)$, we have

$$P^*_\alpha(\boldsymbol{q}, \omega) = \epsilon_0 \sum_\beta \int \mathrm{d}\boldsymbol{q}' \int_{-\infty}^{\infty} \mathrm{d}\omega' \; \chi^*_{\alpha\beta}(\boldsymbol{q}-\boldsymbol{q}', \omega-\omega') E^*_\beta(\boldsymbol{q}', \omega').$$

Making use of $\boldsymbol{P}^*(\boldsymbol{q},\omega) = \boldsymbol{P}(-\boldsymbol{q},-\omega)$ and $\boldsymbol{E}^*(\boldsymbol{q},\omega) = \boldsymbol{E}(-\boldsymbol{q},-\omega)$, we have

$$P_\alpha(-\boldsymbol{q},-\omega) = \epsilon_0 \sum_\beta \int \mathrm{d}\boldsymbol{q}' \int_{-\infty}^{\infty} \mathrm{d}\omega' \, \chi^*_{\alpha\beta}(\boldsymbol{q}-\boldsymbol{q}',\omega-\omega') E_\beta(-\boldsymbol{q}',-\omega').$$

Setting $-\boldsymbol{q}$ to $\boldsymbol{q}$ and $-\omega$ to ω and making a change of integration variables from $\boldsymbol{q}'$ to $-\boldsymbol{q}'$ and from ω' to $-\omega'$, we have

$$P_\alpha(\boldsymbol{q},\omega) = \epsilon_0 \sum_\beta \int \mathrm{d}\boldsymbol{q}' \int_{-\infty}^{\infty} \mathrm{d}\omega' \, \chi^*_{\alpha\beta}(-(\boldsymbol{q}-\boldsymbol{q}'),-(\omega-\omega')) E_\beta(\boldsymbol{q}',\omega').$$

Comparing the above expression for $P_\alpha(\boldsymbol{q},\omega)$ with the original one, we see that

$$\chi^*_{\alpha\beta}(-(\boldsymbol{q}-\boldsymbol{q}'),-(\omega-\omega')) = \chi_{\alpha\beta}(\boldsymbol{q}-\boldsymbol{q}',\omega-\omega').$$

That is,

$$\chi^*_{\alpha\beta}(\boldsymbol{q},\omega) = \chi_{\alpha\beta}(-\boldsymbol{q},-\omega).$$

(2) In terms of its real and imaginary parts, $\chi_{\alpha\beta}(\boldsymbol{q},\omega)$ is written as

$$\chi_{\alpha\beta}(\boldsymbol{q},\omega) = \mathrm{Re}\,\chi_{\alpha\beta}(\boldsymbol{q},\omega) + \mathrm{i}\,\mathrm{Im}\,\chi_{\alpha\beta}(\boldsymbol{q},\omega).$$

Taking the complex conjugation of the above expression yields

$$\chi^*_{\alpha\beta}(\boldsymbol{q},\omega) = \mathrm{Re}\,\chi_{\alpha\beta}(\boldsymbol{q},\omega) - \mathrm{i}\,\mathrm{Im}\,\chi_{\alpha\beta}(\boldsymbol{q},\omega).$$

On the other hand, we have

$$\begin{aligned}\chi^*_{\alpha\beta}(\boldsymbol{q},\omega) &= \chi_{\alpha\beta}(-\boldsymbol{q},-\omega)\\ &= \mathrm{Re}\,\chi_{\alpha\beta}(-\boldsymbol{q},-\omega) + \mathrm{i}\,\mathrm{Im}\,\chi_{\alpha\beta}(-\boldsymbol{q},-\omega)\end{aligned}$$

from which it flows that

$$\begin{aligned}\mathrm{Re}\,\chi_{\alpha\beta}(-\boldsymbol{q},-\omega) &= \mathrm{Re}\,\chi_{\alpha\beta}(\boldsymbol{q},\omega),\\ \mathrm{Im}\,\chi_{\alpha\beta}(-\boldsymbol{q},-\omega) &= -\,\mathrm{Im}\,\chi_{\alpha\beta}(\boldsymbol{q},\omega).\end{aligned}$$

Therefore, the real part of $\chi_{\alpha\beta}(\boldsymbol{q},\omega)$ is an even function of $\boldsymbol{q}$ and ω while its imaginary part is an odd function of $\boldsymbol{q}$ and ω.

29-2 Reflectivity of a simple metal. The optical reflectivity of a solid is given by

$$R(\omega) = \left|[1 - N(\omega)]/[1 + N(\omega)]\right|^2$$
$$= \{[1 - n(\omega)]^2 + k^2(\omega)\}/\{[1 + n(\omega)]^2 + k^2(\omega)\},$$

where $N(\omega) = \sqrt{\epsilon_r(\omega)} = n(\omega) + \mathrm{i}k(\omega)$ is the complex refractive index. Here, the transverse radiation wave is of concern and its wave vector $\boldsymbol{q}$ has been assumed to be zero, $\boldsymbol{q} = 0$. Consider a simple metal and assume that its optical conductivity can be described by the Drude approximation $\sigma(\omega) = (ne^2\tau/m)/(1 - \mathrm{i}\omega\tau)$. The relative dielectric function $\epsilon_r(\omega)$ is related to the optical conductivity $\sigma(\omega)$ through the usual expression $\epsilon_r(\omega) = 1 + \mathrm{i}\sigma(\omega)/\epsilon_0\omega$. We assume that $\omega_p\tau \gg 1$, where ω_p is the plasma frequency, $\omega_p^2 = ne^2/\epsilon_0 m$. We now consider three different ranges of the frequency ω.

(1) Show that, in the Hagen-Rubens regime $0 < \omega \ll \tau^{-1}$, $R(\omega) \approx 1 - (8\omega/\omega_p^2\tau)^{1/2}$.

(2) Show that, in the relaxation regime $\tau^{-1} \ll \omega \ll \omega_p$, $R(\omega) \approx 1 - 2/\omega_p\tau$.

(3) Show that, in the high-frequency regime $\omega \gg \omega_p$, $R(\omega) \approx (\omega_p/2\omega)^4$.

---

(1) In the Hagen-Rubens regime $0 < \omega \ll \tau^{-1}$, the relative dielectric function $\epsilon_r(\omega)$ can be approximated by

$$\epsilon_r(\omega) \approx 1 + \frac{ine^2\tau}{\epsilon_0 m\omega} = 1 + \frac{\mathrm{i}\omega_p^2\tau}{\omega},$$

since $\omega\tau \ll 1$ in this regime. The real and imaginary parts of $\epsilon_r(\omega)$ are then approximately given by

$$\epsilon_r'(\omega) \approx 1, \quad \epsilon_r''(\omega) \approx \frac{\omega_p^2\tau}{\omega}.$$

In terms of ϵ_r' and ϵ_r'', $n(\omega)$ and $k(\omega)$ are given by

$$n(\omega) = \frac{1}{2^{1/2}}\left[\left(\epsilon_r'^2 + \epsilon_r''^2\right)^{1/2} + \epsilon_r'\right]^{1/2},$$
$$k(\omega) = \frac{1}{2^{1/2}}\left[\left(\epsilon_r'^2 + \epsilon_r''^2\right)^{1/2} - \epsilon_r'\right]^{1/2}.$$

Note that $\epsilon_r''(\omega) \gg 1$ in this regime. Making use of the above-obtained approximate expressions of ϵ_r' and ϵ_r'', we have for

$n(\omega)$

$$n(\omega) \approx \frac{1}{2^{1/2}}\left[\left(1+(\omega_p^2\tau/\omega)^2\right)^{1/2}+1\right]^{1/2}$$
$$\approx \frac{1}{2^{1/2}}\left(\omega_p^2\tau/\omega+1\right)^{1/2} \approx \left(\frac{\omega_p^2\tau}{2\omega}\right)^{1/2}.$$

Similarly, we have for $k(\omega)$

$$k(\omega) \approx \left(\frac{\omega_p^2\tau}{2\omega}\right)^{1/2}.$$

Thus, $n(\omega) \approx k(\omega)$ in this regime. We then have for the reflectivity $R(\omega)$

$$R(\omega) \approx \frac{\left[1-(\omega_p^2\tau/2\omega)^{1/2}\right]^2+\omega_p^2\tau/2\omega}{\left[1+(\omega_p^2\tau/2\omega)^{1/2}\right]^2+\omega_p^2\tau/2\omega} \approx \frac{\omega_p^2\tau/\omega-2(\omega_p^2\tau/2\omega)^{1/2}}{\omega_p^2\tau/\omega+2(\omega_p^2\tau/2\omega)^{1/2}}$$
$$= \frac{1-2(\omega/2\omega_p^2\tau)^{1/2}}{1+2(\omega/2\omega_p^2\tau)^{1/2}} \approx 1-4\left(\frac{\omega}{2\omega_p^2\tau}\right)^{1/2} = 1-\left(\frac{8\omega}{\omega_p^2\tau}\right)^{1/2}.$$

(2) In the relaxation regime $\tau^{-1} \ll \omega \ll \omega_p$, the relative dielectric function $\epsilon_r(\omega)$ can be approximated by

$$\epsilon_r(\omega) = 1+\frac{\mathrm{i}\omega_p^2\tau}{\omega}\frac{1}{1-\mathrm{i}\omega\tau} = 1-\frac{\omega_p^2}{\omega^2}\frac{1}{1+\mathrm{i}/\omega\tau} \approx -\frac{\omega_p^2}{\omega^2}\left(1-\frac{\mathrm{i}}{\omega\tau}\right),$$

since $\omega\tau \gg 1$ and $\omega \ll \omega_p$ in this regime. The real and imaginary parts of $\epsilon_r(\omega)$ are then approximately given by

$$\epsilon_r'(\omega) \approx -\frac{\omega_p^2}{\omega^2}, \quad \epsilon_r''(\omega) \approx \frac{\omega_p^2}{\omega^3\tau}.$$

Note that $\epsilon_r''(\omega) \ll |\epsilon_r'(\omega)|$ because $\omega\tau \gg 1$. $n(\omega)$ is approximately given by

$$n(\omega) \approx \frac{1}{2^{1/2}}\left[\frac{\omega_p^2}{\omega^2}\left(1+\frac{1}{\omega^2\tau^2}\right)^{1/2}-\frac{\omega_p^2}{\omega^2}\right]^{1/2} \approx \frac{\omega_p}{2\omega^2\tau}.$$

For $k(\omega)$, we can completely neglect $\epsilon_r''(\omega)$ and approximately have

$$k(\omega) \approx \frac{\omega_p}{\omega}.$$

The reflectivity $R(\omega)$ is then approximately given by

$$R(\omega) \approx \frac{(1-\omega_p/2\omega^2\tau)^2+\omega_p^2/\omega^2}{(1+\omega_p/2\omega^2\tau)^2+\omega_p^2/\omega^2} \approx \frac{-\omega_p/\omega^2\tau+\omega_p^2/\omega^2}{\omega_p/\omega^2\tau+\omega_p^2/\omega^2}$$
$$= \frac{1-1/\omega_p\tau}{1+1/\omega_p\tau} \approx 1-\frac{2}{\omega_p\tau}.$$

(3) In the high-frequency regime $\omega \gg \omega_p$, since $\omega\tau \gg 1$, $\epsilon_r(\omega)$ can be approximated by

$$\epsilon_r(\omega) = 1 + \frac{\mathrm{i}\omega_p^2\tau}{\omega}\frac{1}{1-\mathrm{i}\omega\tau} = 1 - \frac{\omega_p^2}{\omega^2}\frac{1}{1+\mathrm{i}/\omega\tau} \approx 1 - \frac{\omega_p^2}{\omega^2}\left(1 - \frac{\mathrm{i}}{\omega\tau}\right).$$

The real and imaginary parts of $\epsilon_r(\omega)$ are then approximately given by

$$\epsilon_r'(\omega) \approx 1 - \frac{\omega_p^2}{\omega^2}, \quad \epsilon_r''(\omega) \approx \frac{\omega_p^2}{\omega^3\tau}.$$

Since $\epsilon_r''(\omega) \ll \epsilon_r'(\omega)$ in this regime, $n(\omega)$ is approximately given by

$$n(\omega) \approx \frac{1}{2^{1/2}}\left[2\epsilon_r'(\omega)\right]^{1/2} \approx 1 - \frac{\omega_p^2}{2\omega^2}.$$

For $k(\omega)$, we approximately have

$$k(\omega) \approx \frac{1}{2^{1/2}}\left[(1-\omega_p^2/\omega^2)(1+\omega_p^4/\omega^6\tau^2)^{1/2} - 1 + \omega_p^2/\omega^2\right]^{1/2} \approx \frac{\omega_p^2}{2\omega^3\tau}.$$

The reflectivity $R(\omega)$ is then approximately given by

$$R(\omega) \approx \frac{(\omega_p^2/2\omega^2)^2 + \omega_p^4/\omega^6\tau}{(2-\omega_p^2/2\omega^2)^2 + \omega_p^4/\omega^6\tau} \approx \frac{(\omega_p^2/2\omega^2)^2}{(2)^2} = \left(\frac{\omega_p}{2\omega}\right)^4.$$

29-3 Dielectric function of a free electron gas. The relative dielectric function of a free electron gas is given by

$$\epsilon_r(\omega) = \lim_{\gamma\to 0^+}\left(1 - \frac{\omega_p^2}{\omega^2 + \mathrm{i}\gamma\omega}\right),$$

where ω_p is the plasma frequency, $\omega_p^2 = ne^2/\epsilon_0 m$.

(1) Show that $\epsilon_r(\omega)$ satisfies the Kramers-Kronig relations.
(2) Evaluate $\int_0^\infty \mathrm{d}\omega\, \omega\, \mathrm{Im}\,\epsilon_r(\omega)$.
(3) Evaluate $\int_0^\infty \mathrm{d}\omega\ \mathrm{Im}\big[1/\epsilon_r(\omega)\big]$.
(4) Evaluate $\int_0^\infty \mathrm{d}\omega\, \omega\, \mathrm{Im}\big[1/\epsilon_r(\omega)\big]$.

---

(1) The real and imaginary parts of $\epsilon_r(\omega)$ are given by

$$\epsilon_r'(\omega) = 1 - \frac{\omega_p^2}{\omega^2+\gamma^2}, \quad \epsilon_r''(\omega) = \frac{\gamma\omega_p^2}{\omega(\omega^2+\gamma^2)},$$

where γ has been taken implicitly to tend to 0^+. From the above expressions, we see that $\epsilon_r(\infty) = 1$. We first consider the first Kramers-Kronig relation. The integral in the first Kramers-Kronig relation can be evaluated as follows

$$\begin{aligned}\frac{2}{\pi}\,\mathrm{P}\!\int_0^\infty d\omega'\ &\frac{\omega'\epsilon_r''(\omega')}{\omega'^2-\omega^2}\\ &= \frac{2\gamma\omega_p^2}{\pi}\,\mathrm{P}\!\int_0^\infty d\omega'\ \frac{1}{(\omega'^2-\omega^2)(\omega'^2+\gamma^2)}\\ &= \frac{2\gamma\omega_p^2}{\pi(\omega^2+\gamma^2)}\,\mathrm{P}\!\int_0^\infty d\omega'\left(\frac{1}{\omega'^2-\omega^2}-\frac{1}{\omega'^2+\gamma^2}\right)\\ &= -\frac{\omega_p^2}{\omega^2+\gamma^2} = \epsilon_r'(\omega)-1,\end{aligned}$$

where we have made use of $\mathrm{P}\int_0^\infty d\omega'\ (\omega'^2-\omega^2)^{-1} = 0$. Thus, the first Kramers-Kronig relation is satisfied.

For the second Kramers-Kronig relation, we have to be more careful since, in the $\gamma \to 0^+$ limit, $\epsilon_r''(\omega)$ is nonzero only at $\omega = 0$ which is also a singular point. Making use of $\lim_{\gamma\to 0^+} \gamma/(\omega^2+\gamma^2) = \pi\delta(\omega)$, we see that $\omega^2+\gamma^2 = 0$ in the $\gamma \to 0^+$ limit. We thus have $\omega/\gamma = -\gamma/\omega$ for $\epsilon_r''(\omega)$ in the $\gamma \to 0^+$ limit. Performing the integral in the second Kramers-Kronig relation, we have

$$\begin{aligned}-\frac{2\omega}{\pi}\,\mathrm{P}\!\int_0^\infty d\omega'\ \frac{\epsilon_r'(\omega')-1}{\omega'^2-\omega^2} &= \frac{2\omega\omega_p^2}{\pi}\,\mathrm{P}\!\int_0^\infty d\omega'\ \frac{1}{(\omega'^2-\omega^2)(\omega'^2+\gamma^2)}\\ &= -\frac{\omega\omega_p^2}{\gamma(\omega^2+\gamma^2)}.\end{aligned}$$

Making use of $\omega/\gamma = -\gamma/\omega$ for $\epsilon_r''(\omega)$ in the $\gamma \to 0^+$ limit, we have

$$-\frac{2\omega}{\pi}\,\mathrm{P}\!\int_0^\infty d\omega'\ \frac{\epsilon_r'(\omega')-1}{\omega'^2-\omega^2} = \frac{\gamma\omega_p^2}{\omega(\omega^2+\gamma^2)} = \epsilon_r''(\omega).$$

Thus, the second Kramers-Kronig relation is satisfied.

(2) Making use of the above-obtained imaginary part of $\epsilon_r(\omega)$, we have [note that $\epsilon_r''(\omega) = \mathrm{Im}\,\epsilon_r(\omega)$]

$$\int_0^\infty d\omega\ \omega\,\mathrm{Im}\,\epsilon_r(\omega) = \gamma\omega_p^2\int_0^\infty d\omega\ \frac{1}{\omega^2+\gamma^2} = \frac{\pi}{2}\omega_p^2.$$

(3) The imaginary part of the inverse of $\epsilon_r(\omega)$ is given by

$$\begin{aligned}\mathrm{Im}\,\frac{1}{\epsilon_r(\omega)} = \mathrm{Im}\,\frac{1}{\epsilon_r'(\omega)+\mathrm{i}\epsilon_r''(\omega)} &= -\frac{\epsilon_r''(\omega)}{[\epsilon_r'(\omega)]^2+[\epsilon_r''(\omega)]^2}\\ &= -\frac{\gamma\omega_p^2\omega}{(\omega^2-\omega_p^2)^2+\gamma^2\omega^2}.\end{aligned}$$

We thus have

$$\int_0^\infty d\omega\ \mathrm{Im}\,\frac{1}{\epsilon_r(\omega)}$$
$$= -\frac{1}{2}\gamma\omega_p^2\int_0^\infty d(\omega^2)\ \frac{1}{(\omega^2-\omega_p^2)^2+\gamma^2\omega^2}$$
$$= -\frac{\gamma\omega_p^2}{2(\gamma^2\omega_p^2-\gamma^4/4)^{1/2}}\left[\frac{\pi}{2}-\tan^{-1}\frac{-\omega_p^2+\gamma^2/2}{(\gamma^2\omega_p^2-\gamma^4/4)^{1/2}}\right].$$

Taking the $\gamma\to 0^+$ limit, we have

$$\int_0^\infty d\omega\ \mathrm{Im}\,\frac{1}{\epsilon_r(\omega)}\to -\frac{\omega_p}{2}\left(\frac{\pi}{2}-\tan^{-1}\frac{-\omega_p}{\gamma}\right)\to -\frac{\pi}{2}\omega_p.$$

Alternatively, we can first take the $\gamma\to 0^+$ limit and obtain

$$\mathrm{Im}\,\frac{1}{\epsilon_r(\omega)} = -\frac{\gamma\omega_p^2\omega}{(\omega^2-\omega_p^2)^2+\gamma^2\omega^2} = -\frac{\omega_p^2}{\omega}\frac{\gamma}{(\omega-\omega_p^2/\omega)^2+\gamma^2}$$
$$\to -\frac{\pi\omega_p^2}{\omega}\delta(\omega-\omega_p^2/\omega) = -\pi\omega_p^2\delta(\omega^2-\omega_p^2)$$
$$= -\frac{\pi}{2}\omega_p\big[\,\delta(\omega-\omega_p)+\delta(\omega+\omega_p)\,\big].$$

The integral is then trivial to perform. We have

$$\int_0^\infty d\omega\ \mathrm{Im}\,\frac{1}{\epsilon_r(\omega)} = -\frac{\pi}{2}\omega_p\int_0^\infty d\omega\ \big[\,\delta(\omega-\omega_p)+\delta(\omega+\omega_p)\,\big]$$
$$= -\frac{\pi}{2}\omega_p.$$

(4) Making use of the above-obtained expression of $\mathrm{Im}[\epsilon_r(\omega)]^{-1}$ in terms of δ-functions, we can perform this integral trivially as in the above

$$\int_0^\infty d\omega\ \omega\,\mathrm{Im}\,\frac{1}{\epsilon_r(\omega)} = -\frac{\pi}{2}\omega_p\int_0^\infty d\omega\ \omega\big[\,\delta(\omega-\omega_p)+\delta(\omega+\omega_p)\,\big]$$
$$= -\frac{\pi}{2}\omega_p^2.$$

29-4 Static dielectric function of a metal. We can derive the static dielectric function of a metal through considering a small perturbation to the free electron gas in the metal. Under the influence of the small perturbation, the energy of an electron at $\boldsymbol{r}$ is changed by $V(\boldsymbol{r})$ which is small. Study the consequence of the perturbation and derive the static dielectric function of the metal using $\nabla\cdot\boldsymbol{D}(\boldsymbol{r}) = -\epsilon_0\nabla^2\phi_{\rm ext}(\boldsymbol{r}) = \rho_{\rm ext}(\boldsymbol{r})$ and $\nabla\cdot\boldsymbol{E}(\boldsymbol{r}) = -\nabla^2\phi_{\rm tot}(\boldsymbol{r}) = \rho_{\rm tot}(\boldsymbol{r})/\epsilon_0$.

The Fourier transforms of the equations $\nabla\cdot\boldsymbol{D}(\boldsymbol{r}) = -\epsilon_0\nabla^2\phi_{\rm ext}(\boldsymbol{r}) = \rho_{\rm ext}(\boldsymbol{r})$ and $\nabla\cdot\boldsymbol{E}(\boldsymbol{r}) = -\nabla^2\phi_{\rm tot}(\boldsymbol{r}) = \rho_{\rm tot}(\boldsymbol{r})/\epsilon_0$ are

$$\mathrm{i}\boldsymbol{q}\cdot\boldsymbol{D}(\boldsymbol{q}) = \epsilon_0\boldsymbol{q}^2\phi_{\rm ext}(\boldsymbol{q}) = \rho_{\rm ext}(\boldsymbol{q}),$$
$$\mathrm{i}\boldsymbol{q}\cdot\boldsymbol{E}(\boldsymbol{q}) = \boldsymbol{q}^2\phi_{\rm tot}(\boldsymbol{q}) = \rho_{\rm tot}(\boldsymbol{q})/\epsilon_0,$$

where we have made use of

$$\boldsymbol{D}(\boldsymbol{r}) = \sum_{\boldsymbol{q}} \boldsymbol{D}(\boldsymbol{q}) e^{i\boldsymbol{q}\cdot\boldsymbol{r}}, \quad etc.$$

For longitudinal fields, the relative dielectric function is given by

$$\epsilon_r(\boldsymbol{q}) = \frac{D(\boldsymbol{q})}{\epsilon_0 E(\boldsymbol{q})} = \frac{\phi_{\text{ext}}(\boldsymbol{q})}{\phi_{\text{tot}}(\boldsymbol{q})}.$$

In the presence of the small perturbation $V(\boldsymbol{r})$, only electrons near the Fermi surface are disturbed and they move away from the Fermi surface. This brings a change to the number density of electrons. Since the number of electrons per unit volume per unit energy interval near the Fermi surface is given by $g(E_{\text{F}})$ with $g(E_{\text{F}})$ the density of states per unit volume, the number of electrons per unit volume that are relocated within the energy interval $V(\boldsymbol{r})$ about the Fermi energy is given by $g(E_{\text{F}})V(\boldsymbol{r})$. Therefore, the variation in the number density of electrons is given by

$$\delta n(\boldsymbol{r}) = -g(E_{\text{F}})V(\boldsymbol{r}).$$

The variation in the number density of electrons, $\delta n(\boldsymbol{r})$, results in a variation in the electric potential, $\delta\phi(\boldsymbol{r})$, which is to be determined from

$$\nabla^2[\delta\phi(\boldsymbol{r})] = -(-e)\delta n(\boldsymbol{r})/\epsilon_0 = e\delta n(\boldsymbol{r})/\epsilon_0 = -eg(E_{\text{F}})V(\boldsymbol{r})/\epsilon_0.$$

The Fourier transform of the above equation reads

$$-q^2\delta\phi(\boldsymbol{q}) = -eg(E_{\text{F}})V(\boldsymbol{q})/\epsilon_0$$

from which it follows that

$$\delta\phi(\boldsymbol{q}) = \frac{eg(E_{\text{F}})}{\epsilon_0 q^2} V(\boldsymbol{q}).$$

The energy $V(\boldsymbol{q})$ is the $\boldsymbol{q}$ component of the total potential energy of an electron in the external electric field. Let $\phi_{\text{ext}}(\boldsymbol{q})$ denote the electric potential of the external electric field. We then have

$$V(\boldsymbol{q}) = -e\phi_{\text{ext}}(\boldsymbol{q}) - e\delta\phi(\boldsymbol{q}) = -e\phi_{\text{ext}}(\boldsymbol{q}) - \frac{e^2 g(E_{\text{F}})}{\epsilon_0 q^2} V(\boldsymbol{q}).$$

Thus,

$$\phi_{\text{ext}}(\boldsymbol{q}) = -\frac{1}{e}\left[1 + \frac{e^2 g(E_{\text{F}})}{\epsilon_0 q^2}\right] V(\boldsymbol{q}).$$

The relative static dielectric function is then given by

$$\epsilon_r(\boldsymbol{q}) = \frac{\phi_{\text{ext}}}{\phi_{\text{tot}}} = \frac{\phi_{\text{ext}}}{V(\boldsymbol{q})/(-e)} = 1 + \frac{e^2 g(E_{\rm F})}{\epsilon_0 q^2}$$
$$= 1 + \frac{3ne^2}{2\epsilon_0 E_{\rm F} q^2} = 1 + \frac{\kappa_s^2}{q^2},$$

where $\kappa_s = (3ne^2/2\epsilon_0 E_{\rm F})^{1/2}$ is known as *the Fermi-Thomas screening wave vector*.

29-5 Low-frequency response of electrons to an AC field. Consider the following equation of motion of an electron in a material under the influence of an AC field in the x direction

$$m\frac{\mathrm{d}^2x(t)}{\mathrm{d}t^2} + \frac{m}{\tau}\frac{\mathrm{d}x(t)}{\mathrm{d}t} = -eE(t)$$

with $E(t) = E_0\,\mathrm{e}^{-\mathrm{i}\omega t}$.

(1) Solve the equation of motion for the case $\omega \ll 1/\tau$ to derive the low-frequency dielectric function.

(2) Find the phase of the electric polarization relative to $E(t)$ in this frequency range. State whether the electric polarization is leading ahead or lagging behind $E(t)$.

(3) Find the phase of the electric current relative to $E(t)$ in this frequency range. State whether the electric current is leading ahead or lagging behind $E(t)$.

(4) State whether the material is absorptive or not in this frequency range.

---

(1) In consideration that $E(t) = E_0\,\mathrm{e}^{-\mathrm{i}\omega t}$, we set $x(t) = x_0\,\mathrm{e}^{-\mathrm{i}\omega t}$. Inserting the expressions of $E(t)$ and $x(t)$ into the equation of motion, we have

$$-m\omega^2 x_0 - \mathrm{i}\omega\frac{m}{\tau}x_0 = -eE_0.$$

Solving the above equation, we obtain

$$x_0 = \frac{eE_0}{m\omega(\omega + \mathrm{i}/\tau)}.$$

For the case $\omega \ll 1/\tau$, we have

$$x_0 = -\mathrm{i}\frac{e\tau E_0}{m\omega}.$$

The dipole moment of the electron is $p = -ex$ and the electric polarization in the material is

$$P = np = -nex = \mathrm{i}\frac{ne^2\tau}{m\omega}E,$$

where n is the number density of electrons. The electric displacement is then given by $D = \epsilon_0 E + P$. From $D = \epsilon_0 \epsilon_r E$, we obtain the relative dielectric function

$$\epsilon_r = 1 + \mathrm{i}\frac{ne^2\tau}{\epsilon_0 m\omega} = 1 + \mathrm{i}\frac{\omega_p^2\tau}{\omega}.$$

For $\omega \ll 1/\tau$, we have,

$$\epsilon_r \approx \mathrm{i}\frac{\omega_p^2\tau}{\omega}.$$

(2) From the above-obtained result, $P = \mathrm{i}(ne^2\tau/m\omega)E$, we see that P and E has a phase difference of $\pi/2$ and that P is leading ahead of E by a phase of $\pi/2$.

(3) The current density is given by

$$j = -ne\frac{\mathrm{d}x}{\mathrm{d}t} = \mathrm{i}ne\omega x = \frac{ne^2\tau}{m}E.$$

Thus, the electric current is in phase with E.

(4) Since the electric current is in phase with E, there exists power dissipation. Hence, the material is absorptive in this frequency range.

29-6 Dielectric function of an ideal metal. Write down the equation of motion for an electron in an ideal (Drude) metal with damping Γ (rad/s). From this, find expressions for the real and imaginary parts of the dielectric function, $\epsilon'(\omega)$ and $\epsilon''(\omega)$. Sketch $\epsilon'(\omega)$ and $\epsilon''(\omega)$ versus ω.

Assume that the electron is in the electric field $\boldsymbol{E} = \boldsymbol{E}_0 \mathrm{e}^{-\mathrm{i}\omega t}$. Let $\boldsymbol{r}$ be the position of the electron. The equation of motion of the electron is then given by

$$\ddot{\boldsymbol{r}} = -\Gamma\dot{\boldsymbol{r}} - e\boldsymbol{E}/m.$$

In consideration that $\boldsymbol{E} = \boldsymbol{E}_0 \mathrm{e}^{-\mathrm{i}\omega t}$, we must have $\boldsymbol{r} = \boldsymbol{r}_0 \mathrm{e}^{-\mathrm{i}\omega t}$. Inserting these expressions for $\boldsymbol{r}$ and $\boldsymbol{E}$ into the equation of motion yields

$$-\omega^2 \boldsymbol{r}_0 = \mathrm{i}\Gamma\omega\, \boldsymbol{r}_0 - e\boldsymbol{E}_0/m.$$

Thus,

$$\boldsymbol{r}_0 = \frac{e/m}{\omega(\omega + \mathrm{i}\Gamma)}\boldsymbol{E}_0.$$

Because of the oscillation of electron relative to the background of positive charge, an oscillating electric dipole appears

$$\boldsymbol{p} = -e\boldsymbol{r} = -\frac{e^2/m}{\omega(\omega + \mathrm{i}\Gamma)}\boldsymbol{E}.$$

Let n be the number density of electrons. The electric polarization is then given by

$$\boldsymbol{P} = n\boldsymbol{p} = -\frac{ne^2/m}{\omega^2 + \mathrm{i}\Gamma\omega}\boldsymbol{E}.$$

From $\boldsymbol{D} = \epsilon_0\boldsymbol{E} + \boldsymbol{P} = \epsilon\boldsymbol{E}$, we have

$$\epsilon(\omega) = \epsilon_0 - \frac{ne^2/m}{\omega(\omega + \mathrm{i}\Gamma)}.$$

Separating the real and imaginary parts of $\epsilon(\omega)$, we have

$$\epsilon'(\omega) = \epsilon_0 - \frac{ne^2/m}{\omega^2 + \Gamma^2}, \quad \epsilon''(\omega) = \frac{ne^2\Gamma/m}{\omega(\omega^2 + \Gamma^2)}.$$

Note that the relative dielectric function is given by

$$\epsilon_r(\omega) = \frac{\epsilon(\omega)}{\epsilon_0} = 1 - \frac{ne^2/\epsilon_0 m}{\omega(\omega + \mathrm{i}\Gamma)}.$$

The real and imaginary parts of $\epsilon_r(\omega)$ are given by

$$\epsilon_r'(\omega) = 1 - \frac{ne^2/\epsilon_0 m}{\omega^2 + \Gamma^2}, \quad \epsilon_r''(\omega) = \frac{ne^2\Gamma/\epsilon_0 m}{\omega(\omega^2 + \Gamma^2)}.$$

We plot in Fig. 29.1 the real and imaginary parts of the relative dielectric function against ω/ω_p with ω_p the plasma frequency, $\omega_p^2 = ne^2/\epsilon_0 m$. The damping rate $\Gamma = 0.1\omega_p$ has been used. From Fig. 29.1, the Drude peak at $\omega = 0$ is clearly seen in the imaginary part $\epsilon_r''(\omega)$. Beyond the Drude peak, the imaginary part decreases rapidly to zero. The real part has the smallest value at $\omega = 0$ and it approaches to unity as ω increases.

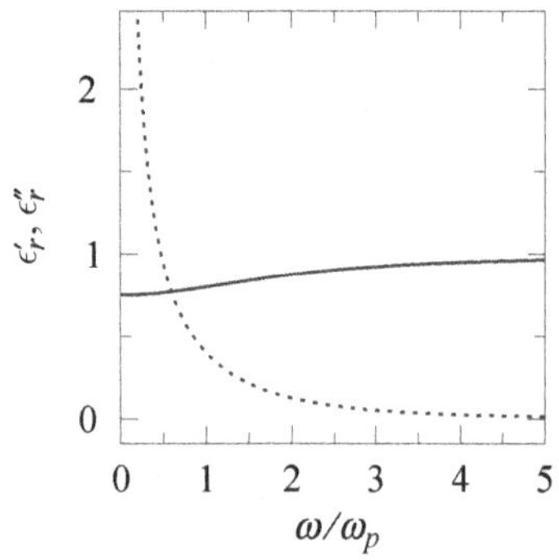

Fig. 29.1 Plots of the real (the solid line) and imaginary (the dotted line) parts of the relative dielectric function as functions of ω/ω_p with $\Gamma = 0.1\omega_p$.

We can see that the real part takes on the smallest value at the frequency where the imaginary part peaks and that the real part takes on the largest value at the frequency where the imaginary part vanishes.

29-7 An isotropic solid modeled by Lorentz oscillators. An isotropic solid medium is model by Lorentz oscillators each of which has electric charge $-e$, mass m, force constant κ, and damping Γ.

(1) Write down the equation of motion for a Lorentz oscillator in which the local electric field $\boldsymbol{E}_{\text{loc}}$ in the medium should be used.

(2) For an isotropic solid medium, how is the local electric field related to the applied electric field $\boldsymbol{E}$ and the electric polarization $\boldsymbol{P}$?

(3) Through examining the equation of motion of $\boldsymbol{P}$, show that the local field leads to an observed resonance frequency that differs from $\omega_0 = (\kappa/m)^{1/2}$.

---

(1) The equation of motion for a Lorentz oscillator is given by

$$\ddot{\boldsymbol{r}} = -\omega_0^2 \boldsymbol{r} - \Gamma \dot{\boldsymbol{r}} - \frac{e}{m}\boldsymbol{E}_{\text{loc}},$$

where $\omega_0 = \sqrt{\kappa/m}$ is the intrinsic frequency of the oscillator and $\boldsymbol{E}_{\text{loc}}$ the local electric field.

(2) The local electric field is the sum of the applied electric field $\boldsymbol{E}$ and the electric field $\boldsymbol{E}_L$ due to all other electrons. The subscript L on $\boldsymbol{E}_L$ is for Lorentz. For an isotropic solid medium, $\boldsymbol{E}_L = \boldsymbol{P}/3\epsilon_0$. The local electric field for an isotropic solid medium is then given by

$$\boldsymbol{E}_{\text{loc}} = \boldsymbol{E} + \boldsymbol{P}/3\epsilon_0.$$

(3) Let n be the number density of electrons. The electric polarization is then given by

$$\boldsymbol{P} = -ne\boldsymbol{r}.$$

Multiplying the equation of motion of $\boldsymbol{r}$ with $-ne$, we obtain the equation of motion of $\boldsymbol{P}$

$$\begin{aligned}\ddot{\boldsymbol{P}} &= -\omega_0^2 \boldsymbol{P} - \Gamma \dot{\boldsymbol{P}} + \frac{ne^2}{m}\boldsymbol{E}_{\text{loc}} \\ &= -\omega_0^2 \boldsymbol{P} - \Gamma \dot{\boldsymbol{P}} + \frac{ne^2}{m}\boldsymbol{E} + \frac{ne^2}{3\epsilon_0 m}\boldsymbol{P} \\ &= -\left(\omega_0^2 - \omega_p^2/3\right)\boldsymbol{P} - \Gamma \dot{\boldsymbol{P}} + \frac{ne^2}{m}\boldsymbol{E},\end{aligned}$$

where $\omega_p = (ne^2/\epsilon_0 m)^{1/2}$ is the plasma frequency. Inserting $\boldsymbol{P} = P_0 \mathrm{e}^{-\mathrm{i}\omega t}$ and $\boldsymbol{E} = E_0 \mathrm{e}^{-\mathrm{i}\omega t}$ into the equation of motion of $\boldsymbol{P}$ yields

$$-\omega^2 \boldsymbol{P} = -\left(\omega_0^2 - \omega_p^2/3\right)\boldsymbol{P} + \mathrm{i}\Gamma\omega\boldsymbol{P} + \frac{ne^2}{m}\boldsymbol{E}.$$

Solving for $\boldsymbol{P}$ from the above equation, we obtain

$$\boldsymbol{P} = \frac{ne^2/m}{\omega_0^2 - \omega_p^2/3 - \omega^2 - \mathrm{i}\Gamma\omega}\boldsymbol{E}.$$

From the above equation, we see that the resonance frequency occurs at $\omega_{\mathrm{res}} = (\omega_0^2 - \omega_p^2/3)^{1/2}$ with the half width at the half maximum (HWHM) approximately given by $(\Gamma\omega_{\mathrm{res}})^{1/2} = \Gamma^{1/2}(\omega_0^2 - \omega_p^2/3)^{1/4}$. Thus, the observed resonance frequency differs from $\omega_0 = (\kappa/m)^{1/2}$.

29-8 Electric susceptibility in the Lorentz oscillator model. In the Lorentz oscillator model, the electric susceptibility as a function of time difference $\tau = t - t'$ is given by

$$\chi(\tau) = \frac{ne^2}{\epsilon_0 m\omega_0}\mathrm{e}^{-\gamma\tau/2}\sin(\omega_0\tau)\theta(\tau).$$

Show that the above expression is a valid expression for the electric susceptibility.

The Fourier components of the electric susceptibility are given by

$$\begin{aligned}
\chi(\omega) &= \int_{-\infty}^{\infty}\mathrm{d}\tau\,\mathrm{e}^{\mathrm{i}\omega\tau}\chi(\tau) = \frac{ne^2}{\epsilon_0 m\omega_0}\int_0^{\infty}\mathrm{d}\tau\,\mathrm{e}^{-(\gamma/2-\mathrm{i}\omega)\tau}\sin(\omega_0\tau)\\
&= \frac{ne^2}{2\mathrm{i}\epsilon_0 m\omega_0}\int_0^{\infty}\mathrm{d}\tau\left[\mathrm{e}^{-(\gamma/2-\mathrm{i}\omega-\mathrm{i}\omega_0)\tau} - \mathrm{e}^{-(\gamma/2-\mathrm{i}\omega+\mathrm{i}\omega_0)\tau}\right]\\
&= \frac{ne^2}{2\mathrm{i}\epsilon_0 m\omega_0}\left(\frac{1}{\gamma/2-\mathrm{i}\omega-\mathrm{i}\omega_0} - \frac{1}{\gamma/2-\mathrm{i}\omega+\mathrm{i}\omega_0}\right)\\
&= \frac{ne^2/\epsilon_0 m}{(\gamma/2-\mathrm{i}\omega)^2+\omega_0^2}.
\end{aligned}$$

Expanding the term $(\gamma/2 - \mathrm{i}\omega)^2$ and neglecting $\gamma^2/4$ which is small in comparison with ω_0^2, we have

$$\chi(\omega) = \frac{ne^2/\epsilon_0 m}{\omega_0^2 - \omega^2 - \mathrm{i}\gamma\omega}.$$

From $\epsilon_r(\omega) = 1 + \chi(\omega)$, we have

$$\epsilon_r(\omega) = 1 + \frac{ne^2/\epsilon_0 m}{\omega_0^2 - \omega^2 - \mathrm{i}\gamma\omega}$$

which is identical with the result for a single Lorentz oscillator. Therefore, the given expression for the electric susceptibility is valid.

29-9 Optical properties of NaCl. Experimental measurements of reflectivity for NaCl reveal that the LO and TO phonon features appear at $\lambda = 38\ \mu$m and 61 μm, respectively.

(1) Estimate the force constant K for the TO phonon mode with only nearest neighbor interactions taken into account.

(2) Compute the magnitude of the lattice polarization contribution to the dielectric constant for crystalline NaCl from the measured value of $\omega_{\rm LO}$ and $\omega_{\rm TO}$ splitting.

(3) Discuss whether $\omega_{\rm LO} - \omega_{\rm TO}$ for NaCl depends on temperature.

---

(1) NaCl has an FCC crystal structure with a two-ion basis as shown in Fig. 29.2. Each ion has six oppositely charged ions as its nearest neighbors. Let the positions of Na$^+$ ions be denoted by $\boldsymbol{R}_{mn\ell} = m\boldsymbol{a}_1 + n\boldsymbol{a}_2 + \ell\boldsymbol{a}_3$ with $\boldsymbol{a}_1 = (\boldsymbol{e}_y + \boldsymbol{e}_z)a/2$, $\boldsymbol{a}_2 = (\boldsymbol{e}_z + \boldsymbol{e}_x)a/2$, and $\boldsymbol{a}_3 = (\boldsymbol{e}_x + \boldsymbol{e}_y)a/2$ the primitive vectors. Here a is the lattice constant.

The origin of each primitive cell is taken to be at an Na$^+$ ion. From Fig. 29.2, it is seen that each primitive cell contains one Cl$^-$ ion whose position within the primitive cell is $(1/2, 1/2, 1/2)a$. Note that the two ions within a primitive cell are not nearest neighbors. However, by choosing a different primitive cell, we can make the two ions within a primitive cell to be nearest neighbors. The displacement of the Na$^+$ ion within the primitive cell at $\boldsymbol{R}_{mn\ell}$ is denoted by $u_{mn\ell,1}$ and that of the Cl$^-$ ion by $u_{mn\ell,2}$.

The six nearest neighboring Cl$^-$ ions of the Na$^+$ ion within the primitive cell at $\boldsymbol{R}_{mn\ell}$ are located relative to $\boldsymbol{R}_{mn\ell}$ at $(\pm 1/2, 0, 0)a$, $(0, \pm 1/2, 0)a$, and $(0, 0, \pm 1/2)a$. These six nearest neighbors are within six other different primitive cells whose positions are, respectively,

$$\begin{aligned}
\boldsymbol{R}_{mn\ell} - \boldsymbol{a}_1 &= \boldsymbol{R}_{m-1,n,\ell}, \\
\boldsymbol{R}_{mn\ell} - \boldsymbol{a}_2 &= \boldsymbol{R}_{m,n-1,\ell}, \\
\boldsymbol{R}_{mn\ell} - \boldsymbol{a}_3 &= \boldsymbol{R}_{m,n,\ell-1}, \\
\boldsymbol{R}_{mn\ell} - \boldsymbol{a}_2 - \boldsymbol{a}_3 &= \boldsymbol{R}_{m,n-1,\ell-1}, \\
\boldsymbol{R}_{mn\ell} - \boldsymbol{a}_3 - \boldsymbol{a}_1 &= \boldsymbol{R}_{m-1,n,\ell-1}, \\
\boldsymbol{R}_{mn\ell} - \boldsymbol{a}_1 - \boldsymbol{a}_2 &= \boldsymbol{R}_{m-1,n-1,\ell}.
\end{aligned}$$

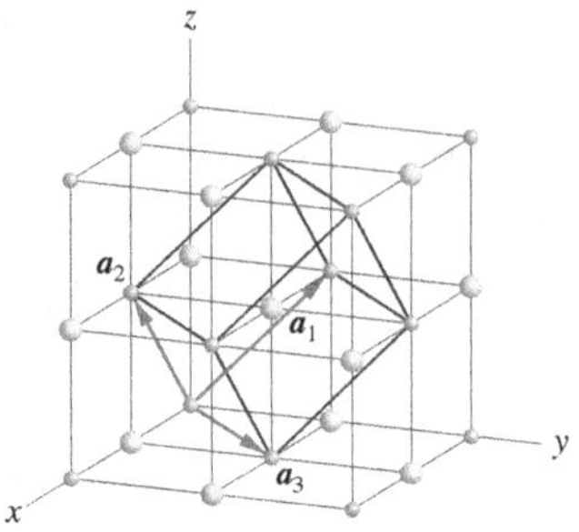

Fig. 29.2 Crystal structure of NaCl. Shown is a full conventional unit cell of Na^+ (small balls) and Cl^- (large balls) ions. Also shown are the primitive vectors and the primitive cell constructed from them.

With only nearest neighbor interactions taken into account, the harmonic lattice potential energy is given by

$$\begin{aligned} U^{\text{harm}} = \frac{1}{4}K \sum_{mn\ell} \Big[& \big(\boldsymbol{u}_{mn\ell,1} - \boldsymbol{u}_{m-1,n,\ell,2}\big)^2 \\ & + \big(\boldsymbol{u}_{mn\ell,1} - \boldsymbol{u}_{m,n-1,\ell,2}\big)^2 \\ & + \big(\boldsymbol{u}_{mn\ell,1} - \boldsymbol{u}_{m,n,\ell-1,2}\big)^2 \\ & + \big(\boldsymbol{u}_{mn\ell,1} - \boldsymbol{u}_{m,n-1,\ell-1,2}\big)^2 \\ & + \big(\boldsymbol{u}_{mn\ell,1} - \boldsymbol{u}_{m-1,n,\ell-1,2}\big)^2 \\ & + \big(\boldsymbol{u}_{mn\ell,1} - \boldsymbol{u}_{m-1,n-1,\ell,2}\big)^2 \Big], \end{aligned}$$

where K is the force constant and one of the two factors of $1/2$ is to remove the double counting. The equations of motion can be set up using $m_\kappa \ddot{\boldsymbol{u}}_{mn\ell,\kappa} = -\boldsymbol{\nabla}_{\boldsymbol{u}_{mn\ell,\kappa}} U^{\text{harm}}$, where $\kappa = 1$ and $\kappa = 2$ are for Na^+ and Cl^- ions, respectively, and m_1 and m_2 are their masses. We have

$$\begin{aligned} m_1 \ddot{\boldsymbol{u}}_{mn\ell,1} = -\frac{1}{2}K\big(& 6\boldsymbol{u}_{mn\ell,1} - \boldsymbol{u}_{m-1,n,\ell,2} - \boldsymbol{u}_{m,n-1,\ell,2} \\ & - \boldsymbol{u}_{m,n,\ell-1,2} - \boldsymbol{u}_{m,n-1,\ell-1,2} \\ & - \boldsymbol{u}_{m-1,n,\ell-1,2} - \boldsymbol{u}_{m-1,n-1,\ell,2}\big), \end{aligned}$$

$$\begin{aligned} m_2 \ddot{\boldsymbol{u}}_{mn\ell,2} = -\frac{1}{2}K\big(& 6\boldsymbol{u}_{mn\ell,2} - \boldsymbol{u}_{m+1,n,\ell,1} - \boldsymbol{u}_{m,n+1,\ell,1} \\ & - \boldsymbol{u}_{m,n,\ell+1,1} - \boldsymbol{u}_{m,n+1,\ell+1,1} \\ & - \boldsymbol{u}_{m+1,n,\ell+1,1} - \boldsymbol{u}_{m+1,n+1,\ell,1}\big). \end{aligned}$$

Expressing the αth component of $\boldsymbol{u}_{mn\ell,\kappa}$, $u_{mn\ell,\kappa,\alpha}$, in terms of its Fourier components, we have

$$u_{mn\ell,\kappa,\alpha}(t) = \sum_{\boldsymbol{k}\omega} \frac{1}{\sqrt{Nm_\kappa}} Q_{\kappa\alpha}(\boldsymbol{k},\omega) \mathrm{e}^{\mathrm{i}(\boldsymbol{k}\cdot\boldsymbol{R}_{mn\ell}-\omega t)}.$$

Inserting the above expansion into the equations of motion, we have

$$\begin{aligned}
&(3K\sqrt{m_2/m_1} - \sqrt{m_1 m_2}\,\omega^2) Q_{1\alpha}(\boldsymbol{k},\omega) \\
&\qquad - \mathrm{e}^{-\mathrm{i}(k_x+k_y+k_z)a/2} K\gamma_{\boldsymbol{k}} Q_{2\alpha}(\boldsymbol{k},\omega) = 0, \\
&- \mathrm{e}^{\mathrm{i}(k_x+k_y+k_z)a/2} K\gamma_{\boldsymbol{k}} Q_{1\alpha}(\boldsymbol{k},\omega) \\
&\qquad + (3K\sqrt{m_1/m_2} - \sqrt{m_1 m_2}\,\omega^2) Q_{2\alpha}(\boldsymbol{k},\omega) = 0,
\end{aligned}$$

where $\gamma_{\boldsymbol{k}} = \cos(k_x a/2) + \cos(k_y a/2) + \cos(k_z a/2)$. Since there exists no coupling between different coordinate components, the two equations for each coordinate component are identical. Thus, each frequency is three-fold degenerate. For each coordinate component, we have the following equation for the determination of frequencies

$$\begin{vmatrix} 3K\sqrt{m_2/m_1} - \sqrt{m_1 m_2}\,\omega^2 & -\mathrm{e}^{-\mathrm{i}(k_x+k_y+k_z)a/2} K\gamma_{\boldsymbol{k}} \\ -\mathrm{e}^{\mathrm{i}(k_x+k_y+k_z)a/2} K\gamma_{\boldsymbol{k}} & 3K\sqrt{m_1/m_2} - \sqrt{m_1 m_2}\,\omega^2 \end{vmatrix} = 0.$$

Solving for ω^2 from the above equation, we obtain

$$\omega_\pm^2(\boldsymbol{k}) = \frac{3(m_1+m_2)K}{2m_1 m_2} \left\{ 1 \pm \left[1 - \frac{4m_1 m_2}{(m_1+m_2)^2} \left(1 - \frac{1}{9}\gamma_{\boldsymbol{k}}^2 \right) \right]^{1/2} \right\}.$$

At $\boldsymbol{k} = 0$, we have $\gamma_{\boldsymbol{k}=0} = 3$. Thus,

$$\omega_+^2(0) = \frac{3(m_1+m_2)K}{m_1 m_2}, \quad \omega_-^2(0) = 0.$$

Therefore, three acoustical branches are degenerate and so are the three optical branches. Then, in the present simple treatment of lattice vibrations in NaCl, the frequencies of the longitudinal and transverse optical modes at $\boldsymbol{k} = 0$, ω_{LO} and ω_{TO} are all given by

$$\omega_{\mathrm{LO,TO}} = \left[\frac{3(m_1+m_2)K}{m_1 m_2} \right]^{1/2}.$$

Of course, $\omega_{\mathrm{LO}} \neq \omega_{\mathrm{TO}}$ in reality as demonstrated by the reflectivity measurements. Identifying ω_{TO} as the frequency corresponding to the feature at $\lambda = 61\ \mu\mathrm{m}$, we have

$$\omega_{\mathrm{TO}} = \frac{2\pi c}{\lambda_2} \approx 3.09 \times 10^{13}\ \mathrm{rad/s},$$

where c is the speed of light in vacuum. Making use of the above-derived expression for $\omega_{\rm TO}$, we have

$$K = \frac{m_1 m_2 \omega_{\rm TO}^2}{3(m_1 + m_2)} \approx 7.37 \text{ N/m},$$

where $m_1 = 22.989\,77$ amu for Na^+ and $m_2 = 35.453$ amu for Cl^- have been used.

(2) For lattice vibrations, only transverse optical modes make contributions to the dielectric function. In the present simple treatment of lattice vibrations in NaCl, the two transverse optical modes at $\boldsymbol{k} = 0$ are degenerate. Applying the Lorentz oscillator model to the doubly-degenerate $\omega_{\rm TO}$, we have

$$\epsilon_r(\omega) = \epsilon_r(\infty) + \frac{2ne^2/\epsilon_0 m}{\omega_{\rm TO}^2 - \omega^2},$$

where the factor of two in the second term arises from the double degeneracy. Note that $\epsilon_r(\infty)$ is actually equal to $\epsilon_r'(\infty)$ since $\epsilon_r''(\infty) = 0$. The factor on the numerator in the second term can be fixed by taking into account the fact that the longitudinal optical mode makes no contribution to the dielectric function. Note that $\omega_{\rm LO} \neq \omega_{\rm TO}$ in a more accurate treatment of lattice vibrations in NaCl. We have

$$\epsilon_r(\omega_{\rm LO}) = \epsilon_r(\infty) + \frac{2ne^2/\epsilon_0 m}{\omega_{\rm TO}^2 - \omega_{\rm LO}^2} = 0$$

from which it follows that

$$\frac{2ne^2}{\epsilon_0 m} = (\omega_{\rm LO}^2 - \omega_{\rm TO}^2)\epsilon_r(\infty).$$

We thus have

$$\epsilon_r(\omega) = \epsilon_r(\infty) + \frac{\omega_{\rm LO}^2 - \omega_{\rm TO}^2}{\omega_{\rm TO}^2 - \omega^2}\epsilon_r(\infty) = \frac{\omega_{\rm LO}^2 - \omega^2}{\omega_{\rm TO}^2 - \omega^2}\epsilon_r(\infty).$$

The second term in the first equality is the lattice contribution to the dielectric function, that is,

$$\epsilon_r^{\rm lattice}(\omega) = \frac{\omega_{\rm LO}^2 - \omega_{\rm TO}^2}{\omega_{\rm TO}^2 - \omega^2}\epsilon_r(\infty).$$

Note that $\epsilon_r(\infty)$ arises from the core electrons.

(3) Since $\omega_{\rm LO}$ and $\omega_{\rm TO}$ are determined from the equations of motion of ions which are independent of temperature, they are temperature independent. Therefore, the difference $\omega_{\rm LO} - \omega_{\rm TO}$ does not depend on temperature.

29-10 Lyddane-Sachs-Teller relation for a crystal with a three-atom basis. Consider a crystal that has a three-atom basis, that is, there are three atoms per primitive unit cell. For this crystal, there are two different transverse optical frequencies $\omega_{\rm TO1}$ and $\omega_{\rm TO2}$. Assume that the dielectric function for this crystal is given by

$$\epsilon_r(\omega) = a + \frac{b_1}{\omega^2 - \omega_{\rm TO1}^2} + \frac{b_2}{\omega^2 - \omega_{\rm TO2}^2},$$

where a, b_1, and b_2 are constants. Establish the Lyddane-Sachs-Teller (LST) relation for this crystal.

There are three constraints on $\epsilon_r(\omega)$. (i) $\epsilon_r(\omega \to \infty) = \epsilon_r(\infty)$; (ii) $\epsilon_r(\omega = 0) = \epsilon_r(0)$; (iii) $\epsilon_r(\omega_{\rm LO}) = 0$. Here $\epsilon_r(\infty)$ and $\epsilon_r(0)$ are known; they are high- and low-frequency dielectric constants, respectively. Making use of these three constraints, we have

$$a = \epsilon_r(\infty),$$

$$\epsilon_r(\infty) - \frac{b_1}{\omega_{\rm TO1}^2} - \frac{b_2}{\omega_{\rm TO2}^2} = \epsilon_r(0),$$

$$\epsilon_r(\infty) + \frac{b_1}{\omega_{\rm LO}^2 - \omega_{\rm TO1}^2} + \frac{b_2}{\omega_{\rm LO}^2 - \omega_{\rm TO2}^2} = 0.$$

The third equation can be cast into a quadratic equation for $\omega_{\rm LO}^2$. We have

$$\omega_{\rm LO}^4 - \left[\omega_{\rm TO1}^2 + \omega_{\rm TO2}^2 - \frac{b_1 + b_2}{\epsilon_r(\infty)}\right]\omega_{\rm LO}^2 + \frac{\epsilon_r(0)}{\epsilon_r(\infty)}\omega_{\rm TO1}^2\omega_{\rm TO2}^2 = 0,$$

where the second equation has been used. The two solutions for $\omega_{\rm LO}^2$ from the above equation are denoted by $\omega_{\rm LO1}^2$ and $\omega_{\rm LO2}^2$. One of the properties of the quadratic algebraic equation $ax^2 + bx + c = 0$ states that the product of the two solutions x_1 and x_2 is equal to c/a. According to this property of the quadratic algebraic equation, we have in the present problem

$$\omega_{\rm LO1}^2\omega_{\rm LO2}^2 = \frac{\epsilon_r(0)}{\epsilon_r(\infty)}\omega_{\rm TO1}^2\omega_{\rm TO2}^2.$$

Rearranging, we have

$$\frac{\omega_{\rm LO1}^2\omega_{\rm LO2}^2}{\omega_{\rm TO1}^2\omega_{\rm TO2}^2} = \frac{\epsilon_r(0)}{\epsilon_r(\infty)}.$$

The above equation is the generalization of the Lyddane-Sachs-Teller relation to the crystal under consideration.

29-11 Absorption coefficient of a nanowire. For a bulk (three-dimensional) semiconductor crystal, the frequency-dependence of the absorption coefficient $\alpha(\omega)$ is given by $\alpha(\omega) \propto (\hbar\omega - E_g)^{1/2}\theta(\hbar\omega - E_g)$. With a similar analysis to that used in deriving the above relation, find the frequency dependence of the absorption coefficient for an infinitesimally thin semiconducting "nanowire".

For a bulk semiconductor crystal, it has been found that $\alpha(\omega)$ is proportional to the joint density of states of electrons in the conduction and valence bands, $\alpha(\omega) \propto g_{cv}(\hbar\omega)$. For a one-dimensional nanowire, we have

$$g_{cv}(\hbar\omega) = \int_{-\infty}^{\infty} \frac{dk}{\pi}\, \delta\big(E_g + \hbar^2k^2/2m_r^* - \hbar\omega\big),$$

where m_r^* is the reduced effective mass, $m_r^* = m_c^* m_v^*/(m_c^* + m_v^*)$. Performing the integral in $g_{cv}(\hbar\omega)$, we obtain

$$\begin{aligned} g_{cv}(\hbar\omega) &= \left(\frac{2m_r^*}{\hbar^2}\right)^{1/2} \frac{1}{2\pi(\hbar\omega - E_g)^{1/2}} \\ &\quad \times \int_{-\infty}^{\infty} dk\, \big[\delta\big(k - (2m_r^*(\hbar\omega - E_g)/\hbar^2)^{1/2}\big) \\ &\qquad + \delta\big(k + (2m_r^*(\hbar\omega - E_g)/\hbar^2)^{1/2}\big)\big]\theta(\hbar\omega - E_g) \\ &= \left(\frac{2m_r^*}{\hbar^2}\right)^{1/2} \frac{1}{\pi(\hbar\omega - E_g)^{1/2}}\theta(\hbar\omega - E_g). \end{aligned}$$

Thus, $\alpha(\omega) \propto (\hbar\omega - E_g)^{-1/2}\theta(\hbar\omega - E_g)$ for a nanowire.

29-12 Model dielectric function for Ge. We model the imaginary part $\epsilon_r''(\omega)$ of the relative dielectric function $\epsilon_r(\omega)$ due to the interband transitions in semiconductor germanium as $\epsilon_r''(\omega) = b$ for $E_1 < \hbar\omega < E_2$ and $\epsilon_r''(\omega) = 0$ otherwise.

(1) Find an expression for the real part $\epsilon_r'(\omega)$ for all ω using the Kramers-Kronig relations, with $\epsilon_r'(\infty)$ taken as unity, $\epsilon_r'(\infty) = 1$. At what photon frequencies does $\epsilon_r'(\omega)$ possess structures? Why these structures appear?

(2) Derive an explicit expression for $\epsilon_r'(0)$ and explain why narrow gap semiconductors tend to have large dielectric constants at $\omega = 0$ using the obtained result.

(3) Prove the sum rule $ne^2/\epsilon_0 m = (2/\pi)\int_0^\infty d\omega\, \omega\epsilon_r''(\omega)$ using the Kramers-Kronig relations, where n is the total carrier density in the semiconductor.

---

(1) The Kramers-Kronig relation for determining the real part of the relative dielectric function from its imaginary part is given by

$$\epsilon_r'(\omega) = \epsilon_r'(\infty) + \frac{2}{\pi}\,\mathrm{P}\int_0^\infty d\omega'\,\frac{\omega'\epsilon_r''(\omega')}{\omega'^2 - \omega^2}.$$

The presence of the first term on the right hand side is because $\epsilon_r(\omega)$ does not go to zero as $\omega \to \infty$, $\epsilon_r(\infty) = \epsilon_r'(\infty) \neq 0$. The Kramers-Kronig relations are obtained by considering $\epsilon_r(\omega) - \epsilon_r'(\infty)$, which leads to the appearance of $\epsilon_r'(\infty)$ on the right hand side of the above expression. Making use of $\epsilon_r'(\infty) = 1$ and the assumed expression of $\epsilon_r''(\omega)$, we have

$$\epsilon_r'(\omega) = 1 + \frac{2b}{\pi}\,\mathrm{P}\int_{E_1/\hbar}^{E_2/\hbar} d\omega'\,\frac{\omega'}{\omega'^2 - \omega^2} = 1 + \frac{b}{\pi}\ln\left|\frac{E_2^2 - (\hbar\omega)^2}{E_1^2 - (\hbar\omega)^2}\right|.$$

Since the logarithmic function has singularities at $\hbar\omega = E_1$ and E_2, the structures in $\epsilon_r'(\omega)$ appear around $\omega = E_1/\hbar$ and $E_2/\hbar$. The presence of these structures is because the imaginary part $\epsilon_r''(\omega)$ also has structures at these frequencies. As a matter of fact, $E_1/\hbar$ corresponds to the onset of the interband absorption while $E_2/\hbar$ corresponds to the maximum frequency for the interband absorption.

(2) Setting $\omega = 0$ in the above-derived result for $\epsilon_r'(\omega)$, we have

$$\epsilon_r'(0) = 1 + \frac{2b}{\pi}\ln\frac{E_2}{E_1}.$$

Note that E_1 corresponds to the band gap, $E_1 \approx E_g = \varepsilon_c - \varepsilon_v$. For narrow gap semiconductors, E_1 is small. Whereas, E_2 does not differ much from that of wide gap semiconductors, which leads to a large value for $\ln(E_2/E_1)$ and consequently results in a large value for the dielectric function at $\omega = 0$. Therefore, narrow gap semiconductors tend to have large dielectric constants at $\omega = 0$.

(3) We prove the sum rule using the following Kramers-Kronig relation for the dielectric function

$$\epsilon_r'(\omega) = \epsilon_r'(\infty) + \frac{2}{\pi}\,\mathrm{P}\int_0^\infty d\omega'\,\frac{\omega'\epsilon_r''(\omega')}{\omega'^2 - \omega^2},$$

where $\epsilon_r''(\omega)$ is generally not the imaginary part of the above-given model dielectric function. We consider the high-frequency limit of $\epsilon_r'(\omega)$. Since $\epsilon_r''(\omega) \to 0$ for $\omega \to \infty$, we can replace the upper limit of the integral by ω_c beyond which $\epsilon_r''(\omega)$ is essentially zero. We then have

$$\epsilon_r'(\omega) = \epsilon_r'(\infty) + \frac{2}{\pi}\,\mathrm{P}\int_0^{\omega_c} \mathrm{d}\omega'\, \frac{\omega'\epsilon_r''(\omega')}{\omega'^2 - \omega^2},$$

For $\omega \gg \omega_c$, we can neglect the ω'^2 term on the denominator of the integrand. Making use of the high-frequency limit of the relative dielectric function

$$\epsilon_r(\omega \to \infty) = \epsilon_r'(\omega \to \infty) = \epsilon_r'(\infty) - \frac{\omega_p^2}{\omega^2},$$

we have

$$\epsilon_r'(\infty) - \frac{\omega_p^2}{\omega^2} = \epsilon_r'(\infty) - \frac{2}{\pi\omega^2}\int_0^{\omega_c} \mathrm{d}\omega'\, \omega'\epsilon_r''(\omega')$$

from which it follows that

$$\frac{2}{\pi}\int_0^{\infty} \mathrm{d}\omega\, \omega\epsilon_r''(\omega) = \omega_p^2 = \frac{ne^2}{\epsilon_0 m},$$

where we have restored the infinite upper integration limit. We have thus proved the sum rule.

29-13 Analysis of reflectance data. The data of reflectance $R(\omega)$ versus $\hbar\omega$ for a solid are given in the following table.

Reflectance data. $\hbar\omega$ is in meV.

$\hbar\omega$	R	$\hbar\omega$	R	$\hbar\omega$	R	$\hbar\omega$	R	$\hbar\omega$	R	$\hbar\omega$	R	$\hbar\omega$	R
0.126	0.98	0.703	0.98	2.611	0.94	4.921	0.27	13.556	0.09	38.155	0.04	130.544	0.00
0.131	0.98	0.723	0.98	2.711	0.93	5.020	0.27	14.058	0.09	39.163	0.04	140.583	0.00
0.141	0.98	0.743	0.98	2.812	0.91	5.222	0.27	14.560	0.1	40.174	0.04	150.635	0.00
0.151	0.98	0.753	0.98	2.912	0.9	5.422	0.26	15.062	0.1	42.175	0.04	160.666	0.00
0.161	0.98	0.763	0.98	3.012	0.88	5.623	0.26	15.564	0.1	44.180	0.03	170.712	0.00
0.171	0.98	0.783	0.98	3.113	0.87	5.823	0.27	16.067	0.09	46.197	0.03	180.749	0.00
0.181	0.98	0.803	0.98	3.213	0.84	6.026	0.27	17.071	0.09	47.195	0.03	200.839	0.00
0.191	0.98	0.823	0.98	3.314	0.81	6.124	0.27	18.075	0.08	48.200	0.03	220.941	0.00
0.201	0.98	0.843	0.98	3.414	0.81	6.225	0.26	19.078	0.07	49.209	0.03	240.999	0.00
0.221	0.99	0.853	0.98	3.515	0.76	6.326	0.26	20.084	0.07	50.202	0.03	261.116	0.00
0.241	0.99	0.863	0.98	3.615	0.69	6.427	0.25	20.585	0.07	51.213	0.03	281.165	0.00
0.261	0.99	0.884	0.98	3.665	0.64	6.626	0.24	21.087	0.07	52.223	0.02	301.234	0.00
0.281	0.98	0.903	0.98	3.715	0.56	6.829	0.23	21.588	0.07	53.228	0.02	351.496	0.00
0.301	0.98	0.924	0.98	3.745	0.5	7.030	0.21	22.090	0.08	54.225	0.02	371.531	0.00
0.321	0.98	0.944	0.98	3.766	0.42	7.230	0.19	22.595	0.09	56.233	0.02	401.742	0.00
0.341	0.98	0.954	0.98	3.816	0.24	7.433	0.18	23.094	0.09	58.232	0.02	502.016	0.00
0.361	0.98	0.964	0.98	3.846	0.17	7.633	0.16	23.597	0.08	60.261	0.02	602.614	0.00
0.382	0.98	0.984	0.98	3.866	0.12	7.830	0.14	24.600	0.07	62.250	0.02	743.284	0.00
0.402	0.98	1.004	0.98	3.897	0.09	8.032	0.11	25.607	0.06	64.275	0.02	803.226	0.00
0.422	0.98	1.105	0.98	3.916	0.06	8.234	0.1	26.106	0.06	66.294	0.02	1004.032	0.00
0.442	0.98	1.205	0.98	3.946	0.06	8.435	0.08	26.608	0.05	68.294	0.02	1506.352	0.00
0.462	0.98	1.305	0.98	3.966	0.06	8.634	0.07	27.112	0.11	70.299	0.02	2008.388	0.00
0.482	0.98	1.406	0.98	3.997	0.07	8.836	0.07	27.618	0.11	72.300	0.02	2510.587	0.00
0.502	0.98	1.506	0.98	4.016	0.09	9.035	0.07	28.117	0.09	74.328	0.01	3012.340	0.00
0.522	0.98	1.607	0.98	4.067	0.12	9.236	0.07	28.621	0.08	76.334	0.01	3514.963	0.00
0.542	0.98	1.707	0.98	4.117	0.15	9.742	0.07	29.123	0.08	78.351	0.01	3614.983	0.00
0.562	0.98	1.807	0.97	4.167	0.17	10.040	0.07	30.131	0.07	80.323	0.01	4017.425	0.00
0.582	0.98	1.908	0.97	4.217	0.19	10.341	0.07	31.133	0.07	85.332	0.01	4519.056	0.00
0.603	0.98	2.008	0.97	4.318	0.22	10.641	0.07	32.137	0.07	90.348	0.01	5020.161	0.00
0.623	0.98	2.109	0.97	4.418	0.24	10.950	0.07	33.138	0.07	95.402	0.01	6026.137	0.00
0.643	0.98	2.209	0.96	4.519	0.25	11.247	0.07	34.147	0.07	100.403	0.00	8032.258	0.00
0.653	0.98	2.309	0.96	4.620	0.26	11.549	0.07	35.150	0.06	105.419	0.00	10040.323	0.00
0.663	0.98	2.410	0.95	4.719	0.26	12.552	0.08	36.150	0.05	110.470	0.00		
0.683	0.98	2.511	0.95	4.820	0.27	13.054	0.08	37.153	0.05	120.523	0.00		

The phase $\varphi(\omega)$ of $r(\omega) = |r(\omega)|e^{i\varphi(\omega)}$ in $r = (n+ik-1)/(n+ik+1)$ is related to its magnitude $|r(\omega)|$ through

$$\varphi(\omega) = -\frac{2\omega}{\pi} \mathrm{P} \int_0^\infty d\omega' \, \frac{\ln(|r(\omega')|/|r(\omega)|)}{\omega'^2 - \omega^2}$$

with $|r(\omega)| = \sqrt{R(\omega)}$.

(1) Determine the refractive index $n(\omega)$ and the extinction coefficient $k(\omega)$.

(2) Find the real and imaginary parts of the relative dielectric function $\epsilon_r(\omega)$.

---

(1) The integral in the expression of $\varphi(\omega)$ is numerically performed by utilizing the GSL C function **`gsl_integration_qawc`** for Cauchy integration and the GSL C function **`gsl_integration_qags`** for non-Cauchy integration. The experimental data for the reflectance are interpolated using the cubic spline interpolation method in performing the integration.
When the value of φ is obtained numerically at a given frequency ω, the refractive index n and extinction coefficient k are evaluated using the expressions

$$n = \frac{1-R}{1-2\sqrt{R}\cos\varphi + R},$$
$$k = \frac{2\sqrt{R}\sin\varphi}{1-2\sqrt{R}\cos\varphi + R}.$$

The obtained values of n and k in such a manner are plotted in Fig. 29.3(b) as functions of $\hbar\omega$. The given experimental data for the reflectance R are plotted in Fig. 29.3(a).

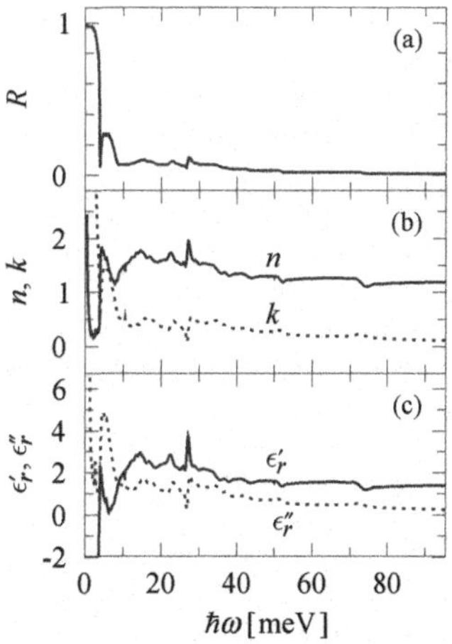

Fig. 29.3 Plots of the reflectance R (a), the refractive index n and extinction coefficient k (b), and the real and imaginary parts of the relative dielectric function ϵ_r, ϵ'_r and ϵ''_r (c), as functions of $\hbar\omega$.

(2) From the values of n and k, the real and imaginary parts of the dielectric function can be obtained from $\epsilon_r' = n^2 - k^2$, $\epsilon_r'' = 2nk$. The computed values of ϵ_r' and ϵ_r'' are plotted Fig. 29.3(c) as functions of $\hbar\omega$.

Chapter 30

Superconductivity

(1) *Zero resistance*
The resistance/resistivity of a superconductor becomes exactly zero below the critical temperature T_c.

(2) *Meissner effect*
A superconductor expels a sufficiently small applied magnetic field completely when it is in the superconducting phase.

(3) *Condensation energy*

$$e_c = f_s(0,0) - f_n(0,0) = -\frac{1}{2}\mu_0 H_c^2(0).$$

(4) *Isotope effect*

$$T_c M^\alpha = \text{const.}$$

(5) *Electron entropy*

$$s_s^e(T,B) - s_n^e(T,B) = \mu_0 H_c(T)\frac{\partial H_c(T)}{\partial T}.$$

(6) *Electron specific heat jump at T_c*

$$\left.\frac{\Delta c_v^e}{c_{vn}^e}\right|_{T=T_c} = \frac{\mu_0}{\gamma}\left[\frac{\partial H_c(T)}{\partial T}\right]^2\Bigg|_{T=T_c}.$$

(7) *Two-fluid model*

$$f_s(T,0) = -\frac{\gamma}{2}\sqrt{1-x_s}T^2 - \frac{\mu_0}{2}x_s H_c^2(0),$$

$$H_c(0) = \left(\frac{\gamma}{2\mu_0}\right)^{1/2} T_c,$$

$$x_s(T) = 1 - \left(\frac{T}{T_c}\right)^4,$$

$$f_s(T,0) = -\frac{\gamma}{4T_c^2}(T^4 + T_c^4),$$

$$c_{vs}^{e} = \frac{3\gamma}{T_c^2} T^3,$$

$$H_c(T) = H_c(0)\left(1 - \frac{T^2}{T_c^2}\right).$$

(8) *London equations*

The first London equation is

$$\boldsymbol{E} = \frac{m}{n_s e^2} \frac{\partial \boldsymbol{j}_s}{\partial t} = \frac{\partial}{\partial t}(\Lambda \boldsymbol{j}_s).$$

The second London equation is

$$\boldsymbol{B} = -\boldsymbol{\nabla} \times (\Lambda \boldsymbol{j}_s).$$

The two London equations can be written as a single equation

$$\boldsymbol{j}_s = -\boldsymbol{A}/\Lambda.$$

The London penetration depth is defined through

$$\nabla^2 \boldsymbol{B} = \lambda_{\mathrm{L}}^{-2} \boldsymbol{B},$$

$$\lambda_{\mathrm{L}} = (\Lambda/\mu_0)^{1/2} = (m/\mu_0 n_s e^2)^{1/2}.$$

(9) *Ginzburg-Landau equation*

$$\frac{(-\mathrm{i}\hbar\boldsymbol{\nabla} + e^* \boldsymbol{A})^2 \psi}{2m^*} + \beta|\psi|^2\psi = -\alpha(T)\psi.$$

(10) *Two types of superconductors*

The Ginzburg-Landau parameter is

$$\kappa = \lambda_{\mathrm{L}}/\xi_{\mathrm{L}}.$$

Type-I superconductors: $\kappa < 1/\sqrt{2}$.

Type-II superconductors: $\kappa > 1/\sqrt{2}$.

(11) *Vortex quantum*

$$\Phi = n\Phi_0, \ \ \Phi_0 = h/e^*.$$

(12) *Electron bound states*

$$\mathscr{E} - 2\varepsilon_{\mathrm{F}} \approx -2\hbar\omega_{\mathrm{D}} e^{-1/N(0)V}.$$

(13) *Pairing symmetry*

S-wave pairing: $S = 0$, $\ell = 0$.

D-wave pairing: $S = 0$, $\ell = 2$.

(14) *BCS wave function*

$$|\Omega\rangle = \prod_{\boldsymbol{k}} \left(u_{\boldsymbol{k}} + v_{\boldsymbol{k}} \hat{c}_{\boldsymbol{k}\uparrow}^{\dagger} \hat{c}_{-\boldsymbol{k}\downarrow}^{\dagger}\right)|0\rangle\,.$$

(15) *Anomalous average*

$$\Delta = V{\sum_{\boldsymbol{k}}}' \langle \hat{c}_{-\boldsymbol{k}\downarrow}\hat{c}_{\boldsymbol{k}\uparrow}\rangle .$$

(16) *BCS Hamiltonian*

$$\hat{H}_{\mathrm{BCS}} \approx {\sum_{\boldsymbol{k}\sigma}}'(\varepsilon_{\boldsymbol{k}} - \mu)\hat{c}^\dagger_{\boldsymbol{k}\sigma}\hat{c}_{\boldsymbol{k}\sigma} + \frac{1}{V}|\Delta|^2 - {\sum_{\boldsymbol{k}}}'\big(\Delta^*\hat{c}_{-\boldsymbol{k}\downarrow}\hat{c}_{\boldsymbol{k}\uparrow} + \Delta\hat{c}^\dagger_{\boldsymbol{k}\uparrow}\hat{c}^\dagger_{-\boldsymbol{k}\downarrow}\big).$$

(17) *Bogoliubov transformation*

$$\hat{\alpha}_{\boldsymbol{k}} = u^*_{\boldsymbol{k}}\hat{c}_{\boldsymbol{k}\uparrow} - v^*_{\boldsymbol{k}}\hat{c}^\dagger_{-\boldsymbol{k}\downarrow},$$
$$\hat{\beta}^\dagger_{\boldsymbol{k}} = v_{\boldsymbol{k}}\hat{c}_{\boldsymbol{k}\uparrow} + u_{\boldsymbol{k}}\hat{c}^\dagger_{-\boldsymbol{k}\downarrow}.$$

(18) *Superconducting density of states*

$$N_s(E) = \frac{g(\varepsilon_{\mathrm{F}})E}{\sqrt{E^2 - \Delta^2}}\theta(E - \Delta).$$

(19) *Self-consistent equation for the superconducting energy gap*

$$\Delta = V{\sum_{\boldsymbol{k}}}' \frac{\Delta}{2E_{\boldsymbol{k}}}\tanh\left(\frac{E_{\boldsymbol{k}}}{2k_{\mathrm{B}}T}\right).$$

(20) *Superconducting energy gap at zero temperature*

$$\Delta(0) \approx 2\hbar\omega_{\mathrm{D}}\mathrm{e}^{-1/N(0)V}.$$

(21) *Superconducting critical temperature*

$$k_{\mathrm{B}}T_c \approx 1.134\hbar\omega_{\mathrm{D}}\mathrm{e}^{-1/N(0)V}.$$

(22) *Gap-T_c ratio*

$$\frac{2\Delta(0)}{k_{\mathrm{B}}T_c} = \frac{2\pi}{\mathrm{e}^\gamma} \approx 3.528.$$

(23) *Specific heat jump at the superconducting phase transition*

$$\frac{\Delta C^e_v\big|_{T_c}}{C^e_{vn}\big|_{T_c}} \equiv \frac{\left(C^e_{vs} - C^e_{vn}\right)\big|_{T_c}}{C^e_{vn}\big|_{T_c}} \approx 1.426.$$

(24) *Critical magnetic field*

$$f_s(T) - f_n(T) = -\mu_0 H^2_c(T)/2,$$
$$H_c(0) = \left[g(\varepsilon_F)/2\mu_0\right]^{1/2}\Delta(0),$$
$$H_c(T \to 0) \approx H_c(0)\left[1 - 1.057(T/T_c)^2\right],$$
$$H_c(T \to T_c^-) \approx 1.74H_c(0)(1 - T/T_c).$$

(25) *Josephson effect*

$$I_s = I_c \sin\Delta\theta, \;\; I_s = I_c \sin(2eVt/\hbar).$$

(26) *Quantum interference*

$$I(\Phi) = I_c(0)\frac{\sin(\pi\Phi/\Phi_0)}{\pi\Phi/\Phi_0}.$$

(27) *Eliashberg equations*

$$\Delta(\mathrm{i}\omega_n) = \frac{1}{Z(\mathrm{i}\omega_n)}\int_0^\infty \mathrm{d}\omega'\,\mathrm{Re}\left\{\frac{\Delta(\omega')}{[\omega'^2-\Delta^2(\omega')]^{1/2}}\right\} \times\left[K_+(\mathrm{i}\omega_n,\omega') - \mu^*\tanh\left(\frac{\beta\omega'}{2}\right)\right],$$

$$Z(\mathrm{i}\omega_n) = 1 - \frac{1}{\mathrm{i}\omega_n}\int_0^\infty \mathrm{d}\omega'\,\mathrm{Re}\left\{\frac{\omega'}{[\omega'^2-\Delta^2(\omega')]^{1/2}}\right\}K_-(\mathrm{i}\omega_n,\omega'),$$

$$K_\pm(\mathrm{i}\omega_n,\omega') = \int_0^\infty \mathrm{d}\omega\,\alpha^2F(\omega)\left[\frac{f(-\omega')+n(\omega)}{\omega'+\mathrm{i}\omega_n+\omega} \pm \frac{f(-\omega')+n(\omega)}{\omega'-\mathrm{i}\omega_n+\omega} \mp \frac{f(\omega')+n(\omega)}{-\omega'+\mathrm{i}\omega_n+\omega} - \frac{f(\omega')+n(\omega)}{-\omega'-\mathrm{i}\omega_n+\omega}\right].$$

30-1 Isotope effect in tin. The following data were obtained for the dependence of critical temperature T_c (in Kelvins) of tin on its average atomic mass M (in the atomic mass unit) [E. Maxwell, Physical Review **86**, 235 (1952)].

M	113.58	116.67	118.05	118.70	119.78	123.01
T_c	3.808 2	3.770 8	3.744 4	3.741 9	3.723 8	3.666 9

(1) Plot T_c versus M, T_c versus $1/\sqrt{M}$, and $\ln T_c$ versus $\ln M$.
(2) Fit the data for $\ln T_c$ and $\ln M$ to a linear relation of the form $\ln T_c = \beta - \alpha \ln M$ using the least-squares fitting method. Comment on the difference between the value of α found here and that predicted by the BCS theory.

---

(1) The plots are given in Figs. 30.1(a), (b), and (c), respectively.
(2) The chi-squared function, χ^2, in the linear least-squares fitting method for the present problem is given by

$$\chi^2 = \sum_{i=1}^{N}(\beta - \alpha\ln M_i - \ln T_{ci})^2,$$

where (M_i, T_{ci}) for $i = 1, 2, \cdots, N$ with $N = 6$ are the given experimental data. The values of the fitting parameters α

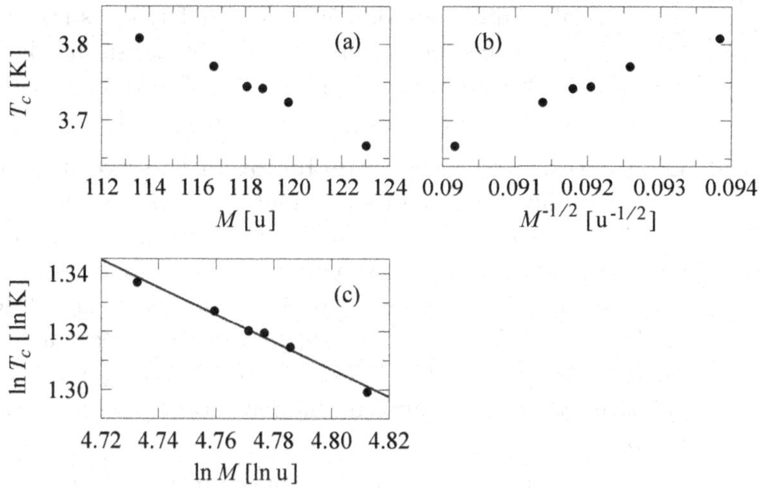

Fig. 30.1 Plots of T_c versus M (a), T_c versus $1/\sqrt{M}$ (b), and $\ln T_c$ versus $\ln M$ (c). The solid line in (c) is a linear-squares fit to the experimental data.

and β can be determined by minimizing χ^2 with respect to α and β. Differentiating χ^2 with respect to α and β, respectively, and setting the obtained derivatives to zero, we have

$$0 = \frac{\partial \chi^2}{\partial \alpha} = -2\sum_{i=1}^{N}(\beta - \alpha \ln M_i - \ln T_{ci}) \ln M_i,$$

$$0 = \frac{\partial \chi^2}{\partial \beta} = 2\sum_{i=1}^{N}(\beta - \alpha \ln M_i - \ln T_{ci}).$$

Solving for α and β from the above two equations yields

$$\alpha = \frac{\sum_i \ln M_i \sum_i \ln T_{ci} - N\sum_i \ln M_i \ln T_{ci}}{N\sum_i \ln^2 M_i - \left(\sum_i \ln M_i\right)^2},$$

$$\beta = \frac{\sum_i \ln^2 M_i \sum_i \ln T_{ci} - \sum_i \ln M_i \sum_i \ln M_i \ln T_{ci}}{N\sum_i \ln^2 M_i - \left(\sum_i \ln M_i\right)^2}.$$

Making use of the given experimental data, we can compute the values for α and β from the above two expressions and we find that $\alpha \approx 0.472\,2$ and $\beta \approx 3.573\,6$. The fitting results are plotted in Fig. 30.1(c) as a solid straight line.

The value of α found here is very close to that predicted by the BCS theory with an absolute difference of 0.0278 and a relative difference of 5.56%. The difference is probably due to the experimental uncertainties in measuring the critical temperatures and sample masses. The results clearly indicate that tin is a weak-coupling superconductor.

30-2 Conductivity of a superconductor in the two-fluid model. In the two-fluid model for a superconductor, it is assumed that there exist both normal and superconducting electrons. The normal electrons obey the Drude-like equation $\mathrm{d}\boldsymbol{j}_n/\mathrm{d}t = (n_n e^2/m)\boldsymbol{E} - \boldsymbol{j}_n/\tau$, where n_n and $\boldsymbol{j}_n$ are respectively the number and current densities of the normal electrons. The superconducting electrons obey the London equation $\mathrm{d}\boldsymbol{j}_s/\mathrm{d}t = (n_s e^2/m)\boldsymbol{E}$, where n_s and $\boldsymbol{j}_s$ are respectively the number and current densities of the superconducting electrons.

(1) Find the frequency-dependent complex conductivity $\sigma(\omega)$ for a superconductor. Use the time dependence of the form $\mathrm{e}^{-\mathrm{i}\omega t}$ for time-dependent quantities and assume that the normal and superconducting fluids respond independently to the electric field.

(2) Show that, in the low-frequency limit, the response of the normal fluid is purely ohmic while the response of the superconducting fluid is purely inductive.

---

(1) Making use of $\boldsymbol{j}_n = \boldsymbol{j}_n(\omega)\mathrm{e}^{-\mathrm{i}\omega t}$, $\boldsymbol{j}_s = \boldsymbol{j}_s(\omega)\mathrm{e}^{-\mathrm{i}\omega t}$, and $\boldsymbol{E} = \boldsymbol{E}(\omega)\mathrm{e}^{-\mathrm{i}\omega t}$, we have

$$-\mathrm{i}\omega\boldsymbol{j}_n(\omega) = \frac{n_n e^2}{m}\boldsymbol{E}(\omega) - \frac{1}{\tau}\boldsymbol{j}_n(\omega),$$

$$-\mathrm{i}\omega\boldsymbol{j}_s(\omega) = \frac{n_s e^2}{m}\boldsymbol{E}(\omega).$$

Solving for $\boldsymbol{j}_n(\omega)$ and $\boldsymbol{j}_s(\omega)$ from the above two equations, we obtain

$$\boldsymbol{j}_n(\omega) = \frac{n_n e^2\tau/m}{1-\mathrm{i}\omega\tau}\boldsymbol{E}(\omega),\ \boldsymbol{j}_s(\omega) = \frac{\mathrm{i}n_s e^2}{m\omega}\boldsymbol{E}(\omega).$$

From $\boldsymbol{j}_n(\omega) = \sigma_n(\omega)\boldsymbol{E}(\omega)$ and $\boldsymbol{j}_s(\omega) = \sigma_s(\omega)\boldsymbol{E}(\omega)$, we obtain the contributions to the conductivity

$$\sigma_n(\omega) = \frac{n_n e^2\tau/m}{1-\mathrm{i}\omega\tau},\ \sigma_s(\omega) = \frac{\mathrm{i}n_s e^2}{m\omega}.$$

The conductivity of the superconductor is then given by

$$\sigma(\omega) = \sigma_n(\omega)+\sigma_s(\omega) = \frac{n_n e^2\tau/m}{1-\mathrm{i}\omega\tau} + \frac{\mathrm{i}n_s e^2}{m\omega} = \frac{e^2}{m}\left(\frac{n_n\tau}{1-\mathrm{i}\omega\tau} + \frac{\mathrm{i}n_s}{\omega}\right).$$

(2) In the low-frequency limit, we have

$$\sigma_n(\omega) \approx \frac{n_n e^2 \tau}{m}, \ \sigma_s(\omega) = \frac{i n_s e^2}{m\omega}.$$

Since $\sigma_n(\omega)$ is real while $\sigma_s(\omega)$ is purely imaginary, we conclude that the response of the normal fluid is purely ohmic while the response of the superconducting fluid is purely inductive.

30-3 Magnetic field inside an infinite superconducting plate. Solve the London equations for an infinite superconducting plate of finite thickness $2t$. Assume that the magnetic field of magnitude B_0 is applied parallel to the plate. Find both the magnetic field and the supercurrent inside the plate. Plot the magnetic field and supercurrent for $2t = \lambda_L$ and $2\lambda_L$.

Take the origin of the Cartesian coordinate system to be at the center of the plate, the y axis perpendicular to the plate, and the z and x axes parallel to the plate and obeying the right hand rule together with the y axis [*cf.* Fig. 30-2]. The magnetic field is taken to be in the positive z direction. Then the magnetic field inside the plate satisfies the following equation

$$\frac{d^2 B_z}{dy^2} = \frac{1}{\lambda_L^2} B_z.$$

The general solution to the above equation is given by

$$B_z(y) = C\cosh(y/\lambda_L) + D\sin(y/\lambda_L).$$

Because of the presence of the geometric symmetry in the coordinate y, we have $B_z(-y) = B_z(y)$. Therefore, $D = 0$. We then have $B_z(y) = C\cosh(y/\lambda_L)$. From the boundary condition $B_z(y = t) = B_0$, we have $C = B_0/\cosh(t/\lambda_L)$. We finally have

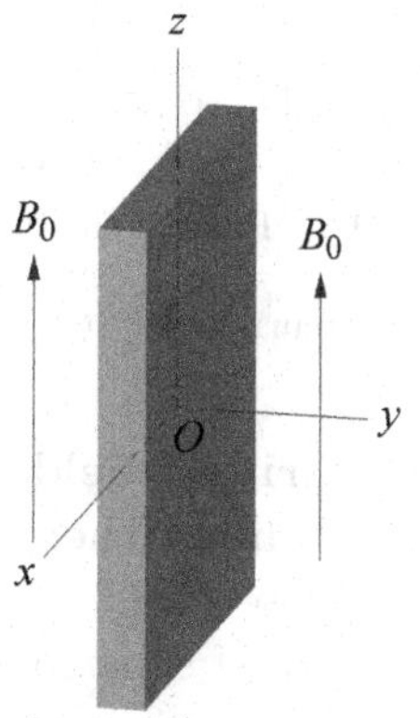

Fig. 30-2: Superconducting plate in a magnetic field.

$$B_z(y) = \frac{\cosh(y/\lambda_L)}{\cosh(t/\lambda_L)} B_0.$$

From Maxwell's equation $\boldsymbol{\nabla}\times\boldsymbol{B} = \mu_0 \boldsymbol{j}_s$, we obtain the supercurrent inside the plate

$$\boldsymbol{j}_s = \frac{1}{\mu_0}\boldsymbol{\nabla}\times\boldsymbol{B} = \frac{B_0 \boldsymbol{\nabla}\times[\cosh(y/\lambda_L)\boldsymbol{e}_z]}{\mu_0\cosh(t/\lambda_L)} = \frac{B_0\sinh(y/\lambda_L)}{\mu_0\lambda_L\cosh(t/\lambda_L)}\boldsymbol{e}_x.$$

Thus, the supercurrent is in the x direction. The plots of the magnetic field and supercurrent for $2t = \lambda_L$ and $2\lambda_L$ are given in Fig. 30.3. From Fig. 30.3, it can be seen that, the larger the ratio $2t/\lambda_L$, the smaller the magnetic field at $y = 0$. For $2t/\lambda_L \gg 1$, $B_z(y = 0) = 0$. It can be also seen that, the larger the ratio $2t/\lambda_L$, the more the supercurrent is distributed near the surfaces. For $2t/\lambda_L \gg 1$, nearly all the supercurrent is distributed on the surfaces.

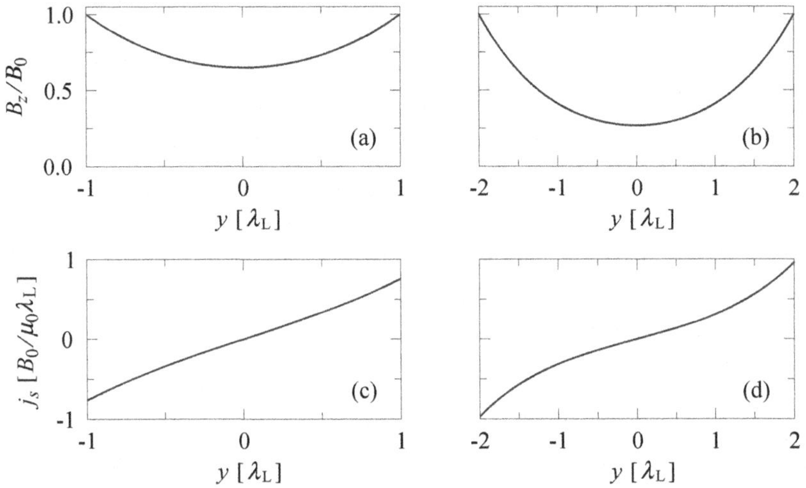

Fig. 30.3 Plots of the magnetic field and supercurrent inside the plate. (a) Magnetic field for $2t = \lambda_L$. (b) Magnetic field for $2t = 2\lambda_L$. (c) Supercurrent for $2t = \lambda_L$. (d) Supercurrent for $2t = 2\lambda_L$.

30-4 Critical field of a type-I superconductor in the Ginzburg-Landau theory. Within the Ginzburg-Landau theory, show that, close to T_c, the temperature dependence of the critical field of a type-I superconductor is given by $H_c(T) = H_c(0)(1 - T/T_c)$. Give an expression for $H_c(0)$ in terms of the penetration depth $\lambda_L(0)$ and the coherence length $\xi_L(0)$ at zero temperature.

We consider the simple case with the order parameter ψ being uniform throughout the superconductor. Then the Helmholtz free energy density of the superconductor in the absence of a magnetic field is given by

$$f_s = f_n + \alpha|\psi|^2 + \frac{1}{2}\beta|\psi|^4.$$

Close to T_c, we have $\alpha \propto (T-T_c)$. We write it as $\alpha = \alpha'(T-T_c)$ with $\alpha' > 0$. The parameter β is a positive constant. We now minimize f_s to find the value for the order parameter. Differentiating f_s with respect to $|\psi|^2$ and setting the result to zero, we have $\alpha + \beta|\psi|^2 = 0$ which leads to $|\psi|^2 = -\alpha/\beta = -\alpha'(T-T_c)/\beta$. The minimum value of the Helmholtz free energy density is then given by $f_s = f_n - \alpha'^2(T-T_c)^2/2\beta$. Comparing this result with $f_s - f_n = -\mu_0 H_c^2(T)/2$, we have

$$H_c(T) = \frac{\alpha'}{(\mu_0\beta)^{1/2}}(T_c - T) = H_c(0)(1 - T/T_c),$$

where $H_c(0) = \alpha' T_c/(\mu_0\beta)^{1/2}$. To express $H_c(0)$ in terms of $\lambda_\mathrm{L}(0)$ and $\xi_\mathrm{L}(0)$, we use the expressions of $\lambda_\mathrm{L}(0)$ and $\xi_\mathrm{L}(0)$ in terms of parameters α' and β. For the penetration depth, we have $\lambda_\mathrm{L}^2 = (m/\mu_0 n_s e^2) = m^*/\mu_0|\psi|^2 e^{*2}$, where we have made use of $m^* = 2m$, $e^* = 2e$, and $n_s = 2|\psi|^2$ to convert single-electron quantities to the corresponding ones for Cooper pairs. Inserting $|\psi|^2 = -\alpha'(T-T_c)/\beta$ into λ_L^2 yields

$$\lambda_\mathrm{L}^2 = \frac{\beta m^*}{\mu_0 e^{*2}\alpha' T_c(1 - T/T_c)}.$$

The coherence length is given by

$$\xi_\mathrm{L}^2 = \frac{\hbar^2}{2m^*|\alpha|} = \frac{\hbar^2}{2m^*\alpha' T_c(1 - T/T_c)}.$$

We see that

$$\lambda_\mathrm{L}^2(0)\xi_\mathrm{L}^2(0) = \frac{\beta\hbar^2}{2\mu_0 e^{*2}\alpha'^2 T_c^2}$$

from which we have

$$\frac{\alpha'}{(\mu_0\beta)^{1/2}} = \frac{\hbar}{2^{1/2}\mu_0 e^* T_c \lambda_\mathrm{L}(0)\xi_\mathrm{L}(0)} = \frac{\Phi_0}{2^{3/2}\pi\mu_0 T_c \lambda_\mathrm{L}(0)\xi_\mathrm{L}(0)}.$$

Finally, in terms of $\lambda_\mathrm{L}(0)$ and $\xi_\mathrm{L}(0)$, $H_c(0)$ is given by

$$H_c(0) = \frac{\alpha'_c}{(\mu_0\beta)^{1/2}} = \frac{\Phi_0}{2^{3/2}\pi\mu_0\lambda_\mathrm{L}(0)\xi_\mathrm{L}(0)}.$$

30-5 Ginzburg-Landau equation and superconducting current density. Using the Ginzburg-Landau equation, show that the superconducting current density $\boldsymbol{j}_s$ vanishes in any direction perpendicular to the surface of a superconductor.

From the Ginzburg-Landau equation, we see that the momentum of a Cooper pair is given by $\hat{p} = (-i\hbar\boldsymbol{\nabla}\psi + e^*\boldsymbol{A}\psi)/m^*$. Since a Cooper pair can not move out of the superconductor through a surface into vacuum, its momentum must be parallel to the surface. Let $\boldsymbol{n}$ denote the normal of the surface. We then have $(-\mathrm{i}\hbar\nabla\psi + e^*\psi\boldsymbol{A})\cdot\boldsymbol{n} = 0$. The supercurrent density in the superconductor is given by

$$\boldsymbol{j}_s = \frac{\mathrm{i}\hbar e^*}{2m^*}\left(\psi^*\nabla\psi - \psi\nabla\psi^*\right) - \frac{e^{*2}}{m^*}|\psi|^2\boldsymbol{A}.$$

Taking the dot-product of $\boldsymbol{j}_s$ at the surface with $\boldsymbol{n}$, we have

$$\begin{aligned}\boldsymbol{j}_s\cdot\boldsymbol{n} &= \frac{\mathrm{i}\hbar e^*}{2m^*}\left(\psi^*\nabla\psi - \psi\nabla\psi^*\right)\cdot\boldsymbol{n} - \frac{e^{*2}}{m^*}|\psi|^2\boldsymbol{A}\cdot\boldsymbol{n}\\ &= -\frac{e^*}{2m^*}\psi^*(-\mathrm{i}\hbar\nabla\psi + e^*\psi\boldsymbol{A})\cdot\boldsymbol{n}\\ &\quad - \frac{e^*}{2m^*}\psi(\mathrm{i}\hbar\nabla\psi^* + e^*\psi^*\boldsymbol{A})\cdot\boldsymbol{n}\\ &= 0,\end{aligned}$$

where we have made use of the condition $(-\mathrm{i}\hbar\nabla\psi + e^*\psi\boldsymbol{A})\cdot\boldsymbol{n} = 0$ and its complex conjugate $(\mathrm{i}\hbar\nabla\psi^* + e^*\psi^*\boldsymbol{A})\cdot\boldsymbol{n} = 0$. Thus, the superconducting current density $\boldsymbol{j}_s$ vanishes in any direction perpendicular to the surface of a superconductor.

30-6 Proximity effect between two planar superconductors. Two planar superconductors 1 and 2 are placed with their flat faces in very good contact. Their critical temperatures are T_{c1} and T_{c2}, respectively, with $T_{c2} > T_{c1}$ and $T_{c2} - T_{c1} \ll T_{c1}$. The system is cooled to a temperature T between T_{c1} and T_{c2} so that only superconductor 2 is superconducting.

(1) Show that the Ginzburg-Landau equation for superconductor 1 can be written as $-\xi_1^2\mathrm{d}^2\varphi/\mathrm{d}x^2 + \varphi + \varphi^3 = 0$, where φ is the dimensionless wave function and ξ_1 is given by $\xi_1 = (\hbar^2/2m^*|\alpha_1|)^{1/2}$.

(2) Making use of the fact that $|\varphi| \ll 1$ in a normal metal so that the cubic term in the above equation can be neglected, show that the wave function decays according to $\varphi = \varphi_0\,\mathrm{e}^{-|x|/\xi_1}$ in superconductor 2, where $x = 0$ is at the interface between the two superconductors and superconductor 2 occupies the $x < 0$ region.

---

(1) In the absence of any magnetic field, the Ginzburg-Landau equation reads for superconductor 1

$$-\frac{\hbar^2}{2m^*}\frac{\mathrm{d}^2\psi}{\mathrm{d}x^2} + \alpha_1\psi + \beta_1|\psi|^2\psi = 0.$$

Since superconductor 1 is in the normal phase, the value of α_1 is greater than 0 for it. We assume that ψ is a real function. We introduce a dimensionless wave function $\varphi = (\beta_1/\alpha_1)^{1/2}\psi$. We then have

$$-\frac{\hbar^2}{2m^*}\frac{\mathrm{d}^2\psi}{\mathrm{d}x^2} = -\frac{\hbar^2\alpha_1^{1/2}}{2m^*\beta_1^{1/2}}\frac{\mathrm{d}^2\varphi}{\mathrm{d}x^2}, \quad \alpha_1\psi+\beta_1|\psi|^2\psi = \frac{\alpha_1^{3/2}}{\beta_1^{1/2}}(\varphi+\varphi^3).$$

The Ginzburg-Landau equation now becomes

$$-\xi_1^2\frac{\mathrm{d}^2\varphi}{\mathrm{d}x^2} + \varphi + \varphi^3 = 0,$$

where $\xi_1 = (\hbar^2/2m^*|\alpha_1|)^{1/2}$. Note that taking the absolute value of α_1 is superfluous since $\alpha_1 > 0$ when superconductor 1 is in the normal phase.

(2) For $|\varphi| \ll 1$, we can neglect the term containing φ^3 in the Ginzburg-Landau equation and have

$$-\xi_1^2\frac{\mathrm{d}^2\varphi}{\mathrm{d}x^2} + \varphi = 0,$$

The general solution to the above equation is of the form $\varphi = Ce^{x/\xi_1} + De^{-x/\xi_1}$. In consideration that $\varphi \to 0$ as $x \to -\infty$, we have $D = 0$. Rewriting C as φ_0, the value of φ at $x = 0$, we have $\varphi = \varphi_0 e^{x/\xi_1} = \varphi_0 e^{-|x|/\xi_1}$.

30-7 Macroscopic quantum wave function of Cooper pairs. We study the following time-dependent Schrödinger equation for the macroscopic quantum wave function $\Psi(\boldsymbol{r},t) = \sqrt{n(\boldsymbol{r},t)}\,\mathrm{e}^{\mathrm{i}\theta(\boldsymbol{r},t)}$ in the electromagnetic fields described by the scalar potential $\phi(\boldsymbol{r},t)$ and the vector potential $\boldsymbol{A}(\boldsymbol{r},t)$, where $n(\boldsymbol{r},t)$ is the number density of Cooper pairs and $\theta(\boldsymbol{r},t)$ the phase of the wave function,

$$\mathrm{i}\hbar\frac{\partial\Psi(\boldsymbol{r},t)}{\partial t} = \frac{1}{2m^*}\big[-\mathrm{i}\hbar\boldsymbol{\nabla} + e^*\boldsymbol{A}(\boldsymbol{r},t)\big]^2\Psi(\boldsymbol{r},t) - e^*\phi(\boldsymbol{r},t)\Psi(\boldsymbol{r},t)$$

with m^* the mass and e^* $(= 2e)$ the magnitude of the charge of a Cooper pair.

(1) Assume that the number density $n(\boldsymbol{r},t)$ is constant in space and time. Derive $-\hbar\partial\theta/\partial t = (\Lambda/2n)j_s^2 - e^*\phi$. Briefly state the physical implication of this result.

(2) Now consider the case with $n(\boldsymbol{r},t)$ being a function of both space and time. Show that $\partial n(\boldsymbol{r},t)/\partial t + \boldsymbol{\nabla}\cdot[n(\boldsymbol{r},t)\boldsymbol{v}_s] = 0$ with $\boldsymbol{v}_s = (\hbar/m^*)\boldsymbol{\nabla\theta} + (e^*/m^*)\boldsymbol{A}$. Give a physical interpretation to this result.

---

(1) For $n(\boldsymbol{r},t)$ being constant in space and time, we have

$$\begin{aligned} \mathrm{i}\hbar\frac{\partial \Psi}{\partial t} &= -\hbar\frac{\partial \theta}{\partial t}\Psi, \\ -\mathrm{i}\hbar\boldsymbol{\nabla}\Psi &= \hbar\Psi\boldsymbol{\nabla}\theta, \\ -\mathrm{i}\hbar\boldsymbol{\nabla}\cdot(-\mathrm{i}\hbar\boldsymbol{\nabla}\Psi) &= \hbar^2\Psi(\boldsymbol{\nabla}\theta)^2 - \mathrm{i}\hbar^2\Psi\nabla^2\theta. \end{aligned}$$

We then have

$$\begin{aligned} (-\mathrm{i}\hbar\boldsymbol{\nabla} + e^*\boldsymbol{A})^2\Psi &= (-\mathrm{i}\hbar\boldsymbol{\nabla} + e^*\boldsymbol{A})(-\mathrm{i}\hbar\boldsymbol{\nabla}\Psi + e^*\boldsymbol{A}\Psi) \\ &= -\mathrm{i}\hbar\boldsymbol{\nabla}\cdot(-\mathrm{i}\hbar\boldsymbol{\nabla}\Psi) + 2e^*\boldsymbol{A}\cdot(-\mathrm{i}\hbar\boldsymbol{\nabla}\Psi) + e^{*2}\boldsymbol{A}^2\Psi \\ &= \left[\hbar^2(\boldsymbol{\nabla}\theta)^2 - \mathrm{i}\hbar^2\nabla^2\theta + 2e^*\hbar\boldsymbol{A}\cdot\boldsymbol{\nabla}\theta + e^{*2}\boldsymbol{A}^2\right]\Psi. \end{aligned}$$

Inserting the above results into the equation for Ψ yields

$$-\hbar\frac{\partial \theta}{\partial t} = \frac{1}{2m^*}\left[\hbar^2(\boldsymbol{\nabla}\theta)^2 - \mathrm{i}\hbar^2\nabla^2\theta + 2e^*\hbar\boldsymbol{A}\cdot\boldsymbol{\nabla}\theta + e^{*2}\boldsymbol{A}^2\right] - e^*\phi.$$

The supercurrent density is given by

$$\boldsymbol{j}_s = \frac{\mathrm{i}\hbar e^*}{2m^*}\left(\Psi^*\boldsymbol{\nabla}\Psi - \Psi\boldsymbol{\nabla}\Psi^*\right) - \frac{e^{*2}}{m^*}|\Psi|^2\boldsymbol{A} = -\frac{n\hbar e^*}{m^*}\boldsymbol{\nabla}\theta - \frac{ne^{*2}}{m^*}\boldsymbol{A}.$$

Squaring $\boldsymbol{j}_s$ and multiplying the resultant equation by $\Lambda/2n = m^*/ne^{*2}$, we have

$$\begin{aligned} \frac{\Lambda}{2n}j_s^2 &= \frac{m^*}{2n^2e^{*2}}\left[\left(\frac{n\hbar e^*}{m^*}\right)^2(\boldsymbol{\nabla}\theta)^2 + \frac{2n^2\hbar e^{*3}}{m^{*2}}\boldsymbol{A}\cdot\boldsymbol{\nabla}\theta + \left(\frac{ne^{*2}}{m^*}\right)^2\boldsymbol{A}^2\right] \\ &= \frac{\hbar^2}{2m^*}(\boldsymbol{\nabla}\theta)^2 + \frac{\hbar e^*}{m^*}\boldsymbol{A}\cdot\boldsymbol{\nabla}\theta + \frac{e^{*2}}{2m^*}\boldsymbol{A}^2 \end{aligned}$$

Making use of the above result in the equation for Ψ, we can rewrite it as

$$-\hbar\frac{\partial \theta}{\partial t} = \frac{\Lambda}{2n}j_s^2 - \frac{\mathrm{i}\hbar^2}{2m^*}\nabla^2\theta - e^*\phi.$$

Comparing the real parts of the two sides of the above equation, we obtain

$$-\hbar\frac{\partial \theta}{\partial t} = \frac{\Lambda}{2n}j_s^2 - e^*\phi.$$

The above equation describes the time variation of the phase of the macroscopic quantum wave function. Note that the first term is the kinetic energy and the second term the potential energy. Then the phase is given by the negative of the integral of the total energy over time, which is just the general principle that the energy of the system dictates its time dependence. Recall that we have generally $\psi(t) = \hat{T}\exp\{-\mathrm{i}\hbar^{-1}\int_0^t dt'\,\hat{H}(t)\}\psi(0)$.

(2) In this case, we have

$$\mathrm{i}\hbar\frac{\partial\Psi}{\partial t}=\mathrm{i}\hbar\left(\frac{1}{2n^{1/2}}\frac{\partial n}{\partial t}+\mathrm{i}n^{1/2}\frac{\partial\theta}{\partial t}\right)e^{\mathrm{i}\theta}=\left(\frac{\mathrm{i}\hbar}{2n}\frac{\partial n}{\partial t}-\hbar\frac{\partial\theta}{\partial t}\right)\Psi,$$

$$-\mathrm{i}\hbar\boldsymbol{\nabla}\Psi=-\frac{\mathrm{i}\hbar}{2n^{1/2}}e^{\mathrm{i}\theta}\boldsymbol{\nabla}n+\hbar n^{1/2}e^{\mathrm{i}\theta}\boldsymbol{\nabla}\theta=\left(-\frac{\mathrm{i}\hbar}{2n}\boldsymbol{\nabla}n+\hbar\boldsymbol{\nabla}\theta\right)\Psi,$$

and

$$\begin{aligned}&-\mathrm{i}\hbar\boldsymbol{\nabla}\cdot(-\mathrm{i}\hbar\boldsymbol{\nabla}\Psi)\\&=-\mathrm{i}\hbar\boldsymbol{\nabla}\left[\left(-\frac{\mathrm{i}\hbar}{2n}\boldsymbol{\nabla}n+\hbar\boldsymbol{\nabla}\theta\right)\Psi\right]\\&=\left[\frac{\hbar^2}{2n^2}(\boldsymbol{\nabla}n)^2-\frac{\hbar^2}{2n}\nabla^2n-\mathrm{i}\hbar^2\nabla^2\theta\right]\Psi+\left(-\frac{\mathrm{i}\hbar}{2n}\boldsymbol{\nabla}n+\hbar\boldsymbol{\nabla}\theta\right)^2\Psi\\&=\left[\frac{\hbar^2}{4n^2}(\boldsymbol{\nabla}n)^2-\frac{\hbar^2}{2n}\nabla^2n-\mathrm{i}\hbar^2\nabla^2\theta+\hbar^2(\boldsymbol{\nabla}\theta)^2-\frac{\mathrm{i}\hbar^2}{n}\boldsymbol{\nabla}n\cdot\boldsymbol{\nabla}\theta\right]\Psi.\end{aligned}$$

The original equation for Ψ now becomes

$$\begin{aligned}&\left(\frac{\mathrm{i}\hbar}{2n}\frac{\partial n}{\partial t}-\hbar\frac{\partial\theta}{\partial t}\right)\Psi\\&=\frac{1}{2m^*}\left(-\mathrm{i}\hbar\boldsymbol{\nabla}+e^*\boldsymbol{A}\right)^2\Psi-e^*\phi\Psi\\&=\frac{1}{2m^*}\left[-\mathrm{i}\hbar\boldsymbol{\nabla}\cdot(-\mathrm{i}\hbar\boldsymbol{\nabla}\Psi)+2e^*\boldsymbol{A}\cdot(-\mathrm{i}\hbar\boldsymbol{\nabla}\Psi)+e^{*2}\boldsymbol{A}^2\Psi\right]-e^*\phi\Psi\\&=\frac{1}{2m^*}\left[\frac{\hbar^2}{4n^2}(\boldsymbol{\nabla}n)^2-\frac{\hbar^2}{2n}\nabla^2n-\mathrm{i}\hbar^2\nabla^2\theta+\hbar^2(\boldsymbol{\nabla}\theta)^2-\frac{\mathrm{i}\hbar^2}{n}\boldsymbol{\nabla}n\cdot\boldsymbol{\nabla}\theta\right.\\&\qquad\left.-\frac{\mathrm{i}\hbar}{n}e^*\boldsymbol{A}\cdot\boldsymbol{\nabla}n+2\hbar e^*\boldsymbol{A}\cdot\boldsymbol{\nabla}\theta+e^{*2}\boldsymbol{A}^2\right]\Psi-e^*\phi\Psi.\end{aligned}$$

Comparing the imaginary parts of the two sides of the above equation, we obtain

$$\begin{aligned}\frac{\hbar}{2n}\frac{\partial n}{\partial t}&=\frac{1}{2m^*}\left[-\hbar^2\nabla^2\theta-\frac{\hbar^2}{n}\boldsymbol{\nabla}n\cdot\boldsymbol{\nabla}\theta-\frac{\hbar}{n}e^*\boldsymbol{A}\cdot\boldsymbol{\nabla}n\right]\\&=\frac{\hbar}{2m^*}\left[-\hbar\nabla^2\theta-\frac{1}{n}\boldsymbol{\nabla}n\cdot(\hbar\boldsymbol{\nabla}\theta+e^*\boldsymbol{A})\right]\\&=\frac{\hbar}{2m^*}\left(-\hbar\nabla^2\theta-\frac{m^*}{n}\boldsymbol{\nabla}n\cdot\boldsymbol{v}_s\right),\end{aligned}$$

where $\boldsymbol{v}_s=(\hbar\boldsymbol{\nabla}\theta+e^*\boldsymbol{A})/m^*$. Upon noticing that $\boldsymbol{\nabla}\cdot\boldsymbol{v}_s=(\hbar/m^*)\nabla^2\theta$ because $\boldsymbol{\nabla}\cdot\boldsymbol{A}=0$ in the Coulomb gauge, we have

$$\frac{\hbar}{2n}\frac{\partial n}{\partial t}=-\frac{\hbar}{2n}(n\boldsymbol{\nabla}\cdot\boldsymbol{v}_s+\boldsymbol{\nabla}n\cdot\boldsymbol{v}_s)=-\frac{\hbar}{2n}\boldsymbol{\nabla}\cdot(n\boldsymbol{v}_s),$$

that is,

$$\frac{\partial n}{\partial t}+\boldsymbol{\nabla}\cdot(n\boldsymbol{v}_s)=0.$$

The above equation is a continuity equation that expresses the conservation of probability.

30-8 Phonon-mediated effective electron-electron attraction. Starting from the electron-phonon interaction Hamiltonian

$$\hat{H}_{\rm int} = \sum_{\boldsymbol{kq}\sigma} \big(M \hat{c}^{\dagger}_{\boldsymbol{k}+\boldsymbol{q},\sigma} \hat{c}_{\boldsymbol{k}\sigma} \hat{a}_{\boldsymbol{q}} + \text{h.c.}\big),$$

show that an effective attraction between two electrons within the Debye energy from the Fermi surface appears in the second-order perturbation theory.

We first develop the second-order time-dependent perturbation theory. Let $|\psi(t)\rangle$ be the state vector of the system of electrons and phonons at time t. The time-dependent Schrödinger equation for $|\psi(t)\rangle$ reads

$$\mathrm{i}\hbar \frac{\mathrm{d}\,|\psi(t)\rangle}{\mathrm{d}t} = \hat{H}\,|\psi(t)\rangle = (\hat{H}_0 + \hat{H}_{\rm int})\,|\psi(t)\rangle\,.$$

The term $\hat{H}_0\,|\psi(t)\rangle$ on the right hand side can be eliminated by setting $|\psi(t)\rangle = \mathrm{e}^{-\mathrm{i}\hat{H}_0 t/\hbar}\,|\phi(t)\rangle$, where $|\phi(t)\rangle$ is the state vector to be solved. This transformation actually brings us into the interaction picture from the Schrödinger picture. Substituting $|\psi(t)\rangle = \mathrm{e}^{-\mathrm{i}\hat{H}_0 t/\hbar}\,|\phi(t)\rangle$ into the Schrödinger equation for $|\psi(t)\rangle$, we obtain an equation for $|\phi(t)\rangle$

$$\mathrm{i}\hbar \frac{\mathrm{d}\,|\phi(t)\rangle}{\mathrm{d}t} = \hat{H}_{\rm int}(t)\,|\phi(t)\rangle,$$

where $\hat{H}_{\rm int}(t) = \mathrm{e}^{\mathrm{i}\hat{H}_0 t/\hbar}\hat{H}_{\rm int}\mathrm{e}^{-\mathrm{i}\hat{H}_0 t/\hbar}$. To solve the above equation, we first convert it into an integral equation by integrating both sides over time t

$$|\phi(t)\rangle = |\phi(0)\rangle + \frac{1}{\mathrm{i}\hbar}\int_0^t \mathrm{d}t'\ \hat{H}_{\rm int}(t')\,|\phi(t')\rangle\,.$$

The above equation for $|\phi(t)\rangle$ can be solved through iteration. Assume that the system is in the state $|\psi(0)\rangle = |\phi(0)\rangle = |\phi_0\rangle$ at time $t = 0$. The zeroth-order solution is given by $|\phi_0\rangle$

$$|\phi^{(0)}(t)\rangle = |\phi_0\rangle\,.$$

Here the order implies the highest power of $\hat{H}_{\rm int}(t)$ in a single term. The first-order solution can be obtained by replacing $|\phi(t')\rangle$ on the right hand side of the integral equation for $|\phi(t)\rangle$ with the zeroth-order solution $|\phi^{(0)}(t')\rangle$. We then have

$$|\phi^{(1)}(t)\rangle = |\phi_0\rangle + \frac{1}{\mathrm{i}\hbar}\int_0^t \mathrm{d}t'\ \hat{H}_{\rm int}(t')\,|\phi_0\rangle\,.$$

The second-order solution can be obtained by replacing $|\phi(t')\rangle$ on the right hand side of the integral equation for $|\phi(t)\rangle$ with the first-order solution $|\phi^{(1)}(t')\rangle$. We then have

$$\begin{aligned}|\phi^{(2)}(t)\rangle &= |\phi_0\rangle + \frac{1}{\mathrm{i}\hbar}\int_0^t \mathrm{d}t'\, \hat{H}_{\mathrm{int}}(t')\,|\phi_0\rangle \\ &\quad + \frac{1}{(\mathrm{i}\hbar)^2}\int_0^t \mathrm{d}t' \int_0^{t'} \mathrm{d}t''\, \hat{H}_{\mathrm{int}}(t')\hat{H}_{\mathrm{int}}(t'')\,|\phi_0\rangle \\ &= \left[1 + \frac{1}{\mathrm{i}\hbar}\int_0^t \mathrm{d}t'\, \hat{H}_{\mathrm{int}}(t') \right. \\ &\quad \left. + \frac{1}{(\mathrm{i}\hbar)^2}\int_0^t \mathrm{d}t' \int_0^{t'} \mathrm{d}t''\, \hat{H}_{\mathrm{int}}(t')\hat{H}_{\mathrm{int}}(t'')\right]|\phi_0\rangle .\end{aligned}$$

We can continue this way up to any arbitrarily high order and obtain an infinite series for $|\phi(t)\rangle$ in $\hat{H}_{\mathrm{int}}$. We do not pursue this any further. We now turn to the computation of the effective electron-electron interaction mediated through phonons by making use of the second-order result obtained in the above. We take

$$|\psi(0)\rangle = |\phi(0)\rangle = |\phi_0\rangle = |i\rangle = |1_{\boldsymbol{k}\sigma}, 1_{\boldsymbol{k}'\sigma'}, 0_{\boldsymbol{k}-\boldsymbol{q},\sigma}, 0_{\boldsymbol{k}'+\boldsymbol{q},\sigma'}; 0_{\boldsymbol{q}}, 0_{-\boldsymbol{q}}\rangle$$

to be the initial state as indicated by i in the ket $|i\rangle$ and take

$$|f\rangle = |0_{\boldsymbol{k}\sigma}, 0_{\boldsymbol{k}'\sigma'}, 1_{\boldsymbol{k}-\boldsymbol{q},\sigma}, 1_{\boldsymbol{k}'+\boldsymbol{q},\sigma'}; 0_{\boldsymbol{q}}, 0_{-\boldsymbol{q}}\rangle$$

as the final state as indicated by f in the ket $|f\rangle$. Each number in the kets $|i\rangle$ and $|f\rangle$ represents the occupation number on the single-particle state indicated by the subscript. The occupation numbers before the semicolon in a ket are for single-electron states while those after the semicolon are for single-phonon states. Note that only those single-particle states that participate in the processes under study are explicitly shown. Those that remain the same in the entire processes are not displayed. When the system goes from the initial state $|i\rangle$ to the final state $|f\rangle$, two electrons make transitions from their initial single-electron states $|\boldsymbol{k}\sigma\rangle$ and $|\boldsymbol{k}'\sigma'\rangle$ to the final single-electron states $|\boldsymbol{k}-\boldsymbol{q},\sigma\rangle$ and $|\boldsymbol{k}'+\boldsymbol{q},\sigma'\rangle$. The conservation of energy requires that

$$\varepsilon_{\boldsymbol{k}} + \varepsilon_{\boldsymbol{k}'} = \varepsilon_{\boldsymbol{k}-\boldsymbol{q}} + \varepsilon_{\boldsymbol{k}'+\boldsymbol{q}}.$$

Our task can be stated as follows: Given that the system is in the initial state $|i\rangle$, compute the probability that the system can be found in the state $|f\rangle$ at $t = \infty$. Because of the particular forms of

$|i\rangle$ and $|f\rangle$, the zeroth- and first-order contributions are identically zero. The first nonzero contribution to the transition probability arises from the second-order term. The state vector at time t is then given by

$$|\psi^{(2)}(t)\rangle = \frac{1}{(\mathrm{i}\hbar)^2} \mathrm{e}^{-\mathrm{i}\hat{H}_0 t/\hbar} \int_0^t \mathrm{d}t' \int_0^{t'} \mathrm{d}t''\ \hat{H}_{\mathrm{int}}(t')\hat{H}_{\mathrm{int}}(t'')|i\rangle\,.$$

The probability amplitude is given by the inner product between $|\psi^{(2)}(t)\rangle$ and $|f\rangle$

$$\langle f|\psi^{(2)}(t)\rangle = \frac{1}{(\mathrm{i}\hbar)^2} \mathrm{e}^{-\mathrm{i}E_f t/\hbar} \int_0^t \mathrm{d}t' \int_0^{t'} \mathrm{d}t''\ \langle f|\,\hat{H}_{\mathrm{int}}(t')\hat{H}_{\mathrm{int}}(t'')|i\rangle,$$

where E_f is the energy of the final state $|f\rangle$. Note that E_f is equal to the energy of the initial state E_i as required as the conservation of energy. It will also follow directly from the computation when the long-time limit is taken. To compute the matrix element in the above equation, we insert in between $\hat{H}_{\mathrm{int}}(t')$ and $\hat{H}_{\mathrm{int}}(t'')$ a complete set of eigenvectors of $\hat{H}_0$ and have

$$\langle f|\,\hat{H}_{\mathrm{int}}(t')\hat{H}_{\mathrm{int}}(t'')|i\rangle = \sum_m \langle f|\,\hat{H}_{\mathrm{int}}(t')|m\rangle\langle m|\,\hat{H}_{\mathrm{int}}(t'')|i\rangle$$

Note that $|i\rangle$ and $|f\rangle$ are eigenvectors of $\hat{H}_0$. From the expressions of $|i\rangle$ and $|f\rangle$, we see that only the following two intermediate states make nonzero contributions

$$\begin{aligned} |m_1\rangle &= |0_{\boldsymbol{k}\sigma}, 1_{\boldsymbol{k}'\sigma'}, 1_{\boldsymbol{k}-\boldsymbol{q},\sigma}, 0_{\boldsymbol{k}'+\boldsymbol{q},\sigma'}; 1_{\boldsymbol{q}}, 0_{-\boldsymbol{q}}\rangle, \\ |m_2\rangle &= |1_{\boldsymbol{k}\sigma}, 0_{\boldsymbol{k}'\sigma'}, 0_{\boldsymbol{k}-\boldsymbol{q},\sigma}, 1_{\boldsymbol{k}'+\boldsymbol{q},\sigma'}; 0_{\boldsymbol{q}}, 1_{-\boldsymbol{q}}\rangle\,. \end{aligned}$$

If the system takes the intermediate state $|m_1\rangle$, then the electron in the single-electron state $|\boldsymbol{k}\sigma\rangle$ makes a transition into the single-electron state $|\boldsymbol{k}-\boldsymbol{q},\sigma\rangle$ by emitting a phonon of wave vector $\boldsymbol{q}$. If the system takes the intermediate state $|m_2\rangle$, then the electron in the single-electron state $|\boldsymbol{k}'\sigma'\rangle$ makes a transition into the single-electron state $|\boldsymbol{k}'+\boldsymbol{q},\sigma'\rangle$ by emitting a phonon of wave vector $-\boldsymbol{q}$. The energy differences between $|i\rangle$ and $|m_{1,2}\rangle$ are

$$\begin{aligned} E_i - E_{m_1} &= \varepsilon_{\boldsymbol{k}} - \varepsilon_{\boldsymbol{k}-\boldsymbol{q}} - \hbar\omega_{\boldsymbol{q}}, \\ E_i - E_{m_2} &= \varepsilon_{\boldsymbol{k}'} - \varepsilon_{\boldsymbol{k}'+\boldsymbol{q}} - \hbar\omega_{-\boldsymbol{q}} \\ &= -\varepsilon_{\boldsymbol{k}} + \varepsilon_{\boldsymbol{k}-\boldsymbol{q}} - \hbar\omega_{\boldsymbol{q}}, \end{aligned}$$

where we have made use of $\omega_{-\boldsymbol{q}} = \omega_{\boldsymbol{q}}$ and the conservation of energy, $\varepsilon_{\boldsymbol{k}} + \varepsilon_{\boldsymbol{k}'} = \varepsilon_{\boldsymbol{k}-\boldsymbol{q}} + \varepsilon_{\boldsymbol{k}'+\boldsymbol{q}}$. The matrix element $\langle f|\,\hat{H}_{\rm int}(t')\hat{H}_{\rm int}(t'')|i\rangle$ can now be evaluated

$$\begin{aligned}
&\langle f|\,\hat{H}_{\rm int}(t')\hat{H}_{\rm int}(t'')|i\rangle \\
&\quad= \langle f|\,\hat{H}_{\rm int}(t')|m_1\rangle\langle m_1|\,\hat{H}_{\rm int}(t'')|i\rangle \\
&\qquad+ \langle f|\,\hat{H}_{\rm int}(t')|m_2\rangle\langle m_2|\,\hat{H}_{\rm int}(t'')|i\rangle \\
&\quad= {\rm e}^{{\rm i}[(E_f-E_{m_1})t'/\hbar-(E_i-E_{m_1})t''/\hbar]}\,\langle f|\,\hat{H}_{\rm int}|m_1\rangle\langle m_1|\,\hat{H}_{\rm int}|i\rangle \\
&\qquad+ {\rm e}^{{\rm i}[(E_f-E_{m_2})t'/\hbar-(E_i-E_{m_2})t''/\hbar]}\,\langle f|\,\hat{H}_{\rm int}|m_2\rangle\langle m_2|\,\hat{H}_{\rm int}|i\rangle \\
&\quad= |M|^2\Big\{\,{\rm e}^{{\rm i}[(E_f-E_{m_1})t'/\hbar-(E_i-E_{m_1})t''/\hbar]} \\
&\qquad\qquad+ {\rm e}^{{\rm i}[(E_f-E_{m_2})t'/\hbar-(E_i-E_{m_2})t''/\hbar]}\,\Big\}.
\end{aligned}$$

Performing the time integrals in $\langle f|\psi^{(2)}(t)\rangle$, we have

$$\begin{aligned}
\langle f|\psi^{(2)}(t)\rangle &= \frac{|M|^2}{({\rm i}\hbar)^2}{\rm e}^{-{\rm i}E_f t/\hbar}\int_0^t {\rm d}t'\int_0^{t'}{\rm d}t''\,\Big\{\,{\rm e}^{{\rm i}[(E_f-E_{m_1})t'-(E_i-E_{m_1})t'']} \\
&\qquad + {\rm e}^{{\rm i}[(E_f-E_{m_2})t'-(E_i-E_{m_2})t'']}\,\Big\}
\end{aligned}$$

$$\begin{aligned}
&= -2{\rm i}|M|^2{\rm e}^{-{\rm i}(E_i+E_f)t/2\hbar}\,\frac{\sin[(E_f-E_i)t/2\hbar]}{E_f-E_i} \\
&\quad\times\left[\frac{1}{E_i-E_{m_1}}+\frac{1}{E_i-E_{m_2}}\right] \\
&\quad+ 2{\rm i}|M|^2{\rm e}^{-{\rm i}(E_{m_1}+E_f)t/2\hbar}\,\frac{\sin[(E_f-E_{m_1})t/2\hbar]}{(E_i-E_{m_1})(E_f-E_{m_1})} \\
&\quad+ 2{\rm i}|M|^2{\rm e}^{-{\rm i}(E_{m_2}+E_f)t/2\hbar}\,\frac{\sin[(E_f-E_{m_2})t/2\hbar]}{(E_i-E_{m_2})(E_f-E_{m_2})}.
\end{aligned}$$

Because of the presence of $E_f - E_i$ in the denominator on the first line, this term is predominant over those on the second and third lines. Therefore, we keep only the term on the first line. We then have

$$\begin{aligned}
\langle f|\psi^{(2)}(t)\rangle &= -2{\rm i}|M|^2{\rm e}^{-{\rm i}(E_i+E_f)t/2\hbar}\,\frac{\sin[(E_f-E_i)t/2\hbar]}{E_f-E_i} \\
&\quad\times\left[\frac{1}{E_i-E_{m_1}}+\frac{1}{E_i-E_{m_2}}\right].
\end{aligned}$$

To see how an effective interaction arises from the second order, although we know that the first-order term makes a zero contribution, we now perform the time integral in it and leave the matrix element $\langle f|\hat{H}_{\rm int}|i\rangle$ as it is (its value is zero). We have

$$\begin{aligned}\langle f|\psi^{(1)}(t)\rangle &= \frac{1}{{\rm i}\hbar}{\rm e}^{-{\rm i}E_f t/\hbar}\int_0^t {\rm d}t'\ \langle f|\,\hat{H}_{\rm int}(t')|i\rangle \\ &= \frac{1}{{\rm i}\hbar}{\rm e}^{-{\rm i}E_f t/\hbar}\int_0^t {\rm d}t'\ {\rm e}^{{\rm i}(E_f-E_i)t'/\hbar}\,\langle f|\,\hat{H}_{\rm int}|i\rangle \\ &= -2{\rm i}{\rm e}^{-{\rm i}(E_i+E_f)t/2\hbar}\,\frac{\sin[(E_f-E_i)t/2\hbar]}{E_f-E_i}\,\langle f|\hat{H}_{\rm int}|i\rangle\,.\end{aligned}$$

The result for $\langle f|\psi(t)^{(2)}\rangle$ can also be put into the above form through introducing an effective interaction Hamiltonian $\hat{H}_{\rm eff}$. We have

$$\langle f|\psi^{(2)}(t)\rangle = -2{\rm i}{\rm e}^{-{\rm i}(E_i+E_f)t/2\hbar}\,\frac{\sin[(E_f-E_i)t/2\hbar]}{E_f-E_i}\,\langle f|\hat{H}_{\rm eff}|i\rangle\,.$$

Comparing this expression with that derived for $\langle f|\psi^{(2)}(t)\rangle$ in the above, we see that the matrix element of the effective interaction Hamiltonian between the initial and final states is given by

$$\begin{aligned}\langle f|\hat{H}_{\rm eff}|i\rangle &= |M|^2\left[\frac{1}{E_i-E_{m_1}}+\frac{1}{E_i-E_{m_2}}\right] \\ &= |M|^2\left[\frac{1}{\varepsilon_{\boldsymbol{k}}-\varepsilon_{\boldsymbol{k}-\boldsymbol{q}}-\hbar\omega_{\boldsymbol{q}}}+\frac{1}{-\varepsilon_{\boldsymbol{k}}+\varepsilon_{\boldsymbol{k}-\boldsymbol{q}}-\hbar\omega_{\boldsymbol{q}}}\right] \\ &= \frac{2|M|^2\hbar\omega_{\boldsymbol{q}}}{(\varepsilon_{\boldsymbol{k}}-\varepsilon_{\boldsymbol{k}-\boldsymbol{q}})^2-(\hbar\omega_{\boldsymbol{q}})^2}.\end{aligned}$$

From the above expression, we see that the effective interaction between electrons is attractive if $\varepsilon_{\boldsymbol{k}}-\varepsilon_{\boldsymbol{k}-\boldsymbol{q}}<\hbar\omega_{\boldsymbol{q}}$. Since the maximum value of $\hbar\omega_{\boldsymbol{q}}$ is $\hbar\omega_{\rm D}$, the effective interaction between electrons is attractive within the shell $\varepsilon_{\rm F}-\hbar\omega_{\rm D}<\varepsilon<\varepsilon_{\rm F}+\hbar\omega_{\rm D}$. We would like to extract $\hat{H}_{\rm eff}$ from the value of $\langle f|\hat{H}_{\rm eff}|i\rangle$ given in the above equation. We first note that $\hat{H}_{\rm eff}$ in the second order represents a two-body interaction between electrons. Therefore, $\hat{H}_{\rm eff}$ must contain two annihilation and two creation operators of electrons. Furthermore, the standard form of operators in a two-body interaction is $\hat{c}^\dagger_{\boldsymbol{k}+\boldsymbol{q},\sigma}\hat{c}^\dagger_{\boldsymbol{k}'-\boldsymbol{q},\sigma'}\hat{c}_{\boldsymbol{k}'\sigma'}\hat{c}_{\boldsymbol{k}\sigma}$. Making use of $\langle f|\,\hat{c}^\dagger_{\boldsymbol{k}+\boldsymbol{q},\sigma}\hat{c}^\dagger_{\boldsymbol{k}'-\boldsymbol{q},\sigma'}\hat{c}_{\boldsymbol{k}'\sigma'}\hat{c}_{\boldsymbol{k}\sigma}|i\rangle=1$, we can extract $\hat{H}_{\rm eff}$ and have

$$\hat{H}_{\rm eff}=\sum_{\boldsymbol{k}\boldsymbol{k}'\sigma\sigma'}\frac{2|M|^2\hbar\omega_{\boldsymbol{q}}}{(\varepsilon_{\boldsymbol{k}}-\varepsilon_{\boldsymbol{k}-\boldsymbol{q}})^2-(\hbar\omega_{\boldsymbol{q}})^2}\hat{c}^\dagger_{\boldsymbol{k}+\boldsymbol{q},\sigma}\hat{c}^\dagger_{\boldsymbol{k}'-\boldsymbol{q},\sigma'}\hat{c}_{\boldsymbol{k}'\sigma'}\hat{c}_{\boldsymbol{k}\sigma}.$$

Note that $\hat{H}_{\text{eff}}$ contains all the interactions among electrons in all possible state configurations. This is necessary because we do not know *a priori* what the initial and final states of electrons are.

30-9 Effect of the Fermi sea. The energy of a Cooper pair is to be determined from $V\sum'_{\boldsymbol{k}}(2\varepsilon_{\boldsymbol{k}} - \mathscr{E})^{-1} = 1$.

(1) We now remove the Fermi sea by setting $k_{\text{F}} = 0$. Show that, in three dimensions, it is no longer true that a bound state solution to the above equation always exists for all $V > 0$. Derive an expression for the critical interaction strength V_c for which there is no bound state for $V < V_c$ and there is a bound state for $V > V_c$.

(2) Study the same problem for the two- and one-dimensional cases.

---

(1) If we remove the Fermi sea, the equation for the energy of a Cooper pair becomes

$$1 = V \sum_{\boldsymbol{k}\ (|\boldsymbol{k}|<k_c)} \frac{1}{\hbar^2 k^2/m - \mathscr{E}},$$

where $k_c = (2m\omega_{\text{D}}/\hbar)^{1/2}$. Converting the summation over $\boldsymbol{k}$ into an integration over $\boldsymbol{k}$, we have

$$1 = \frac{V\mathscr{V}}{2\pi^2}\int_0^{k_c} \mathrm{d}k\, \frac{k^2}{\hbar^2 k^2/m - \mathscr{E}} = \frac{V\mathscr{V}m}{2\pi^2\hbar^2}\int_0^{k_c} \mathrm{d}k\, \frac{k^2}{k^2 - m\mathscr{E}/\hbar^2},$$

where $\mathscr{V}$ is the volume of the crystal. Since a bound state with the Fermi sea removed has a negative energy, we investigate whether a solution exists for $\mathscr{E} < 0$. However, even with the Fermi sea removed, the Fermi wave vector and Fermi energy corresponding to the usual electron gas are still used. Making use of $N(0) = \mathscr{V}mk_{\text{F}}/2\pi^2\hbar^2$ and $\lambda = N(0)V$ and introducing $\kappa = k_c/k_{\text{F}}$, $x = k/k_c$, and $z = (-m\mathscr{E}/\hbar^2 k_c^2)^{1/2}$ for a bound state with $\mathscr{E} < 0$, we have

$$\frac{1}{\lambda\kappa} = \int_0^1 \mathrm{d}x\, \frac{x^2}{x^2 + z^2} = 1 - z\tan^{-1}\left(\frac{1}{z}\right).$$

For the given values of λ and κ, z can be solved from the above equation and the energy of a Cooper pair can be then obtained from $\mathscr{E} = -z^2\hbar^2 k_c^2/m$ if a solution has been found.

For the convenience of studying whether a solution exists, we refer to the right hand side of the above equation as $\eta(z)$

$$\eta(z) = 1 - z\tan^{-1}\left(\frac{1}{z}\right).$$

Then, the equation for the energy of a Cooper pair becomes

$$\eta(z) = 1/\lambda\kappa.$$

The function $\eta(z)$ is plotted in Fig. 30.4 with $1/\lambda\kappa$ also given as a horizontal dashed-line. The intersection of the horizontal dashed-line with the curve for $\eta(z)$ corresponds to the solution to the above equation.

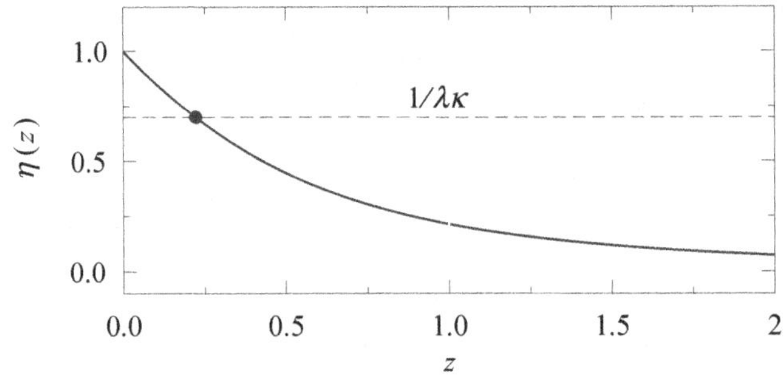

Fig. 30.4 Plot of $\eta(z)$ versus z. The horizontal dashed-line is for $1/\lambda\kappa$. The intersection of the curve for $\eta(z)$ with the horizontal dashed-line for $1/\lambda\kappa$ is marked with a solid circle.

From Fig. 30.4, it is seen that $\max \eta(z) = 1$. Therefore, for a solution (a bound state) to exist, we must have $0 < 1/\lambda\kappa < 1$, *i.e.*,

$$V > V_c \equiv \frac{2\pi^2\hbar^3}{m\mathscr{V}[2m(\hbar\omega_D)]^{1/2}}.$$

Note that there also exists a positive-energy solution if the above condition is satisfied. However, it is not a bound state.

(2) In two dimensions, we have

$$1 = \frac{V\mathscr{A}}{2\pi}\int_0^{k_c} dk\, \frac{k}{\hbar^2k^2/m - \mathscr{E}} = \frac{V\mathscr{A}m}{4\pi\hbar^2}\ln\left|1 - \frac{\hbar^2k_c^2}{m\mathscr{E}}\right|$$
$$= \frac{V\mathscr{A}m}{4\pi\hbar^2}\ln\left|1 - \frac{2\hbar\omega_D}{\mathscr{E}}\right| = \frac{\lambda}{2}\ln\left|1 - \frac{2\hbar\omega_D}{\mathscr{E}}\right|,$$

where $\mathscr{A}$ is the area of the crystal and $\lambda = N_{2D}(0)V$ with $N_{2D}(0) = \mathscr{A}m/2\pi\hbar^2$ the density of states at the Fermi energy for one spin direction in two dimensions. The function $\ln|1 - 2\hbar\omega_D/\mathscr{E}|$ versus $\mathscr{E}/2\hbar\omega_D$ is plotted in Fig. 30.5.

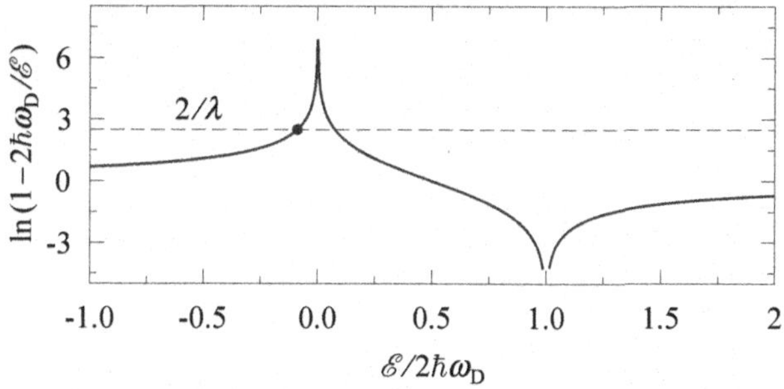

Fig. 30.5 Plot of $\ln|1 - 2\hbar\omega_{\rm D}/\mathscr{E}|$ versus $\mathscr{E}/2\hbar\omega_{\rm D}$. The horizontal dashed-line is for $2/\lambda$. Among the two intersections of the curve for $\ln|1 - 2\hbar\omega_{\rm D}/\mathscr{E}|$ with the horizontal dashed-line for $2/\lambda$, the one with a negative value of $\mathscr{E}$ corresponds to a bound state and is marked with a solid circle.

From Fig. 30.5, we see that two solutions always exist for $V > 0$ in two dimensions. However, only the solution with $\mathscr{E} < 0$ corresponds to a bound state with the energy given by

$$\mathscr{E}_{\rm 2D} = -\frac{2\hbar\omega_{\rm D}}{e^{2/\lambda} - 1}.$$

In one dimension, we have

$$\begin{aligned}1 &= \frac{V\mathscr{L}}{2\pi}\int_{-k_c}^{k_c} {\rm d}k\, \frac{1}{\hbar^2k^2/m - \mathscr{E}} = \frac{V\mathscr{L}m}{\pi\hbar^2}\int_0^{k_c} {\rm d}k\, \frac{1}{k^2 + (-m\mathscr{E}/\hbar^2)} \\ &= \frac{V\mathscr{L}m}{\pi\hbar^2}\frac{1}{(-m\mathscr{E}/\hbar^2)^{1/2}}\tan^{-1}\frac{k_c}{(-m\mathscr{E}/\hbar^2)^{1/2}} = \frac{\lambda}{\kappa}\, z\tan^{-1} z,\end{aligned}$$

where $\lambda = N_{\rm 1D}(0)V$ with $N_{\rm 1D}(0) = m\mathscr{L}/\pi^2\hbar^2k_{\rm F}$ the density of states at the Fermi energy for one spin direction in one dimension, $\kappa = k_c/k_{\rm F}$, and $z = k_c/(-m\mathscr{E}/\hbar^2)^{1/2}$. Thus, the equation for the determination of the energy of a Cooper pair is given by

$$\frac{\kappa}{\lambda} = z\tan^{-1} z.$$

The two sides of the above equation are plotted in Fig. 30.6. From Fig. 30.6, we see that a bound state always exists for $V > 0$ in one dimension. Note that there exists no positive-energy solution in one dimension.

30-10 Size of a Cooper pair. The orbital part of the wave function of a Cooper pair is expanded in terms of the plane waves as $\psi(\boldsymbol{r}) = \mathscr{V}^{-1/2}\sum'_{\boldsymbol{k}} c_{\boldsymbol{k}} e^{{\rm i}\boldsymbol{k}\cdot\boldsymbol{r}}$ with $c_{\boldsymbol{k}}$'s to be determined from $(2\varepsilon_{\boldsymbol{k}} - \mathscr{E})c_{\boldsymbol{k}} = V\sum'_{\boldsymbol{k}'} c_{\boldsymbol{k}'}$. Let $C = \sum'_{\boldsymbol{k}'} c_{\boldsymbol{k}'}$. We then have $c_{\boldsymbol{k}} = C/(2\varepsilon_{\boldsymbol{k}} - \mathscr{E})$.

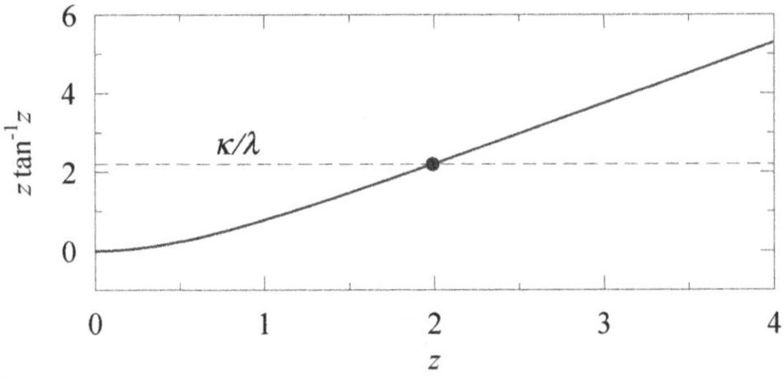

Fig. 30.6 Plot of $z\tan^{-1}z$ versus z. The horizontal dashed-line is for κ/λ. The intersection of the curve for $z\tan^{-1}z$ with the horizontal dashed-line for κ/λ is marked by a solid circle.

(1) Evaluate the expectation value $\langle r^2\rangle$ for a Cooper pair, where $\boldsymbol{r}$ is the relative coordinate between the two electrons. Justify all the approximations made in the computation and express $\langle r^2\rangle$ in terms of the energy gap Δ at zero temperature and the Fermi velocity $v_{\rm F}$.

(2) Let $\xi = \langle r^2\rangle^{1/2}$ be the root-mean-square size of a Cooper pair. Evaluate ξ for Pb, Sn, In, and Al. Their $v_{\rm F}$'s are 1.83, 1.90, 1.74, and 2.03×10^6 m/s, respectively. Their 2Δ's are 2.68, 1.11, 1.05, and 0.42 meV, respectively.

---

(1) $\langle r^2\rangle$ is given by

$$\langle r^2\rangle = \left[\int {\rm d}\boldsymbol{r}\ |\psi(\boldsymbol{r})|^2\right]^{-1}\int {\rm d}\boldsymbol{r}\ r^2|\psi(\boldsymbol{r})|^2.$$

For $\int {\rm d}\boldsymbol{r}\ |\psi(\boldsymbol{r})|^2$, we have

$$\begin{aligned}\int {\rm d}\boldsymbol{r}\ |\psi(\boldsymbol{r})|^2 &= \frac{|C|^2}{\mathscr{V}}{\sum_{\boldsymbol{k}\boldsymbol{k}'}}'\frac{1}{(2\varepsilon_{\boldsymbol{k}}-\mathscr{E})(2\varepsilon_{\boldsymbol{k}'}-\mathscr{E})}\int {\rm d}\boldsymbol{r}\ {\rm e}^{{\rm i}(\boldsymbol{k}-\boldsymbol{k}')\cdot\boldsymbol{r}}\\ &= |C|^2{\sum_{\boldsymbol{k}\boldsymbol{k}'}}'\frac{\delta_{\boldsymbol{k}\boldsymbol{k}'}}{(2\varepsilon_{\boldsymbol{k}}-\mathscr{E})(2\varepsilon_{\boldsymbol{k}'}-\mathscr{E})}\\ &= |C|^2{\sum_{\boldsymbol{k}}}'\frac{1}{(2\varepsilon_{\boldsymbol{k}}-\mathscr{E})^2}.\end{aligned}$$

Converting the summation over $\boldsymbol{k}$ into an integration over the electron energy $\xi = \varepsilon-\varepsilon_{\rm F}$ and making use of $\mathscr{E}\approx 2\varepsilon_{\rm F}-\Delta$, we have

$$\int {\rm d}\boldsymbol{r}\ |\psi(\boldsymbol{r})|^2 = |C|^2N(0)\int_0^{\hbar\omega_{\rm D}}\frac{{\rm d}\xi}{(2\xi+\Delta)^2} = \frac{|C|^2N(0)}{2\Delta},$$

where we have set the density of states to be equal to its value at the Fermi surface and pulled it out of the integral. This is justified in consideration that the integration interval is a narrow region around the Fermi surface since $\hbar\omega_D/\varepsilon_F \ll 1$. For $\int d\boldsymbol{r}\, r^2|\psi(r)|^2$, we have

$$\begin{aligned}\int d\boldsymbol{r}\, r^2|\psi(r)|^2 &= \frac{|C|^2}{\mathscr{V}}\sum_{\boldsymbol{k}\boldsymbol{k}'}{}' \frac{1}{(2\varepsilon_{\boldsymbol{k}}-\mathscr{E})(2\varepsilon_{\boldsymbol{k}'}-\mathscr{E})}\int d\boldsymbol{r}\, r^2 e^{i(\boldsymbol{k}-\boldsymbol{k}')\cdot\boldsymbol{r}} \\ &= \frac{|C|^2}{\mathscr{V}}\sum_{\boldsymbol{k}\boldsymbol{k}'}{}' \frac{1}{(2\varepsilon_{\boldsymbol{k}}-\mathscr{E})(2\varepsilon_{\boldsymbol{k}'}-\mathscr{E})}\int d\boldsymbol{r}\, \left[\nabla_{\boldsymbol{k}}\cdot\nabla_{\boldsymbol{k}'} e^{i(\boldsymbol{k}-\boldsymbol{k}')\cdot\boldsymbol{r}}\right] \\ &= |C|^2\sum_{\boldsymbol{k}\boldsymbol{k}'}{}' \frac{1}{(2\varepsilon_{\boldsymbol{k}}-\mathscr{E})(2\varepsilon_{\boldsymbol{k}'}-\mathscr{E})}\nabla_{\boldsymbol{k}}\cdot\nabla_{\boldsymbol{k}'}\delta_{\boldsymbol{k}\boldsymbol{k}'}.\end{aligned}$$

Implicitly converting the summations over wave vectors into integrations and performing integrations by parts, we obtain

$$\begin{aligned}\int d\boldsymbol{r}\, r^2|\psi(r)|^2 &= |C|^2\sum_{\boldsymbol{k}\boldsymbol{k}'}{}' \delta_{\boldsymbol{k}\boldsymbol{k}'}\nabla_{\boldsymbol{k}}\left(\frac{1}{2\varepsilon_{\boldsymbol{k}}-\mathscr{E}}\right)\cdot\nabla_{\boldsymbol{k}'}\left(\frac{1}{2\varepsilon_{\boldsymbol{k}'}-\mathscr{E}}\right) \\ &= |C|^2\sum_{\boldsymbol{k}}{}'\left[\nabla_{\boldsymbol{k}}\left(\frac{1}{2\varepsilon_{\boldsymbol{k}}-\mathscr{E}}\right)\right]^2.\end{aligned}$$

Evaluating $\nabla_{\boldsymbol{k}}(2\varepsilon_{\boldsymbol{k}}-\mathscr{E})^{-1}$, we have

$$\begin{aligned}\int d\boldsymbol{r}\, r^2|\psi(r)|^2 &= \frac{8\hbar^2|C|^2}{m}\sum_{\boldsymbol{k}}{}'\frac{\varepsilon_{\boldsymbol{k}}}{(2\varepsilon_{\boldsymbol{k}}-\mathscr{E})^4} \\ &= \frac{8\hbar^2 N(0)|C|^2}{m}\int_0^{\hbar\omega_D} d\xi\, \frac{\xi+\varepsilon_F}{(2\xi+\Delta)^4} \\ &\approx \frac{8\hbar^2 N(0)|C|^2\varepsilon_F}{m}\int_0^{\hbar\omega_D} d\xi\, \frac{1}{(2\xi+\Delta)^4} \\ &\approx \frac{4\hbar^2 N(0)|C|^2\varepsilon_F}{3m\Delta^3}.\end{aligned}$$

We then have

$$\langle r^2\rangle = \frac{8\hbar^2\varepsilon_F}{3m\Delta^2} = \frac{4\hbar^2 v_F^2}{3\Delta^2}.$$

The root-mean-square size of a Cooper pair, $\xi = \langle r^2\rangle^{1/2}$, is given by

$$\xi = \frac{2\hbar v_F}{3^{1/2}\Delta}.$$

(2) The values of ξ for Hg, Pb, V, and Nb are given in the following table.

	Pb	Sn	In	Al
$v_{\rm F}$ [10^6 m/s]	1.83	1.90	1.74	2.03
2Δ [meV]	2.68	1.11	1.05	0.42
ξ [nm]	1037.96	2601.93	2518.98	7347.03

30-11 BCS superconducting ground state. The BCS wave function is given by $|\Omega\rangle = \prod_{\boldsymbol{k}}(u_{\boldsymbol{k}} + v_{\boldsymbol{k}}\hat{c}^\dagger_{\boldsymbol{k}\uparrow}\hat{c}^\dagger_{-\boldsymbol{k}\downarrow})\,|0\rangle$.

(1) Show that $|\Omega\rangle$ becomes a state described by the Fermi sphere in the $\Delta \to 0$ limit.

(2) Evaluate the average number of electrons in the superconducting ground state, that is, compute $\langle \hat{n}_{\boldsymbol{k}}\rangle$, where $\hat{n}_{\boldsymbol{k}} = \hat{n}_{\boldsymbol{k}\uparrow} + \hat{n}_{\boldsymbol{k}\downarrow}$ with $\hat{n}_{\boldsymbol{k}\sigma} = \hat{c}^\dagger_{\boldsymbol{k}\sigma}\hat{c}_{\boldsymbol{k}\sigma}$. Show that the root-mean-square fluctuations in the electron number given by $[\,\langle n^2_{\boldsymbol{k}}\rangle - \langle \hat{n}_{\boldsymbol{k}}\rangle^2\,]^{1/2}$ are proportional to the order parameter Δ.

(3) Evaluate the expectation value of the current operator in the superconducting ground-state. Explain the obtained result.

---

(1) From the expressions of $u_{\boldsymbol{k}}$ and $v_{\boldsymbol{k}}$

$$u_{\boldsymbol{k}} = \frac{1}{2}\left[1 + \frac{\xi_{\boldsymbol{k}}}{(\xi^2_{\boldsymbol{k}} + \Delta^2)^{1/2}}\right],\quad v_{\boldsymbol{k}} = \frac{1}{2}\left[1 - \frac{\xi_{\boldsymbol{k}}}{(\xi^2_{\boldsymbol{k}} + \Delta^2)^{1/2}}\right],$$

we see that

$$\lim_{\Delta\to 0} u_{\boldsymbol{k}} = \begin{cases} 0, & |\boldsymbol{k}| < k_F, \\ 1, & |\boldsymbol{k}| > k_F, \end{cases} \quad \lim_{\Delta\to 0} v_{\boldsymbol{k}} = \begin{cases} 1, & |\boldsymbol{k}| < k_F, \\ 0, & |\boldsymbol{k}| > k_F, \end{cases}$$

which implies that all the single-electron states below the Fermi surface are occupied and that all the single-electron states above the Fermi surface are unoccupied. This describes just a fully-occupied Fermi sphere, with all the single-electron states outside the Fermi sphere are unoccupied.

(2) Making use of

$$[\hat{n}_{\boldsymbol{k}\uparrow}, \hat{c}^\dagger_{\boldsymbol{k}\uparrow}\hat{c}^\dagger_{-\boldsymbol{k}\downarrow}] = [\hat{c}^\dagger_{\boldsymbol{k}\uparrow}\hat{c}_{\boldsymbol{k}\uparrow}, \hat{c}^\dagger_{\boldsymbol{k}\uparrow}\hat{c}^\dagger_{-\boldsymbol{k}\downarrow}] = \hat{c}^\dagger_{\boldsymbol{k}\uparrow}\hat{c}^\dagger_{-\boldsymbol{k}\downarrow},$$

$\{\hat{c}_{\boldsymbol{k}\uparrow}, \hat{c}^\dagger_{\boldsymbol{k}\uparrow}\} = 1$, $\{\hat{c}_{-\boldsymbol{k}\downarrow}, \hat{c}^\dagger_{-\boldsymbol{k}\downarrow}\} = 1$, and $\hat{c}_{\boldsymbol{k}\uparrow}|0\rangle = \hat{c}_{-\boldsymbol{k}\downarrow}|0\rangle = 0$, we have

$$\begin{aligned}\langle \hat{n}_{\boldsymbol{k}\uparrow}\rangle &= \langle 0|\big(u_{\boldsymbol{k}} + v_{\boldsymbol{k}}\hat{c}_{-\boldsymbol{k}\downarrow}\hat{c}_{\boldsymbol{k}\uparrow}\big)\hat{c}^\dagger_{\boldsymbol{k}\uparrow}\hat{c}_{\boldsymbol{k}\uparrow}\big(u_{\boldsymbol{k}} + v_{\boldsymbol{k}}\hat{c}^\dagger_{\boldsymbol{k}\uparrow}\hat{c}^\dagger_{-\boldsymbol{k}\downarrow}\big)|0\rangle \\ &= v_{\boldsymbol{k}}\langle 0|\big(u_{\boldsymbol{k}} + v_{\boldsymbol{k}}\hat{c}_{-\boldsymbol{k}\downarrow}\hat{c}_{\boldsymbol{k}\uparrow}\big)\hat{c}^\dagger_{\boldsymbol{k}\uparrow}\hat{c}^\dagger_{-\boldsymbol{k}\downarrow}|0\rangle \\ &= v^2_{\boldsymbol{k}}\langle 0|\hat{c}_{-\boldsymbol{k}\downarrow}\hat{c}_{\boldsymbol{k}\uparrow}\hat{c}^\dagger_{\boldsymbol{k}\uparrow}\hat{c}^\dagger_{-\boldsymbol{k}\downarrow}|0\rangle \\ &= v^2_{\boldsymbol{k}}\langle 0|\hat{c}_{-\boldsymbol{k}\downarrow}(1 - \hat{c}^\dagger_{\boldsymbol{k}\uparrow}\hat{c}_{\boldsymbol{k}\uparrow})\hat{c}^\dagger_{-\boldsymbol{k}\downarrow}|0\rangle \\ &= v^2_{\boldsymbol{k}}\langle 0|\hat{c}_{-\boldsymbol{k}\downarrow}\hat{c}^\dagger_{-\boldsymbol{k}\downarrow}|0\rangle = v^2_{\boldsymbol{k}}\langle 0|1 - \hat{c}^\dagger_{-\boldsymbol{k}\downarrow}\hat{c}_{-\boldsymbol{k}\downarrow}|0\rangle = v^2_{\boldsymbol{k}}.\end{aligned}$$

Note that we also have $\langle 0|\hat{c}^\dagger_{\boldsymbol{k}\uparrow} = \langle 0|\hat{c}^\dagger_{-\boldsymbol{k}\downarrow} = 0$ which is that the Hermitian conjugate of $\hat{c}_{\boldsymbol{k}\uparrow}|0\rangle = \hat{c}_{-\boldsymbol{k}\downarrow}|0\rangle = 0$. We similarly have

$$\begin{aligned}\langle \hat{n}_{\boldsymbol{k}\downarrow}\rangle &= \langle 0|\big(u_{\boldsymbol{k}} + v_{\boldsymbol{k}}\hat{c}_{\boldsymbol{k}\downarrow}\hat{c}_{-\boldsymbol{k}\uparrow}\big)\hat{c}^\dagger_{\boldsymbol{k}\downarrow}\hat{c}_{\boldsymbol{k}\downarrow}\big(u_{\boldsymbol{k}} + v_{\boldsymbol{k}}\hat{c}^\dagger_{-\boldsymbol{k}\uparrow}\hat{c}^\dagger_{\boldsymbol{k}\downarrow}\big)|0\rangle \\ &= v_{\boldsymbol{k}}\langle 0|\big(u_{\boldsymbol{k}} + v_{\boldsymbol{k}}\hat{c}_{\boldsymbol{k}\downarrow}\hat{c}_{-\boldsymbol{k}\uparrow}\big)\hat{c}^\dagger_{-\boldsymbol{k}\uparrow}\hat{c}^\dagger_{\boldsymbol{k}\downarrow}|0\rangle \\ &= v^2_{\boldsymbol{k}}\langle 0|\hat{c}_{\boldsymbol{k}\downarrow}\hat{c}_{-\boldsymbol{k}\uparrow}\hat{c}^\dagger_{-\boldsymbol{k}\uparrow}\hat{c}^\dagger_{\boldsymbol{k}\downarrow}|0\rangle \\ &= v^2_{\boldsymbol{k}}\langle 0|\hat{c}_{\boldsymbol{k}\downarrow}(1 - \hat{c}^\dagger_{-\boldsymbol{k}\uparrow}\hat{c}_{-\boldsymbol{k}\uparrow})\hat{c}^\dagger_{\boldsymbol{k}\downarrow}|0\rangle \\ &= v^2_{\boldsymbol{k}}\langle 0|\hat{c}_{\boldsymbol{k}\downarrow}\hat{c}^\dagger_{\boldsymbol{k}\downarrow}|0\rangle = v^2_{\boldsymbol{k}}\langle 0|1 - \hat{c}^\dagger_{\boldsymbol{k}\downarrow}\hat{c}_{\boldsymbol{k}\downarrow}|0\rangle = v^2_{\boldsymbol{k}},\end{aligned}$$

where we have made use of

$$[\hat{n}_{\boldsymbol{k}\downarrow}, \hat{c}^\dagger_{-\boldsymbol{k}\uparrow}\hat{c}^\dagger_{\boldsymbol{k}\downarrow}] = [\hat{c}^\dagger_{\boldsymbol{k}\downarrow}\hat{c}_{\boldsymbol{k}\downarrow}, \hat{c}^\dagger_{-\boldsymbol{k}\uparrow}\hat{c}^\dagger_{\boldsymbol{k}\downarrow}] = \hat{c}^\dagger_{-\boldsymbol{k}\uparrow}\hat{c}^\dagger_{\boldsymbol{k}\downarrow},$$

$\{\hat{c}_{-\boldsymbol{k}\uparrow}, \hat{c}^\dagger_{-\boldsymbol{k}\uparrow}\} = 1$, $\{\hat{c}_{\boldsymbol{k}\downarrow}, \hat{c}^\dagger_{\boldsymbol{k}\downarrow}\} = 1$, and $\hat{c}_{-\boldsymbol{k}\uparrow}|0\rangle = \hat{c}_{\boldsymbol{k}\downarrow}|0\rangle = 0$. We thus have

$$\langle n_{\boldsymbol{k}}\rangle = 2v^2_{\boldsymbol{k}}.$$

We now compute $\langle n^2_{\boldsymbol{k}}\rangle$. We first note that

$$\begin{aligned}n^2_{\boldsymbol{k}} &= (\hat{n}_{\boldsymbol{k}\uparrow} + \hat{n}_{\boldsymbol{k}\downarrow})^2 = \hat{n}^2_{\boldsymbol{k}\uparrow} + 2\hat{n}_{\boldsymbol{k}\uparrow}\hat{n}_{\boldsymbol{k}\downarrow} + \hat{n}^2_{\boldsymbol{k}\downarrow} \\ &= \hat{n}_{\boldsymbol{k}\uparrow} + 2\hat{n}_{\boldsymbol{k}\uparrow}\hat{n}_{\boldsymbol{k}\downarrow} + \hat{n}_{\boldsymbol{k}\downarrow} = \hat{n}_{\boldsymbol{k}} + 2\hat{n}_{\boldsymbol{k}\uparrow}\hat{n}_{\boldsymbol{k}\downarrow}.\end{aligned}$$

Thus, $\langle n^2_{\boldsymbol{k}}\rangle$ can be obtained from $\langle n_{\boldsymbol{k}}\rangle$ and $\langle \hat{n}_{\boldsymbol{k}\uparrow}\hat{n}_{\boldsymbol{k}\downarrow}\rangle$. For $\langle \hat{n}_{\boldsymbol{k}\uparrow}\hat{n}_{\boldsymbol{k}\downarrow}\rangle$, we have

$$\begin{aligned}\langle \hat{n}_{\boldsymbol{k}\uparrow}\hat{n}_{\boldsymbol{k}\downarrow}\rangle &= \langle 0|\big(u_{\boldsymbol{k}} + v_{\boldsymbol{k}}\hat{c}_{\boldsymbol{k}\downarrow}\hat{c}_{-\boldsymbol{k}\uparrow}\big)\big(u_{\boldsymbol{k}} + v_{\boldsymbol{k}}\hat{c}_{-\boldsymbol{k}\downarrow}\hat{c}_{\boldsymbol{k}\uparrow}\big) \\ &\quad\times \hat{c}^\dagger_{\boldsymbol{k}\uparrow}\hat{c}_{\boldsymbol{k}\uparrow}\hat{c}^\dagger_{\boldsymbol{k}\downarrow}\hat{c}_{\boldsymbol{k}\downarrow}\big(u_{\boldsymbol{k}} + v_{\boldsymbol{k}}\hat{c}^\dagger_{\boldsymbol{k}\uparrow}\hat{c}^\dagger_{-\boldsymbol{k}\downarrow}\big)\big(u_{\boldsymbol{k}} + v_{\boldsymbol{k}}\hat{c}^\dagger_{-\boldsymbol{k}\uparrow}\hat{c}^\dagger_{\boldsymbol{k}\downarrow}\big)|0\rangle \\ &= v_{\boldsymbol{k}}\langle 0|\big(u_{\boldsymbol{k}} + v_{\boldsymbol{k}}\hat{c}_{\boldsymbol{k}\downarrow}\hat{c}_{-\boldsymbol{k}\uparrow}\big)\big(u_{\boldsymbol{k}} + v_{\boldsymbol{k}}\hat{c}_{-\boldsymbol{k}\downarrow}\hat{c}_{\boldsymbol{k}\uparrow}\big) \\ &\quad\times \hat{c}^\dagger_{\boldsymbol{k}\uparrow}\hat{c}_{\boldsymbol{k}\uparrow}\big(u_{\boldsymbol{k}} + v_{\boldsymbol{k}}\hat{c}^\dagger_{\boldsymbol{k}\uparrow}\hat{c}^\dagger_{-\boldsymbol{k}\downarrow}\big)\hat{c}^\dagger_{-\boldsymbol{k}\uparrow}\hat{c}^\dagger_{\boldsymbol{k}\downarrow}|0\rangle \\ &= v^2_{\boldsymbol{k}}\langle 0|\big(u_{\boldsymbol{k}} + v_{\boldsymbol{k}}\hat{c}_{\boldsymbol{k}\downarrow}\hat{c}_{-\boldsymbol{k}\uparrow}\big)\big(u_{\boldsymbol{k}} + v_{\boldsymbol{k}} \\ &\quad\times \hat{c}_{-\boldsymbol{k}\downarrow}\hat{c}_{\boldsymbol{k}\uparrow}\big)\hat{c}^\dagger_{\boldsymbol{k}\uparrow}\hat{c}^\dagger_{-\boldsymbol{k}\downarrow}\hat{c}^\dagger_{-\boldsymbol{k}\uparrow}\hat{c}^\dagger_{\boldsymbol{k}\downarrow}|0\rangle \\ &= v^4_{\boldsymbol{k}}\langle 0|\hat{c}_{\boldsymbol{k}\downarrow}\hat{c}_{-\boldsymbol{k}\uparrow}\hat{c}_{-\boldsymbol{k}\downarrow}\hat{c}_{\boldsymbol{k}\uparrow}\hat{c}^\dagger_{\boldsymbol{k}\uparrow}\hat{c}^\dagger_{-\boldsymbol{k}\downarrow}\hat{c}^\dagger_{-\boldsymbol{k}\uparrow}\hat{c}^\dagger_{\boldsymbol{k}\downarrow}|0\rangle.\end{aligned}$$

We now make use of anticommutations to make further simplifications, starting from $\hat{c}_{\boldsymbol{k}\uparrow}\hat{c}^\dagger_{\boldsymbol{k}\uparrow}$ in the middle. We have

$$\begin{aligned}\langle \hat{n}_{\boldsymbol{k}\uparrow}\hat{n}_{\boldsymbol{k}\downarrow}\rangle &= v_{\boldsymbol{k}}^4 \langle 0|\, \hat{c}_{\boldsymbol{k}\downarrow}\hat{c}_{-\boldsymbol{k}\uparrow}\hat{c}_{-\boldsymbol{k}\downarrow}\hat{c}^\dagger_{-\boldsymbol{k}\downarrow}\hat{c}^\dagger_{-\boldsymbol{k}\uparrow}\hat{c}^\dagger_{\boldsymbol{k}\downarrow}\, |0\rangle \\ &= v_{\boldsymbol{k}}^4 \langle 0|\, \hat{c}_{\boldsymbol{k}\downarrow}\hat{c}_{-\boldsymbol{k}\uparrow}\hat{c}^\dagger_{-\boldsymbol{k}\uparrow}\hat{c}^\dagger_{\boldsymbol{k}\downarrow}\, |0\rangle \\ &= v_{\boldsymbol{k}}^4 \langle 0|\, \hat{c}_{\boldsymbol{k}\downarrow}\hat{c}^\dagger_{\boldsymbol{k}\downarrow}\, |0\rangle = v_{\boldsymbol{k}}^4.\end{aligned}$$

Thus,

$$\langle n_{\boldsymbol{k}}^2\rangle = \langle \hat{n}_{\boldsymbol{k}}\rangle + 2\langle \hat{n}_{\boldsymbol{k}\uparrow}\hat{n}_{\boldsymbol{k}\downarrow}\rangle = 2v_{\boldsymbol{k}}^2 + 2v_{\boldsymbol{k}}^4.$$

The variance of $n_{\boldsymbol{k}}$ is given by

$$\langle n_{\boldsymbol{k}}^2\rangle - \langle n_{\boldsymbol{k}}\rangle^2 = 2v_{\boldsymbol{k}}^2 + 2v_{\boldsymbol{k}}^4 - 4v_{\boldsymbol{k}}^4 = 2u_{\boldsymbol{k}}^2 v_{\boldsymbol{k}}^2 = \frac{\Delta^2}{2E_{\boldsymbol{k}}^2}.$$

Finally, the fluctuations in $n_{\boldsymbol{k}}$ are given by

$$\left[\langle n_{\boldsymbol{k}}^2\rangle - \langle n_{\boldsymbol{k}}\rangle^2\right]^{1/2} = \frac{\Delta}{2^{1/2}E_{\boldsymbol{k}}}.$$

The above result shows that the fluctuations in $n_{\boldsymbol{k}}$ are proportional to the superconducting energy gap Δ.

(3) The paramagnetic current density operator is given by

$$\begin{aligned}\hat{\boldsymbol{j}}_p(\boldsymbol{q}) &= -\frac{e\hbar}{m\mathscr{V}}\sum_{\boldsymbol{k}\sigma}(\boldsymbol{k}+\boldsymbol{q}/2)\,\hat{c}^\dagger_{\boldsymbol{k}\sigma}\hat{c}_{\boldsymbol{k}+\boldsymbol{q},\sigma} \\ &= -\frac{e\hbar}{m\mathscr{V}}\sum_{\boldsymbol{k}}(\boldsymbol{k}+\boldsymbol{q}/2)\,\hat{c}^\dagger_{\boldsymbol{k}\uparrow}\hat{c}_{\boldsymbol{k}+\boldsymbol{q}\uparrow} - \frac{e\hbar}{m\mathscr{V}}\sum_{\boldsymbol{k}}(\boldsymbol{k}+\boldsymbol{q}/2)\,\hat{c}^\dagger_{\boldsymbol{k}\downarrow}\hat{c}_{\boldsymbol{k}+\boldsymbol{q}\downarrow}.\end{aligned}$$

To evaluate $\langle\hat{\boldsymbol{j}}_p(\boldsymbol{q})\rangle$, we need to evaluate $\langle\hat{c}^\dagger_{\boldsymbol{k}\uparrow}\hat{c}_{\boldsymbol{k}+\boldsymbol{q}\uparrow}\rangle$ and $\langle\hat{c}^\dagger_{\boldsymbol{k}\downarrow}\hat{c}_{\boldsymbol{k}+\boldsymbol{q}\downarrow}\rangle$. For $\langle\hat{c}^\dagger_{\boldsymbol{k}\uparrow}\hat{c}_{\boldsymbol{k}+\boldsymbol{q}\uparrow}\rangle$, we have for $\boldsymbol{q}\neq 0$

$$\begin{aligned}\langle\hat{c}^\dagger_{\boldsymbol{k}\uparrow}\hat{c}_{\boldsymbol{k}+\boldsymbol{q}\uparrow}\rangle &= \langle 0|\big(u_{\boldsymbol{k}+\boldsymbol{q}} + v_{\boldsymbol{k}+\boldsymbol{q}}\hat{c}_{-\boldsymbol{k}-\boldsymbol{q}\downarrow}\hat{c}_{\boldsymbol{k}+\boldsymbol{q}\uparrow}\big)\big(u_{\boldsymbol{k}} + v_{\boldsymbol{k}}\hat{c}_{-\boldsymbol{k}\downarrow}\hat{c}_{\boldsymbol{k}\uparrow}\big) \\ &\quad\times \hat{c}^\dagger_{\boldsymbol{k}\uparrow}\hat{c}_{\boldsymbol{k}+\boldsymbol{q}\uparrow}\big(u_{\boldsymbol{k}} + v_{\boldsymbol{k}}\hat{c}^\dagger_{\boldsymbol{k}\uparrow}\hat{c}^\dagger_{-\boldsymbol{k}\downarrow}\big)\big(u_{\boldsymbol{k}+\boldsymbol{q}} + v_{\boldsymbol{k}+\boldsymbol{q}}\hat{c}^\dagger_{\boldsymbol{k}+\boldsymbol{q}\uparrow}\hat{c}^\dagger_{-\boldsymbol{k}-\boldsymbol{q}\downarrow}\big)|0\rangle\,.\end{aligned}$$

For a term in the above average to take on a nonzero value, each annihilation operator in the term must be able to find its conjugate. Because of the presence of $\hat{c}^\dagger_{\boldsymbol{k}\uparrow}\hat{c}_{\boldsymbol{k}+\boldsymbol{q}\uparrow}$, this can not be true for $\boldsymbol{q}\neq 0$. Thus, $\langle\hat{c}^\dagger_{\boldsymbol{k}\uparrow}\hat{c}_{\boldsymbol{k}+\boldsymbol{q}\uparrow}\rangle$ is nonzero only for $\boldsymbol{q}=0$ and we have

$$\begin{aligned}\langle\hat{c}^\dagger_{\boldsymbol{k}\uparrow}\hat{c}_{\boldsymbol{k}+\boldsymbol{q}\uparrow}\rangle &= \langle 0|\big(u_{\boldsymbol{k}} + v_{\boldsymbol{k}}\hat{c}_{-\boldsymbol{k}\downarrow}\hat{c}_{\boldsymbol{k}\uparrow}\big)\hat{c}^\dagger_{\boldsymbol{k}\uparrow}\hat{c}_{\boldsymbol{k}\uparrow}\big(u_{\boldsymbol{k}} + v_{\boldsymbol{k}}\hat{c}^\dagger_{\boldsymbol{k}\uparrow}\hat{c}^\dagger_{-\boldsymbol{k}\downarrow}\big)|0\rangle\,\delta_{\boldsymbol{q}0} \\ &= v_{\boldsymbol{k}}^2\delta_{\boldsymbol{q}0}.\end{aligned}$$

We similarly have

$$\begin{aligned}\langle\hat{c}^\dagger_{\boldsymbol{k}\downarrow}\hat{c}_{\boldsymbol{k}+\boldsymbol{q}\downarrow}\rangle &= \langle 0|\big(u_{\boldsymbol{k}} + v_{\boldsymbol{k}}\hat{c}_{\boldsymbol{k}\downarrow}\hat{c}_{-\boldsymbol{k}\uparrow}\big)\hat{c}^\dagger_{\boldsymbol{k}\downarrow}\hat{c}_{\boldsymbol{k}\downarrow}\big(u_{\boldsymbol{k}} + v_{\boldsymbol{k}}\hat{c}^\dagger_{-\boldsymbol{k}\uparrow}\hat{c}^\dagger_{\boldsymbol{k}\downarrow}\big)|0\rangle\,\delta_{\boldsymbol{q}0} \\ &= v_{\boldsymbol{k}}^2\delta_{\boldsymbol{q}0}.\end{aligned}$$

The average of $\hat{j}_p(\boldsymbol{q})$ is then given by

$$\langle \hat{j}_p(\boldsymbol{q}) \rangle = -\delta_{q0} \frac{2e\hbar}{m\mathscr{V}} \sum_{\boldsymbol{k}} \boldsymbol{k} v_{\boldsymbol{k}}^2 = 0.$$

The zero value for the average of the current density implies that the superconducting ground state of zero momentum does not carry a current. This is because the momentum of each Cooper pair is zero.

30-12 Finite-momentum BCS state. The finite-momentum BCS state is given by

$$|\Omega_{\boldsymbol{K}}\rangle = \prod_{\boldsymbol{k}} (u_{\boldsymbol{k}} + v_{\boldsymbol{k}} \hat{c}^\dagger_{\boldsymbol{k}+\boldsymbol{K}/2\uparrow} \hat{c}^\dagger_{-\boldsymbol{k}+\boldsymbol{K}/2\downarrow}) |0\rangle .$$

(1) Compute the energy gap $\Delta_{\boldsymbol{K}} = V \sum_{\boldsymbol{k}} \langle \hat{c}_{-\boldsymbol{k}+\boldsymbol{K}/2\downarrow} \hat{c}_{\boldsymbol{k}+\boldsymbol{K}/2\uparrow} \rangle$ in $|\Omega_{\boldsymbol{K}}\rangle$.

(2) Evaluate the current carried by the finite-momentum BCS state.

---

(1) The energy gap can be evaluated as follows

$$\begin{aligned}
\Delta_{\boldsymbol{K}} &= V \sum_{\boldsymbol{k}} \langle \hat{c}_{-\boldsymbol{k}+\boldsymbol{K}/2\downarrow} \hat{c}_{\boldsymbol{k}+\boldsymbol{K}/2\uparrow} \rangle \\
&= V \sum_{\boldsymbol{k}} \langle 0 | (u_{\boldsymbol{k}} + v_{\boldsymbol{k}} \hat{c}_{-\boldsymbol{k}+\boldsymbol{K}/2\downarrow} \hat{c}_{\boldsymbol{k}+\boldsymbol{K}/2\uparrow}) \\
&\qquad \times \hat{c}_{-\boldsymbol{k}+\boldsymbol{K}/2\downarrow} \hat{c}_{\boldsymbol{k}+\boldsymbol{K}/2\uparrow} (u_{\boldsymbol{k}} + v_{\boldsymbol{k}} \hat{c}^\dagger_{\boldsymbol{k}+\boldsymbol{K}/2\uparrow} \hat{c}^\dagger_{-\boldsymbol{k}+\boldsymbol{K}/2\downarrow}) |0\rangle \\
&= V \sum_{\boldsymbol{k}} u_{\boldsymbol{k}} v_{\boldsymbol{k}} \langle 0 | \hat{c}_{-\boldsymbol{k}+\boldsymbol{K}/2\downarrow} \hat{c}_{\boldsymbol{k}+\boldsymbol{K}/2\uparrow} \hat{c}^\dagger_{\boldsymbol{k}+\boldsymbol{K}/2\uparrow} \hat{c}^\dagger_{-\boldsymbol{k}+\boldsymbol{K}/2\downarrow} |0\rangle \\
&= V \sum_{\boldsymbol{k}} u_{\boldsymbol{k}} v_{\boldsymbol{k}} = \Delta.
\end{aligned}$$

Thus, the energy gap is independent of $\boldsymbol{K}$ and is the same as that in the zero-momentum BCS state.

(2) As in the previous problem, $\langle \hat{c}^\dagger_{\boldsymbol{k}\uparrow} \hat{c}_{\boldsymbol{k}+\boldsymbol{q}\uparrow} \rangle$ and $\langle \hat{c}^\dagger_{\boldsymbol{k}\downarrow} \hat{c}_{\boldsymbol{k}+\boldsymbol{q}\downarrow} \rangle$ are nonzero only for $\boldsymbol{q} = 0$. For $\langle \hat{c}^\dagger_{\boldsymbol{k}\uparrow} \hat{c}_{\boldsymbol{k}+\boldsymbol{q}\uparrow} \rangle$, we have

$$\begin{aligned}
\langle \hat{c}^\dagger_{\boldsymbol{k}\uparrow} \hat{c}_{\boldsymbol{k}+\boldsymbol{q}\uparrow} \rangle &= \langle 0 | (u_{\boldsymbol{k}-\boldsymbol{K}/2} + v_{\boldsymbol{k}-\boldsymbol{K}/2} \hat{c}_{\boldsymbol{K}-\boldsymbol{k}\downarrow} \hat{c}_{\boldsymbol{k}\uparrow}) \\
&\qquad \times \hat{c}^\dagger_{\boldsymbol{k}\uparrow} \hat{c}_{\boldsymbol{k}\uparrow} (u_{\boldsymbol{k}-\boldsymbol{K}/2} + v_{\boldsymbol{k}-\boldsymbol{K}/2} \hat{c}^\dagger_{\boldsymbol{k}\uparrow} \hat{c}^\dagger_{\boldsymbol{K}-\boldsymbol{k}\downarrow}) |0\rangle \, \delta_{\boldsymbol{q}0} \\
&= v^2_{\boldsymbol{k}-\boldsymbol{K}/2} \delta_{\boldsymbol{q}0}.
\end{aligned}$$

Similarly, for $\langle \hat{c}^\dagger_{\boldsymbol{k}\downarrow} \hat{c}_{\boldsymbol{k}+\boldsymbol{q}\downarrow} \rangle$, we have

$$\begin{aligned}
\langle \hat{c}^\dagger_{\boldsymbol{k}\downarrow} \hat{c}_{\boldsymbol{k}+\boldsymbol{q}\downarrow} \rangle &= \langle 0 | (u_{\boldsymbol{k}-\boldsymbol{K}/2} + v_{\boldsymbol{k}-\boldsymbol{K}/2} \hat{c}_{\boldsymbol{k}\downarrow} \hat{c}_{\boldsymbol{K}-\boldsymbol{k}\uparrow}) \hat{c}^\dagger_{\boldsymbol{k}\downarrow} \hat{c}_{\boldsymbol{k}\downarrow} \\
&\qquad \times (u_{\boldsymbol{k}-\boldsymbol{K}/2} + v_{\boldsymbol{k}-\boldsymbol{K}/2} \hat{c}^\dagger_{\boldsymbol{K}-\boldsymbol{k}\uparrow} \hat{c}^\dagger_{\boldsymbol{k}\downarrow}) |0\rangle \, \delta_{\boldsymbol{q}0} \\
&= v^2_{\boldsymbol{k}-\boldsymbol{K}/2} \delta_{\boldsymbol{q}0}.
\end{aligned}$$

The average of $\hat{j}_p(\boldsymbol{q})$ is then given by

$$\langle \hat{j}_p(\boldsymbol{q})\rangle = -\delta_{\boldsymbol{q}0}\frac{2e\hbar}{m\mathscr{V}}\sum_{\boldsymbol{k}} \boldsymbol{k} v^2_{\boldsymbol{k}-\boldsymbol{K}/2} = -\delta_{\boldsymbol{q}0}\frac{e\hbar}{m\mathscr{V}}\sum_{\boldsymbol{k}}\frac{\xi_{\boldsymbol{k}-\boldsymbol{K}/2}}{E_{\boldsymbol{k}-\boldsymbol{K}/2}}\boldsymbol{k}$$
$$= -\delta_{\boldsymbol{q}0}\frac{e\hbar}{2m\mathscr{V}}\boldsymbol{K}\sum_{\boldsymbol{k}}\frac{\xi_{\boldsymbol{k}}}{E_{\boldsymbol{k}}}.$$

Converting the summation over $\boldsymbol{k}$ into an integration over the electron energy $\xi = \varepsilon - \varepsilon_{\rm F}$, we have

$$\langle \hat{j}_p(\boldsymbol{q})\rangle = -\delta_{\boldsymbol{q}0}\frac{3ne\hbar}{8m\varepsilon_{\rm F}^{3/2}}\boldsymbol{K}\int_{-\hbar\omega_{\rm D}}^{\hbar\omega_{\rm D}} d\xi\,\frac{\xi(\xi+\varepsilon_{\rm F})^{1/2}}{(\xi^2+\Delta^2)^{1/2}}$$
$$= -\delta_{\boldsymbol{q}0}\frac{3ne\hbar}{8m\varepsilon_{\rm F}^{2}}\boldsymbol{K}\int_{0}^{\hbar\omega_{\rm D}} d\xi\,\frac{\xi^2}{(\xi^2+\Delta^2)^{1/2}}$$
$$= -\delta_{\boldsymbol{q}0}\frac{3ne\hbar\Delta^2}{16m\varepsilon_{\rm F}^{2}}\boldsymbol{K}\Big[(\hbar\omega_{\rm D}/\Delta)\sqrt{1+(\hbar\omega_{\rm D}/\Delta)^2}$$
$$- \sinh^{-1}(\hbar\omega_{\rm D}/\Delta)\Big].$$

The above result implies that the current in the BCS state of momentum $\hbar\boldsymbol{K}$ is proportional to $\boldsymbol{K}$.

30-13 Variational approach of the BCS theory. In this problem, we work through the BCS variational computation. Take the BCS wave function $|\Omega\rangle = \prod_{\boldsymbol{k}}(u_{\boldsymbol{k}} + v_{\boldsymbol{k}}\hat{c}^\dagger_{\boldsymbol{k}\uparrow}\hat{c}^\dagger_{-\boldsymbol{k}\downarrow})\,|0\rangle$ as the trial wave function for the superconducting ground state. The Hamiltonian is given by $\hat{H} = \hat{H}_0 + \hat{H}_{\rm int}$, where $\hat{H}_0 = \sum'_{\boldsymbol{k}\sigma}\xi_{\boldsymbol{k}}\hat{c}^\dagger_{\boldsymbol{k}\sigma}\hat{c}_{\boldsymbol{k}\sigma}$ and $\hat{H}_{\rm int} = -V\sum'_{\boldsymbol{k}\boldsymbol{k}'}\hat{c}^\dagger_{\boldsymbol{k}'\uparrow}\hat{c}^\dagger_{-\boldsymbol{k}'\downarrow}\hat{c}_{-\boldsymbol{k}\downarrow}\hat{c}_{\boldsymbol{k}\uparrow}$

(1) Show that $|\Omega\rangle$ is normalized to unity under the condition that $u^2_{\boldsymbol{k}} + v^2_{\boldsymbol{k}} = 1$.

(2) Show that the expectation value of the kinetic energy term in the superconducting ground state is given by $\langle\hat{H}_0\rangle = \sum'_{\boldsymbol{k}} 2\xi_{\boldsymbol{k}}v^2_{\boldsymbol{k}}$.

(3) Show that the expectation value of the interaction term in the superconducting ground state is given by $\langle\hat{H}_{\rm int}\rangle = -V\sum'_{\boldsymbol{k}\boldsymbol{k}'} u_{\boldsymbol{k}}v_{\boldsymbol{k}}u_{\boldsymbol{k}'}v_{\boldsymbol{k}'}$.

(4) Minimize $\langle\hat{H}\rangle = \langle\hat{H}_0\rangle + \langle\hat{H}_{\rm int}\rangle$ with respect to $u_{\boldsymbol{k}}$ or $v_{\boldsymbol{k}}$ under the condition that $u^2_{\boldsymbol{k}}+v^2_{\boldsymbol{k}} = 1$ to determine $u_{\boldsymbol{k}}$ and $v_{\boldsymbol{k}}$. It will be found that it is convenient to introduce $\Delta = V\sum'_{\boldsymbol{k}} u_{\boldsymbol{k}}v_{\boldsymbol{k}}$ and $E_{\boldsymbol{k}} = (\xi^2_{\boldsymbol{k}} + \Delta^2_{\boldsymbol{k}})^{1/2}$.

(5) Derive a self-consistent equation for Δ from $\Delta = V\sum'_{\boldsymbol{k}} u_{\boldsymbol{k}}v_{\boldsymbol{k}}$.

---

(1) Computing $\langle\Omega|\Omega\rangle$, we have

$$\langle\Omega|\Omega\rangle = \langle 0|\prod_{k}(u_k + v_k\hat{c}_{-k\downarrow}\hat{c}_{k\uparrow})(u_k + v_k\hat{c}^\dagger_{k\uparrow}\hat{c}^\dagger_{-k\downarrow})|0\rangle$$
$$= \langle 0|\prod_{k}(u_k^2 + v_k^2\hat{c}_{-k\downarrow}\hat{c}_{k\uparrow}\hat{c}^\dagger_{k\uparrow}\hat{c}^\dagger_{-k\downarrow})|0\rangle$$

$$= \langle 0|\prod_{k}(u_k^2 + v_k^2\hat{c}_{-k\downarrow}\hat{c}^\dagger_{-k\downarrow} - v_k^2\hat{c}_{-k\downarrow}\hat{c}^\dagger_{k\uparrow}\hat{c}_{k\uparrow}\hat{c}^\dagger_{-k\downarrow})|0\rangle$$
$$= \langle 0|\prod_{k}(u_k^2 + v_k^2\hat{c}_{-k\downarrow}\hat{c}^\dagger_{-k\downarrow})|0\rangle = \langle 0|\prod_{k}(u_k^2 + v_k^2 - v_k^2\hat{c}^\dagger_{-k\downarrow}\hat{c}_{-k\downarrow})|0\rangle$$
$$= \prod_{k}(u_k^2 + v_k^2) = \prod_{k} 1 = 1,$$

where we have made use of the facts that $(u_{k'} + v_{k'}\hat{c}_{-k'\downarrow}\hat{c}_{k'\uparrow})$ commutes with $(u_k + v_k\hat{c}_{-k\downarrow}\hat{c}_{k\uparrow})$, $(u_{k'} + v_{k'}\hat{c}^\dagger_{k'\uparrow}\hat{c}^\dagger_{-k'\downarrow})$ commutes with $(u_k + v_k\hat{c}^\dagger_{k\uparrow}\hat{c}^\dagger_{-k\downarrow})$, and that $(u_{k'} + v_{k'}\hat{c}_{-k'\downarrow}\hat{c}_{k'\uparrow})$ commutes with $(u_k + v_k\hat{c}^\dagger_{k\uparrow}\hat{c}^\dagger_{-k\downarrow})$ for $k' \neq k$ so that the product of the two products $\prod_{k'}(u_{k'} + v_{k'}\hat{c}_{-k'\downarrow}\hat{c}_{k'\uparrow})$ and $\prod_{k}(u_k + v_k\hat{c}^\dagger_{k\uparrow}\hat{c}^\dagger_{-k\downarrow})$ can be written as $\prod_{k}(u_k + v_k\hat{c}_{-k\downarrow}\hat{c}_{k\uparrow})(u_k + v_k\hat{c}^\dagger_{k\uparrow}\hat{c}^\dagger_{-k\downarrow})$.

(2) To compute the average of $\hat{H}_0$, we rewrite it as

$$\hat{H}_0 = \sum_{k}{}' \xi_k\hat{c}^\dagger_{k\uparrow}\hat{c}_{k\uparrow} + \sum_{k}{}' \xi_k\hat{c}^\dagger_{-k\downarrow}\hat{c}_{-k\downarrow}.$$

We then have

$$\langle\hat{H}_0\rangle = \langle\Omega|\hat{H}_0|\Omega\rangle$$
$$= \sum_{k}{}' \xi_k \langle 0|(u_k + v_k\hat{c}_{-k\downarrow}\hat{c}_{k\uparrow})\hat{c}^\dagger_{k\uparrow}\hat{c}_{k\uparrow}(u_k + v_k\hat{c}^\dagger_{k\uparrow}\hat{c}^\dagger_{-k\downarrow})|0\rangle$$
$$+ \sum_{k}{}' \xi_k \langle 0|(u_k + v_k\hat{c}_{-k\downarrow}\hat{c}_{k\uparrow})\hat{c}^\dagger_{-k\downarrow}\hat{c}_{-k\downarrow}(u_k + v_k\hat{c}^\dagger_{k\uparrow}\hat{c}^\dagger_{-k\downarrow})|0\rangle,$$

where we have made use of the fact that all the terms in $|\Omega\rangle$ with wave vectors not equal to that of the term in $\hat{H}_0$ yield unity as can be seen from the above computation. Making use of $[\hat{c}^\dagger_{k\uparrow}\hat{c}_{k\uparrow}, \hat{c}^\dagger_{k\uparrow}\hat{c}^\dagger_{-k\downarrow}] = \hat{c}^\dagger_{k\uparrow}\hat{c}^\dagger_{-k\downarrow}$, $[\hat{c}^\dagger_{-k\downarrow}\hat{c}_{-k\downarrow}, \hat{c}^\dagger_{k\uparrow}\hat{c}^\dagger_{-k\downarrow}] =$

$\hat{c}^\dagger_{\boldsymbol{k}\uparrow}\hat{c}^\dagger_{-\boldsymbol{k}\downarrow}$, and $\hat{c}_{\boldsymbol{k}\uparrow}|0\rangle = \hat{c}_{-\boldsymbol{k}\downarrow}|0\rangle = 0$, we have

$$\begin{aligned}
\langle \hat{H}_0\rangle &= 2\sum_{\boldsymbol{k}}{}' \xi_{\boldsymbol{k}} v_{\boldsymbol{k}} \langle 0|\big(u_{\boldsymbol{k}} + v_{\boldsymbol{k}}\hat{c}_{-\boldsymbol{k}\downarrow}\hat{c}_{\boldsymbol{k}\uparrow}\big)\hat{c}^\dagger_{\boldsymbol{k}\uparrow}\hat{c}^\dagger_{-\boldsymbol{k}\downarrow}|0\rangle \\
&= 2\sum_{\boldsymbol{k}}{}' \xi_{\boldsymbol{k}} v_{\boldsymbol{k}}^2 \langle 0|\,\hat{c}_{-\boldsymbol{k}\downarrow}\hat{c}_{\boldsymbol{k}\uparrow}\hat{c}^\dagger_{\boldsymbol{k}\uparrow}\hat{c}^\dagger_{-\boldsymbol{k}\downarrow}|0\rangle \\
&= 2\sum_{\boldsymbol{k}}{}' \xi_{\boldsymbol{k}} v_{\boldsymbol{k}}^2 \langle 0|\,\hat{c}_{-\boldsymbol{k}\downarrow}(1 - \hat{c}^\dagger_{\boldsymbol{k}\uparrow}\hat{c}_{\boldsymbol{k}\uparrow})\hat{c}^\dagger_{-\boldsymbol{k}\downarrow}|0\rangle \\
&= 2\sum_{\boldsymbol{k}}{}' \xi_{\boldsymbol{k}} v_{\boldsymbol{k}}^2 \langle 0|\,\hat{c}_{-\boldsymbol{k}\downarrow}\hat{c}^\dagger_{-\boldsymbol{k}\downarrow}|0\rangle \\
&= 2\sum_{\boldsymbol{k}}{}' \xi_{\boldsymbol{k}} v_{\boldsymbol{k}}^2 \langle 0|(1 - \hat{c}^\dagger_{-\boldsymbol{k}\downarrow}\hat{c}_{-\boldsymbol{k}\downarrow})|0\rangle = 2\sum_{\boldsymbol{k}}{}' \xi_{\boldsymbol{k}} v_{\boldsymbol{k}}^2.
\end{aligned}$$

(3) For the average of the interaction Hamiltonian, it appears that we must separately evaluate the $\boldsymbol{k}' = \boldsymbol{k}$ and $\boldsymbol{k}' \neq \boldsymbol{k}$ terms. However, in a process corresponding to a $\boldsymbol{k}' = \boldsymbol{k}$ term, two electrons are taken out of two single-electron states and they are put into the same single-electron states. Thus, nothing actually happens in such a process and it should not be taken into account. However, the difference between $\boldsymbol{k}$ and $\boldsymbol{k}'$ can become infinitesimally small. Therefore, we first evaluate $\langle H_{\text{int}}\rangle$ under the condition that $\boldsymbol{k}' \neq \boldsymbol{k}$ and then relax this restriction to allow $\boldsymbol{k} \to \boldsymbol{k}'$. Making use of $[\hat{c}_{-\boldsymbol{k}\downarrow}\hat{c}_{\boldsymbol{k}\uparrow}, \hat{c}^\dagger_{\boldsymbol{k}\uparrow}\hat{c}^\dagger_{-\boldsymbol{k}\downarrow}] = \hat{c}_{-\boldsymbol{k}\downarrow}\hat{c}^\dagger_{-\boldsymbol{k}\downarrow} - \hat{c}^\dagger_{\boldsymbol{k}\uparrow}\hat{c}_{\boldsymbol{k}\uparrow}$ and $\hat{c}_{\boldsymbol{k}\uparrow}|0\rangle = \hat{c}_{-\boldsymbol{k}\downarrow}|0\rangle = 0$, we have for $\boldsymbol{k}' \neq \boldsymbol{k}$

$$\begin{aligned}
&\langle \hat{c}^\dagger_{\boldsymbol{k}'\uparrow}\hat{c}^\dagger_{-\boldsymbol{k}'\downarrow}\hat{c}_{-\boldsymbol{k}\downarrow}\hat{c}_{\boldsymbol{k}\uparrow}\rangle \\
&\quad= \langle 0|\big(u_{\boldsymbol{k}'} + v_{\boldsymbol{k}'}\hat{c}_{-\boldsymbol{k}'\downarrow}\hat{c}_{\boldsymbol{k}'\uparrow}\big)\big(u_{\boldsymbol{k}} + v_{\boldsymbol{k}}\hat{c}_{-\boldsymbol{k}\downarrow}\hat{c}_{\boldsymbol{k}\uparrow}\big)\hat{c}^\dagger_{\boldsymbol{k}'\uparrow}\hat{c}^\dagger_{-\boldsymbol{k}'\downarrow}\hat{c}_{-\boldsymbol{k}\downarrow}\hat{c}_{\boldsymbol{k}\uparrow} \\
&\qquad\quad \times \big(u_{\boldsymbol{k}} + v_{\boldsymbol{k}}\hat{c}^\dagger_{\boldsymbol{k}\uparrow}\hat{c}^\dagger_{-\boldsymbol{k}\downarrow}\big)\big(u_{\boldsymbol{k}'} + v_{\boldsymbol{k}'}\hat{c}^\dagger_{\boldsymbol{k}'\uparrow}\hat{c}^\dagger_{-\boldsymbol{k}'\downarrow}\big)|0\rangle \\
&\quad= v_{\boldsymbol{k}}v_{\boldsymbol{k}'}\langle 0|\,\hat{c}_{-\boldsymbol{k}'\downarrow}\hat{c}^\dagger_{-\boldsymbol{k}'\downarrow}\big(u_{\boldsymbol{k}} + v_{\boldsymbol{k}}\hat{c}_{-\boldsymbol{k}\downarrow}\hat{c}_{\boldsymbol{k}\uparrow}\big) \\
&\qquad\quad \times \hat{c}_{-\boldsymbol{k}\downarrow}\hat{c}^\dagger_{-\boldsymbol{k}\downarrow}\big(u_{\boldsymbol{k}'} + v_{\boldsymbol{k}'}\hat{c}^\dagger_{\boldsymbol{k}'\uparrow}\hat{c}^\dagger_{-\boldsymbol{k}'\downarrow}\big)|0\rangle \\
&\quad= v_{\boldsymbol{k}}v_{\boldsymbol{k}'}\langle 0|\big(u_{\boldsymbol{k}} + v_{\boldsymbol{k}}\hat{c}_{-\boldsymbol{k}\downarrow}\hat{c}_{\boldsymbol{k}\uparrow}\big)\big(u_{\boldsymbol{k}'} + v_{\boldsymbol{k}'}\hat{c}^\dagger_{\boldsymbol{k}'\uparrow}\hat{c}^\dagger_{-\boldsymbol{k}'\downarrow}\big)|0\rangle \\
&\quad= u_{\boldsymbol{k}}v_{\boldsymbol{k}}u_{\boldsymbol{k}'}v_{\boldsymbol{k}'}.
\end{aligned}$$

Relaxing the restriction $\boldsymbol{k}' \neq \boldsymbol{k}$, we have

$$\langle H_{\text{int}}\rangle = -V\sum_{\boldsymbol{k}\boldsymbol{k}'}{}' u_{\boldsymbol{k}}v_{\boldsymbol{k}}u_{\boldsymbol{k}'}v_{\boldsymbol{k}'}.$$

(4) From the above results, we have the following average of the Hamiltonian $\hat{H}$

$$\langle \hat{H}\rangle = \langle H_0\rangle + \langle H_{\text{int}}\rangle = 2\sum_{\boldsymbol{k}}{}' \xi_{\boldsymbol{k}}v_{\boldsymbol{k}}^2 - V\sum_{\boldsymbol{k}\boldsymbol{k}'}{}' u_{\boldsymbol{k}}v_{\boldsymbol{k}}u_{\boldsymbol{k}'}v_{\boldsymbol{k}'}.$$

Differentiating $\langle \hat{H} \rangle$ with respect to v_k with the use of $\partial u_k / \partial v_k = -v_k / u_k = -v_k/(1 - v_k^2)^{1/2}$ and setting the result to zero, we obtain

$$4\xi_k u_k v_k - 2(u_k^2 - v_k^2)\left(V \sum_{k'} u_{k'} v_{k'}\right) = 0.$$

Introducing

$$\Delta = V \sum_k u_k v_k$$

and making use of $v_k = (1 - u_k^2)^{1/2}$, we can simplify the above equation into

$$u_k^4 - u_k^2 + \frac{\Delta^2}{4E_k^2} = 0,$$

where

$$E_k = (\xi_k^2 + \Delta^2)^{1/2}.$$

Solving the above equation for u_k, we obtain two roots, $(1 \pm \xi_k/E_k)/2$, for u_k^2. We choose the plus sign for u_k^2 and have

$$u_k^2 = \frac{1}{2}\left(1 + \frac{\xi_k}{E_k}\right).$$

From the condition $u_k^2 + v_k^2 = 1$, we have

$$v_k^2 = \frac{1}{2}\left(1 - \frac{\xi_k}{E_k}\right).$$

(5) From the original equation obtained from $\partial \langle \hat{H} \rangle / \partial v_k = 0$, we have

$$u_k v_k = \frac{\Delta}{2\xi_k}(u_k^2 - v_k^2) = \frac{\Delta}{2E_k}.$$

Substituting the above result into $\Delta = V \sum_k u_k v_k$, we have

$$\Delta = V \sum_k \frac{\Delta}{2E_k}.$$

At zero temperature, Δ is not equal to zero and it can be thus canceled from the two sides. We then have the following self-consistent equation for Δ

$$\frac{1}{V} = \sum_k \frac{1}{2E_k} = \sum_k \frac{1}{2(\xi_k^2 + \Delta^2)^{1/2}}.$$

30-14 Average internal flux density in a triangular vortex lattice. In the mixed state of an isotropic type-II superconductor, the vortex lattice has the triangular symmetry. Show that the average internal flux density is $B = 2\Phi_0/(3^{1/2}a^2)$, where a is the lattice constant.

In a triangular lattice, a vertex of each equilateral triangle is shared by six other equilateral triangles. Thus, each equilateral triangle contains half a vortex on average. Making use of the facts that the magnetic flux in a vortex is equal to Φ_0 and that the area of each equilateral triangle is equal to $3^{1/2}a^2/4$, we see that the average internal flux density is

$$B = \frac{\Phi_0/2}{\sqrt{3}a^2/4} = \frac{2\Phi_0}{3^{1/2}a^2}.$$

30-15 Superconducting instability. The pairing susceptibility is defined as the linear response function of $\hat{\Delta}(\boldsymbol{r}) = \mathscr{V}\big[\hat{\psi}_\uparrow^\dagger(\boldsymbol{r})\hat{\psi}_\downarrow^\dagger(\boldsymbol{r}) + \hat{\psi}_\downarrow(\boldsymbol{r})\hat{\psi}_\uparrow(\boldsymbol{r})\big]$ to the pairing field $V(\boldsymbol{r},t)$ with the interaction Hamiltonian given by

$$\hat{H}_p = \int \mathrm{d}\boldsymbol{r}\; V(\boldsymbol{r},t)\big[\hat{\psi}_\uparrow^\dagger(\boldsymbol{r})\hat{\psi}_\downarrow^\dagger(\boldsymbol{r}) + \hat{\psi}_\downarrow(\boldsymbol{r})\hat{\psi}_\uparrow(\boldsymbol{r})\big].$$

(1) Explain why the Feynman diagram in Fig. 30.7(a) is the first Feynman diagram for the Matsubara pairing susceptibility.

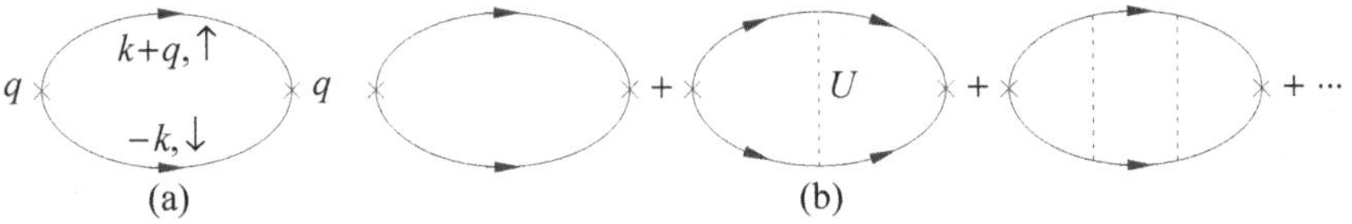

Fig. 30.7 Feynman diagrams for the pairing susceptibility. (a) Lowest-order Feynman diagram. (b) RPA series for the pairing susceptibility.

(2) Evaluate $\chi_p^{(0)}(\boldsymbol{q} = 0, \mathrm{i}\omega_m = 0)$. The divergence at large momenta is to be cut off by the energy $\hbar\omega_c$ relative to the Fermi energy, that is, $|\xi_{\boldsymbol{k}}| < \hbar\omega_c$ with $\hbar\omega_c \ll E_\mathrm{F}$.

(3) Sum the leading divergencies shown in the Fig. 30.7(b) and obtain the RPA result for the pairing susceptibility $\chi_p(\boldsymbol{q}, \mathrm{i}\omega_m) = \chi_p^{(0)}(\boldsymbol{q}, \mathrm{i}\omega_m)/\big[1 + U\chi_p^{(0)}(\boldsymbol{q}, \mathrm{i}\omega_m)\big]$, where we have assumed that the inter-particle interaction is a contact one which is a constant U in k-space.

(4) We now study the behavior of $\chi_p(\boldsymbol{q} = 0, \mathrm{i}\omega_m = 0)$ for a very weak attraction between fermions ($U < 0$ and $|U| \ll \hbar\omega_c$).

Show that, as the temperature decreases, there is always an instability [a divergence in $\chi_p(\boldsymbol{q} = 0, \mathrm{i}\omega_m = 0)$] at some finite temperature T_c. Find an explicit expression for T_c in terms of U and $\hbar\omega_c$.

---

(1) In the linear response theory, the interaction Hamiltonian due to an external field is often put into the form

$$\hat{H}_{\mathrm{ext}} = \frac{1}{\mathscr{V}} \int \mathrm{d}\boldsymbol{r}\ f(\boldsymbol{r},t)\hat{B}(\boldsymbol{r}).$$

Comparing the above expression with the interaction Hamiltonian due to the pairing field, we see that

$$f(\boldsymbol{r},t) = V(\boldsymbol{r},t),\ \ \hat{B}(\boldsymbol{r}) = \mathscr{V}\big[\hat{\psi}_\uparrow^\dagger(\boldsymbol{r})\hat{\psi}_\downarrow^\dagger(\boldsymbol{r}) + \hat{\psi}_\downarrow(\boldsymbol{r})\hat{\psi}_\uparrow(\boldsymbol{r})\big].$$

Note that $\hat{\psi}_\sigma$ has the dimension of $\mathscr{V}^{-1/2}$. According to the linear response theory, we have

$$\langle\hat{\Delta}\rangle(\boldsymbol{q},\omega) = \chi_p(\boldsymbol{q},\omega + \mathrm{i}\delta)V(\boldsymbol{q},\omega).$$

where the response function $\chi_p(\boldsymbol{q},\omega+\mathrm{i}\delta)$ is referred to as *the pairing susceptibility* and is given by

$$\chi_p(\boldsymbol{q},\omega + \mathrm{i}\delta) = \mathrm{i}\int_{-\infty}^{\infty} \frac{\mathrm{d}\omega'}{2\pi}\, \frac{G_R(\boldsymbol{q},\omega')}{\omega - \omega' + \mathrm{i}\delta},$$

where

$$G_R(\boldsymbol{q},\omega) = \int \frac{\mathrm{d}(\boldsymbol{r}-\boldsymbol{r}')}{\mathscr{V}} \int_{-\infty}^{\infty} \mathrm{d}(t-t')\ \mathrm{e}^{-\mathrm{i}[\boldsymbol{q}\cdot(\boldsymbol{r}-\boldsymbol{r}')-\omega(t-t')]} \times G_R(\boldsymbol{r}-\boldsymbol{r}',t-t')$$

with

$$\begin{aligned} &G_R(\boldsymbol{r}-\boldsymbol{r}',t-t') \\ &\quad = -(\mathrm{i}\mathscr{V}^2/\hbar)\theta(t-t')\langle\big[\hat{\psi}_\uparrow^\dagger(\boldsymbol{r},t)\hat{\psi}_\downarrow^\dagger(\boldsymbol{r},t) + \hat{\psi}_\downarrow(\boldsymbol{r},t)\hat{\psi}_\uparrow(\boldsymbol{r},t), \\ &\qquad \hat{\psi}_\uparrow^\dagger(\boldsymbol{r}',t')\hat{\psi}_\downarrow^\dagger(\boldsymbol{r}',t') + \hat{\psi}_\downarrow(\boldsymbol{r}',t')\hat{\psi}_\uparrow(\boldsymbol{r}',t')\big]\rangle. \end{aligned}$$

Analytically continue $\chi_p(\boldsymbol{q},\omega)$ onto the imaginary frequency axis, we have

$$\chi_p(\boldsymbol{q},\mathrm{i}\omega_m) = \mathrm{i}\int_{-\infty}^{\infty} \frac{\mathrm{d}\omega'}{2\pi}\, \frac{G_R(\boldsymbol{q},\omega')}{\mathrm{i}\omega_m - \omega'}.$$

The above expression for $\chi_p(\boldsymbol{q},\mathrm{i}\omega_m)$ is actually its spectral representation with $\mathrm{i}G_R(\boldsymbol{q},\omega)/2\pi$ as its spectral function. We can see that $G_R(\boldsymbol{q},\omega)$ is indeed purely imaginary by taking its complex conjugation. It is well-known that the spectral function of a Green's function is given by the imaginary part of the corresponding retarded Green's function divided by $-\pi$. In the present problem, the spectral function is given by $-\pi^{-1}\,\mathrm{Im}\,\chi_p(\boldsymbol{q},\omega+\mathrm{i}\delta)$ which is equal to $\mathrm{i}G_R(\boldsymbol{q},\omega)/2\pi$. Therefore, $\chi_p(\boldsymbol{q},\mathrm{i}\omega_m)$ is the Fourier transform of the imaginary-time pairing susceptibility $\chi_p(\boldsymbol{r}-\boldsymbol{r}',\tau-\tau')$,

$$\chi_p(\boldsymbol{r}-\boldsymbol{r}',\tau-\tau') = \frac{1}{\beta}\sum_{\boldsymbol{q}m} \mathrm{e}^{\mathrm{i}[\boldsymbol{q}\cdot(\boldsymbol{r}-\boldsymbol{r}')-\omega_m(\tau-\tau')]}\chi_p(\boldsymbol{q},\mathrm{i}\omega_m)$$

with $\chi_p(\boldsymbol{r}-\boldsymbol{r}',\tau-\tau')$ defined as

$$\chi_p(\boldsymbol{r}-\boldsymbol{r}',\tau-\tau') = \mathscr{V}^2\big\langle \hat{T}_\tau\big[\,\hat{\psi}_\uparrow(\boldsymbol{r},\tau)\hat{\psi}_\downarrow(\boldsymbol{r},\tau)\hat{\psi}_\downarrow^\dagger(\boldsymbol{r}',\tau')\hat{\psi}_\uparrow^\dagger(\boldsymbol{r}',\tau')\,\big]\big\rangle .$$

Inserting the expansion of the quantum field operator for free fermions with the spin wave function suppressed

$$\hat{\psi}_\sigma(\boldsymbol{r}) = \frac{1}{\sqrt{\mathscr{V}}}\sum_{\boldsymbol{k}} \mathrm{e}^{\mathrm{i}\boldsymbol{k}\cdot\boldsymbol{r}}\hat{c}_{\boldsymbol{k}\sigma}$$

into $\chi_p(\boldsymbol{r}-\boldsymbol{r}',\tau-\tau')$, we have

$$\chi_p(\boldsymbol{r}-\boldsymbol{r}',\tau-\tau') = \sum_{\boldsymbol{k}_1\boldsymbol{k}_2\boldsymbol{k}_3\boldsymbol{k}_4} \mathrm{e}^{\mathrm{i}[(\boldsymbol{k}_1+\boldsymbol{k}_2)\cdot\boldsymbol{r}-(\boldsymbol{k}_3+\boldsymbol{k}_4)\cdot\boldsymbol{r}']} \times \big\langle \hat{T}_\tau\big[\,\hat{c}_{\boldsymbol{k}_1\uparrow}(\tau)\hat{c}_{\boldsymbol{k}_2\downarrow}(\tau)\hat{c}_{\boldsymbol{k}_3\downarrow}^\dagger(\tau')\hat{c}_{\boldsymbol{k}_4\uparrow}^\dagger(\tau')\,\big]\big\rangle .$$

We first evaluate $\chi_p(\boldsymbol{r}-\boldsymbol{r}',\tau-\tau')$ without considering the interaction between fermions. In this case, the Hamiltonian of the fermion system is given by

$$\hat{H}_0 = \sum_{\boldsymbol{k}\sigma}\xi_{\boldsymbol{k}}\hat{c}_{\boldsymbol{k}\sigma}^\dagger\hat{c}_{\boldsymbol{k}\sigma},$$

where $\xi_{\boldsymbol{k}} = \varepsilon_{\boldsymbol{k}}-\mu = \hbar^2\boldsymbol{k}^2/2m-\mu$. In this case, the imaginary time dependencies of the fermion operators are given by

$$\hat{c}_{\boldsymbol{k}\sigma}(\tau) = \mathrm{e}^{-\xi_{\boldsymbol{k}}\tau/\hbar}\hat{c}_{\boldsymbol{k}\sigma},\quad \hat{c}_{\boldsymbol{k}\sigma}^\dagger(\tau) = \mathrm{e}^{\xi_{\boldsymbol{k}}\tau/\hbar}\hat{c}_{\boldsymbol{k}\sigma}^\dagger.$$

Making use of Wick's theorem, we have

$$\begin{aligned}
&\chi_p^{(0)}(\boldsymbol{r}-\boldsymbol{r}',\tau-\tau')\\
&= \sum_{\boldsymbol{k}_1\boldsymbol{k}_2\boldsymbol{k}_3\boldsymbol{k}_4} \mathrm{e}^{\mathrm{i}[(\boldsymbol{k}_1+\boldsymbol{k}_2)\cdot\boldsymbol{r}-(\boldsymbol{k}_3+\boldsymbol{k}_4)\cdot\boldsymbol{r}']}\\
&\qquad\times \big\langle \hat{T}_\tau\big[\,\hat{c}_{\boldsymbol{k}_1\uparrow}(\tau)\hat{c}_{\boldsymbol{k}_2\downarrow}(\tau)\hat{c}_{\boldsymbol{k}_3\downarrow}^\dagger(\tau')\hat{c}_{\boldsymbol{k}_4\uparrow}^\dagger(\tau')\,\big]\big\rangle_0\\
&= \sum_{\boldsymbol{k}_1\boldsymbol{k}_2\boldsymbol{k}_3\boldsymbol{k}_4} \mathrm{e}^{\mathrm{i}[(\boldsymbol{k}_1+\boldsymbol{k}_2)\cdot\boldsymbol{r}-(\boldsymbol{k}_3+\boldsymbol{k}_4)\cdot\boldsymbol{r}']}\\
&\qquad\times \mathscr{G}_{\uparrow 0}(\boldsymbol{k}_1,\tau-\tau')\mathscr{G}_{\downarrow 0}(\boldsymbol{k}_2,\tau-\tau')\delta_{\boldsymbol{k}_2\boldsymbol{k}_3}\delta_{\boldsymbol{k}_1\boldsymbol{k}_4}\\
&= \sum_{\boldsymbol{k}_1\boldsymbol{k}_2} \mathrm{e}^{\mathrm{i}(\boldsymbol{k}_1+\boldsymbol{k}_2)\cdot(\boldsymbol{r}-\boldsymbol{r}')}\mathscr{G}_{\uparrow 0}(\boldsymbol{k}_1,\tau-\tau')\mathscr{G}_{\downarrow 0}(\boldsymbol{k}_2,\tau-\tau')\\
&= \sum_{\boldsymbol{k}\boldsymbol{q}} \mathrm{e}^{\mathrm{i}\boldsymbol{q}\cdot(\boldsymbol{r}-\boldsymbol{r}')}\mathscr{G}_{\uparrow 0}(\boldsymbol{k},\tau-\tau')\mathscr{G}_{\downarrow 0}(\boldsymbol{q}-\boldsymbol{k},\tau-\tau').
\end{aligned}$$

Thus, the Fourier transform with respect to $\boldsymbol{r} - \boldsymbol{r}'$ is given by

$$\chi_p^{(0)}(\boldsymbol{q}, \tau - \tau') = \sum_{\boldsymbol{k}} \mathscr{G}_{\uparrow 0}(\boldsymbol{k}, \tau - \tau')\mathscr{G}_{\downarrow 0}\boldsymbol{q} - \boldsymbol{k}, \tau - \tau').$$

Inserting

$$\mathscr{G}_{\sigma 0}(\boldsymbol{k}, \tau - \tau') = \frac{1}{\beta} \sum_{n} \mathscr{G}_{\sigma 0}(\boldsymbol{k}, \mathrm{i}\omega_n) \mathrm{e}^{-\mathrm{i}\omega_n(\tau - \tau')}$$

into $\chi_p^{(0)}(\boldsymbol{q}, \tau - \tau')$ yields

$$\begin{aligned}
&\chi_p^{(0)}(\boldsymbol{q}, \tau - \tau') \\
&\quad = \frac{1}{\beta^2} \sum_{\boldsymbol{k}nn'} \mathrm{e}^{-\mathrm{i}(\omega_n + \omega_{n'})(\tau - \tau')} \mathscr{G}_{\uparrow 0}(\boldsymbol{k}, \mathrm{i}\omega_n) \mathscr{G}_{\downarrow 0}(\boldsymbol{q} - \boldsymbol{k}, \mathrm{i}\omega_{n'}) \\
&\quad = \frac{1}{\beta^2} \sum_{\boldsymbol{k}mn} \mathrm{e}^{-\mathrm{i}\omega_m(\tau - \tau')} \mathscr{G}_{\uparrow 0}(\boldsymbol{k}, \mathrm{i}\omega_n) \mathscr{G}_{\downarrow 0}(\boldsymbol{q} - \boldsymbol{k}, \mathrm{i}\omega_m - \mathrm{i}\omega_n).
\end{aligned}$$

Thus, the Fourier transform of $\chi_p^{(0)}(\boldsymbol{q}, \tau - \tau')$ with respect to $\tau - \tau'$ is given by

$$\begin{aligned}
\chi_p^{(0)}(\boldsymbol{q}, \mathrm{i}\omega_m) &= \frac{1}{\beta} \sum_{\boldsymbol{k}n} \mathscr{G}_{\uparrow 0}(\boldsymbol{k}, \mathrm{i}\omega_n) \mathscr{G}_{\downarrow 0}(\boldsymbol{q} - \boldsymbol{k}, \mathrm{i}\omega_m - \mathrm{i}\omega_n) \\
&= \frac{1}{\beta} \sum_{\boldsymbol{k}n} \mathscr{G}_{\uparrow 0}(-\boldsymbol{k}, -\mathrm{i}\omega_n) \mathscr{G}_{\downarrow 0}(\boldsymbol{q} + \boldsymbol{k}, \mathrm{i}\omega_m + \mathrm{i}\omega_n).
\end{aligned}$$

The above expression for $\chi_p^{(0)}(\boldsymbol{q}, \mathrm{i}\omega_m)$ corresponds to the Feynman diagram shown in Fig. 30.7(a).

(2) The summation over n was evaluated in **Problem 26-13**. We have

$$\chi_p^{(0)}(\boldsymbol{q}, \mathrm{i}\omega_m) = \sum_{\boldsymbol{k}} \frac{1 - n_{\mathrm{F}}(\xi_{\boldsymbol{k}}) - n_{\mathrm{F}}(\xi_{\boldsymbol{k}+\boldsymbol{q}})}{\xi_{\boldsymbol{k}} + \xi_{\boldsymbol{k}+\boldsymbol{q}} - \mathrm{i}\hbar\omega_m}.$$

For $\chi_p^{(0)}(\boldsymbol{q} = 0, \mathrm{i}\omega_m = 0)$, we have

$$\begin{aligned}
\chi_p^{(0)}(0,0) &= \sum_{\boldsymbol{k}} \frac{1 - 2n_{\mathrm{F}}(\xi_{\boldsymbol{k}})}{2\xi_{\boldsymbol{k}}} = \sum_{\boldsymbol{k}} \frac{1}{2\xi_{\boldsymbol{k}}} \tanh\left(\frac{\xi_{\boldsymbol{k}}}{2k_{\mathrm{B}}T}\right) \\
&= N(0) \int_{-\hbar\omega_c}^{\hbar\omega_c} \mathrm{d}\xi \, \frac{1}{2\xi} \tanh\left(\frac{\xi}{2k_{\mathrm{B}}T}\right) \\
&= N(0) \int_0^{\hbar\omega_c/2k_{\mathrm{B}}T} \mathrm{d}x \, \frac{\tanh x}{x},
\end{aligned}$$

where $N(0)$ is the density of states at the Fermi surface for one spin direction, $N(0) = \mathscr{V} m k_{\mathrm{F}}/2\pi^2\hbar^2$. The above integral appears in the computation of the critical temperature in the BCS theory and its value is approximately given by

$$\int_0^{\hbar\omega_c/2k_BT} dx\, \frac{\tanh x}{x} \approx \ln \frac{2e^\gamma \hbar\omega_c}{\pi k_B T} \approx \ln \frac{1.134\hbar\omega_c}{k_B T}.$$

We thus have

$$\chi_p^{(0)}(0,0) = N(0) \ln \frac{1.134\hbar\omega_c}{k_B T}.$$

(3) With the interaction between fermions taken into consideration, the Hamiltonian of the fermion system reads

$$\hat{H} = \hat{H}_0 + \hat{H}_1,$$

where $\hat{H}_0$ is given in the above and $\hat{H}_1$ is given by

$$\hat{H}_1 = \frac{1}{2} U \sum_{\boldsymbol{k}\boldsymbol{k}'\boldsymbol{q}\sigma\sigma'} \hat{c}^\dagger_{\boldsymbol{k}+\boldsymbol{q},\sigma} \hat{c}^\dagger_{\boldsymbol{k}'-\boldsymbol{q},\sigma'} \hat{c}_{\boldsymbol{k}'\sigma'} \hat{c}_{\boldsymbol{k}\sigma}$$

with $U < 0$ the effective fermion-fermion coupling constant. The perturbation series for $\chi_p(\boldsymbol{q}, \tau - \tau')$ is given by

$$\begin{aligned}
&\chi_p(\boldsymbol{q}, \tau - \tau') \\
&= \sum_{n=0}^{\infty} (-1)^n \frac{1}{\hbar^n} \sum_{\boldsymbol{k}_1\boldsymbol{k}_2\boldsymbol{k}_3\boldsymbol{k}_4} e^{i[(\boldsymbol{k}_1+\boldsymbol{k}_2)\cdot\boldsymbol{r} - (\boldsymbol{k}_3+\boldsymbol{k}_4)\cdot\boldsymbol{r}']} \int_0^{\beta\hbar} d\tau_1 \cdots \int_0^{\hbar\beta} d\tau_n \\
&\quad \times \left\langle \hat{T}_\tau \left[\hat{c}_{\boldsymbol{k}_1\uparrow}(\tau) \hat{c}_{\boldsymbol{k}_2\downarrow}(\tau) \hat{c}^\dagger_{\boldsymbol{k}_3\downarrow}(\tau') \hat{c}^\dagger_{\boldsymbol{k}_4\uparrow}(\tau') \hat{H}_1(\tau_1) \cdots \hat{H}_1(\tau_n) \right] \right\rangle_{0,\mathrm{dc}},
\end{aligned}$$

where the subscript "dc" indicates that only those terms corresponding to different connected Feynman diagrams are to be retained. In the first order, we have

$$\begin{aligned}
&\chi_p^{(1)}(\boldsymbol{r} - \boldsymbol{r}', \tau - \tau') \\
&= -\frac{U}{2\hbar} \sum_{\boldsymbol{k}_1\boldsymbol{k}_2\boldsymbol{k}_3\boldsymbol{k}_4} \sum_{\boldsymbol{k}\boldsymbol{k}'\boldsymbol{q}\sigma\sigma'} e^{i[(\boldsymbol{k}_1+\boldsymbol{k}_2)\cdot\boldsymbol{r} - (\boldsymbol{k}_3+\boldsymbol{k}_4)\cdot\boldsymbol{r}']} \int_0^{\beta\hbar} d\tau_1 \\
&\quad \times \left\langle \hat{T}_\tau \left[\hat{c}_{\boldsymbol{k}_1\uparrow}(\tau) \hat{c}_{\boldsymbol{k}_2\downarrow}(\tau) \hat{c}^\dagger_{\boldsymbol{k}_3\downarrow}(\tau') \hat{c}^\dagger_{\boldsymbol{k}_4\uparrow}(\tau') \right.\right. \\
&\quad\quad \left.\left. \times\, \hat{c}^\dagger_{\boldsymbol{k}+\boldsymbol{q},\sigma}(\tau_1) \hat{c}^\dagger_{\boldsymbol{k}'-\boldsymbol{q},\sigma'}(\tau_1) \hat{c}_{\boldsymbol{k}'\sigma'}(\tau_1) \hat{c}_{\boldsymbol{k}\sigma}(\tau_1) \right] \right\rangle_{0,\mathrm{dc}} \\
&= -\frac{U}{\hbar} \sum_{\boldsymbol{k}_1\boldsymbol{k}_2\boldsymbol{k}_3\boldsymbol{k}_4} \sum_{\boldsymbol{k}\boldsymbol{k}'\boldsymbol{q}\sigma\sigma'} e^{i[(\boldsymbol{k}_1+\boldsymbol{k}_2)\cdot\boldsymbol{r} - (\boldsymbol{k}_3+\boldsymbol{k}_4)\cdot\boldsymbol{r}']} \int_0^{\beta\hbar} d\tau_1 \\
&\quad \times \mathscr{G}_{\uparrow 0}(\boldsymbol{k}+\boldsymbol{q}, \tau - \tau_1) \delta_{\boldsymbol{k}_1,\boldsymbol{k}+\boldsymbol{q}} \delta_{\sigma\uparrow} \mathscr{G}_{\downarrow 0}(\boldsymbol{k}'-\boldsymbol{q}, \tau - \tau_1) \\
&\quad \times \delta_{\boldsymbol{k}_2,\boldsymbol{k}'-\boldsymbol{q}} \delta_{\sigma'\downarrow} \mathscr{G}_{\downarrow 0}(\boldsymbol{k}', \tau_1 - \tau') \delta_{\boldsymbol{k}_3\boldsymbol{k}'} \mathscr{G}_{\uparrow 0}(\boldsymbol{k}, \tau_1 - \tau') \delta_{\boldsymbol{k}_4\boldsymbol{k}} \\
&= -\frac{U}{\hbar} \sum_{\boldsymbol{k}\boldsymbol{k}'\boldsymbol{q}} e^{i(\boldsymbol{k}+\boldsymbol{k}')\cdot(\boldsymbol{r}-\boldsymbol{r}')} \int_0^{\beta\hbar} d\tau_1\, \mathscr{G}_{\uparrow 0}(\boldsymbol{k}+\boldsymbol{q}, \tau - \tau_1) \\
&\quad \times \mathscr{G}_{\downarrow 0}(\boldsymbol{k}'-\boldsymbol{q}, \tau - \tau_1) \mathscr{G}_{\downarrow 0}(\boldsymbol{k}', \tau_1 - \tau') \mathscr{G}_{\uparrow 0}(\boldsymbol{k}, \tau_1 - \tau').
\end{aligned}$$

Thus, the Fourier transform of $\chi_p^{(1)}(\boldsymbol{r}-\boldsymbol{r}',\tau-\tau')$ with respect to $\boldsymbol{r}-\boldsymbol{r}'$ is

$$\chi_p^{(1)}(\boldsymbol{q},\tau-\tau') = -\frac{U}{\hbar}\sum_{\boldsymbol{k}\boldsymbol{k}'}\int_0^{\beta\hbar} \mathrm{d}\tau_1\, \mathscr{G}_{\uparrow 0}(\boldsymbol{q},\tau-\tau_1)\mathscr{G}_{\downarrow 0}(\boldsymbol{q}-\boldsymbol{k}',\tau-\tau_1)$$
$$\times\, \mathscr{G}_{\downarrow 0}(\boldsymbol{q}-\boldsymbol{k},\tau_1-\tau')\mathscr{G}_{\uparrow 0}(\boldsymbol{k},\tau_1-\tau'),$$

where we have made changes to the dummy summation variables for wave vectors. Inserting the Fourier expansion of $\mathscr{G}_{\uparrow 0}(\boldsymbol{k},\tau)$ into the above equation yields

$$\begin{aligned}\chi_p^{(1)}&(\boldsymbol{q},\tau-\tau')\\ &= -\frac{U}{\hbar\beta^4}\sum_{\boldsymbol{k}\boldsymbol{k}'}\sum_{n_1n_2n_3n_4}\int_0^{\beta\hbar}\mathrm{d}\tau_1\, \mathrm{e}^{-\mathrm{i}(\omega_{n_1}+\omega_{n_2}-\omega_{n_3}-\omega_{n_4})\tau_1}\\ &\quad\times \mathrm{e}^{-\mathrm{i}(\omega_{n_1}+\omega_{n_2})\tau+\mathrm{i}(\omega_{n_3}+\omega_{n_4})\tau'}\mathscr{G}_{\uparrow 0}(\boldsymbol{q},\mathrm{i}\omega_{n_1})\mathscr{G}_{\downarrow 0}(\boldsymbol{q}-\boldsymbol{k}',\mathrm{i}\omega_{n_2})\\ &\quad\times \mathscr{G}_{\downarrow 0}(\boldsymbol{q}-\boldsymbol{k},\mathrm{i}\omega_{n_3})\mathscr{G}_{\uparrow 0}(\boldsymbol{k},\mathrm{i}\omega_{n_4})\\ &= -\frac{U}{\beta^3}\sum_{\boldsymbol{k}\boldsymbol{k}'}\sum_{nn'm}\mathrm{e}^{-\mathrm{i}\omega_m(\tau-\tau')}\mathscr{G}_{\downarrow 0}(\boldsymbol{q}-\boldsymbol{k}',\mathrm{i}\omega_m-\mathrm{i}\omega_{n'})\\ &\quad\times \mathscr{G}_{\uparrow 0}(\boldsymbol{q},\mathrm{i}\omega_{n'})\mathscr{G}_{\downarrow 0}(\boldsymbol{q}-\boldsymbol{k},\mathrm{i}\omega_m-\mathrm{i}\omega_n)\mathscr{G}_{\uparrow 0}(\boldsymbol{k},\mathrm{i}\omega_n).\end{aligned}$$

Thus, the Fourier transform of $\chi_p^{(1)}(\boldsymbol{q},\tau-\tau')$ with respect to $\tau-\tau'$ is

$$\begin{aligned}\chi_p^{(1)}(\boldsymbol{q},\mathrm{i}\omega_m) &= -\frac{U}{\beta^2}\sum_{\boldsymbol{k}\boldsymbol{k}'}\sum_{nn'}\mathscr{G}_{\uparrow 0}(\boldsymbol{q},\mathrm{i}\omega_{n'})\mathscr{G}_{\downarrow 0}(\boldsymbol{q}-\boldsymbol{k}',\mathrm{i}\omega_m-\mathrm{i}\omega_{n'})\\ &\qquad\times \mathscr{G}_{\downarrow 0}(\boldsymbol{q}-\boldsymbol{k},\mathrm{i}\omega_m-\mathrm{i}\omega_n)\mathscr{G}_{\uparrow 0}(\boldsymbol{k},\mathrm{i}\omega_n)\\ &= -U\big[\chi_p^{(0)}(\boldsymbol{q},\mathrm{i}\omega_m)\big]^2.\end{aligned}$$

In the second order, we have

$$\begin{aligned}\chi_p^{(2)}&(\boldsymbol{r}-\boldsymbol{r}',\tau-\tau')\\ &= \frac{U^2}{4\hbar^2}\sum_{\boldsymbol{k}_1\boldsymbol{k}_2\boldsymbol{k}_3\boldsymbol{k}_4}\sum_{\boldsymbol{k}\boldsymbol{k}'\boldsymbol{q}\alpha\alpha'}\sum_{\boldsymbol{p}\boldsymbol{p}'\boldsymbol{g}\gamma\gamma'}\mathrm{e}^{\mathrm{i}[(\boldsymbol{k}_1+\boldsymbol{k}_2)\cdot\boldsymbol{r}-(\boldsymbol{k}_3+\boldsymbol{k}_4)\cdot\boldsymbol{r}']}\int_0^{\beta\hbar}\mathrm{d}\tau_1\int_0^{\beta\hbar}\mathrm{d}\tau_2\\ &\quad\times \big\langle \hat{T}_\tau\big[\hat{c}_{\boldsymbol{k}_1\uparrow}(\tau)\hat{c}_{\boldsymbol{k}_2\downarrow}(\tau)\hat{c}^\dagger_{\boldsymbol{k}_3\downarrow}(\tau')\hat{c}^\dagger_{\boldsymbol{k}_4\uparrow}(\tau')\hat{c}^\dagger_{\boldsymbol{k}+\boldsymbol{q},\alpha}(\tau_1)\hat{c}^\dagger_{\boldsymbol{k}'-\boldsymbol{q},\alpha'}(\tau_1)\\ &\quad\times \hat{c}_{\boldsymbol{k}'\alpha'}(\tau_1)\hat{c}_{\boldsymbol{k}\alpha}(\tau_1)\hat{c}^\dagger_{\boldsymbol{p}+\boldsymbol{g},\gamma}(\tau_2)\hat{c}^\dagger_{\boldsymbol{p}'-\boldsymbol{g},\gamma'}(\tau_2)\hat{c}_{\boldsymbol{p}'\gamma'}(\tau_2)\hat{c}_{\boldsymbol{p}\gamma}(\tau_2)\big]\big\rangle_{0,\mathrm{dc}}.\end{aligned}$$

Making use of Wick's theorem, we have

$$\begin{aligned}\chi_p^{(2)}&(\boldsymbol{r}-\boldsymbol{r}',\tau-\tau')\\ &= \frac{U^2}{\hbar^2}\sum_{\boldsymbol{k}_1\boldsymbol{k}_2\boldsymbol{k}_3\boldsymbol{k}_4}\sum_{\boldsymbol{k}\boldsymbol{k}'\boldsymbol{q}\alpha\alpha'}\sum_{\boldsymbol{p}\boldsymbol{p}'\boldsymbol{g}\gamma\gamma'}\mathrm{e}^{\mathrm{i}[(\boldsymbol{k}_1+\boldsymbol{k}_2)\cdot\boldsymbol{r}-(\boldsymbol{k}_3+\boldsymbol{k}_4)\cdot\boldsymbol{r}']}\int_0^{\beta\hbar}\mathrm{d}\tau_1\int_0^{\beta\hbar}\mathrm{d}\tau_2\\ &\quad\times \mathscr{G}_{\uparrow 0}(\boldsymbol{k}'-\boldsymbol{q},\tau-\tau_1)\delta_{\boldsymbol{k}_1,\boldsymbol{k}'-\boldsymbol{q}}\delta_{\alpha'\uparrow}\mathscr{G}_{\downarrow 0}(\boldsymbol{k}+\boldsymbol{q},\tau-\tau_1)\delta_{\boldsymbol{k}_2,\boldsymbol{k}+\boldsymbol{q}}\delta_{\alpha\downarrow}\\ &\quad\times \mathscr{G}_{\downarrow 0}(\boldsymbol{k},\tau_1-\tau_2)\delta_{\boldsymbol{p}+\boldsymbol{g},\boldsymbol{k}}\delta_{\gamma\downarrow}\mathscr{G}_{\uparrow 0}(\boldsymbol{k}',\tau_1-\tau_2)\delta_{\boldsymbol{p}'-\boldsymbol{g},\boldsymbol{k}}\delta_{\gamma'\uparrow}\\ &\quad\times \mathscr{G}_{\downarrow 0}(\boldsymbol{p},\tau_2-\tau')\delta_{\boldsymbol{k}_3\boldsymbol{p}}\mathscr{G}_{\uparrow 0}(\boldsymbol{p}',\tau_2-\tau')\delta_{\boldsymbol{k}_4\boldsymbol{p}'}.\end{aligned}$$

Performing summations over some of the wave vectors and spin indices using the δ-symbols, we have

$$\begin{aligned}
&\chi_p^{(2)}(\boldsymbol{r}-\boldsymbol{r}',\tau-\tau')\\
&\quad=\frac{U^2}{\hbar^2}\sum_{\boldsymbol{k}\boldsymbol{k}'\boldsymbol{p}\boldsymbol{q}}\mathrm{e}^{\mathrm{i}(\boldsymbol{k}+\boldsymbol{k}')\cdot(\boldsymbol{r}-\boldsymbol{r}')}\int_0^{\beta\hbar}\mathrm{d}\tau_1\int_0^{\beta\hbar}\mathrm{d}\tau_2\,\mathscr{G}_{\uparrow 0}(\boldsymbol{k}'-\boldsymbol{q},\tau-\tau_1)\\
&\qquad\times\mathscr{G}_{\downarrow 0}(\boldsymbol{k}+\boldsymbol{q},\tau-\tau_1)\mathscr{G}_{\downarrow 0}(\boldsymbol{k},\tau_1-\tau_2)\mathscr{G}_{\uparrow 0}(\boldsymbol{k}',\tau_1-\tau_2)\\
&\qquad\times\mathscr{G}_{\downarrow 0}(\boldsymbol{p},\tau_2-\tau')\mathscr{G}_{\uparrow 0}(\boldsymbol{k}+\boldsymbol{k}'-\boldsymbol{p},\tau_2-\tau').
\end{aligned}$$

Thus, the Fourier transform of $\chi_p^{(2)}(\boldsymbol{r}-\boldsymbol{r}',\tau-\tau')$ with respect to $\boldsymbol{r}-\boldsymbol{r}'$ is

$$\begin{aligned}
\chi_p^{(2)}(\boldsymbol{q},\tau-\tau')&=\frac{U^2}{\hbar^2}\sum_{\boldsymbol{k}\boldsymbol{k}'\boldsymbol{p}}\int_0^{\beta\hbar}\mathrm{d}\tau_1\int_0^{\beta\hbar}\mathrm{d}\tau_2\,\mathscr{G}_{\uparrow 0}(\boldsymbol{q}-\boldsymbol{k}',\tau-\tau_1)\\
&\qquad\times\mathscr{G}_{\downarrow 0}(\boldsymbol{k}',\tau-\tau_1)\mathscr{G}_{\downarrow 0}(\boldsymbol{k},\tau_1-\tau_2)\mathscr{G}_{\uparrow 0}(\boldsymbol{q}-\boldsymbol{k},\tau_1-\tau_2)\\
&\qquad\times\mathscr{G}_{\downarrow 0}(\boldsymbol{p},\tau_2-\tau')\mathscr{G}_{\uparrow 0}(\boldsymbol{q}-\boldsymbol{p},\tau_2-\tau').
\end{aligned}$$

Inserting the Fourier expansion of $\mathscr{G}_{\uparrow 0}(\boldsymbol{k},\tau)$ into the above equation yields

$$\begin{aligned}
&\chi_p^{(2)}(\boldsymbol{q},\tau-\tau')\\
&\quad=\frac{U^2}{\beta^4}\sum_{\boldsymbol{k}\boldsymbol{k}'\boldsymbol{p}}\sum_{n_1n_2n_3m}\mathrm{e}^{-\mathrm{i}\omega_m(\tau-\tau')}\mathscr{G}_{\uparrow 0}(\boldsymbol{q}-\boldsymbol{k}',\mathrm{i}\omega_m-\mathrm{i}\omega_{n_1})\\
&\qquad\times\mathscr{G}_{\downarrow 0}(\boldsymbol{k}',\mathrm{i}\omega_{n_1})\mathscr{G}_{\downarrow 0}(\boldsymbol{k},\mathrm{i}\omega_{n_2})\mathscr{G}_{\uparrow 0}(\boldsymbol{q}-\boldsymbol{k},\mathrm{i}\omega_m-\mathrm{i}\omega_{n_2})\\
&\qquad\times\mathscr{G}_{\downarrow 0}(\boldsymbol{p},\mathrm{i}\omega_{n_3})\mathscr{G}_{\uparrow 0}(\boldsymbol{q}-\boldsymbol{p},\mathrm{i}\omega_m-\mathrm{i}\omega_{n_3}).
\end{aligned}$$

Thus, the Fourier transform of $\chi_p^{(2)}(\boldsymbol{q},\tau-\tau')$ with respect to $\tau-\tau'$ is

$$\begin{aligned}
\chi_p^{(2)}(\boldsymbol{q},\mathrm{i}\omega_m)&=\frac{U^2}{\beta^3}\sum_{\boldsymbol{k}\boldsymbol{k}'\boldsymbol{p}}\sum_{n_1n_2n_3}\mathscr{G}_{\uparrow 0}(\boldsymbol{q}-\boldsymbol{k}',\mathrm{i}\omega_m-\mathrm{i}\omega_{n_1})\mathscr{G}_{\downarrow 0}(\boldsymbol{k}',\mathrm{i}\omega_{n_1})\\
&\qquad\times\mathscr{G}_{\downarrow 0}(\boldsymbol{k},\mathrm{i}\omega_{n_2})\mathscr{G}_{\uparrow 0}(\boldsymbol{q}-\boldsymbol{k},\mathrm{i}\omega_m-\mathrm{i}\omega_{n_2})\\
&\qquad\times\mathscr{G}_{\downarrow 0}(\boldsymbol{p},\mathrm{i}\omega_{n_3})\mathscr{G}_{\uparrow 0}(\boldsymbol{q}-\boldsymbol{p},\mathrm{i}\omega_m-\mathrm{i}\omega_{n_3})\\
&=U^2\big[\chi_p^{(0)}(\boldsymbol{q},\mathrm{i}\omega_m)\big]^3.
\end{aligned}$$

Having evaluated the first three terms in the series for $\chi_p(\boldsymbol{q},\mathrm{i}\omega_m)$, we can now write down the whole series. We have

$$\begin{aligned}
\chi_p(\boldsymbol{q},\mathrm{i}\omega_m)&=\chi_p^{(0)}(\boldsymbol{q},\mathrm{i}\omega_m)-U\big[\chi_p^{(0)}(\boldsymbol{q},\mathrm{i}\omega_m)\big]^2\\
&\qquad+U^2\big[\chi_p^{(0)}(\boldsymbol{q},\mathrm{i}\omega_m)\big]^3-\cdots
\end{aligned}$$

which is just the series represented by the Feynman diagrams in Fig. 30.7(b). Summing up the above series, we obtain

$$\chi_p(\boldsymbol{q}, \mathrm{i}\omega_m) = \frac{\chi_p^{(0)}(\boldsymbol{q}, \mathrm{i}\omega_m)}{1 + U\chi_p^{(0)}(\boldsymbol{q}, \mathrm{i}\omega_m)}.$$

(4) For $\boldsymbol{q} = 0$ and $\omega_m = 0$, we have

$$\chi_p(0,0) = \frac{\chi_p^{(0)}(0,0)}{1 + U\chi_p^{(0)}(0,0)}.$$

The instability occurs if the denominator in the above expression vanishes. Let $1 + U\chi_p^{(0)}(0,0) = 0$. Making use of the above-computed value for $\chi_p^{(0)}(0,0)$ and denoting the temperature at which $1 + U\chi_p^{(0)}(0,0) = 0$ by T_c, we have

$$1 + N(0)U \ln \frac{1.134\hbar\omega_c}{k_{\mathrm{B}}T_c} = 0.$$

Solving for $k_{\mathrm{B}}T_c$, we obtain

$$k_{\mathrm{B}}T_c = 1.134\hbar\omega_c\, \mathrm{e}^{-1/N(0)|U|},$$

where we have made use of the fact that $U < 0$. The above expression is just the BCS formula for the superconducting phase transition temperature T_c in the weak-coupling limit.

References

(1) N. W. Ashcroft and N. D. Mermin, *Solid State Physics*, 2nd Edition (Holt, Rinehart, and Winston, 2002).
(2) C. Kittel, *Introduction to Solid State Physics*, 7th Edition (Wiley, 2004).
(3) J. R. Hook and E. R. Hook, *Solid State Physics*, 2nd Edition (Wiley, 1995).
(4) J. S. Blakemore, *Solid State Physics* (Cambridge University Press, 1985).
(5) H. Ibach and H. Lüth, *Solid-State Physics: An Introduction to Principles of Materials Science* (Springer, 1991).
(6) W. Jones and NormN.an H. March, *Theoretical Solid State Physics*, Volumes 1 and 2 (Dover Publications, 1985).
(7) J. M. Ziman, *Principles of the Theory of Solids* (Cambridge University Press, 1979).
(8) C. Kittel, *Quantum Theory of Solids* (Wiley, 1987).
(9) J. Callaway, *Quantum Theory of the Solid State*, 2nd Edition (Academic Press, 1991).
(10) W. A. Harrison, *Solid State Theory* (Dover Publications, 1980).
(11) O. Madelung, *Introduction to Solid-State Theory* (Springer, 1995).
(12) G. D. Mahan, *Many-Particle Physics*, 3rd Edition (Springer, 2000).
(13) A. A. Abrikosov, L. P. Gorkox, and I. E. Dzyaloshinski, *Methods of Quantum Field Theory in Statistical Physics* (Dover Publications, revised Edition, 1975).
(14) S. Doniach and E. H. Sondheimer, *Green's Functions for Solid State Physicists* (World Scientific, 1998).
(15) J. W. Negele and H. Orland, *Quantum Many-particle Systems* (Westview Press, 1998).

(16) A. L. Fetter and J. D. Walecka, *Quantum Theory of Many-Particle Systems* (Dover Publications, 2003).
(17) R. M. White, *Quantum Theory of Magnetism: Magnetic Properties of Materials* (Springer, 2006).
(18) T. Moriya, *Spin Fluctuations in Itinerant Electron Magnetism* (Springer, 1985).
(19) M. Fox, *Optical Properties of Solids* (Oxford University Press, USA, 2010).
(20) M. Tinkham, *Introduction to Superconductivity*, 2nd Edition (Dover Publications, 2004).
(21) P. G. De Gennes, *Superconductivity of Metals and Alloys* (Westview Press, 1999).
(22) J. R. Schrieffer, *Theory of Superconductivity* (Perseus Books, 1999).
(23) A. M. Glazer and G. Burns, *Space Groups for Solid State Scientists*, 2nd Edition (Academic Press, 1990).
(24) P. W. Anderson, *Concepts in Solids* (World Scientific, 1998).
(25) P. W. Anderson, *Basic Notions of Condensed Matter Physics* (Westview Press, 1997).
(26) E. Fradkin, *Field Theories of Condensed Matter Systems* (Sarat Book House, 2007).
(27) R. M. Martin, *Electronic Structure: Basic Theory and Practical Methods* (Cambridge University Press, 2008).
(28) W. A. Harrison, *Electronic Structure and the Properties of Solids* (Dover Publications, 1989).
(29) E. Kaxiras, *Atomic and Electronic Structure of Solids* (Cambridge University Press, 2003).
(30) H. Eschrig, *The Fundamentals of Density Functional Theory* (Edition am Gutenbergplatz, 2003).
(31) D. Sholl and J. A. Steckel, *Density Functional Theory: A Practical Introduction* (Wiley-Interscience, 2009).
(32) J. Weertman and J. R. Weertman, *Elementary Dislocation Theory* (Oxford University Press, USA, 1992).
(33) D. Hull and D. J. Bacon, *Introduction to Dislocations* (Butterworth-Heinemann, 2001).
(34) W. H. Press, S. A. Teukolsky, W. T. Vetterling, and B. P. Flannery, *Numerical Recipes: The Art of Scientific Computing*, 3rd Edition (Cambridge University Press, 2007).
(35) L. Mihály and M. C. Martin, *Solid State Physics: Problems and Solutions* (Wiley-Interscience, 1996).

Index